Handbook of Railway Vehicle Dynamics

Handbook of Railway Vehicle Dynamics

Second Edition

Edited by
Simon Iwnicki
Maksym Spiryagin
Colin Cole
Tim McSweeney

CRC Press
Taylor & Francis Group
Boca Raton London New York

CRC Press is an imprint of the
Taylor & Francis Group, an **informa** business

CRC Press
Taylor & Francis Group
6000 Broken Sound Parkway NW, Suite 300
Boca Raton, FL 33487-2742

© 2020 by Taylor & Francis Group, LLC
CRC Press is an imprint of Taylor & Francis Group, an Informa business

No claim to original U.S. Government works

Printed on acid-free paper

International Standard Book Number-13: 978-1-1386-0285-4 (Hardback)

This book contains information obtained from authentic and highly regarded sources. Reasonable efforts have been made to publish reliable data and information, but the author and publisher cannot assume responsibility for the validity of all materials or the consequences of their use. The authors and publishers have attempted to trace the copyright holders of all material reproduced in this publication and apologize to copyright holders if permission to publish in this form has not been obtained. If any copyright material has not been acknowledged please write and let us know so we may rectify in any future reprint.

Except as permitted under U.S. Copyright Law, no part of this book may be reprinted, reproduced, transmitted, or utilized in any form by any electronic, mechanical, or other means, now known or hereafter invented, including photocopying, microfilming, and recording, or in any information storage or retrieval system, without written permission from the publishers.

For permission to photocopy or use material electronically from this work, please access www.copyright.com (http://www.copyright.com/) or contact the Copyright Clearance Center, Inc. (CCC), 222 Rosewood Drive, Danvers, MA 01923, 978-750-8400. CCC is a not-for-profit organization that provides licenses and registration for a variety of users. For organizations that have been granted a photocopy license by the CCC, a separate system of payment has been arranged.

Trademark Notice: Product or corporate names may be trademarks or registered trademarks, and are used only for identification and explanation without intent to infringe.

Visit the Taylor & Francis Web site at
http://www.taylorandfrancis.com

and the CRC Press Web site at
http://www.crcpress.com

Contents

Preface vii
Editors ix
Contributors xi

Chapter 1 Introduction 1

    Simon Iwnicki, Maksym Spiryagin, Colin Cole and Tim McSweeney

Chapter 2 A History of Railway Vehicle Dynamics 5

    A. H. Wickens

Chapter 3 Design of Unpowered Railway Vehicles 43

    Anna Orlova, Roman Savushkin, Iurii (Yury) Boronenko, Kirill Kyakk, Ekaterina Rudakova, Artem Gusev, Veronika Fedorova and Nataly Tanicheva

Chapter 4 Design of Powered Rail Vehicles and Locomotives 115

    Maksym Spiryagin, Qing Wu, Peter Wolfs and Valentyn Spiryagin

Chapter 5 Magnetic Levitation Vehicles 165

    Shihui Luo and Weihua Ma

Chapter 6 Suspension Elements and Their Characteristics 197

    Sebastian Stichel, Anna Orlova, Mats Berg and Jordi Viñolas

Chapter 7 Wheel-Rail Contact Mechanics 241

    Jean-Bernard Ayasse, Hugues Chollet and Michel Sebès

Chapter 8 Tribology of the Wheel-Rail Contact 281

    Ulf Olofsson, Roger Lewis and Matthew Harmon

Chapter 9 Track Design, Dynamics and Modelling 307

    Wanming Zhai and Shengyang Zhu

Chapter 10 Gauging Issues 345

    David M. Johnson

Chapter 11 Railway Vehicle Derailment and Prevention 373

    Nicholas Wilson, Huimin Wu, Adam Klopp and Alexander Keylin

Chapter 12 Rail Vehicle Aerodynamics 415
   Hongqi Tian

Chapter 13 Longitudinal Train Dynamics and Vehicle Stability in Train Operations 457
   Colin Cole

Chapter 14 Noise and Vibration from Railway Vehicles 521
   David Thompson, Giacomo Squicciarini, Evangelos Ntotsios and Luis Baeza

Chapter 15 Active Suspensions 579
   Roger M. Goodall and T.X. Mei

Chapter 16 Dynamics of the Pantograph-Catenary System 613
   Stefano Bruni, Giuseppe Bucca, Andrea Collina and Alan Facchinetti

Chapter 17 Simulation of Railway Vehicle Dynamics 651
   Oldrich Polach, Mats Berg and Simon Iwnicki

Chapter 18 Field Testing and Instrumentation of Railway Vehicles 723
   Julian Stow

Chapter 19 Roller Rigs 761
   *Paul D. Allen, Weihua Zhang, Yaru Liang, Jing Zeng, Henning Jung,
   Enrico Meli, Alessandro Ridolfi, Andrea Rindi, Martin Heller and Joerg Koch*

Chapter 20 Scale Testing Theory and Approaches 825
   Nicola Bosso, Paul D. Allen and Nicolò Zampieri

Chapter 21 Railway Vehicle Dynamics Glossary 869
   Tim McSweeney

Index 879

Preface

This is the second edition of the handbook. The first edition, published in 2006, has become the established text in this field, is used by many researchers and has over 800 citations. We have completely reviewed all the material and updated much of the text to recognise that some significant new theoretical, numerical and experimental approaches have been developed, and new designs of railway vehicles and their components have been introduced since the publication of the first edition.

There have been rapid developments in many areas, including the application of IT through digitisation and vastly increased access to data. Although many of the key tools and techniques presented in the first edition are still used, most have been modified or updated, and new methods and computer tools have been developed. In this edition, we have included new chapters covering design of powered rail vehicles, aerodynamics of railway vehicles, maglev and the dynamics of the pantograph-catenary system.

We hope that readers find this handbook useful. Railway transport is seeing a resurgence in many countries and can provide efficient passenger and freight operations, but higher demands for safe and reliable operation at higher loads and speeds mean that the dynamic performance of vehicles and their interactions with the track and other infrastructure must be well understood. Engineers and researchers working in this field face significant challenges and the tools and techniques outlined in this handbook will assist in solving the problems faced in designing, operating and maintaining modern railway systems.

Simon Iwnicki
Maksym Spiryagin
Colin Cole
Tim McSweeney

Editors

Simon Iwnicki is professor of railway engineering at the University of Huddersfield in the UK, where he is director of the Institute of Railway Research (IRR). The IRR has an international reputation for its research and support to industry, providing not only valuable practical solutions to specific problems in the industry but also making significant contributions to the understanding of some of the fundamental mechanisms of the wheel-rail interaction on which the safe and economical operation of railways depends. Professor Iwnicki is the editor-in-chief of Part F of the Proceedings of the Institution of Mechanical Engineers (the *Journal of Rail and Rapid Transit*) and co-editor (responsible for railway matters) of the journal *Vehicle System Dynamics*. He was the academic co-chair of the Rail Research UK Association (RRUKA) from 2010 to 2014, and, from 2014 to 2015, he was chair of the railway division of the Institution of Mechanical Engineers. He is a former member of the Scientific Committee of Shift2Rail.

Maksym Spiryagin is a professor of engineering and the deputy director of the Centre for Railway Engineering at Central Queensland University, Australia. He received his PhD in the field of railway transport in 2004 at the East Ukrainian National University. Professor Spiryagin's involvement in academia and railway industry projects includes research experience in Australia, China, Italy, South Korea and Ukraine, involving locomotive design and traction, rail vehicle dynamics, acoustics and real-time and software-enabled control systems, mechatronics and the development of complex mechatronic systems using various approaches (co-simulation, software-in-the-loop, processor-in-the-loop and hardware-in-the loop simulations).

Colin Cole is a professor of mechanical engineering and the director of the Centre for Railway Engineering at Central Queensland University, Australia. His work history includes over 31 years in railway industry and research roles starting in 1984, with six years working in mechanised track maintenance in Queensland Railways. Since then, his experience has included both rolling stock and infrastructure areas. He has worked in railway research for the past 25 years, and his 1999 PhD thesis was on *Longitudinal Train Dynamics*. He has conducted a range of rail projects related to field testing of trains, simulation of dynamics, energy studies, train braking, derailment investigation, railway standards and innovations in measurement and control devices.

Tim McSweeney is an adjunct research fellow at the Centre for Railway Engineering (CRE) at Central Queensland University in Australia. He has over 45 years of experience in the field of railway fixed infrastructure asset management, specialising particularly in track engineering in the heavy-haul environment. He was the senior infrastructure manager overseeing the Bowen Basin Coal Network for Queensland Rail from 1991 until 2001. He then joined the CRE to follow his interest in railway research. Tim is a member of the Railway Technical Society of Australasia and a Fellow of the Permanent Way Institution. Central Queensland University awarded him an Honorary Master of Engineering degree in 2011. He has co-authored 2 books and 30 technical papers and consultancy reports on various aspects of railway engineering and operations.

Contributors

Paul D. Allen is a professor and assistant director of the Institute of Railway Research (IRR) at the University of Huddersfield. As a technical expert, his specialist fields are railway vehicle dynamics and wheel-rail contact mechanics. He completed a PhD on the subject of error quantification of scaled railway roller rigs and led the concept design of the full-scale roller rig at the IRR. His wider research interests include train braking technologies, pantograph-overhead line dynamics and the promotion of innovation in the rail industry.

Jean-Bernard Ayasse is a retired research director. Before joining "The French Institute of Science and Technology for Transport, Development and Networks" (IFSTTAR), France, he worked at the Commissariat à l'Energie Atomique and obtained his PhD from the University of Grenoble in 1970 and a state thesis in 1977 in solid state physics. He is a specialist in numerical simulations in the electromagnetic and mechanical domains. His research field goes from the modelling of linear induction motors to railway dynamics. He is the author of several innovations in the modelling of the wheel-rail contact and of the multibody formalism implemented in the VOCO code.

Luis Baeza is professor and chair at the Technical University of Valencia in Spain. From 2016 to 2018, he was a full professor in the Institute of Sound and Vibration Research at the University of Southampton, UK. His research area comprises various fields of railway technology, including dynamics of railway vehicles and the track, vibration, corrugation of rails and wheel-rail contact mechanics.

Mats Berg is professor and head of the Road and Rail Vehicles Unit at the Royal Institute of Technology (KTH) in Stockholm. Before joining KTH in 1993, he worked at ABB Traction in Västerås and at the University of California at Berkeley. He obtained his PhD from Lund Institute of Technology in 1987. His main research field is vehicle-track interaction, with emphasis on the aspects of structural dynamics, suspension dynamics, track dynamics and wheel-rail wear. Professor Berg has authored many papers and reports in this field and advised several PhD students. He teaches courses on rail vehicle dynamics and general railway engineering in degree programmes as well as for practising engineers of the railway sector (both in Sweden and internationally).

Iurii (Yury) Boronenko is professor and head of the Department of Railcars and Railcar Maintenance at the Petersburg State Transport University in St. Petersburg, Russia. Professor Boronenko is also the director of the Scientific Research Center 'Vagony'. The centre is involved in many practical fields such as monitoring the fleet of freight wagons in Russia, evaluation of the technical condition of railway vehicles, design of new and modification of existing railcars and implementation of repair technologies, as well as in research and consultancy projects for Russian railways and industry. 'Vagony' is the testing centre certified by the Russian Federal Service to test railway vehicles and by the Russian Maritime Register to test containers. Professor Boronenko's special interests include vehicle dynamics and modelling the motion of liquids in tank wagons. For his theoretical and practical contribution in developing railway vehicles, Professor Boronenko became a member of the Transport Academy of Russia.

Nicola Bosso is an associate professor at Politecnico di Torino. He gained his MA degree in 1996 and his PhD in machine design in 2004. After experience at the strategic research group at Fiat Ferroviaria, he joined the railway research group at Politecnico di Torino, where he developed his research in the railway sector, primarily concerning wheel/rail contact, multibody simulation and experimental testing on prototypes and real vehicles. He teaches in several courses at Politecnico di Torino, including rolling stock design.

Stefano Bruni is a full professor at Politecnico di Milano, Department of Mechanical Engineering, where he teaches applied mechanics and dynamics. He is the leader of the 'Railway Dynamics' research group, carrying out research on rail vehicles and their interaction with the infrastructure, with a critical mass of senior research competence. Professor Bruni has authored over 240 scientific papers and has spent a large amount of time lecturing and consulting to industry in Italy and other countries. He has been lead scientist for several research projects funded by the industry and by the European Commission (EC). He is vice president of the International Association for Vehicle System Dynamics (IAVSD) and was chairman of the IAVSD'05 international conference held in Milano in 2005. He is an editorial board member for some international journals in the field of railway engineering.

Giuseppe Bucca is an associate professor of applied mechanics at Politecnico di Milano, Department of Mechanical Engineering, where he teaches applied mechanics and mechatronics. His research activity is focussed on the dynamics and control of mechanical systems, with primary application to railway vehicles. In particular, his main research topics are the theoretical and experimental (laboratory and on-track tests) studies of the dynamical interaction between pantograph and catenary and of the related electromechanical phenomena, and the study of railway and tramway vehicle dynamics focussing on the wheel-rail contact phenomena, passenger comfort and running safety. He has participated and continues to participate in several research projects funded by the EU and in research and technical projects funded by private companies.

Hugues Chollet is researcher at IFSTTAR, France. He graduated from Université de Technologie Compiègne (UTC) in 1984 and obtained a PhD in 1991 at Université Pierre et Marie Curie, Paris 6, on the experimental validation of Kalker's theory for the use in wheel-rail contact. He carries out research and consultancy work on guided transportation systems, dealing with wheel-rail contact fatigue, derailment situations, instabilities, vibration and comfort problems.

Andrea Collina is a full professor at Politecnico di Milano, Department of Mechanical Engineering, where he teaches applied mechanics and dynamics. He is active in the field of railway dynamics, interaction with infrastructure and pantograph-catenary interaction, carrying out both simulation and laboratory and field-testing activities. He has been involved in EU-funded projects. He has authored over 140 scientific papers and acts as consultant to industry in Italy and other countries for railway topics.

Alan Facchinetti is an associate professor of applied mechanics at Politecnico di Milano. He graduated in 2000 and received his PhD in 2004 from that same university before joining the academic permanent staff in 2005. His research focuses on the dynamics, stability and control of mechanical systems, with primary application to railway vehicles and to their interaction with the infrastructure. In this respect, his research activities address the dynamic behaviour of railway vehicles and tramcars, pantograph-catenary interaction, active control, monitoring and diagnostics in railway vehicles, and include the development of numerical and laboratory tools and the design and execution of on-track tests. He has participated and continues to participate in several research projects funded by EU or national grants and in research and technical projects funded by private companies. He is a member of various European Committee for Electrotechnical Standardization working groups dealing with current collection systems.

Veronika Fedorova is a researcher at the Department of Integrated Studies of Track-Train Interaction Dynamics at All-Union Research and Development Center for Transportation Technology (St. Petersburg, Russia). She graduated from Petersburg State Transport University in 2015. She is currently a postgraduate student working on a PhD thesis in the field of railway vehicle dynamics, developing wheel-rail wear and rolling contact fatigue (RCF) simulation models and designing advanced wheel profiles for freight wagons.

Contributors

Roger M. Goodall spent 12 years at British Rail's Research Division in Derby, where he worked on a variety of projects, including Maglev, tilting trains and active railway suspension systems. Roger took up an academic position at Loughborough University in 1982, and he became professor of control systems engineering in 1994. He also has a part-time professorial role at the University of Huddersfield's Institute of Railway Research. His research has been concerned with a variety of practical applications of advanced control, usually for high-performance electro-mechanical systems and, for many years, specifically on active railway vehicle suspensions. Roger has served in a variety of external roles such as a member of the board of the International Association for Vehicle System Dynamics (IAVSD), vice president of the International Federation of Automatic Control (IFAC) and chairman of the IMechE Railway Division. He has been a fellow of the Institution of Mechanical Engineers for a number of years, and he was elected a fellow of the Royal Academy of Engineering in 2007 and a fellow of IFAC in 2017.

Artem Gusev is a researcher at the Department of Integrated Studies of Train-Track Interaction Dynamics at All-Union Research and Development Center for Transportation Technology (St. Petersburg, Russia). He graduated from Petersburg State Transport University in 2014, completed postgraduate studies with a degree in railway rolling stock in 2018 and prepared the thesis for PhD viva examination. He carries out research in the area of improving rolling stock dynamics and developing freight bogies, using the *finite elements method* and computer simulation modelling of railcar movement.

Matthew Harmon is a research associate in the Department of Mechanical Engineering at The University of Sheffield where he studies tribology. He gained his PhD in 2019 which studied the application and mechanisms of lubricants and friction modifiers in the wheel-rail interface. This work developed a number of test methods in the laboratory as well as is developing an in situ method for measuring friction in interfaces. Since finishing his PhD he has continued to study the wheel-rail interface through a variety of projects. His current focus is on the verification of adhesion forecasts and the development of guidance for the rail industry.

Martin Heller is employed at Knorr-Bremse-Systeme für Schienenfahrzeuge, Munich, Germany, and directs the test department with the 'Atlas' Roller Rig. He completed a Dr.-Ing. in mechanical engineering on the subject of mathematical simulation of brake control systems of trains and railway vehicles and has been engaged since then in R&D of railway brakes.

David M. Johnson has a BSc in civil engineering (University of Leeds) and a PhD in mechanical engineering (Imperial College London), where his research and thesis covered 'The Simulation of Clearances between Trains and the Infrastructure'. His early career was at British Rail Research and in the USA, where he specialised in track, structure and soil mechanics disciplines. This led him to form his own business, specialising in technology development, particularly digital engineering systems, through which he developed a number of laser-based measuring technologies and started his interest in gauging. David's company, Laser Rail, specialised in the development of structure measuring technology, analysis of clearances and the management of gauging data. The company is now part of the Balfour Beatty Rail group. His latest venture, a partnership with his son, Colin, has developed gauging analysis technology to another level. Based upon mechanistic modelling of the complete gauging system (the topic of his PhD), DGauge specialises in matching clearance calculations to real life through processes of risk-based analysis (which they have commercially developed as Probabilistic Gauging™), high-definition and complex rail vehicle modelling and three-dimensional simulation, geometric swept envelope analysis of transitional curvature and other bespoke techniques. The Company's Cloud-based gauging service provides near-instant results to a variety of rail vehicle and infrastructure clients, both in the UK and internationally.

Henning Jung, MSc, studied mechanical engineering at the University of Siegen. He is now a research assistant working in the Applied Mechanics group of Professor Claus-Peter Fritzen at the University of Siegen. His research activities are focussed on dealing with the development of modern structural health monitoring systems (SHM) for railway vehicles. Prior to this, he also worked as a research assistant in the field of rolling mill design at Achenbach Buschhütten.

Alexander Keylin is a senior engineer in the Vehicle-Track Interaction group at Transportation Technology Center, Inc. (TTCI) in Pueblo, Colorado. He holds a BSc in mechanical engineering from University of Pittsburgh (2011) and an MSc in mechanical engineering from Virginia Tech (2012). His work involves testing and characterisation of rail vehicles, modelling of special track work, conducting computer simulations of vehicle-track interaction, analysis of ride quality data and development of algorithms for processing of rail profiles and track geometry data.

Adam Klopp is a senior engineer at Transportation Technology Center, Inc. (TTCI) in Pueblo, Colorado, specialising in vehicle-track interaction and train dynamics. He has over 7 years of railroad research and engineering experience at TTCI. His research includes the characterisation, analysis and modelling of rail vehicles and trains. His testing experience includes static and dynamic vehicle and track tests, compressive end load tests and impact tests. He holds a BSc in mechanical engineering from Colorado State University (2012) and an MSc in engineering with emphasis in railroad engineering from Colorado State University-Pueblo (2016).

Joerg Koch is employed at Knorr-Bremse-Systeme für Schienenfahrzeuge, Munich, Germany. He has a university degree as Dipl.-Ing. in Mechanical Engineering. After university, he developed and built a range of test rigs for railway equipment. He has been responsible for the ATLAS test rig since 2010 and was involved in its concept, design, build, commissioning and now its operation.

Kirill Kyakk is executive director of PTK-Engineering LLC (Russia), the freight wagon fleet operating company introducing next-generation freight cars and heavy freight trains on 1520 mm gauge railways. He obtained the PhD in 2007 at Petersburg State Transport University in St. Petersburg, Russia. His field of scientific interest is railcar design theory and system engineering. Dr. Kyakk has 16 years of experience in the railway industry, including leadership and participation in the development of more than 120 new freight wagon models with improved technical characteristics for the railways of Russia, Europe, Africa and the Middle East.

Roger Lewis is a professor of mechanical engineering at the University of Sheffield, where he teaches design and tribology. He received his PhD from that university in 2000, before joining the academic staff in 2002. His research interests are split into three areas: solving industrial wear problems, application and development of a novel ultrasonic technique for machine element contact analysis and design of engineering components and machines. He has worked on a number of projects related to the wheel-rail interface, including understanding fundamental mechanisms and modelling of wheel and rail materials, measurements of wheel-rail interface conditions and rail stress, friction management and understanding of low adhesion. In 2019, he was appointed as a Royal Academy of Engineering Research Chair in 'Wheel/Rail Interface Low Adhesion Management'. In collaboration with the Rail and Safety Standards Board, he will now undertake a 5-year programme of work in this area.

Yaru Liang is a PhD student at the State Key Laboratory of Traction Power, Southwest Jiaotong University, China. She received her bachelor's degree in vehicle engineering at Dalian Jiaotong University in 2010. She then continued her studies for the master's degree course from 2010 to 2012 and later became a PhD student. Her research interests are in vehicle dynamics simulation and roller rig testing.

Contributors

Shihui Luo is a professor at the State Key Laboratory of Traction Power, Southwest Jiaotong University (SWJTU), China. He earned his BEng, MEng and PhD in 1985, 1988 and 1991, respectively, from the Department of Marine Power Machinery Engineering, Shanghai Jiaotong University, China. He then joined SWJTU and started research on vehicle dynamics. During this period, he stayed 1 year in the Duewag Factory, Siemens VT, as a trainee. He teaches in postgraduate courses. He participated in dynamics simulation projects for the Shanghai maglev train in 2002 and has continued research since then for the development of maglev vehicles.

Weihua Ma is a researcher at the Traction Power State Key Laboratory (TPL), Southwest Jiaotong University (SWJTU), China. He was awarded a BEng degree in vehicle engineering from Shandong University, China, in 2002 and a PhD degree in vehicle engineering from SWJTU in 2008. He has worked at TPL since 2008 and completed his postdoctoral fellowship during this period in a joint project of CRRC-SWJTU at Qishuyan Co., Ltd. His research interests include locomotive and heavy-haul train dynamics as well as the maglev train. In recent years, he has focussed on developing maglev for regional applications, with an operational speed range of 160 ~ 200 km/h.

T.X. Mei is a professor in control engineering at the University of Salford, where he leads a research group at the School of Computing, Science and Engineering, carrying out leading-edge research in the area of control and systems study for railway vehicles. Professor Mei has a strong background in railway engineering and substantial expertise in vehicle dynamics and traction control. He has given invited research seminars at an international level and published many papers in leading academic journals and international conferences, which explore the application of advanced control techniques and the use of active components. Professor Mei is one of the most active researchers worldwide in the latest fundamental research into active steering and system integration for railway vehicles and has made significant contributions to several leading-edge research projects in the field. His educational background includes BSc (1982, Shanghai Tiedao), MSc (1985, Shanghai Tiedao), MSc (1991, Manchester) and PhD (1994, Loughborough).

Enrico Meli received his bachelor's degree in mechanical engineering in 2004 and his master degree in mathematical engineering in 2006 from the School of Engineering of the University of Florence. He received his PhD in mechanism and machine theory in 2010 from the School of Engineering of the University of Bologna. In 2014, Dr. Enrico Meli won the prestigious SIR 2014 – Scientific Independence of Young Researchers Project, funded by the Italian Minister for Education, University and Research. He has been an assistant professor at the Department of Industrial Engineering of the University of Florence since 2015. Currently, his main research interests include vehicle dynamics, tribology, turbomachinery, rotor dynamics, robotics and automation. In these fields, Dr. Enrico Meli is the author of over 50 publications in international journals and over 120 publications in Proceedings of International Congresses.

Evangelos Ntotsios is a research fellow at the Institute of Sound and Vibration Research (ISVR), University of Southampton. Before joining the ISVR in 2013, he worked at the School of Architecture, Building and Civil Engineering at Loughborough University and obtained his PhD from the Department of Mechanical Engineering at University of Thessaly (Greece) in 2010. He has participated in EPSRC-funded research projects, including 'MOTIV: Modelling of Train Induced Vibrations' and 'Track to the Future'. His research interests include ground-borne railway noise and vibration as well as structural vibration and system identification.

Ulf Olofsson has been professor in tribology at the Royal Institute of Technology (KTH) since 2006. Before joining KTH, he worked at the Swedish National Testing and Research Institute with tribological material and component testing. He obtained his Licentiate of Engineering degree from Chalmers University of Technology in 1994 and his PhD from the Royal Institute of

Technology in 1996. Dr. Olofsson has 20 years of research experience on the tribology of the wheel-rail contact. His main research interests include interfaces and especially simulation and prediction of friction and wear, mainly applied to problems in mechanical, automotive and railway engineering. New research interests include airborne particles from wear processes such as in disc brakes and the railway wheel to rail contact.

Anna Orlova is deputy director in scientific and technical development for Research and Production Corporation 'United Wagon Company' and CEO of its subsidiary, the All-Union Research and Development Center for Transportation Technology (St. Petersburg, Russia), working on the development of innovative freight wagons and their components for CIS, North American, European and other markets. United Wagon Company carries out the full cycle of research and development of freight wagons, starting from marketing, design and prototype production, production technology development, preliminary and operational testing, development of maintenance strategies and repair technology. Dr. Orlova's special interests include optimisation of running gear parameters for dynamic performance, evaluation of design schemes and the development of simulation models and testing methods. Dr. Orlova is a supervisor of postgraduate students at Petersburg State Transport University and the author of several textbooks on bogie design and multibody dynamics simulation.

Oldrich Polach is an independent consultant and assessor, and an honorary professor at the Technische Universität Berlin. From 2001 to 2016, he was chief engineer dynamics in Bombardier Transportation, Winterthur, Switzerland, responsible for dynamics specialists in Business Unit Bogies Europe. He is a well-recognised expert in railway vehicle dynamics and wheel-rail contact. He acted for 20 years as a member of the working group 'Interaction Vehicle-Track' of the European Committee for Standardisation CEN TC 256 and is accredited by the Railway Federal Authority in Germany for the assessment of railway vehicles. Professor Polach teaches railway vehicle dynamics at the ETH Swiss Federal Institute of Technology Zürich and at the Technische Universität Berlin. He is a member of the editorial boards of the international journals *Vehicle System Dynamics*, the *International Journal of Railway Technology* and the *International Journal of Heavy Vehicle Systems*.

Alessandro Ridolfi is currently an assistant professor within the Department of Industrial Engineering (DIEF) at the University of Florence (UNIFI). He has also been an adjunct professor at Syracuse University in Florence since 2015, teaching dynamics. He graduated in mechanical engineering from UNIFI in 2010 and received his PhD degree in industrial engineering from UNIFI in 2014. At the beginning of his PhD, he worked on railway vehicle localisation and wheel-rail adhesion modelling. His current research interests are underwater and industrial robotics, sensor-based navigation of vehicles, vehicle dynamics and bio-robotics. He is co-author of more than 80 journal and conference papers on vehicle dynamics, robotics and mechatronics topics.

Andrea Rindi received his PhD in 1999 from the University of Bologna, Italy. He is currently associate professor in machine theory with the School of Engineering of the University of Florence. He is cofounder and coordinator of the Laboratory of Mechatronics and Dynamic Modelling (MDM Lab). His current research interests include vehicle dynamics, hardware in-the-loop (HIL) simulation and automation in transport systems.

Ekaterina Rudakova is head of the Department of Integrated Studies of Track-Train Interaction Dynamics at All-Union Research and Development Center for Transportation Technology (St. Petersburg, Russia). She obtained her PhD in 2005 in railway rolling stock from Petersburg State Transport University. The department carries out science-based research in the area of estimation of rolling stock dynamic characteristics, track forces and dynamic loading of railcar elements.

Contributors

Theoretical research, coupled with testing and experimental results, allows to perfect simulation models of railcar movement and computational methods. Dr. Rudakova's special interests include computer simulation modelling, design and developing of bogie suspensions.

Roman Savushkin is a member of the board of directors and CEO of Research and Production Corporation 'United Wagon Company' PJSC (MOEX: UWGN). He received degrees from the Petersburg State Transport University and the University of Antwerp's Management School. Roman Savushkin holds the scientific degree of PhD in technical sciences and is an acting professor at the Russian University of Transport (MIIT). His fields of research include theory, numerical simulation and experimental evaluation of structural strength, durability, dynamic performance and track interaction of railway vehicles; structural fatigue theory and simulation of metal structures; theory and simulation of wheel-rail interaction; theory of casting and welding production processes including automation and robotics; design of railway freight wagons and their major components; and the economics of railway transport. He is the author of multiple papers published in Russian national and international scientific journals.

Michel Sebès is research engineer at IFSTTAR, France. Prior to joining IFSTTAR, he spent 13 years in the service industry, where he carried out studies in the fields of structural mechanics and numerical simulation. He obtained a Master of Science and Engineering degree from the Ecole Centrale de Nantes in 1989 in the field of structural mechanics. His main activities are centred on the development of the VOCO code, dedicated to guided transport dynamics, particularly the last wheel rail contact extensions.

Valentyn Spiryagin received his PhD in the field of railway transport in 2004 at the East Ukrainian National University at Lugansk. His research activities include rail vehicle dynamics, multibody simulation, control systems and vehicle structural analysis. He currently lives in Russia and works as a railway consultant on vehicle dynamics and design, including vehicle structural engineering, mechatronic suspension systems for locomotives, locomotive traction and embedded software development.

Giacomo Squicciarini is a lecturer at the Institute of Sound and Vibration Research (ISVR), University of Southampton. He joined the ISVR in 2012 after obtaining his PhD at Politecnico di Milano studying the acoustics of piano soundboards. His main areas of research are related to railway noise and vibration, including rolling noise, curve squeal and measurement techniques. He is also active in vibroacoustics, structural vibration and musical instrument acoustics. He has written 20 papers in refereed journals and co-authored various conference contributions. He teaches undergraduate and master's students at the University of Southampton.

Sebastian Stichel is professor in Rail Vehicle Dynamics and head of the Department of Aeronautical and Vehicle Engineering at KTH Royal Institute of Technology in Stockholm, Sweden. He is vice chairman of the European Rail Research Advisory Council. Since 2011, he has been director of the KTH Railway Group, a multidisciplinary research centre that deals with most aspects of Railway Technology. He holds a BSc (1989) and an MSc (1992) in vehicle engineering and a PhD (1996) in vehicle dynamics from Technische Universität Berlin. From 2000 to 2010, he was employed at Bombardier Transportation in Sweden, where, from 2003, he headed its Vehicle Dynamics department with employees in Sweden, Germany, UK and France. Professor Stichel has a primary research interest in the dynamic vehicle-track interaction, mainly using multibody simulation; the main concerns are improved ride comfort and reduced wheel and track damage. He is also involved in research on the interaction between pantograph and catenary and active suspension for rail vehicles.

Julian Stow is assistant director at the Institute of Railway Research at the University of Huddersfield. He has 18 years of experience in the railway industry, specialising in rail vehicle dynamics and wheel-rail interface engineering, and has led a wide range of research and consultancy projects for the rail industry of Great Britain in these areas. These include investigating the causes of rolling contact fatigue and other wheel and rail defects, simulation for running acceptance, problem solving on current fleets, safety and maintenance standards development and wheel-rail interface management for existing and new build light rail and metro systems. He is currently responsible for the delivery of a programme of research work under the strategic partnership between the Rail Safety and Standards Board and the University of Huddersfield. Julian is a chartered engineer and a Fellow of the Institution of Mechanical Engineers.

Nataly Tanicheva is a junior professor at the Department of Railcars and Railcar Maintenance at Petersburg State Transport University in St Petersburg, Russia, and a researcher at the Scientific Research Centre 'Vagony', with a special interest in design of rolling stock and its components. She obtained her PhD in 2013 in articulated railway flat wagons from Petersburg State Transport University.

David Thompson is professor of railway noise and vibration at the Institute of Sound and Vibration Research (ISVR), University of Southampton. Before joining the ISVR in 1996, he worked at British Rail Research in Derby, UK, and at TNO Institute of Applied Physics in Delft, The Netherlands, and obtained his PhD from the ISVR in 1990. He has written over 160 papers in refereed journals as well as a book on railway noise and vibration, which has also been translated into Chinese. He is the main author of the TWINS software for railway rolling noise. His research interests include a wide range of aspects of railway noise and vibration as well as noise control, vibroacoustics and structural vibration. He teaches undergraduate- and master-level courses.

Hongqi Tian is a professor of Central South University at Changsha in Hunan, Peoples Republic of China. She is an academician of the Chinese Academy of Engineering. Professor Tian is currently president of Central South University and is a past vice president of the Chinese Academy of Engineering. She was awarded her bachelor's degree and master's degree in railway locomotive vehicles and a PhD degree in fluid mechanics. She has been engaged in the railway science and technology field over several decades. Under Professor Tian's leadership, her research team exploited the research directions of railway vehicle aerodynamics and railway vehicle collision dynamics; established the aerodynamic design theory, technology and method system of high-speed trains; proposed the safety protection technology of train collision; and established the safety protection technology system for railway operations in windy environments. The team developed the 500 km/h moving model rig for the aerodynamic testing of high-speed trains and the actual vehicle impact test platform. These test qualifications are recognised in the relevant field around the world. In addition, the team developed the strong wind monitoring and warning system for the Qinghai-Tibet railway line, designed the shape of the first Chinese high-speed train and proposed the first crashworthiness and energy-absorbing vehicle design in China. The team has participated in the research and construction of all high-speed railway lines in China, including the railway lines of Beijing-Shanghai, Beijing-Guangzhou and the Qinghai-Tibet railway line that is characterised by a frigid plateau.

Jordi Viñolas is dean and professor of the School of Engineering at Nebrija University. He has previously held positions as head of European Projects at Bantec and head of TECNUN (University of Navarra) and CEIT. His scientific interests are focussed on machine dynamics, noise and vibration, railway dynamics and infrastructure. He has published around 60 scientific papers in areas such as vehicle dynamics, rail/vehicle interaction and other topics linked to the performance optimisation of vehicle and machine components. He has directly supervised 16 PhD theses and more than 50 MSc theses. His courses are machine elements design, noise and vibration and also mechanical fatigue analysis. Dr. Viñolas has worked as an evaluator for the European Commission and was one of the promoters of European Rail Research Network of Excellence (EURNEX). He is a member of the Editorial Board of the *IMechE Journal of Rail and Rapid Transit* and of the *International Journal of Rail Transportation*.

A. H. Wickens was educated as an aeronautical engineer at Loughborough and University of London. His 11 years in the aircraft industry culminated as head of Aeroelastics on the Avro Blue Steel missile. He joined British Railways Research in 1962 to carry out research into the dynamics of railway vehicles. He was director of Research from 1971–1983 and director of Engineering Development and Research from 1983–1989. From 1987 to 1990, Alan Wickens was chairman of the Office for Research and Experiments of the International Union of Railways in Utrecht. He was professor of dynamics in the Department of Mechanical Engineering at Loughborough University in 1989–1992 and subsequently Visiting Industrial Professor, where his research interest was in the active guidance and dynamic stability of innovative railway vehicles. He is an honorary member of the Association for Vehicle System Dynamics.

Nicholas Wilson (BSME, Cornell University, 1980) is chief scientist at the Transportation Technology Center, Inc. (TTCI) in Pueblo, Colorado, where he has worked since 1980, specialising in rail vehicle dynamics and wheel-rail interaction. He leads TTCI's Vehicle-Track Interaction group and the team of engineers developing TTCI's NUCARS® multibody vehicle-track dynamic interaction software. Recently, he has been working on flange climb derailment research, derailment investigations of rail vehicles, wheel-rail wear and RCF studies. He has also been working on developing rail vehicle dynamic performance specifications for, and analysing performance of, freight and passenger vehicles and trains to carry high-level radioactive material.

Huimin Wu (PhD, Mechanical Engineering, Illinois Institute of Technology, 2000) retired in May 2018 from her position as a scientist in the Vehicle-Track Interaction group at Transportation Technology Center, Inc. (TTCI) in Pueblo, Colorado. She worked for more than 26 years at TTCI in the simulation, analysis and testing of railway vehicles. Her research carried over into areas including vehicle dynamics, vehicle-track interaction, wheel flange climb derailment criteria, computation methodology of studying wheel-rail contact, NUCARS development, wheel-rail profile design and rail grinding. She has presented papers at international conferences on vehicle-track interaction and published a number of reports on consultancy work carried out for the railways.

Qing Wu is a research fellow at the Centre for Railway Engineering, Central Queensland University (CQU), Australia. His research expertise and interests include parallel computing, 3D train system dynamics, track dynamics and multiobjective optimisations. Dr. Wu has authored more than 50 journal articles and a simulation software. He has conducted a number of railway research and consultancy projects ranging from track dynamics to vehicle and train dynamics. His education background includes a BEng (2010) and MEng (2012) from Southwest Jiaotong University, China, and a PhD from CQU (2016).

Peter Wolfs is a professor of electrical engineering at Central Queensland University, Rockhampton, Australia. His research interests include railway power supply and traction systems, smart-grid technology, distributed renewable resources and energy storage and their impact on system capacity and power quality, the support of weak rural feeders and the remote-area power supply. Prof. Wolfs is a Fellow of Engineers Australia and a senior member of the Institute of Electrical and Electronics Engineers (IEEE).

Nicolò Zampieri is assistant professor at the Department of Mechanical and Aerospace Engineering of the Politecnico di Torino. He received his bachelor's degree and master's degree in mechanical engineering from that university in 2008 and 2010, respectively, and was awarded his PhD in 2014 for the dissertation regarding the development of monitoring systems for railway applications. His research interests are railway vehicle dynamics and monitoring, modelling of wheel-rail/roller contact, wear and RCF. His current research activity concerns the design of test benches for railway applications and the development of specific applications for railway vehicle monitoring. Nicolò Zampieri is co-author of more than 30 scientific publications.

Jing Zeng is professor of railway vehicle system dynamics at the State Key Laboratory of Traction Power, Southwest Jiaotong University (SWJTU), China. He obtained his PhD in Dynamic Simulation of Railway Vehicle Systems at SWJTU in 1991. His expertise covers novel bogie design, dynamic performance simulation and measurement techniques. In recent years, his research has mainly focussed on parameter optimal design, dynamic simulation and laboratory and field tests of high-speed trains. He has won two first-class and one second-class prizes of the State Scientific and Technological Progress Award.

Wanming Zhai is chair professor of railway engineering at Southwest Jiaotong University (SWJTU) in China and is an academician of the Chinese Academy of Sciences. Since 1994, Dr. Zhai has been a full professor and director of the Train and Track Research Institute, which is affiliated to the State Key Laboratory of Traction Power of SWJTU. In 1999, he was appointed Chang Jiang Chair Professor by the Chinese Ministry of Education. Currently, he is the chairman of the Academic Committee of Southwest Jiaotong University. Professor Zhai's research activities are mainly in the field of railway system dynamics, focussing on vehicle-track dynamic interaction and train-track-bridge interactions. He established a new theoretical framework of vehicle-track coupled dynamics and invented new methodologies for solving large-scale train-track-bridge interaction problems. His models and methods have been successfully applied to more than 20 large-scale field engineering projects for the railway network in China, mostly for high-speed railways. He is editor-in-chief of the *International Journal of Rail Transportation* and a trustee member of the International Association for Vehicle System Dynamics. He also serves as the president of the Chengdu Association for Science and Technology, vice president of the Chinese Society of Theoretical and Applied Mechanics and vice president of the Chinese Society for Vibration Engineering.

Weihua Zhang is a distinguished professor of the 'Cheung Kong Scholars' Program of China, winner of the 'National Science Funds for Distinguished Young Scholars' and chief scientist of '973 Program', the National Basic Research Program. In 2012, he was awarded the Guanghua Award of Engineering Technology. His doctoral dissertation was listed among 'National Top 100 Outstanding Doctoral Dissertations' in 2000. Professor Zhang served as an expert on the General Planning Group for Autonomous Innovation & Joint Action Plan for China's High-Speed Trains and an expert of the General Planning Group for the China-standard Electrical Multiple Units (CEMU) Development Program.

Shengyang Zhu is an associate professor at the Train and Track Research Institute, affiliated to the State Key Laboratory of Traction Power, Southwest Jiaotong University (SWJTU) in China. He graduated from SWJTU with a PhD degree in rail transportation engineering in 2015. He had research experience at Rice University, USA, as an award holder from the China Scholarship Council for 18 months since 2012. Dr. Zhu has published more than 30 papers in refereed high-level journals, including 22 papers as the first author or corresponding author, and he has given several keynote or invited presentations in international conferences and seminars. His research interests include a wide range of aspects of train-track interaction as well as track vibration control, track damage mechanisms and structural health monitoring. He has worked on a number of national key projects related to long-term dynamic performance of train and track systems, as well as industry-funded projects related to train induced vibration problems. He supervises graduate students in railway system dynamics, and he is a PhD thesis (international) examiner for some universities.

1 Introduction

Simon Iwnicki, Maksym Spiryagin, Colin Cole and Tim McSweeney

CONTENTS

1.1 Structure of the Handbook 1

The principal aim of this handbook is to present a detailed introduction to the main issues influencing the dynamic behaviour of railway vehicles, and a summary of the history and the state of the art of the analytical and computer tools and techniques that are used in this field around the world. The level of technical detail is intended to be sufficient to allow analysis of common practical situations, but references are made to other published material for those who need more detail in specific areas. The main readership will be engineers working in the railway industry worldwide and researchers working on issues connected with railway vehicle behaviour, but it should also prove useful to those wishing to gain a basic knowledge of topics outside their specialist technical area.

1.1 STRUCTURE OF THE HANDBOOK

The topics covered in this handbook are the main areas that impact the dynamic behaviour of railway vehicles and are intended to present the existing solutions in this area from a multidisciplinary perspective. These include the numerical and tribological analysis of the wheel-rail interface, general railway vehicle design and architecture, suspension and suspension component design, simulation and testing of electrical and mechanical systems, interaction with the surrounding infrastructure and noise and vibration generation. The handbook is international in scope and draws examples from around the world, but several chapters have a more specific focus, where a particular local limitation or need has led to the development of specific techniques or tools.

For example, the chapter on longitudinal train dynamics and vehicle stability expands upon the longitudinal train dynamics chapter in the first edition and mainly uses Australian examples of the issues related to longitudinal dynamics on their heavy-haul lines, where very long trains are used to transport bulk freight. Similarly, the issue of structure gauging largely uses the UK as a case study, because it is there that historic lines through dense-population centres have resulted in a very restricted loading gauge. The desire to run high-speed trains in this situation has led to the use of highly developed techniques to permit full advantage of the loading gauge to be taken.

The history of the field is presented by Alan H. Wickens in Chapter 2, from the earliest thoughts of George Stephenson about the dynamic behaviour of a wheelset through the development of theoretical principles to the application of modern computing techniques. Professor Wickens was one of the modern pioneers of these methods and, as director of research at British Rail Research, played a key role in the practical application of vehicle dynamics knowledge to high-speed freight and passenger vehicles.

Chapters 3 and 4 set out the basic structure of railway vehicles. In Chapter 3, Anna Orlova, Roman Savushkin, Iurii(Yury) Boronenko, Kirill Kyakk, Ekaterina Rudakova, Artem Gusev, Veronika Fedorova and Nataly Tanicheva outline and explain the basic structure of the railway coaches and wagons and the different types of running gear that are commonly used. Chapter 4 covers the design of powered railway vehicles and locomotives. Maksym Spiryagin, Qing Wu, Peter Wolfs and Valentyn Spiryagin explain the type and structure of locomotives in service and the different types of traction system used. Magnetic levitation vehicles are described in Chapter 5 by Shihui Luo and Weihua Ma. MagLev technology has been around for some time but does not yet seem to have achieved full commercialisation. The likely trends are explored in Chapter 5.

Chapter 6 explores the detail of the key suspension components that make up the running gear of typical railway vehicles. Sebastian Stichel, Anna Orlova, Mats Berg and Jordi Viñolas show how these components can be represented mathematically and give practical examples from different vehicles. The key area of any study of railway vehicle behaviour is the contact between the wheels and the rails. An understanding of all the forces that support and guide the vehicle pass through this small contact patch and of the nature of these forces is vital to any analysis of the general vehicle behaviour. The equations that govern these forces are derived and explained by Jean-Bernard Ayasse, Hugues Chollet and Michel Sebès in Chapter 7. They include an analysis of the normal contact that governs the size and shape of the contact patch and the stresses in the wheel and rail, and also the tangential problem, where slippage or creep in the contact patch produces the creep forces which accelerate, brake and guide the vehicle. The specific area of tribology applied to the wheel-rail contact is explained by Ulf Olofsson, Roger Lewis and Matthew Harmon in Chapter 8.

The track on which railway vehicles run is clearly a significant part of the dynamic system, and Wanming Zhai and Shengyang Zhu present the dynamics and modelling of various railway track structures in Chapter 9, as well as the interaction between track and train. Chapter 10 covers the unique railway problem of gauging, where the movement of a railway vehicle means that it sweeps through a space that is larger than it would occupy if it moved in a perfectly straight or curved path. Precise knowledge of this space or envelope is essential to avoid vehicles hitting parts of the surrounding infrastructure or each other. David M. Johnson has developed computer techniques that allow the gauging process to be carried out to permit vehicle designers and operators to ensure safety at the same time as maximising vehicle size and speed, and he explains the philosophies and techniques in this chapter.

The avoidance of derailment and its potentially catastrophic consequences are of fundamental concern to all railway engineers. In Chapter 11, Nicholas Wilson, Huimin Wu, Adam Klopp and Alexander Keylin explain how railway vehicle derailment is prevented. They explore the main causes and summarise the limits that have been set by standards to try to prevent these occurrences and cover the special case of independently rotating wheels and several possible preventative measures that can be taken.

In Chapter 12, Hongqi Tian explains the use of wind tunnels and computational fluid dynamics to improve the understanding of the effects of aerodynamics on the dynamic behaviour of railway vehicles.

Longitudinal train dynamics are covered by Colin Cole in Chapter 13. This is an aspect of vehicle dynamics that is sometimes ignored, but it becomes of major importance in heavy-haul railways, where very long and heavy trains lead to extremely high coupling forces. This chapter also covers rolling resistance and braking systems.

Chapter 14 deals with noise and vibration problems. David Thompson, Giacomo Squicciarini, Evangelos Ntotsios and Luis Baeza explain the key issues, including rolling noise caused by rail surface roughness, impact noise and curve squeal. They outline the basic theory required for a study in this area and also show how computer tools can be used to reduce the problem of noise. The effect of vibrations on human comfort is also discussed, and the effect of vehicle design is considered.

In Chapter 15, Roger M. Goodall and T.X. Mei summarise the possible ways in which active suspensions can allow vehicle designers to provide advantages that are not possible with passive suspensions. The basic concepts from tilting bodies to active secondary and primary suspension components are explained in detail and with examples. Recent tests on a prototype actively controlled bogie are presented, and limitations of the current actuators and sensors are explored before conclusions are drawn about the technology that will be seen in future vehicles.

Computer tools are now widely used in vehicle dynamics, and some specialist software packages allow all aspects of vehicle-track interaction to be simulated. In Chapter 17, Oldrich Polach,

Introduction

Mats Berg and Simon Iwnicki explain the historical development and state of the art of the methods that can be used to set up models of railway vehicles and to predict their behaviour as they run on typical track or over specific irregularities or defects. The material of previous chapters is drawn upon to inform the models of suspension elements and wheel-rail contact, and the types of analysis that are typically carried out are described. Typical simulation tasks are presented from the viewpoint of a vehicle designer attempting to optimise suspension performance, and the key issue of validation of the results of computer models is reviewed.

In Chapter 18, Julian Stow outlines the key aspects of field testing, including the procedures typically used during the acceptance process to demonstrate safe operation of railway vehicles. An alternative to field testing is to use a roller rig on which a vehicle can be run in relative safety, with conditions being varied in a controlled manner. In Chapter 19, Paul D. Allen, Weihua Zhang, Yaru Liang, Jing Zeng, Henning Jung, Enrico Meli, Alessandro Ridolfi, Andrea Rindi, Martin Heller and Joerg Koch summarise the characteristics of the main types of roller rig and the ways in which they are used. Chapter 19 also reviews the history of existing roller rigs, summarising the key details of examples of the main types. Chapter 20 extends the theme to scale testing, which has been used effectively for research into wheel-rail contact. In this chapter, Nicola Bosso, Paul D. Allen and Nicolò Zampieri describe the possible scaling philosophies that can be used and how these have been applied to scaled roller rigs. In Chapter 21, Tim McSweeney provides a glossary of terms relevant to railway vehicle dynamics.

2 A History of Railway Vehicle Dynamics

A. H. Wickens

CONTENTS

2.1 Introduction 5
2.2 Coning and the Kinematic Oscillation 6
2.3 Concepts of Curving 8
2.4 Dynamic Response, Hunting and the Bogie 9
2.5 Innovations for Improved Steering 12
2.6 Carter 13
2.7 Wheel-Rail Geometry 17
2.8 Creep 18
2.9 Matsudaira 19
2.10 The ORE Competition 19
2.11 The Complete Solution of the Hunting Problem 21
2.12 Later Research on Curving 23
2.13 Dynamic Response to Track Geometry 26
2.14 Suspension Design Concepts and Optimisation 26
    2.14.1 Two-Axle Vehicles and Bogies 26
    2.14.2 Forced Steering 27
    2.14.3 Three-Axle Vehicles 28
    2.14.4 Unsymmetrical Configurations 28
    2.14.5 The Three-Piece Bogie 28
    2.14.6 Independently Rotating Wheels 29
    2.14.7 Articulated Trains 29
2.15 Derailment 30
2.16 Active Suspensions 30
2.17 The Development of Computer Simulation 32
2.18 The Expanding Domain of Rail Vehicle Dynamics 33
References 34

2.1 INTRODUCTION

The railway train running along a track is one of the most complicated dynamical systems in engineering. Many bodies comprise the system, and so, it has many degrees of freedom. The bodies that make up the vehicle can be connected in various ways, and a moving interface connects the vehicle with the track. This interface involves the complex geometry of the wheel tread and the railhead and non-conservative frictional forces generated by relative motion in the contact area.

    The technology of this complex system rests on a long history. In the late eighteenth and early nineteenth centuries, development concentrated on the prime mover and the possibility of traction using adhesion. Strength of materials presented a major problem. Even though speeds were low, dynamic loads applied to the track were of concern, and so the earliest vehicles adopted elements of suspension taken from horse carriage practice. Above all, the problem of guidance was resolved by the almost universal adoption of the flanged wheel in the early nineteenth century, the result of empirical development and dependent on engineering intuition.

Operation of the early vehicles led to verbal descriptions of their dynamic behaviour such as Stephenson's description of the kinematic oscillation discussed later. Later in the nineteenth century, Redtenbacher and Klingel introduced the first simple mathematical models of the action of the coned wheelset, but they had virtually no impact on engineering practice. In practice, the balancing of the reciprocating masses of the steam locomotive assumed much greater importance. At this stage, artefacts, not equations, defined engineering knowledge.

A catastrophic bridge failure led to the first analytical model of the interaction between vehicle and flexible track in 1849.

The increasing size of the steam locomotive increased the problem of the forces generated in negotiating curves, and in 1883, Mackenzie gave the first essentially correct description of curving. This became the basis of a standard calculation carried out in design offices throughout the era of the steam locomotive.

As train speeds increased, problems of ride quality, particularly in the lateral direction, became more important. The introduction of the electric locomotive at the end of the nineteenth century involved Carter, a mathematical electrical engineer, in the problem, with the result that a realistic model of the forces acting between wheel and rail was proposed and the first calculations of lateral stability carried out.

Generally, empirical engineering development was able to keep abreast of the requirements of ride quality and safety until the middle of the twentieth century. Then, increasing speeds of trains and the greater potential risks arising from instability stimulated a more scientific approach to vehicle dynamics. Realistic calculations on which design decisions were based were achieved in the 1960s, and, as the power of the digital computer increased, so did the scope of engineering calculations, leading to today's powerful modelling tools.

This chapter tells the story of this conceptual and analytical development. It concentrates on the most basic problems associated with stability, response to track geometry and behaviour in curves of the railway vehicle, and most attention is given to the formative stage in which an understanding was gained. Progress in the last 20 years is not discussed, as the salient points are discussed later in the relevant chapters. As a result, many important aspects such as track dynamics, noise generation and other high-frequency (in this context, above about 15 Hz) phenomena are excluded.

2.2 CONING AND THE KINEMATIC OSCILLATION

The conventional railway wheelset, which consists of two wheels mounted on a common axle, has a long history [1] and evolved empirically. In the early days of the railways, speeds were low, and the objectives were to reduce rolling resistance (so that the useful load that could be hauled by horses could be multiplied) and to solve problems of strength and wear.

The flanged wheel running on a rail existed as early as the seventeenth century. The position of the flanges was on the inside, outside or even on both sides of the wheels and was still being debated in the 1820s. Wheels were normally fixed to the axle, though freely rotating wheels were sometimes used in order to reduce friction in curves. To start with, the play allowed between wheel flange and rail was minimal.

Coning was introduced partly to reduce the rubbing of the flange on the rail and partly to ease the motion of the vehicle round curves. It is not known when coning of the wheel tread was first introduced. It would be natural to provide a smooth curve uniting the flange with the wheel tread, and wear of the tread would contribute to this. Moreover, once wheels were made of cast iron, taper was normal foundry practice. In the early 1830s, the flange way clearance was opened up to reduce the lateral forces between wheel and rail, so that, typically, about 10 to 12 mm of lateral displacement was allowed before flange contact.

A History of Railway Vehicle Dynamics

Coning of the wheel tread was well established by 1821. George Stephenson in his observations on edge and tram railways [2] stated that 'It must be understood the form of edge railway wheels are conical that is the outer is rather less than the inner diameter about 3/16 of an inch. Then from a small irregularity of the railway the wheels may be thrown a little to the right or a little to the left, when the former happens the right wheel will expose a larger and the left one a smaller diameter to the bearing surface of the rail which will cause the latter to lose ground of the former but at the same time in moving forward it gradually exposes a greater diameter to the rail while the right one on the contrary is gradually exposing a lesser which will cause it to lose ground of the left one but will regain it on its progress as has been described alternately gaining and losing ground of each other which will cause the wheels to proceed in an oscillatory but easy motion on the rails'.

This is a very clear description of what is now called the kinematic oscillation, as shown in Figure 2.1.

The rolling behaviour of the wheelset suggests why it adopted its present form. If the flange is on the inside, the conicity is positive, and, as the flange approaches the rail, there will be a strong steering action, tending to return the wheelset to the centre of the track. If the flange is on the outside, the conicity is negative, and the wheelset will simply run into the flange and remain in contact as the wheelset moves along the track. Moreover, consider motion in a sharp curve in which the wheelset is in flange contact. If the flange is on the inside, the lateral force applied by the rail to the leading wheelset is applied to the outer wheel and will be combined with an enhanced vertical load, thus diminishing the risk of derailment. If the flange is on the outside, the lateral force applied by the rail is applied to the inner wheel, which has a reduced vertical load, and thus, the risk of derailment is increased.

As was explicitly stated by Brunel in 1838 [3], it can be seen that, for small displacements from the centre of straight or slightly curved track, the primary mode of guidance is conicity, and it is on sharper curves and switches and crossings that the flanges become the essential mode of guidance.

Lateral oscillations caused by coning were experienced from the early days of the railways. One solution to the oscillation problem that has been proposed from time to time, even down to modern times, was to fit wheels with cylindrical treads. However, in this case, if the wheels were rigidly mounted on the axle, very slight errors in parallelism would induce large lateral displacements that would be limited by flange contact. Thus, a wheelset with cylindrical treads tends to run in continuous flange contact.

In 1883, Klingel gave the first mathematical analysis of the kinematic oscillation [4] and derived the relationship between the wavelength Λ the wheelset conicity λ, wheel radius r_0 and the lateral distance between contact points $2l$ as

$$\Lambda = 2\pi(r_0 l/\lambda)^{1/2} \quad (2.1)$$

Klingel's formula shows that the frequency of the kinematic oscillation increases with speed. Any further aspects of the dynamical behaviour of railway vehicles must be deduced from a consideration of the forces acting, and this had to wait for Carter's much later contribution to the subject.

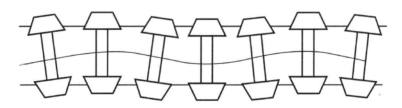

FIGURE 2.1 The kinematic oscillation of a wheelset. (From Iwnicki, S. (Ed.), *Handbook of Railway Vehicle Dynamics*, CRC Press, Boca Raton, FL, 2006. With permission.)

2.3 CONCEPTS OF CURVING

The action of a wheelset with coned wheels in a curve was understood intuitively early in the development of the railways. For example, in 1829, Ross Winans took out a patent that stressed the importance of the axles taking up a radial position on curves [5], a fundamental objective of running gear designers ever since, and W. B. Adams clearly understood the limitations of coning in curves in 1863 [6]. Redtenbacher [7] provided the first theoretical analysis in 1855, which is illustrated in Figure 2.2.

From the geometry in this figure, it can be seen that there is a simple geometric relationship between the outwards movement of the wheel y, the radius of the curve R, the wheel radius r_0, the distance between the contact points $2l$ and the conicity λ of the wheels in order to sustain pure rolling.

The application of Redtenbacher's formula shows that a wheelset will only be able to move outwards to achieve pure rolling if either the radius of curvature or the flangeway clearance is sufficiently large. Otherwise, a realistic consideration of curving requires the analysis of the forces acting between the vehicle and the track. In 1883, Mackenzie [8] gave the first essentially correct description of curving in a seminal paper (which was subsequently translated and published in both France and Germany). His work was suggested by an unintentional experiment in which the springs of the driving wheels of a six-wheeled engine were tightened to increase the available adhesion. The leading wheel mounted the rail when the locomotive approached a curve. Mackenzie gave a numerical but non-mathematical treatment of the forces generated in curving. His discussion is based on sliding friction and neglects coning, so that it is appropriate for sharp curves, where flanges provide guidance. Referring to Figure 2.3, Mackenzie explains 'If the flange were removed from the outer wheel, the engine would run straight forwards, and this wheel, in making one revolution, would run from A to B; but it is compelled by the flange to move in the direction of the line AC, a tangent to the curve at A, so that it slides sideways through a distance equal to BC. If this wheel were loose on the axle, it would, in making a revolution, run along the rail to F; but the inner wheel, in making a revolution, would run from H to K, the centre-line of the axle being KG; so that, if both axles are keyed on the axle, either the outer wheel must slide forwards or the inner wheel backwards. Assuming that the engine is exerting no tractive force, and that both wheels revolve at the speed due to the inner wheel, then the outer wheel will slide forwards from F to G. Take AL equal to BC, and LM equal to FG, the diagonal AM is the distance which the outer wheel slides in making one revolution'.

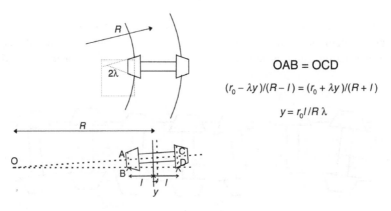

FIGURE 2.2 Redtenbacher's formula for the rolling of a coned wheelset on a curve. (From Iwnicki, S. (Ed.), *Handbook of Railway Vehicle Dynamics*, CRC Press, Boca Raton, FL, 2006. With permission.)

A History of Railway Vehicle Dynamics

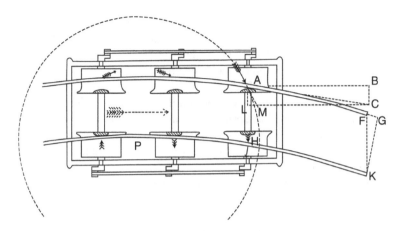

FIGURE 2.3 Forces acting on a vehicle in a curve according to Mackenzie. (From Mackenzie, J., *Proc. Inst. Civil Eng.*, 74, 1–57, 1883; With kind permission from CRC Press: *Handbook of Railway Vehicle Dynamics*, CRC Press, Boca Raton, FL, 2006, Iwnicki, S. [ed.].)

He then applies similar reasoning to the other wheels, assuming various positions for the wheelsets in relation to the rails. Thus, Mackenzie's calculations showed that the outer wheel flange exerts against the rail a force sufficient to overcome the friction of the wheel treads. Previously, centrifugal forces were regarded as the cause of many derailments. He also made the comment that 'the vehicle seems to travel in the direction which causes the smallest amount of sliding', which foresaw a later analytical technique developed by Heumann.

Subsequent work by Boedecker, Von Helmoltz and Uebelacker (described by Gilchrist [9]) was dominated by the need to avoid excessive loads on both vehicle and track, caused by steam locomotives, with long rigid wheelbases traversing sharp curves. Hence, in these theories, the conicity of the wheelsets is often ignored, and the wheels are assumed to be in the sliding regime. The corresponding forces are then balanced by a resultant flange force, or flange forces. This approach culminated in the work of Heumann in 1915 [10] and Porter in 1934 to 1935 [11].

Superelevation of tracks in curves was introduced on the Liverpool and Manchester Railway, and tables giving the relationship between superelevation of the outer rail and maximum speed were available in the 1830s.

2.4 DYNAMIC RESPONSE, HUNTING AND THE BOGIE

It was thought by some early engineers that the track would be so smooth that no vertical suspension would be necessary, but many rail breakages showed that it was necessary to reduce the stresses on the track. William Chapman recognised that, if the weight was spread over several axles, it was necessary to provide flexibility in a plan view to allow the locomotive to follow the curvature of the track. In an 1812 patent with his brother Edward Walton Chapman [12], they suggested a scheme in which the leading wheelset was mounted rigidly in the locomotive body and the trailing wheelsets mounted in a frame pivoted to the locomotive body, in other words, a bogie. It was envisaged that a proportion of the weight of the body would be carried by the bogie through conical rollers, so as to allow rotational freedom of the bogie relative to the body. However, the arrangement was not ideal in that the wheelsets could not adopt a radial position on curves necessary for pure rolling of the wheels. As Marshall [13] comments, what was proposed was either a trailing bogie or a double-bogie locomotive but not a leading bogie with the driving wheels fixed in a frame, which was to become, eventually, the most successful configuration for the steam locomotive. The concept was not widely adopted further at that time, probably because rail materials improved.

Losh and Stephenson in their 1816 patent [14] provided a method of dividing the weight of the locomotive among the wheels, in other words, a form of equalisation. This used steam springs, consisting of pistons mounted on top of the axle bearings moving in cylinders let into the bottom of the boiler. Stephenson built several locomotives with steam springs, starting in 1816.

Though laminated steel springs had become normal practice in road carriages from about 1770, replacing suspension by leather straps, it was not until between 1825 and 1828 that Wood developed adequately strong laminated plate springs, so that Stephenson's steam springs fell out of use. Though the vertical stiffness was a matter for calculation, endplay, clearances and flexibility of the axle-guards introduced variable and unknown lateral and longitudinal flexibility between the wheelset and the vehicle body.

The inception of service on the Liverpool and Manchester Railway in 1829 meant that, for the first time, railway vehicles operated at speeds at which dynamic effects became apparent. The coaches had a very short wheelbase and were reputed to hunt violently at any speed. One measure employed to control this was to close couple the vehicles. The instability of two-axle vehicles was an accepted and often unremarked occurrence throughout their employment on the railways. In the early days of the railways, it had become customary to link together two- and three-axle vehicles not only by couplings but also by side chains to provide yaw restraint between adjacent car bodies in order to stabilise lateral motions.

The two-axle vehicle gave an arguably acceptable ride on well-aligned British lines but suffered riding problems on the curvaceous and roughly aligned American lines and lacked the ability to negotiate irregular track and to maintain contact of all the wheels with the track. As a result, many American engineers promulgated bogie designs from 1826 onwards, so that the rapid adoption of the passenger and freight bogie in American practice took place in the 1830s [15,16]. With two swivelling bogies, good conformity to curved track was obtained. As the ratio of bogie mass to car body mass was low, there could be good stability, provided that adequate yaw restraint between car body and bogie was provided. In practice, because the car body was supported on the bogie by side rollers, the bogie could swivel with little restraint, resulting in a tendency to hunt violently, often leading to derailment. In the late 1830s, a central bearing plate replaced the side rollers, and this became a general practice by the 1850s. These pragmatic measures were intended to allow the bogies to follow sharp curves at low speeds while at the same time preventing bogie hunting on straight track, so as to try and resolve the fundamental conflict between curving and stability.

John B. Jervis made the next major step in the development of the bogie in 1831 to 1832 [17,18]. The two-axle locomotive had deficiencies similar to those of the freight and passenger cars described previously and usually had an unsymmetrical configuration with driving wheels and carrying wheels of different sizes in order to accommodate the layout of boiler, cylinders and driving motion. Jervis's 4–2–0 locomotive had a pivoted leading bogie to provide guidance on curves and the flexibility in plan view to allow good curving. As in the Chapman patent, the pivot was without play and located the bogie horizontally. A proportion of the weight of the locomotive body was carried on rollers running on a track on the side beams of the bogie, which was free to rotate. Jervis's locomotive had a leaf spring primary suspension similar to that used on the rigid two-axle locomotive.

These early bogies had very short wheelbases in order to assist curving. In the 1850s, the bogie wheelbase was increased, which improved stability significantly.

In the 1850s, Bissel proposed a 4–2–0 configuration of locomotive, subsequently patented [19], which removed most of the geometric error inherent in Chapman's and Jervis's designs. Attributing many derailments to this, in his design, the bogie frame was pivoted to the locomotive frame at a point centrally located between the driver and the midpoint of the bogie wheelbase, thus permitting both the locomotive and its bogie to take up a more closely radial attitude in curves. This idea was adopted quite widely when applied to two-wheel trucks in the 2–2–0 configuration. In both cases, a centring device was incorporated. This device provided lateral stiffness between the centre of the bogie and the locomotive frame, that Bissel surmised, quite correctly, would tend to stabilise the locomotive and prevent hunting of the bogie. The centring device was originally formed of double inclined planes but was replaced by the swing bolster, patented by A. F. Smith in 1862 [20], though originally patented in 1841 by Davenport and Bridges for bogies used in passenger cars [21]. It was appreciated that a function of the secondary suspension, connecting the bogie frame to the car body, was to isolate the car body from motions of the bogie.

The next significant advance was the 4–4–0 locomotive, which incorporated the suspension features discussed previously with the addition of equalisation bars on the driving wheels, which provided an effective three-point suspension. Patents by Eastwick [22] in 1837 and Harrison [23] in 1838 initiated the general application of equalisation, which resulted in the invention of many schemes. Equalisation became a common feature of the vertical suspension on both locomotives and carriages in the United States.

In Britain, the 'rigid' 2–2–0 and 2–4–0 configurations persisted throughout the 1840s and 1850s.

The frequency of bogie hunting in America resulted in British engineers distrusting the bogie. For example, Fernihough pointed out the danger of bogie oscillation in his evidence to the Gauge Commission in 1845 [24]. This was one reason why British 4–4–0s with leading bogies did not emerge until between 1855 and 1860, though even then most had the kinematically unsound fixed bogie pivot. The first British 4–4–0s with side play at the pivot were the Adams tank engines of 1863. At first, there was no spring control of the pivot, but later, rubber pads were fitted.

From the earliest days, the bad riding of locomotives was often ascribed, correctly, to the driving motion forces, and this often influenced the choice of configuration. When Stephenson's 'Long Boiler' locomotive was imported into France in 1846, Le Chatelier investigated its bad riding [25]. Le Chatelier, trained at the Ecole Polytechnique, carried out a mathematical analysis that resulted in a comprehensive understanding of the effect of the driving motion forces. His investigations resulted in his book [26] of 1849, in which he derived formulae for the driving motion forces from the geometry of the piston, connecting rod and crank system as a function of the angle of rotation of the driving wheel, considering the forces generated both by steam and by inertia of the reciprocating and rotating parts. This work became accessible to English-speaking engineers in the book by Clark [27, pp. 169–171], which is less formally mathematical.

The impact caused by the lack of balance of the revolving and reciprocating parts of the locomotive could cause severe track damage. Initially, only the lack of balance of the revolving parts was corrected by balance weights attached to the rims of the driving wheels, and Fernihough [27, p. 173] appears to be the first to use weights heavy enough to approach complete balance.

Le Chatelier (and Clark following him) recognised that the lateral motion of locomotives was not solely due to the action of the driving forces. They argued that, at low speeds, any yawing motion would be resisted by the lateral frictional resistance between the wheel and rail, though at higher speeds, the greater disturbing forces would overcome the frictional forces and the motion would become more violent [26, pp. 58–73, and 27, pp. 180–183]. This was an erroneous concept, as the hunting motion actually consisted of a sort of steering oscillation, related to the kinematic oscillation, in which lateral displacement of the wheels was about 90 degrees out of phase with the yaw and in which the frictional forces were quite low. Though erroneous, their ideas represent the first attempt at understanding the hunting oscillation scientifically.

A further aspect of interaction between vehicle and track was revealed by the collapse of Stephenson's bridge across the River Dee at Chester in 1847. At this time, little was known about the dynamic effects of moving loads on bridges. In order to support the enquiry into the accident, Willis carried out a series of experiments on a dynamic test rig at Portsmouth Dockyard. This was followed by further model tests at Cambridge and, in 1849, G.G. Stokes gave the first analysis of the travelling load problem, albeit with severe simplifying assumptions. This was the beginning of a long history of such investigations [28].

Locomotive axle configuration adapted to the growth in power and size demanded by larger and faster trains, but the principles, established entirely by empirical means, remained until the end of steam. Unlike America, in Britain, the short-wheelbase two-axle coach was standard until the 1860s.

After the 860s, three-axle vehicles increasingly replaced two-axle vehicles, often with lateral play on the central axle. The first bogie coaches in Britain were for various railways in 1873 to 1875, and, by the end of the nineteenth century, the bogie configuration of passenger vehicle had become common, though two- and three-axle passenger and freight vehicles lingered on in Europe. As long as engineers did not stray into unconventional arrangements, or incorporated features that caused significant parameter changes, the conventional configuration proved capable of the enormous development seen in the twentieth century.

2.5 INNOVATIONS FOR IMPROVED STEERING

The development of the standard locomotive and bogie vehicle configurations described previously involved the use of two or more wheelsets mounted stiffly in frames, such as a bogie frame or a locomotive mainframe, so that the ability of the wheelsets to take up a radial position in a curve was limited. There is a long history of inventions, and hundreds of patents attempted to ensure that wheelsets are steered, so that they adopt a more or less radial position on curves. A comprehensive account cannot be given here, but Roman Liechty, a well-known authority on the design of vehicles for operation on the curvaceous lines of his native Switzerland, reviewed the historical development of wheelset steering in his book in 1934 [29] and, moreover, gave a valuable insight into the design methods, such as that of Heumann mentioned previously, that were at the disposal of the railway engineer in the 1930s.

Probably, the first vehicle with steered wheelsets was on the Linz-Budweis horse railway in 1827, in which the wheelsets were directly connected by cross-bracing, using rigid joints, as shown in Figure 2.4a [29, p. 25]. This configuration was also used on a number of railways in Germany from 1880 to 1890 but was not successful.

Horatio Allen designed the first locomotive with an articulated mainframe in 1832. Though this had a short career, it probably stimulated several of the articulated designs for the Semmering Contest in 1851. As locomotives became more powerful and hence longer, the problem of steering

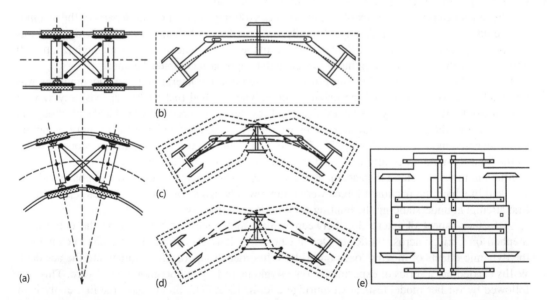

FIGURE 2.4 Innovations for improved steering: (a) direct connections between wheels by cross-bracing, (b) three-axle vehicle, (c) articulation with a steering beam, (d) articulation with linkage steering driven by angle between adjacent car bodies and (e) bogie with steered wheelsets driven by angle between bogie frame and car body. (From Iwnicki, S. (Ed.), *Handbook of Railway Vehicle Dynamics*, CRC Press, Boca Raton, FL, 2006. With permission.)

on the sharp curves typical of railways in mountainous regions became acute. This stimulated the development of various kinds of articulated locomotives, as has been described by Wiener [30]. Many of these designs were highly ingenious such as the Klose locomotive [31], in which the problem of altering the length of the driving motion on each side in order to accommodate radial axles was solved.

According to Liechty [26, p. 29], a three-axle vehicle in which the lateral displacement of the central axle steered the outer axles through a linkage was also tried out on the Linz-Budweis railway in 1826. It was argued that three axles, connected by suitable linkages, would assume a radial position on curves and then re-align themselves correctly on straight track. Other examples of inventions in which wheelsets are connected to achieve radial steering are the three-axle vehicles of Germain (1837), Themor (1844) and Fidler (1868), Figure 2.4b. In these schemes, the outer wheelsets were pivoted to the car body. More refined arrangements due to Robinson (1889) and Faye (1898) were much used in trams on account of the very sharp curves involved in street railways [32].

As an alternative to the use of the bogie, in 1837, W. B. Adams proposed an articulated two-axle carriage. Adams invented a form of radial axle in 1863, which had no controlling force, with the result that, on straight track, there was considerable lateral oscillation of the axle. Phipps suggested the idea of a controlling force, which was subsequently applied by Webb and others [33].

Another form of steering exploited the angle between the bogie and the car body in order to steer the wheelsets relative to the bogie frame, using a linkage, Figure 2.4c. A similar objective is achieved by mounting the outer wheelset on an arm pivoted on the car body and actuated by a steering beam, Figure 2.4d. An alternative approach was to steer the wheelsets by using the angle between adjacent car bodies, Figure 2.4e.

All these developments were based on very simple ideas about the mechanics of vehicles in curves and depended on systems of rigid linkages and pivots. Though the aim of designers was usually radial steering, very often, the design involved a compromise in which only partial steering was achieved. Not surprisingly, in the light of modern knowledge, there is considerable evidence that, when such schemes were built, they exhibited an even wider spectrum of various hunting instabilities than the more conventional mainstream designs. This is probably why many of these inventions failed to achieve widespread adoption. On the other hand, following the publication of his book in 1934, Liechty's designs were applied in the next few years on a number of railways, and the experience gained with these over a period of 40 years was reviewed by Schwanck in 1974 [34], who concluded that wheelset steering is 'one of the most effective means against flange and track wear, and that the expenses resulting from its application will be written off by the resultant savings'.

2.6 CARTER

As already mentioned previously, Le Chatelier made the first attempt to understand the hunting oscillation in the late 1840s, but the first recognition that hunting was the result of an oscillation with growing amplitude seems to have been made by Boedecker in his book of 1887 [35]. Boedecker extended his curving analysis to consider the case of a two-axle vehicle with rigid primary suspension and coned wheels. The wheel-rail forces were represented by Coulomb's law. Using a numerical solution, Boedecker showed that 'each displacement, however small, of the middle of the axle from the centre of the track results in a wavy course of the vehicle of which the amplitude increases with each swing of the axle until limited by the flangeway clearance'.

The model chosen for the wheel-rail forces limited this approach. Carter made the decisive step forward in 1916.

By the end of the nineteenth century, experience had shown that the standard locomotive configuration with a leading guiding bogie was usually safe. Bogie hunting, if occurred on a locomotive or bogie vehicle, usually involved the relatively small mass of the bogie, so that the forces were not dangerous. On the other hand, symmetric configurations, such as the 0–6–0, were used only at low speeds, as at higher speeds, they were subject to riding problems, lateral oscillation and sometimes derailment, as instability involved the relatively large forces generated by the mass of the main frame. Experience had therefore shown that symmetric configurations were best avoided.

This seems to have been forgotten when the first electric locomotives were designed towards the end of the nineteenth century, perhaps because they evolved from trams rather than from steam locomotives, and the operational advantages of a symmetric configuration looked attractive. As a result, the introduction of the symmetric electric locomotive had been accompanied by many occurrences of lateral instability at high speed and, consequently, large lateral forces between the vehicle and track. This was how Carter became involved in the problem.

Up to then, railway engineering and theory followed separate paths. The achievements of railway engineers, in the field of running gear at least, largely rested on empirical development and acute mechanical insight. Mackenzie's work in understanding the forces acting on a vehicle in a curve represents an excellent example. It is perhaps not surprising that the seminal development in railway vehicle dynamics was made not by a mechanical engineer but by an electrical engineer who had been exposed to the new analytical techniques necessary to further the application of electrification.

Carter (1870–1952) read mathematics at Cambridge. After a 4-year spell as a lecturer, he decided to make electrical engineering his career and spent the following 3 years with General Electric at Schenectady, where he was employed in the testing department, working on electric traction. He then returned to England and spent the rest of his career with British Thompson Houston (a company affiliated with General Electric) at Rugby. For most of his career, he was consulting engineer to the company, dealing with problems that were beyond the ordinary engineering mathematics of the day. With his mathematical ability and working at the leading edge of railway electric traction, he was able to bridge the gap between science, theory and railway engineering [36]. After making many significant contributions to electric traction, Carter turned to the mechanical engineering problems of locomotives. The first realistic model of the lateral dynamics of a railway vehicle was that presented by Carter in 1916 [37]. In this model, Carter introduced the fundamental concept of creep and included the effect of conicity. This paper showed that the combined effects of creep and conicity could lead to a dynamic instability.

Carter states that the forces acting between wheels and rails can be assumed to be proportional to the creepages, without reference or derivation in this 1916 paper. Osborne Reynolds had first described the concept of creep in 1874 in relation to the transmission of power by belts or straps, and he pointed out that the concept was equally applicable to rolling wheels [38]. It was Carter's introduction of the creep mechanism into the theory of lateral dynamics that was the crucial step in identifying the cause of 'hunting'.

Carter derived equations of motion for the rigid bogie in which two wheelsets were connected by means of a stiff frame. They consist of the two coupled second-order linear differential equations in the variables lateral displacement y and yaw angle ψ of the bogie, and they are equivalent to

$$m\ddot{y} + 4f(\dot{y}/V - \psi) = Y$$

$$4f\lambda ly/r_0 + I\ddot{\psi} + 4f(l^2 + h^2)\dot{\psi}/V = G \qquad (2.2)$$

where m and I are the mass and yaw moment of inertia of the bogie, f is the creep coefficient, h is the semi-wheelbase of the bogie and V is the forward speed. It can be seen that lateral displacements of the wheelset generate longitudinal creep. The corresponding creep forces are equivalent to a couple, which is proportional to the difference in rolling radii or conicity and which tends to steer the wheelset back into the centre of the track. This is the basic guidance mechanism of the wheelset. In addition, when the wheelset is yawed, a lateral creep force is generated. In effect, this coupling between the lateral displacement and yaw of the wheelset represents a form of feedback, and the achievement of guidance brings with it the possibility of instability. Klingel's solution for pure rolling follows from these equations as a special case in which the wheelset is unrestrained and rolling at low speed.

The theory of dynamic stability had been developed during the nineteenth century by scientists and mathematicians [39]. The behaviour of governors was analysed by Airy in 1840 in connection with design of a telescope. Maxwell analysed the stability of Saturn's rings in 1856 and also derived conditions of stability for governors in 1868. The most significant step forward was Routh's essay for the 1877 Adams Prize [40] that derived comprehensive conditions for stability of a system in steady motion. Routh incorporated discussion of the stability conditions into the various editions of his textbook [41]. However, in England, mechanical engineers were not familiar with these developments. On the other hand, in 1894, in Switzerland, Stodola studied the stability of steam turbine control systems and encouraged Hurwitz to formulate conditions for stability [42] that are equivalent to Routh's criteria. Carter's work was one of the first engineering applications of Routh's work, and it is interesting to note that the pioneering work of Bryan and Williams in 1903 on the stability of aeroplanes using similar methods [43] had only been published a few years before. Significantly, another early application of stability theory was made by another electrical engineer, Bertram Hopkinson, in his analysis of the hunting of alternating machinery, published in 1903 [44]. All these early publications refer to Routh's textbook. In contrast, the stability of the bicycle, another system involving rolling wheels, was analysed by Whipple in 1899 [45], using a solution of the equations of motion.

Moreover, it is interesting that, concurrently with Carter, aeronautical engineers were grappling with the dynamic instabilities of aircraft structures. The first flutter analysis was made in 1916 by Bairstow and Page [46], and the imperatives of aeronautical progress ensured the development of many of the techniques that were to be brought to bear on the 'hunting' problem in the future. By 1927, Frazer and Duncan had laid firm foundations for flutter analysis in [47], and a foundation for the application of matrices in engineering dynamics was laid in the text [48], which eventually was to find application in the railway field.

As Carter's interest was in stability, he considered that the flangeway clearance was not taken up, and he therefore applied Routh's stability theory, not only to electric bogie locomotives but also to a variety of steam locomotives. In his mathematical models, a bogie consists of two wheelsets rigidly mounted in a frame, and locomotives comprise wheelsets rigidly mounted in one or more frames. Following Carter's first paper of 1916, the theory was elaborated in a chapter of his book [49].

So far, Carter had used an approximation to give the value of the creep coefficient (the constant of proportionality between the creep force and the creepage). In 1926, Carter analysed the creep of a locomotive driving wheel by extending the Hertz theory of elastic contact, as presented by Love [50]. He considered the case of creep in the longitudinal direction, treating the wheel as a two-dimensional cylinder [51]. This not only provided an expression for the creep coefficient but also described how the creep force saturated with increasing creepage. The starting point for this was the work of Hertz, the German physicist, who had become interested in 1881 in the theory of compression of elastic bodies as a result of his work on optics. By making some realistic assumptions, he was able to give a theoretical solution for the size of the contact area and the stresses in the two contacting bodies as a function of the normal load between the bodies. This work attracted the attention of not only physicists but also engineers who persuaded Hertz to prepare another version of his paper, including experimental results [52].

Carter's next paper [53] gave a comprehensive analysis of stability within the assumptions mentioned previously. As he was concerned with locomotives, the emphasis of his analyses was on the lack of fore-and-aft symmetry characteristic of the configurations that he was dealing with, and he derived both specific results and design criteria.

His analysis of the 0–6–0 locomotive found that such locomotives were unstable at all speeds if completely symmetric, and he comments that this class of locomotive is 'much used in working freight trains; but is not employed for high speed running on account of the proclivities indicated in the previous discussion'.

Carter analysed the 4–6–0 locomotive both in forward and reverse motions and found that, in forward motion, 'beyond the limits shown (for sufficiently high speed or sufficiently stiff bogie centring spring stiffness) the bogie tends to lash the rails; but being comparatively light and connected with the main mass of the locomotive, the impacts are unlikely to be a source of danger at ordinary speeds'. Two of Carter's stability diagrams, the first of their kind in the railway field, are shown in Figure 2.5. As the system considered has four degrees of freedom (lateral translation and yaw of mainframe and bogies), substitution of a trial exponential solution and expansion of the resulting characteristic equation lead to an eighth-order polynomial. As Carter writes 'expansion of the determinant is long…but not difficult'. Carter examines stability in two ways: firstly by extracting the roots of the polynomial and secondly by Routh's scheme of cross-multiplication [41]. Either method involved tedious and lengthy calculation by hand, and tackling more complex cases 'becomes, more appropriately, an office undertaking'.

In reverse motion, see Figure 2.5b, he found that, beyond a certain value of the centring spring stiffness, 'buckling of the wheelbase tending to cause derailment at a fore-wheel and moreover that the impacts of the flanges on the rail when the locomotive is running at speed are backed by the mass of the main frame and are accordingly liable to constitute a source of danger'. This was Carter's explanation of a number of derailments at speed of tank engines such as the derailment of the Lincoln to Tamworth mail train at Swinderby on 6 June 1928, as discussed in his final paper on the subject [54].

Carter's analysis of the 2–8–0 with a leading Bissel similarly explained the need for a very strong aligning couple for stability at high speed, whilst noting that, in reverse motion, a trailing Bissel has a stabilising effect for a large and useful range of values of aligning couple.

Though Carter had exploited the stabilising influence of elastic elements in his analyses of unsymmetrical locomotive configurations, his brief treatment of symmetric vehicles with two-axle bogies (by now, a common configuration of passenger rolling stock) assumed that the bogies were pivoted to the car body. This case received such brief mention, which is consistent with the fact that railway engineers had, by empirical development, achieved an acceptable standard of ride at the speeds then current. Moreover, as Carter says, 'the destructive effect of the instability is, however, limited on account of the comparatively small mass of the trucks'.

Carter's work expressed in scientific terms what railway engineers had learnt by hard experience that stability at speed required rigid-framed locomotives to be unsymmetrical and unidirectional. A further practical result of his work was a series of design measures, the subject of various patents [55], for the stabilisation of symmetric electric bogie locomotives.

The fact that the analyses, though only involving a few degrees of freedom, required heavy algebra and arithmetic and involved techniques beyond the mechanical engineering training of the day is perhaps one reason why Carter's work was not taken up much sooner.

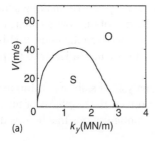
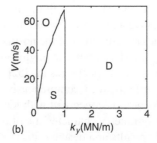

FIGURE 2.5 Carter's stability diagram for the 4–6–0 locomotive in (a) forward motion and (b) reverse motion. k_y is the centring stiffness (recalculated in modern units from [53]). S = stable; O = oscillatory instability; D = divergence. (From Iwnicki, S. (Ed.), *Handbook of Railway Vehicle Dynamics*, CRC Press, Boca Raton, FL, 2006. With permission.)

Thus, while the theoretical foundations had been established, the need for vehicle dynamics was not, and practicing railway engineers were largely sceptical of theory, particularly when the experimental basis was very limited. As a result, the next 20 years saw only a few significant contributions to the science of railway vehicle dynamics.

Rocard [56] in 1935 employed the same form of equations of motion as Carter. In addition to covering much of the ground as Carter, he considered the case of a massless bogie that is connected by a lateral spring to the car body and showed that the system could be stabilised. Rocard also considers the case of the unsymmetric bogie in which the wheelsets have different conicities. He found that the distribution of conicity could be arranged to give stability in one direction of motion but not in both. Rocard [57] states that French National Railways made a successful experiment in 1936.

There were also theoretical contributions by Langer [58] in 1935 and by Cain [59] in 1940 that involved rather severe assumptions, but, in general, papers concerned with bogie design published during this period were purely descriptive, reflecting the negligible role played by analysis in this branch of engineering practice.

However, in 1939, Davies carried out significant model and full-scale experiments with both single axles and two-axle bogies and demonstrated various forms of the hunting instability. This was the beginning of roller rig testing for vehicle dynamics [60]. Notably, he presented equations of motion, including, for the first time, longitudinal and lateral suspension flexibility, later established to be one of the essential features of a realistic model of railway vehicle dynamics. Though Davies felt that the laborious arithmetic involved in solutions of the equations was not justified in view of the assumptions, a particularly valuable contribution was the detailed experimentation and discussion of the importance of worn wheel and rail profiles to vehicle dynamics.

2.7 WHEEL-RAIL GEOMETRY

Carter assumed that the wheel treads were purely conical. In practice, it had been known from the earliest days of the railways that treads wear and assume a hollow form, and rails also wear. It was also known that there was a connection between ride quality and the amount of wheel wear. For example, in 1855, Clark discussed several examples of wheel profiles and how they varied between the various wheels of a locomotive, depending on the history of the distribution of applied forces [27, pp. 181–183]. Conicity varied not only during the running of the locomotive between maintenance but also between examples of the same nominal design, and it was, of course, a function of the particular piece of track on which the locomotive was running. Clark also mentions the high conicity induced by the flat top to the rails caused by the rolling process.

It follows that an important further step in developing a realistic mathematical model was concerned with the treatment of actual wheel and rail profiles. Whilst new wheel profiles were purely coned on the tread, usually to an angle of 1:20, in 1934, Heumann [61] emphasised the important influence of worn wheel and rail profiles on wheelset behaviour in curves. Heumann analysed the effect of the mutual wheel and rail geometry on the variation of the rolling radius as the wheelset is displaced laterally and derived the formula for the effective conicity λ_0 of a wheel-rail combination for small displacements from the central running position, defined as the rate of change of the rolling radius with lateral displacement of the wheelset

$$\lambda_0 = \delta_0 R_w / (R_w - R_r)(1 - r_0 \delta_0 / l) \tag{2.3}$$

where R_w and R_r are the wheel and rail radius of curvature, $2l$ is the distance between the contact points, δ_0 is the slope of the tread at the contact point and r_0 is the rolling radius of the wheelset in the central position. Heumann's expression shows clearly that the effective conicity of a worn wheelset can be much greater than that of the corresponding purely coned wheelset. Moreover, Heumann suggested for the first time that profiles approximating to the fully worn should be used rather than the purely coned treads then standard. He argued that after reprofiling to a coned tread, tyre profiles tend to wear rapidly, so that the running tread normally in contact with the railhead is worn to a uniform profile. This profile then tends to remain stable during further use and is largely independent of the original profile and of the tyre steel. Similarly, railhead profiles are developed, which also tend to remain stable after the initial period of wear is over. Heumann therefore suggested that vehicles should be designed to operate with these naturally worn profiles, as it is only with these profiles that any long-term stability of the wheel-rail geometrical parameters occurs. Moreover, a considerable reduction in the amount of wear would be possible by providing new rails and wheels with an approximation to worn profiles at the outset. Modern wheel and rail profiles are largely based on this concept.

Müller gave a detailed analysis of the wheel-rail contact geometry in 1953 [62] and also tabulated geometric data that was measured for a combination of worn wheels and rails [63]. In the early 1960s, King evaluated the contact conditions between a pair of worn wheels and worn rails [64] and between worn wheels and new rails [65] and on the basis of this work designed a new standard wheel profile, the P8, for British Railways [66]. It was shown in this work that the graph of rolling radius difference versus wheelset lateral displacement was extremely sensitive to the gauge of the track, rail inclination and small variations in profile geometry. As a result, more refined measuring and computational techniques were developed [67], which were subsequently used very widely; see for example [68,69].

The neglect of the effect of wheelset yaw on the wheel-rail geometry is a realistic assumption, except in the case of flange contact at large angles of wheelset yaw. In addition to Müller's pioneering analysis, three-dimensional geometry analyses were developed in the 1970s by Cooperrider et al. [69] and Hauschild [70]. Research on the topic continued; see for example Duffek [71], de Pater [72] and Yang [73].

For linearised analyses of stability and curving, the concept of 'equivalent conicity' was introduced [74]. For a coned wheelset, the equivalent conicity is simply the cone angle of the tread. For a wheelset with worn or profiled treads, the equivalent conicity is defined as the cone angle that for purely coned wheels would produce the same wavelength of kinematic oscillation and is approximately equal to the mean slope of the rolling radius difference versus lateral displacement graph. In other words, the equivalent conicity is a 'describing function', a method of dealing with non-linear control system components introduced by Kochenburger [75] in 1950. The circular arc theory of Heumann is accurate only for extremely small lateral displacements of the wheelset, and the 'equivalent conicity' is equal to the effective conicity for small amplitudes of wheelset lateral displacement only.

2.8 CREEP

Following Carter's analysis of creep, similar results were obtained by Poritsky [76], and in the discussion of [76], Cain [77] pointed out that the region of adhesion must lie at the leading edge of the contact area. A three-dimensional case was solved approximately by Johnson [78], who considered an elastic sphere rolling on an elastic plane. This solution was based on the assumption that the area of adhesion is circular and tangential to the area of contact (which is also circular) at the leading edge. Good agreement with experiment was obtained. The influence of spin about an axis normal to the contact area was first studied by Johnson [79], who showed that spin could generate significant lateral force due to the curvature of the strain field in the vicinity of the contact patch (the couple about the common normal is small and may be safely neglected). Haines and Ollerton [80] considered the general case, where the contact area is elliptical. They confined their attention to creep in the direction of motion and assumed that Carter's two-dimensional stress distribution holds in strips parallel to the direction of motion. Vermeulen and Johnson [81] gave a general theory for the elliptical contact area, based on similar assumptions to those made in [78]. This yielded the relationship between creepage and tangential forces for arbitrary values of the semi-axes of the contact area, but for zero spin.

In parallel with these academic studies, railway engineers sought a simple way of combining the linear creep model with creep saturation, and simple formulae were provided by Levi [82] and Chartet [83], but these were applicable only for the case of zero spin. These were consistent with measurements carried out by Müller on full-scale wheelsets [84]. Later, Shen et al. [85] proposed a heuristic extension to the results of Vermeulen and Johnson to take account of spin, which later was much used in vehicle dynamics studies.

De Pater [86] initiated the complete solution of the problem by considering the case where the contact area is circular and derived solutions for both small and large creepages, without making assumptions about the shape of the area of adhesion. However, this analysis was confined to the case where Poisson's ratio was zero; Kalker [87] gave a complete analytical treatment for the case in which Poisson's ratio was not zero. The agreement between these theoretical results and the experimental results of Johnson is very good. Kalker gave a full solution of the general three-dimensional case in [88], covering arbitrary creepage and spin for the case of dry friction and ideal elastic bodies in Hertzian contact and subsequently gave simpler approximate solution methods [89,90]. Kalker's theory is described in [91]. Recent research has refined and extended the theoretical modelling of creep, as well as provided alternative heuristic models, and this is covered in Chapters 7 and 8.

2.9 MATSUDAIRA

Tadashi Matsudaira studied marine engineering at the University of Tokyo and then joined the aircraft development department of the Japanese Imperial Navy, where he was concerned with the vibration of aeroplanes. After the end of World War II, he moved to the Railway Technical Research Institute of Japanese National Railways to work on railway vehicle dynamics. During the years 1946 to 1957, Japanese National Railways were attempting to increase the speed of freight trains. The short wheelbase two-axle wagons then in use experienced hunting at low speeds and a high rate of derailment. Matsudaira introduced his experience of the flutter problem in aeroplanes (such as the Japanese Imperial Navy's 'Zero' fighter). Then, using both analysis and scale model experiments on roller rigs, he showed that the hunting problem is one of self-excited vibration and not arising from external factors such as uneven rail geometry. In his paper [92], he departed from Carter's model by considering a single wheelset and demonstrated the stabilising effect of elastic restraint. As this paper was in Japanese, it had little impact in the West. Subsequently, Matsudaira introduced into the mathematical model of the two-axle vehicle both longitudinal and lateral suspension flexibilities between wheelset and car body, a crucial step in understanding the stability of railway vehicles, and based on this, he was able to suggest an improved suspension design.

In the 1950s, planning started for the new Tokaido line or Shinkansen, the first purpose-built dedicated high-speed railway. Shima [93] identifies the bogies as one of the key enabling technologies of the Shinkansen. Based on empirical development and simple calculations, surprisingly good results with the conventional bogie vehicle had been obtained up to then, providing that conicities were kept low by re-turning wheel treads, and speeds were moderate, for example, below 160 km/h. The trains for the Shinkansen were to be high-speed, 200 km/h, multiple units in which every bogie is powered. It had been widely assumed that the powered bogies would not run as smoothly as the trailer bogies, but by studying closely the stability of bogies theoretically and experimentally, it was possible to improve the riding quality of the powered bogies up to very high speeds. The analysis of these bogies by Matsudaira and his group led to the choice of suspension parameters, which were subsequently validated by roller rig and track tests.

2.10 THE ORE COMPETITION

In the 1950s, the newly formed Office for Research and Experiments (ORE) of the International Union of Railways (UIC) held a competition for the best analysis of the stability of a two-axle railway vehicle. The specification for the competition, drawn up by Committee C9 under the chairmanship of Robert Levi, emphasised worn wheel and rail profiles and non-linear effects, for it was still widely held, despite Carter's work, that the explanation for instability lay in some way in the non-linearities of the system [94]. The three prize-winning papers (by de Possel, Boutefoy and Matsudaira) [95], in fact, all gave linearised analyses. However, Matsudaira's paper was alone in incorporating both longitudinal and lateral suspension stiffness between wheelsets and frame. Surprisingly, it was awarded only the third prize.

Matsudaira's model has suspension stiffnesses but no suspension damping. It has six degrees of freedom, lateral displacement and yaw of the wheelsets and car body, so that roll of the car body is neglected. Circular arcs approximated worn wheel and rail profiles in order to give an approximation for the effective conicity similar to that of Heumann. Assuming sinusoidal oscillations at the critical state between stable and unstable motion, Matsudaira was able to reduce the order of the characteristic equation. This made it possible to establish the critical speed and the frequency of the oscillation at the critical speed by a graphical method. In this way, Matsudaira avoided the onerous task of calculating the actual eigenvalues for each pair of parameters. In the resulting chart, Figure 2.6, the lines representing an eigenvalue with a zero real part are plotted in the plane of speed versus lateral suspension stiffness. There are four boundaries in the chart on which purely sinusoidal oscillations are possible. Two of these relate to relatively large excursions of the car body and two relate to relatively large excursions of the wheelsets. Matsudaira proposes two approaches to achieve stability for a practical range of speeds. One is to make the lateral stiffness rather large and exploit the lower stable region; the other is to use the upper stable region with the smallest possible value of the lateral stiffness. In this latter case, Matsudaira suggests that the vehicle goes through an unstable region at a very low speed, but the ensuing hunting is not severe.

In fact, an examination of the root locus shows that each root has a positive real part above its critical speed, though the magnitude of the real positive part corresponding to the two lowest critical speeds decreases as the speed increases. The reason that this prescription works for the two-axle vehicle is that inclusion of suspension damping in the lateral direction can, under certain conditions, eliminate the instability at low speeds completely, as was found later.

Another factor that emerged for the first time in de Possel's and Boutefoy's papers, but was neglected in Matsudaira's paper, was that of the gravitational stiffness, the lateral resultant of the resolved normal forces at the contact points between wheel and rail, Figure 2.7. On its own, this effect would be strongly stabilising, but it is largely counteracted by the lateral force due to spin creep (discovered later, see later) for small lateral displacements. The resulting contact stiffness is therefore often ignored for motions within the flangeway clearance, though the correct representation of these forces in the case of flange contact, or in the case of independently rotating wheels, is, of course, vital.

Thus, by the early 1960s, the basic ingredients of an analytical model of the lateral dynamics of a railway vehicle had been identified. Of these, few values of the creep coefficients had been measured and in any case were not under the control of the designer; conicities could only be controlled within a narrow range by re-profiling, and the leading dimensions and number of wheelsets were

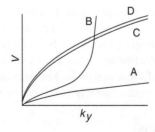

FIGURE 2.6 Stability chart in which the lines represent an eigenvalue with a zero real part. V = vehicle speed; k_y = lateral suspension stiffness; A and B correspond to mainly oscillations of the car body on the suspension and C and D correspond to wheelset oscillations. (From Iwnicki, S. (Ed.), *Handbook of Railway Vehicle Dynamics*, CRC Press, Boca Raton, FL, 2006. With permission.)

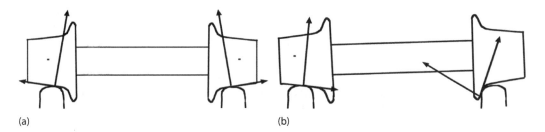

FIGURE 2.7 Normal and lateral tangential forces acting on wheelset in (a) central position and (b) laterally displaced position, illustrating the gravitational stiffness effect. (From Iwnicki, S. (Ed.), *Handbook of Railway Vehicle Dynamics*, CRC Press, Boca Raton, FL, 2006. With permission.)

largely dictated by the proposed duty of the vehicle. But Matsudaira recognised that the designer could vary both the way in which wheelsets were connected and the corresponding stiffness properties, and this pointed the way to future progress.

One of the members of Levi's ORE committee was A. D. de Pater, who considered the hunting problem and formulated it as a non-linear problem [96,97]. Even though severe assumptions were made, interesting theoretical results emerged. In 1964, one of de Pater's students, P. van Bommel, published non-linear calculations [98] for a two-axle vehicle using worn wheel and rail profiles and the creep force-creepage laws proposed by Levi and Chartet. However, the practical relevance of the results was limited because lateral and longitudinal suspension flexibility was not considered.

2.11 THE COMPLETE SOLUTION OF THE HUNTING PROBLEM

In the early 1960s, British Railways, like Japanese National Railways, faced an increasing incidence of derailments of short wheelbase two-axle wagons as freight train speeds increased. After some false starts [9], and the failure to solve the problem by empirical means, a team was formed at Derby to undertake research into railway vehicle dynamics. Based on Carter's work and quickly understanding the significance of Matsudaira's 1960 paper, it was possible to extend the analysis of stability by introducing a new feature, lateral suspension damping, and by reintroducing the 'gravitational stiffness' effect, which de Possel and Boutefoy had already used in [95]. The model used had seven degrees of freedom, lateral displacement and yaw of the wheelsets and car body and roll of the car body. As a result, it was shown that, with a careful choice of lateral suspension damping and the lateral and longitudinal stiffnesses, it was possible to eliminate the low-speed body instability (a strongly contributory factor in wagon derailments) so that the vehicle operating speed was only limited by the wheelset instability. In this work, the application of analytical and both analogue and digital computer techniques marched hand in hand with experimental work on models.

Comprehensive details of the behaviour of a simple elastically restrained wheelset were derived. As the equations of motion are not symmetric and the system is non-conservative, the wheelset is able to convert energy from the forward motion to the energy of the lateral motion [99]. Moreover, the representation and analysis of the wheelset were introduced as a feedback system.

On the full-scale experimental side, measurements of critical speeds and mode shapes, associated with the hunting limit-cycle of a range of vehicles, were made by King [100] and by Pooley [101]. The most striking validation of the theory came from a series of full-scale experiments with two kinds of standard two-axle vehicles, Gilchrist et al. [74]. As the linear critical speeds of these vehicles were low, it was possible to measure the fully developed hunting limit cycle. Quite apart from the highly non-linear suspension characteristics, which were realistically modelled, two major limitations of linear theory were faced. These were creep saturation and wheel-rail geometry. Creep saturation was modelled by a two-part characteristic, Kalker's linear values being taken for the linear part and the creep forces being limited by sliding friction. The wheel-rail geometry was modelled by the use of 'equivalent conicity'. An example of the comparative results by Gilchrist et al. is shown in Figure 2.8, and it demonstrates that both the onset of instability and the fully developed hunting limit cycle were satisfactorily modelled.

So far, in the studies of stability, only the creepage due to longitudinal and lateral relative motion between wheel and rail had been considered, and the relative angular motion about the normal to the contact plane, the spin, had been neglected. When the effect of 'spin creep' was included in the equations [102], the stabilising influence of gravitational stiffness was found to be much reduced and had to be counteracted by an increase in yaw stiffness.

Experimental verification of the stability boundaries predicted by the above prescription was obtained from full-scale roller rig and track testing of the specially constructed vehicle HSFV-1. Some results are shown in Figure 2.9 [103].

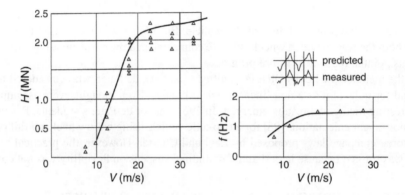

FIGURE 2.8 Measured and predicted lateral forces H and frequency f during hunting of a two-axle vehicle as a function of forward speed V (from [74] in modern units). Inset shows waveforms of H. Predicted results indicated by full lines. (From Iwnicki, S. (Ed.), *Handbook of Railway Vehicle Dynamics*, CRC Press, Boca Raton, FL, 2006. With permission.)

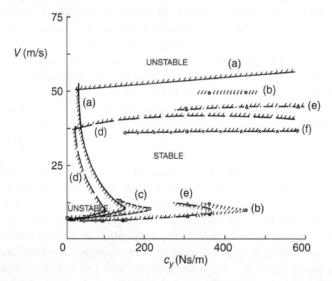

FIGURE 2.9 Comparison between predicted stability boundaries and roller rig, and track test measurements as lateral suspension damping c_y is varied (from [103] in modern units): (a) predicted (equivalent conicity $\lambda = 0.27$, yaw stiffness $k_\psi = 5.88$ MNm), (b) roller rig ($k_\psi = 5.88$ MNm), (c) track tests ($k_\psi = 5.88$ MNm), (d) predicted ($\lambda = 0.27$, $k_\psi = 3.06$ MNm), (e) roller rig ($k_\psi = 5.20$ MNm) and (f) roller rig ($k_\psi = 3.06$ MNm). (From Iwnicki, S. (Ed.), *Handbook of Railway Vehicle Dynamics*, CRC Press, Boca Raton, FL, 2006. With permission.)

A History of Railway Vehicle Dynamics

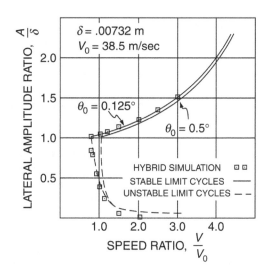

FIGURE 2.10 Limit cycle or bifurcation diagram. δ = nominal flangeway clearance; A = lateral wheelset amplitude of oscillation; V_0 = non-linear critical speed; q_0 = breakaway yaw angle in yaw spring in series with dry friction. (From Cooperrider, N.K. et al., The application of quasilinearisation techniques to the prediction of nonlinear railway vehicle response, In Pacejka, H.B. (Ed.), *The Dynamics of Vehicles on Roads and on Railway Tracks*, Proceedings IUTAM Symposium held at Delft University of Technology, 18–22 August 1975, Swets & Zeitlinger, Amsterdam, the Netherlands, pp. 314–325, 1976; With kind permission from CRC Press: *Handbook of Railway Vehicle Dynamics*, CRC Press, Boca Raton, FL, 2006, Iwnicki, S. [ed.].)

Apart from Van Bommel's work referred to previously, various approaches to the analysis of non-linear hunting motions have been developed. Cooperrider, Hedrick, Law and Malstrom [104] introduced the more formal method of 'quasi-linearisation', in which the non-linear functions are replaced by the linear functions so chosen to minimise the mean-square error between the non-linear and the quasi-linear response. They also introduced the limit-cycle or bifurcation diagram, an example of which is shown in Figure 2.10.

This procedure was extended by Gasch, Moelle and Knothe [105,106], who approximated the limit cycle by a Fourier series and used a Galerkin method to solve the equations. This made it possible to establish much detail about the limit cycle. It is known that apparently simple dynamical systems with strong non-linearities can respond to a disturbance in very complex ways. In fact, for certain ranges of parameters, no periodic solution may exist. Moreover, systems with large non-linearities may respond to a disturbance in an apparently random way. In this case, the response is deterministic but very sensitive to the initial conditions. Such chaotic motions have been studied for railway vehicles by True et al. [107,108]. Many studies specific to the UIC double-link suspension, widely used in European two-axle freight vehicles, have been carried out, for example [109]. Piotrowsky [110] pointed out the effect of dither generated by rolling contact in smoothing dry friction damping.

2.12 LATER RESEARCH ON CURVING

In 1966, Boocock [111] and Newland [112] independently considered the curving of a vehicle by using the same equations of motion used in stability analyses but with terms on the right-hand side representing the input due to curvature and cant deficiency. As the wheelsets are constrained by the longitudinal and lateral springs connecting them to the rest of the vehicle, the wheelsets are not able to take up the radial attitude of perfect steering envisaged by Redtenbacher. Instead, a wheelset will balance a yaw couple applied to it by the suspension by moving further in a radial direction so as to generate equal and opposite longitudinal creep forces, and it will balance a lateral force by yawing further.

For the complete vehicle, the attitude of the vehicle in the curve and the set of forces acting on it are obtained by solving the equations of equilibrium. Newland's model made useful simplifications, but Boocock analysed several configurations, including a complete bogie vehicle, a two-axle vehicle and vehicles with cross-braced bogies. The bogie vehicle had 14 degrees of freedom representing lateral displacement and yaw of the wheelsets, bogie frames and car body. He also included the effects of gravitational stiffness and spin creep. Most important of all, Boocock was able to able to obtain experimental full-scale confirmation of his theory using the two-axle research vehicle HSFV-1, Figure 2.11.

These linear theories are valid only for large radius curves. On most curves, the curving of conventional vehicles involves the same non-linearities due to creep saturation and wheel-rail geometry that were noted in the case of hunting. Elkins and Gostling gave the first comprehensive non-linear treatment of practical vehicles in curves in 1978 [113]. Their treatment covers the movement of the contact patch across the wheel tread through the flange root and on to the flange and its subsequent change in shape, assuming a single point of contact, appropriate for worn or profiled wheels. At this stage, the complication of two-point contact was to be the subject of much future research. As the contact moves across wheel and rail, account is taken of the increasing inclination of the normal force and the lateral creep force generated by spin. They used Kalker's results for the tangential creep forces for arbitrary values of creepage and spin and for a wide range of contact ellipticities, Kalker having issued his results numerically in a table book. Elkins and Gostling installed this table in their computer program so that values could be read by interpolation, as needed. The resulting equations were solved by iterative numerical procedures, of which two alternatives were given. Elkins and Gostling's program required input in numerical form of the wheel and rail cross-sectional profiles, and much research was carried out on the measurement and analysis of profiles.

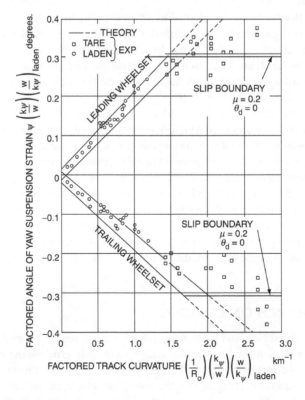

FIGURE 2.11 Comparison between theory and experiment for the steering behaviour of a two-axle vehicle. (HSFV-1, [111], results are factored to account for two loading conditions.) (From Iwnicki, S. (Ed.), *Handbook of Railway Vehicle Dynamics*, CRC Press, Boca Raton, FL, 2006. With permission.)

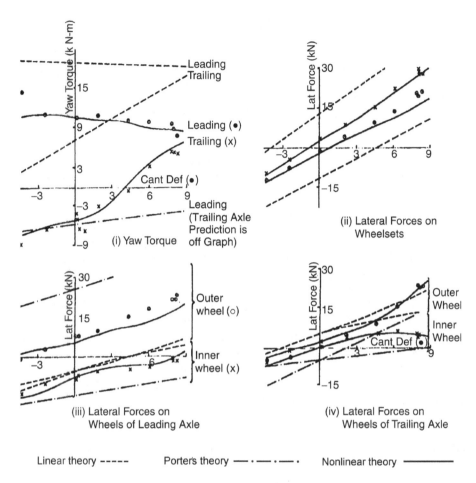

FIGURE 2.12 Curving test results for APT-E for curve radius $R_0 = 650$ m and cant $\phi_0 = 150$ mm. (From Elkins, J.A. and Gostling, R.J., A general quasi-static curving theory for railway vehicles, In Slibar, A. and Springer, H. (Eds.), *Proceedings 5th VSD-2nd IUTAM Symposium held at Technical University of Vienna*, Vienna, Austria, 19–23 September 1977, Swets and Zeitlinger, Lisse, the Netherlands, pp. 388–406, 1978; With kind permission from CRC Press: *Handbook of Railway Vehicle Dynamics*, CRC Press, Boca Raton, FL, 2006, Iwnicki, S. [ed.].)

The accuracy of the predictions of Elkins and Gostling was demonstrated by experiments carried out on HSFV-1 and the tilting train research vehicle APT-E, an example of their results being given in Figure 2.12. Both vehicles were heavily instrumented, including load-measuring wheels [114], allowing individual wheel forces to be measured and the limiting value of friction to be identified. This comprehensive theory encompassed the two extremes envisaged in the earlier theories. Large radius curves, high conicity and a high coefficient of friction produced agreement with the linear theory, and small radius curves and flange contacts produced agreement with Porter's results.

Many wheel-rail combinations experience contact at two points on one wheel for certain values of the lateral wheelset displacement. This commonly occurs, for example, when contact is made between the throat of the flange and the gauge corner of the rail. If the wheels and rails are considered to be rigid, as was done in the case of single-point contact, discontinuities occur in the geometric characteristics such as the rolling radius difference and slope difference graphs. The mathematical aspects of two-point contact in this case have been considered by Yang [73]. However, in this case, the distribution of forces between the points of contact depends on the elasticity in the contact areas, and the formulation of the equations of motion becomes more complicated [115,116].

2.13 DYNAMIC RESPONSE TO TRACK GEOMETRY

Before vehicle dynamics became established, it was engineering practice to carry out a simple static analysis and tests to measure the amount of wheel unloading on track with a defined degree of twist. In 1964, Gilchrist et al. [74] computed the dynamic response of two-axle vehicles to a dipped rail joint and compared the results with experiment. Jenkins et al. [117] analysed the vertical response of a vehicle to a dipped rail joint in 1974 and showed that the response involved two distinct peaks. The first fast transient involves the rail mass and the contact stiffness, and the second slower transient involves the unsprung mass and the track stiffness. Subsequently, quite complex models of track and vehicle have been used to establish transient stresses resulting from geometric defects in both track and wheels.

For the lateral motions of railway vehicles, the excitation terms in the equations of motion (that had been derived for stability analysis) were first derived by Hobbs in 1964 [118]. These were validated by Illingworth, using a model roller rig, in 1973 [119]. A comprehensive approach to the dynamic response to large discrete inputs, including both suspension and wheel-rail contact non-linearities, was carried out by Clark, Eickhoff and Hunt [120] in 1980 and includes full-scale experimental validation. Similar calculations have been carried out on the response of vehicles to switch and crossing work.

In the case where it can be considered that irregularities are distributed continuously along the track, the approach offered by stochastic process theory is appropriate, and this was first applied by Hobbs in 1964 to the lateral motions of a restrained wheelset. The response of complete vehicles became established as an indication of ride quality, and for passenger comfort assessments, existing international standards, which define frequency-weighting characteristics, were extended to lower frequencies to cover the railway case. Gilchrist carried out the first measurements of power spectral density, using a specially developed trolley-based measurement system, in 1967 [121]. Subsequently, extensive measurements of the power spectra of irregularities of track have been made on railways in many countries, resulting in inputs used in design.

With the advent of research into innovative transport systems involving air cushions and magnetic levitation in the late 1960s, the problem of interaction between a flexible track and the vehicle received renewed attention. A review paper by Kortüm and Wormley [122] indicates the progress achieved by 1981 in developing appropriate computer models.

The development of the heavy-haul railway with extremely long trains, often with locomotives attached at various points along the train, introduced serious problems arising from the longitudinal response of trains to hills and to braking. The advent of the digital computer made it possible to develop dynamical models and also to support the development of train-driving simulators.

2.14 SUSPENSION DESIGN CONCEPTS AND OPTIMISATION

When linear theories of the curving of railway vehicles became available, it became possible for the first time to consider the best compromise between the requirements of stability and curving on a numerate basis, enabling the derivation of configurations and sets of suspension parameters. The parameters associated with wheel-rail contact, both geometrical and frictional, are not under the control of the designer or operator, but they can vary over a wide range. It follows that practical designs must be very robust in relation to such parameters. On the other hand, there is enormous scope for vehicle design in terms of the way in which the wheelsets, frames and car bodies in a train are connected, and it became possible to optimise, and even depart from, the conventional configurations of bogie and two-axle vehicle.

2.14.1 Two-Axle Vehicles and Bogies

Considering a two-axle vehicle in which the wheelsets are elastically mounted in a frame or car body, Boocock [111] defined the bending and shear stiffnesses that characterise the elastic properties in plan view. It was shown in [123] that the bending stiffness should be zero to achieve

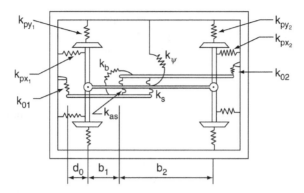

FIGURE 2.13 A generic bogie configuration. (From Kar, A.K. and Wormley, D.N., Generic properties and performance characteristics of passenger rail vehicles, In Wickens, A.H. (Ed.), *Proceedings 7th IAVSD Symposium held at Cambridge University*, 7–11 September 1981, Swets and Zeitlinger, Lisse, the Netherlands, pp. 329–341, 1982; With kind permission from CRC Press: *Handbook of Railway Vehicle Dynamics*, CRC Press, Boca Raton, FL, 2006, Iwnicki, S. [ed.].)

radial steering but would result in dynamic instability at low speeds. The design of a two-axle vehicle with a purely elastic suspension therefore requires a compromise between stability and curving. It was also shown by Boocock that for conventional vehicles in which there are primary longitudinal and lateral springs connecting the wheelsets to a frame, there is a limit to the shear stiffness for a given bending stiffness, and therefore, the stability/curving trade-off is constrained. This limitation is removed if the wheelsets are connected directly by diagonal elastic elements or cross-bracing, or interconnections, which are structurally equivalent. This is termed a self-steering bogie. Superficially, it is similar to the systems of articulation between axles by means of rigid linkages discussed in Section 2.5, but the significant difference is the computed elasticity of the linkages. Self-steering bogies have been applied to locomotives (with benefits to the maximum exploitation of adhesion), passenger vehicles and freight vehicles [124].

Inter-wheelset connections can be provided by means other than elastic elements. For example, the equivalent of cross-bracing can be provided by means of a passive hydrostatic circuit, which has a number of potential design advantages [125]. Hobbs [126] showed that the use of yaw relaxation dampers could provide sufficient flexibility at low frequencies in curves and sufficient elastic restraint at high frequencies to prevent instability, and this was demonstrated on the research vehicle HSFV-1 [127]. Other possibilities were revealed by consideration of generic two-axle vehicles or bogies [128–133], and an example is shown in Figure 2.13, but the added complexity has prevented practical application.

2.14.2 Forced Steering

An alternative to providing self-steering by means of elastic or rigid linkages directly between wheelsets is to use a linkage system that allows the wheelsets to take up a radial position but provides stabilising elastic restraint from the vehicle body. This is the so-called forced steering, as it can be considered that the vehicle body imposes a radial position on the wheelsets. Liechty's work on the earlier equivalent of body steering embodying rigid linkages has already been discussed in Section 2.5, and his later work is discussed in [134,135]. The first analytical studies of forced steered bogie vehicles, taking account of the elasticity of the linkages, were carried out in 1981 by Bell and Hedrick [136] and Gilmore [137], who identified various instabilities, which were promoted by low conicities and reduced creep coefficients. A considerable body of work by Anderson and Smith and colleagues is reported in [138–143], covering the analysis of a vehicle with bogies having separately steered wheelsets. Weeks [144] described dynamic modelling and track testing of vehicles with steered bogies, noting the enhanced sensitivity of this type of configuration to constructional misalignments. Many examples of body steering are in current use.

2.14.3 Three-Axle Vehicles

It was shown in [145] that for a vehicle with three or more axles, it is possible, in principle, to arrange the suspension, so that both radial steering and dynamic stability are achieved, though the margins of stability were small and various instabilities could occur [146,147]. de Pater [148,149] and Keizer [150] followed a slightly different approach with broadly similar results. The maximum damping available when radial steering is achieved is small, so that though cross-bracing is used in various designs of three-axle bogies, a compromise similar to that faced with the two-axle vehicle is necessary.

2.14.4 Unsymmetrical Configurations

As discussed in Section 2.4, the success of the classical steam locomotive configuration depended, in part, on its lack of symmetry and demonstrated that unsymmetrical configurations make it possible to achieve a better compromise between curving and dynamic stability, at least in one direction of motion. But, as Carter showed, additional forms of instability could occur.

A general theory for the stability of unsymmetrical vehicles and for the derivation of theorems relating the stability characteristics in forward motion with those in reverse motion was given in [151,152]. In the case of unsymmetrical two-axle vehicles at low speeds, it was shown that a suitable choice of elastic restraint in the inter-wheelset connections results in static and dynamic stability in forward and reverse motion and that will steer perfectly, without any modification dependent on the direction of motion. However, the margin of stability can be small. Suda [153,154] has studied bogies with unsymmetrical stiffnesses and symmetric conicity, and their development work has led to application in service. Illingworth [155] suggested the use of unsymmetrical stiffness in steering bogies.

2.14.5 The Three-Piece Bogie

The three-piece bogie has evolved over a long period of time [15]. It enjoys wide application throughout the world and presents a very different bogie configuration to the conventional. In this, the bogie frame of a conventional bogie is replaced by two separate side frames that rest directly on the axle boxes through adaptors that allow only rotational freedom. A bolster supports the car body, with a centre plate that allows relative yaw between bogie and body. The bolster is connected to the side frames by a suspension that includes friction wedges and allows some relative motion in all senses except longitudinal. Thus, the side frames have additional degrees of freedom of longitudinal and pitching motions. This results in warping or differential rolling movements of the wheelsets (useful on twisted track) and lozenging of the side frames with respect to the wheelsets in plan view. It follows that the form of construction yields a very low overall shear stiffness. If the warp stiffness is very large, then the bogie behaves like a conventional bogie [156,157]. Quite apart from the low shear stiffness of the configuration, there are many rubbing surfaces, and consequently, the behaviour is highly non-linear, with the result that crabwise motions are common, with resultant asymmetric wear of wheel profiles.

An early study taking account of the effects of unsymmetrical wear is by Tuten, Law and Cooperrider [156]. Since the advent of non-linear techniques of analysis, many studies have been carried out in recent years, exemplified by the work of True [158] and reviewed in [159].

The application of cross-bracing to the three-piece bogie is particularly useful, because the anti-lozenging function of the cross-bracing allows the designer to omit the usual constraints between the bolster and the side-frames and provides a proper lateral suspension at that point. Moreover, if controlled flexibility is provided, by means of an elastomeric pad mounted between the side frames and bearing adaptors, for example, then, in conjunction with cross-bracing, specified suspension stiffnesses can be provided, enabling optimisation of the design. Reduced bending stiffness between the wheelsets improves curving, and the improvement of the primary and secondary lateral suspension enhances ride quality and reduces loads. Crossed-braced three-piece bogies were produced in the 1970s, notably by Scheffel [160] in South Africa, by Pollard [161] in Britain and by List [162] in North America, though more widespread application has been retarded by the increment in first cost.

2.14.6 Independently Rotating Wheels

As mentioned previously, independently rotating wheels have been frequently proposed, as they eliminate the classical hunting problem. The guidance is provided by the lateral component of the gravitational stiffness (reduced by the lateral force due to spin creep), which becomes the flange force when the flangeway clearance is taken up. In the early days, as lateral oscillations were experienced frequently, and as there was concern about large flange forces in curves, various alternatives to the conventional form of wheelset were proposed. Even Robert Stephenson, who was concerned about reducing friction in curves, invented schemes with independently rotating wheels [163,164]. The matter was discussed by the Gauge Commission as early as 1845. Cubitt gave evidence and, in a reply to a question about independently rotating wheels, said that [165] 'I am not sure that we are perfectly right yet about wheels and rails and flanges, we have adopted the old coal railway system designs without trying sufficient experiments'.

Then followed considerable discussion about various combinations of conventional wheelsets and independently rotating wheelsets applicable to locomotives [166], but experiments with them invariably resulted in an unsatisfactory outcome. For example, Nicholas Wood described experience with wheelsets fitted with one loose wheel and said 'the carriages were very liable to get off the road and we had several instances where they had' [167]. Some engineers, such as Crampton and Haswell, believed that wheelsets with cylindrical treads were superior. However, by this time, the coned wheelset had been firmly established as the norm. Nevertheless, experiments with independently rotating wheels have occurred repeatedly over the years, and all have proved nugatory.

More recently, Frederich [168] has surveyed some of the possibilities. Though the kinematic oscillation is eliminated, the self-centring action is slow. Extensive experimental experience has shown that indeed the kinematic oscillation is absent but that one or other of the wheels tends to run in continuous flange contact [169]. Good agreement between calculation and experiment is demonstrated in [170–172].

Elkins [172] and Suda [153,154] showed both by calculation and by experiment that a configuration of bogie, with the trailing axle having independently rotating wheels and the leading axle conventional, significantly improved stability and curving performance and reduced rolling resistance. This provides the possibility of a re-configurable design, as the wheelsets could be provided with a lock that is released on the trailing wheelset (allowing free rotation of the wheels) and locked on the leading wheelset (providing a solid axle). The lock would be switched depending on the direction of motion.

A proposal to incline substantially from the horizontal the axis of rotation of the wheel that would increase the effect of the lateral resultant gravitational force but reduce the amount of spin was put forward by Wiesinger [173]. Jaschinski and Netter [174] studied a generic wheelset model, including the effect of modest amounts of camber.

2.14.7 Articulated Trains

The need to improve curving performance; maximise the use of the clearance gauge; minimise axle loads; and reduce mass, aerodynamic drag and cost has led to many designs in which there is articulation of the car bodies of a vehicle or train, so that the connections between vehicles form an essential part of the running gear, in which the relative motion between the car bodies is used to influence the stability and guidance of the vehicle. Various schemes of articulation have already been discussed in Section 2.5, but modern work shows that the stiffness parameters of the linkages need to be carefully chosen to avoid instability.

The Jacob bogie that is shared by adjacent car bodies is the most common form of articulation. This was used by Gresley [175] in a number of carriage designs between 1900 and 1930. In 1939, Stanier employed a double-point articulation, one at each end of the bogie, which allowed longer car bodies [176]. Application to a new generation of passenger trains in the United States in 1930 to 1950 was less successful, and generally, articulated trains acquired a reputation for bad riding. More recently, articulation of bogie vehicles has been used, most successfully, in high-speed trains such as the French Train a Grande Vitesse [177]. Articulation is used in a wide variety of trams.

The Talgo train combines articulation and independently rotating wheels that are mounted at the rear of a car body, with the front of the body pivoted on the rear of the car in front. The lead vehicle provides guidance, and the rest of the train follows in a stable manner. This scheme was invented by Alejander Goicoechea in 1944 and resulted in several generations of high-speed trains [178], the dynamics of which are discussed in [179].

The articulated Copenhagen S-Tog embodies forced steering of the shared single-axle wheelsets through hydraulic actuators driven by the angle between adjacent car bodies. Extensive design calculations were carried out on this train [180], and its lateral stability is also discussed in [181].

These developments demonstrate that the availability of computational tools in vehicle dynamics has made it possible to depart successfully from established conventional configurations.

2.15 DERAILMENT

The conditions necessary to sustain equilibrium of the forces in flange contact were considered by Nadal in a classical analysis in 1908 [182] that derived his celebrated derailment criterion. Gilchrist and Brickle [183] applied Kalker's theory of creep in a re-examination of Nadal's analysis, and they have shown that Nadal's formula is correct for the most pessimistic case, when the angle of attack is large and the longitudinal creep on the flange is small.

Nadal's criterion is based on a steady-state analysis and assumes that derailment is instantaneous once the critical ratio of horizontal force to lateral force is exceeded. As the wheel actually takes time to climb, Nadal's analysis is conservative. The dynamics of the derailment process was first analysed by Matsui [184]. Sweet et al. [185–188] also investigated the mechanics of derailment and obtained good agreement with model experiments.

2.16 ACTIVE SUSPENSIONS

The concept of an active suspension is to add sensors, a controller and actuators to an existing mechanical system and usually involves feedback action so that the dynamics of the system is modified. The concept of feedback is an ancient one, though the Watts governor was the first widespread application. In railways, there are three principal areas of application of active suspensions – car body tilting systems, secondary suspensions and primary suspensions. The technology of active suspensions is covered in Chapter 15, but in the context of the present chapter, it is relevant to look at the origins of the applications of active suspensions to railway vehicles.

As mentioned previously, superelevation of the track in curves was used at an early date, and tilting the car body to achieve the same effect was demonstrated in monorail systems, such as that at Wuppertal in 1893, with cars suspended from an overhead rail. The developments in dynamics exemplified by Routh's textbook had promoted intense scientific and popular interest in gyroscopes in the 1890s, and various demonstrations were made of monorail systems that exploited gyroscopic stabilisation, that of Brennan in 1906 being perhaps the most successful within the limits of contemporary technology [189]. The dynamics of monorail vehicles has been considered by Ross [190], Arnold and Maunder [191] and Spry and Girard [192].

Bogies with provision for passive pendular suspension of the car body were patented in the United States in 1938 and 1939 [193,194] and put into service in limited numbers on a number of railways [195,196] in the early 1940s. Experiments were carried out on a vehicle with full passive tilting in France in 1957 [197].

The Chesapeake and Ohio Railroad carried out some studies in the 1950s, using as a start a Talgo design as described previously, with passive tilting [198,199]. This work was left undeveloped until the 1960s, when it was applied to the United Aircraft Turbotrain, which first ran in 1968.

The limited dynamic performance of passive tilt systems, together with advances in control system analysis techniques and the availability of sensors and actuators in other fields of technology, stimulated the development of active tilt, starting in 1966.

At British Rail Research, Ispeert carried out a study of active tilting in August 1966 [200], and this led to a proposal for a research vehicle in November 1966; this became the Advanced Passenger Train project in November 1967. An unpowered test vehicle APT-POP with full active tilting first ran in September 1971, and the first run of APT-E was on 25 July 1972; APT-E then went on to demonstrating the advantages of active tilting as well as carrying out crucial validation of the stability and curving calculations described previously [201].

This was paralleled by research in Italy. Franco di Majo carried out a study on high-speed trains in December 1967 for the Italian State Railways, and this was followed by comfort tests in March/April 1969; an experimental vehicle Y0160 was ordered in May 1970, and this first ran in October 1971. Trials with this vehicle over the next 3 years demonstrated the advantages of active tilting and formed the basis of the ETR 401, which entered limited service in 1976 [202].

Though the concept of tilting seems simple, the ramifications of operational service add complexity, and the various technological options are fully discussed in Chapter 15. Following a long period of development, the first commercial service application of active tilt was the Pendolino in Italy in 1988. The operation of tilting trains is now widespread in Europe and Japan.

Some designs of tilt systems incorporated active lateral centring, which acted at low frequencies to counteract the lateral curving forces, thereby enabling the use of a softer lateral suspension. Research on the application of active systems to vertical and lateral vibration isolation and stability augmentation involving the secondary suspension had actually started in the late 1960s. Much early research was motivated by work on air-cushion vehicles [203] and later by the advent of magnetically levitated vehicles. The first full-scale demonstration of an active lateral secondary suspension on a railway vehicle was carried out in 1978 [204], but even though significant benefits in terms of ride quality were demonstrated, the additional cost of the equipment deterred commercial application. Hence, it was not until 2002 that the first service use of an active lateral suspension was employed by Sumitomo for the East Japan Railway Company on their series E2-1000 and E3 Shinkansen vehicles. These pioneering developments are discussed in detail in Chapter 15.

Magnetic suspension relies on active control for stabilisation. Albertson, Bachelet and Graeminger all proposed schemes for magnetic levitation in the 1900s, and in 1938, Kemper [205] demonstrated a model showing the feasibility of a wheel-less train. Subsequently, with the advent of high-power solid-state electronic devices in the mid-1960s and the application of various forms of linear motors, research and development on the topic has flourished. As a result, there has been considerable cross-fertilisation between the dynamics of rail and magnetically levitated vehicles, both in terms of technique and personnel.

Feedback control system methods were used in the stability analysis of the wheelset in 1962 [206]. This approach reveals some of the deficiencies of the wheelset as a guidance element. The unsatisfactory behaviour of independently rotating wheels has already discussed in Section 2.14, so it is natural that the application of active controls should be considered. Bennington [207] used control system techniques to propose an active torque connection between the two wheels of a wheelset, and subsequently, a wheelset with an active torque connection using a magnetic coupling was developed [208]. Various control laws were used with the object of providing a good torque connection between the wheels at low frequencies, so that curving ability is maintained, but at high frequencies, the wheels are more or less uncoupled, so that instability does not arise. Pascal and Petit [209] carried out experiments in which active steering, using freely rotating wheels, was achieved by the yaw moment generated by electromagnets that react against a guide rail. More recently, much research work has been carried out on active enhancement of the dynamic performance of the conventional wheelset and of independently rotating wheels; see Chapter 15. However, the challenge of practical implementation is formidable.

It can be seen that, increasingly, control engineering techniques will exercise a strong influence on the dynamics of railway vehicles – see for example [210] – either by improving the dynamics of vehicles, using the conventional wheelset, or by supporting the development of more innovative systems.

2.17 THE DEVELOPMENT OF COMPUTER SIMULATION

The state of the art of simulation of the dynamics of railway vehicles is discussed in Chapter 17; this section describes the historical background. As discussed previously, owing to the laborious nature of the calculations, the work of Carter in the 1920s and Davies in the 1930s was constrained by the practical limits to computation. On the other hand, there was no pressing need for dynamics calculations, as, in general, empirical development had kept abreast of train speeds.

In the 1960s, attempts to increase the speeds of freight trains resulted in much higher risk of derailment, and the advent of the high-speed train meant that instabilities such as bogie hunting were no longer acceptable. A more numerate approach to dynamics problems on the railways became necessary.

Once the importance of the suspension stiffnesses was recognised, it was possible to carry out the analysis of a restrained wheelset with pencil and paper. This was, perhaps, the most fruitful model with which to understand the mechanics of instability, but a complete vehicle involved the solution of the equations of motion with many degrees of freedom.

The advent of the digital computer and the general-purpose electronic analogue computer provided the means to analyse complete vehicles. Amongst the first to use these computers on a large scale was the aircraft industry, where the solution of widespread flutter and vibration problems occurring in the 1950s was necessary. It is interesting that the first eigenvalue analysis of a complete two-axle vehicle carried out in 1962 by British Rail Research used a flutter routine and computer at English Electric Aviation; the linear equations of motion of a railway vehicle and the aeroelastic flutter equations of an aircraft wing are formally the same, provided that vehicle speed is interpreted appropriately. The work by Gilchrist et al. [74] in 1964, discussed previously, in which the simulation of two-axle vehicles involved some non-linearities, was carried out using an analogue computer.

The next steps in the development of vehicle dynamics required the solution of the equations of motion that were now non-linear. Numerical methods for mathematical analysis date back to Newton, but the mechanisation of these methods by using accounting machines was well established by the 1930s [211]. The advent of the programmable digital computer, with its much larger memory and higher speed of operation, facilitated the step-by-step integration of differential equations in the mid-1940s. With the application of the computer on a large scale in all branches of engineering, interest in numerical methods quickened and efficient methods of computation became available in the form of libraries of standard routines.

In the 1970s, simulations of complex non-linear railway vehicle models, with many degrees of freedom, were developed, which exploited the increase in computer power becoming available. These simulations were based on equations of motion derived manually that were then incorporated into computer programs to solve specific problems such as stability, response to track irregularities and curving.

The derivation of equations of motion, their reduction to first-order differential equations suitable for numerical integration or eigenvalue analysis and the calculation of the parameters in the equations were lengthy and error-prone tasks. Moreover, the application of railway vehicle dynamics was moving from the research laboratory to the industrial design office. Many companies set up groups specialising in bogie and suspension design dependent on dynamics calculations. This motivated the development of complete packages, which covered a range of dynamics calculations, using the same consistent model of the vehicle. In these packages, user-friendly interfaces were provided, which aided and simplified the inputs and outputs of the simulation. Many of these packages are based on the formal methods of multibody dynamics that treat systems of rigid or flexible bodies that are connected together by joints and that may undergo large translational and rotational displacements. Multibody dynamics had its origin in work carried out on satellite dynamics in the 1960s, when they could no longer be considered as single rigid bodies. These packages enabled computer-aided engineering decision-making in an industrial environment supporting, for example, design optimisation, the validation of safety margins and reducing the reliance on safety margins.

It could be considered that by the end of the 1990s, vehicle dynamics simulation had reached a level of maturity. As computer power has increased and vehicle dynamics computations have been used more widely for engineering design purposes, it became necessary to refine the modelling of suspension components (see Chapter 6) and of wheel-rail contact (see Chapter 7).

The state of the art in 1999 is illustrated by the successful results of the Manchester benchmarks for rail simulation [212], in which the performance of five computer packages was compared. It was concluded that the users of one of these packages could be confident that they would achieve similar results with any of the other packages. A similar benchmarking exercise for wheel-rail contact is described in [213].

2.18 THE EXPANDING DOMAIN OF RAIL VEHICLE DYNAMICS

This chapter has reviewed the historical development of ideas about the basic conceptual problems associated with stability, response to track geometry and behaviour in curves of the railway vehicle. It goes from the earliest days of the railways and covers the period in which the most fundamental concepts were formulated. During the nineteenth century and beyond, empirical evolution of configurations that worked generally kept pace with train performance, though departures from the conventional often resulted in problems. These problems attracted the attention of individual and isolated researchers such as Carter in the early twentieth century, but their impact on engineering practice was minimal.

The mid-twentieth-century perception that the railways were outmoded was gradually overtaken by the recognition that they had a future after all, spearheaded by the advent of very high-speed railways in Japan. There was a necessity for a more scientific approach to the problems of vehicle dynamics; dedicated research groups were set up, and the parallel development of the digital computer and electronic instrumentation ensured that rapid progress was made.

The International Association for Vehicle System Dynamics (IAVSD) was formed in 1977 and reflected the fact that there was now a growing international community of vehicle dynamicists. Towards the end of the twentieth century, dynamics simulation was well established as a practical tool in industry as well as in academia. The issues involved in the engineering application of simulation are discussed in [214]. The proceedings of recent IAVSD symposia, the state-of-the-art papers and the following chapters of this handbook reveal the growing range and depth of dynamics studies undertaken in the field of railway dynamics.

Increased computing power and a deeper understanding of non-linear dynamics are making it possible to model, increasingly accurately, vehicles with highly non-linear suspensions such as the three-piece bogie and the standard UIC two-axle suspension.

Higher speeds and axle loads have motivated more detailed modelling of the wheel-rail interface with its complex geometry, and forces generated by large creepages, particularly in switches and crossings and in derailment.

As both traction and guidance forces are provided by the same wheel-rail interface, a systems approach is being followed, in which the design of the drive control system is combined with the needs of guidance.

New problems of interaction between vehicle and track have emerged such as irregular ballast settlement and deterioration, increased levels of rail corrugation and out-of-round wheels. The solution of these problems requires the consideration of structural dynamics of both vehicle and track in the frequency range of about 40 to 400 Hz, together with the analysis of the long-term behaviour of wheel and track components.

The analysis of mechanisms of noise generation, particularly by corrugations and squealing in curves, requires the consideration of an even higher range of frequencies of structural oscillations of vehicles and track, up to 5 kHz or more. At these frequencies, a non–steady-state analysis of the contact forces is needed.

At high speeds, particularly in tunnels, aerodynamic forces are significant not only as a generator of drag but also in affecting the lateral response of vehicles. With the further increases in speed and reductions in mass of car bodies, lateral oscillations have been experienced due to pressure fluctuations caused by unsteady flow separations from the car body surface. This requires simulation in which the vehicle dynamics is combined with the aerodynamics of the flow field with moving boundaries and that is dependent on the car body motion.

Increasing use of electronics and active controls makes it necessary to adopt a mechatronic approach in which the mechanical parts and the electronics are seen as integral parts of the system to be analysed and designed concurrently. Industrial applications demand integration of software tools with design and manufacturing systems.

Though there is much to be understood about the behaviour of apparently simple systems with strong non-linearities, the history of railway vehicle dynamics suggests that, in most cases, the subject rests on a sound conceptual basis, with satisfactory full-scale experimental validation.

REFERENCES

1. A.H. Wickens, The dynamics of railway vehicles—from Stephenson to Carter, *Proceedings of the Institution of Mechanical Engineers, Part F*, 212(3), 209–217, 1998.
2. C.F. Dendy Marshall, *A History of British Railways Down to the Year 1830*, Oxford University Press, Oxford, UK, pp. 147–148, 1938.
3. A. Vaughan, *Isambard Kingdom Brunel: Engineering Knight-Errant*, John Murray, London, p. 102, 1992.
4. J. Klingel, *Uber den Lauf der Eisenbahnwagen auf Gerarder Bahn*, Organ Fortsch, Eisenb-wes 38, pp. 113–123, 1883.
5. R. Winans, Diminishing friction in wheeled carriages used on rail and other roads, *British Patent* 5796, 1829.
6. W.B. Adams, On the impedimental friction between wheel tire and rails with plans for improvement, *Proceedings of the Institution of Civil Engineers*, 23, 411–427, 1864.
7. F.J. Redtenbacher, *Die Gesetze des Locomotiv-Baues*, Verlag von Friedrich Bassermann, Mannheim, Germany, 22, 1855.
8. J. Mackenzie, Resistance on railway curves as an element of danger, *Proceedings of the Institution of Civil Engineers*, 74, 1–57, 1883.
9. A.O. Gilchrist, The long road to solution of the railway hunting and curving problems, *Proceedings of the Institution of Mechanical Engineers, Part F*, 212(3), 219–226, 1998.
10. H. Heumann, Das Verhalten von Eisenbahnfahrzeugen in Gleisbogen, *Organ Fortsch, Eisenb-wes*, 68, pp. 104–108, 118–121, 136–140, 158–163, 1913.
11. S.R.M. Porter, *The Mechanics of a Locomotive on Curved Track*, The Railway Gazette, London, UK, 1935.
12. W. Chapman, E.W. Chapman, Facilitating and reducing the expense of carriage on railways and other roads, *British Patent* 3632, 1812.
13. C.F.D. Marshall, *A History of Railway Locomotives Down to the End of the Year 1831*, The Locomotive Publishing Company, London, pp. 61–76, 1953.
14. W. Losh, R. Stephenson, Improvements in the construction of railways and tramways, *British Patent* 4067, 1816.

15. J.H. White, *The American Railroad Freight Car*, The Johns Hopkins University Press, Baltimore, MD, pp. 164–176, 1993.
16. J.H. White, *A History of the American Railroad Passenger Car*, The Johns Hopkins University Press, Baltimore, MD, pp. 8–20, 1978.
17. J.H. White, *American Locomotives: An Engineering History 1830–1880 (revised and expanded edition)*, The Johns Hopkins University Press, Baltimore, MD, pp. 151–157 and 167–175, 1997.
18. J.H. White, *Introduction of the Locomotive Safety Truck*, United States National Museum Bulletin 228, Smithsonian Institution, Washington, DC, pp. 117–131, 1961.
19. L. Bissel, Truck for locomotives, *U.S. Patent* 17913, 1857.
20. A.F. Smith, Improvement in trucks for locomotives, *U.S. Patent* 34377, 1862.
21. C. Davenport, A. Bridges, Manner of constructing railroad-carriages so as to ease the lateral motions of the bodies thereof, *U.S. Patent* 2071, 1841.
22. A.M. Eastwick, Mode of applying the driving wheels of locomotive-engines, *U.S. Patent* 471, 1837.
23. J. Harrison, Improvement in cars, carriages, trucks etc for railroads, *U.S. Patent* 706, 1838.
24. Report of the Gauge Commissioners with minutes of evidence, question 4373, p. 221, London, UK, 1846.
25. A.H. Wickens, Stephenson's 'Long Boiler' locomotive and the dawn of railway vehicle dynamics, *International Journal for the History of Engineering & Technology*, 87(1), 42–63, 2017.
26. L. Le Chatelier, *Etudes Sur la Stabilite Des Machines Locomotives En Movement*, Mathias, Paris, 1849.
27. D.K. Clark, *Railway Machinery*, Blackie and Sons, London, UK, 1855.
28. S.P. Timoshenko, *A History of the Strength of Materials*, McGraw-Hill, New York, 1953, pp. 173–178.
29. R. Liechty, Das Bogenlaeufige Eisenbahn-Fahrzeug, Schulthess Verlag, Zurich, Switzerland, 1934. Translation: A. H. Wickens, 2008, The motion in curves of railway vehicles—historical development, Loughborough University.
30. L. Weiner, *Articulated Locomotives*, Constable & Co., London, UK, 1930.
31. M. Weisbrod, R. Barkhoff, Die Dampflokmotive -Technik und Funktion - Teil 4 – Eisenbahn Journal, pp. 71–75, 1999.
32. H. Elsner, *Three-Axle Streetcars*, N.J. International, Hicksville, NY, 1994.
33. E.L. Ahrons, *The British Steam Railway Locomotive 1825–1925*, The Locomotive Publishing Company, London, UK, pp. 160–162, 1927.
34. U. Schwanck, Wheelset steering for bogies of railway vehicles, *Rail Engineering International*, 4, 352–359, 1974.
35. Chr. Boedecker, Dir Wirkungen zwischen Rad und Scheine und ihre Einflusse auf den Bewegungswiderstand der Fahrzeuge in der Eisenbahnzugen, Hahn'sche Buchhandlung, Hannover, 1887. Translation: A. Bewley, 1899, The interaction of wheel and rail and its effect on the motion and resistance of vehicles in trains, Lawrence Asylum Press, Madras, 60.
36. K.R. Hopkirk, Frederick William Carter, 1870–1952, *Obituary Notices of Fellows of the Royal Society*, 8(22), 373–388, 1953.
37. F.W. Carter, The electric locomotive, *Proceedings of the Institution of Civil Engineers*, 221, 221–252, 1916.
38. O. Reynolds, On the efficiency of belts or straps as communicators of work, *The Engineer*, 27 November 1874, p. 396.
39. S. Bennett, *A History of Control Engineering, 1800–1930*, Peter Peregrinus, London, UK, 1979, Chapter 3.
40. E.J. Routh, *Stability of a Given State of Motion*, Macmillan, London, UK, 1877. Reprinted: 1975, Taylor & Francis Group, London, UK.
41. E.J. Routh, *Dynamics of a System of Rigid bodies (Advanced Part)*, Macmillan, London, 1860 (1st ed.), 1905 (6th ed.).
42. A. Hurwitz, Über die Bedingungen, unter welchen eine Gleichung nur Wurzeln mit negativen reelen Teilen besitzt, *Mathematische Annalen*, 46, 273–284, 1895.
43. G.H. Bryan, W.E. Williams, The longitudinal stability of aeroplane gliders, *Proceedings of the Royal Society of London*, 73, 100–116, 1904.
44. B. Hopkinson, The 'hunting' of alternating-current machinery, *Proceedings of the Royal Society of London*, 72, 235–252, 1903.
45. F.J.W. Whipple, The stability of motion of the bicycle, *Quarterly Journal of Pure and Applied Mathematics*, 30, 312–348, 1899.
46. L. Bairstow, A. Page, Oscillations of the tailplane and body of an aeroplane in flight, Aeronautical Research Council Reports and Memoranda, 276, Part 2, 1916.

47. R.A. Frazer, W.J. Duncan, The flutter of aeroplane wings, Aeronautical Research Council Reports and Memoranda, 1155, 1928.
48. R.A. Frazer, W.J. Duncan, A.R. Collar, *Elementary Matrices and Some Applications to Dynamics and Differential Equations*, Cambridge University Press, Cambridge, 1938.
49. F.W. Carter, *Railway Electric Traction*, Edward Arnold, London, UK, 1922.
50. A.E.H. Love, *Mathematical Theory of Elasticity*, 2nd ed., Cambridge University Press, Cambridge, UK, pp. 195–198, 1906.
51. F.W. Carter, On the action of a locomotive driving wheel, *Proceedings of the Royal Society of London, Series A*, 112, 151–157, 1926.
52. S.P. Timoshenko, *History of the Strength of Materials*, McGraw-Hill, New York, 1953, p. 348.
53. F.W. Carter, On the stability of running of locomotives, *Proceedings of the Royal Society of London, Series A*, 121, 585–611, 1928.
54. F.W. Carter, The running of locomotives with reference to their tendency to derail, *The Institution of Civil Engineers Selected Engineering Papers*, 1(91), 1930.
55. F.W. Carter, Improvement in and relating to high speed electric or other locomotives, British Patents 128106, 155038, 163185, 1918, 1919, 1920.
56. Y. Rocard, La stabilite de route des locomotives, Actual Sci Ind., 234, Part 1, 1935.
57. Y. Rocard, *General Dynamics of Vibrations*, Crosby Lockwood, London, 1960 (translated from French, first published 1943).
58. B.F. Langer, J.P. Shamberger, Dynamic stability of railway trucks, *Transactions of the American Society of Mechanical Engineers*, 57, 481–493, 1935.
59. B.S. Cain, *Vibration of Road and Rail Vehicles*, Pitman Publishing Corporation, New York, 1940, pp. 149–189.
60. R.D. Davies, Some experiments on the lateral oscillation of railway vehicles, *Journal of the Institution of Civil Engineers*, 11(5), 224–261, 1939.
61. H. Heumann, Lauf der Drehgestell-Radsätze in der Geraden, *Organ Fortschr Eisenb-wes*, 92, 336–342, 1937.
62. C.Th. Müller, Kinematik, Spurführungsgeometrie und Führungsvermögen der Eisenbahnradsatz, *Glasers Annalen*, 77, 264–281, 1953.
63. C.Th. Müller, Wear profiles of wheels and rails, ORE-Report C9/RP6, Office of Research and Experiment (ORE) of the International Union of Railways, Utrecht, the Netherlands, 1960.
64. B.L. King, An evaluation of the contact conditions between a pair of worn wheels and worn rails in straight track, British Rail Research Technical Note DYN/37, Derby, UK, 1996.
65. B.L. King, An evaluation of the contact conditions between a pair of worn wheels and new rails in straight track, British Rail Research Technical Note DYN/42, Derby, UK, 1996.
66. B.L. King, The design of new tyre profiles for use on British Railways, British Rail Research Technical Note DYN/38, Derby, UK, 1996.
67. R.J. Gostling, The measurement of real wheel and track profiles and their use in finding contact conditions, equivalent conicity and equilibrium rolling line. British Rail Research Technical Note TN DA 22, Derby, UK, 1971.
68. A. Nefzger, Geometrie der Berührung zwischen Radsatz und Gleis, *Eisenbahntechnische Rundschau*, 23, 113–122, 1971.
69. N.K. Cooperrider, J.K. Hedrick, E.H. Law, P.S. Kadala, J.M. Tuten, Analytical and experimental determination of nonlinear wheel/rail geometric constraints, U.S. Department of Transportation Report FRA-O&RD 76-244, Washington, DC, 1975.
70. W. Hauschild, Die Kinematik des Rad-Schiene Systems, Doctoral thesis, Technische Universität Berlin, Institut für Mechanik, Berlin, Germany, 1977.
71. W. Duffek, Contact geometry in wheel rail mechanics, In J. Kalousek, R.V. Dukkipati, G.M.L. Gladwell (Eds.), *Proceedings International Symposium on Contact Mechanics and Wear of Rail/Wheel Systems*, 6–9 July 1982, Vancouver, Canada, University of Waterloo Press, pp. 161–179, 1982.
72. A.D. de Pater, The geometric contact between wheel and rail, *Vehicle System Dynamics*, 17(3), 127–140, 1988.
73. G. Yang, Dynamic analysis of railway wheelsets and complete vehicle systems, Doctoral thesis, Delft University of Technology, Faculty for Mechanical Engineering and Marine Technology, Delft, the Netherlands, pp. 42–50, 1993.
74. A.O. Gilchrist, A.E.W. Hobbs, B.L. King, V. Washby, The riding of two particular designs of four-wheeled vehicle, *Proceedings of the Institution of Mechanical Engineers Part 3F*, 180, 99–113, 1965.
75. R.J. Kochenburger, Frequency-response methods for analysis of a relay servomechanism, *Transactions of the American Institute of Electrical Engineers*, 69, 270–284, 1950.

76. H. Poritsky, Stresses and deflections of cylindrical bodies in contact with application to contact of gears and of locomotive wheels, *ASME Journal of Applied Mechanics*, 72, 191–201, 1950.
77. B.S. Cain, Discussion of the Poritsky (1950) reference above, *ASME Journal of Applied Mechanics*, 72, 465–466, 1950.
78. K.L. Johnson, The effect of tangential contact force upon the rolling motion of an elastic sphere upon a plane, *ASME Journal of Applied Mechanics*, 25(3), 339–346, 1958.
79. K.L. Johnson, The effect of spin upon the rolling motion of an elastic sphere upon a plane, *ASME Journal of Applied Mechanics*, 25(3), 332–338, 1958.
80. D.J. Haines, E. Ollerton, Contact stress distributions on elliptical contact surfaces subjected to radial and tangential forces, *Proceedings of the Institution of Mechanical Engineers*, 177, 95–114, 1963.
81. P.J. Vermeulen, K.L. Johnson, Contact of non-spherical elastic bodies transmitting tangential forces, *ASME Journal of Applied Mechanics*, 31(2), 338–340, 1964.
82. R. Levi, Etude relative au contact des roués sur le rail, *Revue General Chemins de Fer*, 54, 81–109, 1935.
83. A. Chartet, La theorie statique de deraillement d'un essieu, *Revue General Chemins de Fer*, 69, 365–386, 1950. Also 71, 442–453, 1952.
84. C.Th. Müller, Dynamics of railway vehicles on curved track, *Proceedings of the Institution of Mechanical Engineers, Part F*, 180(6), 45–57, 1965.
85. Z.Y. Shen, J.K. Hedrick, J.A. Elkins, A comparison of alternative creep-force models for rail vehicle dynamics analysis, In J.K. Hedrick (Ed.), *Proceedings 8th IAVSD Symposium held at Massachusetts Institute of Technology*, 15–19 August 1983, Swets and Zeitlinger, Lisse, the Netherlands, pp. 591–605, 1984.
86. A.D. de Pater, On the reciprocal pressure between two elastic bodies, *Proceedings of a Symposium on Rolling Contact Phenomena held at the General Motors Research Laboratories*, Warren, MI, October 1960, Elsevier, Amsterdam, the Netherlands, 29–74, 1962.
87. J.J. Kalker, The transmission of force and couple between two elastically similar rolling spheres, *Proceedings Koninklijke Nederlandse Akademie van Wetenschappen*, B67, 135–177, 1964.
88. J.J. Kalker, On the rolling of two elastic bodies in the presence of dry friction, Doctoral thesis, Delft University of Technology, Delft, the Netherlands, 1967.
89. J.J. Kalker, Simplified theory of rolling contact, Delft Progress Report Series C1, No 1, 1–10, 1973.
90. J.J. Kalker, A fast algorithm for the simplified theory of rolling contact, *Vehicle System Dynamics*, 11(1), 1–13, 1982.
91. J.J. Kalker, *Three-Dimensional Elastic Bodies in Rolling Contact*, Kluwer Academic Publishers, Dordrecht, the Netherlands, 1990.
92. T. Matsudaira, Shimmy of axles with pair of wheels (in Japanese), *Journal of Railway Engineering Research*, 16–26, 1952.
93. H. Shima, The New Tokaido Line: brief notes on the way the idea of the construction was developed, Paper 10 at Convention on Guided Land Transport, *Proceedings of the Institution of Mechanical Engineers*, 181(7), 13–19, 1966.
94. R. Lévi, Study of hunting movement, Question C9 Preliminary Report, Office of Research and Experiment (ORE) of the International Union of Railways, Utrecht, the Netherlands, 1953.
95. R. de Possel, J. Beautefoy, T. Matsudaira, Essays awarded prizes, Question C9 Interim Report No. 2 Part 2, Office of Research and Experiment (ORE) of the International Union of Railways, Utrecht, the Netherlands, 1960.
96. A.D. de Pater, Etude du mouvement de lacet d'un vehicule de chemin de fer, *Applied Scientific Research, Section A*, 6(4), 263–316, 1957.
97. A.D. de Pater, The approximate determination of the hunting movement of a railway vehicle by aid of the method of Krylov and Bogoljubov, *Applied Scientific Research*, 10, 205–228, 1961.
98. P. van Bommel, Application de la theorie des vibrations nonlineaires sur le problem du mouvement de lacet d'un vehicule de chemin de fer, Doctoral thesis, Technische Hogeschool van Delft, Delft, the Netherlands, 1964.
99. A.H. Wickens, The dynamic stability of railway vehicle wheelsets and bogies having profiled wheels, *International Journal of Solids and Structures*, 1(3), 319–341, 1965.
100. B.L. King, The measurement of the mode of hunting of a coach fitted with standard double-bolster bogies, British Railways Research Department Report E439, Derby, UK, 1963.
101. R.A. Pooley, Assessment of the critical speeds of various types of four-wheeled vehicles, British Railways Research Department Report E557, Derby, UK, 1965.
102. A.H. Wickens, The dynamics of railway vehicles on straight track: Fundamental considerations of lateral stability, *Proceedings of the Institution of Mechanical Engineers, Part 3F*, 180(6), 29–44, 1965.

103. A.E.W. Hobbs, The lateral stability of HSFV-1, British Rail Research Technical Note DYN/53, Derby, UK, 1967.
104. N.K. Cooperrider, J.K. Hedrick, E.H. Law, C.W. Malstrom, The application of quasilinearization techniques to the prediction of nonlinear railway vehicle response, In H.B. Pacejka (Ed.), *The Dynamics of Vehicles on Roads and on Railway Tracks, Proceedings IUTAM Symposium held at Delft University of Technology*, 18–22 August 1975, Swets & Zeitlinger, Amsterdam, the Netherlands, pp. 314–325, 1976.
105. D. Moelle, R. Gasch, Nonlinear bogie hunting, In A.H. Wickens (Ed.), *Proceedings 7th IAVSD Symposium held at Cambridge University*, 7–11 September 1981, Swets and Zeitlinger, Lisse, the Netherlands, pp. 455–467, 1982.
106. R. Gasch, D. Moelle, K. Knothe, The effects of non-linearities on the limit-cycles of railway vehicles, In J.K. Hedrick (Ed.), *Proceedings 8th IAVSD Symposium held at Massachusetts Institute of Technology*, 15–19 August 1983. Swets and Zeitlinger, Lisse, the Netherlands, pp. 207–224, 1984.
107. H. True, Dynamics of a rolling wheelset, *ASME Applied Mechanics Reviews*, 46(7), 438–444, 1993.
108. H. True, Railway vehicle chaos and asymmetric hunting, In G. Sauvage (Ed.), *Proceedings 12th IAVSD Symposium*, Lyon, France, 26–30 August 1991, Swets and Zeitlinger, Lisse, the Netherlands, pp. 625–637, 1992.
109. M. Hoffman, H. True, The dynamics of European two-axle railway freight wagons with UIC standard suspension, *Vehicle System Dynamics*, 46, 225–236, 2008.
110. J. Piotrowsky, Smoothing dry friction damping by dither generated in rolling contact of wheel and rail and its influence in ride dynamics of freight wagons, *Vehicle System Dynamics*, 48, 675–703, 2010.
111. D. Boocock, Steady-state motion of railway vehicles on curved track, *Journal of Mechanical Engineering Science*, 11(6), 556–566, 1969.
112. D.E. Newland, Steering characteristics of bogies, *Railway Gazette*, 124, 745–750, 1968.
113. J.A. Elkins, R.J. Gostling, A general quasi-static curving theory for railway vehicles, In A. Slibar, H. Springer (Eds.), *Proceedings 5th VSD-2nd IUTAM Symposium held at Technical University of Vienna*, Vienna, Austria, 19–23 September 1977, Swets and Zeitlinger, Lisse, the Netherlands, pp. 388–406, 1978.
114. A.R. Pocklington, R.A. Allen, Improved data from load-measuring wheels, *Railway Engineer*, 2(4), 37–43, 1977.
115. H. Netter, G. Schupp, W. Rulka, K. Schroeder, New aspects of contact modelling and validation within multibody system simulation of railway vehicles, In L. Palkovics (Ed.), *Proceedings 15th IAVSD Symposium*, Budapest, Hungary, 25–29 August 1997, Swets and Zeitlinger, Lisse, the Netherlands, 246–269, 1998.
116. J.-P. Pascal, About multi-Hertzian contact hypothesis and equivalent conicity in the case of S1002 and UIC60 analytical wheel/rail profiles, *Vehicle System Dynamics*, 22, 263–275, 1993.
117. H.H. Jenkins, J.E. Stephenson, G.A. Clayton, G.W. Morland, D. Lyon, The effect of track and vehicle parameters on wheel/rail vertical dynamic forces, *Railway Engineering Journal*, 3, 2–16, 1974.
118. A.E.W. Hobbs, The response of a restrained wheelset to variations in the alignment of an ideally straight track, British Railways Research Department Report E542, Derby, UK, 1964.
119. R. Illingworth, The mechanism of railway vehicle excitation by track irregularities, Doctoral thesis, University of Oxford, Oxford, UK, 1973.
120. R.A. Clark, B.M. Eickhoff, G.A. Hunt, Prediction of the dynamic response of vehicles to lateral track irregularities, In A.H. Wickens (Ed.), *Proceedings 7th IAVSD Symposium held at Cambridge University, UK*, 7–11 September 1981, Swets and Zeitlinger, Lisse, the Netherlands, pp. 535–548, 1982.
121. A.O. Gilchrist, Power spectral measurements by TMM 1: Proving trials and three site measurements, British Rail Research Technical Note DYN/67, Derby, UK, 1967.
122. K. Kortüm, D.N. Wormley, Dynamic interactions between traveling vehicles and guideway systems, *Vehicle System Dynamics*, 10, 285–317, 1981.
123. A.H. Wickens, Steering and dynamic stability of railway vehicles, *Vehicle System Dynamics*, 5, 15–46, 1978.
124. H. Scheffel, A new design approach for railway vehicle suspensions, *Rail International*, 5, 638–651, 1974.
125. A.H. Wickens, Improvements in or relating to railway vehicles and bogies, *British Patent* 1179723, 1967.
126. A.E.W. Hobbs, Improvements in or relating to railway vehicles, *British Patent* 1261896, 1972.
127. C.L. Murray, The lateral stability of HSFV1a when fitted with yaw relaxation, British Rail Research Department Technical Note DA 13, Derby, UK, 1969.
128. D. Horak, C.E. Bell, J.K. Hedrick, A comparison of the stability performance of radial and conventional rail vehicle trucks, *ASME Journal of Dynamic Systems, Measurement, and Control*, 103, 181, 1981.

129. A.K. Kar, D.N. Wormley, J.K. Hedrick, Generic rail truck characteristics, In H.P. Willumeit (Ed.), *Proceedings 6th IAVSD Symposium held at Technical University of Berlin*, 3–7 September 1979, Swets and Zeitlinger, Lisse, the Netherlands, pp. 198–210, 1980.
130. A.K. Kar, D.N. Wormley, Generic properties and performance characteristics of passenger rail vehicles, In A.H. Wickens (Ed.), *Proceedings 7th IAVSD Symposium held at Cambridge University*, 7–11 September 1981, Swets and Zeitlinger, Lisse, the Netherlands, pp. 329–341, 1982.
131. T. Fujioka, Generic representation of primary suspensions of rail vehicles, In R.J. Anderson (Ed.), *Proceedings 11th IAVSD Symposium*, Kingston, Canada, 21–25 August 1989, Swets and Zeitlinger, Lisse, the Netherlands, 233–247, 1989.
132. T. Fujioka, Y. Suda, M. Iguchi, Representation of primary suspensions of rail vehicles and performance of radial trucks, *Bulletin of Japan Society of Mechanical Engineers*, 27(232), 2249–2257, 1984.
133. J.K. Hedrick, D.N. Wormley, R.R. Kim, A.K. Kar, W. Baum, Performance limits of rail passenger vehicles: Conventional, radial and innovative trucks, U.S. Department of Transportation Report DOT/RSPA/DPB-50/81/28, 1982.
134. R. Liechty, Studie uber die Spurfuhrung von Eisenbahnfahrzeugen, *Schweizer Archiv für Angewandte Wissenschaft und Technik*, 3, 81–100, 1937.
135. R. Liechty, Die Bewegungen der Eisenbahnfahrzeuge auf den schienen und die dabei auftretenden Kräfte, *Elektrische Bahnen*, 16, 17–27, 1940.
136. C.E. Bell, J.K. Hedrick, Forced steering of rail vehicles: Stability and curving mechanics, *Vehicle System Dynamics*, 10, 357–385, 1981.
137. D.C. Gilmore, The application of linear modelling to the development of a light steerable transit truck, In A.H. Wickens (Ed.), *Proceedings 7th IAVSD Symposium held at Cambridge University*, 7–11 September 1981, Swets and Zeitlinger, Lisse, the Netherlands, pp. 371–384, 1982.
138. J.A. Fortin, R.J. Anderson, Steady state and dynamic predictions of the curving performance of forced steering rail vehicles, In J.K. Hedrick, (Ed.), *Proceedings 8th IAVSD Symposium held at Massachusetts Institute of Technology*, 15–19 August 1983, Swets and Zeitlinger, Lisse, the Netherlands, pp. 179–192, 1984.
139. J.A.C. Fortin, R.J. Anderson, D.C. Gilmore, Validation of a computer simulation forced steering rail vehicles, In O. Nordstrom (Ed.), *Proceedings 9th IAVSD Symposium held at Linköpings University*, Sweden, 24–28 June 1985, Swets and Zeitlinger, Lisse, the Netherlands, pp. 100–111, 1986.
140. R.E. Smith, R.J. Anderson, Characteristics of guided steering railway trucks, *Vehicle System Dynamics*, 17, 1–36, 1988.
141. R.J. Anderson, C. Fortin, Low conicity instabilities in forced steering railway vehicles, In M. Apetaur (Ed.), *Proceedings 10th IAVSD Symposium held at Czech Technical University*, Prague, 24–28 August 1987, Swets and Zeitlinger, Lisse, the Netherlands, pp. 17–28, 1988.
142. R.E. Smith, Forced steered truck and vehicle dynamic modes resonance effects due to car geometry, In M. Apetaur (Ed.), *Proceedings 10th IAVSD Symposium held at Czech Technical University*, Prague, 24–28 August 1987, Swets and Zeitlinger, Lisse, the Netherlands, pp. 423–424, 1988.
143. R.E. Smith, Dynamic characteristics of steered railway vehicles and implications for design, *Vehicle System Dynamics*, 18, 45–69, 1989.
144. R. Weeks, The design and testing of a bogie with a mechanical steering linkage, In M. Apetaur (Ed.), *Proceedings 10th IAVSD Symposium held at Czech Technical University*, Prague, 24–28 August 1987, Swets and Zeitlinger, Lisse, the Netherlands, pp. 497–508, 1988.
145. A.H. Wickens, Stability criteria for articulated railway vehicles possessing perfect steering, *Vehicle System Dynamics*, 7(1), 33–48, 1979.
146. A.H. Wickens, Static and dynamic stability of a class of three-axle railway vehicles possessing perfect steering, *Vehicle System Dynamics*, 6(1), 1–19, 1977.
147. A.H. Wickens, Flutter and divergence instabilities in systems of railway vehicles with semi-rigid articulation, *Vehicle System Dynamics*, 8(1), 33–48, 1979.
148. A.D. de Pater, Optimal design of a railway vehicle with regard to cant deficiency forces and stability behaviour, Delft University of Technology, Laboratory for Engineering Mechanics, Report 751, 1984.
149. A.D. de Pater, Optimal design of railway vehicles, *Ingenieur-Archiv*, 57(1), 25–38, 1987.
150. C.P. Keizer, A theory on multi-wheelset systems applied to three wheelsets, In O. Nordstrom (Ed.), *Proceedings 9th IAVSD Symposium held at Linköpings University*, Sweden, 24–28 June 1985, Swets and Zeitlinger, Lisse, the Netherlands, pp. 233–249, 1986.
151. A.H. Wickens, Steering and stability of unsymmetric articulated railway vehicles, *ASME Journal of Dynamic Systems, Measurement, and Control*, 101, 256–262, 1979.
152. A.H. Wickens, Static and dynamic stability of unsymmetric two-axle railway possessing perfect steering, *Vehicle System Dynamics*, 11, 89–106, 1982.

153. Y. Suda, Improvement of high-speed stability and curving performance by parameter control of trucks for rail vehicles considering independently rotating wheelsets and unsymmetric structure, *Japan Society of Mechanical Engineers International Journal Series III*, 33(2), 176–182, 1990.
154. Y. Suda, High speed stability and curving performance of longitudinally unsymmetric trucks with semi-active control, *Vehicle System Dynamics*, 23(1), 29–52, 1994.
155. R. Illingworth, The use of unsymmetric plan view suspension in rapid transit steering bogies, In J.K. Hedrick (Ed.), *Proceedings 8th IAVSD Symposium held at Massachusetts Institute of Technology*, 15–19 August 1983, Swets and Zeitlinger, Lisse, the Netherlands, 252–265, 1984.
156. J.M. Tuten, E.H. Law, N.K. Cooperrider, Lateral stability of freight cars with axles having different wheel profiles and asymmetric loading, *ASME Journal of Engineering for Industry*, 101(1), 1–16, 1979.
157. A.M. Whitman, A.M. Khaskia, Freight car lateral dynamics—an asymptotic sketch, *ASME Journal of Dynamic Systems, Measurement, and Control*, 106, 107–113, 1984.
158. H. True, On the theory of nonlinear dynamics and its application to vehicle system dynamics, *Vehicle System Dynamics*, 31, 393–421, 1995.
159. S.D. Iwickni, S. Stichel, A. Orlova, M. Hecht, Dynamics of railway freight vehicles, *Vehicle System Dynamics*, 53, 995–1033, 2015.
160. H. Scheffel, Unconventional bogie designs-their practical basis and historical background, *Vehicle System Dynamics*, 26(6–7), 497–524, 1995.
161. M.G. Pollard, The development of cross-braced freight bogies, *Rail International*, 9, 736–758, 1979.
162. H.A. List, An evaluation of recent developments in railway truck design, *IEEE/ASME Joint Railroad Conference Paper 71-RR-1*, New York, 19–21 April 1971.
163. R. Stephenson, Axle-trees, *British Patent* 5325, 1826.
164. R. Stephenson, Axles and bearings of railway vehicles, *British Patent* 6092, 1831.
165. Report of the Gauge Commissioners with Minutes of Evidence, Questions 1628-9, p. 101, London, UK, 1846.
166. Report of the Gauge Commissioners with Minutes of Evidence, Question 1795, p. 115; question 4246, p. 204; questions 6155–6176, p. 315, London, UK, 1846.
167. Report of the Gauge Commissioners with Minutes of Evidence, Question 6158, p. 315, London, UK, 1846.
168. F. Frederich, Possibilities as yet unknown regarding the wheel/rail tracking mechanism, *Rail International*, 16, 33–40, 1985.
169. P. Becker, On the use of individual free rolling wheels on railway vehicles, *Eisenbahntechnische Rundschau*, 19, 457–463, 1970.
170. B.M. Eickhoff, R.F. Harvey, Theoretical and experimental evaluation of independently rotating wheels for railway vehicles, In R. Anderson (Ed.), *Proceedings 11th IAVSD Symposium*, Kingston, Canada, 21–25 August 1989, Swets and Zeitlinger, Lisse, the Netherlands, pp. 190–202, 1990.
171. B.M. Eickhoff, The application of independently rotating wheels to railway vehicles, *Journal of Rail and Rapid Transit*, 205(1), 43–54, 1991.
172. J.A. Elkins, The performance of three-piece trucks equipped with independently rotating wheels, In R. Anderson (Ed.), *Proceedings 11th IAVSD Symposium*, Kingston, Canada, 21–25 August 1989, Swets and Zeitlinger, Lisse, the Netherlands, pp. 203–216, 1990.
173. A.D. de Pater, Analytisch en synthetisch ontwerpen, Public Lecture, Technische Hogeschool van Delft, Delft, the Netherlands, pp. 37–38, 1985.
174. A. Jaschinski, H. Netter, Non-linear dynamical investigations by using simplified wheelset models, In G. Sauvage (Ed.), *Proceedings 12th IAVSD Symposium*, Lyon, France, 26–30 August 1991, Swets and Zeitlinger, Lisse, the Netherlands, pp. 284–298, 1992.
175. D. Jenkinson, *British Railway Carriages of the 20th Century*, Volume 1: The end of an era 1901–22, Patrick Stephens, Wellingborough, UK, 155, 1988.
176. D. Jenkinson, *British Railway Carriages of the 20th Century*, Volume 2: The years of consolidation 1923–53, Patrick Stephens, Wellingborough, UK, 197, 1990.
177. P. Tachet, J.-C. Boutonnet, The structure and fitting out of the TGV vehicle bodies, *French Railway Techniques*, 21(1), 91–99, 1978.
178. M.C. Lopez-Luzzatti, M.G. Eruste, Talgo 1942–2005: De un Sueno a la alta velocidad, Patentes Talgo, Madrid Spain, 2005.
179. J. Carballara, L. Baeza, A. Rovira, E. Garcia, Technical characteristics and dynamic modelling of Talgo trains, *Vehicle System Dynamics*, 46(S1), 301–316, 2008.
180. R.D. Rose, Lenkung und Selbstlenkung von Einzelradsatzfahrwerken am Beispiel des KERF im S-Tog Kopenhagen, In I. Zobory (Ed.), *Proceedings 4th International Conference on Railway Bogies and Running Gears*, Budapest, Hungary, pp. 123–132, 1998.

181. E. Slivsgaard, J.C. Jensen, On the dynamics of a railway vehicle with a single-axle bogie, *Proceedings of the 4th Mini Conference on Vehicle System Dynamics, Identification and Anomalies*, Budapest, University of Technology and Economics, 197–207, 1994.
182. M.J. Nadal, Theorie de la stabilite des locomotives—Part II: mouvement de lacet, *Annales des Mines*, 10, 232–255, 1896.
183. A.O. Gilchrist, B.V. Brickle, A re-examination of the proneness to derailment of a railway wheelset, *Journal of Mechanical Engineering Science*, 18(3), 131–141, 1976.
184. N. Matsui, On the derailment quotient Q/P, Railway Technical Research Institute, Japanese National Railways Report, 1966.
185. L.M. Sweet, A. Karmel, S.R. Fairley, Derailment mechanics and safety criteria for complete rail vehicle trucks, In A.H. Wickens (Ed.), *Proceedings 7th IAVSD Symposium held at Cambridge University*, 7–11 September 1981, Swets and Zeitlinger, Lisse, the Netherlands, pp. 481–494, 1982.
186. L.M. Sweet, J.A. Sivak, Nonlinear wheelset forces in flange contact—Part 1: Steady state analysis and numerical results, *ASME Journal of Dynamic Systems, Measurement, and Control*, 101(3), 238–246, 1979.
187. L.M. Sweet, J.A. Sivak, Nonlinear wheelset forces in flange contact—Part 2: Measurements using dynamically scaled models, *ASME Journal of Dynamic Systems, Measurement, and Control*, 101(3), 247–255, 1979.
188. L.M. Sweet, A. Karmel, Evaluation of time-duration dependent wheel load criteria for wheel climb derailment, *ASME Journal of Dynamic Systems, Measurement, and Control*, 103(3), 219–227 1981.
189. L. Brennan, Means for imparting stability to unstable bodies, *British Patent* 27212, 1903.
190. J.F.S. Ross, *The Gyroscopic Stabilisation of Land Vehicles*, Edward Arnold, London, UK, 1933.
191. R.N. Arnold, L. Maunder, *Gyrodynamics and Its Engineering Applications*, Academic Press, New York, 218–227, 1961.
192. S.C. Spry, A.R. Girard, Gyroscopic stabilization of unstable vehicles: Configurations, dynamics and control, *Vehicle System Dynamics*, 46, 247–260, 2008.
193. W.E. Dorn, P.K. Beemer, Suspension system for railway vehicles, *U.S. Patent* 2,225,242, 1940.
194. W.E. Dorn, Suspension system for railway vehicles, *U.S. Patent* 2,217,033, 1940.
195. Anonymous, Three pendulum type cars being constructed by Pullman, *Railway Mechanical Engineer*, 114, 126, 1940.
196. Anonymous, Pendulum cars for the Santa Fe, Great Northern and Burlington, *Railway Age*, 112, 248–252, 1942.
197. F. Mauzin, M. Chartet, M. Lenoir, A new pendulum type carriage for high speed traffic, *Revue Generale des Chemin de Fer*, 76, 581–593, 1957.
198. K.A. Browne, S.G. Guins, Suspension system for vehicles and the like, *U.S. Patent* 2,893,326, 1959.
199. A.R. Cripe, Radially guided, single-axle, above centre of gravity suspension for articulated trains, *U.S. Patent* 2,954,746, 1960.
200. A.J. Ispeert, A preliminary analysis of passive and active roll control systems for high speed railway vehicles, British Rail Research Technical Note DYN 34, Derby, UK, 1966.
201. A.O. Gilchrist, A history of engineering research on British Railways, Working Paper 10, Institute of Railway Studies and Transport History, York, UK, 40–45, 2009.
202. G.K. Koenig, Gli elettrotreni ad assetto variabile: Storia del progetto FIAT Y 1060: l'ETR. 400 e l'ETR. 450, Valerio Levi editore, Rome, Italy, 1986.
203. D.A. Hullender, D.N. Wormley, H.H. Richardson, Active control of vehicle air cushion suspensions, *ASME Journal of Dynamic Systems, Measurement, and Control*, 94(1), 41–49, 1972.
204. R.M. Goodall, R.A. Williams, A. Lawton, P.R. Harborough, Railway vehicle active suspensions in theory and practice, In A.H. Wickens (Ed.), *Proceedings 7th IAVSD Symposium held at Cambridge University*, 7–11 September 1981, Swets and Zeitlinger, Lisse, the Netherlands, pp. 301–316, 1982.
205. H. Kemper, Schwebende Aufhängung durch elektromagnetische Kräfte: Eine Möglichkeit für eine grundsätzlich neue Fortbewegungsart, Electrotech, *Zeus*, 59, 391–395, 1938.
206. A.H. Wickens, Preliminary analytical study of hunting of an idealised railway vehicle bogie, British Railways Research Department Report E442, Derby, UK, 1963.
207. C.K. Bennington, The railway wheelset and suspension unit as a closed loop guidance control system: A method for performance improvement, *Journal of Mechanical Engineering Science*, 10(2), 91–100, 1968.
208. W. Geuenich, C. Guenther, R. Leo, Dynamics of fiber composite bogies with creep-controlled wheelsets, In J.K. Hedrick (Ed.), *Proceedings 8th IAVSD Symposium held at Massachusetts Institute of Technology*, 15–19 August 1983, Swets and Zeitlinger, Lisse, the Netherlands, pp. 225–238, 1984.
209. J.P. Pascal, J.M. Petit, Dynamique ferroviaire active; vers l'asservissement des be gies, Le Rail, *Juillet-Aout*, 32–35, 1988.

210. R.M. Goodall, H. Li, Solid axle and independently-rotating railway wheelsets—a control engineering assessment, *Vehicle System Dynamics*, 33(1), 57–67, 2000.
211. L.J. Comrie, Inverse interpolation and scientific applications of the National Accounting Machine, *Supplement to the Journal of the Royal Statistical Society*, 3(2), 87–114, 1936.
212. S. Iwnicki (Ed.), *The Manchester Benchmarks for Rail Vehicle Simulation*, Supplement to Vehicle System Dynamics Volume 31, Taylor & Francis Group, Abingdon, UK, 1999.
213. P. Shackleton, S. Iwnicki, Comparison of wheel-rail contact codes for railway vehicle simulation: An introduction to the Manchester contact benchmark and initial results, *Vehicle System Dynamics*, 46(1–2), 129–149, 2008.
214. J. Evans, B. Berg, Challenges in simulation of railway vehicles, *Vehicle System Dynamics*, 47(8), 1023–1048, 2009.

3 Design of Unpowered Railway Vehicles

Anna Orlova, Roman Savushkin, Iurii (Yury) Boronenko, Kirill Kyakk, Ekaterina Rudakova, Artem Gusev, Veronika Fedorova and Nataly Tanicheva

CONTENTS

3.1 General Vehicle Structure, Main Functions and Terminology 43
    3.1.1 Car Bodies 47
    3.1.2 Running Gears, Bogies and Suspensions 49
    3.1.3 Couplers, Automatic Couplers and Draw Bars 51
    3.1.4 Pneumatic Brakes 53
3.2 Design of Car Bodies 54
    3.2.1 Passenger Coaches 54
    3.2.2 Freight Wagons 59
3.3 Running Gears and Components 70
    3.3.1 Wheelsets 71
    3.3.2 Axleboxes and Cartridge Type Bearings 75
    3.3.3 Wheels 77
3.4 Design of Freight Wagons Bogies 79
    3.4.1 Three-Piece Bogies 79
    3.4.2 Primary Suspended H-Frame Bogies 81
    3.4.3 Double-Suspension Bogies 82
    3.4.4 Auxiliary Suspensions (Cross-Anchors, Radial Arms, etc.) 83
3.5 Design of Unpowered Bogies for Passenger Coaches 86
3.6 Design of Inter-Car Connections 91
    3.6.1 Screw Couplers 91
    3.6.2 Automatic Couplers 92
    3.6.3 Draft Gear 95
    3.6.4 Buffers 98
    3.6.5 Inter-Car Gangways 100
3.7 Principles for the Design of Suspensions 103
    3.7.1 Suspension Characteristics in Vertical Direction 103
    3.7.2 In-Plane Suspension Stiffness 106
    3.7.3 Suspension Damping 108
    3.7.4 Car Body to Bogie Connections 110
References 112

3.1 GENERAL VEHICLE STRUCTURE, MAIN FUNCTIONS AND TERMINOLOGY

Unpowered railway vehicles are mainly operated in trains being drawn by a locomotive or several locomotives distributed along the train. This is the basic technology that is equally used for transportation of freight and passengers, providing high energy efficiency. Further, we refer to freight unpowered vehicles as wagons and passenger vehicles as coaches. Typical examples are shown in Figure 3.1.

(a) (b)

FIGURE 3.1 Examples of unpowered rail vehicles: (a) freight wagon and (b) passenger coach. ([b] From Transmashholding, Moscow, Russia. With permission.)

For any railway vehicle, the axle load (mass transmitted onto the rails by one wheelset) is the major characteristic, along with the maximum design speed. In freight transportation, axle loads between 22.5 t and 37.5 t are used, with design speeds below 160 km/h (the bigger the axle load, the smaller the design speed), whereas the tendency for passenger coaches is increasing the design speed while maintaining the axle load as small as possible.

Freight train consists can reach up to 48,000 t in total weight and up to 320 wagons, limited by locomotive capabilities and the landscape of the railway, to achieve maximum efficiency (see heavy-haul freight train example in Figure 3.2). Efficiency of the individual wagon is determined by its mass capacity and volume capacity that indicate the possible maximum mass of the cargo and the absolute maximum volume of the cargo that it can carry. The minimum possible mass of the wagon is its tare mass (when totally empty), and the maximum mass is the fully laden mass (tare plus the capacity) that should not exceed the limitation for the axle load.

The length of passenger trains is usually limited by platform size, and speed and comfort are valued more than the mass of the passengers; thus, coaches are characterised by passenger capacity and carry a lot of specific equipment to provide comfort.

The locomotive hauls the train, while each unpowered vehicle produces the running resistance due to interaction between the wheel and rail, oscillations and damping in the suspension, friction in bearings, air and gravity resistance, longitudinal oscillations between vehicles, etc. To stop the train, brakes are applied to each vehicle individually. The lifecycle of unpowered vehicles also includes loading and unloading operations (with external or built-in devices), shunting manoeuvres during assembling and disassembling of the train and coupling to and uncoupling from the adjacent vehicles. When in the train, the locomotive accelerates or decelerates the vehicles, and they pass straight or curved sections of the railway track, change tracks when passing through switches and crossings, experience wind and interact with bridges and tunnels.

All of these described operational scenarios need to be handled by vehicle design. The main parts of the unpowered railway vehicle are the car body, the bogies and the braking and coupling equipment, as shown in Figure 3.3.

Number of axles: In general, the vehicles are classified by the number of axles (wheelsets); examples are shown in Figure 3.4. Some special transport vehicles for super-heavy cargo can have two bogies with up to 8 axles in each of them. There may be more than two sections in articulated vehicles or one-axle bogies under the articulation. In addition, bogies in articulated vehicles with different numbers of axles can be positioned under the articulation (three-axle) and under the ends of the car body sections (two-axle). Because the axle load is limited, the number of axles is used to distribute the mass of the vehicle over the length of the railway track.

Design of Unpowered Railway Vehicles

FIGURE 3.2 Heavy-haul freight train.

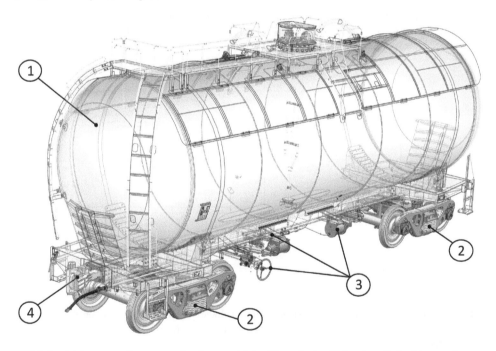

FIGURE 3.3 Main parts of unpowered railway vehicle. (1) car body, (2) bogies, (3) braking equipment and (4) coupling equipment.

The following characteristics are used to describe the geometry of railway vehicles:

Vehicle wheel base: Wheel base of the vehicle is the longitudinal dimension between the centres of outer bogies or between the centres of outer suspensions of the vehicle. For articulated vehicles, as shown in Figure 3.5, besides the wheel base of the whole vehicle, the wheel base of each section can be used, designating the longitudinal distance between the centres of neighbouring bogies.

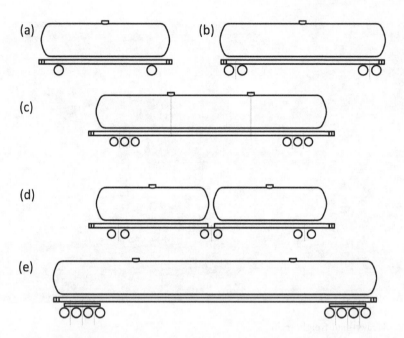

FIGURE 3.4 Types of rail vehicles by the number of axles: (a) 2-axle vehicle, (b) 4-axle vehicle with two 2-axle bogies, (c) 6-axle vehicle with two 3-axle bogies, (d) 6-axle articulated vehicle with three 2-axle bogies and (e) 8-axle vehicle with two 4-axle bogies.

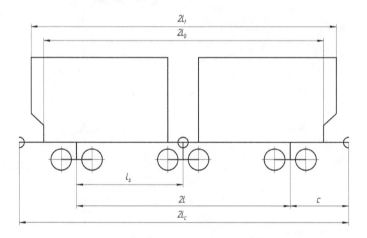

FIGURE 3.5 Dimensions of articulated vehicle: $2L$ – vehicle wheelbase, L_s – section wheel base, $2L_c$ – length between coupler axes, $2L_b$ – length between end beams, $2L_t$ – length of the car body and C – length of the cantilever part.

Length between coupler axes: The length of the vehicle in the train or how much space it takes up in the train is measured between the axes of the coupling devices. This parameter is also important for spacing the loading and unloading devices at the terminal facilities and for distributing the mass of the vehicles in the train over the length of the track. The bearing capacity of bridges that are longer than one vehicle is determined by the permissible load per unit length of the vehicle, that is, fully laden weight of the vehicle divided by its length between coupler axes. To provide safe curving, the length of four-axle vehicles between coupler axes does not exceed 25 m.

Design of Unpowered Railway Vehicles 47

Length of the cantilever part: The difference between the length between coupler axes and vehicle wheel base is double the length of the cantilever part. This parameter becomes important when the vehicles in the train are passing curved sections of the track, because it determines the lateral displacement of the coupling devices.

Car body length: Car body length is its maximum dimension along the track. In most vehicles, car body length equals the length between end beams, but some designs have a protruding top part that increases the loading volume.

Coupler height (above the rail top level): This parameter determines the safety of coupling and prevents uncoupling when the train is moving over summit parts of sloped track. In production, the coupler height is measured in empty vehicles, and it depends on the tolerance for bogie height and tolerances of car body centre sill, as well as the tare weight of the empty car body. In operation, the coupler height is even more variable, as it is dependent on wheel diameter that changes due to turning, suspension deflection that increases when the vehicle is loaded, wear of bogie horizontal surfaces and wear of coupling device surfaces. The railway administrations set safety limits for coupler height in empty and laden vehicle conditions [1,2].

Clearance diagram: The vehicle dimensions in cross-section laterally to the track are limited by the so-called clearance diagram. Two tracks can be laid next to each other, various platforms and other buildings stand near the track and tracks can go over bridges and into tunnels, and on electrified railways, there is the contact wire running over the track. To limit the vertical (top and bottom) and lateral dimensions of the vehicle and to provide safe clearance between the vehicle and stationary structures on straight track and in curves, the clearance diagram was introduced. Conformity of vehicles to clearance diagrams is a basic safety requirement of all railway administrations [3–5].

3.1.1 Car Bodies

During almost 200 years, the evolution of freight wagons and coaches created a huge number of various car body designs for many purposes. Typical examples are shown in Figure 3.6. There are specific freight wagons to carry bulk named gondola wagons (either drop bottom or solid bottom), to carry grain or mineral fertilisers with the capability of discharge though bottom hatches named hopper wagons, to carry liquids named tank wagons, to carry containers named flat wagons, as well as more specific designs.

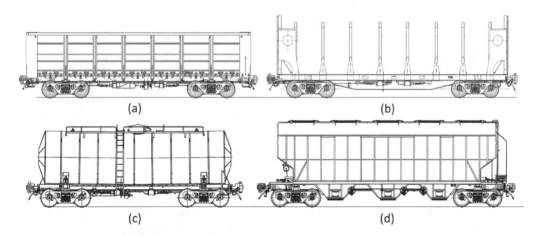

FIGURE 3.6 Various types of freight wagon car bodies: (a) drop-bottom gondola, (b) timber flat wagon, (c) tank wagon and (d) hopper wagon.

FIGURE 3.7 Double deck passenger coach. (From Transmashholding, Moscow, Russia. With permission.)

Passenger coaches differ by either providing travel for sitting or sleeping passengers; there also exist restaurant coaches, laboratory coaches with measurement equipment to monitor railway condition and many other specific types. To increase the passenger capacity of the coach, double decks, as shown in Figure 3.7, can be introduced if the clearance diagram permits.

No matter the variety of car body types, in rail vehicle dynamics, they can be presented either as rigid bodies or as rigid bodies with certain elastic properties. If eigenfrequencies of the car body structure in the empty or laden condition are within the range of the suspension oscillation frequencies (0–20 Hz), then elastic properties of the car body will influence the performance of the vehicle on track and should be taken into account when simulating vehicle dynamics. A typical example of elastic behaviour of a freight wagon is bending oscillations of an 80-foot long container flat [6], as shown in Figure 3.8. The situation of low car body frequencies appears more often for passenger coaches because reduction of tare mass makes the body more flexible. In passenger vehicles, the eigenmodes of the car body not only influence the ride performance but also affect passenger comfort, and standards therefore limit the value of the first eigenmode frequency to be higher than 8 or 10 Hz [7]. Research is being done by Japanese scientists [8] that is aimed to use piezoelectric technology to provide additional damping for eigenmodes of super-light car bodies. Other research tends to implement control in suspensions to dampen the car body's eigenmodes as well as its rigid-body oscillations [9].

The car body is the part that establishes the existence of the railway vehicle in technical terms; it unites all systems, houses cargo or passengers to provide safety of transportation in all operational modes and bears the registration number of the vehicle as the only non-replaceable part. The car body rests on bogies (or suspensions) and interacts with them through car body to bogie connections. Adjacent car bodies are joined in a train via inter-car connections: coupling devices, passages and additional inter-car dampers to improve ride quality. The car body is used to position braking equipment.

Interaction of wagons and cargo is a separate and very important part of vehicle system dynamics. Semi-trailers have their own road suspension and oscillate while the train is in motion [10]. Liquid cargo sloshes inside a tank, changing its behaviour on straight track and in curves [11,12]. Wood or even bulk cargo can displace under the action of longitudinal impacts.

Interaction of vehicle and passengers is also important, producing much research on the perception of oscillations by people [13] and modelling of human bodies in different postures [14].

Design of Unpowered Railway Vehicles

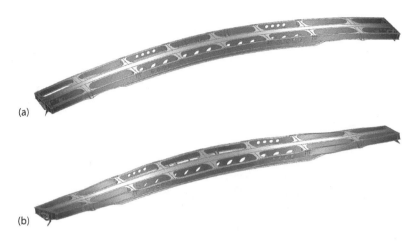

FIGURE 3.8 Typical examples of elastic behaviour of freight flat wagon: (a) bending mode with 11.5 Hz eigenfrequency and (b) torsion mode with 11.8 Hz eigenfrequency.

3.1.2 Running Gears, Bogies and Suspensions

The principal difference between a railway vehicle and other types of wheeled transport is the guidance provided by the railway track. The surface of the rails not only supports the wheels but also guides them in the lateral direction. The rails and the switches change the rolling direction of wheels and thus determine the direction of travel of the railway vehicle. Irregularities on the rails produce oscillations of the vehicles.

The running gear is the system that provides safe motion of the vehicle along the railway track. The running gear includes such components as wheelsets with bearings, the elastic suspension, the brakes and the device to transmit traction and braking forces to the car body. Its main functions are as follows:

- Transmission and equalisation of the vertical load from the wheels of the vehicle to the rails
- Guidance of the vehicle along the track
- Control of the dynamic forces due to motion over track irregularities, in curves, switches and after impacts between the vehicles
- Efficient damping of oscillations
- Application of braking forces

Depending on the running gear, the vehicles may be described as bogied or bogie-less.

In vehicles without bogies, the suspension, brakes and traction equipment are mounted on the car body. The traction and braking forces are transmitted through traction rods or axlebox guides (sometimes known as 'hornblocks') straight onto the wheelset. The typical example is the conventional two-axle wagon (see Figure 3.9), which generates larger forces in tight curves than the equivalent bogie vehicle; therefore, the length of the former is limited.

Running gear mounted on a separate frame that can turn relative to the vehicle body is known as a bogie (or truck). The number of wheelsets that they unite classifies the bogies. The most common type is the two-axle bogie, but three- and four-axle bogies are also encountered. Typical examples are shown in Figure 3.10.

Early bogies simply allowed the running gear to turn in the horizontal plane relative to the car body, thus making it possible for the wheelsets to have smaller angles of attack in curves. In modern bogies, the bogie frame transmits all the longitudinal, lateral and vertical forces between the car body and the wheelsets. The frame also carries braking and traction equipment, suspension and

FIGURE 3.9 Bogie-less two-axle wagon. (From Corporate Media People, Moscow, Russia. With permission.)

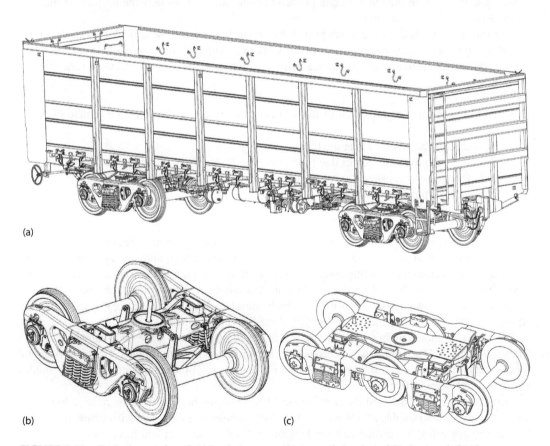

FIGURE 3.10 Typical examples of: (a) bogied wagon, (b) two-axle bogie and (c) three-axle bogie.

Design of Unpowered Railway Vehicles

dampers. It may house tilting devices, lubrication devices for wheel-rail contact and mechanisms to provide radial positioning of wheelsets in curves. Bogied vehicles are normally heavier than two-axle vehicles. However, the design of railway vehicles with bogies is often simpler than for two-axle vehicles, and this may provide reliability and maintenance benefits.

The following characteristics are used to describe the bogies:

Track gauge: To correspond to the track gauge of the railway and to provide appropriate wheelsets.

Axle load and design speed: These are used to check the compatibility of the bogie with the designed vehicle.

(Bogie) wheel base: Wheel base of the bogie is the longitudinal dimension between the centres of its outer wheelsets. This parameter is important for distributing the mass of the vehicles over the length of the track. The bearing capacity of bridges that are shorter than one vehicle is determined by the permissible load per unit length of the vehicle, that is, fully laden weight of the vehicle per one bogie divided by bogie wheel base.

Car body support height over the top of rails: The height is measured between the top of the rail and the centre bowl surface or the surface of the other elements that support the car body. In production, measurements are usually taken in free bogie condition and can vary depending on wheel diameter tolerance, suspension element tolerances, production tolerances of cast or fabricated frames and beams. After the car body is positioned over the bogie, the height reduces, depending on the suspension deflection under the tare weight. The parameter is important for safety to provide tare vehicle coupler height.

Clearance diagram: The bogie dimensions in cross-section laterally to the track are limited by the so-called clearance diagram. Besides requirements that limit the lateral dimensions of the bogies in the same way as the dimensions of the car bodies, there exist special limits on the vertical distance between the bottom of the bogie and on-track devices.

Suspension deflection under empty/laden vehicle conditions: The values of deflection are used to check whether the bogie is appropriate under the chosen vehicle and will provide safety limits of coupler height.

3.1.3 COUPLERS, AUTOMATIC COUPLERS AND DRAW BARS

Various devices are positioned in the end sections of the car body to provide connection between adjacent vehicles in a train, transmitting longitudinal tension and compression forces, providing damping of longitudinal train oscillations and allowing necessary angles of rotation and vertical and lateral displacements for safe passing of curves and humps. These same devices are working when the train is formed out of separate wagons or coaches, facilitating coupling and uncoupling operations when impacts between vehicles need to be handled by the coupling devices.

In general, the coupling devices can work either manually (see screw coupling design in Figure 3.11) or in automatic mode (see automatic coupler design in Figure 3.12). Manual operation is done by the railway staff who lift the screw coupling and connect it to the adjacent vehicle; thus, the screw coupling [15] weight and tension capacity are limited by human capability. Automatic operation is done either by running the vehicle off the sorting yard hump or by moving it by locomotive, with an impact onto the adjacent vehicle. The speed range during sorting is usually below 5 km/h. The special shape of the automatic coupler surfaces [16,17] guides the automatic couplers into their locking position. However, to unlock the automatic couplers, manual force is needed to raise the lock. The introduction of automatic couplers allowed the transmission of much bigger tension as well as compression forces in the train, thus providing for increases in train length and weight. It also speeds up the train sorting operations and provides much higher safety for railway staff who do not need to go between the vehicles during train forming operations.

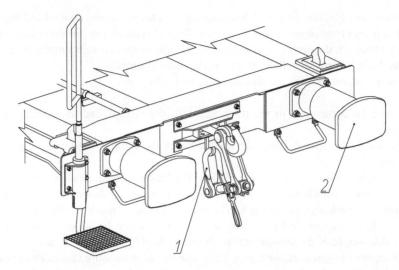

FIGURE 3.11 Wagon with (1) screw coupling and (2) a pair of buffers.

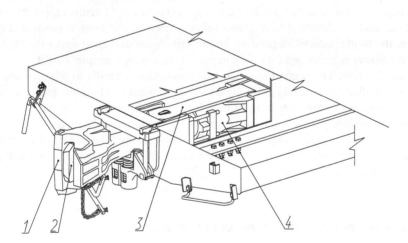

FIGURE 3.12 Wagon with SA-3 automatic coupler and draft gear. (1) coupler body; (2) lock; (3) yoke; and (4) draft gear.

Two adjacent automatic couplers have longitudinal slack (clearance) between the coupling surfaces that results from tolerances and wear in operation. In train starting mode, the slack allows the locomotive to start moving the vehicles one by one and not the whole train consist at once, thus providing energy efficiency. However, in train transition modes from traction to braking, the slack results in impacts between adjacent vehicles that increase accelerations and make ride comfort poor. Therefore, slackless couplers are often used in coaches instead of automatic couplers. In freight wagons, two or more of them can be connected by slackless draw bars if loading and unloading operations allow it.

Vehicles in one train can have different coupler heights due to wear of wheels or deflection of the suspension that can vary due to oscillations in motion. Therefore, the couplers need to have a possibility of relative vertical displacement and are characterised by a safety limit on coupler height difference.

To provide damping of the oscillations in a train, the screw coupling is usually used together with buffers (Figure 3.11) that work only in compression and provide elastic resistance and friction damping [18].

Design of Unpowered Railway Vehicles

In automatic couplers, the special damping device named the draft gear is installed inside the centre sill (Figure 3.12). The draft gear is the damper with initial pre-compression that works in compression when tension or compression forces are applied to the coupler. The yoke transforms both tension and compression forces into compression of the draft gear. The main characteristic of the draft gear is its energy-absorption capacity under the impact conditions. Friction draft gears have the lowest energy capacity; higher capacity is provided by friction-elastomer draft gears, elastomer or even hydraulic ones.

3.1.4 Pneumatic Brakes

The major brake type for unpowered railway vehicles around the world is the automatic pneumatic brake, where the brake control signals are transmitted from the locomotive onto each vehicle in a train by changing the pressure in the pneumatic train line brake pipe. Each vehicle houses braking equipment such as the brake pipe, air distributor, spare tank, brake cylinder, brake rigging, brake shoes and an empty-load device, as shown in Figure 3.13.

When the pressure in the train line brake pipe increases in charging mode or brake release mode, then the air distributor charges the spare tank and connects the brake cylinder with atmosphere. When the brake pipe pressure decreases in brake application mode, then the air distributor connects the spare tank with the brake cylinder. The braking force from the shaft of the brake cylinder is transmitted through the brake rigging onto the brake shoes that press against the wheels in tread brakes. In disc brakes, the force is transmitted by callipers onto the brake pads that press against the brake discs. Friction between the brake shoes and wheel treads, or between the brake pads and discs, decelerates the vehicle.

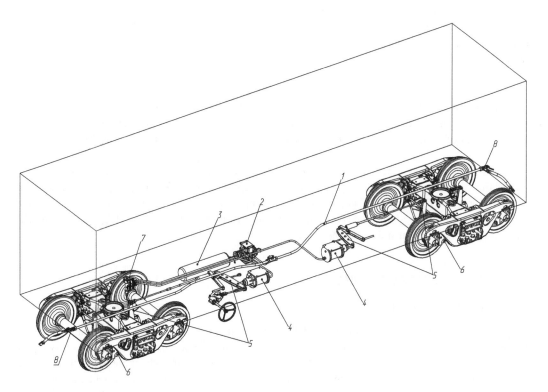

FIGURE 3.13 General composition of braking equipment on a freight wagon. (1) train line brake pipe; (2) air distributor; (3) spare tank; (4) brake cylinder; (5) brake rigging; (6) brake shoes; (7) empty-load device; and (8) end cock.

The term 'automatic' is used for this type of brake equipment because it automatically applies the brake when the vehicles in the train or the train line brake pipe disconnect in the case of an accident.

Pneumatic brake can be applied in normal operational mode to slowly decelerate the train or in emergency braking mode that corresponds to the same situation of an open mainline in the case of an accident.

The following characteristics are used to describe the vehicle brake system:

Shoe force: The physical value of the force applied from the brake shoe to the wheel tread.

Friction coefficient: The friction coefficient between the brake shoe or pad and the wheel tread or brake disc surface. The coefficient is non-linearly dependent on the vehicle speed and the materials of the friction surfaces.

Vehicle stop distance: The distance that the single vehicle will travel from its initial speed to a full stop after the emergency brake is applied or the train line brake pipe disconnects. The stop distance of a single vehicle is measured in tests on straight horizontal track.

Train stop distance: The distance that the train consisting of identical vehicles will travel from its initial speed to a full stop after the emergency brake is applied. Train stop distance is usually bigger than the vehicle stop distance because the pressure decrease wave needs to travel through the train to reach every vehicle. Train stop distance is highly dependent on track gradient.

Skidding: Skidding between the braked wheel and the rail occurs when the brake force is bigger than the wheel-rail traction force. It increases the stop distance and produces wheel and rail defects.

Wheel-rail traction force depends on the actual axle load of the vehicle. This becomes important for wagons because the weight of the empty one can be five times lower than that of the laden one. To avoid skidding, the pressure in the brake cylinder needs to depend on the actual weight of the car body. Special empty-load devices control the brake cylinder pressure, depending on suspension deflection. The anti-skidding devices are sometimes introduced in passenger coaches, especially with disc brakes.

Braking is a source of longitudinal compression forces originating in couplers of vehicles in a train. Non-synchronous braking of vehicles makes the back part of the train run onto the front part, creating the compression forces. This problem is specifically important in long freight trains and is often dealt with by driving them using several locomotives distributed along the train [19].

When assessing derailment safety, it is necessary to simulate the situation when the heavy train passes through the curve and the emergency brake is applied that produces large quasi-static compression forces on the couplers that press the light (empty) vehicle out of the curve [20]. On straight track, the large compression forces can lead to various train buckling modes [21,22], and again, the assessment of derailment stability is necessary.

3.2 DESIGN OF CAR BODIES

3.2.1 Passenger Coaches

To provide the requirements of stiffness, deformations and structural strength, the car bodies for passenger coaches bear the external loads from all parts, such as the frame with the floor, two side walls and two end walls and the roof (see Figure 3.14). Inside the car body, there are structures that do not bear main loads, such as internal walls that split the interior into compartments, toilets, staff rooms, etc.

Splitting the bearing structure of the car body by functions, it is possible to distinguish the top compartment (under the roof), the middle compartment (being actually the space utilised by

Design of Unpowered Railway Vehicles

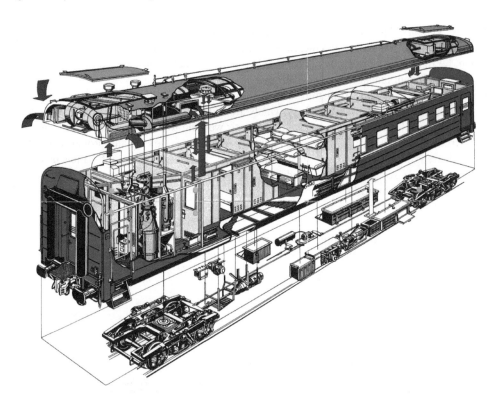

FIGURE 3.14 Main parts of a passenger coach car body.

passengers) and the bottom part (where various items of equipment are installed). The top compartment houses ventilation and conditioning systems, water tanks and tubes. In the bottom part, the frame supports the power generator, batteries, refrigeration units, tanks for used water and other devices providing comfort. In some coaches, the bottom part is open; in others, it is covered with decks or can even be protected by a load bearing structure to increase the car body stiffness.

Most common material for car bodies is low-alloyed steel. In the desire to make the car body lighter, to increase its reliability and durability as well as improve its external look, designers are constantly searching for new materials, designs and technology solutions. In high-speed trains, the aluminium alloys and stainless steel initially found application. Later, various non-metal materials such as plastics, fibreglass and carbon fiber reinforced polymer composites, multi-layer panels and honeycomb panels were implemented.

Aluminium alloys have better material strength and lower density than steel, do not corrode and are easy to cut and shape. Aluminium car bodies can dissipate much energy to deformation in conditions of an accident, thus providing better safety. At the same time, aluminium alloys have an elasticity modulus that is three times lower than that of steel, which makes the structures prone to stability loss under compressive loads and decreases their stiffness and modal oscillation frequencies. To override these problems, the aluminium-bearing structures need bigger sizes of cross-sections than steel ones.

The bearing structures of coaches can be classified into three types:

Reinforced sheets: The main bearing structure are the metal (steel, aluminium or stainless steel) sheets that are reinforced by beams with various shapes of cross-sections. Sometimes, to prevent the sheets from buckling, the corrugated design is used, or the sheets are reinforced with stringers. Beams and sheets are assembled with welding or rivets. Use of welding produces visible deformations of sheets that spoil the visual appearance of the car body.

Carcass systems: The main bearing structures are massive rods with open or closed cross-sections. External sheets just increase the general stiffness of the system. External sheets can be light alloys or plastics that provide a better visual appearance and do not need to bear significant loads.

Panel systems: Loads are transmitted onto the bearing panels that consist of external and internal walls joined together. Such panels are more often produced out of extruded aluminium profiles joined with arc welding or stir welding to form the floor, walls and roof. Extruded panels after stir welding do not have any surface defects and provide high structural strength and good visual appearance. In another case, honeycomb panels (sandwich systems) consisting of external and internal shells joined together with porous polyurethane or foam-aluminium or composite filling are used. Advantages of panels are their ability to damp vibrations and to provide sound and thermal insulation.

Let us consider the car body design with beam-reinforced metal sheets.

The frame (Figure 3.15) consists of the centre sill 1, transverse beam 2, end beam 3, side beams 4, lateral beams 5, lateral floor beams 6 and metal floor sheet 7 that has openings for various service pipes and ducts to pass through.

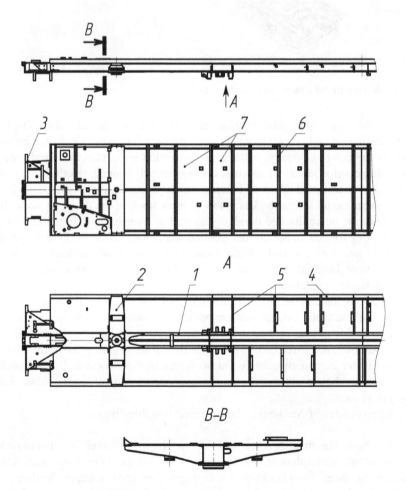

FIGURE 3.15 Passenger coach frame. (1) centre sill; (2) transverse beam; (3) end beam; (4) side beams; (5) lateral beams; (6) lateral floor beams; and (7) metal floor sheet.

Design of Unpowered Railway Vehicles

The roof (Figure 3.16) has uniformly spaced arches 1 that are covered with corrugated sheets 2 having a 2 mm thickness and having bent side sheets 3. The roof has air supply openings 5, hatch 4 and covers 6 and 7 that provide access to various equipment in the top part of the coach.

The side wall (Figure 3.17) is the welded whole-metal structure constructed out of corrugated and plane sheets with thickness of 2–2.5 mm supported by carcass bracing on the inside. The side wall consists of the top belt 1 and bottom belt 2, which are interconnected between each other with fabricated beams 3, and windows 4 and 5 comprising the middle belt. The wall is reinforced with profiled elements of the top beams 6, racks for the doors 7, vertical racks 8 and longitudinal beams 9.

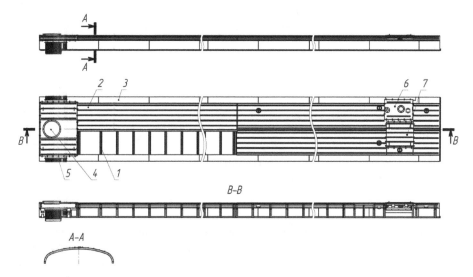

FIGURE 3.16 Passenger coach roof. (1) Arch; (2) corrugated sheet; (3) side sheet; (4) hatch; (5) air supply; (6) and (7) covers.

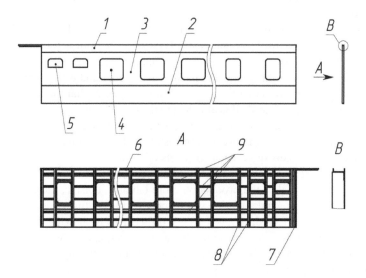

FIGURE 3.17 Passenger coach side wall. (1) Top belt; (2) bottom belt; (3) middle belt with fabricated beams; (4) compartment window; (5) toilet window; (6) top beam; (7) rack for the door; (8) vertical rack; and (9) longitudinal beams.

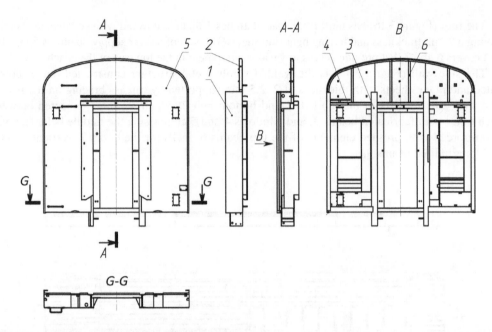

FIGURE 3.18 End wall of the passenger coach. (1) corner racks; (2) roof arc; (3) anti-impact rack; (4) lateral beam; (5) metal sheet; and (6) reinforcement ribs.

The end wall (Figure 3.18) consists of the corner racks 1, roof arc 2 and two anti-impact racks 3, which are welded to the bottom of the frame end beam (see Figure 3.15) and to the lateral beam 4, that altogether provide safety for passengers in the case of accidents. The bearing structures are covered with metal sheets 5, additionally reinforced by ribs 6.

Years of operation have shown that relatively simple reinforced sheet structures of car bodies have one significant disadvantage. Their lower bending mode frequency is often less than 8 Hz, and they can come into resonance with suspension oscillations worsening the ride performance and passenger comfort.

In carcass-type designs, the coach frame consists of the centre sill, with side and transverse beams having box cross-sections. Corrugated sheets used in the bottom and floor further increase the stiffness of the whole system. The side walls of the car body are based on a grid of vertical and horizontal beams. Parts of the beams are narrow and higher and others are wider and thinner. This allows joining them by stamping thinner elements into the higher ones. Altogether, this provides a stiff and stable system. The shell is attached to the carcass with spot welding, thus providing a better visual appearance, as spot welding gives less buckling and deformation.

To increase the roof stiffness, the arches are installed. Longitudinal stiffness of the roof can either be provided with stringers or corrugated sheets of the shell.

General stiffness of a carcass-type car body is higher than that for sheet structures; however, the variability of elements makes the production more labour intensive, increasing the cost of the design. Spot welding needs special attention when applying the coatings.

The typical example of carcass design is the TGV train car body shown in Figure 3.19. The frame is not symmetrical because the example shows the end coach that couples to the locomotive when forming the train. The left part of the car body rests on the bogie through a transverse beam (see 7 on side A), and the right part (side B) rests on an articulated bogie that supports two adjacent car bodies simultaneously. Pay attention to the massive carcass of the side walls. The bottom part of the car body consists of frames joined with longitudinal elements, while the floor and the roof are corrugated sheets. Testing showed that the first bending frequency is higher than 10 Hz.

Design of Unpowered Railway Vehicles 59

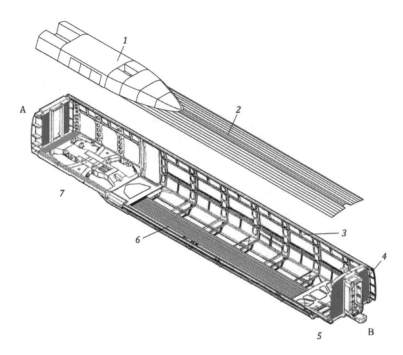

FIGURE 3.19 Design schematic of carcass type car body. (1) roof cowl; (2) roof; (3) side wall; (4) end wall; (5) articulated bogie position; (6) car body frame; (7) individual bogie position; (A) end of the car body facing the locomotive; (B) end of the car body facing the next articulated coach.

The panel type design allows to have an even lighter, stiffer and more reliable car body structure. The main idea is to use large extruded aluminium profiles for the external shell (see Figure 3.20). Extruded profiles are joined with automatic arc welding, followed by grinding the welds or by stir welding. The resulting panels have large bending stiffness and local structural strength to withstand point loads. This property is used to attach internal and external equipment. The main advantages are low production cost and a smaller number of components.

However, the extruded panels are often too strong and do not provide sufficient sound and thermal insulation; in such cases, they are replaced with sandwich or honeycomb panels.

A typical example of panel design is the Talgo train coach (Figure 3.20a). It uses extruded aluminium profiles for the floor and walls. The bottom part of the floor has guides that are used to support the equipment under it. Side walls have windows and doors reinforced with vertical beams.

By the end of the 1990s, much research work was aimed to choose the best design of car bodies for coaches and locomotives. The lightest car bodies with maximum stiffness and structural strength were found to be provided by soldered honeycomb panels that are, at the same time, the most expensive in production. Riveted airplane-type designs were tested as well and did not find practical application. Currently, the double-skin design constructed out of volumetric extruded aluminium profiles seems to have good prospects; however, it is mainly used in high-speed rollingstock.

3.2.2 Freight Wagons

The design of a freight wagon body is determined by its purpose [23–26]. Railway freight wagons are used to transport liquid, loose or bulky product, piece-freight goods, lumber, civil structures, equipment, pipes, wheeled and tracked vehicles, containers and items requiring cooling and maintaining at stable temperatures during transport. To transport each type of cargo, multi-purpose or special

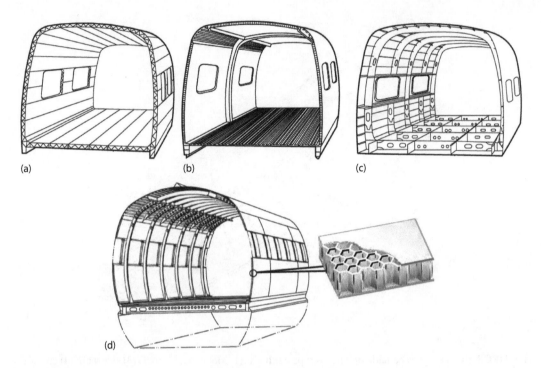

FIGURE 3.20 Schematic of various types of panel car body designs: (a) out of volumetric panels, (b) out of honeycomb elements, (c) riveted airplane type design and (d) honeycomb structure.

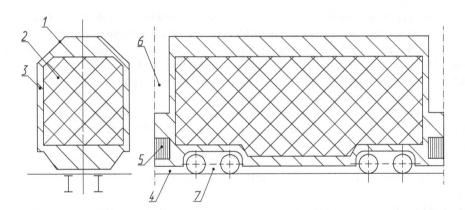

FIGURE 3.21 Using the clearance diagram to determine the car body structure. (1) clearance diagram, (2) transported cargo, (3) space to position the car body structure, (4) bottom clearance diagram, (5) space to position the coupling devices, (6) maintenance zone and (7) space to position the bogies and running gears.

rollingstock is employed. When in transit, the cargo produces impact forces and can cause corrosive and abrasive wear in the vehicle, whereas the wagon body may contaminate or damage the cargo.

The main characteristics of the freight car body are the volume of the carried cargo and the availability of devices for loading, fastening and unloading of the cargo. Since the wagon is designed in a limited clearance diagram and with a minimum possible car length over the couplers, effective volumetric capacity of the car body should tend to reach the volume of the clearance diagram. The space between the cargo and the clearance diagram is used to accommodate the car body structure (see Figure 3.21).

The freight car body can be nominally represented by a box-section beam mounted on two supports, that is, bogies, and subjected to static and dynamic loads from cargo, bogies and adjacent

Design of Unpowered Railway Vehicles

rollingstock transferred through coupling devices. The body structure shall have a minimum weight and a minimum cost with the given strength and stiffness.

The full diversity of freight car body designs can be divided into several basic types.

Enclosed body designs are used in box wagons, hopper wagons, auto carriers, refrigerated wagons, etc. Typical examples are shown in Figure 3.22. A box car body design represents a closed box-section beam. In this type of body, cargo impacts the floor deck by bending the body like a beam lying on two supports, the underframe and the floor deck jointly receive loads in the same fashion as a bottom flange of a nominal beam, the side walls of the car transfer shear stresses and participate in the load transfer in the same fashion as beam webs and the roof with the upper chords

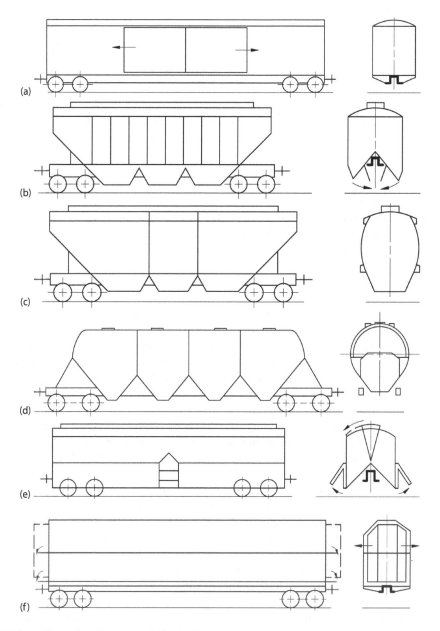

FIGURE 3.22 Wagons with enclosed car body: (a) box wagon, (b) hopper wagon with straight walls, (c) hopper wagon with round walls, (d) tank type hopper wagon, (e) dump wagon and (f) refrigerated wagon.

of the side walls functions like the top flange of the beam. The body can have either flat side walls, with the roof reinforced by posts and horizontal sills, or cylindrical side walls and the roof resting upon—two to three transverse partitions. In the latter case, stiffness and stability of plates are ensured by their cylindrical surface. Cylindrical wall designs have shorter welds, yet thicker plates and greater weight.

The lateral pressure from loose or palletised load is transferred to the sheathing and side wall posts, which redistribute it to the car underframe and roof. The underframe, as a rule, is sufficiently lightweight and can have a centre sill extending the entire length of the car or only two middle sills at the ends, where coupling devices are mounted.

The end wall taking longitudinal dynamic loads from cargo usually consists of a flat sheet and a case frame. Stamped corrugated sheets, light and robust, are also used without additional sills, the function of which is carried out by corrugations.

The main challenges faced in developing enclosed body designs are wide openings of sliding doors, protecting the lightweight structure of the floor decking, causing damage to cargo in transit and being able to withstand concentrated loads from loaders, as well as the connection of hopper car pockets with the underframe and location of discharge hatches.

Open body designs are used in gondola wagons with discharge hatches, solid-bottom gondola wagons and open hopper wagons, as shown in Figure 3.23. Owing to an open-top design, the upper chord of the side wall takes up compressive stress resulting from the bending of the car, for which, generally, a reinforced rectangular hollow section is used. The upper chord of the side wall also resists the load when the car is turned over during unloading on a car dumper and can also be subjected to impact loads from a grapple or excavator bucket. Owing to the open-top design, the side wall is attached to the underframe in a cantilever fashion, and the side wall posts resist the bending moment from the load lateral pressure. The largest bending torque is observed at the bottom of the posts and at the junction with crossbearers. An open-top design with internal transverse trusses is occasionally used, which receives a lateral pressure acting on the side walls and transfers forces from the side walls to the centre sill. Although such a technical solution makes it possible to reduce the body weight, it worsens unloading conditions for bulk cargo, in particular if frozen at low temperatures.

Gravity-based unloading of cargo without the use of a rotary dump or grapple can be carried out through discharge hatches made in the floor of the open body. The hatches can take up the entire area of the body, and the end and side walls can then be made vertical, with the maximum body volume attained. If the hatches only take up a part of the car surface area, the lower portion of end and side walls needs to be inclined to direct cargo into hatches. Although the use of the hatches enables the car to be unloaded at an unloading point without technical means, it increases the weight and cost of the body and reduces the volume of the carried cargo.

The main challenges faced in designing the open body are the provision of fatigue resistance of side wall posts to underframe attachments, strength of the underframe with large hatches under longitudinal dynamic loads and strength of the upper chord of the side wall during mechanised loading and unloading.

Load-carrying underframes are employed in flat wagons serving different intended purposes – for containers, for multi-purpose use and for the transportation of wood, large-diameter pipes, coiled and sheet steel, semi-trailer trucks, etc. The flat wagon body consists of an underframe and, when so required by the intended purpose, end walls, side posts, side and end swivel flaps, end platforms and cargo securing devices. As shown by the examples in Figure 3.24, the load can be placed above the underframe, partially lowered into special niches or placed in a well of the underframe.

The flat wagon underframe can have a load-bearing centre sill, load-bearing side sills or a combination of the two. Since the size of the loading area is limited by the upper part of the gauge profile and the top surface of the underframe, developers try to lower the loading surface of the underframe as much as possible. Therefore, in long-wheelbase flat cars, the centre sill and side sills of the underframe extend downwards to the lower part of the loading gauge profile. If the side

Design of Unpowered Railway Vehicles 63

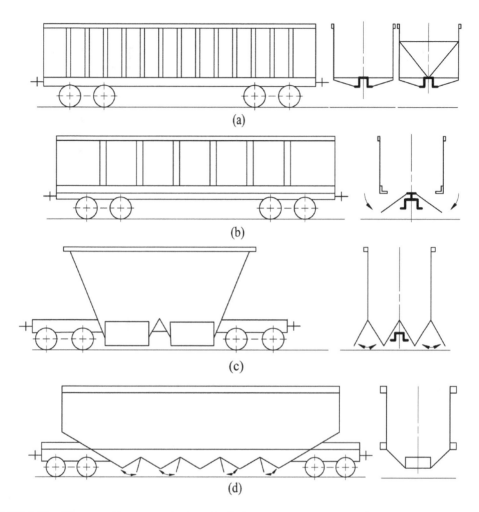

FIGURE 3.23 Wagons with open car body: (a) solid bottom gondola, (b) drop bottom gondola and (c) and (d) open hopper wagons.

sills of the underframe are located above the bogies, the height of the sills is limited, and, in order to obtain the required stiffness and strength, complicated designs comprising plates of varying thickness are used.

The top surface of the underframe may have wooden or metal floor decking used for carrying various cargoes, including wheeled and tracked vehicles, and heavy loose goods, for example, crushed stone.

The flat wagon underframe can be fitted with removable or non-removable end flaps and side posts. They are used for stowing large-diameter pipes and timber. There are underframe designs featuring a middle truss to which pipes or other long cargoes are attached on either side.

Underframes with a deck lowered in the middle are used in well cars for large-size cargoes. Dimensions of the carried cargo do not permit having it placed above the bogies and fit into the loading gauge, so the provision for locating the cargo between the bogies at a minimum height above rail level is made in the car layout. Such an underframe has bends and a minimum height. In order to obtain the necessary strength and especially stiffness, the top and bottom plates of the underframe use the entire width of the car and are connected to one another by several vertical plates.

When designing flat wagon underframes, fatigue resistance and stiffness of sills of a limited height and attachment points connecting the sills into a single structure may pose difficulties.

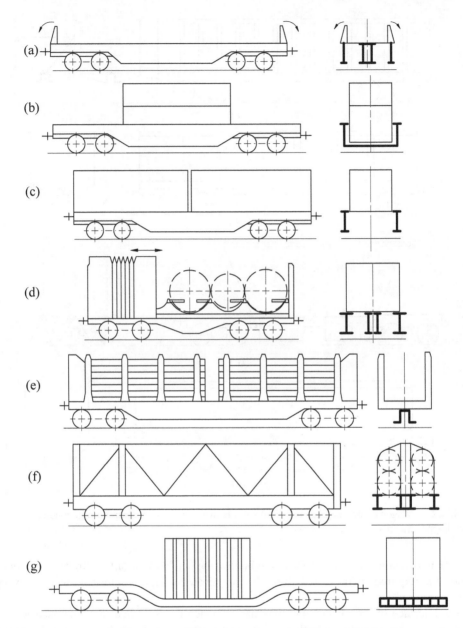

FIGURE 3.24 Wagons with load carrying underframes: (a) general purpose, (b) double stack containers, (c) containers and swap bodies, (d) steel rolls, (e) wood, (f) large radius pipes and (g) special cargo.

Tanks are used for the transportation of liquid and powdery bulk cargoes. The tank body consists of a cylindrical hermetically sealed vessel used to carry the goods and an underframe or two half-frames that receive and transfer loads from couplers and undercarriages to the vessel. The material of the vessel and the thickness of the vessel case are determined according to cargo properties and conditions of carriage. Cargoes can be basically divided into cargoes transported under pressure such as compressed natural gas; cargoes unloaded under pressure, for example, acids and other chemical products; and cargoes transported and unloaded in unpressurised condition such as oil. The case thickness depends on the maximum pressure and can range from 6 to 25 mm. If cargo has a corrosive effect on the walls of the vessel or formation of corrosion products leads to cargo contamination, the vessel is made of corrosion-resistant materials or provided with an internal coating.

In the upper or lower part of the vessel, special fittings for filling and draining cargo and monitoring cargo level and pressure are provided. The top or end face of the vessel is equipped with a hatch, providing access to the vessel, which can also be used for loading non-dangerous goods.

Examples of wagons with tanks are shown in Figure 3.25. The tank may have a separate vessel and underframe or have no centre sill. In a tank wagon with the separate vessel, longitudinal loads from coupling devices are received and transferred by the underframe. The vessel is subjected only to static and dynamic loading from cargo. In a tank without the centre sill, the vessel receives and

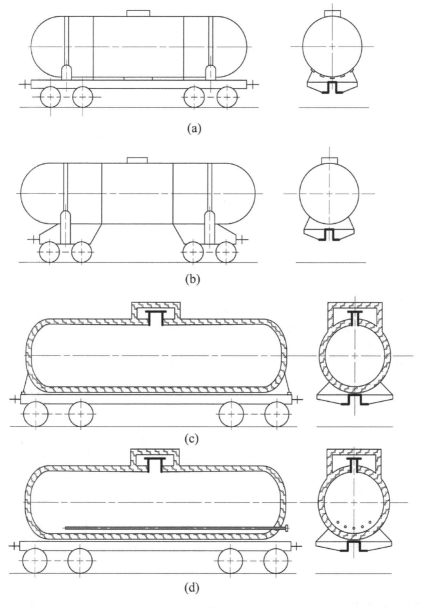

FIGURE 3.25 Wagons with tanks: (a) with a frame, (b) with two sub-frames, (c) with insulation and (d) with insulation and heating.

transfers all loads from couplers and undercarriages. A tank without a centre sill is usually lighter than centre-sill designs; however, the possibility of assembling vessel and underframe separately at different production facilities and then putting them together at the final stage of the rail car assembly attracts manufacturers.

For the transportation of goods solidifying at low temperatures, the vessel is provided with an insulating jacket. A cargo heating system can also be installed inside the tank or on its external surface to facilitate cargo discharge. For the transportation of liquefied gases at low temperatures, cryogenic vessels, that is, thermoses, are employed. Such a vessel consists of an inner case interacting with the cargo and an external case. The space between the cases is exhausted of air to incorporate vacuum insulation.

Tanks intended for the transportation of dangerous goods are equipped with protective emergency devices, as shown in Figure 3.26. Dished end heads of the vessel have shields, and fittings are protected by arcs or a strong shell. A relief valve is used to prevent pressure build-up in the vessel, and fittings are provided with safety shut-off valves for stopping or slowing the flow of cargo when the fitting is damaged.

The main functional characteristics of the tank, which determine the possibility and efficiency of the carriage of one or several goods, are the volume of the vessel, payload capacity, allowable pressure, material of the vessel and design of the loading/discharge fitting.

The most difficult tasks in designing a tank car are to maximise the volume of the vessel within a limited clearance diagram and car length, achieve the minimum weight of the vessel and ensure strength and fatigue resistance of the vessel to underframe attachment points and those of the vessel in the hatch area.

Combined structure car body designs are used for railcars, the loading and unloading of which involve displacement or detachment of body parts. The bodies of dump wagons consist of a main underframe and a separate freight body with side and end flaps, which in unloading turns about the longitudinal axis to an angle of 45°. When the body is turned, the side wall swivels to the position of the floor plane. Unloading can be done on either side of the track. The body is rotated by pneumatic cylinders with the pressure air supply through a separate main line passing along the train consist from the locomotive. This arrangement and other examples of combined structure car bodies are shown in Figure 3.27.

Flat wagons with a rotary flat deck are used to transport steel sheets up to 4.8 m wide. For loading and unloading of sheets, the flat deck is rotated to the horizontal position. In this condition, the width of a transported sheet and that of the flat deck is greater than the width of the vehicle clearance diagram. To bring the flat deck into the service position, the flat deck loaded with sheets is

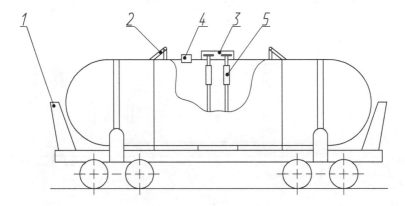

FIGURE 3.26 Protective emergency devices for tank wagons for transportation of dangerous goods. (1) End shields; (2) arcs to protect fittings; (3) strong shell to protect fittings; (4) relief valve; and (5) safety shut-off valves.

Design of Unpowered Railway Vehicles

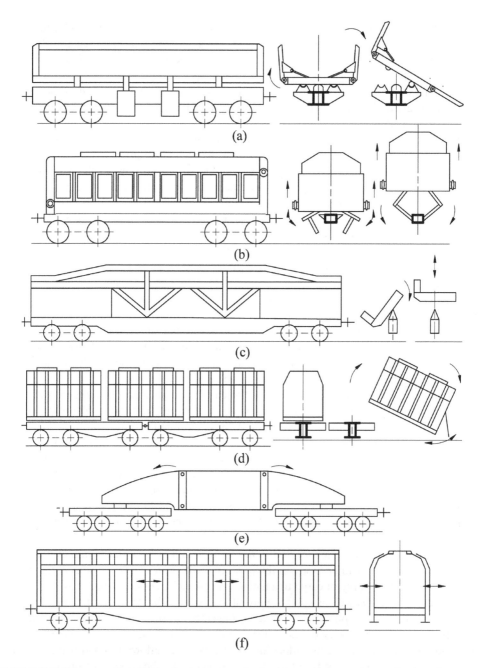

FIGURE 3.27 Wagons with combined structure car body: (a) rotary dump wagon, (b) drop dump wagon, (c) rotary frame wagon, (d) wagon for swap bodies, (e) transporter wagon and (f) sliding walls wagon.

rotated by pneumatic cylinders to an angle of 45° to 55°. In this position, the flat deck and cargo are within the clearance diagram. A similar operation principle is employed for flat wagons designed for the transportation of track switches; the only difference is a hydraulic drive used for rotating the flat deck.

The body of a wagon for the transportation of bulk chemical products with a lifting cargo reservoir comprises an underframe and the cargo reservoir with loading and unloading hatches. For unloading, the train consist with lifting bodies moves at a low speed inside an unloading trestle.

Rollers on the side walls of the cargo reservoir move on inclined ways of the trestle, the reservoir carrying cargo rises and rods connected to the underframe open discharge hatches. In addition, designs of an ore-carrying reservoir detachable from the underframe, which are turned over when the train consist enters the unloading trestles, are available.

Flat wagon underframes with a movable additional second top deck are used to transport automobiles. The height of the top deck is changed by means of a mechanical drive for loading automobiles of different heights.

In order to handle large-size heavy loads, for example, electric transformers, underframes of well cars consisting of two separate parts are used. The two underframe halves are attached to the carried cargo on either side, and, during transportation, cargo transfers loads as an underframe component. In the empty condition, the halves of the underframe are interconnected.

Swap bodies mounted on the flat wagon underframe are removed and separated from the underframe while being loaded or unloaded. Swap bodies can be stored in the loaded or empty condition, separately from the car pending unloading or sending for loading. The model range of swap bodies is customised for transportation, loading and unloading of various goods or one in particular. Such designs are used to transport coal, mineral fertilisers, grain and raw materials of woodworking industry. Since the swap body is not intended for road or water transport, it is developed in a vehicle clearance diagram; therefore, its dimensions and volume are much larger than the volume of a multi-purpose container having the same length.

To haul cargo, it is usually loaded at the shipper's site and unloaded at the consignee's site. The methods and features of loading and unloading determine a cargo vessel design. For the purposes of loading, securing cargo in transit and unloading, the freight car body is equipped with loading and unloading devices, as well as with cargo securing devices.

Loading devices in the form of hatches in the car roof ensure access of cargo during loading and provide integrity of the body and protection of cargo from atmospheric precipitation. The location and size of loading hatches generally correspond to cargo sources on a loading trestle. On some vehicles, the entire roof or parts thereof are shifted or removed to load cargo. When loading cargo with loaders, sliding doors in side walls or sliding side walls are used. In order to protect cargo from atmospheric precipitation, removable hoods or sliding soft tents on arches can be used. In tank cars, a loading device ensures the supply of liquid cargo into the vessel. Auto carriers are loaded and unloaded through doors made in end walls, using stationary ramps and a fold-away crossover platform, allowing automobiles to move through rail cars along the entire consist.

Unloading devices can be implemented in the form of unloading hatches intended for gravity-based unloading. Hatches can be made in the floor, in walls or in special pockets at the bottom of the body, providing the supply of cargo to a hatch. Unloading hatches, as a rule, have a driving unit for opening and closing, as well as seals to prevent load spills. The drive of unloading hatches can be actuated manually, by a pneumatic cylinder connected through an external mechanical rotator, in lifting or turning the body on the unloading trestle. On some types of rail cars, such as an automatic-discharge hopper car, an unloading mechanism that allows to control the volume and direct the discharged cargo is used. Unloading of bulk goods can be done using pneumatic vacuum haulage systems installed at the bottom of the body.

Unloading of tank cars is carried out through fittings and pipelines mounted on the top or bottom of the vessel.

Load-securing devices. Since the car is subjected to dynamic loading during movement and shunting operations, all cargoes other than liquid, loose and bulk shall be attached to the vehicle by special devices. Swivel and fixed fitting lugs are used to fasten containers of different lengths to flat cars. Coiled steel carried on flat cars is stowed between transverse or longitudinal sections.

Automobiles in special auto carriers are attached to the floor by wheel retainers. Pipes are pulled together using soft-coated tie-downs to prevent longitudinal displacement and damage to a polymer coating. Large machines and equipment are secured to special brackets with restraining straps. Wooden or combined decks of box cars and flat cars are fitted with nailed wooden lugs for fixing cargo. To secure timber lading above the body level, wooden stakes are inserted in special rings on the car sides and underframe.

Materials. The most common material used for the production of freight car bodies is a low-alloy steel. Such materials have a high yield strength from 290 to 390 MPa, depending on the thickness, chemical composition, ductility and good weldability, and can be used to manufacture bent and stamped parts. Corrosion resistance of low-alloy steel increases the body service life, when protected by paint-and-varnish coating, to 40 years.

The use of high-strength and more expensive alloy steels in the freight car body makes it possible to reduce the weight only for the elements resisting extreme tensile, compressive or bending loads, mainly under impact conditions. Such elements are centre sills and an end wall frame, and a top sill of the side wall of a gondola wagon. Other body elements receive multi-cycle loads, and, in order to reduce their weight, fatigue resistance characteristics of both the material and welds need to be improved. However, with an increase in the yield strength of steel, fatigue resistance of welds does not improve, or such improvement is insignificant. Therefore, high-strength steels have not become widespread in freight car bodies.

Since pressure vessels of tank cars for handling gases are charged with an internal pressure of up to 30 kPa, and the thickness of their case amounts to 24 mm, the use of a high-strength steel helps to reduce the wall thickness and increase payload capacity.

If the cargo, for example, coal or ore, causes friction during loading and unloading, steels of grades with increased abrasion resistance can be used in body plates that are in contact with the cargo.

Pure steel grade is used for tank car vessels carrying sulphuric acid and for the plates made from this material, and welds are not susceptible to corrosion when in contact with cargo.

Corrosion-resistant steels can be used for bodies that are exposed to corrosive, chemically active cargoes. Since such steels are four to five times more expensive than low-alloy steels, the extent of their application is usually kept at a minimum level to cut back on the cost of the structure. To this end, tank cars only have the vessel made from corrosion-resistant steel, whereas hopper cars only have the body sheathing as such. To reduce the cost, clad steels consisting of a layer of low-alloyed or carbon steel, which is 6 to 12 mm thick, and a layer of corrosion-resistant steel, which is 0.5 to 2 mm thick, are used.

Reducing the vehicle body weight by using aluminium alloys is viewed as a highly promising approach. Weldable aluminium alloys are used in sheathing and the case frame of the bodies of gondola wagons, hopper wagons, box wagons and tank wagons. Since the aluminium alloy structure is lighter but more expensive than steel, the combined body designs are considered optimal. In such bodies, highly loaded elements, for example, underframes, are made of steel, whereas less loaded but sizeable elements, such as sheathing and the case frame of side, end walls and roof are made of light alloys.

Since aluminium materials with a low content of alloying elements are resistant to aggressive chemical products, they are used to manufacture tank car vessels for the transportation of acids and other chemical cargoes.

The use of more expensive light alloys to replace steel is effective for freight car bodies of large volume and carrying capacity. Therefore, aluminium structures are widely used for railway cars with 32 t axle load and on a limited basis on those with an axle load under 25 t. High corrosion resistance to mineral fertilisers and compatibility with food products contribute to improving the effectiveness of Al-alloy. Therefore, aluminium alloys are used for bodies hauling mineral fertilisers and foodstuffs, for example, grain. Coating is not applied on the internal surface of the vessel carrying cargo.

Composite and polymeric materials, for example, fibreglass, are applied in freight car bodies in a limited way, mainly for loading hatch covers of hopper wagons, roof panels of box wagons, as well as outer case embracing thermal insulation. This is due to their high cost and low strength and stiffness.

Low-alloy and carbon steel structures are susceptible to corrosion when exposed to atmospheric humidity, and as a result of the transportation of cargo, exterior and interior surfaces of the vehicle body are adequately coated. To protect the interior surface of the body from chemical products or increase corrosion protection for foodstuffs, special coatings are applied. Vessels carrying acids are provided with internal rubber lining plates installed using an adhesive.

Connections. Electric arc welding is the most commonly used type of connection of steel parts within the vehicle body structure. Welded joints have the same yield strength as a steel plate, yet their fatigue resistance parameters are 2 to 2.5 times less. Therefore, in structural designing, welds are located in areas with reduced cyclic loads, or the weld area is increased with the use of additional parts to reduce local cyclic stresses. For welds of the vessel case designed to carry liquefied gases, weld ductility and high fracture energy shall be obtained throughout the entire range of operating temperatures. Obtaining such qualities for welds of high-strength steel plates can turn out to be problematic in low-temperature environments with temperatures under $-40°C$ and down to $-60°C$.

For Al-alloy structures, both arc welding and friction stir welding are used. Friction stir welding provides a connection with a higher strength and fatigue resistance than those of the parent material, which is also less costly.

Hot riveting is also used to join the parts. Riveted connections are less durable and somewhat more expensive than welded joints, yet they have high robustness and maintainability and are used to fasten parts damaged or worn in service. To connect light-alloy lining sheets with the light alloy or steel case frame, lock bolts are used. The lock bolt connection is strong and has a higher fatigue safety and maintainability but is more expensive than the weld.

Owning to the fact that transported goods have different physical and chemical properties and that conditions of carriage and loading for different cargoes are widely varied, the number of vehicle body designs amounts to several hundreds. Thus, to create new models of rollingstock with a competitive edge over those in service, each car must be designed individually. For the most common types of rail cars, there are conventional time-proven designs in place, and they can be used for new models with technical characteristics similar to those previously produced. However, putting new bogies into operation with increased axle load, as well as the need for a higher payload capacity of a vehicle with a limited axle load, requires a redesign or complete review of the conventional body structure.

3.3 RUNNING GEARS AND COMPONENTS

Traditionally, wheelsets provide running support of railway vehicles and their guidance along the rails. In a traditional wheelset, the left and right wheels are rigidly connected to each other via the axle; the axle and wheels work together and are considered as one unit. The wheelsets are guided by normal and tangential forces between profiled surfaces of wheels and rails that act on either the wheel tread or wheel flange.

Interaction between wheels and rails has proved to be able to efficiently support axle loads up to 37.5 t (Australia, Brazil, South Africa) for freight wagons or running speeds up to 550 km/h (experiments in China and France) for passenger vehicles with light axle loads. At the same time, malfunctions in design or maintenance of wheel-rail interaction can lead to fast degradation of the elements and poor economy or even insufficient safety of the railway system.

Design of Unpowered Railway Vehicles

3.3.1 Wheelsets

A *wheelset* comprises two wheels rigidly connected by a common axle. The wheelset is supported on bearings mounted on the axle journals.

The wheelset provides the following:

- The necessary distance between the vehicle and the track
- The guidance that determines the motion within the rail gauge, including at curves and switches
- The means of transmitting traction and braking forces onto the rails to accelerate and decelerate the vehicle

The design of the wheelset depends on the following:

- The type of the vehicle (traction or trailing)
- The type of braking system used (shoe brake, brake disc on the axle or brake disc on the wheel)
- The construction of the wheel centre and the position of bearings on the axle (inside or outside)
- The desire to limit higher-frequency forces by using resilient elements between the wheel centre and the tyre

The main types of wheelset design are shown in Figure 3.28. Despite the variety of designs, all these wheelsets have two common features: the rigid connection between the wheels through the axle and the cross-sectional profile of the wheel rolling surface, named the wheel profile.

In curves, the outer rail will be of a larger radius than the inner rail. This means that a cylindrical wheel has to travel further on the outer rail than on the inner rail. As the wheels moving on the inner and outer rails must have the same number of rotations per time unit, such motion cannot occur by pure rolling. To make the distances travelled by the two wheels equal, one or both of them will therefore 'slip', thus increasing the rolling resistance and causing wear of wheels and rails. The solution is to machine the rolling surface of wheels to a conical profile with variable inclination angle γ to the axis of the wheelset (see Figure 3.29). The position of the contact point when the wheelset is at a central position on the rails determines the so-called 'tape circle', where the diameter of the wheel is measured. On the inner side of the wheel, the conical profile has a flange that prevents derailment and guides the vehicle once the available creep forces have been exhausted.

An unrestrained wheelset with conical profiles will move laterally in a curve such that the outer wheel is rolling on a larger radius (due to the cone angle) than the inner one. It can be seen that, for each curve radius, only one value of conicity exists that eliminates slip. As different railways have varying populations of curve radii, the shape of wheel profile that provides minimum slip depends on the features of the track. Railway administrations normally specify allowable wheel profiles for their infrastructure and the degree of wear permitted before re-profiling is required [27,28,29].

Figure 3.30 shows several examples of new wheel profiles. For understanding the dynamic behaviour of a railway vehicle, the conicity of the wheel-rail interface is critical. Conicity is defined as the difference in rolling radii between the wheels for a given lateral shift of the wheelset.

Despite the variety of wheel profiles, they have a number of common features. The width of the profile is typically 125 to 135 mm, and flange height for vehicles is typically 28 to 30 mm. The flange inclination angle is normally between 65° and 70°. In the vicinity of the tape circle, the

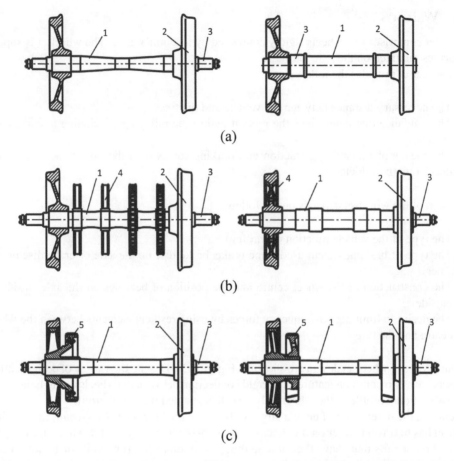

FIGURE 3.28 Main types of wheelset design: (a) with external and internal journals, (b) with brake discs on the axle and on the wheel and (c) traction wheelsets with asymmetric and symmetric positioning of gears. (1) Axle; (2) wheel; (3) journal; (4) brake disc; and (5) tooth gear. (From Iwnicki, S. (Ed.), *Handbook of Railway Vehicle Dynamics*, CRC Press, Boca Raton, FL, 2006. With permission.)

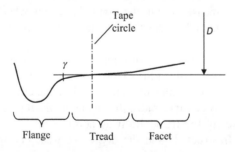

FIGURE 3.29 Main elements of wheel profile. (From Iwnicki, S. (Ed.), *Handbook of Railway Vehicle Dynamics*, CRC Press, Boca Raton, FL, 2006. With permission.)

γ is 1:10 or 1:20 for common rollingstock. For high-speed rollingstock, the γ is reduced to around 1:40 or 1:50 to prevent hunting. It can be seen from Figure 3.30 that the wheel profile has a relief towards the outer side of the wheel. This is intended to lift the outer side of the wheel off the rail and thus ease the motion on switches. Some modern wheel profiles, particularly for passenger rollingstock or heavy haul freight wagons, are not conical but designed instead from a series of radii

Design of Unpowered Railway Vehicles

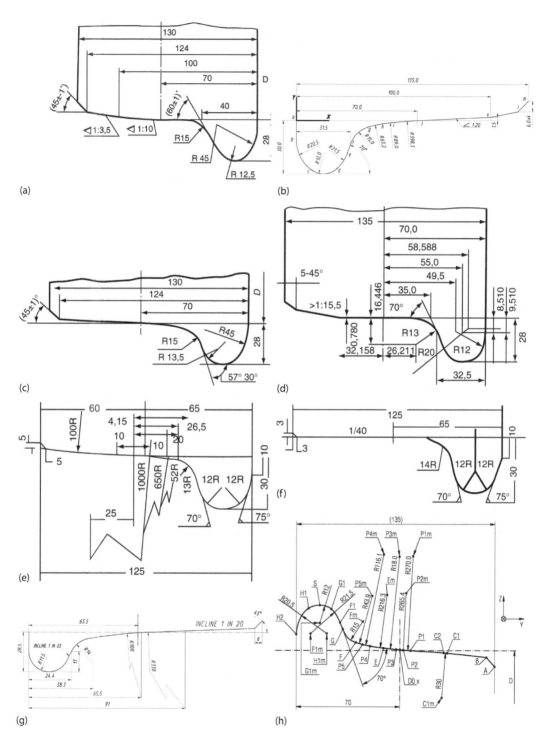

FIGURE 3.30 Common wheel profiles: (a) for freight and passenger railcars (Russia), (b) for freight railcars (China), (c) for industrial rollingstock (Russia), (d) for European freight and passenger railcars, (e) and (f) for high-speed trains (Japan), (g) for freight railcars (India) and (h) for high-speed railcars (Russia). (From Iwnicki, S. (Ed.), *Handbook of Railway Vehicle Dynamics*, CRC Press, Boca Raton, FL, 2006. With permission.)

that approximate a partly worn shape. This is intended to give a more stable shape and prevent the significant changes in conicity that may occur as a conical wheel profile wears. An example of such as profile is the UK P8 wheel profile.

For profiles whose shape is not purely conical (either by design or through wear in service), the term equivalent conicity is applied [30]. This is the ratio of the rolling radius difference to twice the lateral displacement of the wheelset:

$$\gamma_{eq} = \frac{\Delta R}{2y} \tag{3.1}$$

It is important to note that the rolling radius difference is a function of both the wheel and rail shape, and hence, a wheel profile on its own cannot be described as having an equivalent conicity.

As the wheel wears, the shape of the profile in operation may alter significantly from the initial design shape, depending upon a large number of factors. These may include the curvature profile of the route, the suspension design, the level of traction and braking forces applied, the average rail profile shape encountered and the lubrication regime. Tread wear (Figure 3.31) will increase the height of the flange and eventually cause it to strike fishplate bolts, etc. If the tread wear causes the profile to become excessively concave (named hollow wear), damaging stresses may arise at the outer side of the wheel and rail and this is known as false flange damage. Flange wear may lead to increase of the flange angle and reduction of the flange thickness. In extreme conditions this could increase the risk of switch splitting derailments. Wheel profiles are generally restored to their design shape by periodic turning on a wheel lathe. This can normally be carried out without the necessity to remove the wheelset from the vehicle.

It is clear that contact conditions will vary considerably, depending upon the shape of the wheel and rail profiles. This may take the form of single-point, two-point or conformal contact, as shown in Figure 3.32. One-point contact (a) develops between the conical or tread worn wheel profiles and rounded rail profile. Wheels wear quickly towards the local rail shape. With two-point contact (b), the wheel additionally touches the rail with its flange. In this case, the rolling contact has two different radii which causes intensive slip and fast flange wear. Conformal contact (c) appears when the wheel profile and the gauge side of the railhead wear to the extent that their radii in the vicinity of the contact patch become very similar.

Many researchers design the optimal combinations of wheel and rail profiles that will provide low wear of both of them as well as reduce the rolling contact fatigue [29,31,32].

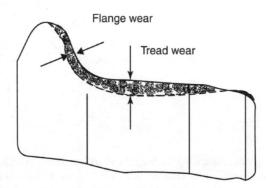

FIGURE 3.31 Tread and flange wear. (From Iwnicki, S. (Ed.), *Handbook of Railway Vehicle Dynamics*, CRC Press, Boca Raton, FL, 2006. With permission.)

Design of Unpowered Railway Vehicles

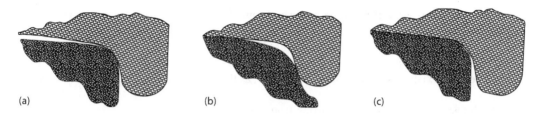

FIGURE 3.32 Possible contact situations between the wheel and the rail: (a) one-point contact, (b) two-point contact and (c) conformal contact. (From Iwnicki, S. (Ed.), *Handbook of Railway Vehicle Dynamics*, CRC Press, Boca Raton, FL, 2006. With permission.)

3.3.2 Axleboxes and Cartridge Type Bearings

The axlebox is the device that allows the wheelset to rotate by providing the bearing housing and also the mountings for the primary suspension to attach the wheelset to the bogie or vehicle frame. The axlebox transmits longitudinal, lateral and vertical forces from the wheelset onto the other bogie elements. Axleboxes are classified according to:

- Their position on the axle depending on whether the journals are outside or inside the wheel
- The bearing type used, either roller or plain bearings

The external shape of the axlebox is determined by the method of connection between the axlebox and the bogie frame and aims to achieve uniform distribution of forces on the bearing. Internal construction of the axlebox is determined by the bearing and its sealing method.

Axleboxes with plain bearings (see Figure 3.33) consist of housing 1; bearing 2, which is usually made out of alloy with low friction coefficient (e.g., bronze or white metal); bearing shell 3 that transmits the forces from the axlebox housing to the bearing and a lubrication device 4 that lubricates the axle journal. Front and rear seals 5 and 6 prevent dirt and foreign bodies entering

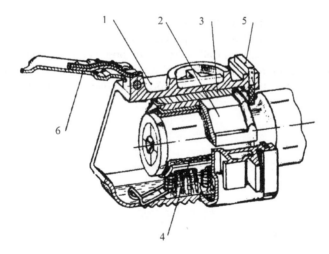

FIGURE 3.33 Construction of axlebox with friction-type bearing. (1) housing; (2) bearing; (3) bearing shell; (4) lubrication device; (5 and 6) front and rear seals. (From Iwnicki, S. (Ed.), *Handbook of Railway Vehicle Dynamics*, CRC Press, Boca Raton, FL, 2006. With permission.)

the axlebox, while front seal 6 can be removed to monitor the condition of the bearing and add lubricant.

Vertical and longitudinal forces are transmitted through the internal surface of the bearing and lateral forces by its faces.

Plain bearing axleboxes are now largely obsolete, as they have several serious disadvantages:

- High friction coefficient when starting from rest
- Poor reliability
- Labour-intensive maintenance
- Environmental pollution

However, from a vehicle dynamic behaviour point of view, axleboxes with plain bearings had certain positive features. In recent years, plain bearing axleboxes that do not require lubrication have been reintroduced on certain types of rollingstock, though their use is still rare.

Axleboxes with roller-type bearings (examples shown in Figure 3.34) are classified according to:

- The bearing type (cylindrical, conical, spherical)
- The number of bearing rows in one bearing (two or one row)
- The fitting method (press-fit, shrink-fit and bushing-fit)

The main factor that determines the construction of axlebox is the way it experiences the axial forces and distributes the load between the rollers.

Cylindrical roller bearings have high dynamic capacity in the radial direction but do not transmit axial forces (Figure 3.34a). Experience in operation of railway rollingstock showed that the faces of rollers can resist lateral forces. However, to do this successfully, it is necessary to regulate not only the diameter but also the length of rollers and the radial and axial clearances.

Conical bearings (Figure 3.34d and e) transmit axial forces through the cylindrical surface due to its inclination to the rotation axis. This makes it necessary to keep the tolerances on roller diameters and clearances almost an order of magnitude tighter than for cylindrical bearings. In addition, conical bearings have higher friction coefficients compared with the radial roller bearings and therefore generate more heat. This produces higher requirements for seals and grease in conical bearings.

Spherical bearings have not been widely applied due to their high cost and lower weight capacity, although they have a significant advantage in providing better distribution of load between the front and rear rows in case of axle bending. Ball bearings are, however, often combined with cylindrical bearings in railway applications to transmit axial forces. High-speed rollingstock often has three bearings in the axlebox – two transmitting radial forces and one (often a ball bearing) working axially (see Figure 3.35).

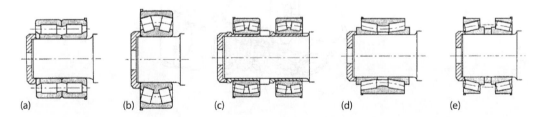

FIGURE 3.34 Constructions of roller bearings: (a) cylindrical double-row, (b) one-row self-alignment, (c) two-row self-alignment, (d) two-row conical and (e) one-row conical (two bearings are shown). (From Iwnicki, S. (Ed.), *Handbook of Railway Vehicle Dynamics*, CRC Press, Boca Raton, FL, 2006. With permission.)

Design of Unpowered Railway Vehicles

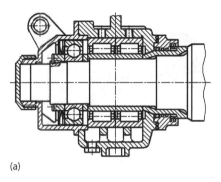
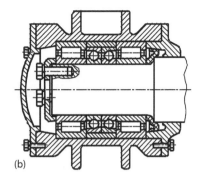

(a) (b)

FIGURE 3.35 Use of spherical bearings: (a) triple bearing of Japanese high-speed trains and (b) triple bearing of French high-speed trains. (From Iwnicki, S. (Ed.), *Handbook of Railway Vehicle Dynamics*, CRC Press, Boca Raton, FL, 2006. With permission.)

Recently, cartridge-type roller bearings gained wide application for higher speeds and axleloads due to many advantages:

- Resistance to combined axial and radial loads that guarantees increased mileages between repairs
- High accuracy in production that eliminates clearances and reduces vibrations at high rotation speeds
- Compact design with smaller external dimensions and mass
- Cartridge-type design that does not need disassembling during maintenance but is replaced as one piece
- Internal seals, grease and clearances are pre-set during the production and thus greasing and adjustment in operation are excluded

The most common cartridge type bearing (Figure 3.36a) consists of the two-row basic bearing: two rows of rollers inside the polymer cages, two cones separated with a spacing ring, one cup on the outside, seals and grease. The basic bearing is a press-fit onto the axle journal having the backing ring with a seal and is then fastened with a washer and bolts (Figure 3.36b). The cup is used as a seat for the adapter.

3.3.3 Wheels

Wheels and axles are the most critical parts of the railway rollingstock. Mechanical failure or exceedance of design dimensions cause derailment. Wheels are classified into solid, tyre and assembly types, as shown in Figure 3.37.

Solid wheels (Figure 3.37a) have three major elements: the tyre, the disc and the hub, and they mainly differ in the shape of the disc.

Tyred wheels (Figure 3.37b) have a tyre fitted to the wheel disc that can be removed and replaced when it reaches its maximum turning limit.

Wheels may have straight, conical, S-shaped, spoked or corrugated-type discs when viewed in cross-section. A straight disc reduces the weight of the construction and can be shaped such that the metal thickness corresponds to the level of local stress. The conical and S-shape discs serve to increase the flexibility of the wheel, therefore reducing the interaction forces between the wheels and the rails. Corrugated discs have better resistance to lateral bending.

The desirability of reducing wheel-rail interaction forces by reducing the unsprung mass has led to development of resilient wheels (Figure 3.37c) that incorporate a layer of material with low

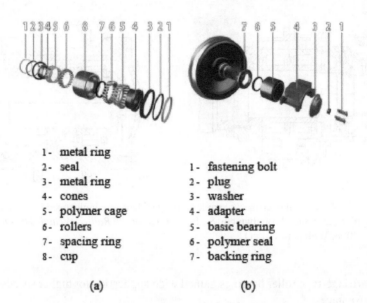

FIGURE 3.36 Cartridge type roller bearing: (a) bearing and (b) its application with adapter.

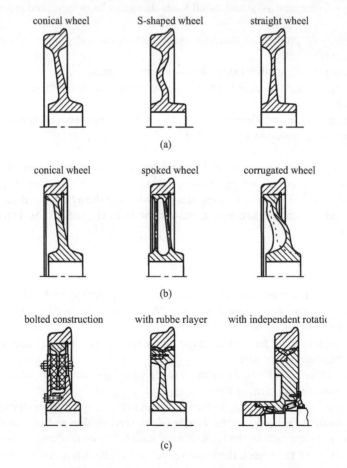

FIGURE 3.37 Major types of railway wheels: (a) solid wheels, (b) tyred wheels and (c) assembly wheels. (From Iwnicki, S. (Ed.), *Handbook of Railway Vehicle Dynamics*, CRC Press, Boca Raton, FL, 2006. With permission.)

elasticity modulus (rubber or polyurethane). These help to attenuate the higher frequency forces acting at the wheel-rail interface.

Improved bearing reliability aroused interest in independently rotating wheels, which provide significant reductions in unsprung mass due to the elimination of the axle. By decoupling the wheels, an independently rotating wheelset inevitably eliminates the majority of wheelset guidance forces. Such wheelsets have found application either on variable gauge rollingstock, providing fast transition from one gauge width to another, or on urban rail transport, where a low floor level is necessary.

3.4 DESIGN OF FREIGHT WAGONS BOGIES

In most cases, freight wagons use two two-axle bogies per vehicle. Articulated freight wagons can have three or more two-axle bogies, with two neighbouring frames resting on one bogie between them.

The majority of freight bogies have single-stage suspension, either between the wheelsets and the bogie frame (similar to passenger bogie primary suspension and often termed 'axlebox' suspension) or between the bogie frame and the bolster (similar to passenger bogie secondary suspension and often termed 'central' suspension). It can be seen from Figure 3.38 that central suspension makes up approximately 6% more of the designs than axlebox suspension. Some wagons use double suspensions similar to passenger bogies to reduce track forces or improve isolation of the load from excess vibrations.

3.4.1 THREE-PIECE BOGIES

Bogies with central suspension are common in the countries of the former USSR, the USA, Canada, China, Australia, South America and most countries in Africa, thus outnumbering axlebox and double-suspension bogies by tens of times. Examples of the classical CNII-H3 (type 18-100) Russian bogie and the Barber bogie from the USA are shown in Figures 3.39 and 3.40, respectively. Such bogies are often termed 'three-piece' bogies [33].

The frame of a three-piece bogie consists of the bolster and two sideframes that are elastically connected by a coil spring and friction wedge-type central suspension system, which, besides other functions, resists asymmetrical loads and holds the bogie frame square in plane. Such suspension allows independent pitch of the sideframes when passing a large vertical irregularity on one rail, allowing the bogie to safely negotiate relatively poor track.

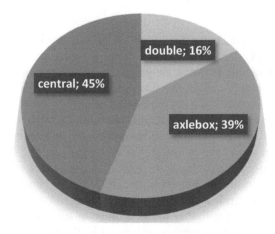

FIGURE 3.38 Proportions of the types of suspension design in freight bogies.

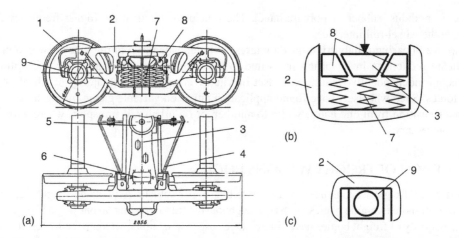

FIGURE 3.39 Type 18–100 three-piece bogie: (a) general view, (b) central suspension schematic and (c) primary 'suspension' schematic. (1) Wheelset; (2) sideframe; (3) bolster; (4) braking leverage; (5) centre bowl; (6) rigid side bearings; (7) suspension springs; (8) friction wedge; and (9) axlebox. (From Iwnicki, S. (Ed.), *Handbook of Railway Vehicle Dynamics*, CRC Press, Boca Raton, FL, 2006. With permission.)

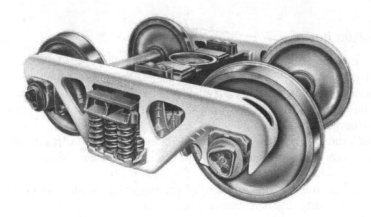

FIGURE 3.40 Barber S-2 bogie. (From Iwnicki, S. (Ed.), *Handbook of Railway Vehicle Dynamics*, CRC Press, Boca Raton, FL, 2006. With permission.)

The vehicle body is connected to the bogie with a flat centre bowl and rigid side bearings having clearance in the vertical direction. When moving on straight track, the car body rocks on the centre bowl and does not touch the side bearings, the gravitational force providing recovery to the central position. In curves, the car body contacts the side bearings.

The central suspension consists of a set of nested coil springs and the wedge arrangement that provides friction damping in the vertical and lateral directions. The inclination of the friction wedges may vary between designs: in the 18–100 bogie, the angle is 45°, whilst in the Barber bogie, it is 35°.

Freight wagons suspensions have to operate under a wide range of load conditions from tare to fully laden, when axle loads can change by more than four times and the load on the spring set more than five times. In the 18–100 bogie, the stiffness of the spring set is independent of the load, which leads to poor ride and increased derailment risk of empty wagons due to small deflections of the springs. For the Barber bogie, a range of suspension spring sets are available for axle loads from 7 to 34 t that include spring sets with bilinear and multi-linear vertical force characteristics.

Design of Unpowered Railway Vehicles

The sideframes of a three-piece bogie rest on the wheelsets. In the 18–100 bogie, the bearing is mounted inside the axlebox, whilst the Barber bogie has an adapter between the cylindrical cartridge-type bearing and the pedestal of the sideframe. Clearances between the adapter (or the axlebox) and the sideframe in the longitudinal and lateral directions allow the wheelsets to move in curves and when passing the large horizontal irregularities. Thus, the axlebox unit does not steer the wheelsets but damps their displacements by friction forces. Owing to the absence of a primary suspension, such bogies have a large unsprung mass, which causes increased track forces on short irregularities or rail joints.

In curves, the three-piece bogies demonstrate the 'lozenging' or 'warping' effects, when the two sideframes adopt a parallelogram position (in plan view). In this instance, the wheelsets cannot adopt a radial position in the curve and generate large angles of attack. This leads to constant contact between the high-rail flange of the leading wheelset and the rail, causing high levels of wear [34].

Since classic three-piece bogies were introduced in the 1930s, there have been many advances in their design that deal with the known disadvantages and are explained in detail in [35–37]:

- Increase in suspension deflection under the tare wagon as well as under the fully loaded one
- Wedge designs that increase the warping stiffness by spatial configuration of inclined surfaces [38]
- Introduction of elastic pads in primary suspension to provide steering of wheelsets and reduce unsprung mass
- Installation of constant contact side bearings that have an elastic element compressed by the weight of the car body and provide yaw and roll resistance to the car body [39].

3.4.2 PRIMARY SUSPENDED H-FRAME BOGIES

The Y25 (and similar bogies such as the Y33) are predominantly used on European freight vehicles [40]. An example of this bogie is shown in Figure 3.41.

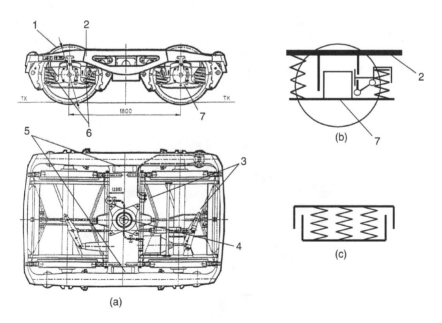

FIGURE 3.41 The Y25 bogie: (a) general view, (b) primary suspension schematic (Lenoir damper) and (c) elastic side bearing schematic. (1) Wheelset; (2) rigid H-shaped frame; (3) braking leverage; (4) centre bowl; (5) side bearings; (6) suspension springs; and (7) axlebox. (From Iwnicki, S. (Ed.), *Handbook of Railway Vehicle Dynamics*, CRC Press, Boca Raton, FL, 2006. With permission.)

The Y25 bogie has a single-stage primary suspension consisting of a set of pairs of nested coil spring (with a bi-linear characteristic for tare/laden ride) and a Lenoir link friction damper (Figure 3.41b) providing vertical and lateral damping. The friction force depends on the vertical load on the spring set, a component of which is transferred to the friction face by the inclined Lenoir link. Derailment safety is improved by the provision of vertical clearance between the inner and outer springs in each pair, giving a lower stiffness in tare than in laden. Whilst improving the ride in both conditions, problems may still arise with the part-laden ride, when the bogie is just resting on the inner 'load' spring, making the suspension relatively stiff for the load being carried.

The bogie has rigid H-shaped frame that consists of two longitudinal beams, one lateral and two end beams, and may be either cast or fabricated. The connection of the vehicle body is different to the three-piece bogies described previously. The centre bowl has a spherical surface to reduce asymmetric forces on the frame and elastic side bearings without clearance resist the body roll motions (Figure 3.41c).

Advances in primary suspended freight bogies [37] include:

- Using progressive rubber springs or bell springs instead of coil springs and Lenoir damper
- Introducing the additional pusher spring into the Lenoir damper or using double Lenoir links on each of the axleboxes to reduce longitudinal stiffness and facilitate curving
- Using wedge type friction damper with four rows of springs

3.4.3 Double-Suspension Bogies

In the 1980s, the UK had developed a novel, track-friendly bogie using passenger vehicle technology. The LTF25 bogie is shown in Figure 3.42 and is described in [41].

The LTF25 bogie was specifically designed to reduce dynamic track forces, and, as part of this, effort was made to reduce the unsprung mass. Small wheels (813 mm diameter) were used and inside axleboxes, giving a 30% reduction in wheelset mass, although this necessitated the use of on-board hotbox detectors. Primary suspension is through steel coil springs, and secondary suspension is through rubber spring elements and hydraulic dampers. The high cost of the LTF25 bogie and concerns about axle fatigue with inboard axleboxes militated against its adoption, but a modified version known as the TF25 bogies (with suspension reduced to only primary) have achieved considerable production success.

FIGURE 3.42 LTF25 bogie. (From Etwell, M.J.W., *J. Rail Rapid Transit*, 204, 45–54, 1990. With permission.)

Design of Unpowered Railway Vehicles

FIGURE 3.43 Double-suspended RC25NT bogie with direct inter-axle linkages. (From Scholdan, D. et al., RC25NT - ein neues, gleisfreundliches Drehgestell für den schweren Güterverkehr (RC25NT - a new, track-friendly bogie for the heavy freight transport), ELH Eisenbahnlaufwerke Halle GmbH Slideshow Presentation delivered at 2011 Modern Rail Vehicles Conference, Graz, Austria, September 2011. With permission.)

The more recent RC25NT bogie shown in Figure 3.43 has horizontally soft rubber bushes in the primary suspension and flexicoil dual rate springs with friction damping via Lenoir link in the secondary suspension [42]. The bogie is equipped with disc brakes and inter-axle linkages that we discuss in the next section.

3.4.4 AUXILIARY SUSPENSIONS (CROSS-ANCHORS, RADIAL ARMS, ETC.)

Many novel bogie designs address the fundamental conflict between stability on straight track and good curving [43]. It is clear from the foregoing discussion that the bogie should maintain stable conditions on straight track but allow the wheelsets to adopt a radial position in curves.

Bogies where the wheelsets adopt or are forced to take an approximately radial position in curves, as shown in Figure 3.44, are called radially steered bogies. Such designs have small angles of attack, which leads to significantly decreased flange wear and lower track forces.

Radially steered bogies fall into two groups: those with forced steering of the wheelsets in curves and those with self-steering of the wheelsets. In the first case, the wheelsets are forced

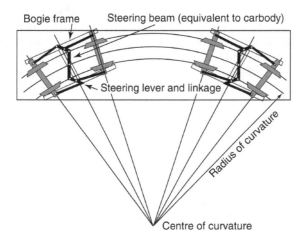

FIGURE 3.44 Radial position of wheelsets in curves having additional inter-axle linkages in the bogies. (From Iwnicki, S. (Ed.), *Handbook of Railway Vehicle Dynamics*, CRC Press, Boca Raton, FL, 2006. With permission.)

TABLE 3.1
Classification of Forced Steering Mechanisms

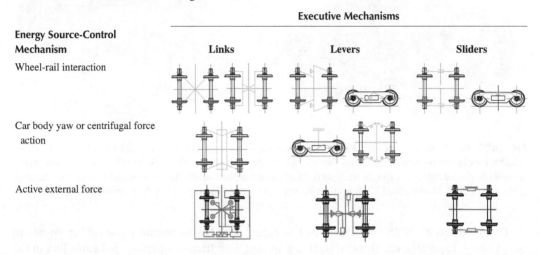

Source: With kind permission from CRC Press: *Handbook of Railway Vehicle Dynamics*, CRC Press, Boca Raton, FL, 2006, Iwnicki, S. [ed.].

to adopt a radial position due to linkages between the wheelsets or linkages from the wheelset to the vehicle body. Various methods of obtaining forced steering for radially steered bogies are shown in Table 3.1. The bogies may be split into three groups depending on the control principle used:

- Wheelsets yawed by the wheel-rail contact forces
- Wheelsets yawed by the relative rotation between the bogie frame and vehicle body (either yaw or roll)
- Wheelsets yawed by an external energy source (electric, hydraulic or pneumatic actuators)

The first two groups in Table 3.1 have passive control systems that change the kinematic motion of the wheelset, depending on the curve radius. Designs where the energy source is provided by the steering force in wheel-rail contact may be considered preferable, as the behaviour of systems relying on interconnection to the car body is dependent upon vehicle speed. Designs where the wheelsets are forced to adopt a radial position by hydraulic, pneumatic or electric actuators (or a combination of these) are called actively controlled bogies. These are considered in detail in Chapter 15.

Three main groups of executive mechanisms are common – those using links between wheelsets, those using an arrangement of levers and those using sliders.

An example of a freight bogie with passive control using diagonal links between the axleboxes designed by Scheffel [44] is shown in Figure 3.45.

The second group of radially steered bogies are those with wheelsets that are self-steering in curves. The design of such bogies is based on selecting the optimum shear and bending stiffnesses [45]. This may be aided by using designs that allow these stiffnesses to be de-coupled.

In conventional suspension arrangements, the bending and shear stiffness are not independent. Decreasing the bending stiffness leads to a reduction of shear stiffness, which means that improving

Design of Unpowered Railway Vehicles

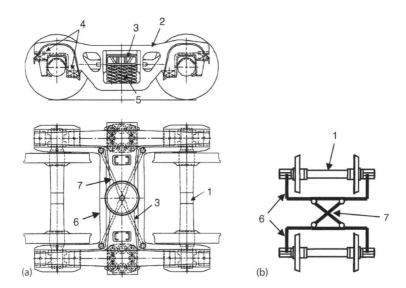

FIGURE 3.45 Scheffel HS bogie with diagonal linkage between wheelsets: (a) general view and (b) the principal scheme of inter-axle linkages. (1) Wheelset; (2) sideframe; (3) bolster; (4) primary suspension; (5) secondary suspension; (6) subframe; and (7) diagonal links. (From Iwnicki, S. (Ed.), *Handbook of Railway Vehicle Dynamics*, CRC Press, Boca Raton, FL, 2006. With permission.)

the curving qualities leads to reduced stability on straight track. Inevitably, therefore, the bogie in-plane stiffness is chosen to give the best compromise between curving and stability.

In order to resolve the curving-stability controversy, Scheffel proposed several arrangements of inter-axle linkages [46], two of which are shown in Figures 3.46 and 3.47.

The Scheffel radial arm bogie is shown in Figure 3.46, and the generalised bogie stiffnesses for it have the following expressions:

Shear stiffness:

$$K_{s\Sigma} = 2k_y + K_s \tag{3.2}$$

where k_y is the lateral stiffness of the inter-axle linkage (per side) and K_s is the shear stiffness provided by the bogie frame.

Bending stiffness:

$$K_{b\Sigma} = 4b^2 k_x + K_b \tag{3.3}$$

where k_x is the longitudinal stiffness of the inter-axle linkage (per side) and K_b is the bending stiffness provided by the bogie frame. $2b$ is the distance between axle journal centres.

Thus, the expressions for $K_{s\Sigma}$ and $K_{b\Sigma}$ contain two independent parameters k_x and k_y that allow optimum shear and bending stiffnesses to be selected. Scheffel bogies have the axle load of 32 t and provide mileage between wheel turning of up to 1.5 million kilometres, thus proving the high efficiency of the design to reduce track forces.

Shown bogie designs are based on the three-piece bogie consisting of a bolster and two sideframes. They retain the advantages of the three-piece bogie when negotiating large track irregularities and carrying the asymmetric loads. However, wheelset steering is provided not by the frame (as in traditional designs) but also by the inter-axle links. In order for the inter-axle links to be effective, the bogie must have low longitudinal and lateral primary suspension stiffnesses. These bogie

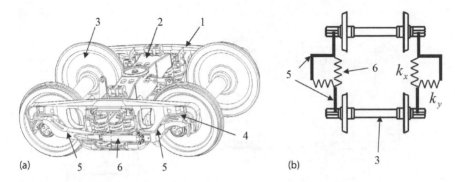

FIGURE 3.46 Bogie construction by Scheffel with radial arm device: (a) general view and (b) principal scheme of inter-axle linkages. (1) Sideframe; (2) bolster; (3) wheelset; (4) primary suspension; (5) two longitudinal arms; and (6) elastic elements between the arms. (From Iwnicki, S. (Ed.), *Handbook of Railway Vehicle Dynamics*, CRC Press, Boca Raton, FL, 2006. With permission.)

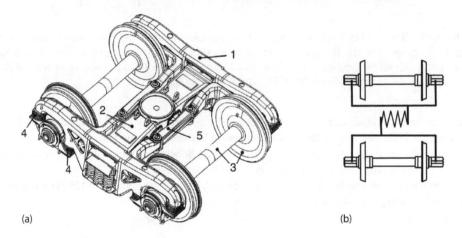

FIGURE 3.47 Scheffel bogie with A-shaped inter-axle linkages: (a) general view and (b) principal scheme of inter-axle linkages. (1) Sideframe; (2) bolster; (3) wheelset; (4) primary suspension; and (5) elastic connection between the sub-frames. (From Iwnicki, S. (Ed.), *Handbook of Railway Vehicle Dynamics*, CRC Press, Boca Raton, FL, 2006. With permission.)

designs are therefore effectively double suspended. Infra-Radial [47] and SUSTRAIL [48] projects have shown that links can be effective with primary suspended bogies as well.

3.5 DESIGN OF UNPOWERED BOGIES FOR PASSENGER COACHES

The most common passenger vehicle designs use a pair of two-axle bogies on each vehicle. However, in articulated trains, for example, the French TGV, two-axle bogies are positioned between the car bodies, whilst the Spanish Talgo trains use single-axle articulated bogies [49].

For passenger bogies, the wheelsets are generally mounted in a rigid H-shaped frame that splits the suspension into two stages. The primary suspension transmits forces from the wheelsets to the bogie frame, and the secondary suspension transmits forces from the bogie frame to the car body.

The principal functions of the primary suspension are guidance of wheelsets on straight track and in curves and isolation of the bogie frame from dynamic loads produced by track irregularities. The secondary suspension provides the reduction of dynamic accelerations acting on the car body that determines passenger comfort. The source of these accelerations is excitation from the track irregularity/roughness profile and the natural oscillations of the bogie frame and car body on their suspension elements. It is particularly important to reduce the lateral influences, to which the passengers are more sensitive, and therefore, the stiffness of secondary suspension in the lateral direction is designed to be as small as possible.

An example of a traditional type of secondary suspension (used on passenger vehicles for over 100 years) is shown in Figure 3.48. The secondary suspension swing consists of the secondary springs and dampers 2 and spring plank 1 that is attached to the bogie frame 3 by swing hangers 4. This arrangement provides low lateral stiffness, and the height of the secondary springs remains comparatively small.

When curving, the bogie should rotate under the car body to reduce track forces, whereas on straight track, it should resist yawing motion. In the case of bogies with swing link secondary suspension, part of the car body mass is transmitted to bolster 5 through bogie centre 6 and part of it is transmitted through side bearings 7. The bogie centre serves as the centre of rotation and transmits the traction forces, whilst the side bearings provide friction damping to the bogie yaw motion. The traction rod usually limits longitudinal displacements of the bolster relative to the bogie frame.

Swing link secondary suspension may be acceptable for speeds up to 200 km/h. Its disadvantage is the large number of wearing parts that require relatively frequent maintenance to prevent deterioration of ride quality.

Modern bogie designs have a smaller number of parts in the secondary suspension and thus reduce maintenance costs. They typically use elastic elements that have a small stiffness in horizontal direction. Examples include the ETR-500 bogie (Figure 3.49), which uses Flexicoil secondary springs, and the Series E2 Shinkansen (Figure 3.50), which uses an air spring secondary suspension.

In such secondary suspension arrangements, the vehicle body may rest on a bolster (as in the swing link bogie) or be directly mounted on the secondary suspension, as in the bolster-less bogies in Figures 3.49 and 3.50. In bolster-less bogies, the traction forces are transmitted through the centre pivot arrangement, and the bogie rotates under the car body by using the flexibility of secondary suspension in longitudinal direction. In such designs, yaw dampers are often fitted longitudinally between the body and the bogie to damp hunting motion on straight track.

Modern bogies are normally equipped with separate secondary dampers to damp oscillations in vertical and lateral directions. Lateral damping is normally achieved with a hydraulic damper, whilst vertical damping may be hydraulic or orifice damping within the air spring.

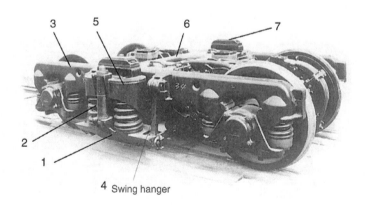

FIGURE 3.48 Bogie with swing link secondary suspension. (1) Spring plank; (2) springs and dampers; (3) frame; (4) swing hangers; (5) bolster; (6) central column support; and (7) side bearing. (From Iwnicki, S. (Ed.), *Handbook of Railway Vehicle Dynamics*, CRC Press, Boca Raton, FL, 2006. With permission.)

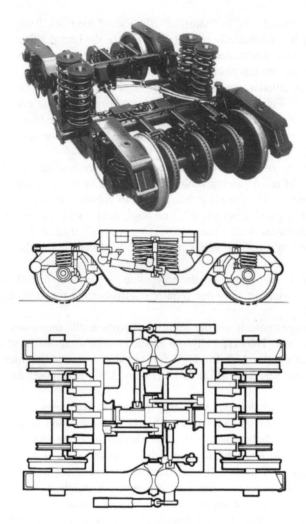

FIGURE 3.49 ETR-500 train bogie (Italy). (From Iwnicki, S. (Ed.), *Handbook of Railway Vehicle Dynamics*, CRC Press, Boca Raton, FL, 2006. With permission.)

FIGURE 3.50 Series E2 Shinkansen bogie (Japan). (From Iwnicki, S. (Ed.), *Handbook of Railway Vehicle Dynamics*, CRC Press, Boca Raton, FL, 2006. With permission.)

Design of Unpowered Railway Vehicles

Various types of elastic elements are used in the primary suspensions of passenger bogies. To achieve high speeds, the longitudinal stiffness of the primary suspension should be high, whereas the lateral stiffness may be lower. In curves, the high longitudinal primary stiffness leads to the increase of contact forces between the wheels and rails, causing increased wear. Similarly, high lateral stiffness may lead to increased dynamic force when negotiating lateral track irregularities. For passenger bogies, it is therefore preferable if the suspension design can provide different stiffnesses in the lateral and longitudinal directions.

The three most common types of primary suspension are those with coil springs and longitudinal traction rods or links (Figure 3.51), coil springs with guide posts (Figure 3.52) and chevron (rubber interleaved) springs (Figure 3.53).

The ETR-460 bogie (Figure 3.51) is an example of a primary suspension using traction links with resilient bushes. The wheelset is guided by two links with spherical joints, and the vertical and lateral loads are mainly reacted by the coil springs.

In the primary suspension of Series 300 Shinkansen bogies, coil springs are used together with cylindrical guide posts containing rubber-metal blocks (Figure 3.52). The springs bear the vertical

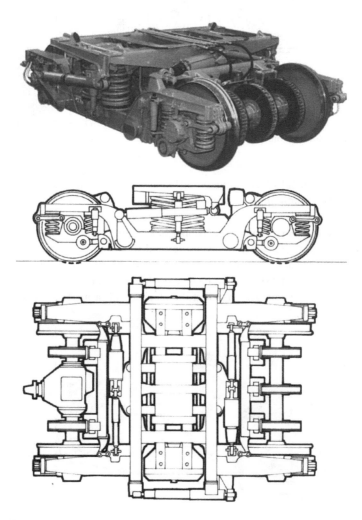

FIGURE 3.51 ETR-460 bogie (Italy). (From Iwnicki, S. (Ed.), *Handbook of Railway Vehicle Dynamics*, CRC Press, Boca Raton, FL, 2006. With permission.)

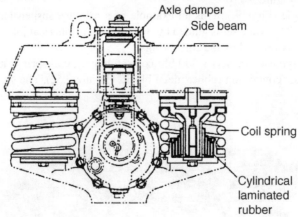

FIGURE 3.52 Series 300 Shinkansen bogie (Japan). (From Iwnicki, S. (Ed.), *Handbook of Railway Vehicle Dynamics*, CRC Press, Boca Raton, FL, 2006. With permission.)

load, whilst the rubber-metal block provides different stiffnesses in the longitudinal and lateral directions. It also acts to damp high-frequency vibrations.

The X-2000 high-speed train bogie primary suspension uses chevron (rubber-interleaved) springs (Figure 3.53). In this type of spring, rubber blocks are separated by steel plates arranged at an inclined position to the vertical. In this way, vertical forces on the spring cause both shear and compression forces in the rubber blocks. Depending on the V-angle and material properties of the chevron spring, the longitudinal stiffness can be made three to six times higher than the lateral stiffness. The disadvantage of such a design is that the mechanical properties are highly dependent on

FIGURE 3.53 The X-2000 bogie with chevron spring primary suspension (Sweden). (From Iwnicki, S. (Ed.), *Handbook of Railway Vehicle Dynamics*, CRC Press, Boca Raton, FL, 2006. With permission.)

temperature, and this may become a significant factor when operating in climates where extremes of temperature are common.

3.6 DESIGN OF INTER-CAR CONNECTIONS

Inter-car connections are intended for performing the following functions:

- Connect cars to locomotives and to each other; keep them at a certain distance from each other; and take up, transmit and soften tension and compression forces during the train movement and manoeuvres
- Constrain relative displacement of cars during the train movement in order to improve running smoothness and passenger comfort, as well as to prevent cars from piling up on each other in case of accidents
- Protect passengers passing between cars from environmental impacts

Devices performing the first function are named draw-and-buff gears. Present-day draw-and-buff gears usually include couplers providing inter-vehicle connection and holding strings of cars together, reversive shock absorbers (draft gears) providing energy absorption and return to the initial state, lugs transmitting loads to the car frame and additional irreversible protective devices for taking up excessively large forces, which are intended to prevent cars from piling up on each other in case of emergency.

3.6.1 SCREW COUPLERS

Screw couplers are used in Europe and in some countries of Asia and Africa. They include a hook with a shock absorber, a turnbuckle and two buffers (Figure 3.54). Transmission of draft forces is carried out from one hook to another via the turnbuckle. Compression forces are transmitted through buffers 4; screw coupler 2 is loose and does not affect transmission of compression forces.

Figure 3.55 represents the screw coupler hook with the shock absorber. Hook 1 is connected to thrust plate 4 by means of pin 2 through shackles 3. Draft gear 6 is located between thrust plates 4 and 5. Thrust plate 5 is fixed on the headstock of the car frame, and when draw forces

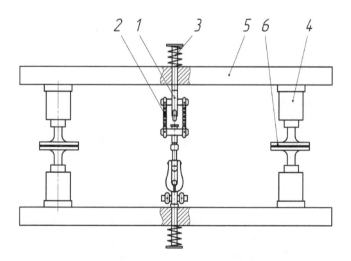

FIGURE 3.54 Principal schematic of screw coupler. (1) hook; (2) turnbuckle; (3) shock absorber; (4) buffer; (5) end beam of the car body; and (6) buffer plate.

FIGURE 3.55 Hook of the screw coupler with the shock absorber. (1) hook; (2) pin; (3) shackles; (4 and 5) thrust plates; (6) draft gear; and (7) slot for pin of the turnbuckle.

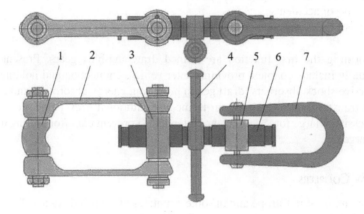

FIGURE 3.56 The turnbuckle. (1) Pin; (2) shackles; (3 and 5) nuts; (4) handle; (6) screw; and (7) coupling link.

occur, draft gear 6 gets compressed, absorbing transmitted forces. Pin 1 of the turnbuckle is installed in slot 7.

The turnbuckle (Figure 3.56) consists of two nuts 3 and 5 that are screwed onto screw 6. On one side, the thread is right-hand; on the other side, the thread is left-hand. The screw itself is turned by means of handle 4. When the handle is turned, the nuts converge or diverge. Bent coupling link 7 that is placed on the hook of the adjoining car is fixed on trunnions of the right nut. On the trunnions of the left nut, two shackles 2 are installed, and pin 1 passing through the slot in the car hook is installed on the ends thereof.

3.6.2 Automatic Couplers

Automatic couplers, in their turn, are divided into non-rigid (allowing reciprocal displacements), rigid (not allowing reciprocal displacements) and semi-rigid (having the possibility of reciprocal displacements only within the clearance limits).

Non-rigid and semi-rigid couplers are applied to freight and passenger cars; rigid couplers are applied to high-speed passenger cars, underground cars and electric trains where electric train-line connections are required. In freight cars, non-rigid automatic couplers SA-3 (for track gauge 1520 mm) and AAR type E (for track gauge 1435 mm, North America, India and Japan) are the most widely used. Their devices are similar, though they have some differences.

The automatic coupler SA-3 (Figure 3.57) includes coupler shank 1, coupler follower 2, draft gear 3, rear 4 and front 5 draft lugs, coupler yoke 6, striker 7 and draft key 8. The front and rear draft lugs as well as the striker are rigidly connected to car centre sill 9. Compression and tension forces are

Design of Unpowered Railway Vehicles

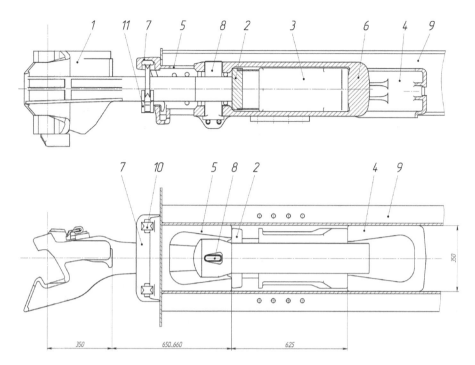

FIGURE 3.57 Automatic SA-3 coupler for 1520 mm gauge railways. (1) automatic coupler body; (2) coupler follower; (3) draft gear; (4) rear lugs; (5) front lugs with striker; (6) yoke; (7) striker; (8) draft key; (9) centre sill; (10) pendulum suspension; and (11) centring beam.

transmitted in the following way. Under compression, the force coupler head 1 is transmitted via its shank onto coupler follower 2 onto draft gear 3 and further onto car center sill 9 via the rear draft lugs 4. Under tension, the force from coupler head 1 via draft key 8 (in some designs, via the pin) is transmitted onto coupler yoke 6 and from the coupler yoke, via draft gear 3, coupler follower 2 and front draft lugs 5, onto car centre sill 9. In the case of a large impact force, at exhaustion of clearance between type F coupler head 1 and the striker, the force is directly taken up by striker 7. In order to provide curving, the coupler is suspended by means of two pendulum-type suspensions 10 and centring beam 11; they ensure the coupler turning when rounding a curve and its return to the initial state.

As for the American non-rigid type E coupler represented in Figure 3.58, forces thereon are transmitted in a similar way; the differences involve the draft key position (horizontal) and the absence of the centring device.

The American semi-rigid automatic coupler is represented in Figure 3.59. Forces thereon are transmitted in a similar way. The differences involve the elevating stop at the bottom of the automatic coupler head, the spherical end of the coupler shank, the vertical position of the pin connecting the automatic coupler to the coupler yoke and the spring-loaded transverse strap that provides the possibility of turning in the vertical plane for the automatic coupler.

Among rigid automatic couplers, couplers using the bell-and-hopper arrangement connection, which minimises the end play, have become the most widely spread (Scharfenberg patent, 1903). There are also other couplers providing connections of electric and pneumatic trainlines: Tomlinson (the USA), wedge lock couplers (Great Britain), the GF-type conical coupler (Belgium and Switzerland) and zero-clearance couplings BSU-3 (Russia).

In recent years, requirements of energy absorption management in case of train collisions (Crash Energy Management) have been imposed for automatic couplers. Several matched energy-absorbing components, which are integrated into both the coupler, and the car body are used for this purpose.

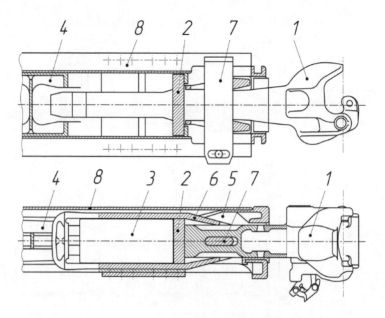

FIGURE 3.58 American non-rigid type E coupler. (1) Coupler body; (2) coupler follower; (3) draft gear; (4) rear lug; (5) front lug; (6) yoke; (7) draft key; and (8) centre sill.

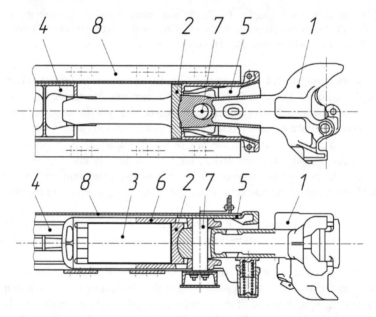

FIGURE 3.59 American type F semi-rigid automatic coupler. (1) Coupler body; (2) coupler follower; (3) draft gear; (4) rear lug; (5) front lug; (6) yoke; (7) draft pin; and (8) centre sill.

Components for non-reversible absorption of energy (deformation tubes) and devices preventing cars from climbing up on each other are installed in addition to draft gears of the reverse type.

In Figure 3.60, the modular structure elements of the rigid automatic Voith Turbo Scharfenberg couplers are presented. The coupler includes coupler head 1 with the locking device and the signal transmission unit, deformation tube 2 and lug 3 with resilient damping device 4.

Design of Unpowered Railway Vehicles

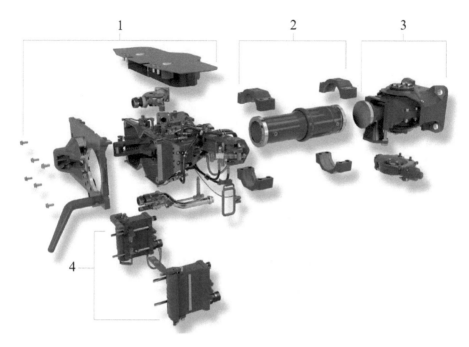

FIGURE 3.60 Rigid automatic Voith Turbo Scharfenberg couplers. (1) Coupler head; (2) deformation tube; (3) lug; and (4) resilient damping device. (From Voith Turbo Scharfenberg GmbH, Salzgitter, Germany. With permission.)

3.6.3 Draft Gear

The draft gear is a device behind the coupler for absorbing longitudinal forces and energy dissipation, which occur during transient modes of a train's movement, impacting in the course of shunting operations, automatic train shunting, etc. The kinetic energy of interacting car is transformed into potential energy of its resilient elements and work of friction forces. Dry friction, viscous friction and internal friction are used for energy dissipation. Draft gears containing only resilient elements are called spring draft gears. Draft gears using dry friction are called friction draft gears. Draft gears using viscous friction are called hydraulic draft gears. Draft gears using internal friction in material are called rubber or polymer draft gears. Draft gears using compressible bulk polymer of high viscosity as the working mechanism are called elastomeric draft gears.

Performance indicators of draft gears are estimated by means of the force characteristic, which is the relationship between the force affecting the draft gear, its displacement x and the speed v. The force characteristic is usually represented in the form of dependences: for the compression stroke ($v > 0$) and for the rebound stroke ($v < 0$) (Figure 3.61).

The work of the force on the route equal to the full travel of the draft gear is called the energy-absorption capacity of the draft gear. It is numerically equal to the area of the diagram under the straight line AB:

$$E = \int_0^x P(x,v)dx$$

After an impact, the draft gear is to return to its initial state. In this case, a part of energy returns. The returned energy is determined by the area under the line CD:

$$E_R = \int_0^{x_{max}} P(x)dx \qquad (3.4)$$

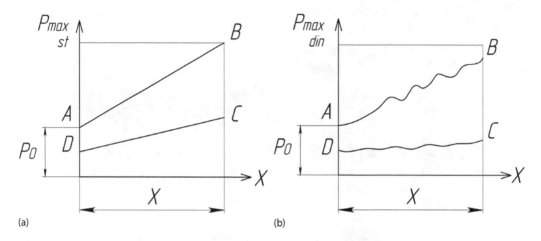

FIGURE 3.61 Friction draft gear force characteristics. (a) static and (b) dynamic.

The irreversibly absorbed energy is characterised by the energy absorption factor:

$$\eta = 1 - \frac{E_R}{E} \qquad (3.5)$$

In addition, the following parameters of the draft gear are used:
X_0 is the full travel;
P_0 is the initial tightening force;
P_{st}^{max} is the maximum resistance force at static loading;
P_{dyn}^{max} is the maximum force at dynamic loading.

Friction draft gears. The wedge thrust is the most common way of pressure generation on friction surfaces (Figure 3.62) [50]. It makes it possible to impose large normal loads on friction surfaces and change them by means of changing the angles of inclination.

Thrust cone 1 takes up forces and transmits them to three wedges 2 providing pressing to hex-shaped nozzle 3 and case 4. The return is carried out by spring 5. The energy-absorbing capacity of such draft gears depends on the breaking-in and wear degree of wedges. That is why, in recent years, ceramic metal plates are applied as friction pairs, and polymer materials are applied as resilient elements in friction draft gears. The design of present-day friction draft gears is shown in Figure 3.63.

Rubber-metal and polymer draft gears. Specific properties of rubber as a material, which combines elastic behaviour and relatively large internal friction, are the cause of rubber-metal draft gears application in passenger cars. Structural designs of rubber-metal draft gears are varied. In trial designs, rubber works in shear and in compression with shear. In most cases, rubber-metal draft gears, wherein rubber works in compression, are applied in actual practice.

The disadvantages of rubber-metal draft gears connected with the specific properties of rubber as an engineering material are partly eliminated in polymer draft gears, wherein resilient elements from polyether thermoplastic elastomer are used, as shown in Figure 3.64. Their application makes it possible to considerably increase the energy-absorbing capacity of draft gears.

Hydraulic and hydraulic-gas draft gears. They use liquid flow through narrow calibrated orifices for energy dissipation. The main difficulty is caused by transmission of quasi-static loads. For this purpose, resilient elements, such as a spring or pressurised gas, are required in the draft gear. Despite the positive test results for their trial models, they have not become widely spread because of the high cost and low reliability. Information related to these draft gears is presented in [51].

Design of Unpowered Railway Vehicles

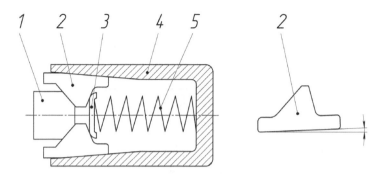

FIGURE 3.62 Schematic of friction draft gear with wedge thrust. (1) thrust cone; (2) wedge; (3) nozzle; (4) case; and (5) spring.

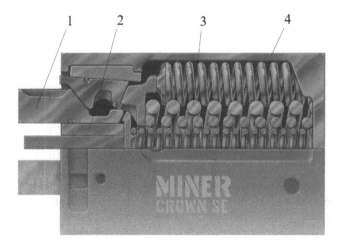

FIGURE 3.63 Miner Crown SE friction draft gear. (1) friction clutch; (2) wrap-around barrier plates; (3) heavy-duty coil spring package; and (4) case.

Elastomeric draft gears. Nowadays, on the strength of their technical and economic parameters, draft gears wherein compressible bulk polymer (elastomer) is used as the working medium are considered the most promising. Elastomers, which have the compressibility of about 15% to 20% at the pressure of 500 MPa and high viscosity, perform the function of a spring and viscous fluid. As for their energy characteristics, they are close to hydraulic draft gears; however, their cost is lower and design is simpler.

> In the course of loading, the piston movement causes pressure increase; as a result of elastomer bulk compression, additional resilient elements become unnecessary for the transmission of static loads and return to the initial state.
> The elastomeric draft gear ZW-73 (Figure 3.65) contains hollow cylindrical case 1 closed with bottom plate 2 on one end. Elastomer shock absorber 3 consisting of cylindrical body 4 filled with elastomer, and rod 5, are partly located inside cylindrical case 1, with the possibility of moving with respect to it. At one end of rod 5, there is piston 6, which divides the working chamber of cylindrical body 4 into two chambers 7 and 8, with the possibility of the elastomer mass flowing from one chamber to the other through calibrated slot 9 between the lateral face of piston 6 and the interior face of cylindrical body 4. Rod 5 rests with its other end on bottom plate 2 of cylindrical case 1. Elastomer

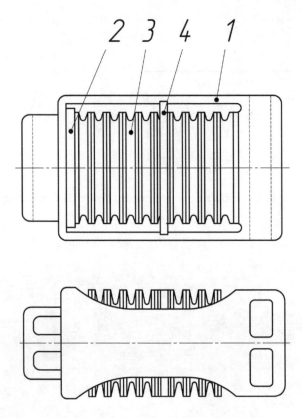

FIGURE 3.64 Rubber-metal or polymer draft gear R-2P. (1) case; (2) thrust plate; (3) rubber-metal elements; and (4) intermediate plate.

shock absorber 3 is pressed by coupler follower 10, which is connected with the draft gear case by means of bolts 11 and mounting strips 12, with the possibility of moving with respect to it.

The draft gear operates as follows: When affected by the compression load transmitted through coupler follower 10 to elastomer shock absorber 3, rod 5 enters the cavity of cylindrical body 4 and compresses the elastomer mass, generating high internal pressure, and energy is absorbed as a result of this process. In the course of impact (dynamic) compression of the shock absorber, energy absorption happens owing to the elastomer mass flowing (throttling) through calibrated slot 9 between the lateral face of piston 6 and the interior face of cylindrical body 4.

3.6.4 Buffers

Buffers are intended for absorption of longitudinal forces; however, depending on the rollingstock type and the coupler type applied, they perform a number of additional functions.

Buffers of freight cars with screw couplers are intended for compressive longitudinal forces absorption and energy dissipation; that is, they perform the same function as draft gears of cars with automatic couplers.

The typical buffer (Figure 3.66) consists of plate 1 and case 2, inside of which the absorbing device is located. Depending on the absorbing device type, buffers can be spring, with helical springs, spring with ring springs, hydraulic, hydraulic gas and elastomeric.

Design of Unpowered Railway Vehicles

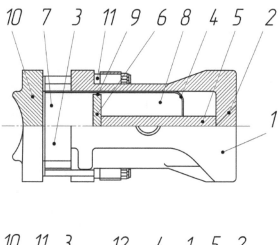

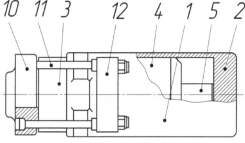

FIGURE 3.65 Elastomeric draft gear ZW-73. (1) case; (2) bottom plate; (3) elastomer shock absorber; (4) body filled with elastomer; (5) rod; (6) piston; (7 and 8) chambers; (9) calibration slot; (10) coupler follower; (11) bolts; and (12) mounting strips.

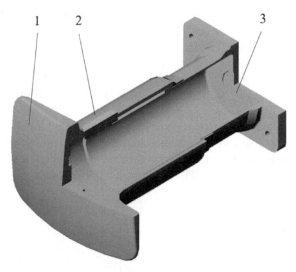

FIGURE 3.66 Typical buffer. (1) Plate; (2) case; (3) foundation.

According to EN 15551 [18], depending on the type A, B, C or L (long strobes), the dynamic energy absorption capacity of buffers is 30 to 72 kJ per each buffer. The full travel of buffers for freight cars is 105 mm, and, for buffers for passenger cars, it is 110 mm; for type L, it is 150 mm.

Wagons for hazardous materials are equipped with crash buffers. In the case of large impact forces (more than 1.5 MN), crash buffers are plastically deformed and do not return to their initial state.

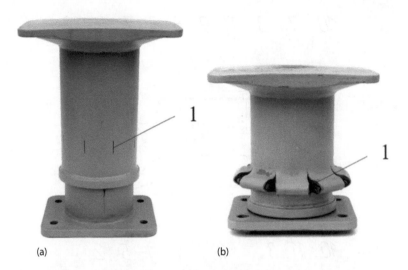

FIGURE 3.67 Innova System and Technologies buffer. (a) general view and (b) after deformation. (1) Slots. (From Innova Systems and Technologies, Arad, Romania. With permission.)

The Innova System and Technologies buffer is shown in Figure 3.67a. Deliberate weakening of the case is made in the form of slots. When affected by large forces, the case gets plastically deformed, as shown in Figure 3.67b, to absorb the impact energy. The energy-absorption capacity of such buffers is 250 to 400 kJ.

Passenger car buffers with automatic non-rigid coupling are intended for thrusting and clearance adjustment in automatic couplers of passenger cars. Their energy-absorption capacity is not large, because the main energy is taken up by the draft gear. Buffers protrude 65 ± 10 mm over the engagement plane of automatic couplers (Figure 3.68). After automatic couplers have been coupled, the tight condition of the automatic coupler and the clearance adjustment result. The tight condition of the automatic coupler reduces the level of longitudinal acceleration when that train movement modes are changed, which is important for maintaining comfort of passengers. Apart from that, the coupler contour wear of automatic couplers reduces, and the danger of uncoupling for automatic couplers decreases [50].

In some designs of transitive platforms (see later), structures of inter-car gangways rest on buffers.

Passenger car buffers with automatic rigid coupling are intended for absorption of longitudinal forces, because there is no necessity of clearance adjustment in automatic couplers. Their design is essentially similar to designs of buffers in freight cars; the differences are a bigger travel (110 mm) and greater energy-absorption capacity. The example of a buffer with a polymer hydraulic absorber is shown in Figure 3.69.

3.6.5 INTER-CAR GANGWAYS

Inter-car gangways are intended for passing from one car to the next one, and they often consist of an enclosure, a sliding frame and a connecting bridge [52].

Inter-car gangways can be with the closed contour, which provides the complete enclosure of the passage, or with an unclosed contour closing only the sides and the top of the passage. Inter-car gangways without the sliding frame, but with rubber tubes, are applied for passenger coaches of 1520 mm track gauge (see Figure 3.70), together with the SA-3 automatic coupler. They consist of three rubber tubes 1 that are fixed onto the car body. In their bottom part, between the pair of

Design of Unpowered Railway Vehicles

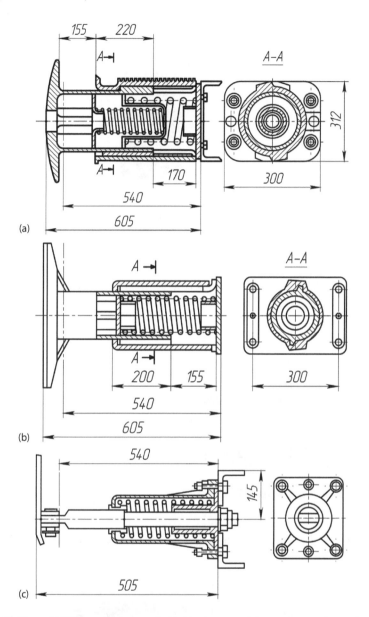

FIGURE 3.68 Buffers of 1520 mm gauge passenger coaches: (a) with two consecutive springs, (b) with one spring and (c) with spindle type damper.

buffers 3, buffer coupling is located, whereon connecting bridge 2 rests, closing the automatic coupler and providing safe passage for passengers. Relative displacements of cars are compensated for by deformation of the rubber tubes.

Inter-car hermetic gangways (Figure 3.71) are used with rigid-type couplers. Relative displacements of the coaches are compensated for by corrugated bellows.

In some high-speed trains, shock absorbers are installed between cars, and they damp reciprocal oscillation of cars. This allows the reducing of input energy from the rail track. The inter-car gangway with inter-car damper of the Talgo train is shown in Figure 3.72.

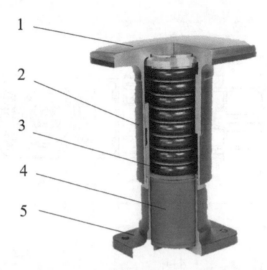

FIGURE 3.69 ETH hydraulic combination shock-absorber from Eisenbahntechnik Halberstadt, Germany. (1) Plate; (2) case; (3) hydraulic-polymer elastic elements; (4) foundation; and (5) mounting.

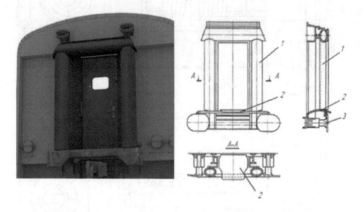

FIGURE 3.70 Inter-car gangway without the sliding frame with rubber tubes. (1) rubber tubes; (2) connecting bridge; and (3) buffers.

FIGURE 3.71 Inter-car hermetic gangways. (1) Frame and (2) corrugated bellows.

Design of Unpowered Railway Vehicles

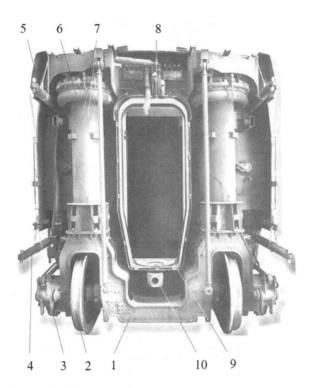

FIGURE 3.72 End wall of Talgo Pendular passenger coach. (1) One-axle bogie frame; (2) wheel; (3) axlebox and gauge changing mechanism; (4) bottom inter-car hydraulic damper; (5) top inter-car hydraulic damper; (6) pneumatic suspension bellow; (7) support column; (8) inter-car gangway; (9) bridge and door; (10) coupler element.

3.7 PRINCIPLES FOR THE DESIGN OF SUSPENSIONS

The parameters of a rail vehicle may be considered optimal if its dynamic characteristics meet three groups of requirements:

- There is sufficient reserve of critical speed with respect to design speed
- Ride quality, track forces and safety factors satisfy the standards on a straight track and in curves for all range of operational speeds
- Wear of friction elements and wheel profiles is within acceptable limits

Experience in the development of rail vehicles shows that, at the preliminary stage, the suspension parameters can be estimated using the simple engineering approaches described later [53]. To make sure that the parameters are optimised, further refinement is usually done using computer simulation.

3.7.1 Suspension Characteristics in Vertical Direction

Suspension should control and damp the motion of both the sprung and unsprung masses in the vehicle to obtain the best possible ride qualities whilst strictly fulfilling the safety requirements and satisfying specific service limitations such as ensuring that the vehicle remains within the clearance diagram.

Bogie elastic elements have various constructions, which, for example, can be cylindrical, rubber, leaf or pneumatic springs. From the dynamic behaviour from the vehicle point of view, the specific construction of the elastic element is not important, but the force characteristic that it provides is significant, that is, the dependence of the vertical load on the element P from its static deflection f: $P = P(f)$.

The static deflection of a suspension with linear characteristics (constant stiffness) is determined by the formula:

$$f_{st} = \frac{P_{st}}{c}, \tag{3.6}$$

where P_{st} is the static load on the suspension, and c is the stiffness of the suspension.

For a linear suspension, there is a dependence between the bounce natural frequency and the static deflection:

$$\omega^2 = \frac{c}{M} = \frac{g}{f_{st}}, \tag{3.7}$$

where M is the sprung mass of the vehicle, and g is the gravity acceleration.

Research has shown that decreasing the suspension vertical stiffness (increasing the static deflection) is favourable for the dynamic performance and force influence of the rail vehicle on the railway track if other conditions do not change.

In general, a low suspension stiffness gives lower acceleration, but practical considerations dictate that there must be a relatively small height difference between tare and laden conditions. To overcome such a limitation, new designs of coupling devices and brake systems are necessary [54]. In addition, the human perception of vibration over a range of frequencies must be considered. For passenger vehicles, the car body bounce frequency is generally in the range of 0.9 to 1.2 Hz, whilst this frequency for freight wagons can rise to 2.5 Hz in laden and up to 4 Hz in the tare conditions.

In order to realise an increased suspension deflection, modern suspensions use non-linear springs to provide optimal stiffness in the vicinity of the static deflection corresponding to the required load.

In suspension elements with variable stiffness (Figure 3.73), the dynamic oscillations appear around an equilibrium position given by static force P_{st}. To estimate the oscillation frequency in this case, the equivalent stiffness and equivalent deflection are used:

$$c_{eq} = \left.\frac{dP}{df}\right|_{P=P_{st}} ; \quad f_{eq} = \frac{P_{st}}{c_{eq}}. \tag{3.8}$$

In a suspension with a stiff force characteristic, as shown by curve b in Figure 3.73, $f_{1eq} < f_{1st}$ and, for a soft characteristic, as shown by curve b, $f_{2eq} > f_{2st}$, where $f_{1st, 2st}$ is the total static deflection.

In general, for the suspension with variable stiffness (bilinear characteristic), the first part of the characteristic has a constant stiffness c_1, chosen to give the required frequency for the tare condition, whilst the second part with constant stiffness c_2 gives the required frequency for the laden wagon (see trace c in Figure 3.73):

In this case:

$$c_{eq} = \begin{cases} c_1, f \leq \Delta \\ c_2, f > \Delta \end{cases} ; \quad f_{eq} = \begin{cases} \dfrac{P_{st}}{c_1}, P_{st} \leq c_1\Delta \\ \dfrac{P_{st}}{c_2}, P_{st} > c_1\Delta \end{cases}, \tag{3.9}$$

where Δ is the deflection corresponding to the breakpoint of the characteristic.

In freight wagon bogies, a non-linear vertical characteristic is usually provided by combinations of cylindrical springs having different free heights. In such bogies, the characteristic can be multi-linear, containing up to four linear pieces.

Design of Unpowered Railway Vehicles

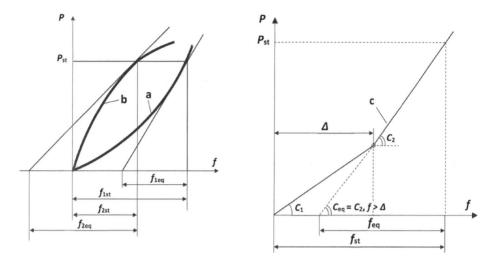

FIGURE 3.73 Non-linear elastic force characteristics of a suspension: stiff (curve a), soft (curve b) and bilinear (trace c). (From Iwnicki, S. (Ed.), *Handbook of Railway Vehicle Dynamics*, CRC Press, Boca Raton, FL, 2006. With permission.)

In theoretical calculations of the suspension force characteristic, there are sharp transition points from one piece of multi-linear characteristic into the other; these are much smoother in the real world, where the springs have free height production tolerances. Therefore, the transition zones in simulation are often smoothed with fillets as well.

To provide structural strength of the elastic elements and no impacts in the suspension, especially with coil springs, they should provide no solid state during the oscillations of a fully laden car body while moving in a train. To consider this situation, the deflection reserve coefficient K_{res} is introduced for suspensions in general:

$$K_{res} \geq 1 + k_d, \quad (3.10)$$

with a limit value that should be bigger than $(1 + k_d)$, where k_d is the maximum possible ratio of dynamic vertical force in car body oscillations to the static vertical force in the suspension [55].

For suspensions in general, the deflection reserve coefficient is the minimum value among all elastic elements. For each of the coil springs in the suspension, the deflection reserve coefficient can be calculated using the formula:

$$K_{res} = 1 + \frac{h_f - h_s - (f_{st} - e)}{f_{eq}}, \quad (3.11)$$

where h_f and h_s are the spring free height and solid height, respectively; e is the height difference between the tallest spring and the chosen spring; f_{st} is the full static deflection of the spring and f_{eq} is the equivalent deflection of the suspension.

To provide the safe operation of coupling devices in the train, the height difference between the longitudinal coupler (or buffer) axes of two neighbouring cars should not exceed the prescribed value. The worst case is calculated from the coupling height of the gross laden car with maximum possible wear of bogie components and the height of the tare vehicle with new bogies (without wear). The difference in the coupler levels is due to static deflection of the suspension under the maximum load, aging of elastic elements and wear of bogie components (e.g., wheel profile wear or wear of centre bowls and side bearings).

In service, the car body roll must also be limited to prevent the risk of overturning on highly canted curves and to ensure that the vehicle remains within the required clearance diagram. Once the maximum allowable roll angle for the vehicle body and the maximum lateral force (centrifugal, wind and lateral components of the interaction force between the vehicles in curves) have been established, the equilibrium equation gives the minimum acceptable vertical stiffness of the suspension.

The final value of vertical stiffness for the suspension is chosen to be the maximum of the minimum values, calculated using the service and design limitations.

3.7.2 In-Plane Suspension Stiffness

Theoretical investigations and experiments show that wheelset stability increases with increasing stiffness of the connection to the bogie frame. However, the character of this dependence is highly non-linear, and the relation between suspension stiffness and the mass and conicity of the wheels influences the critical speed. Increasing the longitudinal stiffness of the primary suspension impairs the guiding properties of the wheelset in curves, whilst increasing the lateral stiffness reduces the ability of the wheelset to safely negotiate large lateral irregularities.

A fundamental conflict therefore exists between the requirement for high-speed stability on straight track and good curving with safe negotiation of track irregularities. The 'in-plane' (lateral and longitudinal) stiffnesses must therefore be selected to give the best compromise for the conditions under which the vehicle will operate.

In order to make a preliminary choice of bogie in-plane stiffness, it is useful to know the relationship between stiffness and the ride quality in an analytical or graphical form. The simplified approach described in [32] is useful as a starting point.

The natural vibration modes shown in Figures 3.74 and 3.75 can be obtained from the linear equations of motion for a two-axle bogie [46].

Analysis of the modes shows:

- For in-phase yaw, as shown in Figure 3.74a, there is a relative lateral displacement between the centres of wheelsets O_1 and O_2 and the bogie centre
- Similar lateral displacements appear for the anti-phase mode, shown in Figure 3.75b
- Relative rotation between wheelset centres O_1 and O_2 occurs only for anti-phase yaw of wheelsets (Figure 3.75a)

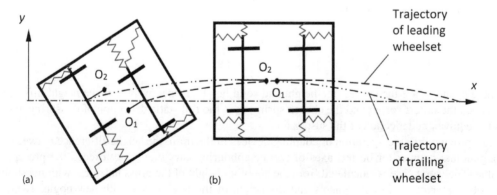

FIGURE 3.74 Wheelset modes for a two-axle bogie: (a) in-phase yaw and (b) in-phase lateral displacement. (From Iwnicki, S. (Ed.), *Handbook of Railway Vehicle Dynamics*, CRC Press, Boca Raton, FL, 2006. With permission.)

Design of Unpowered Railway Vehicles

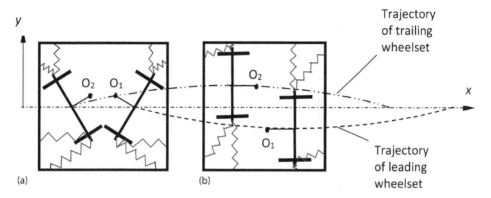

FIGURE 3.75 Wheelset modes for a two-axle bogie. (a) anti-phase yaw and (b) anti-phase lateral displacement. (From Iwnicki, S. (Ed.), *Handbook of Railway Vehicle Dynamics*, CRC Press, Boca Raton, FL, 2006. With permission.)

Thus, two generalised parameters can be introduced for the bogie:

- A stiffness corresponding to relative lateral displacement between the centres of wheelsets, referred to as the shear stiffness (K_s)
- A stiffness corresponding to the relative yaw angle between the wheelsets, referred to as the bending stiffness (K_b)

The conventional representation of bogie shear and bending stiffness is shown in Figure 3.76, represented as translational and torsion springs, respectively. The generalised stiffnesses K_s and K_b have a particular physical meaning. The shear stiffness K_s has a greater influence on the critical speed

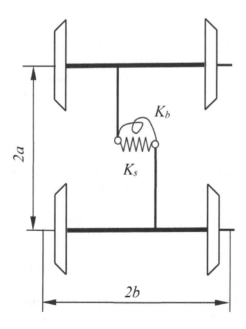

FIGURE 3.76 Representation of the primary suspension using shear and bending stiffness. (From Iwnicki, S. (Ed.), *Handbook of Railway Vehicle Dynamics*, CRC Press, Boca Raton, FL, 2006. With permission.)

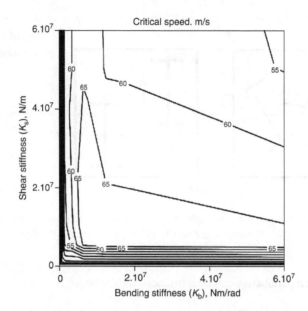

FIGURE 3.77 Critical speed (m/s) as a function of shear and bending stiffness. (From Iwnicki, S. (Ed.), *Handbook of Railway Vehicle Dynamics*, CRC Press, Boca Raton, FL, 2006. With permission.)

of the vehicle, whilst the bending stiffness K_b mainly determines the wheelsets' angles of attack in curves.

The use of shear and bending stiffness to give a simplified representation of the primary suspension without consideration of the bogie frame inertia (Figure 3.76) allows the in-plane bogie stiffnesses to be chosen, without considering its specific design.

Solution of the stability problem [36] shows that the critical speed of a conventional railway vehicle is a function of its shear and bending stiffnesses, as shown in Figure 3.77. The quality of curving can be estimated using the relationship of the wear number (the sum of creep force power for all wheels of the vehicle) to the shear and bending stiffnesses, as shown in Figure 3.78.

These relationships show that the chosen bending stiffness of the bogie should be the minimum that provides the required critical speed, and the shear stiffness should be within the critical speed range for the chosen bending stiffness.

3.7.3 Suspension Damping

Damping is typically provided within the suspension by either friction or hydraulic devices. Some types of elastic elements, such as leaf springs, have sufficient internal friction damping to avoid the necessity of a separate damper.

The selection of the optimum damping levels is a more complicated problem than the choice of suspension stiffness, although damping is less dependent on existing operational limitations. High levels of damping decrease the amplitudes of vibrations in resonance situations but significantly increase the accelerations acting on the vehicle body for higher-frequency inputs such as short-wavelength track irregularities.

Hydraulic dampers are almost universally used for passenger vehicles. Let us consider the simplified case of linear dependence between the damper force and the velocity. In this case, attenuation of vehicle vibrations is determined by the ratio of the real part of the eigenvalue to the corresponding natural frequency. This is termed the damping coefficient and is different for different natural vibration modes:

Design of Unpowered Railway Vehicles 109

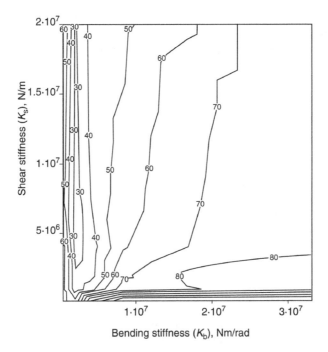

FIGURE 3.78 Sum of four wheels' wear numbers (N) as a function of shear and bending stiffness for a 600 m curve at 30 m/sec. (From Iwnicki, S. (Ed.), *Handbook of Railway Vehicle Dynamics*, CRC Press, Boca Raton, FL, 2006. With permission.)

$$d_i = \frac{1}{2\omega_i} \frac{\{v_i\}^T [B]\{v_i\}}{\{v_i\}^T [M]\{v_i\}}, \qquad (3.12)$$

where $[B]$ and $[M]$ are the damping and inertia matrices of the vehicle multi-body model, respectively; $\{v_i\}$ is the column-vector of the i-th eigenmode and ω_i is the natural frequency of the i-th eigenmode.

Effective damping of the vibrations of railway vehicles is typically obtained with damping coefficients, which lie in the following ranges: 0.2 to 0.3 for vertical oscillations, 0.3 to 0.4 for horizontal oscillations and 0.1 to 0.2 for vehicle body roll.

In freight bogies, friction dampers are commonly used. When making the preliminary choice of parameters, the friction force in the damper is estimated on the basis that the amplitude should not increase in the resonance case.

Assuming that the amplitude of oscillations at resonance increases by $\Delta A'$ during one period and the friction force F acting in the suspension reduces it by $\Delta A''$, the following conditions must apply to prevent the amplitude increasing in the resonant case:

$$\Delta A'' \geq \Delta A'. \qquad (3.13)$$

The equations of oscillation for the system with dry friction under periodic excitation give:

$$F \geq \frac{\pi q}{4} c_{eq}, \qquad (3.14)$$

where q is the estimated amplitude of periodic track irregularity (prescribed in regulations), and c_{eq} is the equivalent stiffness of the suspension.

Estimating the magnitude of the friction force is easier when using a relative friction coefficient that equals the ratio of friction force to the static vertical load:

$$\varphi = \frac{F}{P_{st}} \geq \frac{\pi q}{4 f_{eq}}, \quad (3.15)$$

where $f_{eq} = \frac{P_{st}}{c_{eq}}$.

The relative friction coefficient is a general parameter of the wagon, and the optimal value of friction force depends on the equivalent static deflection of the suspension or, for the case of a non-linear suspension characteristic, the vertical load. For freight wagons, the recommended optimum relative friction coefficient lies within the range of 0.2 to 0.4 for the empty condition and 0.07 to 0.13 for the fully laden condition [36].

3.7.4 CAR BODY TO BOGIE CONNECTIONS

Car body to bogie connections have the following functions:

- Allow the bogie to turn relative to the car body in curves
- Transmit the vertical, traction and braking forces
- Provide additional control of lateral suspension inputs
- Assist in maintaining the stability of the bogie
- Provide longitudinal stability of bogie frames and equal distribution of load over the wheelsets

These functions depend on the type of the rollingstock: traction or trailing, passenger or freight, moderate or high speed. If the vehicle is stable up to the design speed, then introduction of additional yaw resistance torque is not necessary. If the static deflection of the suspension is sufficient, then vertical flexibility in car body to bogie connections may not be necessary.

Designs generally aim to make the bogie to car body connections as simple as possible by the use of a small number of elements and reduction of the number of elements with surface friction. The conventional design uses the centre bowl to centre plate connection (flat or spherical) in the middle, with side bearings (elastic constant contact or rigid with a clearance) by the sides. From the railway vehicle dynamics point of view, it is described by the following parameters:

- Vertical and longitudinal stiffness of the elastic element of the side bearing (zero for rigid side bearings with a clearance)
- Static deflection of the constant contact side bearing under the car body (usually expressed in percentage of car body weight carried by the side bearings)
- Friction coefficient on the side bearing surface
- Friction yaw torque and roll recovery in the centre bowl
- Maximum possible deflection (or clearance for rigid) of the side bearing limited by the bump stop

The weight per one constant contact side bearing is determined by the formula:

$$P_{st} = c_{spr} \cdot f_{st}, \quad (3.16)$$

Design of Unpowered Railway Vehicles

where c_{spr} is the vertical stiffness of the elastic element, and f_{st} is the static deflection of the side bearing.

To provide uniform distribution of car body weight to the bogie, the sum of vertical forces per side bearing is usually limited to not more than 85% of empty car body weight:

$$\eta = \frac{4P_{st}}{M_b g} \cdot 100\%, \tag{3.17}$$

where M_b is the minimum car body mass.

The constant contact side bearings experience not only the static weight of the car body but also the dynamic roll oscillations, providing the recovery torque and thus improving the ride performance [39]. The dependence of mode damping on the side bearing vertical stiffness is presented in Figure 3.79. It appears that the damping is highly dependent on the longitudinal mass moment of inertia of the car body (23,000 kg·m² for the gondola and 131,000 kg·m² for the hopper). With the increase of the vertical stiffness of the constant contact side bearing, the damping of roll oscillations increases. In addition, there is an increase in overturning safety in curves but a decrease in derailment safety on straight track (Figure 3.80). The rational values of the side bearing vertical stiffness that provide the necessary compromise correspond to damping coefficients in the range of 0.25 to 0.3.

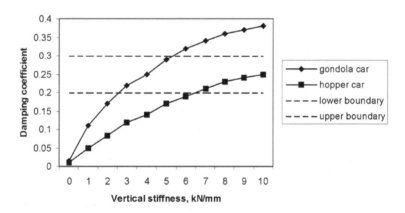

FIGURE 3.79 Dependence of car body roll damping coefficient on side bearing vertical stiffness for gondola car and hopper car.

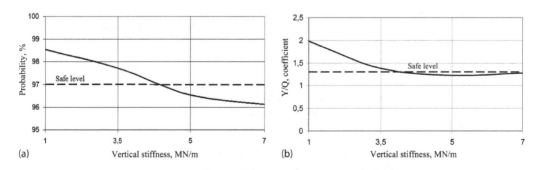

FIGURE 3.80 Impact of side bearing vertical stiffness on the probability that the Y/Q coefficient is greater than the limit value for: (a) the empty wagon and (b) the fully laden wagon.

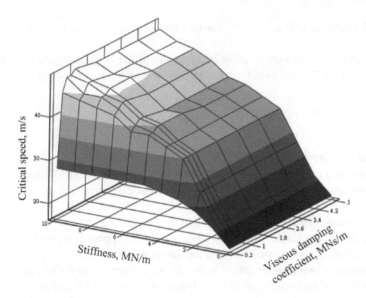

FIGURE 3.81 Dependence of the critical speed on longitudinal stiffness and viscous damping being modelled in series in the side bearing.

To keep down the car body roll angle and provide the structural strength of the side bearing elastic elements, their vertical deflection is usually limited with a bump stop or locking between the cap and the cage. The maximum vertical force on the side bearing appears when the laden wagon passes through sharp curves, with large rail superelevation at maximum possible speed.

The longitudinal stiffness and friction force in the side bearing (that can be viewed in series with elastic element shear deflection limited by the bump stop) influence the critical speed and wheel-rail lateral forces in curving. Increase in the shear longitudinal stiffness of the side bearing (see Figure 3.81) leads to the increase of the critical speed with a larger rate of increase for stiffness below 3.5 MN/m and with a much smaller rate for stiffness above this. Analysis of curving using the nonlinear models showed that, within the possible range of yaw friction torque and the range of side bearing longitudinal stiffness, providing necessary critical speed level, no variation of curving qualities was obtained. However, standards tend to limit the total yaw torque provided by constant contact side bearings down to 12 to 15 MN·m/rad.

REFERENCES

1. AAR, Manual of Standards and Recommended Practices, Section C, Part II, Design, Fabrication, and Construction of Freight Cars, M-1001, Association of American Railroads, Washington, DC, 2015.
2. GOST 32885-2014, Automatic coupler model SA-3 and its components: Design and dimensions, 2015.
3. AAR, Manual of Standards and Recommended Practices, Section C, Car Construction—Fundamentals and Details, Association of American Railroads, Washington, DC, 2014.
4. GOST 9238-2013, Construction and rolling stock clearance diagrams, 2014.
5. European Union Commission, Technical specification for interoperability relating to the subsystem rolling stock—freight wagons, Annex to Commission decision of 28 July 2006 concerning the trans-European conventional rail system, Brussels, Belgium, 2006.
6. A. Afanasiev, K. Kyakk, Improving the method to determine the fatigue limit of frame elements in long flat wagons, *Vagony I Vagonnoe Hoziaystvo*, 1(37), 42–46, 2014.
7. GOST 34093-2017, Passenger cars on locomotive traction: Requirements for structural strength and dynamic qualities, 2018.

8. T. Kamada, R. Kiuchi, M. Nagai, Suppression of railway vehicle vibration by shunt damping using stack type piezoelectric transducers, *Vehicle System Dynamics*, 46(S1), 561–570, 2008.
9. G. Schandl, P. Lugner, C. Benatzky, M. Kozek, A. Stribersky, Comfort enhancement by an active vibration reduction system for a flexible railway car body, *Vehicle System Dynamics*, 45(9), 835–847, 2007.
10. J.R. Ellis, A model of semi trailer vehicles including the roll modes of motion, *Vehicle System Dynamics*, 6(2–3), 124–129, 1977.
11. G.I. Bogomaz, O.M. Markova, Y.G. Chernomashentseva, Mathematical modelling of vibrations and loading of railway tanks taking into account the liquid cargo mobility, *Vehicle System Dynamics*, 30(3–4), 285–294, 1998.
12. S.F. Feschenko, I.A. Lukovsky, B.I. Rabinovich, L.V. Dokuchaev, Methods for determining the added fluid masses in mobile cavities, *Naukova Dumka, Kiev*, 1969.
13. M. Wollström, Effects of vibrations on passenger activities: writing and reading—a literature survey. TRITA-FKT Report 2000:64, Royal Institute of Technology (KTH), Stockholm, Sweden, 2000.
14. P. Carlbom, M. Berg, Passengers, seats and carbody in rail vehicle dynamics, *Vehicle System Dynamics*, 37(S1), 290–300, 2002.
15. EN 15566:2009+A1:2010, Railway applications-railway rolling stock-draw gear and screw coupling, European Committee for Standardization, Brussels, Belgium, 2010.
16. GOST 21447-75, Coupler contour line: dimensions, 1976.
17. AAR, Manual of Standards and Recommended Practices, Section B, Couplers and Freight Car Draft Components, Association of American Railroads, Washington, DC, 2012.
18. EN 15551:2009+A1:2010, Railway applications-railway rolling stock-buffers, European Committee for Standardization, Brussels, Belgium, 2010.
19. M. Spiryagin, Wolfs, C. Cole, V. Spiryagin, Y.Q. Sun, T. McSweeney, *Design and Simulation of Heavy Haul Locomotives and Trains*. CRC Press, Boca Raton, FL, 2017.
20. Y. Boronenko, A. Orlova, A. Iofan, S. Galperin, Effects that appear during the derailment of one wheelset in the freight wagon: Simulation and testing, *Vehicle System Dynamics*, 44(S1), 663–668, 2006.
21. Norms for calculation and design of cars for MPS railways of 1520 mm gauge (not self-propelled), GosNIIV–VNIIZhT, Moscow, 1996.
22. Y.P. Boronenko, A. Orlova, On possibility to operate the articulated two-frame container flat wagon on the Russian railways, In I. Zobory (Ed.), *Proceedings 11th Mini Conference on Vehicle System Dynamics, Identification and Anomalies*, Komaromi Nyomda es Kiado Kft., Budapest, Hungary, pp. 277–284, 2010.
23. Railway Transport Encyclopedia, In N.S. Konarev (Ed.), *Big Russian Encyclopedia*, Scientific Publishing, Moscow, Russia, pp. 48–52, 101–102, 206–207, 354, 456, 491, 1995. (in Russian)
24. American Railway Encyclopedia, Wagons and their Maintenance, All-Union Publishing and Polygraph Union of Railway Ministry, 1961. (in Russian-shortened translation from English)
25. L.A. Shadur, I.I. Chelnokov, L.N. Nikolsky, E.N. Nikolsky, V.N. Koturanov, P.G. Proskurnev, G.A. Kazansky, A.L. Spivakovsky, V.F. Devyatkov, *Wagons: Student Book for Railway Institutes*, L.A. Shadur (Ed.), Moscow: Transport, Russia, 1980. (in Russian)
26. Railway wagons of 1520 mm gauge railways: album and lookbook, 002I-97, PKB TsV, Moscow, Russia, 1998. (in Russian)
27. GOST 10791-2011, All-rolled wheels-specifications, 2012.
28. International Heavy Haul Association, Guidelines to best practices for heavy haul railway operations: Management of the wheel and rail interface, J. Leeper, R. Allen (Eds.), Simmons-Boardman Books, Omaha, NE, 2015.
29. EN 13715:2006+A1, Railway applications-wheelsets and bogies-wheels-tread profile, European Committee for Standardization Brussels, Belgium, 2006.
30. EN 15302:2008+A1:2010, Railway applications - Method for determining the equivalent conicity, European Committee for Standardization, Brussels, Belgium, 2010.
31. I.Y. Shevtsov, Wheel/rail interface optimization, PhD Dissertation, Delft University of Technology, Delft, the Netherlands, 2008.
32. A.H. Wickens, Fundamentals of rail vehicle dynamics: Guidance and stability, Swets & Zeitlinger, Lisse, the Netherlands, 2003.
33. Y. Boronenko, A. Orlova, E. Rudakova, Influence of construction schemes and parameters of three-piece freight bogies on wagon stability, ride and curving qualities, *Vehicle System Dynamics*, 44(S1), 402–414, 2006.
34. A. Orlova, Y. Boronenko, The influence of the condition of three-piece freight bogies on wheel flange wear: Simulation and operation monitoring, *Vehicle System Dynamics*, 48(S1), 37–53, 2010.

35. A. Orlova, Y. Boronenko, Reconsidering requirements to friction wedge suspensions from freight wagon dynamics point of view, *Proceedings 22nd IAVSD Symposium*, 14–19 August 2011, Manchester, UK, CD-ROM, 2011.
36. A.M. Orlova, A.V. Saidova, E.A. Rudakova, A.N. Komarova, A.V. Gusev, Advancements in three-piece freight bogies for increasing axle load up to 27t, In: M. Rosenberger, M. Plöchl, K. Six, J Edelmann (Eds.), *Proceedings 24th IAVSD Symposium held at Graz*, Austria, 17–21 August 2015, CRC Press/Balkema, Leiden, the Netherlands, pp. 1044–1049, 2016.
37. S.D. Iwnicki, S. Stichel, A. Orlova, M. Hecht, Dynamics of railway freight vehicles, *Vehicle System Dynamics*, 53(7), 995–1033, 2015.
38. A. Orlova, E. Rudakova, Comparison of different types of friction wedge suspensions in freight wagons, *Proceedings 8th International Conference on Railway Bogies and Running Gears*, 13–16 September 2010, Budapest University of Technology and Economics, Budapest, Hungary, pp. 41–50, 2010.
39. Y.P. Boronenko, A.M. Orlova, Influence of bogie to car body connection parameters on stability and curving of freight vehicle, *Extended Abstracts 6th International Conference Railway Bogies and Running Gears*, 13–16 September 2004, Budapest University of Technology and Economics, Budapest, Hungary, pp. 23–25, 2004.
40. J. Piotrowski, P. Pazdzierniak, T. Adamczewski, Suspension of freight wagon bogie with the Lenoir friction damper ensuring low wear of wheels and good lateral dynamics of the wagon, *Proceedings of XVIII Conference on Railway Vehicles ('Pojazdy Szynow')*, Katowice, Poland, Vol. I, pp. 199–211, 2008.
41. M.W.J. Etwell, Advances in rail wagon design, *Journal of Rail and Rapid Transit*, 204(1), 45–54, 1990.
42. D. Scholdan, N. Gabriel, W. Kik, RC25NT-ein neues, gleisfreundliches Drehgestell für den schweren Güterverkehr (RC25NT-a new, track-friendly bogie for the heavy freight transport), ELH Eisenbahnlaufwerke Halle GmbH Slideshow Presentation delivered at 2011 Modern Rail Vehicles Conference, September 2011, Graz, Austria.
43. Y.P. Boronenko, A.M. Orlova, E.A. Rudakova, The influence of inter-axle linkages on stability and guidance of freight bogies, *Proceedings 8th Mini Conference on Vehicle System Dynamics, Identification and Anomalies*, Budapest University of Technology and Economics, Budapest, Hungary, pp. 175–182, 2002.
44. H. Scheffel, R.D. Fröhling, P.S. Heynes, Curving and stability analysis of self-steering bogies having a variable yaw constraint, *Vehicle System Dynamics*, 23(S1), 425–436, 1994.
45. H. Scheffel, A new design approach for railway vehicle suspension, *Rail International*, 10, 638–651, 1974.
46. A. Orlova, Y. Boronenko, H. Scheffel, R. Fröhling, W. Kik, Tuning von Güterwagendrehgestellen durch Radsatzkopplungen, *ZEV-Glasers Annalen*, 126, 270–282, 2002.
47. W. Kik, D. Scholdan, J. Stephanides, Project INFRA-RADIAL-bogies for axle loads of 25t-test and simulation, *XXI Century Rolling Stock: Ideas, Requirements, Projects Conference*, St. Petersburg, 2007.
48. S.D. Iwnicki, A. Orlova, P.-A. Jonsson, M. Fartan, Design of the running gear for the SUSTRAIL Freight Vehicle, *Proceedings of IMechE Stephenson Conference: Research for Railways*, 21–23 April 2015, London, UK, pp. 299–306, 2015.
49. L.M. de Oriol, El Talgo Pendular, *Revista Asociacion de Investigacion del transporte*, 53, 1–76, 1983.
50. V.V. Lukin, L.A. Shadur, V.N. Koturanov, A.A. Koholov, P.S. Anisimov, Design and calculation of railcars: Notebook, UMK MPS, Moscow, Russia, 2000. (in Russian)
51. A.P. Boldyrev, B.G. Keglin, Calculation and design of shock absorbers for rolling stock, Mashinostroenie-1, Moscow, Russia, 2004. (in Russian)
52. I.P. Kiselev (Ed.), L.S. Blazhko, A.T. Burkov, N.S. Bushuev, V.A. Gapanovich, V.I. Kovalev, A.P. Ledyaev et al. High speed railway transport: Notebook, FGBOU Educational and Methodical Center for Education in Railway Transport, Moscow, Russia, 2004. (in Russian)
53. V.K. Garg, R.V. Dukkipati, *Dynamics of Railway Vehicle Systems*, Academic Press, Orlando, FL, 1984.
54. A. Orlova, R. Savushkin, A. Sokolov, S. Dmitriev, E. Rudakova, A. Krivchenkov, M. Kudryavtsev, V. Fedorova, Development and testing of freight wagons for 27t per axle loads for 1520 mm gauge railways, In: P.J. Grabe, R.D. Fröhling (Eds.), *Proceedings 11th International Heavy Haul Association Conference*, 2–6 September 2017, Cape Town, South Africa, pp. 1089–1096, 2017.
55. A. Orlova, E. Rudakova, A. Gusev, Reasoning for assignment of permissible minimum value of deflection reserve coefficient for freight cars' bogies, *Proceedings of Petersburg State Transport University*, 14(1), 73–87, 2017. (in Russian)

4 Design of Powered Rail Vehicles and Locomotives

Maksym Spiryagin, Qing Wu, Peter Wolfs and Valentyn Spiryagin

CONTENTS

- 4.1 Introduction 116
- 4.2 Types of Railway Traction Rolling Stock and Their Classification 116
  - 4.2.1 Passenger Traction Rolling Stock 118
    - 4.2.1.1 Light Rail Vehicles 118
    - 4.2.1.2 Passenger Trains and Locomotives 118
    - 4.2.1.3 High-Speed Trains 119
  - 4.2.2 Freight Traction Rolling Stock 119
    - 4.2.2.1 Freight Locomotives 119
    - 4.2.2.2 Heavy-Haul Locomotives 119
  - 4.2.3 Shunting Locomotives 119
  - 4.2.4 Traction Rolling Stock for Special Purposes 119
- 4.3 Motive Power Energy Principles 119
  - 4.3.1 Electric Traction 121
    - 4.3.1.1 Light Rail Vehicles 122
    - 4.3.1.2 Electric Locomotives 125
    - 4.3.1.3 Electric Multiple Units 126
  - 4.3.2 Diesel Traction 128
    - 4.3.2.1 Diesel Locomotives 129
    - 4.3.2.2 Diesel Multiple Units 131
  - 4.3.3 Gas Turbine Traction 132
  - 4.3.4 Hybrid Traction 134
- 4.4 Classification of Main Rail Traction Vehicle Components and Suspension Systems 135
  - 4.4.1 Main Frames and Bodies 135
  - 4.4.2 Bogies 136
  - 4.4.3 Traction Drives 138
  - 4.4.4 Suspension Systems 141
- 4.5 Connection between a Frame/Car Body and Bogies 142
  - 4.5.1 Centre Pivots 142
  - 4.5.2 Traction Rods 143
- 4.6 Traction Systems and Their Classification 144
  - 4.6.1 DC Traction Power 147
  - 4.6.2 AC Traction Power 149
  - 4.6.3 Hybrid Traction Power 150
- 4.7 Brake Systems and Their Components 151
  - 4.7.1 Basic Components of Air Brake Systems 152
  - 4.7.2 Dynamic Brake Systems 153
  - 4.7.3 Electromagnetic Brakes 155
  - 4.7.4 Rail Brakes 156

4.8   Design Classification and Critical Parameters for Vehicle System Dynamics 156
       4.8.1   Classification and Main Design Principles in Traction Studies 156
       4.8.2   Traction/Adhesion and Slip Control Principles 157
       4.8.3   Critical Parameters in Traction Studies 159
                4.8.3.1   Axle Load 159
                4.8.3.2   Tractive and Dynamic Braking Efforts 161
                4.8.3.3   Maximum Adhesion/Traction Coefficient 161
                4.8.3.4   Power Output 161
                4.8.3.5   Maximum Speed 162
References 162

4.1 INTRODUCTION

The history of rail transport development is directly linked with the advent of powered rail vehicles and locomotives and improvements of their designs and their manufacturing. The first locomotive building process can be dated to 1801, with the construction of a high-pressure steam road locomotive, known as the 'Puffing Devil', which had been designed by British inventor Richard Trevithick. Within a few years, an era of steam trams and locomotives commenced, which was only to itself be challenged at the beginning of twentieth century when the first modern competitors of steam locomotives were beginning to appear. By the middle of the twentieth century, all industrialised countries had begun the transition to new, advanced forms of traction, which fully replaced steam locomotives in passenger and freight operations with electric and diesel-powered rail vehicles and locomotives. This chapter provides an introduction to the classification of powered rail vehicles/locomotives and focuses on existing modern designs of their components. It also discusses details of their practical application for vehicle system dynamics studies.

4.2 TYPES OF RAILWAY TRACTION ROLLING STOCK AND THEIR CLASSIFICATION

There are many ways in which railway traction rolling stock can be classified. Details of those generally used are as follows [1,2]:

- By energy source: This divides all traction rolling stock into two groups, namely non-autonomous and autonomous. Non-autonomous rolling stock is provided with energy from a source being outside the powered vehicle. Electric locomotives and electric trains are good examples of non-autonomous rolling stock. Conversely, autonomous rolling stock receives the energy required for its motion process from a power plant that is mounted directly on or inside the vehicle. Steam, diesel and gas turbine locomotives, as well as diesel engine and hybrid transport rail vehicles, comprise this type of rolling stock.
  Advantages of non-autonomous traction are the possibility to realise lower mass and higher power by a rail traction vehicle, significant reduction of effects on the environment during operation of such rail vehicles, and also the possibility of the more efficient use of energy (e.g., regenerative braking in electric locomotives).
  However, autonomous rolling stock has its own advantages such as much lower costs of construction and maintenance of railway infrastructure (absence of a network of electrical supply substations, etc.) and also provides the possibility for working in critical conditions and extraordinary situations (failure of the electrical supply stations/substations, loss of connection in contact conductor wire networks in cases of bad weather conditions such as icing and hurricanes, etc.).

Currently, some works on the development of rolling stock designs with combined energy sources are in progress, which should allow these rail vehicles to work in both autonomous and non-autonomous modes.
- By type of use: Powered rolling stock are commonly divided into categories of passenger or freight transportation (or both can be combined in one), shunting operations and powered vehicles for special purposes (usually rail vehicles for special loads or rolling stock for maintenance activities).
- By track gauge: Many different dimensions for gauge have found wide application throughout the world. For example, the dimensions for commonly used gauges can vary from 1000 to 1676 mm.
- By loading gauge: This covers predefined dimensions (height and width) for rail vehicles, which should allow vehicles to be kept within a specific 'swept envelope' that provides adequate clearance to the surrounding structure outlines (e.g., tunnels, bridges and platforms) in order to ensure safe operation of vehicles through them. The loading gauge for rail vehicles is usually different for different countries and even for individual railways within a country.
- By vertical axle load: Limits in loadings on a track are determined by the structure of individual rail networks, which can potentially use rails with a variety of load-bearing capacity, different types and spacing of sleepers, various track substructure designs and limits on bearing capacity of bridges and other engineering structures.
- By electric power type used in the power traction system of a rail vehicle: Direct current (DC), alternating current (AC) or their combination e.g., an AC-DC locomotive).
- By maximum power of the powered rail traction vehicle: The maximum power of an autonomous locomotive means the effective maximum capacity of the power plant. Non-autonomous locomotives such as electric locomotives typically have a peak power output and are rated by their power delivered at the wheel.
- By maximum operational speed: Regular or high speed.
- By type of car body: Cab unit or hood unit.
- By number of driving cabs: One-cab or two-cab designs.
- By number of units in a vehicle set/train configuration: One, two, ..., multi-unit(s).
- By type of connection between units in a vehicle set/train configuration (single or articulated).
- By types of coupling and absorbing devices (also called draft gear): Couplers are used for the connection of railway transport vehicles in a train, for the transmission of tractive and brake efforts from powered transport vehicles to unpowered ones (e.g., to the wagons and carriages) and for the absorption of the shock loadings, which occur during motion, stops and also during shunting.
- By number of bogies in a rail-powered vehicle.
- By type of motor installation: Independent motorised/rotated wheels, individual motorised axles or group drive of multiple axles by a single traction motor.
- By number of consecutive motorised and non-motorised axles and related details.

According to the International Union of Railways (UIC) classification of axle arrangements [3], the axles within the same bogie (truck) are classified starting from the front end of the locomotive by alphabetical symbols for the number of consecutive motorised axles (A for one, B for two, C for three, etc.) and numerals for the number of consecutive non-motorised axles between motorised axles. The use of a lowercase 'o' as a suffix after the letter indicates that those motorised axles are individually driven by separate traction motors. A prime sign indicates axles that are mounted in a bogie; alternatively, brackets can be used to group letters and numbers describing a particular bogie. For example, Bo'Bo' means a rail traction vehicle with two independent two-axle bogies, all of which have axles/wheelsets individually motorised [3]. According to the system developed by the Association of American Railroads (AAR), a 'minus' sign is used to indicate the separation (non-articulation) of bogies used in one locomotive, and a 'plus' sign indicates articulated connections between bogies of a vehicle or between vehicles in a vehicle set/train configuration. Moreover, the suffix 'o' indicating individual drive axles in the UIC classification is not used. That is, the simplified classification used in the AAR system does not distinguish between drive axles that are individually driven and mechanically linked drive axles. For example, a rail traction vehicle with two independent two-axle bogies which have all axles/wheelsets individually motorised is written as B-B in the AAR system [4].

The description of railway traction vehicles using the classification approaches given previously is presented and discussed in the following sections of this chapter.

4.2.1 Passenger Traction Rolling Stock

Passenger rolling stock is commonly classified with the following groups: Light rail vehicles, trams (also called streetcars), locomotives and motorised carriages (cars), including multiple unit trains.

4.2.1.1 Light Rail Vehicles

It is believed that the term 'light rail vehicle' originally came from Britain in order to describe vehicles used in urban transportation. The formal definition of light rail made by the American Public Transportation Association states [5]: *'An electric railway with a "light volume" traffic capacity compared to heavy rail. Light rail may use shared or exclusive rights-of-way, high or low platform loading and multi-car trains or single cars. Also known as "street cars," "trolley car" and "tramway."'*

A more detailed definition can be formulated from [1,6] as follows:

- A light rail vehicle is a non-autonomous electrically propelled passenger vehicle with steel wheels that runs on steel rails.
- A light rail vehicle receives propulsion power from an overhead line by means of a pantograph or other collector and returns it back to an electrical substation(s) through the rails.
- A light rail vehicle is a vehicle that is able to negotiate sharp curves (radius of 25 metres or less).
- A light rail vehicle is not constructed to structural criteria (so-called 'buff strength'), as are heavy rail vehicles.

4.2.1.2 Passenger Trains and Locomotives

Passenger trains consist of motorised passenger cars or locomotives and unpowered passenger cars that are designed to carry passengers.

The motorised passenger cars are commonly implemented in diesel multiple unit (DMU) or electric multiple unit (EMU) designs for passenger transportation on city, suburban and regional rail networks.

An EMU comes under the non-autonomous category of rolling stock, as it receives energy from an external electrical supply source.

A DMU is an autonomous multiple unit train that has diesel engines as the power plant and usually provides passenger transportation in urban, suburban and inter-regional service areas, which are non-electrified or partially electrified.

In some countries, it is also common to use a passenger locomotive (a standalone unit that does not carry passengers itself and is designed to reach an operational speed up to 200 km/h very quickly) to haul passenger and non-passenger (post service, baggage, dining, etc.) coaching stock. A passenger locomotive can be an autonomous or non-autonomous powered traction vehicle, depending on the railway network where it is in use.

4.2.1.3 High-Speed Trains

A high-speed train consists of motorised passenger cars or locomotives and unpowered passenger cars that are designed to carry passengers at operational speeds in excess of 200 km/h on existing lines and 250 km/h on new railway lines. Similar to an EMU, it usually comes under the non-autonomous category of rolling stock, which receives energy from an external electrical supply source.

4.2.2 Freight Traction Rolling Stock

Freight traction rolling stock consists of the locomotives designed for high tractive effort for hauling large freight and heavy-haul ore or coal trains. Moreover, individual units hauling a train are placed in groups (parallel control) at the head of the train and in various position along the train (distributed control).

4.2.2.1 Freight Locomotives

Freight locomotives are powered traction vehicles designated to transport cargo between the shipper and the intended destination. The main requirements for a freight locomotive are to achieve and sustain the maximum tractive efforts for as long as necessary and to keep to the desired speed.

These locomotives can be an autonomous or non-autonomous powered traction vehicles, and they do not usually have any on-board payload capacity (with the exception of the Cargo Sprinter innovation).

4.2.2.2 Heavy-Haul Locomotives

Unlike freight locomotives, heavy-haul locomotives are designated for hauling the much longer unit trains used to carry very large payloads of bulk products such as coal and iron ore [2]. Like freight locomotives, they can be either autonomous or non-autonomous rail traction vehicles.

4.2.3 Shunting Locomotives

Shunting locomotives (switch engines) perform works in stations related to the forming of trains that they assemble for dispatch or disassemble upon arrival. They usually do not possess the large power capabilities of main-line locomotives and are able to work on track with lesser axle loading.

4.2.4 Traction Rolling Stock for Special Purposes

The rail traction vehicles for special purposes are vehicles that are able to perform special user functions other than transportation of cargo and passengers. For example, there are such vehicles as maintenance of way vehicles, military use rail vehicles and firefighting and rescue vehicles.

4.3 MOTIVE POWER ENERGY PRINCIPLES

Modern traction rolling stock is divided into four groups based on the type of the energy/power supply system:

- Electric
- Diesel
- Gas turbine
- Hybrid

The energy obtained from a power supply system is then transformed to motive power, as shown in Figure 4.1.

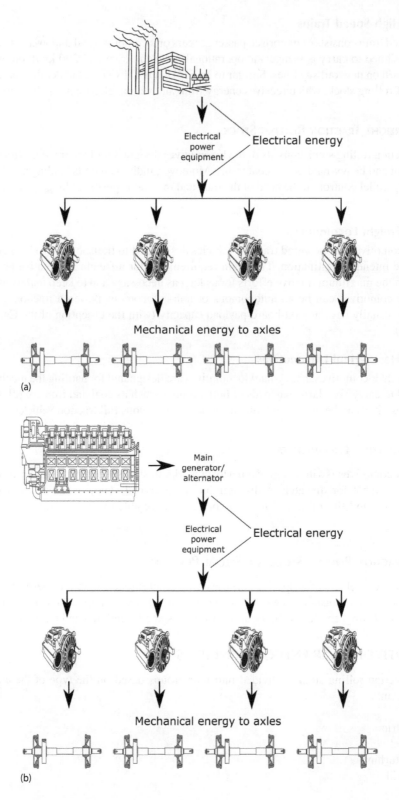

FIGURE 4.1 Example of energy transformation to motive power for (a) four-axle electric locomotive and (b) four-axle diesel-electric locomotive.

4.3.1 Electric Traction

The first practical prototypes of electric rolling stock were created in the 1920s. Industrial production of electric locomotives began in the 1930s, and there was a progressive improvement in both continuous traction power capacity and speed of locomotives until World War II. During the war, the production of electric locomotives was completely suspended. After the war, railway operators in Europe renewed their demand for electric locomotives, and this was connected with the repairing of damaged sections of electric traction railway infrastructure and the creation of new railway corridors.

Many modern electric rail traction vehicles can operate using multiple types of voltages and currents, and this is referred to as multi-system performance (allowing operation on AC and DC and at various voltages). Overhead traction wiring systems are generally used, except for underground railways, where limited clearances generally result in third rail delivery of electricity for trains. In most operational cases, an electric traction vehicle or a locomotive receives electrical power for its motion from an external electrical supply source. The general scheme for the electrical power supply system used in electrified railways is presented in Figure 4.2. The electricity from the power plant is transmitted to traction substations over the high-voltage distribution power lines. The substations perform the transformation of the current in accordance with the parameters required and then supply it through feeder power lines to points along the overhead line equipment for powering electric rail traction vehicles/locomotives through the contact conductor wire. For closed-loop networks, the railway track is equipped with special return feeders, which are connected to the power substations.

Electric rail traction vehicles can be divided into three types based on the types of railway electrification systems:

- DC rail traction vehicles
- AC rail traction vehicles
- Multi-system rail traction vehicles, which can operate with more than one railway electrification system

Electric traction vehicles or electric locomotives consist of the following basic systems: electrical, mechanical, pneumatic and hydraulic.

The car body, main frame, coupling devices, suspension, devices for transmission of tractive and brake efforts, bogies and a system for air cooling and ventilation of the electric traction equipment belong to the mechanical system.

The pneumatic system includes an air compressor, which supplies compressed air through connecting pipelines to the brake system; an automatic control system; reservoirs for storage of the compressed air and control and management systems and instrumentation (valves, manometers, etc.).

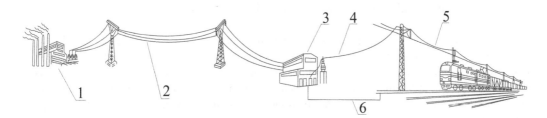

FIGURE 4.2 Example of operational principle for power supply in electrified railway systems: (1) power station; (2) distribution power lines; (3) electrical traction substation; (4) feeder power line; (5) overhead line equipment; (6) return feeder. (From Spiryagin, M. et al., *Design and Simulation of Rail Vehicles*, Ground Vehicle Engineering Series, CRC Press, Boca Raton, FL, 2014. With permission.)

Contact conductor, power transformers, inverters, traction electric motors, auxiliary machines, electrical control and management units and the dynamic and regenerative braking systems are all parts of the electrical equipment.

The hydraulic system includes liquid cooling systems (oil, water, etc.) and also a hydraulic control system and associated instrumentation.

On electric rail traction vehicles, the following types of traction motors can be used:

- Brushed DC electric motors
- AC motors
- Brushless DC electric motors

Traction motors are used in the current designs for the dynamic and regenerative brakes with the purpose of reducing wear of the contact parts of the mechanical and hydraulic brake systems and also for the economy of electrical power consumption.

During dynamic braking, the electrical energy either dissipates as heat from variable resistors or can be returned back to the power supply network if the system design allows.

Leaders in the development and production of electric rail traction vehicles and power equipment were and are the long established concerns such as 'Siemens' and 'Alstom', as well as big firms such as 'AnsaldoBreda', 'ASEA Brown Boveri', 'Bombardier', 'Krauss-Maffei', 'Mitsubishi', 'Kawasaki Heavy Industries', 'Hitachi' and others. Design improvements and modularisation of components and parts for electric rolling stock have required the creation of new technologies in the fields of electrical engineering, aerodynamics, super-lightweight and durable materials, control systems and safety performance on the track.

The advantages of the operation of electric rail traction vehicles are better energy efficiency, greater reliability, higher available power per traction unit, lower maintenance and repair costs. The main disadvantages in comparison with other types of traction rolling stock are the high costs of electrified network infrastructure and its maintenance and also the lack of fully autonomous operation because the electric rail traction vehicle (except an electric hybrid rail traction vehicle) cannot be operated on tracks without overhead line equipment fitted.

4.3.1.1 Light Rail Vehicles

In the case of high-floor light rail vehicles, the equipment is commonly installed under the main frame. An example of the layout scheme of a DC light rail vehicle with a high floor design is shown in Figure 4.3. In light rail vehicles with a low-level floor, which have found wide application due to the ease of entry and exit for passengers (particularly those with physical limitations), the equipment is commonly installed under the roof. An example of the layout scheme of a DC light rail vehicle with a low floor design is shown in Figure 4.4a. The combination of high-floor and low-floor designs in a single light rail vehicle also exists now, as shown in Figure 4.4b.

It is also possible to see an application of modular design in the case of light rail vehicles, as shown in Figure 4.5. This design consists of a car body supported on bogies or supported by other car bodies through special beam and joint connections, as well as motorised and non-motorised bogies. In addition, individually driven wheels have begun to be used in bogies of light rail vehicles instead of conventional wheelsets. Light rail vehicles are commonly equipped with standard pneumatic, electric, rail and/or eddy current brakes.

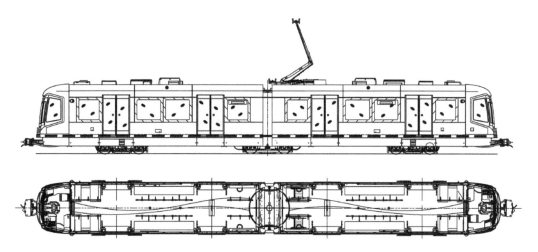

FIGURE 4.3 Layout scheme of a high-floor light rail vehicle; the dashed line in the side elevation view indicates the floor level (manufactured by Siemens, Florin, California, USA).

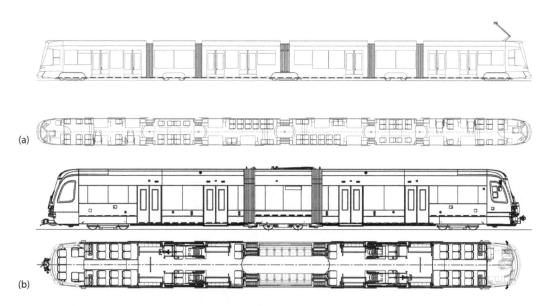

FIGURE 4.4 Layout schemes of light rail vehicles, with the dashed lines in the side elevation views indicating the floor level: (a) 100% low-floor light rail vehicle (manufactured by Stadler Rail Group, Bussnang, Switzerland) and (b) 70% low-floor light rail vehicle (manufactured by Siemens, Sacramento, California, USA).

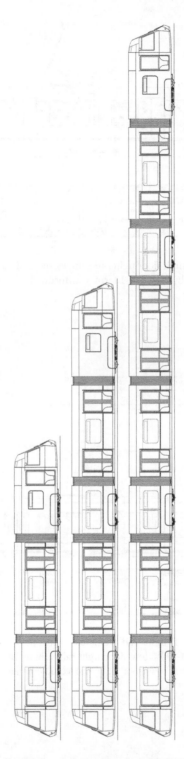

FIGURE 4.5 Example of light rail vehicle configurations (manufactured by CRRC Dalian Locomotive & Rolling Stock Company, Dalian, China).

4.3.1.2 Electric Locomotives

Unlike light rail vehicles, the equipment of electric locomotives is located in the car body of the locomotive. An example of the layout scheme of a DC electric locomotive is shown in Figure 4.6. The DC electric locomotives are different from the AC locomotives because they do not have high-voltage AC electrical power systems. An example of the layout scheme of an AC electric locomotive is shown in Figure 4.7.

All electric locomotives are equipped with both standard pneumatic and electric brakes. Electric locomotives can be designed as one-cab or two-cab versions, and they can also operate as a multiple-unit system (two or more locomotives being controlled by one driver). An example of an electric locomotive with two cabs is shown in Figure 4.8.

Multi-system electric locomotives have the current collection, traction and power equipment required for working with several different combinations of current and voltage. An example of

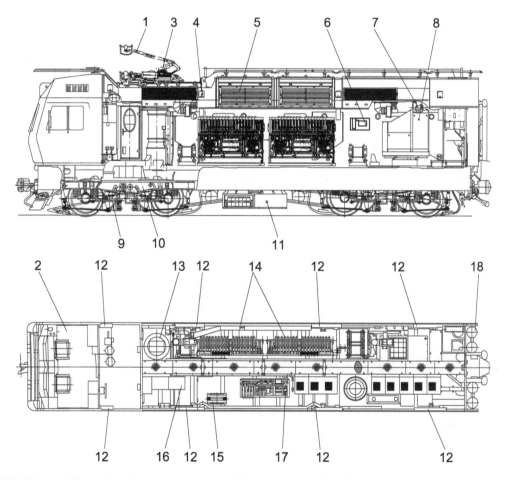

FIGURE 4.6 Example of layout scheme of a direct current electric locomotive (manufactured by Ural Locomotives, Ekaterinburg, Russia): (1) pantograph; (2) driver's compartment (cab); (3) line filter element; (4) dynamic brake grid; (5) circuit breaker; (6) air dryer; (7) auxiliary compressor; (8) main compressor; (9) traction motor; (10) bogie; (11) battery box; (12) sand boxes; (13) traction motor blower; (14) traction blocks 1 and 2; (15) high speed circuit breaker block; (16) locomotive microprocessor and monitor system; (17) traction block 3; (18) brake pneumatic system main reservoirs. (From Spiryagin, M. et al., *Design and Simulation of Heavy Haul Locomotives and Trains*, Ground Vehicle Engineering Series, CRC Press, Boca Raton, FL, 2017. With permission.)

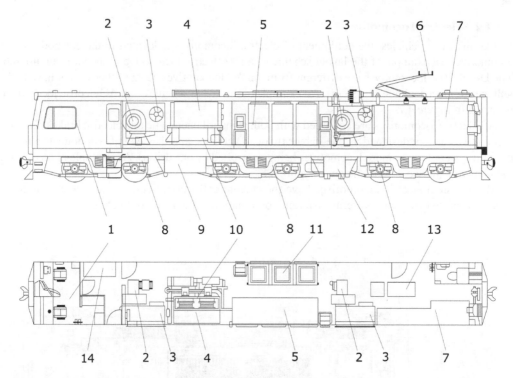

FIGURE 4.7 Example of layout scheme of an alternating current electric locomotive (manufactured by Siemens, Germany): (1) driver's compartment (cab); (2) traction motor blower; (3) inertial air filter boxes; (4) cooling rack; (5) main converter; (6) pantograph; (7) auxiliary switchgear compartment; (8) bogie; (9) battery box; (10) main transformer; (11) brake resistors; (12) air compressor; (13) brake rack; (14) air conditioning unit. (From Spiryagin, M. et al., *Design and Simulation of Heavy Haul Locomotives and Trains*, Ground Vehicle Engineering Series, CRC Press, Boca Raton, FL, 2017. With permission.)

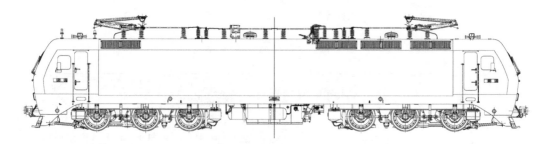

FIGURE 4.8 Example of an electric locomotive design with two cabs and two three-axle bogies (manufactured by CRRC Zhuzhou Electric Locomotive Co. Ltd, Zhuzhou, China).

the layout scheme of a multi-system electric locomotive with four pantographs (two for AC and two for DC) is shown in Figure 4.9.

4.3.1.3 Electric Multiple Units

An electric multiple unit or EMU is a train used for passenger transportation on city, suburban and regional rail networks and also for high-speed passenger trains. The EMUs are one of the primary means of passenger transport and successfully compete over short and medium distances against road and air transport.

Design of Powered Rail Vehicles and Locomotives

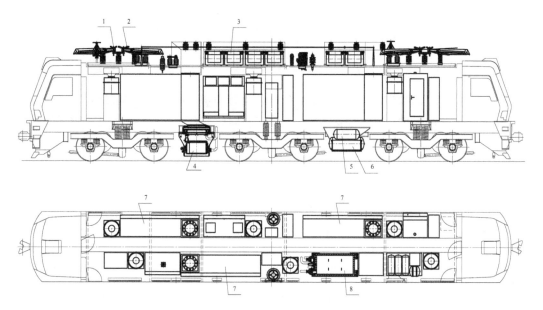

FIGURE 4.9 Layout scheme of a multi-system electric locomotive with two cabs and three two-axle bogies (manufactured by Transmashholding's Novocherkassk Electric Locomotive Plant, Novocherkassk, Russia): (1) AC pantograph; (2) DC pantograph; (3) brake resistors; (4) inductance unit; (5) pneumatic system main reservoirs; (6) battery compartment; (7) traction converters; and (8) transformer.

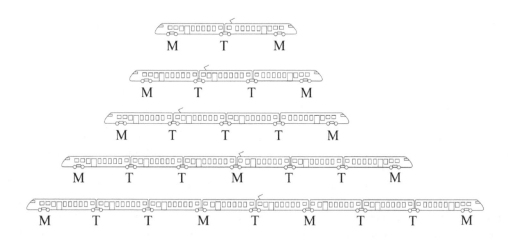

FIGURE 4.10 Examples of EMU train configurations. M – motorised bogie; T – trailer bogie.

Designs of major equipment and other systems used on EMUs are similar to those of electric locomotives. The difference is that an EMU is a powered train that consists of driving, motor and/or trailer cars in a classic design scheme. The driving car can also be a motor car. In some cases, a power car (similar term to an electric locomotive) can also be added to the configuration of such a train as a separate unit. Trailer cars are usually not used for traction equipment; in rare cases, pantographs and brake air compressor units can be installed on them. The EMU trains can have a modular design, often with a shared bogie approach between adjacent cars. An EMU train configuration usually includes from 2 to 16 cars. Cars are equipped with motorised or trailer bogies and also with traction equipment and pantographs. Examples of different EMU train configurations with modular designs are shown in Figure 4.10.

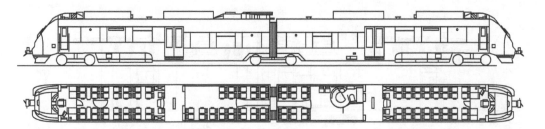

FIGURE 4.11 Example of mixed equipment locations for an EMU (manufactured by Bombardier, Hennigsdorf, Germany).

Unlike electric locomotives where the equipment is located in the car body, the equipment on EMUs is similar to light rail vehicle designs, and it is installed outside of the car bodies (under the car frames or on the roofs). Figure 4.11 shows an example of a typical layout of the roof power equipment installation and two transformers and batteries under the car frame for an EMU double-car configuration.

Suburban EMU trains normally operate at speeds no higher than 180 km/h. Their design should provide good train dynamics under high rates of acceleration and braking, which are associated with the short distances between stations. Therefore, they have increased numbers of driven wheels or wheelsets in their train configurations. Furthermore, they are not only equipped with standard pneumatic and electric brakes but can also be equipped with rail brakes and eddy current brakes.

In high-speed train operations, EMU speeds can reach 400 km/h. Operation at such a speed requires a significant increase in power (e.g., 12500 kW for Train à Grande Vitesse (TGV) trains), as well as the application of new design solutions to ensure reliability and safety. An example of the layout scheme of a high-speed train is shown in Figure 4.12. These types of trains are widely used with active suspension systems to guarantee tilting in curves, better load transfer between elements of the running gear, levelling of the floor, installation of steering bogies and the application of traction control systems for individual driving wheels. To improve the dynamic performance of these trains, it is necessary to reduce the unsprung weight of the running gear. For this purpose, these vehicles are equipped with solid wheels with small diameters up to 600 mm and traction motors with gearboxes hung on the car body. The transfer of torque to the wheelsets is performed by means of drive shafts. Car bodies are manufactured with high usage of light alloy or composite material with fire-resistant properties. As in the case of light rail vehicles, high-speed train designs with a low-level floor have found wide application. Such a design solution provides better train stability at high-speed operation. Close attention is given to the aerodynamic design of high-speed trains. This is due to the presence of significant drag forces, as well as a significant increase in aerodynamic noise and vibrations, which appear at speeds over 200 km/h and can become dominant, exceeding the noise level from the wheels, running gear and traction equipment.

4.3.2 Diesel Traction

A diesel-powered vehicle is the most common autonomous traction vehicle currently running on rail networks. The power plant uses internal combustion engines, usually running on diesel. Engines that run on petrol (gasoline) are not common on railways due to high maintenance costs. The main advantages of diesel power are that such rail traction vehicles are self-contained, are capable of performing operations in any climate zone and can be designed to produce sufficient power and traction to efficiently undertake particular operations on a specific railway network. The main disadvantages include harmful effects on the environment due to the emission of products of combustion and higher costs of maintenance and repair compared with electric traction-powered vehicles.

Design of Powered Rail Vehicles and Locomotives 129

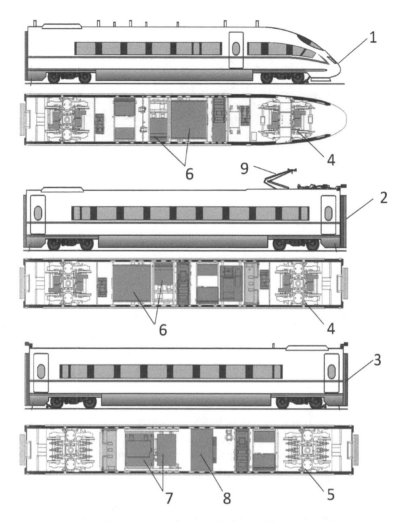

FIGURE 4.12 Layout scheme of high-speed train (manufactured by Tangshan Railway Vehicle Co. Ltd, Tangshan, China): (1) driver car; (2) motor car; (3) trailer car; (4) motor bogie; (5) trailer bogie; (6) traction inverter and cooling equipment; (7) transformer and cooling equipment; (8) auxiliary inverter box; and (9) catenary.

4.3.2.1 Diesel Locomotives

According to their service operations, diesel locomotives may be divided into the following groups:

- Freight locomotives (in some cases, for trains with large total mass and heavy axle loads, they can be designated as heavy-haul locomotives)
- Passenger locomotives
- Mainline or freight-passenger locomotives
- Shunting locomotives (also called switchers)

Diesel locomotives usually consist of the power plant and the four basic systems: mechanical, electrical, pneumatic and hydraulic. Main frame (platform) or monocoque car body designs are used for the transmission of tractive and braking efforts generated by a locomotive to other rail vehicles in the train configuration by means of coupling devices installed on them.

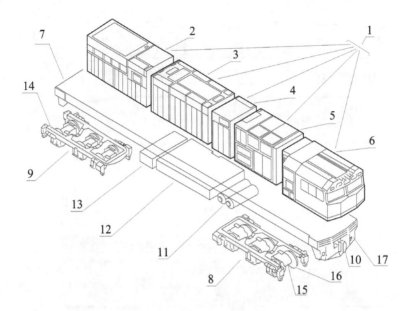

FIGURE 4.13 General layout scheme of a diesel-electric locomotive: (1) car body; (2) radiator compartment; (3) diesel engine compartment; (4) alternator compartment; (5) auxiliary compartment; (6) driver cab; (7) main frame; (8) front bogie; (9) rear bogie; (10) coupler; (11) air reservoirs; (12) fuel tank; (13) batteries; (14) headstock; (14) traction motor; (16) wheelset; and (17) headstock.

The car body of a locomotive with diesel-electric transmission is usually divided into the following areas: operator, auxiliary, alternator, engine and radiator modules. As the frame and modules are placed on the bogies, which have some space between them under the middle of the main frame, the fuel tanks and batteries are commonly installed in that space. An example of such a design scheme is shown in Figure 4.13.

The working principle of the locomotive is to convert the energy of the gases produced by combustion processes in engine cylinders into a pressure force on the pistons, which is then converted into rotational energy of the crankshaft. This energy is transferred to the transmission system (electric, hydraulic or mechanical) and is then transformed into the energy for the traction motors, which deliver the traction through a gearbox or directly to the wheels or wheelsets. The traction, which is realised as a tractive force applied on the rails, is needed for the movement of the locomotive and the wagons coupled to it.

Diesel locomotives can be designed as one-cab or two-cab versions, and they can also operate as a multiple-unit system (two or more locomotives being controlled by one driver). An example of a locomotive with two cabs is shown in Figure 4.14.

An electric transmission system provides optimal tractive and economic operating characteristics for locomotives. Electric power transmissions are characterised by the types of currents used by the main generator (alternator) and the traction motors. They are as follows:

- DC, where both the generator and the traction motors are DC
- AC-DC, where the generator is AC and traction motors are DC
- AC, where both the generator and the traction motors are AC

Mechanical transmission systems are used for locomotives with a low power. Such a transmission is similar to an automotive one, but it has some distinguishing features for the reverse mode of operation.

Design of Powered Rail Vehicles and Locomotives

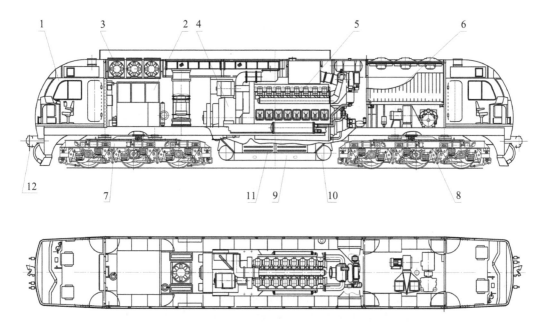

FIGURE 4.14 Example of a diesel-electric locomotive design with two cabs (manufactured by Lugansk Diesel Locomotive Plant, Lugansk, Ukraine): (1) driver cab; (2) high voltage chamber; (3) motor-fans and electro-dynamic brake resistors, traction motor field-weakening resistors, compressor starting and traction generator emergency excitation resistors; (4) alternator; (5) diesel engine; (6) radiator compartment; (7) and (8) bogies; (9) fuel tank; (10) air reservoir; (11) batteries; and (12) coupler.

Hydraulic transmission systems consist of a hydraulic gearbox connected to the crankshaft of the diesel engine and mechanical transmission to the wheelsets. The adjustment of traction torque is performed by changing the flow rate and pressure of the working liquid (oil). In comparison with the electric transmission, the hydraulic transmission does not need non-ferrous metals, and it was widely adopted in the period of electrical copper deficiency during the 1950s and 1960s. However, the hydraulic transmission is a precise machine that requires high-level skills and technical expertise from service personnel, and it also needs high-quality and expensive oils. One more disadvantage of hydraulic transmissions is lower efficiency compared with electric transmissions.

The auxiliary equipment of the locomotive includes the cooling, air supply and fuel supply systems of the diesel engine, the sanding system, fire protection system, electrical auxiliary equipment, associated low-voltage circuits, etc.

Recent years have seen the application of dual-fuel diesel locomotives, allowing the locomotive to run on either diesel or liquefied natural gas. In this case, an additional tank for storing liquefied natural gas on a separate wagon can be attached to the locomotive or mounted on the extended frame of the locomotive.

4.3.2.2 Diesel Multiple Units

Diesel multiple units usually provide passenger transportation in urban, suburban and inter-regional service areas that are non-electrified or partially electrified. The main elements of equipment on DMUs are similar to diesel locomotives.

Similar to EMUs, the DMU trains consist of driving, motor and/or trailer cars in a classic design scheme. The main difference from the electric rolling stock is that, instead of pantographs and electric control circuits of high-voltage equipment, the motor cars have diesel power plants that produce energy that is then transferred/transformed to the traction motors (traction transmission).

Similar transmission types as for diesel locomotives are in use. Diesel multiple units can therefore be divided into three categories:

- Diesel-EMUs
- Diesel-mechanical multiple units
- Diesel-hydraulic multiple units

The electrical transmission has found much wider application in comparison with the others. As for diesel locomotives, hydraulic and mechanical transmissions are generally used with low powered diesel engines.

There are two common locations for the power plant in DMU trains. The first is a traditional one, as shown in Figure 4.15, where the diesel engine is installed in the driving car behind the driver cab. In this case, this compartment has soundproofed insulation on both driver cab and passenger compartment sides. The advantage of such a design is better access to the diesel engine during service or repair works. However, it significantly reduces the size of the passenger compartment.

In order to increase passenger capacity on modern DMU trains, the diesel unit is often placed in the underfloor space between the bogies of the motor or the driving cars, as shown in Figure 4.16. In this case, the power plants are made up of special packages called modules. If there is a failure, then the relevant module is simply replaced by a new one. The engines in these modules are usually flat engines, where pistons move in a horizontal plane. This is necessary for the reduction of the height of the diesel engines.

4.3.3 Gas Turbine Traction

Gas turbine rail traction vehicles are equipped with a gas turbine as a power plant. Gas turbine rail traction vehicles have not found wide application in railway operations, but research and development on their further improvement are still in progress, because such a design solution has

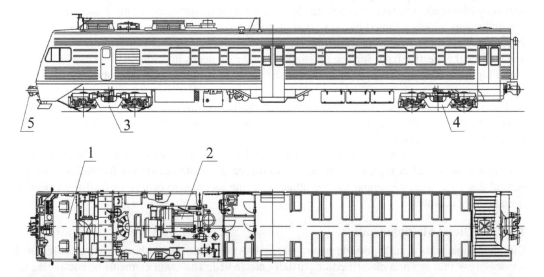

FIGURE 4.15 Example of the traditional arrangement for a motorised passenger car (manufactured by Lugansk Diesel Locomotive Plant, Lugansk, Ukraine): (1) driver cab; (2) diesel generator; (3) motorised bogie; (4) non-motorised bogie; and (5) coupler. (From Spiryagin, M. et al., *Design and Simulation of Rail Vehicles*, Ground Vehicle Engineering Series, CRC Press, Boca Raton, FL, 2014. With permission.)

Design of Powered Rail Vehicles and Locomotives 133

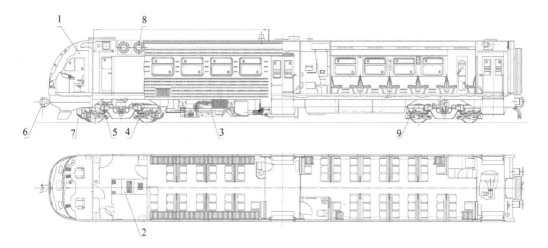

FIGURE 4.16 Example of the underframe power plant location for a motor car (manufactured by Lugansk Diesel Locomotive Plant, Lugansk, Ukraine): (1) driver cab; (2) electric equipment; (3) diesel engine; (4) motorised bogie; (5) traction motor; (6) coupler; (7) sand nozzle; (8) dynamic brake; and (9) non-motorised bogie. (From Spiryagin, M. et al., *Design and Simulation of Rail Vehicles*, Ground Vehicle Engineering Series, CRC Press, Boca Raton, FL, 2014.)

potentially significant advantages in power density and the cost of fuel, as well as provides a significantly simpler design in comparison with diesel traction. Disadvantages include the low value of energy conversion efficiency, large variation in this efficiency across the notch position operating range, high consumption of fuel at idle notch position and increased aerodynamic noise from the operation of the turbine engine.

Gas turbine traction has been considered for freight and heavy-haul train operation scenarios. However, considering the high fuel consumption, most operational scenarios would require the locomotive to be connected with tank wagons for storing fuel oil, liquefied natural gas, pulverised coal or peat to allow an increase in the operating range of the locomotive. This can produce some undesirable effects on longitudinal train dynamics, and such fuel wagons reduce the train payload. An example of the layout scheme of a gas turbine locomotive is shown in Figure 4.17.

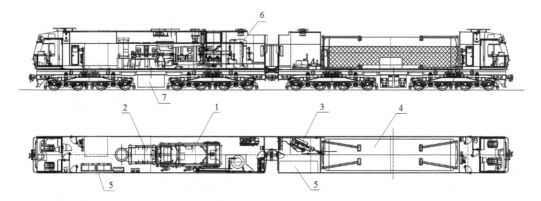

FIGURE 4.17 Layout scheme of a gas turbine locomotive (manufactured by Lyudinovo Diesel Locomotive Plant, Lyudinovo, Russia): (1) gas turbine engine; (2) main generator; (3) cryogenic gas pump; (4) gas tank; (5) high-voltage traction equipment; (6) gas equipment unit; and (7) battery.

4.3.4 Hybrid Traction

By design, the hybrid traction vehicles are similar to diesel and gas turbine traction vehicles. A significant difference is that, in addition to diesel or gas turbine power plants, hybrid locomotives also use electrical energy stored in electric batteries, supercapacitors or flywheels. The charge process of these components occurs during the operation of the diesel generator or gas turbine at idle speed or when the kinetic energy of braking (of both the train and the locomotive) is transformed into electric power. During hauling operations (traction mode), a combination of energies might be used (i.e., drawing simultaneously from the energy storage and the main generator) when additional power is required for acceleration or travelling up long gradients. Further transformation and transmission of energy to the wheels of the locomotive are carried out in a standard manner, as is performed in diesel or turbine-electric locomotives with electric transmission.

The classification of the main systems (mechanical, electrical, hydraulic and pneumatic systems) for hybrid locomotives is similar to that for diesel locomotives. To this classification, the following hybridisation designs can be added:

- Hybrid design with no internal energy storage and only external storage units (hybrid network energy is stored in the energy supply plants or made available to other rail traction vehicles via the overhead line equipment)
- Hybrid construction with internal accumulator units (autonomous hybrid internal energy storage)
- Complex hybrid structures that combine several varieties of these types

An example of a hybrid locomotive design with internal energy storage is shown in Figure 4.18.

At the present stage of hybrid traction technology development, hybrid locomotives are already in operation for shunting services, as well as for suburban and urban passenger traffic. However, they are not used for freight or heavy-haul operations due to limitations of existing

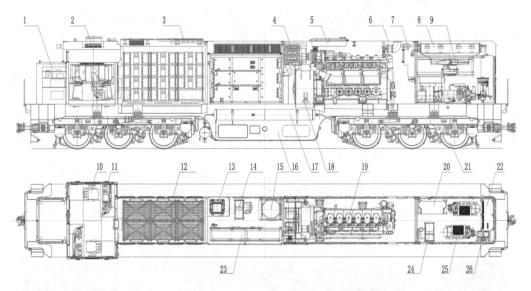

FIGURE 4.18 Layout scheme of a hybrid locomotive design with internal energy storage (manufactured by CRRC Ziyang Locomotive Co. Ltd, Ziyang, China): (1) low voltage electrical cabinet; (2) driver cab air-conditioner; (3) traction battery air-conditioner; (4) air inlet; (5) air outlet; (6) car body ventilator; (7) expansion tank; (8) radiator; (9) cooling fan and motor; (10) driver seat; (11) console; (12) traction battery array; (13) transformer; (14) rear ventilator; (15) dynamic brake cabinet; (16) fuel tank; (17) main reservoir; (18) main generator; (19) diesel; (20) brake cabinet; (21) bogie; (22) auxiliary transformer cabinet; (23) auxiliary convertor; (24) front ventilator; (25) air compressor; and (26) dryer.

Design of Powered Rail Vehicles and Locomotives

energy storage options (large energy storage capacities are required), and this is the reason why such a locomotive or locomotive consist would need to use an additional powered vehicle (booster).

4.4 CLASSIFICATION OF MAIN RAIL TRACTION VEHICLE COMPONENTS AND SUSPENSION SYSTEMS

The study of rail traction vehicle dynamics requires consideration of their design, that is, what components and systems the vehicles have, as well as an understanding of their functionalities and principles of work. For the development of models in multibody dynamics packages, the main components are classified and described from the point of view of rail vehicle dynamicists.

4.4.1 MAIN FRAMES AND BODIES

The car body of the vehicle is designed to accommodate its equipment, personnel and, in the case of the presence of a passenger compartment, the passengers, as well as to cope with the application of external and internal loads.

Depending on the structural approach, car bodies can be divided into two types:

- Those with a main frame (underframe) as the main load-bearing component
- Monocoque construction (stressed skin design)

For the main frame type, all the main loads from the weight of installed equipment as well as traction and braking forces and dynamic and impact loads are received, carried or borne by the strong longitudinal design of the main frame. The side and end walls, the roof and the driver cab are provided solely for the protection of drivers, equipment and passengers from the environment.

The first type mainly uses two styles of car body installed on the main frame:

- Hood unit with external service walkways on both sides (see Figure 4.19)
- Cowl unit with full-width car body and internal service walkway (see Figure 4.20)

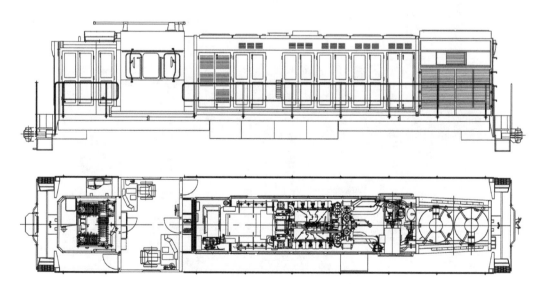

FIGURE 4.19 Example of a hood unit locomotive design (manufactured by Lyudinovo Diesel Locomotive Plant, Lyudinovo, Russia).

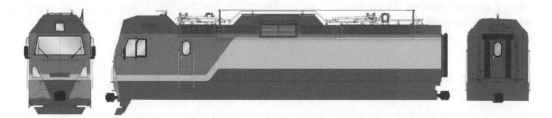

FIGURE 4.20 Example of a cowl unit locomotive design (manufactured by Ural Locomotives, Yekaterinburg, Russia).

Car bodies with a main frame have a simple design with removable body cover elements. This design approach allows reducing the complexity and cost of assembly and maintenance. However, such a design approach introduces a large specific weight, which greatly reduces its competitiveness when creating rolling stock for high speed, or to haul small loads on the tracks.

The monocoque body has rigid link connections between elements such as the frame, roof and side walls, the tightening belt, etc. It enables collaboration of all elements of the design to resist loads acting on it. This also includes skin elements of the body shell such as the wall-covering sheets. Car bodies of this type are produced in the cowl unit style. The advantage of monocoque construction is the high rigidity and low weight. One of the designs of this type is shown in Figure 4.21.

4.4.2 Bogies

Most of the early designs of running gear of powered rail vehicles were without bogies. This was due to the use of crank mechanisms as a traction transmission; these were applied widely in the steam locomotive and do not allow for displacement or rotation of wheel sets in the horizontal plane. A rail vehicle design without bogies does not run well through curves in the track. The widespread introduction of running gear with bogies became possible with the introduction of individually driven wheelsets, and this design was critical for the development of modern rail traction vehicles.

FIGURE 4.21 Example of a monocoque construction design (manufactured by CRRC Zhuzhou Electric Locomotive Co. Ltd, Zhuzhou, China).

The main purpose of the rail vehicle bogie is to improve the dynamic interaction between the running gear and the rails in curved sections of track. In addition, a bogie takes over the support or suspension of the upper-weight structure (above the bogie, i.e., the car body) and redistributes it between the wheels or wheelsets through elastic-damping connections. It also transmits the traction and braking forces to the upper-weight structure and coupling devices.

Depending on the design parameters of a rail traction vehicle (service, weight, length and tractive effort) and restrictions due to loading gauge and axle load, bogies are available in two-axle, three-axle and four-axle design variants. Typical two-axle and three-axle bogie designs are shown in Figures 4.22 and 4.23, respectively.

Recently, single-axle bogies have begun to be considered in the design of EMUs in order to reduce train weight and energy consumption. However, this type of 'bogie' has not yet been considered for being powered as a traction running gear. Future design concepts for such bogies assume the development of fully controlled mechatronic running gear.

The main elements of a bogie are the bogie frame, on which are installed the braking system equipment, elements of the locomotive sanding system, spring suspension, wheelsets with associated assemblies and traction drives or wheel-traction drive assembles.

The conventional bogie, which is a rigid bogie and equipped with solid wheelsets, has some clearance provided in its primary suspension design in order to improve dynamic interaction between wheels and rails that allows small displacements of wheelsets, which results in reduced cornering forces. However, for better dynamic performance and significant wear reduction, it is desirable to reduce the wheelset angles of attack in curved sections of track. For this purpose, steering of the wheelsets within a bogie is a good solution. An alternative bogie design can be equipped with individual drives, where the traction torque from the motor acts on each wheel. In this case, the steering of bogie wheels is performed by special traction drive control strategies, as described in Chapter 15. Design of locomotive traction drives and their connections with spring suspension and wheels will be described in the next section.

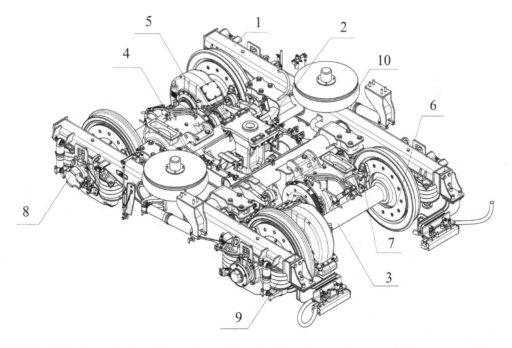

FIGURE 4.22 Two-axle bogie (manufactured by Hyundai Rotem Company, Uiwang, South Korea): (1) bogie frame; (2) centre pivot; (3) wheelset; (4) traction motor; (5) gear box; (6) brake disc; (7) friction air brake mechanism; (8) axle box; (9) primary suspension (coil and rubber springs); (10) secondary suspension (air spring).

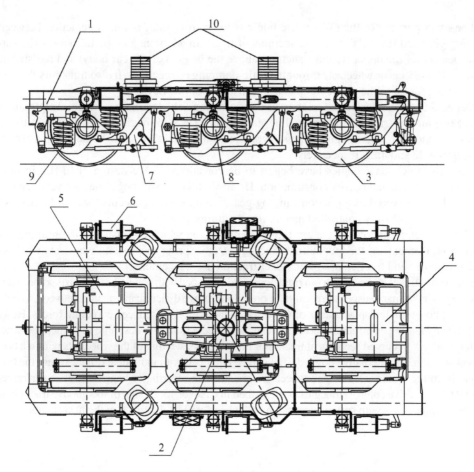

FIGURE 4.23 Three-axle bogie (manufactured by Lugansk Diesel Locomotive Plant, Lugansk, Ukraine): (1) bogie frame; (2) centre pivot; (3) wheelset; (4) traction motor; (5) gear box; (6) brake cylinder; (7) brake lever mechanism; (8) axle box; (9) primary suspension (coil springs); and (10) secondary suspension (side bearers with return devices).

4.4.3 Traction Drives

Traction drives are made up of mechanisms and units engaged in the transfer of kinematic power from the traction motors (electric and hydraulic) or the output shaft of the mechanical gear transmission to the wheelsets or wheels of the powered rail vehicle. Designs of drives are varied and depend on the type and operational service parameters of rail traction vehicles, the selected mode of transmission, the design of wheelsets/wheels and the mounting methods of the traction motor. Traction drive designs can be divided into two types: individual or grouped.

For the individual drive design, the traction torque from the motor acts on one wheelset or one wheel. An example of such a design is shown in Figure 4.24.

For the grouped drive design, the traction torque from the motor or an output shaft of transmission is shared between multiple wheelsets or bogie wheels. The monomotor bogie, which has a grouped drive design, is shown in Figure 4.25.

Design of Powered Rail Vehicles and Locomotives 139

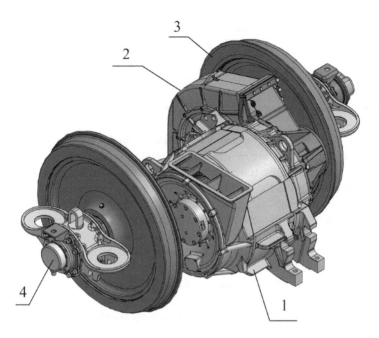

FIGURE 4.24 Example of the individual drive design (manufactured by Ural Locomotives, Yekaterinburg, Russia): (1) traction motor; (2) gear box; (3) wheelset; and (4) axle box.

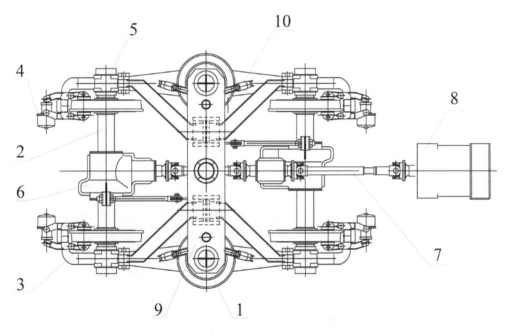

FIGURE 4.25 Example of the grouped drive design (manufactured by Lugansk Diesel Locomotive Plant, Lugansk, Ukraine): (1) air spring; (2) axle; (3) wheel; (4) brake cylinder; (5) axle box; (6) gear box; (7) shaft; (8) body-mounted traction motor; (9) bolster; and (10) damper. (From Spiryagin, M. et al., *Design and Simulation of Rail Vehicles*, Ground Vehicle Engineering Series, CRC Press, Boca Raton, FL, 2014. With permission.)

The design and parameters of traction drives are often dependent on the installation designs of traction motors and associated gearing. Three design variants have found wide application:

- With a nose-suspended traction motor
- With a frame-mounted traction motor
- With a body-mounted traction motor (see Figure 4.25)

Generally, the first of these design variants has traction drives, of which one part is resting on the axle of the wheelset through rolling or slip bearings and the other part is connected through the elastic-damping suspension to the frame of a bogie or the rail vehicle (see Figure 4.26). Torque from the motor is transmitted to the gear box, the driven gear of which is seated firmly on the axle. The advantages of this drive design are the low price and the simplicity of design. It enables the effective transfer of high tractive effort. However, about 60% of the weight of the engine and the traction gear account for unsprung mass in this case; this causes increased dynamic effects of the traction vehicle on the track. This type of suspension is widely used in rail traction vehicles with a relatively low design speed.

The other two design variants are similar because the traction motor is mounted to the bogie frame or the main frame (car body). In such designs, the wheelset receives a torque through mobile and flexible connection elements that provide the necessary freedom of movement of the wheelset or the wheels relative to the traction motor. In this case, unsprung weight is sharply reduced, and this improves the dynamic performance of powered rail vehicles. This type of design is also used in high-speed vehicles. An example of a drive design mounted to the bogie frame for a tram bogie is shown in Figure 4.27.

The wheels can have their own traction drives, providing independent rotation of each of them. In this case, a differential gear is typically used, either with both wheels driven by a single motor or with each wheel driven using its own motor controlled by the principles of differential gearing with the harmonisation of the frequency of rotation of each wheel. Depending on the type of traction gear, the drive can be made with an axial gear in which the driving shaft is perpendicular to the axis of rotation of the wheel or with a radial gear when the axes of the input

FIGURE 4.26 Example of a nose-suspended traction motor design (manufactured by Ural Locomotives, Yekaterinburg, Russia): (1) bogie frame; (2) traction motor; and (3) dog bone (nose link) designated for the suspension of the traction motor from the bogie frame (see Figure 4.24).

Design of Powered Rail Vehicles and Locomotives

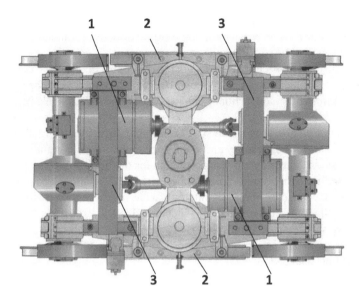

FIGURE 4.27 Bogie mounted drive design (manufactured by Ust-Katav Wagon-Building Plant, Ust-Katav, Russia): (1) traction motor; (2) bogie frame; and (3) motor-mount beam.

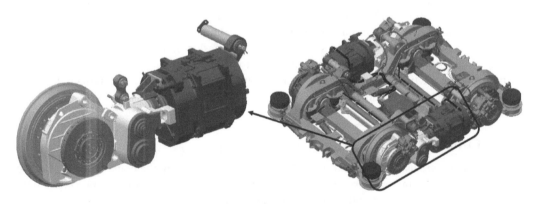

FIGURE 4.28 Drive design of an independent rotating wheel bogie (manufactured by CRRC Dalian Co. Ltd, Dalian, China).

shaft and the wheelset are parallel. Such designs of the latter type are commonly used for light rail vehicles, and an example is shown in Figure 4.28.

4.4.4 Suspension Systems

The spring suspension is necessary for a rail vehicle to reduce its force interaction with the track, which arises from rolling contact on its irregularities, and to minimise and damp the dynamic forces and the natural oscillations of the vehicle in order to reduce their effect on cargo or to provide passengers with a comfortable ride.

Suspension of a rail vehicle can be performed in several stages (one, two or more), and it acts in the horizontal, vertical and transverse planes.

The primary suspension acts in the vertical plane, and it is usually located at the connection points of the wheelset or its axle box with a bogie frame or body, but it can also be located inside the wheelset or the wheels (the so-called elastic wheel).

The secondary suspension is commonly located at the connection points between a bogie frame with the car body, but it may also be incorporated between the elements of the bogie itself.

More detailed information on primary and secondary suspension elements is provided in Chapter 6.

The development of passive suspension systems is currently approaching its practical optimisation limit. The needs for further security and stability of operation with higher traction and braking forces require designers to create suspension systems for traction rolling stock that could provide opportunities for the redistribution of loads under different operational conditions, reduction of the centrifugal forces, change of frequency ranges of vibrations and the possibility of utilisation of the energy from oscillations. These can all be made possible by introducing a complex system that controls the processes in the suspension systems of running rail vehicles. Suspension equipped with a control system is called an active suspension. More detailed information on active suspension designs is provided in Chapter 15.

4.5 CONNECTION BETWEEN A FRAME/CAR BODY AND BOGIES

Connection elements between a frame or a car body and bogies are used for supporting the rail vehicle car body on the bogie frames and the transmission of traction and braking forces from the bogies to the car body, and they can also form parts of the secondary suspension system. Such elements make rotations and displacements of bogies relative to the car body possible within the prescribed limits and their return to the initial position. These include pivot assemblies, side bearings, links and linkages, return devices and flexi-coil suspension. In order to transmit traction and braking forces between them, traction rods are also used. The commonly used connection designs for transmission traction and braking forces are discussed in the following two subsections. Other connection elements are classified and discussed in Chapter 6.

4.5.1 CENTRE PIVOTS

Pivot assemblies are used to transmit traction and braking forces from the bogie to the car body or the main frame of the rail traction vehicles. The pivot assembly is also the point about which a bogie undergoes rotation movement in the horizontal plane relative to the car body.

Pivot assemblies can be divided into two types, which are characterised by their position relative to the centre of wheelset axles or wheels in the horizontal plane:

- With a high location of the pivot point: In this case, the force is transmitted from the bogie to the car body at a point that is located higher than the centre of the wheelset in the horizontal plane; an example of such a design is shown in Figure 4.29.
- With a low location of the pivot point: In this case, the force is transmitted from the bogie to the car body at a point that is located below the centre of the wheelset in the horizontal plane; an example of such a design is shown in Figure 4.30.

When these points have low locations, then a higher value of tractive and brake efforts can be achieved by a rail traction vehicle in comparison with a rail traction vehicle of the same design and configuration that has pivot assemblies with high pivot points.

Design of Powered Rail Vehicles and Locomotives

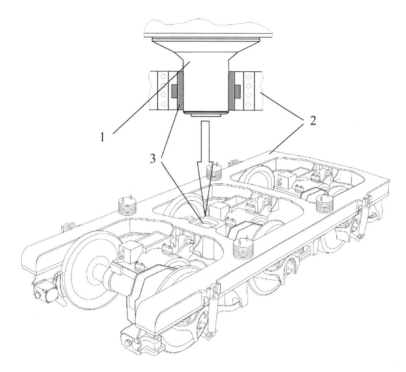

FIGURE 4.29 Example of bogie design with a high location of the pivot point (manufactured by United Group Limited, Newcastle, Australia): (1) pivot pin; (2) bogie frame; and (3) traction centre rubber-steel adaptor.

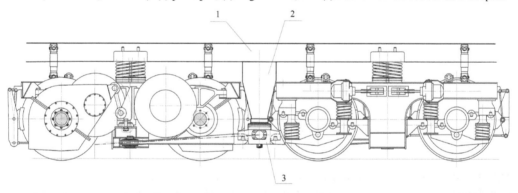

FIGURE 4.30 Low-positioned centre pivot assembly (manufactured by Lugansk Diesel Locomotive Plant, Lugansk, Ukraine): (1) under frame; (2) pivot pin; (3) low-positioned centre pin connection assembly.

4.5.2 Traction Rods

Traction rods are used to transfer traction and braking efforts. An example of the usage of traction rods for the connection of the pivot assembly is shown in Figure 4.31. When a powered rail vehicle is not equipped with pivot assemblies, then the traction rod(s) can directly connect a car body and a bogie. An example of such a connection scheme is shown in Figure 4.32.

For damping of oscillations of traction and brake forces, traction rods can be equipped with absorbing devices; most often, in such cases, rubber and rubber-metal elements or bushings have found wide application.

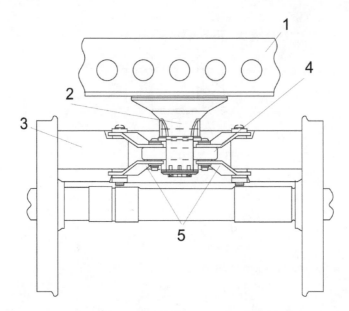

FIGURE 4.31 Example of pivot assembly with the usage of traction rods (manufactured by Electro-Motive Division, McCook, USA): (1) main frame; (2) pivot pin; (3) bogie frame; (4) yoke; (5) traction rod. (From Design and Simulation of Heavy Haul Locomotives and Trains, Ground Vehicle Engineering Series, CRC Press, Boca Raton, FL, 2017. With permission.)

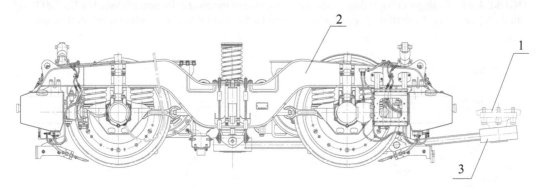

FIGURE 4.32 Example of a connection scheme when a powered rail vehicle is not equipped with a pivot assembly (manufactured by CRRC Zhuzhou Electric Locomotive Co. Ltd, Zhuzhou, China): (1) car body connection point (see both ends of the locomotive design shown in Figure 4.8); (2) bogie frame; and (3) traction rod assembly.

4.6 TRACTION SYSTEMS AND THEIR CLASSIFICATION

For both types of rail traction vehicle (diesel and electric), it is common to classify them as either AC or DC rail traction vehicles based on the type of electricity supplied to their traction motors [7,8]. With any electrical machine, the torque production relies upon either the Lorentz force or reluctance forces. The magnetic component of the Lorentz force, the force on electric charges moving in a magnetic field, dominates in most traction machines. In these machines, the torque is produced by the interaction of currents in two windings, that is, the currents in the field winding and armature in a DC machine or in the stator and the rotor in the induction machine.

To illustrate some of the key principles that are common to both AC and DC traction machines, a simple model is presented. Figure 4.33 shows a single conductor of length l fixed on a cylindrical

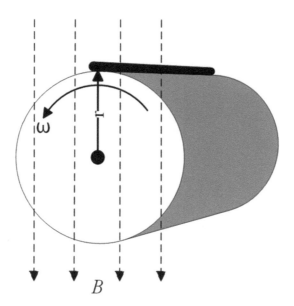

FIGURE 4.33 A current carrying conductor on a rotor in a uniform magnetic field.

rotor or armature of radius r in a uniform magnetic field with a flux density B. The cylinder is free to rotate. This forms an elemental electrical machine that can be used to illustrate some of the key operational limitations for both AC and DC machines.

The equation for force on the current carrying conductor is:

$$F = Bil \tag{4.1}$$

where F is the tangential force, B is the magnetic field strength, i is the current and l is the conductor length.

For both AC and DC traction machines, the force production is proportional to the machine length, the air-gap flux density and the total current flowing in the conductors on the surface of the rotor or armature. The tangential components of these forces are translated into torques by multiplying by the armature or rotor radius. The air-gap flux strength is referred to as the level of machine excitation or magnetisation. A given torque can be achieved with the lowest current in a fully excited or fully magnetised machine. This is operationally desirable, as it results in the lowest conduction losses.

The force equation determines the limit on machine torque. The magnetic field strength will be determined by the magnetic properties of the machine magnetic materials. The saturation level of steel laminations is typically 1.7 T. All machines will have current limits that are largely imposed by the lifetime of the electrical insulation. For continuous operation, the cooling methods employed will determine the maximum allowable losses. Higher currents are possible for short time durations. For short-term operation, the specific heat of the motor materials will determine the rate of temperature rise. All electric drives have a characteristic low-speed region, where the machine torque is adjustable within limits set by the rated maximum field and rated maximum currents. In this region, the machine is operated in a fully excited state, and the torque producing current is controlled to adjust the machine torque. The low-speed region is also termed the constant-torque region. Figure 4.34 shows the generalised torque speed characteristic of an electrical machine. This applies to both AC and DC machines.

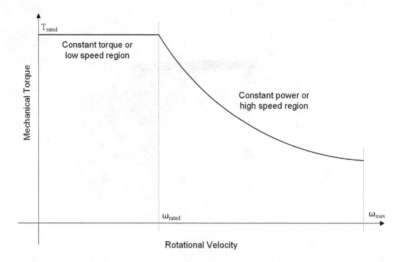

FIGURE 4.34 Torque speed curve for a generalised electrical machine.

As the number and size of the conductors on the surface of a rotor or armature is proportional to the available area, the machine torque varies with the square of the cylinder radius and cylinder length. Torque is proportional to the machine volume. The mechanical power is the product of the machine torque and rotational velocity. The mechanical power is proportional to the machine volume and the rated speed. In a traction application, a gearbox will allow a higher-speed traction machine to have a lower volume to achieve the same mechanical power.

Another key equation that determines the behaviour of electrical traction machines is the generator equation. For an orthogonal arrangement of conductors and flux, the expression for the voltage induced on a conductor moving in a magnetic field is:

$$E = Blv \tag{4.2}$$

where E is the induced voltage, v is the velocity, B is the magnetic field strength and l is the conductor length.

In an electric machine, the conductor velocity is proportional to the cylinder radius and the rotational speed. In any traction application, the allowable machine voltage will be limited by a combination of the voltage rating of the insulation system, the voltage rating of the traction converters and the available voltage from the overhead power system or from the on-board alternator. If the machine is fully excited, there will be a designed rotational speed, ω_{rated} or f_{rated}, where the voltage limit is reached. Beyond that, the machine flux density must reduce either to limit the machine voltages or to maintain the current flows within the machine. Above that speed, the machine flux must reduce, and the motor enters the field-weakening mode. Above the threshold field-weakening speed, ω_{rated}, the flux and the available machine torque fall inversely with speed, as shown in Figure 4.34. The machine power, which is the product of speed and torque, is constant. This region is also termed the constant-power region or the high-speed region.

Finally, Equations 4.1 and 4.2 can be considered together to illustrate the equality of electrical and mechanical power in an ideal machine. The mechanical power delivered by a conductor on an armature or rotor moving at w_{mech} is:

$$P_{mech} = w_{mech}T = w_{mech}rBil \tag{4.3}$$

Design of Powered Rail Vehicles and Locomotives

This is equal to the electrical power:

$$P_{elec} = iE = iBlv = w_{mech} rBil \qquad (4.4)$$

For both AC and DC traction machines, and their electronic controls, the mechanical to electrical efficiencies are in the approximate range of 80%–95%. In machines where fast dynamics are not significant, simple models can be developed based on the equality of the input and output powers and constraints on torque and power in the low-speed constant-torque region and high-speed or constant-power operational regions.

4.6.1 DC Traction Power

In a DC rail traction vehicle, the traction machines require DC supplies for both the armature and field windings [1]. At low speeds, the field currents are maintained at high values. Equation 4.1 shows that high field strengths result in the largest levels of torque production for any level of armature current. At higher speeds, the induced armature voltage, which depends on Equation 4.2, must be limited by reducing the field-winding currents, and the maximum torque production falls. The DC traction machine exhibits the classical torque speed dependency shown in Figure 4.34. This dependency is directly visible in the tractive effort curve of any DC rail traction vehicle.

The following are common DC rail traction vehicle topologies:

- Diesel-electric rail traction vehicles with an AC-DC topology are equipped with a power traction transmission system, the main components of which are main and auxiliary alternators, rectifiers and traction motors; because of size and maintenance advantages, alternators and rectifiers are used in preference to DC generators, and an example of such a topology is shown in Figure 4.35.
- Electric rail traction vehicles with a DC-DC topology commonly have main components of pantograph/s, DC:DC converters, which include choppers, control equipment and traction

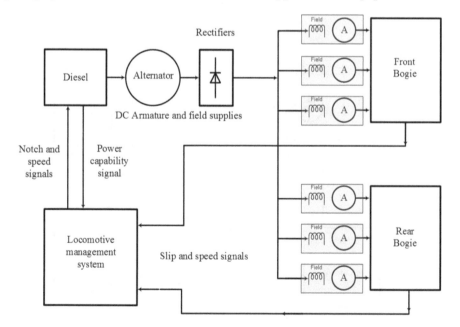

FIGURE 4.35 Example of an electric traction scheme for a diesel-electric rail traction vehicle with an AC-DC topology.

motors; because of the lower overhead voltages used with DC systems, these will be lower rating vehicles such as light rail vehicles and trams, and an example of such a topology is shown in Figure 4.36.
- Electric rail traction vehicles with an AC-DC topology commonly use AC power from an overhead network, but they are equipped with DC traction motors; these rail traction vehicles have main components of pantograph/s, transformers, rectifiers and traction motors, and an example of such a topology is shown in Figure 4.37.

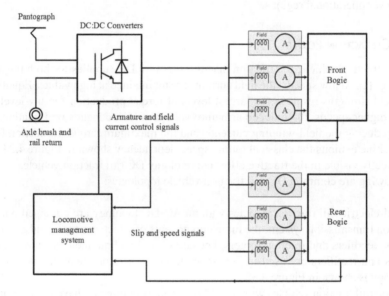

FIGURE 4.36 Example of an electric traction scheme for an electric rail traction vehicle with an DC-DC topology.

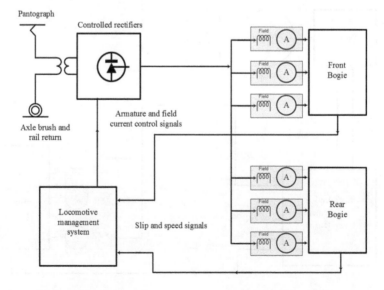

FIGURE 4.37 Example of an electric traction scheme for an electric rail traction vehicle with an AC-DC topology.

Design of Powered Rail Vehicles and Locomotives

4.6.2 AC Traction Power

In an AC rail traction vehicle, the traction machines are controlled by DC-AC inverters [8]. In modern rail traction vehicles, two common inverter control strategies are applied: field-oriented control (FOC) and direct torque control (DTC). Both strategies offer similar high levels of performance. In both cases, a low-speed, constant-torque region occurs, where the machines are operated at the highest flux levels and have their maximal torque-producing ability. At higher speeds, the machine magnetisation must reduce because of operating voltage limitations. This results in a constant-power region of operation. These regions are clearly visible in the tractive effort curves of any AC rail traction vehicle. The following are common AC rail traction vehicle topologies:

- Diesel-electric rail traction vehicles with an AC-DC-AC topology are equipped with a power traction transmission system, the main components of which are main and auxiliary alternators, rectifiers, traction inverters and traction motors; two types of traction system configurations are commonly in use, namely one inverter per bogie and one inverter per wheelset, and a typical scheme of electrical traction with such a topology for a diesel-electric rail traction vehicle with one inverter per bogie is shown in Figure 4.38.
- Electric rail traction vehicles with a DC-AC topology use DC power from an overhead network, but they are equipped with AC traction motors, and their other main components are pantographs and traction inverters; these are typically lower-powered vehicles such as light rail vehicles and trams, and an example of such a topology is shown in Figure 4.39.
- Electric rail traction vehicles with an AC-DC-AC topology have main components of pantographs, transformers, rectifiers, traction inverters and traction motors; an example of such a topology, as may be applied in high-speed trains, is shown in Figure 4.40.

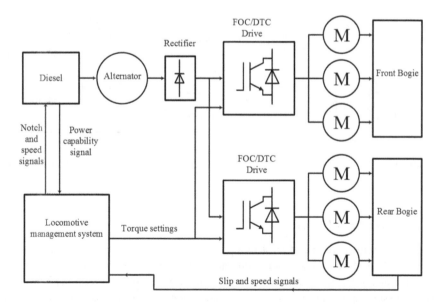

FIGURE 4.38 Example of an electric traction scheme for a diesel-electric rail traction vehicle with an AC-DC-AC topology.

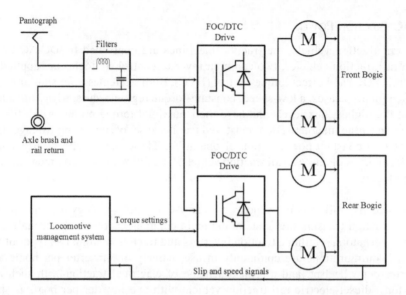

FIGURE 4.39 Example of an electric traction scheme for an electric rail traction vehicle with a DC-AC topology.

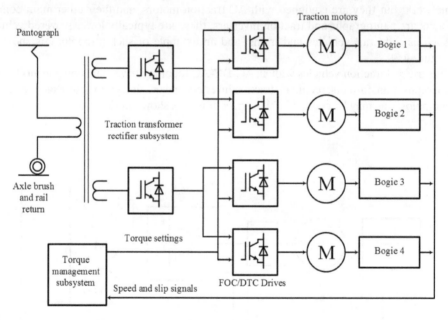

FIGURE 4.40 Example of an electric traction scheme for an electric rail traction vehicle with a AC-DC-AC topology.

4.6.3 Hybrid Traction Power

Hybrid traction systems combine at least two power sources or energy storage systems. The most common railway hybrids are those that incorporate energy storage. For the modern hybrid rail traction vehicle, the electrical traction machine is most often the induction machine, but brushless DC machines are possible for the lower power ranges and offer slightly improved efficiencies. The energy storage units are most frequently batteries [9], flywheels [10] or supercapacitors [9], as shown in Figure 4.41. Amongst the battery technologies, lithium-ion batteries are most frequently used.

Design of Powered Rail Vehicles and Locomotives

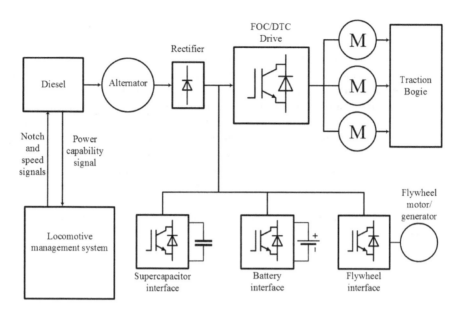

FIGURE 4.41 Series hybrid diesel electric system with flywheel, battery and supercapacitor storage.

A diesel electric system with energy storage has the option to recover braking energy or adjust the loading of the diesel traction engine to optimise fuel consumption or particulate emissions [11]. An electric light rail traction vehicle or tram that includes energy storage may be able to operate in electrified and non-electrified sections of a network [12].

Hybrid traction systems are classified as series or parallel types. In the series hybrid, energy is transferred to the driving wheels via a single pathway. In a parallel hybrid, at least two pathways exist. These are common in hybrid automobiles, where traction energy may be delivered by an internal combustion engine operating in parallel with an electrical machine. As both machines are mechanically coupled to the driving wheels, a degree of complexity exists. Most modern hybrid rail traction vehicles are series hybrids. The traction machines are electric, and these are supplied from an on-board energy source in combination with energy storage. In principle, a train may be configured with some conventional rail traction vehicles and some units with energy storage. In that case, it can be argued that the train, if considered as a single entity, could be viewed as a parallel hybrid.

Hybrid vehicles are further classified as weak and strong hybrids. This is a function of the degree of energy storage. A weak hybrid has limited energy storage that is often used to recover braking energy. If the rail traction vehicle frequently accelerates and brakes, the energy storage may perform a large number of lower-energy cycles. Flywheels and supercapacitors may be best suited to this duty. A strong hybrid has significant energy storage and may be capable of many periods of operation solely drawing upon its reserves of stored energy. A hybrid shunting rail traction vehicle or switcher rail traction vehicle may be a strong hybrid with a relatively large battery that is capable of high peak powers and that is coupled to a relatively small traction engine that is sized to provide the average power demand [11].

4.7 BRAKE SYSTEMS AND THEIR COMPONENTS

Brake systems are used to exert braking force(s) on the wheels or rails that then transfer to the contact patches between the running rails and the wheels in order to maintain or reduce the operational speed or bring the rail vehicle(s) to a full stop, whether operating as an independent unit or as a traction vehicle in a train configuration.

Brake systems are divided into two groups based on the method used to create the resistance force, these being either adhesion or non-adhesion. Adhesion brakes are commonly split into two types of system: frictional and dynamic. In frictional braking systems, energy is usually absorbed by the friction between the wheel and the brake shoes, pads or discs or between the rails and brake shoes in the less common case of rail brakes, with appropriate force loads attached on them. Dynamic brake systems usually work based on principles of transformation of kinetic energy of the train or rail traction vehicle into other types of energy (the main one being electrical) for further recovery processes and utilisation. Non-adhesion brakes are divided into two types of systems. The first uses electromagnetic forces acting on rails to stop or decelerate a rail traction vehicle or a train. The second is focused on the creation of additional aerodynamic resistance forces to stop or decelerate a rail traction vehicle or a train. The latter is more applicable for high-speed rail application scenarios.

Based on the method of the creation of the acting control forces, the brakes are divided into the following types:

- Mechanical
- Pneumatic
- Electric
- Hydraulic
- Magnetic

The brake system of powered rail vehicles can contain several types of brakes at the same time, such as shoes and discs, and vehicles can also be equipped with dynamic, electromagnetic and rail brakes.

4.7.1 Basic Components of Air Brake Systems

The most common braking systems of rail traction vehicles used in rolling stock design are pneumatic systems, which use shoes or wheel disc brakes. The simplified scheme of such a system is shown in Figure 4.42.

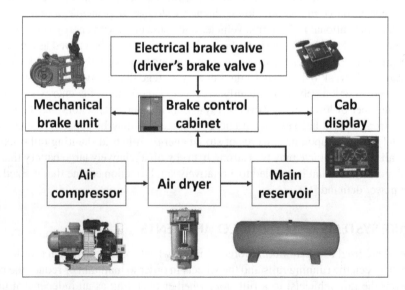

FIGURE 4.42 Basic components of an air brake system.

Design of Powered Rail Vehicles and Locomotives

A typical air brake system includes the following main components:

- Feeding and supply components (e.g., air compressor and air brake pipes which are used to supply cylinders)
- Energy storage components (e.g., main and auxiliary air reservoirs)
- Activation components or actuators (e.g., pneumatic brake cylinders)
- Mechanical system for transferring of braking efforts (e.g., brake rigging connected with brake shoes or disc brakes)

It is necessary to mention that, in the case of shoe braking, the number of pneumatic brake cylinders indicates what type of mechanism is in use. If one cylinder activates several brake shoes located on different wheelsets, then such a mechanism is called a grouped scheme, and if each wheelset uses its own cylinder, then it is called an individual scheme. In the case of disc brakes, the most common type of scheme is an individual design where one cylinder is used to activate the braking discs on one wheelset through friction pads.

In addition, a brake system includes the following elements:

- Control devices and instrumentation (driver's brake valve, emergency stop valves, etc.)
- Transportation elements (e.g., pipes that transport air to acting devices or actuators)

Taking into account that the brake system is one of the critical systems, especially for operational safety, it also has separate pneumatic control blocks and electronic control systems (check of system integrity, vigilance, automatic braking and automatic control, etc.).

4.7.2 Dynamic Brake Systems

The definition of a dynamic braking system covers systems that use the absorption of kinetic energy by means of various effects to control the speed of rail traction vehicles.

The main type of dynamic brake used on rail vehicles is the electric brake. The traction motors of a rail traction vehicle can be used to produce a braking force by converting the kinetic energy of the moving vehicle into electrical energy. Both AC and DC machines are capable of regenerative or dynamic braking. The machines are inherently reversible. There are three major advantages of these electrical braking systems. The kinetic energy can be recovered and reused if desired. The braking effect does not rely upon mechanical friction components, and therefore, wear is avoided. The braking energy is not dissipated as heat in wheels or brake discs, and the restrictions imposed by limited thermal mass and temperature rise do not apply.

In a DC rail traction vehicle, braking is accomplished by providing a power supply to the traction motor fields. The field power requirement is a few percentages of the machine rating, and the power is generally provided by operating an alternator at a low notch setting. If the armature is rotating in the presence of a magnetising field, a voltage is generated. The traction motor is then electrically loaded using braking resistors or a braking grid. With a fixed braking resistor and a fixed level field current, the retarding force varies linearly with velocity. The brake force can be controlled by adjusting the field or by switching braking resistors in or out of circuit. The switching of braking elements causes characteristic fluctuations in the braking torque, as shown in Figure 4.43. In this example, four braking resistor values are available. A feature of DC machine braking is that the brake force will fall to zero as the rail traction vehicle speed approaches zero. For a DC rail traction vehicle with several values of braking resistor, the peak braking torque achievable in the low-speed region is comparable to that available for the tractive effort. Above some speed, the power rating of the braking resistors will be a limitation, and the field current must be reduced to limit the braking power. The capability in this constant-power region is somewhat diminished due to more severe operating conditions at the commutator.

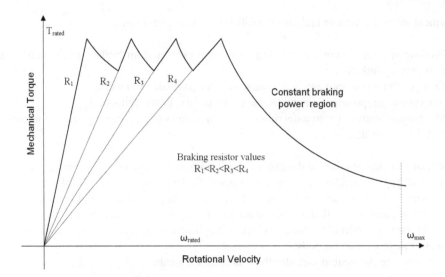

FIGURE 4.43 Dynamic braking for DC rail traction vehicle.

In an AC rail traction vehicle, the traction machine control strategies are either DTC or FOC. These are highly responsive precision torque-regulating systems. Regeneration occurs naturally if the torque commands are set to any negative value. The inverters automatically maintain an appropriate level of magnetisation and adjust the stator current to produce a prescribed torque. In principle, the full torque speed capability of the drive, as depicted in Figure 4.43 for DC rail traction vehicles, is available.

The level of dynamic braking available in an AC rail traction vehicle is limited by the constant-torque and constant-power regions of the drive. If there is sufficient ability to dissipate or store the braking energy on the DC bus bar, there is no significant torque derating relative to the traction case, and full braking force is available down to zero speed. Further, given the high dynamic capability of the inverters, any wheel slip control systems can potentially be applied in the braking mode.

If braking occurs, the mechanical power is transferred automatically to the DC bus bar of the rail traction vehicle. In this case, the DC bus bar voltage will rise. To arrest the voltage rise, energy must be withdrawn from the DC bus bar. This can be dissipated in braking resistors, which are physically similar to the braking resistors in DC rail traction vehicles. However, the braking resistor is controlled with a DC chopper, and, as this is continuously adjustable, there are no fluctuations in braking capability, as with the DC rail traction vehicles. In an AC-DC-AC rail traction vehicle, the DC bus braking power can be transferred to the overhead power system and potentially used by other trains in the network. In a hybrid rail traction vehicle with either AC or DC traction, the braking energy can be stored on board in an energy storage system for later use.

An alternative to the electric dynamic brake is the hydrodynamic brake. This type works based on the creation of the friction forces arising in fluid flow. Such brakes are also often used for high-speed operations. The design of a hydrodynamic brake is most often represented as a water turbine, which is connected through the drive or mounted directly on the wheelset's axle. When braking starts, the turbine is fed with a liquid that begins to circulate on a power circuit; because the fluid has a viscosity, there is resistance, and its circulation is accompanied by heating. The generated heat is dissipated into the environment through the heat-conducting walls of the turbine.

Design of Powered Rail Vehicles and Locomotives 155

4.7.3 ELECTROMAGNETIC BRAKES

Eddy currents are produced within any metal object that has relative motion with a nearby magnetic field. These currents dissipate energy within the metallic object. Conservation of energy requires that these losses be reflected in the mechanical system as retarding forces. As an eddy current brake uses a non-contact braking method, mechanical wear is avoided. In railway applications, two configurations are used. These are linear railhead brake systems or disc brake systems. Both are widely deployed in high-speed trains.

Linear railhead eddy current brakes produce a braking force by applying a strong magnetic field to a section of the railhead. A typical brake cross-section is shown in Figure 4.44. A series of electromagnets are arranged as a linear array that is positioned above the rail. The application of current into the five coils produces magnetic flux, which flows in the direction of the arrows marked as 'B'. The magnetic field moves relative to the rail and induces circulating currents in the railhead. These interact with the applied field to produce a retarding force. The kinetic energy of the vehicle is converted into heat in the railhead. One study has shown that the heat is dissipated in the top 2 mm of the railhead, and, for the cases studied, the highest temperature rises observed were 86°C [13].

Eddy current brakes are most effective in high-speed train applications. Some sample data produced by Knorr Bremse (see [14]) show that, at speeds below 50 km/h, the braking force increases linearly with speed to approximately 8 kN/m of the braking system. This then reduces steadily with further increase in speed, falling to 6 kN/m at 350 km/h. This force and velocity imply a braking power of 583 kW for each metre of the active braking system. One more advantage is that the braking force is independent of the wheel/rail contact conditions.

The linear railhead brake system has some features that may pose design challenges. The linear magnet array must be physically close to the railhead, at least during the braking operation. The air gap is typically 7 mm [13,14]. Another feature of eddy current brakes is the force of attraction between the brake array and the railhead, which can reach 50kN/m at low speeds. A challenge may be to maintain a well-controlled gap despite an attraction force that is somewhat variable.

An eddy current disc brake applies the braking field to a rotating brake disc, and some of the linear railhead brake challenges are avoided. A disadvantage is that the kinetic energy of the vehicle does heat the brake discs, and these will have a thermal capacity limit.

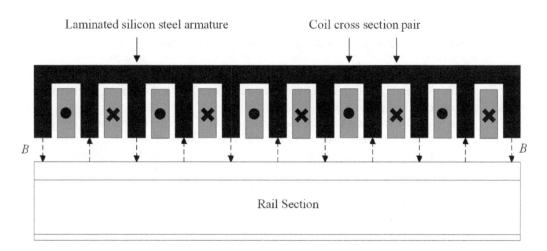

FIGURE 4.44 Typical linear eddy current brake cross-section.

4.7.4 Rail Brakes

This type of brake is based on utilising the force of friction generated by rubbing elements, which bear against the rail. Friction brake elements may operate on the horizontal surface of the top of the head of the rail and/or on the side surface of the rail; the latter has limited application due to the frequent presence of obstructions on this area of the rails. Unlike electromagnetic brakes, which allow rigid fixation on the running gear, rail brakes are installed with an elastic-damping suspension in order to reduce shock and vibration caused by their interaction with the rail.

4.8 DESIGN CLASSIFICATION AND CRITICAL PARAMETERS FOR VEHICLE SYSTEM DYNAMICS

The successful application of a rail traction vehicle design requires a comprehensive understanding of its critical parameters in vehicle dynamics studies. In this section, major design characteristics and parameters for rail traction vehicles are classified and discussed.

4.8.1 Classification and Main Design Principles in Traction Studies

When a rail traction vehicle design is being considered, it is necessary to understand how traction effort is produced at the wheel level and then transferred to vehicle (car body) motion.

Therefore, as the first step, it is necessary to have a detailed view on the classification of bogies, that is, to understand how traction effort is generated and applied to driven axles. In this discussion, the following bogie classification criteria are very important:

- By the number of wheel pairs enclosed in a vehicle frame:
  - According to the UIC classification of axle arrangements, the axles within the same bogie (truck) are classified starting from the front end of the rail vehicle by alphabetical symbols for the number of consecutive driven axles (A for one, B for two, C for three, etc.)
- By the functioning of the axles in the bogie design:
  - Motorised (independently rotating) wheels, when each wheel is driven by its own traction motor
  - A motorised axle, when driven by its own traction motor
  - A non-motorised axle, when the axle is used only to support part of the weight of a rail vehicle
- By the traction drive design:
  - An independently rotating wheel drive design, where the traction torque from a motor acts directly on a single wheel
  - An individual drive design, where the traction torque from a motor acts on a single axle through a gear box; note that, under the UIC classification of axle arrangements, the use of a lower case 'o' as a suffix after the letter indicates that those motorised axles are individually driven by separate traction motors; however, this suffix is not used in the system for axle arrangements developed by the AAR
  - A grouped drive design, where the traction torque from the motor or an output shaft of the transmission is shared between multiple axles; this type of design is uncommon for heavy-haul locomotives

Design of Powered Rail Vehicles and Locomotives

Secondly, it is necessary to classify how the traction effort is transferred or transmitted from a wheel or a wheelset to a bogie frame and from a bogie frame to a car body. In this case, the following bogie classification criteria are very important:

- By connection types between journal housings (axle boxes) and a bogie frame:
  - Connection with pedestal legs (also called a jaw connection)
  - Connection with cylindrical guides and a link arm (also called a traction rod)
  - Connection with radius links
- By the method of transmission of traction and braking force from the bogie to the locomotive car body:
  - Through the pivot point
  - Through the traction rod(s)

All these criteria should be systematised, and a conceptual scheme on the delivering of traction efforts should be finalised before any traction dynamics study for a rail vehicle starts.

4.8.2 Traction/Adhesion and Slip Control Principles

Rail traction vehicles are commonly equipped with four alternative traction control schemes [2,15–18]:

- Locomotive or whole vehicle traction control system: Where the same torque value is given to all traction motors of a rail traction vehicle
- Bogie traction control: The two or three bogies are each equipped with traction control systems that act independently on each bogie, that is, one inverter per bogie (also called group traction control)
- Individual wheelset traction control: The four or six individual axles are each equipped with traction control systems that act independently on each wheelset; that is, each wheelset has its own inverter (also called axle traction control)
- Independently rotating wheel traction control: Where the torque value is given and adjusted by a traction system to each wheel's traction motor of a rail traction vehicle

All these traction controls schemes are commonly limited by creep/wheel slip control principles. The terms 'slip' and 'creep' are often used interchangeably, and, for the estimation of wheel slip (creep), data obtained from angular velocity sensors mounted on wheels or wheelsets is essential. The data measured is used directly in the case of individual axle or independently rotating wheel controls or processed in order to find a minimal angular velocity in the case of whole vehicle or bogie traction controls, for their comparison with the rail vehicle ground speed (referenced or measured). A value of the longitudinal slip (creep) is estimated based on the following relation:

$$s = \frac{wr - V}{V} \tag{4.5}$$

where ω is the wheelset angular velocity, V is the rail vehicle speed and r is the nominal rolling radius of the wheels.

Modern adhesion/traction control systems can use conventional and extended slip control techniques. The basic operational principle of such techniques for rail traction vehicles is that action should be taken so that the estimated slip value remains situated in the stable wheel slip zone, as shown in Figure 4.45a [2,18–26]. The conventional systems usually operate with a predefined

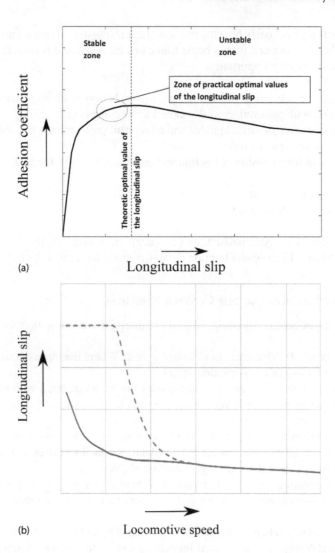

FIGURE 4.45 Examples of the relationships of longitudinal slip (creep) with (a) adhesion coefficient and (b) locomotive linear speed (solid line – conventional wheel slip control; dashed line – extended wheel slip control).

wheel slip threshold; that is, they maintain a nominal level of wheel slip, which is dependent on the rail vehicle speed. Meanwhile, the extended slip control technique should operate at the peak of the adhesion-slip curve, that is, at the point of the theoretic optimal wheel slip, as shown in Figure 4.45a. This is very important for poor adhesion conditions, and various algorithms can be used to detect a reference value (threshold) of the wheel slip. However, it is common that such a traction algorithm operates in the area defined as the practical optimal wheel slip zone indicated in Figure 4.45a. In this case, different adhesion estimation approaches can be used in order to change the value of slip threshold, but it is still limited by the rail vehicle speed, as shown in Figure 4.45b.

All these design system variations should be considered at the vehicle dynamics modelling stage, and a proper technique should be used to apply appropriate wheel torques to traction machines as well as have their limits set by slip control strategies.

Design of Powered Rail Vehicles and Locomotives

4.8.3 Critical Parameters in Traction Studies

4.8.3.1 Axle Load

Axle loads can be specified in two ways. The first method is indicative, with the mass of the rail traction vehicle considered to be carried equally by each axle, giving an average mass per axle, m_{axle}, and a corresponding axle load, F_{axle}, determined by the following formulas:

$$m_{axle} = \frac{m_{tot}}{n} \tag{4.6}$$

$$F_{axle} = m_{axle} \times g = \frac{m_{tot} \times g}{n} \tag{4.7}$$

where m_{tot} is the total rail vehicle mass, kg; g is the gravitational acceleration, equal to 9.81 m/s²; and n is the number of axles in the rail vehicle.

The second method is a more detailed one to determine the variation in load on each individual wheel (and hence on each axle). As an example, let us consider a six-axle locomotive. The calculation approach for the position of the centre of mass for the locomotive car body structure is shown in Figure 4.46. The centre of mass can be found as:

$$x_c = \frac{\sum_{i=1}^{k} M_i}{\sum_{i=1}^{k} W_i} \tag{4.8}$$

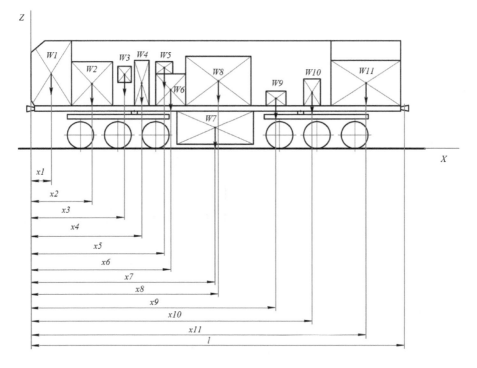

FIGURE 4.46 Calculation scheme for the determination of the position of the centre of mass for the locomotive car body structure in a six-axle locomotive.

where $\sum_{i=1}^{k} M_i$ is the sum of inertia moments of components, and $\sum_{i=1}^{k} M_i$ is the total weight of the locomotive.

In addition, assume that the mass of the car body is bearing on four secondary suspension points for each bogie and that the sprung mass of the car body and bogies is equally distributed on the bogie wheels. Knowing the position of the centre of mass for a locomotive car body for the defined axis coordinate system, as shown by the offsets b and c from the dotted line geometrical centreline axes in Figure 4.47, determines the mass of the locomotive attributed to each wheel, according to the formula:

$$m_{wheel\,j} = \frac{m_c}{12} \pm \frac{m_c \times c}{6 \times S} \pm \frac{m_c \times b}{6 \times S} + \frac{m_{bs}}{6} + m_{bwu\,j} \qquad (4.9)$$

where j is an index for right or left wheel; m_c is the total mass of the car body, kg; b and c are the coordinates of the centre of mass for the locomotive car body, m; S is the distance between nominal rolling radii of left and right wheels of the same axle/wheelset, m; m_{bs} is the sprung mass of a bogie, kg; and m_{bwu} is the unsprung mass of a bogie attributed to each wheel, kg. The signs '+' and '−' are used in cases when the centre of mass either loads or unloads a wheel of the wheelset, respectively, relative to the axis of symmetry. In this example, if c is positive, then the second component in Equation 4.9 should be taken as positive for a right wheel and negative for a left wheel. If b is negative, then the third component in Equation 4.9 should be taken as positive for wheelsets situated between the front of the locomotive and the car body centre of mass and as negative for wheelsets situated behind the car body centre of mass.

Thus, knowing the mass of the locomotive attributable to each wheel, determine the mass of the locomotive attributable to each axle:

$$m_{axle} = m_{wheel\,left} + m_{wheel\,right} \qquad (4.10)$$

Finally, each axle load can be calculated using Equation 4.7.

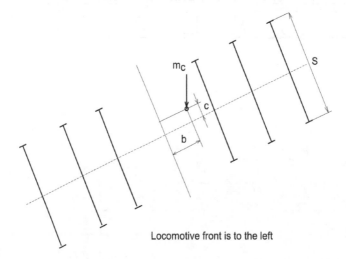

FIGURE 4.47 Calculation scheme for the determination of wheel loads in a six-axle locomotive. (From Spiryagin, M. et al., *Design and Simulation of Heavy Haul Locomotives and Trains*, Ground Vehicle Engineering Series, CRC Press, Boca Raton, FL, 2017. With permission.)

Design of Powered Rail Vehicles and Locomotives

4.8.3.2 Tractive and Dynamic Braking Efforts

The tractive effort exerted by a rail traction vehicle, F_{TE}, is necessary to move its train along the track [27–29].

The maximum possible tractive effort, $F_{TE\,max}$, is limited by the adhesion between wheels and rails, μ, and is defined by:

$$F_{TE\,max} = \mu \times m_{tot} \times g \tag{4.11}$$

Then, the tractive effort able to be realised by a rail traction vehicle, F_{TE}, obeys the following law:

$$F_{TE} \leq \mu \times m_{tot} \times g \tag{4.12}$$

In practice, two characteristics are often used in the description of tractive efforts: starting and continuous. Starting tractive effort is needed to determine how much train weight may be set into motion by a rail traction vehicle. Starting traction is mainly limited by the vehicle weight and the achievable adhesion/traction coefficient between wheels and rails, as shown in Equation 4.11. Continuous mode tractive effort allows for the possibility of an indefinite period of vehicle/train operation. In other words, the continuous tractive effort is designed to determine the vehicle/train weight that can be moved over very long periods of traction operation. This effort is limited by the power and dynamic performance of the traction electric transmission of a rail vehicle.

Equations similar to Equations 4.11 and 4.12 can also be written for dynamic braking effort calculations:

$$F_{DB\,max} = \mu \times m_{tot} \times g \tag{4.13}$$

$$F_{DB} \leq \mu \times m_{tot} \times g \tag{4.14}$$

4.8.3.3 Maximum Adhesion/Traction Coefficient

The value of the maximum adhesion/traction coefficient able to be achieved by a rail traction vehicle can be defined as:

$$\mu_{max} = \frac{F_{TE\,max}}{m_{tot} \times g} \tag{4.15}$$

A maximum realised traction coefficient of more than 40% can be achieved only on a dry track with ideal conditions and optimised traction control algorithms managing the power traction transmission system of a rail traction vehicle.

4.8.3.4 Power Output

The power output of rail traction vehicles is usually defined in units of watts or horsepower. The tangent power in kW, which is applied to the rail, can be calculated as:

$$P_{rail} = \frac{F_{TE} \times V}{3.6} \tag{4.16}$$

where V is the vehicle speed, km/h.

To calculate the horsepower, the ratio of 1 kW = 1.3596 metric horsepower should be used in Equation 4.16, giving rail traction vehicle power in horsepower (hp) as:

$$P_{rail\,hp} = \frac{1.3596 \times F_{TE} \times V}{3.6} \tag{4.17}$$

Effective power of a rail traction vehicle in kW, which equals the power of the power plant for an autonomous rail traction vehicle or the power consumed from the external power supply for a non-autonomous rail traction vehicle, can then be defined as:

$$P_5 = \frac{P_{rail}}{\eta_t \times \eta_a} \quad (4.18)$$

where η_t is the efficiency of the rail vehicle traction transmission, and η_a is a coefficient that takes into account the auxiliary equipment power needs.

4.8.3.5 Maximum Speed

The maximum rail traction vehicle speed, V_{max} (km/h), is commonly limited by a speed characteristic of the traction electric motor, which can be determined by the following formula:

$$V_{max} = \frac{1.8 \times d_{wheel} \times \omega_{m\,max}}{i} = \frac{0.18849 \times d_{wheel} \times n_{m\,max}}{i} \quad (4.19)$$

where d_{wheel} is the wheel diameter, m; $\omega_{m\,max}$ is the maximum angular velocity of rotor/shaft of the traction motor, rad/s; $n_{m\,max}$ is the maximum rotor/shaft speed of the traction motor, rev/min; and i is the gear box ratio, equal to the number of teeth of the gears mounted on the wheelset divided by the number of teeth of the gears mounted on the rotor/shaft of the traction motor. However, it should be noted that the maximum speed can also be further limited by the technical solutions or design of the running gear of a rail traction vehicle.

REFERENCES

1. M. Spiryagin, C. Cole, Y.Q. Sun, M. McClanachan, V. Spiryagin, T. McSweeney, *Design and simulation of rail vehicles, Ground Vehicle Engineering Series*, CRC Press, Taylor & Francis Group, Boca Raton, FL, 2014.
2. M. Spiryagin, P. Wolfs, C. Cole, V. Spiryagin, Y.Q. Sun, T. McSweeney, *Design and Simulation of Heavy haul Locomotives and Trains, Ground Vehicle Engineering Series*, CRC Press, Taylor & Francis Group, Boca Raton, FL, 2017.
3. UIC Leaflet 650, Standard Designation of Axle Arrangement on Locomotives and Multiple-Unit Sets, 5th ed., International Union of Railways, Paris, France, 1983.
4. Association of American Railroads, AAR Manual of Standards and Recommended Practices, Section M – Locomotives and Locomotive Interchange Equipment: RP-5523 Axle Nomenclature Arrangement – Locomotives, AAR, Washington, DC, 2008.
5. American Public Transit Association, Glossary of Transit Terminology, Produced by APTA Governing Boards Committee and compiled by Peggy Glenn, July 1994. Available at: https://www.apta.com/resources/reportsandpublications/Documents/Transit_Glossary_1994.pdf
6. Parsons Brinckerhoff, Inc., Track design handbook for light rail transit, 2nd ed., Transit Cooperative Research Program (TCRP) Report 155, Transportation Research Board, Washington, DC, 2012.
7. R.J. Hill, Electric railway traction, Part 1: Electric traction and DC traction motor drives, *Power Engineering Journal*, 8(1), 47–56, 1994.
8. R.J. Hill, Electric railway traction, Part 2: Traction drives with three-phase induction motors, *Power Engineering Journal*, 8(3), 143–152, 1994.
9. M. Spiryagin, Q. Wu, P. Wolfs, Y. Sun, C. Cole, Comparison of rail traction vehicle energy storage systems for heavy haul operation, *International Journal of Rail Transportation*, 6(1), 1–15, 2018.
10. M. Spiryagin, P. Wolfs, F. Szanto, Y.Q. Sun, C. Cole, D. Nielsen, Application of flywheel energy storage for heavy haul rail traction vehicles, *Applied Energy*, 157, 607–618, 2014.
11. R. Cousineau, Development of a hybrid switcher locomotive – the Railpower Green Goat, *IEEE Instrumentation & Measurement Magazine*, 9(1), 25–29, 2006.

12. Bombardier's battery powered tram sets range record, 3 November 2015, available at: https://www.bombardier.com/en/media/newsList/details. BT-20151103-Bombardiers-Battery-Powered-Tram-Sets-Range-Record-01.bombardiercom.html
13. P.J. Wang, S.J. Chiueh, Analysis of eddy-current brakes for high speed railway, *IEEE Transactions on Magnetics*, 34(4), 1237–1239, 1998.
14. X. Lu, Y. Li, M. Wu, J. Zuo, W. Hu, Rail temperature rise characteristics caused by linear eddy current brake of high speed train, *Journal of Traffic and Transportation Engineering* (English Edition), 1(6), 448–456, 2014.
15. A. Steimel, *Electric Traction – Motive Power and Energy Supply: Basics and Practical Experience*, Oldenbourg Industrieverlag GmbH, Munich, Germany, 2008.
16. L. Liudvinavičius, L.P. Lingaitis, G. Bureika, Investigation on wheel-sets slip and slide control problems of locomotives with AC traction motors, *Maintenance and Reliability*, 4, 21–28, 2011.
17. M. Spiryagin, P. Wolfs, F. Szanto, C. Cole, Simplified and advanced modelling of traction control systems of heavy-haul locomotives, *Vehicle System Dynamics*, 53(5), 672–691, 2015.
18. M. Spiryagin, P. Wolfs, C. Cole, S. Stichel, M. Berg, M. Plöch, Influence of AC system design on the realisation of tractive efforts by high adhesion locomotives, *Vehicle System Dynamics*, 55(8), 1241–1264, 2017.
19. D. Frylmark, S. Johnsson, Automatic slip control for railway vehicles, M.Sc. Thesis, Linköpings Universitet, Linköping, Sweden, 2003.
20. M. Spiryagin, K.S. Lee, H.H. Yoo, Control system for maximum use of adhesive forces of a railway vehicle in a tractive mode, *Mechanical Systems and Signal Processing*, 22(3), 709–720, 2008.
21. M. Spiryagin, C. Cole, Y.Q. Sun, Adhesion estimation and its implementation for traction control of locomotives, *International Journal of Rail Transportation*, 2(3), 187–204, 2014.
22. Y. Yao, H. Zhang, Y. Luo, S. Luo, Theory of stick-slip vibration and its application in locomotives, *Chinese Journal of Mechanical Engineering*, 46(24), 75–82, 2010. (In Chinese).
23. P. Pichlik, J. Zdenek, Overview of slip control methods used in locomotives, *IEEE Transactions on Electrical Engineering*, 3(2), 38–43, 2014.
24. O. Polach, Creep forces in simulations of traction vehicles running on adhesion limit, *Wear*, 258(7–8), 992–1000, 2005.
25. M. Spiryagin, O. Polach, C. Cole, Creep force modelling for rail traction vehicles based on the Fastsim algorithm, *Vehicle System Dynamics*, 51(11), 1765–1783, 2013.
26. M. Spiryagin, I. Persson, Q. Wu, C. Bosomworth, P. Wolfs, C. Cole, A co-simulation approach for heavy haul long distance locomotive-track simulation studies, *Vehicle System Dynamics*, In Press, DOI: 10.1080/00423114.2018.1504088, 2018.
27. Q. Wu, M. Spiryagin, C. Cole, Longitudinal train dynamics: an overview, *Vehicle System Dynamics*, 55(4), 1–27, 2017.
28. C. Cole, M. Spiryagin, Q. Wu, Y.Q. Sun, Modelling, simulation and applications of longitudinal train dynamics, *Vehicle System Dynamics*, 55(10), 1498–1571, 2017.
29. Q. Wu, M. Spiryagin, P. Wolfs, C. Cole, Traction modelling in train dynamics, *Journal of Rail and Rapid Transit*, 233(4), 382–395, 2019.

5 Magnetic Levitation Vehicles

Shihui Luo and Weihua Ma

CONTENTS

- 5.1 The Option of Magnetic Levitation (Maglev) Vehicles 166
  - 5.1.1 Maglev for Whole Speed Ranges 166
  - 5.1.2 A Mega Trend for Higher-Speed Rail Transport 167
  - 5.1.3 Maglev for Higher Speed 168
  - 5.1.4 Maglev for Lower Dynamic Impact 169
  - 5.1.5 Maglev System Dynamic Problems 169
- 5.2 Classification and System Characteristics of Maglev Vehicles 170
  - 5.2.1 Maglev Vehicle Classification 170
  - 5.2.2 Technological Characteristics of Typical Vehicles 171
- 5.3 Dynamics of Moving Loads on Flexible Girder 172
  - 5.3.1 Periodical Excitation between Vehicle and Girder 172
  - 5.3.2 Single Moving Force on Flexible Girder 173
  - 5.3.3 Multiple Moving Forces on Flexible Girder 175
    - 5.3.3.1 Case Study (a) 176
    - 5.3.3.2 Case Study (b) 177
    - 5.3.3.3 Influence of Vehicle Load Distribution 178
- 5.4 The Methodology for Maglev Vehicle Dynamics Analysis 178
  - 5.4.1 Equations of Flexible Girder 179
    - 5.4.1.1 Dynamic Equations of the Girder 179
    - 5.4.1.2 Solution of the Dynamic Equations 179
    - 5.4.1.3 Solving Vibration Mode and Frequency 180
    - 5.4.1.4 Equations of Motion of the Beam Expressed Using Generalised Coordinates 182
  - 5.4.2 Vehicle System Dynamic Equations 183
  - 5.4.3 Levitation Forces 184
    - 5.4.3.1 EMS Levitation Forces 184
    - 5.4.3.2 EDS Levitation Forces 185
    - 5.4.3.3 HTS Levitation Forces 187
  - 5.4.4 System Equations and Solutions 187
- 5.5 Simulation Examples of Maglev by Multibody System 188
  - 5.5.1 Moving Loads on Flexible Beam 188
    - 5.5.1.1 Single Moving Force Passing a Single-Span Beam 188
    - 5.5.1.2 Single Moving Force Passing a Double-Span Beam 189
    - 5.5.1.3 Multiple Moving Forces Passing a Double-Span Beam 189
    - 5.5.1.4 Modelling of Moving Lumped Mass 190
  - 5.5.2 Modelling of EMS Levitation Control Loop 190
  - 5.5.3 Modelling of Vehicle-Girder Interaction 192
- 5.6 Concluding Remarks 193
- References 193

5.1 THE OPTION OF MAGNETIC LEVITATION (MAGLEV) VEHICLES

Traditional railways have been playing a significant role in both short-distance and long-distance mass passenger transport. In the modern commuter environment, traditional railways also face new challenges from different perspectives. For example, urban railways must focus on how to minimise the influence of vibration and noise from trams, subways and light rail vehicles on the environment. Regional railways need to cater to public desire for rapid trips to suburban areas as well as direct connections to downtown areas, with less station transfers. Intercity transport must focus on how to achieve higher speeds so as to minimise travel time and even achieve the potential of diverting passengers from air traffic and decreasing exhaust pollution in the atmosphere. Last but not least, urban transport, regional transport or intercity transport must all develop strategies to achieve better economic values regardless of what type of transport is used.

Since the construction of the very first functional steam locomotive by British engineer Richard Trevithick in 1803, traditional railways have gone through more than 200 years of technology evolution. Some researchers consider that the technological potential of traditional railways has approached its limit. Facing future demands and challenges, maglev trains provide a new option [1].

Maglev trains are defined as trains that are supported and steered via magnetic forces, without contact with the guideway, and are propelled by electric magnetic forces from linear motors.

5.1.1 MAGLEV FOR WHOLE SPEED RANGES

Compared with traditional wheel-rail railways, maglev transport has a significantly wider speed range capability. The operational speed range of traditional railways usually does not exceed 400 km/h, and the speed range can be divided into four classes of regular-speed railways (~120 km/h), speeded-up railways (120 ~ 160 km/h), quasi high-speed railways (160 ~ 200 km/h) and high-speed railways (200 ~ 400 km/h). The available speed range of maglev transport can also be divided into four classes:

- Medium-low speed (MLS) (~200 km/h)
- Medium speed (MS) (200 ~ 400 km/h)
- High speed (HS) (400 ~ 1000 km/h)
- Super-high speed (SHS) (1000 km/h ~ several Mach numbers)

Maglev trains have been commercialised or are approaching the status of commercialisation in the MLS and HS ranges [2–4]. There are five maglev lines around the world that have been put into commercial operations. Among those five lines, only one is in the high-speed range; this being in Shanghai (2003, 29.9 km, 430 km/h), which has used German-developed technology. All the other four lines have medium-low operational speed ranges. They are the Nagoya TKL line in Japan (2005, 8.9 km, 100 km/h), the Incheon Airport line in Korea (2016, 6.1 km, 110 km/h), the Changsha Airport line in China (2016, 18.6 km, 100 km/h) and the Beijing S1 line in China (2017, 10.2 km, 80 km/h). There are also two other maglev lines under construction, namely the route between Tokyo and Nagoya (286 km, 505 km/h), which is part of the Chuo Shinkansen high-speed line and the first section of the Qingyuan tour line in China (9.4 km, 120 km/h).

Medium- and super-high-speed maglev trains are under continuous development; however, commercialisation has not been reported yet. There are different source types for the magnet forces of the medium- and super-high-speed maglev trains: high-temperature superconducting (HTS) maglev [5], electric dynamic suspension (EDS) maglev with on-board permanent magnets [6,7] and electromagnetic suspension (EMS) maglev with a hybrid source of both permanent and electric magnets [8]. It can be seen that, from MLS to SHS ranges, using maglev trains to meet the demand of future rail transport has been a rapidly developing and essential research topic.

5.1.2 A Mega Trend for Higher-Speed Rail Transport

Traditional high-speed railways have achieved commercial operational speeds of 350 km/h; technically, higher speeds up to 400 km/h are also possible. Passenger jets have speeds of approximately 1000 km/h. Even with these high-speed transports that are already available for public, continuously pursuing ground mass transports that have higher speeds are still a mega trend. The motivations are multiple as follows. The development of land-based high-speed mass rail transport is beneficial to promote economic development in large areas along the railway corridor. Further, land-based electrical propulsion technology is more beneficial for decreasing greenhouse gas exhausts.

Each industrial revolution marks a milestone of the development of human civilisation. Technological advancements increased the speed of land-based mass transport, and they also promoted the development of improved productivity. Reciprocally, these developments fostered human's desire to reach places that are further in shorter travelling time. Figure 5.1 chronologically describes the relationship between the advancement of human civilisation and the speed of land-based transport.

The invention and application of steam-powered machines in the seventeenth century doubled the speed of land-based transport compared with the era when traction forces were delivered by livestock. They also provided the foundation for railways whose hauling capacity had to progressively meet the growing demand of intense mass passenger and freight transport. The second industrial revolution, which was represented by electrification, achieved high-speed capability for rail transport. One example is that Japan, in 1964, achieved a train speed higher than 200 km/h on the Tokaido Shinkansen railway. Third industrial revolution was characterised by computerisation and informatisation and resulted in new mass rail transport featuring high-speed maglev trains that increased the speed of land-based transport to about 500 km/h. The twenty-first century has seen the dawn of the fourth industrial revolution, which is characterised by new energy, artificial intelligence, quantum communication, etc. It can be inferred that the pursuit of higher rail transport speeds should remain the same. Judging from previous trends, it could be predicted that the speed of land-based rail transport could be doubled to 1000 km/h. Considering the state of the art of maglev research and its technological characteristics, maglev can very probably be used to achieve this goal.

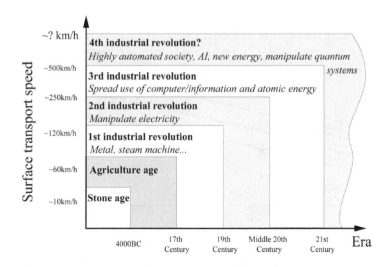

FIGURE 5.1 Relationship between technological advancement and land-based transport speed.

5.1.3 Maglev for Higher Speed

Traditional railways rely on rolling friction between wheels and rails to operate. Therefore, the utilisation level of friction limits the realisation of traction forces of the system. To achieve higher speed, it is essential to use as much as possible of the train's weight to generate friction forces. The ultimate scenario would be for all axles to be motorised axles, and the whole weight of the train is used to generate friction forces. Under this condition, Figure 5.2 plots the propulsion resistance and traction force per unit weight of the train. The intersection of the two plots indicates the possible maximum speed of the train. Note that these plots are merely rough calculations to explain the basic principles and trends.

On wheel-rail high-speed electric trains, the traction motors are mounted on the vehicles. The motors receive electrical energy from ground-based power plants via pantograph contact with overhead catenary system wires. Figure 5.3 shows the relationships among the tension force of the contact wire, train operational speed and the contact stress of the pantograph carbon strip. In this figure, the contact pressure plots are the boundaries that can be used to assess the working conditions of the carbon strip contacts. Areas beneath the plots indicate that the carbon strip can be kept in contact, while the areas above the plots indicate that the carbon strips will become detached from the contact wires [9]. Higher wave speeds in the contact wire are beneficial for keeping in contact. However, there are trade-offs among wave speed limits and contact wire strain, permissible strain and specific electric resistance. The maximum speed of the train is therefore limited by these factors also.

Regardless of what maglev principle is being used, mechanical contact friction does not exist in high-speed maglev transport. Maglev transport operates without adhesion and has broken through

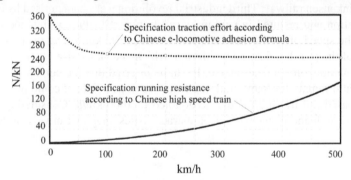

FIGURE 5.2 Speed limitation for trains relying on wheel-rail adhesion.

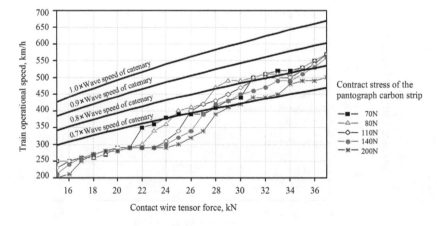

FIGURE 5.3 Speed limitation for trains relying on current collection by pantograph will become detached from the contact wires.

the limit of friction. In the high-speed application, propulsion of maglev is always realised by linear synchronous motors (LSMs) with on-track long stators; the massive amount of electrical tractive energy that is needed to power the train does not need to be transmitted from the ground to the train, and the maglev operation is therefore not limited by current collection. Furthermore, the support between the vehicle and the line is in the form of translation motion instead of rolling; therefore, maglev trains are not subject to the limits of axle journal bearings under high load and high speed. High-speed maglev in Japan and Germany has reached test speeds of 603 km/h and 501 km/h, respectively. Theoretically, much higher speeds can be achieved using maglev [10,11].

5.1.4 Maglev for Lower Dynamic Impact

Compared with traditional wheel-rail trains, the dynamic impacts that maglev trains impose on the infrastructure are smaller. In traditional wheel-rail trains, the total mass of the train is supported by a limited number of wheel-rail contact points and then transmitted to the track. The wheel-rail contact points can be regarded as rigid contacts, and they impose concentrated loads. Regardless of what theory a maglev train is based on, the weight of the maglev train is evenly applied to the infrastructure and the loads imposed are distributed loads. In addition, magnetic levitation systems have characteristics of suspensions [12], and they further attenuate the impacts from the train to the infrastructure.

Maglev trains can be classified into active support (AS) and passive support (PS) according to their different supporting methods; they have different requirements to be provided by the supporting 'track'.

Electromagnetic levitation is a type of AS approach; it uses feedback control to maintain a small gap (8 ~ 12 mm) between the vehicle and the track. This type of support needs to effectively suppress coupled vibrations between the vehicles and the track to maintain good vehicle operational performance. Therefore, it poses very high requirements for the bridge design of the viaduct structure, which will certainly increase the construction cost of the system. For example, even the weakest section of the German Transrapid Versuchsanlage Emsland (TVE) high-speed maglev test line used a 1/4000 deflection-span ratio [13]. As the first commercially applied high-speed maglev line in the world, the deflection-span ratio on the Shanghai high-speed line has reached approximately 1/15000. As the impacts of maglev distributed loads are smaller, the maintenance cost for the lines is significantly lower than that for traditional railway lines. For example, the maintenance cost of the Shanghai high-speed maglev line is about 50% of that for traditional high-speed railways [13].

Electric dynamic suspension maglev and HTS maglev use a PS approach. During high-speed operations, the levitation gap for electric dynamic suspension (EDS) maglev can exceed 200 mm, while that for HTS maglev is about 10 ~ 15 mm. For the PS method, vehicle-track coupled vibrations will not influence the levitation stability of the train. Therefore, large deflection-span ratios can be used for the bridge designs of the viaduct structure, similar to that used in the traditional high-speed railways. The constraint condition that must be satisfied is the requirement for a comfortable ride index during high-speed operations. As the impacts of maglev distributed loads are smaller, the maintenance cost for maglev infrastructure will also be significantly lower than that used for traditional wheel-rail trains.

Figure 5.4 [14] shows a comparison of pier supporting forces that are generated during operations of wheel-rail vehicles and maglev vehicles (electromagnetic maglev using AS). The results show that maglev vehicles that have distributed loads have evidently lowered the dynamic actions on the bridge than the wheel-rail vehicles that have concentrated loads.

5.1.5 Maglev System Dynamic Problems

Maglev lines commonly use viaduct structures. The essence of maglev transport system dynamics is to determine the response of vehicle and girder by simulating moving loads on flexible girders, as well as to consider the interaction between the vehicle and the girder. Therefore, maglev system

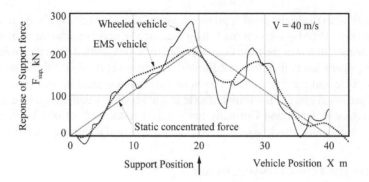

FIGURE 5.4 Supporting forces of viaducts with maglev and wheeled vehicle. (From M. Nagai, M. Iguchi, Vibrational characteristics of electromagnetic levitation vehicles-guideway system, in H.P. Willumeit (Ed.), *Proceedings 6th IAVSD Symposium held at Technical University Berlin*, 3–7 September 1979, Swets and Zeitlinger, Lisse, the Netherlands, 352–366, 1980. With permission.)

dynamics and wheel-rail system dynamics share some common features. The main examples are the multibody system (MBS) modelling principles and methods that are used to model the vehicles as well as flexible girder modelling theory and methods that are used to model the infrastructure. The difference between the maglev and traditional systems is the interface models between their vehicles and infrastructure. The interface between wheel and rail experiences creep forces and normal forces generated from mechanical rolling contact, while the maglev interface force consists of the magnetic attraction or repulsion forces.

The research of maglev system dynamics involves achieving the following objectives:

- From a bridge design perspective, to obtain the dynamic response of the bridge subjected to the passage of trains. This allows the use of dynamic factors to describe the level of bridge response and provide guidance for bridge designs.
- From a vehicle design perspective, to assess and optimise the integration of bridge structure, levitation system and vehicle structure. This is necessary to assure that the train has stable levitation performance, a comfortable ride index and good curve negotiation capability. It is particularly important to decrease the requirements for the infrastructure to a reasonable range so as to increase the economic efficiency of the system.

5.2 CLASSIFICATION AND SYSTEM CHARACTERISTICS OF MAGLEV VEHICLES

5.2.1 Maglev Vehicle Classification

Different from the traditional railway transport, there is no compatibility or universality among different types of maglev transport. For a wheel-rail railway, regardless of whether considering locomotives, passenger vehicles or freight vehicles, all vehicles use the same fundamental mechanism of using wheel-rail rolling contact to achieve support and steering. Also, different vehicles share basically the same bogie and suspension principles. The basic structures of the track infrastructure are relatively universal but with a wide range of customised variations to suit the operational requirements of particular railways.

For maglev trains, the three most basic types of maglev levitation support are EMS, EDS and HTS magnet suspension. They follow different fundamental principles to achieve levitation, steering and propulsion. It is impossible for different types of maglev to use simple universal guideway structures and vehicle structures. Even vehicles that use the same levitation principle but need to be

Magnetic Levitation Vehicles

operated in different speed ranges cannot use the same guideway and vehicle structures. Therefore, maglev trains cannot be classified in a similar way as used for traditional railway trains. For example, wheel-rail vehicles can be simply classified according to vehicle function (locomotive, passenger car, freight car and so on) and follow similar principles to design the vehicles. Similar methods can also be used to assess the dynamic performance of these vehicles.

Maglev vehicles can be classified from the following different perspectives:

- By *levitation principles*: EMS, EDS and HTS
- By *supporting methods*: AS and PS
- By *propulsion methods*: On-board short-stator linear induction motor (LIM) and on-track long-stator LSM
- By *operational speed*: MLS, MS, HS and SHS
- By *materials of conductor*: Normal conductor (NC), HTS and low-temperature superconductor (LTS)
- By *the application of permanent magnet*: No permanent magnet (NPM), vehicle-mounted permanent magnet (VPM) and track-mounted permanent magnet (TPM)

According to the above classifications, a tree map for maglev vehicle technological characteristics can be drawn up, as shown in Figure 5.5. More detailed classifications can be conducted from the perspective of bogie structures, traction configuration and steering approaches. But these detailed classifications are only different applications of the same basic technological characteristics and therefore are not expanded upon in this chapter.

5.2.2 Technological Characteristics of Typical Vehicles

Using the classifications shown in Figure 5.5, main technological characteristics of maglev vehicles can be easily identified. The following typical maglev trains that have reached the development level for commercial operation are used as examples to discuss the main characteristics:

- Series L0 train (Japan): A PS HS maglev train using EDS levitation, with LSM propulsion and on-board LTS magnet
- TR08, TR09 train (Germany): An AS HS maglev train using EMS levitation, with LSM propulsion and NC

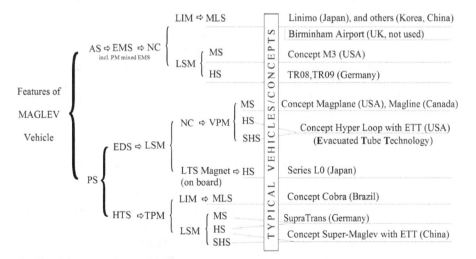

FIGURE 5.5 Classification of maglev vehicles and typical vehicles/concepts.

- Linimo train (Japan) and other commercially applied MLS trains: An AS MLS maglev train using the EMS levitation, with on-board short-stator LIM propulsion and NC

Similar descriptions can also be used for the technological characteristics of other systems that are still being developed:

- Super-Maglev concept vehicle (China): A PS HS or SHS maglev concept using HTS levitation, with LSM propulsion and running on permanent magnet track in a vacuum tube
- Hyperloop concept vehicle (USA): A PS HS or SHS maglev concept using EDS levitation, with LSM propulsion and on-board permanent magnet, running in a vacuum tube
- M3 concept vehicle (USA): An AS MS maglev concept using permanent magnet mixed EMS levitation, with LSM propulsion
- MAG-plane (USA) and MAG-line (Canada) concept vehicle: Same as the hyperloop concept but at MS

5.3 DYNAMICS OF MOVING LOADS ON FLEXIBLE GIRDER

Maglev lines usually use viaduct structures, which comprise a big proportion of the system cost. Under the precondition of satisfying all operational requirements, lowering the requirements for the basic structure of the viaduct as much as possible can help to lower the cost of the whole system. For maglev vehicles that have active suspensions, the vertical dynamics between vehicles and the guideway have evident implications for the levitation stability. For maglev vehicles that have passive suspensions, the dynamics between vehicles and the guideway can influence the ride index of the train but not levitation performance. Therefore, maglev vehicle dynamics need to mainly focus on the vertical interactions between vehicles and the flexible girders. If the generation mechanism of the maglev forces in the interface between vehicles and guideways is neglected (in other words, model the generation mechanism as simple forces), the dynamics of maglev vehicles can then be simplified as the question of moving loads on flexible girders. Focusing on this research question, plenty of effective research has been done in the 1980s, and many important conclusions have been reached [15].

5.3.1 PERIODICAL EXCITATION BETWEEN VEHICLE AND GIRDER

When a train is running at the speed of v on a girder that has a span of l, two periodical excitations are generated. The first is the periodical excitation generated from the vehicle to the girder, while the second is generated from equally spaced piers to the vehicle.

The periodical excitations from vehicles to girders are related to the characteristic length of the vehicle L, and the excitation frequency can be determined as $f_L = v/L$. The characteristic lengths of traditional wheel-rail vehicles have only vehicle spacing, axle spacing and bogie spacing. However, there are more characteristic lengths that need to be considered in maglev trains. Take the TR08 train, which is shown in Figure 5.6, as an example. Various characteristic lengths include vehicle spacing $L_{car} = 25$ m, air spring suspension spacing $L_{as} = 3.096$ m, coil spacing $L_{coil} = 0.258$ m and the length of a half magnet (with 6 coils) $L_{hm} = 1.548$ m, and the spacings for the supports of levitation electromagnets $L_{rb1} = 2.189$ m and $L_{rb2} = 0.907$ m.

Excitation frequencies that originate from various vehicle characteristic lengths linearly increase with the train speed v. These patterns are shown as skew lines in Figure 5.7. Modal frequencies of the girder do not change with train speed. Take the double-span simple beam girder that is used for the Shanghai HS maglev line as an example. Its first four orders of modal frequencies are shown as horizontal lines in Figure 5.7. The intersection point between an excitation frequency skew line and a horizontal line indicates a speed where resonance may be caused for the girder. If only the vehicle spacing characteristic length L_{car} is considered, resonance will not be caused under the speed of 500 km/h. However, the weight of the vehicle is not applied on the girder in a concentrated

Magnetic Levitation Vehicles

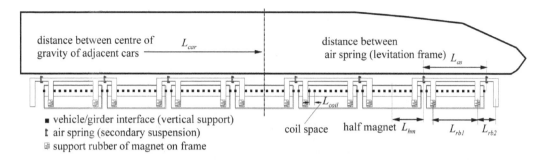

FIGURE 5.6 Characteristic lengths of TR08 train.

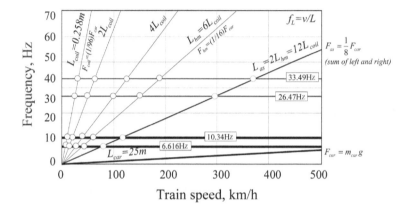

FIGURE 5.7 Periodical excitation applied by TR08 train on flexible girder.

form. Therefore, characteristic lengths L_{as} and L_{hm} and other even shorter characteristic lengths (the corresponding applied forces are also smaller) can still cause resonance of the girder at the speeds of 370 km/h, 295 km/h, 190 km/h and even lower speeds. Under these conditions, the levitation stability and passenger ride index will be influenced. One of the tasks of maglev vehicle dynamics is to select appropriate minimum characteristic lengths. Excessively short characteristic lengths will not significantly influence the vehicle performance but will make the model more complex.

Similarly, when a train is passing over a girder at the speed of v, the periodical excitations from the equally spaced piers to the trains have the frequency of $f_p = v/l$. The first few modal frequencies of the vertical vibrations of the girder will also generate periodical excitations to the vehicles, as shown in Figure 5.8.

5.3.2 Single Moving Force on Flexible Girder

The dynamic response of the girder excited by a single moving constant force is related to the passing speed. When the speed is very low, the deflection of the girder in the middle of the span tends to be the same as that under static load. With the increase of passing speed, the dynamic response becomes intensified and will reach a peak value at a critical speed and then decrease as speed increases further.

For the girder shown in Figure 5.9a, the critical speed is defined for the condition of a moving load as follows [15]:

$$V_{cr} = \frac{\pi}{l}\sqrt{\frac{EI}{\rho a}} \quad (5.1)$$

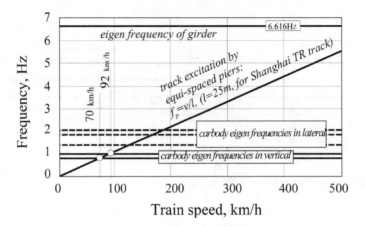

FIGURE 5.8 Periodical excitation applied by elevated girder piers on vehicle.

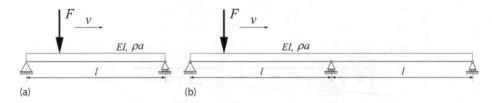

FIGURE 5.9 Single moving force on flexible girders: (a) single-span girder and (b) double-span girder.

where E is the modulus of elasticity, I is moment of inertia, EI is the bending stiffness of the girder, ρ is the density of the girder, and a is the cross-sectional area of the girder. The relationship between the critical speed and the frequency of the first-order bending mode (vertical) is:

$$V_{cr} = 2lf_1 \qquad (5.2)$$

The crossing frequency of the load when passing the girder at the speed of v is defined as:

$$f_c = v/l \qquad (5.3)$$

The ratio of the crossing frequency over the modal frequency f_1 of the girder is defined as the crossing frequency ratio [15], given by:

$$v_c = f_c / f_1 = 2v / V_{cr} \qquad (5.4)$$

Figure 5.10a shows the ratio Y_M of middle-point deflections at different passing speeds for single-span and double-span girders. Figure 5.10b shows the middle-point acceleration under the same conditions. Figure 5.11 [15] shows the middle-point deflection ratio between dynamic condition and static condition at different speeds for various bridge constructions (girders of single span, three spans and continuous multiple spans). The results show that (1) until v_c exceeds 1.0, the acceleration of the double-span girder increases rapidly; (2) at a speed of $v_c \approx 1.33$, the deflection of the single-span girder reaches its maximum value; and (3) approaching a speed of $v_c = 2.0$, both deflection and acceleration of the double-span girder reach their maximum value.

Assume that the maximum speed of a maglev train is 500 km/h and the girder span is 25 m, then the crossing frequency of the vehicle centre of mass (CoM) is $f_c = 5.56$ Hz. In order to keep $v_c < 0.9$,

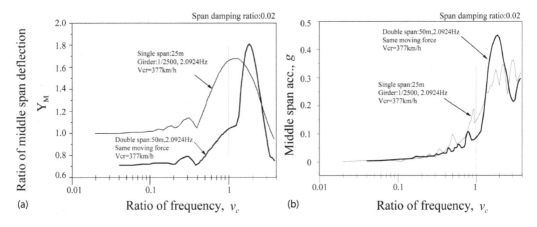

FIGURE 5.10 Response of single- and double-span girders for (a) deflection and (b) acceleration.

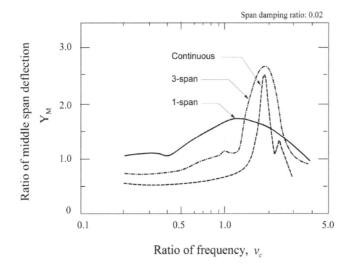

FIGURE 5.11 Deflection response of single- and multiple-span girders. (From W. Kortüm, Vehicle response on flexible track, *International Conference on Maglev Transport Now and For the Future*, 9–12 October 1984, Solihull, UK, IMechE, C405/84, 47–58, 1984. With permission.)

the first-order vertical bending modal frequency of the girder has to satisfy $f_1 > 1.1$ v/l according to Equation 5.4. This is one of the basic requirements of the Shanghai TR08 HS maglev train bridges. Under the condition of maintaining a girder span of 25 m, to increase the first-order vertical bending modal frequency requires an increase of girder stiffness. This in turn requires an increase of the girder cross-sectional area. Consequently, the mass of the girder may also increase, and the strength of the girder will far exceed the needs of the girder as a supporting structure. Understandably, these factors will increase the cost of the girders. Considering that the loads from the vehicles of maglev trains are not single concentrated loads, but are distributed on the line, the girder design requirements proposed by using the rules of Figures 5.10 and 5.11 could be very conservative.

5.3.3 Multiple Moving Forces on Flexible Girder

As shown in Figure 5.7, when maglev trains are passing over the girders of viaduct bridges, the characteristic lengths of the train apply periodical excitations to the flexible girders. If the excitation frequency coincides with the natural frequency of the girders, girder resonances will be caused.

The following analysis takes a maglev train with five vehicles as an example; the dimensions of the vehicles are the same as shown in Figure 5.7. For the convenience of comparisons, the same loads and girder characteristics used in Figure 5.10 are used again here. Note that the characteristics of the girders are relatively soft compared with actual girders used on the Shanghai Maglev line; therefore, this design can be called a 'soft girder'. When a concentrated load F is applied to the middle point of a single-span girder, the deflection at that specific point is $l/2500$, while the deflection to span length ratio is $1/2500$. The following analyses will also compare different cases of multiple moving constant forces negotiating double-span simple girders, as shown in Figure 5.12 (25 m span length and 50 m girder length).

5.3.3.1 Case Study (a)

In the first case study, vehicles in the train are treated as concentrated loads. The deflection and acceleration response at the mid-span of the girder under different speeds of the five moving forces are shown in Figure 5.13. For better comparisons, the corresponding results for the situation of a single constant moving force are also shown in Figure 5.13. The results show that there is one resonance speed for the passage of the five-vehicle train. During the resonance, deflection and acceleration responses of the girder have far exceeded the results of the single constant-moving force situation. The resonance speed of the multiple force situation is different from the crossing speed when the single constant force causes maximum response. Actually, the moving train at this speed causes resonance of the second vertical bending mode of the girder.

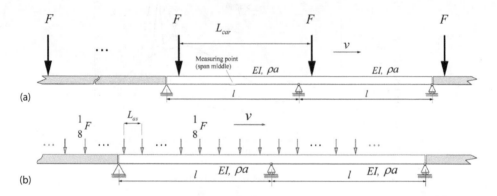

FIGURE 5.12 Multiple equally spaced moving forces on a double-span flexible girder: (a) train consisting of five vehicles, each of which is represented as a concentrated force F and (b) each force F is further distributed on eight air spring positions.

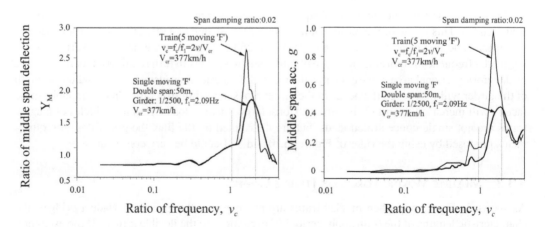

FIGURE 5.13 (Left and Right) Girder mid-span responses in case study (a).

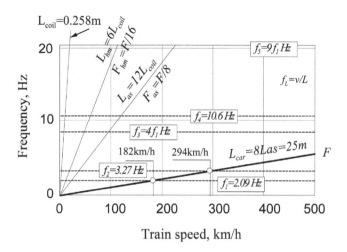

FIGURE 5.14 Possible resonance moving speeds under the action of concentrated forces F.

Figure 5.14 shows the characteristic lengths of the vehicles and the corresponding excitation frequency with different speeds. It also shows the first few orders of the vertical bending modal frequencies of the soft girder. It can be seen that, at the speed of 294 km/h (corresponds to $v_c = 1.56$ in Figure 5.13), the vehicle length excitation frequency is the same as the second-order vertical bending modal frequency. As a result, girder resonance is caused. At the speed of 182 km/h (corresponds to $v_c \approx 1.0$ in Figure 5.13), the vehicle length excitation frequency is the same as the first-order vertical bending modal frequency of the girder, but resonance has not occurred. This is because a short train with only five vehicles is considered; hence, the first modal resonance has not been fully developed. According to these results, the train speed has to be limited to 208 km/h in order to avoid resonance.

5.3.3.2 Case Study (b)

In the second case study, the weight of each vehicle is shared by eight pairs of air springs; the load spacing is $L_{as} = L_{car}/8$, and each load is $F/8$. This case is closer to the reality of maglev trains that impose evenly distributed loads to the track. The comparison of girder response with that of the first case is shown in Figure 5.15. The results show that both the deflections and accelerations caused by maglev distributed loads are very low. According to this result, the allowable train speed can reach $v_c = 3.0$ or even higher; the corresponding train speed is about 570 km/h.

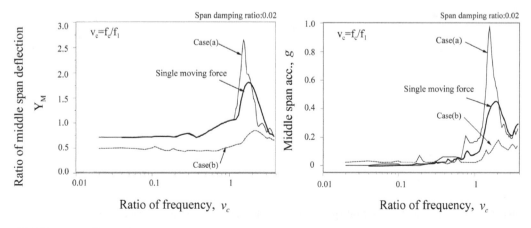

FIGURE 5.15 (Left and Right) Girder mid-span responses in case study (b).

5.3.3.3 Influence of Vehicle Load Distribution

In the following general cases, each of the concentrated forces that used to represent a vehicle is shared by N positions. The spacing of adjacent positions is L_{car}/N; each position takes the load of F/N. Obviously, the case of $N = 1$ corresponds to case study (a) discussed in Section 5.3.3.1, while the case of $N = 8$ corresponds to case study (b) discussed in Section 5.3.3.2. For maglev vehicles, the weight of each vehicle is evenly applied to the track via maglev supporting components. In dynamics analyses, the evenly distributed forces can be simplified and discretised into a certain number of smaller concentrated forces. N indicates the level of discretisation. The responses of a double-span simple girder under different discretisation levels ($N = 1 \sim 8$) are shown in Figure 5.16.

In Figure 5.16, the $N = 1$ case can be regarded as representing each vehicle using one concentrated load. When a long train is treated as a fleet of concentrated loads, the continuous passage of large loads fully excites girder vibrations, which becomes a situation that is extremely detrimental to both infrastructure and vehicle. In reality, a train consists of a limited number of vehicles; in the second simulation case study, previously examined, only five vehicles were considered, and the results were significantly lower than that of long trains. With the increase of vehicle numbers in the train, the girder response will significantly increase. The $N = 2$ case is similar to traditional wheel-rail railways that use two bogies for each vehicle. In this case, a vehicle load is simplified as two concentrated loads that are applied to the track at the positions of the bogies. Once v_c is greater than 1.0, the deflection and acceleration responses of the girder will increase significantly. When N is no smaller than 4, the deflection and acceleration responses of the girder basically stay the same; the response is about 50% lower than that in the case of $N = 2$.

5.4 THE METHODOLOGY FOR MAGLEV VEHICLE DYNAMICS ANALYSIS

The method of simplifying maglev trains to a sequence of forces moving on flexible beams only considers the dynamic response of the beam under nominal loads; it has reasonable rationality and accuracy and can be used to design and assess the maglev line girders. If the dynamics performance of the vehicle is of interest, detailed vehicle dynamics models are needed to replace the moving forces. Using the vehicle models, inertial forces of rigid bodies, suspension forces and damping forces among the bodies, external excitation forces of the vehicle as well as the dynamic response of the rigid bodies subjected to these forces can be presented. Therefore, a whole maglev vehicle-bridge dynamics system includes equations for flexible track and vehicles; these two parts are coupled via the levitation force equations.

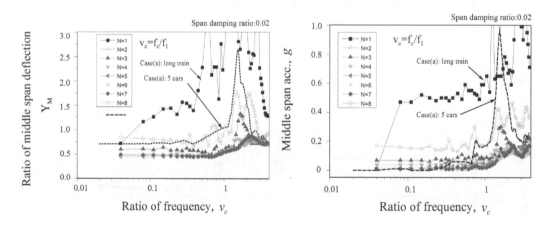

FIGURE 5.16 (Left and Right) Mid-span response under various levels of vehicle load distribution.

Magnetic Levitation Vehicles

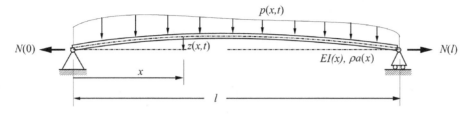

FIGURE 5.17 Beam subjected to static axial force and vertical dynamics forces.

5.4.1 Equations of Flexible Girder

5.4.1.1 Dynamic Equations of the Girder

A flexible girder of a rail transport viaduct bridge is a continuous system that has distributed parameters. The girders have a characteristic that their spans are significantly larger than their cross-sectional measurements. The Euler-Bernoulli beam theory can thus be used to develop the partial differential kinetics equations, and the equations can be solved using separation of variables and modal superposition methods. A very brief summary of the method is given in [15], and a much more detailed description can be found in [16]. For the beam shown in Figure 5.17, $N(x)$ represents the axial forces applied on the beam and a_1 is the structural Rayleigh damping factor, which is proportional to the stiffness. The external stiffness and damping applied on the unit length of the beam are $k(x)$ and $c(x)$, respectively. The structural bending stiffness is $EI(x)$, and the mass of each unit length is $\rho a(x)$. The bending deflection $z(x, t)$ is governed by the following equation:

$$\frac{\partial^2}{\partial x^2}\left[EI(x)\left(\frac{\partial^2 z}{\partial x^2}+a_1\frac{\partial^3 z}{\partial x^2 \partial t}\right)\right]-\frac{\partial}{\partial x}\left(N(x)\frac{\partial z}{\partial x}\right)+\rho a(x)\frac{\partial^2 z}{\partial t^2}+c\frac{\partial z}{\partial t}+kz=p(x,t) \qquad (5.5)$$

The physical meaning of Equation 5.5 is that, at an arbitrary time instant t and an arbitrary position x of the beam, the structural elastic force, structural damping force, geometrical stiffness force, external damping force and external elastic forces that are applied on an incremental section of the beam should be balanced with externally applied forces.

5.4.1.2 Solution of the Dynamic Equations

A system that has continuously distributed parameters has an infinite number of degrees of freedom (DoFs). It cannot be exactly solved but can be discretised into a finite number of DoFs and solved approximately. The methods for discretisation include lumped mass method, finite element method and generalised coordinate or mode superposition method. The following analysis uses the generalised coordinate method to study maglev vehicle-flexible bridge coupled dynamics. Using this method, the deflection of the girder at an arbitrary time instant can be approximated by using the first k orders of vibration mode expressed by $\varphi_i(x)$ and their corresponding generalised coordinates $Z_i(t)$ of the structure:

$$z(x,t)=\sum_{i=1}^{k}\phi_i(x)Z_i(t) \qquad (5.6)$$

Therefore, the dynamics equations of the beam have been converted to vibration modes and generalised coordinates of the beam.

5.4.1.3 Solving Vibration Mode and Frequency

For the case of a uniform beam that is not subjected to any external forces and shows undamped free vibrations, Equation 5.5 can be simplified as:

$$EI\frac{\partial^4 z(x,t)}{\partial x^4} + \rho a \frac{\partial^2 z(x,t)}{\partial t^2} = 0 \tag{5.7}$$

According to the separation of variables method, assume the general expression of the solution as:

$$z(x,t) = \phi(x)Z(t) \tag{5.8}$$

Substituting Equation 5.8 into Equation 5.7, the following equation can be derived:

$$\frac{d^4\phi(x)/dx^4}{\phi(x)} = -\frac{\rho a}{EI} \cdot \frac{d^2 Z(t)/dt^2}{Z(t)} = b^4 = \omega^2 \frac{\rho a}{EI} \tag{5.9}$$

where b is an intermediate parameter and ω is the vibration frequency.

For the ith mode $\phi_i(x)$, there is the relationship $b_i^4 = \omega_i^2 \frac{\rho a}{EI}$. An ordinary differential equation that is related to the vibration mode can be derived from Equation 5.9 as:

$$\frac{d^4\phi(x)}{dx^4} - b^4\phi(x) = 0 \tag{5.10}$$

The general expression of the solution for this equation is:

$$\phi(x) = A_1 \sin bx + A_2 \cos bx + A_3 \sinh bx + A_4 \cosh bx \tag{5.11}$$

Using the four boundary conditions from two ends of the beam, the relationships among four constants $A_1 \sim A_4$ can be determined, as well as the parameter b_i. Substituting b_i into Equation 5.11, the vibration mode $\phi_i(x)$ can be obtained. The corresponding vibration frequency can then be determined via Equation 5.9.

For a beam with a standard fixed end and a hinged end, the vibration modes that correspond to different mode frequencies satisfy the orthogonality conditions:

$$\int_0^l \rho a \cdot \phi_i(x)\phi_j(x)dx = 0, \quad \omega_i \neq \omega_j, \quad i,j = 1,2,\ldots,k \tag{5.12}$$

$$\int_0^l EI \cdot \phi_i''(x)\phi_j''(x)dx = 0, \quad \omega_i \neq \omega_j$$

Example 5.1: Single-Span Simple Beam

As shown in Figure 5.18a, substituting boundary conditions into Equation 5.11 leads to $A_2 = A_3 = A_4 = 0$, $A_1 \sin bl = 0$. In order to make the solutions of Equation 5.11 meaningful, $\sin bl = 0$ should be satisfied. The frequency equation is thus obtained:

$$bl = i\pi, \quad i = 1,2,\ldots,\infty$$

Magnetic Levitation Vehicles

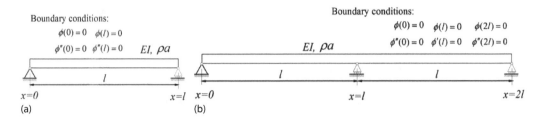

FIGURE 5.18 Boundary conditions of simple supported beams: (a) single-span girder and (b) double-span girder.

According to Equation 5.9, the mode frequencies are obtained as:

$$\omega_i = \frac{\pi^2}{l^2}\sqrt{\frac{EI}{\rho a}} \cdot i^2, \quad i = 1,2,...,\infty$$

Assuming that modal shape coefficient A_1 is 1.0, the vibration modes can then be expressed as:

$$\phi_i(x) = \sin\left(\frac{i\pi}{l} \cdot x\right)$$

Example 5.2: Double-Span Simple Beam

As shown in Figure 5.18b, the beam is divided into two sections and has six boundary conditions. It can be solved by dividing the vibration modes into dissymmetrical modes ($i = 1,3,5,...$) and symmetrical modes ($i = 2,4,6,...$). The dissymmetrical modes can be regarded as the even number modes of a single-span simple beam with the span length of $2l$; therefore:

$$\omega_i = \frac{\pi^2}{(2l)^2}\sqrt{\frac{EI}{\rho a}} \cdot (i+1)^2 = \frac{\pi^2}{l^2}\sqrt{\frac{EI}{\rho a}} \cdot \left(\frac{i+1}{2}\right)^2, \quad i = 1,3,...$$

Assuming the modal shape coefficient is 1.0, the vibration modes can then be expressed as:

$$\phi_i(x) = \sin\left(\frac{(i+1)\pi}{2l} \cdot x\right), \quad i = 1,3,...$$

For the symmetrical modes, only the $x \in (0,l)$ section of the beam needs to be solved; for the $x \in (l,2l)$ section, it satisfies $\phi(x) = \phi(2l - x)$. According to the four boundary conditions of the two ends of the $x \in (0,l)$ section, the constants can be determined as $A_2 = A_4 = 0$, and A_1 and A_3 satisfy the relationships of:

$$\begin{pmatrix} \sin bl & \sinh bl \\ \cos bl & \cosh bl \end{pmatrix} \begin{pmatrix} A_1 \\ A_3 \end{pmatrix} = 0$$

In order to make the solutions meaningful, the determent of the matrix should be equal to zero, and then, the frequency equation can be derived as:

$$\sin bl \cosh bl - \cos bl \sinh bl = 0$$

This is a transcendental equation that can be solved using numerical methods to determine parameter b_i. After this, ω_i, $i=2,4,...$ can be determined via Equation 5.9.

Assuming the modal shape coefficient is 1.0, the vibration modes can then be expressed as ($i = 2,4,\ldots$):

$$\phi_i(x) = \cosh(b_i l)\sin(b_i x) - \cos(b_i l)\sinh(b_i x), \quad x \in (0,l)$$
$$\phi_i(x) = \phi_i(2l-x), \quad x \in (l,2l)$$

$$b_2 = 3.926602/l, \quad b_4 = 7.068582/l,\ldots; \qquad \omega_i^2 = b_i^4(EI/\rho a).$$

5.4.1.4 Equations of Motion of the Beam Expressed Using Generalised Coordinates

Having obtained the vibration modes and frequencies, the equations of motion of the beam can be derived using generalised coordinates. For a uniform beam that is subjected to external loads, if the geometrical stiffness, external stiffness and external damping are not considered, Equation 5.5 can then be simplified as:

$$EI\frac{\partial^4 z(x,t)}{\partial x^4} + a_1 EI\frac{\partial^5 z(x,t)}{\partial x^4 \partial t} + \rho a\frac{\partial^2 z(x,t)}{\partial t^2} = p(x,t) \qquad (5.13)$$

Considering only the first k orders of vibration modes and substituting Equation 5.6 into Equation 5.13, the following equation can be derived:

$$\sum_{i=1}^{k} EI \cdot \frac{d^4\phi_i(x)}{dx^4} \cdot Z_i(t) + \sum_{i=1}^{k} a_1 EI \cdot \frac{d^4\phi_i(x)}{dx^4} \cdot \frac{dZ_i(t)}{dt} + \sum_{i=1}^{k} \rho a \cdot \frac{d^2 Z_i(t)}{dt^2} = p(x,t) \qquad (5.14)$$

Note the existence of Equation 5.9, that is, $d^4\phi_i(x)/dx^4 = b_i^4\phi_i(x)$. Pre-multiplying Equation 5.14 by $\phi_j(x)$ and integrating the product on the whole length of the beam, the following equation can be derived:

$$\sum_{i=1}^{k}\int_0^l \left[EI \cdot b_i^4 \phi_j(x)\phi_i(x)dx\right] \cdot Z_i(t) + \sum_{i=1}^{k}\int_0^l \left[a_1 EI \cdot b_i^4 \phi_j(x)\phi_i(x)dx\right] \cdot \dot{Z}_i(t)$$
$$+ \sum_{i=1}^{k}\left[\int_0^l \rho a \cdot \phi_j(x)\phi_i(x)dx\right] \cdot \ddot{Z}_i(t) = \int_0^l \phi_j(x)p(x,t)dx$$

Using the orthogonal relation of different modes (Equation 5.12) and the relationship expressed by Equation 5.9, the following equation can be derived for the nth order vibration mode ($i = j = n$):

$$\ddot{Z}_n(t) + 2\xi_n \omega_n \dot{Z}_n(t) + \omega_n^2 Z_n(t) = P_n(t)/M_n, \qquad n = 1,2,\ldots,k$$

where $P_n(t) = \int_0^l \phi_n(x)p(x,t)dx$ are the generalised loads, $\xi_n = a_1\omega_n/2$ is the damping ratio, $M_n = \rho a\int_0^l \phi_n^2(x)$ is the generalised mass and k is the number of generalised coordinates that are used to describe the motions of the beam (the first k orders of modes). For maglev vehicle–girder dynamics simulation, the total number of orders that needs to be considered is within 4 to 6, and reasonably good results can be achieved. Using generalised coordinates matrixes, the equations of motion for the beam can be expressed as:

$$\ddot{\mathbf{Z}}(t) = \Lambda \mathbf{Z}(t) + \Lambda_d \dot{\mathbf{Z}}(t) + \mathbf{P} \qquad (5.15)$$

where $\mathbf{Z} = [Z_1, Z_2, \ldots, Z_k]^T$, $\Lambda = \mathrm{diag}(\omega_1^2, \omega_2^2, \ldots, \omega_k^2)$, $\Lambda_d = \mathrm{diag}(2\xi_1\omega_1, 2\xi_2\omega_2, \ldots, 2\xi_k\omega_k)$ and $\mathbf{P} = [P_1/M_1, P_2/M_2, \ldots, P_k/M_k]^T$.

5.4.2 Vehicle System Dynamic Equations

The method to develop equations for maglev vehicles is almost the same as that for traditional wheel-rail vehicles, except that in maglev vehicle equations, the 'magnet wheels' are used instead of wheelsets. The vehicle system is a typical rigid MBS in which flexible deformation of structures is neglected. For a rigid MBS with holonomic constraints, the Newton-Euler method can be used to establish dynamic equations. This section briefly introduces the process according to [17].

For a vehicle with p bodies and q holonomic constraints, the total DoFs are $6p$, and among them, $f = 6p - q$ DoFs are independent. That means the motion of the system is described by f generalised coordinates. It needs f equations with only f independent variables.

For an arbitrary body i, the position of its CoM and the attitude (the rotational orientation) with respect to the inertia reference frame $\mathbf{e}^r$ are described as:

Position: $\mathbf{r}_i^r = \bar{r}_i^{rT} \mathbf{e}^r$

Attitude: $\mathbf{e}^r = A^{rb} \mathbf{e}^b$

where $\bar{r}_i^r = (r_{ix}, r_{iy}, r_{iz})^T$ is a 3 × 1 array representing the CoM position of body i with respect to $\mathbf{e}^r$ in three directions, respectively. A^{rb} is a 3 × 3 matrix representing the rotation tensor of the body fixed frame $\mathbf{e}^b$ with respect to $\mathbf{e}^r$. This matrix is determined either by three Cardan angles or by three Euler angles. Both $\mathbf{e}^r$ and $\mathbf{e}^b$ are 3 × 1 vector arrays, and the elements of the array form a unit vector. These two arrays represent the directions of the inertia reference frame and body fixed frame, respectively. The motion of a free body is described by six general coordinates corresponding to its six DoFs. These coordinates are expressed as a 6 × 1 array $\bar{z} = (r_{ix}, r_{iy}, r_{iz}, \alpha_i, \beta_i, \gamma_i)^T$. Or, in other words, the motion of a body is a function of $\bar{z}$.

From the position of the CoM, its translational velocities and accelerations are derived as:

$$\bar{v}_i = \frac{d\bar{r}_i}{dt} = \frac{\partial \bar{r}_i}{\partial \bar{z}} \frac{d\bar{z}}{dt} = J_{Ti}(\bar{z}) \cdot \dot{\bar{z}} \tag{5.16a}$$

$$\dot{\bar{v}}_i = J_{Ti}(\bar{z}) \cdot \ddot{\bar{z}} + K_{Ti}(\bar{z}) \dot{\bar{z}} \tag{5.16b}$$

where $J_{Ti}(\bar{z})$ is a 3 × 6 array, and $K_{Ti}(\bar{z}) = \partial(J_{Ti}\dot{\bar{z}})/\partial z$ is also a 3 × 6 array. The results of both equations are 3 × 1 arrays corresponding to three directions.

The angular velocity vector $\omega_i^{rb} = \bar{\omega}_i^{rbT} \mathbf{e}^r = (\omega_{ix}^{rb}, \omega_{iy}^{rb}, \omega_{iz}^{rb}) \mathbf{e}^r$ is obtained from the 3 × 3 matrix expression of the vectors, which are determined by the rotation tensor:

$$\tilde{\omega}_i^{rb} = \dot{A}_i^{rb} A_i^{rbT} = \begin{bmatrix} 0 & -\omega_{iz} & \omega_{iy} \\ \omega_{iz} & 0 & -\omega_{ix} \\ -\omega_{iy} & \omega_{ix} & 0 \end{bmatrix}$$

More discussions about rotation velocities and accelerations can be found in Section 2 of [18]. They can be expressed in a form as shown below [17,18]:

$$\bar{\omega}_i^{rb} = J_{Ri}(\bar{z}) \cdot \dot{\bar{z}} \tag{5.17a}$$

$$\dot{\bar{\omega}}_i^{rb} = J_{Ri}(\bar{z}) \cdot \ddot{\bar{z}} + K_{Ri}(\bar{z}) \dot{\bar{z}} \tag{5.17b}$$

where $J_{Ri}(\bar{z})$ is a 3 × 6 matrix, and $K_{Ri}(\bar{z}) = \partial(J_{Ri}\dot{\bar{z}})/\partial z$ is also a 3 × 6 matrix. The results of both equations are again 3 × 1 arrays corresponding to three directions.

In this analysis, the Newton-Euler method is applied to each body in all translational and rotational directions. The linear and angular momentum laws for body i are expressed as:

$$m_i \dot{\bar{v}}_i = \bar{f}_i, \quad i = 1 \sim p \tag{5.18a}$$

$$\bar{I}_i \dot{\bar{\omega}}_i + \tilde{\omega}_i \bar{I}_i \bar{\omega}_i = \bar{l}_i, \quad i = 1 \sim p \tag{5.18b}$$

where m_i is the body mass, $\bar{f}_i$ is the sum of all applied and constraint forces on the body, $\bar{I}_i$ is a 3 × 3 inertia tensor of the body with respect to $\mathbf{e}^r$ and $\bar{l}_i$ is the sum of applied and constraint torques on the body relative to its CoM.

Substituting Equations 5.16 and 5.17 into Equation 5.18, the $6p$ equations of motion for all p bodies in all translational and rotational directions are derived; their matrix form can be written as:

$$\bar{M}(\bar{z},t)\ddot{\bar{z}} + \bar{g}(\bar{z},\dot{\bar{z}},t) = \bar{q}(\bar{z},\dot{\bar{z}},t) \tag{5.19}$$

Some bodies have less than six generalised coordinates, and the whole system has f generalised coordinates. In Equation 5.19, $\bar{z}$ is an $f \times 1$ array, $\bar{M} = [m_i J_{Ti}, \bar{I}_i J_{Ri}]^T$ ($i = 1 - p$) is a $6p \times f$ matrix, J_{Ti} and J_{Ri} are both $3 \times f$ matrixes, $\bar{q} = [\bar{f}_i, \bar{l}_i]^T$ ($i = 1-p$) is a $6p \times 1$ array and $\bar{g}$ is also a $6p \times 1$ array comprising all remaining terms of Equation 5.18. Note that $\bar{g}$ actually consists of the gyroscopic and centrifugal forces.

The term $\bar{q}$ consists of applied forces $\bar{q}^e$ and constraint forces $\bar{q}^z$: $\bar{q} = \bar{q}^e + \bar{q}^z$. For a holonomic constraint, virtual work done by $\bar{q}^z$ should be zero, that is, $\bar{J}^T \bar{q}^z = \bar{0}$, where $\bar{J} = [J_{Ti}^T, J_{Ri}^T]^T$, $i = 1 \sim p$ is a $6p \times f$ matrix. Pre-multiplying Equation 5.19 by $\bar{J}^T$, the constraint forces are then completely eliminated and f equations with f independent variables are thus obtained. That leads to the dynamic equations of the vehicle:

$$M(\bar{z},t)\ddot{\bar{z}} + \bar{g}(\bar{z},\dot{\bar{z}},t) = \bar{q}(\bar{z},\dot{\bar{z}},t) \tag{5.20}$$

where $M(\bar{z},t) = \bar{J}^T(\bar{z},t)\bar{M}(\bar{z},t)$ is an $f \times f$ matrix, and $\bar{q}(\bar{z},\dot{\bar{z}},t) = \bar{J}^T(\bar{z},t)\bar{q}^e(\bar{z},\dot{\bar{z}},t)$ are generalised forces expressed by an $f \times 1$ array.

5.4.3 Levitation Forces

In maglev vehicle dynamics, except for the interacting forces between the track and the vehicle, all other suspension forces and constraint forces are the same as with traditional wheel-rail vehicles. They both adhere to the same suspension dynamics. The following contents will demonstrate the methods to determine vertical levitation forces when using different levitation theories.

5.4.3.1 EMS Levitation Forces

The electromagnetic levitation principle generates a magnetic field by energising the magnets on the vehicle underneath the guideway. It creates magnetic attraction forces that then lift the loads (the vehicles) up towards the guideway. Active controls are used to maintain a small normal magnetic gap, which is approximately 8–12 mm. Each measuring point can relate to an independent controller and a group of coils; this scheme forms a single magnet control. Another approach is to conduct Fourier analysis on the measured signals from two measuring points that are then used to individually adjust coil currents according to the output signals determined

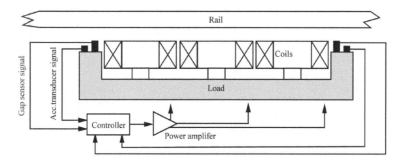

FIGURE 5.19 Principle of EMS levitation.

by using Fourier Synthesis. The second approach forms a modal observer-controller [19], as shown in Figure 5.19.

The levitation force that corresponds to a particular electromagnetic coil is determined by the coil current and air gap. The force can be expressed as $F_m = F_m(i,\delta)$, where $i(t)$ is the coil current and $\delta(t)$ is the levitation gap. Levitation force F_{0m} at a nominal working point (i_0,δ_0) balances the weight of the vehicle. The levitation force at a specific time instant can be regarded as the combination of a nominal levitation force and a variable component: $F_m = F_{0m} + \Delta F_m$. In the case without stray flux, the relationship between $\Delta F_m(t)$, gap increment $\Delta\delta(t)$ and voltage increment $\Delta u(t)$ can be expressed as a first-order linear differential equation [20]:

$$\Delta \dot{F}_m = d_1 \cdot \Delta F_m + d_2 \cdot \Delta\delta + d_3 \cdot \Delta u \tag{5.21}$$

where parameters $d_i (i = 1 \sim 3)$ are all known and determined by the designed parameters of the magnets and the locations of nominal working points (i_0,δ_0). The third item on the right of the equation is the control parameter that is obtained from the outputs of the levitation controller. In linearised systems, $\Delta F_m(t)$ can be considered as a state variable and $\Delta\delta(t)$ depends on state variables. However, in the simulations of non-linear systems, the calculated levitation force in each time step is applied to vehicle and track systems. And the levitation variable can be determined via the following equation [21,22]:

$$\Delta F(t) = k_\delta \Delta\delta(t) - k_i \Delta i \tag{5.22}$$

where k_δ and k_i are linear changing rates of levitation force due to gap and current changes, respectively, around a nominal working point (i_0,δ_0) in a small range. Δi is governed by a differential equation $\Delta \dot{i}(t) = e_1\Delta i(t) + e_2\Delta\dot{\delta}(t) + e_3\Delta u$. k_δ, k_i and e_i, $(i = 1 \sim 3)$ are all known and determined by the design parameters of the magnets and nominal working points (i_0,δ_0). Δu is the control parameter determined by the output of the levitation controller.

5.4.3.2 EDS Levitation Forces

Electric dynamic suspension has some typical characteristics such as on-board high-strength magnetic fields and high moving speed. The basic theory is to utilise the relative movements between the on-board magnetic field (usually superconducting or permanent magnets) and good non-ferromagnetic NCs or coils of the non-magnetic guideway to generate induced currents. The induced currents further generate mirrored magnetic fields. The levitation of vehicle body is achieved by using the repelling forces between the two magnetic fields, as shown in Figure 5.20.

Three magnetic forces exist between the maglev coils and mirrored coils, that is, levitation repelling force, lateral force and drag force. Experiments that were conducted focusing on a large number of coils have indicated that the geometry of the coils does not have any influence on the drag force F_{dD}. The lifting force F_{lD} can be approximated using the following empirical formulas [22]:

$$F_{lD}(v,z,t) = \eta F_{LD}(t) \tag{5.23a}$$

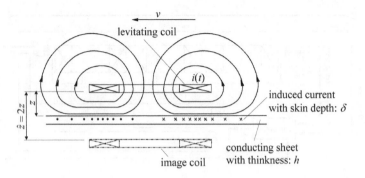

FIGURE 5.20 Principle of EDS levitation.

$$F_{dD} = \frac{w}{v} F_{lD}(v,z,t) \qquad (5.23b)$$

where η is a coil geometry-dependent parameter valued between 0.2 and 0.3; $w = 2/\mu_0\sigma h$ is a constant related to the thickness of the conductor sheet h, absolute magnetic permeability μ_0 and conductivity σ. $F_{LD} = i^2(t)[\mathrm{d}M(x,\hat{z},t)/\mathrm{d}\hat{z}]$, where $i(t)$ is the coil current, $M(x,\hat{z},t)$ is the mutual inductance between the actual and mirrored coils at distance $\hat{z}$ and x is the longitudinal distance to the fixed reference point. The longitudinal velocity is defined as $v = dx/dt$.

There are different approaches [23–26] to determine the magnetic forces when the EDS principle is used. As an example, Figure 5.21 shows the calculated levitation force and drag force of the MLU001 maglev vehicle using the stored energy variation method [23]. The drag force is proportional to speed in the low-speed region and reaches a peak value and then decreases inversely proportional to $\sqrt{v}$. In the high-speed region, it decreases inversely proportional to v, as shown in Equation 5.23b. The lifting force increases monotonically with speed. In the low-speed region, the lifting force is not enough to support the vehicle, and rollingstock wheels are needed to support the operation.

In order to avoid complicated calculations of magnetic field during vehicle-girder dynamic simulation, the magnetic forces can be pre-calculated or experimentally determined. They can then be used as input function arrays. Furthermore, in order to avoid vertical oscillation caused by a negative damping coefficient developed when lifting force becomes greater than the drag force, a small vertical damping should be added to the vehicle model.

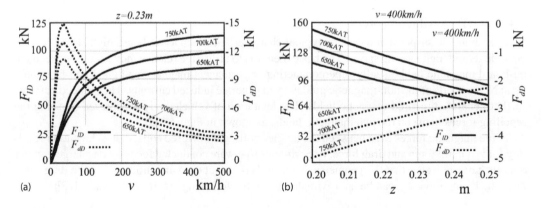

FIGURE 5.21 Levitation and drag forces versus: (a) longitudinal speed and (b) levitation height.

5.4.3.3 HTS Levitation Forces

A high-temperature superconducting bulk (HTSCB) cooled in a non-uniform magnetic field will be pinned in its initial position; in other words, the HTSCB will generate a restoring force if it experiences any change of field gradient. If the applied field is generated by a permanent magnet guideway, as shown in Figure 5.22, the field is non-uniform in guideway cross-section. Any motion disturbance of the HTSCB in the vertical or lateral direction causes changes of the field and its gradient, which lead to a corresponding restoring Lorentz force [27]. Along the guideway, the field is uniform and no field gradient will be experienced by the HTSCB; this leads to zero drag.

The fundamental features of the physics of superconductors are reviewed in [11]. It is pointed out that the levitation force cannot be represented as a position- and velocity-dependent force, but its time evolution involves the entire history of the motion from the moment when the superconductor shows superconducting phenomenon. In addition to this, the levitation and guidance force relate to lots of design parameters such as the size and arrangements of the permanent magnet guideway (PMG) cross-section and the HTSCB in vacuum cryostats. The first prototype HTS passenger vehicle was reported in [28]. A macroscopic modelling of a vehicle application of HTS levitation force and stiffness is reported in [29]. In general, for the dynamic simulation of an HTS vehicle, the levitation and guidance force are suggested to be pre-calculated or experimentally determined. They can then be used as input function arrays in a multi-body simulation model.

5.4.4 System Equations and Solutions

The whole vehicle-girder dynamic system consists of beam equations, vehicle equations and the levitating force laws between the vehicle and the guideway (girder). A general state-space form of the whole system can be expressed as:

$$\underline{\dot{x}} = \underline{A}(t)\underline{x} + \underline{D}\underline{u} + \underline{B}(t)\underline{b}$$

$$\underline{y} = \underline{C}(t)\underline{x}$$

where $\underline{x}$ is the state vector, $\underline{A}(t)$ is the system matrix, $\underline{D}$ is the control input matrix, $\underline{B}(t)$ is the disturbance matrix, $\underline{C}(t)$ is the measurement matrix, $\underline{y}$ is the measurement vector, $\underline{u}$ is the control input vector and $\underline{b}$ is the disturbance vector. Detailed application of the method can be found in [20] with an example for a 2-DoF car model (KOMET) with an EMS levitation control loop. A similar but more general case of maglev vehicle-guideway vertical interaction was reported in [19].

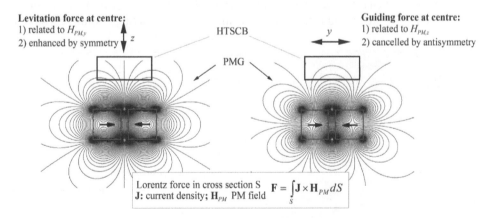

FIGURE 5.22 Principle of HTS levitation and guidance (one side of a track) showing magnetic field contours of the permanent magnet guideway (PMG) in cross-section: (a) lateral component $H_{PM,y}$ and (b) vertical component $H_{PM,z}$.

For non-linear numerical simulations, lots of researches were reported for maglev-girder interactions. Reference [21] has presented an example of an EMS vehicle that shows resonance at standstill (zero travelling speed) or very low speed. A flowchart for numerical simulation of a more comprehensive case, including the foundation-soil system, can be found in [30].

5.5 SIMULATION EXAMPLES OF MAGLEV BY MULTIBODY SYSTEM

Nowadays, an efficient and powerful simulation approach to study a maglev passing over a flexible guideway can be carried out by means of MBS simulation software packages. The development of some MBS software packages and their general purposes were reviewed in [31]. In this section, the MBS modelling method is introduced through typical examples with the help of MBS SIMPACK. The predecessor of SIMPACK was MEDYNA, which has been described in detail in [31]. Other MBS programs may share some similar modelling approaches. There are surely other modelling approaches for different considerations. The examples given here are just for reference.

5.5.1 Moving Loads on Flexible Beam

The simplest case of this type of simulation is by considering only one moving force. A similar modelling approach can also be used when simulating more complex cases such as multiple moving forces, masses and even sprung moving masses with periodical spaces between loads. The beam can be a single span, a double span, or even multi-span.

5.5.1.1 Single Moving Force Passing a Single-Span Beam

The model of a single moving force passing a single-span beam is shown in Figure 5.23a. The modelling steps are as follows:

1. Prepare an interface beam file for MBS main model using the 'beam_nt.exe' (or other program with the same purpose) module. The beam properties will be defined during the process (refer to SIMPACK Documentation: Interfaces).
2. Open a new model, modify the default body into a flexible body by loading the beam interface file built in Step 1 and set the global gravity to zero.
3. Add (define) a movable marker on the flexible beam body. The marker translates only along the x direction according to time excitation (constant velocity).
4. Define a dummy body (see the dark shape with an arrow attached in Figure 5.23). The dummy body has two dependent translation DoFs in the x and z directions. These two translation DoFs are locked with the movable marker defined in Step 3 with a constraint.

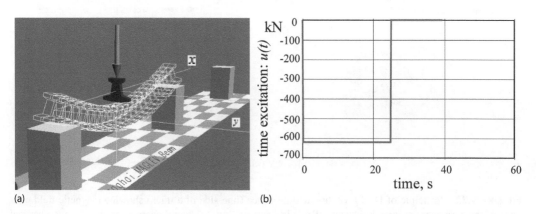

FIGURE 5.23 Modelling of single moving force on single-span beam: (a) model and (b) applied moving force.

Magnetic Levitation Vehicles

5. Define an applied force that has time excitation only in the z direction. The time excitation of this force is defined as having a 'constant second derivative' in order to make the force to be active only when passing over the beam; the time excitation is shown in Figure 5.23b. Set the switch point parameters according to span length and passing velocity, which enables the achievement of this parametric simulation.
6. Define a sensor for the middle marker of the span to record the results, as shown in Figure 5.10, Section 5.3.2.

5.5.1.2 Single Moving Force Passing a Double-Span Beam

This model is shown in Figure 5.24. Compared with the single-span model described in the previous section, the difference is the interface beam file, which is prepared according the properties of a double-span beam. Again, the applied force is active only when passing over the beam.

5.5.1.3 Multiple Moving Forces Passing a Double-Span Beam

This model is shown in Figure 5.25. Based on the single force modelling, the Steps 4 and 5 described in Section 5.5.1.1 are repeated for each applied force. Note that all defined movable markers on a flexible body share the same time excitation for the translational motion in the x direction, but each

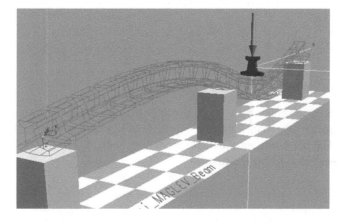

FIGURE 5.24 Single moving force on double-span beam.

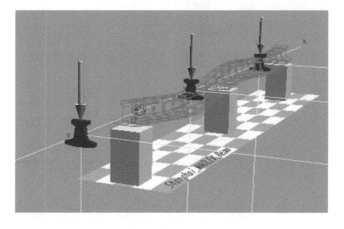

FIGURE 5.25 Multiple moving forces on double-span beam.

force has an individual time excitation. The switch on and off points for this should be set to ensure that each force is active only when it is passing over the beam.

The results shown in Figures 5.15 and 5.16 were obtained using this modelling method.

5.5.1.4 Modelling of Moving Lumped Mass

In dynamic modelling, the flexible beam has a very limited length. In the case of moving forces, the forces can be set to be active only when passing over the beam. For a moving lumped mass, it always causes inertia effects on the flexible beam if a lumped mass does not jump from one beam to another. Using contact pairs, the situation where the lumped mass is able to jump from one beam to another may occur. A simple way to solve this issue here is to configure a multi-section long beam, as shown in Figure 5.26b.

The corresponding model is shown in Figure 5.26a. The modelling is still mainly based on the previous considerations but with some differences as follow:

1. Beam interface file for the MBS main model is defined according to Figure 5.26b.
2. In the main model, add a new body to simulate the lumped mass, which is shown as the dark cube in Figure 5.26a. The mass can be fixed on a dummy body, as shown using the dark shape below the arrow, or the mass can move in the vertical direction with respect to this dummy body in the case of being supported by a force element such as a spring.
3. Global gravity is still zero, and the lumped mass only responds to vibration. Nominal weight is applied by using constant forces that correspond with the values of the weight.

5.5.2 MODELLING OF EMS LEVITATION CONTROL LOOP

A possible control loop as shown in Figure 5.27 can be built using SIMPACK as a substructure and then repeatedly loaded by the main vehicle model or be built via co-simulation between SIMPACK and SIMULINK. The procedure for the direct modelling method is shown in Figure 5.28 (refer to SIMPACK Documentation: Control elements).

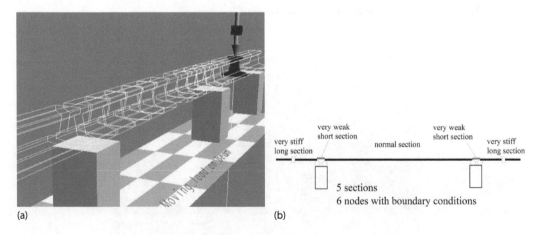

FIGURE 5.26 Modelling of moving lumped mass: (a) model and (b) special configuration of beam.

Magnetic Levitation Vehicles

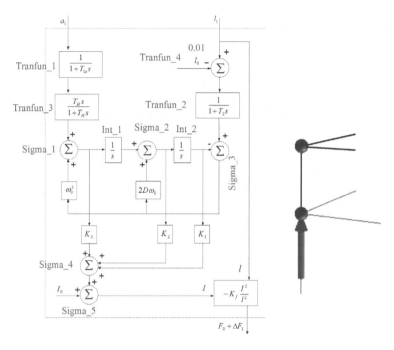

FIGURE 5.27 A possible EMS levitation control loop and substructure model.

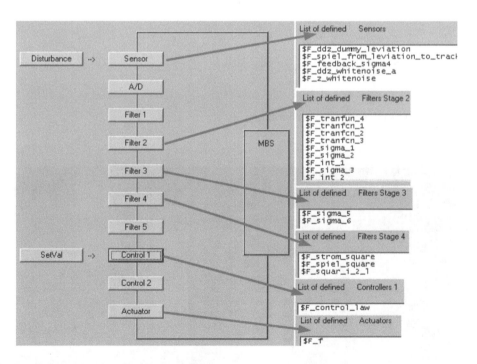

FIGURE 5.28 Procedure of modelling control loop directly using SIMPACK.

5.5.3 Modelling of Vehicle-Girder Interaction

Except for the difference between vehicles and flexible guideways, the building of a vehicle-girder model is very similar to that in conventional wheel-rail vehicle models. In order to represent the forward translation of the whole vehicle without the rolling wheels, a translational platform is considered to support the whole vehicle, as shown in Figure 5.29. The platform can represent the track or rail irregularities of a rigid track layout or just as a forward state if the vehicle runs on a flexible girder. In the latter case, the weight of the vehicle will be supported by moving markers defined on the flexible girder.

The following different models can be built according to whether the control loop is used and whether a rigid or flexible girder track is considered:

- Mechanical model on rigid track with irregularities and arbitrary geometrical layout
- Mechanical model on straight flexible girder
- Mechatronic model on rigid track with irregularities and arbitrary geometrical layout
- Mechatronic model on straight flexible girder, as shown in Figure 5.30

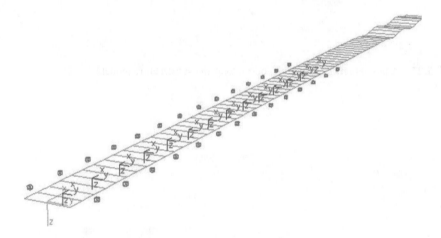

FIGURE 5.29 Forward translational platform.

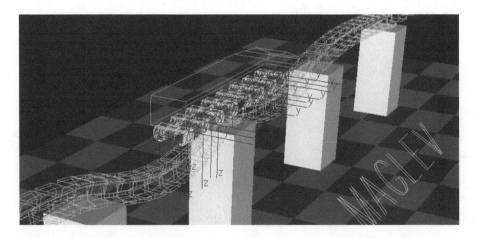

FIGURE 5.30 Mechatronic model of vehicle on a multiple-span flexible girder.

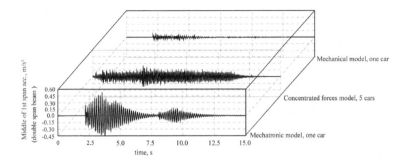

FIGURE 5.31 Different vehicle models passing over flexible girder.

Each type of model can simulate some interesting problems but also has its limitations of course. A comparison among different models is shown in Figure 5.31. It can be seen that resonance trends between vehicle and beam can obviously appear with the mechatronic model but only at a very low speed of 37 km/h, which will not cause resonance by the mechanical model.

5.6 CONCLUDING REMARKS

Nothing is perfect in the world. This philosophy also applies to maglev trains that are using three different basic maglev principles.

For AS EMS levitation vehicles, the most difficult problem is to effectively suppress the interaction between vehicle and flexible girder at a reasonably low cost. Significant efforts have been made by optimising control strategies, but these have still not been sufficient to allow low cost construction of guideways. Just as the bogies of a conventional wheel-rail vehicle have great influences on dynamic interaction between train and track, levitation frames for EMS maglev vehicles should also play a similarly important role to realise low dynamic interaction between EMS maglev vehicles and their elevated guideway. A recent article [32] has reported on its feasibility and gives an optimistic prospect for this type of vehicle.

Passive-support EDS levitation vehicles are suitable for HS applications. However, the magnetic drag forces are still considerable at the present operational speed range of 400 ~ 600 km/h. Evacuated tube technology (ETT) is better for reducing drag forces to realise higher speeds such as those above 1000 km/h, but this technology is still under development, and attention should also be paid to the cost of obtaining satisfactory vacuum condition. For EDS vehicles, running gear with wheels is necessary when operating at low speeds or when at standstill.

For AS HTS levitation vehicles, magnetic drag forces no longer exist, but the levitation capability should be further improved to meet the mass transport requirements, and the levitation frames should be very carefully designed to reduce the dynamic loads. Another issue with this type of vehicle is that they run on rare-earth permanent magnet guideways. Sufficient experience in this regard has not been gained for open environment applications. In an ETT application environment, again at a very high speed, the dynamic loads should not be too high.

REFERENCES

1. J. Klühspies, The Maglev case: Prospects and limitations, *22nd International Conference on Magnetically Levitated Systems and Linear Drives (Maglev 2014)*, 28 September–1 October 2014, Rio de Janeiro, Brazil, 2014.
2. X.M. Wu, W.S. Chang, W.M. Liu, Shanghai MAGLEV train and discussion of technology development, *Chinese Journal of Comprehensive Transportation*, 1, 28–31, 2005.

3. K. Sawada, M. Murai, M. Tanaka, Magnetic levitation (Maglev) technologies, *Japan Railway & Transport Review*, 25, 58–67, 2000.
4. K. Sawada, Outlook of Maglev Chuo Shinkansen, *22nd International Conference on Magnetically Levitated Systems and Linear Drives (Maglev 2014)*, 28 September–1 October 2014, Rio de Janeiro, Brazil, 2014.
5. A.S. chute, The big picture news, *IEEE Spectrum*, 51(7), 20–21, 2014.
6. SpaceX, Hyperloop Alpha, http://www.spacex.com/sites/spacex/files/hyperloop_alpha.pdf, 2013.
7. Virgin Hyperloop One, New chairman, new funding, & new speed records, https://hyperloop-one.com/blog/new-chairman-new-funding-new-speed-records, 2017.
8. R.D. Thornton, T. Clark, B. Perreault, Linear synchronous motor propulsion of small transit vehicles, *Proceedings ASME/IEEE Joint Rail Conference*, 6–8 April 2004, Baltimore, MD, RTD2004-66020, 101–107, 2004.
9. S. Zhang, *Study on the Design Methods of High-Speed Trains*, China Railway Press, Beijing, China, 2009. (in Chinese)
10. Y. Hsu, D. Ketchen, L. Holland, A. Langhorn, D. Minto, D. Doll, Status of the magnetic levitation upgrade to Holloman high speed test track, *20th International Conference on Magnetically Levitated Systems and Linear Drives (Maglev 2008)*, 15–18 December 2008, San Diego, CA, 2008.
11. K.B. Ma, Y.V. Postrekhin, W.K. Chu, Superconductor and magnet levitation devices, *Review of Scientific Instruments*, 74(12), 4989–5017, 2003.
12. R.M. Goodall, R.A. Williams, Dynamic criteria in the design of Maglev suspension systems, *International Conference on Maglev Transport Now and For the Future*, 9–12 October 1984, Solihull, UK, IMechE, C393/84, 77–86, 1984.
13. K.C. Coates, Shanghai's Maglev project: Levitating beyond transportation theory, *Engineering World*, 26–33, 2005.
14. M. Nagai, M. Iguchi, Vibrational characteristics of electromagnetic levitation vehicles-guideway system, In H.P. Willumeit (Ed.), *Proceedings 6th IAVSD Symposium held at Technical University Berlin*, 3–7 September 1979, Swets and Zeitlinger, Lisse, the Netherlands, 352–366, 1980.
15. W. Kortüm, Vehicle response on flexible track, *International Conference on Maglev Transport Now and For the Future*, 9–12 October 1984, Solihull, UK, IMechE, C405/84, 47–58, 1984.
16. R. Clough, J. Penzien, *Dynamics of Structures*, 2nd ed., Computers and Structures Inc., Berkeley, CA, 2003.
17. W. Kortüm, Introduction to system-dynamics of ground vehicles, *Vehicle System Dynamics*, 16(S1), 1–36, 1987.
18. A.A. Shabana, *Dynamics of Multibody Systems*, 3rd ed., Cambridge University Press, New York, 2005.
19. E. Gottzein, L. Miller, R. Meisinger, Magnetic suspension system for high speed ground transportation vehicles, *World Electrotechnical Congress*, 21–25 June 1977, Moscow, USSR, 1678–1724, 1977.
20. R. Meisinger, Vehicle-guideway dynamics of a high-speed MAGLEV train. *Proceedings of International Conference of Cybernetics and Society*, 7–9 October 1980, Cambridge, MA, 1028–1035, 1980.
21. K.J. Kim, J.B. Han, H.S. Han, S.J. Yang, Coupled vibration analysis of Maglev vehicle-guideway while standing still or moving at low speeds, *Vehicle System Dynamics*, 53(4), 587–601, 2015.
22. P.K. Sinha, *Electromagnetic Suspension Dynamics & Control*, Peter Peregrinus Ltd., London, UK, 1987.
23. N. Carbbonari, G. Martinelli, A. Morini, Calculation of levitation, drag and lateral forces in EDS-MAGLEV transport systems, *Archiv für Elektrotechnik*, 71(2), 139–148, 1988.
24. S. Nonaka, T. Hirosaki, E. Kawakam, Analysis of characteristics of repulsive magnetic levitated train using a space harmonic technique, *Electrical Engineering in Japan*, 100(5), 80–88, 1980.
25. Y. Iwasa, High speed magnetically levitated and propelled mass ground transportation, Chapter 6 in S. Foner, B. Schwartz (Eds.), *Superconducting Machines and Devices: Large Systems Applications*, Plenum Press, New York, 347–399, 1974.
26. T. Saijo, Thrust and levitation force characteristics of linear synchronous motors, *Proceedings of International Conference on Maglev & Linear Drives*, 14–16 May 1986, Vancouver, Canada, 157–164, 1986.
27. G.T, Ma, H.F. Liu, J.S. Wang, S.Y. Wang, X.C. Li, 3D modeling permanent magnet guideway for high temperature superconducting Maglev vehicle application, *Journal of Superconductivity and Novel Magnetism*, 22(8), 841–847, 2009.

28. J. Wang, S. Wang, Y. Zeng, H. Huang, F. Luo, et al., The first man-loading high temperature superconducting Maglev test vehicle in the world, *Physica C, Superconductivity and Its Applications*, 378–381, 809–814, 2002.
29. C. Navau, N. Del-Valle, A. Sanchez, Macroscopic modeling of magnetization and levitation of hard type-II superconductors: The critical-state model, *IEEE Transactions on Applied Superconductivity*, 23(1), 8201023, 2013.
30. Y.B. Yang, J.D. Yau, An iterative interacting method for dynamic analysis of the Maglev train-guideway/foundation-soil system, *Engineering Structures*, 33(3), 1013–1024, 2011.
31. W. Kortüm, W. Schiehlen, General purpose vehicle system dynamics software based on multibody formalisms, *Vehicle System Dynamics*, 14(4–6), 229–263, 1985.
32. M. Zhang, S. Luo, C. Gao, W. Ma, Research on the mechanism of a newly developed levitation frame with mid-set air spring, *Vehicle System Dynamics*, 56(12), 1797–1816, 2018.

6 Suspension Elements and Their Characteristics

Sebastian Stichel, Anna Orlova, Mats Berg and Jordi Viñolas

CONTENTS

6.1 Introduction 198
6.2 Elastic Elements 198
    6.2.1 Coil Compression Helical Springs 199
        6.2.1.1 Stiffness Calculations (Spring Rate) 202
        6.2.1.2 Modelling 205
    6.2.2 Rubber Springs 206
        6.2.2.1 One-Dimensional Models 207
        6.2.2.2 Multi-Dimensional Models 209
    6.2.3 Air Springs 209
        6.2.3.1 Effect of Amplitude 211
        6.2.3.2 Effect of the Connection Pipe 212
        6.2.3.3 Effect of the Auxiliary Air Chamber 212
        6.2.3.4 Effect of Orifice Plates in the Connection Pipe 212
        6.2.3.5 Equivalent Mechanical Models 214
        6.2.3.6 Thermodynamic Models 216
        6.2.3.7 Air Spring Models in the Horizontal Plane 218
        6.2.3.8 Definition of Air Spring Model Parameters 219
    6.2.4 Leaf Springs 221
6.3 Dampers 222
    6.3.1 Viscous Dampers 223
    6.3.2 Friction Dampers 225
6.4 Constraints and Bumpstops 228
    6.4.1 Horn Guides 229
    6.4.2 Cylindrical Guides 229
    6.4.3 Beam Links 230
    6.4.4 Constraints Using Radius Links 231
    6.4.5 Constraints Using Trailing (Radial) Arms 232
    6.4.6 Traction Rods 232
6.5 Car Body to Bogie Connections 233
    6.5.1 Flat Centre Bowl and Side Bearings 233
    6.5.2 Spherical Centre Bowl and Side Bearings 234
    6.5.3 Centre Column 234
    6.5.4 Watts Linkage 235
    6.5.5 Pendulum Linkage 236
    6.5.6 Connection of Car Body to Bolsterless Bogie 236
References 238

6.1 INTRODUCTION

The *suspension* is the set of elastic elements, dampers and associated components that connect wheelsets to the car body. Even linkages like traction rods are often regarded as suspension elements. The most common suspension components are coil springs, friction-based components such as leaf springs, rubber springs, air springs and hydraulic dampers. Secondary suspensions' air springs have replaced most of the coil springs due to the advantage of system natural frequencies being almost independent of payload variation and a good high-frequency vibration isolation.

Suspension elements are designed to have certain force-displacement or, in the case of dampers, force-velocity characteristics to guarantee running safety and to provide good ride characteristics to the vehicle. Force characteristics can be linear or non-linear. In linear characteristics, the deflection is proportional to the force, as shown in Figure 6.1a [1]. For non-linear characteristics, the deflection rate increases (or less often for railway applications, decreases) with an increase of the load (Figure 6.1a). The characteristics can include a so-called hysteresis, as shown in Figure 6.1b. Hysteresis means that the force-displacement curve does not follow the same path for the loading and unloading parts of the cycle. The area described inside the curve determines the amount of damping in the suspension element. In Figure 6.1b, a (quasi-)static stiffness is also defined. This stiffness is the tangent stiffness when virtually all spring friction has been released. Suspension elements, as well spring as damper elements, can also be stepwise linear and even have so called 'dead bands'. A typical example is a bumpstop characteristic.

If the bogie has a rigid frame, the suspension usually consists of two stages: a primary suspension connecting the wheelsets to the bogie frame and a secondary suspension between the bogie frame and the bolster or car body (see also Chapter 3). Such bogies are termed double-suspended. Sometimes, typically in freight bogies, only a single-stage suspension is used. Where this occupies the primary suspension position, it is often termed an 'axlebox suspension'. In the secondary suspension position, it may be termed a 'centre suspension'.

See also [2] and Section 17.2.3.

6.2 ELASTIC ELEMENTS

Elastic elements are components that return to their original position when forces causing them to deflect are removed. Elastic elements are used to:

- Equalise the vertical loads between wheels (unloading of any wheel causes a reduction/loss of guidance forces)

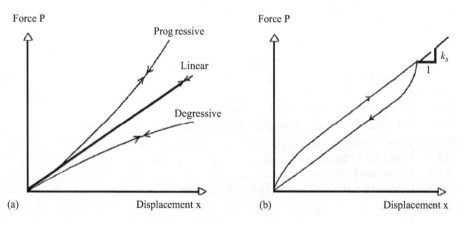

FIGURE 6.1 Force-displacement graphs for quasi-static loading and unloading: (a) elastic behaviour (linear, progressive and degressive) and (b) hysteresis and definition of static stiffness k_s. (From Andersson, E. et al., *Rail Vehicle Dynamics*, Royal Institute of Technology (KTH), Stockholm, Sweden, 2014.)

Suspension Elements and Their Characteristics

- Stabilise the motion of vehicles on track (self-excited lateral oscillations, i.e., hunting of wheelsets is dangerous, as discussed in Chapter 2)
- Reduce the dynamic forces and accelerations due to track irregularities

The capability of elastic elements to provide the above functions is determined by their *force-displacement characteristic*, which is the dependence between the force acting on the elastic element P and its displacement z: $P = P(z)$. The principal types of elastic elements are shown in Table 6.1.

Coil springs and air springs have specific properties and, in what follows, the important design variables of these components will be highlighted. When they are used as part of the suspension, stiffness and damping are definitely two of the important parameters. In spring standards, 'stiffness' is often split into a static one and a dynamic one. Then, great care has to be taken regarding the definition of 'dynamic stiffness'. As an example, a single (vertical) dynamic stiffness value is not sufficient in modelling most air spring systems [2].

For air springs and most coil springs, significant horizontal effects take place, and this means that one-dimensional vertical models are not sufficient. Thus, horizontal forces and bending moments are introduced at the interfaces, with the bodies connected, and the destabilising effect of the compressive preload needs to be considered. For some metal-to-metal interfaces, two-dimensional friction sliding is possible. Spring-dashpot-friction suspension models sometimes resemble the component's physical appearance but can also be a pure mathematical representation where such a direct coupling cannot be identified. Often the suspension models include internal degrees of freedom (or first-order differential equations), that is, to imitate, say, the emergency spring of an air spring system.

6.2.1 Coil Compression Helical Springs

According to EN 13906-1 [3], a helical coil spring is a *'Mechanical device designed to store energy when deflected and to return the equivalent amount of energy when released that offers resistance to a compressive force applied axially made of wire of circular, non-circular, square or rectangular cross-section, or strip of rectangular cross-section, wound around an axis with spaces between its coils'*.

In this subsection, the definition of parameters is the same as in the cited standard:

d (mm)	Nominal diameter of wire (or bar)
D (mm)	Mean diameter of coil
E (N/mm² – MPa)	Young's modulus
F (N)	Spring force
G (N/mm² – MPa)	Shear modulus
L (mm)	Spring length
L_0 (mm)	Spring nominal free length
n	Number of active coils
R (N/mm)	Spring rate
R_Q (N/mm)	Transverse spring rate
s (mm)	Spring deflection
s_Q (mm)	Transverse spring deflection, for the transverse force F_Q
w	Spring index, D/d
η	Spring rate ratio, R_Q/R
λ	Slenderness ratio, L_0/D
υ	Seating coefficient (buckling)
ξ	Relative spring deflection, s/L_0
ρ (kg/dm³)	Density

TABLE 6.1
Principal Types of Elastic Elements

Features	Schematic of Elastic Elements							
	A	B	C	D	E	F	G	H
Application	Attenuation of vibrations and impacts in one direction	Multi-directional attenuation of vibrations	Multi-directional attenuation of vibrations	Multi-directional attenuation of vibrations and impacts	Attenuation of vibrations and impacts in one direction	Multi-directional attenuation of vibrations and impacts	Multi-directional attenuation of vibrations and impacts; various characteristics are possible	Attenuation of vibrations and impacts in one direction
Advantages	Stiffness elastic and damping properties in a single component	Good vibration and noise insulation	Good vibration and noise insulation	Easy production and maintenance, small mass and dimensions	Possible to obtain large deflections, small mass and dimensions	Damping of high-frequency vibrations, stiffness and damping in a single component	Ability to vary stiffness and maintain constant ride height, good noise and vibration insulation	Easy to maintain and repair
Disadvantages	Unpredictable damping	Complicated control of parameters, production and maintenance	Complicated control of parameters, production and maintenance	Inability to vary characteristics in operation	Complicated production, vertical displacements cause longitudinal displacements	Ageing of rubber, inability to vary characteristics in operation	Complicated air supply system and maintenance	Wear, large size

Source: With kind permission from CRC Press: *Handbook of Railway Vehicle Dynamics*, CRC Press, Boca Raton, FL, 2006, Iwnicki, S. [ed.].

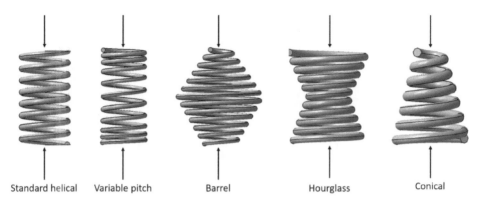

FIGURE 6.2 Shapes of helical springs.

TABLE 6.2
Material Properties for the Calculation of Some Steel Springs

Material	E (MPa)	G (MPa)	ρ (kg/m³)
Spring steel wire according to EN 10270-1	206000	81500	7850
Spring steel wire according to EN 10270-1	206000	79500	7850
Steels according to EN 10089	206000	78500	7850

Compression springs can also have different shapes other than helical, as Figure 6.2 shows.

Different material can be used in the manufacturing of springs; however, steel is the most common. Table 6.2 provides the values of some steels used to manufacture springs.

Depending on the deformations they are subjected to, modelling of coil springs is more difficult than initially expected. In many cases, one has to consider more than just the axial (vertical) stiffness due to the coupling between different directions and the reaction moments caused by the non-axial displacement of the vertical force. In addition, the compressive preload provokes destabilising effects in the transverse plane, and, consequently, some cases require a particular analysis of buckling limits (critical axial load or critical axial deflection), which depend on how the coil spring is able to deform and how its ends are fixed. More details about how different types of seating (and associated seating coefficients, ν) influence the buckling of axially loaded springs are found in EN 13906-1 [3].

The simplest way to model a coil spring is by a single, linear stiffness (which corresponds to the axial direction). Three-dimensional models include three perpendicular springs. Non-linearities associated with bumpstops or vertical non-linearity, which commonly appears in the primary suspension due to inner coil springs with vertical clearance, have to be included. However, these simple models frequently need an upgrade that includes transverse effects to represent adequately realistic coil spring suspensions. Transverse and bending stiffnesses play an important role (e.g., coil springs located in the secondary suspension of a railway vehicle), as transverse shear deformation may compromise vehicle dynamics.

The contribution of coil springs to damping can be neglected as load/unloading curves coincide, showing nearly no energy dissipation/hysteresis. In addition, the dynamic behaviour of coil springs used in railway applications has nearly no dependency on frequency in the range of 0–20 Hz. For higher frequencies, some analytical formulae might be used to find the dynamic response of the spring [4,5]; however, finite element models can also be used. Figure 6.3 [2] compares analytical and numerical solutions for a typical rail vehicle spring and shows that, for frequencies lower than 20 Hz, the coil spring can be modelled using its static properties.

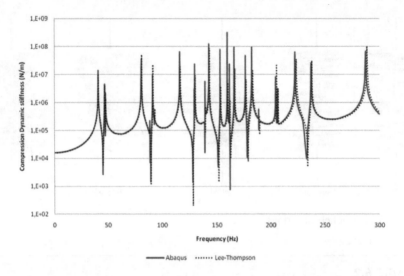

FIGURE 6.3 Analytical (Lee-Thompson [5])) and numerical (Abaqus) solutions for the axial dynamic stiffness of a typical rail vehicle coil spring. (From Bruni, et al., *Vehicle. Syst. Dyn.*, 49, 1021–1072, 2011. With permission.)

6.2.1.1 Stiffness Calculations (Spring Rate)

The European standard EN 13906-1:2013 [3] provides different methods for the calculation and design of cylindrical helical springs made from round wire and bar. It includes transverse buckling and impact loading, stress correction factors over the cross-section of the wire, material property values for the calculation of the spring and a review of the formulas needed in the design process.

Axial stiffness, spring rate R, is estimated using a simple expression that depends on geometric data of the coil spring.

$$R = \frac{Gd^4}{8D^3 n} \tag{6.1}$$

The following equation is also proposed for the transverse spring rate, R_Q; however, the standard notes that the '*transverse spring rate is only constant for short transverse spring deflections, s_Q, for a given length L under compression*'.

$$R_Q = R\,\xi \left[\xi - 1 + \frac{\frac{1}{\lambda}}{\frac{1}{2} + \frac{G}{E}} \sqrt{\left(\frac{1}{2} + \frac{G}{E}\right)\left(\frac{G}{E} + \frac{1-\xi}{\xi}\right)} \tan\left\{\lambda\xi\sqrt{\left(\frac{1}{2} + \frac{G}{E}\right)\left(\frac{G}{E} + \frac{1-\xi}{\xi}\right)}\right\} \right]^{-1} \tag{6.2}$$

where λ is the slenderness ratio L_0/D, and ξ is the relative spring deflection s/L_0.

Based on Timoshenko's work (see [6–10]), the concept for the Haringx model [7] is the division of the spring into small elements consisting of ordinary, linear springs; it assumes a small helix angle. By integration, the relationship between axial and shear stiffness is obtained.

For the coil spring of Figure 6.4 [2] to be in equilibrium when compressed between two non-parallel plates, the equations for axial and lateral reactions at the seats are:

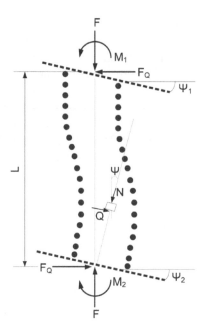

FIGURE 6.4 Spring subjected to axial and shear forces and moments at both ends. (From Bruni, S. et al., *Vehicle. Syst. Dyn.*, 49, 1021–1072, 2011. With permission.)

$$F = \frac{Gd^4(L_0 - L)}{8nD^3} \quad (6.3)$$

$$F_Q = \frac{F(\psi_1 - \psi_2)}{\left(\dfrac{4G}{E} + 2\right)\left(\dfrac{L_0 - L}{qD^2 \tan\dfrac{qL}{2}}\right) - 2} \quad (6.4)$$

$$M_1 = \frac{EGqLd^4}{16(2G+E)nD \tan\dfrac{qL}{2}}\left\{\psi_1 + \frac{F_Q}{F}\left[1 - \left(\frac{4G}{E} + 2\right)\left(\frac{L_0 - L}{qD^2 \sin qL}\right)\right]\right\} \quad (6.5)$$

$$M_2 = M_2 + F_Q L \quad (6.6)$$

where D is the mean coil diameter, n is the number of active coils, d is the nominal diameter of wire, E is Young's modulus, G is shear modulus, L_0 is the unloaded spring length, Ψ_1 and Ψ_2 are the angles of the ends of the spring or seat angles when the spring is seated and q is the so-called *buckling factor*. Buckling refers to the loss of stability of a component and is usually independent of material strength. In this case, it can be calculated as:

$$q = \frac{1}{D}\sqrt{\left[\left(\frac{2G}{E}\right)^2 + \frac{2G}{E}\right]\left[\frac{E}{G}\left(\frac{L_0}{L} - 1\right) + \left(\frac{L_0}{L} - 1\right)^2\right]} \quad (6.7)$$

Krettek and Sobczak worked on the same problem [8]. They estimated different correction factors following extensive tests that relate the axial and shear deformation for a given loading factor, as they found differences between experiments and previously proposed formulae. These correction factors, a_i, depend on the relative axial deformation of the spring and are foreseen for slenderness ratios between 1.5 and 3 as:

$a_1 = 1.9619\xi + 0.6740$	Correction factor for lateral stiffness
$a_2 = 1.3822\xi + 0.6513$	Correction factor for mixed term (lateral/bending coupling)
$a_3 = 0.8945\xi + 0.6690$	Correction factor for bending stiffness and the spring deflection ratio, ξ, is given by: $\xi = \dfrac{s}{L_0} = \dfrac{L_0 - L}{L_0}$

They also provided the following formulae for lateral and bending stiffness $\left(K_y^* \text{ and } K_\psi^*\right)$:

$$K_y^* = a_1 K_y = a_1 \left(\frac{F}{\dfrac{2}{c}\left(1 + \dfrac{F}{R_Q'}\right)\tan\left(\dfrac{cL}{2}\right) - L} \right) \tag{6.8}$$

$$K_\psi^* = a_3 K_\psi = a_3 \left(K_y \frac{R_\psi'}{F}\left[\left(1 + \frac{F}{R_Q'}\right) - \frac{cL}{\tan cL}\right] \right) \tag{6.9}$$

The contribution of bending angle to lateral force and the contribution of lateral displacement to bending moment can be included using a 'mixed term', this being:

$$K_{y\psi}^* = a_2 K_{y\psi} = a_2 \left(K_y \frac{c R_\psi'}{F} \tan \frac{cL}{2} \right) \tag{6.10}$$

where c is a factor that relates the axial load with the bending and shear properties of the spring through the equation:

$$c^2 = \frac{F}{R_\psi'}\left(1 + \frac{F}{R_Q'}\right) \tag{6.11}$$

and

$$R_\psi' = \frac{Ed^4 L}{32nD(v+2)} \tag{6.12}$$

$$R_Q' = \frac{Gd^4 L}{4nD^3}(v+1) \tag{6.13}$$

$$R' = R.L = \frac{Gd^4 L}{4nD^3} \tag{6.14}$$

Figure 6.5 [2] compares the lateral stiffness for a particular coil spring using the different approaches. This coil spring has a slenderness ratio of 1.63; in the case of higher values of λ, the results of the Timoshenko formulae would give less accurate results. The influence of the correction factors proposed by Krettek and Sobczak is also appreciated, which justifies why, in applications where the transverse stability of the spring is an important operational factor, the calculated values should be verified by practical tests, as suggested by [3].

Suspension Elements and Their Characteristics

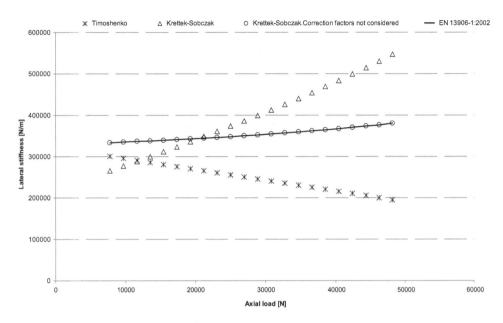

FIGURE 6.5 Lateral stiffness (transverse spring rate) of a coil spring versus axial load, F; coil spring data: $L_0 = 0.32$ m; $d = 0.036$ m; $D = 0.196$ m; and $n = 4.5$. (From Bruni, S. et al., *Vehicle. Syst. Dyn.*, 49, 1021–1072, 2011. With permission.)

6.2.1.2 Modelling

There are two main options to model spring components in general in a multi-body vehicle model [11]. The first is the so-called point-to-point force element (PtP), which exerts only an axial force along the line of action. The second is the compact force element (Cmp), which enables axial, shear forces and reaction moments. In both kinds of elements, linear and non-linear characteristics can be modelled; if necessary, pre-compression can also be accounted for.

What all the force elements have in common is being mass-less. The mass and inertia of the spring component are very small compared with other vehicle bodies. The options are to neglect the masses or to share them among the bodies connected to retain the mass of the actual vehicle.

The PtP elements act along the connecting line of their coupling markers (Mi, Mj), with all their outputs (forces/torques) applied in this direction. An example is shown in Figure 6.6 [2]. At $t = 0$, Mi and Mj are coupled markers defining the line (Mi-Mj) in which the forces are acting. At $t = t_1$, Bi moves towards the final position defining the final position of marker Mi, which is Mi'; Bj does not change its position. Consequently, at t_1, a different line of action is defined (Mi'-Mj); that is, the direction of the acting force has changed, as well as its magnitude, because of the new distance between those markers (|Mi'-Mj|).

Compact spring elements allow the user to take into consideration the three coordinate directions: X, Y and Z. In addition, the stiffness curve of each direction can be different. In this case, the reaction moments are not neglected; therefore, moments are generated from the offset, as indicated in Figure 6.7 [2]. The reaction moments can be distributed between the connected bodies. This element is recommended when bending and reaction moments are important.

Frequently, the coil spring has rubber seats at both ends in order to reduce vibration transmission and improve its seating. Not only does the geometric configuration of the spring and seats affects the force line, but it has also been observed experimentally that the spring seat material has an effect [9].

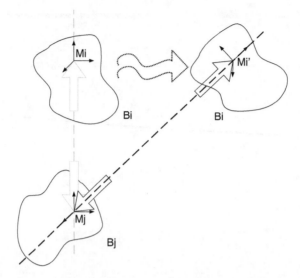

FIGURE 6.6 Point-to-point force element diagram. (From Bruni, S. et al., *Vehicle. Syst. Dyn.*, 49, 1021–1072, 2011. With permission.)

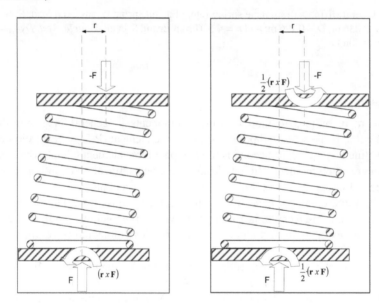

FIGURE 6.7 Compact spring element connecting two bodies: (a) reaction moment only at one end and (b) reaction moment distributed between both ends. (From Bruni, S. et al., *Vehicle. Syst. Dyn.*, 49, 1021–1072, 2011. With permission.)

6.2.2 Rubber Springs

Rubber springs found in rail vehicles usually serve as part of the primary and secondary suspension systems. An example of a rubber spring is the layered rubber-metal spring (chevron) used in the primary suspension; see Figure 6.8 [2]. Sometimes, these components can provide flexibility in all three directions; however, they only give flexibility in the horizontal plane, as shown in Figure 6.8b. Rubber in the secondary suspension mainly shows up as the bellows of air springs, hosting the pressurised air but also affecting the lateral air spring characteristics. Air springs are discussed in Subsection 6.2.3. Rubber is also used as bushes for hydraulic dampers, traction rods, etc.

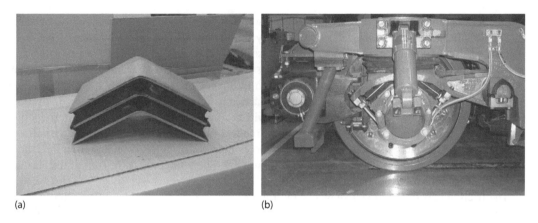

FIGURE 6.8 Example of layered rubber-metal components: (a) single component and (b) part of primary suspension. (From Bruni, S. et al., *Vehicle. Syst. Dyn.*, 49, 1021–1072, 2011. With permission.)

In contrast to coil springs, rubber springs provide damping as well as increased stiffness with increased excitation frequency and decreased amplitude. Rubber spring modelling in the present context mainly focuses on the frequency range of 0–20 Hz and on displacement (deformation) amplitudes typically found in primary and secondary suspensions.

6.2.2.1 One-Dimensional Models

In many cases, the very simple model of a spring and dashpot in parallel is used. This might be acceptable if only a very limited frequency range is studied, but the increase in stiffness with frequency is significant and often not representative for rubber. A common rubber spring model is instead the one shown in Figure 6.9 [2]. Here, a series spring has been added to the dashpot (Maxwell element), which means that the model stiffness will stay in the range of k to $k + k_s$ and monotonically increase with increasing excitation frequency. This better matches the results of rubber spring measurements. The characteristic frequency k_s/c will determine the excitation frequency around which the main transition between the two stiffness levels will take place.

To improve the frequency dependency, additional Maxwell elements may be added in parallel to the model components of Figure 6.9 [12,13]. But such, and more advanced, linear models, in principle, need additional measurements to be justified and more input parameters.

The models described previously imply that, for harmonic excitation, the hysteresis effect will tend to zero when the excitation frequency tends to zero. But such behaviour is generally not supported by measurements. Instead, a certain hysteresis effect remains even for very low frequencies. This can be attributed to internal rubber friction associated with the introduction of carbon black in

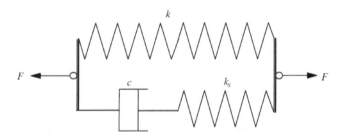

FIGURE 6.9 One-dimensional model often used to represent rubber (springs) with parallel stiffness k, viscous damping c and series stiffness k_s. (From Bruni, S. et al., *Vehicle. Syst. Dyn.*, 49, 1021–1072, 2011. With permission.)

the rubber manufacturing. Some rubber springs also allow for friction sliding between rubber and metal parts. In such cases, the hysteresis can be very significant, also for low frequencies. The internal and external rubber frictions will also increase the rubber spring stiffness, in particular for low-amplitude motions. The trend of higher rubber stiffness for higher frequencies is also due to the lower displacement amplitudes associated with high-frequency motions.

To extend the models described previously to account for the phenomena explained, friction needs to be represented. In this way, the models will become non-linear. One way of introducing the friction is to add another parallel element to the model of Figure 6.9. This approach is visualised by Figure 6.10 [2], where the elastic and viscous forces may, for instance, originate from the k and the series c and k_s components, respectively, of Figure 6.9. The total force is the sum of the elastic, friction and viscous force contributions.

The friction part of the model shown in Figure 6.10 may be represented by a Coulomb friction element with a series spring. As for the viscous part, several such friction components may be added in parallel to imitate a successive stick/slip; see, for instance, Figure 6.11 [2], referring to [14]. Models used for leaf springs may also come into question (Subsection 6.2.4). Other examples are given in [15] and [16]. In [15–18], two-parameter models have been used to represent 'smooth friction'. In [15] and [16], a logarithmic function is used, whereas a fractional expression was used in [17] and [18]. In [19], the friction was included in a non-linear viscous part by a velocity-dependent friction model. Also, a non-linear elastic part was used through a three-parameter polynomial expression. See models for lateral behaviour of air springs (Subsection 6.2.3).

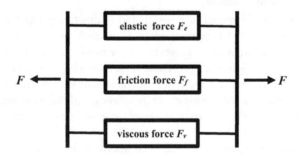

FIGURE 6.10 One-dimensional model principle to represent rubber (springs). (From Bruni, S. et al., *Vehicle. Syst. Dyn.*, 49, 1021–1072, 2011. With permission.)

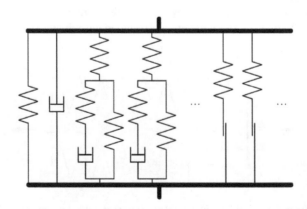

FIGURE 6.11 Example of one-dimensional model for elastomeric components: generalised Zener model from [14]. (From Bruni, S. et al., *Vehicle. Syst. Dyn.*, 49, 1021–1072, 2011. With permission.)

In principle, the parameters of rubber spring models need to be determined by component measurements. Measurements on rubber specimens can also support this process. In the component measurements, the dynamic behaviour at different excitation frequencies and displacement amplitudes needs to be evaluated, often also for different preloads.

Traditionally, rubber (spring) damping is expressed as a loss angle, or the tangent of this angle, describing the phase angle by which displacement lags force at harmonic excitation. For strong frictional rubber behaviour, alternative damping definitions to those discussed previously may be considered, since, at low-frequency, large-amplitude excitation, the frictional hysteresis can be significant, but still, the phase angle remains very small.

6.2.2.2 Multi-Dimensional Models

For two- or three-dimensional rubber spring models, one-dimensional ones are often superimposed. This may also be the case for non-linear one-dimensional models; however, this is not theoretically correct. Aspects such as reference points/levels and possible resisting moments must be considered, as described for coil springs in Subsection 6.2.1.

6.2.3 Air Springs

Air springs, as shown in Figure 6.12, consist of the mounting of a rubber-cord elastic chamber filled with compressed gas (usually air). This type of elastic elements is characterised by its small mass, excellent noise and vibration isolation, as well as its ability to maintain a constant ride height for different vehicle payload conditions. Such springs are found almost universally in the secondary suspension of modern passenger vehicles. Air springs are often arranged in series with a rubber or rubber-interleaved spring to provide some compliance in the suspension if the air spring becomes deflated. Bear in mind that, in situations of deflated air springs, there is a need to consider friction damping and to model the emergency rubber spring, which is precisely designed to tackle this situation. How to model rubber springs was explained in Subsection 6.2.2.

FIGURE 6.12 Typical air spring installation: (a) air spring and emergency spring, and typical air spring installation: (b) auxiliary air chamber (auxiliary volume), with pipe connecting both components and emergency rubber spring.

The operation of a typical air suspension with pressure control to maintain constant ride height is shown in schematic form in Figure 6.13.

In position (a) above, the system is in static equilibrium when the pressure inside the elastic chamber (airbag) 1 provides the prescribed ride height P. To reduce the spring stiffness, the elastic chamber is connected to the surge reservoir (additional volume) 2. When the load increases (position (b)), airbag 1 is compressed and moves valve 5 of the control system 4 down. This causes the compressed air from the main reservoir 6 to be admitted to the air spring system through pipe 9 and orifice 7, thus increasing the pressure. This restores air spring 1 to the equilibrium position (a) again, and control valve 4 stops the flow of air from the main chamber 6 into the airbag. Reduction of the load (position (c)) makes the airbag rise and control valve 5 moves up. In this case, pipe 9 connects to atmosphere 10 through orifice 8 and drops the pressure in airbag 1. The spring height reduces and returns to the equilibrium position again. The surge reservoir 2 and the damping orifice 7 are important features in the operation of the air spring. Increasing the surge reservoir volume leads to a decrease in spring stiffness. Reducing the size of the damping orifice not only increases the damping properties of the spring (by increasing the kinetic energy dissipation) but also increases the stiffness. The lateral stiffness of the pneumatic spring depends on the shape of the elastic chamber.

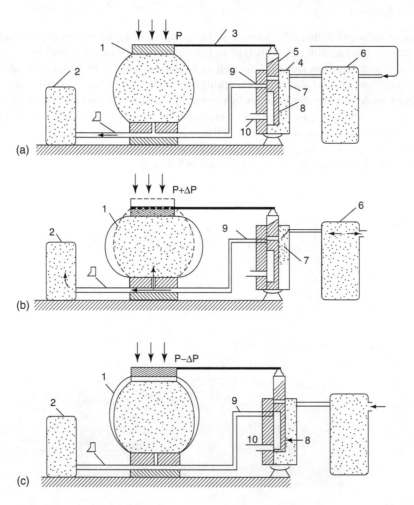

FIGURE 6.13 Schematic showing the operation of a typical air suspension: (a) equilibrium position, (b) upstroke and (c) downstroke. (From Iwnicki, S. (Ed.), *Handbook of Railway Vehicle Dynamics*, CRC Press, Boca Raton, FL, 2006. With permission.)

Air springs are often used in the secondary suspension of passenger railway vehicles, and their modelling has important implications for the accuracy of quasi-static and dynamic multi-body simulations. The overall behaviour of this suspension element can be described in terms of vertical and horizontal behaviours, generally with a weak interaction between the two; however, the vertical preload has an important influence on the lateral behaviour of the suspension.

In the vertical direction, air spring suspensions show behaviour highly dependent on the preload and on the amplitude and frequency of dynamic displacements. Hence, specific models have been defined, which are reviewed in this subsection. This is illustrated in the following discussion, taking some results from reference [20]; see also [21].

In the horizontal plane, air springs represent a particular case of shear springs, and the relationship between shear and rotational deformation and the shear forces and moments reacted by the spring is often not negligible, requiring a three-dimensional modelling approach. When deflated, air springs sit down on a rubber emergency spring. They can be modelled using the rubber and friction elements in two dimensions, as explained in Subsection 6.2.2.

6.2.3.1 Effect of Amplitude

Figure 6.14 shows the vertical dynamic stiffness of the pneumatic system at different frequencies and for two different amplitudes (0.5 mm and 1 mm). The dynamic stiffness has been calculated as the quotient between the modulus of the normal force (approximating it as a harmonic function) and the modulus of the displacement applied to the pneumatic suspension. The following conclusions can be drawn from the analysis of Figure 6.14.

The pneumatic system is clearly non-linear in the medium-frequency range (from 6 to 14 Hz in this case). As can be seen, the dynamic stiffness value in these frequencies varies with oscillation amplitude. However, the stiffness value at high and low frequencies is not related to oscillation amplitude.

As the frequency is reduced, the stiffness value tends towards a fixed value. This value is known as the pneumatic system static stiffness. Likewise, when the frequency is increased, the system tends towards a different fixed value: the dynamic stiffness of the system for high frequencies. As can be seen, the stiffness value for high frequencies is greater than the stiffness value for low frequencies.

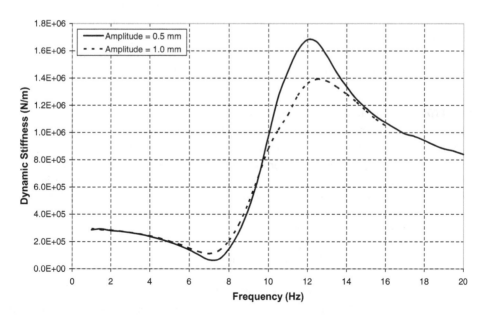

FIGURE 6.14 Dynamic stiffness at different oscillation amplitudes. (From Alonso, A. et al., *Vehicle. Syst. Dyn.*, 48, 271–286, 2010. With permission.)

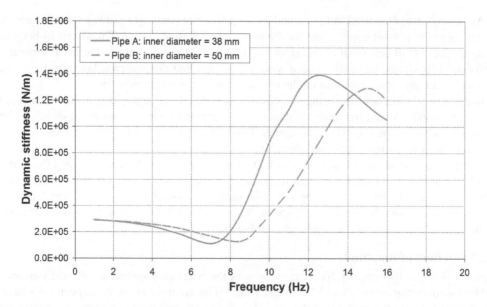

FIGURE 6.15 Effect of varying the pipe diameter interconnecting the air spring with the auxiliary reservoir. (From Alonso, A. et al., *Vehicle. Syst. Dyn.*, 48, 271–286, 2010. With permission.)

Resonance behaviour can be observed at 7.74 Hz. The frequency at which this resonance occurs depends on the volume of the auxiliary chamber and the mass of the air inside the connection pipe.

6.2.3.2 Effect of the Connection Pipe

Figure 6.15 compares the results obtained when the pipe that connects the air spring and the auxiliary air chamber is modified. Pipe A has an internal diameter of 38 mm, and pipe B has 50 mm. Both pipes are of similar shape and length. As can be seen, the stiffness of the system for low and high frequencies is not affected by changing the pipe; however, the medium-frequency range is clearly affected by this change. The position and the maximum and minimum stiffness values are modified: as the pipe diameter increases, the transition range moves towards higher frequencies.

6.2.3.3 Effect of the Auxiliary Air Chamber

With the purpose of checking the effect caused by the chamber, Figure 6.16 compares the results obtained using an auxiliary chamber with the case without the chamber. As can be seen, virtual constant dynamic stiffness is achieved when an auxiliary chamber is not used. This stiffness value is similar to the dynamic stiffness value at high frequencies, obtained when using an auxiliary chamber. Therefore, it can be stated that using an auxiliary chamber enables modification (always decreasing) of the stiffness value of the pneumatic system for low frequencies. On the other hand, when using an auxiliary chamber, a transition is generated between the stiffness at low frequencies and the stiffness at high frequencies. As has been shown in the previous paragraph, the transition rather depends on the characteristics of the connection pipe.

6.2.3.4 Effect of Orifice Plates in the Connection Pipe

To check the effect of losses produced in the connection pipe, Figure 6.17 shows the response of the pneumatic system by using different orifice plates (whose function is to increase the load loss). As can be observed, using these elements enables us to modify the dynamic stiffness

Suspension Elements and Their Characteristics

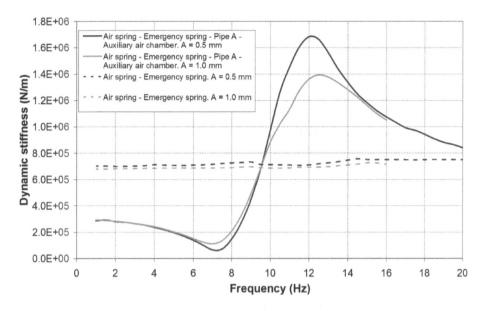

FIGURE 6.16 Effect of the auxiliary chamber on the dynamic stiffness. (From Alonso, A. et al., *Vehicle. Syst. Dyn.*, 48, 271–286, 2010. With permission.)

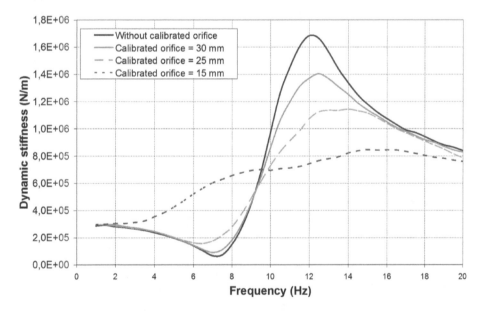

FIGURE 6.17 Effect of orifice plates on the dynamic stiffness of pipe A at an amplitude of 0.5 mm. (From Alonso, A. et al., *Vehicle. Syst. Dyn.*, 48, 271–286, 2010. With permission.)

value in the medium-frequency range completely. As expected, it is important to point out that the introduction of this element does not modify the stiffness value for low or high frequencies. Likewise, it does not modify the frequency at which the transition range maximums and minimums are produced.

Models of the vertical air spring behaviour can be classified into 'equivalent mechanical models' and 'thermodynamic models'. Equivalent mechanical models are based on the use of lumped parameter springs, dashpots and masses. These allow a relatively simple mathematical description of the suspension but generally do not account for the levelling system behaviour and do not provide an estimate of air consumption. Furthermore, these models may not be well suited to consider non-conventional suspension configurations (e.g., cross-piping of the bellows) or active/semi-active suspension control.

Thermodynamic models instead aim at representing the actual mechanical and thermodynamic processes occurring in the air spring suspension, and hence, all parameters in such models have a clear physical meaning. Despite this, tuning may be needed to define the values of model parameters describing the concentrated and distributed losses in the pneumatic circuit.

More details on equivalent mechanical and thermodynamic models are provided in the next two subsections. It is not the intention of this discussion to cover in detail all existing air spring models but instead to focus on some representative ones. In a subsequent subsection, models of the air spring suspension in the horizontal plane are reviewed. The final subsection deals with the definition of model data.

6.2.3.5 Equivalent Mechanical Models

The simplest model of the air spring suspension in the vertical direction consists of a spring with a viscous dashpot in parallel. This model, however, only reproduces the quasi-static stiffness of the suspension, and it is difficult to define a correct value for the damper parameter, because the actual dissipative effects in the suspension are far from linear.

A different model, appropriate in a wider frequency range and known as a 'Nishimura model' [22], is shown in Figure 6.18a. It consists of a spring K_1 representing the bellow, in series with the parallel combination of a spring K_2 representing the compressibility of the air in the surge reservoir (auxiliary reservoir) and a viscous damper C. This damper accounts for dissipations in the surge pipe and from the orifice between the bellows and the reservoir typically present in suspension systems with air damping which could even replace hydraulic dampers. Optionally, a spring K_3 may be added in

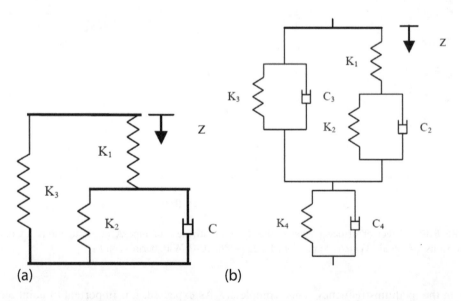

FIGURE 6.18 (a) Oda-Nishimura model (from [22]) and (b) Simpack FE83 model [23]. (From Bruni, S. et al., *Vehicle. Syst. Dyn.*, 49, 1021–1072, 2011. With permission.)

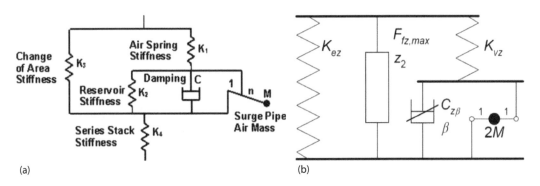

FIGURE 6.19 (a) VAMPIRE vertical model [16] and (b) Berg model in vertical direction [17]. (From Bruni, S. et al., *Vehicle. Syst. Dyn.*, 49, 1021–1072, 2011. With permission.)

parallel to represent the additional stiffness effect due to the change of the effective area with the suspension height [24]. A similar model is the 'linear air-spring element FE83', implemented in Simpack [23]; see Figure 6.18b. This model also includes an additional damper in parallel to spring K_3 and a spring-damper element K_4 and C_4 in series to the rest of the suspension, representing the emergency spring.

The simplest Nishimura model with $K_3 = 0$ is sufficient to define a frequency-dependent behaviour of the air spring: when the frequency of deformation is low, the force in the damper C is negligible, and the model behaves as the series of springs K_1 and K_2, but the damper force increases with frequency until the deformations of the damper and of spring K_2 become negligible and the complete model approaches the behaviour of the single spring K_1. Hence, the Nishimura model allows the reproduction of the transition between the 'low frequency' and 'high frequency' stiffness of the pneumatic suspension but does not include internal suspension resonances, which may occur on account of inertial effects in the air mass contained in the surge pipe. As pointed out in [25,26], this is a relatively small mass, but that is accelerated to high velocities when air is exchanged between the bellows and the reservoir, and therefore, the equivalent inertia can be important. Two models allowing consideration of this effect are the 'VAMPIRE model' [27] and the 'Berg model' [28].

The VAMPIRE vertical model is shown in Figure 6.19a. It consists of a Nishimura model to which a lumped mass M and a series stack stiffness K_4 are added. The K_4 spring represents the emergency spring and, when needed, may be taken to be infinite. The lumped mass M represents the inertia of the air mass in the surge pipe. It is worth remarking that, in the VAMPIRE model, the damping term can follow a square law, which is considered to be more accurate than linear damping. Berg in [28] defines a three-dimensional model, of which only the vertical part is described here; see Figure 6.19b. This is similar to the VAMPIRE vertical model, but it also includes a friction force element and a velocity exponent parameter (β), which in [28] is set to 1.8, supported by measurements.

When compared to the Nishimura model, both the VAMPIRE and Berg models are capable of accounting for an internal resonance of the air spring suspension, leading to a maximum value of the dynamic stiffness in an intermediate frequency range, with the quasi-static and high-frequency stiffness approaching the same values as that for the Nishimura model; see Figure 6.20 for the VAMPIRE model and Figure 6.21 for the Berg model.

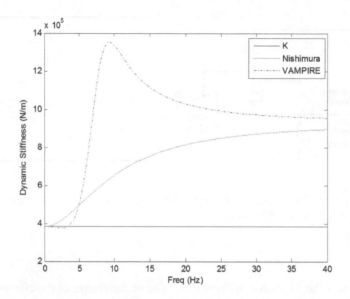

FIGURE 6.20 Comparison of the vertical dynamic stiffness for an air spring system, Nishimura and VAMPIRE models ($K_1 = 7.6\text{E}5$ N/m, $K_2 = 5.4\text{E}5$ N/m, $C = 1.29\text{E}5$ Ns/m, $M = 5.4\text{E-}3$ kg, $\kappa = 209$) [20]. (From Bruni, S. et al., *Vehicle. Syst. Dyn.*, 49, 1021–1072, 2011. With permission.)

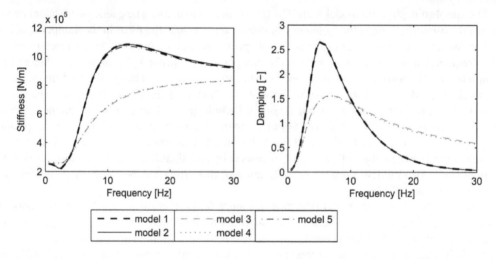

FIGURE 6.21 Dynamic stiffness and damping for different air spring models [29]: model (1) oscillating air mass; model (2) incompressible differential; model (3) incompressible algebraic; model (4) ISO 6358; model (5) Berg's equivalent mechanical model. (From Bruni, S. et al., *Vehicle. Syst. Dyn.*, 49, 1021–1072, 2011. With permission.)

6.2.3.6 Thermodynamic Models

Figure 6.22 shows the main elements that thermodynamic models of an air spring suspension ([23,29–33]) are composed of:

- A model of the bellows and surge reservoir
- A model of the surge pipe connecting the bellows with the reservoir

Additionally, a model of the levelling system can also included, as done in [32] and [33] dependent on the scope of the air spring suspension model.

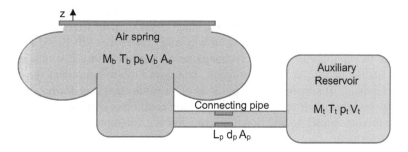

FIGURE 6.22 Thermodynamic model of an air spring suspension. (From Docquier, N. et al., *Vehicle. Syst. Dyn.*, 45, 505–524, 2007.)

The bellows and reservoir are modelled as variable-size and constant-size air volumes, respectively. Their thermodynamic states vary in order to take into account of the boundary conditions applied and of the fluid exchange between the volumes. Some references such as Quaglia and Sorli [30], Nieto et al. [31] and Docquier [33] show examples of different types of bellows, where the measured volume vs height is analysed. All these results depict that although the actual relationship between these parameters is non-linear, it is well approximated by a linearised expression.

The vertical force F generated by the air spring is expressed as:

$$F = p_{rel} A_e \tag{6.15}$$

where p_{rel} is the relative pressure of the air in the bellows and A_e is a geometric parameter called 'effective area'. The value of A_e is obtained either based on the geometry of the bellows or from measurements. Some references, such as [30,31,33], illustrate the relationship between the effective area and the air spring height.

As already pointed out, they show that the effective area can be well approximated either using a constant value or a linear expression as a function of the air spring height. Additionally, a formulation based on mass and energy balances is the base to describe the thermodynamic state of the air volumes in the bellows and surge reservoir. Instead of the energy conservation equation, one can consider a polytropic law of the type:

$$p \left(\frac{V}{M} \right)^k = p_0 \left(\frac{V_0}{M_0} \right)^k \tag{6.16}$$

where V is the volume of the bellows or tank; M is the air mass in the bellows or tank; p is the (absolute) air pressure in the bellows or tank; p_0, V_0 and M_0 are given reference conditions and k is the polytropic exponent whose value depends on assumptions made on the energy balance of the system. This means that $k = 1$ would assume an isothermal transformation and $k = \gamma$ (the specific heat ratio) an adiabatic transformation. The exponent k used in the polytropic law has a significant influence in the suspension behaviour. This matter is specifically addressed in references [33] and [34], where the two possible extreme cases (i.e. zero and infinite heat exchange capacity, which correspond, respectively, to the adiabatic and isothermal transformations) are analysed. When a specific rail vehicle is considered, the calculations show how changes in the heat transfer coefficient conduce to significant differences in wheel unloading while the vehicle negotiates a track twist and also heavily affects the estimate of air consumption of the pneumatic suspension due to passenger going in or out. Note however, that both conditions, zero and infinite heat exchange, take place on a time scale in the range of 10^2 seconds, much larger than the timescales typically associated with vehicle dynamics. This is the reason why other references [29,30,35] use the adiabatic exponent under the assumption of heat exchange being negligible in a fast manoeuvre. A good correlation of simulation results and experimental tests prove this assumption to be correct. On the other hand, reference [31] supports the hypothesis of an isothermal transformation. This is based on the fact that suspension air temperature in working conditions was monitored experimentally by means of a thermocouple; no additional information is given, however, on the timescale of the experiments performed.

The second main component that air spring thermodynamic models have to take into account is the model of the surge pipe, i.e. the pipe that connects the bellow with the reservoir. Different levels of complexity are possible. The model can be as simple as just a simple fluidic resistance, defined according to ISO 6358 and accounting for concentrated and distributed losses, and may include the effect of an orifice introduced in the duct to increase the damping of the suspension such as in [30].

References [29,33,35] propose a more detailed model, the 'incompressible differential' model, which accounts for the inertial effects associated with the oscillation of the fluid in the pipe (the model assumes a one-dimensional incompressible flow in the pipe). In this case, a first-order differential equation is written for the pipe, and one additional state is introduced for each pipe. The multibody software Simpack uses a similar approach which is implemented in its FE82 element [23]: a thermodynamic model which includes an oscillating mass between the bellows and the reservoir (i.e. the mass of the air inside the pipe) and dissipative forces representing losses in the pipe.

Docquier et al. [33] also proposed an 'incompressible algebraic' pipe model which does take into account the inertial effects related to the air confined in the pipe, and is derived from the incompressible differential model. As was the case for the model based on ISO 6358 formulae, this model provides an algebraic equation relating the mass flow rate in the pipe to the pressure drop between the reservoir and the bellows. Figure 6.21 is a good illustration of results obtained with different pipe models. It shows that the incompressible differential model (or equivalent models) is able to account for an internal resonance of the air spring, as is the case for the VAMPIRE and Berg equivalent mechanical models.

It is also possible to use different compressible flow models, such as Docquier in his PhD thesis [33], in order to describe fluid motion in the piping. These models show that by considering fluid flexibility, a second internal resonance of the suspension can exist and, depending on the length of the pipes, it may fall below 20 Hz.

6.2.3.7 Air Spring Models in the Horizontal Plane

The air spring has generally a rather low lateral stiffness. Its value depends on the structural stiffness of the bellows, friction effects at the rubber-metal interface at the upper part of the bellows and the effect of the vertical load, which combined with shear and rotational deformation of the bellows produce shear forces and moments on the end mountings.

A typical way to model the horizontal air spring behaviour is using an elastic force element (with either a linear or a non-linear characteristic), in parallel with an element able to represent both rubber hysteresis and viscoelasticity. In the VAMPIRE model (see Figure 6.23a) it consists of a damper with series stiffness [24,27]. In addition and in order to secure static equilibrium of the air spring, a balancing moment is introduced; the user is allowed to define a specific share of the balancing moment between the upper and lower ends of the air spring which mostly depends on the preload, shear force and spring deformation.

In the case of the Berg model in the horizontal plane [28] three factors are considered: a set of elastic forces plus frictional and viscous contributions. The elastic part consists of two shear forces, a roll moment and a pitch moment applied at the air spring upper end, defined as linear functions of the shear deformation in longitudinal and lateral directions, and of the roll and pitch rotations of the end mountings. The static equilibrium of the air spring provides the

Suspension Elements and Their Characteristics

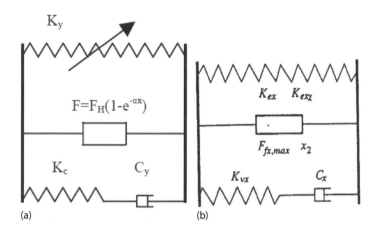

FIGURE 6.23 (a) VAMPIRE model in lateral direction [13] and (b) Berg model in lateral direction [17]. (From Bruni, S. et al., *Vehicle. Syst. Dyn.*, 49, 1021–1072, 2011. With permission.)

values for the forces and moments at the lower end mounting. The friction and viscous forces are defined only in the lateral and longitudinal directions. There are clear similarities when comparing the Berg model and the VAMPIRE model for the lateral direction (Figure 6.23), the differences being the linearity of the elastic term and the different function for the 'smooth' friction.

Facchinetti et al. [35] propose two different models for the air spring suspension. The first one is a linear quasi-static elastic model which takes into account the lateral/roll coupling. The second one is a combination of a typical vertical air spring model (able to represent the frequency dependency of the air spring) and of the quasi-static model.

Simulations are carried out using the second model and the simplified one which neglects the direct roll stiffness term and the shear-roll coupling term. The stiffness parameters are identified from full-scale measurements performed by applying combinations of shear and roll deformation. The results are compared in terms of wheel-rail contact forces, showing that the coupling of lateral and roll deformations have a non-negligible effect on the load transfer effects when the vehicle is negotiating a curve. On the other hand, the frequency-dependent effect only has minor effects on the curving behaviour if slow variations in the load act on the suspension, which is not the case when analysing ride safety in the presence of crosswinds.

6.2.3.8 Definition of Air Spring Model Parameters

Presthus collated in [24] formulae to define the parameters of various equivalent mechanical models, including the Nishimura, Simpack FE83 and Berg models, based on the physical properties of the air spring suspension (effective area, air volumes, concentrated and distributed loss coefficients), and similar expressions are available for the VAMPIRE model. Alonso et al. [20] reported a comparison between the measured and simulated dynamic stiffness of an air spring suspension, showing a good agreement between the two sets of data for different cases of concentrated losses in the surge pipe (no orifice, orifices with various sizes).

In order to calculate the parameters of the Nishimura or VAMPIRE model, the following formulae can be used:

$$K_1 = \frac{p_0 A_e^2 n}{V_{b0}} \qquad (6.17)$$

$$K_1 = \frac{p_0 A_e^2 n}{V_{r0}} = K_1 \frac{V_{b0}}{V_{r0}} \qquad (6.18)$$

$$M = m_s = \rho A_s L_s \qquad (6.19)$$

$$C = C_s \left(\frac{A_e}{A_s}\right)^3 \qquad (6.20)$$

$$\kappa = \frac{A_e}{A_s} \qquad (6.21)$$

where p_0 is the static pressure inside the air spring, A_e is the effective area of the air spring, n is the air polytrophic constant, V_{b0} is the volume of the air spring, V_{r0} is the volume of the auxiliary chamber, ρ is the density of the air inside the pneumatic system, A_s is the area of the pipe, L_s is the length of the pipe and C is a damping coefficient related with the load losses inside the pipe. In order to calculate this parameter, empirical formulae obtained from the literature have been used [36,37]. It is worth mentioning that all the parameters can be defined from geometrical and physical magnitudes, that is, without testing. As can be seen, all three models have different degrees of complexity.

Figure 6.24 shows the experimental results obtained in certain cases, along with their corresponding theoretical predictions provided by the VAMPIRE model; the theoretical results of the VAMPIRE model are very close to the experimental ones. It can be concluded that the VAMPIRE models are able to reproduce the behaviour of the pneumatic system adequately in the frequency range of interest for vehicle dynamics.

For thermodynamic models, the definition of the input parameters is more straightforward, since these are directly represented by the physical parameters of the system. However, the accuracy of both equivalent mechanical and thermodynamic air spring models may be affected by uncertainties in some physical parameters. In particular, the variation of the effective area with

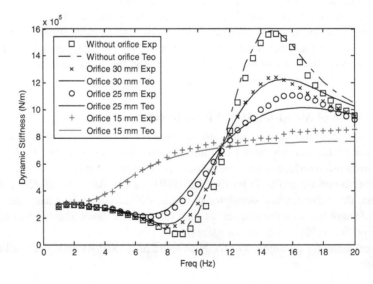

FIGURE 6.24 Theoretical versus experimental comparison at different frequencies, with orifice plates of different diameters (pressure 4 bar, amplitude 0.5 mm). (From Alonso, A. et al., *Vehicle. Syst. Dyn.*, 48, 271–286, 2010. With permission.)

the air spring height can hardly be defined, other than by a direct measurement, and the range of validity of semi-empirical formulae defining the loss coefficients based on the geometry of the surge pipe needs to be carefully considered, especially in the case of complicated geometry of the pipes and orifices.

Based on the examination of the state of the art, the direct measurement currently remains the most frequently used way to define some critical air spring geometric parameters, namely the relationships between the effective area and the volume of the bellows and the height of the air spring; however, attempts have been made, for example, by Qing and Shi [38], to compute some of these parameters based on the analysis of the air spring geometry.

Mathematical models of active and semi-active vertical suspensions can be derived from both the 'equivalent mechanic' and 'thermodynamic' air spring suspension models. As explained in Chapter 15, active and semi-active pneumatic suspensions are used in the secondary suspension stage of passenger railway vehicles and can act either in the vertical direction (active/semi-active air springs) or in the lateral direction (i.e., car body centring system) [39].

A few examples can be cited. Tang [40] proposes an equivalent mechanical model for a semi-active air spring with a controlled variable-size orifice in the surge pipe. The model consists of a modified Nishimura model, in which the effect of the variable size orifice is represented by changing the damping rate C of the viscous damper. The same modelling approach is followed by Sugahara et al. [41], where a model for the vertical vibration of a complete rail vehicle equipped with semi-active air springs and adaptive primary dampers is also defined.

A thermodynamic model of an active air spring suspension is defined by Alfi et al. [42] as a modification of the model shown in Figure 6.22, by modelling one additional constant air volume representing an additional supply reservoir and the servo-valve connecting the additional reservoir with the surge reservoir. The model of the active suspension is compared with experiments performed on a full-scale prototype of the active suspension, showing a good match of numerical simulations with the measurements. The model of the active suspension is then used in combination with a multi-body model of the entire vehicle to investigate the use of the active suspension to improve ride comfort in curves and to increase the vehicle's resistance to overturning in the presence of crosswind.

As far as pneumatic active lateral suspensions are concerned, a thermodynamic model is proposed by Sorli et al. [43], which is integrated into a multi-physics simulator of the lateral dynamics of a Pendolino vehicle; numerical results from the model are compared with on-track measurements, showing good agreement. Conde Mellado et al. [44] propose a model of an active lateral suspension interconnected to the vertical pneumatic secondary suspension; the model is used to compare the performances of the proposed active suspension with a conventional passive one.

6.2.4 Leaf Springs

A *leaf spring* (diagram A in Table 6.1) is an elastic element comprising a number of steel leaves. The leaves work in bending, and the 'fish-bellied' shape of the beam provides smaller spring stiffness. Depending on their design, leaf springs can be closed (diagram A in Table 6.1), elliptical or open. They consist of layered leaves, 1 and 2 having different lengths and held together by a buckle 3. The largest leaf (1) is named the master and other leaves (2) the slaves. Leaf springs also provide damping due to the inter-leaf friction. However, it is difficult to obtain specific desired damping values, and the damping can change considerably due to lubrication or contamination of the rubbing surfaces. A typical application of leaf springs in rail vehicles today is the two-axle freight wagon with single-axle running gear common in Europe. The leaf spring is combined with a link suspension, providing the flexibility in the horizontal direction. There is also a bogie version with this type of suspension that exists in Europe. Both single-stage and two-stage leaf springs are used in this suspension arrangement. The typical characteristics for large displacements can be seen in Figure 6.25a. For small dynamic displacements around a static equilibrium position, the characteristics are as

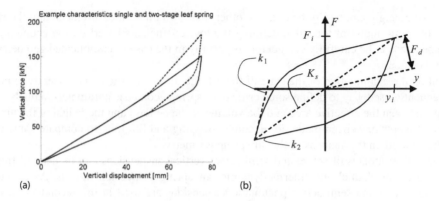

FIGURE 6.25 Typical force-displacement diagrams of leaf spring/link suspension: (a) example of curve for large displacements of leaf spring and (b) example of curve for small displacements around a static equilibrium. (From Bruni, S. et al., *Vehicle. Syst. Dyn.*, 49, 1021–1072, 2011. With permission.)

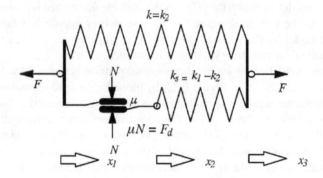

FIGURE 6.26 Model for leaf spring or link suspension, as used by Royal Institute of Technology (KTH) [47]. (From Bruni, S. et al., *Vehicle. Syst. Dyn.*, 49, 1021–1072, 2011. With permission.)

shown in Figure 6.25(b), where k_1 is the stiffness for the leaves sticking together and k_2 is the stiffness when the leaves have started to slide on each other. F_d is the force at the transition from sticking to sliding, and k_s is the difference between k_1 and k_2. Leaf springs are described in the Office for Research and Experiments (ORE) of the International Union of Railways reports [45,46]. A possible simulation model used to describe the dynamic properties of the leaf spring is shown in Figure 6.26 [47]. More advanced models and measurement results for this type of suspension can be found in [48–57]. For more details on friction damping in general, refer to Section 6.3.

6.3 DAMPERS

Damping is usually provided in railway vehicle suspension by the use of viscous or friction damping devices. A damper is the device that controls oscillations in the primary or secondary suspension of the vehicle by energy dissipation.

Dry friction results from the relative slip between two rigid bodies in contact. The friction force can be constant or dependent on the mass of the car body but always acts to resist the relative motion. Friction force is proportional to friction coefficient μ, pressure between surfaces Q and contact surface area S. This dependence can be represented by the following formula:

$$F_{dry\ fric} = -\mu S Q \frac{\dot{z}}{|\dot{z}|} = -F_0 \frac{\dot{z}}{|\dot{z}|}, \qquad (6.22)$$

Suspension Elements and Their Characteristics

where F_0 is the magnitude of friction force, $\dot{z}$ is the relative velocity of motion and $|\dot{z}|$ is the magnitude of velocity. The minus sign denotes that the friction force is always in the opposite direction to the velocity.

Viscous damping develops between two parts separated with a layer of viscous liquid (lubricant) or in devices known as hydraulic dampers, where the viscous liquid flows through an orifice and dissipates the energy. The damping force in the viscous case is proportional to velocity:

$$F_{hydr\,fric} = -\beta\,\dot{z}^n, \tag{6.23}$$

where β is the coefficient, $\dot{z}$ is the velocity of relative motion and n is the power. Depending on the construction of the device and the liquid properties, the power n can be greater than, equal to or less than 1.

If the liquid flow is laminar, then $n \approx 1$, and damping is described as linear viscous damping:

$$F_{lin\,visc\,fric} = -\beta_1 \dot{z}, \tag{6.24}$$

where β_1 is the coefficient, named the damping coefficient for the hydraulic damper.

For $n = 2$, the damping is called turbulent or quadratic:

$$F_{turb\,visc\,fric} = -\beta_2\,|\dot{z}|\,\dot{z}. \tag{6.25}$$

Gases are also viscous. Therefore, driving the gas through a throttle valve (damper orifice) may also produce sufficient force for damping the oscillations of railway vehicles.

Intermolecular damping (hysteresis) originates mainly in rubber and polyurethane elastic elements. In such cases, the damping force is proportional to the oscillation velocity and is inversely proportional to the frequency:

$$F_{molec\,fric} = -\frac{\beta_0}{\omega}\dot{z}. \tag{6.26}$$

Damping of vibrations can also be obtained by other means such as the introduction of active dampers being controlled proportionally to velocity.

6.3.1 Viscous Dampers

Hydraulic dampers are almost universally used in passenger bogies and are sometimes also used in modern freight bogies.

The energy dissipated in a hydraulic damper is related to velocity and therefore to the amplitude and frequency of vibration. Thus, the hydraulic damper is self-tuning to dynamic excitations and provides reliable and predictable damping of vehicle oscillations.

Railway vehicles use the telescopic hydraulic dampers, as shown in Figure 6.27. The hydraulic damper operates by forcing the working fluid through an orifice (flow-control valve) from one chamber into the other, as the vehicle oscillates on the suspension. This produces viscous damping, and the kinetic energy of the oscillations is transformed into heat.

As shown in Figure 6.27, telescopic hydraulic dampers consist of the body 1 with the sealing device, the working cylinder 2 with valves and the shaft 3 with a piston 5 that also has valves (position 6). When the piston moves relative to the cylinder, the working fluid flows through the valves from the chamber over the piston to the chamber under it and back.

The reliability of hydraulic dampers mostly depends on the sealing between the shaft and the body. Occasionally, a malfunction of this unit causes excessive pressure in the chamber over the piston, resulting in leakage of the working fluid.

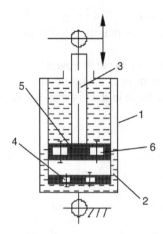

FIGURE 6.27 Telescopic hydraulic damper. (From Iwnicki, S. (Ed.), *Handbook of Railway Vehicle Dynamics*, CRC Press, Boca Raton, FL, 2006. With permission.)

The capability of hydraulic dampers to dissipate energy is characterised by its force versus velocity characteristic, which is the dependence between the resistance force developed in the hydraulic damper, P, and the piston displacement velocity, $\dot{d}$.

The damper characteristic may be either symmetrical, when the resistance forces are the same for extension and compression, or asymmetric. Dampers with symmetric characteristics are typically used in secondary suspensions. In primary suspensions, asymmetric dampers are often used as a wheelset running over a convex irregularity (hill) causes larger forces than negotiating a concave irregularity (valley). As a result, dampers may be designed with an asymmetric characteristic, providing a smaller force in compression than in extension. However, large damping forces in extension can significantly decrease the vertical wheel load, thus increasing the risk of derailment. Therefore, railway dampers are less asymmetric than automobile ones.

Figure 6.28 shows common force characteristics of hydraulic dampers. The force characteristic in Figure 6.28a shows a hydraulic damper with a resistance force proportional to velocity and not exceeding the 'blow-off' (saturation) force. When a predetermined pressure value is reached inside the working chamber, the 'blow-off' valve opens to prevent excessive forces being developed by the damper.

Hydraulic dampers having the characteristic shown in Figure 6.28b have a resistance force proportional to the velocity and the displacement. Such a characteristic is obtained by the provision of specially calibrated needles (or other devices) into the flow-control valve to change its cross-section. The size of the valve cross-section is controlled depending on static deflection of the suspension.

The characteristic shown in Figure 6.28c is typical for devices where the dissipative force is proportional to velocity, but the operation of the 'blow-off' valve is controlled depending on the displacement and velocity of the piston. For the characteristic in Figure 6.28d, the size of the valve cross-section and the saturation limit for the emergency valve are controlled together, depending on the relative displacement and velocity of the piston.

Attachment of hydraulic dampers to the vehicle is usually done by using elastic mountings or bushes to prevent the transmission of high-frequency vibrations. The internal pressure in the damper often gives it elastic properties. Therefore, hydraulic dampers are often modelled as a spring and viscous damper in series.

In some designs, the hydraulic dampers are united with the elastic elements. The schematic of a hydraulic damper integrated into a coaxial rubber-metal spring is shown in Figure 6.29.

For modelling of hydraulic dampers, see [2] and Section 17.2.3.

Suspension Elements and Their Characteristics 225

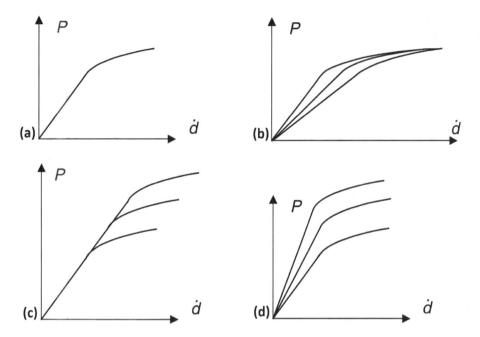

FIGURE 6.28 Common force characteristics of hydraulic dampers. (With kind permission from CRC Press: *Handbook of Railway Vehicle Dynamics*, CRC Press, Boca Raton, FL, 2006, Iwnicki, S. [ed.].)

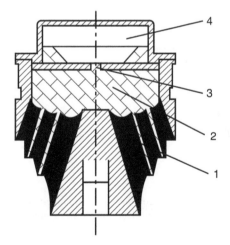

FIGURE 6.29 Hydraulic spring: (1) rubber-metal conical spring; (2) working fluid; (3) flow control valve; (4) compensation reservoir with rubber diaphragms. (From Iwnicki, S. (Ed.), *Handbook of Railway Vehicle Dynamics*, CRC Press, Boca Raton, FL, 2006. With permission.)

6.3.2 Friction Dampers

Friction damping is used in many suspension components in railway vehicles, especially in freight wagons or deflated air springs. In some components, such as rubber and (inflated) air springs, friction damping is only a side effect. Friction dampers are devices that transform the energy of oscillations into heat energy by dry friction. One advantage of such friction components is that they are relatively cheap and almost maintenance free. Another major advantage is that the amount of damping in many arrangements is more or less proportional to the axle load. This is an important property, especially in freight wagons, where a loaded wagon can have up to five times higher axle load than an empty one. A disadvantage is that the efficiency of friction dampers depends on parameters such as the friction coefficient in the contacting surfaces and the flexibility in the damper arrangement. The friction coefficient can vary significantly during operation of a vehicle, which can change the running behaviour of a freight wagon. The uncertainty regarding the parameters also makes it a challenge to simulate the running behaviour of vehicles, which, to a great extent, relies on friction damping.

In most cases, friction in rail vehicle suspensions is modelled as dry Coulomb friction. The disadvantage of the Coulomb model is that it is non-smooth, multi-valued and non-differentiable, which cause numerical problems in simulations; see the force-displacement curve in Figure 6.30a. Therefore, most authors presenting work on simulation of dry friction dampers apply regularisation to avoid the difficulties mentioned previously, for example, [58,59]. One possibility to avoid the problem of a multi-valued function is a linear spring in series with a friction slider; see Figure 6.31. The resulting force-displacement characteristics can be seen in Figure 6.30b.

Depending on their construction, friction dampers may be classified as one of four types, as shown in Table 6.3, namely integrated into the elastic element, telescopic, lever action and spring suspension.

Dampers integrated into an elastic element consist of the barrel 1 and friction wedges 2 that are held in contact by a spring. When the elastic element deforms, the friction forces act on the contacting surfaces between the barrel and the wedges, transforming the kinetic energy into heat.

Telescopic friction dampers consist of the body 1 that contains the piston, with the system of friction wedges 2 clamped by a spring.

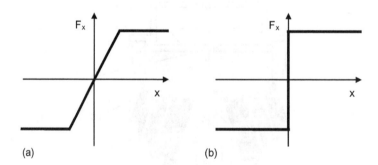

FIGURE 6.30 Force-displacement curve of Coulomb friction model: (a) Coulomb model with friction slider alone and (b) with spring in series, as in Figure 6.31. (From Bruni, S. et al., *Vehicle. Syst. Dyn.*, 49, 1021–1072, 2011. With permission.)

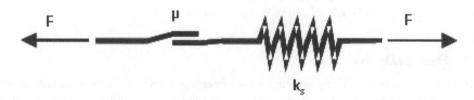

FIGURE 6.31 Friction element with spring in series. (From Bruni, S. et al., *Vehicle. Syst. Dyn.*, 49, 1021–1072, 2011. With permission.)

TABLE 6.3
Classification of Friction Dampers

	Linear Action		Planar Action	Spatial Action
Integrated with Elastic Element	Telescopic	Lever	Integrated in the Suspension	
Constant friction				
Variable friction				

Source: With kind permission from CRC Press: *Handbook of Railway Vehicle Dynamics*, CRC Press, Boca Raton, FL, 2006, Iwnicki, S. [ed.].

Dampers integrated in the suspension are mostly used in three-piece bogies and consist of friction wedges 2 that move relative to side frame 6 and bolster 5. Construction of the dampers (Table 6.3) differs by the position of friction wedges 2 (inside the bolster 5 or inside the frame 6), by the number of springs and their inclination angles, as well as by the design of the friction wedges. For example, the Russian CNII-H3 bogie has wedges 2, with inclined faces contacting with the bolster 5 and pressed to the side frame 6 by springs underneath.

Simultaneous and integrated friction dampers are connected to the springs in the suspension, whereas telescopic dampers are independent devices. Friction dampers may be arranged to produce either constant or variable friction force and can be designed to act in one (linear), two (planar) or three (spatial) directions.

Friction dampers integrated in elastic elements have found wide application in freight bogies in Russia, the USA and many other countries because of their advantages of simplicity of design and fabrication, low cost and easy maintenance. Disadvantages of such dampers include sub-optimal damping in the partially laden condition, the difficulties of controlling friction to the desired design values and changes in friction levels as the surfaces wear or become contaminated in service.

The main advantage of planar and spatial friction dampers is their ability to damp vibrations in several directions and, in certain cases, provide friction-elastic connections between parts of the bogie frame. Such properties allow significant simplification of the bogies while retaining reasonable damping of complex vibrations. They are therefore widely used in freight bogies despite a number of disadvantages, including providing unpredictable friction forces, and the fact that repair and adjustment of friction forces may require lifting the car body and disassembling the spring set.

Telescopic dampers have the advantage of being autonomous, protected from the environment (which reduces the likelihood of contamination of the friction surfaces) and being able to be installed at angles other than vertical, and hence, they can be used to damp vertical or horizontal vibrations of sprung elements of the vehicle. They can be inspected and repaired without lifting the car body. One of the reasons that such telescopic dampers are not widely used in freight vehicles using the

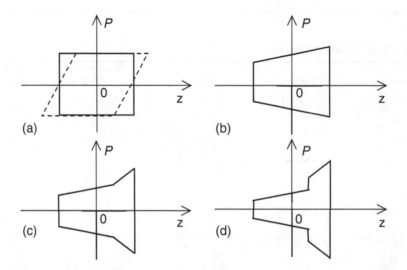

FIGURE 6.32 Typical force characteristics of friction dampers. (a) constant friction damper where the friction force does not depend on deformation on the spring set. (b) Most common friction damper in freight bogies. The friction force depends on the deflection of the suspension. (c) and (d) Multi-mode dampers where the friction forces vary according to the design and depend on spring set deflection. (From Iwnicki, S. (Ed.), *Handbook of Railway Vehicle Dynamics*, CRC Press, Boca Raton, FL, 2006. With permission.)

popular three-piece bogie is that an integrated friction wedge is required to resist warping in vertical and horizontal planes.

In case of the bogies with a rigid frame, friction dampers in the primary suspension must resist wheelset displacements. It is desirable that, in primary suspensions, the damper has an asymmetric characteristic, providing lower damping forces in compression than in extension. Hydraulic dampers are superior in this respect.

Typical force characteristics for friction dampers are presented in Figure 6.32. Different designs of friction damper have varying arrangements for transmitting the normal force to the friction surfaces. Depending on the design, the damper may provide constant or variable friction. In the latter case, such dampers are usually arranged, such that a component of the force in one or more of the suspension springs is transmitted via a linkage or wedge to the friction surfaces.

The force characteristic shown in Figure 6.32a describes a constant friction damper, where the friction force does not depend on deformation of the spring set and is the same for compression and tension. The dashed line shows the characteristic of the same damper, where the friction pairs are elastically coupled. This can occur, for example, as a result of the friction surface having an elastic layer underneath. Force P first deforms the elastic pad, and, when the shear force equals the friction breakout force (i.e., μN), then relative displacement of the friction pair occurs.

The characteristic in Figure 6.32b is common for most friction dampers used on freight bogies. The friction force depends on the deflection of the suspension and is different for tension and compression.

The characteristics in Figure 6.32c and d are typical for multi-mode dampers, where the friction forces vary according to the given design and depend on the spring set deflection in tension or compression.

It can be seen that the variety of force characteristics available from friction dampers allows freight vehicles to be designed with suspensions, providing satisfactory ride qualities.

6.4 CONSTRAINTS AND BUMPSTOPS

Constraints are the devices that limit the relative displacements of bogie units in longitudinal and lateral directions. See also the discussion in [2].

Suspension Elements and Their Characteristics

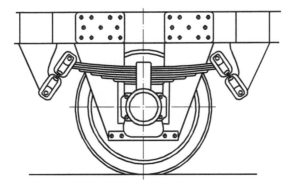

FIGURE 6.33 Axlebox located by horn guides. (From Iwnicki, S. (Ed.), *Handbook of Railway Vehicle Dynamics*, CRC Press, Boca Raton, FL, 2006. With permission.)

6.4.1 Horn Guides

As shown in Figure 6.33, a simple primary suspension design uses horn guides to limit the movement of the axlebox.

This design has several disadvantages, including fast wear of friction surfaces, leading to the increases in clearances; lack of elastic longitudinal and lateral characteristics and increased friction force in the vertical direction in traction and braking modes when the axlebox is pressed against the slides. The design could be improved by the application of anti-friction materials that do not require lubrication and have high resistance to wear.

6.4.2 Cylindrical Guides

These comprise two vertical guides and two barrels sliding along them. Typically, the vertical guides are attached to the bogie frame and the barrels are attached to the axlebox, as shown in Figure 6.34. The barrels are attached to the axlebox through rubber coaxial bushings and

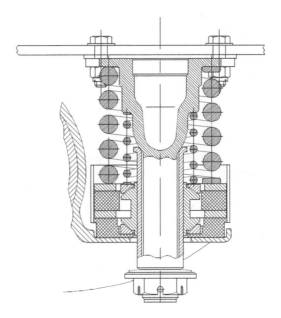

FIGURE 6.34 Connection between the axlebox and bogie frame using cylindrical guides. (From S. Iwnicki (ed.), *Handbook of Railway Vehicle Dynamics*, CRC Press, Boca Raton, FL, 2006. With permission.)

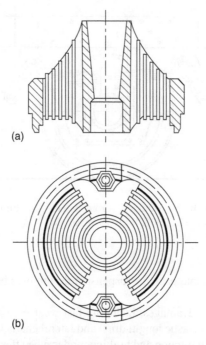

FIGURE 6.35 Two-section rubber-metal block used to connect the axlebox and bogie frame. (From Iwnicki, S. (Ed.), *Handbook of Railway Vehicle Dynamics*, CRC Press, Boca Raton, FL, 2006. With permission.)

therefore provide some flexibility between the wheelset and the bogie frame in the longitudinal and lateral direction. Owing to the axial symmetry of the rubber bushes, the stiffness in the longitudinal and lateral directions is the same, which may limit the provision of optimal suspension characteristics.

The axlebox constraint with cylindrical guides, where the displacement of the axlebox along the guides occurs by shear deformation of multi-layer rubber-metal block, is free from the disadvantages of classical constructions. Such axlebox designs are used on French TGV Y2-30 bogies. In order to obtain the optimum relationship of horizontal and vertical stiffness, this block consists of two longitudinally oriented sections, as shown in Figure 6.35.

6.4.3 Beam Links

The desire to avoid wear led to the development of links in the form of thin elastic beams that hold the wheelset in the longitudinal direction, as shown in Figure 6.36.

When primary suspension springs deflect, the beam links bend, whereas they experience tension or compression for traction and braking. To provide vertical flexibility in such construction, it is necessary for at least one of the links to have longitudinal flexibility. This is achieved by attaching the beam to a longitudinally flexible spring support (the Minden Deutz link) or by attaching the links to the frame through radially elastic joints (as in the primary suspension design of Japanese trains).

The main disadvantage of such designs is the high stress developed around the joints at either end of the beam.

Suspension Elements and Their Characteristics 231

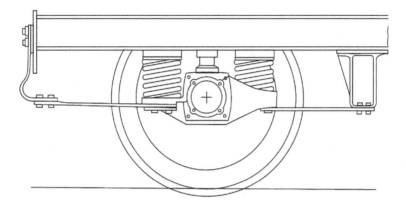

FIGURE 6.36 Connection between axlebox and bogie frame, using beam links. (From S. Iwnicki (ed.), *Handbook of Railway Vehicle Dynamics*, CRC Press, Boca Raton, FL, 2006. With permission.)

6.4.4 Constraints Using Radius Links

The use of rubber-metal bushes avoids surface friction and corresponding wear. The main problem with a radius link arrangement is obtaining linear motion of axleboxes when the links rotate. Alstom designed such an arrangement where the links are positioned on different levels in an anti-parallelogram configuration, as shown in Figure 6.37, and this has found wide application. Links that connect the axlebox to the frame provide linear displacement of its centre. By careful choice of size and the material of the rubber elements, it is possible to obtain required stiffness values in

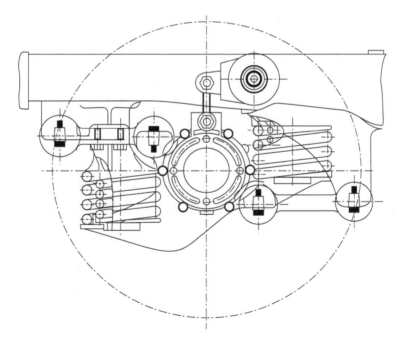

FIGURE 6.37 Radius links positioned at different heights in an anti-parallelogram configuration. (From Iwnicki, S. (Ed.), *Handbook of Railway Vehicle Dynamics*, CRC Press, Boca Raton, FL, 2006. With permission.)

different directions. Owing to the position of the links, lateral displacements do not cause misalignment of the axlebox, therefore providing optimum conditions for the bearings.

Disadvantages of the radius link design include the significant vertical stiffness of the connection due to torsion stiffness of the bushes. Increasing the length of the levers would decrease the vertical stiffness, but this is limited by the space available in the bogie frame.

6.4.5 Constraints Using Trailing (Radial) Arms

Trailing arm suspensions allow the design of shorter and lighter bogie frames. Such designs are now widely used in passenger vehicle primary suspensions such as the Y32 bogie shown in Figure 6.38.

The disadvantages of such designs include the longitudinal displacement of the axleboxes caused by vertical displacement of the suspension and changes in the wheelset attitude due to lateral displacements.

6.4.6 Traction Rods

Traction rods are normally used to transmit longitudinal (traction and braking) forces in either the primary or secondary suspension. They are typically composed of a rod with a rubber 'doughnut' or bushes at each end, as shown in Figure 6.39. They may be of adjustable length to maintain the necessary linear dimensions as wheels or suspension components wear.

FIGURE 6.38 Trailing arm suspension on Y32 bogie. (From Iwnicki, S. (Ed.), *Handbook of Railway Vehicle Dynamics*, CRC Press, Boca Raton, FL, 2006. With permission.)

FIGURE 6.39 Traction rod. (With kind permission from CRC Press: *Handbook of Railway Vehicle Dynamics*, CRC Press, Boca Raton, FL, 2006, Iwnicki, S. [ed.].)

Suspension Elements and Their Characteristics

6.5 CAR BODY TO BOGIE CONNECTIONS

The connection between the car body and bogie must:

- Allow the bogie to turn relative to the car body in curves;
- Transmit the vertical, traction and braking forces;
- Provide additional control of lateral suspension inputs;
- Assist in maintaining the stability of the bogie; and
- Provide longitudinal stability of bogie frames and equal distribution of load over the wheelsets.

These problems are solved differently, depending on the type of the rolling stock – traction or trailing, passenger or freight, moderate or high speed.

If the vehicle is stable up to the design speed, then introduction of additional yaw resistance torque is not necessary. If the static deflection of the suspension is sufficient, then vertical flexibility in connections between the car body and bogie may not be necessary.

Designs generally aim to make the connections between the car body and bogie as simple as possible by utilising a small number of elements and reducing the number of elements with surface friction.

6.5.1 Flat Centre Bowl and Side Bearings

In three-piece freight bogies, the most common connection is the flat circular centre bowl, which is secured by a pin pivot at the centre, with side bearings moved closer to the ends of the bolster, as shown in Figure 6.40. The centre bowl on the bogie interacts with the centre plate on the car body.

The bowl transmits most of the car body weight through its horizontal support surface and the longitudinal and lateral interaction forces through the rim. The pin pivot has large in-plane gaps to the car body and only provides emergency restraint. When the car body rocks on the flat centre plate, a gravitation resistance torque having soft characteristics is produced. The centre plate to centre bowl connection allows the bogie to rotate in curves and creates a friction torque that resists such rotation. Hence, the cylindrical centre bowl provides a connection between the bogie and the car body in all directions.

Such a unit is of simple construction but has several disadvantages. Firstly, clearances exist in the lateral and longitudinal directions. Secondly, relative motion occurs under high contact pressure and hence the surfaces are subject to significant wear. Various solutions can be applied to protect the surfaces from wear, with the use of wear resistant disk and ring being the most popular.

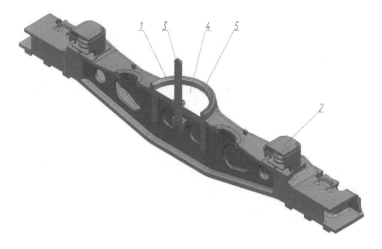

FIGURE 6.40 Flat centre plate with side bearings: (1) centre bowl; (2) side bearing; (3) pivot; (4) wear-resistant disk; (5) wear-resistant ring.

FIGURE 6.41 Variants of constant contact side bearings: (a) with non-metal elastic element and a roller; (b) with coil springs; and (c) with non-metal elastic element.

In classic bogies, the side bearings are metal structures having a clearance between their horizontal surface and the corresponding plates on the car body. In curves, the car body leans on the side bearing, creating additional friction torque that resists bogie rotation and increases wheel-rail forces. When the car body rocks on straight track, the contact surface becomes very small, and high contact pressures can lead to cracks in the centre bowl and the centre plate.

To combat these problems, modern designs use a flat centre bowl combined with constant contact side bearings, as shown in Figure 6.41. This resists car body rock and reduces the load on the centre bowl. Constant contact side bearings can use either coil springs or non-metal elastic elements to support part of the car body weight. Some of them are equipped with a roller that plays the part of the bumpstop without increasing the yaw torque.

6.5.2 Spherical Centre Bowl and Side Bearings

With the arrangement shown in Figure 6.42, the car body rests on the spherical centre bowl and on elastic constant contact side bearings that are necessary, because the centre bowl itself does not produce any roll recovery torque.

The advantage of this design is the lack of clearances in the horizontal plane and no edge contact during car body roll. This results in reduced levels of contact stress and increases the centre bowl service life. Such centre bowls are widely used in Y25 freight bogies, electric trains and underground cars in Russia.

6.5.3 Centre Column

The desire to exclude edge contact and increase the friction torque to resist bogie yaw led to the development of bogies with centre columns, as shown in Figure 6.43. Most of the car body weight

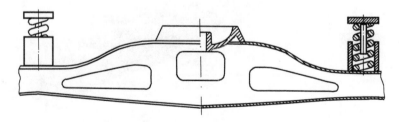

FIGURE 6.42 Spherical centre bowl with constant contact side bearings. (From Iwnicki, S. (Ed.), *Handbook of Railway Vehicle Dynamics*, CRC Press, Boca Raton, FL, 2006. With permission.)

Suspension Elements and Their Characteristics

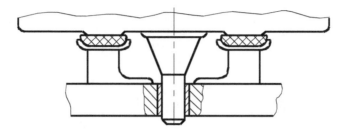

FIGURE 6.43 Centre column and side bearings. (From Iwnicki, S. (Ed.), *Handbook of Railway Vehicle Dynamics*, CRC Press, Boca Raton, FL, 2006. With permission.)

in this case is transmitted to the side bearings, and the car body can only turn relative to the bolster about the vertical axis.

This design is widely used in passenger coaches of the former USSR. The disadvantages include the clearances in longitudinal and lateral directions. The design provides sufficient ride quality only for bogies having low lateral stiffness of the secondary suspension.

6.5.4 Watts Linkage

The Watts linkage arrangement illustrated in Figure 6.44 allows the bogie to rotate while restricting longitudinal and lateral movement. It therefore provides a means of transmitting traction and braking forces. Pivots in the linkage are provided with rubber bushes to prevent the transmission of high-frequency vibrations through the mechanism.

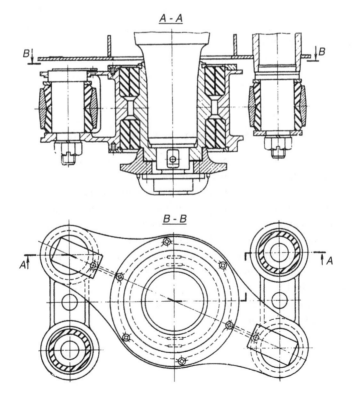

FIGURE 6.44 Watts linkage. (From Iwnicki, S. (Ed.), *Handbook of Railway Vehicle Dynamics*, CRC Press, Boca Raton, FL, 2006. With permission.)

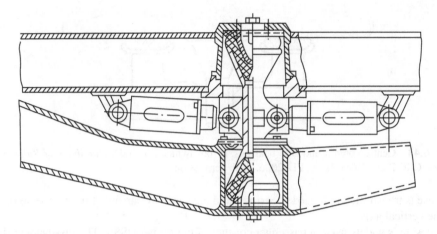

FIGURE 6.45 Pendulum linkage. (From Iwnicki, S. (Ed.), *Handbook of Railway Vehicle Dynamics*, CRC Press, Boca Raton, FL, 2006. With permission.)

6.5.5 Pendulum Linkage

The pendulum linkage consists of a vertical rod connected at each end to the body and bogie frame by conical rubber bushes, as shown in Figure 6.45. The mechanism is held in a central position by two pre-compressed springs. Elastic side supports provide lateral stability to the car body. For the small displacements that are typical of bogie hunting on straight track, the pendulum support provides almost infinite stiffness determined by the initial compression of the springs. When large displacements develop in curves, the support provides low stiffness. Thus, the pendulum support has a soft non-linear characteristic.

The drawbacks of such an arrangement are the rigid connection with a gap in the longitudinal direction, complex tuning requirements for the pre-compressed springs and friction forces in the additional sliding supports.

6.5.6 Connection of Car Body to Bolsterless Bogie

The complexity of the designs described previously accounted for the development of modern bolsterless bogies using either coil springs or air springs.

In such suspensions, the springs can achieve large deflections in shear, providing sufficiently large longitudinal displacements to allow the bogie to rotate in curves, as depicted in Figure 6.46.

The top of the coil springs rests on resilient blocks, arranged to provide a cylindrical joint, with rotation axis perpendicular to the track axis, as shown in Figure 6.47.

A similar approach is used in bogies with secondary air suspension. In this case, the air spring is often arranged in series with a rubber-metal spring to provide some suspension if the air spring deflates. Transmission of longitudinal forces is done through the centre pivot, Watts linkage, traction rods or, in the case of a Y32 bogie, through backstay cables. Bolsterless bogie designs typically achieve reductions in bogie mass of around 0.5–1.0 tonne.

Suspension Elements and Their Characteristics

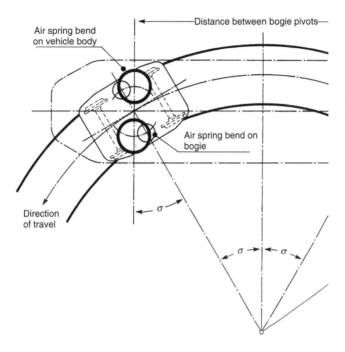

FIGURE 6.46 Schematic showing a bolsterless bogie passing through a curve. (From Iwnicki, S. (Ed.), *Handbook of Railway Vehicle Dynamics*, CRC Press, Boca Raton, FL, 2006. With permission.)

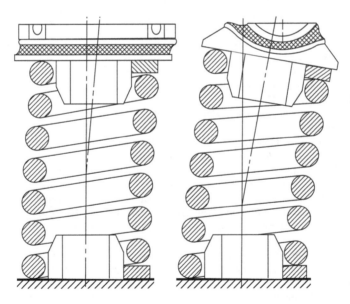

FIGURE 6.47 Spring resting on rubber-metal cylindrical joints. (From Iwnicki, S. (Ed.), *Handbook of Railway Vehicle Dynamics*, CRC Press, Boca Raton, FL, 2006. With permission.)

REFERENCES

1. E. Andersson, M. Berg, S. Stichel, *Rail Vehicle Dynamics*, Royal Institute of Technology (KTH), Stockholm, Sweden, 2014.
2. S. Bruni, J. Vinolas, M. Berg, O. Polach, S. Stichel, Modelling of suspension components in a rail vehicle dynamics context, *Vehicle System Dynamics*, 49(7), 1021–1072, 2011.
3. EN 13906-1:2013, Cylindrical helical springs made from round wire and bars: Calculation and design – Part 1: Compression springs, 2013.
4. W.H. Wittrick, On elastic wave propagation in helical springs, *International Journal of Mechanical Science*, 8(1), 25–47, 1966.
5. J. Lee, D.J. Thompson, Dynamic stiffness formulation, free vibration and wave motion of helical springs, *Journal of Sound and Vibration*, 239(2), 297–320, 2001.
6. S.P. Timoshenko, J.M. Gere, *Theory of Elastic Stability*, 2nd ed., McGraw-Hill, New York, 1961.
7. J.A. Haringx, On highly compressible helical springs and rubber rods, and their application for vibration-free mountings, Philips Research Report 4, Eindhoven, the Netherlands, 1949.
8. O. Krettek, M. Sobczak, Zur Berechnung der Quer- und Biegekennung von Schraubenfedern für Schienenfahrzeuge, *ZEV-Glasers Annalen*, 112(9), 319–326, 1988.
9. S. Nishizawa, M. Ikeda, J. Logsdon, H. Enomoto, N. Sato, T. Hamano, The effect of rubber seats on coil spring force line, SAE 2002 World Congress, Detroit, MI, 4–7 March 2002, Technical Paper 2002-01-0317, Society of Automotive Engineers, Warrendale, PA, 2002.
10. C.M. Wang, J.N. Reddy, K.H. Lee, *Shear Deformable Beams and Plates: Relationships with Classical Solutions*, Elsevier, Oxford, UK, 2000.
11. C. Weidemann, S. Mulski, Modelling helical springs, *SIMPACK News*, 11(1), 7, 2007.
12. J. Nicolin, T. Dellmann, Über die modellhafte Nachbildung der dynamischen Eigenschaften einer Gummifeder, *ZEV-Glasers Annalen*, 109(4), 1985.
13. P.E. Austrell, *Mechanical Properties of Carbon Black Filled Rubber Vulcanizates, Proceedings of Rubbercon'95*, Gothenburg, Sweden, 10–12 May, 1995.
14. S. Bruni, A. Collina, Modelling the viscoelastic behaviour of elastomeric components: An application to the simulation of train-track interaction, *Vehicle System Dynamics*, 34(4), 283–301, 2000.
15. H. Peeken, S. Lambertz, Nichtlineares Materialmodell zur Beschreibung des Spannungs-Dehnungsverhaltens gefüllter Kautschuk-Vulkanisate, *Konstruktion*, 46(1), 9–15, 1994.
16. P. Pfeffer, K. Hofer, Einfaches nichtlineares Modell für Elastomer- und Hydrolager zur Optimierung der Gesamtfahrzeug-Simulation, *ATZ – AutomobiltechnischeZeitschrift*, 104(5), 442–451, 2002.
17. M. Berg, A model for rubber springs in the dynamic analysis of rail vehicles, *Journal of Rail and Rapid Transit*, 211(2), 95–108, 1997.
18. M. Berg, A non-linear rubber spring model for rail vehicle dynamics analysis, *Vehicle System Dynamics*, 30(3–4), 197–212, 1998.
19. P. Allen, A. Hameed, H. Goyder, Measurement and modelling of carbon black filled natural rubber components, *International Journal of Vehicle System Modelling and Testing*, 2(4), 315–344, 2007.
20. A. Alonso, J.G. Giménez, J. Nieto, J. Vinolas, Air suspension characterisation and effectiveness of a variable area orifice, *Vehicle System Dynamics*, 48(S1), 271–286, 2010.
21. L. Mazzola, M. Berg, Secondary suspension of railway vehicles – air spring modelling: Performance and critical issues, *Journal of Rail and Rapid Transit*, 228(3), 225–241, 2014.
22. N. Oda, S. Nishimura, Vibration of air suspension bogies and their design, *Bulletin of JSME*, 13(55), 43–50, 1970.
23. C. Weidemann, Air-Springs in simpack, *SIMPACK News*, December 2003, pp. 10–11.
24. M. Presthus, Derivation of air spring model parameters for train simulation, Master's thesis, Luleå University of Technology, Sweden, 2002.
25. J. Evans, M. Berg, Challenges in simulation of rail vehicle dynamics, *Vehicle System Dynamics*, 47(8), 1023–1048, 2009.
26. B.M. Eickhoff, J.R. Evans, A.J. Minnis, A review of modelling methods for railway vehicle suspension components, *Vehicle System Dynamics*, 24(6–7), 469–496, 1995.
27. DeltaRail Group Ltd, *VAMPIRE Pro User Manual—V 5.02*, Derby, UK, 2006.
28. M. Berg, A three-dimensional airspring model with friction and orifice damping, In R. Fröhling (Ed.), *Proceedings 16th IAVSD Symposium Held in Pretoria*, South Africa, 30 August–3 September 1999, Swets and Zeitlinger, Lisse, the Netherlands, 528–539, 2000.
29. N. Docquier, P. Fisette, H. Jeanmart, Multiphysic modelling of railway vehicles equipped with pneumatic suspensions, *Vehicle System Dynamics*, 45(6), 505–524, 2007.

30. G. Quaglia, M. Sorli, Air suspension dimensionless analysis and design procedure, *Vehicle System Dynamics*, 35(6), 443–475, 2001.
31. A.J. Nieto, A.L. Morales, A. Gonzalez, J.M. Chicharro, P. Pintado, An analytical model of pneumatic suspensions based on an experimental characterization, *Journal of Sound and Vibration*, 313, 290–307, 2008.
32. N. Docquier, P. Fisette, H. Jeanmart, Model based evaluation of railway pneumatic suspensions, *Vehicle System Dynamics* 46(S1), 481–493, 2008.
33. N. Docquier, Multiphysics modelling of multibody systems: Application to railway pneumatic suspensions, PhD thesis, Université Catholique de Louvain, Louvain-la-Neuve, Belgium, 2010.
34. N. Docquier, P. Fisette, H. Jeanmart, Influence of heat transfer on railway pneumatic suspension dynamics, Paper presented at *21st International Symposium on Dynamics of Vehicles on Roads and Tracks (IAVSD'09)*, Stockholm, Sweden, 17–21 August 2009.
35. A. Facchinetti, L. Mazzola, S. Alfi, S. Bruni, Mathematical modelling of the secondary airspring suspension in railway vehicles and its effect on safety and ride comfort, *Vehicle System Dynamics*, 48(S1), 429–449, 2010.
36. M.C. Potter, D.C. Wiggert, *Mechanics of Fluids*, 3rd ed., Brooks/Cole, Pacific Grove, CA, 2002.
37. Crane Co., Flow of fluids through valves, fittings, and pipe: Metric edition, Technical Paper No. 410 M, New York, 1999.
38. O. Qing, S. Yin, The non-linear mechanical properties of an airspring, *Mechanical Systems and Signal Processing*, 17(3), 705–711, 2003.
39. S. Bruni, R. Goodall, T.X. Mei, H. Tsunashima, Control and monitoring for railway vehicle dynamics, *Vehicle System Dynamics*, 45(7–8), 743–779, 2007.
40. J.S. Tang, Passive and semi-active airspring suspension for rail passenger vehicle – theory and practice, *Journal of Rail and Rapid Transit*, 210(2), 103–117, 1996.
41. Y. Sugahara, A. Kazato, R. Koganei, M. Sampei, S. Nakaura, Suppression of vertical bending and rigid-body-mode vibration in railway vehicle carbody by primary and secondary suspension control: Results of simulations and running tests using Shinkansen vehicle, *Journal of Rail and Rapid Transit*, 223(6), 517–531, 2009.
42. S. Alfi, S. Bruni, G. Diana, A. Facchinetti, L. Mazzola, Active control of airspring secondary suspension to improve ride quality and safety against crosswinds, *Journal of Rail and Rapid Transit*, 225(1), 84–98, 2011.
43. M. Sorli, W. Franco, S. Mauro, G. Quaglia, R. Giuzio, G. Vernillo, Features of the lateral active pneumatic suspensions in the ETR 470 high speed train, In J. Adolfsson, J. Karlsen (Eds.), *Proceedings 6th UK Mechatronics Forum International Conference*, 9–11 September 1998, Skövde, Sweden, Pergamon, Oxford, UK, 621–626, 1998.
44. A. Conde Mellado, C. Casanueva, J. Vinolas, J.G. Giménez, A lateral active suspension for conventional railway bogies, *Vehicle System Dynamics*, 47(1), 1–14, 2009.
45. ORE, Flexibility of trapezoidal springs, Technical Report ORE B12/Rp25, 2nd ed., Office of Research and Experiment (ORE) of the International Union of Railways, Utrecht, the Netherlands, 1986.
46. ORE, Technical Report ORE B12/Rp43, Parabolic springs for wagons (design, calculation, treatment), Office of Research and Experiment (ORE) of the International Union of Railways, Utrecht, the Netherlands, 1988.
47. S. Stichel, On freight wagon dynamics and track deterioration, *Journal of Rail and Rapid Transit*, 213(4), 243–254, 1999.
48. ORE, Technical Report ORE B56/Rp2, Part 1: Study of the effect of various simple modifications on the stability of old-type two-axled wagons (first work progress report). Part 2: Attempt to define a general criterion for riding quality, Office of Research and Experiment (ORE) of the International Union of Railways, Utrecht, the Netherlands, 1967.
49. M.R. Joly, Technical Report ORE DT30, Etude de la stabilité transversale d'un véhicule ferroviaire à deux essieux, Office of Research and Experiment (ORE) of the International Union of Railways, Utrecht, the Netherlands, 1974.
50. J.B. Ayasse, et al., Computer simulation of freight vehicles with leaf springs – A comparison between different packages, Technical Report INRETS/RE-01–046-FR, 2001.
51. P.A. Jönsson, Modelling and laboratory investigations on freight wagon link suspension, with respect to vehicle-track dynamic interaction, Licentiate thesis, TRITA AVE 2004:48, Royal Institute of Technology (KTH), Stockholm, Sweden, 2004.
52. P.A. Jönsson, E. Andersson, S. Stichel, Experimental and theoretical analysis of freight wagon link suspension, *Journal of Rail and Rapid Transit*, 220(4), 361–372, 2006.

53. P.A. Jönsson, E. Andersson, S. Stichel, Influence of link suspension characteristics variation on two-axle freight wagon dynamics, *Vehicle System Dynamics*, 44(S1), 415–423, 2006.
54. P.A. Jönsson, S. Stichel, I. Persson, New simulation model for freight wagons with UIC link suspension. *Vehicle System Dynamics*, 46(S1), 695–704, 2008.
55. M. Hoffmann, Dynamics of European two-axle freight wagons, PhD thesis, IMM-PHD-2006–170, Technical University of Denmark, Kongens Lyngby, Denmark, 2006.
56. H. Lange, *Dynamic Analysis of a Freight Car with Standard UIC Single-Axle Running Gear*, TRITA-FKT 1996:34, Royal Institute of Technology (KTH), Stockholm, Sweden, 1996.
57. M. Stiepel, S. Zeipel, Freight wagon running gears with leaf spring and ring suspension, *6th SIMPACK User Group Meeting*, Eisenach, Germany, 9–10 November 2004.
58. F. Xia, Modelling of wedge dampers in the presence of two-dimensional dry friction, *Vehicle System Dynamics* 37(S1), 565–578, 2002.
59. A.B. Kaiser, J.P. Cusumano, J.F. Gardner, Modelling and dynamics of friction wedge dampers in railroad trucks, *Vehicle System Dynamics*, 38(1), 55–82, 2002.

7 Wheel-Rail Contact Mechanics

Jean-Bernard Ayasse, Hugues Chollet and Michel Sebès

CONTENTS

7.1 Introduction 242
    7.1.1 Basic Model of a Wheelset, Degrees of Freedom 243
7.2 The Normal Contact 244
    7.2.1 Hertzian Contact...244
        7.2.1.1 Curvature Ratio *A/B*, Relation with *b/a*...246
        7.2.1.2 Calculation of the Semi-Axis...246
        7.2.1.3 Convexity, Concavity and Radius Sign..247
        7.2.1.4 Particular Case: 2D Contact 248
        7.2.1.5 Contact Pressure 248
    7.2.2 Tables, Polynomial Expressions and Faster Methods....................................248
        7.2.2.1 Continuous Expression of the Tables 248
        7.2.2.2 Analytical Approximation of the Tables..249
    7.2.3 Application to the Railway Field 249
        7.2.3.1 Contact Plane Angle, Conicity...249
        7.2.3.2 Note on Conicity 249
        7.2.3.3 Normal Load...250
        7.2.3.4 Determination of the Longitudinal Curvature of the Wheel............250
        7.2.3.5 Contact Point between the Wheel and Rail Profiles........................250
        7.2.3.6 Direct Comparison of the Antagonist Curvatures 250
        7.2.3.7 One or Several Ellipses? Notion of Contact Jump 251
        7.2.3.8 One or Two Wheel-Rail Couples? 251
7.3 The Tangent Problem...251
    7.3.1 Forces and Couples on a Wheelset...251
        7.3.1.1 Approximations...252
    7.3.2 Tangent Forces: Rolling Friction Simple Models 253
        7.3.2.1 Historical Review...253
    7.3.3 Linear Expressions of the Creep Forces 253
    7.3.4 Definition of Creepages 254
        7.3.4.1 Quasi-Static Creepages 254
        7.3.4.2 Quasi-Static Creepages in the Railway Case...................................254
        7.3.4.3 Dynamic Formulation of the Creepages 256
        7.3.4.4 Damping Terms and Stability 256
        7.3.4.5 Non-Dimensional Spin Creepage 256
    7.3.5 Kalker's Coefficients c_{ij}..257
        7.3.5.1 Values of c_{ij} for Simplified Bogie Models..257
    7.3.6 Creep Forces in the Linear Domain 257
        7.3.6.1 Dependence with the Load 257
        7.3.6.2 Creepages Combinations and Saturation 259
        7.3.6.3 Using Linear Models: The C110, C220 Stiffnesses 259
        7.3.6.4 Reduced Creepages..259

7.3.7 Saturation Laws 259
    7.3.7.1 Vermeulen and Johnson 260
    7.3.7.2 Kalker's Empirical Proposition 260
    7.3.7.3 Exponential Saturation Law: CHOPAYA, Ohyama and Others 260
7.3.8 FASTSIM-Based Contact Models 261
    7.3.8.1 Discretised Ellipse: FASTSIM from Kalker 261
    7.3.8.2 Stresses 262
    7.3.8.3 Linear Contact Forces, Elastic Coefficients 263
    7.3.8.4 Reduced Creepages in FASTSIM 264
    7.3.8.5 Extensions of FASTSIM 265
7.3.9 Pre-Tabulated Methods 265
7.4 Contact Forces in the Railway Context 266
    7.4.1 From the Pure Dicone to the Wheelset with Real Profiles 266
        7.4.1.1 Equivalent Conicity 266
        7.4.1.2 Variable Conicity 266
        7.4.1.3 Profile Measurements: Importance of the Angular Reference ... 266
    7.4.2 Gravitational Stiffness 266
    7.4.3 Flange Contact 267
        7.4.3.1 Flange Contact Jump and Definition of the Nominal Play in the Track 267
        7.4.3.2 Contact Jump and Load Transfer 267
    7.4.4 Using the Contact Angle Function 268
        7.4.4.1 Localisation of Multiple Contacts on the Tread 268
        7.4.4.2 Gravitational Centring Ability in Standard Conditions 269
    7.4.5 The Diverging Effect of Spin, Influence on the Normal Load 269
    7.4.6 Safety Criteria, Nadal's Formula 270
    7.4.7 Independent Wheel, Application to Industrial Mechanisms on Rails 270
    7.4.8 Modelling the Contact Jumps 271
        7.4.8.1 Hertzian Multiple Contacts 271
    7.4.9 Advanced Methods for Non-Hertzian Contacts 272
7.5 Recent Works and Advanced Models 274
Appendices 275
    Appendix 7.1 Kinematic Movement and the Klingel's Formula 275
    Appendix 7.2 Kinematic Hunting and Equivalent Conicity 275
    Appendix 7.3 The Circle Theory 275
    Appendix 7.4 Analysis of Y/Q and Nadal's Criteria 276
Nomenclature 277
References 278

7.1 INTRODUCTION

For more than 150 years, the steel wheel on the steel rail has provided the railway system with an exceptional level of safety. This safety level is so high that the contact mechanism is neglected by many and considered as being a simple slider by most people.

However, the railway engineer's point of view can be very different, especially when taking into account responsibilities in a railway network. The wheel-rail contact is actually a complex and imperfect link. Firstly, it is a place of highly concentrated stresses. Secondly, the simple conical wheel shape makes the wheelset a complex mechanical actuator, incorporating a sensor limited by the transverse play, a force amplifier able to steer dozens of tons and with partially sliding surfaces limiting the forces. The contact surfaces can be described by the Hertzian theory, similarly to those in roller bearings but without protection against dust, rain, sand, leaves or even ballast stones. This is why, Chapter 8 of this handbook is dedicated to wheel-rail tribology.

Wheel-Rail Contact Mechanics

Looking closer, the railway safety can be maintained at a very high level if some precautions are taken. The aim of this chapter is to give to railway engineers a basis for understanding and evaluating the wheel-rail contact situation.

Historically, the first theoretical model of the wheel-rail longitudinal contact force is due to Carter in the USA [1]. More recently, Vermeulen and Johnson in the UK [2] and Kalker in the Netherlands [3] set out the basis for an accurate description.

In parallel, multibody dynamic software developers decided to model the wheel-rail contact with, at first, a constant conicity limited by the two flange contact springs. Then, variable conicity was taken into account, then the spin effect, and later contact 'jumps' representing contact location changes due to the profile combination. These jumps are a major difficulty in the calculation of wheel-rail forces, able to produce unrealistic wheelset jumps if incorrectly managed, and this is a major step towards the non-Hertzian description of the contact.

However, the first step in the wheel-rail contact study is to consider the Hertzian modelisation, starting from a simple model of a wheelset rolling with two contacts on two rails.

7.1.1 Basic Model of a Wheelset, Degrees of Freedom

A real track being sinuous, the longitudinal axis is a curvilinear abscissa frequently denoted by S. The track being considered rigid, the railway wheelset behaviour can be simplified to a plane model (Figure 7.1) limited to two main degrees of freedom:

- The lateral displacement, t_y
- The yaw angle, α

When the behaviour of a wheelset is unstable, the dynamic combination of these two degrees of freedom results in an action called 'hunting'.

The lateral displacement t_y and the yaw angle α must be considered as two small displacements relative to the track. Unless considering derailment, the play in the track gauge will be the limit of the lateral displacement. It is generally about ±8 mm. The yaw angle is limited to the order of two degrees (35 milliradians) for urban railway applications in sharp curves. With this low value, it is possible to consider that $\sin(\alpha) = \alpha$, $\cos(\alpha) = 1$, and other approximations are made.

The other degrees of freedom are constrained; the displacement along OS and the axle rotation speed ω around oy are determined by the longitudinal speed V_x and the rolling radius of the wheel r_0 in the central position, with $V_x = \omega \, r_0$. The wheelset centre of gravity height z and the roll angle around OS are linked to the rails when there is contact on both rails. The speed of the wheelset V is not necessarily parallel to ox or to OS; a lateral speed V_y will be introduced in the creepage definition later.

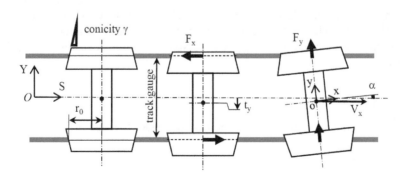

FIGURE 7.1 Wheelset degrees of freedom. (From Iwnicki, S. (ed.), *Handbook of Railway Vehicle Dynamics*, CRC Press, Boca Raton, FL, 2006. With permission.)

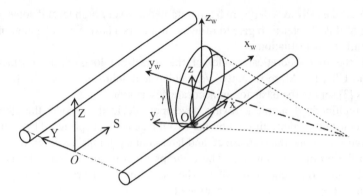

FIGURE 7.2 Rail, wheel and contact frames. (From Iwnicki, S. (ed.), *Handbook of Railway Vehicle Dynamics*, CRC Press, Boca Raton, FL, 2006. With permission.)

The railway wheelset is basically described by two conical, nearly cylindrical wheels (Figures 7.1 and 7.2), linked together with a rigid axle. Each wheel is equipped with a flange, the primary role of which seems to be to avoid derailment. In a straight line, the flanges are not in contact, but the rigid link between the two wheels suggests that the railway wheelset is designed to go straight ahead and will go to flange contact only in curves.

This is the railway dicone or wheelset.

The interface between the wheel and the rail appears to be a small quasi-horizontal contact patch. The contact pressure on this small surface is closer to a stress concentration. The centre of this surface is also the application point of tangential forces (traction and braking F_x, guiding or parasite forces F_y, as shown in Figure 7.1). The knowledge of these forces is necessary to determine the general wheelset equilibrium and its dynamic behaviour.

In order to determine this behaviour and these forces, the first step is to determine some contact parameters: the contact surface, the pressure and the tangential forces. This determination is generally separated into two procedures:

- Determining the normal problem (Hertz theory)
- Determining the tangential problem (Kalker's theory)

7.2 THE NORMAL CONTACT

The study of the contact between bodies is possible today with finite element methods. However, the necessity to calculate as fast as possible millions of contact cases during the millions of time steps of a dynamic simulation leads to the use of analytical methods. This section first describes the classical Hertzian model, and then, some particular considerations aimed at speeding up and softening the numerical calculation are presented. Non-Hertzian models will be briefly described in Section 7.4.9.

7.2.1 Hertzian Contact

Hertz demonstrates [4] that, when two elastic bodies are pressed together in the following conditions:

- Elastic behaviour
- Semi-infinite spaces
- Large curvature radius compared to the contact size
- Constant curvatures inside the contact patch

Wheel-Rail Contact Mechanics

then:

- The contact surface is an ellipse
- The contact surface is considered flat
- The contact pressure is a semi-ellipsoid

The main curvatures of the two semi-spaces are needed for the calculation of the surface dimensions and the pressure distribution. In the railway case, the four main curvatures can be considered to be in perpendicular planes; their directions correspond to the main axes of the frame: xOy (Figures 7.3 and 7.4).

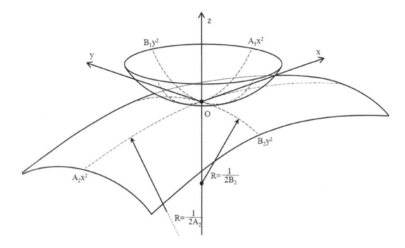

FIGURE 7.3 Hertzian contact: general case. (From Iwnicki, S. (ed.), *Handbook of Railway Vehicle Dynamics*, CRC Press, Boca Raton, FL, 2006. With permission.)

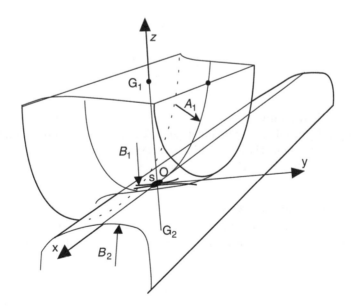

FIGURE 7.4 Hertzian contact: the railway case. (From Iwnicki, S. (ed.), *Handbook of Railway Vehicle Dynamics*, CRC Press, Boca Raton, FL, 2006. With permission.)

Considering the two elastic bodies in contact, they will meet at a single point O, where the normal distance between them is minimal. Near this contact point O, without load, the surface shapes of the bodies are represented by two second-order polynomials:

$$z_1 = A_1 x^2 + B_1 y^2$$
$$z_2 = A_2 x^2 + B_2 y^2 \qquad (7.1)$$

The $A_{1,2}$ and $B_{1,2}$ coefficients are assumed constant in the neighbourhood of the contact point O and linked to the main local curvatures by the second partial differential equations, the first being neglected if described in the contact frame. In the railway case (Figure 7.4), these curvatures and radii are expressed as:

wheel:
$$\frac{d^2 z_1}{dx^2} = 2A_1 \approx \frac{1}{r_n}$$
$$\frac{d^2 z_1}{dy^2} = 2B_1 \approx \frac{1}{R_{wx}}$$

rail:
$$\frac{d^2 z_2}{dy^2} = 2B_2 \approx \frac{1}{R_{rx}} \qquad (7.2)$$

In the railway case, the curvature A_2 is generally neglected, as the rail is straight; the radius is infinite. B_1 and B_2 are deduced from the transversal profiles, A_1 from r_n (the radius of the wheel normal to the contact plane) is itself deduced from the rolling radius of the wheel, as described in Section 7.2.3.4.

7.2.1.1 Curvature Ratio A/B, Relation with b/a

Before being loaded, the vertical relative distance $d(x, y)$ between the two bodies can be written as:

$$z_1 + z_2 = d = Ax^2 + By^2$$

with
$$A = \frac{1}{2r_n} \quad \text{and} \quad B = \frac{1}{2}\left(\frac{1}{R_{wx}} + \frac{1}{R_{rx}}\right) \qquad (7.3)$$

and A and B being strictly positive.

Conventionally, a is the ellipse longitudinal semi-axis length in the Ox direction, and b is that in the transversal Oy direction. The A/B and b/a ratios vary in the same way: if $A > B$, then $b > a$. The equality $A/B = 1$ leads to a circular contact patch, as $a = b$.

7.2.1.2 Calculation of the Semi-Axis

The traditional calculation is based on the determination of the ratio of the semi-axis lengths: $g < 1$, ($g = b/a$ or a/b), function of B/A by using an intermediate parameter, the angle θ defined as:

$$\cos\theta = \frac{|B - A|}{B + A} \qquad (7.4)$$

Wheel-Rail Contact Mechanics

The practical values of the semi-axis lengths a and b, and of δ being the reduction of the distance between the bodies' centres, are given by:

$$a = m \left(\frac{3}{2} N \frac{1-v^2}{E} \frac{1}{A+B} \right)^{1/3}$$

$$b = n \left(\frac{3}{2} N \frac{1-v^2}{E} \frac{1}{A+B} \right)^{1/3}$$

$$\delta = r \left(\left(\frac{3}{2} N \frac{1-v^2}{E} \right)^2 (A+B) \right)^{1/3} \tag{7.5}$$

E is the Young's modulus and v is the Poisson's ratio, assuming the same material for the rail and the wheel. These equations are correct if $a \geq b$. This limitation may be overcome with the solution proposed in Section 7.2.2.

Here m, n and r are non-dimensional Hertz coefficients, which can be calculated by elliptic integrals, tabulated for an engineering use as a function of the ratio $g = n/m$ or the virtual angle θ, as shown in Table 7.1.

πab, being the surface of the ellipse, can be expressed as a function of:

$$ab = mn \left(\frac{3}{2} \frac{1-v^2}{E} \frac{1}{A+B} \right)^{2/3} N^{2/3} \tag{7.6}$$

in which the first term contains the material and geometrical constants, and the second term contains only the load.

7.2.1.3 Convexity, Concavity and Radius Sign

The sign of each radius is important, because one of the calculation methods uses $A - B$ and $A + B$ values to determinate the shape of the ellipse. Each radius is, by definition, positive if the curvature centre is *inside* the body. The wheel rolling radius and, most of the time, the rail transverse radius are positive (convex). But the wheel transverse radius at the contact point can be positive (convex) or negative (concave); see further discussion in Section 7.2.3.6.

TABLE 7.1
Hertz Coefficients ($A/B < 1$)

					$\theta°$					
	90	80	70	60	50	40	30	20	10	0
$g = n/m$	1	0.7916	0.6225	0.4828	0.3652	0.2656	0.1806	0.1080	0.0470	0
m	1	1.128	1.285	1.486	1.754	2.136	2.731	3.816	6.612	∞
n	1	0.8927	0.8000	0.7171	0.6407	0.5673	0.4931	0.4122	0.3110	0
r	1	0.9932	0.9726	0.9376	0.8867	0.8177	0.7263	0.6038	0.4280	0

7.2.1.4 Particular Case: 2D Contact

When the wheel and the rail are conformal, they have the same transversal curvature value with opposite sign, giving $B = 0$ and particular values for m, n and r; see Table 7.1 for $\theta = 0$. In this case, the contact is simplified to two-dimensional (2D) or 'cylindrical contact' in engineering texts.

However, in multibody software packages, the integration procedure could result in perturbations due to the shift between 2D and three-dimensional (3D) expressions. It is better to stay with a finite 3D surface giving a continuous numerical result from time step to time step. A practical way to manage the evolution that may include slender ellipses could be to limit the g value or A/B or b/a ratios to, for example, 1/20. Another solution is proposed in Section 7.2.2.

7.2.1.5 Contact Pressure

With an elliptical pressure distribution and the mean pressure being $N/\pi ab$, the maximal pressure will be simply $\sigma_{max} = 1.5 N / \pi ab$.

In the railway field, the maximal contact pressure is frequently over 1000 MPa. This value is over the elastic limit of most steels determined in traction, but the compression state is more complex than a simple tensile test, and the elastic limit is not reached. The determination of the plastification (which is a limit of the Hertzian hypothesis) must be calculated with a criterion based on the hydrostatic (Von Mises) stress and with the use of a more accurate tool, namely finite elements. However, in a dynamic software package, simplified expressions can estimate appropriate modifications of the contact pressure distribution [5].

7.2.2 TABLES, POLYNOMIAL EXPRESSIONS AND FASTER METHODS

7.2.2.1 Continuous Expression of the Tables

The traditional Hertz table limited to $1 > g > 0$ leads to the necessity of determining if the A/B value is more or less than 1 and to switch between $g = b/a$ or $g = a/b$, while at the same time, $g = n/m$ becomes $g = m/n$, which is somewhat confusing.

This ambiguity can be avoided with a description of the tables in the interval $0 < g < \infty$. Then, in all the cases, $g = b/a = n/m$.

The expressions of the elliptic integrals used to calculate the tabulated values of n and m have the property:

$$n(A/B) = m\left(\frac{1}{A/B}\right) \quad (7.7)$$

The traditional Hertz table can be rewritten as shown in Table 7.2.

With this presentation, A/B can be used directly as the input of the tables, instead of $\cos\theta$.

TABLE 7.2
Hertz Coefficients for θ = 0 to 180°

	$\theta°$										
	0	5	10	30	60	90	120	150	170	175	180
A/B	0	0.0019	0.0077	0.0717	0.3333	1	3.0	13.93	130.6	524.6	∞
$b/a = n/m = g$	0	0.0212	0.0470	0.1806	0.4826	1	2.0720	5.5380	21.26	47.20	∞
m	∞	11.238	6.612	2.731	1.486	1	0.7171	0.4931	0.311	0.2381	0
r	0	0.2969	0.4280	0.7263	0.9376	1	0.9376	0.7263	0.4280	0.2969	0

Wheel-Rail Contact Mechanics

7.2.2.2 Analytical Approximation of the Tables

Rather reliable expressions of the *n/m* and *mn* values as a function of *A/B* [0, ∞] can be proposed with:

$$\frac{b}{a} = \frac{n}{m} \approx \left(\frac{A}{B}\right)^{0.63} \qquad (mn)^{3/2} \approx \left(\frac{1+A/B}{2\sqrt{A/B}}\right)^{0.63} \tag{7.8}$$

These expressions have the advantage of being continuous, simple and fast to calculate.

The proposed exponent 0.63 is a compromise. It is equal to 2/3 when *A/B* is close to 1, but the value 0.63 will better describe the slender ellipses; the difference with the tabulated values is about +/−5% between *b/a* = 1/25 and *b/a* = 25 (see Table 7.3).

7.2.3 APPLICATION TO THE RAILWAY FIELD

7.2.3.1 Contact Plane Angle, Conicity

With a perfectly conical wheel, the contact place on the rail appears to be the point with the same slope as that on the wheel, measured in the *YOZ* frame (see Figure 7.4).

For a perfectly conical wheel of 1:20 slope, and with new rails tilted at 1:20, this give a contact point in the middle of the rail (see Table 7.4).

7.2.3.2 Note on Conicity

There is such a large set of wheel profiles that the main cone of the wheel is not able to describe the wheel-rail contact; for a large part of the time, the contact is not on the conical part of the wheel. The mean value is considered to be the conicity. Its calculation not only considers the wheel profile but also includes the rail profile and the track clearance.

TABLE 7.3
Approximation of the *n/m* Values

						$\theta°$					
	0	5	10	30	60	90	120	150	170	175	180
A/B	0	0.0019	0.0077	0.0717	0.3333	1	3.0	13.93	130.6	524.6	∞
b/a = *n/m*	0	0.0212	0.0470	0.1806	0.4826	1	2.0720	5.5380	21.26	47.20	∞
$(A/B)^{0.63}$	0	0.0193	0.0466	0.1901	0.5005	1	1.9980	5.2564	21.530	51.700	∞

TABLE 7.4
Typical Wheelset Loads, Rail Inclinations and Wheel Cone Values

	Intercity & Freight	Metros	Tramway
Wheel cone	1:20 to 1:40 for High Speed Train (HST)	1:20	1:20 to flat
Rail tilt	1:20 France, England 1:30 Sweden 1:40 Germany	1:20	Flat
Wheelset load	22.5 t per wheelset, but 17 t per wheelset HST	10 t per wheelset	6 t per wheelset

7.2.3.3 Normal Load

On the tread, with a low conicity, the normal load N has practically the same value as the vertical load on the wheel; for example, having a cone angle with a tangent of 1:20 gives:

$$N = Q\cos(\operatorname{atan}(1/20)) = 0.9988\,Q \tag{7.9}$$

However, a distinct reference frame must be considered and is noted as z for the normal direction and x, y for the tangent ones.

7.2.3.4 Determination of the Longitudinal Curvature of the Wheel

This curvature is referred to as 'longitudinal' because the rolling circle of a cylindrical wheel would be in the XOS plane. But the wheel is generally conical, the contact angle γ is not zero, and the A_1 curvature in the xOz plane differs from the rolling radius r_o. The intersection between xOz and the wheel cone is an ellipse, with one focus situated on the wheelset axis, as shown in Figure 7.5, and the curvature radius can be approximated to:

$$\frac{1}{r_n} = \frac{\cos\gamma}{r_o} \tag{7.10}$$

Note than the radius value at the contact point is a little different between the tread and the flange, but this variation (+10 to +15 mm) can be neglected, being of a second order, compared to the $\cos\gamma$ influence. The centred value r_o is taken.

7.2.3.5 Contact Point between the Wheel and Rail Profiles

In order to determine the contact point, the wheel and rail profiles are placed one relative to the other, with the wheelset being centred on the track as a function of the wheelset gauge D_w and rail gauge D_r, and then moved laterally (Figure 7.6a), with the lateral relative displacement, or lateral shift, being a translation noted in the figure as t_y. In this latter position, the minimal vertical distance between the wheel and the rail defines the contact point O. On a real track, the rail gauge D_r will be variable along S.

The minimal distance is simply determined by the minimal vertical distance when the lateral position is fixed. It is of no use to calculate accurately the vertical translation t_z of the rigid wheel profile at this step, because we are interested in the contact situation, not the relative trajectory when the wheel lifts.

7.2.3.6 Direct Comparison of the Antagonist Curvatures

Independently of the minimal distance determination, it can be observed that B cannot have a negative value at the contact point. A simple comparison of the corresponding curvatures (Figure 7.6b) gives information on the possibility of having or not having the centre of a Hertzian contact in this zone.

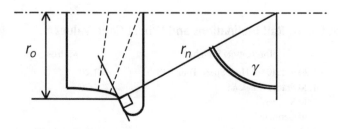

FIGURE 7.5 Determination of the longitudinal curvature r_n from the rolling radius r_o. (From Iwnicki, S. (ed.), *Handbook of Railway Vehicle Dynamics*, CRC Press, Boca Raton, FL, 2006. With permission.)

Wheel-Rail Contact Mechanics

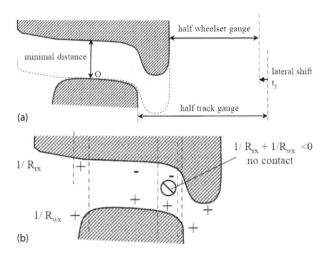

FIGURE 7.6 Contact between wheel and rail: (a) minimal distance determination of contact point and (b) corresponding curvatures indicate zone of no contact. (From Iwnicki, S. (ed.), *Handbook of Railway Vehicle Dynamics*, CRC Press, Boca Raton, FL, 2006. With permission.)

In the transition zone on the wheel between the tread and the flange, contact is frequently impossible, as the sum $\frac{1}{R_{wx}} + \frac{1}{R_{rx}}$ must always be positive to have contact.

However, if the centre of a contact cannot be located in this area, it can be covered partially by the contact ellipses.

7.2.3.7 One or Several Ellipses? Notion of Contact Jump

When the wheel profile is concave, during wheel lateral translation, the contact point seems to 'move' gradually on the surfaces. However, when the curvatures are discontinuous, the contact point sometimes seems to move suddenly or jump, for example, from one side of the 'no contact' zone to the other side, as shown on Figure 7.6b. The word 'jump' used here does not have a dynamic but rather a geometric meaning. This jump is associated with a difficulty in the railway contact calculation: the determination of the load transfer, which itself is able to have a dynamic effect on the wheel. This will be developed in Section 7.4.

Practically, the profile description, the digitisation and smoothing operations will have an influence on the accuracy of the curvatures, on the contact position and on the contact jumps' description.

7.2.3.8 One or Two Wheel-Rail Couples?

As a first step, the study of a single wheel-rail pair is enough to give a large set of information on the contact as a function of the lateral relative displacement t_y.

In a more complex algorithm, it is also possible to take into account the roll effect of the opposite wheel-rail pair of the wheelset, in which case the tables are functions of t_y and roll.

In the same way, the yaw angle has an influence on the longitudinal position of the contact on the wheel, and this can be taken into account with an input of multiple parameters.

Generally, these effects are not very influential on the wheelset equilibrium.

7.3 THE TANGENT PROBLEM

7.3.1 FORCES AND COUPLES ON A WHEELSET

The kinematic representation of the wheelset (Klingel's formula, Appendix 7.2) has been used for a long time to explain the sinusoidal behaviour of a free wheelset rolling without external forces [6], but the situation is different under a real vehicle.

The real wheelset is strongly linked to the vehicle through flexible suspension elements, and these links create significant forces when the wheelset is entering a curve or running on a real track with defects.

The suspension forces find their reaction forces (normal and tangent) at the wheel-rail contact interface, where the tangent components or **creep forces** are related to the relative speed between the two bodies: the creepages.

In the contact coordinate systems, the forces are denoted by:

- N for the normal force
- F_x for the longitudinal creep force
- F_y for the lateral creep force in the contact plane

The F_y forces must be projected on the track plane OY and summed with the projection of the normal forces N to give the guiding force.

7.3.1.1 Approximations

The main couple exerted on a rigid wheelset around OZ is coming from the longitudinal forces F_x on each rail that are separated laterally by the contact distance D_c:

$$M_z = -\left(F_{xg}\frac{D_c}{2} - F_{xd}\frac{D_c}{2}\right) \tag{7.11}$$

The spin creepage, due to the relative rotation of the contact patches around the normal axis to the contact, generates a couple (ϕ in Figure 7.7), but it can be neglected in comparison with the longitudinal forces couple M_z. However, the spin generates a lateral force, which is not negligible when the contact angle becomes higher. This lateral force is described separately or included with the yaw force, depending on the theory used.

The couple $\Sigma\phi$ has a component resisting the forward displacement, which can be neglected in a first approximation.

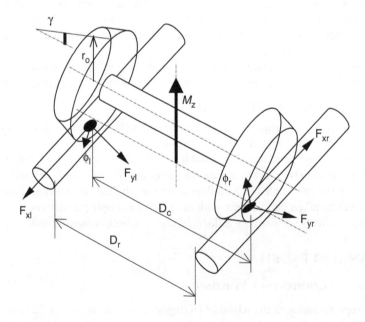

FIGURE 7.7 Wheelset geometry and creep forces. (From Iwnicki, S. (ed.), *Handbook of Railway Vehicle Dynamics*, CRC Press, Boca Raton, FL, 2006. With permission.)

Wheel-Rail Contact Mechanics

7.3.2 Tangent Forces: Rolling Friction Simple Models

The wheel-rail contact is a rolling friction contact. It differs from the sliding friction Coulomb model shown in Figure 7.8a (which can be found at the brake shoes level) with an area of adhesion and an area of slip, which appears progressively as the slip speed increases. The transition is characterised by the initial slope or 'no-slip force', representing the force if the friction coefficient μ was infinite and the continuous 'S' saturation curve (see Figure 7.8b).

7.3.2.1 Historical Review

Following on from Hertz [4], Boussinesq [7] and Cerruti [8], three authors became interested in the beginning of the twentieth century in wheel-rail contact modelisation: Carter and Fromm for longitudinal models [1] and Rocard for the lateral force.

Carter described a simple 2D contact surface, but he was the first to give a rather adequate expression of the force relatively to the creepage in the longitudinal direction. His method to describe the stresses in the adhesive zone (see further discussion at Section 7.3.7.3) has been used until the 1960s. Fromm produced a similar work. Rocard described the linear relationship between the yaw angle and the guiding force, for rubber tyres and for railway wheels, in the lateral direction. He was interested particularly in the equivalent of bogie hunting for automobiles: the shimmy phenomenon.

In the 1960s, more experimental data were available, and the definitive expressions were established mainly by Johnson and Kalker, who give an expression of the creepage stiffness, introducing variable coefficients depending on the b/a ratio of the contact ellipse. This expression is the most common today.

7.3.3 Linear Expressions of the Creep Forces

In the case of a Hertzian contact, the creep forces are a function of the relative speeds between rigid bodies near the contact point, the **creepages**.

The general expressions of the creep forces take into account stiffness coefficients c_{ij} expressed in the linear theory of Kalker [3] by:

$$\begin{aligned} F_x &= -Gabc_{11}v_x \\ F_y &= F_{y\,yaw} + F_{y\,spin} \\ F_{y\,yaw} &= -Gabc_{22}v_y \\ F_{y\,spin} &= -Gabc_{23}c\varphi \qquad c = \sqrt{ab} \end{aligned} \qquad (7.12)$$

where G is the material shear modulus (steel in the railway case), πab is the contact ellipse surface and c_{ij} are coefficients given by Kalker in [9].

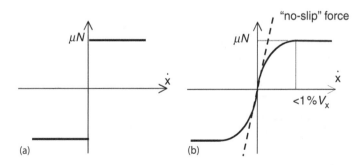

FIGURE 7.8 Simple friction models: (a) Coulomb's model and (b) rolling friction model. (From Iwnicki, S. (ed.), *Handbook of Railway Vehicle Dynamics*, CRC Press, Boca Raton, FL, 2006. With permission.)

7.3.4 Definition of Creepages

7.3.4.1 Quasi-Static Creepages

A general expression for two rolling bodies, as in the case of a roller rig (Figure 7.9), can be given by the projection of the speed vectors on Ox, Oy and Oz:

$$\text{Longitudinal: } v_x = \frac{proj./x\left(\vec{V}_0 - \vec{V}_1\right)}{\tfrac{1}{2}\left\|\vec{V}_0 + \vec{V}_1\right\|} \quad \text{(dimensionless)}$$

$$\text{Lateral: } v_y = \frac{proj./y\left(\vec{V}_0 - \vec{V}_1\right)}{\tfrac{1}{2}\left\|\vec{V}_0 + \vec{V}_1\right\|} \quad \text{(dimensionless)}$$

$$\text{Rotation: } \varphi = \frac{proj./z\left(\vec{\Omega}_0 - \vec{\Omega}_1\right)}{\tfrac{1}{2}\left\|\vec{V}_0 + \vec{V}_1\right\|} \quad (m^{-1}) \quad (7.13)$$

V_0 and V_1 are the absolute speeds at the contact, and $\tfrac{1}{2}\left(\vec{V}_0 + \vec{V}_1\right)$ is the mean speed.

ω_0 and ω_1 are the angular speeds of the two solids, with $\Omega i = Vi/ri$ being projected on the normal to the contact (see Figure 7.9).

7.3.4.2 Quasi-Static Creepages in the Railway Case

The general expressions described previously are useful for the test rigs used in research; however, the railway track case leads to simplified expressions.

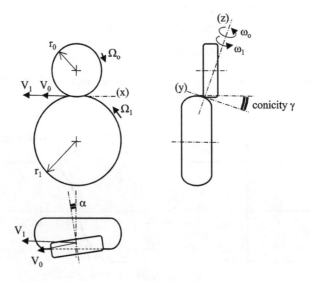

FIGURE 7.9 General geometry and creepages on a roller rig. (From Iwnicki, S. (ed.), *Handbook of Railway Vehicle Dynamics*, CRC Press, Boca Raton, FL, 2006. With permission.)

Wheel-Rail Contact Mechanics

7.3.4.2.1 Longitudinal Creepage ν_x

With small creepage, one may assume:

$$\tfrac{1}{2}(V_x + r\omega) \approx V_x$$

In quasi-static conditions, with r_o being the mean rolling radius, and ν_{xo} the mean longitudinal creepage, at the left:

$$\nu_{xl} = \frac{V_x - r_l\omega}{\tfrac{1}{2}(V_x + r_l\omega)} \approx \frac{V_x - (r_o + \Delta r_l)\omega}{V_x}$$

$$\nu_{xo} = \frac{V_x - r_o\omega}{V_x} \quad (7.14)$$

$$\nu_{xl} = \nu_{xo} - \frac{\Delta r_l}{r_o}\frac{r_o\omega}{V_x}$$

When the wheelset is rolling freely without traction or braking, ν_{xo} may be neglected. The static longitudinal creepage can be described with:

$$\nu_{xl} = -\frac{\Delta r_l}{r_o}$$

In the case of a perfectly conical wheelset with angle γ at left:

$$\Delta r_l = \gamma y$$

then

$$\nu_{xl} = -\gamma \frac{y}{r_o}$$

$$\nu_{xr} = -\nu_{xl}$$

Due to the conicity, the two different rolling radii generate two opposite forces, f_x and $-f_x$. Rather than the cone value, the equivalent conicity γ_e can be used there for simplified models (see Appendix 7.1).

7.3.4.2.2 Lateral Creepage ν_y

The lateral creepage in quasi-static conditions, with small creepages, is simply the yaw angle common to the two wheels:

$$\nu_y = -\alpha \quad (7.15)$$

In this expression, it is assumed that both wheels have a small contact angle γ. Otherwise, the lateral creepage may differ, as given by:

$$\nu_y = -\alpha / \cos \gamma$$

7.3.4.2.3 Spin Creepage φ

In the quasi-static case, the rail speed Ω is equal to zero, and the general expression is simplified.

The spin creepage φ is:

$$\varphi = -\sin\gamma/r_o \qquad (7.16)$$

(γ *is an algebraic value, different for the two wheels*)

This expression shows that the spin has an important value when flanging and a larger value with small radius wheels.

7.3.4.3 Dynamic Formulation of the Creepages

For the dynamic forces at the contact level, a formulation identical to the other dynamic links is established with an elastic term and a damping term.

The expressions of the dynamic creepages contain the speed terms whose sign is opposed to the elastic deformation, which will contribute to damping:

$$\text{For the left wheel: } v_{xl} = -\left(\frac{\Delta r_l}{r_o} + \frac{D_c}{2}\frac{\dot{\alpha}}{\dot{x}}\right)$$

$$\text{For the right wheel: } v_{xr} = -\left(\frac{\Delta r_r}{r_o} - \frac{D_c}{2}\frac{\dot{\alpha}}{\dot{x}}\right)$$

For conical wheels: $v_{xr} = -v_{xl}$

$$v_y = \frac{1}{\cos\gamma}\left(\frac{\dot{y}}{\dot{x}} - \alpha\right)$$

$$\varphi = -\frac{\sin\gamma}{r_o} + \frac{\dot{\alpha}}{\dot{x}}\cos\gamma \quad (\text{2nd term} \ll \text{1st term}) \qquad (7.17)$$

where α is the wheelset yaw angle relative to the rail (radians), r_0 is the mean rolling radius of the wheel, y is the lateral displacement of the wheel relatively to the centred position, $\dot{\alpha}, \dot{y}$ is the relative speed in the track reference system and $\dot{x} = V_x$ is the running speed of the wheelset along the curvilinear abscissa.

7.3.4.4 Damping Terms and Stability

In the creepage, the damping terms $\frac{\dot{y}}{\dot{x}}$ and $\frac{\dot{\alpha}}{\dot{x}}$ are inversely proportional to the forward speed. This means that these terms are reduced as the speed increases, and the wheelset becomes unstable.

7.3.4.5 Non-Dimensional Spin Creepage

The creepages are relative slips and have no dimension.

To become dimensionless, the last spin expression must be multiplied by a distance. The characteristic dimension of the ellipse $c = \sqrt{ab}$ is used to obtain the spin creepage in a homogeneous form [10]:

$$c\varphi = c\left(-\frac{\sin\gamma}{r_o} + \frac{\dot{\alpha}}{\dot{x}}\cos\gamma\right) \qquad (7.18)$$

Wheel-Rail Contact Mechanics

7.3.5 Kalker's Coefficients c_{ij}

The c_{ij} coefficients are given as a function of the b/a ratio of the ellipse, as shown in Table 7.5 (adapted from [10]). Their values are not far from π for b/a, close to 1. Initially, Carter uses the value π. Kalker's coefficients are deduced from the slope of the 'no-slip' force (see Figure 7.8b), computed with Kalker's exact theory [11].

In Table 7.5, the c_{11} and c_{22} values are given for Poisson's ratio values of 0 or 0.25 or 0.5. The typical value of steel is close to 0.27, and the tables must be interpolated.

Polynomial fits are proposed as:

$$c_{11} = 3.2893 + \frac{0.975}{b/a} - \frac{0.012}{(b/a)^2}$$

$$c_{22} = 2.4014 + \frac{1.3179}{b/a} - \frac{0.02}{(b/a)^2} \tag{7.19}$$

$$c_{23} = 0.4147 + \frac{1.0184}{b/a} + \frac{0.0565}{(b/a)^2} - \frac{0.0013}{(b/a)^3}$$

Before using these expressions, it is necessary to limit the b/a ratio to an interval (i.e., 1/25 to 25), the fit becoming wrong for very slender A/B ratios. It is possible to express the c_{ij} directly from the curvature ratio A/B (see Section 7.2.2.2).

7.3.5.1 Values of c_{ij} for Simplified Bogie Models

In some simplified bogie models with equivalent conicity (Appendix 7.1), a constant value is taken for c_{11} and c_{22}. The contact is assumed to be on the tread at all times, with values varying from 3 to 5.

When the flange contact is considered, the range grows to $a/b = 20$ (c_{22} reaches 12.8 when $a/b = 10$). During quasi-static flanging in a curve, an accurate determination is not critical, as the creepages are very large and the creep forces can be considered saturated. But, during a derailment, the lateral creepage diminishes and the c_{22} determination can be influential on the simulated dynamics.

7.3.6 Creep Forces in the Linear Domain

7.3.6.1 Dependence with the Load

The expressions at Equation 7.12, considering a Hertzian contact, can be normalised with the load, implicit in the contact dimensions a and b:

$$a = a_1 N^{1/3} \quad b = b_1 N^{1/3} \quad c_1 = \sqrt{a_1 b_1}$$

giving:

$$ab = a_1 b_1 N^{2/3} \quad abc = a_1 b_1 c_1 N \tag{7.20}$$

where a_1 and b_1 are the semi-axes, and c_1 is the square root of the semi-axes' product for 1 Newton load. Equation 7.12 becomes:

$$F_x = -G a_1 b_1 c_{11} N^{2/3} v_x$$

$$F_{y\ yaw} = -G a_1 b_1 c_{22} N^{2/3} v_y \tag{7.21}$$

$$F_{y\ spin} = -G a_1 b_1 c_{23} c_1 N \varphi$$

The spin force is proportional to the load, and the pressure $N/\pi ab$ is proportional to $N^{1/3}$.

TABLE 7.5
Kalker's Coefficients

	g	C_{11}			C_{22}			$C_{23} = -C_{32}$			C_{33}		
		$\nu = 0$	0.25	0.5	$\nu = 0$	0.25	0.5	$\nu = 0$	0.25	0.5	$\nu = 0$	0.25	0.5
	↓0.0	$\pi^2/4(1-\nu)$			$\pi^2/4 = 2.47$			$\dfrac{\pi\sqrt{g}}{3(1-\nu)}\left\{1+\nu(\ln(16/g)-5)\right\}$			$\pi^2/16(1-\nu)g$		
a/b	0.1	2.51	3.31	4.85	2.51	2.52	2.53	0.334	0.473	0.731	6.42	8.28	11.7
	0.2	2.59	3.37	4.81	2.59	2.63	2.66	0.483	0.603	0.809	3.46	4.27	5.66
	0.3	2.68	3.44	4.80	2.68	2.75	2.81	0.607	0.715	0.889	2.49	2.96	3.72
	0.4	2.78	3.53	4.82	2.78	2.88	2.98	0.720	0.823	0.977	2.02	2.32	2.77
	0.5	2.88	3.62	4.83	2.88	3.01	3.14	0.827	0.929	1.07	1.74	1.93	2.22
	0.6	2.98	3.72	4.91	2.98	3.14	3.31	0.930	1.03	1.18	1.56	1.68	1.86
	0.7	3.09	3.81	4.97	3.09	3.28	3.48	1.03	1.14	1.29	1.43	1.50	1.60
	0.8	3.19	3.91	5.05	3.19	3.41	3.65	1.13	1.25	1.40	1.34	1.37	1.42
	0.9	3.29	4.01	5.12	3.29	3.54	3.82	1.23	1.36	1.51	1.27	1.27	1.27
	1.0	3.40	4.12	5.20	3.40	3.67	3.98	1.33	1.47	1.63	1.21	1.19	1.16
b/a	0.9	3.51	4.22	5.30	3.51	3.81	4.16	1.44	1.59	1.77	1.16	1.11	1.06
	0.8	3.65	4.36	5.42	3.65	3.99	4.39	1.58	1.75	1.94	1.10	1.04	0.954
	0.7	3.82	4.54	5.58	3.82	4.21	4.67	1.76	1.95	2.18	1.05	0.965	0.852
	0.6	4.06	4.78	5.80	4.06	4.50	5.04	2.01	2.23	2.50	1.01	0.892	0.751
	0.5	4.37	5.10	6.11	4.37	4.90	5.56	2.35	2.62	2.96	0.958	0.819	0.650
	0.4	4.84	5.57	6.57	4.84	5.48	6.31	2.88	3.24	3.70	0.912	0.747	0.549
	0.3	5.57	6.34	7.34	5.57	6.40	7.51	3.79	4.32	5.01	0.868	0.674	0.446
	0.2	6.96	7.78	8.82	6.96	8.14	9.79	5.72	6.63	7.89	0.828	0.601	0.341
	0.1	10.7	11.7	12.9	10.7	12.8	16.0	12.2	14.6	18.0	0.795	0.526	0.228

Note: a, b = longitudinal and lateral semi-axis lengths of wheel-rail contact patch ellipse; $g = \min(a/b, b/a)$.

Wheel-Rail Contact Mechanics

7.3.6.2 Creepages Combinations and Saturation

In Equation 7.12, the transversal force due to the spin φ is separated from the force due to the lateral creepage ν_y. The total transversal force is:

$$F_y = F_{y\,\text{yaw}} + F_{y\,\text{spin}} \tag{7.22}$$

In the case of negligible spin, that is, in models considering mainly the tread contact, these expressions can be used, and the lateral force due to the spin can be added to the yaw force, or even neglected.

In the case of combined creepages, when the spin is not negligible, these independent expressions are not adequate because of the non-uniform combined saturation of the shear contact stresses inside the contact area. A model based on the surface description is necessary, the most commonly used being the FASTSIM model developed by Kalker [12].

7.3.6.3 Using Linear Models: The C110, C220 Stiffnesses

Some models need to establish the pre-calculated stiffness coefficients (mean or variable values). These parameters have been standardised by the UIC (French acronym of International Union of Railways) in the two main directions by the expressions:

$$C110 = G\,a\,b\,c_{11}$$

$$C220 = G\,a\,b\,c_{22}$$

It is useful to normalise them with the normal load N:

$$C110s = G\,a_1\,b_1\,c_{11}\,N^{2/3}$$

$$C220 = G\,a_1\,b_1\,c_{22}\,N^{2/3}$$

Note that these expressions are neglecting the spin effect.

7.3.6.4 Reduced Creepages

In the following expressions, it is useful to introduce a reduced parameter, which will be used in a great number of equations:

$$u_x = \frac{Gabc_{11}v_x}{\mu N} \quad u_y = \frac{Gabc_{22}v_y}{\mu N} \quad u_\varphi = \frac{Gabc_{23}c\varphi}{\mu N} \tag{7.23}$$

These parameters are called the no-slip reduced friction forces; they are characteristic of both the stiffness of the contact and the creepage. Reduced creepages enable the solving of the tangent problem in a dimensionless way with only four data inputs, these being the three reduced creepages in Equation 7.23 and the b/a ratio. See Section 7.3.8.4 for an alternate formulation of reduced creepages.

7.3.7 SATURATION LAWS

The first dynamic software packages used the linear stiffnesses expressions in order to calculate the critical speed of a bogie with eigenvalues.

The first time domain resolution programs introduced saturation as an evolution of the linear expressions with analytical formulations.

7.3.7.1 Vermeulen and Johnson

Vermeulen and Johnson's law [2] is a function of a reduced creepage coefficient neglecting the spin, given by $\tau = \sqrt{\tau_x^2 + \tau_y^2}$ that results in:

$$\frac{F}{\mu N} = 1 - (1-\tau)^3 \qquad 0 \le \tau \le 1$$

$$\frac{F}{\mu N} = 1 \qquad \tau > 1 \qquad (7.24)$$

$$\text{with} \quad \tau_x = \frac{Gabc_{11}v_x}{3\mu N} \quad \tau_y = \frac{Gabc_{22}v_y}{3\mu N}$$

$$\text{or} \quad \tau_x = \frac{u_x}{3} \quad \tau_y = \frac{u_y}{3}$$

Vermeulen and Johnson's proposition is close to Carter's. The method by Shen, Hedrick and Elkins [13] improves on Vermeulen and Johnson by adding the spin in the expression of τ_y:

$$\tau_y = \frac{u_y + u_\varphi}{3}$$

7.3.7.2 Kalker's Empirical Proposition

If the expression is limited to the lateral creepage and force modulus, a comparison is possible [9]; using the reduced parameter $\tau_y = u_y / 3$ gives:

$$\frac{F_y}{\mu N} = \left(\frac{3}{2}\tau_y \cos^{-1}\tau_y\right) + \left(1 - (1 + \frac{\tau_y^2}{2})\sqrt{1-\tau_y^2}\right) \qquad 0 \le \tau_y \le 1$$

$$\frac{F_y}{\mu N} = 1 \qquad \tau_y > 1 \qquad (7.25)$$

The expression is more complex if both directions are considered. Separating the two parts of his formulation, Kalker introduces the difference between the forces (e_2) and the slip direction (e_1, see [11]) as the creepage increases.

The original slope $C220_K$ calculated by Kalker is different from Johnson's; both are close to Johnson's experiments.

7.3.7.3 Exponential Saturation Law: CHOPAYA, Ohyama and Others

Ohyama [14] and, later, Ayasse-Chollet-Pascal (under the name CHOPAYA) have proposed a classical exponential saturation, starting from forces measurement in railway conditions:

$$\frac{F}{\mu N} = 1 - e^{-u} \qquad u = \sqrt{u_x^2 + (u_y + u_\varphi)^2} \qquad (7.26)$$

Some other saturation expressions have been proposed, such as the hyperbolic tangent and the arctangent [15]. All these laws (see Figure 7.10) are 'heuristic'; they are fitted on measured data [16], respecting the c_{ij}, but do not correspond to a physical saturation model.

Wheel-Rail Contact Mechanics

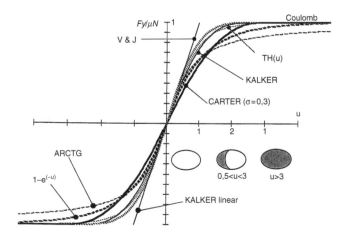

FIGURE 7.10 Heuristic expressions used for the saturation, and physical meaning of the different parts. (From Iwnicki, S. (ed.), *Handbook of Railway Vehicle Dynamics*, CRC Press, Boca Raton, FL, 2006. With permission.)

Neglecting the spin, the general mechanism of saturation can be divided into three steps. The linear zone is a full adhesion surface. The saturated case is slipping everywhere with the dry friction Coulomb value, and the intermediate zone is a partially saturating surface, where the slipping area is always at the rear side of the ellipse.

Qualitatively, when there is no spin, the slip-stick frontier presented in Figure 7.11 looks like a quarter moon propagating from the rear to the front with increasing creepage, ν_x or ν_y. Between these creepages, only the stress directions are different.

The situation is more complex with the presence of spin. To determine the shear stresses quantitatively, it is necessary to use a physically based model. The most important model for railway use is FASTSIM [12] from Kalker. It is presented in the following paragraph.

7.3.8 FASTSIM-BASED CONTACT MODELS

Kalker proposed several methods to solve the contact problem with models based on the surface description; these methods are widely described in his book [11], and only the simplest one is described here briefly. Both CONTACT and FASTSIM algorithms are based on the 'strip theory' originally proposed by Haines and Ollerton, and extended to the three creepages [10] (see Figure 7.11).

CONTACT [17] is a program based on the complete theory of elasticity. It can take into account several body shapes, including the railway case. Several methods are available to calculate the tangent stresses and/or the internal stresses. However, the calculation of one case takes several seconds, and it is limited to half-space elastic bodies.

7.3.8.1 Discretised Ellipse: FASTSIM from Kalker

FASTSIM, originally a Fortran subroutine [12], is based on the 'simplified theory' [18] and, in the original publication, is limited to ellipses, which means that some assumptions are common to Hertz:

- The contact surface is elliptic and flat, and the traction bound defined by the product of the friction coefficient μ with the pressure p_z may be an ellipsoid (consistent with Hertz) or a paraboloid, as proposed in [12] (leading to a more precise computation of the limit between the zone of adhesion and the zone of slip).
- The creepages are estimated at the ellipse centre.
- Kalker's coefficients c_{ij} are constant everywhere in the ellipse, with their values deduced from A/B or b/a.

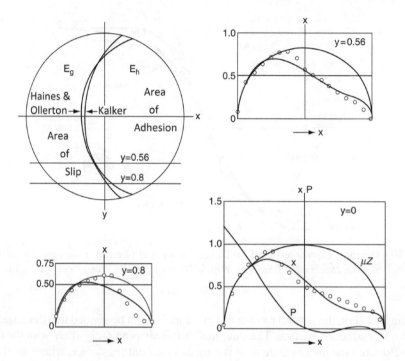

FIGURE 7.11 Comparison of separatrix results from Haines and Ollerton and the proposition from Kalker.

- The elliptic contact surface is divided into independent longitudinal parallel strips of length $2a_i$ and width Δy_i; in the original algorithm [12], the width Δy_i of the strip depends on its relative location with respect to the spin pole (the point where creepage vanishes), keeping it equidistant if the spin pole is out the contact patch.
- All the strips are divided in the same number of elements; the stress calculation is started from the leading edge, from element to element.
- The method is simplified: a local deformation corresponds to a local force.
- The saturation is calculated independently for each element, loaded by the normal force n_{ij}.

Practically, the surface is described by a grid separating parallel strips in the direction of rolling. Owing to the elliptic shape, the elements do not have the same length, and $\Delta x_i = a_i/MX$ (see Figure 7.12). Internal creepages are computed for each element, starting from the central creepage. The pressure and elementary forces are considered in the centre of each element. The pressure is defined by the ellipsoid value at this point.

7.3.8.2 Stresses

The simplified theory is an approximation of the exact one where a Winkler foundation replaces the continuous equations of Cerruti [8]. The xOy reference frame here is the plane containing the ellipse. The unsaturated stress distribution in the x and y directions is the following (simplified or 'linear' theory, [12]):

$$p_x(x,y) = \left(\frac{v_x}{L_1} - y\frac{\varphi}{L_3}\right)(x - a_i) \tag{7.27}$$

Wheel-Rail Contact Mechanics

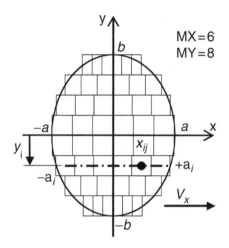

FIGURE 7.12 Strips and elements discretisation with FASTSIM. (From Iwnicki, S. (ed.), *Handbook of Railway Vehicle Dynamics*, CRC Press, Boca Raton, FL, 2006. With permission.)

The first term represents the mean rigid longitudinal slip, and the second term is the spin effect as a local rigid slip at the point (x, y) in the strip, with a_i being the leading edge of this strip.

$$p_y(x,y) = \frac{v_y}{L_2}(x - a_i) + \frac{\varphi}{2L_3}(x^2 - a_i^2) \tag{7.28}$$

L_1, L_2 and L_3 are the elasticity coefficients (or flexibilities) of the contact [12,19]. Their values are deduced by equating the sum of previous expressions over the ellipse with Equation 7.12. The simplified theory matches with the exact theory in a small pure creepage case:

$$L_1 = \frac{8a}{3Gc_{11}} \qquad L_2 = \frac{8a}{3Gc_{22}} \qquad L_3 = \frac{\pi a \sqrt{a/b}}{4Gc_{23}} \tag{7.29}$$

As an alternative to this formulation with three flexibilities, another possibility is to define one equivalent flexibility:

$$L = \frac{|v_x|L_1 + |v_y|L_2 + \sqrt{ab}|\varphi|L_3}{\sqrt{v_x^2 + v_y^2 + ab\varphi^2}}$$

This latter expression is more suitable for high values of combined creepages [18]. In the following, the formulation with three flexibilities is described with a parabolic traction bound.

7.3.8.3 Linear Contact Forces, Elastic Coefficients

The contact forces are the integral distributions of the surface stresses. When there is no saturation:

$$F_x = \iint_{\text{ellipse}} p_x dS = -\frac{8a^2 b}{3L_1} v_x$$

$$F_y = \iint_{\text{ellipse}} p_y dS = -\frac{8a^2 b}{3L_2} v_y - \frac{\pi a^3 b \varphi}{4L_3} \tag{7.30}$$

7.3.8.4 Reduced Creepages in FASTSIM

FASTSIM is composed of two parts; the main program calls the subroutine SR, in which the stresses are computed in loops.

Practically, starting from the physical creepages, the four inputs of SR are the following reduced creepages (see also Section 7.3.6.4):

$$u'_x = \frac{3\pi}{16} \frac{Gabc_{11}}{\mu N} v_x = \frac{3\pi}{16} u_x$$

$$u'_y = \frac{3\pi}{16} \frac{Gabc_{22}}{\mu N} v_y = \frac{3\pi}{16} u_y$$

$$u'_{\varphi y} = 2 \frac{G(ab)^{3/2} c_{23}}{\mu N} \varphi = 2u_\varphi$$

$$u'_{\varphi x} = \frac{b}{a} u'_{\varphi y} \tag{7.31}$$

Here, it can be seen that the numerical coefficient $3\pi/16$ is coming from the L_1 and L_2 parameters.

The SR algorithm is computed in a normalised form (ellipse reduced to a circle of radius one). Preceding values are associated to a parabolic traction bound. In order to switch to an elliptic traction bound, they should be multiplied by 4/3, the ratio between σ_{max} with a parabolic pressure and σ_{max} with an elliptic one (see Section 7.2.1.5).

At a given point of coordinates (x'_{ij}, y'_i) of the circle of radius one, the parabolic traction bound is defined by:

$$\mu p'_n = 1 - x'^2_{ij} - y'^2_i \tag{7.32}$$

On a given strip, starting from the leading edge where shears (p'_{xi0}, p'_{yi0}) are assumed to be zero, the following iterative process is used:

$$p'_{Axij} = p'_{xij-1} - \Delta x'_i (u'_x - y'_i u'_{\varphi x})$$

$$p'_{Ayij} = p'_{yij-1} - \Delta x'_i (u'_y + x'_{ij} u'_{\varphi y}) \tag{7.33a}$$

where p'_A is the shear vector assuming adhesion.

The shear vector p' at element (i, j) is limited by the traction bound:

$$\|p'_{Aij}\| \leq \mu p'_n : \quad p'_{ij} = p'_{Aij} \quad \text{adhesion}$$

$$\|p'_{Aij}\| > \mu p'_n : \quad p'_{ij} = \frac{p'_{Aij}}{\|p'_{Aij}\|} \mu p'_n \quad \text{slip} \tag{7.33b}$$

The vector of physical tangent forces $F = (F_x, F_y)$ is deduced from the summation over the unit circle:

$$F = \mu \frac{2N}{\pi} \Delta y' \sum_{i=1}^{MY} \sum_{j=1}^{MX} p'_{ij} \Delta x'_i \tag{7.34}$$

The stresses are independent from one strip to the other, but they are not independent inside a strip. The calculation is shown on Figure 7.13 on a strip; on the right side, the pure spin case is presented without any longitudinal stress, even from the spin effect, in order to simplify the presentation.

Wheel-Rail Contact Mechanics

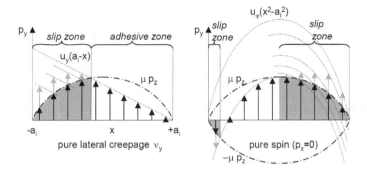

FIGURE 7.13 Internal saturation of the stresses inside FASTSIM showing simplified cases. (From Iwnicki, S. (ed.), *Handbook of Railway Vehicle Dynamics*, CRC Press, Boca Raton, FL, 2006. With permission.)

The calculation begins with a deformation and stress at zero on the leading edge of each strip: $+a_i$. The stress is incremented as a function of the stresses' functions; for example, in the direction Oy (second line of Equation 7.33a), the increment is constant for the lateral creepage but variable for the spin.

When the incremented stress reaches more than μp_z (shadowed arrows on Figure 7.13), it is saturated, and the new step starts from this correct value (black arrows).

The pure spin calculation shows that an adhesive zone is persistent after the centre of the contact. It can disappear with the other creepages.

7.3.8.5 Extensions of FASTSIM

Although intended to handle elliptical contact patches, FASTSIM may be extended to non-Hertzian cases. A contour and the corresponding pressure distribution can be inserted coming from a model other than Hertz. In this case, it is necessary to adapt the creepages' description and to adapt the L_i parameters [20].

Although a calculation of the dissipated power is accessible with the FASTSIM algorithm [18,21], a more robust assessment is often preferred, using the wear factor (also known as T-gamma):

$$F_x v_x + F_y v_y$$

It can be a criterion to compare different vehicles going into the same curve at the same speed. It must be used to give access to wear simulations.

The program CONTACT from Kalker is more adapted to such studies. And the wear coefficients are themselves very dependent on lubrication and slip speed. They are adjusted from experiments.

7.3.9 Pre-Tabulated Methods

As shown in the previous section, tangent forces in any ellipse may be expressed as a function of four inputs: u'_x, u'_y, $u'_{\varphi x}$ and $u'_{\varphi y}$, or alternatively u_x, u_y, u_φ and b/a. As an alternative to FASTSIM, this multivariate function may be pre-tabulated with the exact theory [11], leading to Kalker's USETAB book of tables.

The two software packages NUCARS and VAMPIRE are interpolating large pre-tabulated tables. This pre-tabulating strategy has been recently extended by Piotrowski et al. [22] to a special case of a non-Hertzian patch, the simple double-elliptical contact region (SDEC).

7.4 CONTACT FORCES IN THE RAILWAY CONTEXT

7.4.1 FROM THE PURE DICONE TO THE WHEELSET WITH REAL PROFILES

For a complete understanding of the wheel-rail forces, a description of the cross-sectional profiles of the wheel and rail are required.

7.4.1.1 Equivalent Conicity

The assimilation of these profiles to a cone rolling on a cylinder has been useful for quasi-static curving but cannot be used when the contact is moving on the wheel tread or the rail head or to predict stability or derailment.

The first linearisation was to consider only the wheel tread dicone.

The 'circle theory' (see Appendix 7.3) has been an improvement of the diconic formulation as a more realistic case; both rail and wheel profiles are considered as circles, giving an additional term to the yaw couple. Today, it is easier to take into account the real profiles, and this method has become obsolete.

7.4.1.2 Variable Conicity

The real conicity to be taken into account has a variable value. On the wheel, the contact with the flange is at least a cone at 70°, which supplements the usual 1:20 cone, but in most cases, these two slopes are connected by a concave part, which can frequently be in contact with the rail. This 'intermediate' contact, where the cone angle γ is variable, is very important, both for steering and for stability considerations.

Real profiles, measured with various sorts of apparatus or 'profile-meters', will be exploited in an operational software for the profiles' study.

The theoretical profiles can be described by analytical expressions from the standards; they are generally calculated at some discrete places and stored in a text file as a set of coordinates: $z(y)$. The measured profiles are directly given in this form.

7.4.1.3 Profile Measurements: Importance of the Angular Reference

The profiles are compared in the same reference frame. This implies that the measurement devices used to measure the wheel and the rail must be correctly referenced to each other.

For the rails, the reference line rests on the top of the two rails. It can be achieved by a transversal cylinder.

For the wheel, the reference line is considered to pass on the 'middle of the treads', at a point 70 mm from the inner flange (see Figure 7.6). These two points are not so easy to find. Some devices use the inner flange face, considering that it is a plane. If this face is not correctly machined, it could be a cone, and the profile study can be relatively biased.

It is useful to measure the two wheels of a wheelset and the two rail profiles at the same abscissae, especially in a curve. However, it must be possible to consider a single wheel on a single rail.

7.4.2 GRAVITATIONAL STIFFNESS

A description closer to the real shape of the wheels and of the rails is necessary to approach the principle of the gravitational centring mechanism.

In a first step, the vertical left and right loads Q are considered identical, and the profiles of wheels and rails are considered the same on each side.

When a wheelset is perfectly conical, the horizontal reaction forces $Q\ tg(\gamma)$ to the normal loads N_g and N_d are equal and opposite so long as there is no flange contact (Figure 7.14).

When a wheelset has concave profiles, the normal loads do not stay symmetrical with the lateral displacement. A transversal force appears with the projection of the normal forces in both directions (Figure 7.15).

Wheel-Rail Contact Mechanics

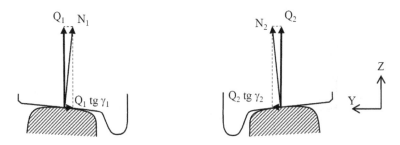

FIGURE 7.14 Conical profiles: compensation of the horizontal components. (From Iwnicki, S. (ed.), *Handbook of Railway Vehicle Dynamics*, CRC Press, Boca Raton, FL, 2006. With permission.)

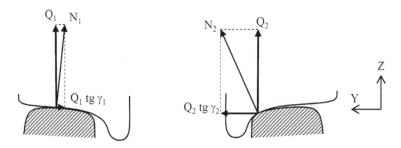

FIGURE 7.15 Progressive centring differential force with concave profiles. (From Iwnicki, S. (ed.), *Handbook of Railway Vehicle Dynamics*, CRC Press, Boca Raton, FL, 2006. With permission.)

7.4.3 Flange Contact

7.4.3.1 Flange Contact Jump and Definition of the Nominal Play in the Track

Even when the profiles are close to two opposite cones, the gravitational effect will appear when flanging occurs (Figure 7.16). The effects of friction and spin are considered in Section 7.4.5.

This main contact jump is a way to define the nominal play in the track by two means:

- For the value where γ is maximal
- For the value where there is the jump

The value is the same only when the flange is a perfectly straight line (Figures 7.14 and 7.16).

7.4.3.2 Contact Jump and Load Transfer

From a geometrical point of view, a rigid jump is possible. However, it is unrealistic from a dynamic point of view and unacceptable from a numerical point of view.

So, a large part of the wheel-rail contact modelisation looks at the load transfer when flanging occurs and more generally when there is a jump between two contact points on the profiles. The load transfer is calculated on the basis of the elastic deformation in the neighbourhood of each contact.

However, the contact stiffness is not the only elasticity to be taken into account, and the track stiffness itself can be used to smoothen the load variation.

The first modelisation of the flange contact has been to consider it as an elastic spring whose reaction comes from the track and the rail beam deformation.

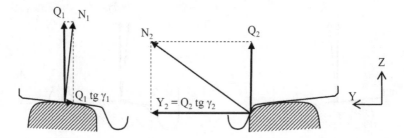

FIGURE 7.16 Gravitational forces and flange contact, without the effects of friction and spin. (From Iwnicki, S. (ed.), *Handbook of Railway Vehicle Dynamics*, CRC Press, Boca Raton, FL, 2006. With permission.)

7.4.4 Using the Contact Angle Function

7.4.4.1 Localisation of Multiple Contacts on the Tread

In order to locate contact jumps, one of the best contact parameters is the contact angle function (CAF see Figure 7.17), that is, the variation of the contact angle with the transversal displacement t_y, which results in the variable conicity. An alternative is the $\delta R/R$ function.

Knowing, for example, the rail profile:

- By integration gives the rolling radius (and the wheel profile in those parts where there is contact)
- By derivation gives the transversal curvature of the wheel

As far as the jumps are concerned, any discontinuity in this function can be considered as a jump. An example is given in Figure 7.17, with the well-known S1002 wheel profile on a UIC60 rail, where it is clearly evident that flange contact is not the only possible discontinuity.

If these jumps are giving ellipses whose centres are too close to each other, the Hertzian assumption of a constant curvature in the contact area is probably no longer valid.

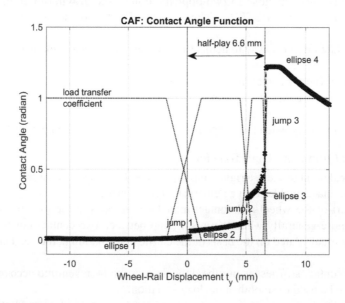

FIGURE 7.17 Jump localisation with the contact angle function (CAF), in the case of an S1002 wheel profile on a UIC60 rail at 1:40. (From Iwnicki, S. (ed.), *Handbook of Railway Vehicle Dynamics*, CRC Press, Boca Raton, FL, 2006. With permission.)

Wheel-Rail Contact Mechanics

7.4.4.2 Gravitational Centring Ability in Standard Conditions

The left wheel and right wheel contact angle values are used to calculate the gravitational forces directly. They are represented in Figure 7.18 with the same positive sign, each one being in its own reference frame, while the equilibrium equation needs to report the angle in the general reference frame.

When there is a large difference between the angle values, the profiles' combination is strongly centring. This is always the case in flange contact. However, the gravitational effect must be effective even around the central position (Figure 7.16 right).

Routes with many tight curves sometimes use vehicles with concave profiles to improve steering. Worn wheels also generally show this behaviour. However, a tendency to instability can be observed with such profiles.

7.4.5 THE DIVERGING EFFECT OF SPIN, INFLUENCE ON THE NORMAL LOAD

When the contact friction is not negligible, Figure 7.14 becomes an incorrect representation. A large value of the contact angle generates a large spin creepage value. A large friction value generates a spin torque in the neighbourhood of the contact, generating a lateral force that is always diverging.

Despite the fact that this torsion value is small, it generates the equivalent of a yaw angle offset.

A second effect of this force will be on the wheelset equilibrium: this spin force, due to friction, is directed mainly upward. Considering the equilibrium between the vertical force Q, this new force

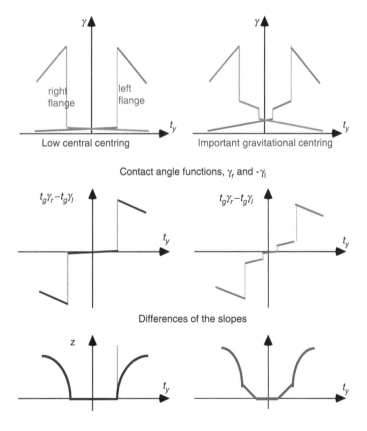

FIGURE 7.18 Gravitational stiffness parameters with conical (left) or concave (right) profiles. (From Iwnicki, S. (ed.), *Handbook of Railway Vehicle Dynamics*, CRC Press, Boca Raton, FL, 2006. With permission.)

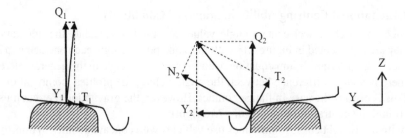

FIGURE 7.19 Gravitational mechanism with the effects of friction and spin. (From Iwnicki, S. (ed.), *Handbook of Railway Vehicle Dynamics*, CRC Press, Boca Raton, FL, 2006. With permission.)

F_{yspin} and the reaction force (Figure 7.19), it is found that the normal force N on a flanging wheel at the equilibrium is reduced by the friction; then, the gravitational effect is also reduced. This combined mechanism is a way to explain some derailments on dry rails. The association with independent wheels increases the friction effect.

7.4.6 Safety Criteria, Nadal's Formula

The Y/Q ratio is used as a safety criterion when flanging. In the real case of an attacking wheel, the spin force when flanging is added to the yaw lateral force, and this reduces the Y guiding force. For a given wheel:

$$Y = -N \sin \gamma + F_y \cos \gamma$$
$$Q = N \cos \gamma + F_y \sin \gamma \tag{7.35}$$

When F_x is negligible, $F_y \approx \mu N$, and $\mu \approx tg\,\mu$. A safe level of Y/Q has been set by Nadal's formula:

$$\left(\frac{Y}{Q}\right)_{max} = tg(\gamma - \mu) \tag{7.36}$$

where γ is the contact angle (radians) and μ is the friction coefficient.

The first way to ensure safety is to give a sufficient contact angle to the flange. The second preference is to reduce the Y force by a good design of the bogie. Another option is to limit the track twist, limiting the diminution of Q when flanging. The last way is to reduce the friction coefficient by lubrication. In Europe, this is commonly carried out at the wheel flange or on the gauge corner of the rail in particular curves.

However, the friction forces are also limited in the lateral direction by the presence of a longitudinal force. The friction coefficient is shared between the two directions (this is the 'friction cone'). In Nadal's formula, this sharing effect is not considered, making the formula adequate for independent wheels. This also means that a rigid wheelset, where the longitudinal forces are important on the attack wheel, will be safer than a wheelset equipped with independent wheels in the same conditions.

7.4.7 Independent Wheel, Application to Industrial Mechanisms on Rails

Independent wheels are mounted similarly to the rear wheelset of a car, on a fixed axle with roller bearings. This solution is more and more popular for tramways with low floors. A large number of mining trucks and many rolling bridges and rolling cranes are also equipped with independent wheels. In the railway domain, some propositions have even been made to use freely steerable independent wheels, like the steering wheelset of a car, using the gravitational phenomenon (Figures 7.15 and 7.16), to naturally steer the wheelset [23].

Except due to friction in the bearings, or when braking, there are no longitudinal forces on the independent wheel contact patch. The wheelset cannot be steered by the dicone effect, and the gravitational centring effect is the only passive mechanism that can be used to centre the wheelset. This implies that it is helpful to adopt concave profiles for the independent wheels. For large industrial cranes rolling with heavily loaded independent wheels, diabolo profiles can be adopted.

In the case where they are purely conical or cylindrical, the wheelset tends to run with one wheel flange permanently against the rail. This is due to the high sensitivity of the steel wheel to any small perturbation of the yaw angle, as the tangent stiffness is very high.

Because there are no longitudinal friction forces, the adhesion is fully available to generate lateral force, and the independent wheels lead more easily to derailment in curves. The diverging force due to the spin must be counterbalanced largely by the gravitational effect and reduced by lubrication.

Many people do not know about the dicone effect and believe that the gravitational mechanism is the main guiding mode of the railway axle. The gravitational mechanism is an effect of the normal forces, while the dicone is an effect of tangential forces, the nature of which is not so easy to explain.

7.4.8 Modelling the Contact Jumps

During the 1970s, Kalker set out the basis for stress calculations in the contact, followed by Knothe and others. However, the difficult problem of contact jumps has not been managed correctly for several years.

Several methods were proposed by using the last parameter calculated from Hertz: the contact deflection δ.

7.4.8.1 Hertzian Multiple Contacts

Sauvage in 1988 proposed to interpenetrate the two initial profiles with the δ values found in monocontact conditions. With circular profiles, the normal relative distance is a parabola; FASTSIM used such parabola in the longitudinal direction, and it is not so far from the Hertzian ellipsoid pressure.

It was found [19,24–26] that the intercepted contour, as shown in the lower part of Figure 7.20, is too large, and a better result is obtained with $\delta/2$. An even better result is obtained with the virtual interpenetration of the longitudinal curvatures.

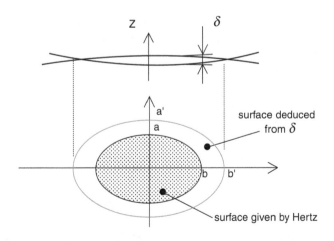

FIGURE 7.20 Interpenetration and contour. (From Iwnicki, S. (ed.), *Handbook of Railway Vehicle Dynamics*, CRC Press, Boca Raton, FL, 2006. With permission.)

When the lateral curvatures are irregular, outside of the Hertzian hypothesis, this intersection profile is no longer a parabola in the lateral direction. In the 1990s, this led different researchers to the idea of deducing the parabola, representing the virtual interpenetration of a main Hertzian ellipse, in order to deduce a second ellipse, then a third one, and so on in an iterative process, with particular criteria to decide how many ellipses are present at the same time. These methods have not been published. At that time, the relevant software packages were too slow to be used for industrial simulations.

In order to accelerate the calculation, Pascal and Sauvage proposed to construct a table, giving the equivalent forces with a single 'equivalent' Hertzian ellipse [27].

At the same time, Ayasse solved the problem by developing analytical equations, giving access to the jump width between two consecutive ellipses (Figure 7.21) [28,29].

The jump width is defined in two parts, as a function of the angles and of the interpenetration:

$$\Delta t_y = \Delta t_{y1} + \Delta t_{y2}$$

$$\Delta t_{y1} = \frac{\cos \gamma 1}{tg|\gamma_2 - \gamma_1|} \frac{\delta_1}{2}$$

$$\Delta t_{y2} = \frac{\cos \gamma 2}{tg|\gamma_1 - \gamma_2|} \frac{\delta_2}{2}$$

(7.37)

Between each mono-contact situation as presented in Figure 7.21, the load transfer is considered to be linear, as presented in Figure 7.17.

When the transfer appears at low angles between two points on the tread, the transfer width Δt_y is large (Figure 7.17 between ellipses 1 and 2), whereas it becomes closer to δ when flanging (Figure 7.17 between ellipses 3 and 4).

Both the Sauvage and Ayasse methods have been applied in different versions of the VOCO family codes.

These methods have common hypotheses, one of which is that, when calculating the forces, the different ellipses are independent. In Figure 7.22, for example, the pressure ellipsoids of the multi-elliptic method are not cumulated. However, when they are, the pressure shape is sometimes not very realist, when there is a large common area between the ellipses. For this reason, more advanced models have been developed.

7.4.9 Advanced Methods for Non-Hertzian Contacts

At the end of the 1980s, the reference model for normal and tangential contact in Hertzian and non-Hertzian cases was the CONTACT software developed by Kalker to simulate his complete theory. This accurate model is generally too slow to be used in a dynamic software loop, so it has been mainly used in the railway domain for the validation of other models. It has been regularly improved by Vollebregt [17].

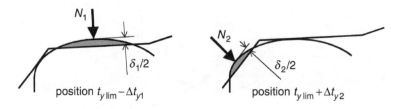

FIGURE 7.21 Multi-Hertzian contact Contact Angle Function (CAF) method: principle of the determination of the jump limits. (From Iwnicki, S. (ed.), *Handbook of Railway Vehicle Dynamics*, CRC Press, Boca Raton, FL, 2006. With permission.)

Wheel-Rail Contact Mechanics

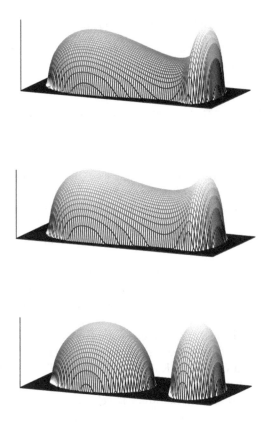

FIGURE 7.22 Contact pressure for S1002 wheel on UIC60 rail at 1:40 in the centred position showing non-Hertzian CONTACT model (top), semi-Hertzian model (middle) and multi-Hertzian model (bottom).

A non-Hertzian simplified method was proposed by Kik and Piotrowski in 1996 in order to make a better stress description in the contact patch and to be used directly in a multibody code [19]. This method is presently used in MEDYNA and ADAMS/Rail and has inspired the wheel-rail contact model used in the Universal Mechanism software. A second publication by the same authors in the Vehicle System Dynamics (VSD) journal includes improvements [30].

With regard to the Hertzian basis, an improved method has been developed by the authors of [22] by considering the curvatures A/B, instead of local b/a values. Even in Hertzian slender cases, the contact contour is very similar to the Hertzian solution, owing to curvature analytical corrections. In non-Hertzian cases, the results are close to Kalker's CONTACT, and the multibody simulations can be made in real time. An advanced model of this type has been proposed from this development for use in the SIMPACK software.

These methods are semi-Hertzian, and the contact is separated into longitudinal strips, as in FASTSIM. The longitudinal pressure can be elliptic. The transversal pressure is non-Hertzian, evaluated in different ways from the indentation contour, and the shear stresses are calculated in a similar way to that used in FASTSIM.

Comparison with the reference model provided by CONTACT is shown in Figure 7.22. For this calculation, the order of magnitude of the elapsed time is about 1 second with CONTACT, while it is less than 0.1 millisecond for the other methods, both semi-Hertzian and multi-Hertzian.

The basis of these methods is contained in a state-of-the-art paper presented at the 19th International Association for Vehicle System Dynamics (IAVSD) Symposium in Milan [31].

7.5 RECENT WORKS AND ADVANCED MODELS

While this chapter is dedicated to the basic quasi-static description of the wheel-rail contact, complementary approaches and complex models have been developed to address particular subjects including:

- Non-stationary problems, during braking, squealing and corrugation
- Fatigue and plasticity

Various families of models can be found.

The first type of model is dedicated to fast simulations, like the simplified multi-contact model presented by Pombo, in order to avoid the problem of convexity [32].

The second type is based on improvements of FASTSIM. Sichani has proposed a method based on Kalker's strip theory [3], using only the FASTSIM algorithm to determine the shear direction in the zone of slip. This method, named Fastrip, produces more accurate results than FASTSIM, with a similar duration of calculation [33].

Another field of research deals with transient formulations for the tangent problem. This is especially required with flexible models of the track and wheelset concerned with a higher frequency range than the traditional cases in railway dynamics. For instance, a non-stationary formulation has been proposed by Alonso and Gimenez [34] and improved by Guiral et al. [35].

The comparisons of experimental creepage-creep force laws with Kalker's theory seem to show some discrepancies. In low creepage values, the slope of the curve is lower than that predicted by Kalker, leading to the concept of reduced Kalker's coefficients. This may be due to the effect of a third body layer interacting between the wheel and the rail. High tractive or braking efforts lead to high creepage values, where the friction limit is velocity-dependent and creepage-dependent. Polach [36] has adapted his own contact theory described in [15] in order to fit with experimental data with high traction forces (see Figure 7.23). Polach's model computes only creep forces. In order to assess shear in the contact patch, FASTSIM may be modified to handle falling friction [37].

In the domain of non-Hertzian models, the STRIPES method has been studied and improved by Sichani [38,39] to propose an even better pressure distribution while still utilising a simple algorithm. Such improvements are facilitating access to real-time simulations.

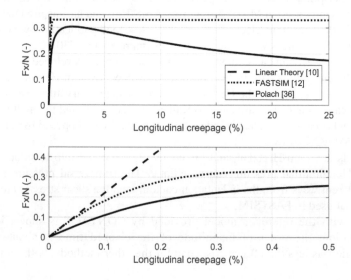

FIGURE 7.23 Creepage-creep force law according to the linear theory [10], the simplified theory [12] and Polach's method [36], respectively.

Wheel-Rail Contact Mechanics

Integral methods are not as general as finite element-based models; however, they are faster. Vollebregt continues to develop the CONTACT software with more functionalities [17]. Toumi [40] developed his own integral method, similar to CONTACT, under the name MIME. These algorithms are sometimes proposed in post- or co-simulation with dynamic software packages.

The most general approach is based on the finite element method. For example, Toumi [41] made an analysis of the sensitivity of the c_{ij} to plasticity at high load under ABAQUS and studied separately the influence on the critical speed of a vehicle.

More complex contact models can be found in the domain of tribology, where granular third bodies are sometimes taken into account. This could be a way to address the displacement of wear debris in the contact area; for the moment, wear modules in the post-processing step of dynamic simulations are based mainly on the energy dissipation estimated in the contact and on the simplified T-gamma method.

APPENDICES

APPENDIX 7.1 KINEMATIC MOVEMENT AND THE KLINGEL'S FORMULA

For this simplified formula, the wheels are considered perfectly conical, rolling on a line representing the rail (Figure 7.6).

The Klingel's formula is the expression of the hunting wavelength [9], without any tangential forces:

$$\lambda = 2\pi \sqrt{\frac{r_o Dc/2}{\gamma}} \quad (A7.1)$$

The amplitude of this sinusoidal movement is an initial condition. However, it is limited by the flange contact.

APPENDIX 7.2 KINEMATIC HUNTING AND EQUIVALENT CONICITY

Even if the wheel profile is not a perfect cone and the rail is not a line, the periodic movement of a free real wheelset in the track will remain close to a sinusoidal movement. This is called kinematic hunting. Its wavelength is a way to determine the equivalent conicity of the wheel-rail profile combination.

The equivalent conicity notion is well known in the railway field, where it was useful at the time of the linearised dynamic models.

APPENDIX 7.3 THE CIRCLE THEORY

From the concave shape of the wheel (R_{wx}) and of the rail (R_{rx}), with the approximation that these radii are constant in the wheelset travel, for a small lateral movement around the central position, Joly shows in [42] that the equivalent conicity γ_e must be used in place of the value γ_o of the cone angle in the central position:

$$\gamma_e = \frac{R_{wx}}{R_{wx} - R_{rx}} \gamma_o \quad (A7.2)$$

If the wheel radius is infinite, the wheel is a cone, and the expression returns to the central value γ_o.

This expression is an improvement of the wavelength expression, in comparison with the cone value. The running safety of the TGV (French HST) was established before 1980 with this conicity determination [42].

With 1:20 tapered wheels, this formula can be used for an equivalent conicity of 0.2. Over this value, the contact probably differs from the initial dicone.

Appendix 7.4 Analysis of Y/Q and Nadal's Criteria

The Nadal's criteria $Y/Q < tg(\gamma-\mu)$ is critical for the evaluation of the safety of a wheelset to avoid derailment. It can be estimated by measurement, by numerical simulation and also by an analytical quasi-static model. The following study establishes such an analytical expression of Y/Q at the contact point, in the track frame, with a complete set of hypotheses.

At the contact point, the tangent plane common to the wheel and the rail makes an angle γ with the track plane. The contact force can be expressed as a normal force N, and there are two tangential forces f_x and f_y in this plane, due to the friction (Figure 7.24).

Projecting these forces in the track frame OYZ, it follows that the forces Y and Q in the directions OY and OZ, respectively, are given by:

$$Y = f_y \cos\gamma - N \sin\gamma$$
$$Q = f_y \sin\gamma + N \cos\gamma \tag{A7.3}$$

Inversely, the forces f_y and N can be expressed by:

$$f_y = Y \cos\gamma + Q \sin\gamma$$
$$N = -Y \sin\gamma + Q \cos\gamma \tag{A7.4}$$

In the particular case when the tangential resultant force is saturated, the additional relation follows:

$$f_x^2 + f_y^2 = (\mu N)^2 \tag{A7.5}$$

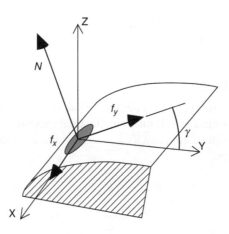

FIGURE 7.24 Normal and tangent forces. (From Iwnicki, S. (ed.), *Handbook of Railway Vehicle Dynamics*, CRC Press, Boca Raton, FL, 2006. With permission.)

where μ is the friction coefficient at the contact point.

Replacing f_y and N in Equation A7.5 by their expressions in Equation A7.4, one obtains the relation:

$$\left(\frac{Y}{Q}\right)^2 \left(\cos^2\gamma - \mu^2 \sin^2\gamma\right) + 2\frac{Y}{Q}\cos\gamma\sin\gamma\left(1+\mu^2\right) + \sin^2\gamma - \mu^2\cos^2\gamma + \left(\frac{f_x}{Q}\right)^2 = 0 \quad (A7.6)$$

A second-order equation in Y/Q whose determinant Δ can be written as:

$$\Delta = \mu^2 \left[1 - \left(\frac{f_x}{\mu Q}\right)^2 \left(\cos^2\gamma - \mu^2\sin^2\gamma\right)\right] \quad (A7.7)$$

It is always positive if the relation in Equation A7.5 concerning the saturation is respected. With a limited development of the first order:

$$\left(\frac{f_x}{\mu Q}\right)^2 \left(\cos^2\gamma - \mu^2\sin^2\gamma\right) \ll 1 \quad (A7.8)$$

And the two roots of the equation can be written as:

$$\frac{Y}{Q} \approx -tg\left(\gamma - \varepsilon\,Atg\,\mu\right) - \frac{\varepsilon}{2\mu}\left(\frac{f_x}{Q}\right)^2 \quad (A7.9)$$

where $\varepsilon = \pm 1$ depends on the choice for $\sqrt{\Delta}$.

This indeterminate form comes from the saturation relation in Equation A7.5; as N is always positive, and if the longitudinal force f_x is considered small, f_y is the main factor responsible for the saturation hypothesis (meaning that the yaw angle is important). It then follows that $f_y = \varepsilon\,\mu\,N$; the value of ε is directly linked to the sign of the force f_y.

The ratio Y/Q for one contact point depends mainly, respecting the hypothesis, on the contact angle and the friction coefficient.

For a single contact in the flange, with a contact angle of 70°, a friction coefficient of 0.35 and neglecting the term f_x, the unfavourable force f_y that facilitates the wheel climb is positive and $\varepsilon = +1$. In this case, the absolute value of Y/Q reaches 1.22, and this value is commonly given as critical for derailment.

Equation A7.9, however, shows that, with a reduced friction coefficient, Y/Q can, with the same other conditions, reach more important values ($Y/Q = 2$ if $\mu = 0.1$). However, this value is difficult to reach in reality because, with a low friction coefficient on the flange, there will certainly be a bi-contact on both flange and tread, in contradiction with the proposed hypothesis. On the other hand, if $\mu = 0.5$, Y/Q decreases to 0.9.

Note that the ratio Y/Q is close to the friction coefficient on the tread, where the contact angle is low. This situation is also found in the case, during derailment, when the flange top is rolling across the tread.

These considerations show that Y/Q, for a single wheel-rail contact, can be expressed analytically in a simple way and with clearly defined approximations and that the limit of 1.2 commonly proposed corresponds to specific values of the contact angle, the friction coefficient and saturation.

NOMENCLATURE

a, b	Longitudinal and lateral semi-axis lengths of wheel-rail contact patch ellipse
$c_{11}\ c_{22}\ c_{23}$	Kalker's coefficients (see Table 7.5)
f_x, f_y	Contact forces in the tangent plane Oxy

m, n, r	Hertz parameters
$p_x p_y p_z$	Contact pressures in normal, transversal and longitudinal directions
r_o	Rolling radius of the wheel, around axis Oy
r_n	Longitudinal radius of the wheel at the contact point
y or t_y	Lateral displacement of the wheel with respect to the rail
A, B	Curvatures at the contact point
D_r	Rail gauge: distance between the inner faces of the rails
D_w	Wheelset gauge: distance between the inner flanges
D_c	Track gauge: distance between the contacts
E	Young's modulus of the material
G	Shear modulus of the material
F_x, F_y	Longitudinal force, lateral force
$L_1 L_2 L_3$	Kalker's elastic coefficients
N	Normal load on a contact patch
$OXYZ$	Reference frame of the track
$Oxyz$	Reference frame at the contact point
Ox	Longitudinal axis, direction of rolling
Oy	Lateral axis, to the left
Oz	Vertical axis, to the top
Q	Vertical load on the wheel-rail contact, $OXYZ$ frame
R_{rx}	Transversal radius of the rail profile
R_{wx}	Transversal radius of the wheel profile
V_x	Longitudinal speed of the wheelset
Y	Lateral load on the wheel-rail contact, $OXYZ$ frame
α	Yaw angle
δ	Relative reduction of distance between elastic bodies, Hertz' theory
φ	Spin creepage
γ	Contact angle, inclination of the profiles at the contact point, for any position of the wheelset
γ_o	Contact angle in the central position of the wheelset
γ_e	Equivalent conicity
μ	Friction coefficient
ν	Poisson's ratio of the material
ν_x	Longitudinal creepage
ν_y	Lateral creepage
ω	Rotation speed of the wheelset around Oy
Ψ	Roll angle of the wheelset

REFERENCES

1. F.W. Carter, On the action of a locomotive driving wheel, *Proceedings of the Royal Society of London, Series A*, 112, 151–157, 1926.
2. P.J. Vermeulen, K.L. Johnson, Contact of non-spherical elastic bodies transmitting tangential forces, *ASME Journal of Applied Mechanics*, 31(2), 338–340, 1964.
3. J.J. Kalker, A strip theory for rolling with slip and spin, *Proceedings Koninklijke Nederlandse Akademie van Wetenschappen*, B70 (No.1, Mechanics), 10–62, 1967.
4. H. Hertz, On the contact of elastic solids, 1881, *Miscellaneous Papers by Heinrich Hertz*, Macmillan & Co., London, UK, 146–162, 1896.
5. M. Sebès, L. Chevalier, J-B. Ayasse, H. Chollet, A fast-simplified wheel-rail contact model consistent with perfect plastic materials, *Vehicle System Dynamics*, 50(9), 1453–1471, 2012.
6. W. Klingel, über den lauf der eisenbahnwagen auf gerader bahn, *Organ für die Fortschritte des Eisenbahnwesens in technischer Beziehung*, 20(4), 113–123, 1883.

7. J. Boussinesq, *Application des potentiels à l'étude de l'équilibre et du mouvement des solides élastiques*, Gauthier-Villars, Paris, France, 1885.
8. V. Cerruti, *Ricerche intorno all'equilibrio dei corpi elastici isotropi*, Reale Accademia dei Lincei, Rome, Italy, 1882.
9. J.J. Kalker, The tangential force transmitted by two elastic bodies rolling over each other with pure creepage, *Wear*, 11(6), 421–430, 1968.
10. J.J. Kalker, On the rolling contact of two elastic bodies in the presence of dry friction, *Doctoral Thesis*, Delft University of Technology, Delft, the Netherlands, 1967.
11. J.J. Kalker, *Three-Dimensional Elastic Bodies in Rolling Contact*, 1st ed., Kluwer Academic Publishers, Dordrecht, the Netherlands, 1990.
12. J.J. Kalker, A fast algorithm for the simplified theory of rolling contact (FASTSIM program), *Vehicle Systems Dynamics*, 11(1), 1–13, 1982.
13. Z.Y. Shen, J.K. Hedrick, J.A. Elkins, A comparison of alterative creep force models for rail vehicle dynamic analysis, In J.K. Hedrick (Ed.), *Proceedings 8th IAVSD Symposium held at Massachusetts Institute of Technology*, 15–19 August 1983, Swets and Zeitlinger, Lisse, the Netherlands, 591–605, 1984.
14. T. Ohyama, Some problems of the fundamental adhesion at higher speeds, *Quarterly Report of RTRI*, 14(4), 181, 1973.
15. O. Polach, A fast wheel-rail forces calculation computer code, In: R. Fröhling (Ed.), *Proceedings 16th IAVSD Symposium held in Pretoria, South Africa*, Swets & Zeitlinger, Lisse, the Netherlands, 728–739, 2000.
16. H. Chollet, Etude en similitude mécanique des efforts tangents au contact roue-rail, Thèse de Doctorat de l'Université Paris 6, 1991. (in French)
17. E.A.H. Vollebregt, User guide for CONTACT, rolling and sliding contact with friction, Technical Report TR09-03, version v19.1, VORtech CMCC, Delft, the Netherlands, 2019.
18. E.A.H. Vollebregt, P. Wilders, FASTSIM2: A second-order accurate frictional rolling contact, *Computational Mechanics*, 47(1), 105–116, 2011.
19. W. Kik, J. Piotrowski, A fast, approximate method to calculate normal load at contact between wheel and rail and creep forces during rolling, In: I. Zobory (Ed.), *Proceedings 2nd Mini Conference on Contact Mechanics and Wear of Rail/Wheel Systems*, Technical University of Budapest, Hungary, 52–61, 1996.
20. J.B. Ayasse, H. Chollet, Determination of the wheel rail contact patch in semi-Hertzian conditions, *Vehicle System Dynamics*, 43(3), 161–172, 2005.
21. B. Soua, Étude de l'usure et de l'endommagement du roulement ferroviaire avec des modèles d'essieu non-rigides, Doctoral thesis, Ecole Nationale des Ponts et Chaussées, 1997. (in French)
22. J. Piotrowski, B. Liu, S. Bruni, The Kalker book of tables for non-Hertzian contact of wheel and rail, *Vehicle System Dynamics*, 55(6), 875–901, 2017.
23. F. Frederich, Possibilities as yet unknown regarding the wheel/rail tracking mechanism, *Rail International*, 16(11), 33–40, 1985.
24. J.P. Pascal, About multi-Hertzian-contact hypothesis and equivalent conicity in the case of S1002 and UIC60 analytical wheel/rail profiles, *Vehicle System Dynamics*, 22(2), 57–78, 1993.
25. J.P. Pascal, G. Sauvage, The available methods to calculate the wheel/rail forces in non Hertzian contact patches and rail damaging, *Vehicle System Dynamics*, 22(3–4), 263–275, 1993.
26. J.P. Pascal, Benchmark to test wheel/rail contact forces, *Vehicle System Dynamics*, 22(S1), 169–173, 1993.
27. J.P. Pascal, G. Sauvage, New method for reducing the multicontact wheel/rail problem to one equivalent rigid contact patch, In: G. Sauvage (Ed.), *Proceedings 12th IAVSD Symposium, Lyon, France*, Swets and Zeitlinger, Lisse, the Netherlands, 1992.
28. J.B. Ayasse, H. Chollet, J.L. Maupu, Paramètres caractéristiques du contact roue-rail, INRETS Report No. 225, 2000. (in French)
29. J.B. Ayasse, H. Chollet, J.S. Fleuret, E. Lévêque, CAF, a generalised conicity criteria for the wheel-rail contact: Example of a switch blade safety study, In: I. Zobory (Ed.), *Proceedings 8th Mini Conference Vehicle System Dynamics Identification and Anomalies*, Budapest University of Technology and Economics, Hungary, 2002.
30. J. Piotrowski, W. Kik, A simplified model of wheel/rail contact mechanics for non-Hertzian problems and its application in rail vehicle dynamic simulations, *Vehicle System Dynamics*, 46(1–2), 27–48, 2008.
31. J. Piotrowski, H. Chollet, Wheel-rail contact models for vehicle system dynamics including multi-point contact, *Vehicle System Dynamics*, 43(6–7), 455–483, 2005.

32. J. Pombo, J. Ambrósio, M. Silva, A new wheel-rail contact model for railway dynamics, *Vehicle System Dynamics*, 45(2), 165–189, 2007.
33. M.Sh. Sichani, R. Enblom, M. Berg, An alternative to FASTSIM for tangential solution of the wheel-rail contact, *Vehicle System Dynamics*, 54(6), 748–764, 2016.
34. A. Alonso, J.G. Giménez, Non-steady state contact with falling friction coefficient, *Vehicle System Dynamics*, 46(S1), 779–789, 2008.
35. A. Guiral, A. Alonso, L. Baeza, J.G. Giménez, Non-steady state modelling of wheel–rail contact problem, *Vehicle System Dynamics*, 51(1), 91–108, 2013.
36. O. Polach, Creep forces in simulations of traction vehicles running on adhesion limit, *Wear*, 258(7–8), 992–1000, 2005.
37. M. Spiryagin, O. Polach, C. Cole, Creep force modelling for rail traction vehicles based on the Fastsim algorithm, *Vehicle System Dynamics*, 51(11), 1765–1783, 2013.
38. M.Sh. Sichani, R. Enblom, M. Berg, A novel method to model wheel-rail normal contact in vehicle dynamics simulation, *Vehicle System Dynamics*, 52(12), 1752–1764, 2014.
39. M.Sh. Sichani, On efficient modelling of wheel-rail contact in vehicle dynamics simulation, *Doctoral Thesis*, KTH Royal Institute of Technology, Stockholm, Sweden, 2016.
40. M. Toumi, Modélisation numérique du contact roue-rail pour l'étude des paramètres influençant les coefficients de Kalker: Application à la dynamique ferroviaire, Doctoral thesis, Université Paris-Est, Paris, France, 2016. (in French)
41. M. Toumi, H. Chollet, H. Yin, Finite element analysis of the frictional wheel-rail rolling contact using explicit and implicit methods, *Wear*, 366–367, 157–166, 2016.
42. R. Joly, Etude de la stabilité transversale d'un véhicule ferroviaire, Revue Francaise de Mécanique, No. 36, 1970.

8 Tribology of the Wheel-Rail Contact

Ulf Olofsson, Roger Lewis and Matthew Harmon

CONTENTS

8.1 Introduction 281
8.2 Contact Conditions for the Wheel-Rail Interface..282
8.3 Surface Damage Mechanisms 285
    8.3.1 Wear..285
    8.3.2 Plastic Deformation 288
    8.3.3 Rolling Contact Fatigue 288
8.4 Friction..290
    8.4.1 Wheel-Rail Friction Conditions...291
    8.4.2 Friction Modification 292
    8.4.3 Adhesion Loss..294
    8.4.4 Increasing Adhesion 295
8.5 Lubrication and Surface Modification...296
    8.5.1 Benefits of Lubrication 296
    8.5.2 Lubrication Methods..296
    8.5.3 Problems with Lubrication 297
    8.5.4 Lubricator System Selection and Positioning 297
8.6 Emission of Sound and Airborne Particles...298
Acknowledgements..299
References..299

8.1 INTRODUCTION

Tribology, the science and technology of friction, wear and lubrication, is an interdisciplinary subject. It can therefore be addressed from several different viewpoints. This chapter focuses on the friction, wear and lubrication of the tiny contact zone (roughly 1 cm^2), where steel wheel meets steel rail, from a mechanical engineer's viewpoint. In contrast to other well-investigated machinery such as roller bearings, the wheel-rail contact is an open system. It is exposed to dirt and particles and natural lubrication, such as high humidity, rain and leaves, all of which can seriously affect the contact conditions and the forces transmitted through the contact. In contrast, the ball-cage contacts in roller bearings are sealed away. The steel rail meets a population of steel wheels from a number of different vehicles, and the form of both the wheels and the rail can change due to wear. In contrast, a roller bearing meets the same rollers, without any form change of the contacting bodies.

A comprehensive overview of the science of tribology is presented in the *ASM Handbook* [1], while a closer examination of the material science field is given by Hutchings [2]. The mathematical modelling aspects of tribology, that is, contact mechanics and fluid film lubrication, are presented by Johnson [3] and Dowson and Higginson [4]. An excellent historical overview of the field is presented by Dowson [5].

In the contact zone between wheel and rail, normal and tangential loads are transmitted. How the steel wheel meets the steel rail and how the size of the forces transmitted in the contact zone influences damage mechanisms, such as wear and surface cracking, are discussed. The contact conditions of the wheel-rail contact interface are covered in Section 8.2.

When two surfaces under load move relative to each other, wear will occur. Wear is often defined as damage to one or both surfaces, involving loss of material. Wear and other surface damage mechanisms are discussed in Section 8.3.

The friction force can be defined as the resistance encountered by one body moving over another body. This definition covers both sliding and rolling bodies. Note that even pure rolling nearly always involves some sliding and that the two classes of motion are not mutually exclusive. Any substance between the contacting surfaces may affect the friction force. The contact conditions may cause the substance to be wiped away quickly, and its effect will be minimal. On the other hand, surface films formed between interposed substances have a major effect on the frictional behaviour. The friction of the wheel-rail contact, as well as causes of friction loss and methods for increasing the friction, is discussed in Section 8.4.

Lubricant application to the wheel-rail contact as well as surface coatings are used to reduce friction and damage due to wear, etc. This is discussed in Section 8.5.

Open systems such as the wheel-rail contact also radiate emissions such as sound and airborne particles from wear products without shielding. This is discussed in the final Section 8.6 of this chapter.

What one should always bear in mind when studying and using tribological data is that friction and wear are system parameters and not material parameters like modulus of elasticity or fracture toughness. This means that frictional and wear data taken from one system, such as a roller bearing, cannot be directly applied to another system, such as the wheel-rail contact. This also highlights the need for a special study of the tribology of the wheel-rail contact interface.

8.2 CONTACT CONDITIONS FOR THE WHEEL-RAIL INTERFACE

In the contact zone between railway wheel and rail, the surfaces and the bulk material must be strong enough to resist the normal (vertical) forces introduced by heavy loads and the dynamic response induced by track and wheel irregularities. The tangential forces in the contact zone must be low enough to permit moving heavy loads with little resistance and, at the same time, must be high enough to provide traction, braking and steering of the trains.

The contact zone (roughly 1 cm^2) between a railway wheel and rail is small compared with their overall dimensions, and its shape not only depends on the rail and wheel geometry, but is also influenced by how the wheel meets the rail, that is, lateral position and angle of wheel relative to the rail, as shown by Le The Hung [6].

It is difficult to make direct measurements of the contact area between wheel and rail. An interesting approach for measuring the contact area for full-scale worn wheel and rail pieces is presented by Marshall et al. [7]. They used an ultrasonic reflection technique, and the results were compared with calculated contact areas, showing good agreement. This work was initially static, but recent work took this approach further by measuring the contact area whilst a wheel rolled over the sensor area, with the sensors mounted under the railhead [8]. The surface topographies of the ultrasonic measured surfaces were measured with a stylus instrument and used as input to a contact mechanics method for rough surfaces (see Björklund et al. [9] for details). Poole [10] used low-pressure air passing through 1-mm-diameter holes drilled into the railhead to measure the contact area as the holes were being blocked by the passing wheel. Measurement of these pressure variations allows studying of the contact area shape under dynamic conditions.

The size and shape of the contact zone where the railway wheel meets the rail can be calculated with different techniques. Traditionally, the Hertz theory of elliptical contacts [3] has been used, implying the following assumptions: the contact surfaces are smooth and can be described by second-degree surfaces, the material model is linear elastic and there is no friction between the contacting surfaces and the contacting bodies are assumed to deform as infinite half spaces.

Tribology of the Wheel-Rail Contact

The half-space assumption puts geometrical limitations on the contact; that is, the significant dimensions of the contact area must be small compared with the relative radii of the curvature of each body. Especially in the gauge corner of the rail profile, the half-plane assumption is questionable, since the contact radius here can be as small as 10 mm. Owing to its simple closed-form solutions, the Hertz method is the most commonly used approach in vehicle dynamics simulation. However, other methods are used for simulation of wear and surface fatigue due to the overestimation of the contact stresses attributed to the non-validity of the half-plane assumption and non-linear material behaviour. Kalker's numerical programme CONTACT [11] still depends on the half-space assumption, but is not restricted to elliptical contact zones. The contact surfaces are meshed into rectangular elements, with constant normal and tangential stresses in each rectangular element. Telliskivi and Olofsson [12] developed a finite element model, including plastic deformation of the wheel-rail contact, using measured wheel and rail profiles as input data. They compared the traditional methods (Hertz and CONTACT) with their detailed finite element solutions of the wheel in contact with the rail gauge (see load case 1 in Figure 8.1a), and the wheel in contact with the rail head (see load case 2 in Figure 8.1b). The results in terms of contact zone shape and size, as well as stress distribution, are presented in Figure 8.2. The results from the two load cases show that the difference in maximum contact pressure between CONTACT/Hertz and the model were small for case 2 when the minimum contact radius was large compared with the significant dimensions of the contact area (half-space assumptions valid). However, in case 1, where the minimum contact radius was small compared with the significant dimensions of the contact area, the difference between the model and CONTACT/Hertz was as large as 3 GPa. Here, the difference was probably due to both the half-space assumption and the material model.

The Stockholm local railway network was the subject of a national Swedish transport programme (the Stockholm test case) [13–16], in which the wear, surface cracks, plastic deformation and friction of rail and wheel were observed for a period of 4 years. The data from the Stockholm test case have been used for validation of different wear models; see [17–19] and also surface crack models [20].

Furthermore, the trains used in this study have been modelled with train dynamic simulation software such as GENSYS [18] and Medyna [21]. A parametric study [18] was performed on curves with different radius, representative of Stockholm local traffic. The results are presented in Figure 8.3 in the form of a contact pressure sliding velocity diagram [14]. A clear difference could be found between the railhead-wheel tread contacts and the rail gauge-wheel flange contacts in terms of sliding velocity and contact pressure. For the railhead-wheel tread contacts, the sliding velocity and the contact pressure were never above 0.1 m/s and 1.5 GPa, respectively. However, for the rail gauge-wheel flange contacts, the maximum sliding velocities reached 0.9 m/s, and maximum contact pressure was observed up to 2.7 GPa. Also shown in Figure 8.3 are simulation results from a curve with 303 m radius, for the Stockholm test case, using the Medyna software. This is a sharp curve with one of the smallest radii in the network, and one can note a very high contact pressure for the first wheel on the leading bogie in contact with the rail gauge. Other examples that modern railway operation leads to high contact stresses that are significantly over the yield strength of the material are presented in Kumar [22] and Cassidy [23].

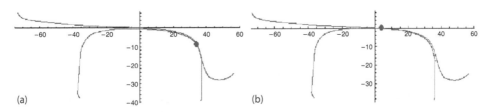

FIGURE 8.1 Contact point location for the two load cases: (a) case 1 and (b) case 2. (From Iwnicki, S. (Ed.), *Handbook of Railway Vehicle Dynamics*, CRC Press, Boca Raton, FL, 2006. With permission.)

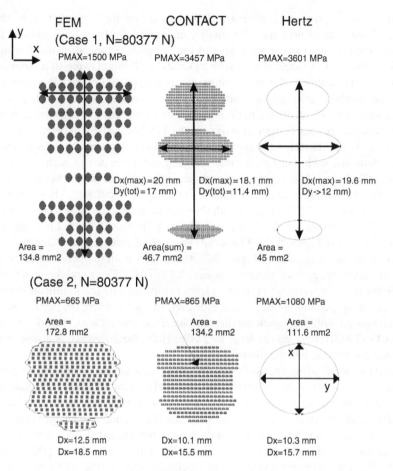

FIGURE 8.2 Comparisons with respect to maximum contact pressure and the contact area between the three different contact mechanics analysis methods. (From Telliskivi, T., and Olofsson, U., *J. Rail Rapid Transit*, 215(2), 65–72, 2001. With permission.)

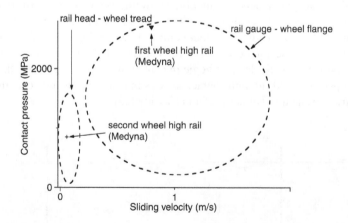

FIGURE 8.3 Sliding velocity contact pressure chart from the Stockholm test case: the elliptical areas show typical regions where railhead-wheel tread and rail gauge-wheel flange contact occur (18); also shown in the figure are simulation results from a small radius curve in the Stockholm test case, using the Medyna software (21). (From Olofsson, U., and Telliskivi, T., *Wear*, 254(1–2), 80–93, 2003. With permission.)

8.3 SURFACE DAMAGE MECHANISMS

The profile change of rail on curves makes a large contribution to track the maintenance cost. The profile change on wheels can also be significant, especially on a curved track. Damage mechanisms such as wear and plastic deformation are the main contributors to profile change. Another growing problem for many railways is rolling contact fatigue [24]. In Europe, there are more than 100 broken rails each year due to rolling contact fatigue. Rail maintenance costs within the European Union were estimated to have totalled 300 million Euro in 1995 [24].

8.3.1 Wear

Wear is the loss or displacement of material from a contacting surface. Material loss may be in the form of debris. Material displacement may occur by the transfer of material from one surface to another by adhesion or by local plastic deformation. There are many different wear mechanisms that can occur between contacting bodies, each of them producing different wear rates. The simplest classification of the different types of wear that produce different wear rates is 'mild wear' and 'severe wear'. Mild wear results in a smooth surface that is often smoother than the original surface. On the other hand, severe wear results in a rough surface that is often rougher than the original surface [25]. Mild wear is a form of wear characterised by the removal of materials in very small fragments. Mild wear is favourable in many cases for the wear life of the contact, as it causes a smooth run-in of the contacting surfaces. However, in some cases, it has been observed that it worsens the contact condition, and the mild wear can change the form of the contacting surfaces in an unfavourable way [26]. Another wear process that results in a smooth surface is the oxidative process characterised by the removal of the oxide layer on the contacting surfaces. In this case, the contact temperature and asperity level influence the wear rate [27]. Abrasive wear caused by hard particles between the contacting surfaces can also cause significant wear and reduce the life of the contacting bodies [28].

In wheel–rail contact, both rolling and sliding occur in the contacting zone. Especially in curves, there can be a large sliding component on the contact patch at the track side of the railhead (gauge corner). Owing to this sliding, progressive wear, as shown in Figure 8.4, occurs in the contact under the poorly lubricated condition that is typical of wheel–rail contact. An observation that can be made on sliding wear is that an increase of the severity of loading (normal load, sliding velocity or bulk temperature) leads at some stage to a sudden change in the wear rate (volume loss per sliding distance). The severe wear form is often associated with seizure. The transfer from mild acceptable wear to severe/catastrophic wear depends strongly on the surface topography. The loading capability of a sliding contact may be increased considerably by smoothing the surface [29]. Chemi-reacted boundary layers imposed by additives in the lubricant can improve the properties of lubricated contacting surfaces and reduce the risk of seizure [30]. Also, as shown by Lewis and Dwyer-Joyce [31], the surface temperature influences the transition from severe to catastrophic wear.

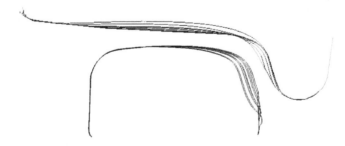

FIGURE 8.4 Progressive change of wheel and rail profiles from the Stockholm test case. (From Iwnicki, S. (Ed.), *Handbook of Railway Vehicle Dynamics*, CRC Press, Boca Raton, FL, 2006. With permission.)

In addition to the contact pressure and the size of the sliding component, natural and applied lubrication strongly influenced the wear rate [14–16] for the full-scale test results from the Stockholm test case. Both lubricated and non-lubricated conditions, as well as seasonal variations, were studied. In addition, two different rail hardnesses were studied in the same test curves. Trackside lubrication reduced the wear significantly, and a lubrication benefit factor of 9 for small radius curves (300 m) was reported. For radius curves of 600 to 800 m, the lubrication benefit factor was about 4. The variation seen in wear rates over the course of a year was probably due to natural lubrication caused by changing the weather conditions. An analysis of the relationship between the weather conditions and the measured rail wear shows that the precipitation has a significant effect on rail wear [16], as shown in Figure 8.5. Waara [32] reports that gauge face wear in a heavy-haul application in northern Sweden can be reduced three to six times with proper full-year lubrication. Engel [33] also reports significant reduction of wear by lubrication, where the lubricant benefit factor was 4 in a twin-disc test. An on-board lubrication system was evaluated by Cantara [34] in a Spanish study. The results were that the flange wear was reduced by a factor of 4.5 for wheels equipped with the on-board lubrication device.

The curve radius of the track has a strong influence on rail wear. The influence strongly depends on the vehicles and their behaviour. In the Stockholm test case, all vehicles were of the same type and passed over all the test sites with the same frequency. In this case, the influence of curve radius can be clearly seen when comparing rail wear rate as the function of curve radius. The rail wear rate seems to increase exponentially for decreasing curve radius [16], as shown in Figure 8.6.

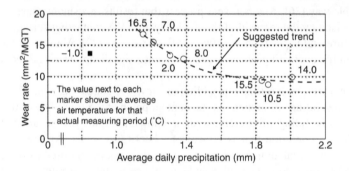

FIGURE 8.5 Rail wear rate versus average daily precipitation (MGT = million gross tonnes of traffic). (From Iwnicki, S. (Ed.), *Handbook of Railway Vehicle Dynamics*, CRC Press, Boca Raton, FL, 2006. With permission.)

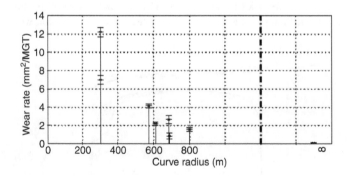

FIGURE 8.6 Wear rate for high rail as function of curve radius in the Stockholm test case (MGT = million gross tonnes of traffic). (From Iwnicki, S. (Ed.), *Handbook of Railway Vehicle Dynamics*, CRC Press, Boca Raton, FL, 2006. With permission.)

Tribology of the Wheel-Rail Contact

For a given situation, a higher steel grade usually reduces rail wear. This effect is shown in Figures 8.7 and 8.8 for two different lubricated high rails with steel grades UIC 900A and UIC 1100, respectively. For a non-lubricated curve of 303 m radius, the ratio between rail wear rates for the 900A-grade rail compared with that of the 1100-grade rail is approximately 2, as detailed in [15]. This can be compared with the lubricant benefit factor that was approximately 9 in this curve. The difference between railhead wear (low sliding velocities and low contact pressure) and rail gauge wear (high contact pressure and high sliding velocities) was seen to be a factor of 10. This is also comparably higher than the rail grade benefit for modern rail steels such as UIC 900A and UIC 1100. This observation that the contact conditions in terms of contact pressure and sliding velocity are more important than the grade of steel (900A and 1100) has also been verified in two-roller tests [14]. However, when Lewis and Olofsson [35] compared rail steel wear coefficients taken from laboratory tests run on twin disc and pin-on-disc machines, as well as those derived from measurements taken in the field, they found that the introduction of more modern rail materials had reduced wear rates by up to an order of magnitude in the last 20 years.

Fully pearlitic rail steels are still the most common and are used by most railways. Pearlite is a lamellar product of eutectoid composition that is formed in steel during transformation under isothermal continuous cooling. It consists of ferrite and cementite. Perez-Uzeta and Beynon [36] have shown that the wear rate of pearlitic rail steel decreases with lower interlamellar spacing between the cementite lamella, giving a corresponding increase in hardness. Steels with a bainitic microstructure are the other main rail steels. They have shown better

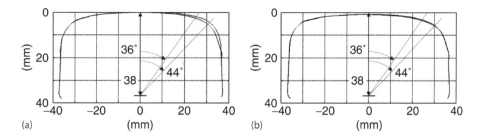

FIGURE 8.7 Results from measurements of high rail at test start and after 2 years of traffic: (a) new rail at test start, (b) worn rail at test start; curve radius 346 m, rail steel grade UIC 900A and curve lubricated during measuring period. (From Olofsson, U., and Nilsson, R., *J. Rail Rapid Transit*, 216, 249–264, 2002. With permission.)

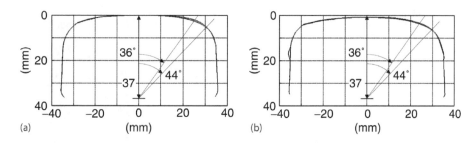

FIGURE 8.8 Results from measurements of high rail at test start and after 2 years of traffic: (a) new rail at test start, (b) worn rail at test start; curve radius 346 m, rail steel grade UIC 1100 and curve lubricated during measuring period. (From Olofsson, U., and Nilsson, R., *J. Rail Rapid Transit*, 216, 249–264, 2002. With permission.)

rolling contact fatigue resistance than pearlitic rail steels. However, the wear resistance of bainitic rail steels is inferior to that of pearlitic rail steels at a fixed tensile strength, as shown by Garnham and Beynon [37] and Mitao et al. [38].

8.3.2 Plastic Deformation

On a straight track, the wheel is in contact with the top of the rail, but the wheel flange, when passing through curves, may be in contact with the gauge corner of the rail. The wheel load is transmitted to the rail through a tiny contact area under high contact stresses. This results in repeated loading above the elastic limit, which leads to plastic deformation. The depth of plastic flow depends on the hardness of the rail and the severity of the curves; it can be as much as 15 mm [39,40]. When a material is subjected to repeat loading, its response depends on the ratio of the amplitude of the maximum stress to the yield stress of the material. When the load increases above the elastic limit, the contact stresses exceed yield and the material flows plastically. After the wheel has passed, residual stresses develop. These residual stresses are protective in nature in that they reduce the tendency of plastic flow in the subsequent passes of the wheel. This, together with any effect of strain hardening, makes it possible for the rail material to support stresses that are much higher than its elastic limit. This process is called elastic shakedown, and the contact pressure limit below which this process is possible is known as the elastic shakedown limit. There is also a plastic shakedown limit. Loads between the elastic and plastic shakedown limit will lead to cyclic plasticity of the rail. If repeated, cyclic plastic deformation takes place. The rail material can cyclically harden, which leads to an increase in the yield stress and reduces the tendency of plastic flow [41]. For loads above the plastic shakedown limit, plastic ratchetting will occur; that is, small increments of plastic deformation accumulate with each pass of the wheel [42]. Plastic ratchetting can be found in curved track as a lip down of the rail gauge corner, as shown in Figure 8.9. Plastic ratchetting is the main cause of headcheck surface cracks [20]. The consequences of ratchetting are wear and the initiation of fatigue cracks as the material accumulates strain up to its limiting ductility. Beyond this limit, failed material can separate from the surface as wear debris or form crack-like flaws, as shown in Figure 8.10.

8.3.3 Rolling Contact Fatigue

Rolling contact fatigue (RCF) cracks on rail can be classified into those that are subsurface-initiated and surface-initiated. Subsurface-initiated cracks are often caused by metallurgical defects. On the other hand, surface-initiated cracks seem to be the result of traffic intensity and axle load. A more specific division can be made into shelling, head checks, tache ovale and squats. *Shelling* (see Grassie and Kalousek [43]) is a subsurface defect that occurs at the gauge corner of the high rail in curves

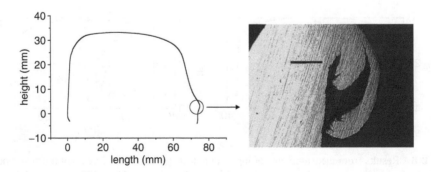

FIGURE 8.9 Lip down of rail from the Stockholm test case showing plastic ratchetting. (From Iwnicki, S. (Ed.), *Handbook of Railway Vehicle Dynamics*, CRC Press, Boca Raton, FL, 2006. With permission.)

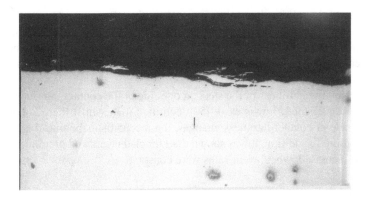

FIGURE 8.10 Micrograph from the Stockholm test case showing wear debris formation and crack-like flaw; length of wear debris = 50 μm. (From Iwnicki, S. (Ed.), *Handbook of Railway Vehicle Dynamics*, CRC Press, Boca Raton, FL, 2006. With permission.)

on railways with high axle loads. An elliptical shell-like crack propagates predominantly parallel to the surface. In many cases, the shell causes metal to spall from the gauge corner. However, when the crack length reaches a critical value, the crack may turn down into the rail, giving rise to the fracture of the rail. *Head checks* (Boulanger et al. [44]) generally occur as a surface-initiated crack on or near the gauge corner in curves. Head checks may branch up towards the surface of the rail, giving rise to spalls. However, for reasons still not clearly understood, cracks can turn down into the rail and, if not detected, cause the rail to break. These events are rare, but are dangerous, since surface cracks tend to form continuously [15]. Frederick [45] discusses the effect of train speed and wheel-rail forces as a result of surface roughness. Furthermore, he discusses whether hard rails or soft rails should be used in curves and the relationship between wear rate and surface crack propagation. The conclusion was that hard rails are more prone to surface cracking. This could be because the wear rate for harder rails is lower. This means that cracks that are initiated are not worn away by the 'natural' wear of the system and so can propagate. However, with a softer rail, the cracks are initiated, but are worn away by wear before they have a chance to propagate through the rail. This suggests that there is a balance to be struck between minimising wear (prolonging rail life) and reducing rolling contact fatigue (increasing safety). This was also seen in the Stockholm test case [15], where UIC 900A rail material was compared against UIC 1100 rail material. Both materials seemed to be similarly sensitive to crack initiation, but the UIC 1100 rail was more sensitive to crack propagation and also more sensitive to the formation of headcheck cracks. More information on the initiation mechanisms and growth of rolling contact fatigue cracks can be found in Beynon et al. [46]. *Tache ovale*, or shatter cracks [43], are defects that develop about 10 to 15 mm below the railhead from cavities, caused by hydrogen. They can occur in the rail or in welds from poor welding practise. Development of *tache ovale* is influenced by thermal or residual stresses from roller straightening. *Squats* [43,44] occur on the railhead in tangent track and in curves of large radius and are characterised by a darkened area on the rail. Squats are surface initiated defects that can initiate from a white etching martensitic layer on the surface of the rail. Other mechanisms of squat formation are linked to longitudinal traction by wheels, which causes plastic ratchetting of the surface layer of material, until a crack develops at the railhead.

Rolling contact fatigue cracks on wheels can be classified as shelling and spalling. *Shelling* is a subsurface rolling contact fatigue defect that occurs on the wheel tread, and the mechanism is similar to the formation of shelling in rails. *Spalling* (Bartley [47]) can be initiated on the wheel tread surface when the wheel experiences gross sliding on rail (braking). Large wheel surface temperatures above the austenisation limit (720°C) can form martensite, a hard, brittle steel phase. This brittle phase will easily fracture during later wheel passes and eventually result in spalling.

Surface coating of the rail has been shown to reduce the advent of RCF cracking in the laboratory [48]. Recent work has also looked at wear rates of clad layers using a twin-disc tester with martensitic stainless steel clad onto two different bulk rail materials [49]. It showed significant reductions in wear rates for clad layers compared with a standard grade rail, even if a softer grade substrate material was used.

The effect of changing hardness of the steel on one side of the contact interface on the other side is important to be understood. Lewis et al. [50] identified that, whilst there is consensus on wear trends when one side of contact hardness changes, the mechanisms behind the trends is currently lacking in explanation. In addition, it was shown that, for clad layers and premium rail grades, when rail hardness was increased, wheel wear rates were constant.

8.4 FRICTION

The friction force can be defined as the resistance encountered by one body moving over another body. This definition covers both sliding and rolling bodies. Note that even pure rolling nearly always involves some sliding and that the two classes of motion are not mutually exclusive. The resistive force, which is parallel to the direction of motion, is called the friction force. If the solid bodies are loaded together, the static friction force is equal to the tangential force required to initiate sliding between the bodies. The kinetic friction force is then the tangential force required to maintain sliding. Kinetic friction is generally lower than static friction.

For sliding bodies, the friction force, and thereby the coefficient of friction (friction force divided by normal force), depends on three different mechanisms in dry and mixed lubricated conditions: deformation of asperities, adhesion of the sliding surfaces and ploughing caused by deterioration particles and hard asperities [51]. For most metal pairs, the maximum value of the coefficient of friction ranges from 0.3 to 1.0 [52]. The ploughing component of the coefficient varies from 0 to 1.0, and the adhesion component varies from 0 to 0.4 [53]. It is generally recognised that friction due to rolling of non-lubricated surfaces over each other is considerably less than dry sliding friction of the same surfaces [54]. For the steel wheel on steel rail contact, the rolling coefficient of friction is of the order of 1×10^{-4}.

As shown in Figure 8.11, the contact area between a wheel and the rail can be divided into stick (no slip) and slip regions. Longitudinal creep and tangential (tractive) forces arise due to the slip that occurs in the trailing region of the contact patch. With increasing tractive force, the slip region increases and the stick region decreases, resulting in a rolling and sliding contact. When the tractive force reaches its saturation value, the stick region disappears, and the entire contact area is in a state of pure sliding. The maximum level of tractive force depends on the capability of the contact patch to absorb traction. This is expressed in the form of the friction coefficient, μ (ratio of tractive force to normal load, N). Normally, wheel-rail traction reaches a maximum at creep levels of 0.01 to 0.02.

The traction versus creep curve can be dramatically affected by the presence of a third body layer in the wheel-rail contact. This could be formed either by a substance applied to increase/decrease friction (friction modifier or lubricant) or by a naturally occurring substance acting to decrease friction (water or leaves etc.). Hou et al. [55] have proposed a frictional model for rolling-sliding contacts separated by an interfacial layer, which is based on three rheological parameters: the shear modulus of elasticity (G), the shear modulus of plasticity (k) and the critical shear stress (τ_c). It shows that the friction is greatly affected by the rheology of the third body, slip distance and load, with the shear stress versus slip distance relationship exhibiting the dominant influence.

There are many other models that have been developed or extended to model the presence of a third body layer. CONTACT [56] includes the third body layer by adapting its elastic properties but has a very high computational cost. FASTSIM [57] is a more basic model, but forms the basis for many others, including Spiryagin [58], Polach [59], Tomberger [60] and the 'Extended Creep Force' (ECF) model [61,62]. The Polach model has been identified as most suitable to model the effects

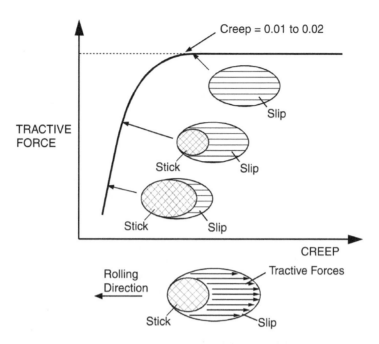

FIGURE 8.11 Relationship between traction and creep in the wheel-rail contact. (From Iwnicki, S. (Ed.), *Handbook of Railway Vehicle Dynamics*, CRC Press, Boca Raton, FL, 2006. With permission.)

of water on friction; it is already implemented in multi-body simulation software, fully published, computationally fast and hence easy to implement. This model takes into account contamination in the contact by describing changes to the initial slope of the traction curve and properties of the interfacial layers. The Spiryagin model allows the friction coefficient to change depending on the slip velocity. The ECF model is an extension of the Tomberger model, but is not widely available. It uses laboratory tests [63] to define the properties of a third body layer. The Water Induced Low Adhesion Creep force model [64] is a recent development to predict the effect of small amounts of water mixed with iron oxide ('wet-rail' syndrome) on the friction in the contact. These models are important for improving inputs to dynamic models of train performance. For example, WILAC has been implemented into the Low Adhesion BRAking Dynamic Optimisation for Rolling stock (LABRADOR) train braking model to improve its capability [65].

8.4.1 Wheel-Rail Friction Conditions

The friction between the wheels and rail is extremely important as it plays a major role in the wheel-rail interface processes such as adhesion, wear, rolling contact fatigue and noise generation. Effective control of friction through the application of friction modifiers to the wheel-rail contact is therefore clearly advantageous; however, the process has to be carefully managed. The aim of friction management is to maintain friction levels in the wheel-rail contact to give [66]:

- *Low friction* in the wheel flange-rail gauge corner contact
- *Intermediate friction* wheel tread-rail top contact (especially for freight trucks)
- *High friction* at the wheel tread-rail top contact for locomotives (especially where adhesion loss problems occur)

Ideal friction conditions in these contact regions for high and low rails are shown in Figure 8.12 [67]. These are similar to values quoted for Canadian Pacific [68].

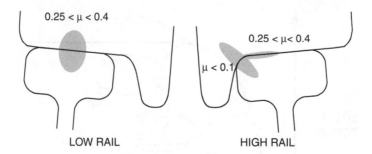

FIGURE 8.12 Ideal friction coefficients in the wheel-rail contact. (From Iwnicki, S. (Ed.), *Handbook of Railway Vehicle Dynamics*, CRC Press, Boca Raton, FL, 2006. With permission.)

Olofsson and Telliskivi [14] compared coefficients of friction measured on track and in the laboratory. For pure non-lubricated sliding tests, the level is roughly the same, varying between 0.5 and 0.6. For full-scale lubricated rail, the coefficient of friction was lower and varied between 0.2 and 0.4. Other results found in the literature support the measured coefficients of friction from the full-scale tests. In another project, the Swedish National Rail Administration studied how leaves on the track influenced the coefficient of friction, using a special friction measurement train [69]. The reported coefficient of friction varied between 0.1 and 0.4. Harrison et al. [70] compared a hand-pushed rail tribometer and a TriboRailer that operated from a companion vehicle. For the hand-pushed tribometer, the coefficient of friction was typically 0.7 under dry conditions and varied between 0.25 and 0.45 under lubricated conditions. The TriboRailer presented lower values of the coefficient of friction. Under dry conditions, the coefficient of friction was roughly 0.5 and varied between 0.05 and 0.3 under lubricated conditions.

To gain the greatest benefit from friction management and to ensure efficient train operation, the coefficient of friction needs to be integrated with the overall wheel-rail management system. It has been noted that, for example, it should be closely tied in with grinding schedules used in the maintenance of wheels and rails [66]. A diagram for the systematic approach to wheel-rail interface research and development is shown in Figure 8.13, which emphasises the consideration of all aspects, including materials, dynamics and friction. Nothing can really be treated in isolation.

8.4.2 Friction Modification

Products can be applied to the wheel-rail contact to generate the required coefficients of friction. These can be divided into three categories [72]:

- *Lubricants* are used to give friction coefficients less than 0.2 at the wheel flange-gauge corner interface
- *Top-of-rail products* with intermediate friction coefficients of 0.2 to 0.4 are used in wheel tread-rail top applications
- *Traction enhancers* are used to increase adhesion for both traction and braking

There is a broad spectrum of top-of-rail products, and there is often confusion in industry and academia about what to call them. Stock et al. [73] define terms to bring clarity to this issue. Top-of-rail products are defined here according to their 'drying' behaviour. Products that stay 'wet' are defined as top-of-rail lubricants, and products that dry out are called top-of-rail friction modifiers. Products that dry out are typically water-based particle suspensions; the water is evaporated in the contact, leaving behind solid particles to mix with the existing third-body layer. Solid friction modifiers do exist; these are made from an easily sheared material and are typically applied to on-board

Tribology of the Wheel-Rail Contact

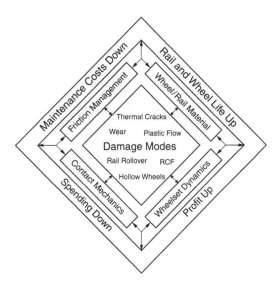

FIGURE 8.13 Systems approach to wheel-rail interface research and development. (From Kalousek, J. and Magel, E., *Railway Track Struc.*, 93, 28–32, 1997; Iwnicki, S. (Ed.), *Handbook of Railway Vehicle Dynamics*, CRC Press, Boca Raton, FL, 2006. With permission.)

trains. Top-of-rail lubricants can be oil based, grease based or a hybrid between oil based and water based [73]. These products still allow contact between wheel and rail surfaces, so a small change in the applied quantity can dramatically change the friction level.

Lubricants can be solid or liquid (greases), the main difference between the two being the thickness of the film that they form in the wheel-rail contact (solid lubricants will give a film of 10 to 30 μm, and grease lubricants will give a film of less than 5 μm) [66]. The primary application of these modifiers is in reducing friction in the wheel flange-rail gauge corner contacts, particularly in curves where the contact conditions can be quite severe. The main focus of the remainder of this section is on low-friction conditions and how to deal with them. Further discussion relating to reduction of friction can be found in the subsequent section on lubrication.

Top-of-rail products are also classified according to their influence on friction after full slip conditions have been reached in the wheel-rail contact, as shown in Figure 8.14 [74]. If friction increases after the saturation point, the modifier has positive friction properties; if friction reduces, the modifier has negative friction properties. Positive friction modifiers can be described as high positive friction or very high positive friction, depending on the rate of increase in friction.

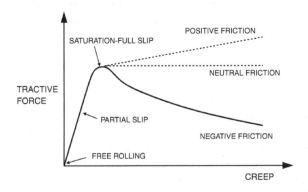

FIGURE 8.14 Behaviour of friction modifiers. (From S. Iwnicki, S. (Ed.), *Handbook of Railway Vehicle Dynamics*, CRC Press, Boca Raton, FL, 2006. With permission.)

8.4.3 ADHESION LOSS

Loss of friction or adhesion between the wheel and rail is particularly important, as this has implications for both braking and traction. Poor adhesion in braking is a safety issue, as it leads to extended stopping distances, and also in traction, as it may lead to reduced acceleration, which will increase the risk of a rear collision from a following train. In traction, however, it is also a performance issue. If a train experiences poor adhesion when pulling away from a station and a delay is enforced, the train operator will incur costs. Similar delays will occur if a train passes over the areas of poor adhesion while in service.

A great deal of research was carried out on adhesion loss in the UK during the 1970s, using both laboratory and field tests [75–79]. This identified the major causes of adhesion as being water (from rainfall or dew), humidity, leaves, wear debris and oil contamination. Oxides have also been shown to cause low adhesion. Recent work with high-viscosity hematite powder/water mixture has produced very low traction coefficients in laboratory tests [80]. This work also updated an adhesion model and showed how very low moisture levels (such as dew in the morning) create a paste on the rail, which causes low adhesion.

Relative humidity has been shown to influence the frictional behaviour of a wide variety of materials [81]. By increasing the relative humidity, an absorbed layer of water molecules can be produced; this can modify frictional behaviour. Relative humidity effects may also produce new chemical reactions on the surface, together with other added substances.

The problems caused by leaves on the line remain prevalent today and each autumn can cause considerable delays to trains on the UK rail network. A review of low adhesion problems caused by leaves [82] identified that wet leaves (caused by dew or rainfall) produced very low traction coefficients of 0.05. It also concluded that the leaf layer was caused by a chemical reaction between the leaves and rail and provided several hypotheses on how the leaf layer bonds to the rail. Laboratory testing with extracts from leaves and analysis of the leaf films [83] found that the low traction levels are likely caused by graphitic carbon, iron oxides and phosphate compounds. Leaves are also a problem in Sweden, where it has been estimated by the Swedish National Railroad Administration that the cost of leaves on the rails is about 10 million Euros annually [69].

Work carried out on Japanese, American and Canadian railways has re-emphasised the effect of the problems outlined above and identified further causes of adhesion loss, such as frost and mud deposited on rails by automobile wheels passing over level crossings [84–86]. This work also showed the varying effects of different types of leaves on adhesion. Oily leaves, such as pine and cedar, caused a larger decrease in adhesion. Tunnels were also highlighted as being a problem, especially where water was leaking onto the track. Full-scale testing has also shown that weather conditions affect both the coefficient of friction and wear rates [14,16].

Most of the work carried out in the UK was at a relatively low speed. Work on adhesion issues related to high-speed lines using both full-scale roller rigs and field measurements has shown that adhesion decreases with train velocity and wheel-rail contact force [87,88].

A number of experimental and theoretical investigations have revealed other significant parameters affecting adhesion. Chen et al. [89] carried out a detailed theoretical investigation of a water-lubricated contact, studying the effect of rolling speed, slip, load, surface roughness and water temperature. The results indicated that the biggest influence on adhesion was the roughness of the wheel and rail surfaces (with adhesion rising with increased roughness). Third body effects due to material generated within the wheel-rail contact and due to externally applied materials have been characterised by Niccolini and Berthier [90] and Hou et al. [55], respectively. These can have a large influence on the adhesion, which is heavily dependent on the rheological properties of the layer formed in the contact. There is only limited data to validate these studies, but they give an important insight into the aspects of the problem that are harder to evaluate in the field.

8.4.4 Increasing Adhesion

While conditions leading to poor adhesion have been well investigated, methods for addressing the problems have not. The main adhesion enhancer used on railway networks worldwide is sand. Sanding is used in train operation to improve adhesion in both braking and traction. In braking, it is used to ensure that the train stops in as short a distance as possible. It usually occurs automatically when the train driver selects emergency braking. Sanding in traction, however, is a manual process. The train driver must determine when to apply the sand and how long the application should last.

The sand is supplied from a hopper mounted under the train. Compressed air is used to blow the sand out of a nozzle attached to the bogie and directed at the wheel-rail contact region (see Figure 8.15). In most systems, the sand is blown at a constant flow rate, but some can provide a variable flow rate. A laboratory test method using a full-scale test facility [91] showed how even moderate cross-winds reduce the effectiveness of sand application and longitudinal winds increased the amount of sand that enters the contact. In addition, the nozzle and hose type could be changed to find the optimum setup for sand application. A method to characterise the size and shape of sand particles and the effect of size and shape on traction characteristics using high-pressure torsion tests has also been developed and could be used to further optimise sand application [92].

While sanding is effective and easy to use, it can potentially cause complex and costly problems relating to both rolling stock and track infrastructure. Sand application has been shown to increase wear rates of both wheel and rail materials by up to an order of magnitude [93,94,86]. Maintenance of sanders and control of sand build-up around track adhesion trouble spots are also issues that require particular attention.

Very high positive friction modifiers to enhance the coefficient of friction to 0.4 to 0.6 are available but are really only in the development stage. There are a number of different products available, but most involve a solid stick of material that is applied directly to the wheel tread. Laboratory test methods have been developed [95,96] to analyse the effect of traction gels. One method measures the effect of the traction gel on a leaf layer, and the other measures its effect on a pectin layer. Electrical isolation and wear can also be measured. These laboratory tests could increase the development rate of traction gels, as the gels can be tested and adapted in the laboratory before being used in the field.

During autumn, when leaf fall occurs, leaf mulch is compressed in the wheel-rail contact and forms an extremely hard layer on the rail surface. This layer can cause adhesion loss problems, as already mentioned, but is also extremely hard to remove. A number of methods are used, including using high-pressure water jets and blasting with Sandite (a mixture of sand and aluminium oxide

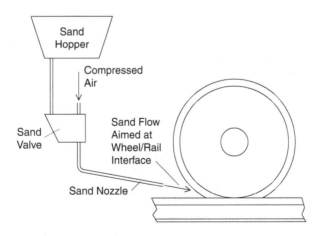

FIGURE 8.15 Sanding apparatus. (From Iwnicki, S. (Ed.), *Handbook of Railway Vehicle Dynamics*, CRC Press, Boca Raton, FL, 2006. With permission.)

particles), and a new system has now been developed that involves using a high-power laser to burn away the layer. In the UK, however, all of these are applied by maintenance trains, which are very few, and gaining track access is extremely difficult. Water jets and Sandite also have knock-on effects, which may be detrimental to the track infrastructure.

8.5 LUBRICATION AND SURFACE MODIFICATION

This section focuses on the problems of high friction coefficients and how to reduce them using lubrication. High friction coefficients are most prevalent at the wheel flange-rail gauge corner contact, particularly in curves. Load and slip conditions are also high, which means that wear and rolling contact fatigue are more likely to occur at these sites. In order to reduce the wear problems, lubrication can be applied to reduce friction and alter the load-bearing capacity. Lubrication, however, is also applied to alleviate other problems, as will be shown. Surface coatings have also been applied to track to address the problem of high friction.

8.5.1 BENEFITS OF LUBRICATION

The benefits of lubrication have been well documented [97,66] and are concerned with the reduction of:

- Wheel flange and rail gauge corner *wear*
- *Energy* consumption
- *Noise* generation

Laboratory [98–100] and field tests [15,32,101] have all shown the wear-reducing benefits of lubrication in the wheel-rail contact.

Fuel savings of around 30% (compared with dry conditions) have been reported for measurements taken on test tracks [102]. Other studies carried out in the field have shown improvements of a similar order of magnitude [103,104].

8.5.2 LUBRICATION METHODS

There are a number of different ways to apply lubricant:

- *Mobile lubricators*: These are basically railway vehicles designed to apply lubricant to the gauge corner of the track.
- *Wayside lubricators*: These are mounted next to the track and apply lubricant to the rail gauge corner. There are three types: mechanical, hydraulic and electronic.
- *On-board lubricators*: These apply grease or solid lubricant or spray oil on to the wheel flange, which is then transferred to the gauge corner of the rail. Complex control systems are used in the application process to avoid the application of lubricant at inappropriate locations.

Mechanical wayside lubricators rely on the wheel making contact with a plunger, which operates a pump. The pump supplies lubricant from a reservoir to a distribution unit. The lubricant is then picked up by the wheel flanges and distributed along the rail. Problems exist because there is only a single circuit, so if a failure occurs, the lubricant supply is ineffective. Mechanical lubricators have a low initial cost because of their simple design but require good maintenance to remain effective. Hydraulic lubricators have been found to be more reliable but have some of the same problems as their mechanical counterparts.

Electronic lubricators use sensors to detect the approach of a train and activate electric pumps to deliver the lubricant. They are inherently more reliable than mechanical or hydraulic lubricators and can also be adjusted away from the track.

On-board lubricators supply lubricant to the wheel flange-rail gauge corner. In most designs, the lubricant is deposited on the wheel flange and spread along the rail; however, in some, the lubricant is directly applied to the rail. Grease or oil spray systems are used; these employ complex control strategies using sensors measuring vehicle speed and track curvature to govern lubricant application. Solid stick lubricators are also available, in which a stick of lubricant is spring loaded against the wheel flange.

On-board systems have a number of advantages over wayside lubricators [67]:

- Reduced safety risk exposure to staff during installation, inspection and maintenance
- Easier inspection and maintenance (carried out in more controlled conditions)
- The rail will continue to receive some friction control protection in the event of the failure of an individual on-board lubricator

Despite these advantages, wayside lubricators will still be a necessity at problem track sites.

8.5.3 Problems with Lubrication

Problems with lubrication systems have been found to be related to both technical and human issues [105]. The main technical problems with wayside lubricators have been highlighted as blocked applicator openings, leaking holes, ineffective pumps and trigger mechanisms and poor choice of lubricant. Human-related problems can result from the technical issues. If over-lubrication occurs and lubricant migrates onto the rail top, adhesion loss can occur. Train drivers may then be tempted to apply sand to compensate and increase friction; however, this will lead to increased wear and could cause the applicators to become blocked. The thought that application of lubricant will lead to wheel slip can also lead train drivers to switch off on-board lubrication systems.

Some of the consequences of poor wayside lubrication have been listed as [68]:

- Wheel slip and loss of braking (and potentially, wheel flats and rail burns)
- Poor train handling
- Prevention of ultrasonic flaw detection
- Wastage of lubricant
- High lateral forces in curves and subsequent increase in wear

Other than adhesion problems, over-lubrication can cause an increase in rolling contact fatigue crack growth on the rail gauge corner [24]. This can be due to pressurisation of the crack, leading to increased growth rates or because reduced wear means that cracks are truncated less. However, full-scale test results from sharp curves show that well-maintained lubrication could reduce both the wear rate and the propagation rate of surface cracks [15].

8.5.4 Lubricator System Selection and Positioning

The effectiveness of a lubrication system is affected by a number of parameters, including the climate, the railway operating conditions, the dispensing mechanism and the maintenance of the lubricating equipment. Clearly, the selection of the most appropriate type of lubricator and lubricant is very important, but also, the positioning of the lubricator is critical to its successful operation.

The key characteristics required of a lubricant are as follows [68]:

- *Lubricity* or the ability of the lubricant to reduce friction (although the effect on wear is of greater importance).
- *Retentivity* or the measure of time over which the lubricant retains its lubricity. Flash temperatures in the wheel-rail contact can be as high as 600°C to 800°C, and these lead to the lubricant in the contact being *burned up*. The retentivity is therefore a function of the loads and creepages seen at the lubrication site, as these dictate the temperature in the contact.
- *Pumpability* or how easily the lubricant can be applied to the track. The temperature is an issue here, as some track locations will experience a wide range across which some lubricants may not maintain their pumpability. Some networks use different lubricants in the winter and summer for this reason.

Laboratory tests have been developed to assess the wear-reducing capacity of lubricants, tackiness, lubricant pick-up and energy-saving potential [98–100,106,107]. These are good for screening and ranking purposes and for selecting those lubricants suitable to take forward for field trials.

Monitoring the effectiveness of lubricants in the field, either during trials or in actual practice, is clearly essential. This will provide information necessary to decide on a lubrication strategy during trials or in maintaining performance once implemented. Friction can be measured using tribometers, either hand-propelled along the track or train-mounted. The hand-propelled equipment is useful for monitoring short stretches of track. Obviously, for long stretches, a train- or vehicle-mounted system is preferable. It has been shown that the benefits of lubrication may take some time to become evident on the installation of a lubrication system [108].

Correct positioning of a wayside lubricator is critical to providing effective lubrication. Each site will require something different, which makes this task quite complex. Controlled field testing has been used to assess the reliability and efficiency of wayside lubricators based on a number of factors related to the lubricant including: waste prevention, burn up, distance covered, washing off by rain or snow and migration to the rail top. These data and factors related to the track, such as length of curve, gradient and applicator configuration, and traffic issues, including direction, types of bogie, axle loads and speeds, have been combined to develop criteria and a model for positioning wayside lubricators [109]. The laboratory tests developed by Temple et al. [107] can help to identify the key pump parameters and allow bespoke settings for different lubricator positions to optimise lubrication. In addition, the tests can be used to assess new designs of lubricator.

Ultimately, however, the most critical element in preserving effective lubrication is maintenance. Once in place, wayside lubricators need regular maintenance to prevent the problems outlined from occurring.

8.6 EMISSION OF SOUND AND AIRBORNE PARTICLES

Both noise and airborne wear particles can be emitted from the open wheel-rail contact system. Noise originating from the wheel-rail contact can radiate both to the wheelset and to a considerable length of the track. Noise emitted from a wheel-rail contact can be categorised into three forms. Rolling noise, which has a broad frequency, occurs frequently on straight track due to the regular surface roughness of the wheel and rail. Impact noise is an intermittent noise that also occurs on straight tracks as a result of discrete rail joints or irregularities. Curve squeal noise can be generated within a narrow frequency band as a railway vehicle negotiates sharp curves. Currently, research into rolling and impact noise concentrates on the effects of surface roughness, rail joints, and wheel and rail profiles [110–113]. These two types of noise can be reduced by such measures as smoothing the wheel and rail surfaces, optimising the wheel and rail profiles and using wheel and rail damping. However, undue smoothing of wheel and rail surfaces can aggravate curve squeal [114]. It was also found in [114] that the orientation of the grinding pattern significantly influenced the onset and level of the generated sound. Speed and curve radius were also found to strongly affect curve squeal [115]. With regard to environmental factors, higher relative humidity is likely to generate curve squeal [116], while liquid friction modifiers significantly reduce noise levels in curves [117]. Sound pressure measurements near the wheel-rail contact can also be used to identify the transfer from mild acceptable wear to severe catastrophic wear; this implies high maintenance cost [114]. Details are provided in [118,119] of an on-board sound pressure measurement system developed; this system identifies in real time the probability for occurrence and the exact location of severe wear in the wheel-rail contact. It calculates the probability of severe wear at the track and transfers this probability online to the traffic control for further actions. This system is today installed on eight underground trains in the Stockholm local network as a maintenance planning aid.

The wheel-rail contact can also emit airborne wear particles. These particles are part of the total worn volume loss of the contact between railway wheels and rail. Researchers have noted high concentrations of airborne particles on underground stations [120] as well as inside train cabins [121]. A literature review showed that the particulate level on an underground subway platform has been commonly reported to be several times higher than the 24-hour limit on ambient PM10 (particles smaller than 10 μm in aerodynamic diameter size) set by the World Health Organization [122]. Dedicated studies of airborne wear particles from the wheel-rail contact are rare. However, results in dedicated laboratory scale test rigs [123,124] indicate transitions to an increase in the number of nanoscale particles at higher contact pressures and increased relative sliding speeds. The increased contact pressure and relative sliding speed simulated the traffic situation in sharp curves. In [125], airborne particle characteristics were investigated in dry contacts and in ones lubricated with rail grease and friction modifiers. The number of particles declined with the grease; the number of ultrafine particles increased with a water-based friction modifier, mainly due to water vaporisation.

Train-based particle emission measurements [126–128] have revealed differences in the number of generated airborne particles, depending on whether the train is operating on straight track or on curved track, with higher values of number concentrations on curved unlubricated track. Here also, a clear dependence of the mechanical braking action on the airborne particulates was noted [126,127]. In [120], it was estimated that half of the airborne particulates measured as PM10 in an underground station trafficked by commuter trains were originated from the mechanical braking action and that the rest were from the wheel-rail contact and the electric power system. Another underground study [128] estimated that the wheel-rail contact generated 1.3 g/rolled km, while the mechanical brakes generated between 0.9 and 1.2 g/rolled km, depending on the type of underground train. In another study, [129], it is even suggested that the number of measured airborne particles is indicative of the occurrence of severe or catastrophic wear on the track, and thus, an on-board particle measurement system can serve as a planning tool for maintenance.

ACKNOWLEDGEMENTS

This work formed part of the activities of the KTH Railway Group at the Department of Machine Design KTH in Stockholm, Sweden. The partners in the KTH Railway Group are Banverket, Green Cargo, SL Infrateknik, Bombardier Transportation, SJ and Traintech Engineering.

REFERENCES

1. ASM Handbook, Volume 18, *Friction, Lubrication, and Wear Technology*, ASM International, Materials Park, OH, 2017.
2. I.M. Hutchings, *Tribology: Friction and Wear of Engineering Materials*, Edward Arnold, London, UK, 1992.
3. K.L. Johnson, *Contact Mechanics*, Cambridge University Press, Cambridge, UK, 1985.
4. D. Dowson, G.R. Higginson, *Elasto-Hydrodynamic Lubrication*, Pergamon Press, Oxford, UK, 1977.
5. D. Dowson, *History of Tribology*, Longman, London, UK, 1979.

6. L-T. Hung, Normal und tangentialspanningsberuchnung beim rollenden kontakt fur rotaionskörper mit nichtellipischen kontaktflächen, Fortschrifts-Berichte VDI 12, Düsseldorf, Germany, 1987.
7. M.B. Marshall, R. Lewis, R.S. Dwyer-Joyce, U. Olofsson, S. Björklund, Ultrasonic characterisation of a wheel/rail contact, In: G. Dalmaz, A.A. Lubrecht, D. Dowson, M. Priest (Eds.), *Proceedings 30th Leeds-Lyon Symposium on Tribology*, 2–5 September 2003, Institut National des Sciences Appliques de Lyon, Villeurbanne, France, 151–158, 2003.
8. L. Zhou, H.P. Brunskill, R. Lewis, M.B. Marshall, R.S. Dwyer-Joyce, Dynamic characterisation of the wheel/rail contact using ultrasonic reflectometry, In: J. Pombo (Ed.), *Proceedings 2nd International Conference on Railway Technology: Research, Development and Maintenance*, 8–11 April 2014, Ajaccio, Corsica, France, Publisher Civil-Comp Press, Stirlingshire, UK, Paper 185, 2014.
9. S. Björklund, U. Olofsson, M. Marshall, R. Lewis, R. Dwyer-Joyce, Contact pressure calculations on rough measured surfaces, *9th International Conference on Metrology and Properties of Engineering Surfaces*, 10–11 September 2003, Halmstad University, Sweden, 2003.
10. W. Poole, The measurement of contact area between opaque objects under static and dynamic rolling conditions, In: G.M.L. Gladwell, H. Ghonem, J. Kalousek (Eds.), *Proceedings International Symposium on Contact Mechanics and Wear of Wheel/Rail Systems*, 8–11 July 1986, University of Rhode Island, Kingston, RI, Publisher University of Waterloo Press, Ontario, Canada, 59–72, 1987.
11. J.J. Kalker, *Three-Dimensional Elastic Bodies in Rolling Contact*, Kluwer Academic Publishers, Dordrecht, the Netherlands, 1990.
12. T. Telliskivi, U. Olofsson, Contact mechanics analysis of measured wheel-rail profiles using the finite element method, *Journal of Rail and Rapid Transit*, 215(2), 65–72, 2001.
13. U. Olofsson, R. Nilsson, Initial wear of a commuter train track, In: S.S. Eskildsen, D.S. Larsen, H. Reitz, E.J. Bienk, C.A. Straede (Eds.), *Proceedings 8th International Conference on Tribology NORDTRIB'98*, 7–10 June 1998, Ebeltoft, Denmark, 1998.
14. U. Olofsson, T. Telliskivi, Wear, plastic deformation and friction of two rail steels – a full-scale test and a laboratory study, *Wear*, 254(1–2), 80–93, 2003.
15. U. Olofsson, R. Nilsson, Surface cracks and wear of rail: A full-scale test on a commuter train track, *Journal of Rail and Rapid Transit*, 216, 249–264, 2002.
16. R. Nilsson, Wheel/rail wear and surface cracks, Licentiate thesis, TRITA-MMK 2003:03, Royal Institute of Technology (KTH), Stockholm, Sweden, 2003.
17. T. Telliskivi, Wheel-rail interaction analysis, TRITA-MMK 2003:21, Doctoral thesis, Royal Institute of Technology KTH), Stockholm, Sweden, 2003.
18. T. Jendel, Prediction of wheel profile wear: Methodology and verification, Licentiate thesis, TRITA-FKT 2000:49, Royal Institute of Technology (KTH), Stockholm, Sweden, 2000.
19. F.A.M. Alwahdi, Wear and rolling contact fatigue of ductile materials, Doctoral thesis, University of Sheffield, Sheffield, UK, 2004.
20. J.W. Ringsberg, Rolling contact fatigue of railway rails with emphasis on crack initiation, Doctoral thesis, Chalmers University of Technology, Göteborg, Sweden, 2000.
21. K. Knothe, A. Theiler, S. Güney, Investigation of contact stresses on the wheel/rail-system at steady state curving, In: R. Fröhling (Ed.), *Proceedings 16th IAVSD Symposium held in Pretoria, South Africa*, 30 August–3 September 1999, Swets & Zeitlinger, Lisse, the Netherlands, 616–628, 2000.
22. S. Kumar, Y.S. Adenwale, B.R. Rajkumar, Experimental investigation of contact stresses between a U.S. locomotive wheel and rail, *Journal of Engineering for Industry*, 105(2), 64–70, 1983.
23. P.D. Cassidy, Variation of normal contact stresses for different wheel/rail profile combinations, Report ERRI D 173/DT 336, European Rail Research Institute, 1996.
24. D.F. Cannon, H. Pradier, Rail rolling contact fatigue: Research by the European Rail Research Institute, *Wear*, 191(1–2), 1–13, 1996.
25. J.A. Williams, Wear modelling: Analytical, computing and mapping: A continuum mechanics approach, *Wear*, 225–229, 1–17, 1999.
26. U. Olofsson, S. Andersson, S. Björklund, Simulation of mild wear in boundary lubricated spherical roller thrust bearings, *Wear*, 241(2), 180–185, 2000.
27. J.L. Sullivan, Boundary lubrication and oxidational wear, *Journal of Physics D: Applied Physics*, 19(10), 1999–2011, 1986.
28. U. Olofsson, G. Svedberg, Low concentration level contaminant-related wear in sliding and rolling contacts, 2nd World Tribology Congress, 3–7 September 2001, Austrian Tribology Society, Vienna, Austria, 2001.
29. S. Andersson, E. Salas-Russo, The influence of surface roughness and oil viscosity on the transition in mixed lubricated sliding steel contacts, *Wear*, 174(1–2), 71–79, 1994.

30. S. Dizdar, Wear transition of a lubricated sliding steel contact as a function of surface texture anisotropy and formation of boundary layers, *Wear*, 237(2), 205–210, 2000.
31. R. Lewis, R.S. Dwyer-Joyce, Wear mechanisms and transitions in railway wheel steels, In: *Proceedings of the Institution of Mechanical Engineers, Part J: Journal of Engineering Tribology*, 218(6), 467–478, 2004.
32. P. Waara, Wear reduction performance of rail flange lubrication, Licentiate thesis, Luleå University of Technology, Luleå, Sweden, 2001.
33. S. Engel, Reibungs- und Erdmudungsverhalten des Rad-Schiene-System mit und ohne Schmierung, Doctoral thesis, Otto von Guericke Universität Magdeburg, Germany, 2002.
34. F. Cantera, Investigation of wheel flange wear on the Santander FEVE Rail – a case study, *Wear*, 162–164, 975–979, 1993.
35. R. Lewis, U. Olofsson, Mapping rail wear regimes and transitions, *Wear*, 257(7–8), 721–729, 2004.
36. A.J. Perz-Unzueta, J.H. Beynon, Microstructure and wear resistance of pearlitic rail steels, *Wear*, 162–164, 173–182, 1993.
37. J.E. Garnham, J.H. Beynon, Dry rolling-sliding wear of bainitic and pearlitic steels, *Wear*, 157(1), 81–109, 1992.
38. S. Mitao, H. Yokoyama, S. Yamamoto, Y. Kataoka, T. Sugiyama, High strength bainitic steel rails for heavy haul railways with superior damage resistance, *Proceedings IHHA'99 STS-Conference on Wheel/Rail Interface*, Moscow, Russia, 1999.
39. K.L. Johnson, The mechanism of plastic deformation of surface and subsurface layers in rolling and sliding contact, In: R. Solecki (Ed.), *Materials Science Forum: The Role of Subsurface Zones in the Wear of Materials*, Trans Tech Publications, Switzerland, 33–40, 1988.
40. C.P. Jones, W.R. Tyfor, J.H. Beynon, A. Kapoor, The effect of strain hardening on shakedown limits of a pearlitic rail steel, *Journal of Rail and Rapid Transit*, 211(2), 131–140, 1997.
41. G. Schleinzer, Residual stress formation during the roller straightening of rails, Doctoral thesis, Montanuniversität Leoben, Austria, 2000.
42. A. Kapoor, K.L. Johnson, Plastic ratchetting as a mechanism of metallic wear, *Proceedings: Mathematical and Physical Sciences*, 445(1924), 367–384, 1994.
43. S.L. Grassie, J. Kalousek, Rolling contact fatigue of rails: Characteristics, causes and treatments, In: *Proceedings 6th IHHA Railway Conference*, 6–10 April 1997, Cape Town, South Africa, 381–404, 1997.
44. D. Boulanger, L. Girardi, G. Galtier, G. Baudry, Prediction and prevention of rail contact fatigue, *Proceedings IHHA'99 STS-Conference on Wheel/Rail Interface*, Moscow, Russia, 1999.
45. C.O. Frederick, Future rail requirements, In: J.J. Kalker, D.F. Cannon, O. Orringer (Eds.), *Rail Quality and Maintenance for Modern Railway Operation*, Kluwer Academic Publishers, Dordrecht, the Netherlands, 3–14, 1993.
46. J.H. Beynon, M.W. Brown, A. Kapoor, Initiation, growth and branching of cracks in railway track, In: J.H. Benyon, M.W. Brown, R.A. Smith, T.C. Linley, B. Tomkins (Eds.), *Proceedings International Conference on Engineering Against Fatigue*, 17–21 March 1997, Sheffield, UK, Publisher Balkema, Rotterdam, the Netherlands, pp. 461–472, 1999.
47. G.W. Bartley, A practical view of wheel tread shelling, 9th International Wheelset Congress, Montreal, Canada, 1988.
48. E.J.M. Hiensch, F.J. Franklin, J.C.O. Nielsen, J.W. Ringsberg, G.J. Weeda, A. Kapoor, B.L. Josefson, Prevention of RCF damage in curved rail through development of the INFRA-STAR two-material rail, *Fatigue and Fracture of Engineering Materials and Structures*, 26(10), 1007–1017, 2003.
49. P. Lu, S.R. Lewis, S. Fretwell-Smith, D.I. Fletcher, R. Lewis, Laser cladding of rail: The effects of depositing material on lower rail grades, In: Z. Li, A. Núñez (Eds.), *Proceedings 11th International Conference on Contact Mechanics and Wear of Rail/Wheel Systems*, 24–27 September 2018, Delft, the Netherlands, 610–617, 2018.
50. R. Lewis, P. Christoforou, W.J. Wang, A. Beagles, M. Burstow, S.R. Lewis, Investigation of the influence of rail hardness on the wear of rail and wheel materials under dry conditions (ICRI Wear Mapping Project), In: Z. Li, A. Núñez (Eds.), *Proceedings 11th International Conference on Contact Mechanics and Wear of Rail/Wheel Systems*, 24–27 September 2018, Delft, the Netherlands, 510–517, 2018.
51. N.P. Suh, H.C. Sin, The genesis of friction, *Wear*, 69(1), 91–114, 1981.
52. H. Czichos, Presentation of friction and wear data, In: P.J. Blau (Ed.), *Friction, Lubrication and Wear Technology, ASM Handbook*, ASM International, Materials Park, Ohio, 18, 489–493, 1992.
53. N.P. Suh, *Tribophysics*, Prentice-Hall, Englewood Cliffs, NJ, 1986.
54. T.A. Harris, *Rolling Bearing Analysis*, 3rd ed., John Wiley & Sons, New York, 1991.

55. K. Hou, J. Kalousek, E. Magel, Rheological model of solid layer in rolling contact, *Wear*, 211(1), 134–140, 1997.
56. E.A.H. Vollebregt, Numerical modeling of measured railway creep versus creep-force curves with contact, *Wear*, 314(1–2), 87–95, 2014.
57. J.J. Kalker, A fast algorithm for the simplified theory of rolling contact, *Vehicle System Dynamics*, 11(1), 1–13, 1982.
58. M. Spiryagin, O. Polach, C. Cole, Creep force modelling for rail traction vehicles based on the Fastsim algorithm, *Vehicle System Dynamics*, 51(11), 1765–1783, 2013.
59. O. Polach, Creep forces in simulations of traction vehicles running on adhesion limit, *Wear*, 258(7–8), 992–1000, 2005.
60. C. Tomberger, P. Dietmaier, W. Sextro, K. Six, Friction in wheel-rail contact: A model comprising interfacial fluids, surface roughness and temperature, *Wear*, 271(1–2), 2–12, 2011.
61. K. Six, A. Meierhofer, G. Müller, P. Dietmaier, Physical processes in wheel-rail contact and its implications on vehicle-track interaction, *Vehicle System Dynamics*, 53(5), 635–650, 2014.
62. A. Meierhofer, A new wheel-rail creep force model based on elasto-plastic third body layers, Doctoral thesis, Graz University of Technology, Austria, 2015.
63. M.D. Evans, R. Lewis, C. Hardwick, A. Meierhofer, K. Six, High pressure torsion testing of the wheel/rail interface, In: H. Tournay (Ed.), *Proceedings 10th International Conference on Contact Mechanics and Wear of Rail/Wheel Systems*, 31 August–2 September 2015, Colorado Springs, CO, Paper 136, 2015.
64. G. Trummer, L.E. Buckley-Johnstone, P. Voltr, A. Meierhofer, R. Lewis, K. Six, Wheel-rail creep force model for predicting water induced low adhesion phenomena, *Tribology International*, 109, 409–415, 2018.
65. H. Alturbeh, R. Lewis, K. Six, G. Trummer, J. Stow, Implementation of the Water Induced Low Adhesion Creep Force Model (WILAC) into the Low Adhesion Braking Dynamic Optimisation for Rolling Stock Model (LABRADOR), In: Z. Li, A. Núñez (Eds.), *Proceedings 11th International Conference on Contact Mechanics and Wear of Rail/Wheel Systems*, 24–27 September 2018, Delft, the Netherlands, 37–43, 2018.
66. S. Zakharov, *Wheel/Rail Performance, Part 3 in Guidelines to Best Practice for Heavy Haul Railway Operations: Wheel and Rail Interface Issues*, International Heavy Haul Association, Virginia Beach, VA, 2001.
67. J. Sinclair, Friction modifiers, in *Vehicle Track Interaction: Identifying and Implementing Solutions*, IMechE Seminar, London, 17 February 2004.
68. M.D. Roney, *Maintaining Optimal Wheel and Rail Performance, Part 5 in Guidelines to Best Practice for Heavy Haul Railway Operations: Wheel and Rail Interface Issues*, International Heavy Haul Association, Virginia Beach, VA, 2001.
69. L. Forslöv, Wheel slip due to leaf contamination, Report TM 1996 03 19, Swedish National Rail Administration, Borlänge, Sweden, 1996. (in Swedish)
70. H. Harrison, T. McCanney, J. Cotter, Recent development in COF measurements at the rail/wheel interface, *Proceedings 5th International Conference on Contact Mechanics and Wear of Rail/Wheel Systems*, Tokyo, Japan, 25–27 July 2000.
71. J. Kalousek, E. Magel, Optimising the wheel/rail system, *Railway Track and Structures*, 93(1), 28–32, 1997.
72. J. Kalousek, E. Magel, Modifying and managing friction, *Railway Track and Structures*, 95(4), 5–6, 1999.
73. R. Stock, L. Stanlake, C. Hardwick, M. Yu, D. Eadie, R. Lewis, Material concepts for top of rail friction management: Classification, characterisation and application, *Wear*, 366–367, 225–232, 2016.
74. D.T. Eadie, J. Kalousek, K.C. Chiddick, The role of high positive friction (HPF) modifier in the control of short pitch corrugation and related phenomena, *Wear*, 253(1–2), 185–192, 2002.
75. A.H. Collins, C. Pritchard, Recent research on adhesion, *Railway Engineering Journal*, 1(1), 19–29, 1972.
76. M. Broster, C. Pritchard, D.A. Smith, Wheel/rail adhesion: It's relation to rail contamination on british railways, *Wear*, 29(3), 309–321, 1974.
77. T.M. Beagley, C. Pritchard, Wheel/rail adhesion: The overriding influence of water, *Wear*, 35(2), 299–313, 1975.
78. T.M. Beagley, I.J. McEwen, C. Pritchard, Wheel/rail adhesion: The influence of railhead debris, *Wear*, 33(1), 141–152, 1975.
79. T.M. Beagley, I.J. McEwen, C. Pritchard, Wheel/rail adhesion: Boundary lubrication by oily fluids, *Wear*, 31(1), 77–88, 1975.

80. B. White, P. Laity, C. Holland, K. Six, G. Trummer, R. Lewis, Iron oxide and water paste rheology and its effect on low adhesion in the wheel/rail interface, In: Z. Li, A. Núñez (Eds.), *Proceedings 11th International Conference on Contact Mechanics and Wear of Rail/Wheel Systems*, 24–27 September 2018, Delft, the Netherlands, 1084–1089, 2018.
81. K. Demizu, R. Wadabayashi, H. Ishigaki, Dry friction of oxide ceramics against metals: The effect of humidity, *Tribology Transactions*, 33(4), 505–510, 1990.
82. K. Ishizaka, S.R. Lewis, R. Lewis, The low adhesion problem due to leaf contamination in the wheel/rail contact: Bonding and low adhesion mechanisms, *Wear*, 378–379, 183–197, 2017.
83. K. Ishizaka, S.R. Lewis, D. Hammond, R. Lewis, Chemistry of black leaf films synthesised using rail steels and their influence on the low friction mechanism, *RSC Advances*, 8(57), 32506–32521, 2018.
84. K. Nagase, A study of adhesion between the rails and running wheels on main lines: Results of investigations by slipping adhesion test bogie, *Journal of Rail and Rapid Transit*, 203(1), 33–43, 1989.
85. C.F. Logston, G.S. Itami, Locomotive friction-creep studies, *Journal of Engineering for Industry*, 102(3), 275–281, 1980.
86. C.W. Jenks, Improved methods for increasing wheel/rail adhesion in the presence of natural contaminants, *Transit Cooperative Research Program, Research Results Digest, No. 17*, Transportation Research Board, Washington, DC, 1997.
87. W. Zhang, J. Chen, X. Wu, X. Jin, Wheel/rail adhesion and analysis by using full scale roller rig, *Wear*, 253(1–2), 82–88, 2002.
88. T. Ohyama, Tribological studies on adhesion phenomena between wheel and rail at high speeds, *Wear*, 144(1–2), 263–275, 1991.
89. H. Chen, T. Ban, I. Ishida, T. Nakahara, Adhesion between rail/wheel under water lubricated contact, *Wear*, 253(1–2), 75–81, 2002.
90. E. Niccolini, Y. Berthier, Progression of the stick/slip zones in a dry wheel/rail contact: Updating theories on the basis of tribological reality, In: D. Dowson, M. Priest, G. Dalmaz, A.A. Lubrecht (Eds.), *Proceedings 29th Leeds-Lyon Symposium on Tribology*, 3–6 September 2002, University of Leeds, UK, Elsevier Tribology Series, 41, 845–853, 2003.
91. S.R. Lewis, S. Riley, D.I. Fletcher, R. Lewis, Optimisation of a railway sanding system for optimal grain entrainment into the wheel-rail contact, *Journal of Rail and Rapid Transit*, 232(1), 43–62, 2018.
92. W.A. Skipper, A. Chalisey, R. Lewis, Particle characterisation of rail sand for understanding tribological behaviour, In: Z. Li, A. Núñez (Eds.), *Proceedings 11th International Conference on Contact Mechanics and Wear of Rail/Wheel Systems*, 24–27 September 2018, Delft, the Netherlands, 886–895, 2018.
93. R. Lewis, R.S. Dwyer-Joyce, Wheel-rail wear and surface damage caused by adhesion sanding, In: G. Dalmaz, A.A. Lubrecht, D. Dowson, M. Priest (Eds.), *Proceedings 30th Leeds-Lyon Symposium on Tribology*, 2–5 September 2003, Institut National des Sciences Appliques de Lyon, Villeurbanne, France, Elsevier Tribology Series, 43, 731–741, 2003.
94. S. Kumar, P.K. Krishnamoorthy, D.L. Prasanna Rao, Wheel-rail wear and adhesion with and without sand for a North American locomotive, *Journal of Engineering for Industry*, 108(2), 141–147, 1986.
95. S.R. Lewis, R. Lewis, J. Cotter, X. Lu, D.T. Eadie, A new method for the assessment of traction enhancers and the generation of organic layers in a twin-disc machine, In: H. Tournay (Ed.), *Proceedings 10th International Conference on Contact Mechanics and Wear of Rail/Wheel Systems*, 31 August–2 September 2015, Colorado Springs, CO, Paper 121, 2015.
96. B. White, R. Lewis, The development of a traction gel assessment method, In: Z. Li, A. Núñez (Eds.), *Proceedings 11th International Conference on Contact Mechanics and Wear of Rail/Wheel Systems*, 24–27 September 2018, Delft, the Netherlands, 1090–1095, 2018.
97. S. Marich, S. Makie, R. Fogarty, The optimisation of rail/wheel lubrication practice in the Hunter Valley, *Proceedings of the Conference on Railway Engineering*, 21–23 May 2000, Railway Technical Society of Australasia, Adelaide, Australia, 4.1–4.13, 2000.
98. P. Clayton, D. Danks, R.K. Steele, Laboratory assessment of lubricants for wheel/rail applications, *Lubrication Engineering*, 45(8), 501–506, 1989.
99. X. Zhao, J. Liu, B. Zhu, C. Wang, Laboratory assessment of lubricants for wheel/rail lubrication, *Journal of Materials Science & Technology*, 13(1), 57–60, 1997.
100. A. Alp, A. Erdemir, S. Kumar, Energy and wear analysis in lubricated sliding contact, *Wear*, 191(1), 261–264, 1996.
101. P. Waara, Lubricant influence on flange wear in sharp railroad curves, *Industrial Lubrication and Tribology*, 53(4), 161–168, 2001.

102. R. Reiff, D. Cregger, Systems approach to best practices for wheel and rail friction control, *Proceedings International Heavy Haul Association Specialist Technical Session Conference*, 14–17 June 1999, Moscow, Russia, 323–330, 1999.
103. R.A. Allen, W.E. Mims, R.C. Rownd, S.P. Singh, Energy savings due to wheel rail lubrication: Seaboard system test and other investigations, *Journal of Engineering for Industry*, 107, 190–196, 1985.
104. J.M. Samuels, D.B. Tharp, Reducing train rolling resistance by on-board lubrication, *Proceedings 2nd Rail and Wheel Lubrication Symposium*, 2–4 June 1987, Memphis, TN, 1987.
105. G. Thelen, M. Lovette, A parametric study of the lubrication transport mechanism at the rail-wheel interface, *Wear*, 191, 113–120, 1996.
106. M. Harmon, B. Powell, I. Barlebo-Larsen, R. Lewis, Development of a grease tackiness test, *Tribology Transactions*, 62, 207–217, 2019.
107. P.D. Temple, M. Harmon, R. Lewis, M.C. Burstow, B. Temple, D. Jones, Optimisation of grease application to railway tracks, *Journal of Rail and Rapid Transit*, 232(5), 1514–1527, 2018.
108. R.P. Reiff, Rail-wheel lubrication, a strategy for improving wear and energy efficiency. *Proceedings 3rd International Heavy Haul Conference*, 13–17 October 1986, Vancouver, Canada, International Heavy Haul Association, 1986.
109. J.J. de Koker, Development of a formula to place rail lubricators, *Proceedings 5th International Tribology Conference*, 27–29 September 1994, Pretoria, South Africa, The South African Institute of Tribology, 1994.
110. B.L. Stoimenov, S. Maruyama, K. Adachi, K. Kato, The roughness effect on the frequency of frictional sound, *Tribology International*, 40(4), 659–664, 2007.
111. D.J. Thompson, C.J.C. Jones, A review of the modelling of wheel/rail noise generation, *Journal of Sound and Vibration*, 231(3), 519–536, 2000.
112. T.X. Wu, D.J. Thompson, On the impact noise generation due to a wheel passing over rail joints, *Journal of Sound of Vibration*, 267(3), 485–496, 2003.
113. D.J. Thompson, On the relationship between wheel and rail surface roughness and rolling noise, *Journal of Sound and Vibration*, 193(1), 149–160, 1996.
114. Y. Lyu, E. Bergseth, U. Olofsson, A. Lindgren, M. Höjer, On the relationships among wheel-rail surface topography, interface noise and tribological transitions, *Wear*, 338–339, 36–46, 2015.
115. M.J. Rudd, Wheel/rail noise: Part II: Wheel squeal, *Journal of Sound and Vibration*, 46(3), 381–394, 1976.
116. X. Liu, P.A. Meehan, Investigation of the effect of relative humidity on lateral force in rolling contact and curve squeal, *Wear*, 310(1–2), 12–19, 2014.
117. D.T. Eadie, M. Santoro, J. Kalousek, Railway noise and the effect of top of rail liquid friction modifiers: Changes in sound and vibration spectral distributions in curves, *Wear*, 258(7), 1148–1155, 2005.
118. M. Höjer, E. Bergseth, U. Olofsson, R. Nilsson, Y. Lyu, A noise related track maintenance tool for severe wear detection of wheel-rail contact, In: J. Pombo (Ed.), *Proceedings 3rd International Conference on Railway Technology: Research, Development and Maintenance*, 5–8 April 2016, Cagliari, Sardinia, Italy, Publisher Civil-Comp Press, Stirlingshire, UK, Paper 146, 2016.
119. Y. Lyu, S. Björklund, E. Bergseth, U. Olofsson, R. Nilsson, A. Lindgren, M. Höjer, A. Lindgren, Development of a noise related track maintenance tool, In: M.J. Crocker, M. Pawelczyk, F. Pedrielli, E. Carletti, S. Luzzi (Eds.), *Proceedings 22nd International Congress on Sound and Vibration*, 12–16 July 2015, Florence, Italy, 2248–2255, 2015.
120. M. Tu, Y. Cha, J. Wahlström, U. Olofsson, Towards a two-part train traffic emissions factor model for airborne wear particles, *Transportation Research Part D: Transport and Environment*, 67, 67–76, 2019.
121. Y. Cha, M. Tu, M. Elmgren, S. Silvergren, U. Olofsson, Factors affecting the exposure of passengers, service staff and train drivers inside trains to airborne particles, *Environmental Research*, 166, 16–24, 2018.
122. S. Abbasi, A. Jansson, U. Sellgren, U. Olofsson, Particle emissions from rail traffic: A literature review, *Critical Reviews in Environmental Science and Technology*, 43(23), 2511–2544, 2013.
123. J. Sundh, U. Olofsson, L. Olander, A. Jansson, Wear rate testing in relation to airborne particles generated in a wheel-rail contact, *Lubrication Science*, 21, 135–150, 2009.
124. H. Liu, Y. Cha, U. Olofsson, L.T.I. Jonsson, P.G. Jönsson, Effect of the sliding velocity on the size and amount of airborne wear particles generated from dry sliding wheel-rail contacts, *Tribology Letters*, 63(3), 30, 2016.
125. S. Abbasi, U. Olofsson, Y. Zhu, U. Sellgren, Pin-on-disc study of the effects of railway friction modifiers on airborne wear particles from wheel-rail contacts, *Tribology International*, 60, 136–139, 2013.
126. S. Abbasi, L. Olander, C. Larsson, U. Olofsson, A. Jansson, U. Sellgren, A field test study of airborne wear particles from a running regional train, *Journal of Rail and Rapid Transit*, 226(1), 95–109, 2012.

127. Y. Cha, S. Abbasi, U. Olofsson, Indoor and outdoor measurement of airborne particulates on a commuter train running partly in tunnels, *Journal of Rail and Rapid Transit*, 232(1), 3–13, 2018.
128. C. Johansson, Källor till partiklar i Stockholms tunnelbana, Report SLB6:2005, SLB-analys, Stockholm, Sweden, 2005. (in Swedish)
129. E. Fridell, A. Björk, M. Ferm, A. Ekberg, On-board measurements of particulate matter emissions from a passenger train, *Journal of Rail and Rapid Transit*, 225(1), 99–106, 2011.

9 Track Design, Dynamics and Modelling

Wanming Zhai and Shengyang Zhu

CONTENTS

9.1 The Railway Track System	308
9.2 Brief Overview of Track Dynamics Modelling	310
9.3 Modelling of Track Dynamics	311
9.3.1 Ballasted Track Model	311
9.3.1.1 Rail Motion	312
9.3.1.2 Sleeper Motion	313
9.3.1.3 Ballast Motion	314
9.3.2 Slab Track Model	315
9.3.2.1 Motions of Slab Track	315
9.4 Interaction between Track and Train	317
9.4.1 Vehicle-Track Coupled Dynamics Model	317
9.4.2 Wheel-Rail Coupling Model	319
9.4.3 Numerical Solution	321
9.4.4 Experimental Validation	322
9.5 System Excitations	323
9.5.1 Models of Impact Loads	323
9.5.1.1 Impact Model of Rail Joints	323
9.5.1.2 Impact Model of Turnouts	323
9.5.1.3 Impact Model of Wheel Flats	325
9.5.2 Models of Harmonic Loads	326
9.5.2.1 Model of Rail Corrugations	326
9.5.2.2 Model of Wheel Polygon	327
9.5.3 Random Track Irregularity	327
9.5.3.1 American Track Spectrum	327
9.5.3.2 German Track Spectrum	328
9.5.3.3 Chinese Track Spectrum	329
9.5.3.4 Comparison of Track Irregularity PSDs for High-Speed Railway	329
9.6 Dynamic Properties of Track Components	331
9.6.1 Dynamic Response due to Impact Loads	331
9.6.2 Random Dynamic Response	332
9.7 Track Design Based on Vehicle-Track Coupled Dynamics	336
9.7.1 Design Scheme	337
9.7.2 Dynamic Simulation and Evaluation	338
9.7.3 Experimental Validation	339
Acknowledgements	341
References	341

9.1 THE RAILWAY TRACK SYSTEM

The railway track is a basic technical structure that is essential for the operation of trains. A traditional railway track structure usually consists of rails, rail pads, rail-to-sleeper fasteners, sleepers, ballast and subgrade, as shown in Figure 9.1. The purpose of a railway track is to guide the trains to run smoothly, to carry the dynamic loads from wheel-rail interaction and to transmit the dynamic loads to the subgrade, bridge deck or tunnel floor. The railway track should be constructed with precise geometry and must be strong and stable enough to ensure train's running safety. In addition, the ballastless track (shown in Figure 9.2) is also now widely applied in high-speed railways and urban rail transit. It employs concrete slabs with good integrity to substitute for the sleepers and discrete ballast bed, with the aim of continuously maintaining the track geometry and, at the same time, substantially decreasing the level of repair and maintenance works required.

The running rails are the main component of railway tracks and function to guide train wheels to move forward along the direction of the track, to carry the large loading pressure of the wheels and to distribute the load over the sleepers, ballast bed and subgrade. Meanwhile, the rails should provide a continuous and smooth rolling surface for the wheels. In this regard, rails should satisfy the following requirements: have enough stiffness to resist elastic deformation induced by dynamic loads, possess a certain ruggedness to avoid damage or fracture due to dynamic loads, have sufficient hardness to prevent fast wheel wear and provide a certain surface roughness at the railhead to realise traction and braking forces of locomotives.

The sleepers, an important component of the track structure, are usually laid laterally on the ballast bed under the rails, carrying the multidirectional pressure from the rails and elastically distributing the resulting loads over the ballast bed. Meanwhile, the sleepers effectively preserve the gauge, level and alignment of the track. The sleepers should have the necessary firmness, elasticity and durability and provide the capacity to resist longitudinal and lateral displacements. The sleepers can be divided into categories of lateral sleepers, longitudinal sleepers and short sleepers according to the construction and laying method. The lateral sleeper is the most commonly used sleeper and is laid perpendicular to the rails. The longitudinal sleeper is only applied in some limited sections with special requirements. Short sleepers are placed separately under each of the running rails and are usually employed in monolithic concrete track-bed structures. Sleeper types can also be categorised based on the primary material from which they are manufactured; timber sleepers, concrete sleepers and steel sleepers are the most common types, but sleepers manufactured from various polymeric composite materials are also now in limited low-axle-load and/or low-speed use. Nowadays, railway tracks in most new and upgraded railway lines use concrete sleepers.

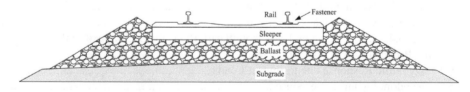

FIGURE 9.1 Traditional ballasted track.

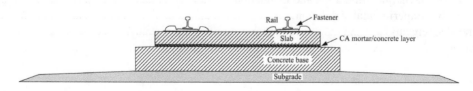

FIGURE 9.2 Typical ballastless track.

The rail pads and fasteners are used to achieve the connection of the rails to the sleepers. The rail pads are placed between the rails and the sleepers to protect the sleepers from wear and impact damage and to provide electrical insulation of the rails. The fasteners are required to have sufficient strength, durability and a certain elasticity, ensuring the reliable connection of the rails and the sleepers and preventing the longitudinal movement of rails with respect to sleepers. From a track dynamics point of view, the rail pads play an important role. They influence the overall track stiffness. A soft rail pad results in a larger deflection of the rails, and the axle load from the train is distributed over more sleepers. In addition, soft rail pads isolate high-frequency vibrations and suppress the transmission of high-frequency vibrations down to the sleepers and further down into the ballast. A stiff rail pad, on the other hand, gives a more direct transmission of the axle load, including the high-frequency load variations, down to the sleepers below the wheels [1].

The ballast bed is the foundation for the sleepers of a ballasted track, on which sleepers are laid at a certain longitudinal spacing interval, aiming to provide the track elasticity for absorbing wheel-rail impacts and vibrations and to provide both lateral and longitudinal resistances for retaining the track stability. To meet the high-speed and high-density operation of trains, ballastless track designs with concrete slabs are being adopted to replace the discrete ballast in modern railway tracks. Compared with the performance of ballasted track, the ballastless track offers better stiffness uniformity, higher running stability and lower maintenance costs. Its main structural types include cast-in-situ monolithic tracks, slab tracks and the ballastless track with sleepers. For example, the China Railway Track System (CRTS) I, II and III slab track systems and the CRTS double-block ballastless track system have been developed and employed in Chinese high-speed railways, as shown in Figure 9.3.

FIGURE 9.3 CRTS series track systems in high-speed railways: (a) CRTS I slab track, (b) CRTS II slab track, (c) CRTS III slab track and (d) CRTS double-block ballastless track. (From Wang, M., *J. Rail Rapid Transit.*, 231(4), 470–481, 2017.)

The subgrade, also called the substructure, is the foundation of the track bed, which carries the static and dynamic loads from the track and train and distributes these loads deep into the underlying ground. The surface layer of the subgrade requires a certain strength to bear train dynamic loads over the long term and should have enough stiffness to limit its elastic deformation to a certain range. In addition, the subgrade should have sufficient depth to ensure that the dynamic stress transmitted to the bottom layer of the subgrade is smaller than the load-bearing capacity of the underlying natural ground. The subgrade is a very important component of the track structure, and not recognising this has been the cause of track failure and poor track quality; for example, uneven subgrade settlement can cause interface damage of slab tracks [3,4] and can disable the proper functioning of sleepers of ballasted tracks [5,6]. The subgrade composed of geomaterial is a potentially weak and unstable part of railway lines. Therefore, subgrade deformation control is critical for the success of subgrade design.

9.2 BRIEF OVERVIEW OF TRACK DYNAMICS MODELLING

The earliest reference related to track dynamics can be traced back to 1867 when Winkler proposed the elastic foundation beam theory, which was soon used for track modelling. In 1926, Timoshenko first studied the dynamic stress of a rail by using the elastic foundation beam model, a classic method that is widely used nowadays. In the 1970s, the rapid development of railway transportation greatly promoted the research into wheel-rail interaction, and remarkable achievements were made in applied mathematics and mechanics models for solving practical engineering problems. It can be said that the test and theoretical research on wheel-rail interaction force at a rail joint, conducted by the British Railways technology research centre at Derby [7,8], moved this research topic forward to a substantially new stage. They established a model for wheel-rail dynamic interaction analysis and first studied the influence of parameters of vehicle and track (such as unsprung mass and track stiffness) on the wheel-rail interaction force. In the model, the track was described as a Euler beam supported by a continuous elastic foundation, the vehicle was simplified as an unsprung mass with the primary suspension and the Hertz non-linear spring model was used for the wheel-rail contact. In 1982, Clark [9] studied the dynamic effect of vehicles running on corrugated rails. The model described the track as a continuous beam supported by discrete elastic points and considered the influence of sleeper vibration, which enables the simulation to more closely replicate the actual support state of the track structure.

At the same time, a number of researchers employed the lumped parameter model, a more simplified model, to simulate track structures that have the characteristics of distributed parameters. By considering the sprung and unsprung masses of vehicles, the lumped parameter model of the wheel-rail system was established to investigate the wheel-rail interactions [10,11]. In the model, the rail is simplified as a concentrated mass block, and the foundation under the rail is treated as a supporting spring and damping element.

With further development of high-speed and heavy-haul railways in the 1990s, especially the booming development of Chinese railways, the vehicle and track dynamic interactions became very crucial topics. In order to meet the requirements of railway engineering applications, in 1990, Zhai [12] started to consider the vehicle and track as an integrated system coupled by the wheel-rail interaction, providing a new theory and a new method for the best matching of designs between modern railway trains and track structures. He established a series of three-dimensional vehicle-track coupled dynamics models that fully consider the lateral dynamic system interactions [13–15] (noting that the previously mentioned references only consider the vertical interactions), to comprehensively investigate the dynamic performance of vehicles running on the elastic track structure and the wheel-rail interaction characteristics. Meanwhile, a number of scholars developed various analysis models [16–29], from the coupled model of a bogie and the track with two-layer distributed parameter model [24] to the integrated vehicle-track dynamic interaction model [19,20,28,29], as well as the model that accounts for the influence of ballast bed vibration [17,25,26]. The overall trend is to consider the dynamic interactions of vehicle and track systems more and more comprehensively.

Track Design, Dynamics and Modelling

After a long period of development, the vehicle-track coupled dynamics theory has been a fundamental method to investigate the interactions between vehicle and track [30]. The vertical vehicle-track coupled dynamics model was widely used in railway engineering dynamics initially [5,6,31–34], and the lateral interaction model is widely applied in this field nowadays [35–43]. The research issues are becoming more and more extensive and complicated, including wheel-rail interactions at turnout sections [35], lateral wheel-rail interaction on curved tracks [40,42], rail corrugations [33,44], vehicle running stability on elastic tracks [37,43], dynamic problems induced by track damage [3–6,45], long-term degradation of vehicle and track systems [31,39,46,47], track vibration assessment and control [48–50], train induced environmental vibrations [34,51,52] and wheel-rail noises [53–55].

9.3 MODELLING OF TRACK DYNAMICS

For the modelling of track dynamics, a number of tools are available, including the finite element method, mode-superposition method, discrete element method, boundary element method and analytical methods. The finite element method can be used to model the most realistic track models with different service states, but it normally requires the most computer capacity [3,56]. The mode-superposition method is often applied for system dynamics analysis in the time domain and possesses high computational efficiency compared with the finite element method [30,50]. The discrete element method is an effective tool to simulate the special morphology and sophisticated mechanism of the ballast bed [57]. The boundary element method is mainly employed in noise prediction of railway systems [55], as well as in modelling boundary conditions to avoid wave reflections at the boundaries [52]. In addition, the analytical methods are normally adopted in track dynamics analysis in the frequency domain, without considering any non-linearities [58]. Here, the introductions of the track dynamics model and its interaction with trains are based on the framework of vehicle-track coupled dynamics theory [13,30] and are parts of the authors' previous research works.

9.3.1 BALLASTED TRACK MODEL

The ballasted track is a traditional structure form in railways around the world and still occupies an important position. For a ballasted track, the rails, sleepers and ballast bed all participate in the system vibration. Accurately simulating the dynamic behaviour of ballasted track, especially the discrete ballast bed, is still a challenging topic in the railway community. The authors have proposed a three-dimensional dynamic model of a typical ballasted track, as shown in Figure 9.4. In the model, the rail is regarded as a Bernoulli-Euler beam discretely supported by three layers of springs and dampers, representing the elasticity and damping of the rail pad and rail to sleeper fasteners, the ballast bed and the subgrade, respectively. The vertical, lateral and torsional degrees of freedom (DoFs) of the rails are considered in the model. The sleeper is treated as a rigid body, and its vertical, lateral and rotational motions are taken into account. Lateral springs and dampers are applied to represent the lateral dynamic properties in the fastener system and to simulate the elasticity and damping property between the sleeper and the ballast in the lateral direction.

For modelling the ballast bed, a five-parameter model of ballast under each rail-supporting point could be adopted [15,59], which is based upon the hypothesis that the load transmission from a sleeper to the ballast approximately coincides with the cone distribution [10]. To consider the continuity and the coupling effects of the interlocking ballast granules, a couple of shear stiffness and shear damping is employed between adjacent ballast masses in the ballast model. To account for the practical situation when adjacent ballast masses overlap each other, the vibrating mass of ballast under a rail support point can be defined as the shadowed area, as shown in Figure 9.5, where h_0 is the height of the overlapping regions, h_b is the depth of ballast, l_s is the sleeper spacing, l_b is the width of sleeper underside and α is the ballast stress distribution angle.

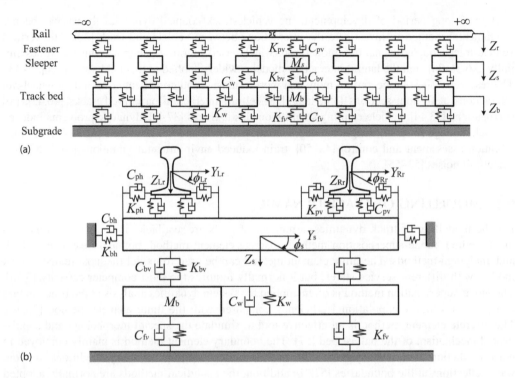

FIGURE 9.4 Three-dimensional dynamics model of ballasted track: (a) side view and (b) end view.

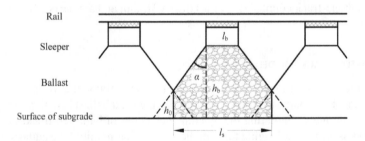

FIGURE 9.5 Modelling of the ballast.

9.3.1.1 Rail Motion

Generally, the Euler beam or Timoshenko beam model could be applied to model the continuous elastic rails. The authors have proven that the wheel-rail interaction force obtained by using the two models do not show a significant difference. Take the Euler beam model, for example; the rail model with infinite length can be simplified as the simply supported beam model with a finite length, considering the influence of the boundary conditions of rail vibration. Because the rail cross-section is small and symmetric about the vertical axis, the torsional centre of the rail cross-section could be assumed to coincide with the shape centre, without any warping deformation. Therefore, the vertical, lateral and torsional vibration of rails under vehicle loads can be expressed in the form of the fourth-order partial differential equations:

$$EI_y \frac{\partial^4 Z_r(x,t)}{\partial x^4} + m_r \frac{\partial^2 Z_r(x,t)}{\partial t^2} = -\sum_{i=1}^{N} F_{svi}(t)\delta(x-x_i) + \sum_{j=1}^{4} P_j(t)\delta(x-x_{wj}) \quad (9.1)$$

Track Design, Dynamics and Modelling

$$EI_z \frac{\partial^4 Y_r(x,t)}{\partial x^4} + m_r \frac{\partial^2 Y_r(x,t)}{\partial t^2} = -\sum_{i=1}^{N} F_{sli}(t)\delta(x-x_i) + \sum_{j=1}^{4} Q_j(t)\delta(x-x_{wj}) \quad (9.2)$$

$$GK \frac{\partial^2 \Phi_r(x,t)}{\partial x^2} + \rho I_0 \frac{\partial^2 \Phi_r(x,t)}{\partial t^2} = -\sum_{i=1}^{N} M_{si}(t)\delta(x-x_i) + \sum_{j=1}^{4} M_{wj}(t)\delta(x-x_{wj}) \quad (9.3)$$

where $Z_r(x, t)$, $Y_r(x, t)$ and $\Phi_r(x, t)$ are the vertical, lateral and torsional displacements of the rail, respectively; m_r is the rail mass per unit length; ρ is the rail density; EI_y and EI_z are the rail bending stiffness about the y-axis and the z-axis, respectively; I_0 is the torsional inertia of the rail, GK is the rail torsional stiffness; $F_{svi}(t)$ and $F_{sli}(t)$ are the vertical and lateral dynamic forces, respectively, at the i-th rail supporting point; $P_j(t)$ and $Q_j(t)$ are the j-th wheel-rail vertical and lateral forces, respectively; $M_{si}(t)$ and $M_{wj}(t)$ are the moments acting on the rails due to the forces $F_{svi}(t)$ and $F_{sli}(t)$ and due to the forces $P_j(t)$ and $Q_j(t)$, respectively; and $\delta(x)$ is the Dirac delta function.

As shown in Figure 9.6, the moments $M_{si}(t)$ and $M_{wj}(t)$ can be determined based on the force relationship acting on the rail cross-section.

$$\begin{cases} M_{si}(t) = [F_{sv2i}(t) - F_{sv1i}(t)]b - F_{sli}(t)a \\ M_{wj}(t) = Q_j(t)h_r - P_j(t)e \end{cases} \quad (9.4)$$

where a is the vertical distance between the rail torsional centre and the lateral force from the fastening system, b is half of the distance between the two vertical forces from the fastening system, h_r is the vertical distance from the rail torsional centre to the lateral wheel-rail force and e is the lateral distance from the rail torsional centre to the vertical wheel-rail force.

9.3.1.2 Sleeper Motion

For improving computational efficiency, the sleeper can be regarded as a rigid body in the track dynamics analysis. The force analysis of the i-th sleeper with vertical, lateral and torsional motions is shown in Figure 9.7, where F_{LrVi} and F_{RrVi} are the vertical forces applied on the i-th sleeper by the left and right rails, respectively; F_{LrLi} and F_{RrLi} are the lateral forces applied on the i-th sleeper by

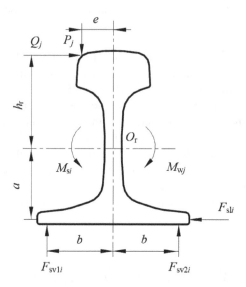

FIGURE 9.6 Force relationship acting on the cross-section of the rail.

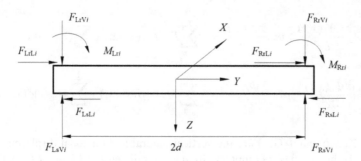

FIGURE 9.7 Force analysis for the sleeper.

the left and right rails, respectively; M_{Lri} and M_{Rri} are the moments applied on the i-th sleeper by the left and right rails, respectively; F_{LsVi} and F_{RsVi} are the vertical counterforces acting on the i-th sleeper from the ballast bed under the left and right rails, respectively; and F_{LsLi} and F_{RsLi} are the lateral counterforces acting on the i-th sleeper from the ballast bed under the left and right rails, respectively. Note that the subscripts Lr and Rr indicate the left and right rails, and the subscripts Ls and Rs mark the left and right sleepers, respectively.

Based on the force analysis of the sleeper, the equations of motion of the sleeper can be obtained as follows:

Vertical motion:

$$M_s \ddot{Z}_s = F_{LrVi} + F_{RrVi} - F_{LsVi} - F_{RsVi} \qquad (9.5)$$

Lateral motion:

$$M_s \ddot{Y}_s = F_{LrLi} + F_{RrLi} - F_{LsLi} - F_{RsLi} \qquad (9.6)$$

Rotational motion:

$$J_s \ddot{\phi}_s = M_{Lri} + M_{Rri} + (F_{RrVi} - F_{RsVi})d - (F_{LrVi} - F_{LsVi})d \qquad (9.7)$$

where M_s is the mass of the sleeper, and J_s is the rotational inertia of the sleeper: $J_s = M_s L_s^2 / 12$.

9.3.1.3 Ballast Motion

In the ballast model, only the vertical vibration is considered, simultaneously accounting for the interactions of ballast mass blocks between the left and right sides and between the front and back sides. The force analysis of the ballast blocks is shown in Figure 9.8, where F_{Rb1i} and F_{Rb2i} are the shear forces acting on the right-side ballast block from the front and back ballast blocks, respectively; F_{Lb1i} and F_{Lb2i} are the shear forces acting on the left-side ballast block from the front and back ballast blocks, respectively; F_{LbRi} is the shear force acting on the left-side ballast block from the right-side ballast block; F_{RbLi} is the shear force acting on the right-side ballast block from the left-side ballast block; and F_{Rbfi} and F_{Lbfi} are the vertical supporting forces acting on the right- and left-side ballast blocks from the subgrade, respectively.

According to the force analysis of the ballast blocks, the equations of vertical motion of the ballast can be written as follows:

Left-side ballast:

$$M_b \ddot{Z}_{Lbi} = F_{LsVi} - F_{Lbfi} - F_{Lb1i} - F_{Lb2i} - F_{LbRi} \qquad (9.8)$$

Track Design, Dynamics and Modelling

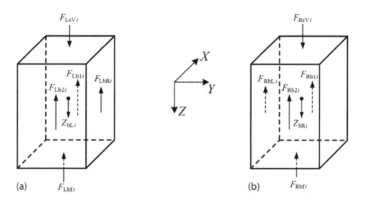

FIGURE 9.8 Force analysis of the ballast blocks under the i-th sleeper: (a) left-side ballast and (b) right-side ballast.

Right-side ballast:

$$M_b \ddot{Z}_{Rbi} = F_{RsVi} - F_{Rbfi} - F_{Rb1i} - F_{Rb2i} - F_{RbLi} \qquad (9.9)$$

9.3.2 Slab Track Model

The slab track is a typical ballastless track that is widely adopted in high-speed railways around the world. It usually consists of the rails, rail pads, fasteners, concrete slabs, adjustment layers, concrete bases, etc. Compared with the modelling of ballasted track, the key difference is the modelling of slabs for simulating the dynamic behaviour of the slab track. The slab is generally considered as a thin plate or a finite element structure in the regular dynamic analysis. The authors developed a three-dimensional dynamic model of the slab track, as shown in Figure 9.9. In this model, the rail is also treated as a continuous Bernoulli-Euler beam resting on rail pads, and the vertical, lateral and torsion motions of the rails are simultaneously taken into account. Equations 9.1–9.3 remain the equations of motion of the rail. The concrete slab and base are described as elastic thin plates supported on a viscoelastic foundation. As shown in Figure 9.9, three layers of discrete springs and dampers represent the elasticity and damping effects of the rail fastener system, the mortar/concrete adjustment layer and the subgrade, respectively. In Figure 9.9b, K_{sv} and C_{sv} are the vertical stiffness and damping of the mortar/concrete layer, respectively, and K_{sh} and C_{sh} are the lateral stiffness and damping of that layer, respectively.

9.3.2.1 Motions of Slab Track

In the model, the track slab is regarded as an elastic thin plate in the vertical direction, because the thickness of the track slab is much smaller than its length and width. As the lateral bending stiffness is quite large, it can be simplified as a rigid body in the lateral direction. The force analysis of the track slab is shown in Figure 9.10, where P_{rVi} and P_{rLi} are the vertical and lateral forces acting on the slab at the i-th rail fastener, respectively; F_{sVj} and F_{sLj} are the vertical and lateral supporting forces acting on the slab, respectively; L_s, W_s and h_s are the length, width and thickness of the slab, respectively; X and Y represent the longitudinal and lateral directions, respectively.

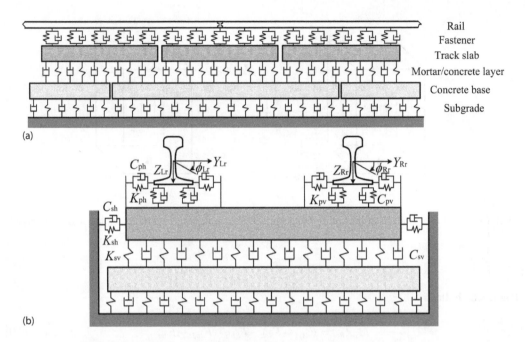

FIGURE 9.9 Three-dimensional dynamics model of slab track: (a) side view and (b) end view.

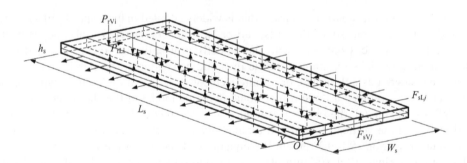

FIGURE 9.10 Force analysis of the track slab.

In general, when the boundaries of a rectangular thin plate are simply supported, fixed, free or elastic boundaries, the series approximation for the combination of bidirectional beam functions can obtain satisfactory results for the solution of the elastic thin plate. In this regard, using the elastic thin plate theory and the combination of bidirectional beam functions, the vertical vibrations of the slab can be written as a series of second-order ordinary differential equations in terms of the generalised coordinates, as expressed by Equation 9.10, which can be solved with the time-stepping integration method.

$$\ddot{T}_{mn}(t) + \frac{C_s}{\rho_s h_s}\dot{T}_{mn}(t) + \frac{D_s}{\rho_s h_s}\frac{B_3 B_2 + 2B_4 B_5 + B_1 B_6}{B_1 B_2} T_{mn}(t)$$
$$= \frac{1}{\rho_s h_s B_1 B_2}\left[\sum_{i=1}^{N_p} P_{rVi}(t) X_m(x_{pi}) Y_n(y_{pi}) - \sum_{j=1}^{N_b} F_{sVj}(t) X_m(x_{bj}) Y_n(y_{bj})\right]$$

(9.10)

where

$$\begin{cases} B_1 = \int_0^{L_s} X_m^2(x)\,dx \\ B_2 = \int_0^{W_s} Y_n^2(y)\,dy \\ B_3 = \int_0^{L_s} X_m''''(x) X_m(x)\,dx \\ B_4 = \int_0^{L_s} X_m''(x) X_m(x)\,dx \\ B_5 = \int_0^{W_s} Y_n''(y) Y_n(y)\,dy \\ B_6 = \int_0^{W_s} Y_n''''(y) Y_n(y)\,dy \end{cases} \quad (9.11)$$

The solution of vertical motions of the slab can be assumed as:

$$w(x,y,t) = \sum_{m=1}^{N_x} \sum_{n=1}^{N_y} X_m(x) Y_n(y) T_{mn}(t) \quad (9.12)$$

Regarding the lateral motions, the slab is considered as a rigid body due to its large bending stiffness in the lateral direction. Thus, the lateral vibrations can be written as:

$$\rho_s L_s W_s h_s \ddot{y}_s = \sum_{i=1}^{N_p} P_{rLi} - 2\sum_{j=1}^{N_l} F_{sLj} \quad (9.13)$$

In Equations 9.10–9.13, $X_m(x)$ and $Y_n(y)$ are the mode functions of the slab with respect to x and y coordinates, respectively; $T_{mn}(t)$ are the generalised coordinates with respect to time and describe the vertical motion of the slab; N_x and N_y are the mode numbers of $X_m(x)$ and $Y_n(y)$, respectively; C_s and D_s are the vertical damping and bending stiffness of the slab, respectively; ρ_s, L_s, W and h_s are the density, length, width and thickness of the slab, respectively; $F_{svj}(t)$ and $F_{sLj}(t)$ are the vertical and lateral dynamic forces at the jth supporting point under the slab, respectively; N_p and N_b are the total number of rail fasteners on the slab and the total number of discrete supporting points under one slab, respectively.

The equations of motion of the concrete base can be easily obtained in a similar manner and are omitted here for brevity. Detailed descriptions of the equations of the vehicle-track coupled dynamics model can be found in [13] and [30].

9.4 INTERACTION BETWEEN TRACK AND TRAIN

9.4.1 Vehicle-Track Coupled Dynamics Model

The vehicle-track coupled dynamics theory has been a fundamental method to investigate the interactions between vehicle and track, and it has wide applications in various practical engineering topics. As mentioned previously, the authors developed a series of three-dimensional vehicle-track coupled dynamics models, including a typical locomotive model, passenger car model, freight car model, ballasted track model and non-ballasted track model. These models have subsequently been modified or improved in many aspects in recent years, for example, non-linear models for track components have been implemented in the vehicle-track coupled dynamics model [4,48], a longitudinal train dynamics model has been proposed in the spatial coupled dynamics analysis [60], a robust non-Hertzian contact method has been introduced for the wheel-rail normal contact analysis [61] and a temporal-spatial stochastic model has been introduced for the coupled system dynamics [62]. Figure 9.11 shows a typical three-dimensional coupled dynamics model of a passenger vehicle and a ballasted track, where the vehicle and ballasted track subsystems are spatially coupled by the wheel-rail interface.

In the vehicle submodel, the car body is supported on two bogies connected by the secondary suspensions. The bogie frames are linked with the wheelsets through the primary suspensions.

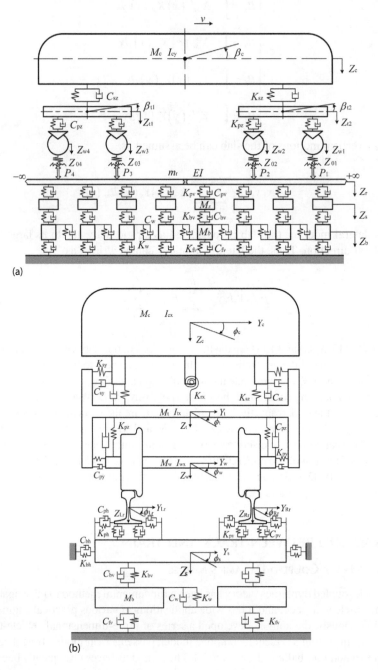

FIGURE 9.11 Three-dimensional coupled dynamics model of a passenger vehicle and a ballasted track: (a) elevation and (b) end view.

Track Design, Dynamics and Modelling

Three-dimensional spring-damper elements are adopted to model the primary and secondary suspensions. Yaw dampers and anti-roll springs are considered in the secondary suspensions. The vehicle is assumed to move along the track with a constant running speed. A total of 35 DOFs of the passenger vehicle are considered in the model, including the vertical displacement Z, the lateral displacement Y, the roll angle Φ, the yaw angle ψ and the pitch angle β with respect to the centre of mass for each component.

The equations of motion of the vehicle subsystem can be easily derived in the form of a second-order differential equations in the time domain:

$$\mathbf{M}_V \mathbf{A}_V + \mathbf{C}_V(\mathbf{V}_V)\mathbf{V}_V + \mathbf{K}_V(\mathbf{X}_V)\mathbf{X}_V = \mathbf{F}_V(\mathbf{X}_V, \mathbf{V}_V, \mathbf{X}_T, \mathbf{V}_T) + \mathbf{F}_{EXT} \qquad (9.14)$$

where $\mathbf{X}_V$, $\mathbf{V}_V$ and $\mathbf{A}_V$ are the vectors of the displacement, velocity and acceleration of the vehicle subsystem, respectively; $\mathbf{M}_V$ is the mass matrix of the vehicle; $\mathbf{C}_V$ and $\mathbf{K}_V$ are the damping and the stiffness matrices that can be related to the current state of the vehicle subsystem for describing suspension non-linearities; $\mathbf{X}_T$ and $\mathbf{V}_T$ are the vectors of displacement and velocity of the track subsystem, respectively; $\mathbf{F}_V$ is the system load vector representing the non-linear wheel-rail contact forces, which is dependent on the motions $\mathbf{X}_V$ and $\mathbf{V}_V$ of the vehicle and $\mathbf{X}_T$ and $\mathbf{V}_T$ of the track; and $\mathbf{F}_{EXT}$ describes external forces.

The track submodel shown in Figure 9.11 represents a typical ballasted track structure; the modelling methodology and equations of motion of the ballasted track subsystem have been mentioned previously. The model allows the input of track parameters varying along the rail longitudinal direction, such as track supporting stiffness. If the analysed track structure is a ballastless track such as the slab track, the above track dynamic model should be changed correspondingly, as shown in Figure 9.9. Therefore, the final equations of the track subsystem, either for the ballasted track structure or for the non-ballasted track structure, could be expressed in terms of the standard matrix form as

$$\mathbf{M}_T \mathbf{A}_T + \mathbf{C}_T \mathbf{V}_T + \mathbf{K}_T \mathbf{X}_T = \mathbf{F}_T(\mathbf{X}_V, \mathbf{V}_V, \mathbf{X}_T, \mathbf{V}_T) \qquad (9.15)$$

where $\mathbf{M}_T$ is the mass matrix of the track structure; $\mathbf{C}_T$ and $\mathbf{K}_T$ are the damping and the stiffness matrices of the track subsystem, respectively; $\mathbf{A}_T$ is the vector of the acceleration of the system; and $\mathbf{F}_T$ is the load vector of the track subsystem that represents the non-linear wheel-rail forces.

9.4.2 Wheel-Rail Coupling Model

The wheel-rail coupling model is the fundamental element to accurately investigate the interaction between wheels and rails, by which the vehicle and the track subsystems are spatially coupled at the wheel-rail interfaces.

In classical vehicle dynamics, the rails are generally assumed to be fixed without any deformation and movement in the early wheel-rail contact model. However, rail motions in the vertical, lateral and torsional directions are now considered in the wheel-rail coupling model for the analysis of the three-dimensional vehicle-track coupled dynamics, as shown in Figure 9.12.

Owing to the yaw angle of the wheel that occurs during the actual train operation process, the wheel-rail contact point is no longer located on profiles of wheel and rail that are normal to the centreline of the track but needs to be determined based on the spatial geometry analysis of the wheel-rail contact relation. In this regard, the trace curve method [63] is adopted to locate the wheel-rail spatial contact points, which are located on a path named the trace curve. Therefore, two-dimensional scanning can be replaced by one-dimensional scanning through a trace curve. Figure 9.13 shows the schematic of the spatial wheel-rail contact geometry, where O_R and B represent the spatial contact point and the centre of the wheel rolling circle, respectively; A represents the intersection between the common normal line at the contact point and the wheelset centreline; and

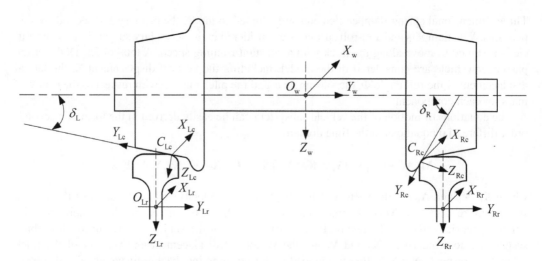

FIGURE 9.12 Wheel-rail coupling model.

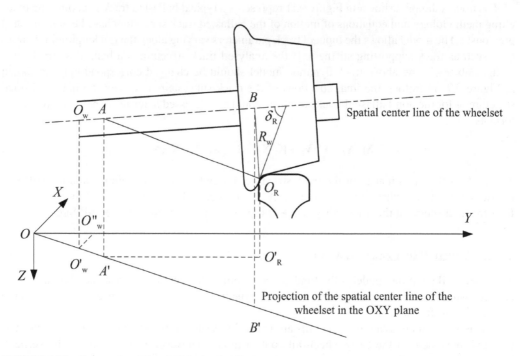

FIGURE 9.13 Schematic of the spatial wheel-rail contact geometry.

O_w is the origin of the wheelset coordinate system. It can be seen from Figure 9.13 that the spatial contact point O_R is located before the main profiles, and OO''_w is the lateral displacement of the wheelset (represented by Y_w).

From the spatial geometry relationship, it can be seen that the contact point O_R is located within three surfaces, namely the rolling circle plane containing the point B, the spherical surface with the centre B and the rolling radius R_w and the plane A-A'-O'_R-O_R. Derived from the equations of three surfaces, the absolute coordinates of the contact point O_R can be obtained as

$$\begin{cases} x = x_B + l_x R_w \tan\delta_R \\ y = y_B - \dfrac{R_w}{1-l_x^2}(l_x^2 l_y \tan\delta_R + l_z m) \\ z = z_B - \dfrac{R_w}{1-l_x^2}(l_x^2 l_z \tan\delta_R - l_y m) \end{cases} \qquad (9.16)$$

where δ_R is the contact angle of wheel tread, and $m = \sqrt{1-l_x^2(1+\tan^2\delta_R)}$; l_x, l_y, l_z are the direction cosines, which are written as

$$\begin{cases} l_x = -\cos\phi_w \sin\psi_w \\ l_y = \cos\phi_w \cos\psi_w \\ l_z = \sin\phi_w \end{cases} \qquad (9.17)$$

The coordinate of the centre B of the rolling circle is

$$\begin{cases} x_B = d_w l_x \\ y_B = d_w l_y + Y_w \\ z_B = d_w l_z \end{cases} \qquad (9.18)$$

where d_w is the horizontal coordinate (in the wheelset coordinate system) of the rolling circle plane of the wheel tread. Therefore, by obtaining the horizontal coordinate of each rolling circle plane step by step, the spatial trace curve of wheel-rail contact points at each time step can be determined.

To calculate the wheel-rail normal contact forces, the non-linear Hertzian elastic contact theory is adopted according to the relative displacements of wheels and rails at contact points. The tangential wheel-rail creep forces are calculated first by use of Kalker's linear creep theory and then modified by Shen-Hedrick-Elkins' non-linear model [64]. Clearly, this spatial wheel-rail coupling model abandons the assumption that the wheel and the rail keep in contact at all times as applied in classical vehicle dynamics. Thus, the current model has the advantage of dealing with the case that the wheel may lose its contact with the rail, as occurs in practical railway operations. This potentially provides a new way to simulate the dynamic process of derailment.

9.4.3 Numerical Solution

For calculating the dynamic response of the vehicle-track coupled system, the efficiency, the accuracy and the stability of the numerical integration method are quite important, owing to the large-scale and non-linear dynamic system. A fast integration algorithm, namely the Zhai method [65], was developed specially to solve the large-scale coupled dynamics equations of vehicle and track. This method has been widely used due to its efficiency and simplicity in the dynamic analysis of the railway vehicle and track. The scheme of this integration method for the calculation of the displacement **X** and the velocity **V** of a system is expressed as follows:

$$\begin{cases} \mathbf{X}_{n+1} = \mathbf{X}_n + \mathbf{V}_n \Delta t + (\tfrac{1}{2}+\psi)\mathbf{A}_n \Delta t^2 - \psi \mathbf{A}_{n-1}\Delta t^2 \\ \mathbf{V}_{n+1} = \mathbf{V}_n + (1+\varphi)\mathbf{A}_n \Delta t - \varphi \mathbf{A}_{n-1}\Delta t \end{cases} \qquad (9.19)$$

where Δt is the time step; the subscripts $n+1$, n and $n-1$ denote the integration time at $(n+1)\Delta t$, $n\Delta t$ and $(n-1)\Delta t$, respectively; and ψ and φ are free parameters that control the stability and numerical dissipation of the algorithm. Usually, a value of 0.5 could be assigned to ψ and φ to achieve good compatibility between numerical stability and accuracy. In order to get a high calculation accuracy, however, the actually adopted effective time step is suggested to be $\Delta t = 1.0 \times 10^{-4}$ s.

9.4.4 Experimental Validation

Field measurements and laboratory tests are often carried out for comparison with simulation results obtained by numerical modelling, these being typical and effective ways to validate the reliability of the proposed theoretical and numerical models. Hence, to demonstrate the validity of the three-dimensional vehicle-track coupled models and the corresponding simulation program TTISIM, many field measurements with a variety of vehicles and track conditions have been conducted, including the speed-up test on the Beijing-Qinhuangdao line, the high-speed running test on the Qinhuangdao-Shenyang line, the wheel-rail dynamic interaction experiment on curved tracks of the Chengdu-Chongqing line, etc. The main results from the first two tests will be presented here and compared with the calculated results obtained from the three-dimensional vehicle-track coupled model.

Figure 9.14a presents the measured lateral wheel-rail force, when the tested vehicle is running at 160 km/h through a curve section with a radius of 1200 m and with a superelevation of 100 mm. Figure 9.14b shows the calculated history of the lateral wheel-rail force. It can be seen from these figures that the lateral wheel-rail force predicted by the proposed vehicle-track coupled model is in reasonable agreement with the field-measured response.

The measured results of the lateral and vertical Sperling indices of the car body are given in Figure 9.15, juxtaposed with the calculated results obtained with the current model. Clearly, the calculated results are close to the measured data.

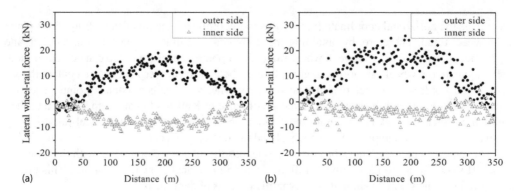

FIGURE 9.14 Comparison of measured and calculated lateral wheel-rail forces on a curved track: (a) measured results and (b) calculated results. (From Zhai, W. et al., *Veh. Syst. Dyn.*, 47(11), 1349–1376, 2009.)

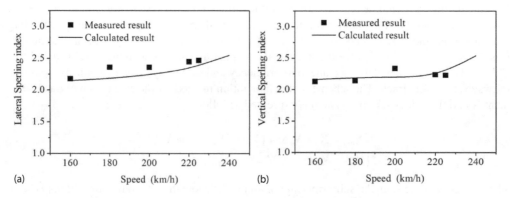

FIGURE 9.15 Comparison of measured and calculated Sperling indices of the car body: (a) lateral Sperling and (b) vertical Sperling. (From Zhai, W. et al., *Veh. Syst. Dyn.*, 47(11), 1349–1376, 2009.)

9.5 SYSTEM EXCITATIONS

9.5.1 MODELS OF IMPACT LOADS

When a wheel passes a dipped joint, dislocation joint or turnout or when a wheel with a flat on the tread moves along a rail, the abrupt change of the instantaneous rotation centre of the wheel leads to a vertical impact velocity to the track, resulting in a sudden impact and vibrations of the wheel-rail system, which disappear instantly when the wheel moves away from these locations or the flat. These excitations are defined as impact excitations; they are often input into the wheel-rail system as impact velocities or as impact displacements.

9.5.1.1 Impact Model of Rail Joints

There are various types of rail joints in practical railway lines, such as rail dislocation joints and rail dipped joints. They all cause sudden impact vibrations of wheel-rail system.

A rail dislocation joint refers to an abnormal joint that has a height difference between the adjacent rail surfaces at the joint. According to the vehicle running direction, rail dislocation joints can be divided into forward and backward dislocation joints. A backward dislocation joint corresponds to the case where the wheel runs from a lower rail surface to a higher one, while a forward dislocation joint is the opposite case.

The rail dipped joint is the most common impact excitation for jointed track lines, occurring when both rail ends dip at the rail joint, as shown in Figure 9.16. The impact velocity of a dipped joint is usually expressed by the product of the joint angles α_1 and α_2 and the vehicle running speed v as

$$v_0 = 2\alpha v = (\alpha_1 + \alpha_2)v \tag{9.20}$$

where 2α is the total angle of the dipped joint.

9.5.1.2 Impact Model of Turnouts

When a vehicle passes through a turnout on the straight track, the vehicle and turnout interactions mainly present vertical impact and vibrations at crossing frogs. When the vehicle runs through the turnout in the curved direction, this mainly presents lateral interactions between the vehicle and the turnout.

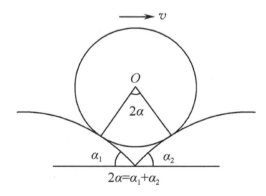

FIGURE 9.16 Schematic of rail dipped joint.

The vertical impact at a turnout crossing mainly occurs at the fixed frog. As shown in Figure 9.17, the fixed frog has running rail gaps, leading to the discontinuous rolling path of the wheels and inducing wheel-frog impacts. When a wheel rolls from the wing rail to the point rail, the wheel rolling radius becomes increasingly smaller as the wheel passes away from the wing rail, which causes a descending of the wheel gravity centre. To avoid a collision between the wheel and the point rail, the rail surface height at this location decreases dramatically and then increases gradually to the normal height. In this way, the wheel gradually rolls back to the original height when it completely runs onto the point rail. In this regard, a vertical impact model of a turnout, as discussed in [13], can be adopted in the track dynamic analysis.

When a vehicle enters the divergent route of a turnout in the facing direction, the wheel will impact the switch rail, as shown in Figure 9.18. The induced lateral impact force is dependent on the

FIGURE 9.17 Turnout with fixed frog.

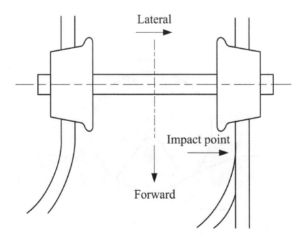

FIGURE 9.18 Impact at curved switch rail.

impact angle β between the wheel and the switch rail, passing speed v and the types of the vehicle and switch rail. Large lateral impact forces can cause severe wear and damage to the turnout switch and vehicle bogie. Meanwhile, the wheel can climb up the switch rail, which produces the potential for derailment.

When a vehicle passes through the divergent route of a turnout in the facing direction, the lateral impact velocity from the wheel to the switch rail can be given as

$$v_{0L} = v \sin \beta \approx v\beta \qquad (9.21)$$

Therefore, when a vehicle passes through the divergent route, the excitation between the wheelset and switch rail could be simply applied on the lateral velocity of the wheelset by an instantaneous impact velocity v_{0L}.

9.5.1.3 Impact Model of Wheel Flats

During train operations, local scratching and spalling of the wheels may occur for various reasons (braking, wheel spin and slip). These phenomena are collectively known as wheel flats and are of the form of a flat section worn on the circumference of the wheel tread, as shown in Figure 9.19, that will induce special dynamic effects during rolling. As the running speed increases, the impact characteristics of wheel flats will have a sudden change at a critical running velocity given by

$$v_{cr0} = \sqrt{\mu R} \qquad (9.22)$$

where μ is the falling acceleration of the wheel, and R is the wheel radius.

At a low running speed $v \leq v_{cr0}$, the wheel impact velocity acting on the track consists of two parts. One is the vertical velocity component of the impact velocity induced by the wheel that rotates around point A and hits the rail. The other part is the instantaneous impact component in the opposite direction of the wheel vertical velocity, which is produced at the moment when the rail hinders the wheel rotating around point B. Therefore, the impact velocity equation can be expressed as

$$v_0 = (1+\gamma)\frac{L}{2R}v \qquad (9.23)$$

where γ is the coefficient for the transformation from rotational inertia to reciprocating inertia, L is the length of the flat and R is the wheel radius. Obviously, the impact velocity at a low speed $v \leq v_{cr0}$ is proportional to the wheel flat length and the running speed and is inversely proportional to the wheel radius.

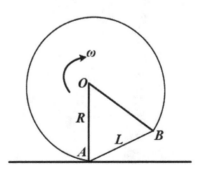

FIGURE 9.19 Schematic of wheel flat.

For a high running speed $v > v_{cr0}$, the impact velocity also consists of two parts, which are the falling speed of the flat part of the wheel circumference onto the rail surface and the vertical component of the wheel centre velocity due to the rotation. Therefore, the impact velocity can be written as

$$v_0 = \frac{L}{v + \sqrt{\mu R}} \left[\mu + \gamma v \sqrt{\frac{\mu}{R}} \right] \tag{9.24}$$

As can be seen, the impact velocity at a high running speed is also proportional with the wheel flat length, while it slowly decreases with an increase of the velocity, and it finally goes to a constant value given by

$$\bar{v}_0 = \lim_{V \to \infty} v_0 = \gamma L \sqrt{\frac{\mu}{R}} \tag{9.25}$$

9.5.2 Models of Harmonic Loads

9.5.2.1 Model of Rail Corrugations

In many cases, track irregularities can be described by a single harmonic wave or multiple harmonic waves. For example, squats that occur at poorly welded rail joints under repeated wheel loads are a single harmonic excitation. For another example, rail corrugations that widely exist in railways worldwide present undulating waves on the rail surface, as shown in Figure 9.20; they are a typical continuous harmonic excitation.

A cosine function can be simply employed to describe the rail corrugations, written as

$$Z_0(t) = \frac{1}{2} a (1 - \cos \omega t) \quad (0 \le t \le \frac{nL}{v}) \tag{9.26}$$

where

$$\omega = \frac{2\pi v}{L} \tag{9.27}$$

where L is the irregularity wavelength, a is the irregularity wave depth and n is the total number of waves.

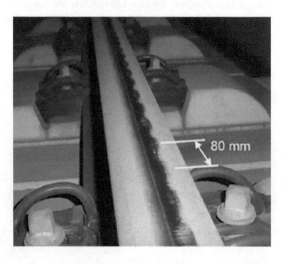

FIGURE 9.20 Rail corrugation in high-speed railways.

9.5.2.2 Model of Wheel Polygon

Wheel polygonisation refers to a periodic radial deviation along the wheel circumference, formed by non-uniform wheel wear, which appears widely in railway operations. Wheel polygon excitation can be represented by the following equation using a Fourier series:

$$Z_0(t) = \sum_{i=0}^{\infty} A_i \sin\left[i\left(\frac{v}{R}\right)t + \varphi_i\right] \quad (9.28)$$

where i is the order of the wheel polygon, A_i is the amplitude of the ith harmonic wave and φ_i is the corresponding phase; A_i and φ_i can be obtained by performing discrete Fourier transforms of the measured radial deviation of the wheel along its circumference. In the equation, A_0 describes the overall deviation of the test data from the radial deviation of the zero position, which is negligible for the vehicle-track system dynamics. In fact, most wheel polygons present the main harmonic components within a 40th-order series, and components with higher order usually have a small amplitude. Meanwhile, owing to the contact filtering effect, the high-order components have negligible effect on the vehicle-track coupled dynamics. Considering the first N-order polygon of the dominant components and ignoring the component A_0, Equation 9.28 can be further written as

$$Z_0(t) = \sum_{i=1}^{N} A_i \sin\left[i\left(\frac{v}{R}\right)t + \varphi_i\right] \quad (9.29)$$

9.5.3 RANDOM TRACK IRREGULARITY

The geometrical state of real-life railways always shows clear randomness, this being caused by many factors, including initial rail bending; wear and damage of rails; non-uniform sleeper spacing and quality; non-uniform gradation, contamination and hardening of the ballast bed; subgrade asymmetrical settlement and stiffness variation; and so on. Through the combined action of all these factors, the randomness of track irregularities is formed. Under the excitation of a random track irregularity, the vehicle-track coupled system will vibrate in a stochastic way, which affects the ride comfort of passengers and the stability of cargo in one way and the fatigue and serviceability of system components in another way. At the same time, the fatigue damage of track structures in turn aggravates the deterioration of the track's geometrical state.

An actual track irregularity is a superposition of random harmonic waves of different wavelengths, phases and amplitudes; it is a complicated random process depending on the location along the track. In general, a power spectral density (PSD) is the most important and most commonly used statistical function for the representation of a random track irregularity, which is usually considered as a stationary stochastic process, hence the so-called track irregularity PSD. Typical track irregularity PSDs around the world are presented later. Their features are clarified through analysis and comparison, so that readers can conveniently choose the appropriate random track irregularity excitation model for their particular dynamics analysis.

9.5.3.1 American Track Spectrum

Based on a large track geometry measurement database, the track irregularity PSDs were obtained by the U.S. Federal Railway Administration (FRA). The track irregularity PSDs were fitted by even rational functions containing cut-off frequencies and roughness constants, in which the wavelengths ranged from 1.524 m to 304.8 m and six track classes were considered [66].

1. Longitudinal level:

$$S_v(\phi) = \frac{A_v \phi_{v2}^2 (\phi^2 + \phi_{v1}^2)}{\phi^4 (\phi^2 + \phi_{v2}^2)} \quad (9.30)$$

2. Alignment:

$$S_a(\phi) = \frac{A_a \phi_{a2}^2 (\phi^2 + \phi_{a1}^2)}{\phi^4 (\phi^2 + \phi_{a2}^2)} \quad (9.31)$$

3. Cross level:

$$S_c(\phi) = \frac{A_c \phi_{c2}^2}{(\phi^2 + \phi_{c1}^2)(\phi^2 + \phi_{c2}^2)} \quad (9.32)$$

4. Gauge:

$$S_g(\phi) = \frac{A_g \phi_{g2}^2}{(\phi^2 + \phi_{g1}^2)(\phi^2 + \phi_{g2}^2)} \quad (9.33)$$

Here, the symbols represent the following: $S(\phi)$, the track irregularity PSD [m²/(1/m)]; ϕ, the spatial frequency of track irregularity (1/m); A, the roughness constants (m); and ϕ_1, ϕ_2, the cut-off frequencies (1/m). The roughness constants and cut-off frequencies of each of the six FRA track classes can be found in [66].

9.5.3.2 German Track Spectrum

The track irregularity PSD for Germany railways is expressed as follows [67]:

1. Longitudinal level:

$$S_v(\Omega) = \frac{A_v \Omega_c^2}{(\Omega^2 + \Omega_r^2)(\Omega^2 + \Omega_c^2)} \quad (9.34)$$

2. Alignment:

$$S_a(\Omega) = \frac{A_a \Omega_c^2}{(\Omega^2 + \Omega_r^2)(\Omega^2 + \Omega_c^2)} \quad (9.35)$$

3. Cross level:

$$S_c(\Omega) = \frac{A_v \cdot b^{-2} \cdot \Omega_c^2 \cdot \Omega^2}{(\Omega^2 + \Omega_r^2)(\Omega^2 + \Omega_c^2)(\Omega^2 + \Omega_s^2)} \quad (9.36)$$

Here, the units of longitudinal level and alignment PSDs are m²/(rad/m); as cross level was measured as the inclination angle, the unit for $S_c(\Omega)$ became 1/(rad/m); Ω is the spatial frequency of track irregularity (rad/m); Ω_c, Ω_r, and Ω_s are the cut-off frequencies (rad/m); A_v and A_a are the

Track Design, Dynamics and Modelling   329

roughness constants (m²·rad/m); and b is half the distance between the two rolling circles (m), normally set as 0.75 m.

4. Gauge:

No equation was provided in [67] for the gauge PSD; it just stipulated that the range of gauge variation be from −3 mm to 3 mm. In general, the PSD expressions for gauge and cross level are similar, so the equation of the gauge PSD can be written as

$$S_g(\Omega) = \frac{A_g \Omega_c^2 \Omega^2}{(\Omega^2 + \Omega_r^2)(\Omega^2 + \Omega_c^2)(\Omega^2 + \Omega_s^2)} \tag{9.37}$$

The relevant roughness coefficients and cut-off frequencies are listed in [67].

9.5.3.3 Chinese Track Spectrum

To date, the track irregularity PSD standard system to represent various track geometry states is not completely formed in China. But relevant departments have already carried out many studies into track irregularity PSDs, and PSD formulas of various track types are available according to the measurement results. For example, [68] provides the PSD for Chinese conventional main lines, which reflects the track geometry state after the speed-up renovation of the existing main lines with a maximum operational speed of up to 160 km/h, and [69] covers the short-wavelength (0.01 ~ 1 m) longitudinal level PSD of track with the Chinese 50 kg/m rail. The track irregularity PSDs of Chinese high-speed railways [70,71] are introduced here as follows.

Ballastless track is widely used in Chinese high-speed railways with operational speeds between 300 km/h and 350 km/h, and ballasted track is frequently employed for high-speed railways with operational speeds between 200 and 250 km/h. The track irregularity PSDs of these high-speed railways have been based on their typical operational measurement data during the time period since their opening. Piecewise power function fitting was adopted for the PSDs, and each wavelength segment of the PSD curves had the same expression, as follows:

$$S(f) = \frac{A}{f^n} \tag{9.38}$$

where the unit of $S(f)$ is mm²/(1/m); f is the spatial frequency(1/m); and A and n are the fitting coefficients. The fitting coefficients of the mean ballastless and ballasted track irregularity PSDs for Chinese high-speed railways, as well as their corresponding cut-off spatial frequencies and wavelengths, can be found in [70–72].

Previous study has shown that track irregularity PSDs estimated from a large amount of irregularity data approximately obeyed the χ^2 distribution with 2 DOFs. For different track geometrical states, the percentile PSDs can be estimated according to the mean track irregularity PSDs of the high-speed railway; their transformation coefficients K are given in Table 9.1.

9.5.3.4 Comparison of Track Irregularity PSDs for High-Speed Railway

With regard to the track irregularity PSDs under high-speed operation, the most typical standard should be the German track irregularity PSDs, where the low-disturbance PSD is used for operational speeds above 250 km/h. In China, the ballastless and ballasted track irregularity PSDs of the high-speed railway have been published (see Equation 9.38). These are compared as follows.

The alignment and longitudinal level of the aforementioned three categories of high-speed railway track irregularity PSDs are compared in Figure 9.21; the wavelength range is from 1 m to 200 m. It is worth noting that, for the Chinese ballastless track irregularity PSDs, the effective wavelength ranges of the alignment and longitudinal level PSDs are between 2 m and 200 m, and

TABLE 9.1
Transformation Coefficients between the Mean and Percentile Track Irregularity PSDs of Chinese High-Speed Railway

Percentile (%)	10.0	20.0	25.0	30.0	50.0	60.0	63.2	70.0	75.0	80.0	90.0
Transformation coefficient, K	0.105	0.223	0.288	0.357	0.693	0.916	1.000	1.204	1.386	1.609	2.303

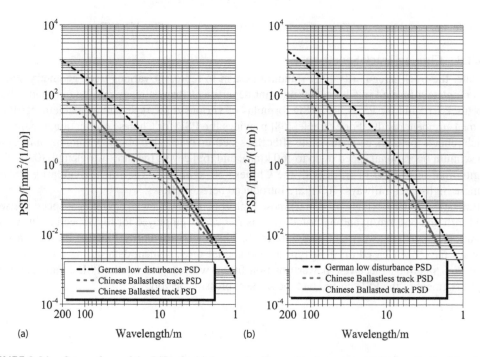

FIGURE 9.21 Comparison of the PSDs for high-speed railway: (a) alignment and (b) longitudinal level.

they are suitable for ballastless track with operational speeds from 300 km/h to 350 km/h. For the Chinese ballasted track irregularity PSDs, the effective wavelength ranges of the alignment and longitudinal level PSDs are between 2 m and 100 m; they are suitable for ballasted track with operational speeds from 200 to 250 km/h.

For the alignment (Figure 9.21a), the ballastless track irregularity PSD of the Chinese high-speed railway is generally better than the German low-disturbance track irregularity PSD in the whole wavelength range of 2 m to 200 m. In this wavelength range, the Chinese ballast PSD is slightly poorer than the Chinese ballastless PSD, but it is still generally better than the German low-disturbance PSD. As longer wavelength alignment irregularities have critical effects on the ride comfort of high-speed trains, it can be deduced that the lateral ride stability will be better under the excitation of Chinese ballastless or ballasted track irregularities.

For the longitudinal level (Figure 9.21b), the Chinese ballastless track irregularity PSD is generally better than the German low-disturbance PSDs over the whole wavelength range of 2 to 200 m, especially for wavelengths between 10 and 100 m. The Chinese ballasted track PSD is slightly poorer than the Chinese ballastless track PSD, but it is obviously better than the German low-disturbance PSD in the wavelength range of 2 to 100 m, especially for wavelengths between 10 and 60 m. Similarly, it can be deduced that the vertical passenger ride comfort will be better under the excitation of Chinese ballastless or ballasted track irregularities.

9.6 DYNAMIC PROPERTIES OF TRACK COMPONENTS

9.6.1 Dynamic Response due to Impact Loads

Without loss of generality, the rail dipped joint is taken as an example to explain its vertical impact vibration characteristics. Figure 9.22 shows the time history of the vertical wheel-rail force when a high-speed vehicle passes through a rail dipped joint of a slab track at a speed of 300 km/h. As can be seen, the wheel-rail force has two peaks in the full time history, which are called P_1 and P_2 forces, respectively, according to the definition of British Railways [8]. The P_1 force is a high-frequency impact force with the frequency of more than 500 Hz, which is roughly equivalent to the frequency of Hertz contact vibration between the vehicle unsprung mass and the rail mass. The P_1 force is generated rapidly following the appearance of the impact source. Generally, it occurs about 0.5 ms after the impact starts. Such a fast dynamic effect cannot be transmitted to the vehicle and the track substructures, and it disappears rapidly. Therefore, it is directly borne by the wheel and rail, causing unfavourable effects of damage to the wheel and rail. By contrast, the P_2 force is a low- and medium-frequency force induced by the impact source. Owing to its low-frequency dynamic effect (generally between 30 Hz and 100 Hz), longer duration (usually occurs about 3 ms after the impact starts) and slower variations, it can be fully transmitted to track substructures, thus having an important role in the degradation of track structures and infrastructure.

Figure 9.23 shows the dynamic displacement of track structures induced by the rail joint impact. It can be seen that the vertical dynamic displacement curve of track structures varies slowly with low frequencies. In Figure 9.23a, the maximum displacement of the rail occurs around 6 ~ 8 ms after the impact, which indicates that the vertical displacement of track structures is closely related to the P_2 force curve. Further, it can be seen from Figure 9.23b that the induced vibration is transmitted from the rail to the track slab, with the vibration amplitude decreasing during this process.

The accelerations of track structures under the vertical impact excitation are presented in Figure 9.24. As can be seen, the impact acceleration decreases rapidly from the rail to the slab. The rail acceleration varies drastically with high frequencies, which is mainly related to the P_1 force. Conversely, the slab acceleration varies slowly, and its frequency has some relationship with the

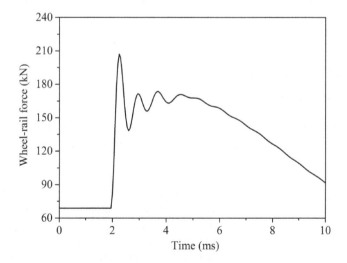

FIGURE 9.22 Vertical wheel-rail force under impact excitation.

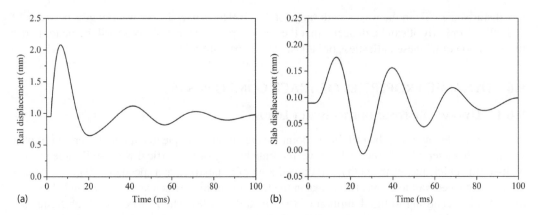

FIGURE 9.23 Dynamic displacement of track structures under impact excitation: (a) rail displacement and (b) slab displacement.

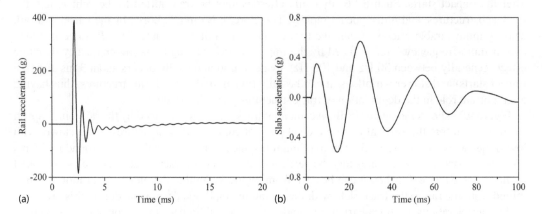

FIGURE 9.24 Accelerations of track structures under impact excitation: (a) rail and (b) slab.

P_2 force. It should be pointed out that the frequency and amplitude of the acceleration of track components are related to the track structure parameters, such as track mass, stiffness and damping. They could be different for different types of tracks, but their qualitative laws will be consistent.

9.6.2 RANDOM DYNAMIC RESPONSE

To introduce the characteristics of random dynamic response of the vehicle-track coupled system, a high-speed vehicle with a running speed of 300 km/h is adopted in the following calculations. The CRTS III slab track is applied. The ballastless track irregularity PSDs of the Chinese high-speed railway are employed with the superimposing of the short wavelength of 0.01 ~ 1 m in the longitudinal level.

Figures 9.25 and 9.26 show the vertical and lateral car body acceleration PSDs, respectively, under the track irregularities of longitudinal level, alignment, cross level and gauge. It can be found that the main frequency range of vertical car body acceleration is 0.5 to 40 Hz. The first main frequency is around 0.7 Hz, which is related to the natural vibration frequency of the car body's vertical suspension. The other main frequencies are around 4, 10 and 25 Hz, which reflect car body vertical bending, wheel and rail excitations, etc. The main frequency range of lateral car body acceleration is 0.5 to 2 Hz, which corresponds to the natural vibration frequency of the suspension

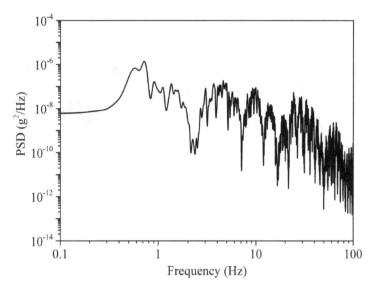

FIGURE 9.25 Vertical car body acceleration PSD.

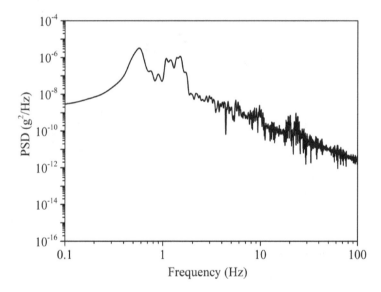

FIGURE 9.26 Lateral car body acceleration PSD.

system. Similar to the vertical car body acceleration, small peak frequencies around 5, 10 and 23 Hz can be found for the lateral car body acceleration.

Figures 9.27 and 9.28 present the vertical and lateral wheel-rail force PSDs, respectively. As can be seen in Figure 9.27, the vertical wheel-rail force PSD has several main frequencies, namely 0.6, 1.1, 35, 160 and 260 Hz. The low frequencies around 1 Hz may be induced by car body vibrations, the medium frequency of 35 Hz probably reflects the coupled resonant vibrations of the wheel and track and the high frequencies over 160 Hz could be attributed to the high-frequency Hertzian contact vibration with local wheel-rail deformations. It can also be seen from Figure 9.28 that the main frequencies of the lateral wheel-rail force PSD are mostly distributed in the low-frequency range below 100 Hz, and its main frequencies are 0.6, 1.1, 7 and 20 Hz. It is worth pointing out that the

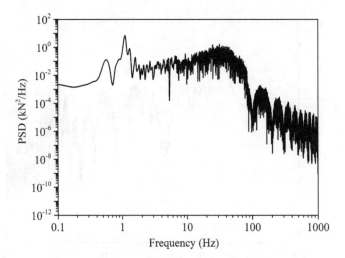

FIGURE 9.27 Vertical wheel-rail force PSD.

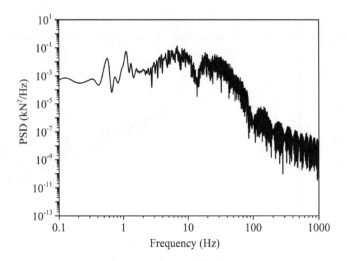

FIGURE 9.28 Lateral wheel-rail force PSD.

frequency component of wheel-rail forces that reflect the wheel-rail coupled vibrations occupies a considerable part of this low-frequency range. Therefore, to understand the basic characteristics of vehicle-track coupled dynamics is of great theoretical significance to propose appropriate technical measures for reducing wheel-rail dynamic interactions.

Figures 9.29 and 9.30 show the vertical and lateral rail acceleration PSDs, respectively. Clearly, medium- and high-frequency vibrations of tens, hundreds, even thousands of Hertz can be found in the vertical and lateral rail vibrations. The main frequency ranges for the lateral rail acceleration are 30 to 100 Hz and 400 to 700 Hz. From the comparisons between the lateral and the vertical rail acceleration PSDs, it can be found that the main frequency of the vertical rail acceleration is higher than that of the lateral rail acceleration, and the vertical PSD acceleration is also larger.

Track Design, Dynamics and Modelling

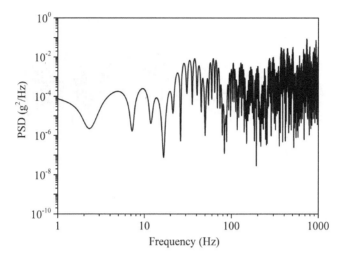

FIGURE 9.29 Vertical rail acceleration PSD.

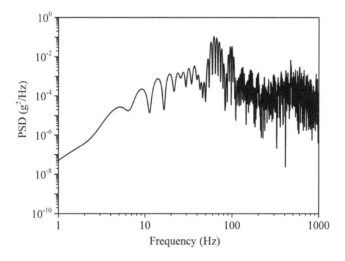

FIGURE 9.30 Lateral rail acceleration PSD.

This is because the rail stiffness in the vertical direction is larger than that in the lateral direction, and the vertical wheel-rail force is often much larger than the lateral force. The root cause of rail high-frequency vibrations is that the vertical short-wave irregularity on the rail surface excites the high-frequency vibration mode of the rail.

Figures 9.31 and 9.32 show the vertical and lateral slab acceleration PSDs, respectively. As can be seen, the main frequencies of vertical slab acceleration are distributed in the ranges of 30 to 120 Hz and 200 to 400 Hz. In addition, for the lateral slab vibration, its main frequency range is 25 to 100 Hz. The vibration characteristics of vertical and lateral slab vibration are obviously different, which is mainly because the slab is subjected to different directional forces from the rails in the vertical and lateral directions.

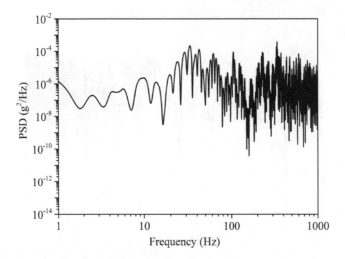

FIGURE 9.31 Vertical slab acceleration PSD.

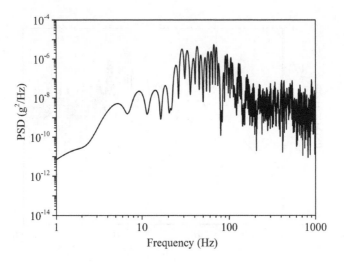

FIGURE 9.32 Lateral slab acceleration PSD.

9.7 TRACK DESIGN BASED ON VEHICLE-TRACK COUPLED DYNAMICS

During the design process of railway tracks, vehicle-track coupled dynamics analysis has proven to be an effective and important way to evaluate dynamic performance of railway tracks and the dynamic interaction with trains. Thompson [73] designed a tuned damping device to reduce the component of railway rolling noise radiated by using a theoretical model "TWINS" for wheel-rail rolling noise prediction. Bezin et al. [74,75] proposed an innovative slab track system composed of a two-layer steel-concrete structure, which was developed based on the numerical analysis of the vehicle-track system interaction and its comparison with respect to typical ballasted track. Zhai [13] performed the optimisation of the line design and safety pre-evaluation of the Fuzhou-Xiamen shared passenger and freight railway by comprehensively investigating the running safety and stability of the track-vehicle system, as well as the dynamic interaction between the vehicle and the track. In addition, he carried out complete dynamic simulations for the running safety and ride comfort analysis of high-speed trains passing through the horizontal alignment and vertical profile sections of four proposed line-selection schemes in the Shiziyang area Guangzhou-Shenzhen-Hong Kong high-speed railway. Through safety evaluation and scheme comparison, the best scheme option was able to be put forward. To elucidate the application of vehicle-track coupled dynamics in railway track design, an example of developing a vibration-attenuation track (VAT) for urban rail transit is introduced here.

As is known, low-frequency vibrations (<20 Hz) of underground railway tracks induced by wheel-rail interaction have become important environmental issues, as they result in undesirable effects on nearby buildings and cause disturbance and annoyance to people. One of the effective means for reducing vibrations from underground railway tracks is the use of floating-slab tracks (FSTs) [58,76–78]. An FST basically consists of concrete slabs supported on resilient elements such as rubber bearings, glass fibre and steel springs. Although a large amount of literature has reported its effectiveness, it remains controversial for underground railways due to the interactions with the supporting foundations at lower frequencies [79–83]. Therefore, modern railway designers face challenging requirements for finding an effective solution to control low-frequency vibrations of underground railways.

In this regard, based on vehicle-track coupled dynamics and passive vibration isolation theory, the authors [49,50,56] have developed a novel VAT, capable of mitigating low-frequency vibrations for urban rail transit systems after comprehensive research works combining analytical, numerical and experimental studies. This new track has been successfully applied to practical engineering, which provides valuable contributions to the reduction of environmental vibrations induced by railways.

9.7.1 Design Scheme

The designed VAT is capable of not only passively attenuating low-frequency vibrations of track slabs but also reducing vibration transmission to the underlying subgrade in real time. The VAT is developed by attaching discrete dynamic vibration absorbers (DVAs) to the centre line of a discontinuous FST, as shown in Figure 9.33.

Here, the low-frequency vibration energies of the track are intended to be absorbed by the DVAs though the linked rubber elements. This is achieved by obtaining optimal design parameters of the DVAs for attenuating low-frequency vibration modes of the FST and thus realising an opposite vibration phase of the FST and the DVAs.

The optimal parameters of DVAs incorporated in the VAT for suppressing the ith mode vibration of its primary system FST can be given as follows, based on the fixed-point theory [84,85]:

$$k_i = m_i \frac{K_i}{M_i} \frac{1 - 2Z_i \zeta_i (1 + \mu_i) - 2Z_i^2}{(1 + \mu_i)^2 + \mu_i \frac{Z_i}{\zeta_i}} \quad (9.39)$$

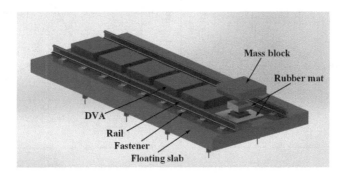

FIGURE 9.33 Prototype of the proposed vibration-attenuation track.

$$c_i = 2m_i \sqrt{\frac{K_i}{M_i}} \sqrt{\frac{3\mu_i}{8(1+\mu_i)^3}} \qquad (9.40)$$

where μ_i is the mass ratio between the ith DVA and the FST; m_i, k_i and c_i are the mass, stiffness and damping of the ith DVA, respectively; and ζ_i and Z_i are the modal damping ratio of the DVA and the FST, respectively. From passive vibration isolation theory, it is known that the attenuation level of the primary system is directly proportional to the mass ratio of the DVA with respect to the primary system, meaning that a larger mass ratio of the DVA or a smaller modal mass of the primary system would lead to a better vibration attenuation performance of the primary system. Therefore, the DVA should be attached at the modal antinode location of the ith mode; if considering the DVAs for suppressing other mode vibrations, then the appropriate DVA location for suppressing the ith mode vibration should not only be the modal antinode location of the ith node but also simultaneously the modal node position of other modes.

9.7.2 Dynamic Simulation and Evaluation

The vibration attenuation performance of the VAT is investigated under train dynamic loads by establishing a three-dimensional coupled dynamic model of a metro vehicle-VAT-subgrade system. Figure 9.34a shows the comparison of the slab acceleration in the time domain between the FST and the actual VAT. It can be seen that much smaller accelerations of the VAT are induced by train dynamic loads compared with that of the FST, due to the fact that vibration energies of the VAT slab are transmitted to the attached DVAs through the linked rubber mats and consumed by its damping elements. The comparison of the root-mean-square (RMS) acceleration levels of slabs in the one-third octave band is shown Figure 9.34b. Clearly, the RMS acceleration levels of the VAT slab display quite discernible reductions at the frequencies of 9 to 16 Hz, attributed to the mitigation of the first-order mode vibrations by the attached DVAs.

To further illustrate the vibration attenuation performance of the VAT in terms of the vibration transmission to the substructures, the comparison of vibration transmitted from the slab to the subgrade is plotted in Figure 9.35. As can be seen in Figure 9.35a, the acceleration amplitudes in the time domain are much reduced for the VAT compared with that of the FST. In addition, as shown in Figure 9.35b, the RMS acceleration levels of the subgrade show a decrease of around 15 dB for the VAT at the frequency of 10 Hz. This is because vibration energies are effectively consumed by the DVAs, which largely reduce the vibration transmission to the subgrade.

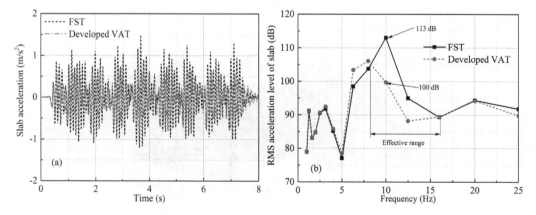

FIGURE 9.34 Comparison of slab acceleration in (a) the time domain and (b) the one-third octave band.

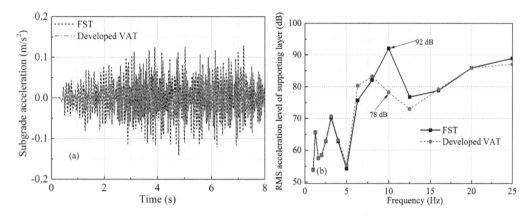

FIGURE 9.35 Comparison vibrations transmitted to the subgrade in (a) the time domain and (b) the one-third octave band.

The effect of DVAs on the wheelset acceleration is shown in Figure 9.36. It can be seen that the DVAs slightly reduce wheelset vibrations at frequencies of around 10 to 13 Hz, whereas the DVAs have negligible influence on the car body and bogie vibrations.

9.7.3 Experimental Validation

To validate the effectiveness of the VAT, a full-scale VAT was developed by attaching the DVAs to appropriate positions of the full-scale discontinuous FST system, with a slab length of 9.06 m, a slab height of 0.495 m and a slab width of 3.15 m, which was designed by Train and Track Research Institute at Southwest Jiaotong University, China, as shown in Figure 9.37. The optimal design parameters of the VAT were determined based on passive vibration isolation theory with the help of laboratory modal testing of the full-scale FST. Thus, by installing the corresponding DVAs on the FST, the proposed VAT can be formed in the laboratory.

Full-scale dynamic tests of the VAT and the FST under harmonic loads have been carried out in the laboratory, as shown in Figure 9.37. Note that the DVAs are adopted to mitigate the first-order mode vibration of the primary system of the VAT with the frequency of 10.19 Hz, so harmonic loads with frequency from 5 to 15 Hz are applied by the loading system.

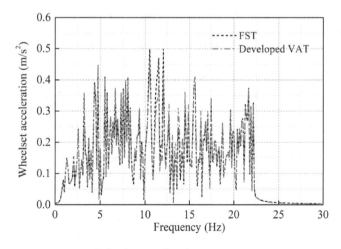

FIGURE 9.36 Frequency contents of wheelset acceleration.

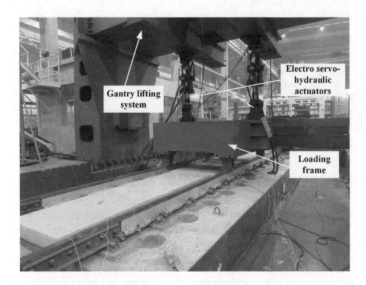

FIGURE 9.37 Full-scale dynamic tests of the VAT and the FST under harmonic loads.

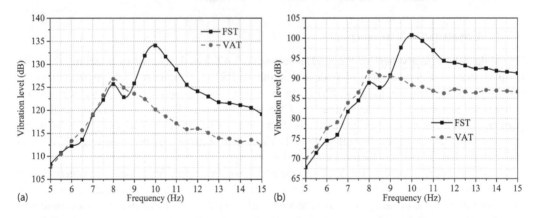

FIGURE 9.38 Comparison of vibration level between the VAT and the FST: (a) slab and (b) supporting layer.

Figure 9.38 shows the vibration level of the VAT slab and supporting layer under the harmonic loads, together with results yielded by the FST. As can be seen in Figure 9.38a, the slab vibration levels of the developed VAT at around 10 Hz are substantially reduced to small proportions because of vibration absorption of the DVAs at low frequencies. It can be seen from Figure 9.38b that the resonance peak vibrations of the VAT supporting layer at around the natural frequency are also significantly mitigated, while, for the FST they are not effectively isolated but transmitted to the supporting layer.

To capture the vibration transmission characteristics, the comparison of the vibration transfer loss from the slab to the supporting layer between the VAT and the FST is provided in Figure 9.39. As it can be seen in the shaded area, the transfer loss of the VAT is much larger than that of the FST above 9 Hz, which has a maximum value of 13 dB at the frequency of 10 Hz. This illustrates that slab vibrations transmitted to the substructures are much reduced for the VAT because of the absorption of slab vibrations by the DVAs.

Therefore, it can be concluded from the train-track-subgrade dynamic analysis that VAT can effectively and passively attenuate low-frequency vibrations at the frequency of 9 and 16 Hz and thus significantly reduce vibration transmissions to the supporting layer and the subgrade.

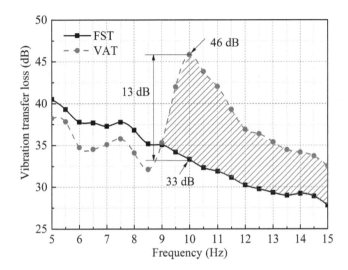

FIGURE 9.39 Comparison vibrations transfer loss from the slab to the subgrade.

ACKNOWLEDGEMENTS

The authors would like to thank their colleagues from the Train and Track Research Institute at Southwest Jiaotong University for their help in preparing this chapter.

REFERENCES

1. S. Iwnicki, *Handbook of Railway Vehicle Dynamics*, CRC Press, Boca Raton, FL, 2006.
2. M. Wang, C. Cai, S. Zhu, W. Zhai, Experimental study on dynamic performance of typical non-ballasted track systems using a full-scale test rig, *Journal of Rail and Rapid Transit*, 231(4), 470–481, 2017.
3. S. Zhu, C. Cai, Interface damage and its effect on vibrations of slab track under temperature and vehicle dynamic loads, *International Journal of Non-Linear Mechanics*, 58, 222–232, 2014.
4. S. Zhu, C. Cai, W. Zhai, Interface damage assessment of railway slab track based on reliability techniques and vehicle-track interactions, *Journal of Transportation Engineering*, 142(10), 2016.
5. A. Lundqvist, T. Dahlberg, Load impact on railway track due to unsupported sleepers, *Journal of Rail and Rapid Transit*, 219(2), 67–77, 2005.
6. J.Y. Zhu, D.J. Thompson, C.J.C. Jones, On the effect of unsupported sleepers on the dynamic behaviour of a railway track, *Vehicle System Dynamics*, 49(9), 1389–1408, 2011.
7. D. Lyon, The calculation of track forces due to dipped rail joints, wheel flats and rail welds, *The Second ORE Colloquium on Technical Computer Programs*, Derby, UK, 1972.
8. H.H. Jenkins, J.E. Stephenson, G.A. Clayton, G.W. Morland, D. Lyon, The effect of track and vehicle parameters on wheel/rail vertical dynamic forces, *Railway Engineering Journal*, 3(1), 2–16, 1974.
9. R.A. Clark, P.A. Dean, J.A. Elkins, S.G. Newton, An investigation into the dynamic effects of railway vehicles running on corrugated rails, *Journal of Mechanical Engineering Science*, 24(2), 65–76, 1982.
10. D.R. Ahlbeck, H.C. Meacham, R.H. Prause, The development of analytical models for railroad track dynamics, In A.D. Kerr (Ed.), *Railroad Track Mechanics & Technology*, Pergamon Press, Oxford, UK, 1978, pp. 239–263.
11. Y. Sato, Abnormal wheel load of test train, *Permanent Way* (Tokyo), 14(53), 1–8, 1973.
12. W.M. Zhai, Vertical vehicle–track coupled dynamics, PhD thesis, Southwest Jiaotong University, Chengdu, China, 1991. (in Chinese)
13. W. Zhai, *Vehicle–Track Coupled Dynamics Theory and Application*, Springer, Singapore, 2019.
14. W.M. Zhai, C.B. Cai, S.Z. Guo, Coupling model of vertical and lateral vehicle/track interactions, *Vehicle System Dynamics*, 26(1), 61–79, 1996.
15. W.M. Zhai, X. Sun, A detailed model for investigating vertical interaction between railway vehicle and track, *Vehicle System Dynamics*, 23, 603–615, 1994.

16. T.X. Wu, D.J. Thompson, Vibration analysis of railway track with multiple wheels on the rail, *Journal of Sound and Vibration*, 239(1), 69–97, 2001.
17. C. Andersson, J. Oscarsson, J.C.O. Nielsen, Dynamic train/track interaction including state-dependent track properties and flexible vehicle components, *Proceedings 16th IAVSD Symposium held in Pretoria, South Africa*, 30 August–3 September 1999, Swets and Zeitlinger, Lisse, the Netherlands, 47–58, 2000.
18. L. Auersch, Vehicle-track-interaction and soil dynamics, *Vehicle System Dynamics*, 28(S1), 553–558, 1998.
19. R.D. Frohling, Low frequency dynamic vehicle/track interaction: Modelling and simulation, *Vehicle System Dynamics*, 28(S1), 30–46, 1998.
20. G. Diana, F. Cheli, S. Bruni, A. Collina, Interaction between railroad superstructure and railway vehicles, *Vehicle System Dynamics*, 23(S1), 75–86, 1994.
21. J. Drozdziel, B. Sowinski, W. Groll, The effect of railway vehicle-track system geometric deviation on its dynamics in the turnout zone, *Proceedings 16th IAVSD Symposium held in Pretoria, South Africa*, 30 August–3 September 1999, Swets and Zeitlinger, Lisse, the Netherlands, 641–652, 2000.
22. S. Gurule, N. Wilson, Simulation of wheel/rail interaction in turnouts and special track work, *Proceedings 16th IAVSD Symposium held in Pretoria, South Africa*, 30 August–3 September 1999, Swets and Zeitlinger, Lisse, the Netherlands, 143–154, 1999.
23. K.L. Knothe, S.L. Grassie, Modeling of railway track and vehicle/track interaction at high frequencies, *Vehicle System Dynamics*, 22(3/4), 209–262, 1993.
24. J.C.O. Nielsen, Train/track interaction: Coupling of moving and stationary dynamic systems, PhD Thesis, Chalmers University of Technology, Gothenburg, Sweden, 1993.
25. J. Oscarsson, Dynamic train-track-ballast interaction with unevenly distributed track properties, *Vehicle System Dynamics*, 37(S1):385–396, 2002.
26. J. Oscarsson, T. Dahlberg, Dynamic train/track/ballast interaction: Computer models and full-scale experiments, *Vehicle System Dynamics*, 28(S1), 73–84, 1998.
27. K. Popp, H. Kruse, I. Kaiser, Vehicle-track dynamics in the mid-frequency range, *Vehicle System Dynamics*, 31(5–6), 423–464, 1999.
28. B. Ripke, K. Knothe, Simulation of high frequency vehicle-track interactions, *Vehicle System Dynamics*, 24(S1), 72–85, 1995.
29. Y.Q. Sun, M. Dhanasekar, A dynamic model for the vertical interaction of the rail track and wagon system, *International Journal of Solids and Structures*, 39(5), 1337–1359, 2002.
30. W. Zhai, K. Wang, C. Cai, Fundamentals of vehicle-track coupled dynamics, *Vehicle System Dynamics*, 47(11), 1349–1376, 2009.
31. E. Kabo, J.C.O. Nielsen, A. Ekberg, Prediction of dynamic train–track interaction and subsequent material deterioration in the presence of insulated rail joints, *Vehicle System Dynamics*, 44(S1), 718–729, 2006.
32. R.U.A. Uzzal, A.K.W. Ahmed, S. Rakheja, Analysis of pitch plane railway vehicle-track interactions due to single and multiple wheel flats, *Journal of Rail and Rapid Transit*, 223(4), 375–390, 2009.
33. Z. Wen, X. Jin, X. Xiao, Z. Zhou, Effect of a scratch on curved rail on initiation and evolution of plastic deformation induced rail corrugation, *International Journal of Solids and Structures*, 45(7–8), 2077–2096, 2008.
34. Y.B. Yang, Y-S. Wu, Transmission of vibrations from high speed trains through viaducts and foundations to the ground, *Journal of the Chinese Institute of Engineers*, 28(2), 251–266, 2005.
35. S. Alfi, S. Bruni, Mathematical modelling of train–turnout interaction, *Vehicle System Dynamics* 47(5), 551–574, 2009.
36. J.L. Escalona, H. Sugiyama, A.A. Shabana, Modelling of structural flexibility in multibody railroad vehicle systems, *Vehicle System Dynamics*, 51(7), 1027–1058, 2013.
37. E.D. Gialleonardo, F. Braghin, S. Bruni, The influence of track modelling options on the simulation of rail vehicle dynamics, *Journal of Sound and Vibration*, 331(19), 4246–4258, 2012.
38. I. Kaiser, Refining the modelling of vehicle–track interaction, *Vehicle System Dynamics*, 50(S1), 229–243, 2012.
39. K. Popp, K. Knothe, C. Popper, System dynamics and long-term behaviour of railway vehicles, track and subgrade: Report on the DFG priority programme in Germany and subsequent research, *Vehicle System Dynamics*, 43(6–7), 485–538, 2005.
40. P.T. Torstensson, J.C.O. Nielsen, Simulation of dynamic vehicle–track interaction on small radius curves, *Vehicle System Dynamics*, 49(11), 1711–1732, 2011.

41. R.U.A. Uzzal, A.K.W. Ahmed, R.B. Bhat, Modelling, validation and analysis of a three-dimensional railway vehicle–track system model with linear and nonlinear track properties in the presence of wheel flats, *Vehicle System Dynamics*, 51(11), 1695–1721, 2013.
42. W.M. Zhai, K.Y. Wang, Lateral interactions of trains and tracks on small-radius curves: Simulation and experiment, *Vehicle System Dynamics*, 44(S1), 520–530, 2006.
43. W.M. Zhai, K.Y. Wang, Lateral hunting stability of railway vehicles running on elastic track structures, *Journal of Computational and Nonlinear Dynamics*, 5(4), 041009, 2010.
44. Z. Wen, X. Jin, Effect of track lateral geometry defects on corrugations of curved rails, *Wear*, 259(7–12), 1324–1331, 2005.
45. Xiao X, Jin X, Wen Z. Effect of disabled fastening systems and ballast on vehicle derailment. *Journal of Vibration and Acoustics*, 129(2), 217–229, 2007.
46. S. Zhu, C. Cai, Stress intensity factors evaluation for through-transverse crack in slab track system under vehicle dynamic load, *Engineering Failure Analysis*, 46, 219–237, 2014.
47. S. Zhu, Q. Fu, C. Cai, P.D. Spanos, Damage evolution and dynamic response of cement asphalt mortar layer of slab track under vehicle dynamic load, *Science China – Technological Sciences*, 57(10), 1883–1894, 2014.
48. S. Zhu, C. Cai, P.D. Spanos, A nonlinear and fractional derivative viscoelastic model for rail pads in the dynamic analysis of coupled vehicle-slab track system, *Journal of Sound and Vibration*, 335, 304–320, 2015.
49. S. Zhu, J. Wang, C. Cai, K. Wang, W. Zhai, J. Yang, H. Yan, Development of a vibration attenuation track at low frequencies for urban rail transit, *Computer-Aided Civil and Infrastructure Engineering*, 32(9), 713–726, 2017.
50. S. Zhu, J. Yang, H. Yan, L. Zhang, C. Cai, Low-frequency vibration control of floating-slab tracks using dynamic vibration absorbers, *Vehicle System Dynamics*, 53(9), 1296–1314, 2015.
51. G. Kouroussis, D.P. Connolly, O. Verlinden, Railway-induced ground vibrations: A review of vehicle effects, *International Journal of Rail Transportation*, 2(2), 69–110, 2014.
52. W. Zhai, K. Wei, X. Song, M. Shao, Experimental investigation into ground vibrations induced by very high speed trains on a non-ballasted track, *Soil Dynamics and Earthquake Engineering*, 72, 24–36, 2015.
53. D.J. Thompson, E. Latorre Iglesias, X. Liu, J. Zhu, Z. Hu, Recent developments in the prediction and control of aerodynamic noise from high-speed trains, *International Journal of Rail Transportation*, 3(3), 119–150, 2015.
54. Z. Xu, Prediction and control of wheel-rail noise for rail transit, PhD thesis, Southwest Jiaotong University, Chengdu, China, 2004. (in Chinese)
55. X. Yang, Theoretical analysis and control studies in wheel-rail noises of high-speed railway, PhD thesis, Southwest Jiaotong University, Chengdu, China, 2010. (in Chinese)
56. S. Zhu, J. Yang, C. Cai, Z. Pan, W. Zhai, Application of dynamic vibration absorbers in designing a vibration isolation track at low-frequency domain, *Journal of Rail and Rapid Transit*, 231(5), 546–557, 2017.
57. X. Zhang, C. Zhao, W. Zhai, Dynamic behavior analysis of high-speed railway ballast under moving vehicle loads using discrete element method, *International Journal of Geomechanics*, 17(7), 04016157-1, 2016.
58. M.F.M. Hussein, P.A. Costa, The effect of end bearings on the dynamic behaviour of floating-slab tracks with discrete slab units, *International Journal of Rail Transportation*, 5(1), 38–46, 2017.
59. W.M. Zhai, K.Y. Wang, J.H. Lin, Modelling and experiment of railway ballast vibrations, *Journal of Sound and Vibration*, 270(4–5), 673–683, 2004.
60. P. Liu, W. Zhai, K. Wang, Establishment and verification of three-dimensional dynamic model for heavy-haul train–track coupled system, *Vehicle System Dynamics*, 54(11), 1511–1537, 2016.
61. Y. Sun, W. Zhai, Y. Guo, A robust non-Hertzian contact method for wheel-rail normal contact analysis, *Vehicle System Dynamics*, 56, 1899–1921, 2018. doi:10.1080/00423114.2018.1439587.
62. L. Xu, W. Zhai, A new model for temporal–spatial stochastic analysis of vehicle–track coupled systems, *Vehicle System Dynamics*, 55(3), 427–448, 2007.
63. G. Chen, W.M. Zhai, A new wheel/rail spatially dynamic coupling model and its verification, *Vehicle System Dynamics*, 41(4), 301–322, 2004.
64. Z. Shen, J. Hedrick, J. Elkins, A comparison of alternative creep-force models for rail vehicle dynamic analysis, *Proceedings 8th IAVSD Symposium*, held at Massachusetts Institute of Technology, Cambridge, MA, 15–19 August 1983, Swets and Zeitlinger, Lisse, the Netherlands, 591–605, 1984.

65. W.M. Zhai, Two simple fast integration methods for large-scale dynamic problems in engineering, *International Journal of Numerical Methods in Engineering*, 39(24), 4199–4214, 1996.
66. V.K. Garg, R.V. Dukkipati, *Dynamics of Railway Vehicle Systems*, Academic Press Canada, Toronto, Canada, 1984.
67. P. Meinke, A. Mielcarek, Design and evaluation of trucks for high-speed wheel/rail application, In W.O. Schiehlen (Ed.), *Dynamics of High-Speed Vehicles*, Springer Verlag, Vienna, Austria, 1982, pp. 281–331.
68. Railway Construction Institute, Study on the track irregularity power spectral density of Chinese main lines, Report TY-1215, China Academy of Railway Sciences, Beijing, China, 1999.
69. L. Wang, Random vibration theory of rail/track structure and its application in the rail/track vibration isolation, PhD thesis, China Academy of Railway Sciences, Beijing, China, 1988.
70. National Railway Administration of China, TB/T 3352–2014 PSD of ballastless track irregularities of high-speed railway, China Railway Publishing House, Beijing, China, 2014.
71. China Railway, *Q/CR 508–2016 PSD of Ballasted Track Irregularities of High-Speed Railway*, China Railway Publishing House, Beijing, China, 2016.
72. W. Zhai, P. Liu, J. Lin, K. Wang, Experimental investigation on vibration behaviour of a CRH train at speed of 350 km/h, *International Journal of Rail Transportation*, 3(1), 1–16, 2015.
73. D.J. Thompson, C.J.C. Jones, T.P. Waters, D. Farrington, A tuned damping device for reducing noise from railway track, *Applied Acoustics*, 68, 43–57, 2007.
74. Y. Bezin, D. Farrington, A structural study of an innovative steel-concrete track structure, *Journal of Rail and Rapid Transit*, 224(4), 245–257, 2010.
75. Y. Bezin, D. Farrington, C. Penny, B. Temple, S. Iwnicki, The dynamic response of slab track constructions and their benefit with respect to conventional ballasted track, *Vehicle System Dynamics*, 48(S1), 175–193, 2010.
76. Z.G. Li, T.X. Wu, Modelling and analysis of force transmission in floating-slab track for railways, *Journal of Rail and Rapid Transit*, 222(1), 45–57, 2008.
77. G. Lombaert, G. Degrande, B. Vanhauwere, B. Vandeborght, S. Francois, The control of ground-borne vibrations from railway traffic by means of continuous floating slabs, *Journal of Sound and Vibration*, 297(3–5), 946–961, 2006.
78. K.E. Vogiatzis, G. Kouroussis, Prediction and efficient control of vibration mitigation using floating slabs: Practical application at Athens metro lines 2 and 3, *International Journal of Rail Transportation*, 3(4), 215–232, 2015.
79. F. Cui, C.H. Chew, The effectiveness of floating slab track system, part 1: Receptance methods, *Applied Acoustics*, 61(4), 441–453, 2000.
80. D. Ding, W. Liu, K. Li, X. Sun, W. Liu, Low frequency vibration tests on a floating slab track in an underground laboratory, *Journal of Zhejiang University - SCIENCE A - Applied Physics & Engineering*, 12(5), 345–359, 2011.
81. H. Saurenman, J. Phillips, In-service tests of the effectiveness of vibration control measures on the BART rail transit system, *Journal of Sound and Vibration*, 293(3–5), 888–900, 2006.
82. L. Schillemans, Impact of sound and vibration of the North-South high-speed railway connection through the city of Antwerp Belgium, *Journal of Sound and Vibration*, 267(3), 637–649, 2003.
83. W. Zhai, P. Xu, K. Wei, Analysis of vibration reduction characteristics and applicability of steel-spring floating-slab track, *Journal of Modern Transportation*, 19(4), 215–222, 2011.
84. J.P. Den Hartog, *Mechanical Vibrations*, 4th ed., McGraw-Hill, New York, 1956.
85. K. Seto, *Dynamic Vibration Absorbers and its Application*, China Machine Press, Beijing, China, 2013. (in Chinese)

10 Gauging Issues

David M. Johnson

CONTENTS

- 10.1 Philosophy and History of Gauging 346
  - 10.1.1 Gauges 347
    - 10.1.1.1 Static Gauges 347
    - 10.1.1.2 Geometric or Swept Gauges 347
    - 10.1.1.3 Pseudo-Kinematic Gauges 347
    - 10.1.1.4 Kinematic Gauges 347
    - 10.1.1.5 UIC Gauges 348
  - 10.1.2 Swept Envelopes 349
    - 10.1.2.1 Kinematic Envelopes 349
    - 10.1.2.2 Dynamic Envelopes 350
- 10.2 Components of Gauging 350
  - 10.2.1 Structure 350
    - 10.2.1.1 Shape 350
    - 10.2.1.2 Accuracy of Measurement 350
    - 10.2.1.3 Choice of Measuring System 351
    - 10.2.1.4 The Need for Measurement 352
  - 10.2.2 Track 352
    - 10.2.2.1 Track Position 352
    - 10.2.2.2 Track Geometry 352
  - 10.2.3 Vehicle 353
    - 10.2.3.1 Geometric Considerations 353
    - 10.2.3.2 Dynamic Considerations 354
    - 10.2.3.3 Vehicle Tolerances 358
    - 10.2.3.4 Tilting Trains 359
    - 10.2.3.5 Comparison of BASS 501 and Multibody Simulation Methods 362
- 10.3 The Gauging System 364
  - 10.3.1 Vehicle-Track Interaction 364
  - 10.3.2 Track-Structure Interaction 365
    - 10.3.2.1 Track Tolerances 365
  - 10.3.3 Structure-Vehicle Interaction 366
    - 10.3.3.1 Clearances 366
    - 10.3.3.2 Stepping Distances 367
  - 10.3.4 Standards 367
  - 10.3.5 Advanced Analysis 368
  - 10.3.6 Digital Systems 369
  - 10.3.7 Risk Analysis 369
- References 371

10.1 PHILOSOPHY AND HISTORY OF GAUGING

Gauging is the name given to the techniques used to ensure that rail vehicles fit through the railway infrastructure and pass by each other in safety. Increasingly, there is emphasis on maximising the capacity of the railway corridor through a more thorough understanding of the gauging system and reducing conservatism in the processes that ensured adequate space was available when the railways were first built [1].

This chapter is intended to provide readers with an insight into the techniques used in Britain, where an infrastructure of up to 200 years old is now required to deliver the capability of running large, intermodal freight trains and passenger trains of increased capacity and comfort at higher speeds that these railways were not designed for. Internationally, capacity constraints are also real, but infrastructure that has been built later invariably requires less incremental change to cope with the near-standardised loads now being transported. Virtually, every country has its own methodologies through which loading gauge is managed, and it is beyond the scope of this chapter to provide anything but a brief look at gauging principles and approaches that form the basis of most of these gauging practices. Beyond Britain, western Europe has generally adopted a standard Union Internationale des Chemins de Fer (UIC) approach, which is described briefly for information.

Railways were originally built to gauges—vehicles to a maximum vehicle (or load) gauge and infrastructure (structures) to a minimum structure gauge. A clearance was included between the vehicle gauge and the structure gauge to allow for unknowns or those items that were known but had not been included in the gauge.

At the turn of the nineteenth century, the UK Board of Trade (whose job was to monitor rail traffic) had registered 127 different load gauges from (private) railway companies. No load gauge was universal, except, perhaps, the smallest. Many railway administrations still work by these simple gauging methods; indeed, the methodology used in much of Europe is a derivative of the earlier fixed-gauge approach. Plenty of the original railway infrastructure built to accommodate these load gauges still exists, but the trend is to increase vehicle size. The challenge is to develop new gauging methodologies that enable this to happen. The original methods provide a good starting point. A seminal publication on which modern practice is founded is the *Railway Safety Principles and Guidance* [2].

British engineers, forced to make increasingly better use of small Victorian era (predominantly arched) infrastructure, have been at the forefront of developing gauging systems that analyse the vehicle-infrastructure interaction on a case-by-case basis to minimise the cost of upgrade works needed to run these larger passenger coaches and bigger freight loads.

This saw the development of what is now known as the absolute gauging process, where the space between vehicles (or vehicle gauges) is determined relative to measurements of the actual infrastructure, rather than to an infrastructure gauge.

New developments also introduced additional factors that had to be taken into account. Early railways used short wagons, and their swept envelopes were not significantly different from their static size. The introduction of long coaches rather than short carriages generated a new vehicle-infrastructure interaction. Overthrow complicated the basic interface between mechanical engineer and civil engineer, as it related to both the curve geometry and the arrangement of the vehicle. Railways that had worked well with short vehicles now exhibited weaknesses in certain situations, having restricted clearances on curves. The first significant development in the trend towards gauging analysis was the adjustment of gauges to include vehicle overthrows associated with curvature of the track.

Increasingly, the understanding of vehicle dynamics has led to techniques that predict suspension movements (and hence the local swept envelope) in response to curvature and speed. Those techniques, largely developed by British Rail Research, became invaluable in the acceptance processes for air-suspension rolling stock (with implicitly softer suspension) in the 1980s. Although it became

Gauging Issues

increasingly possible to calculate vehicle movements with precision, gauging standards were slow to react to these improved methodologies and, for a while, failed to allow all of the benefits that the techniques could offer.

Significant advances in vehicle dynamics and the introduction of computerised techniques have allowed tolerances, clearances and 'unknowns' to be defined in a more robust manner. Tolerances may now be calculated accurately, and clearances may be provided for the fewer remaining unknown or incalculable effects. An important factor is that, as unknowns are understood, they may be removed from mandated clearances and analysed as appropriate tolerances. Conservatism is thus being progressively removed from the system.

Modern gauging technology is far removed from the simple pen-and-paper solutions of 100 years ago. This chapter aims to give an insight into the factors considered and calculations performed in modern gauging methods.

In simple terms, gauging has moved on from being the technique for simply deciding whether something will fit to what can be done to enable something to fit.

10.1.1 GAUGES

10.1.1.1 Static Gauges
In what may be described as 'simple gauging', the mechanical engineer built vehicles to a 'vehicle gauge', being the maximum permissible cross-section of the train, and the civil engineer ensured that structures were always larger than the 'structure gauge'. A separation between the two, known as clearance, allowed for any variations of track position and the suspension movements of the vehicle. These are known as static gauges.

10.1.1.2 Geometric or Swept Gauges
Geometric or swept gauges represented a development of Static Gauges where the vehicle size (swept envelope) was substantially affected by the geometry of the track. On curves, vehicles sweep a larger path than on straight track, a phenomenon known as 'overthrow'. The amount depends on the tightness of the curve, the vehicle bogie (or axle) centres and the overall length. In the immediate post-nationalisation period in Britain (approximately 1951 onwards), 'national gauges' for rail passenger vehicles (known as C1) and freight vehicles (W5) were defined, based upon the vehicle gauges used by the majority of component railway companies absorbed into British Railways (BR). C1 and W5 gauges are geometric gauges, requiring knowledge of both vehicle parameters and curve geometry in order to calculate the clearance to a structure. A clearance of 150 mm (6 inches) was usually allowed, comprising 100 mm for potential vehicle movement on its suspension and 50 mm for potential track positional and geometric errors.

10.1.1.3 Pseudo-Kinematic Gauges
A pseudo-kinematic gauge is where maximum vehicle movements are included in the gauge. It is common for light rail and metro systems to use a vehicle gauge, which includes all suspension movement for particular vehicles (this is sometimes known as a red-line kinematic gauge). The system used across Europe is a further development of this, using a reference profile to define a notional boundary between train and infrastructure under certain, prescribed limits, beyond which both vehicle builder and infrastructure controller must make adjustment. Pseudo-kinematic gauges work well for new infrastructure but lead to the restriction of vehicle size as softer suspension is introduced.

10.1.1.4 Kinematic Gauges
It should be noted that the swept envelope of a vehicle is, in reality, a series of swept envelopes, since some parts of a vehicle move more than others (depending on where the cross-section of the vehicle

is, in relation to bogie or axle centres), and some cross-sections may have projections. In particular, the cross-section of a vehicle at the bogie/axle positions will exhibit minimal throw, whereas the section located in the centre of the vehicle will have maximum throw towards the inside of a curve. Similarly, the ends of a vehicle will have maximum throw towards the outside of a curve. A kinematic gauge (examples being those that define the GB/SNCF/SNCB Class 373 Eurostar vehicle [3] and the W12 freight gauge) is a union of different kinematic vehicle dimensions defining the largest envelope under given operating conditions. W12, for instance, consists of many thousands of gauge diagrams appropriate to curve radius, installed cant and speed.

Gauges refer not only to a cross-sectional profile but also to a set of rules that must be applied. A basic understanding of the gauge definitions will show that the clearance required for safe operation is intrinsically linked to the derivation of the gauge and the parameters considered.

10.1.1.5 UIC Gauges

The Union Internationale des Chemins de Fer or UIC gauging methods were originally defined in the 505 series of leaflets and use reference gauges shown in Figure 10.1 as the basis for gauging [4–6]. These methods are now described in the European Standard EN 15273 [7]. The method dates back to 1913 and was developed as a hand-calculated technique that contained a number of simplifications. The infrastructure gauge is an obstruction gauge—the conservatism ensures that contact is not physically possible. From a vehicle's perspective, this reference profile defines a base gauge into which the vehicle must fit under certain defined conditions.

The conditions under which vehicles must be contained within the reference dimensions are (inclusively):

- Horizontally, on a 250 m radius curve
- Under 50 mm of installed cant and 50 mm of developed cant deficiency
- With full horizontal suspension travel
- Stationary

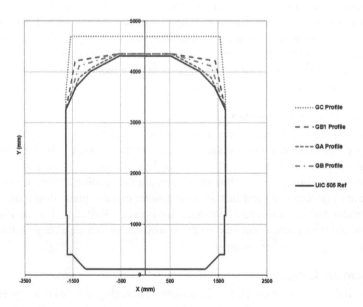

FIGURE 10.1 Basic UIC reference profiles.

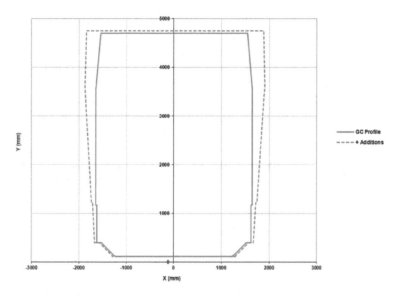

FIGURE 10.2 Example UIC GC reference profile with infrastructure additions.

Note that this is not a vehicle gauge in the British sense. It represents a vehicle snapshot, which requires further analysis in applying this to the infrastructure. In particular, the following effects (known as infrastructure additions) must be taken into account:

- 'Authorised projections' on curves and throws on curves between 150 and 250 m radius (note that UIC vehicles are not designed to run on curves of less than 150 m radius)
- The effect of track quality associated with speed
- Gauge widening
- Roll from cant excess or deficiency above 50 mm
- Track alignment tolerances
- Static and dynamic cross-level error
- Possible vehicle loading asymmetry
- Vertical curvature

Figure 10.2 shows a developed infrastructure gauge for a specific condition (1000 m horizontal curve, 150 mm installed cant, 160 km/h developing a cant deficiency of 150 mm, good track and a 1000 m vertical curve). Similarly, reductions are made to the profile to provide a developed vehicle gauge. Under UIC rules, it is not necessary to provide any clearance between these envelopes; there is sufficient conservatism in the calculation process to ensure that vehicle-structure contact can never occur. However, the UIC process by this very conservatism makes poor use of the space envelope that would normally be used in Britain.

10.1.2 Swept Envelopes

10.1.2.1 Kinematic Envelopes

In the late 1970s and 1980s, cost engineering became prevalent in Britain, particularly in the area of track maintenance. A given ride quality can be achieved by maintaining high-quality track geometry or by providing softer vehicle suspensions. The former solution is particularly expensive, since, as track quality is raised, the cost of maintaining it increases exponentially. The new generation of rollingstock then being commissioned could readily be given suspension capable of providing adequate ride comfort on poorer track. Air suspension provided this mechanism but at the expense of having greater dynamic movement (movement associated with the speed of the vehicle). The methodologies described would have meant that the infrastructure would have required enlargement to maintain clearance. However, it was recognised that by relating kinematic movement to operating environment, the locations where enlarged infrastructure was required could be minimised. A publication known as 'Design Guide BASS 501' [8] provided a methodology whereby the 'kinematic envelope' of a vehicle (the space required by a given vehicle, moving at speed) at a specific location could be manually calculated from a number of input parameters. The techniques used are quasi-static, equating dynamic conditions to stationary forces, and are generally conservative. Nevertheless, the techniques were very successful in allowing larger trains to operate on restrictive infrastructure at a minimal cost. In particular, a derivation of the technique has allowed tilting trains to be designed for Britain that would otherwise have been of a non-viable cross-section if traditional gauging rules were applied.

The BASS 501 method originated as a hand-calculation method but achieved widespread use with the development of computerised gauging software in the 1990s such as ClearRoute™.*

10.1.2.2 Dynamic Envelopes

More recently, with the advent of powerful computers, multibody simulation (MBS) systems such as VAMPIRE™[†] or SimPack™[‡] have replaced the empirical rules developed in BASS 501 with more robust estimations of vehicle dynamic movement, derived from modelling a vehicle's suspension whilst running on digital models of benchmark track. Using such simulations, together with more advanced gauging software such as PhX rail™,[§] enables many of the allowances used to accommodate unknown values in the BASS 501 method to be replaced by known values and thus improve the accuracy of the clearance calculation. Such simulations have paved the way for the analysis of modern rollingstock, which may require longitudinal variation of dynamic movements to be analysed, rather than simply considering consistent movement along a vehicle's length.

10.2 COMPONENTS OF GAUGING

A full gauging model requires the interactions between structure, track and vehicle to be understood. Before examining these interactions, an understanding of the behaviour of the individual components is required.

10.2.1 STRUCTURE

10.2.1.1 Shape

In Britain, clearance issues mainly relate to arch bridges, tunnels and platforms. Containers, particularly, provide an obvious 'square peg in a round hole' challenge when trying to run the former through the latter. Overbridges generate height restrictions, and platforms generate width restrictions. All obstacles in the vicinity of the train must be measured.

10.2.1.2 Accuracy of Measurement

Accuracy of structure measurement is becoming increasingly significant. As analysis methodologies improve, conservatism in infrastructure measurement that results from inaccurate measurement becomes less acceptable. In particular, whilst it may be possible to define the swept envelope

---

* ClearRoute™ was developed by Laser Rail Ltd and is the trademark of Balfour Beatty Rail.
† VAMPIRE™ is the trademark of Resonate Ltd.
‡ SimPack™ is the trademark of Dassault Systemes Deutschland GmbH.
§ PhX rail™ is the trademark of DGauge Ltd.

Gauging Issues

FIGURE 10.3 Network rail high-speed-structure gauging train.

of a vehicle to within a few millimetres, the accuracy of many structure-measuring techniques may be as much as 50 mm. In order to maximise infrastructure capacity, it is important that an opportunity is not lost through poor measurement accuracy.

Early measuring systems (with relatively low accuracy) consisted of simple measuring frames or used triangulation of distances measured using tape measures to a point on the structure from each of the two running rails. These systems have largely been superseded by higher-accuracy laser-based measuring systems, which are deployed either manually or mounted on measuring trains (see Figure 10.3).

The accuracy of a measuring system should consider the following:

- The accuracy of the measuring system itself (usually quoted as 2 standard deviations or 95% certainty to a Gaussian distribution)
- The accuracy by which the measurements relate to the (usually rail) datums
- The accuracy of ancillary measurements, such as cant and curvature, which are required by the gauging process

It has been customary to consider a 'composite accuracy' for any measuring system, determining an Root Mean Squared (RMS) value that would be removed from any calculated clearance. More sophisticated analyses, such as that performed using PhX rail™, apply accuracies to the appropriate part of the gauging model, which enables a level of conservatism to be removed.

10.2.1.3 Choice of Measuring System

The choice of an appropriate measuring system would normally be based upon the need for accuracy, safety and the relative cost of deployment, the latter of which would consider the following:

- Capital cost
- Maintenance cost
- Operating cost

- Possession/protection cost
- Train path cost
- Data-processing cost
- Effective working shift length

10.2.1.4 The Need for Measurement

Measurements of every structure on a route are required to determine the suitability of a route to carry particular traffic. After an initial measurement, repeat measurements are required from time to time to ensure the following:

- The shape of the structure has not changed (e.g., due to geological effects)
- The track-structure relationship remains within the bounds under which gauging calculations are performed

In both cases, an initial structure measurement (which largely defines a route's capability) would only be replaced if the structure shape was found to have changed or if the *design* position of the track had been changed. Structure measurements should not be replaced as a result of track maintenance or movement within authorised bounds.

10.2.2 TRACK

10.2.2.1 Track Position

Knowledge of the position of the track and the amount by which it may move are vital in accurate gauging calculations. Track position controls vehicle position. Track position may vary as a result of traffic loading, effects of weather and, most importantly, the movement that is allowed for maintenance and alignment purposes. Some tracks are better restrained than others. Slab track is generally very stable. Ballast gluing can reduce lateral movement (although it would not affect vertical movement significantly). Strutting sleepers against platforms will generally reduce lateral movement towards the platform. Track fixity is discussed later in this chapter.

10.2.2.2 Track Geometry

As we try to model vehicle behaviour more accurately, irregularities of track geometry become increasingly important. In basic gauging, the only geometric input comes from the curve radius used in the calculation of throws. In the more complex models, the full spectrum of track geometry must be considered.

Track geometry is essentially the variation of lateral and vertical track position in relation to the longitudinal position. On perfect, straight track, there is no track geometry. However, track is neither constantly straight nor perfect and consists of straights, curves and track irregularities. This is generally referred to as 'design geometry' and 'roughness'. These parameters are handled separately, in that they have different effects on vehicles. However, in practice, the boundary between geometry and roughness is difficult to define. The key issue is that of wavelength. Generally, design geometry is of long wavelength and defects are of short wavelength. However, it is possible for deviations to be longer than design alignments. In particular, some transition curves are very short and would, in other circumstances, be considered to be irregularities.

In Figure 10.4, the shape on the top would be considered to be normal curved track. It is a summation of the true design curvature (shown by the shape in the middle) and the track irregularity (shown bottom right). If the exact (design) curvature is known, then the irregularity can be calculated easily. In practice, it is unusual (in Britain) to know the exact design curvature, and thus, approximations must be performed to extract the two shapes. This process is known as filtering. The use of a high-pass filter (one that lets high frequency, short wavelength through) would produce the roughness from the measured, compound profile. This process is used to extract track roughness from track topography in order to determine when to maintain the track. The curvature of the track

Gauging Issues

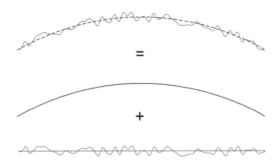

FIGURE 10.4 Superposition of track design alignment and irregularity. (From Iwnicki, S. (ed.), *Handbook of Railway Vehicle Dynamics*, CRC Press, Boca Raton, FL, 2006.)

can be calculated by using a low-pass filter. This process is sometimes referred to as regression, the regressed alignment being the underlying geometric shape of the track.

The filter used makes a significant difference in the separation of track roughness and track geometry. A commonly used filter is a four-pole Butterworth filter, which has a relatively sharp cut-off frequency. However, this filter, on its own, creates a spatial phase shift (where there is a longitudinal movement of peaks and troughs), which may be corrected by performing a reverse pass by using the same filter.

It is important to understand how to use track geometry data in gauging calculations. It will become clear that the curvature data and the roughness data are used for different purposes. However, measuring of track provides a single curvature reading, the nature of which is dependent on how it is measured and processed.

High-speed systems tend to use inertial geometry measurement and sample at frequent intervals. Such systems are principally used for track's quality recording but have been adapted for use on gauging systems. Inertial systems are best at measuring high rates of change of curvature. Since sharper curves provide the greatest input to gauging calculations, these systems may be used, provided the roughness (which they can also measure well) is removed. If unfiltered geometry is used, there is a risk of under- or overcalculating overthrows and double-counting dynamic effects.

Manual systems tend to sample infrequently, usually every 10 to 20 m. As such, they tend to pick up more general curvature, without significant effect from track geometry errors. However, it should be noted that track faults will have an effect on the measured curvature. The true 'design' curvature may need to be extracted using methods such as 'Hallade', a filtering technique using a combination of mathematics and human skill that is used to determine optimal lateral track alignment.

10.2.3 Vehicle

10.2.3.1 Geometric Considerations

The axles of a railway vehicle form the end points of a chord placed on curved track. The body represents an extension of this chord. As the vehicle traverses the chord, the centre of the vehicle is thrown towards the inside of the curve, and the end of the vehicle is thrown towards the outside of the curve. The overthrow effect increases with vehicle length and tighter curvature. A bogie is simply a vehicle with centre throw only. Vertical curvature is not generally an issue on main-line railways but is often considered on metro and light rail systems.

The equations for calculating throw are shown later. The simplified equations ignore some small angle effects, leading to marginal inaccuracy, but are useful for quick calculations.

If we consider Figure 10.5, the overthrow at a point on a vehicle body is the difference between the radial distance from the track centreline to the point and the lateral distance from the vehicle centreline to the point (W_o or W_i). This is calculated with the vehicle stationary.

Consider a vehicle with bogie centres B and a bogie axle semi-spacing of a_o (the actual axle spacing is $2a_o$).

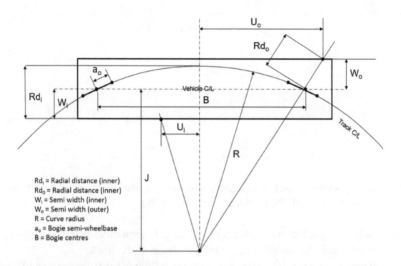

FIGURE 10.5 Curve overthrow diagram.

The inner overthrow of a point U_i from the centre of the vehicle is:

$$R - W_i - \sqrt{[U_i^2 + (J - W_i)^2]}a$$

The outer overthrow of a point U_o from the centre of the vehicle is:

$$\sqrt{[U_o^2 + (J + W_o)^2]} - R - W_o$$

where $J = \sqrt{[R^2 - a_o^2 - B^2/4]}$

The simplified equations are:

$$\text{Inner throw} = 125\,(B^2 - (B - 2x)^2)/R\ [\text{body}] + 500\,(a_o^2/R)\ [\text{bogie}]$$
$$\text{Outer throw} = 125\,[(B + 2x)^2 - B^2]/R\ [\text{body}] - 500\,(a_o^2/R)\ [\text{bogie}]$$

where $x = B/2 - U_o$

Note: [body] is the overthrow of the vehicle body in relation to the bogie pivots.
[bogie] is the inwards overthrow of the pivot of the bogie.

10.2.3.2 Dynamic Considerations

A vehicle moves on its suspension in relation to applied lateral and vertical forces. In simple models, the total amount of lateral and vertical suspension travel (limited by bump stops) is used to determine the required clearance, and no relationship is assumed. As stated earlier, there is a low risk in this approach, but it makes poor use of infrastructure space.

In more sophisticated analyses, the non-linear suspension characteristics are used to determine the amount of suspension movement according to vehicle's running conditions.

The dynamic movement of the vehicle defines the position adopted by the vehicle resulting from the forces applied, and allowances for uncertainty. These can be summarised as follows:

- Lateral, vertical, roll, yaw and pitch movements due to curving forces
- Lateral, vertical, roll, yaw and pitch movements due to motion

- Drops due to loading and suspension condition
- Lateral, vertical and roll due to tilting and active suspensions

10.2.3.2.1 Movement from Curving Forces

The curving forces acting on a vehicle are generally expressed as cant deficiency or cant excess.

The force applied to a vehicle is related to the vehicle speed (V km/h) according to the following approximate formula (on conventional 1435 mm standard gauge track).

$$\text{Cant (deficiency or excess) (mm)} = 11.82\, V^2/\text{radius (m)} - \text{Installed cant (mm)}$$

Figure 10.6 shows cant on the horizontal axis (cant excess to the left and cant deficiency to the right) and vehicle sway on the vertical axis (positive is sway to the outside of the curve). It can be seen that as cant increases (or decreases), the sway in the appropriate direction (inwards or outwards) also increases (or decreases).

It is normal to validate the relationship using a 'sway test' where the vehicle is progressively tilted and the resultant sway curve is calculated.

10.2.3.2.2 Movement due to Track Roughness

Track is imperfect. As a result of unevenness of the rail alignments, there will be local variations of lateral alignment, vertical alignment and the cant that the vehicle sees. Since this is a dynamic phenomenon, the effect on the vehicle suspension is likely to depend on a number of factors, in particular the roughness of the track and the mass/inertia system of the train. These movements are considered differently in quasi-static methods (such as BASS 501) and dynamic methods such as MBS.

BASS 501 empirically considers inputs due to track roughness as a component of the cant applied to the vehicle, known as the equivalent cant (being the sum of actual cant experienced and other roll-inducing effects expressed as cants), using a parameter known as K_{speed}. This parameter defines a linear relationship between the notional force applied to the vehicle (expressed as a cant) and the speed of a vehicle. Typically, K_{speed} is around 0.5, meaning that,

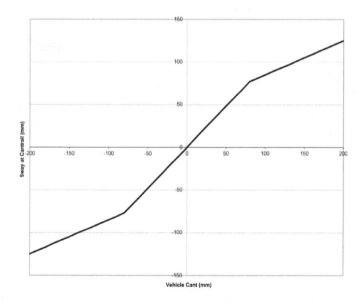

FIGURE 10.6 Typical relationship between applied cant and sway.

at 100 km/h, the rolling force seen by the vehicle, acting on the suspension, is equivalent to an additional ±50 mm of cant above that caused by curving. It should be emphasised that equivalent cant is an input to the suspension relationship given in the previous section. Although the relationship between speed and equivalent cant is assumed to be linear, the resulting suspension movement is unlikely to be so.

In addition, the vehicle responds to vertical track irregularities. Accordingly, in BASS 501, upward and downward movements of the vehicle calculated on the remaining suspension travel for given load cases are defined. This is known as 'dynamic drop', although the term 'dynamic lift' is appropriate for upward movements. Both cases need to be considered simultaneously, since these define the 'bounce' of the vehicle. A number of techniques have been used to limit these according to the true amount of suspension travel available once roll drop is considered (by relating it to equivalent cant) and to linearise the value with speed.

In MBSs, the dynamic movement is calculated by considering the reaction of the spring/damper system to both the speed of the vehicle and the track irregularities. Such movements are presented statistically as a standard deviation of movements in each plane. Typically, mean movement (generated by curving forces) is supplemented by 2.12 standard deviations of dynamically generated movement. Note that the distribution of movement is not generally normal (Gaussian), as it is likely to contain movements beyond the normally expected distribution tails.

It is worth noting that the use of mean + 2.12 standard deviations of movement is merely an empirical rule to determine a credible maximum movement that may be used in the gauging calculations, without introducing extreme conservatism. In practice, a vehicle will move more than this, but not frequently, and simultaneously with all other conservative assumptions in the calculation process. This is explored later in this chapter, under risk considerations.

10.2.3.2.3 Critical Speeds

Speed must also be considered in relation to the maximum movements of a vehicle that are generated in service. A vehicle will usually be designed to run at a maximum 'line speed'. This is generally limited by cant deficiency, based upon passenger comfort. The faster a conventional vehicle travels around curves, the more it will sway towards the outside of the curve, limited only by suspension travel. However, we must consider the possibility that the vehicle may travel at reduced speed or may even be stationary. The maximum static force on a vehicle to the inside of a curve occurs when stationary, due to an excess of cant. It is frequently assumed that this is the worst case for inside curve clearances.

In practice, a vehicle moving slowly on high-installed cant will oscillate around the mean position, defined by the cant excess as it travels, and will thus sway more to the inside of a curve when moving slowly than when it is stationary. As speed increases, the mean position moves towards the outside of the curve because of the increase in curving force due to that speed. However, the oscillations will not necessarily increase at the same rate, and there will exist a speed at which the peak oscillations to the inside occur. This is known as an inside critical speed. There may be a similar outside critical speed if a vehicle's suspension is non-linear.

In quasi-static analysis, the linearity assumptions related to K_{speed} allow this inside critical speed to be derived from K_{speed} and curve radius and is known as 'trundle speed':

$$V_{trundle} = K_{speed} R / 23.64$$

Note that tangent track has infinite radius, and hence, vehicles running on this exhibit maximum movement to the inside at line speed.

In MBSs, the non-linearities of the system require analysis of movements at all speeds in order to determine maximum movements. Advanced software, such as PhX rail™, considers the critical speeds in all directions for every point on a vehicle body, which also enables the longitudinal components of movement to be considered.

10.2.3.2.4 Effect of Loading

Different relationships occur for different suspension loading and failure conditions. In particular, Figure 10.7 shows relationships defined as inflated and deflated, referring to airbag condition. The possibility of airbag failure (or accidental isolation) is usually considered in analysing clearances. As can be seen from the graph, deflation of airbags results in a stiffer vehicle, but one that may have a 'locked-in' suspension lateral movement (see also time-related effects discussed later). At lower cants, and particularly on lower parts of the vehicle body, this lateral offset may be the worst defining movement of the vehicle. Note also that there is likely to be some hysteresis, since this locked-in movement requires a force in the opposite direction for it to break out.

Obviously, the suspension performance relates to many factors, the principal ones being the passenger (or freight) load and the condition of the springs. The normal conditions analysed are as follows:

- *Tare or tare inflated*: The condition where the vehicle is likely to be tallest, since the springs are least compressed
- *Laden or crush inflated*: This condition results in the greatest sways, due to the mass of passengers or load and higher centre of gravity.
- *Crush deflated*: This condition has a larger locked-in lateral component but smaller sways, leading to a greater risk to low structures (such as platforms). It is also the lowest position in which a vehicle may operate.
- *Tare deflated*: This is an unusual condition in that it is only likely to occur on delivery or depot routes, where there are no passengers. Its principal use is in clearing low structures for delivery where a tare inflated vehicle may not pass.

On coil sprung vehicles, failure of springs is considered unlikely, and hence, these conditions are not generally analysed.

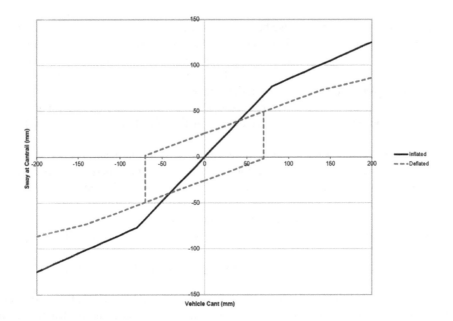

FIGURE 10.7 Relationship between vehicle cant and sway for vehicles with air suspension.

10.2.3.2.5 Time Factors

Time has not generally been considered in terms of gauging, but increasingly, the effects are becoming significant in the context of accurate analysis. Two particular issues should be considered:

- *Air suspension*: These systems generally have self-levelling valves that compensate for vehicle asymmetry caused by loading and curving forces, tending to compensate for roll on curves. The time constants for such systems are generally long and thus unlikely to have a noticeable effect on normal, at-speed analyses. However, where vehicles are moving very slowly or are stationary, the compensation effect may need to be considered.
- *Locked-in suspension movement*: Where air suspensions are run in deflated failure mode, hysteresis, as shown on the earlier suspension characteristics graph, is gradually 'shaken out' by normal track oscillations over a short period of time. Such effects can only be considered by advanced dynamic prediction methods, as described later.

10.2.3.2.6 Vehicle Height

A vehicle has a nominal, static height. This would normally be in the tare condition, with airbags inflated if the vehicle has them. When loaded (passengers or freight), the suspension is compressed, depending on the loading, resulting in a lowered static height for this condition. Loading is defined by strict rules, and there may be different operating conditions associated with different loadings. The static height is also reduced in the case of airbag deactivation or failure. Occasionally, over-inflation of airbags is considered.

10.2.3.3 Vehicle Tolerances

Commonly considered tolerances are as follows:

- *Uncompensated wheel wear*: This is the amount of wheel wear that can develop before being compensated by shimming of the suspension. A worn vehicle will be lower than a new vehicle, and thus, this parameter must be considered when analysing lower-body clearances (platforms, etc.).
- *Suspension creep*: With age, rubber suspension components compress (creep) and thus lower the body. This parameter must also be considered when assessing lower-body clearances.
- *Body build tolerance (BOD)*: This parameter represents tolerances in the building processes and must be added to the static shape of the vehicle. Construction methods sometimes produce smaller tolerances at the vehicle solebar, and thus, it is possible to have a varying BOD for different positions on the vehicle.
- *Height-setting tolerance*: This value affects both the maximum and minimum height of the vehicle and depends on the accuracy with which the static height of the vehicle air suspension may be set.
- *Air bag compensation*: A self-levelling system on air suspensions means that, over a period of time, the suspension gradually corrects for load imbalance. This is particularly noticeable on canted platforms, either as passengers embark or disembark or as the cant excess is gradually compensated. Being a relatively slow process, this parameter, on its own, is unlikely to be an issue. Where a vehicle stops on a curve, the effect is to reduce sway. However, there may be a tolerance of operation of the self-levelling valves, and these are occasionally considered.

In BASS 501, or simple MBS analyses, it is necessary to consider the following:

- *Vehicle yaw*: This is a lateral movement of the end of the vehicle in relation to the centre. It is not strictly a tolerance but is occasionally included as one. The further the sections are from the centre of the vehicle, the more progressively the vehicle yaw affects them.

Usually, this is only considered if their effect is of significance in relation to the clearance regime under which the vehicle is operating.
- *Vehicle pitch*: This is the vertical equivalent of yaw, and the same considerations apply.

In defining dynamic movements, it is usual to refer to specific reference points of significance on a vehicle. Typically, these would be as follows:

- *Cantrail*: A notional line drawn along the vehicle, which for passenger stock represents the upper limit of the body side and the start of the roof contour. This height represents a combination of semi-width and sways likely to present the greatest risk of infringement to the arch bridges prevalent in the British infrastructure.
- *Waist*: The widest part of the vehicle (statically), which is likely to present the greatest risk of infringement to passing vehicles (with the possible exception of tilting trains).
- *Step*: Traditionally, the part of the vehicle designed to come into close proximity to the infrastructure (platforms).

Using BASS 501, the sways and drops of cantrail, waist, step and, occasionally, yaw damper may be calculated. In MBSs, movements may be defined as lateral, vertical and roll movements about a notional datum or (more accurately) at two datum points.

10.2.3.4 Tilting Trains

The basic relationships described are valid for most normal vehicles. However, tilting trains have a more complex relationship. In this case, not only do the inputs from curving forces need to be considered, but also the effect of the active suspension (which principally operates in roll) should be taken into account. The relationship between tilt angle, cant deficiency and speed varies between trains and is non-linear. In general, tilting trains tend to behave conventionally below a cut-in speed known as the tilt threshold speed or on cant excess. Vehicles with tilt locked out or in tilt failure behave as conventional vehicles, although the positional error of the body in the latter case will also need to be considered.

Figure 10.8 shows an idealised relationship for a tilting train suspension, showing the effects of speed and cant on the sway at the cantrail. The horizontal axis shows cant deficiency as positive and

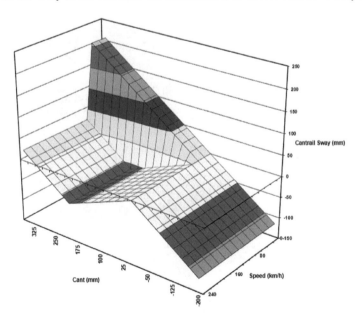

FIGURE 10.8 Relationship between speed, cant and sway for a tilting train at the cantrail.

cant excess as negative. A positive sway on the vertical axis is towards the outside of the curve. It can be seen that at low speeds (less than 50 km/h on this particular vehicle) and at cant excess (negative cant), a similar relationship to that of a conventional train can be seen (i.e., a simple linear relationship, where increased cant deficiency results in a greater sway to the outside of the curve). In the tilt active area, it can be seen that the tilt system causes the vehicle to lean inwards, where a normal vehicle would lean outwards (i.e., under cant deficiency). As the tilt movement is used up, the vehicle again begins to move outwards at high cant deficiencies. If the point being considered is above the tilt centre (normally the case with the cantrail), then the characteristic inward tilt can be seen.

If the point is below the tilt centre (usually the case with a vehicle footstep), then tilt angle is additive to the roll caused by cant deficiency. This is closer to the performance of a conventional train, but with additional roll. Vehicles whose tilt centre is high (i.e., those where the bodies are suspended from a high-level suspension) behave like conventional trains, but with significantly greater sways in the lower sector. These vehicles sway more at the footstep than at the cantrail (see Figure 10.9).

For tilting trains, the relative angles of primary and secondary suspensions must also be considered, since the lateral force of curving at high tilt angles may cause a significant compression of the secondary suspension, which is now not operating truly vertically. This component is known as 'compression drop'.

The number of operational cases of tilting trains is higher than the four conventional cases, since the possibility of failure of the tilt system or other parts of the active suspension must be considered. Also, an effect known as 'tilt lag' means that, as the train travels onto and off transition curves, the active suspension is slightly delayed in its response for electromechanical reasons. For some trains, this can be as much as six degrees. There is thus an entire range of situations that may occur individually or concurrently, including the following:

- Tare or laden condition (various loading factors)
- Inflated or deflated air suspension
- Active, active lagged, passive (locked) or failed tilt system
- Active or failed active suspension components

This may result in many potential situations requiring analysis.

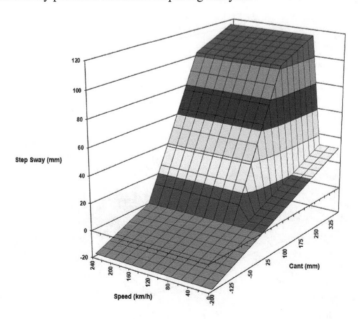

FIGURE 10.9 Relationship between speed, cant and sway for a tilting train at the footstep.

10.2.3.4.1 Effects of Speed

In the analysis of tilting trains, the speed of the vehicle has been included into the basic relationship between applied force and suspension movement. However, this results in more complicated analyses than those of conventional vehicles, where sway is directly related to speed.

10.2.3.4.2 Critical Speeds

The maximum dynamic envelope of a tilting train cannot be determined using the simple concepts of maximum and trundle speeds, as used for BASS 501 calculations, because:

- High-speed cases can lead to sways to the inside greater than low-speed cases.
- Points above the tilt centre behave differently to points below it, where more conventional rules apply.
- Worst sway does not necessarily mean worst drop, since this depends on the tilt system geometry.

Consider the graph of cantrail sway in relation to speed (Figure 10.8). If the vehicle is operating below the tilt threshold speed, then the worst sway to the outside occurs at maximum speed. If operating above tilt threshold speed, then the outward critical speed is dependent upon cant deficiency (also dependent on speed). At low cant deficiencies, the critical speed is likely to be the tilt threshold speed. But above the limit of tilt compensation, critical speed can again be the maximum speed. There are a number of intermediate combinations of speed and cant deficiency leading to critical speeds.

Crucial parts of the train need to be assessed over an entire speed spectrum to ensure that all combinations of sway and drop over the speed range are considered.

10.2.3.4.3 Time Factors

Tilting and active suspensions, by their nature, have delayed responses either by design or due to the time required to provide a measured response to inputs. The most common form of tilting suspensions measure cant deficiency and curvature on a leading bogie and calculate the required tilt demand from this, which is applied to the leading vehicle and subsequently to trailing vehicles. In order to avoid false responses to track irregularities and ensure that there is only a response to true curving forces, a delay period (normally no more than 1 second) is provided, during which no tilt is applied to the lead vehicle. This is progressively less pronounced on trailing vehicles, where the time lag is less. This effect is known as tilt precedence. A further effect is that the tilt system may not be able to respond at the same rate as transition curves develop. Figure 10.10 illustrates the tilt lag phenomenon.

In Figure 10.10, the horizontal axis shows the position of a train entering into a curve, which starts 100 m into the diagram. The solid line shows a linear cant transition for this curve, in degrees. On this particular curve, the maximum cant is six degrees (approximately 150 mm), and the transition is 100 m long. At 50 m/s, this represents a cant gradient of 75 mm/s, typical of a tilting train at its enhanced speed. The dashed line shows the response of the tilt system. The system does not respond for the first 50 m of the curve (1 s at 50 m/s) and then responds at a rate of two degrees per second. The dotted line shows the imbalance between tilt required and tilt achieved. A maximum tilt lag of four degrees develops in this particular scenario.

Tilt lag means that, in some cases, the use of conventional gauging models will not provide adequate clearance assessment. In these cases, it is necessary to use 'lead-lag' models, where the kinematic envelope of the vehicle is expanded to include this error. Tilt lag refers to the error that develops as a vehicle moves onto a transition, and tilt lead (technically a misnomer) refers to the opposite error that develops as a vehicle moves off a transition. In the latter case, it is important to note that the effect can occur on tangent track, where simple analyses would normally be performed.

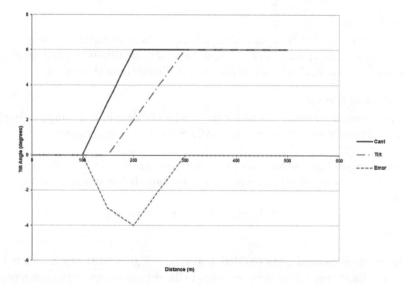

FIGURE 10.10 Tilt-lag relationship.

10.2.3.5 Comparison of BASS 501 and Multibody Simulation Methods

BASS 501 has a number of inherent simplifications:

- It assumes that all lateral movement of the vehicle is from roll generated by cant forces. Pure lateral irregularities are not considered dynamically.
- It assumes a linear relationship between speed and equivalent cant leading to sway. In practice, this relationship is non-linear as a result of harmonic responses of the spring/mass system,
- It assumes that all generated sway is 'upper sway', where the roll centre is low.
- The locked-in movements predicted by quasi-static analysis are rapidly shaken out by dynamic movement of the vehicle.

Using MBSs, it is possible to perform full dynamic simulations of the vehicle and thus consider true vehicle behaviour within the bounds of a simulation that is now considered to be extremely accurate. The process is as follows:

- A range of 'real' track data is assembled, according to the probable range of track roughness that will be experienced.
- The vehicle is run, in a variety of suspension conditions, over a full range of applied cants and speeds. It is not necessary to consider radius, since the cant deficiency or excess drives the behaviour.
- A series of lookup tables are produced, defining lateral, vertical and roll performance of the vehicle or suite of vehicles. Different configurations of the same vehicle can behave differently, and there can be a different behaviour when running in different directions.
- The curving behaviour is calculated (as explained in vehicle/track interaction).

The process of defining the relationships between track inputs and vehicle body dynamic behaviour is statistically based. Track inputs are provided from a variety of 'typical' track geometries appropriate to the speed of the vehicle. In general, lower-speed (and lower-quality) track provides greater dynamic inputs to the vehicle. The resultant body movements generated at each combination of applied cant and speed are summarised statistically as mean and standard deviation of lateral, vertical and roll movements, onto which a certainty limit is applied. The definition of appropriate track quality indices, combined with the maintenance regime of the railway, is an important factor in determining the level of risk in the gauging calculations.

Figure 10.11 shows the sway predicted at the cantrail by the different models of the same vehicle on the vertical axis. The horizontal axes show the inputs of cant and speed into the model.

The following observations can be made:

- Sways predicted by both methods of analysis are similar, indicating a generally good correlation between the techniques.
- The non-linearity associated with speed is clearly visible. In some cases, BASS 501 over-predicts and, in others, it under-predicts.
- Generally, BASS 501 over-predicts sways.
- The significant conservatism of BASS 501 at low cants results from the shakeout of hysteresis due to the movement of the vehicle that is not considered in the quasi-static analysis.

It should be noted that, whilst Figure 10.11 shows axes of speed and cant, these are not independent of each other, a factor that should be considered in understanding the effect of speed. On a given curve, increasing speed will have the effect of increasing cant deficiency, whilst decreasing speed will reduce this or generate cant excess. The dependency between cant and speed is a squared relationship, as described previously.

Such dynamic systems can provide vehicle movement information associated with particular track geometry, and in real time. However, real-time gauging is flawed in that it takes no account of the spectrum of track geometry that may develop as track deteriorates, or to which it is maintained. In particular, a track defect that causes a vehicle to sway away from a structure (providing clearance) could cause a gauging infringement if removed.

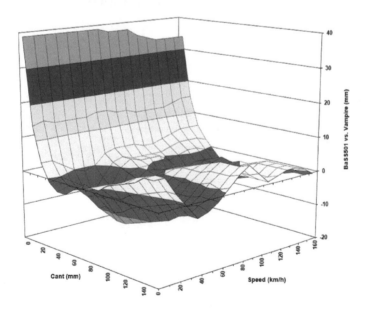

FIGURE 10.11 Comparison of cantrail sways for the same vehicle modelled by BASS 501 and multibody simulation.

10.3 THE GAUGING SYSTEM

10.3.1 Vehicle-Track Interaction

The primary interface between vehicle and track occurs at the wheel-rail interface. The wheelset has freedom to move within the rails, limited by flange contact. The size of the gap depends upon the gauge of the track, the wheel flange wear and the rail sidewear. The various gauging models handle this interface in various ways:

- In simple analyses, this interface is ignored, since its effect is small.
- In basic analyses, such as BASS 501, it is assumed that all possible combinations of wide gauge, wheel flange wear and rail sidewear occur simultaneously. Typical values would assume an 8-mm flange-rail gap, 3 mm of wheel flange wear and 6 mm of rail sidewear. Owing to the curving nature of bogies, this latter value is usually reduced to 3 mm, since it would be very unlikely for all wheelsets to be running to either the outside or the inside rail (as demonstrated later). Thus, a global value of 14 mm is often used.
- In more sophisticated analyses, the nature of curving is considered and is used to correct the centreline position of the vehicle for the curving behaviour of the bogies or wheelsets. It has been found that, in cant-deficient situations (where there is insufficient installed cant to balance curving force), the bogies (and hence the vehicle) move towards the outside of the curve. Figures 10.12 and 10.13 show the approximate bogie behaviour as curve radius varies. On tangent (straight) track, the wheelsets assume a mean position running centrally between the rails, and there is generally no offset, although asymmetric running on straight track has been observed on some flexible-frame bogies. As radius progressively tightens, the angle of attack increases as the leading wheelset moves towards the outside rail. The trailing wheelset continues to follow a path more centrally between the rails. At approximately 500 m radius, the leading wheelset will come close to flange contact. This is the point of maximum outward bogie movement. As radius further decreases, the angle of attack increases further by the trailing wheelset moving towards the inside rail. An extremely sharp curve would cause the trailing wheelset to come into flange contact with the inner rail,

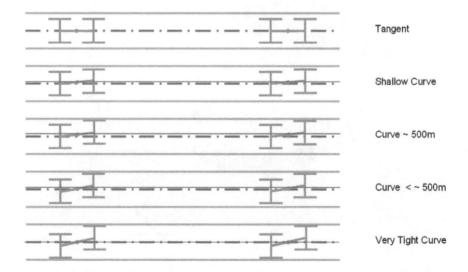

FIGURE 10.12 Approximate relationship between curve radius and wheelset movement.

Gauging Issues

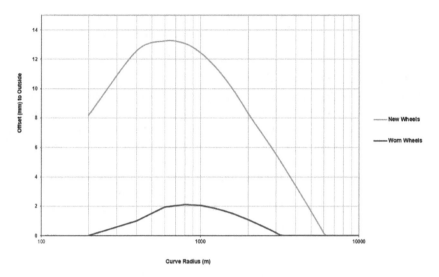

FIGURE 10.13 Approximate relationship between curve radius and bogie pivot offset.

giving a resultant zero offset. In practice, radii this severe will not be encountered. It must be emphasised that the exact relationship of offset to curve radius is complex and vehicle-specific. It requires complex modelling software, such as VAMPIRE™ to generate the exact relationship.

It is necessary to consider a spectrum of operating conditions to determine the maximum and minimum wheelset movements at different radii. This will include modelling various conditions of worn wheel profile. In the gauging analysis, the maximum outward wheelset movement (with new wheels) will be applied to cant-deficient cases, and the minimum movement (with worn wheels) will be applied to cant-excess cases.

The UIC rules consider the behaviour of wheel-rail interaction and, in particular, gauge widening and bogie alignment as part of the structure-vehicle relationship.

10.3.2 Track-Structure Interaction

10.3.2.1 Track Tolerances

The relationship between track position and structures is a significant factor to consider in gauging. The following components of track tolerance must be considered.

10.3.2.1.1 Lateral Track Positional Tolerance

Sometimes known as the 'track alignment error', this relates to the possible movement of the track over its maintenance cycle. Normal ballasted track is generally maintained by tamping, with lateral slues being applied to correct geometric errors. In normal circumstances, track is maintained within a tolerance of ±25 mm. Datum plates are used to provide guidance to machine operators on track position, and overhead contact wire registration (if present) normally requires the track to be maintained to this tolerance. Where the track is non-electrified and datum plates are not present, care must be taken to ensure (by more frequent measurement) that track remains within positional tolerance. Normal ballasted track is known as low fixity. High-fixity track, such as slabs, may be held to much tighter tolerances (even zero). Ballast gluing and strutting tracks against platforms are considered of medium fixity, with a nominal lateral tolerance of ±15 mm.

It should be noted that, whilst track is likely to move within the tolerances, it is usual to consider track tolerances as an allowance, which may be used to maintain the track within an allowable position regime.

10.3.2.1.2 Vertical Track Positional Tolerance

Track level deteriorates under the effect of traffic and time. In general, the settlement of ballasted track is logarithmic in nature. Rapid settlement occurring immediately after maintenance becomes more linear as quality deteriorates towards the end of a maintenance cycle. Over a maintenance cycle, of 1 to 2 years, depending on track condition and quality, it would be expected that track would settle around 25 mm from its highest level. Unfortunately, it is difficult to apply this tolerance without knowing the position within the maintenance cycle. Just maintained track will settle by up to 25 mm, but track that is just about to be maintained could be lifted by 25 mm. By assuming track to be at a position that could be lifted 15 mm and lowered by 10 mm provides a regime that statistically covers a large part of the maintenance cycle.

10.3.2.1.3 Cross-Level Error

On low-fixity track, it is generally assumed that absolute cross level may vary by ±20 mm in relation to that required, as a result of differential settlement or measuring errors. It is considered that half of this value (±10 mm) would be long-wavelength (i.e., a drift) and half (±10 mm) would be short wavelength (i.e., part of the track roughness). The long-wavelength (static) component affects steady-state curving forces and vehicle position. The short-wavelength component affects dynamic performance of the vehicle. In applying this error, it is usual to consider the long-wavelength component in relation to track fixity. High fixity (slab track) may be laid to such precision that there is no long wavelength error, and a zero value may be used. However, it is unusual to reduce the short-wavelength cross-level error significantly below ±10 mm. This latter value is usually included in the vehicle model (although it is a track parameter) and is implicitly related to K_{speed} in BASS 501 calculations and to the track geometry files in dynamic simulations.

10.3.2.1.4 Sidewear

On tight curves, rail sidewear tends to occur. Its formation can be slowed by lubrication, and it is generally a high-rail problem. However, it serves to widen gauge and affects the vehicle positioning on the rails. The amount of sidewear included in analyses depends on whether it can develop (unusual on straight track) and what the maintenance intervention level is. A normal sidewear limit is 6 to 9 mm. However, as discussed earlier, the amount of movement that this can generate in the vehicle is generally less.

10.3.3 STRUCTURE-VEHICLE INTERACTION

10.3.3.1 Clearances

Clearance is required for a variety of reasons. Historically, clearance provided the safe boundary between vehicle and structure where there were significant unknowns in each, which has included suspension movements, tolerances and inaccuracies in the measurement of structures. Clearance provides space to allow for aerodynamic effects and for safe walkways.

As vehicle behaviour and system tolerances are better understood, a differentiation between what is calculable and what remains unknown is possible. Unfortunately, this has not always led to a relaxation of clearances, as tolerances are extracted, which increasingly leads to conservatism and smaller trains. Modern trains, with air suspension and about which the behaviour is well understood, tend to be smaller inside than their predecessors whilst occupying what appears to be a larger swept envelope.

Pressures on the infrastructure, especially in the face of an increasing need to move larger intermodal freight containers (notably 2900 mm high × 2500–2600 mm wide ISO boxes), require

Gauging Issues

clearances to be specified frugally if rail is to survive in the increasingly competitive environment offered by road transport where larger paths routinely exist.

Clearance is about risk management. The larger the clearance provided, the smaller the risk and thus the need for control measures is minimised. Modern standards specify clearance according to risk regime, where the available clearance dictates what control measures are required [9]. Typically, actual clearances greater than 100 mm are defined as normal, whereas below this, reduced and special reduced clearances (the latter being clearances >0 mm) require increasingly rigorous control measures. Control measures involve processes to control track position such as slab track and glued ballast. The regime of inspection is also important, ensuring that tight structures are inspected more regularly than those that are well clear of the track.

The UIC rules require a reference profile to be enlarged (infrastructure) or reduced (vehicle) for various effects. Clearance between the developed reference profiles is not specifically mandated.

10.3.3.2 Stepping Distances

Whilst not strictly a clearance issue, stepping distances are an integral component of gauging analyses and are generally the most difficult to resolve. To provide an adequate clearance to a moving vehicle, whilst still providing safe passenger access and egress to stationary vehicles, involves considering opposite, worst-case scenarios.

Clearance analysis involves calculating the tightest reasonable clearance that may develop. Stepping calculation involves determining the maximum distance between a platform edge and vehicle step that may develop. In the latter case, it is customary to consider the static (thrown) position of a vehicle in relation to the platform edge. In Britain, the Health and Safety Executive's Railway Inspectorate [2] requires maximum stepping distances of:

- Lateral 275 mm
- Vertical 230 mm
- Diagonal 350 mm

This is known as the 'stepping triangle', although it does not conform to Pythagoras' rule.

The values required are theoretical, taking no account of many tolerances that affect the actual stepping distance. In particular, air suspension system performance (i.e., with self-levelling valves), installed cant and track tolerances can have a significant effect on the stepping distances measured in relation to those calculated. However, the Her Majesty's Railway Inspectorate Guidance values do provide a sensible benchmark for static values.

Improvement of stepping distances is likely to be a characteristic of increasing regulation. Increasingly, dynamic methods of analysing the platform-train interface (PTI) are being developed.

10.3.4 STANDARDS

In Britain, the legal framework behind gauging is covered by the European Technical Standards for Interoperability. European Standard BS-EN15273 [7] covers the gauging methodologies across Europe, which include those used in Britain. At the time of writing, BS-EN15273 requires supplementing with British Railway Group Standards, which presently provide the most comprehensive interface standards but lack the descriptive methodologies that the Euronorm incorporates. Railway Group Standards (RGS) are published by the Rail Safety and Standards Board (RSSB), and the following are appropriate for gauging:

- GM/RT2173 Issue 1 (Dec 2015): Requirements for the Size of Vehicles and Position of Equipment. This standard defines the requirements for those building, maintaining or operating trains.

- GI/RT7073 Issue 1 (Dec 2015) [8]: Requirements for the Position of Infrastructure and for Defining and Maintaining Clearances. This standard defines the requirements for constructing and maintaining the railway infrastructure.
- GI/RT7016 Issue 5 (Mar 2014): Interface between Station Platforms, Track and Trains. This standard covers the stepping gap between trains and platforms.
- GE/RT8073 Issue 3 (Dec 2015): Requirements for the Application of Standard Vehicle Gauges. This standard defines the standard vehicle gauges used on the British railway network and the rules for their application.
- RIS-2773-RST Issue 1 (Mar 2013): Format for Vehicle Gauging Data. This railway industry standard (RISs are presently voluntary, and Railway Group Standards (RGSs) are mandatory) defines a standard format by which gauging characteristics of rollingstock should be presented, such that common descriptors are used. This is particularly important when computerised methods of analysis are undertaken. The format prescribed is suitable for describing common railway vehicles but cannot fully describe some of the more advanced technology presently in use. It should be emphasised that this represents a standard for the transfer of information, not the computer models themselves, which is a popular misconception, as human intervention is required to ensure that specific behaviour, as defined in notes, can be correctly interpreted.

In addition to standards, RSSB also publishes guidance and research. Of particular note are the following:

- GE/GN8573 Issue 4 (Dec 2015): Guidance on Gauging and Platform Stepping Distances. This guidance note provides background information and sample calculation methods associated with the various standards.
- The Vehicle/Structures Systems Interface Committee (V/S SIC) Guide to British Gauging Practice was produced by the author under RSSB Research Project T926 and provides a simple guide to the subject of gauging for intended practitioners.
- The Archive of Documents Relating to Gauging Issue 2 (July 2009) was produced by the author as a repository of information relating to the pre-privatised railway, from which much of the current practice has been drawn.

Further standards are published by the National Infrastructure Operator (Network Rail), which relate solely to the responsibilities of the infrastructure operator. One notable standard is

NR/L2/TRK/3203 Issue 1 (Sept 2011): Structure Gauge Recording, which defines the requirements to be used when measuring and recording structure measurement data.

10.3.5 Advanced Analysis

Computerised systems, such as ClearRoute™, began to replace traditional methods of structure gauging, which used drawings, in the early 1990s. Techniques for electronic measurement of structures and emerging analysis methods turned gauging from a necessary safety check into a process whereby the economic operations of railways could be improved by translating wasted infrastructure space into a usable commodity.

However, as a result of the requirement to provide safety and recognising that analysis at this time could only be rudimentary, due to the lack of detailed understanding of the gauging system, allowances were required as a proxy for those parameters in the system that could not be defined or were, indeed, unknown.

As a result of this, a number of issues emerged:

- Rail vehicles known to safely operate (under 'grandfather rights') could not be demonstrated to be running to mandated clearances. Indeed, in some situations, simulations demonstrated that such vehicles would not physically fit the infrastructure through which they

were actually running. A direct result of this was that new trains became smaller than the trains they replaced in order to reduce acceptance risk.
- Clearances became misunderstood and were interpreted literally, rather than being correctly considered to be a value under benchmark conditions. This has led to unnecessary infrastructure works, where, for example, calculated clearances are 1 mm less than a prescribed 100-mm limit.

Accordingly, a new generation of analysis tools has emerged, typified by PhX™ rail, where a systems approach has been adopted to incorporate mechanistic models of all gauging system components (foundation, track, rails, wheelsets, bogies, bodies, pantographs and the infrastructure) and to allow each to be developed, as knowledge emerges. In the last 25 years, development of understanding, combined with a revolution in computer technology, means that, using modern systems, the ability to model accurately in relation to real life and to make competent decisions based upon those results has increased dramatically.

10.3.6 Digital Systems

Whilst it could be argued that computerised gauging technology was founded on digital computers, to do so would belittle the effect that the digital revolution is having on the world and on engineering in particular.

Within the analytical engines associated with modern gauging software, many of the processes used in the digital broadcasting industry are used to great effect in filtering and transforming the huge quantities of data contained in, for example, point cloud infrastructure, and in using high-definition vehicle models to remove the simplifications of early vehicle profiles, which falsely intrude beyond a true swept envelope.

Perhaps, the biggest development is the use of Cloud-based systems, typified by RouteSpace®,* where the computing and data hosting elements of previously PC-based software systems are hosted in a virtual world. Apart from the obvious advantages of not requiring high-specification personal computers with installed software (with implicit problems associated with corporate IT regulation), the use of common data allows intelligence sharing and a dramatic reduction in duplication of processing effort on project works.

A particular effect of the introduction of such technology has been the significantly wider use of gauging information in, for example, the cascade of rollingstock from one operation to another, where previously, the costs of such an investigation would have been prohibitive.

A likely consequence of increasing availability of gauging information at reduced cost will be a shift from inefficient (of space) gauging methods (seen as cheap), such as vehicle gauges, towards the more efficient absolute gauging methods.

10.3.7 Risk Analysis

Railways are risk averse, and many engineering systems have evolved to ensure that safety is never compromised. Gauging is a classic example, where the size of a swept envelope of a vehicle is determined to be 'with everything going wrong'. This is beyond safety and leads to an inefficient use of the infrastructure, smaller trains and unnecessary costs. It is not tolerated in other, more progressive engineering-led industries.

The process traditionally used is typified by aggregating all tolerances, allowances and movements in a constructive manner, without considering whether all could occur simultaneously.

---

* RouteSpace® is the registered trademark of DGauge Ltd.

Consider the case of two passing trains. Each will be considered with the following:

- Body movements towards each other at the point of passing
- Each track being at its extreme position of movement towards each other
- Each track having a maximum installed cant error towards each other
- All tolerances and allowances being aggregated to reduce the calculated passing clearance

In reality, such events are extremely unlikely, although, technically, possible. Recognition is given to this in empirical ways:

- Maximum vehicle movements are considered to be mean + 2.12 standard deviations, rather than the peak movements of around mean + 4.8 standard deviations that have been measured in practice. Empirically, 2.12 standard deviations have been determined as an RMS value generated from a sine wave.
- Measurement accuracy is usually taken as mean + 2 standard deviations, as quoted by suppliers.
- Track tolerances may 'discounted' by 25% to reflect the unlikelihood of both tracks moving towards each other.
- Required clearances to platforms are reduced by 50% where deflated airbags are involved, to reflect the unlikeliness of such an event.
- Required clearances between failed tilting trains are reduced from 100 to 1 mm to reflect the unlikeliness of occurrence of such an event.

In short, the industry recognises risk but does not apply it in a mathematically competent manner.

In recent times (since the end of the last century), recognition has been given to the use of uncertainty theory in the calculation of clearances. Using such theory, every component of the gauging system is assigned a bias and an uncertainty, the latter being determined according to the statistical probability of deviation from the bias. The resultant calculations consider aggregation of parameters in a statistically correct manner (i.e., they are not simply added together) and result in a level of probability (and hence risk) of a particular clearance arising over the system lifetime.

Consider a structure whose clearance to a vehicle, using conventional calculations, is 100 mm. Figure 10.14 shows an illustrative probabilistic assessment of the *likely* clearance. This has been

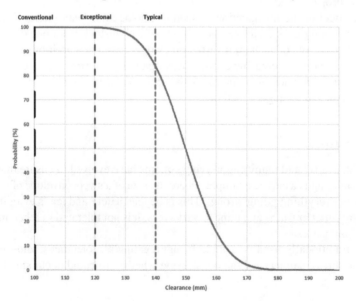

FIGURE 10.14 Clearance probability distribution.

undertaken using a 'Monte-Carlo' method, performing 10,000 calculations, each using a different combination of tolerances (according to each tolerance's individual probability distribution). In agreement with the central limit theorem, despite each tolerance having an individual probability distribution, given sufficient simulations, the resultant combined probability distribution will be Gaussian (normal). Accordingly, any clearance below 120 mm (mean − 3 standard deviations) would be extremely unlikely, or exceptional, and clearances would typically not be below 140 mm (mean − 1 standard deviation).

Whilst work to determine realistic probability thresholds is ongoing, the use of mean − 3 standard deviation values (c. 99.7%) in defining a normal/exceptional clearance limit boundary would release significant amounts of infrastructure space in many circumstances, without introducing a threat to safety.

Systems such as Probabilistic Gauging®* are presently available to implement such strategies. However, it should be noted that most situations do not require the sophistication of such analysis. The advantages of uncertainty analysis are as follows:

- It can explain how some trains running, presently not deemed clear, are able to run and the relative level of risk incurred.
- It can be used to quantify 'engineering judgment' in circumstances where statutory normal clearances cannot be achieved, but where, for a minor and manageable increased risk, expensive correction/monitoring regimes can be avoided.
- It can reduce the costs associated with the introduction of new rollingstock where substandard clearances exist with current rollingstock, yet standard clearances may be mandated for new rollingstock of near-identical size.

It should be emphasised that uncertainty methods are a subset of the absolute gauging methods described earlier, but rather than using pre-determined levels of probability for components, they use a mathematically robust method of applying them to the gauging system.

REFERENCES

1. *Strategic Rail Authority*, Gauging policy, London, UK, 2005.
2. Health and Safety Executive, Railway safety principles and guidance, *Her Majesty's Railway Inspectorate*, HSE Books, UK, 1996.
3. European Passenger Services Ltd, General gauging procedure for Class 373 trains—Mechanical Systems Engineer Report 42, British Rail, London, UK, 1995.
4. UIC Code 505-1, Railway transport stock—rolling stock construction gauge, 10th edition, UIC, Paris, France, 2006.
5. UIC Code 505-4, Effects of the application of the kinematic gauges defined in the 505 series of leaflets on the positioning of structures in relation to the tracks and of the tracks in relation to each other, 4th edition, UIC, Paris, France, 2007.
6. UIC Code 505-5, History, justification and commentaries on the elaboration and development of UIC leaflets of the series 505 and 506 on gauges, 3rd edition, UIC, Paris, France, 2010.
7. Euronorm BS EN 15273, *Railway Applications—Gauges*, British Standards Institution, London, UK, 2017.
8. Bogie and Suspension Section, Kinematic envelope and curve overthrow calculations (Design Guide BaSS501), Department of Mechanical and Electrical Engineering, British Railways Board, Derby, UK, 1985.
9. Railway Group Standard GI/RT7073 Issue 1, Requirements for the position of infrastructure and for defining and maintaining clearances, Rail Safety and Standards Board, London, UK, 2015.

---

* Probabilistic Gauging® is the registered trademark of DGauge Ltd.

11 Railway Vehicle Derailment and Prevention

Nicholas Wilson, Huimin Wu, Adam Klopp and Alexander Keylin

CONTENTS

11.1	Introduction	374
11.2	History and Statistics	374
11.3	Railway Vehicle Derailment Mechanisms and Safety Criteria	376
	11.3.1 Flange Climb Derailment	376
	11.3.1.1 Wheel Climb Process	376
	11.3.1.2 Flange Climb Safety Criteria	378
	11.3.2 Application of Flange Climb Derailment Criteria	386
	11.3.2.1 Flange Climb due to Low Flange Angle	386
	11.3.2.2 Increase of Flange Length Can Increase Flange Climb Distance Limit	386
	11.3.2.3 Flange Climb due to High Coefficient of Friction at Wheel-Rail Interface	387
	11.3.2.4 Flange Climb of Independently Rotating Wheels	388
	11.3.2.5 Low-Speed Flange Climb Derailment	389
	11.3.2.6 Vertical Wheel Unloading Criteria	389
	11.3.3 Derailments due to Gauge Widening and Rail Rollover	390
	11.3.3.1 The AAR Chapter 11 Rail Roll Criterion	391
	11.3.3.2 The Gauge-Widening Criterion	392
	11.3.3.3 Effect of Hollow-Worn Wheels on Gauge Widening and Rail Roll Derailment	393
	11.3.4 Derailment due to Track Panel Shift	395
	11.3.4.1 Causes of Track Panel Shift	395
	11.3.4.2 Track Panel Shift Criterion	397
11.4	Causes of Railway Vehicle Derailments	398
	11.4.1 Derailment Caused by Vehicle Lateral Instability	398
	11.4.2 Derailment Caused by Vehicle Body Resonance	399
	11.4.3 Derailment Caused by Longitudinal In-Train Forces	400
	11.4.4 Derailment Caused by Vehicle Overspeed	401
11.5	Assessment and Prediction of Derailment	401
	11.5.1 Assessment of Wheel-Rail Contact Parameters	401
	11.5.2 Dynamic Simulation of Vehicle-Track Interaction	402
	11.5.3 Track Tests	402
11.6	Prevention of Derailment	403
	11.6.1 Wheel-Rail Profile Optimisation	404
	11.6.1.1 Addressing Wheel Flange Angle	404
	11.6.1.2 Removing Hollow-Worn Wheels	404
	11.6.1.3 Wheel Taper and Effective Conicity	404
	11.6.1.4 Contact Conditions at the Wheel-Rail Interface	405

11.6.2	Independently Rotating Wheels	405
11.6.3	Installation of Guardrails or Restraining Rails on Sharp Curves	406
11.6.4	Vehicle Dynamic Response and Optimising the Vehicle Suspension	406
	11.6.4.1 Bogie Lateral and Yaw Stiffness	406
	11.6.4.2 Vertical and Roll Stiffness and Wheel Load Equalisation	406
	11.6.4.3 Bogie Warp and Rotational Resistance	407
11.6.5	Vehicle Loading	407
11.6.6	Train Marshalling and Handling	408
11.6.7	Lubrication and Wheel-Rail Friction Modification	408
11.6.8	Track Geometry Inspection and Maintenance	408
11.6.9	Switches and Crossings	409
11.6.10	System Monitoring	409
References		409

11.1 INTRODUCTION

Railway vehicle derailment can cause significant casualties and property loss. Avoidance of derailment is vital to railways for both safety and economic reasons.

Railway vehicle derailments are the result of wheels running off the rails, which are intended to provide the support and guidance. The reason for wheels running off rails can be very complicated; however, the final scenario of derailment can result in wheels climbing off the rail, rail gauge widening or rail rollover that causes wheels to fall between rails. Therefore, any conditions that may reduce the lateral guidance provided by the rails can increase the risk of derailment.

Note that the derailments discussed in this chapter relate only to the cause of losing lateral constraint at the wheel and rail interface. Derailments due to other causes, such as vehicle component failure, are not covered in this chapter.

11.2 HISTORY AND STATISTICS

Derailment has always been one of the major concerns for railway operations since the first day of wheels running on rails. The essential feature of wheels running on rails creates a unique challenge for railways to ensure that wheels stay on the rail. The high speed and heavy-haul operations developed in the past century demand a stricter control of vehicle lateral guidance. Railway technologies have advanced significantly in recent years, and safety levels are high compared with the early days or compared with other modes of transport. Unfortunately, derailments still occur. An internet search results in numerous examples of derailment incidents worldwide, caused by a variety of factors. Among those examples is a report that describes a derailment on 10 November 1881, at Carnforth, United Kingdom [1]. This derailment resulted in four passengers being injured, three carriages damaged and some track damage. It was concluded that the event was caused by a signalman having shifted the facing points before the entire train had passed over them.

Some of the most catastrophic derailments in recent history include the 2013 derailment of a runaway crude oil train in Lac-Mégantic, Canada (47 fatalities) [2]; the 2013 derailment of a high-speed train near Santiago de Compostela, Spain, caused by exceedance of train speed in a curve (80 fatalities) [3]; and the 2014 derailment of a Metro train in Moscow, Russia, caused by an improperly installed turnout (24 fatalities) [4].

Railroads in the United States began reporting accidents to the federal government following the passage of the Report Act of 1910 [5], after which an accident/incident database was established. Derived from this database, Figure 11.1 illustrates the significant reduction in derailment rates on freight railways in the United States between 1977 and 2017 [6].

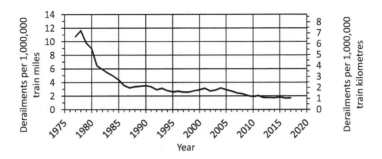

FIGURE 11.1 Derailment rates in the United States between 1977 and 2017. (From FRA Office of Safety Analysis, Train accidents and rates, Available at: https://safetydata.fra.dot.gov/OfficeofSafety/publicsite/query/TrainAccidentsFYCYWithRates.aspx.)

A recent study [7] ranked the top causes of freight train derailments in the United States between 2001 and 2010 (Table 11.1). On main lines, the most common causes by far were broken rails or welds, followed by track geometry defects, bearing failures, broken wheels, train handling and wide gauge. On sidings and in yards, turnout defects and their improper use were found to be additional common causes.

Statistics of freight train derailments [8] show significant differences between frequency and causes of derailment in continental Europe, Russia, the United Kingdom and the United States. For example, while rail failure is the top cause of derailments in the United States, they are less common in Europe than axle failures (the top cause of derailments in Europe), wheel failures, uneven loading, wide gauge and other track geometry defects. One of the likely reasons for this discrepancy is the higher axle loads and higher train tonnages in the United States compared with Europe.

TABLE 11.1
Top Causes of Freight Train Derailments in the United States between 2001 and 2010

	Main Line		Siding		Yard	
Rank	Cause Group	%	Cause Group	%	Cause Group	%
1	Broken rails or welds	15.3	Broken rails or welds	16.5	Broken rails or welds	16.4
2	Track geometry (excluding wide gauge)	7.3	Wide gauge	14.2	Use of switches	13.5
3	Bearing failure	5.9	Turnout defects – switches	9.7	Wide gauge	13.5
4	Broken wheels	5.2	Switching rules	7.7	Turnout defects – switches	11.1
5	Train handling (excluding brakes)	4.6	Track geometry (excluding wide gauge)	7.2	Train handling (excluding brakes)	6.7
6	Wide gauge	3.9	Use of switches	5.8	Switching rules	6.2
7	Obstructions	3.5	Train handling (excluding brakes)	3.5	Track geometry (excluding wide gauge)	3.6
8	Buckled track	3.4	Lading problems	2.3	Miscellaneous track and structure defects	3.4
9	Track-train interaction	3.4	Roadbed defects	2.1	Track-train interaction	3.1
10	Other axle or journal defects	3.3	Miscellaneous track and structure defects	2.1	Other miscellaneous	3.0

Source: Liu, X., Saat, R., and Barkan, C., *J. Transp. Res. Board*, 2289, 154–163, copyright © 2012 National Academy of Sciences U.S.A. Reprinted by Permission of Sage Publications, Inc.

Derailments due to component failure are self-explanatory, and their prevention has to do more with metallurgy and mechanical design than vehicle dynamics. However, many other causes such as track geometry defects and track-train interaction directly relate to poor wheel-rail interaction. This necessitates the study of derailment as a special topic in rail vehicle dynamics.

11.3 RAILWAY VEHICLE DERAILMENT MECHANISMS AND SAFETY CRITERIA

Railway derailments due to loss of the lateral guidance at the wheel-rail interface may be classified into four major mechanisms based on the ways that wheel-rail lateral constraints are lost: wheel flange climb, gauge widening, rail rollover and track panel shift.

A derailment review by Blader discussed the mechanisms of these types of derailment and some related test methods [9].

11.3.1 Flange Climb Derailment

Wheel flange climb derailments are caused by wheels climbing onto the top of the railhead, then further running off the field side of the rail. Wheel climb derailments generally occur in situations where the wheel experiences a high lateral force, combined with circumstances where the vertical force is reduced on the flanging wheel. The high lateral force is usually induced by a large wheelset angle of attack. The vertical force on the flanging wheel can be reduced significantly on bogies having poor vertical wheel load equalisation, such as when negotiating rough track and large track twist or when the vehicle is experiencing roll resonances. The forces between the wheel and the rail are explained in more detail in Chapter 7.

Flange climb derailments generally occur on curves. The wheels on the outer rail usually experience a base level of lateral force to vertical force ratio (L/V or Y/Q) that is mainly related to the following:

- Curve radius
- Wheel-rail profiles
- Bogie suspension characteristics
- Vehicle speed
- Track irregularities

These factors combine to generate a base wheelset angle of attack, which in turn generates the base level of lateral curving force.

A significantly misaligned bogie is likely to induce higher wheelset angle of attack. Furthermore, any track irregularities and dynamic discontinuities may lead to an additional increase of the wheel L/V force ratio. When this ratio exceeds the limit that the wheel can sustain, flange climb occurs.

Wheel climb derailments can also occur on tangent track when track irregularities and vehicle lateral dynamic motion are severe, such as during vehicle hunting (lateral instability) and aggressive braking.

11.3.1.1 Wheel Climb Process

The lateral velocity of a wheel due to its rotational velocity is given by:

$$V_t = -\omega r \sin(\psi) \qquad (11.1)$$

where V_t is the lateral velocity of a wheelset, ω is angular velocity, r is rolling radius and ψ is wheelset angle of attack. Figure 11.2 shows a plan view of a wheelset with a yaw angle relative to the track. This angle, which is known more commonly as the angle of attack, contributes to the lateral creepage through a component of the wheelset's rotational velocity.

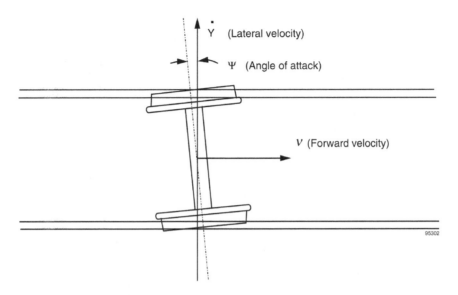

FIGURE 11.2 Wheelset angle of attack. (From Wu, H., Elkins, J., Investigation of wheel flange climb derailment criteria, Report R-931, Association of American Railroads, Washington, DC, 1999. With permission.)

If the wheelset has a lateral velocity in addition to the component of lateral velocity due to its rotation, the net lateral velocity of the wheelset at the contact zone, assuming the angle of attack to be small, ($\psi = \sin \psi$), is given by:

$$V_y = \dot{y} - \omega r \psi \qquad (11.2)$$

The lateral creepage is defined as the wheel-rail relative lateral velocity divided by the forward velocity and adjusted for the contact angle δ.

$$\gamma_y = \left(\psi - \frac{\dot{y}}{V} \right) \sec(\delta) \qquad (11.3)$$

The term ($\psi - \dot{y}/V$) is commonly known as the effective angle of attack and is a function of the wheelset lateral velocity. It is clear that, if the wheelset is moving towards flange contact with a positive angle of attack, the lateral velocity tends to reduce the effective angle of attack.

Since the term $\sec(\delta)$ always has a positive value during flange climb, the direction of the lateral creepage is dependent on the sign of the term ($\psi - \dot{y}/V$). The lateral creepage equals zero when ψ equals $\dot{y}/V$. The lateral creepage changes direction when $\psi < \dot{y}/V$. The spin creepage also affects the lateral creep force. The direction of the lateral creep force depends on the resultant of the contribution of both the lateral and spin creepages.

As Figure 11.3 shows, the process of the wheel flange climbing up the gauge face and onto the head of the rail may be illustrated in three phases. A single point of contact is assumed in this description.

In phase 1, under the influence of a lateral force, the wheel moves to the right towards flange contact. This produces a lateral creep force F_C acting on the wheel, which is opposing flange climb. In phase 2, as the flange contact angle increases, the wheelset lateral velocity decreases. As a result, the lateral creepage and creep force reverse direction due to the change in sign of the effective angle of attack. During this phase, the lateral creep force is assisting the wheel to climb. Figure 11.4 illustrates the forces acting between both wheels on the wheelset and the rails during phase 2 for a wheelset running

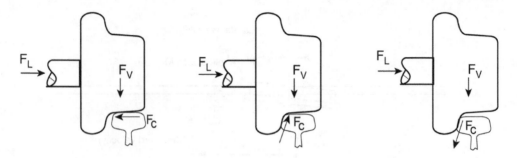

FIGURE 11.3 Process of derailment. (From Wu, H., Elkins, J., Investigation of wheel flange climb derailment criteria, Report R-931, Association of American Railroads, Washington, DC, 1999. With permission.)

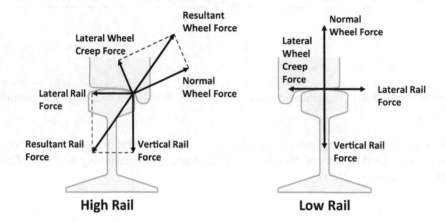

FIGURE 11.4 Forces acting between wheels and rails from a wheelset in flange contact with large angle of attack.

with a large angle of attack. Note that the low rail wheel generates a lateral force that acts to push the wheelset towards flange contact and that the reaction forces on both rails act to spread the track gauge.

After the maximum contact angle has been passed, the wheelset lateral velocity increases and the wheelset lateral displacement increases rapidly. As a result of the changing wheelset lateral velocity, the effective angle of attack approaches zero and then changes sign. Consequently, the lateral creepage and creep force also reverse direction and, once again, the lateral creep force opposes the climbing motion of the wheel, as shown in phase 3 (Figure 11.3).

11.3.1.2 Flange Climb Safety Criteria

Wheel flange climb derailment phenomena have been investigated for more than 100 years. Several flange climb safety criteria have been proposed. These criteria have been used by railway engineers (globally or locally) as guidelines for safety certification testing of railway vehicles. The following are examples of commonly used published criteria:

- The Nadal single-wheel L/V limit criterion
- The Weinstock axle-sum L/V limit criterion
- The high-speed passenger distance limit (5 ft, or 1.52 m), Code of Federal Regulations, United States
- The combined 50-millisecond wheel climb time limit and 3-feet (0.91 m) distance limit, Association of American Railroads (AAR), United States

Railway Vehicle Derailment and Prevention

- The L/V time duration criterion proposed by Japanese National Railways (JNR)
- The L/V time duration criterion proposed by Electromotive Division of General Motors (EMD)

The first two are related to the L/V ratio limits. The rest are related to the time or distance limits, which are applied to limit the exceeding duration of the L/V ratio limit, in either time or distance scale. The wheel climb would be very likely to occur if both the L/V ratio criterion and duration limit are exceeded. Brief descriptions are provided in below subsections for each criterion listed previously.

11.3.1.2.1 The Nadal Single-Wheel L/V Limit Criterion

The Nadal single-wheel L/V limit criterion, proposed by Nadal for the French railways [10], has been used throughout the railway community, including in the Euronorms for railway vehicles [11] and US Code of Federal Regulations [12]. Nadal established the original formulation for limiting the L/V ratio in order to minimise the risk of derailment. Nadal assumed that the wheel was initially in two-point contact, with the flange point leading the tread point. He concluded that the wheel material at the flange contact point was moving downwards relative to the rail material due to the wheel rolling about the tread contact. He further theorised that wheel climb occurs when the downward motion ceases with the friction saturated at the contact point. Based on Nadal's assumption and a simple equilibrium of the forces between a wheel and rail at the single point of flange contact, as illustrated in Figure 11.5, Equation 11.4 can be derived.

$$\begin{cases} F_3 = V\cos\delta + L\sin\delta = V\left(\cos\delta + \dfrac{L}{V}\sin\delta\right) \\ F_2 = V\sin\delta - L\cos\delta = V\left(\sin\delta - \dfrac{L}{V}\cos\delta\right) & \text{when } (V\sin\delta - L\cos\delta) < \mu F_3 \\ F_2 = \mu F_3 & \text{when } (V\sin\delta - L\cos\delta) \geq \mu F_3 \end{cases} \quad (11.4)$$

From Equation 11.4, the L/V ratio can be expressed as:

$$\frac{L}{V} = \frac{\tan\delta - \dfrac{F_2}{F_3}}{1 + \dfrac{F_2}{F_3}\tan\delta} \quad (11.5)$$

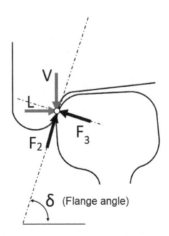

FIGURE 11.5 Forces at flange contact location. (From Iwnicki, S. (ed.), *Handbook of Railway Vehicle Dynamics*, CRC Press, Boca Raton, FL, 2006.)

Nadal's famous L/V ratio limiting criterion, given by Equation 11.5, was proposed for the saturated condition $F_2/F_3 = \mu$.

$$\frac{L}{V} = \frac{\tan\delta - \mu}{1 + \mu\tan\delta} \quad (11.6)$$

If the maximum contact angle is used, this equation gives the minimum wheel L/V ratio at which flange climb derailment may occur for the given contact angle and friction coefficient μ. In other words, below this L/V value, flange climb cannot occur.

Figure 11.6 plots Equation 11.6 for the coefficient of friction range between 0.1 and 1.0.

It indicates that the larger the maximum contact angle, the higher the L/V ratio limit required for flange climb. For the same contact angle, Figure 11.6 also indicates that the lower the friction coefficient, the higher the L/V ratio limit required for flange climb.

Nadal's formula relies on a simplified two-dimensional analysis of wheel-rail force and does not account for the wheelset's angle of attack [13]. At low angles of attack, lateral creep force at a flanging wheel is substantially reduced, such that the condition $F_2/F_3 = \mu$ is not true, and Nadal's formula can become overly conservative [14].

To demonstrate the effect of wheelset angle of attack (defined in Figure 11.2) on wheel L/V ratio limit, Figure 11.7 displays an example of a single-axle wheel climbing. These are the results from simulations using NUCARS®, Transportation Technology Center, Inc.'s (TTCI) rail vehicle dynamic simulation software and flange climb tests conducted using the AAR's track loading vehicle [15]. The wheels used have a flange angle of 75 degrees. In this example, wheel climb will not occur for an L/V ratio level below the solid line for a specified angle of attack. Figure 11.7 also indicates that, for large wheelset angles of attack (about 10 mrad in Figure 11.7), derailments occurred at Nadal's value. However, for smaller and negative angles of attack, the L/V ratio required for derailment increased considerably.

It should be noted that, with regard to Figure 11.5, 'maximum contact angle' and 'maximum wheel flange angle' are identical. However, when considering the wheel-rail contact as a three-dimensional problem, the maximum contact angle of the wheel and rail surfaces may be lower than the maximum flange angle by 0 to 0.3 degrees (1–1.5 degrees at very large angles of attack). This means that, in some cases, at large angles of attack, Nadal's limit calculated using maximum flange angle may be slightly nonconservative [13]. Therefore, when accurate wheel and rail contact geometry, wheelset angle of attack and contact point positions are known (usually during analysis and computer simulations), it is recommended to use contact angle rather than flange angle when calculating Nadal's limit.

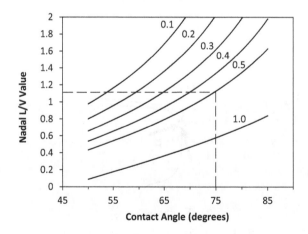

FIGURE 11.6 Relationship of limiting wheel L/V ratio and maximum wheel-rail contact angle. (Adapted from Blader, F.B., A review of literature and methodologies in the study of derailments caused by excessive forces at the wheel/rail interface, Report R-717, Association of American Railroads, Chicago, IL, 1990. With permission.)

Railway Vehicle Derailment and Prevention

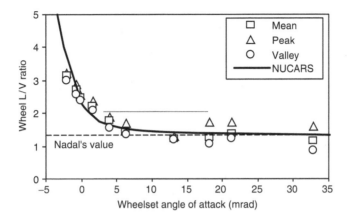

FIGURE 11.7 Effect of wheelset angle of attack on wheel L/V ratio limit. (Adapted from Shust, W.C., Elkins, J.A., Kalay, S., El-Sibaie, M., Wheel-climb derailment tests using AAR's track loading vehicle, Report R-910, Association of American Railroads, Washington, DC, 1997. With permission.)

In summary, Nadal's criterion agrees with situations when a large angle of attack is experienced and is conservative for small angles of attack. It does not consider the effects of friction coefficient of the nonflanging wheel on the flanging wheel climbing, which will be discussed in a later section. It assumes flange-climbing derailment is instantaneous once the L/V limit has been exceeded. Both field tests and simulations have also proved that wheel flange climb derailments would only occur when the L/V ratio limit has been exceeded for a certain distance limit or time duration limit.

11.3.1.2.2 Weinstock Criterion

In 1984, Weinstock proposed a less conservative wheel flange climb criterion [14]. This criterion evaluates incipient derailment by summing the absolute values of L/V on the two wheels on the same axle, known as the 'axle sum L/V' ratio.

When analysing wheel-rail forces obtained from computer simulations of wheel climb derailments, Weinstock observed the following:

- As a wheelset's angle of attack increases, the L/V ratio on the flanging wheel during an incipient derailment decreases and approaches Nadal's limit
- At the same time, the L/V ratio on the nonflanging wheel increases and approaches the wheel-rail friction coefficient because, at large angles of attack, the contact patch at the wheel tread-rail crown region is close to saturation, longitudinal creep force is small compared with the lateral creep force and contact angle is small (generally 0–3 degrees); the lateral creep force on the nonflanging wheel therefore acts to push the wheelset into hard flange contact on the opposite wheel, as illustrated in Figure 11.4
- Consequently, for flange climb derailments occurring at positive wheelset angles of attack, the sum of L/V ratios on both wheels stays approximately constant and equals the Nadal's limit at the flanging wheel plus the wheel-rail friction coefficient on the nonflanging wheel

Based on the above, Weinstock suggested that wheel climb derailment will not occur as long as the sum of L/V ratios on the flanging and nonflanging wheel is less than the sum of Nadal's limit for the flanging wheel and the friction coefficient at the nonflanging wheel-rail interface:

$$\left|\frac{L_F}{V_F}\right| + \left|\frac{L_{NF}}{V_{NF}}\right| < \frac{\tan\delta_F - \mu_F}{1 + \mu_F \tan\delta_F} + \mu_{NF} \qquad (11.7)$$

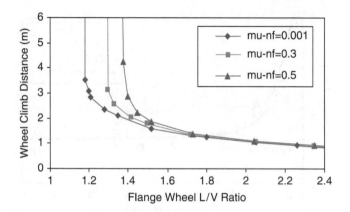

FIGURE 11.8 Effect of nonflanging wheel friction coefficient (5 mrad wheelset angle of attack, 75-degree flange angle). (Adapted from Wu, H., Elkins, J., Investigation of wheel flange climb derailment criteria, Report R-931, Association of American Railroads, Washington, DC, 1999. With permission.)

As Equation 11.7 suggests, the Weinstock criterion converges with that of Nadal as the coefficient of friction on the nonflanging wheel approaches zero, since there is no contribution of L/V value from the nonflanging wheel. A flange climb derailment study conducted by Wu and Elkins [16,17] showed that the L/V ratio limit for the flanging wheel increases with the increase of friction coefficient on the nonflanging wheel. As Figure 11.8 shows, wheel climb will not occur for an L/V ratio level less than the asymptotic line for each friction coefficient level on the nonflanging wheel.

The Weinstock criterion retains the advantage of simplicity. It can be measured with an instrumented wheelset, which measures the values of L/V ratio on both wheels on an axle. It is not only more accurate than Nadal's criterion, but it also has the merit of being less sensitive to errors or variations in the coefficient of friction. Figure 11.9 compares the Nadal and Weinstock criteria variation with the coefficient of friction at the wheel-rail interface (in the case of identical coefficient of friction on both rails).

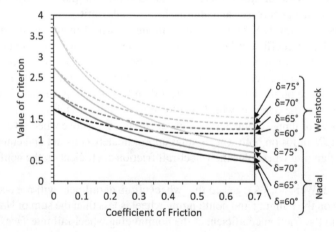

FIGURE 11.9 Comparison of Nadal and Weinstock criteria. (From Blader, F.B., A review of literature and methodologies in the study of derailments caused by excessive forces at the wheel/rail interface, Report R-717, Association of American Railroads, Chicago, IL, 1990; Iwnicki, S. (ed.), *Handbook of Railway Vehicle Dynamics*, CRC Press, Boca Raton, FL, 2006.)

Although Weinstock's criterion is less conservative than Nadal's, it is nevertheless conservative at small (approximately <5 mrad) and especially at negative angles of attack [15]. Because it is less conservative than Nadal, the Weinstock criterion is useful as an aid to evaluate risk of derailment. Derailment risk is likely to be higher if both Nadal and Weinstock limits are exceeded than if only the Nadal limit is exceeded.

11.3.1.2.3 Duration-Based Criteria

While investigating the duration of the single wheel L/V criterion necessary for derailment, researchers at the JNR proposed a modification to Nadal's criterion [9,18]. They suggested that, for a duration of lateral thrust (lateral force impulse) less than 50 msec, such as the one that might be expected during flange impacts while hunting, the allowable value of the L/V criterion should be increased, as shown in Figure 11.10. The analytical expression for the JNR criterion is given as:

$$\frac{L}{V} = \pi \left\{\frac{i_B}{G}\right\} \left\{\frac{\tan(\delta) - \mu}{1 + \mu \tan(\delta)}\right\} \sqrt{\frac{hP_W}{gP}} \left\{\frac{1}{T}\right\} \qquad (11.8)$$

where G is lateral distance between contact points, i_B is the radius of gyration of the axle about longitudinal axis through contact point, h is the height of wheel flange, g is gravitational acceleration, P is wheel load, P_W is wheel load due to unsprung mass and T is the time duration of lateral force thrust.

An even less conservative approach was proposed by the EMD [9,19]. Its L/V criterion is also shown in Figure 11.10.

11.3.1.2.3.1 AAR's Wheel Climb Duration Limit Based on the JNR and EMD research, and considerable experience in on-track testing of freight vehicles, a 0.05-second (50 msec) time duration was adopted by the AAR for certification testing of new freight vehicles in accordance with Chapter 11 of its Manual of Standards and Recommended Practices [20]. Subsequent research [15,16,21,22] has demonstrated that a distance-based limit should also be applied. The Chapter 11 criterion states that the single-wheel L/V and axle sum L/V ratios are 'not to exceed the indicated value for a period greater than 50 milliseconds and a distance greater than 3-feet per instance', as illustrated in Figure 11.11. This combined limit is also applied to the AAR's vertical wheel unloading criteria.

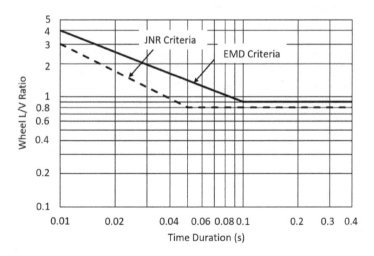

FIGURE 11.10 The JNR and EMD flange climb duration criteria. (Adapted from Koci, H.H., Swenson, C.A., Locomotive wheel-rail loading – A systems approach, In: Proceedings Heavy Haul Railways Conference, 18–22 September 1978, Perth, Australia, The Institution of Engineers Australia, Paper F.3, 1978. With permission.)

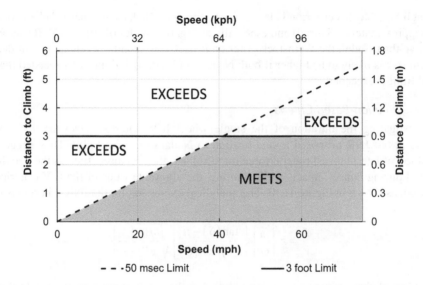

FIGURE 11.11 The AAR combined time and distance criterion. (From AAR, Manual of standards and recommended practices, Section C, Part II, Volume 1, Chapter 11, Association of American Railroads, Washington, DC, 2015. With permission.)

This combined time and distance duration limit has since been widely adopted by test engineers throughout North America for both freight and passenger vehicles that have adopted a wheel flange angle of 75 degrees.

11.3.1.2.3.2 Federal Railroad Administration's Wheel Climb Duration Limit A flange climb distance limit of 5 ft (1.5 m) was adopted by the US Department of Transportation's Federal Railroad Administration (FRA), for Class 6 (and higher) high-speed track standards [12]. This distance limit appears to have been based partly on the results of the joint AAR/FRA flange climb research conducted by TTCI [16] and also on experience gained during the testing of various commuter rail and passenger vehicles. Further discussion on the applicability of this criterion is included in discussion on low-speed flange climb derailment in Section 11.3.2.5.

11.3.1.2.3.3 TTCI's Proposed Wheel Climb Distance Criterion TTCI has proposed flange climb criteria for North American freight wagons using the AAR-1B wheel profile with a 75-degree flange angle at speeds below 80 km/h in curving [16,17]. These criteria encompass two limits: the single wheel L/V limit and the L/V distance limit. The distance limit is the maximum distance that the single-wheel L/V limit can be exceeded without risk of flange climb derailment. It is possibly the first time that the wheelset angle of attack has explicitly been included in the flange climb criterion. Figure 11.12 shows the simulation results of the L/V distance limits under different wheelset angles of attack. The test and simulation results showed that the distance limit is a function of wheelset angle of attack.

The following are the proposed criteria. Since measurements of angle of attack are usually difficult in track tests, the criterion is given in two forms: one for the use of simulations in terms of wheelset angle of attack, and one for the use of track tests in terms of track curvature (in degrees). Figures 11.13 and 11.14 graphically display the criteria as a function of angle of attack.

Note that both the wheel L/V ratio limit and distance limit will converge to a constant value as the wheelset angle of attack reaches a certain level.

The TTCI conducted research to update the proposed criterion and further develop flange climb derailment L/V ratio and distance criteria for application to the North American freight

Railway Vehicle Derailment and Prevention

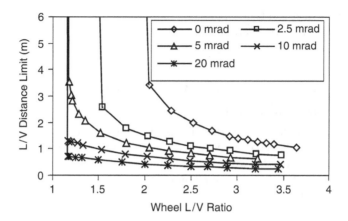

FIGURE 11.12 Effect of wheelset angle of attack on L/V distance limit, $\mu = 0.5$, no longitudinal creepage. (Adapted from Wu, H., Elkins, J., Investigation of wheel flange climb derailment criteria, Report R-931, Association of American Railroads, Washington, DC, 1999. With permission.)

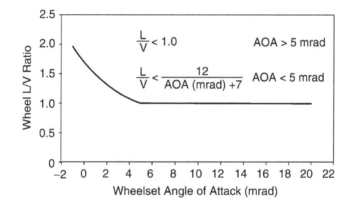

FIGURE 11.13 Proposed single wheel L/V criterion as a function of angle of attack. (Adapted from Wu, H., Elkins, J., Investigation of wheel flange climb derailment criteria, Report R-931, Association of American Railroads, Washington, DC, 1999. With permission.)

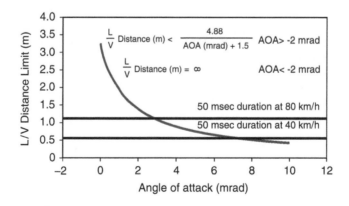

FIGURE 11.14 Proposed L/V distance limit as a function of angle of attack. (Adapted from Wu, H., Elkins, J., Investigation of wheel flange climb derailment criteria, Report R-931, Association of American Railroads, Washington, DC, 1999. With permission.)

railroads and North American transit operation. The updated criteria have a more general form for application to the variety of wheel profile designs used by different freight vehicles and transit systems [21,22].

11.3.2 Application of Flange Climb Derailment Criteria

11.3.2.1 Flange Climb due to Low Flange Angle

Figure 11.15 provides two examples of wheel flange angles. One is a wheel profile with a 75-degree flange angle and the other has a 63-degree flange angle. Referring to Figure 11.6, at a friction coefficient of 0.5 (representing the dry wheel-rail contact condition), the limiting L/V value is 1.13 for wheels with a 75-degree flange angle, according to the Nadal criterion, and 0.73 for wheels with a 63-degree flange angle. Clearly, wheels with low flange angles have a higher risk of flange climb derailment.

For historic reasons, some railway systems have adopted relatively low wheel flange angles in the range of 63 to 65 degrees. New systems now generally start with a wheel profile having a flange angle of 72 to 75 degrees. Note that systems with very sharp curves such as those used by streetcars and trams frequently have low flange angles due to the large wheelset angles of attack, causing high wheel flange and rail gauge face wear rates when steeper flange angles are used.

A wheel profile with a higher flange angle can reduce the risk of flange climb derailment and can have much better compatibility with any new designs of vehicle/bogie that may be introduced compared with wheels with lower flange angles. Also, with a higher L/V ratio limit, high flange angles will tolerate greater levels of unexpected track irregularity.

In the *Track Design Handbook for Light Rail Transit* [23], a wheel flange angle of 70 degrees was proposed based on Professor Heumann's design. The American Public Transit Association (APTA) now recommends a minimum flange angle of 72 degrees (including manufacturing tolerances) to be maintained for a minimum length of 0.1 inch (2.54 mm) [24].

11.3.2.2 Increase of Flange Length Can Increase Flange Climb Distance Limit

The flange length is defined as a measure of wheel flange contour with an angle above a given degree, as explained in Figure 11.16. The flange angle may not be a constant in the length F_L but must be above a specified value.

A concept of increasing flange length to increase the flange climb distance limit was proposed by Wu and Elkins [16] and further validated by Wilson et al. [25]. They concluded that increasing flange length would increase flange climb distance appreciably at a lower angle of attack (approximately 5 mrad) and produce only a small increase in climb distance at a higher angle of attack.

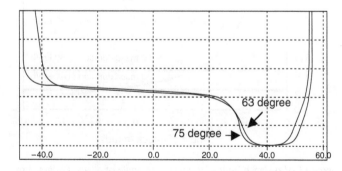

FIGURE 11.15 Wheels with 75- and 63-degree flange angle. (From Iwnicki, S. (ed.), *Handbook of Railway Vehicle Dynamics*, CRC Press, Boca Raton, FL, 2006. With permission.)

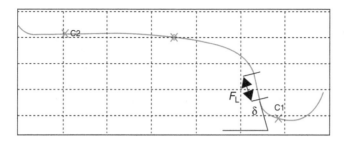

FIGURE 11.16 Definition of flange length. (From Iwnicki, S. (ed.), *Handbook of Railway Vehicle Dynamics*, CRC Press, Boca Raton, FL, 2006. With permission.)

11.3.2.3 Flange Climb due to High Coefficient of Friction at Wheel-Rail Interface

Flange climb derailments have been reported to occur at curves or switches in maintenance yards when the vehicles were just out of the wheel truing machines. This type of derailment is probably caused by the wheel surface roughness after wheel truing. Figure 11.17 compares a wheel surface just after truing and the surface after many miles of running. The left wheel in Figure 11.17 was trued by the milling-type machine with very-clear-cutting traces on the surface, the middle one was trued by the lathe-type machine with shallower-cutting traces and the right wheel was back from operation with a smooth surface but a flat spot on the tread.

Generally, the coefficient of friction for dry and smooth steel-to-steel contact is about 0.5. The effective friction coefficient for the rough surface could be much higher. For example, if the coefficient reaches 1.0, the L/V limit, as shown in Figure 11.6, would be 0.5 for a 75-degree flange angle and 0.3 for a 63-degree flange angle. Therefore, the rough surface produced by wheel truing could significantly reduce the L/V limit for flange climb. A low flange angle would further increase the derailment risk. Addressing the final surface finish and wheel-rail lubrication after reprofiling are two possible remedies to improve the potential adverse outcomes of rough surface condition after reprofiling.

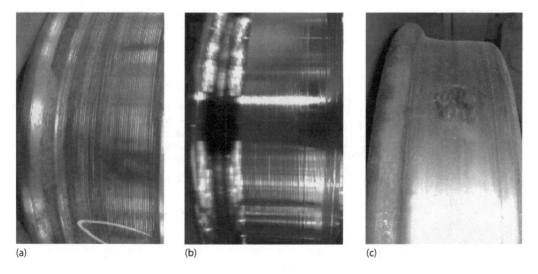

FIGURE 11.17 Comparison of wheel surface roughness: (a) surface after wheel truing from milling type machine, (b) surface after wheel truing from lathe-type machine and (c) surface of wheel with a flat spot from operation in service. (From Iwnicki (ed.), *Handbook of Railway Vehicle Dynamics*, CRC Press, Boca Raton, FL, 2006. With permission.)

11.3.2.4 Flange Climb of Independently Rotating Wheels

Wheels mounted on a solid axle must rotate at the same speed. To accommodate running in curves, a taper is usually provided on the wheel tread. The wheelset shifts sideways, as shown in Figure 11.18, to allow the outer wheel to run with a larger rolling radius than the inner wheel. The resulting longitudinal creep forces at the wheel-rail interfaces on wheels of the same axle form a moment that steers the bogie around curves (Figure 11.18). Previous flange climb studies have indicated that, as the ratio of longitudinal force to vertical force increases, the wheel L/V ratio required for derailment also increases (Figure 11.19). Therefore, the Nadal flange climb criterion can be relaxed based on the level of longitudinal force.

In simple terms, the longitudinal steering forces can be viewed as 'using up' some of the available wheel-rail friction. This reduces the effective friction coefficient for flange climbing, increasing the L/V ratio required for flange climb.

Independently rotating wheels can rotate at different speeds and therefore produce no longitudinal forces to form a steering moment. This can lead to higher wheelset angles of attack, consequently

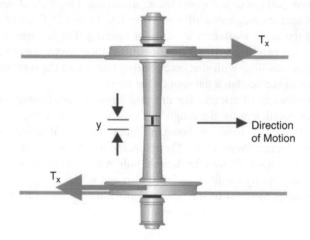

FIGURE 11.18 Steering moment formed by wheel longitudinal forces due to different rolling radius on two wheels. (From Iwnicki, S. (ed.), *Handbook of Railway Vehicle Dynamics*, CRC Press, Boca Raton, FL, 2006. With permission.)

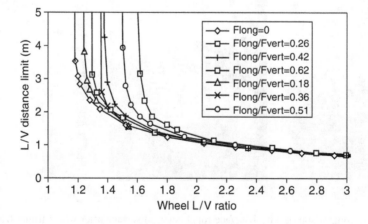

FIGURE 11.19 Effect of wheel L/V ratio on wheel climb distance (5 mrad wheelset angle of attack) [16]. (Adapted from Wu, H., Elkins, J., Investigation of wheel flange climb derailment criteria, Report R-931, Association of American Railroads, Washington, DC, 1999. With permission.)

resulting in higher lateral forces (before reaching saturation), higher L/V ratios and increased wheel and rail wear. In addition, since there are no longitudinal forces (the line of $f_{long} = 0$ in Figure 11.19), the wheel-rail friction acts entirely in the lateral direction, resulting in the shortest distance to climb and greater flange climb risk.

The conservative nature of the Nadal criterion was discussed in Section 11.3.2.1. However, for independently rotating wheels, any L/V values that exceed the Nadal limit would cause wheel flange climb, because there is no relaxation from the effect of longitudinal force and friction coefficient level on the nonflanging wheel. Therefore, independently rotating wheels have less tolerance to track irregularities and switch points that may suddenly increase wheel lateral forces or reduce vertical forces.

In summary, vehicles with independently rotating wheels need to be carefully designed to control flange climb and wheel wear. Additional control mechanisms, such as linkages and active control systems, can be used to steer the wheelset on curves and track perturbations. Without such control mechanisms, the wheel-rail profiles, vehicle-track maintenance and wheel-rail friction will need to be much more strictly controlled and monitored to prevent wheel flange climb.

11.3.2.5 Low-Speed Flange Climb Derailment

Low-speed flange climb derailments can occur when vehicles with stiff suspension systems negotiate sharp curves with track twist defects. Certain passenger vehicles are designed with stiff suspension systems suitable for high-speed operating environments; however, the suspension systems may perform poorly in low-speed environments over tracks with sharp curves and track twist. In these situations, a combination of high lateral curving forces and vertical wheel unloading can lead to flange climb derailments.

Several standards exist in the industry [11,26] to ensure passenger vehicle bogies distribute vertical wheel loads properly, which reduces the risk of flange climb derailment. The FRA has also issued a low-speed derailment safety advisory [27] with recommended measures to help the industry evaluate the low-speed curving performance of passenger vehicles in the presence of track twist defects. The safety advisory states that suspension systems should 'prevent wheel climb while negotiating, at a minimum, a 12 degree curve with a coefficient of friction representative of dry track conditions (i.e., 0.5) and 3-inch track warp variations with the following wavelengths: 10, 20, 40, and 62 ft' (i.e., 76-mm track warp variations at wavelengths of 3.0, 6.1, 12.2 and 18.9 m). The advisory also recommends that L/V ratios on any wheel during the evaluation should be limited to the 5-feet (1.52 m) FRA single-wheel L/V ratio criterion [12].

Current criteria for assessing flange climb derailment risk use duration-based climb assessment windows, such as the FRA's 5-feet single wheel L/V ratio limit [12] and the AAR's Chapter 11 3-feet/50-msec L/V ratio limit [20]. Wilson et al. [22] compared the use of 50-msec time-based and 3-feet distance-based flange climb criteria and concluded that the time criterion was more conservative at speeds below 40 mph (64.4 km/h), while the distance criterion was more conservative at speeds above this (Figure 11.11). Similarly, the FRA's 5-feet assessment window for single-wheel L/V ratio, which was intended for use at high speeds, may not be conservative enough for evaluating flange climb derailment risk at low speeds. The FRA and TTCI are currently working with the APTA to develop a new standard for assessing low-speed curving performance of passenger vehicles over track twist defects. A part of this effort is to determine the appropriate duration-based climb assessment windows to be used in the new standard.

11.3.2.6 Vertical Wheel Unloading Criteria

Flange climb frequently occurs when the vertical load on a wheel is decreasing at the same time as other forces are acting to push the wheelset sideways into flange contact, increasing the lateral load and L/V ratios. The loss of vertical wheel load, which can often be caused by track irregularities and/or dynamic response of the vehicle such as vehicle body rocking motions, can lead to a situation where even a small absolute increase in lateral load could result in flange climb derailment.

A lateral load of 9 kN and vertical load of 10 kN are more dangerous conditions than a lateral load of 90 kN and vertical load of 100 kN, even though L/V ratios are identical, because in the former case, a relatively small additional lateral load would cause flange climb.

Therefore, vertical wheel unloading criteria are commonly specified in addition to L/V ratio criteria:

$$\frac{\Delta Q}{Q} = \frac{Q_{stat} - Q_{min}}{Q_{stat}} \qquad (11.9)$$

where Q_{stat} is static vertical wheel load and Q_{min} is minimum vertical wheel load over a defined distance or time period. A maximum wheel unloading of 90% is specified in the AAR Chapter 11 performance standards [20], and the FRA [12] specifies 85% unloading. The corresponding L/V ratio duration limits are also used.

These criteria can also detect other dangerous conditions, such as incipient vehicle rollover due to rocking motion or crosswind forces.

11.3.3 Derailments due to Gauge Widening and Rail Rollover

Derailments caused by gauge widening usually involve a combination of wide track gauge and large lateral rail deflections (rail roll), as shown in Figure 11.20. When a wheelset operates with a large angle of attack, large lateral forces from the wheels act to spread the rails in curves, as illustrated in Figure 11.4. Both rails may experience significant lateral translation and/or railhead roll, which often cause the nonflanging wheel to drop between rails. Figure 11.21 was produced based on a photo of an actual derailment caused by rail rollover, in which the nonflange wheel fell between rails and the outer rail was rolled over. Frequently, the inner rail will roll over due to contact with hollow-worn wheels, as discussed in Section 11.3.3.3.

When a bogie experiences poor steering, the wheelsets may experience high angles of attack in curves, resulting in large lateral forces exerted on the rails. The poor steering can be caused by inadequate suspensions (generally indicated by low warp or skew stiffness), high bogie turning resistance, misaligned axles, poor wheel and rail profile compatibilities [28] and wheels having

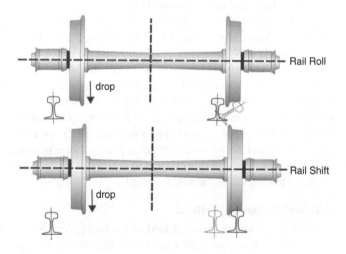

FIGURE 11.20 Gauge widening derailment. (Adapted from Blader, F.B., A review of literature and methodologies in the study of derailments caused by excessive forces at the wheel/rail interface, Report R-717, Association of American Railroads, Chicago, IL, 1990. With permission.)

Railway Vehicle Derailment and Prevention

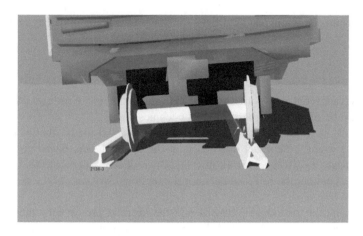

FIGURE 11.21 Nonflange wheel falls between the rails, while the outer rail is rolled over. (From Iwnicki, S (ed.), *Handbook of Railway Vehicle Dynamics*, CRC Press, Boca Raton, FL, 2006. With permission.)

significant tread hollowing. The dynamic forces caused by track lateral perturbations and curve entry/exit transition spirals can intensify the lateral force level to deflect the rail further.

Bogies with high turning resistance, misaligned axles and lateral track perturbations can also cause these large angles of attack and the consequent gauge spreading forces in tangent track.

Locomotives produce high traction forces on rails. Since their bogies are long and can result in large angles of attack, six-axle locomotives (now common for heavy-haul operations) have been considered as an important cause of gauge widening and rail rollover derailments.

Rail gauge face wear is another cause for gauge widening and is further discussed in Section 11.3.3.2.

11.3.3.1 The AAR Chapter 11 Rail Roll Criterion

The AAR Chapter 11 rail roll criterion is established by using the L/V force ratio. The rail is assumed to rotate about the rail base corner under the load, as shown in Figure 11.22. The roll moment about the pivot point is given by:

$$M = Vd - Lh \tag{11.10}$$

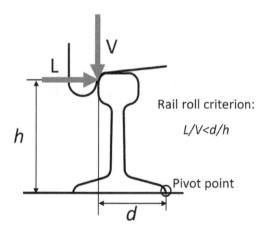

FIGURE 11.22 Illustration of rail roll criterion. (Adapted from Blader, F.B., A review of literature and methodologies in the study of derailments caused by excessive forces at the wheel/rail interface, Report R-717, Association of American Railroads, Chicago, IL, 1990. With permission.)

Under an equilibrium condition, just before the rail starts to roll, M approaches zero, and then,

$$\frac{L}{V} = \frac{d}{h} \tag{11.11}$$

This L/V ratio is considered as the critical value to evaluate the risk of rail roll. When the L/V ratio is larger than the ratio of d/h, the risk of rail roll becomes high.

The critical L/V ratio for rail roll can vary from above 0.6 for contact at the gauge side to approximately 0.2 when the contact position is at the far-field side, based on the dimension of rails. This is because the distance d in Figure 11.22 is reduced. Note that this L/V ratio is calculated assuming that neither the rail fasteners nor the torsional stiffness of the rail section provides any restraint.

When considering the torsional rigidity of rail, the restraint provided by the rail fasteners and the vertical force applied to the rail by the adjacent wheels, a criterion that only considers the forces due to a single wheel may be too conservative for predicting the stability of the rail. Therefore, the limiting criterion has counted the combined forces from all wheels on the same side of the bogie. Hence, a bogie (truck) side L/V ratio is defined by:

$$\left(\frac{L}{V}\right)_{bogie} = \frac{\text{sum of Lateral forces on bogie side}}{\text{sum of Vertical forces on bogie side}} \tag{11.12}$$

In Chapter 11 of the *AAR Manual of Standards and Recommended Practices* [20] and the FRA standards for Class 6 (and higher) high-speed track [12], the bogie side L/V ratio has been limited to below 0.6 in (15.2 mm) the vehicle yaw and sway tests. A duration limit of 6 ft (1.83 m) is used by the AAR. The FRA uses a more conservative duration limit of 5 ft (1.52 m).

11.3.3.2 The Gauge-Widening Criterion

The gauge-widening criterion is related to the wheel and rail geometries and their relative positions, as illustrated in Figure 11.23.

When the wheel drops between the rails, as in Figure 11.20, the geometry of wheel and rail must meet the following expression:

$$G \geq B + W + f \tag{11.13}$$

where G, B, W and f are the rail gauge distance, wheel back-to-back space, wheel width and flange thickness, respectively.

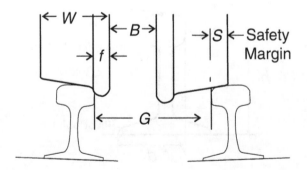

FIGURE 11.23 Wheel and rail geometry related to gauge-widening derailment. (Adapted from Blader, F.B., A review of literature and methodologies in the study of derailments caused by excessive forces at the wheel/rail interface, Report R-717, Association of American Railroads, Chicago, IL, 1990. With permission)

Railway Vehicle Derailment and Prevention

Therefore, a safety margin (S), expressed in Equation 11.14, represents the minimum overlap of wheel and rail required on the nonflanging wheel when the flanging wheel contacts the gauge face of the rail. In this circumstance, the instantaneous flangeway clearance on the flanging wheel is zero.

$$(B + W + f) - G > S \quad (11.14)$$

In general, the wheel back-to-back space (B) is a constant for a solid axle and so is the wheel width (W). However, the flange thickness (f) is gradually reduced as the wheel wears. The track gauge variations are influenced by multiple factors. As discussed in the previous section, rail roll and the lateral movement of rail due to weakened fasteners can widen the gauge. As shown in Figure 11.24, rail gauge wear can also contribute to gauge widening with wear of about 8 mm.

The North American standard AAR-1B freight wagon interchange wheel profile is taken as an example. The back-to-back spacing of the AAR-1B is 1350 mm, the wheel width is 145 mm and the flange thickness is 35 mm. With a standard gauge of 1435 mm, the safety margin is 95 mm. A maximum 31.5-mm gauge widening (include the rail gauge wear measured under an unloaded condition) from the standard value is allowed for a freight vehicle operating in the speed range of 40 to 60 km/h, if a maximum of 15 mm of wheel flange wear is allowed. Under this extreme condition, the overlapping is reduced to 48.5 mm. Therefore, any lateral shift and rotation of rail under the loaded condition can further reduce the overlapping to increase the risk of the wheel falling between the rails, especially on poorly maintained track.

11.3.3.3 Effect of Hollow-Worn Wheels on Gauge Widening and Rail Roll Derailment

Wheel hollowing is defined as the vertical difference in rolling radius between the end of the tread and the minimum point around the middle of the tread. The value is found by placing a horizontal line through the highest point on the end of the wheel tread. The wheel tread hollow tends to form a false flange at the end of the tread, as Figure 11.25 illustrates.

Figure 11.26 shows a likely contact condition of a measured hollow-worn wheel on a measured low rail. With wide gauge, or with combined wide rail gauge and thin wheel flange, the false flange of the hollow wheel is likely to contact the top of the low rail towards the field side on curves. Referring to the rail rollover criterion stated in Equation 11.11, the value of d would be quite low under this condition, leading to a low ratio of d/h. Therefore, any bogie (truck) side L/V ratio larger than this d/h ratio would increase the risk of rail rollover or put excessive forces on the fasteners. Figure 11.27 shows an example of loose spikes caused by repeated contact towards the field side of the rail.

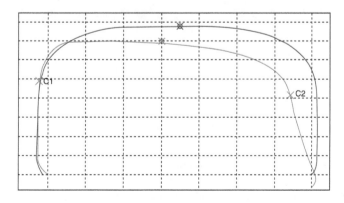

FIGURE 11.24 Gauge-widening caused by rail gauge wear. (From Iwnicki, S. (ed.), *Handbook of Railway Vehicle Dynamics*, CRC Press, Boca Raton, FL, 2006. With permission.)

FIGURE 11.25 Hollow-worn wheel. (From Iwnicki, S. (ed.), *Handbook of Railway Vehicle Dynamics*, CRC Press, Boca Raton, FL, 2006. With permission.)

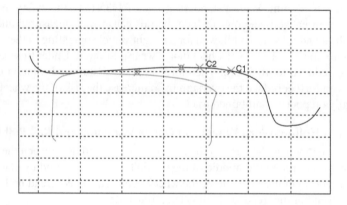

FIGURE 11.26 False flange contact of hollow-worn wheel. (From Iwnicki, S. (ed.), *Handbook of Railway Vehicle Dynamics*, CRC Press, Boca Raton, FL, 2006. With permission.)

FIGURE 11.27 Example of loose rail fasteners. (From S. Iwnicki, S. (ed.), *Handbook of Railway Vehicle Dynamics*, CRC Press, Boca Raton, FL, 2006. With permission.)

The contact between the false flange of a hollow wheel and the field side of the inner rail in a curve can lead to an adverse rolling radius difference condition between the inside wheel and the outside wheel in curves. Under certain conditions of wheel-rail flange lubrication, this can cause the wheelset steering forces to reverse, increasing the wheelset angles of attack and lateral forces, causing increased gauge widening and rail rollover [29,30].

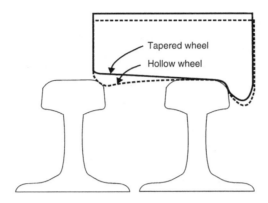

FIGURE 11.28 False flange of a hollow-worn wheel applies a roll force to the stock rail on a riser-less switch point. (From Iwnicki, S. (ed.), *Handbook of Railway Vehicle Dynamics*, CRC Press, Boca Raton, FL, 2006. With permission.)

Switches are consistently one of the most common track-related causes of derailments. Some incidents of rail rollover derailment in switches are expected to be directly related to hollow-wheel profiles. At switch points without risers, a vehicle with hollow-worn wheels may cause rail rollover [31].

For a vehicle making a trailing point move, the wheel stays on the switch point, with the false flange hanging down below the level of the stock-rail running surface. As the wheel approaches the point where the two railheads converge, the false flange can strike the side of the stock rail (see Figure 11.28). The high force produced from the false flange-stock rail interaction may either cause a stock rail to roll out or, in severe cases, cause a wheel to climb.

11.3.4 Derailment due to Track Panel Shift

Track panel shift is the cumulative lateral displacement of the track panel, including rails, sleeper plates and sleepers over the ballast (see Figure 11.29). A small shift of these components may not immediately cause the loss of guidance to bogies. However, as the situation gradually deteriorates to a certain level, wheels could lose guidance and drop to the ground at some speed. The derailments caused by track panel shift usually result in one wheel falling between the rails and the other falling outside of the track.

11.3.4.1 Causes of Track Panel Shift

Track panel shift is a lateral misalignment phenomenon primarily caused by repeated lateral axle loads. Tracks that possess low resistance to lateral force, such as poorly laid track, newly laid track and newly maintained track, often show separations between the track panel and ballast. Track

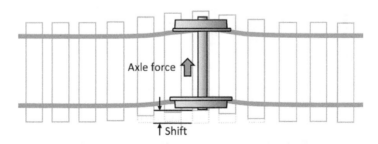

FIGURE 11.29 Lateral track panel shift. (From Iwnicki, S. (ed.), *Handbook of Railway Vehicle Dynamics*, CRC Press, Boca Raton, FL, 2006.)

panels could shift under large lateral forces under such conditions. The capacity of a track panel to resist lateral movement is measured by lateral track strength and stiffness. Soft subgrade may also allow the panel to shift more freely.

Track panel shift has become increasingly important, as both speeds and loads increase and more continuous welded rail is placed in use. The increase in speed may result in an increase in the unbalanced forces on curves or poorly aligned track by curving considerably above the balance speed. At high imbalance speeds, the wheelsets of many bogie designs stop generating gauge spreading forces, and, instead, both wheels generate wheel-rail forces that act to force the rails outwards in a curve leading to panel shift.

At balance speed, all centrifugal forces are balanced by the force of gravity due to superelevation in the track. Thus, for a bogie that is free to rotate relative to the car body, the sum of all lateral forces in the track plane generated from the wheelsets of a vehicle will approximately equal zero. For moderate to sharp curves, this usually results in the left and right wheels generating forces in opposite directions that act to spread the rails (Figure 11.4). The magnitude of the gauge spreading forces depends on the axle angles of attack.

As the vehicle speed increases above the balance speed, the centrifugal force becomes greater than the forces due to gravity. This increased force is reacted by lateral forces generated at the wheel-rail interface. For most H-frame bogies in moderate curves, the wheel-rail forces tend to be saturated on the lead axle due to the magnitude of the angle of attack. This means that the trailing axle must therefore generate the additional lateral force to counteract the increasing centrifugal force. Figure 11.30 shows these effects for a simulation of an H-frame bogie.

As this lateral force increases, at very high speeds, this can result in the trailing axle changing from generating small gauge spreading forces to having a large net lateral force acting to displace

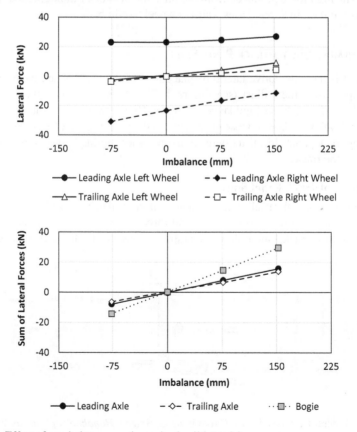

FIGURE 11.30 Effect of overbalance speed on wheel-rail lateral forces.

the entire track outward. At extremely high speeds, the lead axle will also begin to generate very large net axle lateral force. Note that the effects shown in Figure 11.30 are dependent on the bogie and suspension design, wheelset steering ability, wheel-rail contact conditions, track curvature and superelevation and must be evaluated for each vehicle design and its operating conditions.

The escalation in load can increase the magnitude of lateral force at the wheel-rail interface. Continuous welded rail can increase the risk of panel shift (track buckling) due to the longitudinal force caused by temperature change. Aggressive acceleration and braking can also induce large forces to cause panel shift on track with poor lateral resistance [32].

Knothe and Böhm [33] commented that the BB 9104 locomotive nearly came to a catastrophe when it achieved the world record of 331 km/h in 1955. The test locomotive caused a strong sinusoidal alignment fault of the track. The evidence suggests that there were two causes. Firstly, the track was tamped just before the record test. The strength of track to resist lateral displacements was reduced by this maintenance operation. Secondly, the locomotive was unstable and thus was exerting high lateral forces on the dislodged ballast bed.

11.3.4.2 Track Panel Shift Criterion

Lateral track strength indicates the capability of track to resist track buckling and retain lateral alignment under traffic. Track buckling is defined as the lateral deformation of track due to high compressive rail force in the longitudinal direction at a temperature above the rail neutral temperature. Track panel shift normally accumulates gradually; however, when the critical load level is exceeded, panel shift increases rapidly with the number of repeated load applications.

The definition of the critical lateral load can be based on either accumulated or incremental deformation after each load application. Note that, with each load pass, the increment of total deformation (elastic plus residual) remains constant for a stable track. Below critical loads, the elastic deformation remains constant and the residual deformation tends to zero. Both elastic and residual deformation increment will grow with each load pass over a segment of unstable track.

Research by the French National Railways and subsequent studies by other agencies suggest that the limiting lateral axle load can be defined in a general expression shown in Equation 11.15 for preventing excessive track panel shift [9,34–38]:

$$L_c = aV + b \tag{11.15}$$

where L_c is the critical lateral load and V is the vertical axle load. Table 11.2 lists some of the suggested values of a and b.

A multiplying factor proposed by Ahlbeck and Harrison considers the effect of track curvature and temperature in determining the lateral force limit [9,36] and is given by:

$$A = 1 - \frac{S\Delta}{22300}(1 + 0.46D) \tag{11.16}$$

where S is the area of rail section (square inches), Δ is the temperature change (degrees Fahrenheit) and D is the degree of track curvature.

The lateral track strength and track panel shift study conducted by Li and Shust [34] concluded that vertical axle load has a major effect on the resistance of a track panel to lateral deflection. The lateral track strength and stiffness would only be valid for a given vertical load. In their in-motion panel shift test, the critical lateral axle force for causing panel shift is approximately 15% to 30% higher for track with concrete sleepers than wooden sleeper track. Longitudinal rail forces have a relatively small effect on lateral track strength; 560-kN change in rail longitudinal forces, corresponding to approximately 30 degrees Celsius change in rail temperature, caused less than 10% change in lateral strength.

TABLE 11.2
Suggested Values of *a* and *b* for Track Panel Shift Criterion

Source	Values	Applicable Conditions
Prud'homme (original) [35]	$a = 0.33$	Uncompacted ballast and wooden sleepers
	$b = 10$ kN	
	$a = 0.33$	Well-compacted ballast and concrete sleepers
	$b = 15$ kN	
Prud'homme (revised) [9,35]	$a = 0.28$	Uncompacted ballast and wooden sleepers
	$b = 8.5$ kN	
	$a = 0.28$	Well-compacted ballast and concrete sleepers
	$b = 12.75$ kN	
Ahlbeck and Harrison [9,36]	$a = 0.4$	Uncompacted ballast on wooden sleepers
	$b = 9.96$ kN	
	$a = 0.7$	Compacted ballast on wooden sleepers
	$b = 24.6$ kN	
Li and Shust [34]	$a = 0.5$	Compacted ballast on wooden sleepers
	$b = 26.7$ kN	
US DOT/FRA [12]	$a = 0.4$	Vehicle performance limit on high-speed track
	$b = 22.2$ kN	
EN 14363:2005 [11]	$a = 0.67$	Performance criteria for locomotives, power cars, multiple units, and passenger coaches
	$b = 10$ kN	
	$a = 0.57$	Performance criteria for freight wagons
	$b = 8.5$ kN	
Kish et al. [38]	$a = 0.28$	High-speed track, concrete ties, continuously welded rail, no initial alignment defects
	$b = 19.8$ kN	
	$a = 0.28$	Same as above, but slightly misaligned track condition
	$b = 17.7$ kN	

In the United States, the FRA Track Safety Standards prescribe a net axle L/V limit defined similar to the Prud'homme criterion, with $a = 0.4$ and $b = 22.2$ kN, regardless of track type, not to be exceeded for a continuous distance of more than 5 ft (1.52 m) [12].

11.4 CAUSES OF RAILWAY VEHICLE DERAILMENTS

The four railway derailments mechanisms discussed in the previous sections (wheel flange climb, gauge widening, rail rollover and track panel shift) describe the specific manner in which vehicles lose lateral guidance at the wheel-rail interface and derail. The following sections describe different conditions that can initiate the chain of events leading to derailment and produce the derailment mechanism.

11.4.1 DERAILMENT CAUSED BY VEHICLE LATERAL INSTABILITY

On tangent track, the wheelset generally oscillates around the track centre due to any vehicle and track irregularities, as shown in Figure 11.31. This movement occurs because vehicle and track are never absolutely smooth and symmetric. This self-centring capability of a wheelset is induced by the coned shape of the wheel tread (see Chapter 2, Section 2.2). However, as speed is increased, if the wheelset conicity is high, the lateral wheelset movement, as well as the associated bogie and vehicle body motion, can cause oscillations with large amplitude and a well-defined wavelength. The lateral movements are limited only by the contact of the wheel flanges with the rail.

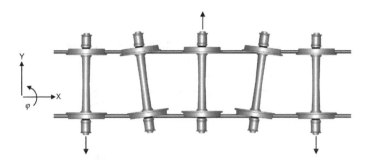

FIGURE 11.31 Wheelset oscillation around the track centre. (From Iwnicki, S. (ed.), *Handbook of Railway Vehicle Dynamics*, CRC Press, Boca Raton, FL, 2006. With permission.)

This vehicle dynamic response is also termed vehicle hunting and can produce high lateral forces to damage track and cause derailments.

Derailments caused by vehicle hunting can have derailment mechanisms of all the four types discussed in the previous sections. The high lateral force induced from hunting may cause wheel flange climbing on the rail, gauge widening, rail rollover, track panel shift or combinations of these. The safety concerns for this type of derailment, usually occurring at higher speeds, make it an important area of study.

Hunting predominantly occurs in empty or lightweight vehicles. The critical hunting speed is highly dependent on the vehicle/track characteristics. When vehicle hunting occurs, the displacements of wheelset are generally large, alternatively flanging from one side of the vehicle/bogies to the other. Considering the wheel-rail geometry and the creep force saturation, the vehicle/track system under hunting conditions should be treated as nonlinear. Investigation of the critical speed for such a system with nonlinearities is to examine the vehicle dynamic response to a disturbance, using a numerical solution of the equations of motion [39].

Vehicle simulation computer models, which include the processes to solve these equations of motion, are often used to predict the hunting speed. Track tests are also generally required to either validate the hunting speed predicted by modelling or ensure that the system operating speed is below the hunting onset speed.

The effective conicity of wheel-rail contact has a considerable influence on the vehicle hunting speed. As wheelset conicity increases, the onset critical speed of hunting decreases. For this reason, it is important when designing wheel and rail profiles to ensure that, for a specific bogie/vehicle, the critical hunting speed is above the operating speed. Methods for evaluating wheel tread taper and wheel-rail conicity have been documented by the APTA for the US transit industry [40].

11.4.2 Derailment Caused by Vehicle Body Resonance

Derailment can occur due to excessive vehicle body movement under transient forces, especially when a vehicle body experiences resonance. In extreme cases, vertical (pitch or bounce) resonance can cause a vehicle body to separate from bogies; somewhat more commonly, roll resonance can cause overturning (for overturning due to quasi-steady forces, see Section 11.4.4).

Safety against such derailment can be assessed by measuring the following:

- Maximum vertical and lateral vehicle body acceleration [11,12,20,41,42]
- Maximum vehicle body roll angle [20,41,43]
- Maximum wheel unloading [12,20,41,43]

11.4.3 DERAILMENT CAUSED BY LONGITUDINAL IN-TRAIN FORCES

Longitudinal in-train forces develop due to interactions between adjacent, coupled vehicles in a train. The development of these forces is affected by factors such as train configuration, track topography, train handling and vehicle characteristics. In-train forces can be steady state or transient in nature. Steady state in-train forces act on vehicles over a long period of time, including train resistance, steady locomotive traction and constant grade forces. Transient in-train forces act over a shorter duration and can have significant effects on train dynamics. Transient in-train forces can include rapid changes in locomotive tractive effort, braking and undulating grade forces, which can contribute to slack action in a train [44]. Slack action events include run-ins due to buff (compressive) forces and run-outs due to draft (tension) forces (see Chapter 13). Buff and draft forces due to slack action in trains have been measured as high as 1000 to 2000 kN, depending on the train and operating conditions [45,46,47].

High longitudinal in-train forces can increase the risk of derailment, given certain vehicle and track conditions. When trains negotiate horizontal track curves, the couplers of adjacent vehicles develop an angular offset relative to the track centreline due to the geometry of the coupled vehicles and lateral coupler swing. The angularity of the couplers causes a portion of the buff and draft forces to be transferred laterally through the vehicles to the wheel-rail interface rather than longitudinally down the length of the train [48].

Two types of derailments related to longitudinal in-train forces are jackknifing and stringlining.

Jackknifing derailments typically occur in curves when high buff forces and high coupler angles in a bunched train cause adjacent vehicles to derail towards the outside of the curve and fold up in a fashion similar to a jackknife. Although jackknifing on curved track is the more common scenario, derailments due to jackknifing on tangent track have also occurred [49]. Figure 11.32 illustrates the general scenario in which jackknifing may occur on curved track.

Stringlining derailments are similar in nature to jackknifing derailments but are related to high draft forces rather than high buff forces. Stringlining derailments occur in curves when longitudinal draft forces and the associated lateral components are high enough to stretch and straighten the train, causing vehicles to derail towards the inside of the curve like a tightly pulled string. Figure 11.33 illustrates the general scenario in which stringlining may occur. High draft forces in trains can also cause draft system failures and train separations, which can also contribute to derailments [50].

The mechanism of derailment in these coupler force-related scenarios can be any of those discussed earlier in this chapter (wheel climb, rail rollover, gauge widening or track panel shift). Another possibility is vertical vehicle jackknifing, leading to centre-plate lift-off and separation of a vehicle body from its bogies [51].

FIGURE 11.32 Jackknifing derailment due to high buff forces and body/coupler geometry in a curve.

FIGURE 11.33 Stringlining derailment due to high draft forces and body/coupler geometry in a curve.

Such derailments are more likely to occur in long, high-tonnage freight trains with heavy buff and draft loads, exacerbated by improper train handling. Contributing factors include undulating terrain, vehicles of substantially differing lengths and use of locomotives with no coupler alignment control.

11.4.4 Derailment Caused by Vehicle Overspeed

Overspeed derailments occur when a train enters a curve at an excessive speed, usually due to operator error or brake system failure. While this type of derailment is relatively rare, its consequences can be catastrophic [52].

Risk of overspeed derailment can be assessed by a quasi-static analysis of uncompensated centrifugal acceleration and crosswind force [53]:

$$Q_{min} = \frac{G}{2} - G\frac{p_c\left(\frac{sv^2}{gR} - h\right)}{s^2} - H_w \frac{p_w}{s} \tag{11.17}$$

where Q_{min} is the vertical wheel force on inner rail, G is the static axle load, p_c is the height of vehicle's centre of gravity, h is cant (superelevation), s is track width (distance between centres of the two rails), g is gravitational acceleration, R is curve radius, H_w is crosswind force per axle and p_w is the effective height of the point of application of crosswind force. The cant deficiency term, $(sv^2/(gR) - h)$, designates the difference between the actual cant in a curve and a theoretical cant that would be required to cancel the centrifugal force for a vehicle travelling at a speed v.

The safety criterion against overspeed derailment is Q_{min} staying above some maximum value, for example, 10% of its static load (see Section 11.3.2.6).

Equation 11.17 is the simplest type of overturning analysis and does not account for displacement of vehicle's gravity centre due to suspension deflection, bogie vibration, etc. More complex forms of the equation such as Kuneida's formula and Japan's Railway Technical Research Institute (RTRI) detailed equation [54] account for these and other factors.

11.5 ASSESSMENT AND PREDICTION OF DERAILMENT

Three types of approach are usually used for assessing and predicting the risks of derailment or diagnosing the causes of derailments: assessment of wheel-rail parameters, dynamic simulations of vehicle-track interaction and vehicle performance track tests. In many cases, all three approaches are applied. A wide range of assessment methods are used worldwide [55]. The following subsections outline some common methods used in North America and Europe.

11.5.1 Assessment of Wheel-Rail Contact Parameters

Assessment of wheel-rail contact parameters may predict the risks of derailment that are the result of unfavourable wheel-rail contact. Since the wheel and rail are operating as a mechanical system involving two-body contact and interaction, this assessment of wheel-rail parameters should include wheels operating on the line where rail condition parameters have been measured (or designed for the new line condition). The contact parameters that may affect derailment include the following:

- Maximum contact angle and length of flange (related to flange climb)
- Rolling radius difference on curves (related to flange climb)
- Effective conicity (related to vehicle lateral instability on tangent track)
- Rail gauge (related to gauge widening)

- Contact positions (related to rail rollover)
- Wheel-rail contact conformity (affects bogie steering and level of lateral forces)
- Rolling radius difference of two wheels on a same axle (affects bogie steering)
- Level of wheel tread hollowing (related to rail rollover, also affects wheel climb and vehicle instability under certain conditions and rolling radius difference)
- Significant wear of wheel flange and rail gauge that can increase wheelset lateral movement and may reduce the effectiveness of restraining rail on curves

A comprehensive view of wheel-rail contact at a system level is important to reveal the overall patterns of the contact. For example, thousands of wheels with different profiles (due to different levels of wear or resulting from different bogie performances) contact a section of rail at different positions and could produce different levels of contact stress. Therefore, the performances of most wheel-rail pairs are of interest in system assessment.

However, derailments are often related to the behaviours of individual wheels and rails that possess undesired shapes. In diagnosing the causes of derailment, the wheel profiles on derailed vehicles and rail profiles at derailment sections must be measured and analysed to study the contributions from the wheel-rail contact geometry.

11.5.2 Dynamic Simulation of Vehicle-Track Interaction

Since the 1980s, computer simulations have been extensively used to study vehicle-track interaction. They are also useful tools to diagnose the causes of derailments. For some derailments, the explicit causes cannot be simply identified. Some derailments cannot even be repeated by track tests that appear to have similar conditions as the derailed vehicles and tracks. The advantages of a computer modelling study are that parameters of vehicle/track can be conveniently (also cost-effectively) varied to investigate the effects of either single parameters or combinations of multiple parameters on derailments. Consequently, if the simulation reveals derailment risk, modifications to vehicle and track for preventing derailments can also be defined by a parametric study.

The flange climb risk can be evaluated in the simulation using the flange climb criteria discussed in Section 11.3.1, including the L/V ratio limit and the exceeding of distance (or time) limit. Simulation of derailments caused by gauge widening and panel shift under dynamic load requires more advanced vehicle/track simulation models that include the capability of assessing rail roll under the load, which results in changing wheel-rail contact condition, and a detailed track model to describe the structure below the rail.

The accuracy of the description of vehicle and of track parameters is crucial for simulation to reflect the actual responses of vehicle and track. For example, a sudden change of wheel-rail forces caused by a large track lateral irregularity combined with wheels with low flange angle could lead to derailment. Therefore, only when this irregularity and the shape of the wheels are accurately described in the model with other vehicle/track parameters will the derailment scenario be reproduced by the simulation. When evaluating a system, it is important to examine the vehicles under the worst track conditions possibly allowed in the system, in order to ensure safe operation under those conditions. When examining newly designed (or modified) vehicles, a certain limit of wear of vehicle elements and wheel profiles should also be considered in the simulations.

11.5.3 Track Tests

On-track tests are generally required for the acceptance of new designs or modifications of vehicles. Track tests are also often conducted for diagnosing performance problems (including derailment) caused by vehicle or track conditions, or a combination of both. These vehicles may have been examined by computer simulation. However, precisely describing every element and parameter in the vehicle/track system is very difficult, especially for those nonlinear elements such as friction elements, damping elements and gaps/stops. Modelling can reveal the trends or probable performances with regard to derailment, while track tests will demonstrate the actual performance under different test conditions for the test unit. Therefore, computer simulations and track tests are often combined efforts in derailment evaluations or diagnoses.

Chapter 11 of the *AAR Manual of Standard and Recommended Practices* describes the regimes of vehicle performance to be examined and the required test conditions [20]. The test regimes in that document include the following:

- Hunting (vehicle lateral instability)
- Constant curving
- Spiral
- Twist, roll
- Pitch
- Yaw, sway
- Dynamic curving (negotiating a curved track with specified alignment and surface irregularities)

Usually, the test unit is instrumented with a number of gauges to measure forces, accelerations and displacements at critical locations on the test unit, depending on the test objective. Instrumented wheelsets have been applied in recent years to more accurately determine wheel-rail interaction forces. The wheel L/V ratio limit (or axle L/V ratio limit) and the exceeding distance (or time) limit are used in AAR Chapter 11 tests to evaluate vehicle curving performance. Wheel unloading is also an important criterion used in the Chapter 11 test. It limits the level of minimum vertical wheel force to no less than 10% of static load.

The FRA has also implemented a requirement for testing of new passenger trains that run at speeds above 90 mph (144 km/h) for passenger vehicles and 80 mph (128 km/h) for freight wagons [12] and is now working to develop testing and analysis requirements standards for operations of passenger vehicles at speeds below 90 mph (144 km/h).

Euronorm EN 14363 [11] requires track tests under all combinations of the following test conditions:

- Test zones: Straight track, large radius curves and intermediate (400 m $\leq$ R $\leq$ 600 m) and small radius curves (250 m $\leq$ R < 400 m)
- Rail inclination: 1:20 and 1:40
- Loading condition: Empty and loaded

The parameters assessed during the track tests include wheel-rail forces and L/V ratios, plus vehicle body and bogie frame accelerations. Safety against derailment is assessed by single wheel L/V ratio, limited to 0.8 in curves with R $\geq$ 250 m and to 1.2 in transition curves.

The Rail Safety and Standards Board (RSSB) of the United Kingdom has issued railway groups standards relevant to engineering acceptance. These standards describe the permissible track forces for railway vehicles [56,57]. Other countries and railway operators issue similar standards and regulations [55].

11.6 PREVENTION OF DERAILMENT

Many derailments could be prevented if factors that can lead to derailment, for example, vehicle- or track-related issues and human factors, could be identified and adequate preventive actions could be taken at an early stage.

Many of the four types of derailments discussed in the previous sections have a common cause of high lateral force at the wheel-rail interface. Therefore, any conditions that lead to high lateral forces or lead to a lowering of the ability of the system to sustain the force should be corrected.

Some general preventive methods are introduced in the following subsections. Still, the wide range of vehicle types and track conditions dictate that any methods adopted by a system to prevent derailment must be carefully assessed by considering the specific vehicle and track conditions in that system to ensure the effectiveness of the methods.

11.6.1 Wheel-Rail Profile Optimisation

As discussed in previous sections, the wheel and rail profile contact geometry can have a significant effect on the forces generated at the wheel-rail interface that lead to derailments. The following sections describe different aspects of wheel profile optimisation and remediation to decrease the risk of derailments. The optimisation starts with the careful design of wheel and rail profiles, taking into account vehicle type, mass and operating conditions such as route curvature, track structure design and operating speeds.

11.6.1.1 Addressing Wheel Flange Angle

To prevent flange climb derailment, the maximum wheel flange angle should be sufficiently high to increase the allowed L/V ratio limit. For a new wheel profile design, a higher flange angle should be emphasised. A wheel flange contact angle above 70 degrees is commonly recommended. For many railway systems, the wheels naturally wear to angles of 75 degrees or more. However, for systems with very sharp curves and consequent very high axle angles of attack, such as streetcars, the natural worn angle will be less.

Should flange climb derailments be a concern to an existing system that has adopted wheels with a low flange angle, a transition to a higher flange angle might be considered. However, this transition needs to be carefully planned according to the capacity of wheel truing and rail grinding in the system [21].

Flange angle changes with wear, sometimes significantly. For example, the flange angle on the British Rail P5 wheel profile was found to be increasing rapidly with wear. For this reason, British Standard GM/RT2466 allows 60-degree flange angles on freshly profiled P5 wheelsets [58], and Euronorm EN 14363 requires taking 'previous service experience' of increasing flange angle into account when calculating Nadal's criterion [11].

11.6.1.2 Removing Hollow-Worn Wheels

Removing significantly hollow-worn wheels from the system may reduce the risk of gauge widening and rail rollover derailment, as described in Section 11.3.3.3. Hollow wheels can also reduce the rolling radius difference required in curving and increase lateral instability on tangent track [59,60]. The current AAR interchange standard for freight cars in North America limits wheel hollowing to 5 mm [61]. A 4-mm hollow-wheel removing limit has been recommended by the TTCI for the North American interchange operation. In the future, the aim would be to eventually remove the wheels with 3-mm hollow tread from the service [62].

11.6.1.3 Wheel Taper and Effective Conicity

Higher effective conicity values may improve steering in curves, but they also decrease the lateral stability of the vehicle on tangent track and lower the onset speed of hunting (see Section 11.4.1). The remediation starts with the careful design of wheel and rail profiles, taking into account vehicle type, mass and operating speed. For example, wheel profiles designed for high-speed rolling stock are generally designed with lower conicity (see Chapter 3, Section 3.3), which improves lateral stability on tangent track at the expense of curving performance.

There have been multiple studies on the optimisation of wheel profiles for hunting stability, lateral forces, derailment coefficient in curves, wear and ride index, etc., using various computer simulation methods [63].

With any wheel profile design, it is important to remember that effective conicity may significantly increase with wheel wear [64,65].

Railway Vehicle Derailment and Prevention

11.6.1.4 Contact Conditions at the Wheel-Rail Interface

Contact conditions at the wheel-rail interface can have a significant impact on wheel-rail wear and vehicle performance. The position and number of contact points between a wheel and rail are determined by the shapes of the wheel and rail profiles. Three common contact conditions that can occur in curves are single-point contact, two-point contact and conformal contact [66].

As a solid-axle wheelset negotiates a curve, it shifts sideways and generates a rolling radius difference, which, in turn, produces longitudinal creep forces that allow the wheelset to steer through the curve (see Figure 11.18).

11.6.1.4.1 Single-Point Contact

When the outer wheel of an axle contacts the high rail at a single point, usually on the tread or in the flange root, the wheelset is able to generate the required rolling radius difference and longitudinal forces to steer through the curve. The longitudinal steering forces produced from the rolling radius difference 'use up' some of the available wheel-rail friction at the contact locations. On the outside (high) rail, this interaction reduces the lateral force in the contact plane that contributes to flange climb, as discussed in Section 11.3.2. On the inside (low) rail, this interaction reduces the lateral force that acts to push the wheelset into hard flange contact.

11.6.1.4.2 Two-Point Contact

Two-point contact occurs when a single wheel contacts the rail at two distinct locations. Severe two-point contact occurs when contact exists at one point on the wheel tread and another point on the flange of the wheel. In curves where the radius is smaller than 2000 ft (610 m), flange contact between outside wheel and rail becomes more likely [66]. When the outer wheel contacts the high rail on the tread and the flange of the wheel, the contact conditions tend to generate competing rolling radius differences that reduce the longitudinal forces and the steering ability of the wheelset. On the outside (high) rail, the contact point on the flange can increase the lateral force in the contact plane, which can contribute to flange climb. On inside (low) rail, the interaction increases the lateral force that acts to push the wheelset into hard flange contact and exacerbates the problem.

The presence of flange lubrication can reduce the longitudinal forces generated on the flange contact point and may cause the forces generated on the tread contact point (which has a smaller rolling radius difference) to dominate, thereby reducing the overall steering forces generated by the wheelset.

11.6.1.4.3 Conformal Contact

Conformal contact occurs when the outer wheel and rail profiles wear and take on similar shapes with similar curve radii. Wheels and rails with conformal contact in curves tend to experience lower lateral forces and rolling resistance, but wheels and rails that are too conformal can also contribute to problems such as rolling contact fatigue [66]. In general, wheelsets with strong two-point contact between the tread and flange of the outside wheel and high rail tend to produce higher lateral forces and higher wear rates in tight curves than wheelsets with single-point contact or conformal contact.

11.6.2 Independently Rotating Wheels

As discussed in Section 11.3.2.4, independently rotating wheels tend to run with a larger angle of attack in curves than the conventional coupled wheelsets and can generate greater lateral forces that increase the risk of wheel climb. Therefore, independent rotating wheels require more carefully designed wheel profiles and control mechanisms for curving. Elkins [67] and Suda et al. [68] have proposed self-steering bogies equipped with independent rotating wheels on the trailing axle only. In recent years, the concept of active control has been studied [69,70]. As the yaw angle and lateral motion of wheels can be accurately controlled, more applications of independent rotating wheels can be expected.

11.6.3 Installation of Guardrails or Restraining Rails on Sharp Curves

Restraining rails and guardrails have been frequently applied in transit operations on sharp curves to prevent flange climb derailment (or to reduce gauge wear on the high rail) [23,66,71]. The restraining rails/guardrails are generally installed inside of the low (inner) rail, as shown in Figure 11.34. In extremely sharp curves, restraining rails are sometimes installed on both the inside and outside rails.

The clearance between the low rail and the restraining rail is critical for the effectiveness of restraining rails. Too tight a clearance may reduce wheelset rolling radius difference required for bogie curving by limiting the flange contact on the high rail. Overwide clearance may cause complete loss of the restraining function.

Wear at the wheel flange back and the contact face of the restraining rail can vary the clearance between the low rail and the restraining rail. The wheel flange and high rail gauge wear can affect the amount of wheelset lateral shift on curves. Note that track lateral geometry irregularities, including alignment and gauge variations, can also affect the performance of restraining rails.

11.6.4 Vehicle Dynamic Response and Optimising the Vehicle Suspension

11.6.4.1 Bogie Lateral and Yaw Stiffness

Lateral stiffness of the bogie refers to the resistance of the two wheelsets to lateral movement with respect to each other and with respect to the bogie frame and is primarily a function of the lateral stiffness of the primary suspension. Yaw stiffness refers to the resistance of the wheelsets to rotation around the vertical axis and is primarily the function of the longitudinal stiffness of primary suspension.

Lateral stability of the vehicle requires a combination of lateral and yaw stiffness of the bogie, which can be fine-tuned to achieve desired vehicle characteristics [72,73]. Low lateral and/or yaw stiffness can cause lateral instability on tangent track, which may result in a derailment (see Section 11.4.1). On the other extreme, excessive yaw stiffness leads to poor steering by making it harder for wheelsets to assume a radial position in a curve. This leads to higher lateral forces and L/V force ratios on flanging wheels in curves, which increase the risk of flange climb.

Depending on the design, lateral and yaw stiffness of a bogie can decrease significantly during its life cycle. For example, severe wear on bearing adapters and adapter pads on three-piece bogies allows for excessive longitudinal and lateral motion of the bearing adapters inside the pedestals, which leads to lateral instability [64].

11.6.4.2 Vertical and Roll Stiffness and Wheel Load Equalisation

Wheel load equalisation is the ability of a bogie to maintain constant vertical wheel loads when negotiating track with transition curves and uneven track. Good equalisation characteristics reduce vertical wheel unloading, decreasing lateral/vertical force ratios at the wheel-rail interface and reducing the risk of flange climb derailment.

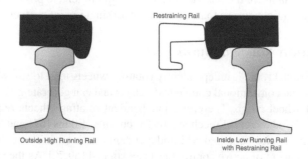

FIGURE 11.34 Restraining rail. (From Iwnicki, S. (ed.), Handbook of Railway Vehicle Dynamics, CRC Press, Boca Raton, FL, 2006. With permission.)

Railway Vehicle Derailment and Prevention

In Europe, wheel load equalisation performance of a bogie is tested on a twist test rig in accordance with Euronorm EN 14363 [11]. The APTA has developed a standard for testing minimum wheel load equalisation characteristics of passenger bogies [26]. For freight bogies in North America, wheel load equalisation is addressed in AAR standards on trackworthiness criteria [41].

11.6.4.3 Bogie Warp and Rotational Resistance

Warp stiffness is a measure of resistance of a bogie to lozenging (parallelogram distortion in the horizontal plane). Low warp stiffness and damping of bogies are generally undesirable, because they can cause lateral instability. This condition can occur due to low shear stiffness and insufficient damping of a bogie's secondary suspension, which can gradually develop as bogie components such as friction dampers in three-piece bogies are worn down.

A bogie's resistance to yaw rotation with respect to the vehicle body affects curving performance as well as lateral stability. High resistance, such as that caused by excessive friction on centre bowls or side bearers of freight bogies [74], leads to higher lateral forces in curves. On the other hand, low rotational resistance decreases lateral stability.

Passenger vehicles and locomotives often use a combination of centre linkage mechanism and hydraulic yaw dampers to simultaneously ease rotation in curves and preserve lateral stability of tangent track. Most freight wagons use dry friction forces to control bogie rotation, creating a trade-off between curving ability and lateral stability of the vehicle [75].

11.6.5 Vehicle Loading

Vertical wheel load imbalance (skew loading) increases a risk of flange climb derailment by affecting a bogie's rotational resistance and wheel load equalisation [41,76]. Load imbalance can be caused by an eccentric cargo loading, vehicle or bogie frame twist, broken suspension elements or air spring load imbalance [77].

One mitigation method is to follow the limits for vehicle skew loading using wheel load detectors. The loading guidelines of the International Union of Railways (UIC) [78,79] specify the maximum longitudinal and lateral loading skew ratios:

$$\frac{E_1}{E_2} \leq \begin{cases} 2 \text{ (vehicles with 2 axles)} \\ 3 \text{ (vehicles with > 2 axles)} \end{cases} \tag{11.18}$$

$$\frac{R_1}{R_2} \leq 1.25 \tag{11.19}$$

where E_1 and E_2 are static vertical loads under the two bogies, and R_1 and R_2 are static vertical loads under the two wheels of a single axle.

A study performed under the European D-RAIL project suggested a 1.35 safety limit for a bogie-side lateral skew ratio, a 1.7 limit for a single axle lateral skew ratio and a 3.0 limit for a longitudinal skew ratio [74]. In addition, a diagonal skew ratio of a tare vehicle was defined as the ratio of the sum of vertical loads on the left front and right rear wheels to the loads on the right front and left rear wheels:

$$\varphi_{\text{diag}} = \max\left[\frac{Q_{11r} + Q_{12r} + Q_{21l} + Q_{22l}}{Q_{11l} + Q_{12l} + Q_{21r} + Q_{22r}}, \frac{Q_{11l} + Q_{12l} + Q_{21r} + Q_{22r}}{Q_{11r} + Q_{12r} + Q_{21l} + Q_{22l}}\right] \tag{11.20}$$

It was recommended that vehicles with diagonal skew ratios over 1.3 must be inspected for chassis twist. Vehicles with a ratio over 1.7 must be stopped and taken out of service immediately.

11.6.6 Train Marshalling and Handling

Risks of derailment due to longitudinal train forces can be reduced by minimising steady-state and transient buff-draft forces by a combination of [44]:

- Careful train handling to minimise in-train forces and slack action
- Proper placement of remote locomotives in trains using distributed power
- Avoidance of concentrating loaded wagons at the rear end of the trains behind empty wagons
- Placement of the wagons with the least amount of slack on the head end of the train and those with the most amount of slack at the rear whenever possible

Following additional risk mitigation strategies are directed at limiting coupler angles:

- Avoiding using locomotives with nonaligning couplers or limiting their number in a train
- Restricting the placement of long-vehicle/short-vehicle combinations and other certain vehicle types within the train, as recommended in the *AAR Train Make-Up Manual* [44]
- Taking coupler angularity effects into account when designing vehicles; the *AAR Manual of Standards and Recommended Practices*, Section C, Part II, Chapter 2 [80] specifies the methodology for determining the minimum permitted radius of track curvature as a function of vehicle dimensions, including body and coupler lengths, and, for interoperable rolling stock, the maximum allowable bogie side L/V ratio may not exceed 0.82 in a 10-degree ($R = 175$ m) curve under a draft load of 200,000 lbs (890 kN)
- Requiring minimum lengths of straight track between reverse curves [81,82]

11.6.7 Lubrication and Wheel-Rail Friction Modification

Friction plays an important role in the wheel-rail interface (see Chapters 7 and 8). It affects many wheel-rail interaction scenarios. With respect to derailment, proper lubrication at the wheel-rail interface can reduce wheel lateral forces, because wheel lateral creep force (F_{lat}) saturates at a lower level ($F_{lat} = \mu N$, where N is the contact normal force). Therefore, the potential of wheel climbing and gauge widening is reduced. As illustrated in Figure 11.6, the limiting wheel L/V ratio for wheel climb increases with the decreases in the friction coefficient at the contact surface. However, wheel-rail lubrication is not recommended as a solution to a derailment problem. This is because lubrication systems are not fail-safe. If the lubrication method fails, the system can quickly revert to a high-friction condition with increased risk of flange climb.

In recent years, top-of-rail friction management, with an intermediate range of coefficient of friction (from 0.2 to 0.4), has been tested [83,84]. Application of top-of-rail friction modifiers is intended to reduce rolling resistance and corrugation and eliminate wheel squeal. Top-of-rail friction management can also be expected to reduce vehicle lateral instability on tangent track.

Application of wheel-rail lubrication and friction modifiers must be carefully managed. Misapplication can cause a variety of issues such as increased wheel and rail wear, increased bogie warping, rail rollover and gauge widening and the possibility of derailment [85–87].

11.6.8 Track Geometry Inspection and Maintenance

Severe lateral and vertical track irregularities and gauge widening are significant causes of derailment. On many railways, track geometry recording vehicles are used regularly to survey the track. The frequency of survey varies greatly (between monthly and yearly) according to track category [12]. Track is categorised based on a combination of maximum speed and annual gross tonnage. Regular visual inspection is often carried out at much shorter intervals (between once and twice a week). Regulations of track geometry for different regions and railways may be different. However, maintenance is normally required when the track geometry deviations exceed the specified limits.

Note that newly laid or newly maintained track requires special attention due to its low lateral strength. Speed restrictions may be required for a period to let the track settle.

11.6.9 Switches and Crossings

Switches and crossings present special challenges:

- Tight curve radii and/or sharp entry angles in switches lead to high lateral wheel-rail forces, especially in facing point movements on the diverging route [74].
- Cant deficiency and/or track twist in switches contributes to vertical wheel unloading.
- The profile of switch rails makes flange climb derailment more likely compared with stock rail [74]; switch points are subject to significant wear, decreasing the wheel-rail contact angle and lowering the safety threshold against flange climb derailment [88].
- Discontinuity in rolling surface, inherent in the design of most frogs and crossings, may cause the wheel to enter the wrong flange way and result in a wheel drop derailment; this is mitigated by the use of restraining rails in the design of frogs and crossings.
- Field-side switch point protectors, sometimes installed at switch points to decrease rail wear by pushing the wheel flange laterally towards the centre of the track, are known to have caused wheel climb derailments by providing a grip surface for the chamfer on the field side of the wheels; this particular risk is mitigated by specifying the geometry of wheel chamfers [88].

Because switches and crossings pose a higher derailment risk than plain track, maintenance tolerances for track geometry in switches may need to be tighter [74].

11.6.10 System Monitoring

Derailments usually occur from a combination of unfavourable vehicle and track conditions. Normally, only a single vehicle with a particular problem derails in a section of track with an adverse condition. In comparison, a bogie with a particular defect usually does not derail in every section of track. A program of system monitoring can be implemented to detect any derailment-related vehicle and track problems, including wheel and rail profiles, in their early stages [89]. Corrective action then can be promptly taken to prevent the occurrences of derailment.

REFERENCES

1. C.S. Hutchinson, The report of the derailment of a passenger train at Carnforth, December 1881. Available at: http://www.railwaysarchive.co.uk/docsummary.php?docID=1453.
2. Transportation Safety Board of Canada, Runaway and main-track derailment: Montreal, Maine & Atlantic Railway freight train MMA-002, Lac-Mégantic, Quebec, 6 July 2013, Railway investigation report R13D0054, 2013.
3. J.M. Shultz, M.P. Garcia-Vera, C.G. Santos, J. Sanz, G. Bibel, C. Schulman, G. Bahouth, Y.D. Guichot, Z. Espinel, A. Rechkemmer, Disaster complexity and the Santiago de Compostela train derailment, *Disaster Health*, 3(1), 11–31, 2016.
4. V. Barinov, Faulty switch pointed at sentences. Metro staff members and a contractor's representative are convicted in the first Metro disaster trial (in Russian), Kommersant 206, p. 5, November 10, 2015. Available at: https://www.kommersant.ru/doc/2850084.
5. Office of Railroad Safety, FRA guide for preparing accident/incident reports, Report No. DOT/FRA/RRS-22, Federal Railroad Administration, Washington, DC, 2011.
6. FRA Office of Safety Analysis, Train accidents and rates, Available at: https://safetydata.fra.dot.gov/OfficeofSafety/publicsite/query/TrainAccidentsFYCYWithRates.aspx.

7. X. Liu, M.R. Saat, C.P.L. Barkan, Analysis of causes of major train derailment and their effect on accident rates, *Journal of the Transportation Research Board*, 2289(1), 154–163, 2012.
8. M. Robinson, P. Scott, B. Lafaix, G. Kozyr, A. Zarembski, G. Vasic, F. Franklin, B. Gilmartin, A. Schoebel, B. Ripke, Summary report and database of derailment incidents, D-RAIL Report SCP1-GA-2011-285162, 2012.
9. F.B. Blader, A review of literature and methodologies in the study of derailments caused by excessive forces at the wheel/rail interface, Report R-717, Association of American Railroads, Chicago, IL, 1990.
10. J. Nadal, Locomotives à vapeur, Collection Encyclopédie Scientifique, Bibliothèque de Mecanique Appliquée et Genie, Vol. 186, Paris, France, 1908.
11. EN 14363:2005(E), Railway applications – testing for the acceptance of running characteristics of railway vehicles – testing of running behaviour and stationary tests, European Committee for Standardization, Brussels, Belgium, 2005.
12. United States Department of Transportation, Code of Federal Regulations, Title 49 – Transportation, Part 213 – Track Safety Standards, Federal Railroad Administration, Washington, DC, 2011.
13. J.J. O'Shea, A.A. Shabana, Further investigation of wheel climb initiation: Three-point contact, *Journal of Multi-body Dynamics*, 231(1), 121–132, 2017.
14. H. Weinstock, Wheel climb derailment criteria for evaluation of rail vehicle safety, In: *Proceedings ASME Winter Annual Meeting*, 9–14 December 1983, New Orleans, LA, Paper 84-WA/RT-1, 1984.
15. W.C. Shust, J.A. Elkins, S. Kalay, M. El-Sibaie, Wheel-climb derailment tests using AAR's track loading vehicle, Report R-910, Association of American Railroads, Washington, DC, 1997.
16. H. Wu, J. Elkins, Investigation of wheel flange climb derailment criteria, Report R-931, Association of American Railroads, Washington, DC, 1999.
17. J. Elkins, H. Wu, New criteria for flange climb derailment, In: *Proceedings IEEE/ASME Joint Railroad Conference*, 4–6 April 2000, Newark, NJ, American Society of Mechanical Engineers, New York, 1–7, 2000.
18. T. Matsudaira, Dynamics of high speed rolling stock, *Japanese National Railways RTRI Quarterly Report*, Special Issue, 1963.
19. H.H. Koci, C. A. Swenson, Locomotive wheel-rail loading – A systems approach, In: *Proceedings Heavy Haul Railways Conference*, 18–22 September 1978, Perth, Australia, The Institution of Engineers Australia, Paper F.3, 1978.
20. AAR, Manual of standards and recommended practices, Section C, Part II, Volume 1, Chapter 11, Association of American Railroads, Washington, DC, 2015.
21. H. Wu, X. Shu, N. Wilson, Flange climb derailment criteria and wheel/rail profile management and maintenance guidelines for transit operations, *Transit Cooperative Research Program Report 71*, Volume 5, Washington, DC, 2005.
22. N. Wilson, X. Shu, H. Wu, J. Tunna, Distance-based flange climb L/V criteria, *TTCI Technology Digest*, TD-04-012, Pueblo, CO, 2004.
23. Parsons Brinckerhoff Quade & Douglas, Inc., Transit Cooperative Research Program Report 57: Track design handbook for light rail transit, National Academy Press, Washington, DC, 2000.
24. APTA PR-M-S-015-06, Standard for wheel flange angle for passenger equipment, American Public Transportation Association, Washington, DC, 2007.
25. N. Wilson, X. Shu, K. Kramp, Effects of independently rolling wheels on flange climb derailment, In: *Proceedings ASME International Mechanical Engineering Congress*, 13–19 November 2004, Anaheim, CA, Paper IMECE2004-60293, 2004.
26. APTA PR-M-S-014-06, Wheel load equalization of passenger railroad rolling stock, Revision 1, American Public Transportation Association, Washington, DC, 2017.
27. United States Department of Transportation, FRA Safety Advisory 2013-02, Low-speed, wheel-climb derailments of passenger equipment with "stiff" suspension systems, Federal Railroad Administration, Washington, DC, 2013.
28. S.E. Mace, D.A. DiBrito, R.W. Blank, S.L. Keegan, M.G. Allran, Effect of wheel profiles on gage widening behaviour, In: *Proceedings ASME/IEEE Joint Railroad Conference*, 22–24 March 1994, Chicago, IL, Institute of Electrical and Electronics Engineers, New York, NY, 51–56, 1994.
29. S. Mace, R. Pena, N. Wilson, D. DiBrito, Effect of wheel-rail contact geometry on wheelset steering forces, *Wear*, 191(1–2), 204–209, 1996.
30. H.M. Tournay, H. Wu, T. Guins, The influence of hollow-worn wheels on the incidence and costs of derailment, Report R-965, Association of American Railroads, Transportation Technology Center, Inc., Pueblo, CO, 2004.
31. S. Singh, D.D. Davis, Effect of switch-point risers on turnout performance, *TTCI Technology Digest*, 98–018, Pueblo, CO, 1998.

32. A. Kish, G. Samavedam, Track buckling prevention: Theory, safety concepts, and applications, Report No. DOT/FRA/ORD-13/16, Federal Railroad Administration, Washington, DC, 2013.
33. K. Knothe, F. Bohm, History of stability of railway and road vehicles, *Vehicle System Dynamics*, 31(5–6), 283–323, 1999.
34. D. Li, W. Shust, Investigation of lateral track strength and track panel shift using AAR's track loading vehicle, Report No. R-917, Association of American Railroads, Transportation Technology Center, Pueblo, CO, 1997.
35. A. Prud'homme, La résistance de la voie aux efforts transversaux exercés par le matériel roulant, *Revue Général des Chemins de Fer*, 731–766, 1967.
36. D.R. Ahlbeck, H.D. Harrison, An evaluation of the Canadian LRC train from the wayside track load measurements on curved and perturbed tangent track, Battelle Columbus Laboratory Report to AMTRAK, May 1977.
37. G. Samavedam, F. Blader, D. Thomson, Track lateral shift: fundamentals and state-of-the-art review, Report No. DOT/FRA/ORD-96/03, Federal Railroad Administration, Washington, DC, 1996.
38. A. Kish, G. Samavedam, D. Wormley, New track shift safety limits for high-speed rail applications, In: *Proceedings 5th World Congress on Railway Research*, 25–29 November 2001, Cologne, Germany, PB011756, 2001.
39. A.H. Wickens, *Fundamentals of Rail Vehicle Dynamics: Guidance and Stability*, Swets & Zeitlinger, Lisse, the Netherlands, 2003.
40. APTA PR-M-S-017-06, *Standard for the Definition and Measurement of Wheel Tread Taper*, American Public Transportation Association, Washington, DC, 2007.
41. AAR, Manual of standards and recommended practices, Section D, Trucks and truck details, Specification M-976: Truck performance for rail cars, Association of American Railroads, Washington, DC, 2013.
42. Railway Group Standard GM/RT2141 Issue 3, Resistance of railway vehicles to derailment and rollover, Rail Safety and Standards Board, London, UK, 2009.
43. H. Weinstock, H.S. Lee, R. Greif, *A Proposed Track Performance Index for Control of Freight Car Harmonic Roll Response*, Transportation Research Record 1006, Transportation Research Board, Washington, DC, 9–16, 1985.
44. P.M. Lovette, J. Thivierge, Train make-up manual, Report R-802, Association of American Railroads Research and Test Department, AAR Technical Center, Chicago, IL, 1992.
45. K. Koch, Tank car operating environment study – Phase I, Report No. DOT/FRA/ORD-07/22, Federal Railroad Administration, Washington, DC, 2007.
46. K. Koch, Measurement of revenue service coupler force environment, 286,000-pound gross rail load boxcar equipped with cushioning units, Report R-1006, Association of American Railroads, Washington, DC, 2014.
47. K. Koch, Measurement of revenue service coupler force environment, bi-level Autorack car equipped with cushioning units, Report R-1010, Association of American Railroads, Washington, DC, 2015.
48. W.R. McGovern, W.B. Egan, D.A. Harrison, P.L. Montgomery, R.R. Newman, D.G. Orr, J. Thivierge, C.J. Whitmore, L.L. Valko, Train derailment cause finding, Report R-522, Association of American Railroads, Washington, DC, 1990.
49. Transportation Safety Board of Canada, Main-track derailment, Canadian National train M37631-30, Pickering, Ontario, 30 March 2010, Railway investigation report R10T0056, 2013.
50. Transportation Safety Board of Canada, Main-track train derailment: Canadian National train number M36231-20, Brighton, Ontario, 21 March 2009, Railway investigation report R09T0092, 2009.
51. M. El-Sibaie, Dynamic buff and draft testing techniques, Report No. DOT/FRA/ORD-92/31, Association of American Railroads, Transportation Test Center, Pueblo, CO, 1992.
52. A. Matsumoto, Y. Michitsuji, K. Tanifuji, Train-overturned derailments due to excessive speed – analysis and countermeasures, In: M. Rosenberger, M. Plöchl, K. Six, J. Edelmann (Eds.), *Proceedings 24th IAVSD Symposium, The Dynamics of Vehicles on Roads and Tracks, Graz, Austria*, 17–21 August 2015, Taylor & Francis Group, London, UK, 1549–1554, 2016.
53. C. Esveld, *Modern Railway Track*, 2nd ed., MRT-Productions, Delft, the Netherlands, 2001.
54. Y. Kurihara, A. Oyama, K. Doi, Y. Yasuda, Introduction of new methods for train operation control in strong winds, *JR East Technical Review*, 27, (17–22), 2013.
55. N.G. Wilson, R. Fries, M. Witte, A. Haigermoser, M. Wrang, J. Evans, A. Orlova, Assessment of safety against derailment using simulations and vehicle acceptance tests: A worldwide comparison of the state-of-the-art assessment methods, *Proceedings 22nd IAVSD Symposium*, 14–19 August 2011, Manchester, UK, CD-ROM, 2011.

56. British Railways Board, Permissible track forces for railway vehicles (GM/TT0088), 1993.
57. Network Rail, Commentary on permissible track forces for railway vehicles (GM/RC2513), 1995.
58. Railway Group Standard GM/RT2466 Issue 1, Railway wheelsets, Rail Safety and Standards Board, London, UK, 2003.
59. K. Sawley, C. Urban, R. Walker, The effect of hollow-worn wheels on vehicle stability in straight track, *Wear*, 258(7–8), 1100–1108, 2003.
60. K. Sawley, H. Wu, The formation of hollow-worn wheels and their effect on wheel/rail interaction, *Wear*, 258(7–8), 1179–1186, 2003.
61. AAR, 2018 *Field Manual of the AAR Interchange Rules*, Association of American Railroads, Washington, DC, 2018.
62. K. Sawley, S. Clark, The economics of removing hollow wheels from service, *TTCI Technology Digest*, TD-99-034, Pueblo, CO, 1999.
63. B. Liu, T.X. Mei, S. Bruni, Design and optimisation of wheel-rail profiles for adhesion improvement, *Vehicle System Dynamics*, 53(3), 429–444, 2016.
64. N. Wilson, H. Wu, H. Tournay, C. Urban, Effects of wheel/rail contact patterns and vehicle parameters on lateral stability, *Vehicle System Dynamics*, 48(S1), 487–503, 2009.
65. O. Polach, D. Nicklisch, Wheel/rail contact geometry parameters in regard to vehicle behaviour and their alteration with wear, In: *10th International Conference on Contact Mechanics (CM2015)*, 30 August – 3 September 2015, Colorado Springs, CO, 2015.
66. X. Shu, N.G. Wilson, Track-Related Research, Volume 7: Guidelines for guard/restraining rail installation (TCRP Report 71), National Academies Press, 2010.
67. J.A. Elkins, Independently rotating wheels: a simple modification to improve the performance of the conventional three-piece Truck, In: *Proceedings Ninth International Wheelset Congress*, Montreal, Canada, 1988.
68. Y. Suda, R. Nishimura, N. Kata, A. Matsumoto, Y. Sata, H. Ohno, M. Tanimoto, E. Miyauchi, Self-steering trucks using unsymmetrical suspension with independently rotating wheels – comparison between stand tests and calculations, In: R. Fröhling (Ed.), *Proceedings 16th IAVSD Symposium held in Pretoria*, South Africa, 30 August – 3 September 1999, Swets & Zeitlinger, Lisse, the Netherlands, 180–190, 2000.
69. T.X. Mei, R.M. Goodall, Wheelset control strategies for a two-axle railway vehicle, In: R. Fröhling (Ed.), *Proceedings 16th IAVSD Symposium held in Pretoria*, South Africa, 30 August – 3 September 1999, Swets & Zeitlinger, Lisse, the Netherlands, 653–654, 2000.
70. S. Shen, T.X. Mei, R.M. Goodall, J. Pearson, G. Himmelstein, A study of active steering strategies for railway bogies, In: M. Abe (Ed.), *Proceedings 18th IAVSD Symposium held in Atsugi, Kanagawa, Japan*, 24–30 August 2003, Taylor & Francis Group, London, UK, 182–291, 2004.
71. X. Shu, N.G. Wilson, Use of restraining/guard rail study, Transit Cooperative Research Program Research Results Digest 82, Washington, DC, 2007.
72. H. Tournay, Supporting technologies – vehicle track interaction, In: *Guidelines to Best Practices for Heavy Haul Railway Operations: Wheel and Rail Interface Issues*, International Heavy Haul Association, Virginia Beach, VA, 2001.
73. C.O. Phillips, H. Weinstock, *The Reduction of Wheel/Rail Curving Forces on U.S. Transit Properties*, Transportation Research Record 1071, Transportation Research Board, Washington, DC, 6–15, 1986.
74. A. Ekberg, B. Pålsson, D. Sala, D. Nicklisch, E. Kabo, F. Braghin, P. Allen, P. Shackleton, T. Vernersson, M. Pineau, Development of the future rail freight system to reduce the occurrences and impact of derailment, D-RAIL Report SST.2011.4.1-3, 2013.
75. H. Wu, Effects of truck center bowl lubrication on vehicle curving and lateral stability, Report R-959, Association of American Railroads, Transportation Technology Center, Inc., Pueblo, CO, 2002.
76. G. Astin, T. Andersen, Assessment of freight train derailment risk reduction measures: A3 -functional and performance assessment, Report ERA/2010/SAF/S-03, BA000777/04, European Railway Agency, Valenciennes, France, 2011.
77. R. Sarunac, P. Klauser, Effects of secondary suspension imbalance on wheel-climb derailment (Part 2 of 2), Interface, *The Journal of Wheel/Rail Interaction*, 2014.
78. UIC, *Loading Guidelines – Volume 1 – Principles*, 2nd ed., International Union of Railways, Paris, France, 2018.
79. UIC, *Loading Guidelines – Volume 2 – Goods*, 2nd ed., International Union of Railways, Paris, France, 2018.
80. AAR, *Manual of Standards and Recommended Practices*, Section C, Part II, Chapter 2, Association of American Railroads, Washington, DC, 2015.

81. C.L. Gatton, Minimum tangent length between reverse curves for slow speed operation, Report R-228, Association of American Railroads, Chicago, IL, 1976.
82. AREMA, *Practical Guide to Railway Engineering*, 2nd ed., American Railway Engineering & Maintenance of Way Association, Lanham, MD, 2013.
83. S. Zakharov, Wheel/rail performance, In: *Guidelines to Best Practices for Heavy Haul Railway Operations: Wheel and Rail Interface Issues*, International Heavy Haul Association, Virginia Beach, VA, 2001.
84. D.T. Eadie, M. Santoro, J. Kalousek, Railway noise and the effect of top of rail liquid friction modifiers: Changes in sound and vibration spectral distributions, *Wear*, 258(7–8), 1148–1155, 2005.
85. K.J. Laine, N.G. Wilson, Effect of track lubrication on gage spreading forces and deflections, Report R-712, Association of American Railroads, Transportation Technology Center, Inc., Pueblo, CO, 1989.
86. W. Pak, R.W. Gonsalves, Effect of experimental rail profile grinding, flange lubrication, and wheel profile on the steering behaviour of 100-ton freight car trucks, Report S742-82, Canadian Pacific Ltd., 1982.
87. H. Wu, S. Cummings, Investigation of root causes of rail rollover derailments and prevention remedies, Report R-1017, Association of American Railroads, Transportation Technology Center, Inc., Pueblo, CO, 2016.
88. H. Wu, S. Cummings, Causes of locomotive wheel climb at switch point protectors, In: *10th International Conference on Contact Mechanics* (CM2015), 30 August – 3 September 2015, Colorado Springs, CO, 2015.
89. E. Magel, NYCT's Maintenance innovation using research, In: *2018 APTA Rail Conference*, 10–13 June 2018, Denver, CO, American Public Transportation Association, 2018.

12 Rail Vehicle Aerodynamics

Hongqi Tian

CONTENTS

12.1 Terminology 416
    12.1.1 Terminology and Definition 416
    12.1.2 Symbols and Units.. 419
12.2 Introduction.. 419
12.3 Research Methods of Aerodynamics for Rail Vehicles 422
    12.3.1 Wind Tunnel Test.. 422
    12.3.2 Moving-Model Test 424
    12.3.3 Real-Vehicle Test 425
    12.3.4 Numerical Simulation... 426
12.4 Properties of Exterior Flow Structure of Rail Vehicles 431
    12.4.1 Distribution Law and Development Characteristics of Flow Field around Trains....... 431
    12.4.2 Characteristics of Vortex Structure in Flow Field around Trains 433
    12.4.3 Characteristics of Wake Flow of the Train.. 434
    12.4.4 Distribution and Development Characteristics of Flow Field around Train Running in Transient Conditions.. 436
        12.4.4.1 Flow Field Characteristics at Rail Vehicle Crossing 436
        12.4.4.2 Flow Field Characteristics of Vehicles under Crosswind 436
        12.4.4.3 Single Train Passing through Tunnel 436
        12.4.4.4 Trains Passing Each Other in Tunnel 438
12.5 Properties of Aerodynamic Load on Rail Vehicles 439
    12.5.1 Influence Factors of Aerodynamic Load on Rail Vehicles.................................... 439
        12.5.1.1 Single Vehicle Running in Open Air without Environmental Wind.... 439
        12.5.1.2 Vehicles Crossing in Open Air without Environmental Wind 440
        12.5.1.3 Single Vehicle Passing through Tunnel 441
        12.5.1.4 Vehicles Crossing Inside Tunnel.. 441
        12.5.1.5 Single Vehicle Running in Open Air with Crosswind........................... 441
        12.5.1.6 Vehicles Crossing with Crosswind 442
    12.5.2 Distributive Load of Aerodynamic Pressure on Rail Vehicle Surface.................. 442
        12.5.2.1 Single Vehicle Running in Open Air without Environmental Wind..... 442
        12.5.2.2 Vehicles Crossing in Open Air without Environmental Wind 443
        12.5.2.3 Pressure Distribution for Vehicles Running in Tunnel..........................444
        12.5.2.4 Pressure Distribution for Vehicles Running in Open Air with Environmental Wind 444
    12.5.3 Aerodynamic Forces and Moments of Rail Vehicles 446
        12.5.3.1 Aerodynamic Drag of Rail Vehicles..446
        12.5.3.2 Aerodynamic Lift of Rail Vehicles..447
        12.5.3.3 Aerodynamic Lateral Force of Rail Vehicles 448
        12.5.3.4 Aerodynamic Pitching, Yawing and Overturning Moment of Rail Vehicles... 448
    12.5.4 Data Entry of Aerodynamic Load for Vehicle Dynamic Response 451

12.6 Evaluation of Aerodynamic Indexes for Rail Vehicles 452
    12.6.1 Aerodynamic Load on Rail Vehicle 452
        12.6.1.1 Aerodynamic Load in Open Air 452
        12.6.1.2 Aerodynamic Load in Railway Tunnels 452
        12.6.1.3 Safe Distance for Human Body 453
    12.6.2 Rail Vehicle Stability under Strong and Rapidly Changing Wind 454
References 454

12.1 TERMINOLOGY

12.1.1 Terminology and Definition

Aerodynamic loads include six components, which are vectors defined in a Cartesian coordinate system attached to the train. The schematic of aerodynamic loads of a train is shown in Figure 12.1. The point of origin O of the coordinate system can be set at different positions on the train according to the requirement of the aerodynamic analysis. As shown in Figure 12.1, the values of the drag force (F_x), lateral force (F_y) and lift force (F_z) along the X, Y and Z directions, respectively, are positive. The Mx, My and Mz moments follow the right-hand rule of the Cartesian coordinate system. In addition, pressure on the train surface is positive when pointing perpendicularly towards the acting surface.

Relevant terminologies in rail vehicle aerodynamics are defined as follows:

Rail vehicle: Individual vehicle of a train, which could be a locomotive, passenger vehicle or freight vehicle.
Head car: When the train is moving forward, the first vehicle of the train is the head car, which could be a locomotive, traction vehicle or control vehicle.
Tail car: When the train is moving forward, the last vehicle of the train is the tail car.
Middle car: All vehicles marshalled between the head and tail car are referred to as middle cars.
Streamlined head: A three-dimensional geometric profile of the head car that can make the air flow smoothly around its surface, to reduce drag force.
Streamlined head length: Longitudinal length of the streamlined structure of the head car.
Blunt head: A three-dimensional geometric profile whose front surface is nearly perpendicular to the incoming airflow, which leads to significant stagnation effect of air flow and extensive flow separation.
Rail vehicles crossing: Two trains operate in opposite directions on two adjacent rail tracks.
Train-induced airflow: The airflow in the vicinity of the train that is induced by train operation in the absence of environmental wind.

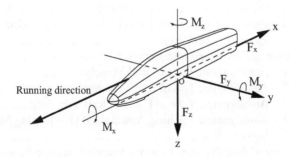

FIGURE 12.1 Schematic of aerodynamic forces and moments of a train.

Environmental wind: The wind with a certain angle between the wind's direction and train's moving direction.
Crosswind: The wind perpendicular to the train's moving direction.
Low-speed flow: Low-speed flow is flow with $Ma < 0.3$.
Subsonic flow: Subsonic flow is flow with $0.3 \leq Ma < 0.8$.
Compressible flow: Flow in which the effects brought about by the change of fluid density cannot be neglected; it generally refers to the flow of $Ma \geq 0.3$.
Incompressible flow: Flow in which the effects brought about by the change of fluid density can be neglected; it generally refers to the flow of $Ma < 0.3$.
Unsteady flow: Flow in which parameters such as velocity, pressure and density change with time.
Viscous flow: Flow considering fluid viscosity.
Inviscid flow: Flow without considering fluid viscosity.
Laminar flow: In fluid dynamics, laminar flow occurs when a fluid flows in parallel layers, with no disruption between the layers.
Turbulent flow: In fluid dynamics, turbulent flow is a pattern of fluid motion characterised by chaotic changes in pressure and flow velocity.
Boundary layer: Thin layer of viscous fluid that forms near the wall when the fluid flows around solid objects.
Laminar boundary layer: Boundary layer in which the flow is laminar.
Turbulent boundary layer: Boundary layer in which the flow is turbulent.
Boundary layer thickness: The distance between the solid wall and the locations where the flow velocity is 0.99 times that of the free stream velocity along the normal direction of the solid wall.
Stagnation point: The point where the velocity of the airflow is zero.
Wake flow: The recirculating flow full of vortices behind a moving or stationary object caused by flow separation or turbulence.
Vortex: Flows rotating around a hypothetical axis.
Streamline: A curve on which any point is tangential to the local velocity vector in the flow field.
Reference point: Origin of local reference coordinate system for vehicle dynamic analysis (see Table 12.1).

TABLE 12.1
Symbols and Units

Symbol	Units	Description	Remarks
A	m²	Surface area	
a	m/s²	Acceleration	
B		Train/tunnel blockage ratio	$B = S_{tr}/S_{tu}$
b	m	Width of vehicle	
c	m/s	Speed of sound	
C_F		Coefficient of aerodynamic force	$C_F = 2F/S\rho V^2$
C_p		Pressure coefficient	$C_p = (p-p_o)/q$
ΔC_p		Coefficient of pressure amplitude	$\Delta C_p = 2(p_{max}-p_{min})/\rho V^2$
Δp		Pressure amplitude	$\Delta p = p_{max}-p_{min}$

(Continued)

TABLE 12.1 (Continued)
Symbols and Units

Symbol	Units	Description	Remarks
C_{Fx}		Coefficients of aerodynamic forces F_x, F_y, F_z	$C_{Fx} = F_x/qS$
C_{Fy}			$C_{Fy} = F_y/qS$
C_{Fz}			$C_{Fz} = F_z/qS$
C_{Mx}		Coefficients of aerodynamic moments M_x, M_y, M_z	$C_{Mx} = M_x/qSL$
C_{My}			$C_{My} = M_y/qSL$
C_{Mz}			$C_{Mz} = M_z/qSL$
Cx		Coefficient of aerodynamic drag for entire train	
c_p	J/kg·K	Specific heat capacity at constant pressure	
c_v	J/kg·K	Specific heat capacity at constant volume	
D	m	Diameter	$D = 2r$
D_L	m	Line spacing	
F	N	Aerodynamic force	
F_x	N	Aerodynamic forces in respective coordinate directions	
F_y			
F_z			
F_x^V	N	Aerodynamic forces acting on reference point for rail vehicle dynamics	
F_y^V			
F_z^V			
g	m/s²	Gravitational acceleration	
h	m	Height	
L	m	Characteristic length	
L_{loco}	m	Length of locomotive	
L_{tr}	m	Length of train	
L_{tu}	m	Length of tunnel	
Ma		Mach number	
M_x	N·m	Aerodynamic rolling moment	
M_y	N·m	Aerodynamic pitching moment	
M_z	N·m	Aerodynamic yawing moment	
M_x^V	N·m	Aerodynamic rolling moment acting on reference point	
M_y^V	N·m	Aerodynamic pitching moment acting on reference point	
M_z^V	N·m	Aerodynamic yawing moment acting on reference point	
m	kg	Mass	
p	Pa	Pressure	
p_o	Pa	Reference static pressure	
p_{max}	Pa	Maximum pressure	
p_{min}	Pa	Minimum pressure	
q	Pa	Dynamic pressure	$q = \rho V^2/2$
Re		Reynolds number	$Re = VL/\upsilon$
r	m	Radius	$r = D/2$
S	m²	Cross-sectional area	
S_{tr}	m²	Cross-sectional area of train	
S_{tu}	m²	Cross-sectional area of tunnel	
T	K	Absolute temperature	
T_f		Tunnel factor	
t	s	Time	
Δt	s	Time interval	
V	m/s	Air speed relative to the train	

(Continued)

Rail Vehicle Aerodynamics

TABLE 12.1 (Continued)
Symbols and Units

Symbol	Units	Description	Remarks
v_r	m/s	Relative speed of two trains	
v_B	m/s	Gust speed	
v_{tr}	m/s	Train speed (Figure 12.2)	km/h may also be used – this shall be indicated
v_w	m/s	Wind speed (Figure 12.2)	
x		Cartesian coordinates (Figure 12.2)	x: Origin is halfway between the axles of the inner wheels
y			y: Origin is track centreline
z			z: Origin is top of rail
β	°	Yaw angle (Figures 12.2 and 12.36)	
γ		Slenderness ratio	$\gamma = L_n/l$
δ	m	Boundary-layer thickness	
κ		Ratio of specific heats	$\kappa = c_p/c_v$
ν	m²/s	Kinematic viscosity	
ρ	kg/m³	Density	

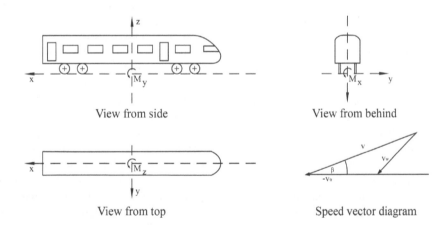

FIGURE 12.2 Schematic of coordinate system.

12.1.2 SYMBOLS AND UNITS

All parameters in Table 12.1 are expressed as SI units. The shape features of the vehicle in Figure 12.2 are based on the European Norm (EN) Standard 14067-1 [1].

12.2 INTRODUCTION

Rail vehicle aerodynamics is a subject that studies the aerodynamic performances concerned with the interaction of a vehicle and its surrounding environment, formation mechanisms of associated forces induced by the relative motion between vehicle and air, and optimisation of the vehicle aerodynamic performance [2]. It follows the fundamental theories of fluid mechanics and classical aerodynamics. As a branch of industrial aerodynamics, rail vehicle aerodynamics is the application and extension of aerodynamics onto the field of rail traffic.

Various aerodynamic problems arise from objects travelling in atmosphere environments with high speed. During the investigation of these problems, the subject of aerodynamics is established from fluid mechanics, which possesses strong engineering applicability. In the early stage, the rapid development of aerodynamics was driven by aerospace engineering. However, aircraft-based aerodynamics is not able to satisfy the requirements of other fields. Unlike automobiles or aircraft, rail vehicles are long and large objects that travel on ground at high speed. The aerodynamic issues arising from the operation of rail vehicles are unique and need to be investigated and solved specifically, such as the pressure impact induced by the crossing of two rail vehicles, the pressure fluctuation caused by vehicles operating through tunnels, as well as the influence of wind on vehicle operation in the open environment. The aerodynamic effects induced by rail vehicle operation become more prominent with the increase of operating speed [2]. Thus, the potential for ongoing increase of the operating speed of rail vehicles relies on the development of rail vehicle aerodynamics.

The increase of rail vehicle speed is the symbol of the development of railway science and technology, and the objective is ever pursued. Since Japan industrialised the world's first high-speed railway in 1964 with a speed of 210 km/h, high-speed trains have been rapidly developed in Britain, France and Germany. The world record of an experimental speed of high-speed train was 574.8 km/h created by France's TGV-A in 2007. And the operational speed of Germany's third-generation ICE high-speed train has reached 330 km/h.

At the beginning of the twenty-first century, China began to develop its high-speed railway system and independently established the Qin-Shen passenger railway (the maximum experimental speed of the train was 321.5 km/h). Thereafter, the world's longest and largest high-speed railway network was formed with a series of high-speed railways being constructed for an operational speed of 350 km/h, including Beijing-Tianjin, Shanghai-Nanjing, Beijing-Shanghai, Beijing-Guangzhou and Harbin-Dalian (maximum crossing speed of each train is 420 km/h). Meanwhile, the Shanghai high-speed maglev line was built (double tracks, maximum operational speed is 430 km/h and maximum crossing speed is 430 km/h against 430 km/h). In addition, the research on a vacuum-tube super-high-speed train with a speed over 1000 km/h has been launched [3]. With the increase of operational speed, breakthroughs have been achieved in the research of rail vehicle aerodynamics.

The shape of high-speed trains is closely related to rail vehicle aerodynamics and directly affects the aerodynamic performances of the train. People pursue the optimal shape of a high-speed train, aiming at improving the aerodynamic performance of the train. Therefore, the shape of the train evolves with the continuous increase of train speed and with the deepening of the research on train aerodynamics. For example, in order to develop the Shinkansen high-speed railway in Tokaido 50 years ago, Japan began to study the aerodynamics-related train shape to solve the aerodynamic problems that arise from the increase of train speed. Initially, the shape design of the train concentrated on the reduction of aerodynamic drag. Therefore, the head of the train was designed as a bullet shape, such as the initial 0-series train, which became known as the 'bullet train'. Besides the dramatic increase of aerodynamic drag with the increase of train speed, many other aerodynamic problems affecting the traffic safety and surrounding environment emerged, as trains are long objects operating on ground. Experts have been working on the improvement of the aerodynamic performance of high-speed trains through shape design and optimisation. The development of new technologies propels the evolution of the shape of rail vehicles. Further increase of train speed, reduction of energy consumption, improvement of passenger comfort and protection of the environment have been achieved through the applications of new technologies on high-speed trains all over the world.

In addition to the wheel-rail high-speed trains, major technological innovations have been made in the research and development of maglev high-speed trains and trains operating in a low vacuum tube. The research of rail vehicle aerodynamics has developed from the low-speed wheel-rail system to subsonic and even supersonic maglev systems.

The negative effects of aerodynamic loads may lead to vehicle damage or device malfunction. The inappropriate design of air conditioning inlets and outlets could cause insufficient ventilation, which results in poor air quality in the passenger compartment and equipment overheating. When high-speed vehicles are crossing, a transient pressure impact will act on the vehicle body. This impact may lead to vehicle wall deformation and sound wave blast that can eventually cause vehicle damage.

The construction of high-speed railways involves tunnel groups and high-rise bridges. These structures create the need for analysing the aerodynamic coupling of the moving vehicle and a tunnel, the moving vehicle and tunnel groups, as well as the moving vehicle and bridges. The flow fields for vehicles operating in the open air and in tunnels exhibit significant differences. When vehicles are operating in the open air, the compressed air in the front region of the head car can expand freely into the atmosphere. Therefore, the pressure change around the train is small. In most scenarios for aerodynamic studies of railway vehicles, incompressible flow is assumed. However, compressible fluid flow should be assumed for vehicle crossing scenarios.

The vehicle entering a tunnel can be compared to a piston moving into a cylinder, where the air flow around the moving object is restricted by the surrounding walls. The heavily compressed air in front of the head car can induce a compression wave that propagates through the tunnel. When the compression wave reaches the exit of the tunnel, a micro-pressure wave will be formed outside of the exit. The micro-pressure wave may cause a sound blast that pollutes the environment. When the tail car enters the tunnel, the flow field at the rear of the tail car is opposite to that of the front car. Therefore, an expansion wave will be created. This expansion wave, along with the compression wave, will propagate throughout the tunnel at the speed of sound. Compressibility of air should be considered for trains passing through the tunnel. The propagation and reflection of expansion and compression waves inside of the tunnel contribute to alternating pressure changes in the pressure field. The alternating pressure will have a negative influence on the ancillary facilities in the tunnel, the vehicle structure and passenger comfort. This negative influence will become more severe in vehicle crossing scenarios. Moreover, the drag force for vehicles operating in tunnels is higher than that in the open air. The various scenarios mentioned above contribute to the subject of vehicle-tunnel coupling aerodynamics.

The aerodynamic performance of railway vehicles is subjected to more severe conditions under strong wind, where the drag force, lift force and lateral force increase drastically, and the lateral stability of vehicles is potentially compromised. A strong crosswind may lead to the overturning of vehicles. Under special wind environments such as a long-span bridge, high-rise bridge and embankments, the change of the surrounding flow field becomes more obvious, and the aerodynamic forces will significantly increase. When trains are passing through curved lines, the superposition of aerodynamic lateral force, aerodynamic lift force and centrifugal force will increase the risk of vehicle overturning. To ensure the safe operation of vehicles in windy areas, systematic investigations on the aerodynamic characteristics of vehicles must be carried out for the determination of the speed limit and the safety control system under windy conditions.

Based on the research scenarios mentioned above, a series of research subjects related to vehicle aerodynamics have been proposed for further investigation. The aerodynamic characteristics of railway vehicles can be studied using numerical simulations, wind tunnel tests, moving model tests and field tests. The research objectives of a vehicle aerodynamic study involve the following contents:

- The investigated train types include rail-wheel high-speed trains, maglev trains, metro, subway and freight trains.
- The speed level of trains includes five speed levels, ranging from 160 to 600 km/h.
- The investigated vehicle shapes include blunt-headed vehicles, straight side walls, vehicle bottom structures with no-bottom skirt-baffle structure and streamlined vehicles.
- The investigated flow field includes both low-speed flow for rail-wheel vehicles and subsonic or supersonic flows for maglev vehicles.

- The research objects include train body, pantograph and human body aerodynamics.
- Research scope covers from vehicle aerodynamics to vehicle-environment coupling aerodynamic characteristics, including vehicle-tunnel (tunnel groups) coupling aerodynamic characteristics, vehicle-bridge (embankment) coupling aerodynamics, vehicle aerodynamic characteristics under strong crosswind and vehicle aerodynamic characteristics under environments of wind, sand, rain and snow.
- Research scope includes vehicle aerodynamic characteristics and mechanisms, vehicle aerodynamic theories and applications, plus numerical simulations, experiments and evaluation for vehicle aerodynamic characteristics.

In summary, vehicle, tunnel, track and the environmental parameters can influence the aerodynamic properties of the rail vehicle system, which consequently impact the dynamic characteristics of vehicles. The aerodynamic forces and moments required in a vehicle dynamics study can be obtained through wind tunnel tests, moving model tests, field tests and numerical simulations under different parameter conditions. The data entry from vehicle aerodynamic analysis to vehicle dynamic analysis will be established. Corresponding assessment criteria for aerodynamic load on dynamic analysis will be proposed to ensure vehicle's operating safety and reduce the negative impacts of vehicle aerodynamics on the environment.

12.3 RESEARCH METHODS OF AERODYNAMICS FOR RAIL VEHICLES

A comprehensive suite of methods for aerodynamics of rail vehicles includes wind tunnel tests, moving-model tests, real-vehicle tests and numerical simulations.

12.3.1 Wind Tunnel Test

A wind tunnel is a specially designed tunnel that can generate manually controllable air flow by using a powerful fan system or other means. Based on the theory of relative motion and similarity principles, a wind tunnel can be used to conduct aerodynamic tests for rail vehicles.

Typically, wind tunnels are classified by the range of flow speed or its Mach (Ma) number in the test section, including low-speed wind tunnel ($Ma < 0.3$), subsonic wind tunnel ($0.3 \leq Ma < 0.8$), transonic wind tunnel ($0.8 \leq Ma < 1.4$), supersonic wind tunnel ($1.4 \leq Ma < 5$) and hypersonic wind tunnel ($Ma > 5$). For rail vehicles, the low-speed and subsonic wind tunnels are appropriate if $Ma < 0.5$. Two types of wind tunnels are used for rail vehicles: open-circuit wind tunnel (Figure 12.3) and closed-circuit wind tunnel (Figure 12.4). To reduce ground effects, the wind tunnel for rail vehicle tests should be equipped with a special ground device, using a conveying belt or ground with suction system.

The wind tunnel test for aerodynamic characteristics of rail vehicles is based on the theory of relative motion and similarity laws of fluid flow [4]. In the test section of the wind tunnel, the scaled vehicle model and other scaled objects, such as tracks and embankment, are placed. The air flow generated by the fan system becomes the low turbulent flow with the required speed, density and

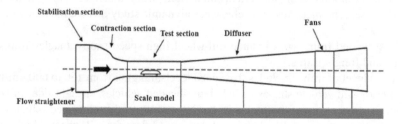

FIGURE 12.3 Open-circuit wind tunnel.

Rail Vehicle Aerodynamics

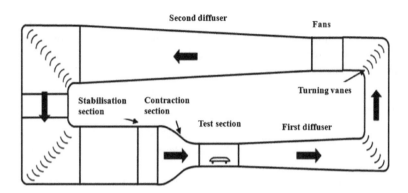

FIGURE 12.4 Closed-circuit wind tunnel.

pressure through the acceleration section and stabilisation section. When the air flow passes the scale model, the measured aerodynamic characteristics of the scale test model are in good accordance with the characteristics of real vehicles.

The aerodynamic forces acting on the moving rail vehicle are caused by the inertia, viscosity, elasticity and gravity of air. The main contributions of aerodynamic forces acting on a rail vehicle are as follows:

- *Inertial force*: Proportional to $\rho l^2 v_{tr}^2$, where ρ is the density of air, l is the characteristic length of the rail vehicle and v_{tr} is the rail vehicle speed
- *Viscous force*: Proportional to $\mu v_{tr} l$, where μ is the dynamic viscosity of air
- *Gravity force*: Proportional to $\rho l^3 g$, where g is the gravitational acceleration
- *Elastic force*: Proportional to $\rho l^2 c^2$, where c is the sound speed of air

In the wind tunnel test of a rail vehicle, the aerodynamic loads on the scale test model and the real full-size vehicle are identical if the following similarities are satisfied:

- *Geometrical similarity*

Reynolds number (Re) similarity as defined by:

$$Re = \frac{\rho v_{tr} l}{\mu} \tag{12.1}$$

The drag, lift, lateral forces, yaw, pitching and rolling moments of the test model are measured by the balance with six components installed inside the scale test model. The balance instrument for aerodynamic loads should be selected according to following principles:

- The components of the balance are the same as the number of forces and moments required to be measured.
- The measuring range of the balance should cover the maximum value of aerodynamic forces and moments.
- The interference of components of the balance should be small.
- The sensitivity, repeatability and accuracy of the balance should be high.

The pressure distribution on the surface of the scale model is measured by pressure sensors and a pressure scanner. The requirements of pressure measurements are almost the same as the requirements of forces measured by the balance. The distribution of pressure sensors is non-uniform, and more sensors should normally be placed on the high-pressure gradient area.

The velocity field and pressure field near the scale model are also important quantities that can be measured by anemometers and barometers.

12.3.2 Moving-Model Test

Similar to a field test, the moving-model test propels the scale model to obtain the aerodynamic characteristics of rail vehicles. This test method is different from a wind tunnel test. However, the moving model test and the wind tunnel test complement each other for the aerodynamic testing of rail vehicles. There exists a typical moving model rig at Central South University of China, as shown in Figure 12.5.

An unpowered high-speed test model is launched on the test track, which can easily reproduce the unsteady compressible flow field of trains crossing, trains passing through a tunnel and vehicles with relative motions to surrounding objects. Based on the similarities of flow for the test model, the tunnel model and other surrounding objects, the aerodynamic characteristics of rail vehicles in different scenarios can be precisely obtained by a moving-model test.

To ensure the representation of the model test results to a real-vehicle test, a moving-model test should obey the following similarity laws:

- Geometrical similarity: Proportionally scale the rail vehicle, tunnel and other objects [5].
- Reynolds number (*Re*) similarity: If the model scale is greater than 1/25, the *Re* number has little influence on the results of the test [6].
- Mach number (*Ma*) similarity: Since the air would be rapidly compressed as vehicles are crossing or passing through a tunnel [7], the moving model test should also match the speed of the real vehicle test.

$$Ma = \frac{v_{tr}}{c} \tag{12.2}$$

It should be noted that the identical *Re* number for the real vehicle is quite hard to achieve in a moving-model test. A more practical and easier way is to make sure the test conditions satisfy the similarity parameters in the self-simulation region. Figure 12.6 [8] shows that the pressure coefficient measured in a moving-model test is no longer changing with *Re* number after exceeding the critical value of 3.6×10^5.

In a moving model test, the following quantities can be measured:

- Aerodynamic drag force
- Pressure waves of vehicles crossing in open air or in a tunnel
- Pressure on surfaces of vehicle, tunnel and surrounding objects

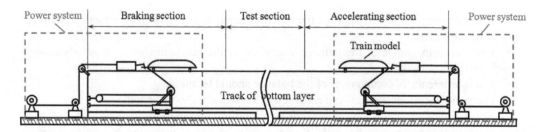

FIGURE 12.5 Schematic diagram of moving model rig at Central South University of China.

Rail Vehicle Aerodynamics

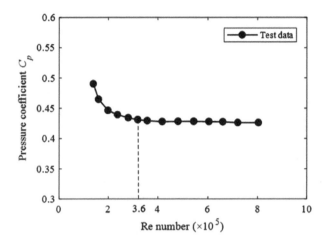

FIGURE 12.6 Pressure coefficients at different *Re* numbers in real-vehicle test. (From Tian, H.Q., *Train Aerodynamics*, China Railway Press, Beijing, China, 2007. With permission.)

- Micro-pressure wave of tunnel portals
- Velocity field and pressure field of flow near vehicle
- Crosswind aerodynamic characteristics

12.3.3 Real-Vehicle Test

The real-vehicle test is the aerodynamic test on the real railway for the real rail vehicle. The realistic aerodynamic characteristics of operating rail vehicles can be reflected in a real-vehicle test. It is an important approach to study and assess the aerodynamic performances of rail vehicles.

In a real-vehicle test, the following aerodynamic or other physical quantities are often measured:

- Aerodynamic drag
- Pressure waves of vehicles crossing in open air and tunnel
- Pressure on surfaces of vehicles, tunnel walls and surrounding objects
- Micro-pressure wave of tunnel portals
- Flow field near vehicles (velocity and pressure fields)
- Sound pressure level of aerodynamic noise
- Internal flow field inside passenger cabin
- Flow field of air conditioner and ventilation ducts for on-board devices

For the measured quantities mentioned above, the pressure distribution on the rail vehicle surface is the primary one for a real-vehicle test. Typical layouts of pressure sensors on the surface of a head car and middle car are presented in Figures 12.7 and 12.8, respectively.

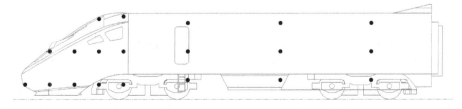

FIGURE 12.7 Typical layout of pressure sensors on head car for crosswind safety test.

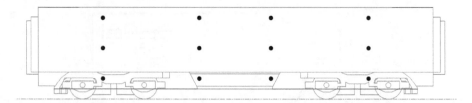

FIGURE 12.8 Layout of pressure sensors on vehicle surface for crosswind safety test.

12.3.4 Numerical Simulation

Numerical simulation is a virtual test method of the aerodynamics of a rail vehicle. It is achieved through the computational fluid dynamics (CFD), which acquires information of flow field by numerically solving the governing equations of the fluid flows.

The aerodynamics of rail vehicles is governed by the Navier-Stokes (N-S) equations. Two forms of N-S equations, compressible and incompressible forms, can describe all aerodynamics of rail vehicles:

- Incompressible N-S equations: For rail vehicles in open air with or without crosswind

$$\frac{\partial u_i}{\partial x_i} = 0 \quad (12.3)$$

$$\frac{\partial u_i}{\partial t} + u_j \frac{\partial u_i}{\partial x_j} = -\frac{1}{\rho}\frac{\partial p}{\partial x_i} + \mu \frac{1}{\rho}\frac{\partial^2 u_i}{\partial x_j \partial x_j} \quad (12.4)$$

where u_i is the component of air velocity; x_i is the x, y, z coordinate; p is the pressure; μ is the dynamic viscosity of air; and ρ is the density of air.

- Compressible N-S equations: For rail vehicles passing through a tunnel and passing each other

$$\frac{\partial \rho}{\partial t} + \frac{\partial (\rho u_i)}{\partial x_i} = 0 \quad (12.5)$$

$$\frac{\partial (\rho u_i)}{\partial t} + \frac{\partial (\rho u_j u_i)}{\partial x_j} = \frac{\partial p}{\partial x_i} + \frac{\partial \tau_{ij}}{\partial x_i} \quad (12.6)$$

$$\frac{\partial}{\partial t}(\rho E) + \frac{\partial}{\partial x_i}[u_i(\rho E + p)] = \frac{\partial}{\partial x_j}\left(K \frac{\partial T}{\partial x_j} + u_i \tau_{ij}\right) \quad (12.7)$$

$$p = R\rho T \quad (12.8)$$

where the shear stress $\tau_{ij} = \mu(\partial u_i/\partial x_j + \partial u_j/\partial x_i - 2/3 \delta_{ij} \partial u_i/\partial x_i)$, T is the absolute temperature; the total energy $E = e + u_i u_i/2$, where e is the internal energy calculated by $e = c_v T$; c_v is the specific heat capacity; K is the thermal conductivity of air; and R is the gas constant.

Theoretically, the flow field and aerodynamics forces of rail vehicles can be obtained by solving the above-mentioned N-S equations with proper boundary conditions. However, no analytical solutions exist for these N-S equations for complex flow near rail vehicles. The most widely used numerical method for rail vehicle aerodynamics is the finite-volume method. Figure 12.9 shows the basic framework for CFD numerical analysis for rail vehicles.

Rail Vehicle Aerodynamics

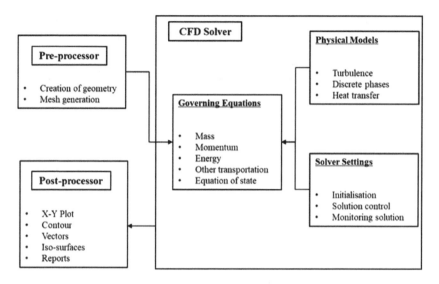

FIGURE 12.9 Framework of CFD analysis for aerodynamics of rail vehicles.

The highly convective effects in airflow would influence the stability and accuracy of numerical simulations. The commonly suggested spatial discretions of convective terms are first-order upwind scheme, second-order upwind scheme, Quadratic Upwind Interpolation of Convective Kinematics (QUICK) scheme, Roe scheme, etc.

In addition, the incompressible N-S equations have the pressure-velocity coupling issues causing the instability of the numerical method. The commonly used pressure-velocity coupling schemes are Semi-Implicit Method for Pressure-Linked Equations (SIMPLE), SIMPLE-Consistent (SIMPLEC), Pressure Implicit with Splitting of Operator (PISO), Fractional step, etc.

The air flow induced by rail vehicles is turbulent with a Reynolds number above 10^5. Proper turbulence modelling is necessary in numerical simulation for rail vehicles. The walls, that is, surfaces of the rail vehicle, have significant effects on the turbulent flows. In boundary layer theory, the near-wall region is made up of a viscous sublayer, a buffer layer and a fully turbulent layer, as shown in Figure 12.10. Figure 12.11 shows the relationship between distance from the wall and fluid velocity and the comparison with experiments. In Figure 12.11 from [9], y^+ on the x-axis is a dimensionless normal distance from the wall and U^+ is a dimensionless velocity. It has been found that the relationship between y^+ and U^+ is linear in the viscous sublayer and logarithmic in the turbulence layer. The semi-empirical wall functions adopted can largely save the computational effort, such as standard wall functions, scalable wall functions, non-equilibrium wall functions and enhanced wall functions.

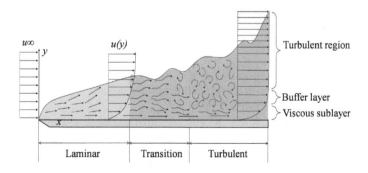

FIGURE 12.10 Turbulent boundary layer development.

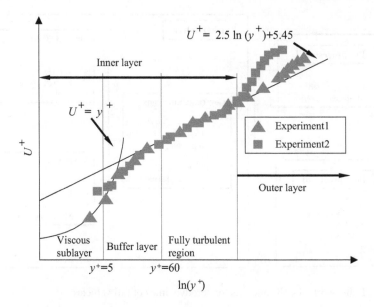

FIGURE 12.11 Relationship between y^+ and U^+ in turbulent boundary layer. (From ANSYS Fluent Theory Guide, ANSYS Inc., 2019. With permission.)

In CFD simulation for rail vehicles, the simulation algorithms for turbulence are simply overviewed as set out in the dot points below, and more details can be found in CFD textbooks [10,11]:

- Large Eddy Simulation (LES): Exactly treats large-scale eddies, quite expensive computationally, solvable on supercomputer, recommended first layer mesh height $y^+ < 5$.
- Reynolds Averaged Navier-Stokes (RANS) equations.
  Standard, Re-Normalisation Group (RNG), and Realisable k-ε models: Two-equations models, good for all industrial flows, with wall function, good convergence, high-Re model with first layer mesh height $30 < y^+ < 300$.
  Standard and SST k-ω model: Two-equations models, sensitive to initial condition for standard k-ω, accurate and reliable for SST k-ω, no wall function, low-Re model with recommended first layer mesh height $y^+ \sim 1$.
- Detached Eddy Simulation (DES): Wall function for near walls, LES for other regions, less expensive than LES, more precise than RANS.

Considering the costs and efficiency, the recommended turbulence models for CFD simulation of rail vehicles are RNG k-ε model, Realisable k-ε model and SST k-ω model.

To obtain the convincible and accurate aerodynamic results, the mesh or grids must be carefully generated with smoothness and good quality according to the selected turbulence model. The basic guidelines for CFD mesh for rail vehicles are as follows:

- y^+ Value of first-layer grid should satisfy the requirement of the selected turbulence model. The estimation of the first-layer cell height for certain y^+ values can use the online calculator at http://www.pointwise.com/yplus/.
- There should be at least five layers of cells with growth rate less than 1.2 in the boundary layer region.
- The last layer of prism cells should have an aspect ratio less than 5 to their neighbouring cells.
- Global mesh size growth rate is recommended to be less than 1.2.

Rail Vehicle Aerodynamics

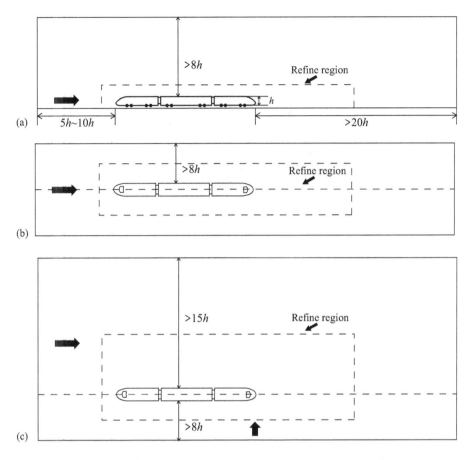

FIGURE 12.12 Computational domain for numerical simulation of rail vehicle aerodynamics: (a) side view, (b) top view without crosswind and (c) top view with crosswind.

- A mesh refinement box surrounding the rail vehicle is highly recommended; see Figure 12.12.
- The three-car train model with head, middle and tail cars is preferred. If no crosswind effect, a train model with head and half middle cars is also acceptable.
- For vehicles in open air, the computational domain should be large enough to minimise the influences of boundary conditions. The computational domain for a three-vehicle train is illustrated in Figure 12.12a and b. For simulation with crosswind, the top view of the computational domain is shown in Figure 12.12c.
- Cartesian grid is recommended. The suggested meshing strategy is hexahedron with hanging nodes for spatial mesh. Figure 12.13 presents the generated mesh by the above strategy.

The boundary conditions in a numerical simulation must be as close as possible to the real physics of the real-vehicle test and wind tunnel model test. The corresponding boundary conditions used in a rail vehicle simulation are:

- *Velocity inlet*: Flow direction is against train, flow velocity magnitude is train speed, velocity profile is uniform or parabolic distribution.
- *Pressure outlet*: Pressure value is ambient pressure, 101.325 kPa.

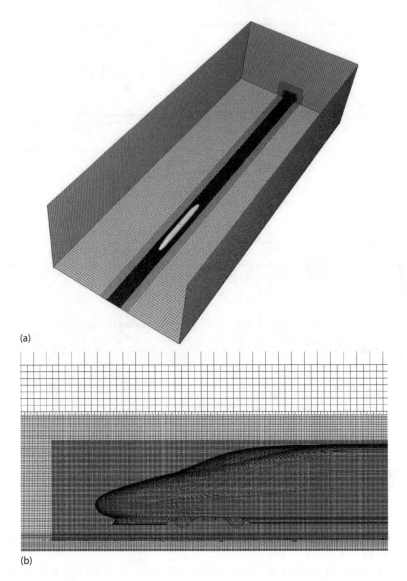

FIGURE 12.13 Cartesian grid for high-speed train: (a) three-dimensional view of mesh and (b) side elevation of head car mesh.

- *No-slip wall*: Stationary wall condition is applied on train model surfaces and tunnel, moving wall condition is applied to ground for the aerodynamic ground effects and rotating wall condition would be applied to wheels.
- *Symmetry*: Top surface of air domain may be applied with symmetry boundary condition to simulate the infinite height of the atmosphere

For simulation of trains crossing and a train passing through a tunnel, the mesh has to be sliding or regenerated due to the motion of rail vehicles. The key for the successful simulation of these problems is the sliding mesh technique. This technique provides adjacent zones of mesh with the flexibility to slide relative to one another. The computational domains of two trains crossing in a tunnel using sliding mesh are plotted in Figure 12.14. Figure 12.15 illustrates the zones and interfaces of the mesh for two trains crossing in a tunnel. Zone 1 is the fixed mesh for air. Zone 2 and zone 3 are two sliding zones of mesh surrounding each of the two moving trains, respectively.

Rail Vehicle Aerodynamics

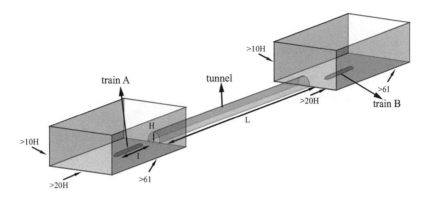

FIGURE 12.14 Computational domains of sliding meshes for two trains crossing in tunnel.

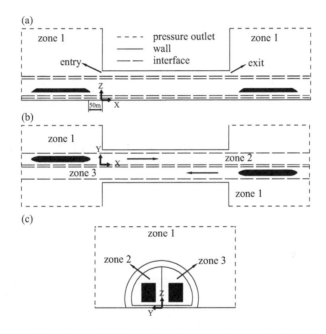

FIGURE 12.15 Zones and interfaces of sliding meshes for two trains crossing in tunnel: (a) side view, (b) top view and (c) cross-section view.

12.4 PROPERTIES OF EXTERIOR FLOW STRUCTURE OF RAIL VEHICLES

12.4.1 Distribution Law and Development Characteristics of Flow Field around Trains

In order to study the general distributions of flow fields around trains, a group of high-speed trains is selected, and the results of flow fields around trains are plotted in Figures 12.16 and 12.17 [12]. Figure 12.16 presents the flow fields around a high-speed train in horizontal section. Figure 12.16a presents the velocity distributions, and Figure 12.16b presents the pressure distributions with isobaric lines. Figure 12.17 shows the flow field around the high-speed train in longitudinal section. Figure 12.17a presents velocity distributions and Figure 12.17b shows pressure distributions.

Analysis of Figures 12.16 and 12.17 shows that the overall variation of the flow fields around the train is as follows: the flow fields around the head, the tail, the lower part and the connecting parts of the train and the wake flow change greatly. The flow field around the regular section of the train changes little [12].

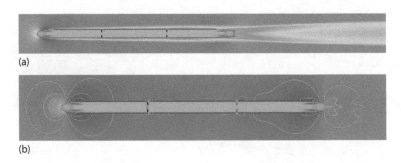

FIGURE 12.16 Flow field distributions around high-speed train (horizontal profile): (a) velocity of field flow and (b) pressure of flow field. (From Tian, H.Q. et al., *J. Cent. South Univ.*, 22, 747–752, 2015. With permission.)

FIGURE 12.17 Flow field distributions around high-speed train (longitudinal profile): (a) velocity of flow field and (b) pressure of flow field. (From Tian, H.Q. et al., *J. Cent. South Univ.*, 22, 747–752, 2015. With permission.)

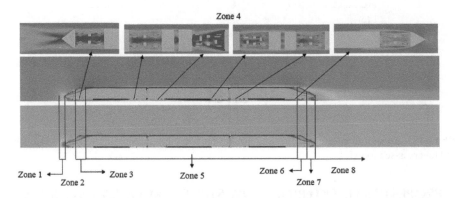

FIGURE 12.18 Flow field zones around high-speed train.

In order to visualise the distribution of the flow fields around the train and determine the special position of the pressure and velocity field change, the flow field can be divided into eight flow zones according to the study of Tian [12], as shown in Figure 12.18, which lay the foundation for the follow-up flow field control.

These eight flow zones can be described as follows:

Zone 1: Airflow stagnation with high pressure
The front zone of the head car is opposing the incoming flow. In this zone, the air flow is stagnated. The closer the airflow is to the nose tip of the head, the closer the flow speed is to zero. In front of the head car, an air stagnation zone with relative velocity of nearly

Rail Vehicle Aerodynamics

zero forms, which is the area with the lowest velocity and highest pressure in the whole flow field, and the air pressure coefficient at the nose tip of the head car is approximately equal to 1.

Zone 2: Airflow-accelerated and pressure-reduced zone

In the zone of the streamlined head, the velocity of airflow increases and the pressure decreases and gradually changes from positive pressure to negative pressure.

Zone 3: Zone I with high-speed airflow and low pressure

In the region of transitioning from the head of the train to the body, the velocity of air flow is greater than that of the free stream. The more drastically the curvature changes from head to body, the greater is the velocity difference. In this region, the pressure decreases sharply to the minimum of the whole flow field. The highest velocity and the lowest pressure in the whole flow field are achieved in this region.

Zone 4: Turbulence zone

In the zones of the lower parts and connecting parts of the vehicles, the changes of airflow velocity and pressure are larger, and the flow field is relatively more disordered.

Zone 5: Steady-flow zone

In the zone of a uniform cross-section of the body of the train, the variation of air velocity and pressure is very small.

Zone 6: Zone II with high-speed airflow and low pressure

In the region of transitioning from the body to the tail of the train, the velocity of air flow is again greater than that of free stream. The more drastically the curvature changes in the transition from the body to the tail, the greater is the speed difference. Because of the same shape of the head and tail, the velocity difference in the high-speed and low-pressure zone II is slightly smaller than that in zone I. This region is the region with the second-highest velocity and lowest pressure in the whole flow field.

Zone 7: Zone of airflow deceleration and pressure rise

In the zone of the streamlined head at the tail car, the airflow velocity decreases and the pressure rises, which result in flow deceleration and pressure boost effects.

Zone 8: Zone of wake flow

In the zone behind the tail car, the flow field has the following characteristics:
- There is a long variable flow zone, the length of which can be equal to, or even longer than, the length of the train
- The intensity of the wake flow varies greatly, and this is the region with the greatest variation in velocity and pressure of the flow field around the train
- With the development of the wake, the wake intensity decreases gradually with the increase of distance from the tail of the train, but this change is relatively light, and the wake flow effects are therefore quite significant.

There are no clear interfaces between these eight types of air flow zones, and each zone cannot exist independently. From the viewpoint of aerodynamics, as a subsonic air flow problem, the disturbance generated by the train motion quickly spreads throughout the whole field. The front and rear flow fields of moving trains interact and interfere with each other.

12.4.2 Characteristics of Vortex Structure in Flow Field around Trains

When the train moves, the airflow around the train will produce many transient irregular vortices, which will develop and detach from the surface of the train and cause the transient velocity fluctuation of the flow field around the train. The unique distribution of these spiral vortices is called the vortex structure of the flow field around the train.

Figure 12.19 shows the distribution of the transient flow field around the high-speed train. According to this figure, the flow field around the high-speed train is a complex structure.

FIGURE 12.19 Distribution of flow field around high-speed train (horizontal profile).

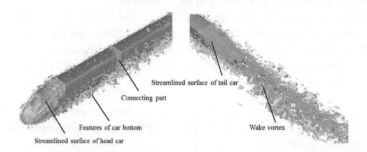

FIGURE 12.20 Distribution and generation of vortices around trains. (From Tian, H.Q. et al., *J. Cent. South Univ.*, 22, 747–752, 2015. With permission.)

In order to study the characteristics of vortices around trains, the Q (vorticity) iso-surface is used to show the distribution structure and source of vortices around trains, as shown in Figure 12.20 [12]. The head of the train is on the left of the figure, and the tail of the train is on the right.

Analysis of Figure 12.20 shows the following distribution laws of vortices around the train [12,13]:

- For trains, the vortices around the head, tail, bottom, vehicle junctions and wake flow are densely distributed. There are very few vortices around the regular section of the train body.
- In the head and tail of the train, because of the continuous change of its cross-section, vortices form, develop and detach from the surface of the train. This causes transient velocity fluctuation of the flow field around the train and results in a dense distribution of vortices in the head and tail of the train.
- A large number of vortices gather around the bottom of the train due to the complex structure of the bogies and other structures, the continuous ground effect and the thick boundary layer at the lower side of the train. Moreover, the concentration of vortices is much larger than that of other parts.
- In the connection part of the vehicle, because of the gap between the two connected vehicles, the vortices in this zone are densely distributed.
- In the wake flow, there are a series of long diffusive vortices due to wake and continuous ground effect. The length of the wake region is tens to hundreds of metres.

According to the above analysis, the vortices mainly locate around the surface structures with complex abrupt changes and large curvature changes. The distributions of vortices in the head, tail, bottom, connecting parts and wake area of trains are dense.

12.4.3 Characteristics of Wake Flow of the Train

Figure 12.21 shows the transient vorticity structures at the rear of a high-speed train. Figure 12.21a displays the transient pressure iso-surface in the wake region. Figure 12.21b shows the transient velocity distribution and streamlines at 0.3H behind the rear of the train. Figure 12.21c shows the transient velocity distribution and particle trajectory at 0.6 H behind the rear.

Figure 12.22 [12] shows the development of the time-averaged vortex structure at the rear section of a high-speed train, which is 0.03 H, 0.17 H, 0.5 H and 0.85 H away from the tail car, respectively.

Rail Vehicle Aerodynamics

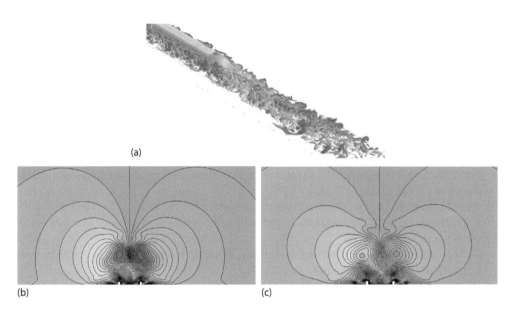

FIGURE 12.21 Transient wake structure of high-speed train: (a) transient pressure iso-surface in wake region, (b) vertical plane velocity distribution and streamlines of wake 0.3 H behind tail car and (c) vertical plane velocity distribution and particle trajectory of wake 0.6 H behind tail car.

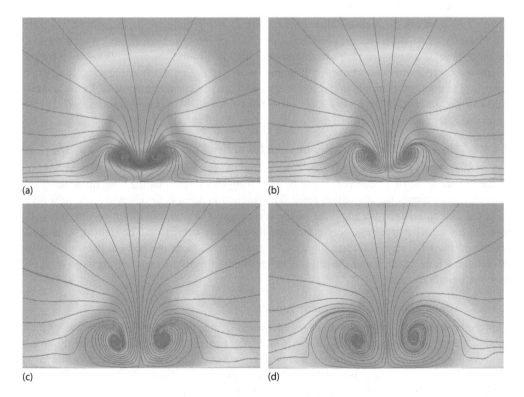

FIGURE 12.22 Development of time-averaged vortices structure in the wake flow of the tail car: (a) 0.03 H away from the train tail, (b) 0.17 H away from the train tail, (c) 0.5 H away from the train tail and (d) 0.85 H away from the train tail. (From Tian, H.Q. et al., *J. Cent. South Univ.*, 22, 747–752, 2015. With permission.)

Based on the analysis of Figures 12.21 and 12.22, combined with a large number of other experimental and numerical results, the spatial structure of the rear vortices of a high-speed train is obtained as follows:

- As analysed in the previous section, there is a long series of transient vortices in the wake region of high-speed trains, as shown in Figure 12.21a.
- Firstly, two main vortices are formed at the tail car. The core of the two main vortices is symmetrically distributed. There are also a large number of relatively small vortices nearby, as shown in Figure 12.21b.
- From the time-averaged vortex structure, the two main vortices are very symmetrical, as shown in Figure 12.22.
- After leaving the tail car, the two main vortices enter the wake flow. On both sides of the symmetrical plane of the train tail, irregular fluctuations of the two main vortices occur, as shown in Figure 12.21a.
- On the vertical plane 0.6 H behind the tail car, it is found that the two main vortices have begun to dissipate, as shown in Figure 12.21c.
- With the development of the vortex structure, two symmetrical vortices are gradually expanding. And far behind the rear of the train, the vortex structure has been diffusing symmetrically and slowly, as shown in Figure 12.22.

12.4.4 Distribution and Development Characteristics of Flow Field around Train Running in Transient Conditions

12.4.4.1 Flow Field Characteristics at Rail Vehicle Crossing

Figure 12.23 presents the pressure distributions of two rail vehicles when crossing. The reference time frame is set to start, that is, $t = 0$ s, as the nose of the two head trains crosses each other. The movement of a high-speed train disturbs the airflow around the train. When two trains are crossing each other, this disturbance will intensify. In particular, when the head or tail of a train passes beside another train, air pressure changes and a transient pressure wave forms around the surface of the other train. In a very short time, the peak values of positive and negative pressure appear one after another.

12.4.4.2 Flow Field Characteristics of Vehicles under Crosswind

Figure 12.24 shows the pressure distributions around the train at the wind flow velocity of 66 m/s and the impact angle of 27 degrees. It can be seen from this figure that the nose tip of the head car of the train is no longer the stagnation point. The stagnation point is deviated to the windward side, where the flow velocity is zero and the positive pressure is the largest. After passing through this point, the air velocity gradually accelerates and the pressure decreases. The train is under positive pressure on the windward side, and the air flow is separated from the body at the transition point between the windward side and the top of the train body, forming a negative pressure zone on the top of the train body.

12.4.4.3 Single Train Passing through Tunnel

When the train enters the tunnel at high speed, the air flow is blocked by the restriction of the tunnel wall, which causes the static air at the front of the train to be compressed violently, resulting in a sudden increase of air pressure and a compression wave. The rear of the train moves forward, forming a low-pressure zone and generating expansion waves. Figure 12.25 shows the pressure distributions on the train surface and tunnel wall when a train passes through a tunnel at a speed of 350 km/h.

Rail Vehicle Aerodynamics

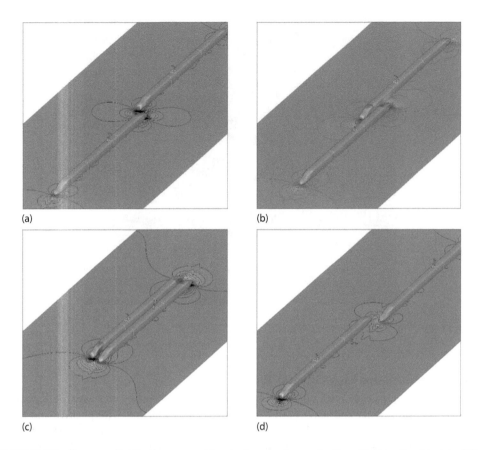

FIGURE 12.23 Pressure distributions around two trains crossing each other: (a) at $t = 0$ s, (b) at $t = 0.093$ s, (c) at $t = 0.417$ s and (d) at $t = 0.833$ s.

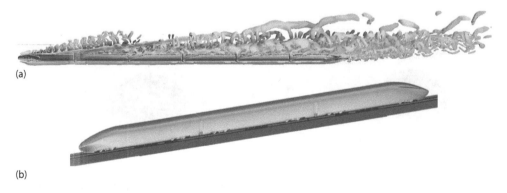

FIGURE 12.24 Vorticity and pressure distributions around train in crosswind: (a) top view of vorticity distribution and (b) side view of pressure distribution on surface of vehicles.

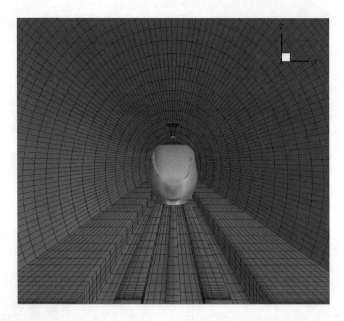

FIGURE 12.25 Pressure distributions around single train passing through tunnel.

12.4.4.4 Trains Passing Each Other in Tunnel

When two trains are passing each other in a tunnel, compression waves and expansion waves will form when the trains enter and leave both ends of the tunnel, together with positive and negative air pressure waves. Various kinds of waves are repeatedly reflected and superimposed during the propagation in tunnels. Figure 12.26 shows the pressure distributions around two trains while passing each other in the tunnel. Figure 12.26a shows the pressure distributions at the moment of the two rail vehicles crossing. When the compression waves are superimposed, the air pressure in the tunnel is positive, but the train on the right side is subjected to negative pressure from the crossing air pressure wave. Therefore, negative pressure appears on the car body. Figure 12.26b shows the pressure distributions after the two trains have crossed and the air pressure begins to decrease under

(a) (b)

FIGURE 12.26 Pressure distributions around trains while passing each other in tunnel: (a) as the two trains are crossing and (b) after the two trains have crossed. (From Tian, H.Q., *Train Aerodynamics*, China Railway Press, Beijing, China, 2007. With permission.)

Rail Vehicle Aerodynamics

the influence of expansion waves. When the length of train is longer, the compression wave and expansion wave will repeat more than once during the rail vehicle crossing.

12.5 PROPERTIES OF AERODYNAMIC LOAD ON RAIL VEHICLES

12.5.1 Influence Factors of Aerodynamic Load on Rail Vehicles

12.5.1.1 Single Vehicle Running in Open Air without Environmental Wind

The influence factors of a single vehicle running in open air without environment wind are as follows [8]:

- Vehicle speed
- The geometric parameters of the vehicle head, including the streamlined head length (Figure 12.27), the longitudinal shape profile of the head car (Figure 12.28), the shape outline of the streamlined head in top view (Figure 12.29) and the cross-sectional area profile throughout the head car (Figure 12.30)

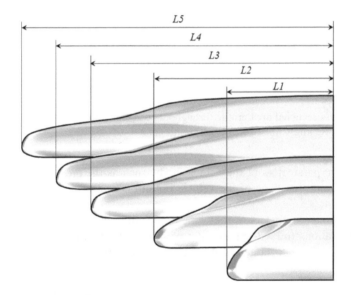

FIGURE 12.27 Schematic of the streamlined head.

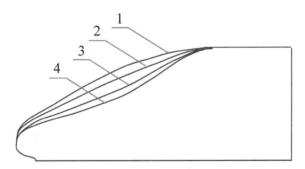

FIGURE 12.28 Schematic of the longitudinal shape profile. (From Tian, H.Q., *Train Aerodynamics*, China Railway Press, Beijing, China, 2007. With permission.)

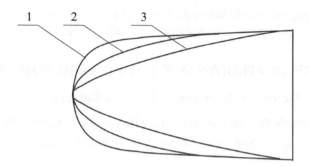

FIGURE 12.29 Schematic of the shape outline of the streamlined head in top view. (From Tian, H.Q., *Train Aerodynamics*, China Railway Press, Beijing, China, 2007. With permission.)

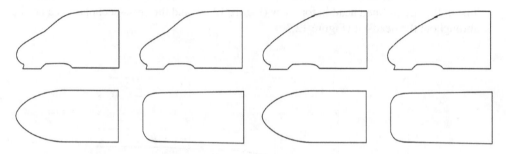

FIGURE 12.30 Cross-sectional area profile throughout head car. (From Tian, H.Q., *Train Aerodynamics*, China Railway Press, Beijing, China, 2007. With permission.)

- The geometric parameters of the vehicle body, such as the cross-sectional shape of the vehicle body and the shape of the vehicle bottom (shape of the bottom cover and skirt board)
- The marshalling parameters of the vehicle, including marshalling length and marshalling configurations (mixed marshalling of passenger train and freight vehicle, mixed marshalling of single-layer and double-layer container trains and the coupling of the two trains)

12.5.1.2 Vehicles Crossing in Open Air without Environmental Wind

The factors that affect the vehicle aerodynamics when vehicles are crossing in the open air without environmental wind are listed as follows:

- The speed of each vehicle
- The geometric parameters of the vehicle head, including the streamlined head length, the longitudinal shape profile of the head car, the shape outline of the streamlined head in top view and the cross-sectional area profile throughout the head car
- The geometric parameters of the vehicle body, such as the cross-sectional shape of the vehicle body and the shape of the vehicle bottom (shape of the bottom cover and skirt board)
- The marshalling parameters of the vehicle, including marshalling length and marshalling configurations (mixed marshalling of passenger train and freight vehicle, mixed marshalling of single-layer and double-layer container trains and the coupling of the two trains)
- Line spacing (i.e., distance between tracks)

Rail Vehicle Aerodynamics

12.5.1.3 Single Vehicle Passing through Tunnel

The factors that affect the vehicle aerodynamics when a single vehicle is passing through a tunnel are:

- Vehicle speed
- The geometric parameters of the vehicle head, including the streamlined head length, the longitudinal shape profile of the head car, the shape outline of the streamlined head in top view and the cross-sectional area profile throughout the head car
- The geometric parameters of the vehicle body, such as the cross-sectional shape of the vehicle body and the shape of the vehicle bottom (shape of the bottom cover and skirt board)
- The marshalling parameters of the vehicle, including marshalling length and marshalling configurations (mixed marshalling of passenger train and freight vehicle, mixed marshalling of single-layer and double-layer container trains and the coupling of the two trains)
- The parameters of the tunnel and its ancillary facilities, which includes the length of the tunnel, the cross-sectional shape of the tunnel, the parameters of the tunnel portal (portal shape, slope value, aperture ratio, etc.) and the parameters of the horizontal passage and shaft (sizes, positions and quantities)
- The line spacing in a double-track tunnel

12.5.1.4 Vehicles Crossing Inside Tunnel

The factors that affect the vehicle aerodynamics when vehicles are crossing in the tunnel are listed as follows:

- The speed of each vehicle
- The geometric parameters of the vehicle head, including the streamlined head length, the longitudinal shape profile of the head car, the shape outline of the streamlined head in top view and the cross-sectional area profile throughout the head car
- The geometric parameters of the vehicle body, such as the cross-sectional shape of the vehicle body and the shape of the vehicle bottom (shape of the bottom cover and skirt board)
- The marshalling parameters of the vehicle, including marshalling length and marshalling configurations (mixed marshalling of passenger train and freight vehicle, mixed marshalling of single-layer and double-layer container trains and the coupling of the two trains)
- The parameters of the tunnel and its ancillary facilities, which includes the length of the tunnel, the cross-sectional shape of the tunnel, the parameters of the tunnel portal (portal shape, slope value, aperture ratio, etc.) and the parameters of the horizontal passage and shaft (sizes, positions and quantities)
- The line spacing in a double-track tunnel

12.5.1.5 Single Vehicle Running in Open Air with Crosswind

The factors that affect the aerodynamics of rail vehicles running in the open air with a crosswind are listed as follows:

- Vehicle speed
- The geometric parameters of the vehicle head, including the streamlined head length, the longitudinal shape profile of the head car, the shape outline of the streamlined head in top view and the cross-sectional area profile throughout the head car
- The geometric parameters of the vehicle body, such as the cross-sectional shape of the vehicle body and the shape of the vehicle bottom (shape of the bottom cover and skirt board)

- The marshalling parameters of the vehicle, including marshalling length and marshalling configurations (mixed marshalling of passenger train and freight vehicle, mixed marshalling of single-layer and double-layer container trains and the coupling of the two trains)
- The wind speed and yaw angle
- The line and environmental conditions, such as embankments, cuttings and bridges
- Wind protection facility parameters, such as shape, height and connection mode

12.5.1.6 Vehicles Crossing with Crosswind

The factors affecting the aerodynamics of the rail vehicles crossing in the open air with a crosswind are listed as follows:

- The speed of each vehicle
- The geometric parameters of the vehicle head, including the streamlined head length, the longitudinal shape profile of the head car, the shape outline of the streamlined head in top view and the cross-sectional area profile throughout the head car
- The geometric parameters of the vehicle body, such as the cross-sectional shape of the vehicle body and the shape of the vehicle bottom (shape of the bottom cover and skirt board)
- The marshalling parameters of the vehicle, including marshalling length and marshalling configurations (mixed marshalling of passenger train and freight vehicle, mixed marshalling of single-layer and double-layer container trains and the coupling of the two trains)
- The wind speed and yaw angle
- The road and environmental conditions, such as embankments, cuttings and bridges
- Wind protection facility parameters, such as shape, height and connection mode
- Line spacing

12.5.2 Distributive Load of Aerodynamic Pressure on Rail Vehicle Surface

12.5.2.1 Single Vehicle Running in Open Air without Environmental Wind

For a single vehicle running in open air without environmental wind, the pressure distribution along the longitudinal shape profile of the vehicle is shown in Figure 12.31.

The pressure coefficient value on surface of the vehicle can be calculated using the following formula:

$$C_p = \frac{p - p_0}{\frac{1}{2}\rho v_{tr}^2} \qquad (12.9)$$

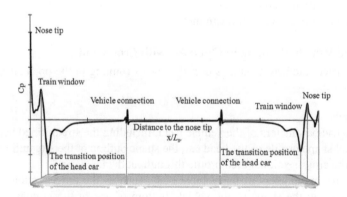

FIGURE 12.31 Pressure distribution along the longitudinal shape profile of the vehicle.

Rail Vehicle Aerodynamics

where ρ is air density (kg/m³), v_{tr} is the speed of the rail vehicle, C_p is the pressure coefficient, p is the pressure and p_0 is the reference static pressure. According to Figure 12.31, the pressure coefficient has the following characteristics:

- The stagnation point is located at the nose tip of the head car, where the velocity of the air is approximately 0. The maximum pressure coefficient on the train surface appears at the stagnation point with a value of approximately 1.0.
- The maximum negative pressure of the vehicle is related to the type of the vehicle.
- The pressure on the vehicle body is basically negative, except for the bottom of the vehicle.

12.5.2.2 Vehicles Crossing in Open Air without Environmental Wind

A representative plot of transient pressure waves induced by two vehicles crossing is shown in Figure 12.32. The two data plots are obtained from symmetric points from both sides of the vehicles.

The pressure wave in Figure 12.32 has the following characteristics:

- At the beginning of vehicle crossing, a pressure wave will be formed with a positive pressure peak, followed by a negative pressure peak. This is known as the head pressure wave. The pressure amplitude of the head pressure wave is defined as ΔP_1. At the end of crossing, a pressure wave will be formed with a negative pressure peak, followed by a positive pressure peak. This is known as the tail pressure wave. The pressure amplitude is defined as ΔP_2.
- When the vehicles are crossing in the open air without environmental wind, $\Delta P_1 > \Delta P_2$.
- When a moving vehicle is crossing with a stationary vehicle, the pressure amplitude for measurement points that are located on the moving vehicle is approximately 0. However, the pressure change on the stationary vehicle is similar to that shown in Figure 12.32. The formula for pressure amplitude calculation is:

$$\Delta p_{tr1} = \alpha_{pv0} v_{tr2}^2 \tag{12.10}$$

where v_{tr2} is the speed of the moving vehicle, and α_{pv0} is a coefficient related to the speed and the pressure wave. The amplitude of the pressure wave is proportional to the square of the speed of the moving vehicle.

- When two vehicles are passing by each other with the same speed, the relationship between the maximum pressure amplitude Δp and the vehicle speed is:

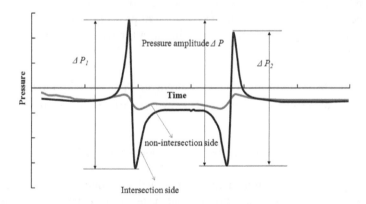

FIGURE 12.32 Transient pressure waves induced by vehicles crossing in open air.

$$\Delta p = \frac{1}{4}\alpha_{pv1}v_{tr}^2 \qquad (12.11)$$

$$v_{tr} = v_{tr1} + v_{tr2} = 2v_{tr1} = 2v_{tr2}$$

where v_{tr} is the relative speed of the two crossing vehicles, and α_{pv1} is a coefficient related to the vehicle speed and the pressure wave.
- When two vehicles are crossing in the open air without environmental wind under different speeds, the maximum pressure amplitude for the measurement point can be calculated using the following formula:

$$\Delta p_{tr1} = \alpha_{pv12}v_{tr2}^2 + 2\alpha_{pv12}v_{tr1}v_{tr2} - \sigma_{pv12}v_{tr1}^2 \qquad (12.12)$$

where α_{pv12} and σ_{pv12} are coefficients related to vehicle speed and pressure wave, and $\alpha_{pv12} > 0$, $\sigma_{pv12} > 0$, $2\alpha_{pv12} > \alpha_{pv12} > \sigma_{pv12}$. Thus, the maximum pressure amplitude on the vehicle with low speed is larger than that of the vehicle running with higher speed.
- When two vehicles are passing by each other, the induced pressure amplitude is:

$$\Delta p = \frac{1}{2}\rho v_{tr}^2 \alpha_1 e^{-m \cdot D_L e^{mb} + n} \qquad (12.13)$$

where D_L is the line spacing, and b is the width of the vehicle. α_1, m and n are coefficients related to the pressure wave, line spacing and the width of the vehicle.

According to Equation 12.13, there is an exponential relationship between the pressure amplitude and the line spacing.

12.5.2.3 Pressure Distribution for Vehicles Running in Tunnel

A compression wave will be formed in front of a vehicle as the head car enters a tunnel. This compression wave will then propagate in the tunnel at approximately the speed of sound. When the compression wave reaches the other side of the tunnel, an expansion wave will be reflected back into the tunnel. An expansion wave will be formed as the tail car enters a tunnel. This expansion wave will then propagate in the tunnel. When the expansion wave reaches the other side of the tunnel, a compression wave will be reflected back into the tunnel. The compression wave and expansion wave induced by a train passing through the tunnel will lead to significant pressure changes on the surface of the vehicle. These pressure waves will be continuously reflected at the tunnel entrance and tunnel exit and propagate in the tunnel. The superposition of pressure waves of the same kind will increase the pressure, while the superposition of pressure waves of different kinds will decrease the pressure. The propagation of pressure waves and pressure changes as vehicles are passing through a tunnel and crossing in a tunnel are shown in Figures 12.33 and 12.34, respectively.

12.5.2.4 Pressure Distribution for Vehicles Running in Open Air with Environmental Wind

In the circumstances of vehicles running in open air with environmental wind present, the following aerodynamic effects are observed:

- The nose tip of the train head is no longer the stagnation point. The pressure coefficient of the nose tip is less than 1.0.
- When a single vehicle is running in the open air with environmental wind, the pressure on the windward side of the vehicle is positive and the pressure on the leeward side is negative; this leads to a strong side force in the direction of the crosswind.

Rail Vehicle Aerodynamics

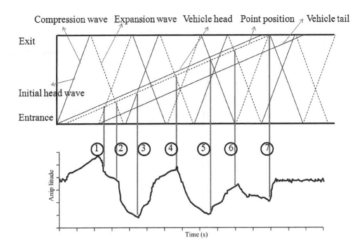

FIGURE 12.33 Pressure changes and propagation of pressure waves induced by a vehicle passing through a tunnel. ① Initial tail wave, ② First reflection of head wave, ③ Second reflection of head wave, ④ Third reflection of head wave, ⑤ Third reflection of tail wave, ⑥ Fifth reflection of head wave and ⑦ Exit wave.

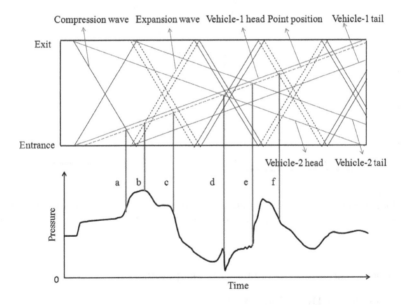

FIGURE 12.34 Pressure changes and propagation of pressure waves induced by vehicles crossing in a tunnel. (a) Initial head wave of vehicle-2, (b) initial tail wave of vehicle-1, (c) initial tail wave of vehicle-2, (d) pass of vehicle-2 head, (e) pass of vehicle-2 tail and (f) second reflection of tail wave of vehicle-2.

- The negative pressure on the top of the vehicle will induce an upward lift force. The upward lift force is larger than the resultant force on the bottom, so that an upward lift force can always be expected. Examples of the pressure distribution on the cross section located in the centre of the middle car are shown in Figure 12.35.
- For vehicles crossing in the open air with crosswind, the pressure change characteristics of the crossing vehicles are that the positive pressure of the pressure wave in Figure 12.32 increases sharply for the vehicle that is crossing on the windward side, and the negative pressure increases sharply for the vehicle that is crossing on the leeward side.

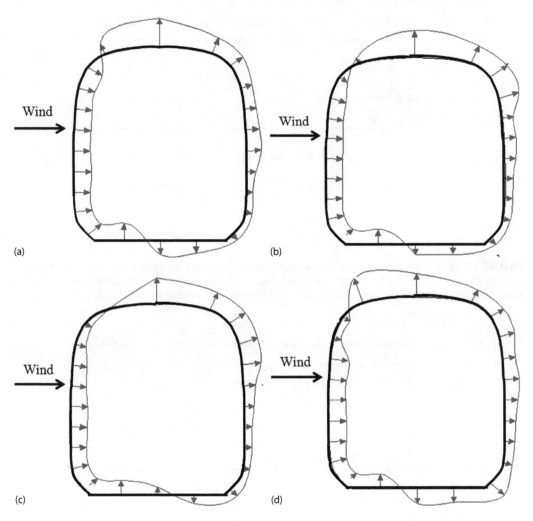

FIGURE 12.35 Diagrams of pressure distribution on cross section of middle car when vehicles are runnin-gin open air with environmental wind and with different streamlined head lengths of (a) 4 m, (b) 7 m, (c) 9 mand (d) 15 m.

12.5.3 Aerodynamic Forces and Moments of Rail Vehicles

12.5.3.1 Aerodynamic Drag of Rail Vehicles

The drag force of a train is related with the drag coefficient as:

$$F_x = \frac{1}{2}\rho S C_{Fx} V^2 \quad (12.14)$$

where ρ is air density, S is the cross-sectional area of the vehicle, C_{Fx} is the drag coefficient and V is the air speed relative to the vehicle. Based on this definition, the aerodynamic drag of the vehicle is proportional to the square of the flow velocity. A correlation term T_f can be added into Equation 12.14 for vehicles operating in a tunnel:

$$F_x = \frac{1}{2}T_f \rho S C_{Fx} V^2 \quad (12.15)$$

Rail Vehicle Aerodynamics

The term T_f is related with blockage ratio (train cross-sectional area over tunnel cross-sectional area), vehicle type, vehicle length, tunnel length and vehicle velocity, etc. For streamlined cars, T_f ranges between 1.4 and 1.8; for blunt-headed cars, T_f ranges between 2 and 3.

The drag coefficient of a train can be reduced by increasing the streamline length of the head car. The relationship between streamline length of the head car and drag coefficient for vehicles operating in open air without environmental wind is shown in Table 12.2.

The drag coefficient is also related with the yaw angle. The yaw angle (β) term is illustrated in Figure 12.36, where v_{tr} is the train speed vector and v_w is the wind speed vector. The angle between v_{tr} and V is defined as yaw angle. The drag coefficient of a streamlined head car under different wind angles is shown in Table 12.3.

12.5.3.2 Aerodynamic Lift of Rail Vehicles

The aerodynamic lift force of the train is defined as:

$$F_z = \frac{1}{2}\rho S C_{Fz} V^2 \tag{12.16}$$

TABLE 12.2
Effect of Head Car Streamline Length on Drag Coefficient

Streamline Length (m)	3.5	5.5	7.5	≥10
C_{Fx}	0.17 ~ 0.25	0.16 ~ 0.22	0.13 ~ 0.19	0.11 ~ 0.17

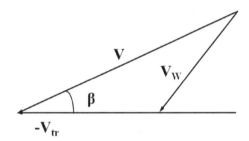

FIGURE 12.36 Speed vector diagram.

TABLE 12.3
Drag Coefficient of Streamlined Head Car for Different Yaw Angles

	Streamline Length (m)	
Yaw Angle (β)	3.5	7.5
0	0.18 ~ 0.27	0.13 ~ 0.19
10	0.20 ~ 0.30	0.19 ~ 0.28
20	0.13 ~ 0.19	0.21 ~ 0.31
30	−0.01 ~ −0.02	0.08 ~ 0.12
40	−0.42 ~ −0.63	−0.16 ~ −0.24
50	−0.74 ~ −1.12	−0.76 ~ −1.14
60	−0.69 ~ −1.03	−0.83 ~ −1.25
70	−0.56 ~ −0.84	−0.48 ~ −0.72
80	−0.64 ~ −0.96	−0.43 ~ −0.65
90	−0.58 ~ −0.87	−0.32 ~ −0.48

TABLE 12.4
Lift Coefficient of Streamlined Head Car for Different Yaw Angles

Yaw Angle (β)	Streamline Length (m)	
	3.5	7.5
0	0	0
10	−0.75 ~ −0.50	−0.93 ~ −0.62
20	−2.88 ~ −1.92	−3.16 ~ −2.10
30	−4.61 ~ −3.08	−5.64 ~ −3.76
40	−6.31 ~ −4.20	−7.64 ~ −5.09
50	−6.31 ~ −4.21	−8.59 ~ −5.72
60	−4.58 ~ −3.05	−7.08 ~ −4.72
70	−4.15 ~ −2.76	−6.07 ~ −4.04
80	−3.93 ~ −2.62	−5.62 ~ −3.75
90	−3.61 ~ −2.40	−4.50 ~ −3.00

where F_z is the lift force of the vehicle, and C_{Fz} is the lift coefficient. The lift force of a vehicle is a strong function of crosswind. Yaw angle can be utilised to relate the lift coefficient with both the environmental wind and vehicle speed. When the yaw angle increases from 0 degrees to 90 degrees, the lift coefficient tends to increase and then decrease after a certain yaw angle. This transition angle may vary with different car shapes. The lift coefficients of two representative trains with different streamline lengths are shown in Table 12.4.

12.5.3.3 Aerodynamic Lateral Force of Rail Vehicles

The lateral force of a train is defined as:

$$F_y = \frac{1}{2}\rho S C_{Fy} V^2 \quad (12.17)$$

where F_y is the lateral force of the vehicle, and C_{Fy} is the lateral coefficient. The aerodynamic lateral force on a vehicle is mainly caused by the crosswind. In most of the studies, yaw angle is used instead of crosswind vector for convenience of analysis. The lateral coefficients with respect to different yaw angles and streamline length of head cars are shown in Table 12.5.

When two vehicles are crossing, a lateral force pulse will be formed. The magnitude of the lateral force with respect to time is shown in Figure 12.37.

In Figure 12.37, Δt is the crossing time of two vehicles, which is defined as $\Delta t = L_{tr}/v_r$. In this definition, L_{tr} is the vehicle length and v_r is the relative speed between the two vehicles. The line spacing in a railway system can significantly influence the lateral coefficient during vehicle crossing. Table 12.6 shows the relationship between line spacing (D_L) and lateral coefficient (C_{Fy}).

Meanwhile, the lateral coefficient during vehicle crossing is also related with the streamline length of the vehicle. Table 12.7 shows the effect of streamline length on lateral coefficient.

12.5.3.4 Aerodynamic Pitching, Yawing and Overturning Moment of Rail Vehicles

The pitching moment M_x, yawing moment M_y and rolling moment M_z of trains are defined as:

$$M_x = \frac{C_{Mx}\rho V^2 LS}{2} \quad (12.18)$$

Rail Vehicle Aerodynamics

TABLE 12.5
Lateral Coefficient of Streamlined Head Car for Different Yaw Angles

Yaw Angle (β)	Streamline Length (m)	
	3.5	7.5
0	0	0
10	0.89 ~ 1.34	0.86 ~ 1.29
20	2.27 ~ 3.40	2.03 ~ 3.05
30	4.31 ~ 6.46	3.56 ~ 5.35
40	7.20 ~ 10.79	5.44 ~ 8.17
50	8.31 ~ 12.46	6.92 ~ 10.39
60	7.75 ~ 11.63	6.81 ~ 10.21
70	7.04 ~ 10.56	5.88 ~ 8.82
80	6.63 ~ 9.94	5.61 ~ 8.41
90	6.06 ~ 9.10	5.35 ~ 8.02

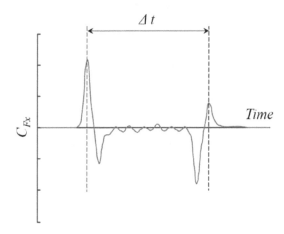

FIGURE 12.37 Lateral coefficient during train crossing.

TABLE 12.6
Lateral Coefficient for Vehicles Crossing for Different Line Distances (D_L)

D_L	4.2	4.6	4.8	5	5.1	5.6	6.1
C_{Fy}	1.17 ~ 1.76	1.03 ~ 1.54	0.96 ~ 1.44	0.9 ~ 1.35	0.87 ~ 1.31	0.72 ~ 1.08	0.63 ~ 0.94

TABLE 12.7
Effect of Streamline Length on Lateral Coefficient

Streamline Length (m)	3.5	5.5	7.5	≥10
C_{Fy}	1.16 ~ 1.74	1 ~ 1.5	0.92 ~ 1.38	0.88 ~ 1.32

$$M_y = \frac{C_{M_y} \rho V^2 LS}{2} \quad (12.19)$$

$$M_z = \frac{C_{M_z} \rho V^2 LS}{2} \quad (12.20)$$

where C_{Mx}, C_{My} and C_{Mz} are the pitching moment coefficient, yawing moment coefficient and rolling moment coefficient, respectively.

The moments of trains can strongly influence the aerodynamic properties of the railway system. The pitching moment, yawing moment and rolling moment of head cars with different streamline lengths and yaw angles are shown in Tables 12.8 and 12.9, respectively. The moment centre is defined in Figure 12.2.

TABLE 12.8
Pitching Moment, Yawing Moment and Rolling Moment of Head Cars with a Streamline Length of 3.5 m

Yaw Angle (β)	C_{Mx}	C_{My}	C_{Mz}
0	0	−0.27 ~ −0.18	0
10	−0.89 ~ −0.59	−0.22 ~ −0.15	−2.26 ~ −1.51
20	−2.21 ~ −1.47	0.21 ~ 0.31	−4.18 ~ −2.78
30	−4.12 ~ −2.75	1.18 ~ 1.77	−4.64 ~ −3.09
40	−6.99 ~ −4.66	1.49 ~ 2.23	−5.45 ~ −3.63
50	−7.98 ~ −5.32	2.02 ~ 3.03	−7.25 ~ −4.84
60	−7.47 ~ −4.98	0.70 ~ 1.05	−4.11 ~ −2.74
70	−6.84 ~ −4.56	−0.07 ~ −0.05	−0.88 ~ −0.59
80	−6.47 ~ −4.31	−0.36 ~ −0.24	−0.11 ~ −0.07
90	−5.96 ~ −3.97	−0.57 ~ −0.38	1.00 ~ 1.51

TABLE 12.9
Pitching Moment, Yawing Moment and Rolling Moment of Head Cars with a Streamline Length of 7.5 m

Yaw Angle (β)	C_{Mx}	C_{My}	C_{Mz}
0	0	−0.40 ~ −0.27	0
10	−0.70 ~ −0.47	−0.61 ~ −0.40	−2.73 ~ −1.82
20	−1.68 ~ −1.12	0.37 ~ 0.56	−4.31 ~ −2.87
30	−2.95 ~ −1.97	1.91 ~ 2.86	−5.26 ~ −3.51
40	−4.53 ~ −3.02	2.75 ~ 4.13	−5.81 ~ −3.87
50	−5.69 ~ −3.79	4.13 ~ 6.20	−7.67 ~ −5.11
60	−5.62 ~ −3.75	3.61 ~ 5.41	−5.47 ~ −3.64
70	−4.88 ~ −3.25	0.11 ~ 0.16	−0.52 ~ −0.35
80	−4.67 ~ −3.12	−1.52 ~ −1.02	0.60 ~ 0.89
90	−4.43 ~ −2.96	−0.04 ~ −0.03	1.82 ~ 2.74

12.5.4 Data Entry of Aerodynamic Load for Vehicle Dynamic Response

The vehicle dynamics of each car of rail vehicle is often described using multi-body dynamics, including wheel set, primary suspension, bogies, secondary suspension and train body. Aerodynamic force components and moments should be entered with respect to the reference point of the vehicle body into the calculation of vehicle dynamics. In general, there are two ways to obtain the required loads:

- In aerodynamic simulation of a rail vehicle, directly use the coordinate system in vehicle dynamics and output aerodynamic moments with respect to the reference point of the vehicle body.
- In a wind tunnel test or existing aerodynamic loads data, the moment centre of measured aerodynamic moments usually differs from the reference point of the vehicle body.
- Figure 12.38 shows the different positions of aerodynamic moment centre and reference point of vehicle dynamics. The following equations are useful to convert previously measured moments to reference point of train body:

$$M_x^V(t) = M_x(t) + F_y(t)d_z + F_z(t)d_y$$
$$M_y^V(t) = M_y(t) + F_x(t)d_z + F_z(t)d_x \quad (12.21)$$
$$M_z^V(t) = M_z(t) + F_x(t)d_y + F_y(t)d_x$$

where $M_i(t)$ is the aerodynamic moment with respect to the i-th axis acting on the reference point of vehicle dynamics at time t, $M_i^V(t)$ is the moment with respect to the i-th axis acting on the reference point of vehicle dynamics at time t, $F_i(t)$ is the component of aerodynamic forces in the i-th direction at time t and d_i is the i-th component of distance to the aerodynamic moment centre.

Under different running conditions, the contributions of aerodynamic forces and moments to the vehicle dynamic response are as follows:

- When a rail vehicle runs in open air with constant speed, the lift and lateral aerodynamic forces are almost zero and can be ignored in vehicle dynamics.
- When rail vehicles cross each other or pass through a tunnel, only the lateral force and rolling moment are needed.
- When a rail vehicle runs in environmental wind, all aerodynamic forces and moments are required for vehicle dynamics analysis.

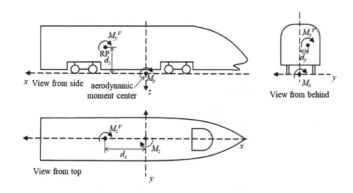

FIGURE 12.38 Positions of aerodynamic moment centre and reference point (RP) of vehicle dynamics.

12.6 EVALUATION OF AERODYNAMIC INDEXES FOR RAIL VEHICLES

12.6.1 Aerodynamic Load on Rail Vehicle

12.6.1.1 Aerodynamic Load in Open Air

- *Drag force*: The maximum allowable air resistance coefficients at different train running speeds are shown in Table 12.10.
- *Lift force*: When there is no environmental wind, the aerodynamic lift force of the high-speed train head car should be close to zero and shall not affect the normal operation of the train.
- *Pressure wave induced by trains crossing*: The maximum value of the pressure wave generated during high-speed train crossing should be less than 6000 Pa.

12.6.1.2 Aerodynamic Load in Railway Tunnels

- *Micro-pressure wave at the tunnel entrance*: When there is a building or special environmental requirement at the entrance of a tunnel, buffering structures should be installed, as shown in Table 12.11.
- *Pressure comfort for railway tunnels*: Table 12.12 lists the standards for human comfort under the air pressure change environment in different countries. In the table, P_{bwmax} is the pressure change amplitude, dP_{bw}/dt is the pressure change rate inside the vehicle and $[P_{bw}]_s$ is the pressure change amplitude within a certain period of time.
- *Train wind*: The maximum allowable train wind speed for platform passengers and track workers is 14 m/s. If the train wind speed is larger than 14 m/s, track workers cannot enter into a tunnel during train operations.
- *Body rigidity*: When high-speed trains cross in the open air or pass through a tunnel, trains with the running speed ranging from 200 km/h to less than 350 km/h shall be assessed

TABLE 12.10
Recommended Upper Limit of Train Air Resistance Coefficient

Rail Vehicle Configuration	Train Running Speed (km/h)				
	200	250	300	350	
Head (with pantograph)	0.25	0.22	0.20	0.19	Values of coefficient of aerodynamic drag for entire train C_x
Head (without pantograph)		0.18	0.17	0.16	
Tail (with pantograph)	0.28	0.26	0.22	0.21	
Tail (without pantograph)		0.19	0.18	0.17	
Middle (with pantograph)		0.16	0.15	0.15	
Middle (without pantograph)	0.10	0.09	0.09	0.09	

TABLE 12.11
Standard for Tunnel Buffer Configuration

Distance from Building to Entrance	Special Environmental Requirements for Building?	Datum Mark	Micro-Pressure Wave Critical Value Standard
<50 m	Yes	Building	As required
<50 m	No	Building	≤20 Pa
≥50 m	Yes	20 m from tunnel entrance	<50 Pa

TABLE 12.12
Pressure Comfort Design Criteria for Railway Tunnels

Country	Criterion	Operating Conditions
China	$dP_{bw}/dt \leq 200$ Pa/s: excellent $[P_{bw}]_s \leq 800$ Pa/3s: good $[P_{bw}]_s \leq 1250$ Pa/3s: qualified $[P_{bw}]_s > 1250$ Pa/3s: not qualified	China Railway High-speed (CRH) train
Japan	$P_{bwmax} < 1$ kPa $dP_{bw}/dt < 200$ Pa/s	High-speed operations: 210, 240, 270 km/h Sealed rolling stock Tight-bore double-track tunnels
UK	$[P_{bw}]_s < 700$ Pa/1.7s $dP_{bw}/dt < 410$ Pa/s	Moderate to high-speed operations: 160, 200 km/h Unsealed rolling stock Tight-bore double-track tunnels
USA	$[P_{bw}]_s < 3$ kPa/3s $[P_{bw}]_s < 4$ kPa/4s	Low-speed operations: 80–100 km/h Unsealed rolling stock Tight-bore tunnels Regular commuter customers
Germany	$P_{bwmax} < 1$ kPa $dP_{bw}/dt < 300 \sim 400$ Pa/s	High-speed operations: 240, 280 km/h Sealed rolling stock Large-bore tunnels

according to the maximum internal and external pressure difference of 4000 Pa. A high-speed train whose operating speed level is equal to or higher than 350 km/h is assessed according to the maximum internal and external pressure difference of 6000 Pa.

12.6.1.3 Safe Distance for Human Body

- *Aerodynamic loads on track workers at the line side*: A full-length train running in the open air at 300 km/h, or at its maximum operating speed $v_{tr,max}$ if lower than 300 km/h, shall not cause an exceedance of the air speed $u_{2\sigma}$ at the trackside, as set out in Table 12.13, at a height of 0.2 m above the top of rail and at a distance of 3.0 m from the track centre, during the passage of the whole train (including the wake).
- *Aerodynamic loads on passengers on a platform*: A full-length train running in the open air at a reference speed $v_{tr} = 200$ km/h, or at its maximum operating speed $v_{tr,max}$ if lower than 200 km/h, shall not cause the air speed $u_{2\sigma}$ to exceed 15.5 m/s at a height of 1.2 m above the platform and at a distance of 3.0 m from the track centre, during the whole train passage (including the wake).
- *Pressure load at trackside*: A full-length train, running in the open air at 250 km/h, or at its maximum operating speed $v_{tr,max}$ if lower than 250 km/h, shall not cause the maximum peak-to-peak pressure changes to exceed a value $\Delta P_{2\sigma}$, as set out in Table 12.14, over the

TABLE 12.13
Maximum Permissible Trackside Air Speed

Maximum Train Speed $v_{tr,max}$ (km/h)	Trackside Maximum Permissible Air Speed (Limit Values for $u_{2\sigma}$ in m/s)
From 190 to 249	20
From 250 to 300	22

TABLE 12.14
Maximum Permissible Trackside Pressure Changes

Reference Train Speed	Maximum Permissible Pressure Change $\Delta P_{2\sigma}$
250 km/h	795 Pa
Maximum speed if <250 km/h	720 Pa

range of heights from 1.5 to 3.3 m above the top of rail and at a distance of 2.5 m from the track centre, during the whole train passage (including the passing of the head, middle and tail cars).

12.6.2 Rail Vehicle Stability under Strong and Rapidly Changing Wind

For rail vehicle stability under strong and rapidly changing wind, the following requirements should be met:

- Vehicle overturning coefficient: ≤0.8
- Body overturning coefficient: ≤0.8
- Bogie overturning coefficient: ≤0.4
- Vehicle derailment coefficient: ≤0.8
- Wheel load reduction: ≤0.8
- Body lateral offset: ≤80 mm
- Body transverse acceleration: ≤2.5 m/s
- Body rolling angle: ≤2.5°
- Body lift-weight angle: ≤0.8°
- Operation stability index: 2.75
- Wheel lateral force: ≤48 kN
- Vibration amplitude of pantograph pan: <150 mm
- Catenary offset: <450 mm
- Catenary pillar offset: <25 mm
- Catenary-pantograph dynamic uplifting amount: ≤120 mm
- Environmental wind speed: ≤50 m/s (for high speed train)

REFERENCES

1. EN 14067-1, Railway applications – aerodynamics – Part 1: Symbols and units, 2 European Committee for Standardization, Brussels, Belgium, 2003.
2. H.Q. Tian, Development of research on aerodynamics of high-speed rails in China, *Engineering Sciences*, 17(4), 30–41, 2015. (in Chinese)
3. H.Q. Tian, Study evolvement of train aerodynamics in China, *Journal of Traffic and Transportation Engineering*, 6(1), 1–9, 2006. (in Chinese)
4. J.R. Bell, D. Burton, M. Thompson, A. Herbst, J. Sheridan, Wind tunnel analysis of the slipstream and wake of a high-speed train, *Journal of Wind Engineering and Industrial Aerodynamics*, 134, 122–138, 2014.
5. D. Zhou, H.Q. Tian, J. Zhang, M.Z. Yang, Pressure transients induced by a high speed train passing through a station, *Journal of Wind Engineering and Industrial Aerodynamics*, 135, 1–9, 2014.
6. EN 14067-5, Railway applications – aerodynamics – Part 5: Requirements and test procedures for aerodynamics in tunnels, European Committee for Standardization, Brussels, Belgium, 2013.
7. M.S. Howe, Mach number dependence of the compression wave generated by a high-speed train entering a tunnel, *Journal of Sound and Vibration*, 212(1), 23–36, 1998.

8. H.Q. Tian, *Train Aerodynamics*, China Railway Press, Beijing, China, 2007. (in Chinese)
9. ANSYS Fluent theory guide, ANSYS Inc., Canonsburg, PA, 2019.
10. H.K. Versteeg, W. Malalasekera, *An Introduction to Computational Fluid Dynamics: The Finite Volume Method*, 2nd ed., Pearson Education Limited, Harlow, UK, 2007.
11. J.Y. Tu, G.H. Yeoh, C.Q. Liu, *Computational Fluid Dynamics: A Practical Approach*, 2nd ed., Butterworth-Heinemann, Kidlington, UK, 2012.
12. H.Q. Tian, S. Huang, M.Z. Yang, Flow structure around high-speed train in open air, *Journal of Central South University*, 22(2), 747–752, 2015.
13. S. Huang, H. Hemida, M.Z. Yang, Numerical calculation of the slipstream generated by a CRH2 high-speed train, *Journal of Rail and Rapid Transit*, 230(1), 103–116, 2016.

13 Longitudinal Train Dynamics and Vehicle Stability in Train Operations

Colin Cole

CONTENTS

13.1 Introduction 457
13.2 Modelling Longitudinal Train Dynamics 458
    13.2.1 Train Models 458
    13.2.2 Modelling Input Forces 460
        13.2.2.1 Locomotive Traction and Dynamic Braking 460
        13.2.2.2 Propulsion Resistance 466
        13.2.2.3 Curving Resistance 469
        13.2.2.4 Gravitational Components 470
        13.2.2.5 Pneumatic Brake Models 470
    13.2.3 Modelling Vehicle Connections 473
        13.2.3.1 Equipment Overview 474
        13.2.3.2 Draft Gear Overview 475
        13.2.3.3 Modelling Fundamentals 476
        13.2.3.4 Modelling for Wagon Connections 477
        13.2.3.5 Methods of Measuring Draft Gear Characteristics 493
13.3 Wagon Dynamic Responses to Longitudinal Train Dynamics 494
    13.3.1 Overview 494
    13.3.2 Fast Methods for Assessing Interactions on Train Routes 497
        13.3.2.1 Lateral Coupler Angles 497
        13.3.2.2 Wheel Unloading, Wheel Climb and Rollover on Curves due to Lateral Components of Coupler Forces 501
        13.3.2.3 Rail Vehicle Body and Bogie Pitch due to Coupler Impact Forces 504
        13.3.2.4 Rail Vehicle Lift-Off due to Vertical Components of Coupler Forces 509
13.4 Longitudinal Train Dynamics and Train Crashworthiness 512
    13.4.1 Locomotive or Power Car Crumple Zones 512
    13.4.2 End Car Crumple Zones 512
    13.4.3 Vertical Collision Posts 512
13.5 Longitudinal Passenger Comfort 514
References 517

13.1 INTRODUCTION

This chapter has been designed to provide a hands-on guide to both understanding and analysing longitudinal train dynamics. It is specifically focussed on the longitudinal dynamics of heavy-haul trains and adds new insights to previous work by the author [1–4], along with recent modelling work relevant to passenger trains [5]. It is 12 years since the first edition and the draft gear section is now expanded. New to the chapter is material concerning lateral wagon stability in the presence of lateral coupler force components. Passenger comfort and crashworthiness sections are also updated.

Longitudinal train dynamics is defined as the motions of rail vehicles in the direction along the track. It therefore includes the motion of the train as a whole and any relative motions between vehicles allowed due to the looseness and travel permitted by spring and damper connections between vehicles. In the railway industry, the term 'slack action' is used for the relative motions of vehicles in a train due to the correct understanding that these motions are primarily allowed by the free slack and deflections allowed in wagon connections. Coupling 'free slack' is defined as the free movement allowed by the sum of the clearances in the vehicle connection. In the case of autocouplers, these clearances consist of clearances in the autocoupler knuckles and draft gear assembly pins. In older rolling stock connection systems, such as draw hooks and buffers, free slack is the clearance between the buffers measured in tension. Note that a system with draw hooks and buffers could be preloaded with the screw link to remove free slack. Longitudinal train dynamics therefore has implications for driver/crew comfort and freight product damage, vehicle stability, rolling stock design and rolling stock metal fatigue [1].

The study and understanding of longitudinal train dynamics were initially motivated by the desire to reduce longitudinal vehicle dynamics in passenger trains and, in so doing, improve the general comfort of passengers. The practice of 'power braking', which is the seemingly strange technique of keeping locomotive power applied while a minimum air brake application is made, is still widely used on passenger trains. Power braking is also sometimes used on partly loaded mixed freight trains to keep the train stretched during braking and when operating on undulating track. Interest in train dynamics in freight trains increased as trains became longer, particularly for heavy-haul trains, as evidenced in published technical papers. In the late 1980s, measurement and simulation of in-train forces on such trains in Australia were reported by Duncan and Webb [6]. The engineering issues associated with moving to trains of double the existing length were reported at the same time, also in Australia, in a paper by Jolly and Sismey [7]. A further paper focussed on train-handling techniques on the Richards Bay Line gave the South African experience [8]. The research at this time was driven primarily by the occurrence of fatigue cracking and tensile failures in wagon structures and autocouplers. From these studies, an understanding of the force magnitudes was developed, along with an awareness of the need to limit these forces with appropriate driving strategies [1–8].

More recent research into longitudinal train dynamics was started in the early 1990s, not motivated at this time by equipment failures and fatigue damage but by derailments. The direction of this research was concerned with the linkage of longitudinal train dynamics to increases in wheel unloading. It stands to reason that, as trains get longer and heavier, the in-train forces get larger. As coupler forces become larger, so too will the lateral and vertical components of these forces resulting from the angles between connected couplers on horizontal and vertical curves. At some point, these components will adversely affect vehicle stability. The first known work published that addressed this issue was that of El-Sibaie in 1993 [9], which looked at the relationship between lateral coupler force components and wheel unloading. Further modes of interaction were reported and simulated in 1999 by McClanachan et al. [10], detailing vehicle body and bogie pitch. This work has been further developed with methods of assessing the effect of longitudinal train dynamics on lateral and vertical train dynamics, using whole trip simulations [11]. Approaches to modelling and simulation of train dynamics have been given more prominence in a recent international benchmarking exercise involving several research groups around the world [12,13].

13.2 MODELLING LONGITUDINAL TRAIN DYNAMICS

13.2.1 Train Models

The longitudinal behaviour of trains is a function of train control inputs from the locomotive, train brake inputs, track topography, track curvature, rolling stock and bogie characteristics and vehicle connection characteristics.

The longitudinal dynamic behaviour of a train can be described by a system of differential equations. For the purposes of setting up the equations for modelling and simulation, it is usually assumed that there is no lateral or vertical movement of the vehicles. This simplification of the system is employed by all known commercial rail-specific simulation packages and by texts such as Garg and Dukkipati [14]. The governing differential equations can be developed by considering the generalised three-mass train shown in Figure 13.1. It will be noticed that the in-train vehicle, whether locomotive or wagon, can be classified as one of only three connection configurations, lead (shown as m_1), in-train and tail. All vehicles are subject to retardation and grade forces. Traction and dynamic brake forces are added to powered vehicles.

In Figure 13.1, a = vehicle acceleration, m/s²; c = damping constant, Ns/m; k = spring constant, N/m; m = vehicle mass, kg; v = vehicle velocity, m/s; x = vehicle displacement, m; F_g = gravity force components due to track grade, N; F_r = sum of retardation forces, N; and $F_{t/db}$ = traction and dynamic brake forces from a locomotive unit, N. Allowing for locomotives to be placed at any train position and extending equation notation for a train of any number of vehicles, a general set of equations can be written as:

For the lead vehicle:

$$m_1(1.0 + J_{f1})a_1 + f_{wc}(v_1, v_2, x_1, x_2) = F_{t/db1} - F_{r1} - F_{g1} \qquad (13.1)$$

For the ith vehicle:

$$m_i(1.0 + J_{fi})a_i + f_{wc}(v_i, v_{i-1}, x_i, x_{i-1}) + f_{wc}(v_i, v_{i+1}, x_i, x_{i+1}) = F_{t/dbi} - F_{ri} - F_{gi} \qquad (13.2)$$

For the nth or last vehicle:

$$m_n(1.0 + J_{fn})a_n + f_{wc}(v_n, v_{n-1}, x_n, x_{n-1}) = F_{t/dbn} - F_{rn} - F_{gn} \qquad (13.3)$$

where f_{wc} is the non-linear function describing the full characteristics of the vehicle connection; $J_f = \sum J/R^2$ is the equivalent translational mass inertia corresponding to the equivalent sum of rotating inertias $\sum J$ due to wheelsets, gearboxes and drive motors in each vehicle; and R is the wheel radius. In heavy-haul analysis, the J_f term is usually neglected, because it is typically 1%–2% of loaded wagon mass and 10%–15% of locomotive mass. As these train systems are large, the effect on the whole train acceleration responses is only ~ 1%–2%, noting that the effect of locomotive rotating inertia is typically less than 0.5% of train mass. In much smaller trains, passenger trains and high-speed trains, the rotary inertia of the locomotive or power car will be proportionately larger and more significant to results.

By including the term $F_{t/db}$ in each equation, and thus on every vehicle, the equations can be applied to any locomotive placement or system of distributed power. $F_{t/db}$ is set to zero for unpowered

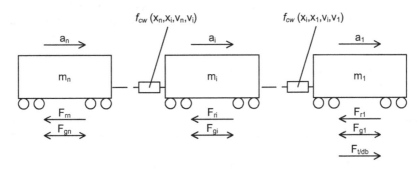

FIGURE 13.1 Generalised three-mass train model. (From Spiryagin, M. et al., *Design and Simulation of Rail Vehicles, Ground Vehicle Engineering Series*, CRC Press, Boca Raton, FL, 2014. With permission.)

vehicles. It will be noted in the model in Figure 13.1 that the grade force can be in either direction. The sum of the retardation forces, F_r, is made up of rolling resistance, curving resistance or curve drag, air resistance and braking (excluding dynamic braking, which is more conveniently grouped with locomotive traction in the $F_{t/db}$ term). Rolling and air resistances are usually grouped as a term known as propulsion resistance, F_{pr}, making the equation for F_r as follows:

$$F_r = F_{pr} + F_{cr} + F_b \tag{13.4}$$

where F_{pr} is the propulsion resistance, F_{cr} is the curving resistance and F_b is the braking resistance due to pneumatic braking.

Note that a positive value of F_g is taken as an upward grade, that is, a retarding force.

Solution and simulation of the above equation set is further complicated by the need to calculate the force inputs to the system, that is, $F_{t/db}$, F_r and F_g. The traction-dynamic brake force term, $F_{t/db}$, must be continually updated for driver control adjustments and any changes to locomotive speed. The retardation forces, F_r, are dependent on air braking forces, vehicle velocity, track curvature and rolling stock design. Gravity force components, F_g, are dependent on track grade and therefore on the position of the vehicle on the track. Approaches to the non-linear modelling of the vehicle connections and modelling of each of the force inputs are included and discussed in the following sections.

13.2.2 Modelling Input Forces

Force inputs to each vehicle can be considered using the following single vehicle equation:

$$m_1 a_1 = F_{t/db} - F_{r1} - F_{g1} \tag{13.5}$$

The following sections briefly explain locomotive traction and braking, resistance forces and grades.

13.2.2.1 Locomotive Traction and Dynamic Braking

Locomotive traction and dynamic braking have evolved over many years, and several systems exist. There is a tradition of notch levels for traction control, in which each level approximates to an almost constant power level over most of the operating range. There is the exception that, at low speeds, motors become current limited rather than power limited.

In diesel locomotives, a tradition of eight notches for the throttle control emerged based on a three-valve fuel control. In fully electric locomotives, there are designs with the number of notches being 8, 10, 16 and 32. As designs have become complex, it is best to base models upon manufacturers' locomotive performance curves. Typical traction curves for a diesel-electric locomotive are shown in Figure 13.2.

While a reasonable fit to the published power curves may be possible with a simple equation of the form $P = F_{t/db} * v$, it may be necessary to modify this model to reflect further control features or reflect changes in efficiency or thermal effects at different train speeds. Enhanced performance at higher speeds is achieved on some locomotives by adding a motor field-weakening control [15]. It can be seen that accurate modelling of locomotives, even without considering the electrical modelling in detail, can become quite complicated. In all cases, the performance curves should be sourced and as much precise detail as possible should be obtained about the control features to ensure that a suitable model is developed.

A key parameter in any discussion about tractive effort is wheel-rail adhesion or the coefficient of friction. Prior to enhancement of motor torque control, a wheel-rail adhesion level of ~0.20 could be expected. Modern locomotive traction control systems are delivering higher values of adhesion reaching ~0.35 in daily operation, with manufacturers claiming up to 0.52 in published performance curves. It needs to be remembered that a smooth control system can only deliver an adhesion level up to the maximum set by the coefficient of friction for the wheel-rail conditions.

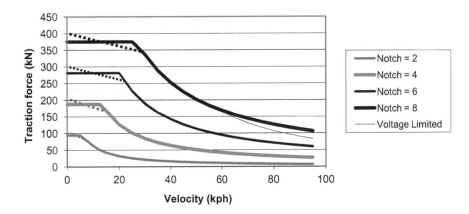

FIGURE 13.2 Typical tractive effort performance curves for diesel-electric locomotive. (From Cole, C., Longitudinal train dynamics, Chapter 9 in *Handbook of Railway Vehicle Dynamics*, S. Iwnicki (Ed.), CRC Press, Boca Raton, FL, 239–278, 2006. With permission.)

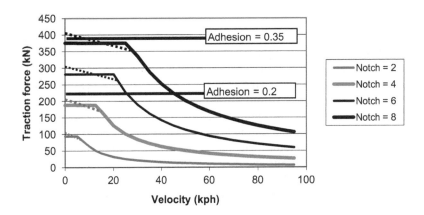

FIGURE 13.3 Tractive effort performance curves showing the effect of adhesion levels. (From Cole, C., Longitudinal train dynamics, Chapter 9 in *Handbook of Railway Vehicle Dynamics*, S. Iwnicki (Ed.), CRC Press, Boca Raton, FL, 239–278, 2006. With permission.)

Wheel-rail conditions in frost and snow could reduce adhesion to as low as 0.1. Figure 13.3 superimposes adhesion levels on the curves from Figure 13.2, showing how significant adhesion is as a locomotive performance parameter.

It is typical for locomotive manufacturers to publish both the maximum tractive effort and the maximum continuous tractive effort. The maximum continuous tractive effort is the traction force delivered at full throttle notch after the traction system has heated to maximum operating temperature. As the resistivity of the windings increases with temperature, motor torque, which is dependent on current, decreases. As traction motors have considerable mass, considerable time is needed for the locomotive motors to heat, and performance levels drop to maximum continuous tractive effort. A typical thermal derating curve for a modern locomotive is shown in Figure 13.4.

Manufacturers' data from which performance curves such as shown in Figure 13.2 are derived can usually be taken to be maximum rather than continuous values. If the longitudinal dynamics problem under study has severe grades and locomotives are delivering large traction forces for long periods, it will be necessary to modify the simple model represented in Figure 13.2 with a further model adding these thermal effects.

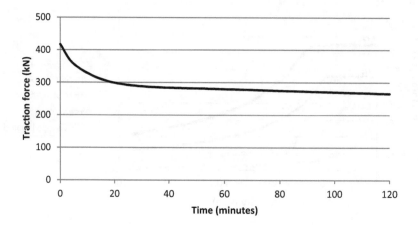

FIGURE 13.4 Tractive effort thermal derating curve. (From Cole, C., Longitudinal train dynamics, Chapter 9 in *Handbook of Railway Vehicle Dynamics*, S. Iwnicki (Ed.), CRC Press, Boca Raton, FL, 239–278, 2006. With permission.)

The use of dynamic brake as a means of train deceleration has continued to increase as dynamic brake control systems have been improved. As shown in Figure 13.5, early systems gave only a 'peaked' forces characteristic and were not well received by train drivers. As the effectiveness was so dependent on velocity, the use of dynamic brake gave unpredictable results unless a mental note was made of locomotive velocity and the driver was aware of what performance to expect. Extended-range systems, which involved switching resistor banks, greatly improved dynamic brake usability on diesel-electric locomotives, and more recent locomotive packages have provided large regions of maximum retardation at steady force levels (see Figure 13.5). Performance at low speeds is limited by the motor field. The performance of the dynamic brake is limited at higher speeds, firstly by current limits and then by both current and voltage limits to give a power limit curve. Commutator limits also noticeably further limit torque on some designs. Designs have continued to extend the full dynamic brake force capability to lower and even lower speeds.

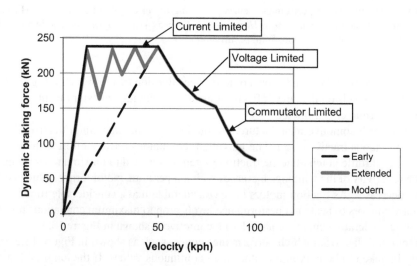

FIGURE 13.5 Dynamic brake characteristics. (From Cole, C., Longitudinal train dynamics, Chapter 9 in *Handbook of Railway Vehicle Dynamics*, S. Iwnicki (Ed.), CRC Press, Boca Raton, FL, 239–278, 2006. With permission.)

Recent designs have achieved the retention of maximum dynamic braking force down to 2 km/h. Dynamic brake can be controlled as a continuous level, or at discrete control levels, depending on the locomotive design.

Modelling of traction and dynamic brake forces can be achieved by:

- Lookup table approach
- Equation-based approach
- Performance curve interpolation approach
- Co-simulation approach

13.2.2.1.1 Lookup Table Approach

With this approach, the generation of a force-time array is based on the data point sets usually measured from a locomotive traction system. This data can be provided as an input to a train dynamics model to replicate and study the train dynamics with accurately determined traction and dynamic forces. This is useful when the data is available and when the characteristics of the traction system are uncertain, and it is an alternative to using locomotive control information with a traction/dynamic braking model, as per Figures 13.3 and 13.5.

13.2.2.1.2 Equation-Based Approach

Traction modelling has been presented in [1–4,16]. Assuming linear distribution of throttle notch levels, the traction characteristics for each notch position are defined in Table 13.1. In this table, F_{TE} is the tractive effort of a locomotive, N; V is the linear speed of a locomotive, m/s; N_{TE} is the notch position from 1 to Nn; P_{max} is the maximum traction power of a locomotive, W; Te_{max} is the maximum starting tractive effort of a locomotive, N; and factors q_1 and q_2 are small correction factors for changes in performance at different speeds to allow fine adjustments to better fit actual performance curves.

The conditions to deliver dynamic braking characteristics are defined in Table 13.2. In this table, N_{DB} is the braking notch position from 0 to 8; F_{DB} is the dynamic braking effort, N; $V_{DB\,min}$ and $V_{DB\,max}$ are minimum and maximum locomotive speeds, respectively, for maximum dynamic braking effort delivered at full dynamic brake (DB), m/s; $P_{DB\,max}$ is the maximum dynamic braking power, W; and DB_{max} is the maximum dynamic braking effort, N. An example of the traction and braking force curves for a locomotive traction model using this approach is shown in Figure 13.6 [17].

13.2.2.1.3 Performance Curve Interpolation Approach

Traction and dynamic brake modelling can also be obtained from using the performance curves directly. These curves will appear the same as those in Figure 13.6 but will exist as a series of points rather than equations. Performance data for each throttle notch and DB level are interpolated for the specific speed.

TABLE 13.1

Equations to Generate Traction Characteristics (Equation 13.6)

Traction Notch	Condition	Tractive Effort (N)
General		
$N_{TE} = 1 \sim Nn$	$F_{TE}*v < (N_{TE}/Nn)^2*P_{max}$	$F_{te} = (N_{TE}/Nn)*Te_{max} - q_1*V$
$N_{TE} = 1 \sim Nn$	$F_{TE}*v \geq (N_{TE}/Nn)^2*P_{max}$	$F_{te} = (N_{TE}/Nn)^2*P_{max}/v \pm q_2*V$
Diesel-Electric		
$N_{TE} = 1 \sim 8$	$F_{TE}*v < (N_{TE}/8)^2*P_{max}$	$F_{te} = (N_{TE}/8)*Te_{max} - q_1*V$
$N_{TE} = 1 \sim 8$	$F_{TE}*v \geq (N_{TE}/8)^2*P_{max}$	$F_{te} = (N_{TE}/8)^2*P_{max}/v \pm q_2*V$

TABLE 13.2
Equations to Generate Dynamic Brake Characteristics (Equation 13.7)

DB Type	Dynamic Brake Force	Limit Controls or Conditions
Modern AC Type	$F_{DB} = (N_{DB}/N_{DBmax})*DB_{max} - q_1*V$ If $(F_{DB} > F_{DB(LSL)})$ $\{F_{DB} = F_{DB(LSL)}\}$ If $(F_{DB} > F_{DB(PL)})$ $\{F_{DB} = F_{DB(PL)}\}$	Low Speed Limit $F_{DB(LSL)} = V/V_{DB\,min}*DB_{max}$ Power Limit $F_{DB(PL)} = P_{DB\,max}/V$
DC Extended Range Type	$F_{DB} = f_{1,2,3}(N_{DB}/N_{DBmax})*DB(N_{DB}, V)$[a,b]	
DC Early Type	$F_{DB} = (N_{DB}/N_{DBmax})(V/V_{DB\,min})*DB_{max}$ If $(F_{DB} > F_{DB(CL)})$ $\{F_{DB} = F_{DB(CL)}\}$ If $(F_{DB} > F_{DB(PL)})$ $\{F_{DB} = F_{DB(PL)}\}$	Current Limit $F_{DB(CL)} = DB_{max}$ Power Limit $F_{DB(PL)} = P_{DB\,max}/V$
DC Very Early Type	$F_{DB} = (N_{DB}/N_{DBmax})(V/V_{DB\,min})*DB_{max}$ If $(F_{DB} > F_{DB(PL)})$ $\{F_{DB} = F_{DB(PL)}\}$	Power Limit $F_{DB(PL)} = P_{DB\,max}/V$

[a] The proportion function $f_{1,2,3}(N_{DB}/N_{DBmax})$ defines the proportionality of notch levels for different speed ranges.
[b] $DB(N_{DB}, V)$ is the general equation describing dynamic brake force in terms of notch levels and speed.

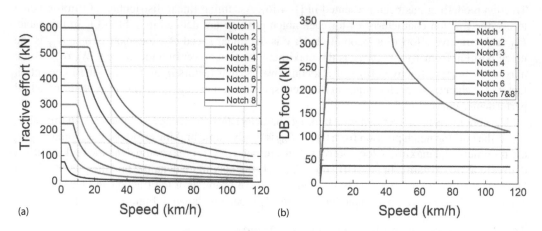

FIGURE 13.6 Heavy-haul locomotive performance curves: (a) tractive effort and (b) dynamic braking. (From Spiryagin, M. et al., *Vehicle Syst. Dyn.*, 55(4), 450–463, 2017. With permission.)

13.2.2.1.4 Co-Simulation Approach

This approach applies the co-simulation modelling technique between two software packages, namely a longitudinal train dynamics simulator and a multibody software package, as described in [18–20]. The concept scheme of the co-simulation process is shown in Figure 13.7. This approach allows the detailed modelling of the electrical control and traction motor characteristics and the detailed modelling of the locomotive wheel-rail contacts to be included. The resulting simulation improves the accuracy of tractive effort results in the longitudinal train dynamics simulator, because the effects of adhesion/traction control of locomotives and processes at the wheel/rail interface are taken into account. As indicated in [3], the traction/adhesion control systems used on rail traction vehicles should achieve an optimal adhesion between wheels and rails and avoid any potential damage caused by exceeding the maximum allowable traction torque applied to the wheelset. As a result, it requires the implementation of adhesion control strategies inside of the multibody software package. This approach gives the ability to comprehensively study specific traction and braking issues, rail and wheel stresses and wear.

Longitudinal Train Dynamics and Vehicle Stability in Train Operations   465

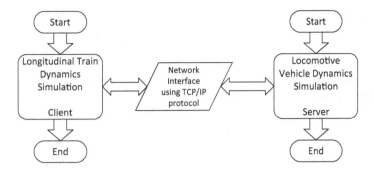

FIGURE 13.7 Design concept of the co-simulation approach.

In [19], the simulations have been performed, as shown in Figure 13.7, for two cases that differ only in their co-simulation data exchange modelling approaches, namely uni-directional and bi-directional. In the first case, the longitudinal train dynamics simulator provides all data required to a locomotive model in the multibody software, but it does not update its own values of traction efforts. In the second case, data exchanges occur in both directions that allow the longitudinal train dynamics simulator to update a value of tractive effort at each co-simulation time step. The locomotive multibody model contains a subroutine with a simplified traction system, based on the bogie traction control strategy of one inverter per bogie and the application of a proportional-integral (PI) controller [4].

In addition, lubricated friction conditions were introduced on the curves in the study in [19]. The published results for this approach show changes in results of the tractive efforts of a locomotive, caused by the activation of the traction control algorithm due to the increase in slips, connected with the change of friction conditions at the wheel-rail interface during running on the lubricated curve. Changes typical of those published in [19] are drawn in Figure 13.8. In addition, the uni-directional co-simulation approach does not allow the multibody locomotive model to achieve the values of tractive effort delivered in the longitudinal train dynamics simulator. This confirms that the wheel-rail contact conditions will have a significant influence on the results delivered in longitudinal train dynamics simulators. Further work by Wu and Spiryagin et al. [20] explores different wheel contact models and the addition of sanding.

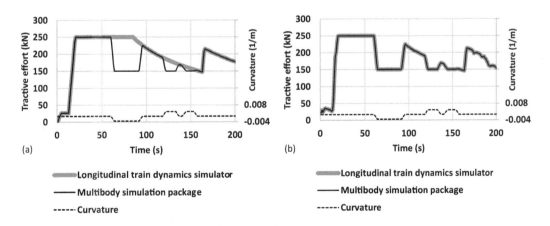

FIGURE 13.8 Locomotive tractive effort in the time-domain during co-simulation studies using (a) uni-directional data exchange approach and (b) bi-directional data exchange approach.

13.2.2.2 Propulsion Resistance

Propulsion resistance is usually defined as the sum of rolling resistance and air resistance for each vehicle. In most cases, increased vehicle drag due to track curvature is considered separately. The variable shapes and designs of rolling stock and the complexity of aerodynamic drag mean that the calculation of rolling resistance is still dependent on empirical formulas. Typically, propulsion resistance is expressed in an equation of the form of $R = A + BV + CV^2$. There are many different equations [16,21–24]. An instructive collection of propulsion resistance formulas has been assembled in Tables 13.3 and 13.4 from [16,21,22]. All equations are converted to SI units and expressed as Newtons per tonne mass. A graphical representation of the various outcomes for freight rolling stock is provided in Figure 13.9. While there is a tradition in literature to quote resistance in terms of force per tonne mass, this practice can be misleading, as the air resistance term has no dependence on mass. In the case of the modified Davis equation formulations, the equation '$K_a[2.943 + 89/m_a + 0.0305V + 1.718k_{ad}V^2/(m_a n)]$', when multiplied by vehicle mass, becomes '$2.943m_a n K_a + 89n K_a + 0.0305m_a n K_a V + 1.787 K_a k_{ad} V^2$', with only the first and third terms proportional to mass. The second term is proportional to the number of axles, and the third term is

TABLE 13.3
Empirical Formulas for Freight Rolling Stock Propulsion Resistance (Equation 13.8)

Description	Propulsion Resistance Equations and Factors
Original Davis equation (AAR-RP548-2001) [22]	$R = A + 129.0/m_a + BV + CLHV^2/(m_a n)$ N/tonne A = 6.376 (typical) [16], 7.358 [20] 3.433–10.448 [24] B = 0.092 locomotives, 0.138 freight cars C = 0.045 lead locomotives, 0.0103 trail locomotives [23,24] C = 0.0094 freight cars (typical) C = 0.0079 loaded coal gondola [23–24] C = 0.0225 empty coal gondola [23–24] LH = frontal area in square metres
Modified Davis equation (USA) [14]	$R = K_a[2.943 + 89/m_a + 0.0305V + 1.718k_{ad}V^2/(m_a n)]$ N/tonne K_a = 1.0 for pre-1950, 0.85 for post-1950, 0.95 container on flatcar, 1.05 trailer on flatcar, 1.05 hopper cars, 1.2 empty covered auto racks, 1.3 for loaded covered auto racks, 1.9 empty, uncovered auto racks k_{ad} = 0.07 for conventional equipment, 0.0935 for containers and 0.16 for trailers on flatcars, giving: $K_a[2.943 + 89/m_a + 0.0305V + 0.122V^2/(m_a n)]$ $K_a[2.943 + 89/m_a + 0.0305V + 0.279k_{ad}V^2/(m_a n)]$
Modified Davis equation (AAR-RP548-2001) [19]	$R = 6.376 + 323/m_a + 0.046V + CV^2/(m_a n)$ N/tonne C = 0.096 [22] C = 0.669 lead loco [23,24] C = 0.153 trail loco [23,24] C = 0.077 loaded coal gondola [23,24] C = 0.220 empty coal gondola [23,24]
French locomotives	$R = 0.65m_a n + 13n + 0.01m_a nV + 0.03V^2$ N/tonne
French standard UIC vehicles	$R = 9.81(1.25 + V^2/6300)$ N/tonne
French Express Freight	$R = 9.81(1.5 + V^2/(2000…2400))$ N/tonne
French 10 t/axle	$R = 9.81(1.5 + V^2/1600)$ N/tonne
French 18 t/axle	$R = 9.81(1.2 + V^2/4000)$ N/tonne
German Strahl formula	$R = 25 + k(V + \Delta V)/10$ N/tonne k = 0.05 for mixed freight trains, 0.025 for block trains
Broad Gauge (i.e., 1.676 m)	$R = 9.81[0.87 + 0.0103V + 0.000056V^2]$ N/tonne
Broad Gauge (i.e., ~1.0 m)	$R = 9.81[2.6 + 0.0003V^2]$ N/tonne

TABLE 13.4
Empirical Formulas for Passenger Rolling Stock Propulsion Resistance (Equation 13.9)

Description	Propulsion Resistance Equations
Original Davis equation (AAR-RP548-2001) [22]	$R = A + 129.0/m_a + BV + CLHV^2/(m_a n)$ N/tonne
	$A = 6.376$
	$B = 0.092$ passenger cars, 0.138 multiple units, 0.092 Talgo type ultra-lightweight trains
	$C = 0.0064$ passenger cars, 0.045 multiple units (lead car), 0.0067 multiple units (trailing car), 0.045 Talgo type ultra-lightweight trains (lead car), 0.0064 Talgo type ultra-lightweight trains (trailing car)
	LH = Frontal area in square metres
French passenger on bogies	$R = 9.81(1.5 + V^2/4500)$ N/tonne
French passenger on axles	$R = 9.81(1.5 + V^2/(2000 \ldots 2400))$ N/tonne
French TGV	$R = 2500 + 33V + 0.543\,V^2$ N/tonne
German Sauthoff formula freight (intercity express, ICE)	$R = 9.81[1 + 0.0025V + 0.0055*((V + \Delta V)/10)^2]$ N/tonne
Broad gauge (i.e., 1.676 m)	$R = 9.81[0.6855 + 0.02112V + 0.000082V^2]$ N/tonne
Narrow gauge (i.e., ~1.0 m)	$R = 9.81[1.56 + 0.0075V + 0.0003V^2]$ N/tonne

proportional to area and drag coefficient. Szanto gives a good discussion of these terms in [24]. It follows for the equations in Tables 13.3 and 13.4 that, if the mass component is not explicitly separated in the air resistance term, the formula can only be used for the specific rolling stock with which it is associated.

In Tables 13.1 and 13.2, B is a flange resistance factor; C is an air resistance factor; V is the velocity in km/h; m_a is mass supported per axle in tonnes; n is the number of axles; K_a is an adjustment factor depending on rolling stock type; k_{ad} is an air drag constant depending on car type; and ΔV is the headwind speed, usually taken as 15 km/h. The effect of headwind can be incorporated into any formula by adding it to the train velocity. The Association of American Railroads (AAR)'s RP-548 [22] suggests that a headwind of 16 km/h be added for schedule calculations.

Even with the number of factors described in Tables 13.3 and 13.4, the effects of many factors are not, and usually cannot be, accurately considered. The frontal area facing air resistance changes depending on whether the vehicle is leading the train and the amount of inter-wagon distance. It will be noted that the air resistance term is highly dependent on vehicle shape and frontal and side areas; see Table 13.3 and Figure 13.9a and b. A clear indication of this can be seen in Table 13.3 in regard to leading and trailing locomotives, noting the air drag coefficients of $C = 0.045$ and 0.0103, respectively, in the original Davis equation. Open hoppers and open box wagons will have significantly higher drag when empty of product and have different drag to equivalent wagons with covers. In Table 13.3 and Figure 13.9b, it will be noted that empty coal gondola wagons are of similar drag to intermodal operations for trailers on flatcars. Loaded hopper wagons have much less drag, as they approximate to a closed-in wagon shape. As railways continue to optimise rolling stock and minimise wasted load space, there is less space between wagons, which will lead to less drag, noting that the loaded coal gondolas show less resistance than the general ARR RP-548 equation in Figure 13.9b. Szanto [24] noted that Lai and Barkan [25] indicated that drag progressively decreased from the lead of a train until about the 10th wagon and then remained approximately constant. The study [25] supports the observation that the differing drag from frontal and side areas will also be highly dependent on inter-wagon spaces (as illustrated by the effect of open hopper or gondola wagons). There are also formulations of train propulsion that treat the train as a whole rather than on a vehicle-by-vehicle basis, as in [26]. These equations have not been included as, for the study of longitudinal train dynamics, it is necessary to formulate the response of each wagon individually.

If the air drag components could be formulated more accurately, it would be possible to make the drag coefficients dependent on train position (Table 13.4).

Despite the complicated formulas already available, issues such as the rail vehicle design, effect of headwinds, crosswinds and tail winds; different rolling resistances; and the in-train position introduce many uncertainties to the calculations.

The dynamicist should therefore be aware that considerable differences between calculations and field measurements are probable.

Starting resistance can also a consideration if the trains have limited free slack and the mass-to-power ratio is large. This usually only arises in short trains, so it is not mentioned in

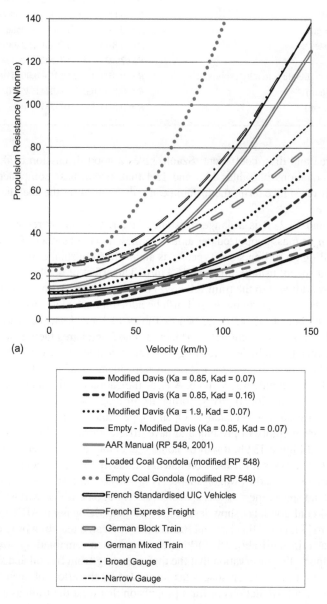

FIGURE 13.9 Comparison of freight rolling stock characteristics: (a) propulsion resistance equations.
(*Continued*)

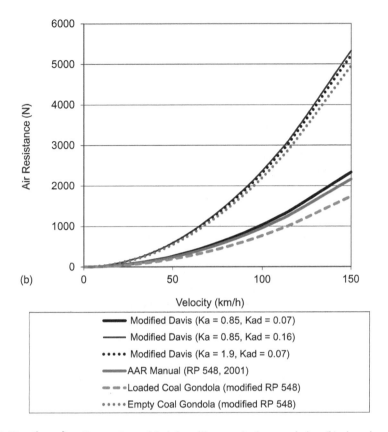

FIGURE 13.9 (Continued) Comparison of freight rolling stock characteristics: (b) air resistance terms.

heavy-haul data. The consideration of starting resistance has become uninteresting in the general study of freight train dynamics, as wagon connections always have coupling slack and, in larger trains, the locomotives have more than sufficient power to move the train, as, of course, the wagons start progressively rather than as a single block. In the consideration of shorter trains and in some railway standards, a higher starting resistance is sometimes noted.

Modern methods to determine the resistance were discussed in the PhD thesis of Lukaszewicz [26], and some 16 different equations are presented in [1–3,23,24]. Of those 16 equations, only 4 recognise a higher resistance at starting, with values ranging from 25 N/tonne to 60 N/tonne, with the effect vanishing at low speeds in the region less than 5 km/h; see Figure 13.10a. Values for starting resistance in [27] for wagons with roller bearings were stated from different sources as 3.5 N/kN and 7.0 N/kN, which translate to the range of 34.3–68.7 N/tonne. Equating the velocity terms of the modified Davis equation to zero (start-up condition), this requires a multiplier factor M in the range of 5.46–10.92.

In a recent paper on tipplers [28], choosing the multiplier factor at M = 5.0 gave reasonable results; see Figure 13.10.

13.2.2.3 Curving Resistance

Curving resistance calculations have similarity to propulsion resistance calculations in that empirical formulas must be used. Vehicle steering design and condition, operating cant (superelevation) condition (cant deficiency, balanced or excess cant), rail profile, rail lubrication and curve radius will all affect the resistance imposed on a vehicle on the curve. All of these factors can vary significantly,

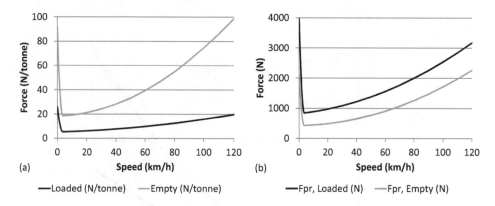

FIGURE 13.10 Propulsion resistance, including starting resistance, for loaded and empty wagons: (a) resistance per tonne mass and (b) total resistance force for a wagon. (From Cole, C. et al., *Vehicle Syst. Dyn.*, 55(4), 534–551, 2017. With permission.)

but for the purposes of longitudinal train simulation, it is usual to estimate curving resistance by a function relating only to curve radius. The equation commonly used, as detailed in [16], is:

$$F_{cr} = 6116/R \tag{13.10}$$

where F_{cr} is in Newtons per tonne of vehicle mass and R is curve radius in metres.

Rail flange lubrication is thought to be capable of reducing curving resistance by 50%. The curving resistance of a vehicle that is stationary on a curve is thought to be approximately double the value given by Equation 13.10 [1].

13.2.2.4 Gravitational Components

Gravitational components, F_g, are added to longitudinal train models by simply resolving the weight vector into components parallel and at right angles to the vehicle body chassis. The parallel component of the vehicle weight becomes F_g. On a grade, a force will either be added to or subtracted from the longitudinal forces on the vehicle; see Figure 13.11 [1].

13.2.2.5 Pneumatic Brake Models

The modelling of the brake system requires the simulation of a fluid dynamic system that must run in parallel with the train simulation. The output from the brake pipe simulation is the brake cylinder force, which is converted by means of rigging factors and shoe friction coefficients into a retardation force that is one term of the sum of retardation forces F_r.

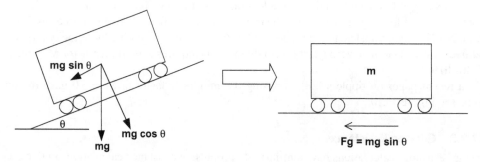

FIGURE 13.11 Modelling gravitational components. (From Cole, C., Longitudinal train dynamics, Chapter 9 in *Handbook of Railway Vehicle Dynamics*, S. Iwnicki (Ed.), CRC Press, Boca Raton, FL, 239–278, 2006. With permission.)

Modelling of the brake pipe and triple valve systems is a subject in itself and will therefore not be treated in this chapter beyond characterising the forces that can be expected and the effect of these forces on train dynamics.

The majority of freight rolling stock still utilise pneumatic control of the brake system. The North American system differs in design from the British/Australian system, but both apply brakes sequentially starting from the point at which the brake pipe is exhausted. Both systems depend on the fail-safe feature, whereby the opening of the brake valve in the locomotive or the fracture of the brake pipe allows loss of brake pipe pressure and results in application of brakes along the train. The particular valves used on each vehicle to apply the brakes work on the same principle but will vary slightly in function and capabilities. The British/Australian system tends to name these valves 'triple valves', while they are known as 'AB valves' in North America and 'distributor valves' in Europe [1].

Irrespective of the particular version of pneumatically controlled brakes, the key issue is that the pneumatic control adjustments made to the brakes via the brake pipe take time to propagate along the train. Since the first triple-valve systems were introduced in the late 1800s, many refinements have been progressively added to ensure or improve brake control propagation. As the control is via a pressure wave, the system is limited to sonic speed, which is 350 m/s for sound in air (noting 318 m/s at –20°C and 349 m/s at 30°C). Allowing for losses in brake equipment, a well-designed system can achieve signal propagation at speeds typically in the range of 250–300 m/s. For short trains of 20 wagons (each 15-m long, ~300-m long train), this gives quite reasonable performance. As trains have increased in length, in particular for heavy-haul applications (lengths of 1.6 to 4.0 km), brake control signal propagation can take several seconds. Some simulated data examples of a brake system emergency application in a long train are given in Figure 13.12 [1].

It is the demand for better braking in these longer trains that is the primary driver for recent adoption of electronically controlled pneumatic (ECP) brakes, which can apply all train brakes

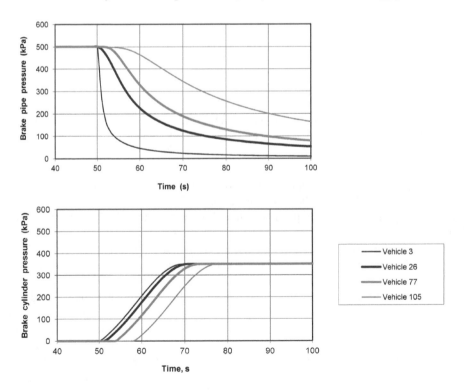

FIGURE 13.12 Simulation results of brake pipe and cylinder responses – emergency application. (From C. Cole, Longitudinal train dynamics, Chapter 9 in *Handbook of Railway Vehicle Dynamics*, S. Iwnicki (Ed.), CRC Press, Boca Raton, FL, 239–278, 2006. With permission.)

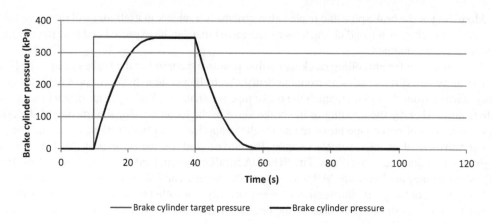

FIGURE 13.13 Simulation results – simplified triple-valve cylinder model. (From M. Spiryagin, M. et al., *Design and Simulation of Rail Vehicles, Ground Vehicle Engineering Series*, CRC Press, Boca Raton, FL, 2014. With permission.)

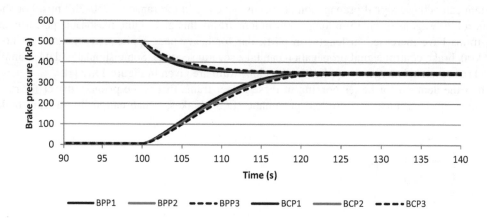

FIGURE 13.14 Simulation results – simplified three-vehicle train air brake model (BPP = brake pipe pressure, and BCP = brake cylinder pressure). (From Spiryagin, M. et al., *Design and Simulation of Rail Vehicles, Ground Vehicle Engineering Series*, CRC Press, Boca Raton, FL, 2014. With permission.)

almost simultaneously. Using a simplified triple-valve model for illustrative purposes, a characteristic curve for a brake cylinder is shown in Figure 13.13 [2].

This simplified model is then implemented in a three-vehicle train brake pipe model, as shown in Figure 13.14.

Note that, for a three-vehicle model, delays are minimal, and the delays shown in Figure 13.14 are exaggerated. To better illustrate the delay issue, the model is simulated again, assuming that the third vehicle is 750 m away; results are shown in Figure 13.15 [2].

As shown in Figure 13.15, the cylinder fill rates for the brake cylinder at the tail of the train are now limited by the control target provided by the brake pipe rather than the fill rates allowed by the chokes in the triple-valve systems. This problem tends to limit the maximum length of brake pipe systems. Much effort has been directed to tuning and improving brake pipe systems over many years to achieve responses like those presented in Figure 13.12 rather than in Figure 13.15 [2].

As train lengths have continued to increase, this issue has been a significant reason for the interest in ECP braking for long heavy-haul trains. An ECP brake system with almost no control delay will give results similar to those in Figure 13.16; control delays are removed, but note that brake cylinders still have a fill time [2].

Longitudinal Train Dynamics and Vehicle Stability in Train Operations

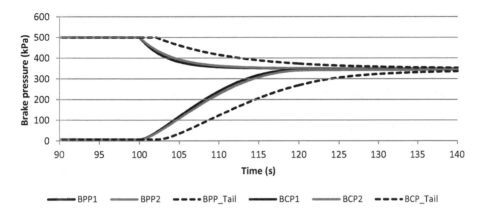

FIGURE 13.15 Simulation results – simplified train air brake model for 750-m brake pipe (BPP = brake pipe pressure, and BCP = brake cylinder pressure). (From Spiryagin, M. et al., *Design and Simulation of Rail Vehicles, Ground Vehicle Engineering Series*, CRC Press, Boca Raton, FL, 2014. With permission.)

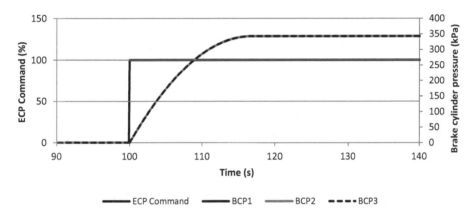

FIGURE 13.16 Simulation results – simplified train ECP brake model for 750-m brake pipe. (From Spiryagin, M. et al., *Design and Simulation of Rail Vehicles, Ground Vehicle Engineering Series*, CRC Press, Boca Raton, FL, 2014. With permission.)

13.2.3 Modelling Vehicle Connections

Perhaps the most important component in any longitudinal train simulation is the connection element between vehicles. While autocouplers are dominating many freight systems, the use of draw hooks and buffers still exists in some systems. Draw hooks and buffers are being phased out wherever possible to improve passenger safety, particularly in older passenger rolling stock. Important differences exist between draw hook and buffer connections and autocouplers, although principles are similar. For draw hook and buffer systems:

- There may or may not be a cushioning device in the draw hooks; in cases where there are only cushioning devices in the buffers, they operate only in compression (buff).
- Free slack in a buffer and draw hook connection are variable; the draw hook link can be tightened or left slack, having different effects on train handling and passenger comfort.

- There is a high degree of asymmetry in the tensile (draft) and compressive (buff) characteristic and structural components.

Conversely, for autocouplers:

- Equal cushioning capability is available in both tensile (draft) and compressive (buff) directions.
- Free slack is determined by design and wear; couplers and knuckles are manufactured with a certain clearance, usually 10–20 mm, with tighter versions available for passenger cars (tightlock types).
- There is a lower degree of asymmetry in the tensile (draft) and compressive (buff) characteristic and structural components.

While draw hooks and buffers will not be dealt with in this chapter, the principles of draft gear operation and modelling are directly applicable to the cushioning devices used. It should also be noted that the legacy of the different features of draw hooks and buffers results in a range of different draft gear and autocoupler designs for freight and passenger applications

13.2.3.1 Equipment Overview

The most common wagon connection arrangements are illustrated in the schematics in Figure 13.17.

As shown in Figure 13.17a, buffers and draw hooks utilise spring-damper units (designated 'DG' in the figure) in the buffers and sometimes in the draw hook. An interesting aspect of the design is that the link can be tightened to give a preloaded slackless connection. Conversely, it can also be left loose, and this allows very large slack values. Autocouplers, as shown in Figure 13.17b, are used extensively in heavy-haul trains and freight trains in the Americas, Australia, China and Russia. Autocouplers can be designed with or without interlocking lugs, as shown in Figure 13.17c. The presence of the locking lugs increases lateral stability and prevents both lateral buckling and relative vertical movement. As heavy-haul trains have become progressively longer, methods of controlling slack have become a priority. Drawbar systems, usually retaining draft gears, as shown in Figure 13.17d, are in common use in heavy-haul trains. The most common configuration is the permanently coupler wagon pair (tandem wagon). There is also some use of 4-wagon sets and occasional examples of 8- and 10-wagon sets in use in Australia and North America. Drawbars can be used with either slackless or energy-absorbing draft gear packages. Early practice seems to favour retaining full capacity dry friction type draft gear packages at the drawbar connections. This is still the practice in the very heavy 160-tonne iron-ore wagons in Australia. Some more recent lighter wagons in coal service utilise small short-pack draft gear units at the drawbar connections. These short packs are quite stiff and provide only short compression displacements; they utilise only polymer or elastomer elements (no friction damping).

Slackless draft gear packages, as in Figure 13.17e, are sometimes used in drawbar coupled wagons or integrated into shared bogie designs. The design of slackless packages is that the components are arranged to continually compensate for wear to ensure that small connection clearances do not get larger as the draft gear components wear. A simple falling wedge is the usual design principle. Slackless packages have been deployed in North American train configurations such as the trough train [29] and bulk product unit trains [30] and are also found in intermodal container carrying fleets. The advantage of slackless systems is found in reductions in longitudinal accelerations and impact forces of up to 96% and 86%, respectively, as reported in [26]. Disadvantages lie in the inflexibility of operating permanently coupled wagons and the reduced numbers of energy absorbing draft gear units in the train. It is usual when using slackless coupled wagon sets that the autocouplers at each end are equipped with heavier-duty energy-absorbing draft gear units. The reduced capacity of these train configurations to absorb impacts can result in accelerated wagon body fatigue or even impact related failures during shunting impacts.

Longitudinal Train Dynamics and Vehicle Stability in Train Operations

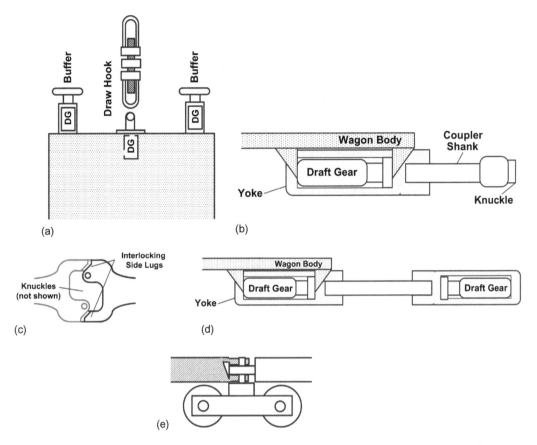

FIGURE 13.17 The most common types of inter-wagon connections: (a) draw hooks and buffers, (b) conventional autocoupler and draft gear, (c) autocoupler with interlocking lugs, (d) drawbars and draft gear and (e) slackless coupling.

13.2.3.2 Draft Gear Overview

Draft gears deploying friction damping, usually used in freight rolling stock, are shown in Figures 13.18 and 13.19.

Full-length draft gears deploying just elastomer damping, usually used in passenger rolling stock, are shown in Figures 13.20 [31]. Note the asymmetry provided in Figure 13.20b and c.

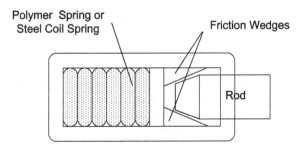

FIGURE 13.18 Friction-type draft gear unit. (From C. Cole, Longitudinal train dynamics, Chapter 9 in *Handbook of Railway Vehicle Dynamics*, S. Iwnicki (Ed.), CRC Press, Boca Raton, FL, 239–278, 2006. With permission.)

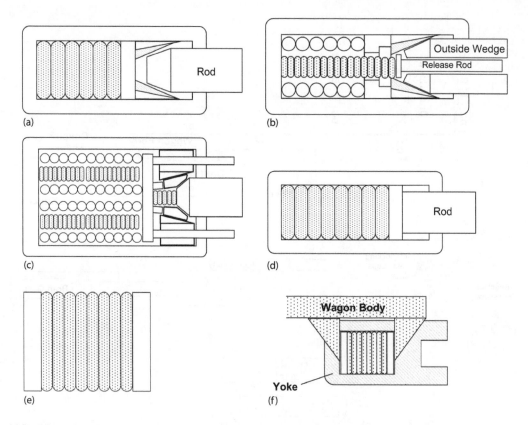

FIGURE 13.19 Various draft gear units (freight rolling stock applications): (a) angled surfaces for increased wedge force, (b) release spring type, (c) friction wedge and plate type, (d) elastomer type, (e) short pack and (f) short pack arrangement. (FIGURE 13.19(c)&(d) From Cole, C. et al., *Vehicle Syst. Dyn.*, 55(10), 1498–1571, 2017. With permission.)

13.2.3.3 Modelling Fundamentals

Before considering draft gear modelling in detail, it is useful to consider the wagon connection as a whole unit, as all aspects must be considered to get an accurate model. A significant issue is that the wagon body structure, draw gear assembly, coupler bar and knuckles or draw hooks might not be significantly stiffer than the draft gear units themselves. A freight-type draft gear might have a mean overall stiffness (F/x) of 25 MN/m at full deflection, while the locked or limiting stiffness of the system reported in [6] was 80 MN/m. This means that draft gear models must include this stiffness into the model. Furthermore, there will be differences in the local structural stiffness of draw gear components due to the asymmetry of the load paths. In compression, load is transferred through the coupler head and bar and to the wagon body via the end clevis of the yoke and draft gear. Conversely, a less stiff tensile load path occurs through the knuckle, coupler head, coupler bar and the full length of the yoke. A further feature of the draft gear design is that the draft gear unit itself is always loaded in compression; hence, in cases of unloading where damping and spring forces are of opposite signs, the damping force cannot exceed the spring force. This is illustrated in the hypothetical diagrams of a linear spring damper in Figure 13.21 and with the addition of structural stiffness in Figure 13.22. This model is now structurally similar to a wagon connection model, but it only has a hypothetical linear spring damper. Draft gear modelling is developed in the next section.

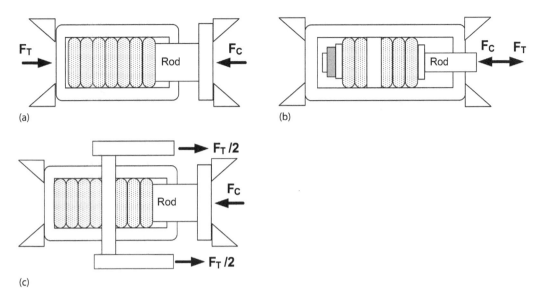

FIGURE 13.20 Various draft gear units (passenger rolling stock applications): (a) single pack, (b) balanced type and (c) floating plate. (From Cole, C. et al., Bosomworth, modelling issues in passenger draft gear connections, In: M. Rosenberger, M. Plöchl, K. Six, J. Edelmann (Eds.), *Proceedings 24th IAVSD Symposium held at Graz*, Austria, 17–21 August 2015, CRC Press/Balkema, Leiden, the Netherlands, 985–993, 2016. With permission.)

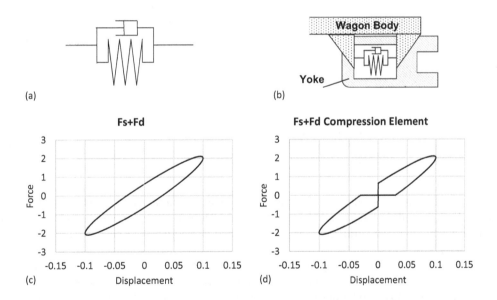

FIGURE 13.21 Implications of the compression only stiffness damper element: (a) spring damper model, (b) spring damper as a compression element, (c) spring damper F-x cross plot and (d) spring damper compression element F-x cross plot.

13.2.3.4 Modelling for Wagon Connections

It was noted from the recent longitudinal train dynamics benchmark study [17] that the most common approach to draft gear modelling was still the use of lookup tables for both upper and lower curves, noting that it is essential to also handle the non-smooth problem of transitioning between upper and lower curves. This can be achieved with some care and be representative of the actual

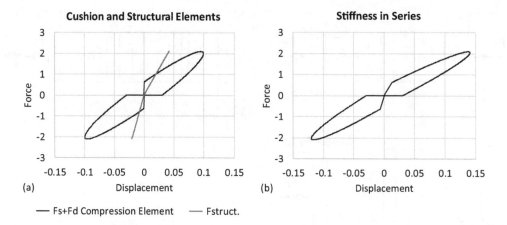

FIGURE 13.22 Asymmetric structural stiffness added to a compression-only stiffness damper element: (a) stiffness components and (b) combined series model.

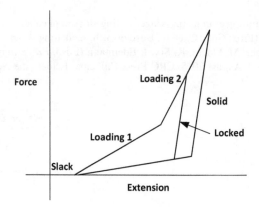

FIGURE 13.23 Piecewise linear vehicle connection model, as proposed in (6). (From C. Cole, Longitudinal train dynamics, Chapter 9 in *Handbook of Railway Vehicle Dynamics*, S. Iwnicki (Ed.), CRC Press, Boca Raton, FL, 239–278, 2006. With permission.)

system, as the transition stiffness can be made equal to the structural stiffness. This approach dates back to Duncan and Webb [6], as shown in Figure 13.23, and solves the problem of near-infinite stiffness evident in typical draft gear drop hammer data in the transition between upper and lower curves; see Figure 13.24.

In the case of friction modelling, the transition can be confidently assumed to be instantaneous as the friction force simply changes direction; however, transition states for polymer draft gears are not so certain. Previous modelling from Cole et al. [1–4] presented an approach to friction draft gear in the form of:

$$F_c = F_s(x) * f(v,x) \qquad (13.11)$$

where $F_s(x)$ is a lookup table or continuous function describing the friction, and $f(v, x)$ is the friction-damping component that is dependent on force, loading rate and friction. This modelling structure is applied to draft gears of the design shown in Figures 13.17 and 13.18a and b. A similar approach was developed for polymer draft gears [5] (such as shown in Figure 13.19d–f):

$$F_c = F_s(x) \pm F_d(v,x) \qquad (13.12)$$

Longitudinal Train Dynamics and Vehicle Stability in Train Operations

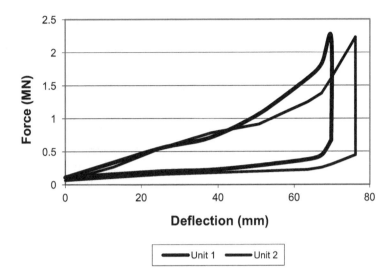

FIGURE 13.24 Typical published draft gear response data – drop-hammer tests. (From Cole, C., Longitudinal train dynamics, Chapter 9 in *Handbook of Railway Vehicle Dynamics*, S. Iwnicki (Ed.), CRC Press, Boca Raton, FL, 239–278, 2006. With permission.)

where $F_s(x)$ is a lookup table or continuous function describing the friction and $F_d(v,x)$ is the damping hysteresis capturing viscous and/or internal friction components. More complicated friction draft gear designs, as shown in Figure 13.19c, can also be modelled in the same way as developed by Qing et al. [32].

Another aspect of the different designs is the treatment of preload. Using the approach proposed for all characteristics being represented in either the form $F_c = F_s(x) * f(v,x)$ or $F_c = F_s(x) \pm F_d(v,x)$, preload can be managed by applying a preload displacement to the spring function $F_s(x)$. (i.e., x can be defined as $x = x_{dg} + x_p$, where x is the displacement of the draft gear from its undeflected position, x_{dg} is the deflection visible in the draw gear pocket and x_p is the deflection due to the preload imposed by the pocket dimensions). Preload will vary depending on pocket dimensions. It will be noted that the friction draft gear in Figure 13.20a and b is shown without preload, whilst a preload is shown in the short pack in Figure 13.20c and d. The subject of draft gear preload has so far been given little attention in literature and, indeed, in this chapter. An interesting study on the effects of preload is given in [33].

An overall schematic of the wagon connection model made up of two sets of components from Figure 13.17b is shown in Figure 13.26. Modelling the coupler slack is straightforward, a simple dead zone. Modelling of the steel components, including body stiffness, can be provided by linear stiffnesses. The stiffness of the yoke, coupler, etc., can significantly change the connection characteristic, as shown in Figure 13.25b. These stiffnesses should be added in series to the draft gear characteristic. Note that, if the wagon connection characteristic is measured directly using inter-wagon distance, all the draw gear structural stiffnesses are already included.

As found by Duncan and Webb [6], there is also an overall stiffness of the rolling stock that must be considered. Work by Duncan and Webb from test data measured on long unit trains identified cases where the draft gear wedges locked and slow sinusoidal vibration was observed. The behaviour was observed in distributed power trains when the train was in a single stress state. The stiffness corresponding to the fundamental vibration mode observed was defined as the locked stiffness of the vehicle connections. The locked stiffness for the freight trains tested in [6] was nominally a value in the order of 80 MN/m. The term 'locked stiffness' came from the particular observation that, when these vibrations were observed, the friction wedges in the draft gears remained locked. To apply the concept more generally, the stiffness can be thought of as the 'limiting stiffness'.

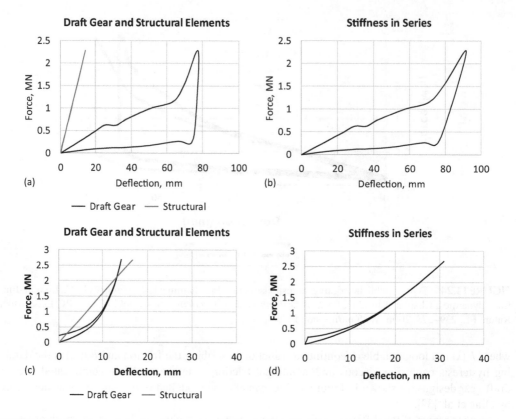

FIGURE 13.25 Typical draft gear characteristics combined in series with draw gear structural stiffness: (a) stiffness components – friction draft gear, (b) friction draft gear – structural stiffness model, (c) stiffness components – short pack polymer draft gear and (d) short pack draft gear – structural stiffness model.

The limiting stiffness is the aggregated effect of all the stiffnesses of the structural components and connections added in series, which include the components such as the coupler shank, knuckle, yoke, draft gear structure and the vehicle body. It may also include any pseudo-linear stiffness due to wagon bulk loading, gravity and bogie steering force components, whereby a longitudinal force is resisted by gravity, as a vehicle is lifted or forced higher on a curve. The limiting stiffness of a long train may therefore vary for different rail vehicle loadings and on-track placement.

The limiting stiffness of the system must be incorporated into the train model in some way. This has been successfully done for many years by applying it as limiting stiffness in the wagon connection model, as done by Duncan and Webb [6] and Cole et al. [1–4]. Note that this is the overall stiffness of the vehicle, so it is not the same stiffness that is used to refine the draw gear models, as shown in Figure 13.25.

Rail vehicle connection modelling can be simplified to a combined draft gear package model equivalent to two draft gear units and including one spring element representing the limiting stiffness; see Figure 13.27. The actual implementation of the model in Figure 13.26 is an 'If–Then' model that works with all stiffnesses arranged in series, while the draft gear has not reached the hard limit. When the model reaches the hard limit, only the limiting stiffness is used. An important issue in longitudinal modelling for heavy-haul trains is that the hard limit must be correctly modelled, as full draft gear deflections are possible in many studies.

The 'Combined Draft Gear Model' in Figure 13.27 can be considered the general model structure for all variations in draft gear type, including friction damped draft gears, polymer draft gears and slackless draft gears. The model is converted to a drawbar model by setting gap element

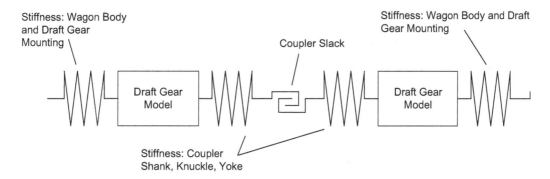

FIGURE 13.26 Rail vehicle connection model components. (From C. Cole, Longitudinal train dynamics, Chapter 9 in *Handbook of Railway Vehicle Dynamics*, S. Iwnicki (Ed.), CRC Press, Boca Raton, FL, 239–278, 2006. With permission.)

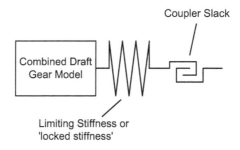

FIGURE 13.27 Rail vehicle connection model. (From C. Cole, Longitudinal train dynamics, Chapter 9 in *Handbook of Railway Vehicle Dynamics*, S. Iwnicki (Ed.), CRC Press, Boca Raton, FL, 239–278, 2006. With permission.)

(coupler slack) to zero and can be adjusted for wear by increasing the gap element. The variations are shown in Table 13.5.

The following models using this approach are presented in the following sub-sections:

- Friction draft gear (wedges)
- Friction draft gear (wedges and plates)
- Polymer (single pack, floating plate and balanced)

13.2.3.4.1 Friction Draft Gear (Wedges)

The lookup table model and locked stiffness wagon connection modelling, as developed by Duncan and Webb [6] (see Figure 13.23), was further improved in [34]. This work referred specifically to friction wedge-type draft gears, as shown in Figure 13.18. The difficulty presented by the work by Duncan and Webb [6] is that draft gear units, and so, the mathematical models used to represent them, differ depending on the regime of train operation expected. Clearly, if extreme impacts were expected in simulation due to shunting or hump yard operations, a draft gear model representing drop hammer test data would be appropriate. Conversely, if normal train operations were expected, a vehicle connection model, as proposed in Figure 13.23, would be appropriate. It was noted in [34] that the stiffness of the draft gear units for small deflections varied by typically five to seven times the stiffness indicated by the drop-hammer test data in Figure 13.24. It is therefore evident that, for mild inter-vehicle dynamics (i.e., gradual

TABLE 13.5
Wagon Connection Model Types

Draft Gear Type

Friction and/or polymer draft gears and autocouplers

Friction and/or polymer draft gears and drawbar

Slackless connection

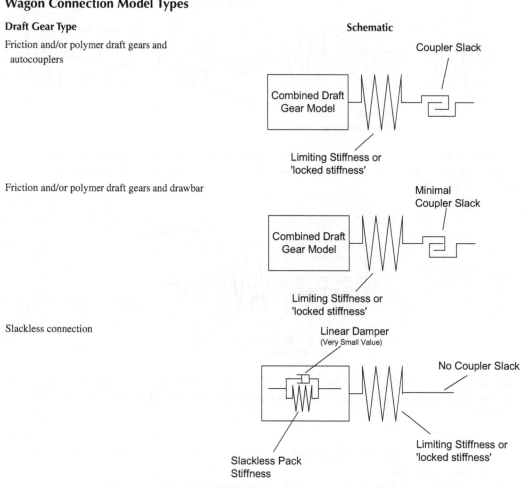

loading of draft gear units), the static friction in the wedge assemblies can sometimes be large enough to keep draft gears locked. A model incorporating the wedge angles and static and dynamic friction was therefore proposed in [34] and published in detail in [1–4]. Assuming symmetry, the draft gear package was considered as a single-wedge spring system, as shown in Figure 13.28. The rollers provided on one side of the compression rod can be justified in that the multiple wedges are arranged symmetrically around the outside of the rod in the actual unit. It will be realised that different equilibrium states are possible, depending the direction of motion, wedge angles and surface conditions. The free-body diagram for increasing load (i.e., compressing) is shown in Figure 13.28. The state of the friction force $\mu_1 N_1$ on the sloping surface can be any value between $\pm\mu_1 N_1$. The fully saturated cases of $\mu_1 N_1$ are drawn on the diagram. If there is sliding action in the direction for compression, then only the Case 1 friction component applies. Case 2 applies if a pre-jammed state exists. In this case, the rod is held in by the jamming action of the wedge. If the equations are examined, it can be seen that, for certain wedge angles and coefficients of friction, wedges are self-locking, and this is sometimes observed in operation.

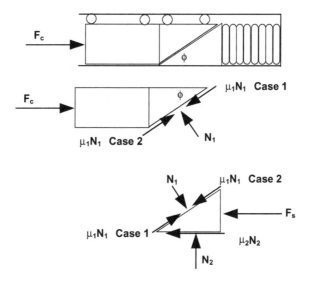

FIGURE 13.28 Free-body diagram of a simplified draft gear rod-wedge-spring system. (From Cole, C., Longitudinal train dynamics, Chapter 9 in *Handbook of Railway Vehicle Dynamics*, S. Iwnicki (Ed.), CRC Press, Boca Raton, FL, 239–278, 2006. With permission.)

Examining the rod:

Case 1: $$F_c = N_1(\sin\phi + \mu_1\cos\phi) \qquad (13.13)$$

Case 2: $$F_c = N_1(\sin\phi - \mu_1\cos\phi) \qquad (13.14)$$

For self-locking, N_1 remains non-zero when F_c is removed, therefore:

$$\sin\phi = \mu_1\cos\phi$$

that is, if $\sin\phi < \mu_1\cos\phi$, then a negative force F_c is required to extend the rod.

From this inequality, it can be seen that, for self-locking:

$$\tan\phi < \mu_1 \qquad (13.15)$$

The self-locking relationship between wedge angle and friction coefficient can therefore be plotted as shown in Figure 13.29.

Further insight can be gained if the equations relating the wedge forces to the coupler force, and draft gear spring force are developed, again assuming saturated friction states and the direction shown in Case 1 for $\mu_1 N_1$, giving:

$$F_c = F_s(\mu_1\cos\phi + \sin\phi) / [(\mu_1 - \mu_2)\cos\phi + (1 + \mu_1\mu_2)\sin\phi] \qquad (13.16)$$

If it is assumed that $\mu_1 = \mu_2$ and that both surfaces are saturated, then the equation reduces to:

$$F_c = F_s\,(\mu\cot\phi + 1) / (1 + \mu^2) \qquad (13.17)$$

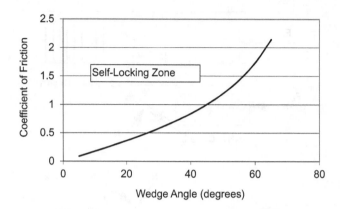

FIGURE 13.29 Friction wedge self-locking zone. (From Cole, C., Longitudinal train dynamics, Chapter 9 in *Handbook of Railway Vehicle Dynamics*, S. Iwnicki (Ed.), CRC Press, Boca Raton, FL, 239–278, 2006. With permission.)

The other extreme of possibility is when there is no impending motion on the sloping surface due to the seating of the rod and wedge, and then, the value assumed for μ_1 is zero, and Equation 13.17 reduces to:

$$F_c = F_s \tan\phi / [\tan\phi - \mu_2] \qquad (13.18)$$

If the same analysis is repeated for the unloading case, a similar equation results:

$$F_c = F_s \tan\phi / [\tan\phi + \mu_2] \qquad (13.19)$$

At this point, it is convenient to define a new parameter, namely friction wedge factor, as follows:

$$Q = F_c / F_s \qquad (13.20)$$

Using the new parameter, the two relationships given by Equations 13.17 and 13.18 are plotted for various values of ϕ in Figure 13.30. There is always difficulty in the estimation of the friction coefficient(s) due to the variable nature of the surfaces.

Surface roughness and wear ensure that the actual coefficients of friction can vary, even on the same draft gear unit, resulting in different responses to drop-hammer tests. It is also difficult to estimate the function that describes the transition zone between static and minimum kinetic friction conditions and the velocity at which minimum kinetic friction occurs. For simplicity and a first approximation, a piecewise linear function can be used, as shown in Figure 13.31a. This approach to modelling has been used quite successfully since [34].

The friction coefficient μ in [34] is given by:

$$\begin{aligned} \mu &= \mu_s & \text{for } v = 0 \\ \mu &= \mu_k & \text{for } v \geq V_f \end{aligned} \qquad (13.21)$$

where $\mu(v)$ can be any continuous function linking μ_s and μ_k. Figure 13.31a shows a piecewise linear approximation, and Figure 13.31(b) shows an exponential function as per [32], giving:

$$\mu = h_2 + h_3 \exp[h_4\, abs(v)] \qquad (13.22)$$

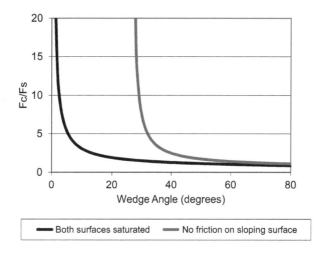

FIGURE 13.30 Friction wedge factor for $\mu = 0.5$. (From Cole, C., Longitudinal train dynamics, Chapter 9 in *Handbook of Railway Vehicle Dynamics*, S. Iwnicki (Ed.), CRC Press, Boca Raton, FL, 239–278, 2006. With permission.)

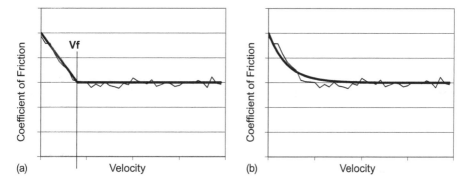

FIGURE 13.31 Approximations of wedge friction coefficient: (a) piecewise linear and (b) exponential. (From Cole, C., Longitudinal train dynamics, Chapter 9 in *Handbook of Railway Vehicle Dynamics*, S. Iwnicki (Ed.), CRC Press, Boca Raton, FL, 239–278, 2006. With permission.)

where μ is the coefficient of friction, v is the relative velocity of two bodies, h_2 represents the kinetic friction, $h_2 + h_3$ represents the static friction and h_4 controls the transition between the static and kinetic friction.

Key data for the model therefore become the wedge angle ϕ, kinetic friction velocity V_f, static friction coefficient μ_s, kinetic coefficient of friction μ_k and the spring force F_s. If the assumption is taken that there is no impending motion on the sloping wedge surface and that the $\mu_1 N_1$ term is small, Equations 13.18 and 13.19 can be used as a starting point for a draft gear model. Alternatively, the more complex Equations 13.16 and/or 13.17 could be used.

Results of this modelling approach are shown in Figure 13.32.

It will be noted that the effect of polymer springs has not been discussed in any detail for the friction type draft gears, although the stiffness characteristic obviously reflects that of an elastomer spring for some models. The reason for this omission is that the effect of the friction wedges on the hysteresis is much larger than the elastomer hysteresis, and so, the damping from the elastomer is just lumped in with the friction. This assumption has proved satisfactory when comparing simulation responses with experimental data. It should be noted that there is no reason that a polymer

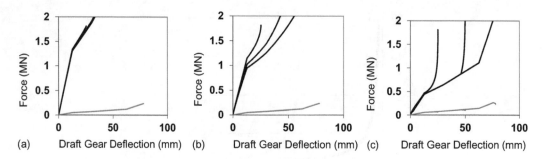

FIGURE 13.32 Vehicle connection model response: (a) slow loading at 0.1 Hz, (b) mild impact loading at 1 Hz and (c) shunt impact at 10 Hz.

modelling of the spring could not be added into the spring function $F_s(x)$, using the polymer modelling introduced later in the chapter.

That is

$$F_{cp} = F_s(x) \pm F_d(v,x) \qquad (13.23)$$

and then

$$F_c = F_{cp} * f(v,x) \qquad (13.24)$$

Despite the possibility, there is no known publication of this more complex approach, perhaps validating the adequacy of existing modelling and/or highlighting the difficulty of measurements for validation.

13.2.3.4.2 Friction Draft Gear (Wedges and Plates)

A more complex friction-type draft gear is one that utilises both wedges and plates to achieve two different stages in damping, as shown in Figure 13.33.

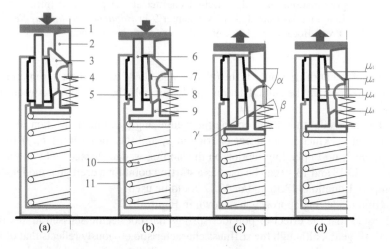

FIGURE 13.33 Double-stage friction draft gear (32): (a) loading stage 1, (b) loading stage 2, (c) unloading stage 1 and (d) unloading stage 2. (1) Follower; (2) central wedge; (3) wedge shoe; (4) release spring; (5) outer stationary plate; (6) movable plate; (7) lubricating metal; (8) inner stationary plate; (9) spring seat; (10) main springs; (11) housing. (From Wu, Q. et al., *Vehicle Syst. Dyn.*, 53(4), 475–492, 2015. With permission.)

This draft gear consists of three sub-systems: friction clutch, springs and housing. The friction clutch is further divided into two groups, namely the wedge group and the plate group. The wedge group includes the central wedge, wedge shoes, inner stationary plates and the spring seat. The plate group includes inner and outer stationary plates and movable plates. The spring sub-system includes the release spring and main springs; the main springs further consist of an outer spring, an inner spring and four corner springs. All springs in this type of draft gear are coil steel springs.

Conventionally, the working processes of a draft gear are generally described as a loading process (draft gear being compressed) and an unloading process (draft gear being released). This description applies fully for single-stage draft gears but is not accurate for double-stage draft gears, as the double-stage design provides a certain amount of clearance between the follower and the tops of movable plates when the draft gear is fully released (see Figure 13.33a). There is also clearance between the spring seat and the bottoms of movable plates when the draft gear is fully compressed, as shown in Figure 13.32b. Given the existence of these clearances, there are two stages in both the loading and unloading processes, and different stages give different force versus displacement characteristics [32].

$$F_{dg,i} = \Psi_i F_{sm} - (\Psi_i - 1) F_{sr}, \quad (i = 1,4) \tag{13.25}$$

$$\Psi_1 = \frac{1 + \tan(\beta + \arctan(\mu_3))\tan(\gamma + \arctan(\mu_1))}{1 - \tan(\alpha + \arctan(\mu_2))\tan(\gamma + \arctan(\mu_1))} \tag{13.26}$$

$$\Psi_2 = \Psi_1 + \frac{2(1 - \mu_1 \tan(\gamma))\mu_4 (\Psi_1 - 1)}{\mu_1 + \tan(\gamma)} \tag{13.27}$$

$$\Psi_3 = \frac{1 + \tan(\beta - \arctan(\mu_3))\tan(\gamma - \arctan(\mu_1))}{1 - \tan(\alpha - \arctan(\mu_2))\tan(\gamma - \arctan(\mu_1))} \tag{13.28}$$

$$\Psi_4 = \frac{(\tan(\gamma) - \mu_1)\Psi_3}{\tan(\gamma)(1 - 2\mu_1\mu_4 + 2\mu_1\mu_4\Psi_3) + 2\mu_4\Psi_3 - 2\mu_4 - \mu_1} \tag{13.29}$$

where $i = 1,4$ corresponds to the L1 stage, L2 stage, U1 stage and U2 stage, respectively; $F_{dg,i}$ is the draft gear force; Ψ_i is the corresponding force coefficient; and F_{sm} and F_{sr} are the main spring force and the release spring force, respectively. The friction model can be expressed as:

$$\mu = h_1 + h_2 \exp(-h_3 v_r) \tag{13.30}$$

where μ is the coefficient of friction (COF); v_r is the relative velocity of the two adjacent objects; and h_1, h_2 and h_3 are parameters that need to be determined experimentally or empirically. These parameters play specific roles in this model: h_1 represents the kinetic friction; the sum of h_1 and h_2 represents static friction; and h_3 is used to adjust the transition from static friction to kinetic friction.

The locked stiffness [6] can be regarded as the system stiffness when the draft gear is locked, that is, when the follower velocity is zero or nearly zero, and is implemented as a limit of draft gear stiffness:

$$\left| \frac{F_{dg}(t) - F_{dg}(t - \Delta t)}{x_t - x_{t-\Delta t}} \right| \leq k \tag{13.31}$$

where t and Δt indicate time and step size, respectively; $x_t, x_{t-\Delta t}$ are draft gear deflections of the current and previous time step, respectively; and k is the locked stiffness. The locked stiffness applies for all draft gear behaviours, which means that it also governs the discontinuities of preload and stage changes. In [6], the locked stiffness was set as 80 kN/mm.

Five sets of experimental data and simulations are shown in Figure 13.34. The experimental data were measured from wagon impact tests, as reported in a different format in [32]. The measured cases were that of one loaded wagon weighing 100 tonnes, impacting another identical

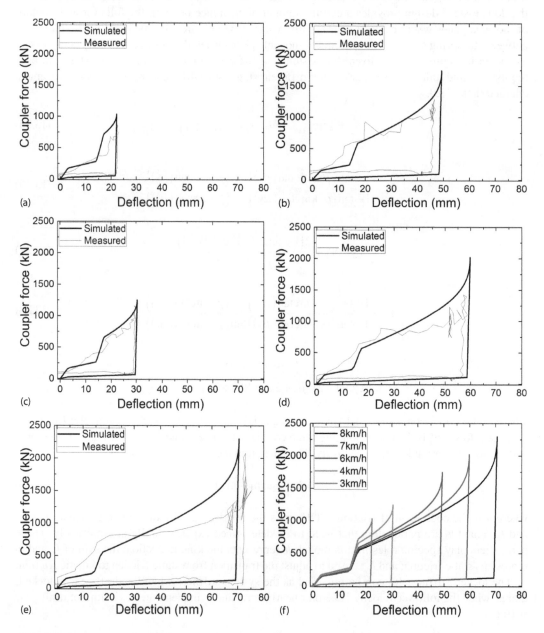

FIGURE 13.34 Characteristics of a double-stage draft gear for (a) impact speed 3 km/h, (b) impact speed 4 km/h, (c) impact speed 6 km/h, (d) impact speed 7 km/h, (e) impact speed 8 km/h and (f) simulation overview. (From Cole, C. et al., *Vehicle Syst. Dyn.*, 55(10), 1498–1571, 2017. With permission.)

stationary loaded wagon at various speeds on a section of tangent flat track. Simulations have achieved general agreement with the tests in terms of maximum forces, maximum deflections and envelope patterns.

Illustrating the difficulty in modelling friction, it is still noted that the modelled forces do not show the softening effect that is evident in the data in the examples in Figure 13.34d and e. A further modification was proposed in [32] to adjust μ_1 to account for the effect of different bushing materials μ_1:

$$\mu_1 = \left[h_1 + h_2 exp(-h_3 v_r) \right] * M_u(v_{f0}, x_{fd}) \tag{13.32}$$

Where v_{f0} is the initial impact velocity of two adjacent vehicles, M_u is the modification factor expressed by a lookup table that has a pattern, as shown in Figure 13.35. The friction μ_1 was adjusted to reflect the characteristic of steel sliding on copper.

Using the model provided in [32], the softening behaviour was simulated, as shown in Figure 13.36.

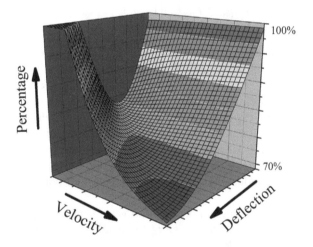

FIGURE 13.35 Friction modification factor. (From Wu, Q. et al., *Vehicle Syst. Dyn.*, 53(4), 475–492, 2015. With permission.)

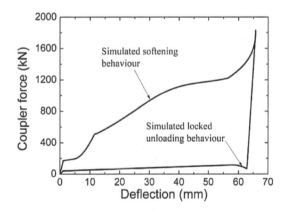

FIGURE 13.36 Simulated softening behaviour. (From Cole, C. et al., *Vehicle Syst. Dyn.*, 55(10), 1498–1571, 2017. With permission.)

13.2.3.4.3 Polymer Draft Gears

For the development of a model for a draft gear with elastomer elements, it is necessary to introduce strongly non-linear characteristics. This step is complicated, as the description of the behaviour is based only on experimental research, and data is always limited. Various approaches have been published [35–37], but difficulties usually arise in obtaining sufficient data to fully describe or tune such models. The method proposed by [35] assumes firstly a higher-order polynomial based on the static load characteristic, as given in Equation 13.33.

$$F_{stat}(x) = \sum_{i=1}^{6} a_{i-1} \cdot x^i \qquad (13.33)$$

The variable x is the draft gear deflection, and v is the associated velocity. The 'a' coefficients in Equation 13.33 for the approximated static load characteristic are determined from experimental results.

$$F(x) = q_l F_{preload} + F_{stat}(x)\left(1 + b_l \left(x - x_p\right)^{c_l(|v|/v_{max})}\right)^L \qquad (13.34)$$

The static characteristic is then scaled for loading and unloading and different loading rates (velocities) by Equation 13.34, where $F_{preload}$ is the preload force, x_p is the initial (preload) displacement in the model, x_{max} is maximum possible displacement, v_{max} is maximum expected velocity, q_l is the preload factor and b_l and c_l are model variables to introduce dynamic components that are obtained from experimental data analysis. L is the selection variable for loading and unloading. Four different curves can be generated for draft and buff and load and unload states. The approach from [35–37] allows development of models when a characteristic curve is known from experimental data. It does not give insight into how the damping processes work, and it cannot be developed unless a comprehensive database of both deflections and velocities is available.

A new approach to the modelling of polymer draft gears is given in [5] and is reproduced here. Sixth-order polynomials can be fitted to each of the loading curves and unloading curves [35]. This gives a set of equation in the form of:

$$F_n(x) = \sum_{i=1}^{6} a_{i-1} \cdot x^i \qquad (13.35)$$

where $n = 1,2,3,4$; 1 = draft loading; 2 = draft unloading; 3 = buff loading; 4 = buff unloading.

Tension and compression were treated separately so as to allow asymmetric designs such as balance-type and floating-plate-type draft gears to be included. It can be reasoned that, in the absence of movement, the force supported by the polymer springs must approximate to a mean or median value between loading and unloading curves. The mean curve or weighted mean curve can be obtained as follows:

$$F_m = f_1 \sum_{i=1}^{6} a_{i-1} \cdot x^i + f_2 \sum_{i=1}^{6} b_{i-1} \cdot x^i \quad \text{where} \quad f_1 + f_2 = 1.0 \qquad (13.36)$$

where F_m is the mean or weighted mean of the sixth-order polynomials that approximate the experimental data, and coefficients a and b are for the sixth-order polynomial that approximate loading and unloading of the experimental data. Factors f_1 and f_2 are determined to suit the polymer characteristic.

The damping component in a polymer draft gear cannot be considered as purely viscous. As noted in [38], the dynamic component often needs to be modelled by an exponential relationship. In [38] the relationship is proposed as:

$$F_d(x,v) = C_1 v [1-\exp(-h_3 x^2)] + C_2 v \qquad (13.37)$$

where F_d is the damping force component, C_1 and C_2 are two viscous damping coefficients and h_3 is a tuning parameter for the x position. Note that the model is both velocity and displacement dependent [32–34].

A more complicated and more generalised model is developed in [5] by Cole et al.

$$F_{dg}(x_a,v) = F_m(x_a) \pm \left\{ C_1 + C_2 \left|\frac{x_{dg}}{x_{dg(max)}}\right|^{h_1} \left|\frac{v}{v_{(max)}}\right|^{h_2} \right\} \left\{ 1 - \exp\left[-\lambda \left|\frac{v}{v_{(max)}}\right|^n \right] \right\} \qquad (13.38)$$

where F_{dg} is the draft gear force, x_{dg} is the measured draft gear deflection, x_a is the draft gear polymer spring deflection from its free length, $x_{dg(max)}$ is the measured hard limit, v is the velocity (or loading rate), v_{max} is an estimated maximum velocity of loading rate, C_1 and C_2 are two viscous damping coefficients, h_1 is a tuning parameter for the x position, h_2 is a tuning parameter for the velocity (or loading rate) and λ and n are tuning parameters for the non-linear damping. The effects of the various parameters are presented in [5].

The modelling response results in Figure 13.37 show the importance of the modelling parameters. The actual differences in the polymer model responses are quite significant. After tuning, the model was used to give a number of simulation results to examine sensitivities to parameters in a train-operating context. Different draft gear rubber types and their responses are illustrated in Figure 13.38.

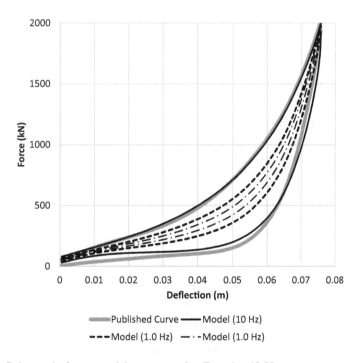

FIGURE 13.37 Polymer draft gear model response using Equation 13.38.

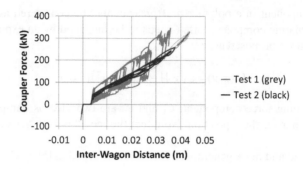

FIGURE 13.38 Model responses with Test 1 (grey) showing non-linear internal friction behaviour, and Test 2 (black) showing mainly viscous behaviour.

Note that the approach of applying preload to the mean force curve differs slightly from the approach in Equation 13.32. The approach of applying preload as an initial displacement, however, has other advantages and can be used with utility when considering the several variants of polymer draft gears, namely single pack, balanced and floating plate units, as per Figure 13.20.

Single pack – pre-load case:

$$\text{Let } x_a = x_{dg} + x_p \qquad (13.39)$$

where x_a is the deflection of the spring from its free length, x_d is the measured deflection of the draft gear and x_p is the value that gives $F_m(x_p) = F_{preload}$.
Both x_a and x_{dg} are used in Equation 13.38.
Balanced – pre-compression case:
In the pre-compression case, the zones of influence of the parallel stiffnesses must first be defined. This means that the pre-compression in buff and draft must be determined by considering the deflections of the springs on the buff and draft sides individually, namely $x_{pre-compression(buff)}$ and $x_{pre-compression(draft)}$. These deflections must be determined, such that:

$$F_{dg}(buff)(x_{pre-compression(buff)}) = F_{dg}(draft)(x_{pre-compression(draft)}) = F_{pre-compression}. \qquad (13.40)$$

There are then three cases of evaluation depending on the displacement range in which the draft gear is working. Using Equation (13.39) and substituting into Equation (13.38):

$$x_a \geq |x_{pre-compression(buff)}| \quad \text{then } F_{dg}(draft)(x_{dg} + x_{pre-compression(draft)})$$

$$x_a \leq -|x_{pre-compression(draft)}| \quad \text{then } F_{dg}(buff)(x_{dg} - |x_{pre-compression(buff)}|)$$

$$x_a > -|x_{pre-compression(draft)}| \text{ and } x_a < |x_{pre-compression(buff)}|$$

$$\text{then } F_{combined} = F_{dg}(draft)(x_{dg} + x_{pre-compression(draft)}) + F_{dg}(buff)(x_{dg} - |x_{pre-compression(buff)}|) \qquad (13.41)$$

where F_{dg} is evaluated as per Equation 13.38 and $F_{combined}$ is the special case where the buff and draft springs act as parallel stiffnesses.
Floating plate – mixed pre-load and pre-compression case:
The floating plate case is handled by combining the above approaches, utilising the preload equations for buff or compression forces and pre-compression equations for the draft or tensile forces.

13.2.3.5 Methods of Measuring Draft Gear Characteristics

Several methods are used to determine draft gear characteristics. An important distinction should be made between tests that are designed to qualify the draft gear product for service versus measurements to understand responses to train dynamics in service. Tests for product qualification, by definition, must be focussed on the most extreme case of train dynamics expected and meet crashworthiness requirements. Examples of these types of tests are the well-known drop hammer tests and wagon impact tests [39]. Not surprisingly, authors of [4,6] and [28] have noted that actual draft gear behaviour in train service can be quite different. Such is the case as there is some dependence in most draft gear designs to the loading rate, and the inter-wagon relative velocities in normal operation are much smaller than those used in the product qualification tests. Common methods of determining in-service draft gear characteristics involve instrumentation of connected wagons or laboratory tests. The use of in-service draft gear at multiple in-train locations during normal running was the method used by both Duncan and Webb [6] and later Cole et al. [28] to achieve measurements that were used to train dynamics modelling. More recently, a high-speed servo controlled hydraulic laboratory rig has been developed at the Centre for Railway Engineering to apply a range of loading rates that allow characterisation of draft gear characteristics, as shown in Figures 13.39 and 13.40. The first stage of the equipment allows displacement feed rates up to 1.0 m/s and forces up to 3.5 MN. The machine tests a single draft gear. Noting that drop-hammer tests require up to 3.3 m/s (single draft gear) and wagon impacts are up to 9.0 m/s (for the daft gear pair wagon connection), the machine is not designed for product qualification, but it is an excellent tool for studying in-train service dynamics.

FIGURE 13.39 Draft gear characterisation test rig.

FIGURE 13.40 Stub sill, drawbar and draft gear – closeup view.

The different testing methods lead to an important discussion about draw gear structural stiffness. The data from draft gear drop hammer tests will include almost no deflections in surrounding frames, etc., and so can be relied upon to give an accurate representation of the response of just the draft gear itself. It therefore follows that structural deflections should always be added as a series stiffness to drop-hammer data, as shown in Figure 13.22, to get an indication of the performance in the inter-wagon connection in service. Conversely, wagon impact test data will include these deflections, as will the data from programs of measuring inter-wagon displacements and forces of trains in service. For the laboratory tests from the rig in Figure 13.39, the deflections of all wagon and test frame components can be measured and calculated. The contribution of the structural stiffness of the draw gear and wagon stub sill assembly can therefore be separated out. Direct measurements of the draft gear force and deflection using an instrumented drawbar are also taken to give an accurate representation of the draft gear; see Figure 13.40.

13.3 WAGON DYNAMIC RESPONSES TO LONGITUDINAL TRAIN DYNAMICS

13.3.1 Overview

The long tradition of analysing train dynamics and vehicle dynamics separately is strongly entrenched in both software and standards. Train dynamics tends to be concerned only with longitudinal dynamics, while vehicle dynamics tends to focus on just one vehicle (or a small number of vehicles) and on vertical and lateral dynamics. The assumption that coupler angles are so small that the consequential vertical and lateral force components can be ignored does not necessarily hold as trains become heavier and longer and coupler forces become larger. Some possibilities for wagon instabilities were examined in [11], namely wheel unloading due to the lateral components of coupler forces and wagon lift due to mismatches in coupling height. There are three issues when considering the effect of longitudinal train dynamics on wagon dynamics:

- The design test
- The operational problem of determining when dangerous lateral or vertical components can occur
- The time-consuming nature of fully detailed wagon simulations

The Australian Standard AS/RISSB 7509 [40] and, no doubt, other such standards give a wagon overturning test. The test must achieve a certain lateral force without overturning, and 90% wheel unloading is allowed. The test is carried out at low speed and so does not consider the effects of curve geometry, track irregularity, wagon steering and speed. The test is therefore a manufacturing 'go' or 'no go' test. Stability, as defined by this type of test, can be improved by reduced coupler angles, lower coupler height and greater wagon mass. There is no known similar test for pitch stability. A more instructive indicator of lateral stability can be found in vehicle roadworthiness standards in reference to lateral wind loads. The presence of wind loads on a curve is not much different to the presence of lateral coupler loads. Wind loads, however, are covered in standards, and wagon stability can be assessed considering the wagon comprehensively, including all vehicle inputs: curve geometry, track irregularity, wagon steering, speed and lateral forces. Excited from the longitudinal direction, bogie and wagon pitch can be observed in the presence of large impact forces and large localised longitudinal accelerations.

It should be noted from Figures 13.41 and 13.42 that the wagon responses can be quite complex and warrant investigation with comprehensive modelling, using fully detailed wagon models and the coupler forces obtained from longitudinal train simulation. Note that, in the cases of the forces presented in Figures 13.41 and 13.42, there is no reason why the two types of interactions cannot be superposed. Impact forces could occur whilst a wagon is on a curve. Likewise, large steady forces can also occur on a curve, combining the instabilities illustrated in Figures 13.41c and 13.42c.

Determining whether such instances can occur on a given route and train design is compounded by the problem that train dynamics forces are dependent on non-unique running situations. Unless automated driverless trains are implemented, each train trip each day can be slightly different. Heavy-haul trains, because of their uniformity in a system and because they are optimised to

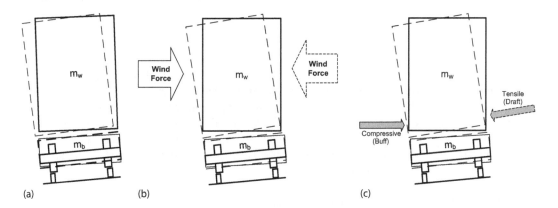

FIGURE 13.41 Wagon stability cases on curves: (a) normal curving, no external forces, full line = cant deficient case, dashed line = cant excess case, (b) curving with lateral wind forces, full line = cant deficient + wind case, dashed line = cant excess + wind case and (c) curving with lateral wind forces, full line = cant deficient + compressive coupler force case, dashed line = cant excess + tensile coupler force case.

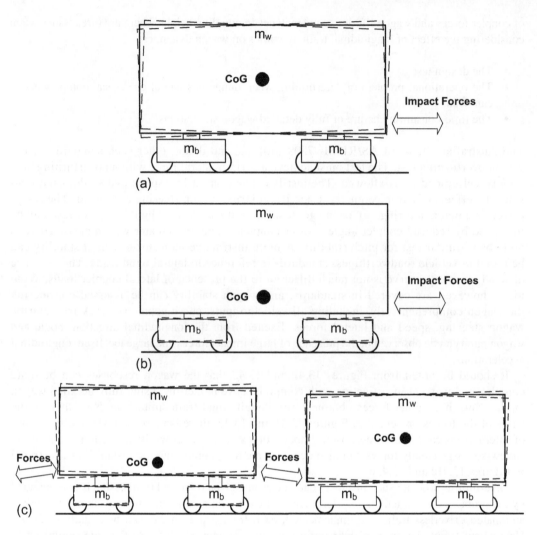

FIGURE 13.42 Wagon body, bogie pitch and jackknifing stability cases: (a) loaded wagon – often predominantly wagon body pitch, (b) empty wagon – often predominately bogie pitch and (c) loaded-empty combination – jackknifing mode.

give maximum delivery of payload per locomotive, often can have high levels of daily uniformity. However, signalling conditions, speeds and stop-start locations can still change on any day. More consistency will be seen in areas of the track that require near-full traction or braking capabilities, which, by necessity, will have almost uniform operations. More variation can be observed on severely undulating tracks and older infrastructure with steeper grades and curves, and coupler forces can change significantly if controls are applied at slightly different locations. Mitigations of this problem for lateral forces include the following:

- Removing sharper curves from the infrastructure
- Shorter couplers, shorter wagons
- Lower coupling points
- Making the wagons heavier
- Careful driving strategies

Similarly, for vertical instability, possible mitigations are as follows:

- Do not permit empty wagons in loaded trains
- Longer couplers
- Lower centre of gravity wagon bodies
- Reduced slack systems (drawbars and slackless system)
- Careful driving strategies

Clearly, several of these factors conflict with other operational needs. For mitigating lateral forces, only the first and last of the above-listed items are practical. The second item (shorter couplers/wagons) conflicts with vertical stability needs. Similarly, for vertical instability, only the last two items are practical. The use of drawbars and slackless systems has added benefit that, if only one wagon in the multi-wagon group is empty, the vehicle accelerations are less than that of a single empty wagon.

The following sections explain some modelling methods that can be used in conjunction with longitudinal train simulations to quickly locate risk areas in a train operation and route. It is emphasised that these methods give estimates assuming perfect track surface and should not be used to indicate safe operation. There are two points of difficulty:

- The train simulation result of a given run is not unique; to be sure of the solution, a set of 'N' train simulations is required over a route to exhaust all possibilities.
- The localised track irregularities and wagon steering condition must be considered; derailment modes can be complex.

To indicate problem areas, guidance can be taken from the wagon dynamics standards. In the case of the AS/RISSB 7509 standard [40], a wagon must negotiate curves without steady wheel unloading exceeding 60% and transients exceeding 80%. In the case of wind loads, no wheel unloading should exceed 90%. Given these mandated requirements, a useful indicator that might result from the following approximate analysis could be that no wheel unloading exceeding 90% would be permitted in the presence of lateral forces. This means that the actual permissible coupler forces will be limited to different levels depending on the track contributions to wagon dynamics already made by the track geometry, track irregularities and speed. It may therefore be necessary to limit lateral force contributions to levels of wheel unloading as low as 10% for transients and up to 30% for steady forces (corresponding to the expected maximum dynamic responses allowed for track geometry and irregularities – steady 60% and transients 80%).

13.3.2 Fast Methods for Assessing Interactions on Train Routes

13.3.2.1 Lateral Coupler Angles

Coupler angles can be calculated by the equations provided in the AAR manual [41] or the technique developed by Simson [42] and utilised in many locomotive traction-steering studies [43,44]. This technique is easier to apply than the AAR method and has no significant error penalty unless used for very sharp curves (error less than 0.1% at $R = 100$ m). The method also allows different curvatures to be applied for movement through curve transitions and is easier to implement in a train simulation context. The method uses the same assumptions as the AAR calculation and makes the further assumption that the two railway vehicles are coupled and are curving normally together, ignoring any offset tracking and/or suspension misalignment at each bogie (the bogie pivot centre is assumed to be located centrally between the rails). The coupler pins are located at some distance overhanging the bogie centre distance. Figure 13.43a shows the configuration of the two vehicles, with the angles between them being θ and the angle of the coupler with respect to vehicle 1 being ϕ. The angle ϕ of the coupler can be determined from the radius of curvature, the lengths between

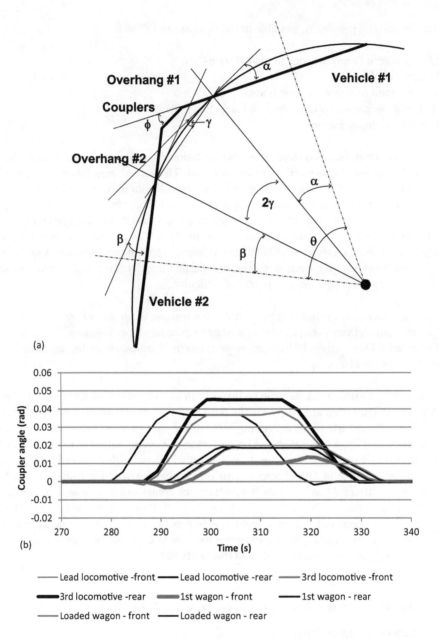

FIGURE 13.43 (a) Vehicle configuration during curving; and (b) coupler angles for various vehicle combinations/locations. (From Spiryagin, M. et al., *Design and Simulation of Rail Vehicles, Ground Vehicle Engineering Series*, CRC Press, Boca Raton, FL, 2014. With permission.)

the adjacent vehicle bogie centres and the overhang distance to the coupler pin and the coupler length. These define the angles α, β and γ, which are the chord angles to the arc for two vehicles and between the adjacent vehicle bogie centres, as shown in Figure 13.43a.

The relationship between θ and the chord arc angles is given by:

$$\theta = \alpha + \beta + \left(2^*\gamma\right) \tag{13.42}$$

where $\alpha = \arcsin(BC_1/R_0)$; $\beta = \arcsin(BC_2/R_0)$; $\gamma = \arcsin(L/2/R_0)$

and where BC_i equals the half length between bogie centres of vehicle i, L equals the chord length between the adjacent vehicle bogie centres and R_0 is the radius of the track curve. By taking small angle approximations, the above expressions can be simplified to:

$$\alpha = BC_1/R_0; \quad \beta = BC_2/R_0; \quad \gamma = L/2/R_0 \tag{13.43}$$

As the arc of the curve must be common, it is not necessary to restrict this calculation to a single radius R_0, so the approach can be used to evaluate coupler angles in curve transitions as follows:

$$\alpha = BC_1/R_{Veh1}; \quad \beta = BC_2/R_{Veh2}; \quad \gamma = L/2/R_L \tag{13.44}$$

Similarly, L can be approximated by using a small-angle assumption as:

$$L = Ov_1 + Ov_2 + Cpl_1 + Cpl_2 \tag{13.45}$$

where Ov_i equals the overhang length of vehicle i (Ov_i = half the coupler pin centre to centre distance less half the bogie centre to centre distance) and Cpl_i equals the coupler length of vehicle i.

The coupler angle ϕ can be approximated by the equation:

$$\phi = \left(L*(\alpha+\gamma)-Ov_2*\theta\right)/D \tag{13.46}$$

where D is the combined length of the two couplers: $D = Cpl_1 + Cpl_2$.

Lateral coupler angles will differ for variations in vehicle length, overhang length and coupling length in the train. In heavy-haul trains, the dimensions of wagons are more uniform, and the dimensions can be standardised to just a few cases. In most heavy-haul trains, just two vehicle lengths need to be analysed for locomotives and wagons, as shown in Table 13.6. A few interesting observations can be made from the table:

- Increasing wagon length increases lateral coupler angles
- Increasing coupler length increases lateral coupler angles
- Unequal coupler pin distances from the bogie give large variations in coupler angles in long/short connections such as locomotive-to-wagon connections

It is also interesting to note where these connection combinations might occur in a train.

Typical cases on curves (excluding transitions) for a head-end train are:

- *Lead locomotive*: No coupling at the front, no lateral force at the front, equal coupling angle coupling at the rear if there are multiple locomotives and unequal coupling angles at the rear between the locomotive and the first wagon if there is only one locomotive
- *Second and further locomotives*: Equal coupling angles at the front between two locomotives, and unequal coupling angles at the rear between the locomotive and the first wagon
- *First wagon*: Unequal coupling angles at the front between locomotive and wagon, and equal coupling angles at the rear between identical wagons
- *In-train wagons*: Equal coupling angles front and rear between identical wagons

For a train with remote-controlled locomotives, the following cases are added:

- *Single remote locomotive*: Unequal coupling angles at both the front and the rear between the locomotive and the two connecting wagons

TABLE 13.6
Coupler Angles for Various Vehicle Combinations on a 300-m Radius Curve

Dimensions[a]	Lead Vehicle			Trailing Vehicle			Angle on Lead Vehicle		
	B_1	C_1	Cpl_1	B_2	C_2	Cpl_2	Angle on	Radians	Degrees
Datum									
Short wagon-short wagon	8.5	9.6	0.8	8.5	9.6	0.8	Wagon	0.0187	1.07
Locomotive-short wagon	16	21	0.8	8.5	9.6	0.8	Locomotive	0.0806	4.62
	8.5	9.6	0.8	16	21	0.8	Wagon	−0.0243	−1.39
Longer Wagons									
Long wagon- long wagon	10	13.4	0.8	10	13.4	0.8	Wagon	0.0250	1.43
Locomotive-long wagon	16	21	0.8	10	13.4	0.8	Locomotive	0.0651	3.73
	10	13.4	0.8	16	21	0.8	Wagon	−0.0025	−0.14
Longer Couplers									
Short wagon- short wagon	8.5	9.6	1.2	8.5	9.6	1.2	Wagon	0.02	1.14
Locomotive-short wagon	16	21	1.2	8.5	9.6	1.2	Locomotive	0.0677	3.87
	8.5	9.6	1.2	16	21	1.2	Wagon	−0.0087	−0.50

[a] All dimensions in metres. B is bogie centre to centre distance; C is coupler pin to pin distance; Cpl is coupler length.

- *Lead locomotive in a remote group*: Unequal coupling angles between the locomotive and the wagon, and equal coupling angles at the rear between two locomotives
- *Single pusher locomotive in a remote group*: Unequal coupling angles at the front between the locomotive and the wagon, no coupling at the rear and no lateral force at the rear

Examples of angles from the various locomotive/wagon combinations are shown in Figure 13.43b. Where couplings are of 'like' vehicles, the angles are equal, as expected. As angles are calculated as the vehicles move through the curve, a small overthrow 'kick' can be seen in all curves. When one vehicle has a longer bogie overthrow than another, the last locomotive and the first wagon, for example, larger and smaller angles than those on matching wagons can occur. In many configurations, the largest angle in the train occurs on the locomotive at the connection between the locomotive (or locomotive group) and the wagons. In such cases, the smallest angle occurs on the connecting wagon. If the mismatch is large enough, the wagon coupling can even be straight or opposite to the direction of the curve. As locomotives almost always have longer overthrow than wagons, the maximum coupler angle in the train is usually one of the locomotive-to-wagon connections. As the minimum wagon angle is also at this connection, the maximum lateral force components on wagons (which can be expected near locomotives) will actually occur at the connection between the first and second wagons in the rake; hence, the second wagon usually has the greatest risk of overturning [3].

One fairly obviously conclusion is that lateral instability risks on curves due to coupler forces can be reduced by either or both reducing coupler forces and by removing sharper curves. As the maximum coupler force is determined by the physical strength of couplers and knuckles, it follows that there is a minimum curve radius, above which all lateral coupler force components are manageable. Using the short-short wagon case from Table 13.6 and assuming a maximum coupler force of 1.8 MN, an empty wagon mass of 22 tonnes, standard gauge and a coupling height of 0.89 m, the wheel unloading due to coupler forces alone can be tabulated as shown in Table 13.7.

Returning to the discussion in Section 13.3.1, a conservative approach would be to limit curve radii to greater than 1100 m, ensuring that coupler forces of 1.8 MN do not add more than 10%

TABLE 13.7
Wheel Unloading of an Empty 22-Tonne Wagon Subject a Coupler Force of 1.8-MN Coupler Force on Curves

Radius (m)	Coupler Angle (Degrees)	Lateral Coupler Force at 1.8 MN (kN)	Wheel Unloading (%)
100	3.209	101	112
200	1.604	50	56
300	1.070	34	37
400	0.802	25	28
500	0.642	20	22
600	0.535	17	19
700	0.458	14	16
800	0.401	13	14
900	0.357	11	12
1000	0.321	10	11
1100	0.292	9	10
1200	0.267	8	9
1300	0.247	8	9
1400	0.229	7	8
1500	0.214	7	7
1600	0.201	6	7
1700	0.189	6	7
1800	0.178	6	6
1900	0.169	5	6
2000	0.160	5	6

wheel unloading to the wagon dynamics. Alternatively, it may be established that the particular wagons have better margins due to design features and better track standards, or the operating standard might allow a greater aggregate of wheel unloading. It is possible that operations and standards might allow a 20% contribution from coupler forces or 1.8 MN, allowing sharper curves (~600 m).

13.3.2.2 Wheel Unloading, Wheel Climb and Rollover on Curves due to Lateral Components of Coupler Forces

The methodology for calculation of coupler angles and associated forces is as follows:

- Calculate front and rear coupler angles on all vehicles using curvature data and vehicle dimensions
- This is completed using the equations described previously, but with the refinement of allowing changes in the overhang distances Ov_1 and Ov_2 in response to draft gear deflections, as measured in the train simulations, and allowing changes in the sum of $Cpl_1 + Cpl_2$ to incorporate the effect of coupling slack
- Combine these angles with coupler forces to get lateral force components at the coupler pins; this is done simply as:

$$F_{lateral} = F_{coupler} * \phi \tag{13.47}$$

- Use moments to translate these forces to the bogies, noting that the forces are not equal during transitions; this parameter is designated as lateral forces from couplers, being

$$F_{lfb} = [F_1 * (C + B) - F_2 * (C - B)] / (2B) \tag{13.48}$$

$$F_{lrb} = [F_2*(C+B) - F_2*(C-B)]/(2B) \tag{13.49}$$

where F_{lfb} and F_{lrb} are the lateral forces at the front and rear bogies, F_1 is the front lateral coupler force component, F_2 is the rear lateral coupler force component, C is the coupler pin half distance and B is the bogie centre half distance

- Match the sign convention of longitudinal forces, considering lateral forces as:
  - Positive if associated with tensile forces: these forces pull the vehicle toward the centre of the curve (stringlining effect)
  - Negative if associated with compressive coupler forces: These forces push the vehicle away from the centre of the curve (buckling effect)
- Add vehicle centripetal forces to the lateral forces, assuming equal distribution of mass between front and rear bogies and using bogie curvature and superelevation; this parameter is designated as total quasi-static bogie lateral force, being

$$F_{lfb\_TL} = m_w/2\,(g*\sin(\psi) - V_w^2/\,abs(R)*\cos(\psi)) + F_{lfb} \tag{13.50}$$

$$F_{lrb\_TL} = m_w/2\,(g*\sin(\psi) - V_w^2/\,abs(R)*\cos(\psi)) + F_{lrb} \tag{13.51}$$

where F_{lfb_TL} and F_{lrb_TL} are the total quasi-static lateral force at the front and rear bogies, m_w is the vehicle mass, V_w is the vehicle velocity, R is the curve radius and ψ is the track cant angle

- Taking moments about each rail, the total quasi-static bogie lateral force could be used to calculate quasi-static vertical forces on each side of each bogie, again assuming equal distribution of mass between front and rear bogies and using bogie curvature and superelevation; this parameter is designated as quasi-static bogie vertical force, being

$$\begin{aligned}F_{vfhr\_TV} =\;& m_w/2*(g*\cos(\psi)/2 - g*\sin(\psi)*H_{cog}/d_c) \\ &+ m_w/2*V_w^2/\,abs(R)*(\sin(\psi)/2 + \cos(\psi)*H_{cog}/d_c) - F_{lfb}*h_c/d_c\end{aligned} \tag{13.52}$$

$$\begin{aligned}F_{vflr\_TV} =\;& m_w/2*(g*\cos(\psi)/2 + g*\sin(\psi)*H_{cog}/d_c) \\ &+ m_w/2*V_w^2/\,abs(R)*(\sin(\psi)/2 - \cos(\psi)*H_{cog}/d_c) + F_{lfb}*h_c/d_c\end{aligned} \tag{13.53}$$

$$\begin{aligned}F_{vrhr\_TV} =\;& m_w/2*(g*\cos(\psi)/2 - g*\sin(\psi)*H_{cog}/d_c) \\ &+ m_w/2*V_w^2/\,abs(R)*(\sin(\psi)/2 + \cos(\psi)*H_{cog}/d_c) - F_{lrb}*h_c/d_c\end{aligned} \tag{13.54}$$

$$\begin{aligned}F_{vrlr\_TV} =\;& m_w/2*(g*\cos(\psi)/2 + g*\sin(\psi)*H_{cog}/d_c) \\ &+ m_w/2*V_w^2/\,abs(R)*(\sin(\psi)/2 - \cos(\psi)*H_{cog}/d_c) + F_{lrb}*h_c/d_c\end{aligned} \tag{13.55}$$

where F_{vfhr_TV} and F_{vflr_TV} are the total quasi-static vertical forces at the front high and low rails, respectively; F_{vrhr_TV} and F_{vrlr_TV} are the total quasi-static vertical forces at the rear high and low rails, respectively; H_{cog} is the height of the vehicle centre of mass above the rail; d_c is the distance between wheel-rail contact points (~track gauge + 0.07 m); and h_c is the height of the vehicle coupler above the rail.

Having derived the total quasi-static bogie lateral force, it is also possible to calculate quasi-static bogie L/V, but the vertical forces calculated cannot be used to give bogie-side L/V, because the lateral force components cannot be separated into right and left rail components. To prevent confusion, it is not recommended that this parameter be used, as it is very different from other definitions of L/V ratio.

To provide context for an example of the effects of coupler angles, results from a train simulation are shown in Figure 13.44. Coupler angles, coupler lateral forces and lateral and vertical forces at the bogies are shown in Figures 13.45 through 13.48.

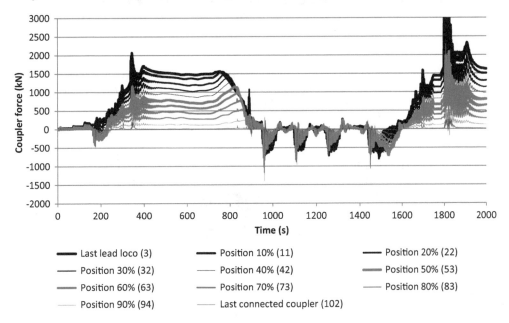

FIGURE 13.44 Simulation results of coupler forces – 100 wagons, head end power. (From Spiryagin, M. et al., *Design and Simulation of Heavy Haul Locomotives and Trains*, CRC Press, Boca Raton, FL, 2016. With permission.)

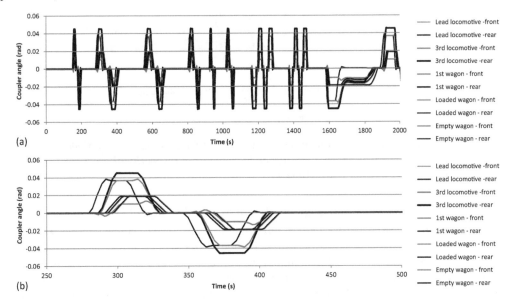

FIGURE 13.45 Simulation results of coupler angles – locomotive and wagon connections: (a) coupler angles and (b) zoom-in view. (From Spiryagin, M. et al., *Design and Simulation of Rail Vehicles, Ground Vehicle Engineering Series*, CRC Press, Boca Raton, FL, 2014. With permission.)

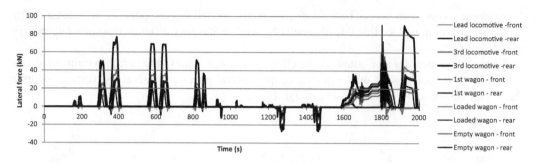

FIGURE 13.46 Simulation results – lateral forces due to couplers – locomotive and wagon connections (positive forces corresponding to tensile coupler stress). (From Spiryagin, M. et al., *Design and Simulation of Heavy Haul Locomotives and Trains*, CRC Press, Boca Raton, FL, 2016. With permission.)

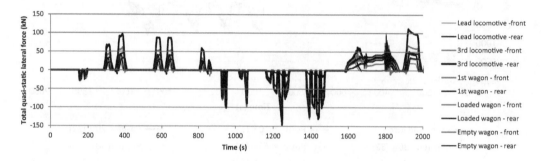

FIGURE 13.47 Simulation results of total quasi-static lateral forces – locomotives, loaded wagons and empty wagons (positive forces corresponding to force toward the low rail). (From Spiryagin, M. et al., *Design and Simulation of Heavy Haul Locomotives and Trains*, CRC Press, Boca Raton, FL, 2016. With permission.)

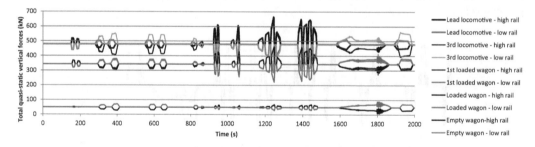

FIGURE 13.48 Simulation results of total quasi-static vertical forces – locomotives, loaded wagons and empty wagons. (From Spiryagin, M. et al., *Design and Simulation of Heavy Haul Locomotives and Trains*, CRC Press, Boca Raton, FL, 2016. With permission.)

13.3.2.3 Rail Vehicle Body and Bogie Pitch due to Coupler Impact Forces

Typically, for rail vehicle dynamics studies, modelling is undertaken of single vehicles with a longitudinal constraint. The models involve full modelling of the wheel-rail contact patch and of the 11 masses and up to 62 degrees of freedom. For the consideration of vehicle body and bogie pitch, longitudinal forces and accelerations need to be known. A rail vehicle model that is computationally economical is also required, so that it can be undertaken for whole train trip simulations (e.g., >100 km). A simplified model is therefore desirable. The vehicle pitch behaviour is modelled, as in [11], with three pitch motions and three vertical motions; see Figure 13.49 for a dynamic model of a typical wagon.

Longitudinal Train Dynamics and Vehicle Stability in Train Operations

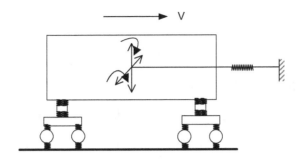

FIGURE 13.49 Schematic of a typical wagon dynamic model. (From Spiryagin, M. et al., *Design and Simulation of Heavy Haul Locomotives and Trains*, CRC Press, Boca Raton, FL, 2016. With permission.)

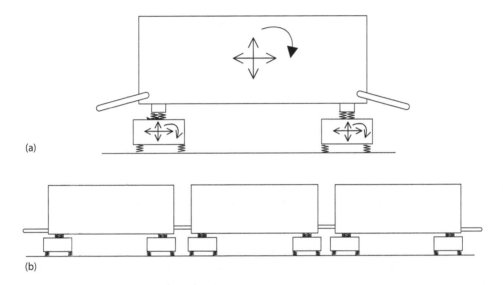

FIGURE 13.50 Simplified rail vehicle pitch models implemented with longitudinal simulation: (a) single vehicle and (b) three-vehicle model. (From Spiryagin, M. et al., *Design and Simulation of Heavy Haul Locomotives and Trains*, CRC Press, Boca Raton, FL, 2016. With permission.)

As only pitch and vertical motions are being modelled, the model can be further simplified by joining the bolster to the car body and modelling the bogie side frames and wheelsets as one mass. As some of the dynamic parameters are already calculated in the train simulation, the modelling of each vehicle can be reduced to just six equations, three describing vertical motions of the vehicle body and the two bogies and three describing the pitch rotations; see Figure 13.50a. It is also necessary to consider the effects of coupling heights and vertical force components between vehicles. If vehicles are of the same type and load, these components will be small, but bogie pitch motions will result in angles and vertical components. To ensure that correct interaction occurs at the couplings, three vehicles are included (see Figure 13.50b), and only the results from the middle vehicle are used. The other two vehicles couple to points that are at a fixed height above the rail.

The modelling equations for the simplified model are reproduced from reference [11]:

$$F_{zwb} = -F_{c1}*(z_{wb} - z_n - C*\sigma_{wb})/(2C_{pl}) - F_{c2}*(z_{wb} - z_p + C*\sigma_{wb})/(2C_{pl})$$
$$+ F_{s1} + F_{s2} + F_{s3} + F_{s4} + F_{d1} + F_{d2} + F_{d3} + F_{d4} - abs(m_{wb}*g\cos(\lambda))$$

(13.56)

$$F_{zfb} = -F_{s1} - F_{s2} - F_{d1} - F_{d2} + F_{wrc1} + F_{wrc2} - m_{fb} * g \quad (13.57)$$

$$F_{zrb} = -F_{s3} - F_{s4} - F_{d3} - F_{d4} + F_{wrc3} + F_{wrc4} - m_{rb} * g \quad (13.58)$$

$$\begin{aligned} M_{wb} = &-F_{s1}*(B+l_s) - F_{s2}*(B-l_s) + F_{s3}*(B-l_s) + F_{s4}*(B+l_s) \\ &- F_{d1}*(B+l_d) - F_{d2}*(B-l_d) + F_{d3}*(B-l_d) + F_{d4}*(B+l_d) \\ &+ F_{c1}*(z_{wb} - z_n - C*\sigma_{wb})/(2C_{pl})*C - F_{c2}*(z_{wb} - z_p + C*\sigma_{wb})/(2C_{pl})*C \\ &- F_{c1}*(h_{cg} + C*\sigma_{wb}) + F_{c2}*(h_{cg} - C*\sigma_{wb}) + m_{fb}*h_{cg}*a_w \\ &+ m_{rb}*h_{cg}*a_w + h_{cg}*(m_w + m_{fb} + m_{rb})*g*\sin(\lambda) \end{aligned} \quad (13.59)$$

Note that the moment from the longitudinal reaction at the centre bowl connection is calculated from the bogie inertia term $m_{fb}*h_{cg}*a_w + m_{rb}*h_{cg}*a_w$. This is also done in Equations 13.60 and 13.61:

$$\begin{aligned} M_{fb} = &-F_{wrc1}*A + F_{wrc2}*A - F_{s1}*l_s + F_{s2}*l_s - F_{d1}*l_d + F_{d2}*l_d \\ &+ m_{fb}*h_b*a_w + F_{brake}/R_w \end{aligned} \quad (13.60)$$

$$\begin{aligned} M_{rb} = &-F_{wrc3}*A + F_{wrc4}*A - F_{s3}*l_s + F_{s4}*l_s - F_{d3}*l_d + F_{d4}*l_d \\ &+ m_{rb}*h_b*a_w + F_{brake}/R_w \end{aligned} \quad (13.61)$$

where A is the axle centre half length; B is bogie centre half length; C is coupler pin centre half length; Cpl is coupler length; F_{zwb}, F_{zfb} and F_{zrb} are the vertical force on the vehicle body, front bogie and rear bogie, respectively; F_{c1} and F_{c2} are the front and rear coupler force, respectively; F_{s1} and F_{s2} are the spring force from the front and rear halves of the two spring nests in the front bogie, respectively; F_{s3} and F_{s4} are the spring force from the front and rear halves of the two spring nests in the rear bogie, respectively; F_{d1} and F_{d2} are the damper force from the front and rear wedges in the front bogie, respectively; F_{d3} and F_{d4} are the damper force from the front and rear wedges in the rear bogie, respectively; F_{wrc1}, F_{wrc2}, F_{wrc3} and F_{wrc4} are the total vertical wheel-rail contact force per axle on wheelsets 1, 2, 3 and 4, respectively; F_{brake} is the bogie braking force; H_b is height of the coupling line above the bogie CoG; M_{wb}, M_{fb} and M_{rb} are the moments about the pitch axis on the vehicle body, front bogie and rear bogie, respectively; R_w is the wheel radius; a_w is longitudinal acceleration of the vehicle obtained from the train simulation; h_{cg} is height of vehicle body centre of mass (CoM) above coupling line; l_s is distance to force centroid of spring half nest; l_d is distance to line of action of wedge dampers; m_{wb} is mass of the vehicle body; m_{fb} is mass of the front bogie; m_{rb} is mass of the rear bogie; z_{wb} is height of the centre of mass of the vehicle body; z_n is height of the centre of mass of the vehicle body connecting to the front; z_p is height of the centre of mass of the vehicle body connecting to the rear; σ_{wb} is the pitch angle of the vehicle body; σ_{fb} is the pitch angle of the front bogie; σ_{rb} is the pitch angle of the rear bogie; and λ is the track grade angle.

To provide context for an example of wagon body pitch, a train simulation result is shown in Figure 13.51. A hypothetical heavy-haul train is simulated with all wagons loaded. To induce wagon body pitch, a minimum brake application is applied. Compressive coupler forces are induced, as shown in Figure 13.52. Details of the coupler forces at the seventh wagon are shown in Figure 13.53, axle forces in Figure 13.54 and 'zoom-in' on axle forces showing body pitch in Figure 13.55.

Similarly, a hypothetical heavy-haul train is simulated with all wagons empty. To induce wagon bogie pitch, a minimum brake application is applied. Details of the coupler forces at the seventh wagon are shown in Figure 13.56, axle forces in Figure 13.57 and 'zoom-in' on axle forces showing bogie pitch in Figure 13.58.

Longitudinal Train Dynamics and Vehicle Stability in Train Operations

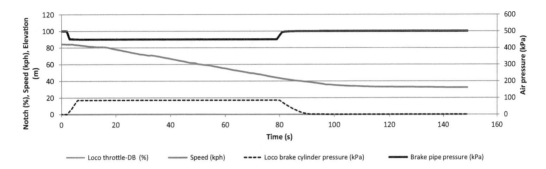

FIGURE 13.51 Simulation results of the body pitch case of operational data – loaded train. (From Spiryagin, M. et al., *Design and Simulation of Rail Vehicles, Ground Vehicle Engineering Series*, CRC Press, Boca Raton, FL, 2014. With permission.)

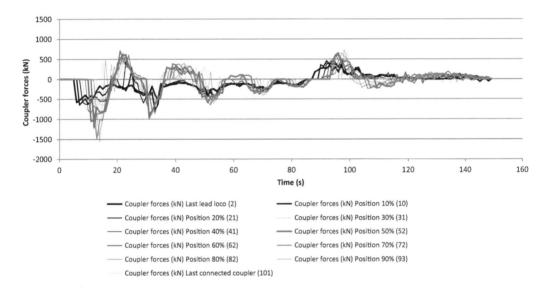

FIGURE 13.52 Simulation results of coupler forces – loaded train. (From Spiryagin, M. et al., *Design and Simulation of Rail Vehicles, Ground Vehicle Engineering Series*, CRC Press, Boca Raton, FL, 2014. With permission.)

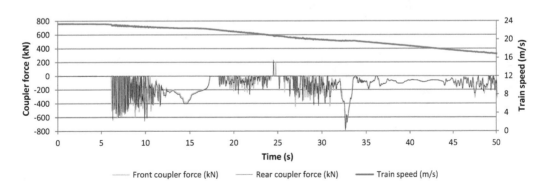

FIGURE 13.53 Simulation results of selected coupler force and operational data, wagon #7 – loaded train. (From Spiryagin, M. et al., *Design and Simulation of Rail Vehicles, Ground Vehicle Engineering Series*, CRC Press, Boca Raton, FL, 2014. With permission.)

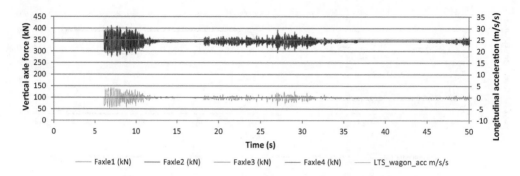

FIGURE 13.54 Simulation results of selected axle force data, wagon #7 – loaded train. (From Spiryagin, M. et al., *Design and Simulation of Rail Vehicles*, Ground Vehicle Engineering Series, CRC Press, Boca Raton, FL, 2014. With permission.)

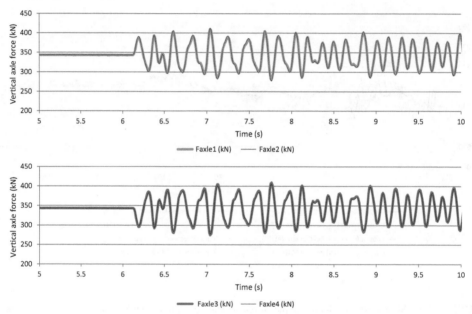

FIGURE 13.55 Simulation results of the body pitch case showing zoom-in on axle force data, wagon #7 – loaded train. (From Spiryagin, M. et al., *Design and Simulation of Rail Vehicles*, Ground Vehicle Engineering Series, CRC Press, Boca Raton, FL, 2014. With permission.)

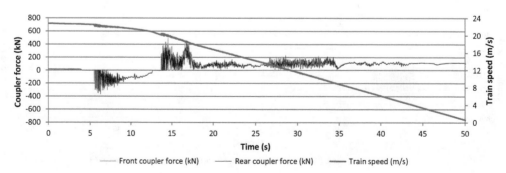

FIGURE 13.56 Simulation results of selected coupler force and operational data, wagon #7 – empty train. (From Spiryagin, M. et al., *Design and Simulation of Rail Vehicles*, Ground Vehicle Engineering Series, CRC Press, Boca Raton, FL, 2014. With permission.)

Longitudinal Train Dynamics and Vehicle Stability in Train Operations

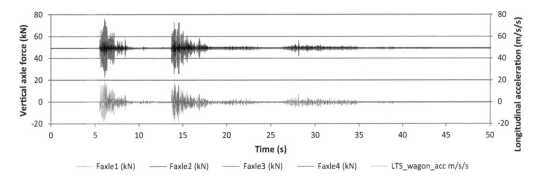

FIGURE 13.57 Simulation results of selected axle force data, wagon #7 – empty train. (From Spiryagin, M. et al., *Design and Simulation of Rail Vehicles, Ground Vehicle Engineering Series*, CRC Press, Boca Raton, FL, 2014. With permission.)

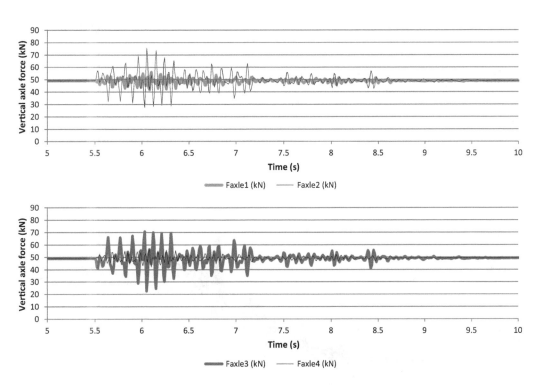

FIGURE 13.58 Simulation results of the bogie pitch case showing zoom-in on axle force data, wagon #7 – empty train. (From Spiryagin, M. et al., *Design and Simulation of Rail Vehicles, Ground Vehicle Engineering Series*, CRC Press, Boca Raton, FL, 2014. With permission.)

13.3.2.4 Rail Vehicle Lift-Off due to Vertical Components of Coupler Forces

Vehicle lift can more easily occur if there is a mismatch in coupling heights. The more severe case for vertical force components from coupling vehicles with different coupling heights, either empty/loaded combinations or vehicles of different types, can also be handled by the three-vehicle model approach. A schematic of this case is shown in Figure 13.59. It is assumed that the

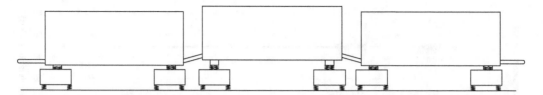

FIGURE 13.59 Simplified vehicle pitch model implemented as a three-vehicle model. (From Spiryagin, M. et al., *McSweeney, Design and Simulation of Heavy Haul Locomotives and Trains*, CRC Press, Boca Raton, FL, 2016. With permission.)

effect of a slight pitch angle on the adjacent vehicles will have no significant effect on the vehicle under study. For the following analysis, the couplers are assumed to be collinear, that is, either interlocked autocoupler connections or drawbars. Note that autocouplers without interlocking lugs or buffer and draw hook connections allow vertical misalignment of the connection. These older-style connections have a more complex set of possibilities. In the case of the autocouplers without interlocking lugs, much depends on whether the couplers are contacting pocket limits. Several different load paths can arise.

To provide context, for example, of wagon lift-off instability, train simulation results for coupler forces are shown in Figure 13.60, which involves coupler tension, so the empty wagon is effectively pulled downward by the couplers, increasing wheel loads, as shown in Figure 13.61. This situation increases wagon stability. The second example in Figure 13.62 is the opposite, involving coupler compression. In this case, the wagon is lifted off the track by the couplers, as shown in Figure 13.63, and severe wheel unloading occurs. If such events are severe enough, complete wheel lift-off and consequent jackknifing can occur.

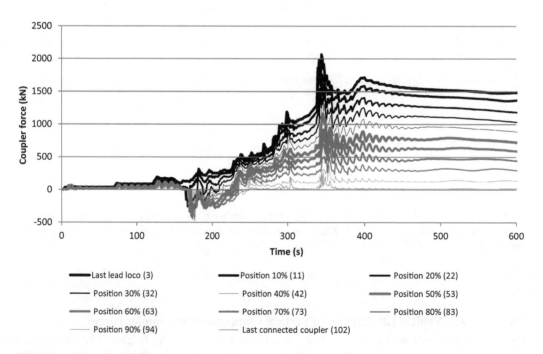

FIGURE 13.60 Simulation results of coupler forces – loaded train, traction case. (From Spiryagin, M. et al., *Design and Simulation of Heavy Haul Locomotives and Trains*, CRC Press, Boca Raton, FL, 2016. With permission.)

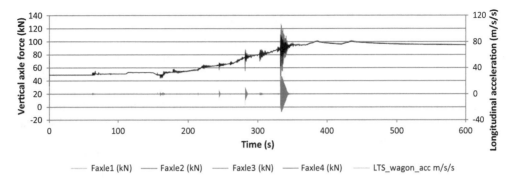

FIGURE 13.61 Simulation results of selected axle force and acceleration data, empty wagon #7 – loaded train, traction case. (From Spiryagin, M. et al., *Design and Simulation of Heavy Haul Locomotives and Trains*, CRC Press, Boca Raton, FL, 2016. With permission.)

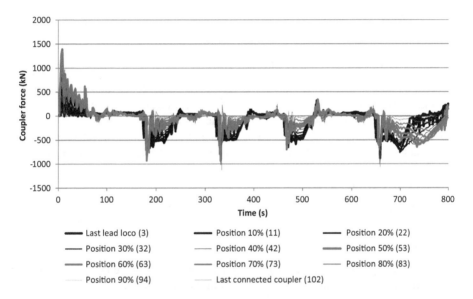

FIGURE 13.62 Simulation results of coupler forces – loaded train, braking case. (From Spiryagin, M. et al., *Design and Simulation of Heavy Haul Locomotives and Trains*, CRC Press, Boca Raton, FL, 2016. With permission.)

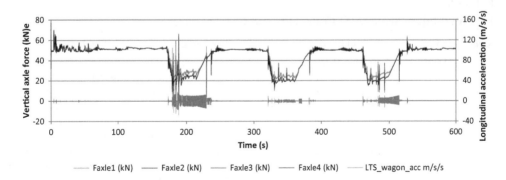

FIGURE 13.63 Simulation results of selected axle force and acceleration data, empty wagon #7 – loaded train, braking case. (From Spiryagin, M. et al., *Design and Simulation of Heavy Haul Locomotives and Trains*, CRC Press, Boca Raton, FL, 2016. With permission.)

13.4 LONGITUDINAL TRAIN DYNAMICS AND TRAIN CRASHWORTHINESS

Longitudinal dynamics is a significant factor in crashworthiness of passenger trains and locomotive cabins. Standards and specifications will differ depending on the expected running speeds and country of operation. In Australia, it is a requirement that energy-absorption elements within draft gears will minimise the effects of minor impacts. The minimum performance of draft gears is the requirement to accommodate an impact at 15 km/h [45]. The code of practice also requires that cars include unoccupied crumple zones between the headstock and bogie centres to absorb larger impacts by plastic deformation.

Design requirements of crashworthiness are focused on improving the chances of survival of car occupants. There is high demand, as noted by Sun et al. [46], for high-energy crush or crumple zones. There are three areas of locomotive and car design related to longitudinal dynamics that require attention and are being mandated by safety authorities in most countries, these being:

- Locomotive or power car crumple zones
- End car crumple zones
- Vertical collision posts

13.4.1 Locomotive or Power Car Crumple Zones

A key issue in passenger and driver survival is the reduction of decelerations. Head injury criteria (HIC) and chest injury decelerations published by the American Association for Automotive Medicine and extracted from [47,48] are given in Table 13.8.

Key to limiting deceleration is the design of crush and crumple zones. References [46,47] give characteristics of idealised crush zones for locomotives; details are shown in Figure 13.64.

13.4.2 End Car Crumple Zones

Similarly, details for passenger cars are shown in Figure 13.65.

13.4.3 Vertical Collision Posts

The requirement is based on the scenario of a wagon becoming uncoupled or broken away and then climbing the next car; see Figure 13.66. The chassis of the raised wagon being much stronger than the passenger car upper structure can easily slice through the car, causing fatalities and horrific injuries.

TABLE 13.8
Abbreviated Injury Scale (AIS) Code, HIC and Chest Deceleration

AIS Code	HIC	Head Injury	Chest Deceleration (g)	Chest Injury
1	135–519	Headache or dizziness	17–37	Single rib fracture
2	520–899	Unconscious less than 1 h; linear fracture	38–54	Two to three rib fractures; sternum fracture
3	900–1254	Unconscious 1 to 6 h; depressed fracture	55–68	Four or more rib fractures; 2–3 rib fractures with hemothorax or pneumothorax
4	1255–1574	Unconscious 6 to 24 h; open fracture	69–79	Greater than four rib fractures with hemothorax or pneumothorax; flail chest
5	1575–1859	Unconscious more than 24 h; large hematoma	80–90	Aorta laceration (partial transection)
6	>1860	Non-survivable	>90	Non-survivable

Longitudinal Train Dynamics and Vehicle Stability in Train Operations 513

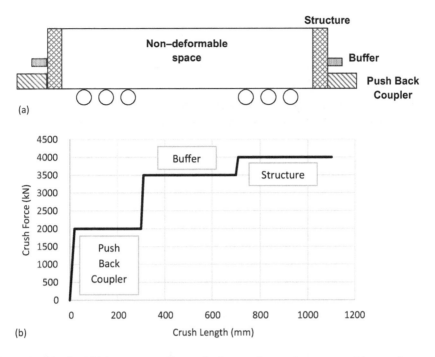

FIGURE 13.64 Idealised high-energy crush zone for locomotives and power cars: (a) crumple zone structure details and (b) idealised crush zone characteristic.

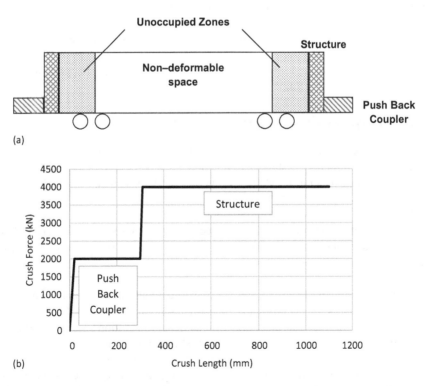

FIGURE 13.65 Idealised low-energy crush zone for inter-car connections: (a) crumple zone structure details and (b) idealised crush zone characteristic.

FIGURE 13.66 Collision illustrating wagon climb. (From Cole, C., Longitudinal train dynamics, Chapter 9 in *Handbook of Railway Vehicle Dynamics*, S. Iwnicki (Ed.), CRC Press, Boca Raton, FL, 239–278, 2006. With permission.)

FIGURE 13.67 Passenger car showing placement of vertical collision posts. (From Cole, C., Longitudinal train dynamics, Chapter 9 in *Handbook of Railway Vehicle Dynamics*, S. Iwnicki (Ed.), CRC Press, Boca Raton, FL, 239–278, 2006. With permission.)

Design requirements to improve occupant survival include the provision of vertical collision posts that must extend from the chassis or underframe to the passenger car roof; see Figure 13.67. Standards will differ depending on the expected running speeds and country of operation. The specification in Australia for operation on the Defined Interstate Rail Network [45] requires the following forces to be withstood without the ultimate material strength being exceeded:

- A total longitudinal force of 1100 kN distributed evenly across the collision posts, with the force applied 1.65 m above the rail level
- A horizontal shear force of 1300 kN applied to each individual post applied at a level just above the chassis or underframe

13.5 LONGITUDINAL PASSENGER COMFORT

Ride comfort measurement and evaluation are usually focussed on accelerations in the vertical and lateral directions. In surveying standards, it is noted that some passenger comfort standards do not refer to longitudinal accelerations at all. In general, standards can be classified as follows:

- Standards that do not consider longitudinal accelerations
- Standards that set the limits for longitudinal accelerations as the same as lateral accelerations
- Standards that explicitly include longitudinal acceleration recommendations for ride comfort

Ride comfort has traditionally been the domain of vertical and lateral vehicle dynamics, so it is logical that many standards follow this same focus. It should also be noted that, with the move to multiple unit suburban trains (rather than mainline passenger operation), vehicles have distributed power, blended and/or electronic braking, advanced slip controls and permanently coupled cars. It follows that, for these applications, longitudinal dynamics may not be such an issue. The same could be stated for high-speed trains. In such cases, if longitudinal comfort is considered at all, it will be considered in the aggregate of comfort parameters. Furthermore, methods typically use root mean square (r.m.s.) values or statistical methods such as those described in ISO 10056, which takes the 95th percentile of weighted r.m.s. values over intervals of 5 seconds. These approaches would not give strong indications of individual events and jerks. Examples of this approach are the European-type standards European Standards (EN) 12299:2009 [49] and earlier International Union of Railways (UIC) Leaflet 513 [50].

The UIC approach integrates longitudinal accelerations into a single parameter. Vertical, lateral and longitudinal accelerations are measured and weighted with appropriate filters. Root mean square values of accelerations taken over 5-second time blocks are calculated. The test data sample is of 5-minute duration. The 95th percentile point in each event distribution is then used to calculate a single parameter. The equation for the simplified method (where measurements are taken on the vehicle floor) is as follows:

$$N_{MV} = 6\sqrt{(a_{XP95})^2 + (a_{YP95})^2 + (a_{ZP95})^2} \qquad (13.62)$$

where a_{XP} is acceleration in the longitudinal direction,
a_{YP} is acceleration in the lateral direction and
a_{ZP} is acceleration in the vertical direction.

A further equation is available for standing passengers, this time using 50 percentile points from event distributions.

$$N_{VD} = 3\sqrt{16(a_{XP50})^2 + 4(a_{YP50})^2 + (a_{ZP50})^2 + 5(a_{ZP95})^2} \qquad (13.63)$$

Ride criteria using the above index parameters are:

- $N < 1$ Very Comfortable
- $1 < N < 2$ Comfortable
- $2 < N < 4$ Medium
- $4 < N < 5$ Uncomfortable
- $N > 5$ Very uncomfortable

There is also a difference in comfort standards depending on the orientation of the human body in the train. Several standards, including the one adopted in the Sydney Suburban System in Australia, apply the lateral comfort limits to longitudinal accelerations if the passengers are orientated on longitudinal seating, rather than traditional lateral seating. Standards that take this approach include:

- AS/RISSB 7513.3, Australia [51]
- ISO 2631 mechanical vibration and shock: Evaluation of human exposure to whole-body vibration [52]

Both these standards use weighted frequency responses and/or r.m.s methods.

The only standards located that give explicit levels for longitudinal accelerations are the RTRI Indexes [53–55] from Japan. Good ride comfort in this approach requires longitudinal accelerations to be below 0.025 g for frequencies 4–15 Hz and accelerations to increase to 0.06 g at 40 Hz. These figures, if translated to jerk, would be as shown in Table 13.9.

The nature of longitudinal dynamics is that trains are only capable of quite low steady accelerations and decelerations due to the limits imposed by adhesion at the wheel-rail interface. Cleary, the highest acceleration achievable will be that of a single locomotive giving a possible ~0.3 g, assuming 30% wheel-rail adhesion and driving all wheels. Typical train accelerations are of course much lower of the order 0.1–1.0 m/s² [21]. Braking also has the same adhesion limit, but rates are limited to values much lower to prevent wheel locking and wheel flats. Typical train deceleration rates are of the order 0.1–0.6 m/s [21]. The higher values of acceleration and deceleration in the ranges quoted correspond to passenger and suburban trains. The only accelerations that contribute to passenger

TABLE 13.9
RTRI Good Ride Comfort Limits Translated to Jerk Values

RTRI Rating	Frequency, Hz	Acceleration, g	Acceleration, m/s²	Jerk, m/s³	X^a, mm
1 – Good	4	0.025	0.245	6.2	0.39
	15	0.025	0.245	23.1	0.03
	40	0.060	0.589	147.9	0.01
5 – Very poor	4	0.070	0.687	17.3	1.09
	15	0.070	0.687	64.7	0.08
	40	0.200	1.962	493.1	0.03

Source: Railway Technical Research Institute, *Evaluating Riding Comfort of Railway Vehicles*, Tokyo, Japan, 2003, Webpage available at: http://www.rtri.or.jp/sales/english/topics/simulation/research09_confort.html.

a Displacement amplitude of the corresponding sinusoidal oscillation.

discomfort or freight damage arise from coupler impact transients. The nature of these events are irregular, so frequency spectral analysis and the development of ride indexes are inappropriate in many instances. It is more appropriate to examine maximum magnitudes of single-impact events.

For comparison, the maximum acceleration limits specified by various standards that can be applied to longitudinal comfort are plotted in Figure 13.68.

The levels permitted for 1-m exposure for fatigue-decreased proficiency boundary (FDPB) and for the reduced comfort boundary (RCB), as per AS 2670 [56], are plotted in Figure 13.68 to compare with peak or maximum criteria found in other standards. Further insight is gained if the longitudinal oscillations are assumed to be sinusoidal; displacement levels associated with these acceleration levels are also plotted. The displacement amplitudes permitted for various frequencies are plotted in Figure 13.69.

In Australia, the now-outdated Railways of Australia's *Manual of Engineering Standards and Practices* [57] included calculations of ride index only for vertical and lateral directions. The only reference to longitudinal comfort was a peak limit of 0.3 g (2.943 m/s), applying to accelerations for all three directions. The 0.3-g limit applied over a bandwidth of 0 to 20 Hz, thereby describing maximum longitudinal oscillation accelerations and displacements in the range of 75–0.2 mm in the range of vibration frequencies from 1 Hz to 20 Hz, as shown by Figures 13.68 and 13.69. The code of practice for the Defined Interstate Rail Network [45] more specifically excludes evaluation of longitudinal comfort, with peak accelerations specified only for vertical and lateral dynamics.

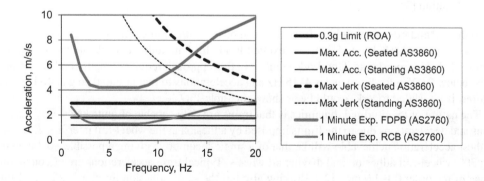

FIGURE 13.68 Passenger comfort acceleration limits. (From Cole, C., Longitudinal train dynamics, Chapter 9 in *Handbook of Railway Vehicle Dynamics*, S. Iwnicki (Ed.), CRC Press, Boca Raton, FL, 239–278, 2006. With permission.)

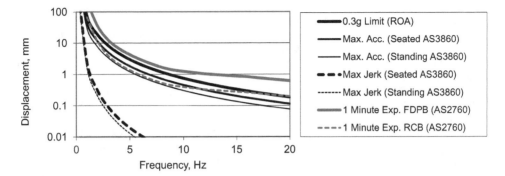

FIGURE 13.69 Passenger comfort displacement limits. (From Cole, C., Longitudinal train dynamics, Chapter 9 in *Handbook of Railway Vehicle Dynamics*, S. Iwnicki (Ed.), CRC Press, Boca Raton, FL, 239–278, 2006. With permission.)

Both standards refer to Australian Standard AS2670 [56], stating that vibration in any passenger seat shall not exceed either the RCB or the FDPB of AS2670 in any axis. Although not specifying calculations in the railway standards, criteria for longitudinal comfort can be drawn from the general Australian standard, AS2670. Another Australian standard that is useful when considering longitudinal comfort issues is AS3860 (Fixed Guideway People Movers) [58]. This standard gives maximum acceleration limits for sitting and standing passengers. It also gives maximum values for 'jerk' the time derivative of acceleration. A jerk limit slightly lower than the Australian standard AS3860 of 1.5 m/s^3 is also quoted by Profillidis [21].

REFERENCES

1. C. Cole, Longitudinal train dynamics, Chapter 9 in *Handbook of Railway Vehicle Dynamics*, S. Iwnicki (Ed.), CRC Press, Boca Raton, FL, 239–278, 2006.
2. M. Spiryagin, C. Cole, Y.Q. Sun, M. McClanachan, V. Spiryagin, T. McSweeney, *Design and Simulation of Rail Vehicles*, CRC Press, Boca Raton, FL, 2014.
3. M. Spiryagin, P. Wolfs, C. Cole, V. Spiryagin, Y.Q. Sun, T. McSweeney, *Design and Simulation of Heavy Haul Locomotives and Trains*, CRC Press, Boca Raton, FL, 2016.
4. C. Cole, M. Spiryagin, Q. Wu & Y.Q. Sun, Modelling, simulation and applications of longitudinal train dynamics, *Vehicle System Dynamics*, 55(10), 1498–1571, 2017.
5. C. Cole, M. Spiryagin, Q. Wu, C. Bosomworth, Practical modelling and simulation of polymer draft gear connections, *Proceedings First International Conference on Rail Transportation*, 10–12 July 2017, Chengdu, China, American Society of Civil Engineers, Reston, VA, 413–423, 2018.
6. I.B. Duncan, P.A. Webb, The longitudinal behaviour of heavy haul trains using remote locomotives, *Proceedings 4th International Heavy Haul Railway Conference*, 11–15 September 1989, Brisbane, Australia, The Institution of Engineers, Canberra, Australia, 587–590, 1989.
7. B.J. Jolly, B.G. Sismey, Doubling the length of coals trains in the Hunter Valley, *Proceedings 4th International Heavy Haul Railway Conference*, 11–15 September 1989, Brisbane, Australia, The Institution of Engineers, Canberra, Australia, 579–583, 1989.
8. R.D. Van Der Meulen, Development of train handling techniques for 200 car trains on the Ermelo-Richards Bay line, *Proceedings 4th International Heavy Haul Railway Conference*, 11–15 September 1989, Brisbane, Australia, The Institution of Engineers, Canberra, Australia, 574–578, 1989.
9. M. El-Sibaie, Recent advancements in buff and draft testing techniques, *Proceedings 5th International Heavy Haul Railway Conference*, 6–11 June 1993, Beijing, China, 146–150, 1993.
10. M. McClanachan, C. Cole, D. Roach, B. Scown, An investigation of the effect of bogie and wagon pitch associated with longitudinal train dynamics, In: R. Fröhling (Ed.), *Proceedings 16th IAVSD Symposium held in Pretoria*, South Africa, 30 August–3 September 1999, Swets & Zeitlinger, Lisse, the Netherlands, 374–385, 2000.

11. C. Cole, M. Spiryagin, Y.Q. Sun, Assessing wagon stability in complex train systems, *International Journal of Rail Transportation*, 1(4), 193–217, 2013.
12. Q. Wu, M. Spiryagin, C. Cole, C. Chang, G. Guo, A. Sakalo, W. Wei et al., *International Benchmarking of Longitudinal Train Dynamics Simulators: Results*, Vehicle System Dynamics, 56(3), 343–365, 2018.
13. Q. Wu, M. Spiryagin, C. Cole, Longitudinal train dynamics: An overview. *Vehicle System Dynamics*, 54(12), 1688–1714, 2016.
14. V.K. Garg, R.V. Dukkipati, *Dynamics of Railway Vehicle Systems*, Academic Press, Orlando, FL, 1984.
15. H.I. Andrews, *Railway Traction: The Principles of Mechanical and Electrical Railway Traction*, Elsevier Science Publishers, New York, 1986.
16. W.W. Hay, *Railroad Engineering*, 2nd ed., John Wiley & Sons, New York, 1982, 1982.
17. M. Spiryagin, Q. Wu, C. Cole, International benchmarking of longitudinal train dynamics simulators: Benchmarking questions, *Vehicle System Dynamics*, 55(4), 450–463, 2017.
18. M. Spiryagin, Q. Wu, Y.Q. Sun, C. Cole, I. Persson, Locomotive studies utilizing multibody and train dynamics. *Proceedings Joint Rail Conference*, 4–7 April 2017, Philadelphia, PA, ASME, Paper No. JRC2017–2221, 2017.
19. M. Spiryagin, Q. Wu, Y.Q. Sun, C. Cole, I. Persson, Advanced co-simulation technique for the study of heavy haul train and locomotive dynamics behaviour, *Proceedings First International Conference on Rail Transportation*, 10–12 July 2017, Chengdu, China, American Society of Civil Engineers, Reston, VA, 79–87, 2018.
20. Q. Wu, M. Spiryagin, P. Wolfs, C. Cole, Traction modelling in train dynamics, *J Journal of Rail and Rapid Transit*, 233(4), 382–395, 2018.
21. V.A. Profillidis, *Railway Engineering*, 2nd ed., Ashgate Publishing, Aldershot, UK, 2000.
22. Association of American Railroads, Standard RP-548: locomotive rating and train/track resistance (revised 2001), in *AAR Manual of Standards and Recommended Practices, Section M, Locomotives and Locomotive Interchange Equipment*, Washington, DC, 2017.
23. American Railway Engineering and Maintenance-of-Way Association, *AREMA Manual for Railway Engineering*, Lanham, MD, 2015.
24. F. Szanto, Rolling resistance revisited, Conference on Railway Excellence, 16–18 May 2016, Melbourne, Australia, Railway Technical Society of Australasia, 2016.
25. Y-C. Lai, C.P.L. Barkan, Options for improving the energy efficiency of intermodal freight trains, *Transportation Research Record: Journal of the Transportation Research Board*, 1916, 47–55, 2005.
26. P. Lukaszewicz, Energy consumption and running time for trains: Modelling of running resistance and driver behaviour based on full scale testing, PhD thesis, Royal Institute of Technology (KTH), Stockholm, Sweden, 2001.
27. The 5AT Group, Steaming Ahead with Advanced Technology, Locomotive and train resistance, Advanced Steam Traction Trust, Stockport, UK, 2012, Available under Rolling Resistance at: https://www.advanced-steam.org/5at/technical-terms-2/
28. C. Cole, M. Spiryagin, C. Bosomworth, Examining longitudinal train dynamics in ore car tipplers, *Vehicle System Dynamics*, 55(4), 534–551, 2017.
29. G.P. Wolf, K.C. Kieres, Innovative engineering concepts for unit train service: the slackless drawbar train and continuous center sill trough train, *Proceedings 4th International Heavy Haul Railway Conference*, 11–15 September 1989, Brisbane, Australia, The Institution of Engineers, Canberra, Australia, 124–128, 1989.
30. G.W. Bartley, S.D. Cavanaugh, The second generation unit train, *Proceedings 4th International Heavy Haul Railway Conference*, 11–15 September 1989, Brisbane, Australia, The Institution of Engineers, Canberra, Australia, 129–133, 1989.
31. C. Cole, M. Spiryagin, Q. Wu, C. Bosomworth, Modelling issues in passenger draft gear connections, In: M. Rosenberger, M. Plöchl, K. Six, J. Edelmann (Eds.), *Proceedings 24th IAVSD Symposium Held at Graz*, Austria, 17–21 August 2015, CRC Press/Balkema, Leiden, the Netherlands, 985–993, 2016.
32. Q. Wu, M. Spiryagin, C. Cole, Advanced dynamic modelling for friction draft gears, *Vehicle System Dynamics*, 53(4), 475–492, 2015.
33. Q. Wu, C. Cole, M. Spiryagin, W. Ma, Preload on draft gear in freight trains, *Journal of Rail and Rapid Transit*, 232(6), 1615–1624, 2018.
34. C. Cole, Improvements to wagon connection modelling for longitudinal train simulation, In: W. Oghanna (Ed.), *Proceedings Conference on Railway Engineering*, 7–9 September 1998, Capricorn International Resort, Central Queensland University, Australia, 187–194, 1998.
35. A.G. Belousov, Development of design and numerical modelling of friction-polymer draft gears, PhD thesis, Bryansk State University, Bryansk, Russia, 2006.

36. M. Berg, A non-linear rubber spring model for rail vehicle dynamics analysis, *Vehicle System Dynamics*, 30(3–4), 197–212, 1998.
37. E. Dumont, W. Maurer, Modelling elastomer buffers with DyMoRail, *Proceedings 10th International Modelica Conference*, 10–12 March 2014, Lund, Sweden, 923–927, 2014.
38. Q. Wu, X. Yang, C. Cole, S. Luo. Modelling polymer draft gear. *Vehicle System Dynamics*, 54(9), 1208–1225, 2016.
39. Association of American Railroads, *Manual of Standards and Recommended Practices*, Section B, Couplers and Freight Draft Gear Components, Washington, DC, 2003.
40. AS/RISSB 7509, *Rolling Stock: Dynamic Behaviour*, Rail Industry Safety & Standards Board (RISSB), Australia, 2017.
41. Association of American Railroads, *AAR Manual of Standards and Recommended Practices*, Section C – Part II, Design, Fabrication, and Construction of Freight Cars, M-1001, Section 2.1 Design Data, pp. C-II-9 – C-II-34, Washington, DC, 2019.
42. S. Simson, Three axle locomotive bogie steering, simulation of powered curving performance: passive and active steering bogies, PhD Thesis, Central Queensland University, Rockhampton, Australia, 2009. Available at: http://hdl.cqu.edu.au/10018/58747.
43. S.A. Simson, C. Cole, Idealized steering for hauling locomotives, *Journal of Rail and Rapid Transit*, 221(2), 227–236, 2007.
44. S.A. Simson, C. Cole, Simulation of traction and curving for passive steering hauling locomotive, *Journal of Rail and Rapid Transit*, 222(2), 117–127, 2008.
45. RCP-6102, *Locomotive Hauled Cars, Part 4 Passenger Cars, Code of Practice for the Defined Interstate Rail Network – Volume 5: Rollingstock*, 6–12, Department of Transport and Regional Services, Canberra, Australia, 2002.
46. Y.Q. Sun, C. Cole, M. Dhanasekar, D.P. Thambiratnam, Modelling and analysis of the crush zone of a typical Australian passenger train, *Vehicle System Dynamics*, 50(7), 1137–1155, 2012.
47. Y.Q. Sun, M. Spiryagin, C. Cole, Rail passenger vehicle crashworthiness simulations using multibody dynamics approaches, *Journal of Computational and Nonlinear Dynamics*, 12(4), 041015, 1–11, 2017.
48. D.C. Tyrell, K.J. Severson, B.P. Marquis, Analysis of occupant protection strategies in train collisions, In: *Crashworthiness and Occupant Protection in Transportation Systems, Proceedings ASME International Mechanical Engineering Congress and Exposition*, 12–17 November 1995, San Francisco, CA, 539–557, 1995.
49. BS EN 12299:2009, *Railway Applications: Ride Comfort for Passengers – Measurement and Evaluation*, British Standards Institution, London, UK, 2009.
50. UIC Leaflet 513, *Guidelines for Evaluating Passenger Comfort in Relation to Vibration in Railway Vehicles*, UIC International Union of Railways, Paris, France, 2003.
51. AS/RISSB 7513.3, *Railway Rolling Stock: Interior Environment – Part 3: Passenger Rolling Stock*, Rail Industry Safety and Standards Board (RISSB), Australia, 2014.
52. ISO 2631-2, *Mechanical Vibration and Shock: Evaluation of Human Exposure to Whole-Body Vibration – Part 2: Vibration in Buildings* (1–80 Hz), International Organization for Standardization, Geneva, Switzerland, 2003.
53. Railway Technical Research Institute, *Evaluating Riding Comfort of Railway Vehicles*, Tokyo, Japan, 2003, Webpage available at: http://www.rtri.or.jp/sales/english/topics/simulation/research09_confort.html.
54. Y. Sugahara, A. Kazato, R. Koganei, M. Sampei, S. Nakaura, Suppression of vertical bending and rigid-body-mode vibration in railway vehicle car body by primary and secondary suspension control: results of simulations and running tests using Shinkansen vehicle, *Journal of Rail and Rapid Transit*, 223(6), 517–531, 2009.
55. T. Tomioka, T. Takigami, Reduction of bending vibration in railway vehicle carbodies using carbody-bogie dynamic interaction. *Vehicle System Dynamics*, 48(S1), 467–486, 2010.
56. AS 2760, *Evaluation of Human Exposure to Whole Body Vibration*, Standards Association of Australia, Sydney, Australia, 2001.
57. Railways of Australia, Passenger cars: passenger comfort index, Section 12.10 in Manual of Engineering Standards and Practices, pp. 12.37–12.40, 1992.
58. AS 3860, *Fixed Guideway People Movers*, Standards Association of Australia, Sydney, Australia, 1991.

14 Noise and Vibration from Railway Vehicles

David Thompson, Giacomo Squicciarini, Evangelos Ntotsios and Luis Baeza

CONTENTS

14.1 Introduction 522
    14.1.1 Importance of Noise and Vibration 522
    14.1.2 Basics of Acoustics 523
    14.1.3 Sources of Railway Noise and Vibration 524
14.2 Rolling Noise 525
    14.2.1 Mechanism of Rolling Noise Generation 525
    14.2.2 Surface Roughness 526
    14.2.3 Wheel Dynamics 527
    14.2.4 Track Dynamics 529
    14.2.5 Wheel-Rail Interaction 531
    14.2.6 Noise Radiation 532
    14.2.7 Overall Model 533
14.3 Reducing Rolling Noise 533
    14.3.1 Controlling Surface Roughness 534
    14.3.2 Wheel-Based Solutions 534
        14.3.2.1 Wheel Damping 534
        14.3.2.2 Wheel Design 535
    14.3.3 Track-Based Solutions 535
        14.3.3.1 Low-Noise Track 535
        14.3.3.2 Slab Tracks 537
    14.3.4 Local Shielding and Barriers 538
14.4 Impact Noise and Vibration 538
    14.4.1 Introduction 538
    14.4.2 Wheel Flats 539
    14.4.3 Rail Joints, Switches and Crossings 541
    14.4.4 Reducing Impact Noise 543
14.5 Curve Squeal 543
    14.5.1 Mechanisms of Squeal Noise Generation 543
    14.5.2 Reducing Squeal Noise 546
14.6 Aerodynamic Noise 547
    14.6.1 Sources of Aerodynamic Noise 547
    14.6.2 Measurement Methods 547
    14.6.3 Prediction Methods 547
    14.6.4 Control of Aerodynamic Noise 548

14.7	Other Sources of Noise		549
	14.7.1	Engine Noise	549
	14.7.2	Fans and Air-Conditioning	549
14.8	Vehicle Interior Noise		549
	14.8.1	Vehicle Interior Noise Levels	549
	14.8.2	Measurement Quantities for Interior Noise	549
	14.8.3	Airborne Transmission	550
	14.8.4	Structure-Borne Transmission	553
	14.8.5	Prediction of Interior Noise	553
14.9	Ground-Borne Vibration and Noise		553
	14.9.1	Introduction	553
	14.9.2	Vibration from Surface Railways	554
	14.9.3	Ground-Borne Noise from Trains in Tunnels	558
	14.9.4	Controlling Low-Frequency Ground Vibration	560
		14.9.4.1 Reducing the Track Geometric Unevenness	560
		14.9.4.2 Rolling Stock Modifications	560
		14.9.4.3 Ground-Based Measures	562
	14.9.5	Controlling Ground-Borne Noise	563
		14.9.5.1 Soft Resilient Fastening Systems	564
		14.9.5.2 Under-Sleeper Pads, Ballast Mats and Booted Sleepers	566
		14.9.5.3 Floating-Slab Tracks	566
14.10	Vibration Comfort on Trains		567
	14.10.1	Introduction	567
	14.10.2	Assessment of Vibration Comfort in Trains	567
	14.10.3	Effects of Vehicle Design	568
		14.10.3.1 Bogie Vehicles	568
		14.10.3.2 Secondary Suspension	569
		14.10.3.3 Primary Suspension	571
		14.10.3.4 Traction Mechanisms	571
		14.10.3.5 Carbody Structure	572
References			572

14.1 INTRODUCTION

14.1.1 IMPORTANCE OF NOISE AND VIBRATION

Environmental noise is an issue that has seen increased awareness in recent years. Noise is often cited as a major factor contributing to people's dissatisfaction with their environment. While this noise exposure is usually due mainly to road traffic, trains also contribute significantly in the vicinity of railway lines. Road vehicles and aircraft have long been the subject of legislation that limits their noise emissions. The European Union has therefore introduced noise limits for new rail vehicles, implemented as part of the Technical Specifications for Interoperability (TSIs) [1]. They state noise limits for new trains under both static and running conditions.

In contrast with exterior noise, the noise inside a vehicle (road or rail) is not generally the subject of legislation, apart from the noise inside the driver's cab. For road vehicles, noise is actually used as a major factor to distinguish vehicles from their competitors and to attract people to buy a particular vehicle. As rail vehicles are for mass use, interior noise is subject instead to specifications from the purchasing organisation. These are usually limited to ensuring that problems are eliminated and that the vehicles are fit for their purpose.

Railway operations also generate vibration that is transmitted through the ground into neighbouring properties. This can lead either to feelable vibration (in the range 2–80 Hz) or to low-frequency rumbling noise (20–250 Hz). Vibration is also transmitted into the vehicle itself, affecting passenger comfort.

14.1.2 BASICS OF ACOUSTICS

The field of acoustics is too large to cover in detail here. This chapter therefore gives only a very brief overview of some basic quantities. For further details, the interested reader is referred to textbooks on the subject [2,3].

Sound consists of audible fluctuations in pressure, usually of the air. These propagate as waves with a wave speed, denoted by c_0, of about 340 m/s in air at 20°C. Simultaneously, fluctuations in air density and particle motion also occur. To express the magnitude of a sound, the root mean square (rms) sound pressure is usually used:

$$p_{rms} = \left(\frac{1}{T} \int_{t_1}^{t_1+T} p^2(t) \, dt \right)^{1/2} \tag{14.1}$$

where $p(t)$ is the instantaneous sound pressure at time t, and T is the averaging time. Much use is made of frequency analysis, whereby sound signals are decomposed into their frequency content (e.g., using Fourier analysis). The normal ear is sensitive to sound in the frequency range 20–20,000 Hz (the upper limit reduces with age and with noise exposure) and to a large range of amplitudes (around six orders of magnitude). Because of these large ranges, and to mimic the way the ear responds to sound, logarithmic scales are generally used to present acoustic data. Thus, amplitudes are expressed in decibels. The *sound pressure level* is defined as

$$L_p = 10 \log_{10} \left(\frac{p_{rms}^2}{p_{ref}^2} \right) = 20 \log_{10} \left(\frac{p_{rms}}{p_{ref}} \right) \tag{14.2}$$

where the standard value for the reference pressure p_{ref} is 2×10^{-5} Pa. Frequencies (expressed in cycles/sec or Hz) are also generally plotted on logarithmic scales, with *one-third octave bands* being a common form of presentation. The frequency range is divided into bands that are of equal width on a logarithmic scale. The centre frequencies of each band can be given by $10^{(N/10)}$, where N is an integer, the band number, although by convention, these frequencies are rounded to standard values. Bands 13–43 cover the audible range.

The total sound emitted by a source is given by its power, W, which in decibel form is given as the *sound power level*:

$$L_W = 10 \log_{10} \left(\frac{W}{W_{ref}} \right) \tag{14.3}$$

where the standard value for the reference power, W_{ref}, is 10^{-12} watt. The power is generally proportional to the square of the sound pressure, so that a 1 dB increase in sound power level leads to a 1 dB increase in sound pressure level at a given location. However, sound pressure also depends on the location, usually reducing as the distance between the source and the receiver increases. For a compact point source, this reduction is 6 dB per doubling of distance, while for a line source, it is 3 dB per doubling. Other quantities can also be expressed in decibels following the pattern of Equations 14.2 and 14.3.

It should be realised that sound generation is often a very inefficient process. The proportion of the mechanical power of a typical machine that is converted into sound is typically in the range 10^{-7} to 10^{-5}. Sound is generated by a variety of mechanisms, but the two main types are:

- *Structural vibrations*: The vibration of a structure causes the air around it to vibrate and transmit sound, for example, a drum, a loudspeaker, wheels and rails.
- *Aerodynamic fluctuations*: Wind, particularly turbulence and flow over solid objects, also produces sound, for example, jet noise, turbulent boundary layer noise, exhaust noise and fan noise.

While the acceptability of sound levels and signal content varies greatly between individuals, it is important to include some approximation to the way the ear weights different sounds. Several weighting curves have been devised, but the A-weighting (Figure 14.1) is the most commonly used. This approximates the inverse of the equal loudness curve at about 40 dB. As the ear is most sensitive around 1–5 kHz, and much less sensitive at low and high frequencies, more prominence is given to this central part of the spectrum. The overall sound level is often quoted as an A-weighted value, meaning that this weighting curve is applied to the spectrum before calculating the total or applied as a filter to the time-domain signal.

Another overall measure of the magnitude of a sound is the *loudness*. Strictly, this is a subjective quantity, but there are ways of estimating a loudness value from a one-third octave band spectrum [4]. However, this is less commonly used than the A-weighted decibel. It should be borne in mind that an increase of 10 dB is perceived as a 'doubling of loudness', while a change of less than 3 dB is normally imperceptible.

14.1.3 Sources of Railway Noise and Vibration

In the case of railway noise, both types of mechanism mentioned previously apply. Aerodynamic noise is important for high-speed operation and is generated by unsteady air flow, particularly over the nose, inter-carriage joints, bogie regions, louvres and roof-mounted equipment such as

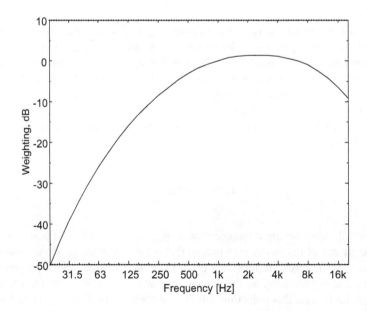

FIGURE 14.1 The A-weighting curve. (From Iwnicki, S. (ed.), *Handbook of Railway Vehicle Dynamics*, CRC Press, Boca Raton, FL, 2006. With permission.)

Noise and Vibration from Railway Vehicles

pantographs; this is described in Section 14.6. However, mechanical sources of noise dominate the overall noise for speeds up to about 300 km/h.

The most important mechanical noise source from a train is generated at the wheel-rail contact. Rolling noise is caused by vibration of the wheel and track structures, induced at the wheel-rail contact point by vertical irregularities in the wheel and rail surfaces; this is described in Sections 14.2 and 14.3. A similar mechanism leads to noise due to discontinuities in the wheel or rail surface (impact noise); see Section 14.4. Squeal noise occurs in sharp curves and is induced by unsteady friction forces at the wheel-rail contact (Section 14.5). Noise from traction equipment and fans is covered briefly in Section 14.7. Finally, ground-borne vibration and noise are caused by track and wheel irregularities and by the movement of the set of axle loads along the track; see Section 14.9. Transmission of noise to the vehicle interior is discussed in Section 14.8 and ride comfort is discussed in Section 14.10. A fuller coverage of these topics can be found in [5].

14.2 ROLLING NOISE

14.2.1 Mechanism of Rolling Noise Generation

As indicated previously, rolling noise is usually the dominant source of noise from moving trains at speeds below about 300 km/h. The sound level increases with speed V at a rate of approximately $30 \log_{10} V$, that is, a 9 dB increase for a doubling of speed. It can be attributed to components radiated by the vibration of both wheels and track. This vibration is caused by the combined surface 'roughness' at their interface, as illustrated in Figure 14.2.

The relative importance of the components of sound radiation from the wheel and track depends on their respective designs as well as on the train speed and the wavelength content of the surface roughness. In most cases, both sources (wheel and track) are significant. As the noise radiation depends on the roughness of both the wheel and track, it is possible that a rough wheel causes a

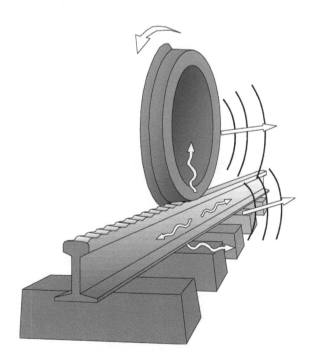

FIGURE 14.2 Schematic view of how rolling noise is generated at the wheel-rail interface. (From Iwnicki, S. (ed.), *Handbook of Railway Vehicle Dynamics*, CRC Press, Boca Raton, FL, 2006. With permission.)

high noise level that is mainly radiated by the track vibration or vice versa. It is therefore difficult to assign noise contributions solely to the vehicle or infrastructure.

14.2.2 Surface Roughness

Irregularities with wavelengths between about 5 mm and 500 mm cause the vibration of relevance to noise. When a wavelength λ (in m) is traversed at a speed V (in m/s) the associated frequency generated (in Hz) is given by

$$f = \frac{V}{\lambda} \tag{14.4}$$

The corresponding amplitudes range from over 100 μm at long wavelengths to much less than 1 μm at short wavelengths. Examples of wheel roughness spectra are shown in Figure 14.3 [6]. These are given in decibels relative to 1 μm (using a definition equivalent to Equation 14.2):

$$L_r = 10\log_{10}\left(\frac{r_{rms}^2}{r_{ref}^2}\right) = 20\log_{10}\left(\frac{r_{rms}}{r_{ref}}\right) \tag{14.5}$$

where r is the roughness amplitude and expressed in one-third octave bands over wavelength. Each line represents the average of between 50 and 100 wheels.

In the TSI Noise [1] and ISO 3095 [7], a limit curve is included for the roughness of a test track. For the characterisation of new vehicles, the rail roughness should be less than this limit curve, which is also shown in Figure 14.3. This represents good-quality track. The purpose of this in the context of [1,7] is to ensure that variations in rail roughness from one site to another do not significantly affect the measurement. The measurement method for rail roughness is defined in [8].

The wheel-rail contact does not occur at a point but over a small area; see Chapter 7. The contact patch is typically 10- to 15-mm long and of a similar width. When roughness wavelengths are short compared with the contact patch length, their effect on the wheel-rail system is attenuated.

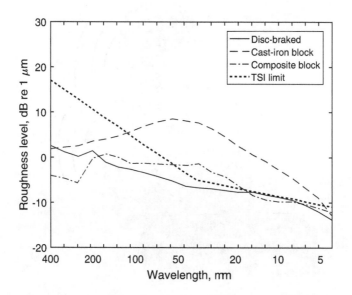

FIGURE 14.3 Typical wheel roughness spectra (data from [6]); also shown is the maximum rail roughness allowed for vehicle noise measurements according to TSI Noise [1] and ISO 3095 [7].

This effect is known as the contact filter. This is significant from about 1–1.5 kHz upwards for a speed of 160 km/h and at lower frequencies for lower speeds.

In early analytical models for this effect [9], the extent of correlation of the roughness across the width of the contact had to be assumed, since very detailed roughness data were not available. The following approximate formula can be used [5]:

$$|H(k)|^2 = \left(1 + \frac{\pi}{4}(ka)^3\right)^{-1} \tag{14.6}$$

where $k = 2\pi/\lambda$ is the roughness wavenumber in the longitudinal direction, λ is the wavelength and the length of the contact patch in the rolling direction is $2a$. Later, Remington developed a numerical discrete point reacting spring (DPRS) model [10]. This model is intended to be used with roughness measurements obtained on multiple parallel lines a few millimetres apart. Figure 14.4 shows the average results obtained using a series of such measurements in combination with the DPRS model [5]. This is compared with the results from Equation 14.6, which confirms the validity of the analytical model at low frequencies but indicates that the filtering effect is less severe at high frequencies than what the analytical model suggests.

14.2.3 Wheel Dynamics

A railway wheel is a lightly damped, resonant structure, which when struck rings like a bell, a structure that it strongly resembles. As with any structure, the frequencies at which it vibrates freely are called its natural frequencies, and the associated vibration pattern is called the mode shape.

Wheels are usually axisymmetric (although the web is sometimes not). Their normal modes of vibration can therefore be described in terms of the number of diametral node lines – lines at which the vibration pattern has a zero. A flat disc, to which a wheel can be approximated, has out-of-plane modes that can be described by the number of nodal diameters, n, and the number of nodal circles, m. A flat disc also has in-plane radial modes and circumferential modes, each with n nodal diameters. In-plane modes with nodal circles occur for railway wheels above 6 kHz.

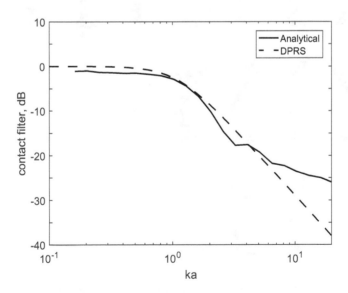

FIGURE 14.4 Contact filter effect from numerical DPRS model (average result from six wheels) and analytical result from Equation 14.6.

A railway wheel differs from a flat disc, having a thick tyre region at the perimeter and a thick hub at the centre connecting the wheel to the axle. A railway wheel is also not symmetric about a plane perpendicular to its axis. The tyre region is asymmetric due to the flange, and the web is usually also asymmetric, at least on wheels designed for tread braking, with the curved web being designed to allow for thermal expansion. An important consequence of this asymmetry is that radial and out-of-plane (axial) modes are coupled.

The finite element method (FEM) can be used quite effectively to calculate the natural frequencies and mode shapes of a railway wheel. Figure 14.5 shows an example of results for an Union International Union of Railways (UIC) 920-mm-diameter freight wheel [11]. The cross-section

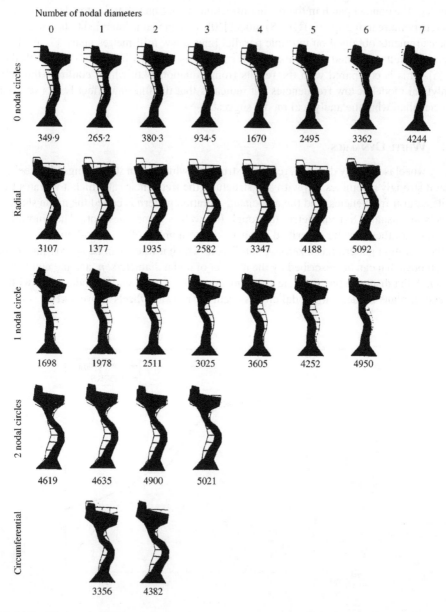

FIGURE 14.5 Modes of vibration and natural frequencies (in Hz) of UIC 920-mm freight wheel calculated using finite elements. (From Thompson, D.J., and Jones, C.J.C., *J. Sound Vib.*, 231, 519–536, 2000. With permission.)

Noise and Vibration from Railway Vehicles

through the wheel is shown, along with an exaggerated form of the deformed shape in each mode of vibration. Each column contains modes of a particular number of nodal diameters, n. The first row contains axial modes with no nodal circle. These have their largest out-of-plane vibration at the running surface of the wheel. These modes are usually excited in curve squeal (see Section 14.5) but are not excited significantly in rolling noise. The second and third rows contain radial modes and one-nodal-circle axial modes. Owing to the asymmetry of the wheel cross-section, and their proximity in frequency, these two sets of modes are strongly coupled; that is, they both contain axial *and* radial motion. It is these modes that are most strongly excited by roughness during rolling on a straight track, due to their radial component at the wheel-rail contact point.

The modes shown in Figure 14.5 are those of the wheel alone, constrained rigidly at the inner edge of its hub. The first column of modes, $n = 0$, is in practice coupled to extensional motion in the axle, and the second set, $n = 1$, is coupled to bending motion in the axle. As a result of this coupling with the axle, which is constrained by the roller bearings within the axleboxes, these sets of modes experience greater damping than the modes with $n \geq 2$. The latter do not involve deformation of the axle and are therefore damped only by material losses; their modal damping ratios are typically about 10^{-4} [5].

In order to couple the wheel to the track in a theoretical model, the frequency response functions of the wheel at the interface point are required. These may be expressed in terms of receptance, the vibration displacement due to a unit force as a function of frequency. Alternatively, mobility, the velocity divided by force, or accelerance, the acceleration divided by force, can be used. Such frequency response functions of a structure can be constructed from a modal summation. For each mode, the natural frequency f_{mn} is written as a circular frequency $\omega_{mn} = 2\pi f_{mn}$. Then, the response at circular frequency ω, in the form of a receptance α_{jk}^W is

$$\alpha_{jk}^W = \sum_{n,m} \frac{\psi_{mnj}\,\psi_{mnk}}{m_{mn}(\omega_{mn}^2 - \omega^2 + 2i\zeta_{mn}\omega\omega_{mn})} \tag{14.7}$$

where:

ψ_{mnj} is the mode shape amplitude of mode m, n at the response position
ψ_{mnk} is the mode shape amplitude of mode m, n at the force position
m_{mn} is the modal mass of mode m, n, a normalisation factor for the mode shape amplitude
ζ_{mn} is the modal damping ratio of mode m, n
i is the square root of -1

Figure 14.6 shows the radial point mobility of a wheelset calculated using the normal modes from an Finite Element Method (FEM), as shown in Figure 14.5. This is based on Equation 14.7 multiplied by $i\omega$ to convert from receptance to mobility. At low frequencies, the magnitude of the mobility is inversely proportional to frequency, corresponding to mass-like behaviour. Around 500 Hz, an anti-resonance trough appears, and above this frequency, the curve rises in stiffness-like behaviour until a series of sharp resonance peaks are reached at around 2 kHz. These peaks correspond to the axial one-nodal-circle and radial sets of modes identified in Figure 14.5.

14.2.4 Track Dynamics

The dynamic behaviour of track is described in detail in Chapter 9. A typical track mobility is also shown in Figure 14.6. This is predicted using a model based on a continuously supported rail, which neglects the effects of the periodic support. A broad peak at around 100 Hz corresponds to the whole track vibrating on the ballast. At the second peak, at about 500 Hz, the rail vibrates on the rail pads. The frequency of this peak depends on the rail pad stiffness. Above this frequency, bending waves propagate in the rail and can be transmitted over quite large distances.

The degree to which these waves are attenuated, mainly due to the damping effect of the pads and fasteners, affects the noise radiation from the rail. The method of measuring these track

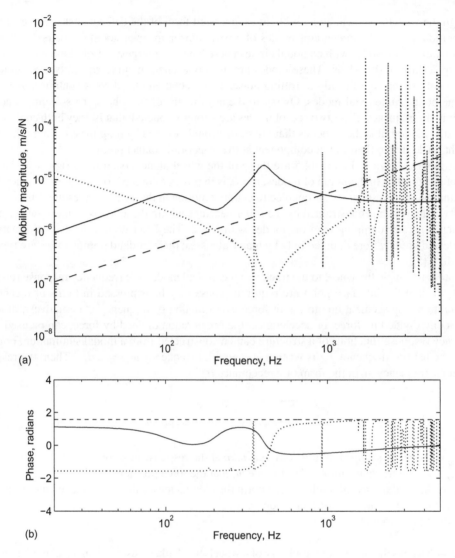

FIGURE 14.6 Vertical mobilities of the wheel-rail system, showing ······ radial mobility of UIC 920-mm freight wheel, —— vertical mobility of track with moderately soft pads and ——— contact spring mobility. (a) Magnitude (b) Phase. (From Thompson, D., *Railway Noise and Vibration: Mechanisms, Modelling and Means of Control*, Oxford, UK, 2009. With permission.)

decay rates is standardised in [12]. Figure 14.7 shows measured decay rates of vertical vibration for three different rail pads installed in the same track. The results for the middle value of pad stiffness correspond to the mobility in Figure 14.6. The vertical bending waves are strongly attenuated in a region between 300 and 800 Hz that depends on the pad stiffness. This peak in the decay rate corresponds to the region between the two resonance peaks in Figure 14.6. Here, the sleeper mass vibrates between the pad and ballast springs and acts as a 'dynamic absorber' to attenuate the propagation of waves in the track. Also shown in Figure 14.7 is the limit curve from ISO 3095 [7]; it can be seen that only the result for the stiff pad exceeds this limit. The attenuation of lateral waves is generally smaller than that for the vertical direction, and a different limit applies.

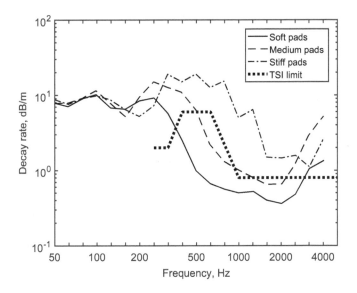

FIGURE 14.7 Measured decay rate of vertical vibration along the track for three different rail pads; also shown is the TSI limit [7].

14.2.5 WHEEL-RAIL INTERACTION

The wheel and rail are coupled dynamically at their point of contact. Between them, local elastic deflection occurs to form the contact patch, which can be represented as a contact spring. Although this spring is non-linear (see Chapter 6), for small dynamic deflections, it can be approximated by a linearised stiffness, k_H [13]. This is shown as contact spring mobility ($= i\omega/k_H$) in Figure 14.6.

The coupled wheel-rail system is excited by the roughness, which forms a relative displacement input; see Figure 14.8. Here, the forward motion of the wheel is ignored, and the system is replaced by one in which the wheel is static and the roughness is pulled between the wheel and rail ('moving irregularity model'). Considering only coupling in the vertical direction, from equilibrium of forces and compatibility of displacements, the vibration amplitude of the wheel (u_W) and rail (u_R) at a particular frequency can be written as

$$u_W = \frac{\alpha_W \, r}{\alpha_W + \alpha_R + \alpha_C}; \quad u_R = \frac{-\alpha_R \, r}{\alpha_W + \alpha_R + \alpha_C} \quad (14.8)$$

where r is the roughness amplitude and α_W, α_R and α_C are the vertical receptances of the wheel, rail and contact spring, respectively. Clearly, where the rail receptance has a much larger magnitude than that of the wheel or contact spring, $u_R \approx -r$; that is, the rail is pushed down at the amplitude of the roughness. From Figure 14.6, this can be expected between about 100 Hz and 1000 Hz. Changing the rail receptance in this frequency region has little effect on the rail vibration at the contact point (although the changes may affect the decay rates).

In practice, coupling also exists in other directions as well as the vertical, notably the lateral, direction [14]. This modifies Equation 14.8 to yield a matrix equation, but the principle remains the same.

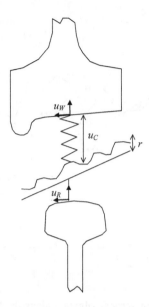

FIGURE 14.8 The wheel-rail contact showing excitation by roughness of amplitude r. (From Iwnicki, S. (ed.), *Handbook of Railway Vehicle Dynamics*, CRC Press, Boca Raton, FL, 2006. With permission.)

14.2.6 Noise Radiation

The vibrations of the wheel, rail and sleepers all produce noise. In general, the sound power W_{rad} radiated by a vibrating surface of area S can be expressed as [2]:

$$W_{rad} = \rho_0 c_0 S \sigma \langle \overline{v^2} \rangle \tag{14.9}$$

where $\langle \overline{v^2} \rangle$ is the spatially averaged mean-square velocity normal to the vibrating surface, ρ_0 is the density of air, c_0 is the speed of sound in air and σ is a frequency-dependent factor called the radiation efficiency. Thus, components radiate large amounts of noise if their vibration is large and/or their surface area is large and/or their radiation efficiency is high. The radiation efficiency is usually close to unity at higher frequencies and much smaller at low frequencies (where the radiating object is small compared with the wavelength of sound). Predictions of this factor can be obtained using numerical methods such as the boundary element method or, for simple cases, analytical models. Models for the wheel radiation are presented in [15] and for the rail radiation are presented in [16].

Figure 14.9 shows predictions of the noise from wheels, rails and sleepers during the passage of a pair of similar bogies. This is shown in the form of the average sound pressure level at a location close to the track (3 m from the nearest rail). The wheel is the most important source of noise at high frequencies, above about 1.6 kHz. From Figure 14.6, it can be seen that this corresponds to the region in which many resonances are excited in the radial direction. Between about 400 Hz and 1600 Hz, the rail is the dominant source of noise. Here, the rail vibrates at the amplitude of the roughness. The support structure affects the rate of decay with distance and hence the spatially averaged velocity. At low frequencies, the sleeper radiates the largest component of noise. Here, the rail and sleeper are well coupled and have similar vibration amplitudes, but the sleeper has a larger area and a radiation efficiency close to unity, whereas that of the rail reduces below 1 kHz.

Noise and Vibration from Railway Vehicles

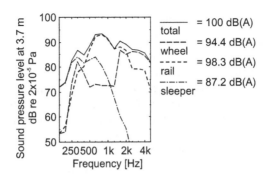

FIGURE 14.9 Predicted noise components from the wheels, rails and sleepers for 920-mm diameter freight wheels at 100 km/h on a track with moderately soft rail pads. (From Iwnicki, S. (ed.), *Handbook of Railway Vehicle Dynamics*, CRC Press, Boca Raton, FL, 2006. With permission.)

Although the details of Figure 14.9 are specific to this combination of wheel and track design, train speed and roughness spectrum, it is generally the case that the most important source is formed by the sleepers at low frequencies, the rails in the mid frequencies and the wheels at high frequencies. As speed increases, the energy in the noise spectrum shifts towards higher frequencies, leading to a greater importance of the wheel in the overall level.

14.2.7 Overall Model

The complete model for rolling noise that has been described in Sections 14.2.1 to 14.2.6 has been implemented in a software package, Track-Wheel Interaction Noise Software (TWINS) [17], which is widely used in the railway industry. This is a frequency-domain model based on the moving irregularity formulation. It produces estimates of sound power and sound pressure spectra in one-third octave bands and allows the user to study the effect of different wheel and track designs on noise.

This model has also been the subject of extensive validation [18,19]. Comparisons between predictions and measurements for three track types, three wheel types and four speeds gave overall sound levels that agreed within about ±2 dB [18]. These predictions were updated in [19] along with new measurements for a range of novel constructions. Revisions to the software have improved agreement slightly. Agreement in one-third octave bands had a larger spread of around ±4 dB, but this was at least partly due to uncertainties in the measured roughness inputs.

14.3 REDUCING ROLLING NOISE

From the theoretical understanding, it is clear that rolling noise can be reduced by:

- Controlling the surface roughness
- Minimising the vibration response of wheels and tracks by adding damping treatments, by shape optimisation of wheels or rails or by introducing vibration isolation
- Preventing sound radiation, for example, by using local shielding measures

In each case, attention must be given to the presence of multiple sources. If more than one source is important, overall reductions will be limited, unless all sources are controlled. For example, if there are initially two sources (wheel and track) that contribute equally and one of them is reduced by 10 dB without affecting the other, the overall reduction will be limited to 2.5 dB.

14.3.1 CONTROLLING SURFACE ROUGHNESS

From the vehicle designer's point of view, the main feature affecting the wheel roughness is the braking system. Traditional tread brakes, in which cast-iron brake blocks act on the wheel tread, lead to the development of high levels of roughness on the wheel running surfaces due to the formation of local hot spots. This can be seen from Figure 14.3, the greatest differences in roughness being at the peak at around 6-cm wavelength. This high roughness in turn leads to higher levels of rolling noise. With the introduction of disc-braked vehicles, for example, the Mk III coach in the UK in the mid-1970s, it became apparent that disc braking can lead to quieter rolling stock. The difference in rolling noise between the Mk III and its tread-braked predecessor, the Mk II, was about 10 dB, mainly due to the difference in roughness. Modern passenger rolling stock is mostly disc-braked for reasons of braking performance, and this brings with it lower noise levels than older stock.

However, environmental noise is usually dominated by freight traffic. Freight vehicles are generally noisier and often run at night when environmental noise limits are tighter. For freight traffic in Europe, a number of factors have meant that cast-iron brake blocks have remained the standard until recently. These include cost, the longevity of wagons (typically 50 years) and, most importantly, the UIC standards for international operation, which required the use of such brakes. However, since 1999, the UIC has been pursuing an initiative to replace cast-iron blocks with alternative materials [20]. The idea is to introduce blocks made of a composite material that do not produce hot spots and therefore leave the wheel relatively smooth. So-called K-blocks and some types of LL-blocks can give a noise reduction of typically 8–10 dB(A) compared with cast-iron blocks on a TSI-compliant track [21]. On track with higher roughness levels, the improvements are more modest.

Rail corrugation is also a source of increased noise. A corrugated track can be up to 20 dB noisier than a smooth one for disc-braked wheels. Grinding of the rail for acoustic purposes is carried out in, for example, Germany to maintain special low noise sections of track [22].

14.3.2 WHEEL-BASED SOLUTIONS

14.3.2.1 Wheel Damping

One means of reducing the amount of noise radiated by the wheels is to increase their damping. Impressive reductions in the reverberation of wheels can be achieved by simple damping measures. However, a wheel in rolling contact with the rail already has, in effect, considerably more damping than a free wheel, since vibration energy flows from the wheel into the track. To improve the rolling noise performance, the added damping must exceed this effective level of damping already present, which is one to two orders of magnitude higher than that of the free wheel.

Various devices have been developed to increase the damping of railway wheels by absorbing energy from their vibrations, thereby reducing the noise produced. Examples are shown schematically in Figure 14.10. These include multi-resonant absorbers (Figure 14.10a), which have been used in Germany since the early 1980s and are fitted to many trains, including the ICE-1. Noise reductions of 5–8 dB have been claimed for speeds of 200 km/h [23]. Another commercial form of damper involves multiple layers of overlapping plates known as the shark's fin damper (Figure 14.10b). Färm [24] found reductions of 1–3 dB(A) overall, associated with wheel noise reductions of 3–5 dB(A). Constrained layer damping treatments (Figure 14.10c) consist of a thin layer of visco-elastic material applied to the wheel and backed by a thin stiff constraining layer (usually metal). Such a treatment was used on the Class 150 Diesel Multiple Unit (DMU) in the UK in the late 1980s and was applied to the whole vehicle fleet to combat a particularly severe curve squeal problem excited by contact between the wheel flange and the check rail. By careful design, sufficient damping can be achieved using constrained layer damping to make significant reductions in rolling noise as well [25,26].

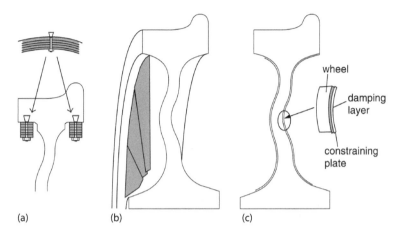

FIGURE 14.10 Various wheel damping devices used on railway wheels: (a) tuned resonant devices, (b) shark's fin dampers, and (c) constrained layer damping. (From Iwnicki, S. (ed.), *Handbook of Railway Vehicle Dynamics*, CRC Press, Boca Raton, FL, 2006. With permission.)

14.3.2.2 Wheel Design

Reductions in the wheel component of radiated noise can also be achieved by careful attention to the wheel cross-sectional shape. In recent years, manufacturers have used theoretical models such as TWINS [17] to assist in designing wheels for low noise.

As an example of the difference that the cross-sectional shape can have, three wheels are shown in Figure 14.11. Wheel (a) is a German Intercity wheel, (b) is a UIC standard freight wheel (cf. Figures 14.5 and 14.6) and (c) was designed several years ago by the Technical University of Berlin on the basis of scale model testing [23]. Figure 14.11 also shows the predicted noise components from the wheel in each case. The track component of noise (not shown) is not affected by these changes and remains the dominant source up to 1 kHz.

These results show that a straight web (wheel (c)) is beneficial compared with a curved web (wheel (b)). This is because the radial and axial motions are decoupled for a straight web. However, it is not always possible to use straight webs if tread brakes are used, as the curve is included in the web to allow thermal expansion. Wheel (a) is particularly noisy, the main difference between this and wheel (c) being the transition between the inside of the tyre and the web and the web thickness. Increasing the web thickness and particularly the transition between the tyre and web are effective means of reducing noise but also lead to increased unsprung mass. Wheels with profiles similar to (a) have shown appreciable rolling noise reductions by the addition of absorbers, whereas wheels such as (b) have shown much smaller reductions.

Another aspect of wheel design that can be used to reduce noise is the diameter. Smaller wheels have higher natural frequencies, so it is possible by reducing the diameter to move many of the resonances out of the range of excitation (i.e., above about 5 kHz) [27]. The upper frequency is somewhat increased for a smaller wheel due to a shift in the contact patch filter, but this effect is much less significant than the shift in natural frequencies. The trend in recent years towards smaller wheels for other reasons is therefore advantageous for noise. This also reduces the unsprung mass. However, if the wheel size is reduced too much, the track noise will increase due to the reduction in the contact filter effect.

14.3.3 TRACK-BASED SOLUTIONS

14.3.3.1 Low-Noise Track

To achieve significant reductions in overall noise, it is usually not sufficient to deal only with the wheel noise. There must be a corresponding reduction in noise from the track vibration. Two very

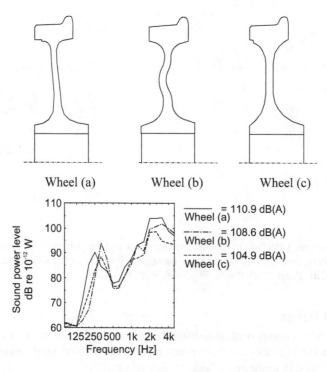

FIGURE 14.11 Top: three wheel designs. Bottom: TWINS predictions of wheel sound power from three types of wheel at a train speed of 160 km/h for the same roughness spectrum in each case, typical of a disc-braked wheel. (From Iwnicki, S. (ed.), *Handbook of Railway Vehicle Dynamics*, CRC Press, Boca Raton, FL, 2006. With permission.)

important parameters of the track that affect its noise emission and that are related to one another are the stiffness of the rail pad and the decay rate of vibration along the rail. A stiff rail pad causes the rail and sleeper to be coupled together over a wide frequency range. Conversely, a soft pad isolates the sleeper for frequencies above a certain threshold. The lower the stiffness of the rail pad, the lower this threshold frequency. Soft rail pads therefore effectively isolate the sleepers and the foundation from the vibration of the rail, reducing the component of noise radiated by vibration of the sleepers. Part of the designed role of the rail pad is to protect the sleeper and ballast from high-impact forces. For this reason, softer rail pads have become more commonplace in recent years. Unfortunately, softer rail pads also cause the vibration of the rail to propagate with less attenuation (see Figure 14.7). As a greater length of rail vibrates with each wheel, this means more noise is generated by the rail, as shown in Figure 14.12. There is thus a compromise to be sought between the isolating and attenuating properties of the rail pad [28].

The EU-funded research project Silent Track successfully developed and demonstrated low-noise technology for the track [5]. The most successful element was a rail damper. Multiple blocks of steel are fixed to the sides of the rail by an elastomer and tuned to give a high damping effect in the region of 1 kHz. This allows a soft rail pad to be used, to give isolation of the sleepers, whilst minimising the propagation of vibration along the rail [29]. Figure 14.13 shows the noise reduction achieved in the field tests. In this case, a low-noise wheel was used for comparison to minimise the effect of the wheel on the total noise, but, even so, some wheel noise was present at high frequencies. The overall reduction in track noise was approximately 6 dB. Various rail dampers have been developed over the last 20 years. A standardised test method was therefore developed to allow their performance to be evaluated on the basis of laboratory tests on a short length of rail [30].

Noise and Vibration from Railway Vehicles

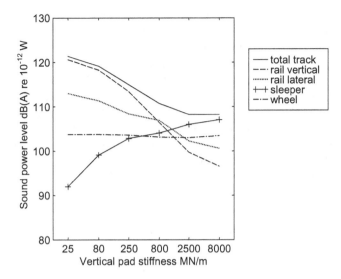

FIGURE 14.12 Effect of rail pad stiffness on predicted components of rolling noise. (From Iwnicki, S. (ed.), *Handbook of Railway Vehicle Dynamics*, CRC Press, Boca Raton, FL, 2006. With permission.)

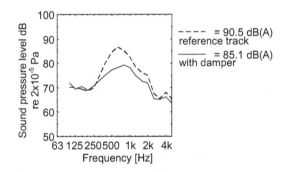

FIGURE 14.13 Measured noise reduction from Silent Track rail damper during the passage of a low-noise wheel at 100 km/h. (From Thompson, Jones, Waters and Farrington, A tuned damping device for reducing noise from railway track, Applied Acoustics, 68(1), 43–57, 2007. With permission.)

Tests with an 'optimum' pad stiffness have been less successful. Although the effect of pad stiffness has been clearly demonstrated in field tests, the optimum for noise radiation is too stiff to be acceptable for other reasons, particularly track damage protection. Stiff pads are also believed to lead to a higher likelihood of corrugation growth, which, in the long term, has a negative effect on the noise. The analysis of the acoustic performance of pads with different stiffnesses is further complicated by their load-dependent characteristics and other factors such as temperature variation [31].

14.3.3.2 Slab Tracks

Tracks mounted on concrete slabs have become more commonplace in the last few years, notably on high-speed lines. Such tracks are generally found to be noisier than conventional ballasted track, typically by around 3 dB. This can be attributed to two features of such tracks. Firstly, they tend to be fitted with softer rail fasteners in order to introduce the resilience normally given by the ballast; this leads to lower track vibration decay rates. Secondly, they have a hard sound-reflecting surface, whereas ballast has an absorptive effect. Although the latter only affects the overall noise by around 1 dB, it has been shown recently that the difference in ground absorption below the rail also affects the radiated sound power from the rail [32].

A number of mitigation measures have been introduced, in which absorbent material is added to the upper surface of the slab. This has the effect of reducing the reflections of sound from the slab surface. Where it is also possible to introduce some shielding of the rail noise, for example, by an integrated mini-barrier, additional attenuation is possible. Such treatments have been found to reduce noise levels from slab track back to those of ballasted track.

For street running trams, a number of embedded rail systems are used. At first sight, an embedded rail might be expected to be silent, as the rail is mostly hidden and therefore should not produce sound. In practice, the rail head is visible, and both the rail head and the embedding material around it vibrate and produce sound [33]. Embedded rail systems offer the possibility of higher rail attenuation rates, due to the damping effect of the embedding material around the rail. They can also be constructed with relatively soft supports and therefore offer the potential to produce good vibration isolation.

14.3.4 LOCAL SHIELDING AND BARRIERS

Conventional noise barriers at the trackside are used widely in many countries. Reductions of 10–20 dB are achievable, depending on the height of the barriers, but they are expensive and visually intrusive, especially if taller than about 2 m. Cost-benefit studies have shown that noise reduction at source can be cost-effective compared with barriers or, in combination, can allow the use of lower barriers for the same overall effect [34].

The efficiency of a barrier is improved by placing it as close as possible to the source. In the Silent Freight project, it was demonstrated that, at least for certain types of wheel, a shield mounted on the wheel covering the web can reduce the noise. A more general solution is to place an enclosure around the bogie. If used in combination with low barriers very close to the rail, reductions of up to 10 dB can be achieved [35].

Bogie shrouds and low barriers were also tested in the Silent Freight and Silent Track projects [5], but in this case, the objective was to find a combination that satisfied international gauging constraints. Unfortunately, this meant that the overall reduction was limited to less than 3 dB, owing to the inevitable gap between the top of the barrier and the bottom of the shroud [36]. There are many other practical difficulties in enclosing the bogies, such as ventilation for the brakes and access for maintenance. Nevertheless, such vehicle-mounted screens are common on trams.

14.4 IMPACT NOISE AND VIBRATION

14.4.1 INTRODUCTION

In the previous sections, noise due to random irregularities on the railhead and wheel tread has been considered. In addition to this, larger discrete features occur on the running surfaces such as rail joints, gaps at points and crossings, dipped welds and wheel flats. These cause high interaction forces and, consequently, noise. In some cases, loss of contact can occur between the wheel and rail, followed by large impact forces. Noise from such discrete features is often referred to as impact noise. Whereas rolling noise can be predicted using a linearised contact spring, in order to predict impact forces and noise, the non-linear contact stiffness must be included (see Chapter 6).

Early models for impact noise were essentially empirical [37]. To predict impact forces, time-domain models incorporating the non-linearities in the contact zone have been used, for example, by Clark et al. [38] and Nielsen and Igeland [39]. These models contain large numbers of degrees of freedom to represent the track. Nevertheless, they are limited to a maximum frequency of around 1500 Hz. In order to model impact noise up to around 5 kHz, simplified models of the wheel and rail have been used in a time-stepping model in order to determine the effects of the non-linearities [40]. These are then used in a hybrid approach with the TWINS model [17] to predict the noise radiation.

14.4.2 Wheel Flats

A wheel flat is an area of the wheel tread that has been worn flat. This usually occurs when the brakes lock up under poor-adhesion conditions at the wheel-rail contact due, for example, to leaves on the railhead in the autumn. Wheels with flats produce high levels of noise and impact loading of the track, which can lead to the damage of track components (see Chapter 9). Typically, flats can be around 50-mm long and in extreme cases up to 100 mm. After their initial formation, flats become 'worn', that is, rounded at their ends due to the high load concentration on the corners. A worn flat of a given depth is longer than the corresponding 'new' flat.

Wheel flats introduce a relative displacement input to the wheel-rail system in the same way as roughness. Figure 14.14 shows examples of the calculated response of the wheel-rail system to a new wheel flat of depth 2 mm (length 86 mm) for a nominal wheel load of 100 kN. The model used here represents the wheel by a mass and spring and the track by a simple state-space model fitted to the track mobility [40].

When the indentation (relative displacement input due to the wheel flat) appears between the wheel and rail, the wheel falls and the rail rises. Since the wheel and rail cannot immediately follow the indentation due to their inertia, the contact force is partly unloaded. At a train speed of 30 km/h, see Figure 14.14a, full unloading occurs first. After the relative displacement input reaches its maximum, the contact force increases rapidly until it reaches its peak. The peak force in this example is about four times as large as the static load. As the speed increases, contact is lost for longer periods during the unloading phase. At 80 km/h, see Figure 14.14b, a second loss of contact can be seen to occur. However, the second impact is much smaller than the first one. Comparisons with measured impact forces [41] suggest that the simplified geometry used here leads to over-estimates of the contact force. Measured wheel flat profiles are required to give more accurate predictions.

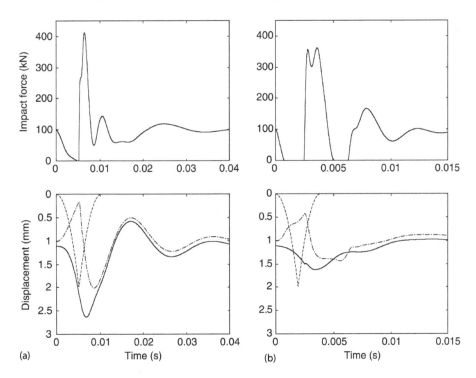

FIGURE 14.14 Predicted wheel-rail interaction force and displacements of wheel and rail due to 2-mm newly formed wheel flat at train speed of (a) 30 km/h and (b) 80 km/h, showing —— wheel displacement, – · – · rail displacement, and ···· relative displacement excitation. (From Wu, T.X., and Thompson, D.J., *J. Sound Vib.*, 251, 115–139, 2002. With permission.)

It is not possible to use the contact force obtained from the impact model and apply it directly within the TWINS model, because the predicted interaction force is very sensitive to details of the wheel and track dynamics used in its prediction. With a modal wheel model, the force spectrum will have strong dips at the wheel's natural frequencies. The wheel response has only shallow peaks, just above the natural frequencies. The interaction with the track thereby introduces apparent damping to the wheel. A hybrid approach has therefore been developed [40], whereby an equivalent roughness spectrum is derived. The equivalent roughness spectrum can then be used as the input to a more detailed linear frequency-domain model, such as the TWINS model, to predict the noise due to the impact.

Example results are given in Figure 14.15a. This shows the sound power due to one wheel and the associated track vibration for a 2-mm-deep new wheel flat at different speeds for a 100-kN wheel load. Results correspond to the average over one whole wheel revolution. Figure 14.15b shows, for comparison, corresponding results for roughness excitation due to a moderate roughness (tread-braked wheel roughness). As the speed increases, the noise at frequencies above about 200–400 Hz increases in both cases. The increase in rolling noise with increasing speed is greater than that due to the flat. For the wheel flats considered here, the noise generated exceeds that due to the tread-braked wheel roughness at all speeds and in all frequency bands; however, the noise due to roughness increases more rapidly with speed, so that, at sufficiently higher speeds, it can be expected to dominate. For corrugated track, the noise due to roughness would exceed that due to these wheel flats at 120 km/h.

Figure 14.16 shows a summary of the variation of the overall A-weighted sound power level with train speed. The predicted noise level due to conventional roughness excitation increases at a rate of approximately $30 \log_{10} V$, where V is the train speed, whereas the noise due to flats increases at an average of around $20 \log_{10} V$ once loss of contact occurs. For example, loss of contact was found to occur for the newly formed 2-mm-deep flat at speeds above 30 km/h and for a rounded 2-mm-deep flat above 50 km/h. This variation with speed indicates that the radiated sound due to wheel flats continues to increase with increasing speed, even though loss of contact is occurring.

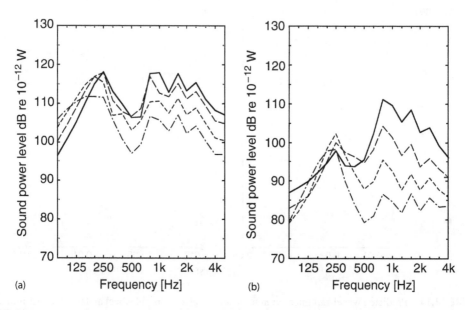

FIGURE 14.15 Sound power level due to wheel and track for (a) 2-mm new wheel flat and (b) rolling noise from moderate roughness, for speeds of — · — · 30 km/h, ····· 50 km/h, - - - 80 km/h, and ——— 120 km/h. (From Wu, T.X., and Thompson, D.J., *J. Sound Vib*, 251(1), 115–139, 2002. With permission.)

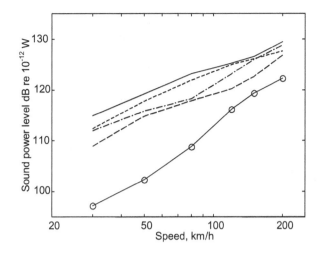

FIGURE 14.16 Sound power radiated by one wheel and the associated track vibration, showing — — — 1-mm rounded flat, ······ 2-mm rounded flat, — · — · 1-mm new flat, ——— 2-mm new flat, o ——— rolling noise due to tread-braked wheel roughness. (From Wu, T.X., and Thompson, D.J. *J. Sound Vib.*, 251(1), 115–139, 2002. With permission.)

Impact noise from wheel flats is found to depend on the wheel load. The increase in noise between a load of 50 kN and 100 kN is about 3 dB. In contrast, the rolling noise due to roughness is relatively insensitive to wheel load.

14.4.3 RAIL JOINTS, SWITCHES AND CROSSINGS

In a similar way to wheel flats, rail joints provide discrete inputs to the wheel-rail system that induce quite large contact force variations. Rail joints can be characterised by a gap width and a step height (either up or down). Moreover, the rail often dips down to a joint on both sides. Such dips are also present at welds and are usually characterised in terms of the angle at the joint.

A similar approach has been used, as previously mentioned, to study the effects of rail joints [42,43]. The sound radiation was calculated using the same hybrid method as for the wheel flats. It was found, for realistic parameter values, that the gap width is insignificant compared with the step height and dip angle.

Results are shown in Figure 14.17a for un-dipped rail joints in the form of the total A-weighted sound power emitted by the wheel and rail during 1/8 sec. The results for a step-down joint are found to be virtually independent of the step height (only results for one value are shown) and also change very little with train speed. However, for step-up joints, both the peak contact force and the sound power level increase with step height and with train speed. The sound power level from a single joint has a speed dependence of around $20\log_{10}V$.

In Figure 14.17b, results are given for dipped joints with no height difference. Here, a dip of 5 or 10 mm is considered as a quadratic function over a length of 0.5 m either side of the joint. A dip of 5 mm corresponds to a joint angle of 0.04 radians, which is large, although within a typical range, and a dip of 10 mm corresponds to 0.08 radians, which is severe. The 10-mm dip produces a similar noise level to a 1-mm step-up un-dipped joint, although for speeds above 120 km/h, the noise level from the dip joint becomes independent of train speed.

Figure 14.18 shows the predicted noise for joints with both dipped rails and steps. The noise radiation generally increases with speed, regardless of whether loss of contact occurs. For the 5-mm dip, the noise level increases by 8 dB when the step height increases from 0 to 2 mm. For the step-down joints, the noise level is higher than that without a step, although at higher speeds, the dip has

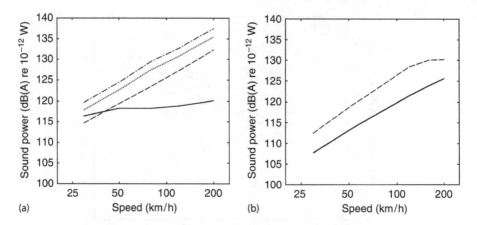

FIGURE 14.17 A-weighted sound radiated by one wheel and the associated track vibration during 0.125 second due to a wheel passing over different rail joints with 7-mm gap for (a) flat rail joints, showing — — — 1-mm step-up, ······ 2-mm step-up, · — · - 3-mm step-up, ——— 2-mm step-down, and (b) dipped rail joints with no height difference, showing ——— 5-mm dip, — — — 10-mm dip. (From Iwnicki, S. (ed.), *Handbook of Railway Vehicle Dynamics*, CRC Press, Boca Raton, FL, 2006. With permission.)

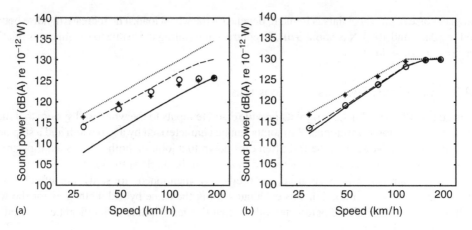

FIGURE 14.18 A-weighted sound power radiated by one wheel and the associated track vibration during 0.125 second due to a wheel passing over rail joints with 7-mm gap for (a) 5-mm dip, and (b) 10-mm dip, showing ······ 2-mm step-up, – – – 1-mm step-up, — no height difference, * 2-mm step-down, o 1-mm step-down. (From Iwnicki, S. (ed.), *Handbook of Railway Vehicle Dynamics*, CRC Press, Boca Raton, FL, 2006. With permission.)

more effect than the step. The results for the 10-mm dip are similar for both step-up and step-down joints, indicating the dominance of the dip in this case.

To compare these results with typical rolling noise results, the time base of the joint noise should be adjusted to the average time between joints. This shows [42] that rolling noise due to the tread-braked roughness considered above is similar to the average noise due to 5-mm dipped joints with no height difference (Figure 14.17b). With a height difference of 2 mm, the average noise predicted from the joints increases to almost 10 dB greater than the rolling noise. Moreover, since the time between rail joints decreases as train speed increases, it is also found that the average noise level from joints increases at about $30 \log_{10} V$, similar to rolling noise.

14.4.4 Reducing Impact Noise

To reduce impact noise, it is clearly desirable to remove the cause if this is possible. Wheel flats can be largely prevented by installation of wheel-slide protection equipment. Monitoring equipment is now widely used to identify wheels with flats, to allow them to be removed from service as quickly as possible for reprofiling. On main lines, jointed track has been mostly replaced by continuously welded rail in the last 40 years, although inevitably, joints remain such as expansion joints, track-circuit insulating joints and switches and crossings. Even so, measures such as swing-nose crossings allow the impact forces, and thus noise, to be minimised. Attention should also be given to ensuring that welded rail joints are as levelled as possible by using rail-straightening equipment.

Countermeasures that are effective for rolling noise, such as those discussed in Section 14.3, can be expected to work equally well for impact noise. This includes, for example, wheel damping, wheel shape optimisation, rail damping and local shielding.

14.5 CURVE SQUEAL

14.5.1 Mechanisms of Squeal Noise Generation

Railway vehicles travelling around tight curves can produce an intense squealing noise. This is a particular problem where curved track exists in urban areas, and it has been found to be annoying to both residents and railway passengers. An extensive review of curve squeal can be found in [44].

When a railway wheelset in a bogie traverses a curve, it is unable to align its rolling direction tangentially to the rail (14.19). Owing to this misalignment, lateral sliding occurs together with the natural rolling of the wheel. In sharp curves, this leads to large creep forces at the wheel-rail interface; see also Chapter 7. The leading outer wheel tends to be in flange contact, with the resultant lateral force acting inwards to ensure that the wheelset remains on the track. Longitudinal and spin creep forces also act as shown in Figure 14.19.

The presence of the creep force due to the sliding at the wheel-rail interface is the main reason for the unstable dynamic behaviour, leading to squeal noise. Two different mechanisms have been

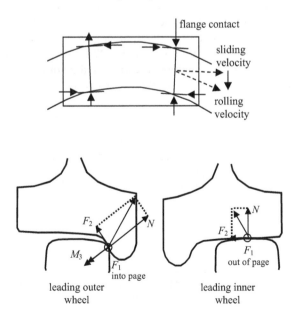

FIGURE 14.19 (Bottom half and top half) Schematic view of forces acting on wheels of a bogie in a curve; N is normal load, F_2 is lateral creep force, F_1 longitudinal creep force and M_3 is spin moment. (From Iwnicki, S. (ed.), *Handbook of Railway Vehicle Dynamics*, CRC Press, Boca Raton, FL, 2006. With permission.)

proposed to explain how the creep force can generate squeal. Originally, attention was focussed on the dependence of the 'creep curve' on the sliding velocity (also known as the Stribeck effect [45]), while, more recently, researchers have also focussed on instabilities due to coupling between normal and tangential directions.

Figure 14.20 shows a typical creep curve relating creep force to creepage. At low values of creepage, the magnitude of the creep force increases linearly. At high values of creepage, the force becomes saturated, with a maximum value of $\mu_0 N$, where μ_0 is the friction coefficient and N is the normal load. In practice, however, the friction coefficient μ is not a constant. It is usually recognised that 'dynamic' or 'sliding' friction coefficients are smaller than 'static' ones. In fact, the friction coefficient depends on the sliding velocity, decreasing as the velocity increases. Thus, as creepage increases beyond the saturation point, the creep force once more reduces in amplitude; see Figure 14.20b.

By analogy with a damper, which gives a reaction force that is proportional to the relative velocity, the falling creep curve can be considered as a 'negative damping'. Thus, the reaction force decreases as the relative velocity increases. Since wheel modes have very low levels of damping (see Section 14.2.3), if this negative damping exceeds a certain level, it causes instability of the wheel modes, making them prone to 'squeal'. In this case, unstable self-excited vibration occurs, and the vibration amplitude increases exponentially until it is bounded to a limit cycle by the non-linear effects in the creep force. The critical value of the structural damping can be calculated as:

$$\zeta_{\lim,n} = -\frac{N}{2m_n \omega_n V_0} \frac{d\mu}{d\gamma_L} \qquad (14.10)$$

with N being the normal load, m_n the modal mass of wheel mode n, ω_n its natural frequency, V_0 the rolling velocity of the train and γ_L the creepage in the lateral direction. For structural damping values smaller than the critical limit, self-excited vibration can occur in the wheel for the mode considered. In this approach, single-wheel modes can be excited, and the squealing frequency would be coincident with the natural frequency of the mode. Equation 14.10 highlights the importance of the slope of the friction curve for this type of excitation. However, the actual trend of the creep curve is usually not known, and assumptions need to be made for modelling purposes.

As an alternative to the negative damping effect, mode-coupling phenomena have attracted the interest of various researchers in the last decade; see [46,47] for examples. This type of instability arises from the coupling between two vibration directions and involves two adjacent modes. In the case of squeal, the non-conservative nature of the friction force can transfer energy between the

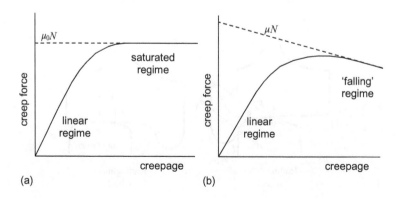

FIGURE 14.20 Typical creep force-creepage relationships for (a) constant friction coefficient and (b) velocity-dependent friction coefficient. (From Iwnicki, S. (ed.), *Handbook of Railway Vehicle Dynamics*, CRC Press, Boca Raton, FL, 2006. With permission.)

normal and tangential directions, leading to this type of instability, which is also known as 'flutter'. As a result, two adjacent modes will tend to merge, and the vibration of the system can build up at a frequency between those of the two vibration modes. The squealing frequency would therefore not necessarily correspond to any of the natural modes of the free wheel. Again, the non-linear nature of the creep force will result in a limit cycle. This type of instability does not require a negative slope of the friction curve. In its most simplified form, it can be illustrated by means of a two-degree-of-freedom system coupled to a moving belt through a simple Coulomb model for friction [48]. This is schematically shown in Figure 14.21, where the mass m and springs of stiffness k_1 and k_2 represent the structural properties of the system and k_H represents the contact stiffness of the wheel-track system.

For mode coupling, it is not easily possible to define threshold values to determine the presence of curve squeal. In fact, this depends on combinations of various parameters such as the friction coefficient, the modal damping ratios and the relative amplitude of the mode shapes in the different directions. A necessary condition [49], however, is that both modes have components in the normal and tangential directions at the contact point.

Various efforts have been made to demonstrate which of the two mechanisms is behind curve squeal, and arguments can be made in favour of one or the other. In various laboratory measurement campaigns [50,51], it was found that the creep curve in the lateral direction, measured in an average sense, showed clear falling trends, and this could be correlated to the presence of squeal. In addition, the measured squealing frequencies can often be attributed to single wheel modes. At the same time, it has also been demonstrated that numerical models relying solely on mode coupling can show good qualitative correlation with field tests [46], and, on some occasions, a clear frequency shift was found between the measured squealing frequency and the wheel's natural frequencies [49]. It is therefore likely that both mechanisms are important, with the possibility for one of the two to be predominant, depending on specific situations.

Several observations indicate that the highest squeal noise amplitude is often generated by the leading inner wheel of a four-wheeled bogie or two-axle vehicle. This noise can be associated with the instability mechanisms listed previously. The fundamental frequency of such squeal noise tends to correspond to a natural frequency of the wheel and is often in the range of 200–2000 Hz. The wheel modes excited in this case tend to be axial modes with no nodal circle, and their maximum amplitude is at the wheel tread (see first row of Figure 14.5).

Contact between the wheel flange and the rail, which occurs at the leading outer wheel (and possibly the trailing inner wheel) in sharp curves, has been initially found to reduce the likelihood of squeal in some cases [52]. For example, Remington concluded from laboratory experiments that flange contact reduces the level of squeal noise [53]. However, curve squeal has also been found for the outer wheels and was associated with flange contact [54–56], either with the outer running rail, with check rails (or grooved heads in tramways), or with wing rails in switches and crossings. Compared with squeal measured at the inner wheel, the cases found at the outer wheels had, in general, higher fundamental frequencies and were more intermittent in nature. In these cases,

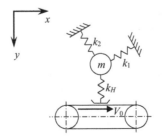

FIGURE 14.21 Simplified model representing mode-coupling mechanism.

additional factors have been studied and found to play a role in curve squeal. These are the presence of two contact points [57], spin creepage [58] and longitudinal creepage [59,60].

Theoretical models for curve squeal have been developed by various authors. Rudd [61] (see also Remington [53]) indicated that instability of the lateral friction force was the most likely cause of squeal and gave a simple model. Fingberg [62] and Périard [63] extended this basic model by including improved models of the wheel dynamics, the friction characteristic and the sound radiation from the wheel. Time-domain calculations allow the squeal magnitude to be predicted as well as the likelihood of squeal to be determined. Heckl [64] has also studied squeal using a simplified model and provided experimental validation using a small-scale model wheel.

De Beer et al. [50] extended these models, based on excitation by unstable lateral creepage, to include feedback through the vertical force as well as through the lateral velocity. Their model consists of two parts: a first part, in the frequency domain, can be used to determine instability and to predict which mode is most likely to be excited, and a second part, in the time domain, calculates the amplitude of the squeal noise. This model has been extended further to allow for an arbitrary contact angle and to include lateral, longitudinal and spin creepage [59]. This allows it to be applied to flange squeal as well as squeal due to lateral creepage.

Chiello et al. [65] developed a curve squeal model that accounts for both axial and radial dynamics to predict growth rates and unstable frequencies. The effect of mode coupling instability was discussed, but it was found that only a creep curve with negative slope would result in squeal.

Brunel et al. [66] introduced a transient model for curve squeal and found that even a positive friction law can lead to instability and limit cycle. This feature was associated with the coupling between the normal and lateral dynamics. More recently, Glocker et al. [46] and Pieringer [47] demonstrated that models with a constant friction coefficient can show squeal at relevant frequencies and give good qualitative comparisons with measured data.

14.5.2 Reducing Squeal Noise

In discussing solutions for curve squeal, it is of little value to quote decibel reductions. The nature of the instability is such that effective measures are those that eliminate the squeal rather than reducing it. Thus, noise barriers are generally ineffective against squeal noise. Curve squeal tests are also extremely unreproducible due to a high sensitivity to parameters such as temperature, humidity, train speed, track geometry, wheel and rail wear.

Known solutions for curve squeal include lubrication using either grease or water or the application of friction modifiers that reduce the difference between static and sliding friction coefficients. If lubricants are used, it must be ensured that they do not lead to loss of adhesion, as this could compromise safety. Grease is therefore only applied to the rail gauge corner or wheel flange. Although this location may not be the primary cause of squeal noise, this can nevertheless reduce the occurrence of squeal by modifying the curving behaviour. Water sprays have also been used effectively in a number of locations.

Friction modifiers act by reducing or eliminating the falling friction characteristic without reducing the level of friction. These can be applied either to the track at the entrance to a curve or on the vehicle. They have been shown to be very effective in eliminating squeal and can be applied to the top of the railhead without compromising traction or braking [67].

Wheel damping treatments are also known to reduce the occurrence of squeal. In this case, a small increase in the level of damping can be effective in eliminating squeal. In addition to the forms of damping discussed in Section 14.3.2, ring dampers have been used as a simple means of increasing the damping of a wheel [68].

Effective solutions can also be sought in the design of vehicles for curving in order to reduce the creepages (see also Section 17.6.5). Unfortunately, this is often in conflict with the design of bogies for stability at high speed.

14.6 AERODYNAMIC NOISE

14.6.1 Sources of Aerodynamic Noise

The sound power from aerodynamic sources increases more rapidly with speed than that from mechanical sources. Aerodynamic sources can be classified as monopole, dipole or quadrupole [2]. Most aeroacoustic sources on a train are of a dipole type, such as the tones generated by vortex shedding from a cylinder and turbulence acting on a rigid surface. For such sources, the sound power increases according to the sixth power of the flow speed V, which when expressed in decibels gives a rate of increase of 60 $\log_{10}V$ [69]. The noise from free turbulence, such as jet noise, has a quadrupole source type and a speed dependence of 80 $\log_{10}V$ [70]. Moreover, the frequency content also changes with speed, shifting towards higher frequencies with increasing speed; consequently, the A-weighted sound pressure level will increase at a greater rate than what these theoretical values suggest.

As a consequence of the higher speed dependence, compared with 30 $\log_{10}V$ typical of rolling noise, aerodynamic noise sources will become predominant in the overall noise above a certain speed. This transition speed is often considered to be around 300 km/h [71], although more recent results suggest that it may be as high as 370 km/h in some situations [72].

Where noise barriers are placed alongside the track, the rolling noise may be attenuated by 10–15 dB, whereas the aerodynamic sources from the upper part of the train, and particularly the pantograph, remain exposed. This causes aerodynamic noise to become important at lower speeds in such situations. Aerodynamic sources are also important for interior noise in high speed trains, particularly the upper deck of double-deck trains, where rolling noise is less noticeable.

Important aerodynamic sources are found to fall into two main categories [71]. Dipole-type sources are generated by air flow over structural elements and cavities, including the bogies, the recess at the inter-coach connections, the pantograph and electrical isolators on the roof and the recess in the roof in which the pantograph is mounted. In addition, the flow over the succession of cavities presented by louvered openings in the side of locomotives is a source of aerodynamic noise, the form of which depends on the length and depth of the cavity. In the second category, which may have a dipole or quadrupole nature, noise is created due to the turbulent boundary layer.

14.6.2 Measurement Methods

An important tool for determining the location of aerodynamic sources on a moving train is an array of microphones, mounted at some distance from the track [73]. An example is shown in Figure 14.22. Using a technique known as beamforming, the sound coming from a certain direction can be obtained by combining the signals from these microphones after applying a suitable delay to each channel. For a moving source, a technique is also required to follow the source motion and to remove the Doppler shift. The output from the beamforming is a map of sound level over the surface of the train from which the locations of the main sources can be identified. However, it is more challenging to quantify the strengths of the sources [72].

Laboratory measurements can also be made in a wind tunnel [72]. This should have a low background noise and should be treated with anechoic boundaries to prevent sound reflections. For practical reasons, testing in wind tunnels often relies on making measurements of individual components or using reduced scale models. In the latter case, scaling laws based on non-dimensional quantities such as the Reynolds number must be applied to derive the corresponding result at full scale. Nevertheless, care is required as, in certain cases, such as the flow over a circular cylinder, the noise level does not follow a linear trend with Reynolds number [74].

14.6.3 Prediction Methods

Computational fluid dynamics (CFD) techniques have become popular in recent years. This is a vast and rapidly changing field that cannot be covered in detail here. Despite the rapid development,

FIGURE 14.22 An array of 90 microphones deployed at the trackside for identifying noise sources on a moving train.

however, it is still the case that conventional CFD methods cannot be used to calculate the aerodynamic noise from a train, so that the most promising developments are based on improved models of components or sub-assemblies [72]. In addition, Lattice Boltzmann methods have also been used recently, which are well suited to handle complex arbitrary geometries; see [75] for example.

An alternative approach is to use a semi-empirical component-based method. This has been used in particular to predict noise from the pantograph [76]. This uses a database of measured spectra from cylinders that are normalised using various non-dimensional parameters. The overall noise spectrum is assembled from the various components by neglecting flow interactions between them.

14.6.4 Control of Aerodynamic Noise

A review of countermeasures for the control of aerodynamic noise is included in [72].

These have focussed particularly on the pantograph, owing to its important location on the roof of the train. Significant noise reductions can be achieved by simplifying the design to a single-arm concept with larger main struts and by eliminating, as far as possible, any small components that generate high-frequency sound. Adding discrete holes to a cylinder can suppress the vortex shedding peak as long as the acoustic resonance of the holes does not match this peak. Roof-mounted shields alongside the pantograph are also included on some high-speed train designs.

The bogie region is also an important source of aerodynamic noise, especially as there are many more bogies on a train than pantographs. The leading bogie usually has the highest noise levels, as it is subjected to a higher in-flow velocity [77]. Attention to the shape of the train nose, particularly the underside and 'snowplough' region, can help to alleviate this [75]. In addition, fairings on the side of the bogies can help to smoothen the flow and to shield the noise [77].

14.7 OTHER SOURCES OF NOISE

14.7.1 ENGINE NOISE

Power units on trains are generally either electric or diesel. Noise from diesel locomotives is mostly dominated by the engine and its intake and exhaust. Space restrictions often limit the ability to silence the exhaust adequately, although in modern locomotives, this has been given serious attention. On electrically powered stock, and on diesels with electric transmission, the electric traction motors and their associated cooling fans are a major source of noise. Most sources of noise from the power unit are largely independent of vehicle speed, depending rather on the tractive effort required. The whine due to traction motors is an exception to this.

14.7.2 FANS AND AIR-CONDITIONING

Fans are an important source of noise on modern trains. In addition to traction motor cooling fans, the ventilation and air-conditioning systems contain fans, which produce significant noise. The most efficient solution for fan noise is to replace axial fans by radial ones, which can give noise reductions of 10 dB(A) [78], but this is not always possible.

The TSI Noise [1] places limits on standstill noise and starting noise as well as pass-by noise at constant speed. These situations are often dominated by the noise from engines and auxiliaries, principally from fans.

14.8 VEHICLE INTERIOR NOISE

14.8.1 VEHICLE INTERIOR NOISE LEVELS

All the noise sources discussed previously are also of relevance to interior noise in trains [79]. Noise is transmitted from each of these sources to the interior by both airborne and structure-borne paths, with structure-borne transmission often dominant at low frequencies and airborne transmission dominant at higher frequencies. The noise from the wheel-rail region is often the major source. In addition, on vehicles with under-floor diesel engines, noise from the engine can be significant. Noise from the air-conditioning system can also require consideration in rolling stock where this is present. There is often very limited space in which to package the air-conditioning unit and ducts.

Example spectra are given in Figure 14.23. These results show that modern high-speed trains are quieter at 300 km/h than a conventional 'rail car' at a lower speed [80]. In tunnels, the noise levels will increase considerably in the mid-frequency range due to a greater contribution from the walls, windows and roof [81].

14.8.2 MEASUREMENT QUANTITIES FOR INTERIOR NOISE

Conventionally the A-weighted sound pressure level has been used to specify acceptable levels inside vehicles, as indicated in the standard [82]. Internationally agreed limits are 68 dB(A) in second class and 65 dB(A) in first class [81]. However, as seen previously, the spectrum of noise inside trains contains considerable energy at low frequency. This low-frequency sound energy can be a source of human fatigue but is not effective in masking speech, for which noise in the range 200–6000 Hz is most effective [2].

Passenger requirements for noise inside a train vary from one person to another [83]. Clearly, it is desirable that the noise should not interfere with conversation held between neighbours. However, particularly for a modern open-saloon-type vehicle, silence would also not be the ideal. There should be sufficient background noise so that passengers talking do not disturb other passengers further along the vehicle (people talking loudly into mobile phones are a particular source of annoyance).

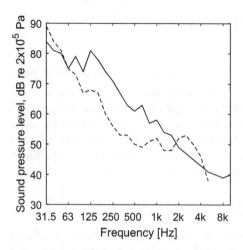

FIGURE 14.23 One-third octave band spectra measured inside French rolling stock (redrawn from [80]), showing —— railcar at 160 km/h (72 dB(A)), and – – – TGV at 300 km/h (64 dB(A)). (From Iwnicki, S. (ed.), *Handbook of Railway Vehicle Dynamics*, CRC Press, Boca Raton, FL, 2006. With permission.)

According to [84], for example, the interior noise level should be at least 60 dB(A) to avoid disturbance by other passengers. Various alternative quantities exist that can be used to define acceptable environments. These include the B-weighted level, preferred speech interference level, loudness level, alternative noise criteria (NCA), noise ratings (NR) and room criteria (RC) [85].

The interior sound level varies considerably within a vehicle. Figure 14.24 shows some example measured results where a loudspeaker has been placed at one end of an open-saloon vehicle. This was a British Rail Mk II coach dating from the 1960s, although the interior dated from the 1990s. The solid line shows the relative sound pressure level along a line down the central gangway at the height of the headrests. Results are shown in three example one-third octave bands. At low frequencies, strong modal patterns are observed due to the long acoustic wavelength. At higher frequencies, considerable decay in the sound level is observed along the coach due to the absorptive properties of the seats, carpets, etc. Additional attenuation is seen at the middle of the coach, where two glass partial screens were present at either side of the door.

Also shown are measured results at positions in front of each seat headrest. The seats were arranged in groups of four, with tables between them. At low frequencies, these measurements follow the same pattern as the gangway measurements, but at higher frequencies, considerable differences can be seen between adjacent seated positions. These spatial variations may be experienced by passengers in the vehicles; the 500-Hz frequency band, for example, is quite important for speech interference. It can also be expected that differences will occur between left and right ear positions at an individual seat, leading to binaural effects. Clearly, in a running vehicle, the source positions will differ from this, but these results serve to illustrate the general trends that can be expected.

14.8.3 Airborne Transmission

Airborne sound transmission into the vehicle occurs due to acoustic excitation of the vehicle floor, walls, windows, doors and roof. The acoustic performance of a panel can be measured by placing it between two reverberant rooms and measuring the difference in sound pressure level between the two rooms [2]. The sound reduction index (or transmission loss) is the difference between the incident intensity level and the transmitted intensity level, which can be derived from such a measurement after allowing for the size of the panel and the absorption in the receiver room.

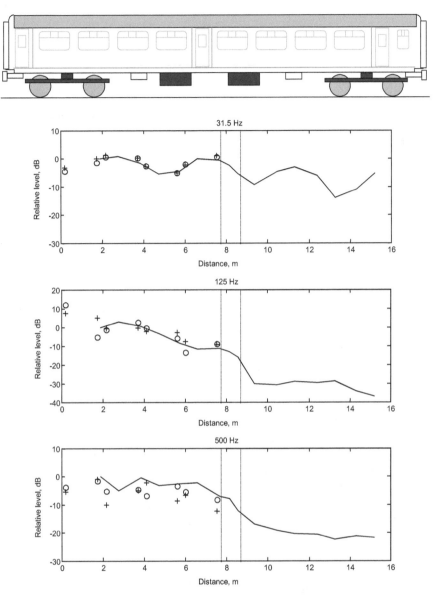

FIGURE 14.24 Relative internal sound levels in selected 1/3 octave bands in ex-BR Mk II coach due to a sound source located at left-hand end, showing ——— measured above floor along central gangway, o measured 0.05 m from aisle seat headrests, and + measured 0.05 m from window seat headrests; all measurements at 1.05 m above floor level, and positions of partial screens indicated by dotted lines. (From Iwnicki, S. (ed.), *Handbook of Railway Vehicle Dynamics*, CRC Press, Boca Raton, FL, 2006. With permission.)

A typical sound reduction index of a homogeneous panel is shown in Figure 14.25. Generally, the sound reduction index of panels is dominated by the 'mass law' behaviour in a wide frequency range. At high frequencies, the coincidence region occurs where the wavelengths in the structure and air are similar. Here, a dip in the sound reduction index occurs, the extent of which depends on the damping. The mass law behaviour extends from the first resonance of the panel up to just below the critical frequency. In this region, the bending stiffness of the panel and its damping have no effect on the sound transmission (see [2] for more details).

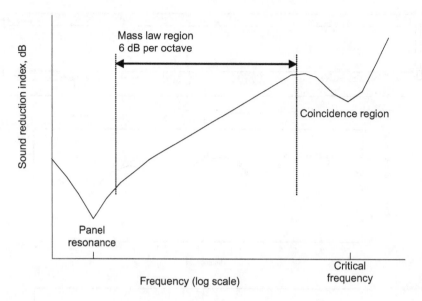

FIGURE 14.25 Typical sound reduction index of a homogeneous panel due to a diffuse incident field.

The use of lightweight constructions such as extruded aluminium and corrugated steel leads to a low sound reduction index. This follows from the mass law, which states that the sound reduction index reduces by 6 dB for a halving of panel mass. However, such structures tend to have a performance that is even worse than what the mass law would suggest, owing to the presence of an extended frequency region over which coincidence effects occur. For example, Figure 14.26 shows measurements of the sound reduction index from a 60-mm-thick extruded aluminium floor of a railway vehicle with a 3-mm-wall thickness, taken from [86] (similar results are also found in [87]). Also shown is the 'field incidence' mass law estimated for a homogeneous panel of similar mass [2].

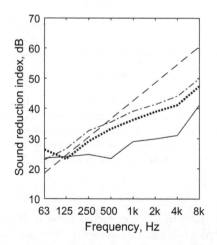

FIGURE 14.26 Octave band sound reduction index of extruded aluminium floor, showing —— measured on bare floor panel, ···· field incidence mass law for 30 kg/m^2, — — — measured for bare floor panel plus 12-mm suspended wooden deck, and — · — · measured for damped floor panel plus 12-mm suspended wooden deck (data from [86]). (From Iwnicki, S. (ed.), *Handbook of Railway Vehicle Dynamics*, CRC Press, Boca Raton, FL, 2006. With permission.)

Clearly, the extruded panel exhibits a much lower sound reduction index than this. It can be brought closer to the mass law behaviour by the use a suspended inner floor and by adding damping treatments to the extruded section.

14.8.4 STRUCTURE-BORNE TRANSMISSION

In addition to the airborne path, considerable sound power is transmitted to the vehicle interior through structural paths. This originates from the wheel-rail region as well as from under-floor diesel engines where these are present. Structure-borne engine noise can be reduced significantly in many cases by applying good mounting practice [79]. The mount stiffness must be chosen considering the frequency characteristics of the engine. An incorrect choice of stiffness can lead to amplification rather than attenuation of transmitted vibration. Flanking paths via pipes and hoses should also be avoided.

Structure-borne noise from the wheel-rail contact is transmitted via the bogie frame through the primary and secondary suspensions as well as a host of other connections such as dampers, traction bars, etc. The dynamic stiffness of these elements are frequency dependent, often with internal resonances, and should be characterised carefully.

14.8.5 PREDICTION OF INTERIOR NOISE

Deterministic methods such as FEM may be applied at low frequencies to predict the vehicle interior noise; see [88] for example. Owing to the regular geometry, an analytical model of the interior may also be used to construct the interior acoustic field on the basis of simple room modes [89].

However, at high frequencies, the number of modes becomes prohibitive for such approaches. The preferred analysis method for frequencies above about 250 Hz is therefore Statistical Energy Analysis (SEA). This can be used in both predictive mode [86,90] and experimental mode [91]. However, in predictive mode, it is not straightforward to define the coupling loss factors between the various sub-systems, especially where aluminium extrusions [87,92,93] or other inhomogeneous constructions are used [94]. Moreover, as SEA is a statistical method, it provides an average result and cannot account easily for the spatial variations in sound field, such as seen in Figure 14.24.

14.9 GROUND-BORNE VIBRATION AND NOISE

14.9.1 INTRODUCTION

Vibration generated at the wheel-rail interface is also transmitted through the ground. Excitation occurs due to the passage of individual wheel loads along the track (quasi-static excitation) and due to dynamic interaction forces caused by irregularities of the wheels and track. This vibration is transmitted into nearby buildings, where it may cause annoyance to people or malfunctioning of sensitive equipment. People may either perceive the vibration directly, as low-frequency 'feelable whole body' vibration (between 2 Hz and 80 Hz) or, indirectly, as re-radiated noise caused by vibration of the floors and walls at higher frequencies (around 20–250 Hz).

The highest levels of low frequency ground-borne vibration are usually produced by freight trains at sites with soft soil. Heavy axle-load freight traffic, travelling at relatively low speeds, causes high-amplitude vibration at the track that excites surface-propagating waves in the ground. This type of vibration often has significant components at very low frequency (below 10 Hz) and may interact with the frequencies of buildings 'rocking' or 'bouncing' on the stiffness of their foundations in the soil. This phenomenon is especially associated with soft soil conditions, where it is found that significant levels of vibration may propagate up to as much as 100 m from the track. At these frequencies, the vibration is perceived in the building as 'whole body' vibration that can be felt. This is usually assessed under the principles of ISO 2631-1 [95]. High levels of vibration cause annoyance and, possibly, sleep disturbance. Complaints are often expressed in terms of concern over possible damage to property, although, for the levels of vibration normally encountered from trains, such concern is unlikely to be borne out when assessed against the criteria for building damage, for example, BS 7385 [96] and DIN 4150 [97].

Passenger trains also may cause significant levels of vibration, particularly electric multiple units with high unsprung masses that dynamically interact with the rails due to the unevenness of the wheels and the track. For the frequency range 2–250 Hz and a train speed range of 10–100 m/s (36–360 km/h), the corresponding wavelengths of the vertical unevenness lie within the range 0.04–50 m. When considering the rail surface, at wavelengths less than about 1 m, this vertical unevenness is most commonly caused by irregular wear or corrugation of the rail contact surface, whereas at much longer wavelengths, it is due to undulations in the track bed. On the wheels, short wavelength unevenness is again caused by wear, whereas discrete wavelengths up to about 3 m are present due to out-of-roundness. In addition, dynamic forces are generated as impacts as the wheels traverse switches and crossings or badly maintained rail joints [98,99]. However, at sites of mixed traffic, it is usually the case that a few freight trains, perhaps running at night, are identified as the worst cases, and it is these that dominate the assessment of potential annoyance.

High-speed passenger trains sometimes travel at speeds approaching the speed of propagation of vibration in the ground and embankments. This has been the concern of track engineers for some years because of the large displacements that can be caused in the track support structure, in electrification masts, etc. This is particularly important on soft soil, where the wave speeds in the ground are relatively low, and, in some cases, it may cause vibration that exceeds the limit for safety and stability. Although the occurrences of this are comparatively rare, the topic has attracted considerable attention amongst researchers recently because of the expansion of the network of high-speed railways; see [100–102].

Trains that run in tunnels also cause vibration that is transmitted to nearby buildings. This has higher frequency content than vibration from surface tracks/trains and generally has lower amplitudes. Although no direct airborne noise can be heard, vibration at the low end of the audible frequency range, from about 20 Hz to 250 Hz, may excite bending in the floors and walls of a building, which then radiates noise directly into the rooms. This rumbling noise may be found to be all the more annoying, because the source cannot be seen and no screening remedy is possible.

14.9.2 Vibration from Surface Railways

All grounds are stratified on some scale, and this layered structure of the ground has important effects on the propagation of surface vibration in the frequency range of interest. Typically, grounds have a layer of softer 'weathered' soil material that is only about 1–3 m deep on top of stiffer soil layers or bedrock, depending on the geology of the site. In such a layered medium, vibration propagates parallel to the surface via a number of wave types or 'modes'. These are often called the Rayleigh waves of different order ('R-waves') and the Love waves. The Rayleigh waves are also called P-SV waves, since they involve coupled components of dilatational deformation and vertically polarised shear deformation. Here, the name P-SV wave is preferred, and the term Rayleigh wave is reserved for the single such wave that exists in a homogeneous half-space. The Love waves are decoupled from these and only involve horizontally polarised shear deformation; they are also known as SH waves. Since the vertical forces in the track dominate the excitation of vibration in the ground, the SH waves are not strongly excited. They are not considered further in the present discussion.

To illustrate, measured examples of P-SV surface waves are shown in Figure 14.27 from two sites in the UK. Such three-dimensional plots are produced by measuring the transfer function along the ground surface over different distances and then applying a Fourier transform in the spatial domain to express the response in terms of wavenumber ($k = 2\pi/\lambda$, where λ is the wavelength) at each frequency [103]. The phase velocity of each wave at a particular frequency is given as the ratio of the frequency over the wavenumber ($c = \omega/k$). Each peak in the diagram represents a wave type associated with a cross-sectional mode of the soil.

Noise and Vibration from Railway Vehicles

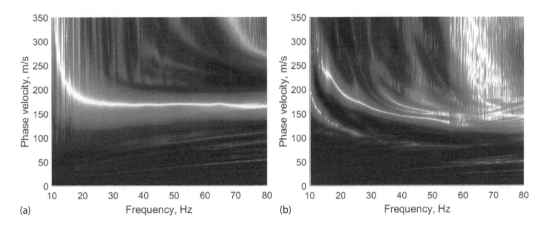

FIGURE 14.27 Phase velocity of P-SV wave modes measured for two sites in the UK: (a) near the Birmingham-Wolverhampton tram line and (b) near the Brighton-Portsmouth main line.

For the first site in Figure 14.27a, at low frequencies (below 15 Hz), the fundamental surface wave has a quite high phase velocity, corresponding to Rayleigh waves of the underlying stiff ground layers. At high frequencies, the phase velocity converges asymptotically to a value of about 170 m/s, corresponding to the Rayleigh wave velocity in the top layer. For the second site in Figure 14.27b, a surface wave is seen between 10 Hz and 30 Hz, with a wave speed of about 100 m/s that tends asymptotically to a value of about 80 m/s at higher frequencies. However, above 20 Hz, the main activity corresponds to a higher wave speed of about 130 m/s.

The layered arrangement and properties of the ground can vary considerably from one site to another and even within a single site. Theoretical ground models are therefore essential for understanding the physics of ground-borne vibration propagation. The majority of these models assume that the ground consists of horizontal, isotropic and homogeneous infinite layers (layered half-space). Apart from a small zone in the immediate vicinity of the track, it can be assumed that the strain levels in the soil remain relatively low during the passage of a train, so that a linear elastic constitutive behaviour can be assumed.

Examples of the wave propagation of the P-SV waves in such a layered half-space are shown in Figure 14.28. These are the characteristic curves of the waves that propagate at the surface of a ground modelled as a typical soft layer of 'weathered' soil overlying a stiffer substratum of material. These results are calculated using a semi-analytical model [104]. The soft soil is modelled as a 2-m-deep layer with a shear wave (S-wave) speed of 118 m/s and dilatational wave (P-wave) speed of 360 m/s, and the substratum is a half-space with shear wave speed of 245 m/s and dilatational wave speed of 1760 m/s. Damping is included in both materials as a loss factor of 0.1.

Since the example ground is not homogeneous, the P-SV waves are dispersive; the phase velocity and the attenuation coefficient of the waves vary with the frequency. Figure 14.28a presents the dispersion diagram for the example ground (only the propagating P-SV modes are shown) in which the wavenumber is plotted as a function of frequency. Each line of the diagram represents a wave type associated with a cross-sectional mode of the layered soil. The corresponding phase velocities are shown in Figure 14.28b.

For this example set of soil parameters, at very low frequency, only a single mode exists, and this has a wave speed close to that of the shear waves in the substratum. Around 15 Hz, a quarter wavelength of the shear wave fits in the depth of the weathered material. Above this frequency, waves can propagate in the upper layer, with little influence of the underlying soil. At higher frequency, as the wavelengths of shear waves become small compared with the depth of the weathered material layer, the wavenumber of the slowest wave converges towards that of the Rayleigh wave in a half-space of the layer material. Higher-order propagating wave types 'cut on' at frequencies of 18 Hz, 35 Hz and 74 Hz.

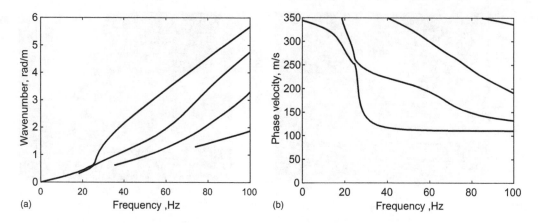

FIGURE 14.28 Characteristic curves for propagating P-SV waves of the example ground: (a) dispersion diagram and (b) phase velocity plot.

In Figure 14.29, the maximum rail response (displacement) is shown for a unit constant load moving with different speeds. The results are given for the same example ground for two different track forms, with the properties given in Table 14.1. An important increase in vibration occurs as the load speed approaches a certain 'critical' value. Comparing the rail vibration level and the values of the peak-response load speed for the two different track forms, it can be seen that for the slab track, the critical speed occurs at a higher speed, and, for all speeds below the critical speed, the maximum rail response is lower than for the ballasted track. A more detailed discussion of the critical speed and the influence of the track parameters on this are given in [105,106].

The wave field for a single constant load moving along the ballasted track at different speeds is shown in Figure 14.30. For a load speed below that of any of the waves in the ground, the displacement 'bowl' under the single load is indicated in Figure 14.30a by the positive (upward) displacement under the track. Little effect is observed only a few metres away, although close to the track, the passage of the quasi-static displacement will be observed. Figure 14.30b shows what happens when the load travels at a speed of 115 m/s near to the 'critical' speed. Larger displacements occur at the loading point and displacements are observable at greater distances along the track than for the

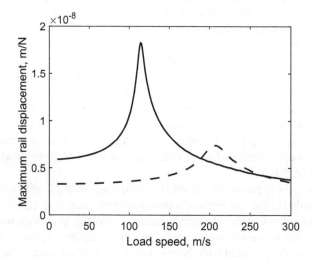

FIGURE 14.29 Maximum rail displacement versus load speed, showing results for ──── ballasted track and for ─ ─ ─ slab track, with parameters in Table 14.1.

TABLE 14.1
Parameters Used for the Ballasted Track and the Slab Track in Figure 14.29

Rail and railpads	Mass of rail per unit length	60 kg/m
	Bending stiffness of rail	6.3×10^6 Nm^2
	Loss factor of the rail	0.01
	Rail pad stiffness	3.5×10^8 N/m^2
	Rail pad loss factor	0.15
Sleepers and ballast	Mass of sleepers per unit length of track	490 kg/m
	Mass of ballast per unit length of track	3300 kg/m
	Ballast stiffness per unit length of track	1.775×10^8 N/m^2
	Loss factor of ballast	1.0
	Effective contact width of railway and ground	2.7 m
Slab	Mass of slab per unit length of track	3720 kg/m
	Bending stiffness of slab	1775×10^6 Nm^2
	Loss factor of slab	0.015
	Effective contact width of railway and ground	3.4 m

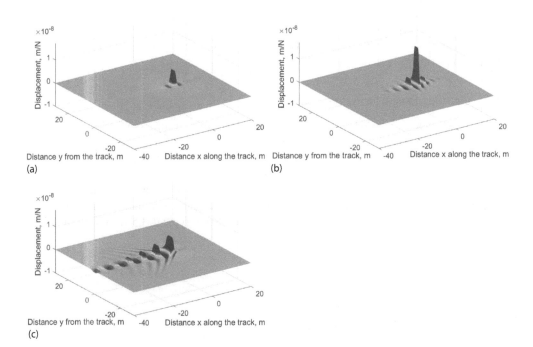

FIGURE 14.30 Displacement pattern in the moving frame of reference for a single non-oscillating axle load on the track moving at: (a) 83 m/s (below the wave speeds in the ground), (b) at 114 m/s (close to the critical speed for this track/ground system) and (c) at 150 m/s.

lower load speed. Waves start to extend sideways from the track. The effects of a further increase of speed to 150 m/s are shown in Figure 14.30c. At this speed, in excess of the P-SV wave speed of the layer material (see Figure 14.28b), a number of waves are created in the track behind the load, and propagating waves may be seen travelling with significant amplitude away from the track.

If multi-body models representing the vehicles of a train are coupled to the model for the track-ground system, a theoretical model that predicts the total vibration field can be produced [107–112]. A number of models in which the vibration excited by the moving axle loads of the whole train and

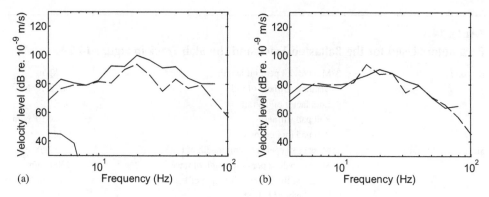

FIGURE 14.31 Vertical velocity level at Steventon for train speed of 170 km/h (47 m/s): (a) at 12 m from the track and (b) at 20 m from the track, showing —— predicted total level, ···· predicted level due to quasi-static loads, and — — measured level. (From Triepaischajonsak, N. et al., *J. Rail Rapid Transit.*, 225, 140–153, 2011. With permission.)

by the irregular vertical profile of the track for all the axles of a train have been validated by comparison with measured vibration for a number of sites, for example, [110,113].

Examples of vibration spectra at the trackside are shown in Figure 14.31, which compares measurements and predictions from a semi-analytical model [114]. Here, the ground is a fairly soft clay and track is on a shallow embankment; the train speed was 47 m/s.

14.9.3 Ground-Borne Noise from Trains in Tunnels

To illustrate the vibration propagation from tunnels, the results from a coupled two-dimensional (2D) FEM/boundary element (FE-BE) model developed in [115] are presented. This was later extended to a so-called 2.5D model, in which the variations in displacement in the third direction are allowed for using a wavenumber transform [116].

Figure 14.32 shows the exaggerated instantaneous particle displacement at a number of points in the ground to illustrate the wave pattern radiating away from an oscillating load applied at the base of a circular tunnel at high frequency. The tunnel has a typical 3.5 m outer radius in this

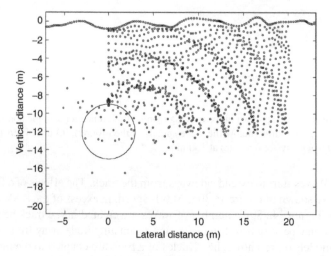

FIGURE 14.32 Predicted vibration field around an unlined tunnel at 100 Hz. (From Jones, C.J.C., *Proc. Inst. Civil Eng. Transp.*, 153, 121–129, 2002. With permission.)

case, without the concrete lining, and is 15 m deep at the rail [115]. Ground properties typical of a deep clay formation have been used, namely, an S-wave speed of 610 m/s and a P-wave speed of 1500 m/s (implying a Poisson's ratio of 0.4); the density of the material has been assumed to be 1700 kg/m³, and the damping loss factor is set to 0.15. Boundary elements are used to represent the ground-tunnel interface and the ground surface from +50 m to −20 m relative to the vertical centreline of the tunnel. The boundary elements allow a ground of infinite extent to be represented [115].

It can be seen that a relatively simple pattern of cylindrical wave fronts radiates towards the surface at greater distances from the tunnel. The strongest component of deformation in these waves is shear. At this frequency (100 Hz), the greatest amplitudes of response on the ground surface are at a distance of about 15–20 m from the tunnel alignment rather than directly above it.

Figure 14.33 shows the calculated response on the surface of the ground to an oscillating load at 60 Hz; in this case, unlike Figure 14.30, the load is in a tunnel. The response is calculated using the model presented in [117] for the soil and tunnel properties taken from [118]. As in Figure 14.32, the highest levels of vibration can be seen to be about 15 m to the side of the tunnel alignment, with the propagation pattern beyond this showing decaying circular wave fronts, whilst the vibration field above the tunnel is more complicated.

Again, by coupling multi-body vehicle models to the track/tunnel/ground system, the total vibration field can be produced. Figure 14.34 shows the total (quasi-static and dynamic) vibration level predicted using the coupled 2.5D FE-BE model [118]; the results are compared with the measured response from a site location in London on a conventional underground line. The response is given at the tunnel invert and on the ground surface. The predictions show a good agreement with the measurements. The fluctuations below 10 Hz are not observed in the measurements, probably due to the influence of background noise in the measurements. At the ground surface, the quasi-static component of vibration is negligible, and the ground vibration shown in the prediction is entirely due to the dynamic component. The response shows a peak at around 63–80 Hz, which is due to the resonance frequency of the vehicle unsprung mass on the track stiffness. Compared with Figure 14.31, it can be seen that the response on the ground surface is 20–30 dB smaller than that for a surface railway. Although this is unlikely to produce feelable vibration, it can nevertheless produce significant ground-borne noise in neighbouring buildings.

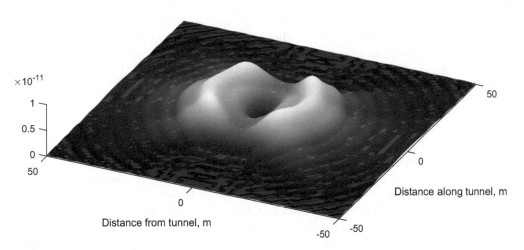

FIGURE 14.33 Vertical response amplitude of the surface of the ground to a load oscillating at 60 Hz; the tunnel lies at a depth of 20 m.

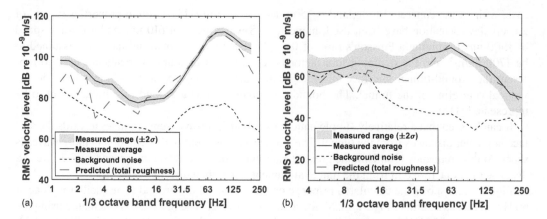

FIGURE 14.34 Comparison of measured and predicted vibration for train passage at 42 km/h (11.7 m/s): (a) on tunnel invert and (b) at the ground surface. (From Jin, Q. et al., *J. Sound Vib.*, 422, 373–389, 2018. With permission.)

14.9.4 Controlling Low-Frequency Ground Vibration

Ground-borne vibration from railways can be controlled at different levels: at the source (train-track-soil interaction), in the transmission path or at the receiver (building). In the present discussion, mitigation measures at the receiver, such as base isolation of buildings and box-in-box arrangement of rooms, will not be considered further. The focus is placed on the dominant mechanism of vibration excitation and the interaction between the track and the ground, as well as on the transmission path through the ground.

In general, the treatment of low-frequency vibration, which involves surface waves with longer wavelengths and larger penetration depths, is considered more difficult, less effective and less economical than the mitigation of higher-frequency ground vibration leading to ground-borne noise.

14.9.4.1 Reducing the Track Geometric Unevenness

Long-term track usage may cause differential ballast/soil settlement and lead to an increase of the long-wavelength geometric unevenness of the track. If the dynamic component of excitation dominates the vibration, reducing the amplitude of long-wavelength components of the vertical track profile should help to reduce the vibration levels in the low-frequency range. This can be achieved by track realignment, in particular for the case of ballasted tracks by tamping the ballast. However, this has no effect on vibration very close to the track, which is dominated by the quasi-static loads. In addition, well-maintained turnouts (switches and crossings) and the removal of wheel flats will reduce large-impact forces and lead to substantial reductions of dynamic loads and further track deterioration.

14.9.4.2 Rolling Stock Modifications

It has been observed at mixed traffic sites that a particular train service may give rise to the main complaints of vibration. Thus, there is potential for vibration mitigation by reducing the dynamic vehicle loads through modification of the rolling stock characteristics.

Concentrating on the dynamic excitation, modifications to the dynamic properties of the vehicles can lead to reduction of the ground vibration. In [119], the effect of the vehicle parameters on railway-induced ground vibration was studied. It was shown that the parameters of most influence are the unsprung mass and the stiffness of the primary suspension, with higher vibration levels occurring for heavier unsprung masses and for stiffer primary suspensions. However, for the range of vehicle parameters investigated, the effect on the levels of ground-borne vibration is relatively small.

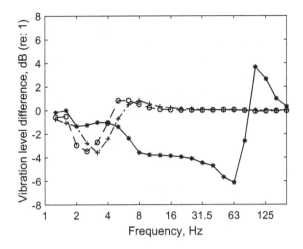

FIGURE 14.35 One-third octave spectra of the vibration level difference 16 m from the track between the nominal train model and the model with (*) 1200-kg wheelset mass, (o) 3000-kg bogie mass, and (+) 425 kN/m primary suspension stiffness.

Similar findings are shown in Figure 14.35, where the reduction in ground-borne vibration at 16 m from the track due to reduction of the unsprung mass, the bogie mass and the primary suspension stiffness is given as a level difference in decibels, relative to a nominal train model. The results for this example are given in one-third octave bands and obtained using the model presented in [112] and the train parameters for the nominal model given in Table 14.2 (a four-vehicle train model was used).

It can be seen that, by reducing the unsprung mass from 1800 kg to 1200 kg per wheelset, the levels of ground-borne vibration can be reduced for the whole range of feelable vibration (between 2 Hz and 80 Hz), with a maximum reduction of 6 dB at 63 Hz. However, above 80 Hz, in the frequency range of ground-borne noise, the reduction of the unsprung mass could lead to an increase of vibration. The modifications to the bogie mass and primary suspension stiffness can also lead to small reductions between 2 Hz and 5 Hz.

TABLE 14.2
Parameters Used for the Nominal Vehicle Model Used in Figure 14.35

Carbody	Mass	26,200 kg
	Pitching moment of inertia	2×10^6 kgm^2
	Overall vehicle length	20 m
Bogie	Mass	5,000 kg
	Pitching moment of inertia	6,000 kgm^2
	Half distance between bogie centres	7.1 m
Wheelset	Mass	1,800 kg
	Half distance between axles	1.3 m
	Total axle load	140.1 kN
Primary suspension	Vertical stiffness per axle	850 kN/m
	Vertical viscous damping per axle	20 kNs/m
Secondary suspension	Vertical stiffness per bogie	600 kN/m
	Vertical viscous damping per bogie	20 kNs/m

The effect of the unsprung mass is especially important for the cases of freight wagons with friction-damped suspensions used to carry materials such as cement, where the suspensions may seize up and result in high vibration levels. Modern passenger trains often have lower unsprung masses than those from prior to the 1990s. This is done in order to reduce track damage, and it has been achieved by avoiding electric motors to be hung directly on the axle. Instead, the motors are hung from the bogie, and vibration isolation is included in the transmissions. However, the means for reducing the unsprung mass significantly beyond this, for the sake of ground vibration reduction, are probably limited.

Regarding the suspension stiffness, some bogies exist for freight wagons that are designed to reduce the track forces by using two-stage suspensions, and these wagons have also been observed to cause much less vibration than conventional freight vehicles at vibration sensitive sites and for frequencies below 10 Hz. It should also be noted that changes in the geometrical parameters (e.g., bogie and axle distance or two-axle wagons) affect the axle and bogie passage frequencies and result in shifts in the one-third octave band spectrum of the quasi-static and the dynamic response [119].

14.9.4.3 Ground-Based Measures

To prevent vibration transmission in the ground, open trenches and 'in-filled' trenches (buried walls) can be considered as possible solutions, as shown in Figure 14.36. For a homogeneous ground, both measurements and simulations [120] show that, for an open trench to be effective, the depth should be more than about half the Rayleigh wavelength. However, for a layered ground, the performance is also influenced by the depth and stiffness of the soil layers. For a layered ground with a soft weathered layer above a stiffer substratum, significant reductions can be achieved if the trench cuts through the upper layer. The width of the open trench has been shown [120] to have a relatively small effect on the benefit. For practical reasons, trenches are usually filled with a soft barrier material. However, this leads to a significant reduction in their performance, as vibration can be transmitted through the barrier material as well as be diffracted beneath it [120].

A stiff wave barrier consisting of a concrete slab wall, a row of steel or concrete piles or a sheet pile wall embedded in the ground can also be effective in the mitigation of vibration [121,122]. In this case, the stability of the barrier is easy to achieve, and the installation of the barrier can be more straightforward. Such mitigation screening measures are more effective at sites with a soft soil.

As an alternative to vibration screening, the vibration may be reduced by installing stiff inclusions in the soil under or near the track [123,124]. This method can be achieved by stiffening the soil locally or replacing it with concrete, as shown in Figure 14.37.

Heavy masses or walls made of concrete or stone gabion baskets next to the track can lead to a reduction in vibration levels above a certain frequency. This frequency can be estimated as a mass-spring system resonance frequency, which is determined by the vertical dynamic stiffness of the soil

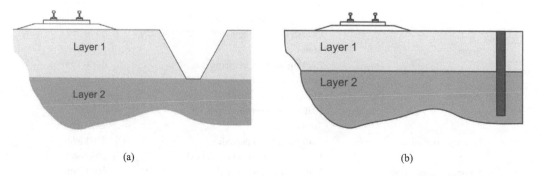

FIGURE 14.36 Examples of vibration transmission prevention using: (a) an open trench and (b) a buried wall.

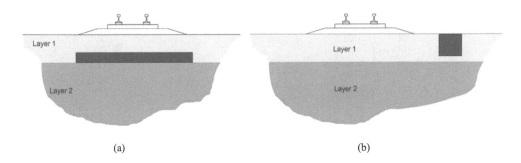

FIGURE 14.37 Examples of wave-impeding blocks: (a) under the track and (b) next to the track.

and the mass of the block or wall [125]. Therefore, by increasing the mass of the block or wall, the mass-spring resonance frequency is reduced and the performance at lower frequencies is improved. This is the case also for sites with soft soil, where the mass-spring resonance is low because of the low dynamic stiffness of the soil. At frequencies higher than the mass-spring resonance, and provided that the stiffness difference between the soil and the installed wall is large enough, the propagation of the surface waves is restricted and the incident waves are scattered, resulting in a reduction of the transmitted wave field. These measures could be designed in conjunction with noise barriers, such as gabion wall noise barriers, in order to obtain feasible mitigation solutions for both ground-borne vibration and airborne noise.

Ground improvement methods are widely used in civil engineering, mainly to strengthen foundations to support the infrastructure or to avoid ground liquefaction during earthquakes. Such ground improvement methods can also be used as vibration mitigation methods at sites with soft soil. Stiffening of the subgrade under the track improves the bearing capacity of soft soils and helps in avoiding excessive track settlements. Various techniques can be applied to achieve the desired subgrade stiffening, for example, vibro-compaction, jet grouting or excavation and replacement by a new material such as concrete. Such methods can reduce the vertical track settlement and thus decrease the long-wavelength track unevenness that creates low-frequency dynamic vibration effects.

Ground and ballast improvements using geosynthetics such as geogrids that can be placed under the ballast or as a reinforcement material inside the ballast layer can also improve the vibration performance of the track. These methods, that are mainly used for stabilising the track on soft soils and for reducing the ballast degradation, can reduce the track unevenness and therefore reduce the dynamic vibration excitation at low frequencies. Over the past years, geogrids have been used mainly in combination with other ground improvement or vibration mitigation methods, and they are relatively convenient; they are easy to install on new tracks but can also be applied on existing railway tracks during ballast renewal.

14.9.5 Controlling Ground-Borne Noise

As ground-borne noise is dominated by higher frequencies than feelable vibration, the main way in which it is controlled is by introducing soft or 'resilient' elements in the track to provide some degree of vibration isolation. The principle of vibration isolation is illustrated in Figure 14.38, using a simple mass-spring system. The ratio of the force transmitted to the foundation to that applied to the mass is called the transmissibility. At very low frequency, this ratio is unity; the whole force is transmitted as it would be in the static case. At the natural frequency of the system f_n, the force is increased. Above $\sqrt{2}$ times the natural frequency, the transmissibility reduces to below unity and continues to decrease with increasing frequency. Here, a hysteretic damping model has been used (constant damping loss factor η) that reflects the behaviour of elastomeric materials. The effect of changes to the damping in the support, the support stiffness and the supported mass is also shown in Figure 14.38.

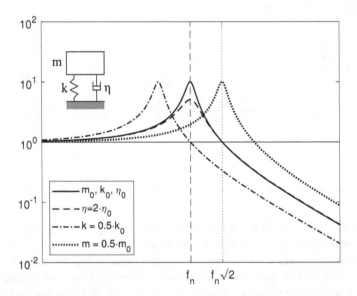

FIGURE 14.38 The force transmissibility showing results for: —— a hysteretically damped single-degree-of-freedom system, — — increased damping loss factor, — · — reducing the spring stiffness, and ···· reducing the mass.

It can be seen that the amplitude of the resonance is dependent on the damping in the support but that the degree of vibration isolation at higher frequencies is not (for this damping model). When reducing the support stiffness, the natural frequency of the system is reduced and a greater degree of vibration isolation is seen at higher frequencies. Conversely, when reducing the mass of the system, the natural frequency is shifted to higher frequencies. In the railway context, the spring represents the stiffness of resilient elements in the track and the mass represents that of the wheelsets and the parts of the track above the resilient elements.

Vibration isolating designs for tracks are commonplace in modern underground railway systems to reduce ground-borne noise, and the subject is an important part of track design. They can also be used for surface railways; however, the insertion loss achieved will be less than that for a track in tunnel, unless a high impedance foundation, for example, a concrete raft, is introduced beneath the track. In order to isolate the track dynamically, resilient elements can be included at different levels in the track structure. The lower the stiffness of the support, the lower the natural frequency of the system and the greater the degree of vibration isolation at higher frequencies. The choice of support stiffness is, however, limited by the allowable vertical and lateral static displacements under the axle load of the train.

Figure 14.39 shows some of the basic design concepts for vibration isolating track designs, described in more detail later. The rail pad is neglected here; it has a stiffness higher than that of the resilient element in each case but is possibly still significant in the behaviour of the track design for the relevant frequency range. Moreover, the geometric track unevenness is assumed as the main source of vibration-generation mechanism. Adding resilience in the track system may also lead to smoothening of the variations in the support stiffness. In this way, the resilient track systems may lead to a reduction of vibration at relatively low frequencies [126]. An important reduction in perceived unevenness may also occur when the unevenness source is the track bed beneath the resilient element, whereas an increase is possible otherwise [127].

14.9.5.1 Soft Resilient Fastening Systems

The ballast layer forms a resilient component for a conventional ballasted track. For this reason, ballastless tracks (slab tracks) with normal pad stiffness give rise to increased vibration transmission compared with ballasted track. Designs of soft fastening systems are used to rectify this. There are

Noise and Vibration from Railway Vehicles

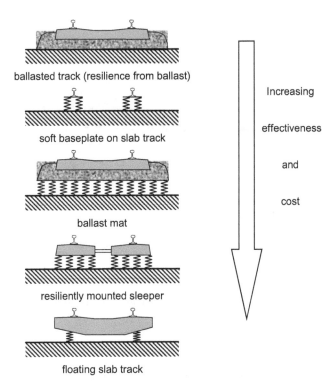

FIGURE 14.39 Design concepts for vibration isolating tracks. (From Iwnicki, S. (ed.), *Handbook of Railway Vehicle Dynamics*, CRC Press, Boca Raton, FL, 2006. With permission.)

many designs of soft fastening systems, with the standard test loadings for rails being the limiting factor to which the vertical stiffness of the system can be lowered.

The most common designs are the soft baseplate systems that allow the rail support stiffness to be reduced to about 12 MN/m per fastener. They are mostly used on slab track but can also be installed on top of sleepers in ballasted track. A typical soft baseplate design is the two-stage system consisting of a relatively stiff rail pad between the rail and a metal plate, beneath which a thicker soft elastomeric pad is used. The baseplate is much wider than the rail foot to prevent excessive rail roll and resultant gauge spreading under the lateral forces of the vehicle, particularly during curving.

There are also different systems available for achieving a low vertical stiffness and, at the same time, limiting the lateral displacement of the railhead. These can achieve vertical stiffness of less than 10 MN/m per fastener; however, a potential undesirable effect is that they may lead to an increase in rolling noise if the vibration decay rates are too low.

Figure 14.40 shows the predicted level difference in ground-borne vibration at 16 m from the track due to reductions of the rail fastening stiffness. This is given in decibels, relative to a nominal surface railway model presented in [112], using a railpad stiffness of 350 MN/m and the parameters given in Tables 14.1 and 14.2. The values of the stiffness used for the comparison are selected to represent a soft railpad case (120 MN/m), a standard baseplate system case (30 MN/m), a soft two-stage baseplate system (12 MN/m) and a very soft fastening system with 9 MN/m. It can be seen that the level difference is positive in the range of feelable vibrations (below 80 Hz), with the reduction of the fastening stiffness corresponding to an increase in maximum ground vibration levels. For higher frequencies, vibration reduction is predicted, which should be the largest at the vehicle-track resonance frequency of the original track system.

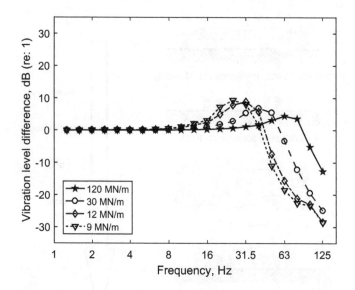

FIGURE 14.40 Vibration level difference in one-third octave pad due to reductions of the rail fastening system.

14.9.5.2 Under-Sleeper Pads, Ballast Mats and Booted Sleepers

Under-sleeper pads and ballast mats can be used for ballasted tracks to lower the stiffness of the ballast layer and therefore the vehicle-track resonance frequency. Booted sleepers perform in the same way as under-sleeper pads, but they are used on slab tracks. Since for all these cases the soft material is installed below the sleeper, the sleeper mass helps to lower the coupled vehicle-track resonance frequency.

Under-sleeper pads are easy to install during a sleeper renewal operation, since they are delivered already fixed to the bottom of the sleeper. Ballast mats can be laid on tunnel inverts or a prepared subgrade and have the additional advantage that the extra mass of the ballast is above the spring in the resonant system. However, if a ballast mat is too soft, there is a risk of making the ballast layer unstable under the vibration of passing trains and therefore compromising ride quality and increasing maintenance costs. For the case of the booted sleepers, these are usually bi-block sleepers, and their design and installation are integrated with the slab track.

14.9.5.3 Floating-Slab Tracks

Floating-slab tracks are used to control vibration and ground-borne noise from underground trains. The track is mounted on a concrete slab that rests on rubber bearings, glass fibre or steel springs. With this design, the highest possible mass is added above the track spring to form a system with a very low resonance frequency.

Floating-slab tracks are typically designed as part of the tunnel structure. In addition to the greater construction cost of the track form itself, great expense can come from any increase in the diameter of the tunnel that has to be made to accommodate sufficient mass for the floating slab. The slab may be cast in situ, resulting in a continuous length of concrete, or may be constructed from pre-cast sections. The continuous slab design usually has a lower deflection for a given resonance frequency and makes maximum use of the tunnel space but is harder to design in such a way that the slab mounts can be replaced.

Noise and Vibration from Railway Vehicles

14.10 VIBRATION COMFORT ON TRAINS

14.10.1 Introduction

The level of vibration in vehicles is a major influence on the perception of the quality of rail travel in comparison with other forms of transport. Vibration in the frequency range from about 0.5–80 Hz causes discomfort as 'whole body' vibration, and frequencies below this may cause nausea. The wavelengths in the vertical and lateral profiles of the track that give rise to this vibration are between about 1 m and 70 m, depending on the train speed. Of course, the comfort of passengers is an important reason for the routine monitoring and maintenance that are central to track management for all railways.

14.10.2 Assessment of Vibration Comfort in Trains

It is important to understand how measured vibration levels in vehicles are used to assess the likely reaction of passengers. A comprehensive background on this subject is given in reference [128]; here, only an indicative overview is given.

The most commonly accepted principles of vibration perception assessment are laid out in the international standard ISO 2631-1 [95] and also in BS 6841 [129]. These set out terms for consideration of health, comfort, incidence of motion sickness and effects on human activities. Frequency weightings or 'filters' are defined that reflect human sensitivity to vibration in a similar way to the A-weighting (Figure 14.1) that is used for sound. Some of these are shown in Figure 14.41. In the assessment of ride comfort, the filter W_b is used in BS 6841 to weight rms vibration in the vertical (spinal) direction for both seated and standing passengers, and filter W_d is used for the two

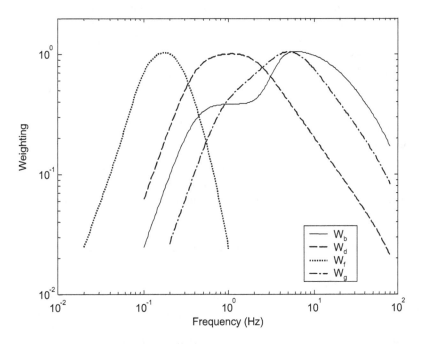

FIGURE 14.41 Some of the frequency weightings for whole-body vibration defined in BS 6841 [129]. (From Iwnicki, S. (ed.), *Handbook of Railway Vehicle Dynamics*, CRC Press, Boca Raton, FL, 2006. With permission.)

components of lateral vibration. Note that there is a difference between ISO 2631 and BS 6841 in that the ISO standard uses a slightly different weighting for vertical vibration, W_k. However, W_b is used more in the railway industry, as is recognised in ISO 2631-4 [130]. Vibration in the frequency range 0.5–80 Hz is considered. It is measured, as appropriate, on the seat surface between the cushion and a subject or on the carriage floor. Since measurements on the seat are dependent on the seated person, measurements should be carried out for a sample of subjects. Vibration on the seat back can also be important and is evaluated using other frequency weightings.

When considering the effects of vibration on human activity, weighting W_g is used for the vertical direction rather than W_b. For assessing the likelihood of vibration to cause motion sickness, weighting W_f is used for vertical vibration and the lower frequency range of 0.1–0.5 Hz is considered. No guidance is given in the standard on the influence of other components of vibration on motion sickness.

Meters and vibration analysis equipment are available that implement the frequency weighting filters, thereby evaluating the overall weighted levels of vibration. To combine the effects of vibration entering the body at the seat, seat back and the floor in different directions, the root sum of squares of these overall levels can be used.

It is for the rolling stock purchaser to set acceptable limits for the vibration measured in this way according to the type of rolling stock, taking into account factors such as the duration of journeys, number of standing passengers, line geometry standard and vehicle speed. In practice, the standards that are set vary from one railway to another.

There are a small number of single-value indicators of ride quality. One is defined by BS 6841, which allows the measurement of weighted accelerations in 12 components on the seat, seat back and floor. These overall levels are then multiplied by 'axis multiplying factors' to give 'component ride values', and these may be combined to give an overall ride index.

Another ride quality indicator for 'average comfort', N_{MV}, is defined by EN 12299 [131]. This uses overall accelerations in the vertical and two horizontal component directions, weighted as in BS 6841 and ISO 2631, but the 95th percentile values of 60 separate 5-second measurements are taken. The measure therefore becomes sensitive to rare events of high acceleration. N_{MV} is evaluated as six times the root sum of squares of these values. Values of N_{MV} are then rated in five bands from 'very comfortable' ($N_{MV} < 1.5$) to 'very uncomfortable' ($N_{MV} > 4.5$). Although it is suggested that all European railways should adopt this measure of 'average comfort', its complexity is a barrier to its acceptance in practice.

14.10.3 Effects of Vehicle Design

An important aspect of passenger comfort in railway vehicles is determined by the vibration induced by irregularities of the track. This vibration is transmitted to the carbody through the suspension, and therefore, the design of the suspension plays an important role in controlling this vibration.

14.10.3.1 Bogie Vehicles

At present, almost all rail vehicles are mounted on bogies. This element generally includes two levels of suspension: the primary suspension, between the wheelsets and the bogie frame, and the secondary suspension, between the bogie frame and the carbody. In addition, there are various linkages between the wheelset and the bogie and between the bogie and the carbody; these include the traction mechanisms, which are stiffer than the suspensions, and transmit the forces associated with traction and braking. Unlike in automotive vehicles, in most railway vehicles, the suspensions hardly restrict any degree of freedom between the elements that they join, so that all relative displacements and rotations between carbody, bogie frames and wheelsets are possible.

The bogie has a number of functions. They were originally adopted to facilitate curve negotiation; they also determine the dynamic stability of the vehicle, and the bogie and suspension contribute to ride quality by filtering the vibration transmitted to the carbody from the track irregularities. This filtering effect is required in both the vertical and horizontal planes. However, since

Noise and Vibration from Railway Vehicles

the suspension is normally built by means of passive elements, there are frequency ranges in which the vibration in the carbody is amplified by the suspension, in the same way as the transmissibility shown in Figure 14.38. In a suitable design, the unavoidable amplification associated with resonances should be outside the frequency range in which humans are most sensitive to vibration, that is, about 2–20 Hz, as seen from the W_b and W_g weighting curves in Figure 14.41. Oscillations at very low frequency produce motion sickness, as identified by the W_f weighting. Consequently, the main resonances of the carbody mass on the secondary suspension are generally arranged to be around 1 Hz.

Frequently, the role of the primary suspension is associated with the dynamic stability and curve negotiation, whereas the secondary suspension influences the vibration comfort of the passengers. Typically, the strategy followed is that the bogie and the carbody vibrations are dynamically uncoupled; this makes it possible to separate the comfort and steering tasks. The uncoupling is achieved because the mass of the carbody is very large with respect to that of the bogie frame, while the vertical primary suspension is stiffer than that of the secondary suspension. Consequently, there are carbody vibration modes (in which the carbody moves and the bogies almost do not) and bogie modes (in which the opposite occurs).

14.10.3.2 Secondary Suspension

In a conventional rail vehicle, there are five vibration modes that characterise the dynamics of the carbody, which are as follows:

- Vertical oscillation of the carbody
- Pitch rotation
- Yaw rotation
- Two lateral modes that couple the roll rotation with the lateral displacement (see Figure 14.42)

Owing to the uncoupling from the dynamics of the bogie, the natural frequencies of these carbody modes depend on the masses and moments of inertia of the carbody and the stiffness and location of the secondary suspension (they may also depend on the existence of anti-roll and traction mechanisms). If the plane of the secondary suspension contains the carbody centre of mass, the two lateral vibration modes shown in Figure 14.42 become a lateral displacement of the carbody (without rotation) and a pure roll rotation. This can be technically very costly, since it may reduce the space

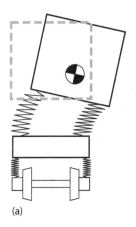
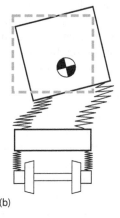

(a) (b)

FIGURE 14.42 Rear view of the vehicle showing two mode shapes that involve coupled lateral displacement and roll rotation: (a) mode with lower natural frequency and (b) mode with higher natural frequency.

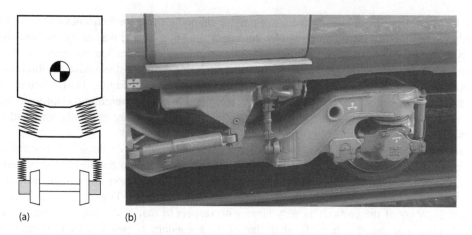

FIGURE 14.43 Alternative secondary suspension strategies: (a) inclined suspension and (b) anti-roll mechanism.

available in the carbody, but a similar effect can also be achieved if the secondary suspensions are inclined or by means of anti-roll mechanisms (both strategies are illustrated in Figure 14.43). If roll and lateral displacement are uncoupled, several beneficial effects are achieved, including the improvement of the passengers' subjective comfort.

As discussed previously, the natural frequencies of the carbody modes should be located in the range between 0.8 Hz and 1.5 Hz. The lowest natural frequency usually corresponds to the vertical mode, and the highest one corresponds to the mode that contains the roll rotation. The lateral mode can potentially have frequencies as low as or lower than the vertical mode, but stiffer lateral suspensions may be required due to gauge limitations in curves and other constraints. If the lateral stiffness of the secondary suspension is very high, the filtering effect is lost, and, moreover, the critical speed for hunting may become too low.

It is convenient if half the longitudinal distance between the two pairs of secondary suspensions coincides with the carbody radius of gyration for pitch motion; this contributes to the reduction of the vibration level. This design is feasible in conventional vehicles, but it is practically impossible in articulated trains or vehicles with Jacobs-type bogies (those in which the bogie is located between two adjacent vehicles).

The fact that the height of the carbody is greater than its width makes its lateral-axis moment of inertia higher than the vertical-axis one, which may prevent stable yaw oscillations of the carbody (that can be coupled with the horizontal dynamics of the bogie). At a certain speed range, the Klingel frequency (frequency of the kinematic oscillation of the bogie) and the carbody horizontal-mode frequency are close; this can produce large oscillations of the carbody if the lateral damping of the secondary suspension is too low.

The secondary suspensions should be as linear as possible in order to facilitate the filtering of the vibration from the bogie. The dampers should be linear and behave symmetrically in the traction and compression directions. If the vehicle is equipped with coil springs for the secondary suspensions, they can have internal modes of vibration at relatively low frequencies (surge frequencies), and, as they are lightly damped, vibration above the cut-off frequency of the secondary suspension can be induced, thus exciting structural modes of the carbody. Air springs can avoid this problem; additionally, they allow the natural frequencies of the carbody to be kept constant independent of the vehicle mass. Owing to their non-linear properties, friction dampers can have a negative effect on the vibration comfort of the vehicle.

Many vehicles travelling at speeds above 120 km/h are equipped with yaw dampers (an example can be seen in Figure 14.43b). Yaw dampers have a non-linear behaviour, with a very high initial slope in the force-velocity curve, until saturation (which is reached at very low velocity, less than 10 mm/s). If the initial slope is too high, the vibration comfort will be adversely affected, leading to a stick-slip phenomenon in the damper, as found in bogies that have friction yaw dampers.

14.10.3.3 Primary Suspension

The stiffness of the primary suspension is generally higher than that of the secondary suspension. The bogies have a similar set of modes of vibration to those of the carbody, but at higher frequencies. The stiffness between the wheelset and the bogie frame must be high to guarantee the running stability of the vehicle and the transmission of the braking/traction forces. In order to produce a cut-off effect, the primary suspension stiffness and damping in the vertical direction must be linear. The required longitudinal stiffness, which is much higher than the lateral one, can hardly be achieved by means of springs; usually, rubber springs and mechanisms (bars with bushings at the ends) are adopted. The traction and braking forces between the bogie frame and the carbody are transmitted through a mechanism or traction lever. The traction lever produces a high stiffness between the carbody and bogie frame in the longitudinal direction.

The natural frequencies of the bogie frame on the primary suspension are typically around 10 Hz. As this is much higher than the cut-off frequency of the secondary suspension, the dynamics of the bogie and the carbody are uncoupled, and the largest displacements of the bogie at its resonances are not transmitted to the carbody.

14.10.3.4 Traction Mechanisms

The traction mechanisms can affect the modes and the transmission of forces from the bogie to the carbody. Some examples are assessed below.

The sketch in Figure 14.44 shows a bogie that has radial arm primary suspensions; an example of this kind of suspension can be also seen in Figure 14.43b. The arms and the bogie frame can be considered as a four-bar mechanism, and the centre of rotation of the bogie frame is at the intersection of the arm lines (which are drawn in the figure in dashed lines). The position of the centre of rotation of the pitch motion should be at the location of the traction lever; otherwise, the pitch mode of vibration could transmit forces to the carbody through the traction lever.

The design in Figure 14.45 represents a bogie that has a Watt mechanism at each primary suspension. The Watt four-bar mechanism is formed by the two rods that have the same lengths and the axlebox. The wheelset axle must be in the middle between the joints of the axlebox. By considering the relative motion with respect to the bogie frame, the displacement of the centre of the axlebox is approximately a vertical straight line (the precise curve is a lemniscata). Consequently, the high-frequency vibration of the axle box centre of mass does not produce any horizontal displacement of the wheelset that may generate horizontal inertial forces that would be transmitted to the bogie frame through the rods. On the other hand, the pitch mode shape of this bogie consists of rotation around the centre of mass. This is only true if the traction lever is in the centre of mass of the bogie frame. If the centre of mass and the traction lever do not match, then the pitch mode produces horizontal forces that can be transmitted through the traction lever.

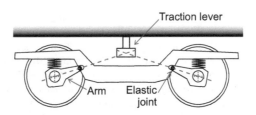

FIGURE 14.44 Bogie with arm linkage.

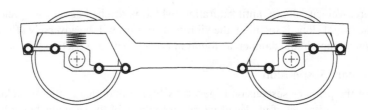

FIGURE 14.45 Bogie with Watt mechanism.

14.10.3.5 Carbody Structure

Finally, the vibration level inside the vehicle body is also affected by its low-order structural resonances. Typically, both vertical and lateral first-order bending modes occur at frequencies around 10 Hz, and resonances of the floor occur at frequencies just above this. The excited amplitude of these should be kept to a minimum by structural design of the carbody, avoiding the coincidence of important modes and using damping treatments. The carbody structural modes should be kept above about 10 Hz to avoid the frequency range to which humans are most sensitive. This is one reason for the trend towards light, stiff materials such as aluminium extrusions in the manufacture of rolling stock.

An additional trend is the introduction of vibration-isolated 'walking' floors in passenger coaches. This is primarily aimed at reducing vibration in the audible frequency range that is important for the interior noise environment (Section 14.8) but can be effective in reducing vibration above about 20 Hz [132].

The carbody vibration, in all three directions, is felt by passengers through the seat and the seat back. Seat dynamics must also therefore be considered. The coupled system of seat and human body exhibits a vertical resonance typically between 4 Hz and 6 Hz and at a similar frequency for fore-aft vibration due to the stiffness of the backrest. With the very soft seats used on some old rolling stock, these resonances can cause the vibration at the floor level to be made worse for the passenger, rather than better, by the soft seat.

REFERENCES

1. European Union Commission, Commission Regulation No 1304/2014 of 26 November 2014 on the technical specification for interoperability relating to the subsystem 'rolling stock – noise', *Official Journal of the European Communities*, 12.12.2014 L356/421–437, Brussels, Belgium, 2014.
2. F. Fahy, D.J. Thompson (Eds.), *Fundamentals of Sound and Vibration*, 2nd edition, CRC Press, Boca Raton, FL, 2015.
3. L.E. Kinsler, A.R. Frey, A. Coppens, J.V. Sanders, *Fundamentals of Acoustics*, 4th edition, Wiley, New York, 2000.
4. ISO 532:1975, *Acoustics: Method For Calculating Loudness Level*, International Organization for Standardization, Geneva, Switzerland, 1975.
5. D. Thompson, *Railway Noise and Vibration: Mechanisms, Modelling and Means of Control*, Elsevier, Oxford, UK, 2009.
6. G. Squicciarini, M.G. Toward, D.J. Thompson, C.J.C. Jones, Statistical description of wheel roughness, In: J. Nielsen, D. Anderson, P. E. Gautier, M. Iida, J. Nelson, T. Tielkes, D.J. Thompson, D. Towers, P. de Vos (Eds.), *Noise and Vibration Mitigation for Rail Transportation Systems, Proceedings 11th International Workshop on Railway Noise*, 9–13 September 2013, Uddevalla, Sweden, Notes on Numerical Fluid Mechanics and Multidisciplinary Design, 126, 651–658, 2015.
7. ISO 3095:2013, *Acoustics: Railway Applications—Measurement of Noise Emitted by Railbound Vehicles*, International Organization for Standardization, Geneva, Switzerland, 2013.
8. EN 15610:2009, *Railway Applications: Noise emission—Rail Roughness Measurement Related to Rolling Noise Generation*, European Committee for Standardization, Brussels, Belgium, 2009.

9. P.J. Remington, Wheel/rail noise – part IV: Rolling noise, *Journal of Sound and Vibration*, 46(3), 419–436, 1976.
10. P.J. Remington, J. Webb, Estimation of wheel/rail interaction forces in the contact area due to roughness, *Journal of Sound and Vibration*, 193, 83–102, 1996.
11. D.J. Thompson, C.J.C. Jones, A review of the modelling of wheel/rail noise generation, *Journal of Sound and Vibration*, 231, 519–536, 2000.
12. EN 15461:2008+A1, *Railway Applications: Noise Emission – Characterization of the Dynamic Properties of Track Selections for Pass by Noise Measurements*, European Committee for Standardization, Brussels, Belgium, 2010.
13. T.X. Wu, D.J. Thompson, Theoretical investigation of wheel/rail non-linear interaction due to roughness excitation, *Vehicle System Dynamics*, 34, 261–282, 2000.
14. D.J. Thompson, Wheel-rail noise generation – part I: Introduction and interaction model, *Journal of Sound and Vibration*, 161(3), 387–400, 1993.
15. D.J. Thompson, C.J.C. Jones, Sound radiation from a vibrating railway wheel, *Journal of Sound and Vibration*, 253(2), 401–419, 2002.
16. D.J. Thompson, C.J.C. Jones, N. Turner, Investigation into the validity of two-dimensional models for sound radiation from waves in rails, *Journal of the Acoustical Society of America*, 113(4), 1965–1974, 2003.
17. D.J. Thompson, B. Hemsworth, N. Vincent, Experimental validation of the TWINS prediction program for rolling noise, part 1: Description of the model and method, *Journal of Sound and Vibration*, 193(1), 123–135, 1996.
18. D.J. Thompson, P. Fodiman, H. Mahé, Experimental validation of the TWINS prediction program for rolling noise, part 2: Results, *Journal of Sound and Vibration*, 193, 137–147, 1996.
19. C.J.C. Jones, D.J. Thompson, Extended validation of a theoretical model for railway rolling noise using novel wheel and track designs, *Journal of Sound and Vibration*, 267, 509–522, 2003.
20. P. Hübner, The action programme of UIC, CER and UIP 'Abatement of railway noise emissions on goods trains', *Journal of Sound and Vibration*, 231, 511–517, 2000.
21. UIC, Real noise reduction of freight wagon retrofitting, MD-AF20120302, Union Internationale des Chemins de Fer, Paris, France, 2013.
22. B. Asmussen, H. Onnich, R. Strube, L.M. Greven, S. Schröder, K. Jäger, K.G. Degen, Status and perspectives of the "Specially Monitored Track", *Journal of Sound and Vibration*, 293, 1070–1077, 2006.
23. G. Hölzl, A quiet railway by noise optimised wheels, *ZEV+DET Glasers Annalen*, 188, 20–23, 1994. (in German)
24. F. Färm, Evaluation of wheel dampers on an intercity train, *Journal of Sound and Vibration*, 267, 739–747, 2003.
25. C.J.C. Jones, D.J. Thompson, Rolling noise generated by wheels with visco-elastic layers, *Journal of Sound and Vibration*, 231, 779–790, 2000.
26. S. Cervello, G. Donzella, A. Pola, M. Scepi, Analysis and design of a low-noise railway wheel, *Journal of Rail and Rapid Transit*, 215(3), 179–192, 2001.
27. D.J. Thompson, Wheel/rail noise generation, part II: Wheel vibration, *Journal of Sound and Vibration*, 161, 401–419, 1993.
28. N. Vincent, P. Bouvet, D.J. Thompson, P.E. Gautier, Theoretical optimization of track components to reduce rolling noise, *Journal of Sound and Vibration*, 193, 161–171, 1996.
29. D.J. Thompson, C.J.C. Jones, T.P. Waters, D. Farrington, A tuned damping device for reducing noise from railway track, *Applied Acoustics*, 68, 43–57, 2007.
30. G. Squicciarini, M.G.R. Toward, D.J. Thompson, Experimental procedures for testing the performance of rail dampers, *Journal of Sound and Vibration*, 359, 21–39, 2015.
31. G. Squicciarini, D.J. Thompson, M.G.R. Toward, R.A. Cottrell, The effect of temperature on railway rolling noise, *Journal of Rail and Rapid Transit*, 230, 1777–1789, 2016.
32. X. Zhang, G. Squicciarini, D.J. Thompson, Sound radiation of a railway rail in close proximity to the ground, *Journal of Sound and Vibration*, 326, 111–124, 2016.
33. C.M. Nilsson, C.J.C. Jones, D.J. Thompson, J. Ryue, A waveguide finite element and boundary element approach to calculating the sound radiated by railway and tram rails, *Journal of Sound and Vibration*, 321, 813–836, 2009.
34. J. Oertli, The STAIRRS project, work package 1: A cost-effectiveness analysis of railway noise reduction on a European scale, *Journal of Sound and Vibration*, 267, 431–437, 2003.
35. R.R.K. Jones, Bogie shrouds and low barriers could significantly reduce wheel/rail noise, *Railway Gazette International*, 150(7), 459–462, 1994.

36. R. Jones, M. Beier, R.J. Diehl, C. Jones, M. Maderboeck, C. Middleton, J. Verheij, Vehicle-mounted shields and low trackside barriers for railway noise control in a European context, *Proceedings of Inter-Noise 2000*, Nice, France, 2372–2379, 2000.
37. I.L. Vér, C.S. Ventres, M.M. Myles, Wheel/rail noise, part II: Impact noise generation by wheel and rail discontinuities, *Journal of Sound and Vibration*, 46, 395–417, 1976.
38. R.A. Clark, P.A. Dean, J.A. Elkins, S.G. Newton, An investigation into the dynamic effects of railway vehicles running on corrugated rails, *Journal of Mechanical Engineering Science*, 24(2), 65–76, 1982.
39. J.C.O. Nielsen, A. Igeland, Vertical dynamic interaction between train and track: Influence of wheel and track imperfections, *Journal of Sound and Vibration*, 187, 825–839, 1995.
40. T.X. Wu, D.J. Thompson, A hybrid model for the noise generation due to railway wheel flats, *Journal of Sound and Vibration*, 251, 115–139, 2002.
41. A. Johansson, J. Nielsen, Railway wheel out-of-roundness: Influence on wheel-rail contact forces and track response, *Proceedings 13th International Wheelset Congress*, Rome, Italy, 2001.
42. T.X. Wu, D.J. Thompson, On the impact noise generation due to a wheel passing over rail joints, *Journal of Sound and Vibration*, 267, 485–496, 2003.
43. T.X. Wu, D.J. Thompson, A model for impact forces and noise generation due to wheel and rail discontinuities, *Proceedings 8th International Congress on Sound and Vibration*, Hong Kong, China, 2905–2912, 2001.
44. D.J. Thompson, G. Squicciarini, B. Ding, L. Baeza, A state-of-the-art review of curve squeal noise: Phenomena, mechanisms, modelling and mitigation, In: D. Anderson, P.-E. Gautier, M. Iida, J. Nelson, D.J. Thompson, T. Tielkes, D. Towers, P. de Vos, J.C.O. Nielsen (Eds.), *Noise and Vibration Mitigation for Rail Transportation Systems: Proceedings of the 12th International Workshop on Railway Noise*, vol. 139, pp. 1–28. Springer, Heidelberg, 2018.
45. B. Armstrong, C. Canudas de Wit, Friction modeling and compensation, chapter in *The Control Handbook*, CRC Press, Boca Raton, FL, 1369–1382, 1996.
46. C. Glocker, E. Cataldi-Spinola, R.I. Leine, Curve squealing of trains: Measurement, modelling and simulation, *Journal of Sound and Vibration*, 324(1), 365–386, 2009.
47. A. Pieringer, A numerical investigation of curve squeal in the case of constant wheel/rail friction, *Journal of Sound and Vibration*, 333(18), 4295–4313, 2014.
48. N. Hoffmann, M. Fischer, R. Allgaier, L. Gaul, A minimal model for studying properties of the mode-coupling type instability in friction induced oscillations, *Mechanics Research Communications*, 29(4), 197–205, 2002.
49. B. Ding, G. Squicciarini, D.J. Thompson, An assessment of mode-coupling and falling-friction mechanisms in railway curve squeal through a simplified approach, *Journal of Sound and Vibration*, 423, 126–140, 2018.
50. F.G. De Beer, M.H.A. Janssens, P.P Kooijman, Squeal of rail bound vehicles influenced by lateral contact position, *Journal of Sound and Vibration*, 267(3), 497–507, 2003.
51. X. Liu, P.A. Meehan, Investigation of the effect of lateral adhesion and rolling speed on wheel squeal noise, *Journal of Rail and Rapid Transit*, 227(5), 469–480, 2013.
52. J.R. Koch, N. Vincent, H. Chollet, O. Chiello, Curve squeal of urban rolling stock – part 2: Parametric study on a 1/4 scale test rig, *Journal of Sound and Vibration*, 293(3), 701–709, 2006.
53. P.J. Remington, Wheel/rail squeal and impact noise: What do we know? What don't we know? Where do we go from here? *Journal of Sound and Vibration*, 116, 339–353, 1987.
54. D. Curley, D.C. Anderson, J. Jiang, D. Hanson, Field trials of gauge face lubrication and top-of-rail friction modification for curve noise mitigation, In: J. Nielsen, D. Anderson, P.-E. Gautier, M. Iida, J. Nelson, T. Tielkes, D.J. Thompson, D. Towers, P. de Vos (Eds.), *Noise and Vibration Mitigation for Rail Transportation Systems, Proceedings 11th International Workshop on Railway Noise*, 9–13 September 2013, Uddevalla, Sweden, Notes on Numerical Fluid Mechanics and Multidisciplinary Design, 126, 449–456, 2015.
55. J. Jiang, R. Dwight, D. Anderson, Field verification of curving noise mechanisms, In: T. Maeda, P.-E. Gautier, C. Hanson, B. Hemsworth, J. Nelson, B. Schulte-Werning, D.J. Thompson, P. de Vos (Eds.), *Noise and Vibration Mitigation for Rail Transportation Systems, Proceedings of the 10th International Workshop on Railway Noise*, 18–22 October 2010, Nagahama, Japan, Notes on Numerical Fluid Mechanics and Multidisciplinary Design, 118, 349–356, 2012.
56. R. Corradi, P. Crosio, S. Manzoni, G. Squicciarini, Experimental investigation on squeal noise in tramway sharp curves, In: *Proceedings 8th International Conference on Structural Dynamics*, Leuven, Belgium, 3214–3221, 2011.

57. G. Squicciarini, S. Usberti, D.J. Thompson, R. Corradi, A. Barbera, Curve squeal in the presence of two wheel/rail contact points, In: J. Nielsen, D. Anderson, P.-E. Gautier, M. Iida, J. Nelson, T. Tielkes, D.J. Thompson, D. Towers, P. de Vos (Eds.), *Noise and Vibration Mitigation for Rail Transportation Systems, Proceedings 11th International Workshop on Railway Noise*, 9–13 September 2013, Uddevalla, Sweden, Notes on Numerical Fluid Mechanics and Multidisciplinary Design, 126, 603–610, 2015.
58. I. Zenzerovic, W. Kropp, A. Pieringer, Influence of spin creepage and contact angle on curve squeal: A numerical approach, *Journal of Sound and Vibration*, 419, 268–280, 2018.
59. A. Monk-Steel, D. Thompson, Models for railway curve squeal noise, 7th International Conference on Recent Advances in Structural Dynamics, Southampton, UK, 2003.
60. D.J. Fourie, J. Petrus, P. Gräbe, S. Heyns, R.D. Fröhling, Experimental characterisation of railway wheel squeal occurring in large-radius curves, *Journal of Rail and Rapid Transit*, 230, 1561–1574, 2016.
61. M.J. Rudd, Wheel/rail noise – part II: Wheel squeal, *Journal of Sound and Vibration*, 46, 381–394, 1976.
62. U. Fingberg, A model of wheel-rail squealing noise, *Journal of Sound and Vibration*, 143, 365–377, 1990.
63. F.J. Périard, Wheel-rail noise generation: Curve squealing by trams, PhD thesis, Technische Universiteit Delft, The Netherlands, 1998.
64. M.A. Heckl, Curve squeal of train wheels, part 2: Which wheel modes are prone to squeal? *Journal of Sound and Vibration*, 229, 695–707, 2000.
65. O. Chiello, J.B. Ayasse, N. Vincent, J.R. Koch, Curve squeal of urban rolling stock – part 3: Theoretical model, *Journal of Sound and Vibration*, 293(3), 710–727, 2006.
66. J.F. Brunel, P. Dufrénoy, M. Naït, J.L. Muñoz, F. Demilly, Transient models for curve squeal noise, *Journal of Sound and Vibration*, 293(3), 758–765, 2006.
67. D.T. Eadie, M. Santoro, W. Powell, Local control of noise and vibration with Keltrack friction modifier and Protector trackside application: An integrated solution. *Journal of Sound and Vibration*, 267, 761–772, 2003.
68. P. Wetta, F. Demilly, Reduction of wheel squeal noise generated on curves or during braking, *Proceedings 11th International Wheelset Congress*, Paris, France, 301–306, 1995.
69. N. Curle, The influence of solid boundaries upon aerodynamic sound, *Proceedings of the Royal Society of London: Series A: Mathematical and Physical Sciences*, 231, 505–514, 1955.
70. M.J. Lighthill, On sound generated aerodynamically: I – general theory, *Proceedings of the Royal Society of London: Series A: Mathematical and Physical Sciences*, 211, 564–587, 1952.
71. C. Talotte, Aerodynamic noise, a critical survey, *Journal of Sound and Vibration*, 231, 549–562, 2000.
72. D.J. Thompson, E. Latorre Iglesias, X. Liu, J. Zhu, Z. Hu, Recent developments in the prediction and control of aerodynamic noise from high-speed trains, *International Journal of Rail Transportation*, 3(3), 119–150, 2015.
73. B. Barsikow, Experiences with various configurations of microphone arrays used to locate sound sources on railway trains operated by DB-AG, *Journal of Sound and Vibration*, 193, 283–293, 1996.
74. E. Latorre Iglesias, D.J. Thompson, M.G. Smith, Experimental study of the aerodynamic noise radiated by cylinders with different cross-sections and yaw angles, *Journal of Sound and Vibration*, 361, 108–129, 2016.
75. E. Masson, N. Paradot, E. Allain, The numerical prediction of the aerodynamic noise of the TGV POS high-speed train power, In: T. Maeda, P.-E. Gautier, C. Hanson, B. Hemsworth, J. Nelson, B. Schulte-Werning, D.J. Thompson, P. de Vos (Eds.), *Noise and Vibration Mitigation for Rail Transportation Systems, Proceedings of the 10th International Workshop on Railway Noise*, 18–22 October 2010, Nagahama, Japan, Notes on Numerical Fluid Mechanics and Multidisciplinary Design, 118, 437–444, 2012.
76. E. Latorre Iglesias, D.J. Thompson, M.G. Smith, Component-based model to predict aerodynamic noise from high-speed train pantographs, *Journal of Sound and Vibration*, 394, 280–305, 2017.
77. F. Poisson, Railway noise generated by high-speed trains, In: J. Nielsen, D. Anderson, P.-E. Gautier, M. Iida, J. Nelson, T. Tielkes, D.J. Thompson, D. Towers, P. de Vos (Eds.), *Noise and Vibration Mitigation for Rail Transportation Systems, Proceedings 11th International Workshop on Railway Noise*, 9–13 September 2013, Uddevalla, Sweden, Notes on Numerical Fluid Mechanics and Multidisciplinary Design, 126, 457–480, 2015.
78. F. Poisson, P.E. Gautier, The railway noise reductions achieved in the Silence project, *Proceedings of Acoustics'08*, Paris, France, 2631–2636, 2008.
79. A.E.J. Hardy, R.R.K. Jones, Control of the noise environment for passengers in railway vehicles, *Journal of Rail and Rapid Transit*, 203(2), 79–85, 1989.

80. L. Guccia, Passenger comfort: General issues, *Presented at 6th International Workshop on Railway Noise*, Ile des Embiez, France, 1998.
81. R. Wettschureck, G. Hauck, Geräusche und Erschütterungen aus dem Schienenverkehr (Noise and ground vibration from rail traffic), In: M. Heckl, H.A. Müller (Eds.), *Taschenbuch der Technischen Akustik*, 2nd edition, Springer Verlag, Berlin, Germany, 1995.
82. ISO 3381:2005, *Railway Applications: Acoustics – Measurement of Noise Inside Railbound Vehicles*, International Organization for Standardization, Geneva, Switzerland, 2005.
83. A.E.J. Hardy, Railway passengers and noise, *Proceedings of the Institution of Mechanical Engineers*, 213(3), 173–180, 1999.
84. L. Willenbrink, Noise inside and outside vehicles and from railway lines, *Proceedings of Inter-Noise 73*, Copenhagen, Denmark, 362–371, 1973.
85. A.E.J. Hardy, Measurement and assessment of noise within passenger trains, *Journal of Sound and Vibration*, 231, 819–829, 2000.
86. N.J. Shaw, The prediction of railway vehicle internal noise using statistical energy analysis techniques, MSc thesis, Heriot-Watt University, Edinburgh, UK, 1990.
87. T. Kohrs, Structural acoustic investigation of orthotropic plates, Diploma thesis, Technische Universität Berlin, Germany, 2002.
88. A. Bracciala, C. Pellegrini, FEM analysis of the internal acoustics of a railway vehicle and its improvements, *Proceedings World Congress on Railway Research*, 16–19 November 1997, Florence, Italy, PB002515, 1997.
89. F. Létourneux, S. Guerrand, F. Poisson, Low-frequency acoustic transmission of high-speed trains: Simplified vibroacoustic model, *Journal of Sound and Vibration*, 231, 847–851, 2000.
90. J. Zhang, X. Xiao, X. Sheng, C. Zhang, R. Wang, X. Jin, SEA and contribution analysis for interior noise of a high speed train, *Applied Acoustics*, 112, 158–170, 2016.
91. K. De Meester, L. Hermans, K. Wyckaert, N. Cuny, Experimental SEA on a highspeed train carriage, *Proceedings of ISMA21*, Leuven, Belgium, 151–161, 1996.
92. P. Geissler, D. Neumann, SEA modelling for extruded profiles for railway passenger coaches, *Proceedings of Euro-Noise 98*, Munich, Germany, 189–194, 1998.
93. G. Xie, D.J. Thompson, C.J.C. Jones, A modelling approach for the vibroacoustic behaviour of aluminium extrusions used in railway vehicles, *Journal of Sound and Vibration*, 293, 921–932, 2006.
94. D. Backström, Analysis of the sound transmission loss of train partitions, MSc thesis, Royal Institute of Technology (KTH), Stockholm, Sweden, 2001.
95. ISO 2631-1:1997, *Mechanical Vibration and Shock: Evaluation of Human Exposure to Whole Body Vibration – part 1: General Requirements*, International Organization for Standardization, Geneva, Switzerland, 1997.
96. BS 7385-2:1993, *Evaluation and Measurement for Vibration in Buildings: Guide to Damage Levels from Groundborne Vibration*, British Standards Institution, London, UK, 1993.
97. DIN 4150-3:2016-12, *Vibration in Buildings – part 3: Effects on Structures*, Deutches Institut für Normung, Berlin, Germany, 2016.
98. E. Kassa, J.C.O. Nielsen, Dynamic interaction between train and railway turnout: Full-scale field test and validation of simulation models, *Vehicle System Dynamics*, 46 (S1), 521–534, 2008.
99. G. Kouroussis, D.P. Connolly, G. Alexandrou, K. Vogiatzis, Railway ground vibrations induced by wheel and rail singular defects, *Vehicle System Dynamics*, 53(10), 1500–1519, 2015.
100. V.V. Krylov, Generation of ground vibration by superfast trains, *Applied Acoustics*, 44, 149–164, 1995.
101. C. Madshus, A.M. Kaynia, High speed railway lines on soft ground: Dynamic behaviour at critical train speed, *Journal of Sound and Vibration*, 231, 689–701, 2000.
102. X. Sheng, C.J.C. Jones, M. Petyt, Ground vibration generated by a load moving along a railway track, *Journal of Sound and Vibration*, 228, 129–156, 1999.
103. S. Nazarian, M.R. Desai, Automated surface wave method: Field testing, *Journal of Geotechnical Engineering, Proceedings of the ASCE*, 119(7), 1094–1111, 1993.
104. E. Kausel, J. Roësset, Stiffness matrices for layered soils, *Bulletin of the Seismological Society of America*, 71, 1743–1761, 1981.
105. X. Sheng, CJC Jones, D.J. Thompson, A theoretical study on the influence of the track on train-induced ground vibration, *Journal of Sound and Vibration*, 272, 909–936, 2004.
106. P.A. Costa, A. Colaço, R. Calçada, A.S. Cardoso, Critical speed of railway tracks: Detailed and simplified approaches, *Transportation Geotechnics*, 2, 30–46, 2015.
107. X. Sheng, C.J.C. Jones, D.J. Thompson, A theoretical model for ground vibration from trains generated by vertical track irregularities, *Journal of Sound and Vibration*, 272, 937–965, 2004.

108. X. Sheng, C.J.C. Jones, D.J. Thompson, Prediction of ground vibration from trains using the wavenumber finite and boundary element methods, *Journal of Sound and Vibration*, 293, 575–586, 2006.
109. L. Auersch, The excitation of ground vibration by rail traffic: Theory of vehicle-track-soil interaction and measurements on high-speed lines, *Journal of Sound and Vibration*, 284(1–2), 103–132, 2005.
110. G. Lombaert, G. Degrande, J. Kogut, S. François, The experimental validation of a numerical model for the prediction of railway induced vibrations, *Journal of Sound and Vibration*, 297(3–5), 512–535, 2006.
111. P.A. Costa, R. Calçada, A.S. Cardoso, Track–ground vibrations induced by railway traffic: In-situ measurements and validation of a 2.5D FEM-BEM model, *Soil Dynamics and Earthquake Engineering*, 32, 111–128, 2012.
112. E. Ntotsios, D.J. Thompson, M.F.M. Hussein, The effect of track load correlation on ground-borne vibration from railways, *Journal of Sound and Vibration*, 402, 142–163, 2017.
113. X. Sheng, C.J.C. Jones, D.J. Thompson, A comparison of a theoretical model for vibration from trains with measurements, *Journal of Sound and Vibration*, 267, 621–635, 2003.
114. N. Triepaischajonsak, D.J. Thompson, C.J.C. Jones, J. Ryue, J.A. Priest, Ground vibration from trains: Experimental parameter characterization and validation of a numerical model, *Journal of Rail and Rapid Transit*, 225, 140–153, 2011.
115. C.J.C. Jones, D.J. Thompson, M. Petyt, A model for ground vibration from railway tunnels, *Proceedings of the Institution Civil Engineers, Transportation*, 153(2), 121–129, 2002.
116. X. Sheng, C.J.C. Jones, D.J. Thompson, Modelling ground vibration from railways using wavenumber finite- and boundary-element methods, *Proceedings of the Royal Society A*, 461, 2043–2070, 2005.
117. M.F.M. Hussein, S. François, M. Schevenels, H.E.M. Hunt, J.P. Talbot, G. Degrande, The fictitious force method for efficient calculation of vibration from a tunnel embedded in a multi-layered half-space, *Journal of Sound and Vibration*, 333(25), 6996–7018, 2014.
118. Q. Jin, D.J. Thompson, D.E.J. Lurcock, M.G.R. Toward, E. Ntotsios, A 2.5D finite element and boundary element model for the ground vibration from trains in tunnels and validation using measurement data, *Journal of Sound and Vibration*, 422, 373–389, 2018.
119. J.C.O. Nielsen, A. Mirza, S. Cervello, P. Huber, R. Müller, B. Nelain, P. Ruest, Reducing train-induced ground-borne vibration by vehicle design and maintenance, *International Journal of Rail Transportation*, 3(1), 17–39, 2015.
120. D.J. Thompson, J. Jiang, M.G.R. Toward, M.F.M. Hussein, E. Ntotsios, A. Dijckmans, P. Coulier, G. Lombaert, G. Degrande, Reducing railway-induced ground-borne vibration by using open trenches and soft-fi Red barriers, *Soil Dynamics and Earthquake Engineering*, 88, 45–59, 2016.
121. P. Coulier, A. Dijckmans, J. Jiang, D.J. Thompson, G. Degrande, G. Lombaert, Stiff wave barriers for the mitigation of railway induced vibrations, In: J. Nielsen, D. Anderson, P.-E. Gautier, M. Iida, J. Nelson, T. Tielkes, D.J. Thompson, D. Towers, P. de Vos (Eds.), *Noise and Vibration Mitigation for Rail Transportation Systems, Proceedings 11th International Workshop on Railway Noise*, 9–13 September 2013, Uddevalla, Sweden, Notes on Numerical Fluid Mechanics and Multidisciplinary Design, 126, 539–546, 2015.
122. A. Dijckmans, A. Ekblad, A. Smekal, G. Degrande, G. Lombaert, Efficacy of a sheet pile wall as a wave barrier for railway induced ground vibration, *Soil Dynamics and Earthquake Engineering*, 84, 55–69, 2016.
123. D.J. Thompson, J. Jiang, M.G.R. Toward, M.F.M. Hussein, A. Dijckmans, P. Coulier, G. Degrande, G. Lombaert, Mitigation of railway-induced vibration by using subgrade stiffening, *Soil Dynamics and Earthquake Engineering*, 79, 89–103, 2015.
124. P. Coulier, S. François, G. Degrande, G. Lombaert, Subgrade stiffening next to the track as a wave impeding barrier for railway induced vibrations, *Soil Dynamics and Earthquake Engineering*, 48, 119–131, 2013.
125. A. Dijckmans, P. Coulier, J. Jiang, M.G.R. Toward, D.J. Thompson, G. Degrande, G. Lombaert, Mitigation of railway induced ground vibration by heavy masses next to the track, *Soil Dynamics and Earthquake Engineering*, 75, 158–170, 2015.
126. A. Johansson, J.C.O. Nielsen, R. Bolmsvikc, A. Karlstrom, R. Lunden, Under sleeper pads: Influence on dynamic train-track interaction, *Wear*, 265, 1479–1487, 2008.
127. R. Verachtert, H.E.M. Hunt, M.F.M. Hussein, G. Degrande, Changes of perceived unevenness caused by in-track vibration countermeasures in slab track, *European Journal of Mechanics A/Solids*, 65, 40–58, 2017.
128. M.J. Griffin, *Handbook of Human Vibration*, Academic Press, London, UK, 1990.
129. BS 6841:1987, *Measurement and Evaluation of Human Exposure to Whole-Body Mechanical Vibration and Repeated Shock*, British Standards Institution, London, UK, 1987.

130. ISO 2631-4:2001+A1, Evaluation of human exposure to whole body vibration – part 4: Guidelines for the evaluation of the effects of vibration and rotational motion on passenger and crew comfort in fixed-guideway transport systems, International Organization for Standardization, Geneva, Switzerland, 2010.
131. EN 12299, *Railway Applications: Ride Comfort of Passengers – Measurement and Evaluation*, European Committee for Standardization, Brussels, Belgium, 2009.
132. M. Wollström, *Internal Noise and Vibrations in Railway Vehicles: A Pilot Study*, TRITA-FKT Report 1998:44, Royal Institute of Technology (KTH), Stockholm, Sweden, 1998.

15 Active Suspensions

Roger M. Goodall and T.X. Mei

CONTENTS

- 15.1 Introduction 580
- 15.2 Basics of Active Suspensions 580
  - 15.2.1 Concepts 580
  - 15.2.2 Active and Semi-Active 581
  - 15.2.3 Design Considerations 581
- 15.3 Tilting Trains 584
  - 15.3.1 Concept and Equations 584
  - 15.3.2 Mechanical Configurations and Requirements ... 585
    - 15.3.2.1 Passive Tilt 585
    - 15.3.2.2 Direct Tilt 585
    - 15.3.2.3 Bolster-Based Tilting 586
  - 15.3.3 Control: Strategies and Assessment 586
    - 15.3.3.1 Control Approaches 587
    - 15.3.3.2 Assessment of Controller Performance ... 588
  - 15.3.4 Summary of Tilting 591
- 15.4 Active Secondary Suspensions 591
  - 15.4.1 Concepts and Requirements 591
  - 15.4.2 Configurations 591
  - 15.4.3 Control Strategies 592
    - 15.4.3.1 Sky-Hook Damping 592
    - 15.4.3.2 Softening of Suspension Stiffness 594
    - 15.4.3.3 Low-Bandwidth Controls 595
    - 15.4.3.4 Modal Control Approach 595
    - 15.4.3.5 Model-Based Control Approaches 596
    - 15.4.3.6 Actuator Response 596
    - 15.4.3.7 Semi-Active Control 597
  - 15.4.4 Technology 597
    - 15.4.4.1 Sensing 597
    - 15.4.4.2 Actuators 597
  - 15.4.5 Examples 598
    - 15.4.5.1 Servo-Hydraulic Active Lateral Suspension ... 598
    - 15.4.5.2 Shinkansen/Sumitomo Active Suspension ... 599
    - 15.4.5.3 Swedish Green Train Project 600
- 15.5 Active Primary Suspensions 600
  - 15.5.1 Concepts and Requirements 600
  - 15.5.2 Configurations and Control Strategies 601
    - 15.5.2.1 Solid-Axle Wheelsets 601
    - 15.5.2.2 Independently Rotating Wheelsets 603
    - 15.5.2.3 Integrated Control Design 604
    - 15.5.2.4 Assessment of Control Performance 604

	15.5.3	Technology 605
		15.5.3.1 Sensing and Estimation Techniques 605
		15.5.3.2 Actuators 606
		15.5.3.3 Controllers and Fault Tolerance 606
	15.5.4	Examples 607
		15.5.4.1 Bombardier Mechatronic Bogie 607
		15.5.4.2 Independently Driven Wheels 607
15.6	Overall Summary and Long-Term Trends 609
Nomenclature 610
References 610

15.1 INTRODUCTION

It is clear from the preceding chapters that railway vehicle dynamics has developed principally as a mechanical engineering discipline, but an important technological change is starting to occur through the use of active suspension concepts. Advanced control on rail vehicles has been common for many decades in the power electronic control of traction systems, and it is now firmly established as the standard technology, which has yielded substantial benefits, but its application to suspensions is much more recent. Although the term 'active suspension' is commonly taken to relate to providing improved ride quality; in fact, it is a generic term that defines the use of actuators, sensors and electronic controllers to enhance and/or replace the springs and dampers that are the key constituents of a conventional, purely mechanical, 'passive' suspension; as such, it can be applied to any aspect of the vehicle's dynamic system.

15.2 BASICS OF ACTIVE SUSPENSIONS

Vehicle dynamicists have been aware of active suspensions for some time, with major reviews having been undertaken in 1975, 1983, 1997 and 2007 [1–4], but so far, they have only found substantial application in tilting trains, which can now be thought of as an established suspension technology. However, there are two other major categories: active secondary suspensions for improved ride quality, and active primary suspensions for improved running stability and curving performance. The sections that follow in this chapter deal with these three categories in turn: tilting, active secondary and active primary suspensions, but first, there are a number of general principles and considerations that need to be explained.

15.2.1 CONCEPTS

The general scheme of an active suspension is shown in diagrammatic form in Figure 15.1. The input/output relationship provided by the suspension, which in the passive case is determined solely by

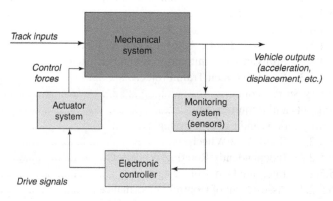

FIGURE 15.1 Generalised active suspension scheme. (From Iwnicki, S. (Ed.), *Handbook of Railway Vehicle Dynamics*, CRC Press, Boca Raton, FL, 2006. With permission.)

Active Suspensions

the values of masses, springs, dampers and the geometrical arrangement, is now dependent upon the configuration of sensors and actuators and upon the control strategy in the electronics (almost invariably now involving some form of software processing). For all the three categories, it will be seen that the introduction of active control enables performances to be achieved that are either not possible or extremely difficult with a passive suspension.

15.2.2 Active and Semi-Active

The greatest benefits can be achieved by using fully controllable actuators with their own power supply, such that the desired control action (usually a force) can be achieved irrespective of the movement of the actuator. Energy can flow from or to the power supply, as required, to implement the particular control law. This is known as a 'full-active' suspension, but it is also possible to use a 'semi-active' approach in which the characteristic of an otherwise-passive suspension component can be rapidly varied under electronic control; see Figure 15.2. Semi-active suspensions usually use controllable dampers of some kind, although the concept is not restricted to dampers.

One benefit of the semi-active approach compared with full-active is its simplicity, because a separate power supply for the actuator is not needed. The disadvantage of a semi-active damper is that the force remains dependent upon the speed of damper movement, which means that large forces cannot be produced when its speed is low, and, in particular, it cannot develop a positive force when the speed reverses, because it is only possible to dissipate energy, not inject it. Figure 15.3 clarifies the limitation by showing areas on the force-velocity diagram that are available for a semi-active damper based upon its minimum and maximum levels, whereas an actuator in a full-active system can cover all four quadrants of the diagram. This limitation upon controllability restricts the performance of a semi-active suspension to a significant degree [5].

Closely related is an option known variously as semi-passive, adjustable passive or adaptive passive, in which the characteristics are varied on the basis of a variable that is not influenced by the dynamic system being controlled (e.g., as a function of vehicle speed).

15.2.3 Design Considerations

For designing active suspension systems such as these, an important difference arises compared with passive suspensions. A conventional suspension is designed with as accurate a model as possible so that the computer simulation can predict the on-track performance effectively. The designer

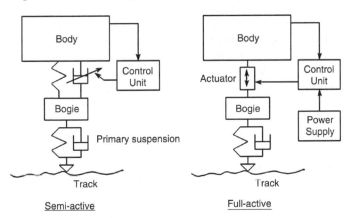

FIGURE 15.2 Semi-active and full-active control. (From Iwnicki, S. (Ed.), *Handbook of Railway Vehicle Dynamics*, CRC Press, Boca Raton, FL, 2006. With permission.)

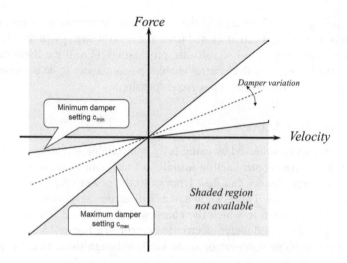

FIGURE 15.3 Force-velocity diagram for semi-active damper. (From Iwnicki, S. (Ed.), *Handbook of Railway Vehicle Dynamics*, CRC Press, Boca Raton, FL, 2006. With permission.)

then adjusts the values of the suspension components based upon well-understood expectations for the particular vehicle configuration until the required performance is achieved. However, for an active suspension, it is important to distinguish between the design model and the simulation model: the former is a simplified model used for synthesis of the control strategy and algorithm, whereas the latter is a full-complexity model to test the system performance, i.e., as used for conventional suspensions [6]. The importance of having an appropriately simplified design model is less critical when 'classical' control design techniques are being used, although even here, key insights arise with simplified models. The real issue arises when modern model-based design approaches are being used, either for the controller itself or for estimators to access difficult or impossible-to-measure variables, in which case the controller and/or estimator assumes a dynamic complexity equal to or greater than that of the design model. Since a good simulation model of a railway vehicle will usually have more than a hundred states, a controller based upon this model would at best be overly complex to implement, at worst impossible because some of the states may be uncontrollable or unobservable.

There are formal methods for reducing the model complexity, but engineering experience will often provide a suitable abstraction. For example, there is a relatively weak coupling between the vertical and lateral motions of rail vehicles, and, depending on the objectives, only selected degrees of freedom need to be included in the design model. Common simplifications are based around a vehicle model that is partitioned into side-view, plan-view and end-view models: the side-view model is concerned with the bounce and pitch degrees of freedom and can be used for active vertical suspensions; the plan-view model deals with the lateral and yaw motions and can be used for active lateral suspensions and active steering/stability control; the end-view model covers the bounce, lateral and roll motions and can be used for the design of tilting controllers.

It is, of course, essential that such modelling software can support the integration of the controller into the mechanical system. This can be achieved within a single package, but there is a strong argument for distinct but well-integrated software, i.e., one of the many MBS dynamics packages in combination with a control design package such as Matlab/Simulink®. Ideally, there should be a number of interface possibilities: controllers designed using the simplified design model can be exported into the MBS package for simulation purposes; equally, it is often valuable to be able to export a complex but linearised model from the MBS package for further controller evaluation, using the targeted analytical tools provided for controller design, and finally, running the two

Active Suspensions

packages simultaneously in a co-simulation mode is also important because this avoids the need for conversion and export. Co-simulation is now a mature and efficient capability of many vehicle dynamics design packages.

In practice, the design process employing both simulation and design models may be quite complex, and Figure 15.4 provides an example. This shows that the design model should be verified by comparison with the full complexity simulation model. Then, the control law developed using the design model can be tested and checked by co-simulation or by incorporation into the MBS software. Control design packages such as Matlab provide code generation, which can be tested with the MBS model, and also the code running within the target processing system can be bench tested, e.g., using time or frequency domain signals and compared with the input-output characteristics of the controller in Matlab.

A final point is illustrated by Figure 15.5, which emphasises the multi-objective nature of the design process. There are a variety of input types and output variables that must be considered, and each output will be affected by different combinations of inputs. The design will require an optimisation involving constraints. For example, an active secondary suspension design must minimise the frequency-weighted accelerations on the vehicle body, without exceeding the maximum suspension deflection; an active primary suspension must optimise the curving performance whilst maintaining minimum levels of running stability on straight track, etc.

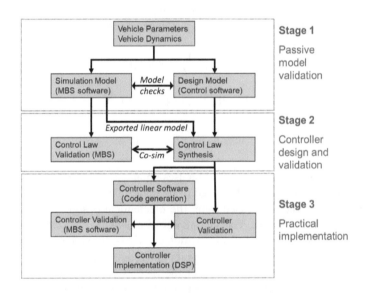

FIGURE 15.4 An example control design and simulation modelling process.

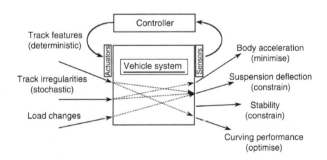

FIGURE 15.5 Multi-objective design optimisation process. (From Iwnicki, S. (Ed.), *Handbook of Railway Vehicle Dynamics*, CRC Press, Boca Raton, FL, 2006. With permission.)

15.3 TILTING TRAINS

The earliest proposals for tilting trains go back into the first half of the twentieth century, but in Europe, it was not until the 1960s and 1970s that experimental developments were aimed towards producing operational trains for prestigious high-speed routes. These emerged as the Talgo Pendular in Spain (1980), the APT in the UK, the LRC in Canada (1982), the first ETR 450 Pendolino trains in Italy (1988) and the X2000 in Sweden (1990) [7]. A similar pattern occurred in Japan, although the developments there were aimed at the regional/narrow-gauge railways rather than the high-speed Shinkansen. The 1990s saw tilting mature into a standard railway technology, with applications extending throughout most of Europe and Japan, and all the major rail vehicle manufacturers now offer and supply tilting trains for regional and high-speed applications.

15.3.1 Concept and Equations

Tilting trains take advantage of the fact that the speed through curves is principally limited by passenger comfort and not by either the lateral forces on the track or the risk of overturning, although these are constraints that cannot be ignored. Tilting the vehicle bodies on curves reduces the acceleration experienced by the passenger, which permits higher speeds and provides a variety of operational benefits. The principles and basic equations related to tilting are relatively straightforward and are given by Equations 15.1 and 15.2 for non-tilting and tilting, respectively:

Steady-state acceleration experienced by the passengers (passive, non-tilting):

$$\ddot{y}_{ss} = g\,\sin\theta_{passive} = \frac{V^2}{R} - g\,\sin(\theta_{cant}[-\theta_R]) \qquad (15.1)$$

Steady-state acceleration experienced by the passengers (tilting):

$$\ddot{y}_{ss} = g\,\sin\theta_{active} = \frac{V^2}{R} - g\,\sin(\theta_{cant} + \theta_{tilt}) \qquad (15.2)$$

These can also be presented in a manner that focuses upon the operational advantages.

There are two primary decisions that need to be made. The first is what maximum tilt angle is to be provided (θ_{tilt}), a decision based upon the mechanical design of the vehicle, especially taking gauging issues into account. The second decision is what cant deficiency the passengers should experience on a steady curve (θ_{active}), which clearly is of primary importance to comfort. Given these two decisions, and the cant deficiency that applies for the passive (non-tilting) case ($\theta_{passive}$), it is possible to derive an equation for the increase in speed offered by tilt. Note that, although the curve radius and the acceleration due to gravity appear in the basic acceleration equations, they disappear when the equation is dealing with the fractional or percentage speed increase:

$$\text{speed increase} = \frac{V_{active} - V_{passive}}{V_{passive}} = \left\{ \sqrt{\frac{\sin(\theta_{cant} + \theta_{tilt} + \theta_{active})}{\sin(\theta_{cant} + \theta_{passive})}} - 1 \right\} \times 100\% \qquad (15.3)$$

Although, in principle, the cant deficiency could be fully compensated by the tilting action, i.e., to make $\theta_{active} = 0$, in practice, this is not sensible either from the operational or the ride comfort viewpoint. It is possible to recognise this by introducing a 'cant deficiency compensation factor' (K_{CD}), an important design parameter in the tilt controller, the choice of which will be discussed later.

$$K_{CD} = 1 - \frac{\theta_{active}}{(\theta_{active} + \theta_{tilt})}, \text{ i.e., } \frac{\theta_{tilt}}{\theta_{active} + \theta_{tilt}} \qquad (15.4)$$

Consider some examples: Track cant is usually around 6°, and typically, 6° of cant deficiency is applied for a non-tilting train. Applying 9° of tilt and a cant deficiency of 6° for the tilting train, the calculation indicates a speed-up of 32%, with a compensation factor of 60%. In this particular case, the passengers nominally experience the same comfort level on curves (although the passive vehicle will usually roll out by a small angle, typically less than 1°, so, in practice, tilting will give a small reduction in the curving acceleration). Another example might be where the tilting cant deficiency is reduced to 4.5°, perhaps to offer an improved ride comfort; using a slightly smaller tilt angle of 8°, the speed-up falls to 24%, with a compensation factor of 64%.

Therefore, speeds on curves may theoretically be increased by around 30% or more with tilting trains. However, the performance on curve transitions as well as the steady curves are important from the comfort viewpoint, and the comfort level can be predicted using a method described by a European standard [8]. It is an empirically based method in which the percentage of passengers (P_{CT}) who are likely to feel uncomfortable during the curve is determined from the lateral acceleration, the lateral jerk and the body roll velocity experienced during the transition. Details of the method are given in the quoted reference, including the way in which the three measurements should be made.

There is also the issue of motion sickness, which is related to, but distinct from, the aforementioned transition response. In contrast to the curve transition comfort level, which may be considered on a curve-by-curve basis, motion sickness is a cumulative effect, which comes as a consequence of a number of human factors, the exact nature of which is not fully understood. The effect is aggravated on highly curvaceous routes with rapid transitions [9], and fully compensating for cant deficiency ($K_{CD} = 1$) is particularly detrimental for motion sickness, probably because passengers sense the rotation and see that they are on a curve but do not feel any lateral acceleration. Hence, the degree to which the curving acceleration is compensated for by the tilting action is an important factor, but once this has been optimised, the only other mitigation measure is operating at lower speed.

Tilting trains still introduce motion sickness to a significant minority of the travelling public, and Section 4.2 in [7] includes a survey on the problem but concludes *'The laboratory tests cannot, therefore, explain the motion sickness experienced in tilting trains. It is likely that combinations of motions, in particular translation combined with rotation, contribute to the motion sickness experienced in tilting trains as these combinations have proven to be highly effective in provoking motion sickness in laboratories.'* It seems probable that motion sickness is essentially an emergent property of tilting that cannot be fully resolved.

15.3.2 Mechanical Configurations and Requirements

Broadly speaking, four mechanical arrangements are possible to provide the tilting action.

15.3.2.1 Passive Tilt

The first is passive or pendular tilt, in which the secondary suspension is raised to around roof level in the vehicle; the vehicle centre of gravity is then substantially below the suspension, and the body naturally swings outwards, reducing the lateral curving acceleration experienced by the passengers. This is a technique pioneered in the Talgo trains – the air springs are raised by means of vertical pillars at the vehicle ends, an arrangement made much easier by the articulated configuration of the trains.

15.3.2.2 Direct Tilt

A second approach is to achieve tilt directly by applying active control to the secondary roll suspension. One method that has been tried in both Europe and Japan is to apply differential control to the air springs, but this may cause a dramatic increase in air consumption and generally has not found favour, although one Japanese development has achieved it by transferring air between the air springs, using a hydraulically actuated pneumatic cylinder [10]. The alternative method of direct control of the roll suspension is by means of an active anti-roll bar (stabiliser), and this is applied in Bombardier's regional talent trains. This uses the traditional arrangement consisting of a transversely mounted torsion tube on the bogie with vertical links to the vehicle body, except that one of the links is replaced by a hydraulic actuator and thereby applies tilt via the torsion tube.

15.3.2.3 Bolster-Based Tilting

The previous two arrangements are very much minority solutions, because most implementations use a tilting bolster to provide the tilt action. An important distinction is where this bolster is fitted, compared with the secondary suspension, which leads to the third and fourth of the arrangements. With the tilting bolster above the secondary suspension, the increased curving forces need to be reacted by the secondary lateral suspension; since a stiffer lateral suspension is not consistent with the higher operating speed of a tilting train, in practice, either an increased lateral suspension movement or some form of active centring method is needed to avoid reaching the limits of travel.

The final arrangement has the tilting bolster below the secondary suspension, thereby avoiding the increased curving forces on the lateral suspension, and this is probably the most common of all schemes, the necessary rotation being achieved either using a pair of inclined swing links or using a circular roller beam. Typical schemes with inclined swing links and with a roller beam are shown in Figure 15.6.

Actuators to provide tilt action have seen significant development since the early days of tilt. Some early systems were based upon controlling the air springs (i.e., intrinsically pneumatic actuation), but it was more normal to use hydraulic actuators because these tend to be the natural choice for mechanical engineers. However, experiments with electro-mechanical actuators in the UK in the 1970s [11], in Switzerland in the 1980s and in Germany in the 1990s paved the way for a progressive change away from the hydraulic solution. Electric motors controlled by solid-state power amplifiers drive screws fitted with high-efficiency ball or roller nuts to convert rotary to linear motion. They are less compact than hydraulic actuators at the point of application, but overall, they provide significant system benefits, and they are now employed in the majority of new European tilting trains. Interestingly, Japanese tilting technology has tended to use pneumatic actuators.

15.3.3 CONTROL: STRATEGIES AND ASSESSMENT

This section explains some of the essential control approaches that are possible to achieve effective tilting action and then discusses how the performance of particular controllers can be assessed.

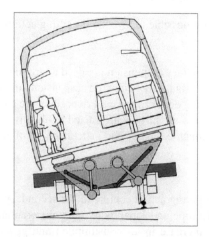 

FIGURE 15.6 Tilt below secondary suspension schemes. (From Iwnicki, S. (Ed.), *Handbook of Railway Vehicle Dynamics*, CRC Press, Boca Raton, FL, 2006. With permission.)

15.3.3.1 Control Approaches

The most intuitive control approach is to put an accelerometer on the vehicle body to measure the lateral acceleration that the tilt action is required to reduce, yielding the 'nulling' controller shown in Figure 15.7. The accelerometer signal is used to drive the actuator in a direction that will bring it towards zero, i.e., a classical application of negative feedback. Implementation of the required value of K_{CD} can be achieved with a modification of the basic nulling controller to give a partial tilt action by including a measure of the tilt angle in the controller, as shown by the dotted arrow on the figure. However, there is a difficulty with this scheme due to interaction with the lateral suspension; the roll and lateral modes of the vehicle body are strongly coupled in a dynamic sense, and it can be shown that, if the loop bandwidth is low enough not to interfere with the lateral suspension, it is then too slow-acting on the curve transition.

Figure 15.8 shows the next solution: the dynamic interaction problem can be avoided by putting the accelerometer on a non-tilting part or, in other words, the bogie. This will then tell how much tilt is needed to reduce the lateral acceleration on the vehicle body and can be multiplied by the factor K_{CD}, which determines what proportion of the lateral acceleration is to be compensated. $K_{CD} = 1$ gives 100% compensation, not a good idea for motion sickness reasons, and typically, 60% or 70% compensation is used (as mentioned previously). This 'tilt angle command signal' then provides the input to a feedback loop, which uses a measurement of the tilt angle.

Unfortunately, there is still a problem, because the accelerometer on the bogie is not only measuring the curving acceleration but also the pure lateral accelerations due to track irregularities. With the accelerometer on the vehicle body, these accelerations are reduced by the secondary suspension, but they are much larger when the accelerometer is on the bogie. Consequently, it is necessary to add a low-pass filter to reduce the acceleration signals caused by the track irregularities; otherwise, there is too much tilt action on straight track, resulting in a worse ride quality. However, to apply sufficient filtering, too much delay is introduced at the start of the curve, so the full lateral curving acceleration is felt for a short time, even though it reduces to an acceptable level once properly on the curve.

Figure 15.9 shows the next step: the signal from the vehicle in front is used to provide precedence, carefully designed so that the delay introduced by the filter compensates for the precedence time corresponding to a vehicle length. In effect, this scheme is what most European tilting trains now use; sometimes, roll and/or yaw gyros are used to improve the response, and normally, a single command signal is generated from the first vehicle and transmitted digitally with appropriate time delays down the train.

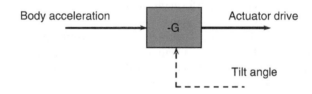

FIGURE 15.7 Nulling tilt controller.

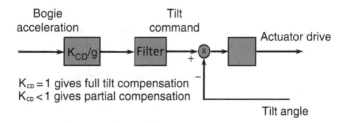

FIGURE 15.8 Command-driven tilt controller.

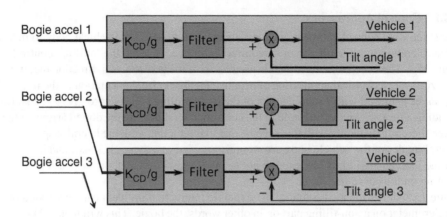

FIGURE 15.9 Precedence tilt controller.

The signal from the bogie-mounted accelerometer is essentially being used to generate an estimate of the true cant deficiency of the track's design alignment, the difficulty being to exclude the effects of the track irregularities.

Of course, the first vehicle does not benefit from the precedence effect, and an obvious longer-term development is to feed the vehicle controllers with signals from a database that defines the track, instead of from the accelerometer. Both the position of the vehicle along the track and the curve data contained in the database need to be known accurately for this approach to work effectively, but it is likely that such systems will become the norm in the future.

Japanese tilting trains often use a balise on the track ahead of the curve to initiate the tilting action, a technique that helps to mitigate the relatively slow response of the pneumatic tilt actuators.

15.3.3.2 Assessment of Controller Performance

It is clear that what happens in the steady curve is important, but also the dynamic response during the transition must be considered. In an ideal tilt control strategy, the tilt angle of the body should rise progressively, perfectly aligned both with the onset of curving acceleration and the rising cant angle, and the difficulties in achieving this kind of response have been explained previously. Since the principal benefit of tilt is to be able to operate at higher speeds without degradation in passenger comfort, there are two issues from a design point of view: how well does the tilting vehicle perform on straight track, and how well does it perform on curve transitions?

The accelerometer-based control strategies mean these two issues must in practice be traded off against each other; if the tilt action is fast to give good transition performance, in general, the straight track ride quality may be degraded. Qualitatively, a good tilt controller responds principally to the deterministic track inputs and as much as possible ignores the random track irregularities. In order to assess different tilt control strategies in an objective manner, it is necessary to define appropriate criteria and conditions.

The straight track performance can be dealt with using a criterion of degrading the lateral ride quality by no more than a specified margin compared with the non-tilting response, a typical value being 7.5%. Note that, for assessing the tilt controller performance, this comparison must be made at the higher speed. Of course, a comparison of ride quality with a lower-speed vehicle is also needed, but achieving a satisfactory ride quality at elevated speeds will require either an improved suspension or a better-quality track, i.e., not a function of the tilt controller.

The curve transition response has to be separated into two aspects. Firstly, the fundamental tilting response, measured by the P_{CT} factors, as described previously, must be as good as a passive vehicle at lower (non-tilting) speed; otherwise, the passenger comfort will inevitably be diminished, no matter how effective the tilt control is. It is possible therefore to introduce the idea of 'ideal tilting', where the tilt action follows the specified tilt compensation perfectly, defined on the basis of

Active Suspensions

the fundamental tilt system parameters – the operating speed (increase), maximum tilt angle and the cant deficiency compensation factor. This combination of parameters can be optimised using the previously mentioned P_{CT} factor approach for deterministic inputs, in order to choose a basic operating condition, and this will give 'ideal' P_{CT} values (one for standing and one for sitting).

Consider, for example, the ideal transition responses for passive and tilting trains shown in Figures 15.10 and 15.11, where the transition length gives a relatively long time of 3.2 seconds for the passive vehicle, and both cant and cant deficiency are 6°. The passive response also includes the effect of a 'passive roll-out' of 1°, but this is obviously vehicle-dependent. Figure 15.11 shows the corresponding acceleration, jerk and roll velocity graphs for the tilting condition. The acceleration is slightly decreased and the jerk level slightly increased, but the most notable change is

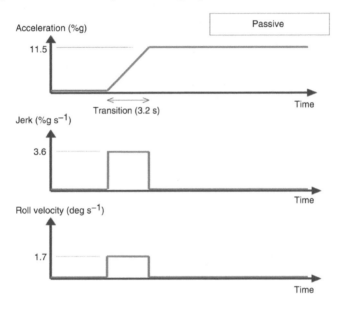

FIGURE 15.10 'Ideal' transition response for non-tilting train. (From Iwnicki, S. (Ed.), *Handbook of Railway Vehicle Dynamics*, CRC Press, Boca Raton, FL, 2006. With permission.)

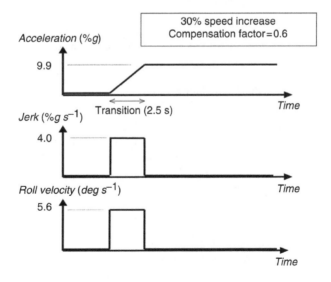

FIGURE 15.11 Ideal transition response for tilting train. (From Iwnicki, S. (Ed.), *Handbook of Railway Vehicle Dynamics*, CRC Press, Boca Raton, FL, 2006. With permission.)

the 3.4 times increase in roll rate, partly due to the tilting action, but enhanced by the reduced transition time arising from the higher speed. These results are for a particular tilting condition, 30% higher speed with a compensation factor of 0.6, but similar assessments can, of course, be made for other conditions.

Figure 15.12 shows the results of P_{CT} calculations undertaken with speed-up factors between 15% and 35% and compensation factors K_{CD} from 40% to 80%, where the dotted horizontal lines show the values for the slower non-tilting train. In this case, when there is a relatively slow transition, increasing the compensation factor improves the comfort level, although this is not necessarily the case with faster transitions. However, Figure 15.13 shows that a larger tilt angle is required. Further discussion of this trade-off can be found in [12].

The other consideration is that it is necessary to quantify the additional dynamic effects that are caused by the suspension/controller dynamics as the transitions to and from the curves are encountered, which can be quantified as the deviations from the 'ideal' response mentioned in the previous paragraph. These deviations relate to both the lateral acceleration and the roll velocity, although the former is likely to be the main consideration. The performance in this respect will depend upon detailed characteristics of the controller such as the filter in the command-driven scheme and the tuning parameters in the tilt angle feedback loop. It is clear that the deviations need to be minimised, but at present, there is no information regarding their acceptable size, although the values derived for a normal passive suspension can be used as a guide.

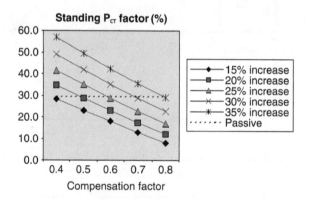

FIGURE 15.12 P_{CT} factors versus K_{CD} (for standing passengers). (From Iwnicki, S. (Ed.), *Handbook of Railway Vehicle Dynamics*, CRC Press, Boca Raton, FL, 2006. With permission.)

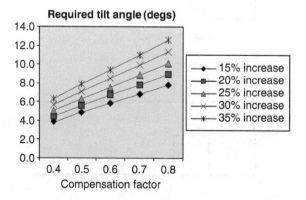

FIGURE 15.13 Tilt angle requirement versus K_{CD} (for standing passengers). (From Iwnicki, S. (Ed.), *Handbook of Railway Vehicle Dynamics*, CRC Press, Boca Raton, FL, 2006. With permission.)

15.3.4 Summary of Tilting

Although tilting seems in many ways to be a rather simple concept, it requires considerable care in practice and has taken many years to introduce reliable operational performance, and tilting controllers still need adjustment for specific route characteristics. It is likely that the state of the art will continue to be developed in the years to come.

15.4 ACTIVE SECONDARY SUSPENSIONS

15.4.1 Concepts and Requirements

For the secondary suspensions, active controls improve the vehicle dynamic response and provide a better isolation of the vehicle body from the track irregularities than the use of only passive springs and dampers. Active control can be applied to any or all of the suspension degrees of freedom but, when applied in the lateral direction, will implicitly include the yaw mode and in the vertical direction will include the pitching mode. Controlling in the roll direction is of course equivalent to tilting, which is essentially a particular form of active secondary suspension but of sufficient importance to have its own section. The improved performance can be used to deliver a better ride quality, but this is not directly cost-beneficial, and so, it will normally be used to enable higher train speed whilst maintaining the same level of passenger comfort. The other possibility is to provide the same ride quality on less-well-aligned track, in which case the cost-benefit analysis needs to take account of the reduced track maintenance cost.

15.4.2 Configurations

Active secondary suspensions can be used in the lateral and/or vertical directions, and a number of actuator configurations are possible as illustrated in Figure 15.14.

Actuators can be used to replace the passive suspensions, as shown in Figure 15.14a, and the suspension behaviour will be completely controlled via active means. In practice, however, it is more beneficial that actuators are used in conjunction with passive components. When connected in parallel, as illustrated in Figure 15.14b, the size of an actuator can be significantly reduced, as the passive component will be largely responsible for providing a constant force to support the body mass of a vehicle in the vertical direction or quasi-static curving forces in the lateral direction. On the other hand, fitting a spring in series with the actuator, as shown in Figure 15.14c, helps with the high-frequency problem

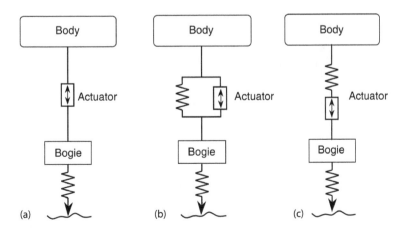

FIGURE 15.14 Active secondary suspension actuator configurations: (a) actuator alone, (b) actuator in parallel with spring, and (c) actuator in series with spring. (From Iwnicki, S. (Ed.), *Handbook of Railway Vehicle Dynamics*, CRC Press, Boca Raton, FL, 2006. With permission.)

caused by the lack of response in the actuator movement and control output at high frequencies (see Section 15.4.3.6), and, in practice, a combination of a parallel spring for load carrying and a series spring to help with the high-frequency response is the most appropriate arrangement. The stiffness of the series spring depends upon the actuator technology; a relatively high value can be used for technologies such as hydraulics that have good high-frequency performance and a softer value can be used for other technologies, which means that achieving a high bandwidth is more problematic.

The other option is to use actuators mounted between adjacent vehicles, although the improvement of ride quality is less significant, and, in general, the design problem is more difficult because the complete train becomes strongly coupled in a dynamic sense via the actuators.

15.4.3 Control Strategies

15.4.3.1 Sky-Hook Damping

Different control approaches are possible for active suspensions. A high-bandwidth system, which deals with the random track inputs caused by irregularities, can be used to improve suspension performance largely through the provision of damping to an absolute datum. The principle of absolute damping is depicted in Figure 15.15a, where a damper is connected from the mass to the sky; hence, the term 'sky-hook' damping is used. For practical implementations, the principle of the sky-hook damping can be realised by an arrangement shown in Figure 15.15b. The feedback measurement is provided from a sensor mounted above the suspension on the body, and the control demand is fed to the actuator, which is placed between the vehicle body and the bogie.

A comparison between the passive and the sky-hook damping of a simple (one-mass) system illustrates very well the potential advantages of the active concept. For a passive damper, a higher level of modal damping can only be achieved at the expense of increased suspension transmissibility at high frequencies, as shown in Figure 15.16. For the sky-hook damper, however, the high-frequency responses are independent of the damping ratio, and the transmissibility is significantly lower than that of the passive damping at all frequencies concerned. This is also the consequence of applying optimal control, as described in [13].

$$F_a = -C_s \cdot \frac{dz}{dt} \qquad (15.5)$$

where C_s is sky-hook damping coefficient and F_a is the actuator force.

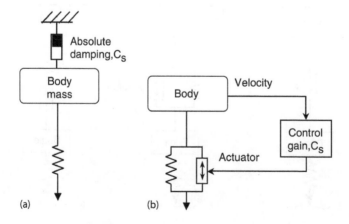

FIGURE 15.15 Sky-hook damping: (a) concept and (b) practical implementation arrangement. (From Iwnicki, S. (Ed.), *Handbook of Railway Vehicle Dynamics*, CRC Press, Boca Raton, FL, 2006. With permission.)

Active Suspensions

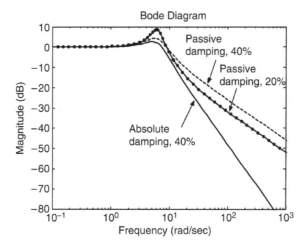

FIGURE 15.16 Comparison of passive and absolute damping. (From Iwnicki, S. (Ed.), *Handbook of Railway Vehicle Dynamics*, CRC Press, Boca Raton, FL, 2006. With permission.)

This yields the transfer function for a simple single-mass suspension as follows:

$$\frac{z}{z_t} = \frac{K}{K + sC_s + s^2 M} \tag{15.6}$$

where K [N/m] and M [kg] are the spring constant and mass, respectively.

The equivalent transfer function for the passive suspension with a conventional damper having a coefficient C [Ns/m] is

$$\frac{z}{z_t} = \frac{K + sC}{K + sC + s^2 M} \tag{15.7}$$

from which it can be seen that the high-frequency response is proportional to $1/f$ for the passive suspension compared with $\alpha\ 1/f^2$ for the active sky-hook suspension, the overall effect of which was seen on Figure 15.16.

Sky-hook damping gives a profound improvement to the ride quality for straight track operation; however, it creates large deflections at deterministic features such as curves and gradients. Although this can be accommodated in the control design, e.g., by filtering out the low frequency components from the measurements, which are largely caused by track deterministic features [14], it is recognised that reducing the deterministic deflections to an acceptable level will compromise the performance achievable with 'pure' sky-hook damping. In fact, the absolute velocity signal that is required for sky-hook damping will usually be produced by integrating the signal from an accelerometer, and so, in practice, it will also be necessary to filter out the low-frequency components in order to avoid problems with thermal drift in the accelerometer – a typical scheme is shown in Figure 15.17. In practice, the integrator and high-pass filter will normally be combined to provide a 'self-zeroing' integration effect.

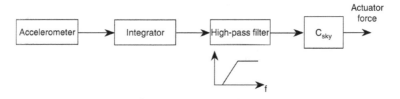

FIGURE 15.17 Scheme for practical implementation of sky-hook damping.

Whilst the use of a high-pass filter can eliminate the quasi-static suspension deflections due to the large quasi-state force of the sky-hook damping on gradients or curves, it is less effective in reducing the transient suspension travel on track transitions, and, in the selection of the filter cut-off frequency, there is a difficult trade-off between the ride quality improvement of the vehicle body and the maximum movement of the suspension.

There are a number of possible solutions proposed to overcome the problem. The complementary filtering approach, as shown in Figure 15.18, uses a relative damping force at the low frequency range in addition to the sky-hook damping at high frequencies, which results in a much improved trade-off. There are also Kalman filter-based strategies, where the effect of the track deterministic input can be minimised or the track features are directly estimated [15]. A typical trade-off comparison between different control approaches is given in Figure 15.19, in this case for the vertical suspension of a vehicle running onto a gradient [14].

15.4.3.2 Softening of Suspension Stiffness

Another strategy is to create a softer suspension by controlling the actuator to cancel part of the suspension force produced by the passive stiffness. The control equation is of a simple form, as shown in Equation 15.8, but note that positive feedback is used to reduce the overall stiffness to a value of $(K-K_s)$. The corresponding transfer function is not given because it is a trivial change to what was given for the passive suspension.

$$F_a = +K_s \cdot (z - z_t) \tag{15.8}$$

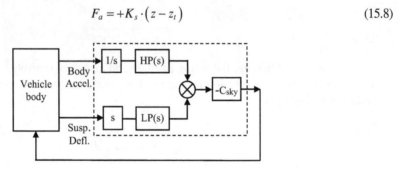

FIGURE 15.18 Complementary filters. (From Iwnicki, S. (Ed.), *Handbook of Railway Vehicle Dynamics*, CRC Press, Boca Raton, FL, 2006. With permission.)

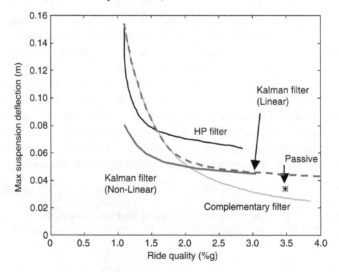

FIGURE 15.19 Trade-off between ride quality and suspension deflection. (From Iwnicki, S. (Ed.), *Handbook of Railway Vehicle Dynamics*, CRC Press, Boca Raton, FL, 2006. With permission.)

15.4.3.3 Low-Bandwidth Controls

Active secondary suspensions can also be used to provide a low-bandwidth control, which is similar to tilting controls in that the action is intended to respond principally to the low-frequency deterministic track inputs. In low-bandwidth systems, there will be passive elements that dictate the fundamental dynamic response, and the function of the active element is associated with some low-frequency activity. A particular use of the concept is for maintaining the average position of the suspension in the centre of its working space, thereby minimising contact with the mechanical limits of travel and enabling the possibility of a softer spring to be used [16,17]. This is a powerful technique for the lateral suspensions, because curving forces are large, and without centring action, there may sometimes be significant reductions in ride quality whilst curving.

The idea of active levelling (or centring for a lateral suspension) can be achieved using the equation:

$$F_a = -K_L \cdot \int (z - z_t) dt \quad (15.9)$$

The suspension transfer function becomes:

$$\frac{z}{z_t} = \frac{K_L + K \cdot s + C \cdot s^2}{K_L + K \cdot s + C \cdot s^2 + M \cdot s^3} \quad (15.10)$$

The integral action changes it from second to third order, the effect of which is less obvious, but it can readily be shown that the suspension deflection $(z - z_t)$ is zero in response to an acceleration input from the track, and it is this characteristic that corresponds to the self-levelling effect.

15.4.3.4 Modal Control Approach

For a conventional railway vehicle with two secondary suspensions between the body frame and the two bogies, it is possible to use local control for each suspension, i.e., the measurement from the sensor(s) mounted above either of the bogies is fed to the controller, which controls the actuator on the same bogie. However, the tuning of control parameters may be problematic, as interactions between the two controllers via the vehicle body will be inevitable. To overcome the difficulty, a centralised controller for both suspensions may be used to enable independent control of the body modes.

Figure 15.20 shows how the lateral and yaw modes of a vehicle body can be separately controlled by using active suspensions in the lateral direction, and a similar scheme can be applied to actuators

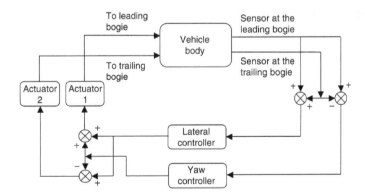

FIGURE 15.20 Modal control diagram. (From Iwnicki, S. (Ed.), *Handbook of Railway Vehicle Dynamics*, CRC Press, Boca Raton, FL, 2006. With permission.)

in the vertical direction to control the bounce and pitch modes. The output measurements from the two bogies are decomposed to give feedback signals required by the lateral and yaw controllers, respectively, and the output signals from the two controllers are then recombined to control two actuators at the two bogies accordingly. In this way, it is possible to apply different levels of control, in particular to reduce the suspension frequency and add more damping to the yaw (or pitch) mode, which is less susceptible to the low-frequency deterministic inputs.

15.4.3.5 Model-Based Control Approaches

Increased system complexity also encourages the use of mathematically rigorous design approaches such as optimal control, which enables a trade-off between ride quality and suspension deflection to be formally defined and optimised [18]. Equation 15.11 gives a typical cost function, which is minimised in the design of an optimal controller to reflect the suspension design problem. Suitable choices of the weighting factors q_1, q_2 and r (on the body acceleration a_b, suspension deflection x_b and actuator force F_a) enable an appropriate design trade-off to be achieved.

$$J = \int \left(q_1 \cdot a_b^2 + q_2 \cdot x_d^2 + r \cdot F_a^2 \right) dt \tag{15.11}$$

15.4.3.6 Actuator Response

In order to implement the control laws, e.g., those listed in the previous subsection, it is necessary to have force control. However, very few actuator types inherently provide a force, and so, an inner force feedback loop is required, but it is important to appreciate that dynamics of this actuator force loop need to be significantly faster than is immediately obvious. The physical explanation can be seen from Figure 15.21, which is a generalised scheme of a force-controlled actuator.

The force command to the actuator would be generated by an active suspension controller, not shown here because it is useful to consider what happens even with a zero force command, which should, in principle, leave the suspension response unchanged compared with the passive suspension. The track input will impact upon the dynamic system, and this will cause actuator movement, which the force control loop must counteract in order to keep its force as close as possible to zero. Remembering that the actuator will be connected across the secondary suspension, its movements at low frequencies will be small as the vehicle follows the intended features of the track but relatively large at high frequencies as the suspension provides isolation by absorbing the track irregularities. How well the actuator generates the force required of it in the presence of the high-frequency movement depends upon the characteristics of the actuator, and it is not possible to generalise. A more detailed analysis reveals that a force loop bandwidth in the region of 20 Hz will still yield noticeable degradations in the acceleration on the suspended mass at around 4 Hz, but this analysis is a detailed control engineering issue and beyond the scope of this handbook. Studies of this problem can however be found in [19].

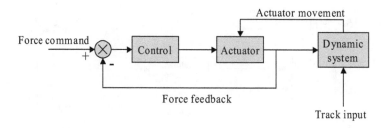

FIGURE 15.21 Actuator force control. (From Iwnicki, S. (Ed.), *Handbook of Railway Vehicle Dynamics*, CRC Press, Boca Raton, FL, 2006. With permission.)

15.4.3.7 Semi-Active Control

The basis of controlling a semi-active system is to replicate as far as possible the action of sky-hook damping [5,20], but using variable/controllable dampers rather than actuators. Most semi-active control strategies are based upon achieving the demanded force as closely as possible, but the actual damper setting is constrained to be between C_{min} and C_{max}, and Figure 15.22 shows the control concept. To achieve operation in the upper left and lower right quadrants of the force-velocity diagram of Figure 15.3, for example, which would require a negative damper setting, the semi-active controller will simply apply C_{min}. As with full-active sky-hook damping, this would potentially create large deflections on response to deterministic features; of course, a semi-active damper cannot create the necessary forces, but pre-filtering, as shown in Figure 15.17, is still required to ensure an effective control law.

Extra performance benefits are realised by adopting a modal approach similar to that shown in Figure 15.20, but achievable improvements in ride quality depend upon both the minimum damper setting and the speed of response of the control action – a switching response time of the damper setting significantly less than 10 ms is needed to ensure effective implementation.

Semi-active control generally works in the 'active' mode only half of the time and is able to deliver about half the performance improvement of a full-active system. This may be changed and performance improvement stretched closer to full-active by using some form of adaptive control strategies to extend substantially the time of the system operating in the active mode [21].

15.4.4 Technology

Technology of control concerned with the practicalities of implementation, the controller, sensors and actuators is an important issue. Satisfactory performances and costs are obviously essential, but more critically, the safety and reliability requirements must be met before any applications can be considered.

15.4.4.1 Sensing

Sensing for the active control of secondary suspensions typically involves the use of inertial sensors (e.g., accelerometers and gyroscopes), and those sensors are normally installed on vehicle frames where the vibrations due to the vehicle responding to track irregularities are largely isolated by the suspensions and the working environment is less hostile than for sensors mounted on the bogie frames or the wheelsets. Therefore, the issue of costs and reliabilities should not be a major concern.

15.4.4.2 Actuators

The provision of high-reliability actuation of sufficient performance (including the issues highlighted in Section 15.4.3.6) is one of the main challenges in active suspensions. Capital cost of the

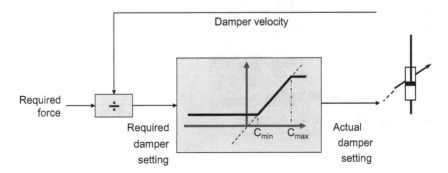

FIGURE 15.22 Controller for semi-active damper. (From Iwnicki, S. (Ed.), *Handbook of Railway Vehicle Dynamics*, CRC Press, Boca Raton, FL, 2006. With permission.)

total system is certainly important, but ease of installation, maintainability and maintenance cost, reliability and failure modes must all have essential inputs into the process of choosing and procuring the actuator system.

Actuator technologies that are possible for active suspensions are servo-hydraulic, servo-pneumatic, electro-mechanical and electro-magnetic. Servo-hydraulic actuators themselves are compact and easy to fit, but, when the power supply is included, the whole system tends to be bulky and inefficient, and there are important questions relating to maintainability. Pneumatic actuators are a possibility, particularly since the air-springs fitted to many railway vehicles can form the basic actuator, but the compressibility of air leads to inefficiency and limited controllability. Electro-mechanical actuators offer a technology with which the railway is generally familiar, and the availability of high-performance servo-motors and high-efficiency power electronics is a favourable indicator. However, they tend to be less compact, and the reliability and life of the mechanical components need careful consideration. Electro-magnetic actuators potentially offer an extremely high reliability and high-performance solution, but they tend to be very bulky and have a somewhat limited travel.

15.4.5 EXAMPLES

15.4.5.1 Servo-Hydraulic Active Lateral Suspension

The first full-scale demonstration of an active railway suspension was an active lateral secondary suspension using hydraulic actuators [22]. An actuator was fitted in parallel with the lateral secondary air suspension at each end of the vehicle, as can be seen in Figure 15.23a. The performance obtained from a comprehensive series of tests is shown in Figure 15.23b, from which it can be seen that a large improvement in ride quality was obtained – a 50% reduction compared with the passive suspension.

The controller used a modal structure shown in Figure 15.24 that provided independent control of the vehicle's lateral and yaw suspension modes using the complementary filter technique. Although hydraulic actuators provide a high bandwidth when used in normal applications; fast-acting force control loops (not shown in the diagram) were included to overcome the difficulty outlined in the Section 15.4.3.6 and to ensure adequate high-frequency performance. Even with these inner loops, it can be seen from Figure 15.23b that there is a small degradation above 3 Hz compared with the passive response.

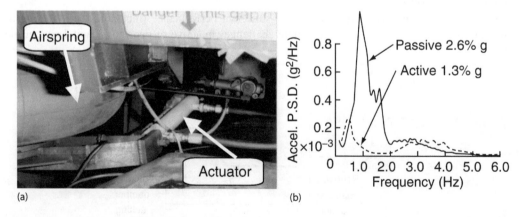

FIGURE 15.23 Servo-hydraulic active lateral suspension: (a) typical set-up and (b) experimental performance results. (From Iwnicki, S. (Ed.), *Handbook of Railway Vehicle Dynamics*, CRC Press, Boca Raton, FL, 2006. With permission.)

Active Suspensions

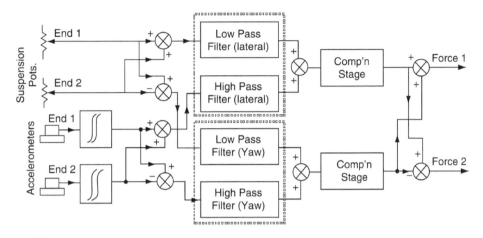

FIGURE 15.24 Controller for servo-hydraulic active lateral suspension. (From Iwnicki, S. (Ed.), *Handbook of Railway Vehicle Dynamics*, CRC Press, Boca Raton, FL, 2006. With permission.)

15.4.5.2 Shinkansen/Sumitomo Active Suspension

The first commercial use of an active suspension was developed by Sumitomo for the East Japan Railway Company on their series E2-1000 and E3 Shinkansen vehicles, introduced in 2002 [23].

The main objective of the control was the lateral vibration, i.e., closely related with riding comfort, the aim being to reduce by more than half the lateral vibration in the frequency range from 1 Hz to 3 Hz. A pneumatic actuator system was adopted, which has the advantages of easy maintenance and low cost and is installed in parallel with a secondary suspension damper (see Figure 15.25). The damper is electronically switched from a soft setting when active control is enabled to the normal harder setting for passive operation.

An H-infinity controller was designed to provide robust vibration control using measurements from body-mounted accelerometers. It provides independent control of the yaw and lateral/roll modes, with the yaw controller driving the two actuators in opposition direction, and the lateral/roll controller driving them in the same direction. Figure 15.26 is a diagram of the overall control scheme.

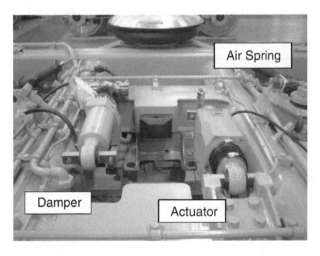

FIGURE 15.25 Actuator installation in bogie. (From Iwnicki, S. (Ed.), *Handbook of Railway Vehicle Dynamics*, CRC Press, Boca Raton, FL, 2006. With permission.)

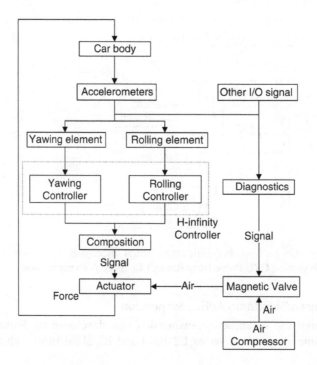

FIGURE 15.26 Overall scheme of control algorithm. (From Iwnicki, S. (Ed.), *Handbook of Railway Vehicle Dynamics*, CRC Press, Boca Raton, FL, 2006. With permission.)

It was shown that improvements of between 5 dB and 9 dB in acceleration level were achievable (44%–64% reduction); initially, it was a problem to achieve this kind of improvement in tunnel sections, and it was necessary to design a special controller that was switched in for use in tunnels.

15.4.5.3 Swedish Green Train Project

The latest experimental study of active secondary suspensions has been carried out in a collaborative project between KTH Royal Institute of Technology and Bombardier Transportation to demonstrate the ride quality improvement in both lateral and vertical directions at an acceptable cost level to enable future implementation [24,25].

On-track tests have been performed with a two-car Regina train, using electro-hydraulic actuators together with sky-hook damping control and a hold-off-device function to actively control the secondary suspensions. The evaluated measurement results show that the active suspension system significantly reduces the lateral/vertical dynamic car-body motions and the lateral quasi-static displacements between car-body and bogies in curves; this improves the ride comfort and allows higher speeds, particularly in curves.

15.5 ACTIVE PRIMARY SUSPENSIONS

15.5.1 CONCEPTS AND REQUIREMENTS

Although active control could be applied to vertical primary suspensions, in fact there seems little to be gained from such an application. The main area of interest therefore relates to controlling the wheelset kinematics through the active primary suspensions. The important issue here is the trade-off between running stability (critical speed) and curving performance, which is difficult with a

Active Suspensions

passive suspension, as has been outlined earlier. Various methods of passive mechanical steering to create radial alignment of the wheelsets on curves have been attempted with some improvement. The idea of using active control for wheelset steering is relatively new, although significant progress has recently been made, mainly theoretically but also experimentally [4,26] for two types of railway wheelset. As has been explained, a solid-axle wheelset consists of two coned or otherwise profiled wheels joined rigidly together by a solid-axle, which has the advantage of natural curving and self-centring but when unconstrained exhibits a sustained oscillation in the lateral plane, often referred to as 'wheelset hunting'. The structure of an independently rotating wheelset is very similar to that of solid-axle wheelset except that two wheels on the same axle are allowed to rotate freely. The release of the rotational constraint between the two wheels significantly reduces the longitudinal creepage on curves, but it loses the ability of natural curving and centring.

The control objectives for active primary suspensions are largely related to the wheelset configurations. For the solid-axle wheelset, the controller must produce a stabilisation effort for the kinematic mode, and it must also ensure desirable performance on curves. For the independently rotating wheelset, there is a weak instability mode that needs to be stabilised. But, more critically, a guidance control must be provided to avoid the wheelset running on flanges.

15.5.2 Configurations and Control Strategies

15.5.2.1 Solid-Axle Wheelsets

A number of actuation schemes are possible for implementing active steering. One of the obvious options is to apply a controlled torque to the wheelset in the yaw direction. This can be achieved via yaw actuators, as shown Figure 15.27a, or, in practice, very likely by means of pairs of longitudinal actuators. Alternatively, actuators may be installed onto a wheelset in the lateral direction, as shown in Figure 15.27b, but a drawback of the configuration is that the stabilisation forces also cause the ride quality on the vehicle to deteriorate.

The control development for active primary suspensions ranges from separate design for stability and steering to integrated design approaches, as presented later.

15.5.2.1.1 Stability Control – Solid-Axle Wheelset

The focus is on the stabilisation of the kinematic oscillation associated with a railway wheelset, but the control is ideally achieved in a way that it does not interfere with the natural curving and centring of the wheelset. One effective control technique is the so-called active yaw damping, where a yaw torque from an actuator, as shown in Figure 15.27a, is proportional to the lateral velocity of the wheelset [27]. The stabilising effect of the control technique can be shown using a linearised

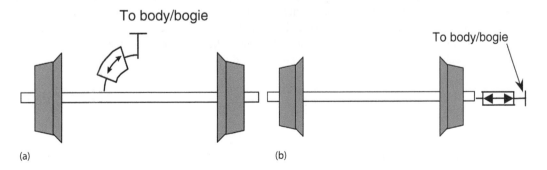

FIGURE 15.27 Actuation configurations for active steering: (a) yaw actuators and (b) lateral actuators. (From Iwnicki, S. (Ed.), *Handbook of Railway Vehicle Dynamics*, CRC Press, Boca Raton, FL, 2006. With permission.)

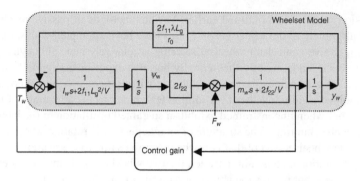

FIGURE 15.28 Active yaw damping. (From Iwnicki, S. (Ed.), *Handbook of Railway Vehicle Dynamics*, CRC Press, Boca Raton, FL, 2006. With permission.)

wheelset model given in Figure 15.28. It is clear from the figure that an unstable mode exists and that the inclusion of the active control loop produces positive damping to the mode. Alternatively, the stabilisation may be achieved by providing a yaw torque proportional to the absolute yaw angle motion of the wheelset (i.e., the concept of sky-hook stiffness), which would also eliminate the adverse effect of passive suspensions on curves [28]. It can also be shown that an alternative and equally effective control method is to apply a lateral force proportional to the yaw velocity of the wheelset, a technique known as active lateral damping [27]. Those control techniques are difficult to realise using conventional passive components but relatively straight-forward to implement with active means using sensors, controllers and actuators.

15.5.2.1.2 Steering Control – Solid-Axle Wheelset

When the stabilisation is obtained passively, or there are (passive) elements in the system that interfere with the natural curving action of the solid-axle wheelset, a steering action may be actively applied to provide a low-bandwidth control that will eliminate, or at least reduce, the adverse effect on curves. Ideally, an active steering is required to achieve equal longitudinal creep between the wheels on the same axle (or zero force if no traction/braking) and equal creep forces in the lateral direction between all wheelsets of a vehicle. The first requirement is obviously to eliminate unnecessary wear and damage to the wheel-rail contact surfaces. The second requirement is concerned with producing and sharing equally the necessary lateral force to balance the centrifugal forces caused by the cant deficiency.

A number of steering strategies are possible [29]. It can readily be shown that perfect steering can be achieved if the angle of attack for two wheelsets (in addition to the radial angular position) can be controlled to be equal and the bogie to be in line with the track on curves. This idea can be implemented by controlling the position of each actuator, such that the wheelset forms an appropriate yaw angle with respect to the bogie. As indicated in Equations 15.12 and 15.13, the required yaw angle is determined by the track curve radius (R), cant deficiency (defining the necessary lateral force F_c for each wheelset), the creep coefficient (f_{22}) and semi-wheelbase (l_x).

$$\phi_{leading} = \sin^{-1}\left(\frac{F_c}{2f_{22}}\right) - \sin^{-1}\left(\frac{l_x}{R}\right) \approx \frac{F_c}{2f_{22}} - \frac{l_x}{R} \tag{15.12}$$

$$\phi_{trailing} = \sin^{-1}\left(\frac{F_c}{2f_{22}}\right) + \sin^{-1}\left(\frac{l_x}{R}\right) \approx \frac{F_c}{2f_{22}} + \frac{l_x}{R} \tag{15.13}$$

Alternately, a yaw torque can be applied such that it cancels out the effect of the longitudinal stiffness of the primary suspension, which otherwise forces the wheelsets away from the pure

Active Suspensions

rolling. As long as the cancellation occurs at frequencies significantly lower than that of the kinematic mode, the steering strategy will not compromise the stability. This can be realised by either measuring the relative yaw angle between the individual wheelset and the bogie and compensating for the primary forces or by controlling the forces and/or moments of the primary suspension [30,31].

15.5.2.2 Independently Rotating Wheelsets

For the independently rotating wheelset, control may be provided in a way similar to that for the solid-axle wheelset (i.e., control effort in the yaw or lateral directions, as illustrated in Figure 15.27), but there is also the possibility of controlling the wheelset via an active torsional coupling between the two wheels, as shown in Figure 15.29a. A more radical approach proposed is to remove the axle from the wheelset and to have two wheels mounted onto a wheel frame, as shown in Figure 15.29b. It is then possible to apply a lateral force between the frame and the wheels to steer the wheel angle directly via a track rod, much like the steering of a car.

15.5.2.2.1 Stability Control – Independently Rotating Wheelset

An independently rotating wheelset can still be unstable, even though the torsional constraint between the two wheels on the same axle is removed – a very effective measure that significantly reduces the longitudinal creep forces at the wheel-rail interface. The instability of an independently rotating wheelset has been reported in [27,32], and it is caused by the need of a longitudinal creep (albeit small) to rotate the wheels. However, the instability is much weaker compared with the kinematic oscillation of a solid-axle wheelset, and a high level of damping can be attained with either a passive yaw damper or an active yaw moment control [33,34]. The latter may be achieved by applying a yaw torque proportional to the lateral acceleration of the wheelset.

15.5.2.2.2 Guidance Control – Independently Rotating Wheelset

For the independently rotating wheelset, a different kind of steering action is required. The longitudinal creep is no longer an issue, which is solved by the introduction of the extra degree of freedom in the relative rotation between the two wheels. However, a guidance control becomes necessary to ensure that the wheelset will follow the track, without running on flanges. To provide the necessary guidance action, it is obvious that the relative displacement between the wheelset and the track (i.e., the wheel-rail deflection) is the natural choice of feedback, and the control design should then be straightforward. Sensing possibilities for the measurement vary from electro-magnetic and eddy current to video imaging or optical techniques, but the potentially high cost and low reliability are the main obstacles for practical applications. Instead, angles between adjacent vehicles as well as

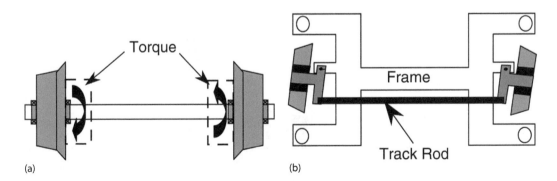

FIGURE 15.29 Alternative actuation configurations for independently rotating wheels: (a) active torsional coupling and (b) track rod for frame-mounted wheels (without an axle). (From Iwnicki, S. (Ed.), *Handbook of Railway Vehicle Dynamics*, CRC Press, Boca Raton, FL, 2006. With permission.)

the vehicle body yaw rate have been used as an indirect measurement of the track curvature, and a steering action is applied to control the wheelset yaw motion [35,36].

Another guidance method is to control the relative rotational speed between the two wheels [33]. Although there is no 'hard' connection between the two wheels on an independently rotating wheelset axle, a control action can be formulated, such that the actuator will steer the axle to achieve the zero-speed difference or a speed bias defined by the track curvature. This approach adds a damping effect between the two wheels via the active means. However, it does not result in the stiff connection of the solid-axle wheelset, which forces the two wheels to be at the same angular position (rather than velocity) at all times.

15.5.2.3 Integrated Control Design

The approach to design separate controllers for the stability and steering/guidance is a pragmatic solution, and the integration of the two parts is, in general, not a problem, as the two functions can be separated in the frequency domain. On the other hand, modern model-based control techniques provide a more effective means to deal with the multi-objective nature of a complex control problem, although the control structures tend to be more dynamically complex. H_2 optimal controls have been proposed to either maintain the natural curving of solid-axle wheelset or to provide the missing curving action for independently rotating wheels [37]. Also, robust H_∞ controls have been studied to tackle the problem of parameter variations, such as the conicity and creepage deviating from their nominal values during operation for a solid-axle wheelset [38] or an independently rotating wheelset [39]. The stability can be guaranteed in the design process, and the focus is then on the other key issues such as curving performance, uncertainty, sensing and actuation requirements [38,39].

15.5.2.4 Assessment of Control Performance

At low speeds, the performance of active primary suspensions is measured by the reduction of creep forces and wear at the wheel-rail interface compared with passive suspensions, and the focus is primarily on curved tracks, where severe wear/noise may occur in passive vehicles. Many proposed active steering schemes deliver similar performances on constant curves, although the responses in transitions will be somewhat affected by different control designs, which are less critical, as track transitions are generally short. Compared with passive suspensions or even the radial steering (where wheelsets are mechanically forced to take a radial angle on curves), actively steered wheelsets provide significant performance improvements, as shown in Figure 15.30. The data has been obtained from a railway bogie with conventional solid-axle wheelset(s) and with much softer passive suspensions [29]. The creep forces produced in non-active cases would be much worse for vehicles with stiffer suspensions. $F_x (w_1)$ is the longitudinal creep force of the leading wheelset of the bogie, $F_x (w_2)$ is that of the

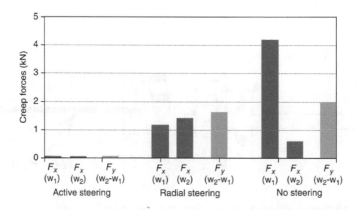

FIGURE 15.30 Steering performance comparison. (From Iwnicki, S. (Ed.), *Handbook of Railway Vehicle Dynamics*, CRC Press, Boca Raton, FL, 2006. With permission.)

trailing wheelset and $F_y (w_2 − w_1)$ is the difference in lateral creep forces between two wheelsets. Note that, while the longitudinal creep is undesirable, except for traction purposes, and should be reduced as much as possible, a certain level of the creep in the lateral direction will be inevitable in order to produce a force to balance that due to the cant deficiency on curves. Therefore, the steering performance in the lateral direction is best assessed by examining the difference in the lateral forces, and a zero difference will indicate that track shifting forces at the two wheelsets are well balanced.

The performance of the active primary suspensions at high speeds is concerned with the running stability and the level and speed of control effort required to control the wheelset kinematic mode and to cope with high-frequency track irregularities. Those factors are affected more by wheelset and actuator configurations than by specific control strategies. In general, the solid-axle wheelset is much more demanding than the independently rotating wheels, as the latter arrangement allows the free rotating of the two wheels and is hence more readily adaptable to track positions.

15.5.3 Technology

Technology of control becomes far more demanding for the primary suspensions compared with that for the secondary suspensions. The performance requirements are expected to be higher in order to be able to effectively handle the unstable kinematic modes of the wheelset. Also, the control of the railway wheelset is concerned with the vehicle stability and track following and is therefore critical to safety. Furthermore, the sensors and actuators will be working in much harsher vibration conditions, because they need to be mounted on the bogies or even wheelsets.

15.5.3.1 Sensing and Estimation Techniques

A large variety of suitable sensors is available, and the key aspects here relate to the conflict between the control requirements and practical issues such as the reliability and cost. In general, the sensing for the active control of primary suspensions is more problematic than that for the secondary suspensions. The measurement of the wheelset movements, in particular those relative to the track, is highly desirable to control the wheelset effectively, but mounting effective sensors on the wheelset is extremely difficult and costly because of the harsh vibration environment.

To enable a practical and cost-effective implementation of the active control schemes, model-based estimation techniques such as Kalman filters provide a very valuable alternative to the direct measurement. Figure 15.31 shows the principle of a model-based estimator. The measured output from the sensors is compared with the output from a mathematical model of the vehicle, and any deviations will produce a corrective action via the gain matrix to compensate for inaccuracies in the model and/or sensors. Estimated state variables, or some of the variables, are then used as the feedback signals for the controller, as shown by the dotted line in the figure. The use of only inertial sensors on the wheelsets and bogies/body was first proposed and proved to give excellent results [40],

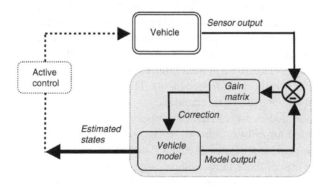

FIGURE 15.31 Block diagram of a model-based estimator. (From Iwnicki, S. (Ed.), *Handbook of Railway Vehicle Dynamics*, CRC Press, Boca Raton, FL, 2006. With permission.)

but it is also possible to remove the sensors from the wheelsets and replace them with bogie-based displacement sensors to provide the primary suspensions deflection [41].

The use of more sophisticated equipment may become economically feasible in the future to directly measure some essential feedback signals and/or track features, e.g., by using a track database and Global Positioning System, whereby the estimation may be simplified and its robustness improved.

15.5.3.2 Actuators

Decisions on the actuator technology for the active primary suspensions will be influenced by practicality as well as functionality. If the stability is provided by passive means and active control is used as an 'add-on' for steering to improve the performance on curved track, the actuator response will not be critical for the low-bandwidth control, and the choice of actuator technology will be determined by other practical considerations at a system's level. In this case, control configurations/laws should be carefully considered to avoid the need for large steering forces to overcome the stiffness of passive suspensions [30]. When the control is used to provide the essential stability, which would be able to take full advantage of active technology, fast-acting and high-bandwidth responses from the actuators will be necessary to overcome the kinematic instability of the wheelsets. This rules out the use of pneumatic actuators, and the practical options include hydraulic or electro-mechanical actuators. In this case, the control effort will be primarily to deal with the wheelsets reacting to track irregularities, which can demand significantly large forces at high operational speeds and/or in poor track conditions, although the movement of the actuators is normally very small [37]. Careful considerations should therefore be given in the development of the actuator subsystem to not only meet the specific control requirements but also optimise the power/size of the actuators, e.g., the optimisation of some key parameters, such as gear ratio in an electro-mechanical actuator can greatly reduce the actuator forces and lower overall power consumption [42].

15.5.3.3 Controllers and Fault Tolerance

The availability of remarkable quantities of computing power means that the controller is unlikely to be a limiting factor in the implementation, although issues such as reliability and ruggedness cannot be ignored.

More importantly, the issues of safety and reliability will have to be addressed satisfactorily. Any new technology must demonstrate that it can cope with component fault(s) without compromising passenger safety and satisfy the requirement that any component fault would not lead to the system failure. On the other hand, reliability and availability are of great importance to rail operators in order to maintain an effective operation of a rail system. Therefore, any active steering scheme must also meet the necessary standards of reliability.

Traditionally, mechanical components are used for the wheelset stabilisation, and they are generally accepted as 'safe'. Safety is ensured by having all safety-critical mechanical components designed as far as is practicable not to fail. This is achieved through a combination of conservative design, careful quality control during manufacture and rigorous maintenance procedures during operation. However, failure modes in sensors and electronics are less definable, so it becomes necessary to reconsider the approach. Having a proven mechanical backup for an active system, which takes over in the case of an electronic system failure, is one solution, but this is not appropriate in the longer term, because it will detract from benefits. The alternative is a fault-tolerant active system based upon functional and/or analytical redundancy.

Many fault-tolerance studies are carried out, especially in the aerospace and process industry, but very little is reported for railway vehicles, except several studies on fault-tolerant sensing and fault detection for the provision of measurement data. Fault tolerance for actuators can be more beneficial because the cost of those units, and hence potential savings, can be far more significant, which is particularly so if the reliance on hardware redundancy is reduced by intelligent management of actuators at the overall system level [43]. This is an important area for the development of both standards and technological solutions in the future.

Active Suspensions

15.5.4 Examples

15.5.4.1 Bombardier Mechatronic Bogie

This example presents an implementation and full-size experiment of active control for railway wheelsets, the first example of its kind in the world [44]. Figure 15.32 shows a photograph of the actively controlled bogie, which is a modified version of a Bombardier VT612 bogie. The bogie has a soft primary suspension and no secondary yaw dampers. In fact, the only stiffness in the longitudinal direction is due to the shear stiffness of the vertical suspensions. Removing the secondary yaw dampers offers significant advantages in terms of the vehicle's weight and comfort; however, stability and, consequently, high-speed operation are significantly compromised by that removal. Without active control, the modified bogie can reach a critical speed of around 90–100 km/h.

Active control is applied by means of two electrically driven actuation mechanisms that apply independent yawing actions to each wheelset. Two alternating current servo-motors act through gearboxes, from which steering linkage mechanisms transfer the control action to the wheelsets. Control strategies for the stability and steering are designed separately [29,44], but the two are brought together through an integration process to ensure there are no adverse interactions. Additional measures in the control loops are needed for reasons of practicality, such as sensing and actuation, which are particularly important for the stability control, owing to the requirement of a high-bandwidth control. Figure 15.33 shows the stability controller for one of the wheelsets, where an inner loop is added to ensure a fast dynamic response of the actuator to the torque demand from the stability control loop. The controller for the second wheelset is the same.

A fully actively controlled bogie was tested on a full size roller rig in Munich, Germany. Extensive stability tests and track field tests were performed and the controller successfully operated at speeds in excess of 300 km/h. Figure 15.34 shows results for both active control and the passive vehicle, and illustrates clearly the effectiveness of active stabilisation.

15.5.4.2 Independently Driven Wheels

Control of the independently rotating wheelset may be achieved by implementing actively a torsional coupling between the two wheels, as illustrated in Figure 15.29a, but this may be more conveniently obtained in practice by using independently driven wheels, as presented in the following example.

FIGURE 15.32 Actively controlled bogie. (From Iwnicki, S. (Ed.), *Handbook of Railway Vehicle Dynamics*, CRC Press, Boca Raton, FL, 2006. With permission.)

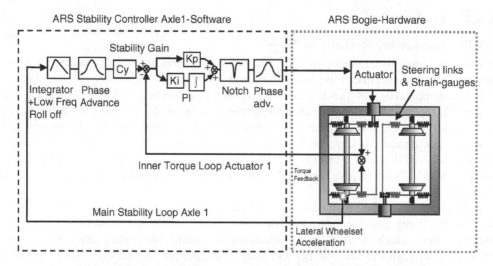

FIGURE 15.33 Stability control loop. (From Iwnicki, S. (Ed.), *Handbook of Railway Vehicle Dynamics*, CRC Press, Boca Raton, FL, 2006. With permission.)

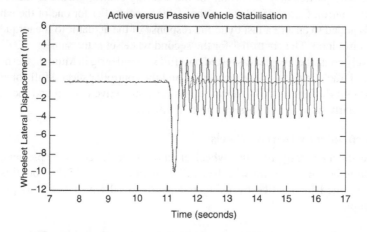

FIGURE 15.34 Rig stability test result – passive control (solid line) and active control (dotted line). (From Iwnicki, S. (Ed.), *Handbook of Railway Vehicle Dynamics*, CRC Press, Boca Raton, FL, 2006. With permission.)

An innovative form of permanent magnet motor fitted inside of a railway wheel, as shown in Figure 15.35a, has been developed by Stored Energy Technology (SET, UK), specifically for light rail or tram applications [45,46]. The wheel motors are developed to not only deliver the necessary tractive effort in both motoring and re-generative modes but also to provide steering control by driving the wheels on two sides of a vehicle independently [45]. For the experimental demonstration of the concept, four of the wheel motors are mounted onto one of the bogies on a modified tram vehicle shown in Figure 15.35b, replacing two conventional solid-axle wheelsets powered by 'old' dc traction motors.

The control scheme for each wheel pair on either side of a vehicle is shown in Figure 15.36. The provision of tractive effort is shared equally between all the motors, and a differential control of the wheel motors is used to control actively the steering and curving actions for the vehicle, as independently rotating wheels do not have the track-following ability that is inherent with the conventional solid-axle wheelset. Trial runs were successfully carried out at a site in Blackpool, UK, although only low-speed (up to 32 km/h) operations were permitted because of speed restrictions on the site.

Active Suspensions   609

FIGURE 15.35 Independently driven wheels: (a) permanent wagon wheel motor and (b) modified tram with wheel motors.

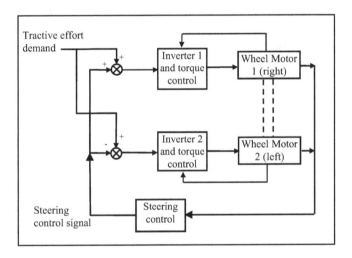

FIGURE 15.36 Control scheme for each wheel pair.

15.6 OVERALL SUMMARY AND LONG-TERM TRENDS

This chapter has covered the range of possible active railway vehicle suspension systems, from present-day tilting trains through to more speculative options, some of which are little more than theoretical possibilities at the moment. Whether the active (secondary) suspensions are being used for improved passenger comfort, or active primary suspensions are being used to control the wheels and wheelsets, it should be clear to the reader that the use of active elements enables substantial performance improvements, improvements that are not possible with purely mechanical or passive solutions.

Active railway suspensions therefore represent an emerging and important technology, offering the railway industry a large variety of commercial and operational opportunities, although there are of course a number of major technical challenges. It is almost inevitable that the concepts will become progressively incorporated into railway vehicles, although it is less clear how quickly this will happen. However, many other industrial devices and systems that have started to replace or enhance mechanically based products using electronic control concepts have never looked back, and so, the already-established tilting technology is almost certainly just at the starting point for active railway suspensions.

NOMENCLATURE

a_b	Body acceleration [m/s^2]
C	Coefficient of a passive damper [Ns/m]
C_s	Sky-hook damping coefficient [Ns/m]
F_w	Lateral control force
F_x	Longitudinal creep force
F_y	Lateral creep force
T_w	Yaw control torque
y_w	Wheelset lateral displacement
ψ_w	Wheelset yaw angle
F_a	Actuator force [N]
F_c	Lateral force on curved track [N]
f_{22}	Lateral creep coefficient [N]
J	Optimisation index
K	Spring constant [N/m]
K_{CD}	Cant deficiency compensation factor
K_L	Control gain of active levelling
K_s	Control gain of suspension stiffness softening
l_x	Semi-wheelbase [m]
M	Body mass [kg]
q_1	Weighting factor for optimisation
q_2	Weighting factor for optimisation
R	Curve radius (m)
r	Weighting factor for optimisation
V_{active}	Vehicle speed in active case [m/s]
$V_{passive}$	Vehicle speed in passive case [m/s]
x_d	Suspension deflection [m]
z	Body displacement [m]
z_t	Track displacement [m]
θ_{active}	Cant deficiency in active case
θ_{cant}	Cant angle of the track
$\theta_{passive}$	Cant deficiency in passive case
θ_{tilt}	Tilt angle
θ_R	Passive vehicle angle of outwards roll
$\varphi_{leading}$	Required yaw angle at the leading wheelset [rads]
$\varphi_{trailing}$	Required yaw angle at the trailing wheelset [rads]

REFERENCES

1. J.K. Hedrick, D.N. Wormley, Active suspensions for ground transport vehicles—A state of the art review, In: B. Paul, K. Ullman, H. Richardson (Eds.), *Mechanics of Transportation Suspension Systems*, ASME, AMD, 15, 21–40, 1975.
2. R.M. Goodall, W. Kortüm, Active controls in ground transportation—A review of the state-of-the-art and future potential, *Vehicle System Dynamics*, 12(4–5), 225–257, 1983.
3. R.M. Goodall, Active railway suspensions: Implementation status and technological trends, *Vehicle System Dynamics*, 28(2–3), 87–117, 1997.
4. S. Bruni, R.M. Goodall, T.X. Mei, H. Tsunashima, Control and monitoring for railway vehicle dynamics, *Vehicle System Dynamics*, 45(7–8), 743–779, 2007.
5. H.R. O'Neill, G.D. Wale, Semi-active suspension improves rail vehicle ride, *Computing & Control Engineering Journal*, 5(4), 183–188, 1994.

6. A.C. Zolotas, R.M. Goodall, Modelling and control of railway vehicle suspensions, In: M.C. Turner, D.G. Bates (Eds.), *Mathematical Methods for Robust and Nonlinear Control*, Springer-Verlag, London, UK, 373–411, 2007.
7. R. Persson, R.M. Goodall, K. Sasaki, Carbody tilting–technologies and benefits, *Vehicle System Dynamics*, 47(8), 949–981, 2009.
8. BS EN 12299:2009, *Railway Applications. Ride Comfort for Passengers. Measurement and Evaluation*, British Standards Institution, London, UK, 2009.
9. J. Förstberg, Ride comfort and motion sickness in tilting trains. Human responses to motion environments in train experiment and simulator experiments, Doctoral thesis, TRITA-FKT 2000:28, Royal Institute of Technology (KTH), Stockholm, Sweden, 2000.
10. Y. Nishioka, A. Tamaki, K. Isshihara, N. Okabe, O. Torii, Tilt control system for railway vehicles using air springs, *Sumimoto Search*, 56, 44–48, 1994.
11. K.W. Pennington, M.G. Pollard, The development of an electro-mechanical tilt system for the Advanced Passenger Train, *Proceedings IMechE Conference on Electric versus Hydraulic Drives*, 27 October 1983, London, UK, C299/83, 21–28, 1983.
12. R.M. Goodall, A.C. Zolotas, J. Evans, Assessment of the performance of tilt system controllers, *Proceedings IMechE Railway Technology Conference*, 21–23 November 2000, Birmingham, UK, C580/028/2000, 231–239, 2000.
13. D.C. Karnopp, Active and passive isolation of random vibration, In: J.C. Snowdon, E.E. Ungar (Eds.), *Isolation of Mechanical Vibration Impact and Noise*, ASME, AMD, 1(1), 64–86, 1973.
14. H. Li, R.M. Goodall, Linear and non-linear skyhook damping control laws for active railway suspensions, *Control Engineering Practice*, 7(7), 843–850, 1999.
15. T.X. Mei, H, Li, R.M. Goodall, Kalman filters applied to actively controlled railway vehicle suspensions, *Transaction of Institute of Measurement and Control*, 23(3), 163–181, 2001.
16. C. Casini, G. Piro, G. Mancini, The Italian tilting train ETR460, *Proceedings IMechE International Railway Conference on Better Journey Time-Better Business*, 24–26 September 1996, Birmingham, UK, 297–305, 1996.
17. D.H. Allen, Active bumpstop hold-off device, *Proceedings IMechE Conference Railtech 94*, 24–26 May 1994, Birmingham, UK, 1994.
18. R.A. Williams, Active suspensions classical or optimal? *Vehicle System Dynamics*, 114(1–3), 127–132, 1985.
19. R.M. Goodall, J.T. Pearson, I. Pratt, Actuator technologies for secondary active suspensions on railway vehicles, *Proceedings International Conference on Railway Speed-up Technology for Railway and Maglev Vehicles, STECH'93*, 22–26 November 1993, Yokohama, Japan Society of Mechanical Engineers, Vol. 2, 377–382, 1993.
20. R. Goodall, G. Freudenthaler, R. Dixon, Hydraulic actuation technology for full-and semi-active railway suspensions, *Vehicle System Dynamics*, 52(12), 1642–1657, 2014.
21. H. Hammood, T.X. Mei, Gain-scheduling control for railway vehicle semi-active suspension, In: M. Spiryagin, T. Gordon, C. Cole, T. McSweeney (Eds.), *Proceedings 25th IAVSD Symposium held at Central Queensland University*, Rockhampton, Queensland, Australia, 14–18 August 2017, CRC Press/Balkema, Leiden, the Netherlands, 893–899, 2017.
22. R.M. Goodall, R.A. Williams, A. Lawton, P.R. Harborough, Railway vehicle active suspensions in theory and practice, In A.H. Wickens (Ed.), *Proceedings 7th IAVSD Symposium held at Cambridge University*, UK, 7–11 September 1981, Swets and Zeitlinger, Lisse, the Netherlands, 301–316, 1982.
23. M. Tahara, K. Watanabe, T. Endo, O. Goto, S. Negoro, S. Koizumi, Practical use of an active suspension system for railway vehicles, *Proceedings International Symposium on Speed-up and Service Technology for Railway and Maglev Systems, STECH'03*, 19–22 August 2003, Tokyo, Japan Society of Mechanical Engineers, A503, 225–228, 2003.
24. A. Orvnäs, S. Stichel, R. Persson, On-track tests with active lateral secondary suspension: A measure to improve ride comfort, *ZEV Rail Glasers Annalen*, 132(11–12), 469–477, 2008.
25. A. Qazizadeh, R. Persson, S. Stichel, On-track tests of active vertical suspension on a passenger train, *Vehicle System Dynamics*, 53(6), 798–811, 2015.
26. Y.H. Cho, Advanced bogie keeps LRVs ahead of the curve, *International Railway Journal*, 58(9), 74–75, 2018.
27. R.M. Goodall, H. Li, Solid axle and independently-rotating railway wheelsets-a control engineering assessment of stability, *Vehicle System Dynamics*, 33(1), 57–67, 2000.

28. T.X. Mei, R.M. Goodall, Stability control of railway bogies using absolute stiffness: Sky-hook spring approach, *Vehicle System Dynamics*, 44(S1), 83–92, 2006.
29. S. Shen, T.X. Mei, R.M. Goodall, J.T. Pearson, G. Himmelstein, A study of active steering strategies for railway bogie, In: M. Abe (Ed.), *Proceedings 18th IAVSD Symposium held in Atsugi, Kanagawa, Japan*, 24–30 August 2003, Taylor & Francis Group, London, UK, 282–291, 2004.
30. G. Shen, R.M. Goodall, Active yaw relaxation for improved bogie performance, *Vehicle System Dynamics*, 28(4–5), 273–289, 1997.
31. J. Pérez, J.M. Busturia, T.X. Mei, J. Vinolas, Combined active steering and traction for mechatronic bogie vehicles with independently rotating wheels, *Annual Reviews in Control*, 28(2), 207–217, 2004.
32. B.M. Eickhoff, The application of independently rotating wheels to railway vehicles, *Journal of Rail and Rapid Transit*, 205(1), 43–54, 1991.
33. T.X. Mei, R.M Goodall, Practical strategies for controlling railway wheelsets with independently rotating wheels, *ASME Journal of Dynamic Systems, Measurement, and Control*, 125(3), 354–360, 2003.
34. T.X. Mei, J.W. Lu, On the interaction and integration of wheelset control and traction system, In: M. Abe (Ed.), *Proceedings 18th IAVSD Symposium held in Atsugi, Kanagawa, Japan*, 24–30 August 2003, Taylor & Francis Group, London, UK, 123–132, 2004.
35. H. Hondius, Microprocessors harnessed to optimise radial steering, *Railway Gazette International*, 151(5), 315–316, 1995.
36. J. Pederson, New S-trains for Copenhagen, *European Railway Review*, 1(4), 29–33, 1995.
37. T.X. Mei, R.M. Goodall, In: R. Fröhling (Ed.), *Proceedings 16th IAVSD Symposium held in Pretoria, South Africa*, 30 August–3 September 1999, Swets & Zeitlinger, Lisse, the Netherlands, 653–664, 2000.
38. S.M.M. Bideleh, T.X. Mei, V. Berbyuk, Robust control and actuator dynamics compensation for railway vehicles, *Vehicle System Dynamics*, 54(12), 1762–1784, 2016.
39. T.X. Mei, R.M. Goodall, Robust control for independently-rotating wheelsets on a railway vehicle using practical sensors, *IEEE Transactions on Control Systems Technology*, 9(4), 599–607, 2001.
40. T.X. Mei, R.M. Goodall, LQG and GA solutions for active steering of railway vehicles, *IEE Proceedings-Control Theory and Applications*, 147(1), 111–117, 2000.
41. J.T. Pearson, R.M. Goodall, T.X. Mei, S. Shen, A. Zolotas, Kalman filter design for a high speed bogie active stability system, *Proceedings UKACC Control 2004 Conference*, 6–9 September 2004, University of Bath, Bath, UK, CD-ROM, ID-128, 2004.
42. L. Weerasooriya, T.X. Mei, Active wheelset control—actuator dynamics and power requirements, *1st International Conference on Rail Transportation*, 10–12 July 2017, Southwest Jiaotong University, Chengdu, China, 2017.
43. M. Mirzapour, T.X. Mei, Control re-configuration for actively steered wheelsets in the event of actuator failure, In: M. Rosenberger, M. Plöchl, K. Six, J. Edelmann (Eds.), *Proceedings 24th IAVSD Symposium held at Graz, Austria*, 17–21 August 2015, CRC Press/Balkema, Leiden, the Netherlands, 1051–1060, 2016.
44. J.T. Pearson, R.M. Goodall, T.X. Mei, S. Shen, C. Kossmann, O. Polach, G. Himmelstein, Design and experimental implementation of an active stability system for a high speed bogie, In: M. Abe (Ed.), *Proceedings 18th IAVSD Symposium held in Atsugi, Kanagawa, Japan*, 24–30 August 2003, Taylor & Francis Group, London, UK, 43–52, 2004.
45. T.X. Mei, K.W. Qu, H. Li, Control of wheel motors for the provision of traction and steering of railway vehicles, *IET Power Electronics*, 7(9), 2279–2287, 2014.
46. J. Stow, N. Cooney, R.M. Goodall, R. Sellick, The use of wheelmotors to provide active steering and guidance for a light rail vehicle, *Proceedings of IMechE Stephenson Conference: Research for Railways*, 25–27 April 2017, London, UK, 2017.

16 Dynamics of the Pantograph-Catenary System

Stefano Bruni, Giuseppe Bucca, Andrea Collina and Alan Facchinetti

CONTENTS

16.1	Introduction	614
16.2	Problems and Issues in the Design of Pantographs and Catenaries	614
	16.2.1 Generalities on Overhead Contact Line Systems	616
	16.2.1.1 Simple Contact Wire	616
	16.2.1.2 Standard Catenary Wire	617
	16.2.1.3 Conductor Bar Overhead Contact System	619
	16.2.2 Main Parameters Influencing the Quality of Current Collection	619
	16.2.3 Multiple Pantograph Operation	620
16.3	Numerical Simulation of Pantograph-Catenary Interaction	621
	16.3.1 Catenary Models	622
	16.3.1.1 Lumped Parameter Models	622
	16.3.1.2 Finite Difference Models	622
	16.3.1.3 Finite Element Models	624
	16.3.1.4 Model of Droppers	624
	16.3.1.5 Catenary Damping	625
	16.3.1.6 Static Position of the Catenary	626
	16.3.2 Pantograph Models	626
	16.3.2.1 Lumped Parameter Models	627
	16.3.2.2 Multibody Models	627
	16.3.2.3 Hybrid Models	627
	16.3.3 Pantograph-Catenary Contact	628
	16.3.4 Software for the Simulation of Pantograph-Catenary Interaction	628
16.4	Measurement, Testing and Qualification of Pantographs	630
	16.4.1 Laboratory Testing of Pantographs	630
	16.4.1.1 Test Rigs for Pantograph Testing	630
	16.4.1.2 Type and Routine Tests	631
	16.4.1.3 Dynamic Characterisation Tests	631
	16.4.1.4 Hardware-in-the-Loop Simulation of Pantograph-Catenary Interaction	632
	16.4.1.5 Calibration of the Measuring System of Contact Force	634
	16.4.2 Qualification and Track Testing of Pantographs and Catenaries	635
	16.4.2.1 Tethered Tests	636
	16.4.2.2 Contact Force Measurement for the Qualification of Current Collection Quality	637
	16.4.2.3 Arcing Detection	638
	16.4.2.4 Contact Wire Uplift and Overhead Contact Line Motion	638

16.5	Wear, Damage and Condition Monitoring in Pantograph-Catenary Systems	638
	16.5.1 Wear of Contact Wire and Contact Strip	638
	16.5.2 Damage of Catenary and Pantograph	642
	16.5.3 Condition Monitoring of Pantograph-Catenary System	642
16.6	Long-Term Trends	644
	16.6.1 Active/Semi-Active Pantograph Control	644
	16.6.2 Instrumented Pantographs for the Continuous Monitoring of the Overhead Contact Line Status	646
	16.6.3 Analysis of Pantograph Aerodynamics Using Computational Fluid Dynamics	646
	16.6.4 Current Collection from an Overhead Contact Line in Heavy Road Vehicles	647
References		647

16.1 INTRODUCTION

Pantograph-catenary interaction is a dynamic problem involving the coupled vibration of one part of the vehicle, the pantograph, with the overhead contact line (OCL), often referred to as the 'catenary' due to its characteristic shape. The sources of excitation for this phenomenon are manifold, but excitations are mainly due to the spatially variable stiffness of the OCL, which is subjected to the uplift force applied by the pantograph. As a consequence, the pantograph head performs a nearly periodic vertical displacement while moving in contact with the catenary and generates dynamic forces affecting the contact force exchanged with the catenary. When the speed of the pantograph is sufficiently high, these effects become so serious that they may impair the proper functioning of the pantograph-catenary dynamic coupling; see Section 16.2.

The pantograph-catenary interaction is a critical issue for electrified railway systems, because it has important repercussions on the maximum service speed at which the electric rolling stock can be operated, and it sets some limits on the dynamic behaviour of a railway vehicle as well as on the reliability, availability and maintainability of the system.

This chapter aims to provide an introductory description of the main problems affecting the design of the pantograph and of the overhead equipment, introduce the modelling and simulation techniques presently in use to investigate pantograph/catenary interaction problems together with the measurement and testing techniques used for the qualification of pantographs and catenaries, describe the main types of damage and failure affecting pantographs and catenaries and outline future trends to improve current collection and reduce damage and life cycle costs.

16.2 PROBLEMS AND ISSUES IN THE DESIGN OF PANTOGRAPHS AND CATENARIES

The transfer of electric current from the OCL to the train is operated through the sliding contact created between the contact wire (CW) and some collector strips mounted on the head of one or more pantographs. During the operation of the train, a proper contact pressure has to be ensured between the pantograph and the OCL, in order to ensure the regular flow of electric current through the sliding contact. To reach this goal, the pantograph is designed as an articulated frame (typically a four-bar linkage) driven by a force applied to the lower member by a mechanical spring or an air bellow, so that the desired pre-load (contact force in static condition) is applied to the OCL. The purpose of the articulated frame is to follow the OCL dynamical uplift in response to the force exchanged at the contact points, in the low-frequency range, namely at span passing frequency, equal to the ratio between train forward speed and span length. In addition, the pantograph head is fitted with a suspension, so that the movements of the head and of the articulated frame are decoupled in the frequency range above 4–5 Hz, reducing the dynamic component of the contact force in this frequency range.

Dynamics of the Pantograph-Catenary System

Dynamic interaction between the pantograph and the OCL causes a dynamic fluctuation of the contact force with respect to its mean value. To ensure proper operation of the pantograph-catenary couple, the fluctuation of the contact force shall not be too large compared with the mean contact force, a requirement typically expressed by the following condition:

$$\sigma_F \leq 0.3\, F_m \qquad (16.1)$$

where F_m and σ_F are the mean value and standard deviation of the contact force.

Current collection involves several physical aspects, including mechanical, aerodynamic, electrical and thermal issues, all interacting with each other. The scheme in Figure 16.1 is an attempt to summarise the main factors affecting current collection and the main issues that can become critical in the operation of the pantograph-OCL couple; the boxes below the sketch of the OCL list the main disturbance effects that are responsible for increased dynamic fluctuation of the contact force, whilst the boxes above the sketch list the effects of contact force fluctuation on the operation of the pantograph-catenary couple. High values of the contact force result in increased mechanical wear of the contacting surfaces, risk of fatigue and other mechanical damage in the OCL and excessive uplift of the CW at the suspension with the risk of interference between the pantograph and the OCL. Conversely, excessively low values of the contact force result in increased electrical resistance of the sliding contact and therefore cause poor quality of current collection, accompanied by electrical arcing, which may cause thermal damage of the CW and pantograph head.

Ultimately, the correct operation of the pantograph-OCL system requires that the contact force is kept within a range of acceptable values, as stated in Equation 16.1. To meet this requirement, proper design and operation of the pantograph-catenary couple are needed. The contact force exchanged by these two bodies depends firstly on the pre-load applied and is also affected by the steady aerodynamic forces acting on the pantograph, which, in turn, depend on train speed and pantograph orientation. In some pantographs for high-speed operation, the pre-load can be adjusted depending on train speed and pantograph orientation to compensate for the steady aerodynamic effects, thereby realising the same mean value of the contact force for different operating conditions. For pantographs with two collectors, another important effect of steady aerodynamic forces is the unbalance of the mean contact force between the two collectors. This effect heavily depends on the kinematics of the suspension of the two collectors and requires that proper measures are adopted to avoid an imbalance in excess of 20%.

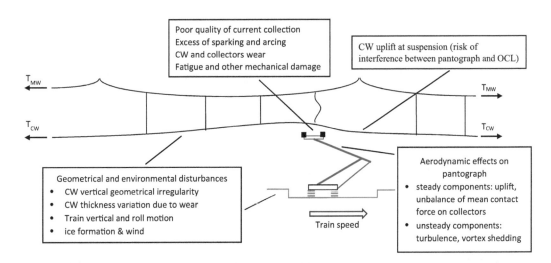

FIGURE 16.1 Scheme of the main issues related to current collection.

16.2.1 GENERALITIES ON OVERHEAD CONTACT LINE SYSTEMS

Different OCL designs are in use, depending on the type of operation being addressed and on environmental issues. All OCL designs include at least one CW that comes in contact with the collector strip(s), realising the sliding contact for the transfer of the feeding current. The CW is staggered in the horizontal plane, usually in a range from 40 cm to 60 cm, to spread wear across the collector strip.

Some frequently used OCL types are shown in Figure 16.2: simple CW ('trolley wire'), CW supported by a messenger wire (MW) through droppers ('standard catenary wire') and the latter with the addition of an auxiliary stitch wire. So-called 'rigid catenaries', not shown in the figure, are used for special purposes (e.g., tunnels) and consist of a CW clamped to a relatively stiff metallic bar. Note that third-rail electrified systems are not covered in this chapter.

16.2.1.1 Simple Contact Wire

Simple CW (or trolley wire) is mainly used for trolley bus and tramway operations. The line is made by a tensioned CW, divided into spans, suspended from supporting poles or stranded by means of an arrangement usually termed a 'delta' suspension, owing to its geometrical appearance; see Figure 16.3. The purpose is to avoid the presence of a stiff point at the suspension. The delta suspension can be realised by a simple portion of shaped CW or by a stranded wire. It is clear from the structure of the system that the highest deformability is at midspan, whereas the lowest one corresponds to the connection with the delta suspension.

The mechanical tension of the CW for this system is in the range of 8–12 kN, with span length in the range 20–35 m. The tension of the CW is, in most cases, not regulated, but solutions with tension applied through a weight and pulley system are also in use, especially when longer span lengths are adopted. The choice of the configuration of regulated/unregulated tension has a clear consequence on the effect of thermal variation on the sag of the CW at midspan; regulated tension enables the maintenance of a constant sag despite temperature changes, whereas the configuration with unregulated tension suffers from uncontrolled changes in the sag arising from thermal expansion or contraction of the CW. In the case of regulated tension, the delta suspension is made of a rope

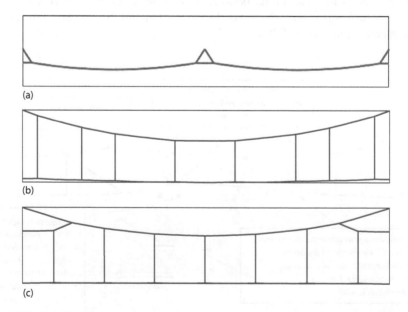

FIGURE 16.2 Main schemes of overhead contact lines: (a) single trolley wire, (b) standard catenary wire, and (c) catenary wire with auxiliary stitch wire.

Dynamics of the Pantograph-Catenary System

FIGURE 16.3 Example of delta suspension with Kevlar rope and longitudinal allowable displacement; the tensioning device is also visible.

(Kevlar is also used) and a pulley, on which the rope can roll to let the mechanical tension on the CW be transmitted across the suspension along the spans. In order to ease the stagger configuration, the delta suspension is inclined, and often, a stagger arm is included. Operating speeds for this OCL system are generally not exceeding 60–80 km/h.

16.2.1.2 Standard Catenary Wire

In its simplest form, the standard catenary wire or, in short, the catenary is composed of a CW suspended to an MW through droppers. The name catenary comes from the shape of the MW that supports its own weight and the weight of the CW. The MW is in turn connected to a suspension, delimiting the span length, while the CW is staggered by means of a steady arm, free to rotate, whose function is to keep the geometry of the stagger, at the same time minimising the introduction of a local stiffness effect in the vertical direction. The steady arm also helps in regulating the vertical position of the CW at the suspension. This OCL system enables the use of longer spans compared with the simple catenary, usually in the range 50–65 m, although special solutions with lengths up to 70 m exist [1].

The main configuration of the standard catenary wire (MW, CW and droppers) can be implemented according to a number of variants, including the use of multiple CWs and sometimes multiple MWs. In particular, standard catenaries for direct current (DC) lines require a larger total section of the conductors and hence often include two CWs and sometimes also two MWs, so that, practically, the system is made of two catenaries in parallel.

In the same way as for the trolley wire, for the standard catenary wire also, the variation of vertical stiffness along the span is the primary source of excitation of the vibration of the pantograph head. In order to reduce the spatial variation of stiffness, a special design of the catenary can be adopted. For instance, an auxiliary wire can be introduced in the suspension area to which the droppers across the suspension are connected instead of being directly connected to the MW. In this way, the local increase of CW stiffness in the suspension area can be mitigated. Owing to its shape, the suspension, including an auxiliary wire, is also termed the 'Y suspension' or 'stitch wire' suspension.

The adoption of a stitch wire solution has advantages and drawbacks. On the one hand, the 'Y' suspension reduces the stiffness variation along the line and the local stiffness under each suspension point. But, on the other hand, it also requires more maintenance effort, considering that the tension in the auxiliary wire is not regulated, which makes this design more sensitive to temperature variation. As a matter of fact, the choice between the standard design and the one incorporating the stitch wire suspension has been so far determined by geographic and historical reasons; the 'Y' suspension is mainly found in central European networks such as in Switzerland, Austria and Germany, whereas the standard suspension is used in France and Italy. Both designs are proved to allow high-speed operation.

The mechanical tension of the CW ranges from 10 kN for low-speed to 15 kN for medium-speed and up to 20–30 kN for high-speed lines. To achieve higher values of the mechanical tension, copper alloys (CuAg and CuMg) have to be used to provide the required mechanical strength and creep behaviour. The tension of the MW is usually lower than the tension of the CW; in this way, the local increase of the CW bending stiffness in correspondence with the location of each of the droppers can be mitigated. Droppers, initially seen just as the connection between CW and MW, become a key issue for a proper operation as speed increases. Rigid (hook) droppers, made of a solid circular section, although much used in the past, are nowadays used only in low-speed catenaries, whilst in the majority of cases, droppers are made of a single stranded wire clamped or hinged at its extremities. The electrical conductivity of the dropper is ensured by the clamps or by additional electrical connections at dropper's ends. Dynamical behaviour of droppers during a pantograph's transit involves dynamic buckling that can occur on the first mode (see Figure 16.4a) or on a higher mode (see Figure 16.4b). Special elastic droppers have been developed recently for high-speed application,

FIGURE 16.4 Excitation of different buckling shape of the dropper during pantograph's transit at 300 km/h: (a) first buckling mode and (b) higher buckling mode.

Dynamics of the Pantograph-Catenary System

to be installed close to the suspensions. The aim is to avoid the transition from a tensioned status to a slackened one, resulting from dynamic buckling, which occurs in standard droppers, as clearly seen in Figure 16.4.

16.2.1.3 Conductor Bar Overhead Contact System

This system is traditionally named the rigid overhead contact system, or 'rigid catenary' in short, although it does not actually include any component in the shape of a catenary. It is a solution composed of an aluminium extruded bar that clamps a CW into its lower part. The section is designed to realise the maximum flexural stiffness in the vertical plane and to clamp the CW, so that it is held clear of the lower edge, as shown in Figure 16.5.

The CW is inserted in the bar by means of a special tool during the mounting procedure. The bars are usually manufactured in 12-m-long modules, which are joined together by bolts and nuts to form a continuous section of several hundred metres. The span length varies according to the speed of operation and type of suspension, ranging from 8 m to 12 m. Longer spans can be used for low speed (80 km/h), whereas, for high speeds (120 km/h), shorter spans should be adopted. Implementations of this design are mainly found in tunnels, because this system has a reduced vertical encumbrance compared with the standard catenary, leading to important savings in the construction costs where a smaller height of the tunnel section is required. Another advantage of this design is the lower risk of mechanical failure compared with a standard catenary wire; this is again particularly attractive for installations in tunnels, where inspection and maintenance activities may require increased effort. The main drawbacks of this system are the initial and installation costs and the limitation in the top speed. Regarding the latter point, recent advances made by leading manufacturers have enabled a significant increase in the speed of operation, resulting in the possibility to use this system in new tunnels connecting some European regional high-speed networks such as in the Rhine-Alpine area.

16.2.2 Main Parameters Influencing the Quality of Current Collection

The first and most important parameter in the case of wired OCL is the mechanical tension of the CW. Increased CW tension raises the 'intrinsic' stiffness of the CW, thus reducing the effect of stiffness variations related to span passing and dropper passing. Thus, it can be said the tension of the CW, as a matter of fact, qualifies the speed range of the OCL system. Experimental evidence supporting this statement is provided by the results of line tests reported in [2], as well

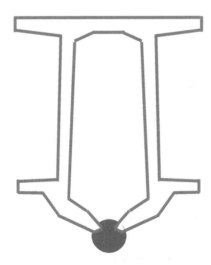

FIGURE 16.5 Schematic view of the section of a conductor bar OCL; the CW is clamped at the section's lower end.

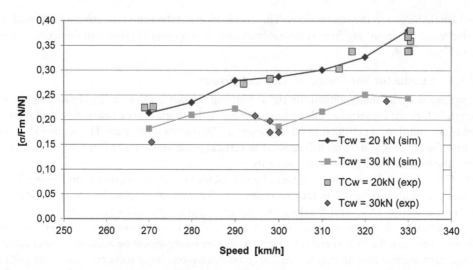

FIGURE 16.6 Standard deviation of contact force (σ) versus train speed for C270 catenary with 20 kN and 30 kN tension of contact wire and 16 kN of messenger wire for 57 m span length (sim: simulation with contact wire irregularity, exp: measurements on the Torino-Milano HS line).

as by tests performed during several testing campaigns on the Italian high-speed line considering two different values of the CW tension; see Figure 16.6.

The importance of this parameter is also witnessed by European Standard EN 50319 (the basis for the International Electrotechnical Commission (IEC) International Standard IEC 60913), which states that the maximum operating speed of a wired OCL is 70% of the speed of the transversal travelling wave in the CW, which is in turn determined by the tension in the wire.

Another parameter affecting the performance of the OCL is the mechanical tension of the MW. Its effect is related to two main issues:

- MW tension affects the stiffness of the catenary at midspan and in the region close to the suspension.
- MW tension also affects the natural frequencies of the span, so it is relevant to multiple current collection, as will be explained in Subsection 16.2.3.

Some guidelines for setting this parameter are also given in [2] in the light of wave reflection effects related to droppers.

A third important parameter in the design of the catenary system is the spacing of droppers. This is strictly related to the mechanical tension in both CW and MW. Generally speaking, it can be said that a higher mechanical tension of the CW allows for a larger spacing of the droppers. Finally, the quality of catenary maintenance has an important effect on the dynamic fluctuations of the contact force. Significant differences have been found during line tests performed under the same nominal conditions, originated by variations in the quality of the vertical alignment of the CW, and the same can be found by simulation; see Figure 16.7. The lesson that can be learnt is that proper maintenance is important to ensure the maximum performance for the pantograph-catenary couple.

16.2.3 MULTIPLE PANTOGRAPH OPERATION

Before closing this section, some considerations are proposed concerning multiple pantograph operation. The operation of multiple pantographs in the same train is sometimes required due to various reasons, e.g., electrical multiple unit (EMU) trains, in which motor units are fed independently, or short individual train sets coupled together in double or triple units or, finally, the need to use

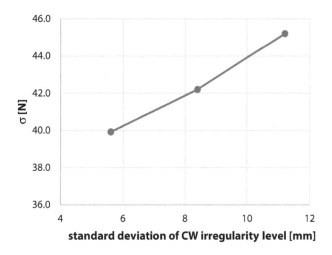

FIGURE 16.7 Simulation results for variation of standard deviation of contact force (σ) for C270 catenary with 20 kN contact wire tension versus various levels of contact wire irregularity for pantograph speed of 300 km/h.

two pantographs to collect enough power from the OCL, as in the case of the Italian high-speed ETR500 train in DC operation.

Multiple pantograph operation introduces diverse problems, with the effects depending on train speed and on the spacing of pantographs. At low speeds, say 60–100 km/h, the effect of multiple pantograph collection mainly consists of an increased CW uplift, which can be an issue for the transit under the suspension points. This effect is particularly significant in those cases where the distance between the pantographs is shorter than the length of one span of the catenary, which means two pantographs can be in a same span of the OCL.

A second typical case is concerned with the operation of multiple pantographs in the low- to medium-speed range (up to 120–140 km/h), with distances between the pantographs comparable to the span length. In this case, problems may arise when the speed of the train maximises the introduction of energy in the vibration of the OCL. In this regard, the most critical condition takes place when the distance between pantographs is close to the length of the span, and, at the same time, the span passing frequency approaches one half of the first natural frequency of the span.

Finally, high-speed operation of multiple pantographs requires that the distance between pantographs is much larger than the length of one span, in the order of 200 m. In this case, the effect of the trailing pantograph on the leading one is generally negligible, but the interaction of the trailing pantograph with the catenary is negatively affected by the vibration of the CW triggered by the leading pantograph. This effect depends on the time delay between the motion of the OCL induced by the transit of the first pantograph and the transit of the trailing one, and it should be emphasised that an increase of the distance between the pantographs does not necessarily correspond to an improvement of current collection quality on the trailing pantograph.

16.3 NUMERICAL SIMULATION OF PANTOGRAPH-CATENARY INTERACTION

Early attempts to formulate numerical models for pantograph-catenary interaction date back to the late 1960s and early 1970s; see [3] as an example. Since then, numerical tools for the modelling and simulation of pantographs and of the overhead equipment have evolved substantially, enabling the accurate and detailed simulation of several phenomena and effects relevant to the operation of pantograph-catenary coupling.

Essentially, a mathematical model of pantograph-catenary interaction is obtained from three main ingredients: a mathematical model of the OCL, a model of the pantograph as a multibody

(MB) vibrating system and a mathematical description of the contact between the pantograph head and the catenary. The available modelling approaches for these three components of the system are outlined in Sections 16.3.1 to 16.3.3.

16.3.1 Catenary Models

From the point of view of OCL vibration, two main factors affecting pantograph-catenary interaction are the spatial stiffness variation of the overhead wire along each span and the propagation of flexural waves in the CW. These effects can be reproduced by numerical models at different levels of accuracy and complexity.

16.3.1.1 Lumped Parameter Models

Early approaches to the modelling of railway catenaries were based on the use of a 1 degree of freedom (DoF) system having variable stiffness, as shown in Figure 16.8 [4]. The periodically varying stiffness can be obtained from the catenary elasticity function $E(x)$, defined as:

$$E(x) = \frac{F}{z_{st}(x)} \quad (16.2)$$

where F is the intensity of a point force applied at position x along the CW in the vertical direction and $z_{st}(x)$ is the static uplift of the CW produced by the application of the force. An example of catenary flexibility, derived from the pantograph-catenary interaction benchmark [5], is shown in Figure 16.9.

The lumped mass in Figure 16.8 represents the mass of a small portion of the CW and hence is often negligible compared with the mass of the pantograph head.

The 1-DoF model with periodically varying stiffness is very simple and allows for in-depth studies of the effect of periodic variation of the catenary stiffness on the interaction with the pantograph. However, it cannot represent wave propagation in the CW.

16.3.1.2 Finite Difference Models

The simplest model of the OCL capable of representing wave propagation is a semi-analytical model in which the CWs and MWs are represented as a tensioned string or a tensioned beam, neglecting the curvature of the MW, as shown in Figure 16.10. A set of n_d droppers is assumed to connect the two wires at discrete locations, and a set of n_p forces is considered acting on the CW, each one representing the effect of one pantograph. Lumped springs, dashpots and masses represent the suspensions and registration arms at the masts.

The equation of motion for each wire is then formulated as:

$$\rho A \frac{\partial^2 w}{\partial t^2} + \beta \frac{\partial w}{\partial t} + EI \frac{\partial^4 w}{\partial x^4} - T \frac{\partial^2 w}{\partial x^2} = -\rho A g + \sum_{j=1}^{n_d} f_{d,j} \delta(x - x_{d,j}) + \sum_{j=1}^{n_p} f_{p,j} \delta(x - x_{p,j}) \quad (16.3)$$

where w is the vertical displacement of the wire, ρA is the line density of the wire, β is a damping coefficient, EI is the wire bending stiffness, T is the wire tension and g is gravitational acceleration. The third term on the left-hand side of Equation 16.3 represents the effect of the wire's bending

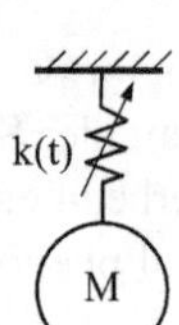

FIGURE 16.8 Model of the OCL as a 1-DoF system with periodically varying stiffness.

Dynamics of the Pantograph-Catenary System

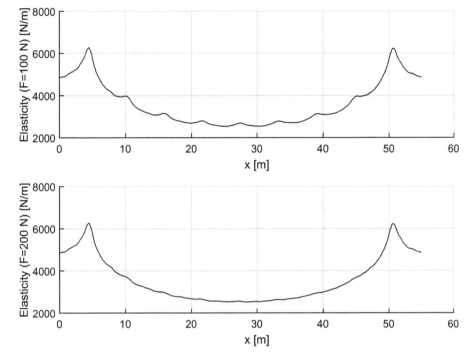

FIGURE 16.9 Elasticity of the catenary used in the pantograph-catenary benchmark. Top: $F = 100$ N, bottom: $F = 200$ N. (From Bruni, S. et al., *Veh. Syst. Dyn.*, 53, 412–435, 2015.)

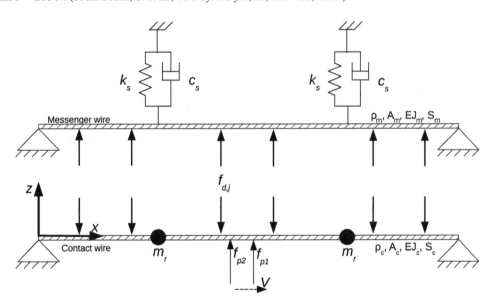

FIGURE 16.10 Model of the catenary as two tensioned beams connected by droppers. (From Facchinetti, A. and Bruni, S., *J. Sound Vib.*, 331, 2783–2797, 2012. With permission.)

stiffness and is sometimes neglected, especially for the MW [6,7]. The last term on the right-hand side of the equation represents the forces applied by n_p pantographs and therefore is only included in the equation of motion of the CW.

A discretisation in space of the equations of motion for the MW and CW can be obtained using the displacements w_i of the wire at discrete locations $x = x_i$ as unknowns and approximating the partial derivatives involving the space coordinate x by means of finite differences (FDs) of the discrete set of displacements w_i. The details of this procedure can be found in [7]; this results in a set of second-order differential equations in the form:

$$\mathbf{M}\ddot{\mathbf{x}}_c + \mathbf{C}\dot{\mathbf{x}}_e + \mathbf{K}\mathbf{x}_c = \mathbf{F}_g + \mathbf{F}_d\left(\mathbf{x}_c, \dot{\mathbf{x}}_c\right) + \mathbf{F}_p\left(\mathbf{x}_c, \dot{\mathbf{x}}_c, \mathbf{x}_p, \dot{\mathbf{x}}_p, t\right) \tag{16.4}$$

where $\mathbf{x}_c$ is a vector collecting the unknown displacements of the catenary at the discrete locations $x = x_i$; $\mathbf{M}$, $\mathbf{C}$ and $\mathbf{K}$ are the mass, damping and stiffness matrices of the catenary, respectively; $\mathbf{F}_g$ is a constant vector representing gravitational forces, $\mathbf{F}_d$ is the vector of forces representing the droppers (see Subsection 16.3.1.4 below) and $\mathbf{F}_p$ is a vector representing the forces applied by the pantographs (see Subsection 16.3.3).

16.3.1.3 Finite Element Models

The discretisation of catenary equations based on FDs is simple and efficient but is not well suited to modelling complex geometries, such as three-dimensional (3D) schemes of the catenary resulting from the effect of the stagger [5,8] and curved catenaries [9], or to consider specific features of the OCL, such as neutral section with insulators, transitions to tunnels, rigid catenaries and the related transitions. In these cases, mathematical models based on the finite element method (FEM) are generally preferred. According to this method, the MW and CW are generally modelled as tensioned Euler-Bernoulli beam elements [10–12]. The FEM also enables the use of detailed models for the steady arms [9,10,13]. The droppers are represented by a set of point forces depending non-linearly on the motion of the nodes at which they are attached. More details about the model of the droppers is provided at point 4 below. The forces applied by the pantographs on the CW are represented by a set of moving point forces, whose value depends on the motion of the CW and pantograph head, based on a model of pantograph-catenary interaction as described in Subsection 16.3.3.

The equations of motion for the OCL resulting from the application of the FEM are in the same form as the ones resulting from the use of FDs, see Equation 16.4, but the way in which the matrices and vectors appearing in the equation are defined is different. As far as the mass and stiffness matrices are concerned, their expression depends on the formulation used for the application of the FEM. The standard linear formulation for the 2-node Euler-Bernoulli beam with tension [14,15] is the most widely used, but attempts were made to use 4-node elements. The use of advanced Finite Element formulations was recently proposed to consider large rotations and large deformations in the catenary [16–18].

The damping matrix $\mathbf{C}$ is usually defined according to the Rayleigh damping (or proportional damping) assumption:

$$\mathbf{C} = \alpha\mathbf{M} + \beta\mathbf{K} \tag{16.5}$$

where $\mathbf{M}$ and $\mathbf{K}$ are the mass and stiffness matrices obtained from the FE model, and α and β are two constant coefficients, see Subsection 16.3.1.5 below.

16.3.1.4 Model of Droppers

The slackening of the droppers results in non-linear effects strongly influencing the dynamics of the pantograph-catenary system [10,11]. Therefore, the use of a suitable non-linear model of the droppers is a key point in the modelling of pantograph-catenary interaction. The simplest and most frequently used approach to modelling the slackening effect consists of considering the droppers as bilinear springs with zero tension in slackening [5]. Alternatively, the force-displacement characteristic curve obtained from laboratory tests can be used [10]; see example in Figure 16.11. In this latter case, the transition from tension to slackening is more gradual, so the choice between using this model or the simpler bilinear one may have some effect on simulation results.

Dynamics of the Pantograph-Catenary System

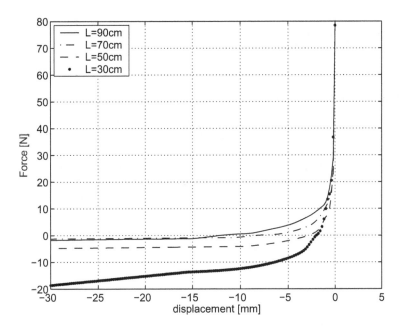

FIGURE 16.11 Experimental force-displacement characteristic curve for single wire droppers with different lengths. (From Collina, A. and Bruni, S., *Veh. Syst. Dyn.*, 38, 261–291, 2002. With permission.)

It has to be noted that the axial stiffness in tension of the droppers is inversely proportional to their length. Therefore, different values of the non-zero stiffness of the bilinear model or different characteristic curves need to be used for droppers located at different positions along the span, depending on their length. Figure 16.11 also shows that droppers with shorter length (30 cm) have a non-zero stiffness in the compression region. The mass of the dropper and of the clamps is normally considered by introducing two lumped masses attached to the MW and CW.

16.3.1.5 Catenary Damping

Catenaries are lightly damped systems; therefore, the results of pantograph-catenary interaction simulations can be sensitive to the assumptions made in the catenary model regarding structural damping. This issue is particularly important when the case of multiple pantograph collection is considered. Indeed, whilst a single pantograph, or the leading pantograph in the case of multiple collection, encounters an unperturbed catenary, the trailing pantographs are affected by the residual oscillation generated in the catenary by the passage of any other pantograph moving ahead of them. In [5] it is concluded, based on comparison of results from different simulation codes, that the modelling of catenary damping is a present challenge in the simulation of pantograph-catenary interaction.

Realistic values of the damping coefficients can be defined from the measurement of catenary vibration during and after the passage of a train. In investigation performed for three catenary sections on a conventional railway line in Norway [19], the identified values of the damping ratio were generally below 1% and below 0.25% at frequencies higher than 5 Hz. The decreasing trend of the damping ratio with frequency obtained from these experiments could be well interpolated according to Rayleigh damping using coefficients $\alpha = 0.062$ s^{-1} and $\beta = 6.13\text{e-}06$ s. In [5], the use of Rayleigh damping coefficients $\alpha = 0.0125$ s^{-1} and $\beta = 1.0\text{e-}04$ s is proposed, these values having been identified on a high-speed catenary in Italy. There is a significant difference between the coefficients of the Rayleigh damping model proposed in [19] and in [5], but it has to be noted that the catenaries in Norway and Italy on which measurements were taken have totally different designs. It should also be remarked that, given the decreasing trend of the damping ratio with frequency, the importance of the second term in the expression of Rayleigh damping becomes negligible, which explains the large difference observed in the identified values for coefficient β.

16.3.1.6 Static Position of the Catenary

The static position of the catenary under the effect of gravity and of the static tensioning of the wires needs to be defined to correctly initialise the simulation of train passage. Solving the static problem for the catenary is a quite peculiar problem. When a catenary system is installed in a railway line, the tension of the wires is set to specified values by proper tensioning devices, while, at the same time, the position of the CW is set by adjusting the length of the droppers and the steady arms, so to meet a prescribed pre-sag in the vertical plane and polygonalisation in the horizontal plane. Hence, when the static problem is solved for a catenary model, not only the loads (gravity, tension in the wires) but also the static position of the catenary at some selected locations are prescribed, whereas the unknowns of the problem are the lengths of the wires and droppers in their undeformed configuration. Examples of procedures for solving the above formulated static problem are proposed in [13] and [20].

16.3.2 Pantograph Models

The pantograph, as shown in Figure 16.12, is a mechanical system formed by a frame and a head carrying the collector strips that are in contact with the overhead equipment. The head is connected by an elastic suspension to the frame, which is an articulated system formed by a lower arm, an upper arm, a coupling rod and a guiding rod. An actuation system, normally realised by a pneumatic bellows, is used to move the frame for lifting and lowering the head. The actuation system also applies a static force, known as pre-load, pressing the collector strips against the CW.

Real pantographs feature several non-linear effects related to their kinematics and to the presence of non-linear force elements such as dry friction forces in the joints and state-dependent actuation forces. However, their low-frequency dynamics can still be represented to an acceptable level of

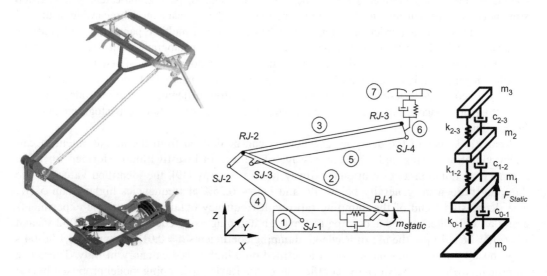

FIGURE 16.12 Typical pantograph (left) and its multibody (centre) and lumped mass (right) models. (From Ambrósio, J. et al. *Veh. Syst. Dyn.*, 53, 314–328, 2015. With permission.)

accuracy using relatively simple models known as 'lumped parameter' (or 'lumped mass') models. Alternatively, more detailed MB models can be used. Finally, 'hybrid models' are an extension of LP models, considering in addition the effect of collector strip flexibility.

16.3.2.1 Lumped Parameter Models

The LP models of pantographs consist of multiple masses connected by springs and viscous dashpots. Each mass can only move in the vertical direction, so the LP model neglects possible lateral and roll movements of the pantograph (especially of the head) that may arise due to flexibility in the joints and in the arms. The number of masses used to model the pantograph is either 2 or 3. In two-mass models, the lower mass represents the kinematic movement of the frame (assuming a rigid motion of the lower and upper arms), whereas the upper mass represents the vertical motion of the pantograph head. In three-mass models, one intermediate mass is introduced to represent, in a simplified way, the effect of bending flexibility in the upper arm. Figure 16.12 on the right shows an example of a three-mass model for the pantograph.

Despite their simplicity, LP models allow an accurate description of pantograph dynamics in a frequency range up to 20 Hz, provided their parameters are defined so that the frequency response function (FRF) of the LP model fits the measured FRF of the pantograph that can be obtained from a laboratory test, as described in Section 16.4. The procedure used to define the parameters of the LP model based on comparison with the experimental FRF of the pantograph coming from an experiment is referred to as 'model calibration'. Given that the FRF of the pantograph is affected by the height of operation, different sets of parameters need to be defined if significantly different heights of operation of the pantograph are to be considered in the simulation.

16.3.2.2 Multibody Models

An alternative to simple LP models is represented by detailed MB models of the pantograph. The MB models represent each body in the pantograph based on its shape and material properties according to a general multibody systems (MBS) modelling procedure [9,19,21,22]. The assumption of rigid body movements is often made for the different bodies composing the pantograph, so that geometry parameters and mass properties need to be defined for each body only. Sometimes, however, MBS models also consider the flexibility of the upper frame and/or collector strips [21,22].

Compared to LP models, MBS models are much more complex and require a larger effort to be defined and to run numerical simulations but offer significant advantages under some circumstances. Firstly, MBS models do not require model calibration, as is required to properly define a LP model. Therefore, MBS models can be used in the design phase of a new pantograph to perform kinematic and dynamic analyses from the starting phase of the design process of the pantograph [6]. Secondly, MBS models are capable of considering different working heights for the pantograph and inherently account for kinematic non-linearities.

16.3.2.3 Hybrid Models

Hybrid pantograph models are an extension of the LP models presented in Subsection 16.3.2.1, in which the effect of collector strip bending deformability is considered to extend the frequency range of the model. In hybrid models, a multi-mass model is adopted to describe the motion of the lower and upper arms, whereas the motion of the flexible collector strips is described as a linear combination of the collector's modal shapes under free-boundary conditions [10,23]. In this way, the low computational effort of LP models is maintained thanks to the reduced number of DoFs involved and to the fact that linear equations of motion are obtained, but the frequency range of validity of the model is significantly extended from approximately 20 Hz for an LP model to 100 Hz and more for a hybrid model.

16.3.3 PANTOGRAPH-CATENARY CONTACT

Two approaches are mainly followed to model the sliding contact between the CW and the collector strips:

- The use of a sliding joint formulation imposing a kinematic constraint on the relative motion of the two bodies at their contact point
- The use of a penalty method

The sliding joint constraint is formulated in terms of non-linear algebraic equations; therefore, the resulting equations of the pantograph-catenary system are in the form of a system of differential-algebraic equations (DAEs) [15]. Lagrange multipliers are introduced in the differential equations of motion of the catenary and of the pantograph(s), representing the contact force exchanged at each contact point. Given that the contact between the pantograph head and the catenary is unilateral, the sliding contact constraint must be de-activated when the contact force vanishes; see for example the procedure described in [7].

According to the penalty method [10,24], sometimes called 'elastic contact' [18], a virtual penetration of the two bodies at their contact point is evaluated, and a unilateral contact force is defined as a function of the penetration and sometimes of the speed of penetration, thus approximately enforcing the sliding joint kinematic constraint. This is often done assuming the compressive contact force to depend linearly on the penetration and speed of penetration, so that the contact is realised by means of a sliding unilateral spring + dashpot element. The penalty method has a simple mathematical formulation, does not require to solve a system of DAEs and its implementation is very efficient from a computational point of view. However, the values of the stiffness and damping parameters defining the contact element need to be carefully set to avoid excessive violation of the sliding contact kinematic constraint, which would lead to inaccurate results in the simulation of pantograph-catenary dynamics. A detailed description of how the contact stiffness and damping parameters should be chosen in order to ensure a proper accuracy of the penalty method is provided in [10].

16.3.4 SOFTWARE FOR THE SIMULATION OF PANTOGRAPH-CATENARY INTERACTION

Table 16.1 provides a necessarily non-exhaustive overview of existing software for the simulation of pantograph-catenary interaction. The table is expanded and adapted from the one in [5]. The table lists software from 16 institutions, in alphabetical order: Alstom, Deutsche Bahn (DB), Instituto Superior Técnico (IST), Korean Rail Research Institute (KRRI), Royal Institute of Technology (KTH), Norwegian University of Science and Technology (NTNU), Politecnico di Milano (POLIMI), Railway Technical Research Institute (RTRI), Société Nationale des Chemins de fer Français (SNCF), SouthWest Jiaotong University (SWJTU), University of Castilla-La Mancha (UCLM), University of Illinois at Chicago (UIC), Università di Salerno (UNISA) University of Huddersfield (UoH), Universidad Pontificia Comillas (UPCo) and Universitat Politècnica València (UPV).

Table 16.1 shows that most of the software packages use the FEM to model the catenary, are capable of 3D analyses and consider the non-linearity due to the slackening of droppers. The table also shows that pantograph models are largely based on LP models, although a significant number of software packages also allow the use of MB models, and only 7 software packages out of 16 allow to consider the flexibility of the contact strips. The modelling of the sliding contact between the collectors and the CW is prevailingly based on the penalty method.

TABLE 16.1
List of Software for the Numerical Simulation of Pantograph-Catenary Interaction

General				Catenary			Pantograph				Sliding Contact	
Institution	Software Name	FEM/FD	2D/3D	Element Types	Droppers	Lumped Mass	Flexibility Contact Strips	Multibody	Multiple Pantos		Penalty Method	Constraint
ALSTOM	INPAC	FEM	3D	EBB	Bar element with slackening	Yes	No	No	Yes		Yes	No
DB	PROSA	FD	2D	Beams, strings	Piecewise linear with slackening	Yes	Yes	Yes	Yes		No	Yes
IST	PantoCat	FEM	3D	EBB, TB	Beam elements with slackening	Yes	Yes	Yes	Yes		Yes	No
KRRI	SPOPS	FEM	2D[a]	EBB	Mass-spring-damper with slackening	Yes	No (rolling considered)	No	Yes		Yes	Yes
KTH	CaPaSIM	FEM	3D	EBB, bars	Bar elements with slackening	Yes	Yes	No	Yes		Yes	No
NTNU	RailCatNOR	FEM	3D	TB	Bar elements with slackening	Yes	No	No	Yes		Yes	No
POLIMI	PCaDA	FEM	3D	EBB	Non-linear visco-elastic with slackening	Yes	Yes	No	Yes		Yes	No
RTRI	Gasen-do FE	FEM	3D	EBB	Bar elements with slackening	Yes	No	No	Yes		Yes	No
SNCF	OSCAR	FEM	3D	EBB	Non-linear with slackening	Yes	Yes	Yes	Yes		Yes	No
SWJTU	PCRUN	FEM	3D	EBB	Spring element	Yes	No	Yes	Yes		Yes	No
UCLM	INDICA3	FEM	3D	EBB	Bar elements with slackening	Yes	Yes	No	Yes		Yes	Yes
UIC	SIGMA/SAMS	FEM and FD	3D	ANCF	Non-linear spring-damper	Yes	Yes	Yes	Yes		Yes	Yes
UNISA	ADA	FEM	3D	ANCF	Non-linear spring-damper	Yes	No	Yes	No		Yes	No
UoH	TOPCAT	FEM	3D	EBB	Non-linear beam elements	Yes	No	No	Yes		Yes	No
UPCo	CANDY	FEM	2D	CRB, bars	Bar elements with slackening	Yes	No	No	Yes		Yes	No
UPV	PACDIN	FEM[a]	3D	ANCF, bars	Bar elements with slackening	Yes	No	No	Yes		Yes	No

[a] Change of lateral position of the contact point due to stagger can be considered.

Abbreviations: FEM, finite element method; FD, finite differences; EBB, Euler-Bernoulli beam; TB, Timoshenko beam; CRB, co-rotational beam; ANCF, absolute nodal coordinate formulation.

16.4 MEASUREMENT, TESTING AND QUALIFICATION OF PANTOGRAPHS

Physical tests are an important part of the design and qualification process for pantographs and catenaries. Depending on their scope, tests can be performed in a laboratory or on track. In this section, the most common measurements and tests performed on the pantograph and on the catenary are described.

16.4.1 Laboratory Testing of Pantographs

16.4.1.1 Test Rigs for Pantograph Testing

Common laboratory tests for the evaluation of pantograph dynamics can, in principle, be performed utilising multi-purpose test facilities, provided these are able to apply vertical excitation to the pantograph collector(s). Nevertheless, dedicated test rigs specifically designed for pantograph dynamic tests have been developed and are in use at some laboratories.

As an example, Figure 16.13 shows the test rig for the evaluation of pantograph dynamics currently in use in the laboratories of Politecnico di Milano, Dipartimento di Meccanica [25]. The rig includes a fixed frame supporting a transversal actuation system, consisting of a slider resting on linear ball guides and moved by an AC motor with a ball screw, and a vertical actuation system mounted on the slider and composed of two independent hydraulic actuators, allowing the application of independent excitations to the collectors of the pantograph.

The electric motor for the lateral actuation is operated in position control mode and used to position the vertical actuators along the collectors of the pantograph under test or to reproduce the catenary stagger, with ±350 mm maximum amplitude. The vertical hydraulic actuators have a maximum stroke of 52 mm and a baseband bandwidth up to 100 Hz (only for very small amplitudes). They are operated in stroke control mode and can be fed with pre-defined waveform or with external signals. Two load cells are interposed between the vertical actuators and the pantograph to measure the force applied on each collector.

Another example of a dedicated test rig is that of DB Systemtechnik in Munich [26]. The main difference is that the test rig in this case is composed of a portal that can perform

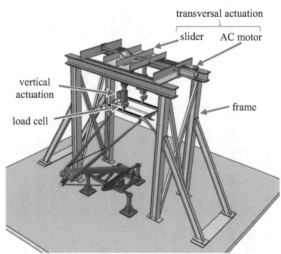

FIGURE 16.13 Pantograph dynamic test rig at Politecnico di Milano.

vertical slow movement with high amplitude (from a height of 0.8 m to 2.2 m above the base plate) in order to test a pantograph in all its working range and to simulate slow changes of the catenary height.

A very similar layout is adopted for the test rig at the State Key Laboratory of Traction Power at SWJTU in Chengdu [27]. A special feature characterising the pantograph test bench at Chengdu and not included in the other two test rig examples is the possibility to apply a given motion to the pantograph base, aimed at simulating the effects of train roof movements. In particular, it is possible to give the pantograph base ± 50 mm vertical displacements and $\pm 3°$ roll rotations, with frequency up to 30 Hz.

All the mentioned test rigs use hydraulic actuators to apply the vertical excitation to the pantograph collector/s. In this respect, the test rig developed by the University of Birmingham and described in [28] presents a distinctive feature, being completely actuated by means of electric drive, also for the vertical excitation of the pantograph. Another peculiarity of this last test rig is that it was designed to be installed in a railway maintenance depot, to test pantograph mounted on train roof.

Test rigs are usually adopted to perform type tests and routine tests prescribed by specific standards, to characterise the dynamic behaviour of pantographs, to reproduce pantograph-catenary interaction in the laboratory and to calibrate measuring systems to be used in line tests.

16.4.1.2 Type and Routine Tests

Existing standards specify different kinds of laboratory tests that a pantograph shall undergo before it is put into service. The European Standard EN 50206 (the basis for the International Standard IEC 60494), in the two versions related to main-line vehicles and metro and light rail vehicles, distinguishes between type tests that shall be performed on a single pantograph of a given design and routine tests to be performed in order to verify that the properties of a product correspond to those measured during the type tests.

Focusing on operating dynamic tests in the laboratory, EN 50206 requires to perform raising and lowering endurance tests (type tests), considering the operation between the housed position and the upper operating position (10,000 cycles) and operation within the working range (75,000 cycles), in this case at fixed speed and with disconnected damper, if present.

Moreover, as type and routine tests, the static contact force of the pantograph shall be measured during a continuous slow motion between the upper and lower operating positions, with disconnected damper, in order to verify that the hysteresis associated with the system is limited and the deviation from nominal static contact force inside the working range is within the limits prescribed by the standard. The same test shall be performed at ambient temperature as a routine test and at extreme temperatures and humidity levels as type tests.

The pantograph shall also undergo tests to verify its resistance when submitted to transversal vibrations.

16.4.1.3 Dynamic Characterisation Tests

Pantograph test rigs are also adopted to characterise the dynamic behaviour of a pantograph, both to verify that it meets the design target and to calibrate/validate numerical models, in particular LP models, used for the numerical simulation of pantograph-catenary interaction.

The dynamic behaviour of the pantograph is usually represented in terms of FRFs between the vertical force applied to the collector/s and the pantograph motion, i.e., the motion of the collector(s) and of the upper and lower frame.

In order to obtain such FRFs, the pantograph is instrumented with suitable transducers, generally accelerometers. Figure 16.14 shows a typical arrangement in terms of positioning of accelerometers for a pantograph equipped with independent collectors.

The pantograph is then excited by applying imposed vertical motion to the collector(s). Swept sine or sinusoidal inputs at different frequencies are adopted in order to introduce adequate energy in

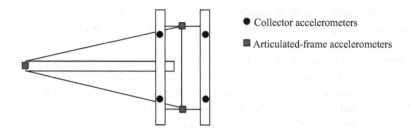

FIGURE 16.14 Position of acceleration transducers for pantograph dynamic characterisation.

the system. Particular care should be paid when selecting the amplitude of the input motion. In fact, the representation of the system by means of FRFs involves the reduction of the pantograph to a linear system, whereas significant non-linearities can be present, especially in the low-frequency range. The excitation amplitudes at different frequencies should thus be chosen consistently with the amplitudes expected in real operation.

To this end, different approaches are adopted, such as applying (i) an excitation amplitude inversely proportional to the frequency, (ii) an excitation amplitude that produces a dynamic force with almost constant amplitude over the considered frequency range, and (iii) different fixed amplitudes in different frequency ranges, as in the example shown in Figure 16.15.

Figure 16.15b shows that the response of the pantograph at low frequencies is significantly different, depending on the chosen excitation amplitude. Two natural frequencies are clearly visible in the 0–10 Hz frequency range, their value and that of the related damping being dependent on the excitation amplitude, whereas a third natural frequency exists around 12 Hz, related to the upper-frame deformability, although its contribution is hardly visible when considering the FRF evaluated on the collector motion. The pantograph behaviour can therefore be represented using a three-mass model, provided that the FRFs associated with the relevant excitation amplitudes are considered for the different frequency ranges. For the sake of completeness, the FRF of the identified LP model is also displayed in Figure 16.15b by the solid line.

It is worth mentioning that the behaviour at very low frequency is significantly affected by the response of the pneumatic system used to feed the pantograph, so this system should be reproduced in the laboratory setup as close as possible to the one adopted in operation.

16.4.1.4 Hardware-in-the-Loop Simulation of Pantograph-Catenary Interaction

Another possible use of a pantograph test rig is the hardware-in-the-loop (HIL) simulation of pantograph-catenary interaction. In this case, the actuation system of the test rig is used to mimic the motion of the catenary, so as to obtain an interaction between the real pantograph and a virtual catenary. To this end, two different approaches can be considered, namely an 'open-loop HIL' approach and a 'closed-loop HIL' approach, the second one resulting in the hybrid simulation of interaction previously mentioned in Subsection 16.3.2.3.

In the 'open-loop HIL' approach, the actuation system is driven by an excitation signal evaluated prior to the tests (Figure 16.16a), representing the vertical motion of the catenary. The excitation signal can either be calculated directly from track measurement results or evaluated by means of numerical simulation of pantograph-catenary interaction. In the latter case, it is possible to fine-tune the pantograph model on the basis of laboratory results and generate a new excitation signal by means of simulation, thus achieving an 'off-line' closure of the loop, as represented by the dashed line path in Figure 16.16a.

In the 'closed-loop HIL' testing, also known as 'hybrid simulation' (see Figure 16.16b), the real pantograph is coupled to a numerical model of the catenary [7,29,30]. The forces generated by the pantograph at each collector are measured by load cells and sent to a real-time board, where they are used to compute the dynamic displacements of the CW at the positions of the contact points

Dynamics of the Pantograph-Catenary System

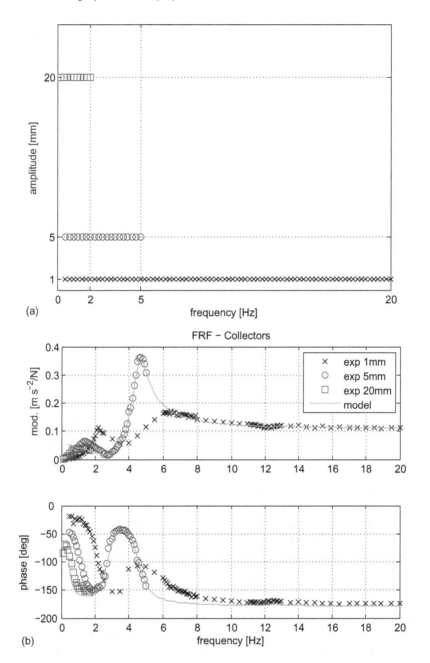

FIGURE 16.15 Example of pantograph experimental FRFs: (a) excitation amplitudes at various frequencies and (b) FRFs between applied force and collector acceleration.

according to a numerical model of the flexible catenary; the displacements computed by the real-time board are then sent as references to the control system, to be applied on the pantograph head by one or more actuators.

The two approaches present advantages and disadvantages that are somehow complementary. Considering the open-loop HIL approach, the principal drawback is that a real interaction is not actually reproduced, so that it is difficult to compare different pantographs or analyse

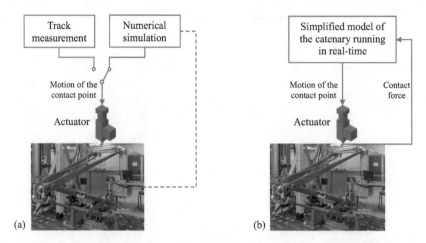

FIGURE 16.16 HIL simulation of pantograph-catenary interaction: (a) open-loop HIL and (b) closed-loop HIL.

the influence of pantograph modifications on the quality of pantograph-catenary interaction. In order to partially overcome this limitation, the pantograph is usually not directly coupled with the actuators, but linking springs reproducing the flexibility of the overhead line are interposed between the actuator and the pantograph [31]. On the other hand, since the simulation of the pantograph-catenary interaction is performed offline, prior to the test, there are no limitations on the complexity of the numerical tool adopted to simulate the pantograph-catenary interaction.

The closed-loop HIL approach allows to reproduce a true interaction between the real pantograph and the virtual catenary but requires the adoption of a simplified numerical model of the catenary compatible with the requirements for real-time execution. Aiming at getting the best trade-off between accuracy and computational efficiency of the catenary model, the use of an Eulerian approach in combination with special absorbing boundary layers is proposed in [30] to efficiently model the dynamics of the catenary, whilst Gregori et al. [32] proposed to split the numerical integration of the catenary model in an 'offline stage' and a fast 'online stage', thus enabling the real-time solution of a relatively large system of equations for the catenary at the price of performing an offline pre-processing stage.

16.4.1.5 Calibration of the Measuring System of Contact Force

Test rigs are used for the calibration of systems to measure the contact force between pantograph and catenary.

The most common measuring setup shown in Figure 16.17 consists of load cells placed between the collector and its own suspensions and accelerometers placed nearby and used to evaluate and compensate for the inertial contribution of the collector in the vertical direction.

The contact force F_c shown in Figure 16.17 then results from the forces measured by the load cells $F_{lc,i}$, the compensation of the inertial contribution evaluated on the basis of the measured accelerations a_j, and the compensation of the aerodynamic lift on the collector $F_{corr,\ aero} = -F_L$ (the lift force F_L acting between the load cells and the contact point), as given by:

$$F_c = \sum_i^{n_{lc}} b_i F_{lc,i} + \sum_j^{n_a} m_j \, a_j + F_{corr,aero} \qquad (16.6)$$

The influence coefficients b_i are evaluated by means of static calibration, applying increasing and decreasing forces in different positions along the collector. This also provides an estimation of the

Dynamics of the Pantograph-Catenary System

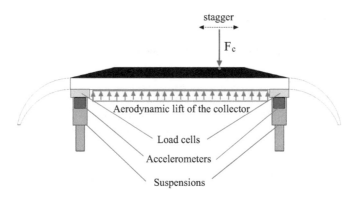

FIGURE 16.17 Principle of the measuring setup for the contact force.

contact point position. The mass coefficients m_j that represent the portion of mass that is associated to each of the accelerometers are evaluated by means of suitable dynamic calibration performed in the laboratory.

The whole measurement system is then tested to check that the accuracy of the measured force is within the requirement provided by the standard (EN 50317 from which IEC 62846 is derived). This is done by evaluating the transfer function between the reference force measured by the load cells of the test rig and the one estimated by the measurement system shown in Figure 16.17, applying sinusoidal excitation to the collector at different frequencies from 0.5 Hz to 20 Hz. The magnitude of the transfer function should be as close as possible to unity, with negligible phase, and allows checking the accuracy of the first two terms on the right side of Equation 16.6, whereas the compensation of the aerodynamic lift is evaluated through specific line tests, as described in the following subsection.

It is worth remarking that the validity of the contact force evaluation is limited to the 0–20 Hz range (25 Hz in some cases), since Equation 16.6 is based on the assumption that the collector can be considered as a rigid body, whilst deformability effects at higher frequencies become non-negligible on the collector (the first natural frequency of the collector usually falls in the 40–70 Hz range).

16.4.2 Qualification and Track Testing of Pantographs and Catenaries

The qualification of a pantograph or of an OCL is a complex process, involving technical and operational issues and composed of several phases, from design to laboratory testing of the pantograph and simulation of pantograph-OCL dynamics in order to demonstrate the feasibility of the operation.

The systems are finally checked by testing the pantograph on the rolling stock in a series of trial tests on track. Qualification may take place at national or international level, adopting national or international rules and standards.

With reference to the European Technical Specifications for Interoperability (TSI), which refer to the relevant EN/IEC standards, the following parameters and the associated line tests for their assessment are usually taken into account to evaluate the dynamics of the interacting systems:

- Mean contact force exchanged between pantograph and OCL as a function of train speed
- Contact wire uplift at suspension
- Dynamic component of the contact force exchanged between the collectors and the CW, in terms of standard deviation of the contact force
- Percentage of contact losses or arcing

The latter two are usually addressed as alternative methods to evaluate the quality of current collection.

The most common measurements and tests performed on track are summarised in the following subsections.

16.4.2.1 Tethered Tests

Especially for medium-high speed applications, aerodynamics has a significant influence on pantograph performances. Tethered tests are thus performed to characterise the aerodynamic behaviour of the pantograph.

In these tests, the pantograph collectors are restrained by means of vertical ropes to prevent contact with the OCL (Figure 16.18), with the restraint height set as close as possible to the CW. The tension of the ropes is measured by means of load cells placed at the lower extremity (Figure 16.18) and corresponds to the force that the pantograph would apply to the OCL, composed of the static preload and the global aerodynamic uplift of the pantograph, the latter resulting from the contribution of the aerodynamic forces acting on the different elements of the pantograph [33,34]. The tests are performed with the vehicle running at different speeds to obtain the global aerodynamic uplift as a function of speed.

It is worth remarking that the orientation of the pantograph, its position along the train and the geometry of the train roof in terms of fairings or protruding elements all have a significant influence on the aerodynamic behaviour of the pantograph and on the resulting aerodynamic uplift. As a consequence, the obtained results are relevant only for the considered configuration(s).

Tethered tests are also necessary to evaluate the term for the compensation of the aerodynamic lift on the collector that appears in Equation 16.6. To this aim, the pantograph undergoing the test shall be instrumented for contact force measurement, and the lift force on the collector can be derived as the difference between the tension in the retaining ropes and the force measured by the load cells in the collector suspension.

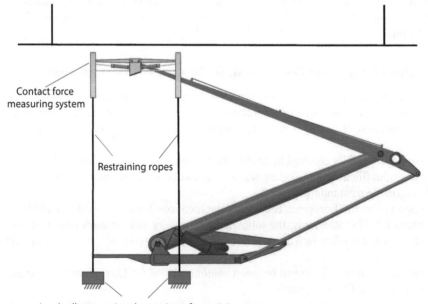

FIGURE 16.18 Configuration for tethered tests.

16.4.2.2 Contact Force Measurement for the Qualification of Current Collection Quality

The assessment of pantograph-catenary interaction is commonly performed by measuring the contact force at different vehicle speeds by means of the measuring technique previously described and by evaluating the mean value and the standard deviation of the contact force.

As far as the mean contact force is concerned, the TSI and EN 50367 (from which IEC 62486 is derived) provide upper and lower limits, depending on the type of power supply (DC 1.5 kV, DC 3 kV or AC). The measured mean force shall fall within the limit curves reported in Figure 16.19, for the entire range of operation speeds of the pantograph-catenary couple. This requirement is introduced to guarantee the interoperability between different systems: the OCL is designed to accept mean forces within the proper range, and the pantograph is designed and regulated to apply mean forces within the same range. The limit curves are currently defined up to a maximum speed of 250 km/h for DC applications and 320 km/h for AC ones. For higher speeds, national rules are applied or special rules need to be established.

On the other hand, the standard deviation of the contact force is used to evaluate the entity of the dynamic component of the force and the quality of current collection. To this end, the assessment criterion provided by the TSI and EN 50367 is the same as presented in Section 16.2:

$$\sigma_F \leq 0.3\, F_m \tag{16.7}$$

A more relaxed requirement is provided by EN 50119 that does not account for interoperability aspects:

$$F_m - 3\sigma_F > 0 \tag{16.8}$$

EN 50119 also provides limits, depending on the power supply (DC or AC), in the form of maximum permissible contact forces that the OCL shall be designed to accept.

The assessment shall be carried out for all the pantographs, in the case of multiple operation, or considering the worst performing pantograph, which can be determined by means of numerical simulation. Moreover, different speeds up to the maximum design speed shall be considered, taking into account that the most critical speed might not be the maximum one, especially in the case of multiple operations, where critical speeds depend on the spacing of the pantographs.

It is worth remarking that all the described evaluations and assessments are performed considering the contact force signal low-pass filtered at 20 Hz, which is the limit of validity of the measurement principle.

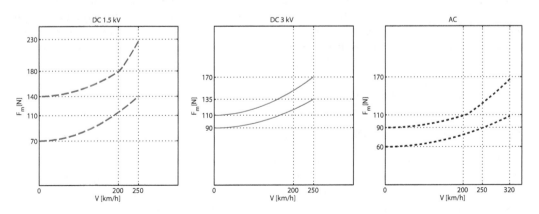

FIGURE 16.19 Mean contact force limits for DC and AC systems as a function of speed according to EN 50367.

16.4.2.3 Arcing Detection

The TSI and EN 50367 address the measurement of arcing as an alternative to the evaluation of the standard deviation of the contact force for the assessment of current collection quality. To this aim, an acceptable percentage of arcing is defined depending on the type of power supply and on the vehicle speed. The conditions of the tests to be performed, in terms of speeds and pantograph(s) to be considered, are the same as already mentioned for dynamic contact force measurement.

The main advantage of this alternative method is that it is not invasive, being based on the optical measurement of the duration of ultraviolet emission due to electric arcing [35], whereas the contact force measurement system unavoidably modifies the aerodynamic shape and mass of the pantograph collector and its suspension, even if the measuring setup is designed by paying particular care to produce the minimum impact on the pantograph.

It is worth mentioning that the two assessment alternatives are not proven to give exactly the same results even though a certain consistency was found when the two methodologies were considered at the same time.

The optical method can be used for both AC and DC current systems, while a method to detect contact loss based on the measurement and analysis of pantograph voltage drop is proposed in [36] just for the DC current system.

16.4.2.4 Contact Wire Uplift and Overhead Contact Line Motion

The analysis of interaction performance is generally completed by measuring catenary motion. In particular, the standards require the verification of the CW uplift at suspensions. This evaluation is needed to verify that prescribed clearances between the pantograph and the infrastructure are respected, namely to avoid possible engagements between the collector(s) and the OCL steady and registration arms, with consequent risks of de-wirement. In this case also, the most critical situation usually corresponds to multiple pantograph operations.

Besides the CW uplift required for assessment and qualification purposes, other measurements of OCL motion can be performed to better characterise and monitor the dynamic behaviour of the OCL; e.g., in [37], a dedicated wireless device is developed and proposed to measure wire accelerations at any location of interest.

16.5 WEAR, DAMAGE AND CONDITION MONITORING IN PANTOGRAPH-CATENARY SYSTEMS

Failures occurring with the pantograph-OCL couple strongly affect the reliability and regularity of railway operation, and the costs incurred by the maintenance of the OCL have a significant share in the total maintenance cost of a railway line. For these reasons, it is extremely important to analyse the causes of wear and damage in both the pantograph and the OCL and to find measures to effectively diagnose faults and improve the long-term behaviour of the overhead line. In this section, the main phenomena that influence the life of OCL and pantograph and that can lead to failure on both subsystems, namely wear of CW and contact strips and general damages of OCL and pantograph, are analysed, and possibilities for the health monitoring of pantograph-OCL couple are outlined.

16.5.1 Wear of Contact Wire and Contact Strip

Electro-mechanical wear is one of the factors that heavily affect the service life of both the OCL and the collector strip and has a direct impact on the maintenance costs. Research in this field aims at the reduction of wear levels on both collector strips and CWs to decrease their life cycle costs. The analysis of the sliding contact between the pantograph contact strip and the CW shows that very complex electromechanical phenomena occur during the dynamic interaction of the two subsystems. These phenomena cause wear on both strips and wire; see Figure 16.20.

Dynamics of the Pantograph-Catenary System

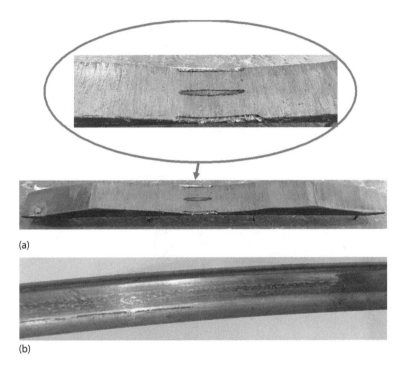

FIGURE 16.20 Examples of effect of wear on (a) contact strip and (b) contact wire.

The analysis and the prediction of wear rate starting from the physical-chemical characteristics of contacting body materials are not possible yet. For this reason, most of the research works on the wear of the couple strip-wire are performed experimentally, mainly with laboratory tests on specimens of collector strip and CW. Field tests are less frequent, as they are extremely expensive and require long-term measurements to observe wear progress in the CW.

In order to perform laboratory tests, specific test benches (see as examples [38–44]) were developed to reproduce the electro-mechanical phenomena occurring at the contact between collector strip and CW (Figure 16.21). Most of these are composed of a rotating disk to which the CW is connected, and the specimen contact strip is pressed against the wire specimen. The sliding speed is reproduced by the peripheral speed of the disk, while the staggering of the CW is obtained by moving the collector strip specimen radially or shaping the CW in the horizontal plane, according to an out-of-round configuration.

Laboratory tests provide results in terms of wear of both CW and collector strips as a function of the main operative parameters, such as the sliding speed, the electrical current, the contact force and the characteristics of the two contacting materials.

The analysis of laboratory tests allows to evaluate the performances in terms of wear rate of materials of wire and especially of strip. In particular, one important research subject is the development of materials and material combinations for both strip and wire, in order to enable the collection of large electrical currents, required especially for DC current collection, while maintaining acceptable wear rates of the two contacting elements.

In modern current collection systems, carbon-based collector strips are generally preferred to metal-based ones (e.g., made of copper or steel), as this solution increases the service life of the CW, although at the price of a shorter life of the contact strips. Given that the costs for the maintenance of the CW are much higher than those related to the frequent replacement of the contact strips, this choice results in an overall improvement of the maintenance costs for the entire pantograph-OCL system. For applications requiring high values of collected electrical current, plain carbon strips are replaced by copper-impregnated carbon strips, with percentage of impregnation up to 50%. This limit

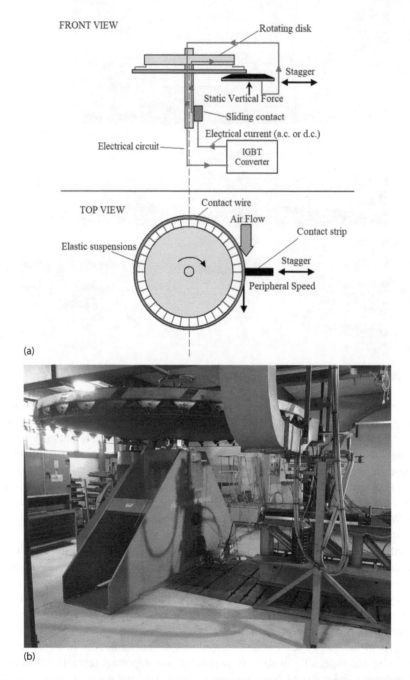

FIGURE 16.21 Test bench developed at Politecnico di Milano: (a) general scheme and (b) photograph of setup.

of impregnation is generally due to the limit on the mass of the strip: high values of strip's mass have a negative effect on the dynamics of pantograph-OCL interaction due to the increase of inertial effects.

The CW material is pure copper (generally electrolytic copper or Cu-ETP), or, in some cases, copper alloys (e.g., CuAg, CuMg and CuSn) are typically used to improve the mechanical behaviour of the CW, especially in terms of creep under the tension applied to the wire.

Laboratory tests highlight that the wear of both strip and wire is the result of three main contributions: (i) a mechanical contribution, due to friction; (ii) an electrical contribution, due to power dissipation related to the Joule effect; (iii) an electric arc contribution, due to the heat generated by the electrical arcs that occur when contact loss occurs. To extend the service life of the CW and collector strips, all these three sources of wear can be addressed. The replacement of metal strips with carbon-based strips in railway operation allowed to heavily reduce the mechanical contribution to the wear; laboratory tests showed the significant difference of friction coefficient values resulting from the use of metal strips and carbon based strips. In [44], some results of friction coefficient values obtained by laboratory tests performed by using the test bench of Politecnico di Milano are reported for different strips; mean values of friction coefficient for metal strips can reach 0.76, while mean values of friction coefficients for carbon-based strips are typically in the range 0.2–0.3.

An interesting phenomenon observed during laboratory tests is the 'current lubrication' [39,44,45]; an increase of electrical current at the contact results in a reduction of the friction effect; i.e., the flow of electrical current through the contact has an effect similar to introducing a friction modifier.

The electrical contribution to wear is strongly related to the electrical current values, which in DC current collection can be up to 1400 A per strip, and to the electrical contact resistance. This last parameter depends on the electrical characteristics of wire and strips (this is why impregnated carbon strips are used – copper impregnation improves the strip's conductivity properties) and on the dynamics of pantograph-catenary interaction. The contact force, indeed, heavily influences the electrical contact resistance, which increases when the contact force decreases. In [45], a relationship for the electrical contact resistance R_c as a function of contact force F_c is defined based on laboratory tests for the carbon-based strip-pure copper CW couple:

$$R_c(F_c) = 0.013 + 0.9 \cdot \exp\left(\frac{F_c - 14}{11}\right) \tag{16.9}$$

Therefore, when the contact force is low, the electrical contribution to the wear becomes very important. Finally, if the contact force goes down to zero and contact loss occurs, electrical arcs are generated, causing very high temperatures at the contact (greater than 3000°C) and consequent wear on both strip and wire (electrical arc contribution). The effect of electrical arcs on wear can be extremely harmful to the life of strips and wire.

In [45], an empirical formula for the CW normal wear rate (NWR), defined as the ratio between the worn volume of wire (mm³) and the sliding distance (km), is introduced to take into account the effect of the three contributions to wear:

$$\text{NWR} = \underbrace{k_1 \cdot \left(\frac{1}{2}\cdot\left(1+\frac{I_c}{I_0}\right)\right)^{-\alpha} \cdot \left(\frac{F_c}{F_0}\right)^{\beta} \cdot \frac{F_c}{H}}_{\text{Mechanical contribution}} + \underbrace{k_2 \cdot \frac{R_c(F_c)\cdot I_c^2}{H\cdot V}(1-u)}_{\text{Electrical contribution}} + \underbrace{k_3 \cdot u \cdot \frac{V_a \cdot I_c}{V \cdot H_m \cdot \rho}}_{\text{Electrical arc contribution}} \tag{16.10}$$

where k_1, k_2 and k_3 are coefficients found by fitting the equation to experimental data obtained on the laboratory test bench shown in Figure 16.21.

The authors in [45] propose to combine the empirical formula for the wear analysis with a numerical model for the dynamical interaction between pantograph and catenary, where the latter is used to provide a statistical distribution of the contact force F_c, which is fed into Equation 16.10, providing as output the corresponding estimated statistical distribution of the CW wear. In this way, not just the mean values of the main parameters but also the variation of these parameters related to the quality of current collection are introduced in the wear model, so that it is possible to assess the extent to which wear rate can be reduced by improving pantograph-catenary interaction.

Even though several important results have been obtained up to now in this field, many items still need to be investigated. One of the main aims of current research works is the development of predictive models that would be very useful for the planning of the maintenance operation and, therefore, convenient for the reduction of the maintenance costs.

16.5.2 DAMAGE OF CATENARY AND PANTOGRAPH

The expected service life of a high-speed catenary (AC current collection) exceeds 50 years. This means that, despite wear being the main phenomenon affecting maintenance costs for the OCL and for the pantograph, other effects related to fatigue and mechanical damage also need to be considered, especially to improve the reliability of operation. Research has recently addressed the issue of bending fatigue in the CW [46–49], with the aim of analysing in-service failures and finding appropriate countermeasures. Fatigue phenomena can rarely be diagnosed in advance of the final failure, because it is difficult to detect crack propagation in the material. Moreover, the unpredictability of damage related to the bending fatigue of the wire has an important impact on the consequences produced on the railway traffic; catenary failure causes traffic disruption and large costs due to the sudden intervention required for the restoration of normal conditions.

Wear and fatigue are in competition with each other, as wear tends to remove surface cracks, thereby mitigating the occurrence of fatigue. Of course, this effect depends on the rates of wear evolution and crack propagation; if wear evolution is faster than crack propagation, the occurrence of fatigue is very low, while, in the opposite case, fatigue phenomena are likely to occur.

Researchers from RTRI developed a specific test bench to study fatigue failure in the catenary CW [47]. The test bench was designed to reproduce realistic operational conditions by using a sample from a real CW. A tensile load was applied to the wire sample to reproduce the effect on the stress of the tensile load in the real operational conditions, and bending stresses were applied according to the three-point bending scheme.

Researchers at SNCF presented a method to analyse fatigue phenomenon, which replicated failure in particular sections of CW by using numerical models. As an example, the case of a failure that occurred in the CW section where a junction claw was present is analysed in [48]. By means of a numerical model for the study of dynamical interaction between pantograph and catenary, the authors highlighted the worst effect of the added mass of junction claw on the contact force, which caused large bending strain in the wire.

Research on CW fatigue highlights the complex nature of this phenomenon, which results from the combined effect of pantograph-catenary interaction, operational conditions and the worn status of the wire. Indeed, wear reduces the wire section and therefore the strength of the CW.

The continuing improvements in the reduction of CW wear are likely to increase the future occurrence of fatigue failures in the wire; for this reason, the interest on this subject will increase in future years.

In addition to the catenary, damage can also occur in the pantograph. As an example, a high wear rate of collector strips may produce contact on its aluminium part (or strip carrier), causing the breaking of the strip. This occurrence may have catastrophic consequences similar to the break of a catenary CW. For this reason, nowadays, the collector strips are typically equipped with the automatic dropper device. This system is able to automatically lower the pantograph, avoiding that damaged or completely worn collector strips continue to be in contact with the catenary CW.

Other possible structural damage on the pantograph head may be produced by defects of CW and/or incorrect catenary laying.

In some cases, dampers and/or springs that are part of the pantograph may be subject to damage, causing a deterioration of current collection quality, which may produce high levels of wear on both strips and wires and, in some worse cases, serious damage of the wire and of the pantograph.

16.5.3 CONDITION MONITORING OF PANTOGRAPH-CATENARY SYSTEM

Monitoring the health state of the pantograph and of the OCL is an effective means to support maintenance planning/scheduling, reduce life cycle costs and reduce in-service failure. To monitor the status of pantograph and catenary and of their components, several means and approaches are available. The traditional way to assess the condition of both pantograph and catenary is the visual and manual inspection; as an example, CW and contact strip thickness verification is performed on a regular time base. Obviously, the effectiveness of monitoring activity depends on the skills of the local maintenance crew.

Another possibility to verify the status of the catenary, in particular in terms of measurement of CW geometry (vertical and lateral position) and thickness of the wire's section, is by means of special inspection vehicles or trains. In this case, sophisticated transducers are used to directly measure the geometrical parameters of the CW as a function of the position along the line. Depending on the class of the line (local, regional, high speed), the inspection is performed at time intervals ranging from 2 months up to 1 year. Such special systems for the inspection need to be typically used by skilled crew who manages the overall measurement process, and the results are analysed by specialised teams belonging to the maintenance management department of the infrastructure manager.

A new emerging option to assess the pantograph and catenary status is the use of trains in revenue service equipped with an instrumented pantograph with simple measurement setups and transducers (typically accelerometers). The condition monitoring of the catenary is performed by using the dynamic response of the pantograph as an indicator to detect defects along the line or to assess the general OCL's status in terms of CW vertical irregularity [50]. In principle, information about the pantograph's status can also be obtained by comparing the dynamic response of different pantographs along the same line. The use of commercial trains allows a daily data flow, and it represents the prerequisite for the development of condition-based monitoring techniques, off-line trend analysis and condition-based maintenance.

Line tests clearly show that the pantograph-catenary contact force and the acceleration of the collector strips are strongly correlated either in terms of peak values or in terms of standard deviation [51]. For this reason, the signal of the collector strip's acceleration can be used for condition-monitoring purposes [51,52]. For the pantograph-catenary application, owing to the problems related to the electrical insulation of sensors, fibre optics sensors are suitable, and optical accelerometers have been developed for this purpose. The advantages of using optical accelerometers are not only related to the electrical insulation but also to the easy measurement setup, especially in terms of maintenance of the system. Thanks to these features, it is possible to envisage the installation of condition-monitoring units based on fibre optic accelerometers on-board trains performing revenue service. In this way, the loss of information inherent in the use of a setup measuring only accelerations, instead of one also including the contact force measurement, can be compensated by the larger amount of data acquired from several trains running under the same line. A cross-reference analysis of the data from different trains allows strengthening of the identification of local and distributed defects of the OCL.

One main difference between the condition monitoring approach and the standard approach based on specialised inspection train is the very large amount of acquired data in terms of frequency of inspections and the number of trains that collect these data. While on the one hand, the availability of more data enables the performance of a more robust analysis, on the other hand, it poses problems in terms of data management. To overcome this problem, real-time processing of the data on-board the train is generally proposed. Suitable algorithms allow to reduce the large amount of raw data to a manageable, but still significant, set of parameters representing the status of the line [51].

To drive OCL maintenance activities, the above-mentioned condition-based monitoring indexes (synthetic parameters) must be related to time, GPS coordinates or milestone position along the line. The latter can be evaluated on-board the train when performing real-time analysis, and on-going research and experimental activities are nowadays aimed at improving the accuracy of algorithms and hardware for train positioning. Assuming a monitoring system based on acceleration measurements is installed on some trains performing regular revenue services, a large set of data can be generated and made available for monitoring the condition of the infrastructure by, for example, trend analysis and/or automated feature searching in large data sets, enabling the early detection and localisation of catenary defects. The comparison between daily monitored condition parameters and reference data enhances the detection of defects and the analysis of their long-term development and reveals sudden changes in the status of the line, eventually allowing appropriate countermeasures to be taken before a serious failure occurs.

Finally, it should be mentioned that other techniques have also been proposed for the monitoring of the pantograph-catenary system, such as an infrared camera to detect the temperature along the strip [53] and digital cameras [54] to infer relevant aspects regarding the status of pantograph and overhead line condition.

16.6 LONG-TERM TRENDS

Improved solutions for the operation of pantograph and OCL systems are actively sought for, and there is a strong potential for enabling better current collection, reducing total life cycle costs and enhancing the reliability of operation. In this section, some research and innovation trends that are likely to affect the state of art of pantograph-catenary interaction are outlined.

16.6.1 Active/Semi-Active Pantograph Control

Active control of the pantograph has been the subject of extensive research in the past, although the use of active control technology in real applications is still limited. The simplest implementation of active pantograph control is to regulate the mean value of the pantograph-OCL contact force through a control system that adjusts the air pressure in the bellows of the pantograph, depending on train speed. This is a quite simple feed-forward solution now in use in some high-speed trains in which a control unit drives a pressure control valve according to train speed and direction of travel. Using this arrangement, a desired trend of mean contact force with train speed can be realised [55]. It is a relatively simple yet efficient way to optimise the overall thrust of the pantograph over the entire range of operational speed and to compensate for varying pantograph orientation when the train performs bi-directional services, without resorting to spoilers or other aerodynamic devices that can be sensitive to operating conditions, such as entry into tunnels, and are also often sensitive to errors and long-term drift in the orientation of winglets. Figure 16.22 compares two configurations for the regulation of the air pressure in the air spring of the pantograph, showing the resulting mean contact force (F_m in the diagrams). Figure 16.22a shows the situation for the configuration with a step-wise regulation performed with two pneumatic valves set at two different fixed levels, while Figure 16.22b shows the results, on the same pantograph, with a continuous regulation of the pressure of the air as a function of train speed. In the latter, it is clearly evident that the mean contact force can more regularly follow the TSI curve.

Besides adjusting the mean contact force with train speed, the active control of pantograph-catenary contact force by a feedback regulator has been considered. Solutions proposed so far mostly address a frequency band limited to the span-passing frequency, i.e., below 2 Hz. This can be obtained by regulating the pressure in the air bellows, using an actuation system placed in the suspension of the pantograph head or using a wire system that applies the control force directly on the pantograph head. Recently, the use of semi-active control is being considered, taking advantage from the availability on the market of simple and robust adjustable dampers that can be used as replacements of the standard passive dampers in the frame or in the pantograph head.

A detailed discussion of different actuation concepts, control strategies and potential benefits of active pantograph control is provided in [56]. Some prototype demonstrators have been built and tested in the laboratory [57–59], but active pantographs are still not used in trains performing revenue service.

The main hurdles to the use of active pantographs are as follows:

- Actuation issues, where pneumatic actuation may be inadequate to realise the desired pass band of the controlled pantograph, electro-mechanical actuation is subject to electromagnetic disturbance, and hydraulic actuation requires that the train is equipped with a hydraulic circuit normally not required for other use on the vehicle

Dynamics of the Pantograph-Catenary System

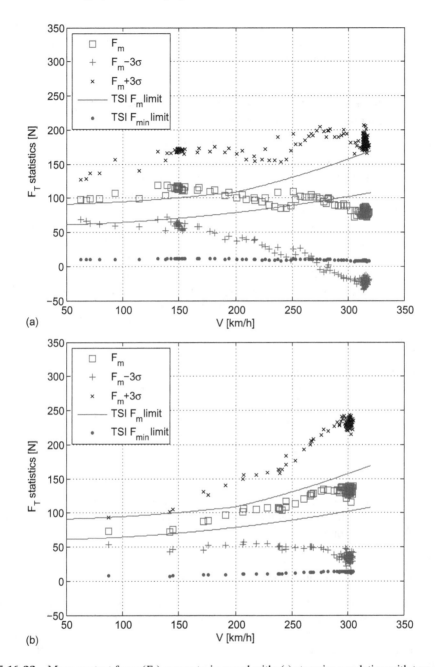

FIGURE 16.22 Mean contact force (F_m) versus train speed with: (a) stepwise regulation with two sequence pneumatic valves and (b) continuous regulation with controlled pressure valve (results from trial test on ETR500-Y1 test train along the Milano-Bologna HS line).

- For feedback control schemes, sensor issues and problems with accurately and reliably measuring quantities such as the contact force and the movement of the pantograph head
- Robustness of the controlled pantograph with respect to harsh environmental disturbances (electromagnetic and aerodynamic actions, vibrations, extreme weather and temperature conditions)

Despite the need to address the issues listed above, the benefits of operating actively controlled pantographs are potentially very high, as the additional costs implied by the control and actuating hardware would be easily paid back by improved current collection, enhanced interoperability (ability of the pantograph to operate correctly under different OCL systems), reduced wear and damage. Therefore, progress in this area can be envisaged in the near future.

16.6.2 Instrumented Pantographs for the Continuous Monitoring of the Overhead Contact Line Status

In recent years, the use of advanced methods and hardware for the condition monitoring of railway assets has become more and more popular. In this perspective, pantographs installed on trains performing standard operations can be equipped with sensors (accelerometers in most cases) to provide information useful for the continuous monitoring of the OCL. The main idea is that pantograph vibration caused by the interaction with the catenary can be measured and used to detect the presence of defects in the OCL, either in terms of singularities (e.g., badly adjusted arms, missing droppers) or in terms of excessive irregularity in the CW geometry (due, e.g., to thermal effects or to a failure in the tensioning device). Raw vibration data are processed on-board the train and used to synthesise suitable condition monitoring indicators. These indicators are then related to the position of the pantograph along the line, stored on-board and then transferred to a central data base, for subsequent long-term and trend analysis. The development of instrumented pantographs and of a system for the continuous monitoring of the OCL poses some challenges listed as follows:

- The sensors installed on the pantograph need to be provided with suitable electrical insulation and the effect of electro-magnetic disturbance shall be minimised – the use of optical based sensors can be a solution to this problem
- Real-time analysis of raw data needs to be performed by a suitable processing unit, and the evaluation of short-time root mean square values of collector accelerations has been proposed as a suitable condition indicator for the purpose of monitoring the OCL [51]

The use of data fusion methods is advisable to consider both measurements performed by in-service trains and measurements of OCL geometry performed by inspection trains. In this way, the accuracy and reliability of detection of faults in the OCL can be improved, and the overall process of maintenance of the OCL optimised.

16.6.3 Analysis of Pantograph Aerodynamics Using Computational Fluid Dynamics

The effect of aerodynamic forces acting on the pantograph becomes more and more important as speed increases, so that aerodynamic forces become a very important issues for very-high-speed trains. The aerodynamic optimisation of the pantograph is usually performed relying mostly on line tests. Wind tunnel tests are often used in the early stage of this process to reduce the number of line tests. With the development of computational fluid dynamics (CFD) tools, it is possible to perform a preliminary evaluation of pantograph aerodynamics by numerical means, reducing dramatically the effort required for the aerodynamic optimisation of the pantograph.

One notable benefit of CFD calculations is that this method allows to evaluate not only the overall aerodynamic force acting on the pantograph but also the contributions to this force caused by the different parts of the pantograph, e.g., the head and the lower/upper arms in the frame, so that the shape of each one of these parts can be optimised in view of minimising, e.g., the effect of train speed on the mean contact force and the unbalance between the force applied by the two collectors. First attempts to correlate wind tunnel test measurements and CFD calculations provided promising results [33,34], showing the potential for the use of this method.

16.6.4 CURRENT COLLECTION FROM AN OVERHEAD CONTACT LINE IN HEAVY ROAD VEHICLES

Current collection from an OCL using one or more pantographs has been so far confined to use in railway-related applications. With the increased attention to the greening of road transport, the same system is being considered to feed road vehicles with electric power, while they are travelling on long-distance roads, extending the range of fully electric heavy road vehicles (trucks). An OCL suitable for this use is composed of a feeding catenary and a return catenary, and two pantographs must be installed on the truck's cabin. Specific hybrid vehicles are of course needed to operate under such a system. Several issues still need to be addressed in view of the practical application of this concept and are mainly related to safety, traffic management under different conditions (i.e., platooning and overtaking) and control of pantograph lateral position with respect to the CW.

This is an interesting topic in which the cooperation between networks (road and electrical power) enables the creation of a new transportation system, whose impact still needs to be assessed. Pilot plants to demonstrate the feasibility of this application have been established in Sweden, Germany and the USA (in the Los Angeles area), and a working group has been recently established by the European Committee for Electrotechnical Standardization (Cenelec) to analyse the regulatory issues related to this new system.

REFERENCES

1. H.J. Schwab, S. Ungvari, Development and design of new overhead contact line systems, *Elektrische Bahnen*, 104(5), 238–248, 2006.
2. F. Kiessling, R. Puschmann, A. Schmieder, E. Schneider, *Contact Lines for Electric Railways: Planning, Design, Implementation, Maintenance*, 3rd ed., Publicis Publishing, Erlangen, Germany, 2018.
3. P.R. Scott, M. Rothman, Computer evaluation of overhead equipment for electric railroad traction, *IEEE Transactions on Industry Applications*, 1A-10, 573–580, 1974.
4. T.X. Wu, M.J. Brennan, Dynamic stiffness of a railway overhead wire system and its effect on pantograph-catenary system dynamics, *Journal of Sound and Vibration*, 219(3), 483–502, 1999.
5. S. Bruni, J. Ambrosio, A. Carnicero, Y.H. Cho, L. Finner, M. Ikeda, S.Y. Kwon, J-P. Massat, S. Stitchel, M. Tur, W. Zhang, The results of the pantograph-catenary interaction benchmark, *Vehicle System Dynamics*, 53(3), 412–435, 2015.
6. G. Poetsch, J. Evans, R. Meisinger, W. Kortüm, W. Baldauf, A. Veitl, J. Wallaschek, Pantograph/catenary dynamics and control, *Vehicle System Dynamics*, 28(2–3), 159–195, 1997.
7. M. Arnold, B. Simeon, Pantograph and catenary dynamics: A benchmark problem and its numerical solution, *Applied Numerical Mathematics*, 34(4), 345–362, 2000.
8. J. Benet, N. Cuartero, F. Cuartero, T. Rojo, P. Tendero, E. Arias, An advanced 3D-model for the study and simulation of the pantograph catenary system, *Transportation Research Part C*, 36, 138–156, 2013.
9. J. Ambrósio, J. Pombo, P. Antunes, M. Pereira, PantoCat statement of method, *Vehicle System Dynamics*, 53(3), 314–328, 2015.
10. A. Collina, S. Bruni, Numerical simulation of pantograph-overhead equipment interaction, *Vehicle System Dynamics*, 38(4), 261–291, 2002.
11. Y.H. Cho, Numerical simulation of the dynamic responses of railway overhead contact lines to a moving pantograph, considering a nonlinear dropper, *Journal of Sound and Vibration*, 315(3), 433–454. 2008.
12. J. Ambrósio, J. Pombo, M. Pereira, P. Antunes, A. Mósca, A computational procedure for the dynamic analysis of the catenary-pantograph interaction in high-speed trains, *Journal of Theoretical and Applied Mechanics*, 50(3), 681–699, 2012.
13. M. Tur, E. García, L. Baeza, F.J. Fuenmayor, A 3D absolute nodal coordinate finite element model to compute the initial configuration of a railway catenary, *Engineering Structures*, 71, 234–243, 2014.
14. D.C. Cook, D.S. Malkus, M.E. Plesha, *Concepts and Applications of Finite Element Analysis*, 3rd ed., John Wiley & Sons, New York, 1989.
15. F. Cheli, G. Diana, *Advanced Dynamics of Mechanical Systems*, Springer International, Cham, Switzerland, 2015.
16. J-H. Seo, H. Sugiyama, A.A. Shabana, Three-dimensional large deformation analysis of the multibody pantograph/catenary systems, *Nonlinear Dynamics*, 45(2), 199–215, 2005.

17. J.L. Escalona, H. Sugiyama, A.A. Shabana, Modelling of structural flexibility in multibody railroad vehicle systems, *Vehicle System Dynamics*, 51(7), 1027–1058, 2013.
18. S. Kulkarni, C.M. Pappalardo, A.A. Shabana, Pantograph/catenary contact formulations, *ASME Journal of Vibration and Acoustics*, 139(1), 011010, 2016.
19. P. Nåvik, A. Rønnquist, S. Stitchel, Identification of system damping in railway catenary wire systems from full-scale measurements, *Engineering Structures*, 113, 71–78, 2016.
20. A. Collina, S. Bruni, A. Facchinetti, A. Zuin, PCaDA statement of methods, *Vehicle System Dynamics*, 53(3), 347–356, 2015.
21. J. Ambrósio, F. Rauter, J. Pombo, M.S. Pereira, A flexible multibody pantograph model for the analysis of the catenary-pantograph contact, In: K. Arczewski, W. Blajer, J. Fraczek, M. Wojtyra (Eds.), *Multibody Dynamics*, Volume 23, Computational Methods in Applied Sciences Series, Springer, Dordrecht, the Netherlands, 2011.
22. J.-P. Massat, E. Balmes, J-P Bianchi, G. Van Kalsbeek, OSCAR statement of methods, *Vehicle System Dynamics*, 53(3), 370–379, 2015.
23. A. Collina, A. Lo Conte, M. Carnevale, Effect of collector deformable modes in pantograph-catenary dynamic interaction, *Journal of Rail and Rapid Transit*, 223(1), 1–14, 2009.
24. N. Zhou, W. Zhang, Investigation on dynamic performance and parameter optimization design of pantograph and catenary system, *Finite Elements in Analysis and Design*, 47(3), 288–295, 2011.
25. A. Facchinetti, S. Bruni, Hardware-in-the-loop hybrid simulation of pantograph-catenary interaction, *Journal of Sound and Vibration*, 331(12), 2783–2797, 2012.
26. J. Deml, W. Baldauf, A new test bench for examinations of the pantograph-catenary interaction, *Proceedings 5th World Congress on Railway Research*, 25–29 November 2001, Cologne, Germany, PB011874, 2001.
27. A. Facchinetti, S. Bruni, W. Zhang, Rolling stock dynamic evaluation by means of laboratory tests, *International Journal of Railway Technology*, 2(4), 99–123, 2013.
28. T. Xin, C. Roberts, P. Weston, E. Stewart, Condition monitoring of railway pantographs to achieve fault detection and fault diagnosis, *Journal of Rail and Rapid Transit*, OnlineFirst, 2018. doi:10.1177/0954409718800567.
29. W. Zhang, G. Mei, X. Wu, Z. Shen, Hybrid simulation of dynamics for the pantograph-catenary system, *Vehicle System Dynamics*, 38(6), 393–414, 2002.
30. A. Schirrer, G. Aschauer, E. Talic, M. Kozek, S. Jakubek, Catenary emulation for hardware-in-the-loop pantograph testing with a model predictive energy-conserving control algorithm, *Mechatronics*, 41, 17–28, 2017.
31. S. Bruni, A. Facchinetti, M. Kolbe, J-P. Massat, Hardware-in-the-loop testing of pantograph for homologation, *Proceedings 9th World Congress on Railway Research*, 22–26 May 2011, Lille, France, PB003802, 2011.
32. S. Gregori, M. Tur, E. Nadal, J.V. Aguado, F.J. Fuenmayor, F. Chinesta, Fast simulation of the pantograph-catenary dynamic interaction, *Finite Elements in Analysis and Design*, 129, 1–13, 2017.
33. M. Carnevale, A. Facchinetti, L. Maggiori, D. Rocchi, Computational fluid dynamics as a means of assessing the influence of aerodynamic forces on the mean contact force acting on a pantograph, *Journal of Rail and Rapid Transit*, 230(7), 1698–1713, 2016.
34. M. Carnevale, A. Facchinetti, D. Rocchi, Procedure to assess the role of railway pantograph components in generating the aerodynamic uplift, *Journal of Wind Engineering and Industrial Aerodynamics*, 160, 16–29, 2017.
35. O. Bruno, A. Landi, M. Papi, L. Sani, Phototube sensor for monitoring the quality of current collection on overhead electrified railways, *Journal of Rail and Rapid Transit*, 215(3), 231–241, 2001.
36. P. Masini, M. Papi, G. Puliatti, Virtual acquisition system for experimentation in pantograph-catenary interaction, *Computers in Railways VI*, WIT Press, Southampton, UK, 827–836, 1998.
37. P. Nåvik, A. Rønnquist, S. Stitchel, A wireless railway catenary structural monitoring system: Full-scale case study, *Case Studies in Structural Engineering*, 6, 22–30, 2016.
38. J. Wilde, A. Rukwied, K. Becker, U. Resch, B-W. Zweig, LCC optimisation in the design process of catenaries of electrical train systems, *Proceedings World Congress on Railway Research*, 16–19 November 1997, Florence, Italy, PB002546, 1997.
39. H. Nagasawa, S. Aoki, K. Kato, Application of precipitation hardened copper alloy to contact wire, *Proceedings World Congress on Railway Research*, 16–19 November 1997, Florence, Italy, PB002498, 1997.
40. D. Klapas, F.A. Benson, R. Hackam, Simulation of wear in overhead current collection systems, *Review of Scientific Instruments*, 56(9), 1820–1828, 1985.

41. K. Becker, U. Resch, B.W. Zweig, Optimizing high-speed overhead contact lines, *Elektrische Bahnen*, 92(9), 243–248, 1994.
42. S. Kubo, K. Kato, Effect of arc discharge on wear rate of Cu-impregnated carbon strip in unlubricated sliding against Cu trolley under electric current, *Wear*, 216(2), 172–178, 1998.
43. H. Zhao, G.C. Barber, J. Liu, Friction and wear in high speed sliding with and without electrical current, *Wear*, 249(5–6), 409–414, 2001.
44. G. Bucca, A. Collina, A procedure for the wear prediction of collector strip and contact wire in pantograph-catenary system, *Wear*, 266(1–2), 46–59, 2009.
45. G. Bucca, A. Collina, Electromechanical interaction between carbon-based pantograph strip and copper contact wire: A heuristic wear model, *Tribology International*, 92, 47–56, 2015.
46. C. Yamashita, A. Sugahara, Influence of mean stress on contact wire fatigue, *Quarterly Report of RTRI*, 47(1), 46–51, 2006.
47. A. Sugahara, Preventing fatigue breakage of contact wires, Railway Technology Avalanche No. 24, 140, 2008.
48. J.P. Massat, T.M.L. Nguyen Tajan, H. Maitournam, E. Balmes, Fatigue analysis of catenary contact wires for high speed trains, *Proceedings 9th World Congress on Railway Research*, 22–26 May 2011, Lille, France, 2011.
49. G. Zhen, Y. Kim, L. Haochuang, J-M. Koo, C-S. Seok, K. Lee, S-Y. Kwon, Bending fatigue life evaluation of Cu-Mg alloy contact wire, *International Journal of Precision Engineering and Manufacturing*, 15(7), 1331–1335, 2014.
50. A. Collina, F. Fossati, M. Papi, F. Resta, Impact of overhead line irregularity on current collection and diagnostics based on the measurement of pantograph dynamics, *Journal of Rail and Rapid Transit*, 221(4), 547–559, 2007.
51. M. Carnevale, A. Collina, Processing of collector acceleration data for condition-based monitoring of overhead lines, *Journal of Rail and Rapid Transit*, 230(2), 472–485, 2016.
52. F. Tanarro, V. Fuerte, OHMS-real-time analysis of the pantograph-catenary interaction to reduce maintenance costs, *5th IET Conference on Railway Condition Monitoring and Non-Destructive Testing*, Derby, UK, 29–30 November 2011.
53. A. Landi, L. Menconi, L. Sani, Hough transform and thermo-vision for monitoring pantograph-catenary system, *Journal of Rail and Rapid Transit*, 220(4), 435–447, 2006.
54. L.G.C. Hamey, T. Watkins, S.W.T Yen, Pancam: In-service inspection of locomotive pantographs, *Proceedings 9th Biennial Conference of the Australian Pattern Recognition Society on Digital Image Computing Techniques and Applications*, 3–5 December 2007, Glenelg, Australia, 493–499, 2007.
55. Patent NA2010A000058, Sistema di comando controllo attivo e diagnostico per pantografo ferroviario, (command system for active control and diagnostics for railway pantographs), 30 November 2010.
56. R.M. Goodall, S. Bruni, A. Facchinetti, Active control in railway vehicles, *International Journal of Railway Technology*, 1(1), 57–85, 2012.
57. A. Collina, Facchinetti, F. Fossati, F. Resta, An application of active control to the collector of an high-speed pantograph: simulation and laboratory tests, *Proceedings 44th IEEE Conference on Decision and Control and European Control Conference*, 12–15 December 2005, Seville, Spain, 4602–4069, 2005.
58. B. Allotta, L. Pugi, F. Bartolini, Design and experimental results of an active suspension system for a high-speed pantograph, *IEEE/ASME Transactions on Mechatronics*, 13(5), 548–557, 2008.
59. W. Baldauf, R. Blaschko, W. Behr, C. Heine, M. Kolbe, Development of an actively controlled, acoustically optimised single arm pantograph, *Proceedings 5th World Congress on Railway Research*, 25–29 November 2001, Cologne, Germany, 2001.

17 Simulation of Railway Vehicle Dynamics

Oldrich Polach, Mats Berg and Simon Iwnicki

CONTENTS

17.1	Introduction	652
17.2	Modelling Vehicle-Track Interaction	653
	17.2.1 Vehicle Models	654
	17.2.2 Vehicle Models – Body Components	655
	17.2.3 Vehicle Models – Suspension Components	658
	17.2.4 Track Models	662
	17.2.5 Wheel-Rail Contact Models	663
17.3	Simulation Methods	663
	17.3.1 Multibody Systems and Equations of Motion	663
	17.3.2 Solution Methods	664
	17.3.2.1 Eigenvalue Analysis	664
	17.3.2.2 Stochastic Analysis	665
	17.3.2.3 Time-Stepping Integration	665
	17.3.2.4 Quasistatic Solution Method	666
17.4	Computer Simulation Tools	666
	17.4.1 Historical Development	666
	17.4.2 Multibody Simulation Tools	667
	17.4.2.1 VAMPIRE	667
	17.4.2.2 Simpack	669
	17.4.2.3 Universal Mechanism	671
	17.4.2.4 Benchmarking	673
17.5	Dynamics in Railway Vehicle Engineering	674
	17.5.1 Railway Vehicle Engineering Process	674
	17.5.2 Task and Applications of Vehicle Dynamics	675
	17.5.3 Model Validation	677
	17.5.3.1 Introduction	677
	17.5.3.2 Features of Testing and Model Validation in Railway Vehicle Dynamics	678
	17.5.3.3 Model Validation in the Context of the Acceptance of Running Characteristics	680
17.6	Typical Railway Vehicle Dynamics Computation Tasks	682
	17.6.1 Introduction	682
	17.6.2 Eigenbehaviour	682
	17.6.2.1 Eigenvalue Analysis	682
	17.6.2.2 Simulation of Eigenbehaviour	684

17.6.3	Stability Analysis	685
	17.6.3.1 Introduction	685
	17.6.3.2 Linearised Stability Analysis	685
	17.6.3.3 Non-Linear Stability Analysis	691
17.6.4	Running on Track with Irregularities	697
	17.6.4.1 Definition of Running Behaviour, Ride Characteristics and Comfort	697
	17.6.4.2 Ride Characteristics	698
	17.6.4.3 Ride Comfort	700
17.6.5	Curving	702
	17.6.5.1 Assessment of Curving Properties	702
	17.6.5.2 Running Safety	704
	17.6.5.3 Track Loading and Wear	705
	17.6.5.4 Curving Optimisation by Using Bogies with Wheelset Steering	708
17.6.6	Examples of Other Investigations	712
	17.6.6.1 Influence of Crosswind	712
	17.6.6.2 Interaction between Vehicle and Traction Dynamics	713
17.7 Conclusions		715
Nomenclature		716
References		717

17.1 INTRODUCTION

With ever more powerful computers, simulation of complex mechanical systems has become a routine activity. A computer model of a railway vehicle can be constructed and run on typical or measured track in a virtual environment, and a wide range of possible designs or parameter changes can be investigated. Outputs from the simulation can provide accurate predictions of the dynamic behaviour of the vehicle and its interaction with the track. Optimisation of suspensions and other parts of the system can be carried out, and levels of forces and accelerations can be checked against standards to ensure safe operation and a comfortable ride.

Excitations of the vehicle model are usually made at each wheelset. Typical inputs are vertical and lateral irregularities and deviations in gauge and cross level. These can be idealised discrete events, such as dipped joints or switches, or can be measured values from a track-recording coach. Additional forces may be specified such as crosswind loading or powered actuators.

Figure 17.1 shows a schematic summary of the main aspects of the simulation process; see further discussion in [1].

The computer tools have developed from routines and programs used by researchers and engineers to solve specific problems, and the theoretical basis of the mathematical modelling used is nowadays mature and reliable; programs originally written by research institutes have been developed into powerful, validated and user-friendly packages.

This chapter covers the basic methods used in setting up a model of a typical railway vehicle and the types of analysis that can be carried out once the equations of motion have been produced. The historical development of the major simulation packages now used is also briefly covered. Typical analysis tasks such as modal analysis and running stability, simulation of ride on track with irregularities and analysis of curving behaviour are then covered in detail, with examples of typical application loads. The main methods used for the assessment of the simulation results are also covered under each of these headings.

Simulation of Railway Vehicle Dynamics

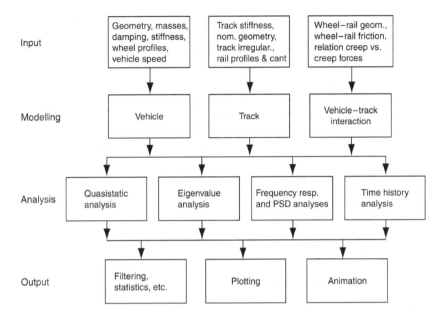

FIGURE 17.1 A flow chart showing the process of computer simulation of vehicle-track interaction. (From Andersson, E. et al., *Rail Vehicle Dynamics*, Royal Institute of Technology (KTH), Stockholm, Sweden, 2014. With permission.)

17.2 MODELLING VEHICLE-TRACK INTERACTION

The potential and success of vehicle-track dynamic simulations very much depend on how well the system is mathematically modelled and fed with pertinent input data. The system modelling involves several fields of mechanics and, owing to its many complexities, is an engineering challenge. The choice of models for the system and its components depends on several aspects, mainly:

- Purpose of simulations, including requested output quantities and their accuracy
- Frequency range of interest
- Access to appropriate simulation packages
- Access to relevant model data
- Time and funding available

The first aspect above should reflect that quite coarse models may be sufficient for preliminary studies and that, for instance, the secondary suspension may not need to be that carefully modelled if only wheel-rail forces, etc., are of interest.

Regarding frequencies, the traditional frequency range of interest is 0–20 Hz. This low-frequency range covers the fundamental dynamics of the vehicle-track system. Obviously, this frequency range and modelling must be extended if noise issues are to be studied (see Chapter 14). However, higher frequencies are also of interest for evaluation of vibrations (ride comfort, etc.) and wheel-rail forces (fatigue, etc.).

Today's simulation packages offer many dynamics modelling options, but often, the engineering guidelines on when to use the various options are limited. Depending on their experience, different engineers may therefore make different model choices. Also, the possibility to add user-defined models will affect this choice.

Some component models may be very advanced and accurate in themselves, but lack of appropriate input data strongly reduces their applicability. Such models often require measurements on the component in question or on similar ones.

Finally, there are also restrictions on available staff time, calendar time and economical resources, which in practice constrain the modelling possibilities.

In conclusion, the art of modelling requires engineering experience and judgement. It also requires a significant amount of relevant and reliable technical information on the vehicle-track system at hand.

In this section, some basic guidelines on modelling vehicle-track systems and their components are given. The focus is on the vehicle modelling, but issues on track and wheel-rail contact modelling are also raised.

17.2.1 Vehicle Models

Railway vehicles consist of many components, and, for vehicle-track dynamic simulations, there is a need to represent the mechanical properties of the main components by proper mathematical models. A main subdivision of the vehicle components can be made into body components and suspension components.

The dominating body components are typically the car body, bogie frames and wheelsets, and they essentially hold the vehicle mass (weight). Thus, the inertia properties of the bodies are of primary interest. However, in many railway applications, the body structural flexibility also needs to be considered. This especially holds for car bodies.

The main suspension components are various physical springs and dampers whose forces are essentially related to the displacements and velocities at the components. Traction rods, bumpstops, anti-roll bars, trailing arms, linkages, etc., also belong to this group of components.

This subdivision of vehicle components relates the vehicle modelling to the mechanics field of multibody dynamics (MBD) or multibody systems (MBS). This also implies that most of the vehicle degrees of freedom (DoFs), or equations of motion, are assigned to the motions of the vehicle bodies.

In dealing with railway vehicles, the body motions are often divided into large desirable motions, making the vehicle travel 'from A to B', and small undesirable motions. The large motions are known for a given track design geometry (curves, etc.) and a given vehicle speed profile. For the large motions of the different vehicle bodies, the nominal positions of the bodies within the vehicle are also known.

For vehicles running on tangent track, this subdivision of motions is simple, since the large motions are given by the speed profile alone. However, for curve negotiation, the kinematics can be quite complicated. Figure 17.2 illustrates the principle of how the kinematics of vehicle bodies can be handled.

In Figure 17.2, an inertial or earth-fixed reference system (coordinate system) I_1–I_2–I_3 is first introduced. Then, the large vehicle body motions are mainly represented by a track-following reference system, X–Y–Z, following the nominal track centreline with the speed of the vehicle. Possible track cant and gradient are also considered through the orientation of the track-following system. The small body motions can be related to a body-following reference system x–y–z, for instance, located at the nominal position of the body centre of gravity and with the same orientation as the system X–Y–Z. Provided the unknown body translations and rotations are small, linear kinematics is sufficient in formulating the system equations of motion.

Simulation of Railway Vehicle Dynamics

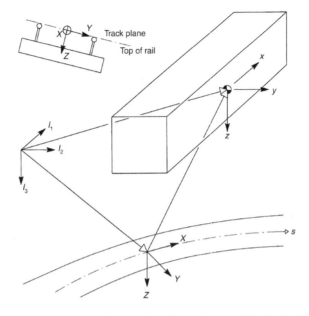

FIGURE 17.2 Reference systems for the kinematics of a railway vehicle body: Inertial system I_1–I_2–I_3, track-following system X–Y–Z and body-following system x–y–z. (From Iwnicki, S. (Ed.), *Handbook of Railway Vehicle Dynamics*, CRC Press, Boca Raton, FL, 2006. With permission.)

17.2.2 Vehicle Models – Body Components

The car body usually holds the main part of the vehicle mass. The mass properties of the car body steel/aluminium structure are not difficult to calculate; a Computer Aided Design (CAD) or finite element (FE) analysis can give the mass, centre of gravity position and mass moments of inertia of the car body structure quite easily. However, in most passenger vehicle applications, the mass of the car body structure is less than half the car body mass. For a coach, the additional mass is a result of the interior layout and interior/exterior equipment. For powered vehicles, especially locomotives, the car body also carries some of the traction equipment. Often, it is quite hard to keep accurate records of all the pertinent masses and positions. Usually, the mass of this equipment is merged to the metal structure model. In many vehicle-track dynamics simulations, the payload, i.e., passengers or goods, also needs to be considered. The corresponding mass might also be combined with the car body structure model.

In most coach applications, it is important to consider the car body structural flexibility so that car body vibrations, and their negative effect on ride comfort are represented. Figure 17.3 exemplifies how car body accelerations increase in amplitude and frequency when the car body structural flexibility is modelled in the vehicle-track simulations. Frequencies of around 10 Hz are usually prominent, and for vertical accelerations, this can cause significant discomfort, as people are most sensitive to vertical accelerations of 8–10 Hz [2,3]. In fact, root mean square (rms) values of comfort-weighted vertical accelerations may be doubled due to the car body structural flexibility. Slender car bodies in vehicles running at fairly high speed on relatively poor track are most prone to this kind of acceleration amplification.

In railway vehicle dynamics simulations, it is desirable to represent the car body structural flexibility by only a limited number of DoFs in addition to the six rigid body DoFs (longitudinal, lateral, vertical, roll, pitch and yaw motions). The most common way of finding such a representation is

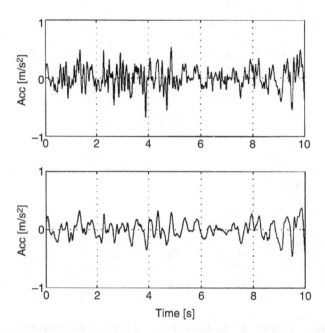

FIGURE 17.3 Influence of car body structural flexibility on car body accelerations (the accelerations are vertical and evaluated at the mid-car body). (From Carlbom, P., Structural flexibility in a rail vehicle car body—Dynamic simulations and measurements, TRITA-FKT Report 1998:37, Royal Institute of Technology (KTH), Stockholm, Sweden, 1998. With permission.)

to use the eigenmodes of the body found from an eigenvalue analysis of the free body. Then, six rigid body modes are required, and a number of structural modes, preferably those with the lowest eigenfrequencies, can also be included. Figure 17.4 shows an example with the four lowest structural modes for a coach car body. Note that the first structural mode has an eigenfrequency of about 10 Hz [1], cf. the discussion on ride discomfort above.

The car body vibrations are also promoted by a low damping of the car body structure. However, in coaches, insulation and other non-metallic interior materials can increase the relative damping to approximately 2%. In fact, passengers can also provide the system with some additional damping, say up to 4% [4]. If possible, the damping for the structural modes should be determined through laboratory tests. In car body modelling, equipment and passengers may be modelled as separate bodies attached to the car body structure. Explicit introduction of seats and passengers into the modelling also promotes the evaluation of ride comfort on the seat cushion and not only on the floor.

When it comes to the bogie frame modelling, it should first be kept in mind that the most common bogie type worldwide, the three-piece bogie, consists of three bodies and not only of a single bogie frame body. Also, some bogies are equipped with a bolster beam to provide the bogie with significant yaw motion possibilities relative to the car body. Such beams may be modelled as separate bodies.

For a bogie frame, seen as a one-piece metal structure, the assumption of essentially rigid body behaviour is fair in most applications. Sometimes, the torsional flexibility about a longitudinal axis is considered, representing an improved negotiation of twisted tracks. Low-floor trams, without traditional wheel axles, need especially designed bogie frames, and, even if the speeds are low, the frame flexibility probably needs to be considered in the bogie frame modelling for the dynamic simulations. The bogie frame flexibility can be determined through FE analysis and may be represented in the MBD simulations by a limited set of eigenmodes of the free frame body. Such analysis can also give the mass and mass moments of inertia of the bogie frame.

Simulation of Railway Vehicle Dynamics

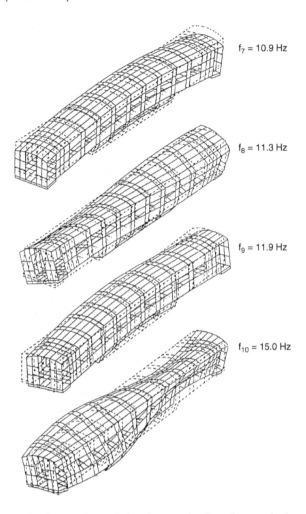

FIGURE 17.4 Four example eigenmodes and eigenfrequencies for a free car body, with dashed lines indicating the undeformed car body and (simple) finite element mesh – modes 1 to 6 are rigid body modes, i.e., f_1 to $f_6 = 0$ Hz. (From Andersson, E. et al., Rail vehicle dynamics, Royal Institute of Technology (KTH), Stockholm, Sweden, 2014. With permission.)

Different braking equipment, for instance brake cylinders, is often attached to the bogie frame. Such equipment may be merged with the bogie frame in the modelling process. For powered bogies, the traction motors are usually mainly supported by the bogie frame. In certain applications, the motors may be considered as separate bodies suspended in the bogie frame. The traction gear housing, etc., is also often partly supported by the bogie frame.

Last but not least, among the main vehicle bodies are the wheelsets. A plain wheelset normally consists of one solid unit with two wheels on a common wheel axle. The wheel diameter is often 0.7–1.0 m, but both smaller and larger wheels exist. The axle diameter is usually 0.15–0.20 m. A typical mass for a plain wheelset is 1000–1500 kg. Additional mass can be introduced through brake discs, mounted on the axle or wheels, and traction gear (wheel axle gear wheel, part of the gear housing, etc.). The axle boxes, or journal-bearing boxes, also add some mass, but they must not be considered in calculating the wheelset mass moment of inertia for pitch motion. In fact, they are sometimes treated as separate bodies in the modelling.

Wheelsets are often modelled as rigid bodies, but wheel axle structural flexibility may affect the vehicle-track interaction. For instance, the axle torsion can cause running instability, and the axle bending can significantly alter the dynamic part of the wheel-rail forces [5]. Expressed in terms of lowest eigenfrequencies for a free wheelset, they may even be below 50–60 Hz in torsion and bending, respectively. More normally, these lowest frequencies are in the range of 60–80 Hz. For wheel diameters above about 1 m, the wheels also contribute significantly to the wheelset structural flexibility, even for frequencies below 100 Hz [6]. An example of eigenmodes and eigenfrequencies for such a slender wheelset is shown in Figure 17.5. Here, the first and second bending modes in the vertical plane are shown along with two modes for which the wheels deform in an umbrella-like fashion.

For slender wheelsets, the structural flexibility should be reflected in the MBD modelling. As for the bodies above, a limited number of the lowest eigenfrequencies and eigenmodes for a free wheelset is usually a fair representation of the wheelset flexibility. The relative damping of these modes is typically below 1%.

17.2.3 Vehicle Models – Suspension Components

Common suspension components in railway vehicles include coil springs, leaf springs, rubber springs, air springs, friction dampers and hydraulic dampers (cf. Chapter 6). They play important roles in reducing bogie frame and car body accelerations as well as dynamic wheel-rail forces. They also allow for proper curve negotiation, but too soft a suspension causes problems with the vehicle gauging. Bumpstops and anti-roll bars may therefore be introduced to mitigate such problems; the bars also reduce quasistatic lateral accelerations on the car body floor. In the present context, traction rods may also be considered as suspension components, although their main task is to transfer longitudinal forces between bogie and car body during acceleration or retardation. In addition, trailing arms and various linkages should be incorporated here.

However, in this chapter, only passive suspension components are considered, and the reader should refer to Chapter 15 for details of active suspensions. Thus, the suspension modelling here assumes that the suspension forces and moments are related to motions only at the interfaces with the connected bodies in question. However, note that the static (vertical) force of an air spring can be altered through changing the air pressure, without modifying the air spring height.

Static forces, or preloads, due to dead weight of bodies and payload are carried mainly by coil springs, air springs, leaf springs and rubber springs. For coil springs and air springs, this normally

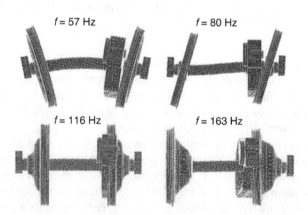

FIGURE 17.5 Four example eigenmodes and eigenfrequencies for a free powered wheelset with 1.3 m wheel diameter – the two lowest torsional modes are not shown but have the eigenfrequencies of 48 and 362 Hz. (From Chaar, N. and Berg, M., Experimental and numerical modal analyses of a loco wheelset, in Abe, M. (Ed.), *Proceedings 18th IAVSD Symposium held in Atsugi*, Kanagawa, Japan, Taylor & Francis, London, UK, 597–606, 2004. With permission.)

leads to compressive forces, whereas leaf springs and rubber springs are also subjected to shear forces and bending moments. In vehicle-track dynamic simulations, the vehicle body motions are usually given as motions relative to the static equilibrium on tangent and horizontal track.

The static behaviour of the springs above can be determined through component measurements by slowly loading and unloading these components (for air springs, the air pressure is also changed accordingly). The vertical (axial) tangent stiffness of a single coil spring is virtually independent of the static load, whereas the air spring stiffness increases almost linearly with increasing preload. Rubber and leaf spring stiffnesses also often increase with increasing preload.

When unloading the coil spring, the corresponding force-displacement graph will almost coincide with that of the loading phase; thus, the energy dissipation or hysteresis is very small. In contrast, the leaf spring undergoes significant hysteresis due to the sliding motions between the leaves. The air spring and rubber springs also experience some hysteresis due to internal friction-like mechanisms of the rubber parts. The hysteresis mentioned is due to friction rather than viscous effects, since it will appear no matter how slowly the loading and unloading are realised.

For coil springs and air springs, the compressive preload gives rise to destabilising effects in the horizontal plane. An increasing preload will give reduced horizontal (shear) stiffness for the coil springs and less-increasing horizontal stiffness for the air springs. To mitigate the destabilising effects of coil springs, two or three springs may be introduced side by side or inside each other. In the latter case, the inner springs may not be activated at low preloads. This design is often used in freight wagons and provides a resulting progressive vertical stiffness.

During each simulation, the suspension preloads are normally assumed constant. For the suspension springs proper, static stiffnesses need to be defined, preferably based on static tests, as indicated above, or on appropriate calculations. Owing to curve negotiation and track irregularities, the springs will deform and the suspension forces need to be re-calculated in the simulations.

A reasonable starting point for suspension spring modelling is to assume models consisting of linear spring elements. The simplest three-dimensional model is one of three perpendicular linear springs, but the shear effect of coil springs and air springs calls for models that consider the component height and compressive preload (see, e.g., [7,8]). Non-linear characteristics due to, for instance, clearances in coil spring sets or bumpstops need special consideration.

Possible friction effects, which would require the use of non-linear models, should be dealt with. The stick-slip motions between leaves of leaf springs are probably the most obvious example of friction effects. Associated linkages of International Union of Railways (UIC) running gear also experience combined rolling-sliding motions [9]. The most common friction model is that suggested by Coulomb. For a one-dimensional case, such a model in series with a linear spring, together with a linear spring in parallel, produces a resulting force-displacement, as exemplified in Figure 17.6. In this way, a parallelogram with an area of $A = 4\mu N(x_o - \mu N/k_s)$ corresponding to the energy dissipation per cycle is obtained. For increasing displacement amplitude x_o, this dissipation will increase, whereas the stiffness $S = F_o/x_o$ will decrease. Note that these two quantities and the graph itself are independent of the excitation frequency [1].

The model shown in Figure 17.6 might be sufficient in some situations but can only represent two distinct friction releases per cycle (at upper left and lower right corners). However, leaf springs in particular instead have a successive friction release, and the corners mentioned would be smoothed out. Also, rubber springs and air springs show similar behaviour, although with smaller hysteresis. A 'smooth friction' model is therefore of interest. For instance, [10] and [11] suggest such a model for leaf springs and rubber springs, respectively. Figure 17.7 shows a comparison of non-smooth and smooth friction models. The latter is in better agreement with measurements and has still only three input parameters, such as the non-smooth model.

If we, for a given displacement amplitude, increase the frequency of harmonic excitation, the force-displacement graphs of both leaf springs and coil springs show very little difference. This implies that models of the type described above should be sufficient. However, for rubber springs and air springs, the viscous effects are significant, thus suggesting frequency-dependent models.

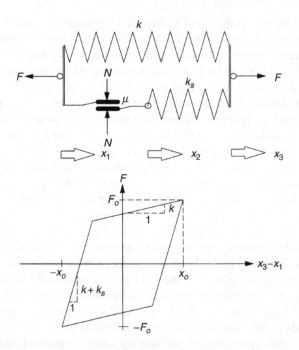

FIGURE 17.6 Simple friction model and corresponding force-displacement graph at harmonic displacement excitation with amplitude x_o, where $x_o > \mu N/k_s$. (From Andersson, E. et al., *Rail Vehicle Dynamics*, Royal Institute of Technology (KTH), Stockholm, Sweden, 2014. With permission.)

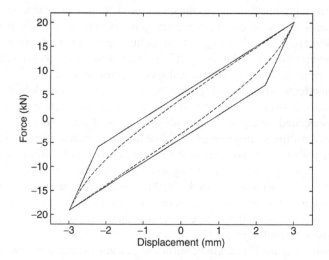

FIGURE 17.7 Force versus displacement graphs at harmonic displacement excitation for the model in Figure 17.6 (solid line) and for a model with smooth friction (dashed line). (From Iwnicki, S. (Ed.), *Handbook of Railway Vehicle Dynamics*, CRC Press, Boca Raton, FL, 2006. With permission.)

The classic model of a linear spring in parallel with a linear viscous damper (dashpot) is sufficient here, provided that the frequency range of interest is very limited. For larger frequency ranges, this model provides a too strong frequency dependence, giving a very significant stiffness and damping at high frequencies. One common way to overcome this is to equip the model dashpot with a series spring, i.e., using a model as in Figure 17.6 but replacing the friction element with a dashpot. In this way, the maximum stiffness will be $k + k_s$, as the excitation frequency ω tends to infinity, and the maximum energy loss per cycle is $\pi k_s (x_o)^2/2$ at $\omega = k_s/c$, with c being the damping rate.

However, for frequencies above $4k_s/c$, the model's energy loss per cycle is already halved, a property not found from tests on rubber. To resolve this, more parallel sets of dashpot with series spring may be added [12], but then, they require a number of additional input parameters.

A proper choice of viscous model can describe the frequency dependence of a rubber spring reasonably well. However, the amplitude dependence, mainly due to friction, should also be represented as indicated above. In [11], a simple rubber spring model, although still representing both frequency and amplitude dependence, is suggested. In fact, comparing the stiffness of a large-displacement and low-frequency case with that of a small-displacement and higher-frequency case, reveals that the stiffness increase for the latter case can very well be greater owing to the smaller amplitude than to the higher frequency.

For the vertical dynamics of air springs with auxiliary air volume, the models above are usually still not sufficient. The viscous effects are rather quadratic than linear with respect to velocity, introducing another non-linearity into the air spring model. Moreover, an inertia effect will arise due to the very high accelerations of the air in the surge pipe connecting the two volumes. For small air spring displacements and no so-called orifice damping of the pipe, this inertia effect is very pronounced and thus needs to be modelled. Examples of such air spring models are [8] and [13]. Figure 17.8 shows an example of the very strong frequency dependence of an air spring system, with the air bag subjected to a small-displacement harmonic excitation. This dependence is shown for three preloads, with both simulation and measurement results. The figure first shows a decrease in stiffness, followed by a significant stiffness increase at about 8 Hz. This effect can be dramatic for the vertical ride comfort [14].

Hydraulic dampers may be modelled as linear or piecewise linear viscous dampers. For dampers with a high damping ratio (i.e., the damping force changes very rapidly with damper velocity), it may be important to consider the inherent stiffness of the damper assembly as well, including rubber end bushings and internal structural and oil stiffnesses. This is particularly true for yaw dampers, which usually have a high damping ratio. Dampers normally have a force-limiting blow-off level, which must also be considered.

Flexibility in damper brackets, etc., must be considered as well, in particular for dampers with high damping rates. This is as important as the internal oil stiffness. It is often desirable that brackets and attachment points are very stiff in order not to reduce damper efficiency. Hydraulic dampers usually have some frictional effects as well.

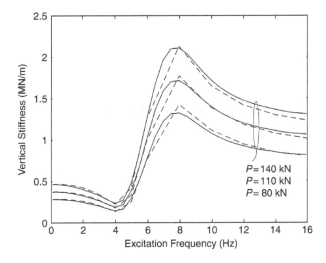

FIGURE 17.8 Vertical air spring stiffness as a function of excitation frequency for three preloads P and displacement amplitude 2 mm for simulations (solid lines) and measurements (dashed lines). (From Berg, M., An airspring model for dynamic analysis of rail vehicles, Report TRITA-FKT 1999:32, Royal Institute of Technology (KTH), Stockholm, Sweden, 1999. With permission.)

Friction damping through plane surfaces sliding against each other during general two-dimensional motion is not simple to model, but one example is shown in [15]. Moreover, the friction characteristics vary with the status of wear, humidity and possible lubrication.

Except for the air spring vertical dynamics, the suspension models do not usually include inertia effects. However, the mass of each suspension component should be split 50/50 to the connected bodies. This is necessary in order not to underestimate the total vehicle mass in the modelling.

For more details on suspension modelling, see [16] and Chapter 6. A final remark on suspension modelling is the need for linearisation of non-linear models when linear types of analyses are to be carried out.

17.2.4 Track Models

Track flexibility, track geometry design (track layout) and track geometry quality (track irregularities) must be modelled and described for simulation of vehicle-track interaction.

Figure 17.9 shows a simple moving track model for lateral, vertical and roll flexibilities of the track. Such a simple model may be sufficient for analysis of the vehicle's interaction with the track [17]. However, this track model is, in most cases, too simple for analysing track behaviour in more detail, cf. Chapter 9. Stiffnesses and damping may have linear or non-linear characteristics. Actual numerical data reflect the track construction, i.e., type of rails, fastenings, rail pads, sleepers and ballast, as well as vehicle static axle loads. The track flexibility will, above all, influence track forces, but under certain circumstances, dynamic running stability will also be affected.

The track design geometry is defined by circular curve radii and lengths, lengths and types of transition curves, track cants, etc.

Track irregularities are either given as lateral and vertical deviations of the track centre line (from track design geometry) together with deviations in track cant and track gauge or by lateral and vertical deviations of the top centre of the left and right rails.

It is important that the track irregularities are representative for a longer section of track and that they are also representative of the worst condition to be considered. A statistical analysis of track data is, therefore, very often needed.

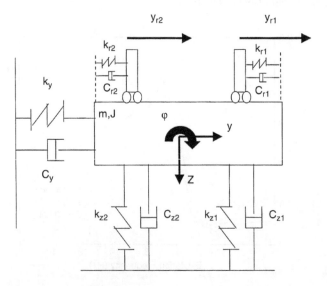

FIGURE 17.9 Example of simple model for track flexibility in cross section – one piece of track is assumed to follow each wheelset (moving track). (From Claesson, S., Modelling of track flexibility for rail vehicle dynamics simulations, M.Sc. Thesis, Report TRITA AVE 2005:26, Royal Institute of Technology (KTH), Stockholm, Sweden, 2005. With permission.)

Simulation of Railway Vehicle Dynamics

17.2.5 WHEEL-RAIL CONTACT MODELS

The characteristics of the wheel-rail contact are of crucial importance for the dynamic interaction between vehicle and track. Therefore, since the 1960s, much research has been devoted to the fundamental issues of wheel-rail contact (see also Chapters 7 and 8). The wheel-rail contact is, as a rule, very non-linear. This is true not only for the so-called contact geometry functions but also for creep forces as functions of creepages and spin.

The stiff wheel-rail contact, including Hertz's contact stiffness, and also the very high damping rates resulting from creep forces provide high system eigenvalues and therefore make it necessary to run time-domain simulations (by means of numerical integration) with time steps in the order of only 0.1–1 msec. Thus, these simulations are quite computer intensive and time-consuming. Advanced wear simulations, see [18], are certainly computer intensive.

A wheel-rail model consists principally of a wheel-rail geometry module (including results from contact geometry functions), a creep/spin calculation procedure and a creep force generator. The theories are described in Chapter 7.

The contact geometry functions are usually calculated in a pre-processor program. Input parameters are wheel and rail geometry (profiles), rail inclination, track gauge and wheelset flange back-to-back spacing. Normally, these quantities, apart from track gauge, are assumed to be constant over the whole simulated section; however, there may also be variations along the track, and, for instance, the possibility of *one-point contact* and *two-point contact* has to be considered. Simulations must automatically change from one state to another, depending on the actual wheel-rail conditions for each individual wheel.

Since wheel and rail parameters (profile shapes, gauge and rail inclination) change over time and over different track sections, many combinations and possibilities have to be included in the model and also have to be systematically investigated. A special case is negotiation of switches and crossings.

Quasistatic and dynamic vehicle-track performance are also dependent on wheel-rail friction, normally varying from approximately 0.6 (very dry rails) to 0.1 or 0.2 (wet or lubricated rails); see also Chapter 8. This must also be considered in the models. In particular, the worst case is usually running stability and wheel-rail wear under dry rail conditions, cf. Section 17.6.

17.3 SIMULATION METHODS

17.3.1 MULTIBODY SYSTEMS AND EQUATIONS OF MOTION

The basis of a computer simulation model is a set of mathematical equations that represent the system being modelled. These are known as the equations of motion and are usually second-order differential equations that can be combined into a set of matrices, e.g., the linear system equation:

$$[\mathbf{M}]\ddot{q} + [\mathbf{C}]\dot{q} + [\mathbf{K}]q = F \tag{17.1}$$

where:
  [**M**] is masses and mass inertia moments matrix
  [**C**] is damping matrix
  [**K**] is stiffness matrix
  q is vehicle motion vector
  F is input force vector

Modern computer simulation packages of the type used to model railway vehicle behaviour can usually prepare the equations of motion automatically with a user interface handling the input of the vehicle and track parameters. This can be via a graphical user interface (GUI) or by entering sets of coordinates and other data describing all the important aspects of the bodies and suspension components.

The vehicle is represented by a network of bodies representing the main parts of the system connected to each other by interconnecting elements. As mentioned earlier, this is called an MBS, and the complexity of the system can be varied to suit the simulation and the results required. Each of the bodies can be considered to have six DoFs, three translational and three rotational. But physical constraints may mean that not all of these movements are possible, and the system can be simplified and the number of DoFs reduced accordingly.

Masses and mass moments of inertia for all bodies and properties such as stiffness and damping for the interconnections need to be specified. Points on the bodies, or 'nodes', are defined as connection locations, and dimensions are specified for these. Springs, dampers, links, joints, friction surfaces or wheel-rail contact elements can often be selected from a library and connected between any of the nodes. All of these interconnections may include non-linearities such as occur with rubber or air spring elements or as in damper blow-off valves or bumpstop contact. Non-linearities also occur at the wheel-rail contact point due to creepage and flange contact. Owing to the presence of these non-linearities, the full equations of motion cannot normally be solved analytically. It is sometimes possible to linearise the equations of motion, but otherwise, a numerical method must be used to integrate the equations at small time intervals over the simulation period, the results at each point being used to predict the behaviour of the system at the next time step. Most simulation packages have several 'solvers', which carry out the numerical simulation, but care must be taken in selecting the most appropriate solver for the particular conditions.

The bodies in an MBS system are usually rigid but can be flexible with given modal stiffness and damping properties if required by the simulation. Modal properties can be measured or simulated using finite FE tools. Some dynamic simulation packages include the possibility of modelling flexible bodies as beams or combinations of beams, and linking to or incorporation of FE models is also usually possible. Examples from some typical simulation packages are given later.

Railway vehicles run on track with irregularities, and, to represent this, inputs are usually made at each wheelset. These can be idealised discrete events representing, for example, dipped joints or switches and crossings, or can be measured values from a real section of track taken from a track-recording vehicle. Most railway administrations use track-recording coaches running on the network and collecting track data at regular intervals. Additional forces, such as wind loading and powered actuators, may be specified. Depending on the purpose of the simulation, a wide range of outputs, e.g., displacements, accelerations and forces, at any point can be extracted.

17.3.2 Solution Methods

The method of solving the equations of motion will depend on the inputs to the model and the required output. Four of the main methods widely used in simulating vehicle dynamic behaviour are explained in the following sections.

17.3.2.1 Eigenvalue Analysis

All systems with mass and stiffness can vibrate, and these vibrations occur most naturally at certain frequencies called modal frequencies and in certain patterns called mode shapes. If the equations of motion are linear (or can be linearised for certain equilibrium positions of the bodies or amplitudes of vibration), then an eigenvalue analysis can be carried out to determine the modal frequencies and mode shapes. This is also known as modal analysis, and it may be useful for a vehicle designer or operator to have a knowledge of these modes to allow unwanted vibrations to be reduced.

Owing to the creep forces that are present at the wheel-rail interface, the railway vehicle can be subject to self-exciting oscillations. These will occur as the vehicle is moving along the track, and their characteristics will depend on its forward velocity. Below a certain speed, oscillations set up by a small disturbance will tend to die away or remain of very small amplitude, their energy being dissipated by the damping present. Above this speed, however, a similar disturbance will cause oscillations that grow until limited by the wheel flanges striking the rails.

Simulation of Railway Vehicle Dynamics

This unstable behaviour is called hunting and can result in damage or derailment. The speed above which this can occur is called the critical speed for the vehicle. An eigenvalue analysis can be used to give information about the stability of vibrations at each mode, and this is useful in establishing the critical speed of a vehicle, above which hunting instability will occur, but caution should be exercised in using this method for establishing the critical speed of a vehicle, as it relies on the linearised equations of motion. In particular, the wheel-rail interface is highly non-linear even over small displacements, and the linearised conicity parameter, which must be used for a linear analysis, cannot fully represent the situation [19,20]. An alternative method is to carry out a time-domain simulation using the full non-linear equations of motion and observing the rate at which the oscillations of vehicle motion (especially wheelset lateral displacement) decay after a disturbance.

17.3.2.2 Stochastic Analysis

In a stochastic analysis, the inputs and outputs are described statistically. Inputs correspond to vertical and lateral track irregularities and deviations in gauge and cross level and usually take the form of a describing function, i.e., a function of the spectral density of the amplitude of the particular parameter against frequency of the irregularity. The parameters of the describing function can be obtained from the measured track data.

The stochastic analysis method is useful for evaluating the general lateral or vertical behaviour of a vehicle to a particular type of track. Responses of various vehicles can then be compared or the effect of minor changes to the design of a vehicle evaluated. The equations of motion can only be linear, and this type of solution method cannot be used to show response to discrete inputs such as bad track joints.

This is an ideal method to use when vehicle ride is of interest. The frequency spectra of the output can easily be analysed against available recommended levels. Caution must be exercised when using this technique for lateral motion and lateral ride, as the non-linearities in this case are severe.

17.3.2.3 Time-Stepping Integration

The most powerful method available for simulating the dynamic behaviour of a vehicle is to solve the equations of motion fully at each of a series of very small time steps. All the non-linearities of the system can be considered and the equations updated accordingly at each time step. A wide range of numerical methods is available for this type of simulation, e.g., the Runge Kutta techniques are widely used. The size of each step must be small enough to ensure that the solution does not become unstable, but the penalty of using a smaller time step is, of course, a longer simulation time. Some solvers use a varying time step, which is automatically adjusted to suit the current state of the simulation. The fastest simulators are now able to solve the equations of motion at faster than real time, even for simulations involving complex suspensions or multiple vehicles. This type of simulation is sometimes called 'dynamic curving', as it is most suitable when a vehicle negotiates a series of curves of differing radii or a curve of changing radius.

Each term of the matrix equation of motion is set out separately in a subroutine, and, at each time step, every term is evaluated. This means that the stiffness or damping coefficient of each suspension element in the model can be calculated with reference to the relevant displacements or velocities. Typical suspension non-linearities encountered are bumpstops or multistage suspension elements such as dampers.

At each time step, the equations of motion are set up and all the suspension non-linearities are evaluated. The creepages and creep forces between the wheels and rails are evaluated, and the resulting accelerations at each body for each DoF are calculated. The displacements and velocities are calculated through the integration routine and stored, the elapsed time is increased and the complete calculation step is repeated. The whole process keeps stepping until the pre-set maximum time or distance is reached.

This solution method is very powerful because of the ease with which it can accommodate non-linearities in the equations of motion.

17.3.2.4 Quasistatic Solution Method

This is a special case in the overall study of the behaviour of a rail vehicle. When a vehicle is negotiating a curve of constant radius at a constant speed, the wheelsets and bogies will take up a certain fixed attitude after a short period of transient motion (provided the vehicle is stable at this speed). The aim of this analysis is to predict the steady-state attitude of the vehicle relative to the track and the resulting wheel-rail and suspension forces. The method is known as 'steady-state curving'. No track irregularities or other varying inputs can be considered when using this technique.

17.4 COMPUTER SIMULATION TOOLS

Using modern computer packages, it is possible to carry out realistic simulation of the dynamic behaviour of railway vehicles. The theoretical basis of the mathematical modelling used is now mature and reliable, and programs often originally written by research institutes have been developed into powerful, validated and user-friendly packages.

17.4.1 Historical Development

In analysing the contact between a railway wheel and a rail, the first step is to establish the location and the size and shape of the contact patch (or patches). As the cross-sectional profiles of the wheel and the rail can be quite complex shapes, most computer simulation packages have a pre-processor, which puts the wheel and rail profiles together for a given wheelset and track and establishes where the contact will occur. A description of the cross-sectional profiles is prepared from the designs or measured using a device such as the widely used 'Miniprof'. More details of these methods are given in Chapter 7.

The classical theory of contact was developed by Hertz [21] in 1882, when he was a 24-year-old research assistant at the University of Berlin. He demonstrated that the contact area between two non-conformal bodies of revolution would be elliptical and established a method for calculating the semi-axes of the ellipse and the pressure distribution within the contact patch. The Hertz theory is theoretically restricted to frictionless surfaces and perfectly elastic solids, but it still provides a valuable starting point for most contact problems and is included in most computer programs that deal with wheel-rail contact.

Some software packages use Hertz theory to establish elliptical contact patches around the contact point. The normal load on the contact point is required, and the calculation may be iterative to allow the correct load distribution between the contact points to be found. In tread contact, the radii of curvature are only changing slowly with position and the contact patch is often close to elliptical in shape. However, if the radii are changing sharply or the contact is very conformal, the contact patch may be quite non-elliptical, and the Hertz method does not produce good results. For example, in 1985, Knothe and Hung [22] set out a numerical method for calculating the tangential stresses for non-elliptical contact.

Multi-Hertzian methods split the contact patch into strips, with Hertz contact being calculated for each strip. Some iteration may be required to establish the correct normal load distribution across the whole contact patch if the contact is being treated as a constraint. Pascal and Sauvage [23] developed a method using an equivalent ellipse, which first calculated the multi-Hertzian contact and then replaced it with a single ellipse, which gives equivalent forces. In the methods developed by Kik and Piotrowski [24], an approximate one-step method is used, and some results for a S1002 wheel profile and a 60E1 rail (formerly UIC60) are shown in Figure 17.10, published in [25]. In the semi-Hertzian methods developed by Ayasse et al. [26], the contact is treated as Hertzian for the longitudinal curvatures (along the rail) and non-Hertzian for the lateral curvatures (across the rail).

In 1916, Carter [27] introduced the concept of creepage or microslip between the wheel and the rail and the corresponding creep force that was generated. A fuller treatment of creep forces was

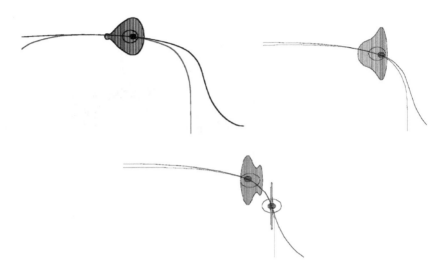

FIGURE 17.10 Non-elliptical contact patches. (From Iwnicki, S.D. (Ed.), *The Manchester Benchmarks for Rail Vehicle Simulation*, Swets & Zeitlinger, Lisse, The Netherlands, 1999. With permission.)

reported in 1964 by Vermeulen and Johnson [28], and in 1967, Kalker [29] provided a full solution for the general three-dimensional case with arbitrary creepage and spin. Haines and Ollerton divided the contact area into strips parallel to the direction of rolling and predicted areas of adhesion and slip in the contact patch [30].

Heuristic methods for predicting creep force were developed initially by Vermeulen and Johnson [28], based on a cubic equation for creep force as it nears saturation. The method developed by Shen et al. [31] is widely used and is very fast, but results are approximate and become less accurate when spin (relative rotation about the normal axis) is high. Polach [32] has developed a method that works well at high levels of creepage and spin and also includes the falling value of the coefficient of friction as slip velocity increases.

17.4.2 MULTIBODY SIMULATION TOOLS

The early packages used text-based interfaces, where vehicle parameters were listed in a particular order or using key words to provide the input to the simulation. User-friendly graphical interfaces were added and packages developed to allow engineers to test the effects of making changes to any part of the system and to animate the output.

A large number of computer codes have been developed by railway organisations to assist in the design of suspensions and the optimisation of track and vehicles. Some of these have been combined into general purpose packages, and three examples from those currently in widespread use are given here, although this is, by no means, a comprehensive list.

17.4.2.1 VAMPIRE

In the UK, British Rail Research developed a number of programs to analyse different aspects of railway vehicle dynamic behaviour. The programs were first used to successfully develop a new high-speed two-axle freight vehicle with much improved stability and then formed the basis for the work on the advanced passenger train. In the late 1980s, the programs were combined into one package called VAMPIRE, which has been continuously developed into the software tool used worldwide today (see Figure 17.11). The VAMPIRE software is now developed and supported by Resonate Group Limited, UK.

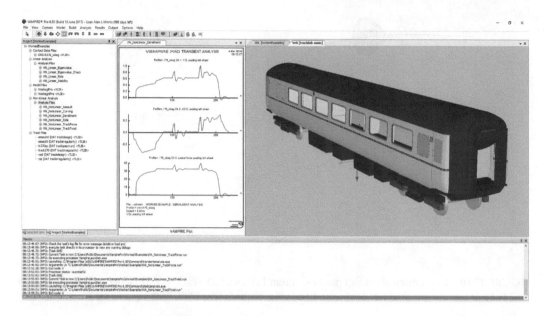

FIGURE 17.11 The VAMPIRE graphical user interface.

Designed from the outset as a dedicated railway vehicle dynamics package, VAMPIRE is significantly faster than other general MBS packages for both model creation and simulation, allowing large numbers of runs to be completed or entire routes to be simulated in a matter of minutes. The VAMPIRE software has a fully interactive front end that enables input files and vehicle models to be constructed quickly and reliably. This is supported by VAMPIRE projects that allow calculation files to be grouped and managed logically, enabling them to be easily manipulated, executed and analysed.

The development of VAMPIRE is focussed on its ease of use and simulation speed; this allows customers to easily pick up the software and get results quickly. The speed of VAMPIRE has enabled the development of automation, where whole calculation processes can be captured and replayed from a single command. The process also allows third-party programs to be executed and for VAMPIRE to be embedded in other applications. The automation and speed of VAMPIRE have opened up a range of new applications.

For example, a project to develop a decision support tool involved performing over 950,000 simulations equating to 114,000 km of track. Another interesting use for the automation process is to analyse track quality from the basis of simulated vehicle response rather than relying solely on traditional track exceedances [33,34]. This requires fast, repeatable, automated processing and simulation of recorded track data. The speed of VAMPIRE enables it to be used during the track-recording process on-board a track-recording vehicle.

VAMPIRE also allows users to embed their own code into the transient solver for improved flexibility. This has been used to simulate novel suspension designs and to evaluate creep laws such as FASTSIM (cf. Chapter 7) and Polach [32] with sophisticated vehicle models.

Currently, VAMPIRE is being developed to simulate Longitudinal 'In-Train' dynamics problems, with the solver able to model traction and braking forces, gradients, curve resistance and propulsion resistance.

Simulation of Railway Vehicle Dynamics

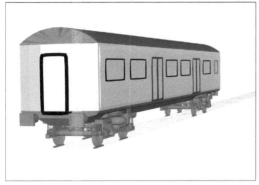 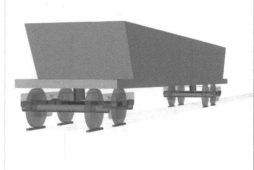

FIGURE 17.12 Typical Simpack models.

17.4.2.2 Simpack

SIMULIA Simpack Rail started as a joint development between the German Aerospace Center DLR and Siemens Transport in 1993. Major rail vehicle manufacturers, e.g., Bombardier and Alstom, very quickly made use of the software for their railway applications. Since that time, a large amount of experience went into the product. After a quite successful history, Simpack Rail is now a standard tool for rail simulations worldwide.

Simpack Rail allows the modelling and analysis of any kind of rail guided system. The most common railway vehicles simulated include high speed, intercity, local trains, locomotives, trams and freight cars, as well as special vehicles, e.g., roller coaster and suspended rails (see Figure 17.12). The software is used for almost any railway application, including the simulation of forces on wheels and rails, derailment, profile wear, passenger comfort, switches and crossings, crosswind, gauging, crash, pantograph-catenary interaction and homologation. The software highlights are the easy assembly of multicar trains from submodels, independently treated wheel, cartographic and measured tracks, unlimited flexibility in modelling, straightforward introduction of linear or non-linear flexible parts and track segments, the large variety of suspension elements, as well as the accurate rail-to-wheel contact. The newest and most valuable rail feature released in the past years by Simpack is the new discrete elastic rail-wheel contact (DEC). The DEC allows a very detailed prediction of the stress distribution within the contact patch while living up to the high demand in performance of the MBS applications (see Figure 17.13).

The standard vehicle components are available, like wheelsets, independent wheels and profiles, as well as a large library of suspension elements, e.g., shear springs, dampers, friction elements, bearings and air spring systems (see Figure 17.14). Arbitrary track layouts of arbitrary complexity are easily entered with inertia-fixed and elastic track components, including additional track irregularities. The Simpack Rail Track feature has turnouts, crossings and switches as common applications and also infrastructure modelling of catenary and overhead line structures in contact with the pantograph. Different wheel-rail contact methods are available with different degrees of complexity and insights into the normal and shear stress distributions of the slip and stick regions within each contact patch.

The Simpack FlexTrack Module allows the introduction of flexible track sections into the track line design to simulate the dynamics forces on bridges, turnouts and crossings (see Figure 17.15).

Simpack Rail Wear (see Figure 17.16) allows users to predict the wear (material removal) of railway wheels and/or rails. The amount of material removed from the wheel or rail after the simulated run is estimated by using the contact patch positions/sizes and contact forces/creepages.

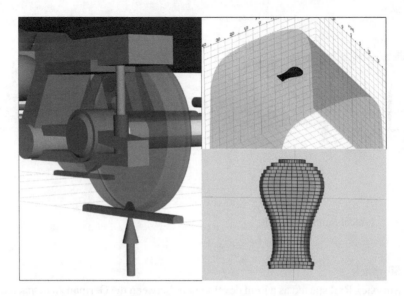

FIGURE 17.13 The wheel-rail interface in Simpack.

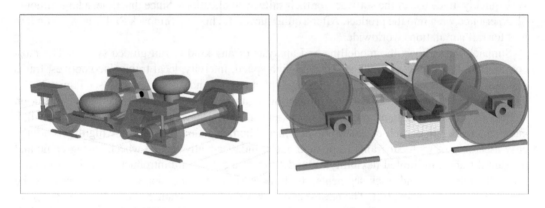

FIGURE 17.14 Simpack bogie models.

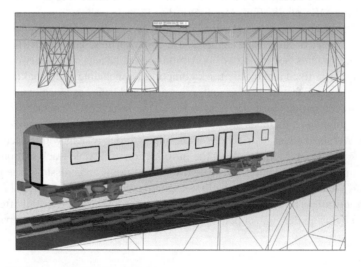

FIGURE 17.15 The Simpack FlexTrack module.

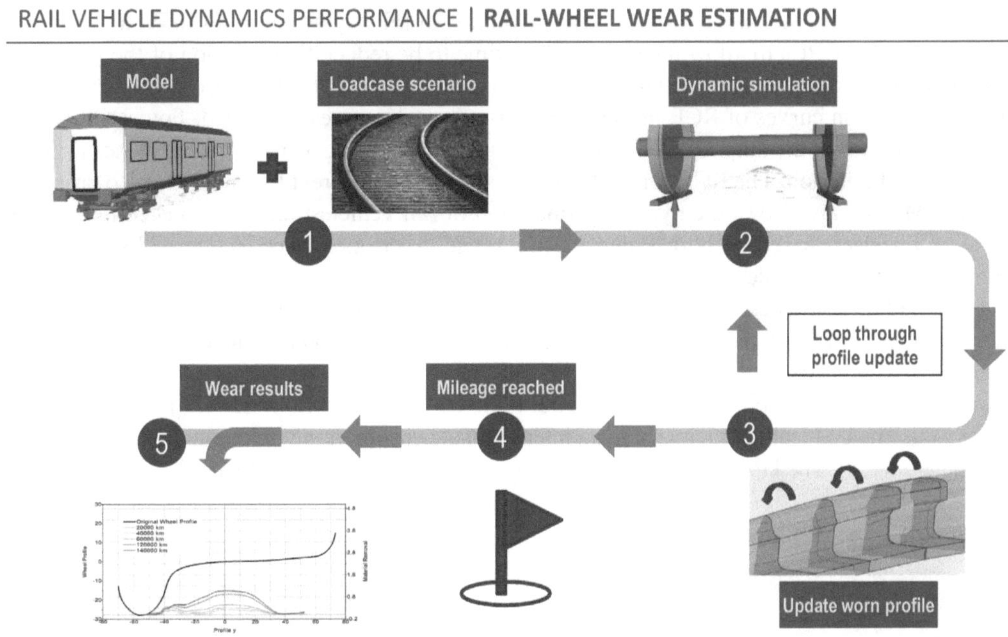

Simulation of Railway Vehicle Dynamics

RAIL VEHICLE DYNAMICS PERFORMANCE | **RAIL-WHEEL WEAR ESTIMATION**

FIGURE 17.16 Simpack rail wear.

17.4.2.3 Universal Mechanism

The Universal Mechanism (UM) software was developed at Bryansk State Technical University (Russia) at the end of the 1980s. UM is a general-purpose MBS software tool, but one of the main directions for the program's practical usage is the simulation of dynamics of rail vehicles. At present, the UM software provides a wide range of techniques and tools for simulation of dynamics of rail vehicles and trains (see Figure 17.17).

UM allows the inclusion of flexible bodies in the model of a railway vehicle. The flexible bodies can be imported from third-party Finite Element Analysis (FEA) software such as ANSYS, MSC. NASTRAN, etc. The CAE-based durability analysis to predict the fatigue damage of parts of railway vehicles has been implemented.

Methods for computation of wheel-rail interaction can be applied to a massless rail (FASTSIM, cf. Chapter 7) and rigid or flexible rail (model of Kik and Piotrowski [24], CONTACT add-on by VORtech CMCC).

FIGURE 17.17 String of 3D vehicles in a train in UM.

The UM software includes a tool for wear prediction of railway wheel profiles. Parallel calculation on multicore CPUs to allow the computation time to be reduced. Simulation of the process of accumulation of rolling contact fatigue (RCF) damage in railway wheels is also available [35]. An approach based on curves of RCF in the form of stress-cycle curves is applied. Four RCF criteria are given for choice: the amplitude of maximum shear stress, Dang Van's criterion, Sines criterion and combined criterion. The RCF curves for wheel steels with different hardness are also proposed.

The UM software allows the dynamic simulation of rail vehicles, taking into account the flexibility of wheelsets based on both Euler and Lagrange descriptions (see Figure 17.18). Roller rig simulation is also available in UM.

A detailed flexible track model includes flexible rails, fasteners, sleepers and sleeper foundation. The UM software allows the analysis of models and takes into account vehicle and flexible bridge interaction.

Simulation of longitudinal dynamics of trains in UM is available in 1D [36], 3D and mixed modes [37]. The UM team participated in an international benchmarking of longitudinal train dynamics simulators [38]. The UM graphical user interface is illustrated in Figure 17.19. A fast solver based

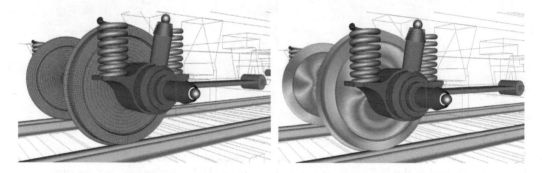

FIGURE 17.18 FE modelling of flexible wheelset in UM – shading of the wheel on the right is in accordance with the equivalent stresses.

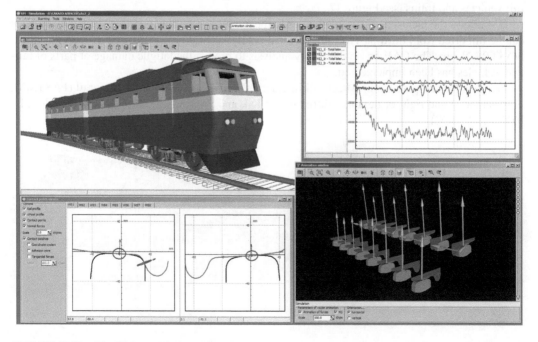

FIGURE 17.19 The UM graphical user interface.

Simulation of Railway Vehicle Dynamics

on the Park method with parallelisation on multicore processors makes possible real-time simulation of 1D and mixed train models, which allows the use of UM in 1D and 3D train simulators [39].

17.4.2.4 Benchmarking

Owing to the high level of complexity of the software codes developed for simulation of railway vehicle dynamics, there is a high level of interest in comparing the results of the different codes for certain test cases.

An early benchmark was proposed by the European Railway Research Institute (ERRI) [40], based on a passenger coach. In the exercise initiated at the Herbertov Workshop on 'MBS Applications to Problems in Vehicle System Dynamics' [41] in 1990 and reported on by Kortüm and Sharp [42], the computer codes that were able to handle wheel-rail contact were asked to simulate a single wheelset and a bogie. The wheelset benchmark was proposed by Pascal, and participants were required to calculate the lateral deflection of the specified wheelset in response to a lateral force of 20 kN and to find the level of lateral force at which the wheelset would derail. In the bogie benchmark defined by Kik and Pascal [43], participants were required to predict the behaviour of the bogie in a vehicle running on straight and curved track at several speeds. Not all codes participated fully in the exercise, but some interesting results were shown.

In the Manchester Benchmarks agreed at the International Workshop on 'Computer Simulation of Rail Vehicle Dynamics' in 1997 [44], with results published in [25], two simple vehicles and four matching track cases were defined to allow comparison of the capabilities of computer simulation packages to model the dynamic behaviour of railway vehicles. One of the aims of this benchmark was to try to encourage railway organisations to accept simulations carried out using any reliable computer simulation package and not to insist on one particular tool. Simulations were carried out with five of the major packages (VAMPIRE, GENSYS, Simpack, ADAMS/Rail-MEDYNA and NUCARS), and the results and statements of methods were presented.

A number of outputs were requested for each of the track cases, each of which was designed to test a particular potential vehicle problem. One of the most useful indicators of derailment potential is the lateral/vertical (L/V or Y/Q) force ratio at each wheel. In a curve, the derailment usually takes place at the outer wheel, and Figure 17.20 shows the L/V ratio for the outer wheel on the first wheelset for one of the benchmark vehicles [25]. The peak value occurs at a dip designed to test vehicle suspension and shows that all five packages give good agreement on the nearness to derailment of this vehicle.

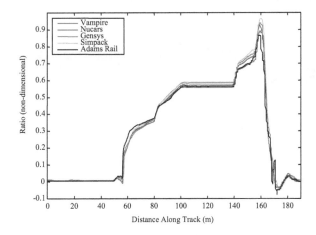

FIGURE 17.20 Example L/V ratio comparisons from the Manchester Benchmarks (vehicle 1, track case 1, 4.4 m/s). (From Iwnicki, S.D. (Ed.), *The Manchester Benchmarks for Rail Vehicle Simulation*, Swets & Zeitlinger, Lisse, The Netherlands, 1999. With permission.)

17.5 DYNAMICS IN RAILWAY VEHICLE ENGINEERING

Computer analyses and simulations of vehicle dynamics constitute an integral part of engineering processes during the development and design of new and modified railway vehicles. Virtual prototyping computer tools have taken a big step forward in recent decades. The simulation of railway vehicle system dynamics can be coupled with other simulations from structural mechanics, aerodynamics, controls, electrotechnics, etc. [45–47] to a virtual development process. Modern simulation packages provide powerful and important analysis and design tools that are well suited to the concurrent engineering process demands in the railway industry.

Despite the significant progress made in the virtual development process, very few publications exist that describe the dynamics methodology used in the railway industry (e.g., [48–50]). Although a wide range of methods is applied for dynamic analyses in general industrial practice, some of the methods are specific to the development of railway vehicles. In Section 17.5.1, the railway vehicle development engineering process and the role of vehicle dynamics in same are described. Section 17.5.2 provides an overview concerning the aim and methods of dynamic calculations applied during the engineering process. Section 17.5.3 shows the recent progress of the methodologies used to validate the simulation models, particularly regarding the simulations in a vehicle acceptance process. Section 17.6 presents in more detail the typical tasks and methods applied during the engineering process of railway vehicle dynamics computations. A description is supplied concerning the aim of the methods, important influencing parameters, typical input parameters and output values. The methods are illustrated with examples from industrial application.

17.5.1 RAILWAY VEHICLE ENGINEERING PROCESS

The phases of the engineering process of railway vehicles as a function of time and knowledge of input parameters are illustrated in Figure 17.21. The engineering process of railway vehicle dynamics commences with a feasibility study and concept analysis phase. The major emphasis of engineering work is concentrated on the optimisation and verification design phases. The process continues with the realisation, test and qualification phases. Warranty and field support issues also benefit from vehicle dynamics analyses, if required. Service experience provides feedback concerning the vehicle dynamics behaviour predicted during engineering.

Railway vehicle dynamics simulations are already applied at an early stage for concept investigation and feasibility study. The models applied here are normally based on other similar vehicles, and most of the input parameters are estimated. Input parameter uncertainty is covered by parameter variation and sensitivity tests.

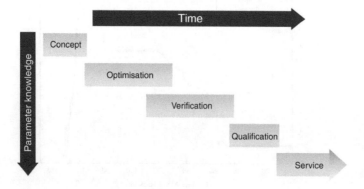

FIGURE 17.21 Engineering process of railway vehicle design. (From Iwnicki, S. (Ed.), *Handbook of Railway Vehicle Dynamics*, CRC Press, Boca Raton, FL, 2006. With permission.)

Simulation of Railway Vehicle Dynamics

The main part of the dynamic calculations takes place during vehicle design. In the optimisation (preliminary) design phase, the topology of the vehicle is specified, but the input parameters are based on a first estimation only. The parameters of suspension and other coupling elements have to be optimised based on the target values given by the standards and vehicle specification. All limit and target values have to be checked and design changes introduced if necessary. Output values required for other engineering activities such as movements and loads are calculated to support the engineering process.

In the verification (detail) design phase, the vehicle topology is frozen. The vehicle is modelled in detail to extensively study the vehicle's behaviour. The parameters of elements are specified and, if possible, also verified by measurements on prototype elements. Parasitic effects such as vertical and lateral parasitic stiffness of traction rods, anti-roll bars and other elements have to be taken into consideration. The structural elasticity also has to be considered if an influence can be expected on the stiffness parameters, damper performance or vehicle behaviour.

The aim of the verification phase is to prove conformity with the limiting values stated in the norms and regulations, as well as the vehicle specification. The verification calculation should provide proof of whether the vehicle will pass the acceptance test. The verification calculation report is often used to support the acceptance test but can also be used to reduce the extent of type tests. Final movement and load collective calculations are applied to finalise the analyses of other specialists. A failure analysis and a sensitivity analysis can be provided as part of the verification calculations.

At a later stage, the vehicle acceptance and qualification tests are supported by vehicle dynamics calculations, if necessary. Computer simulations can allow the reduction of the amount of testing or also the replacement of some tests under certain conditions. During the tests, the predicted and measured vehicle performance can be compared and the modelling improved to gain experience for future projects.

Following the vehicle acceptance tests, the vehicles go into service. Although the engineering process is completed, feedback from service promotes the improvement of modelling and simulation in future projects. Requirements for improvement arising during service can lead to warranty issues, which in turn may necessitate vehicle dynamics simulation support.

17.5.2 Task and Applications of Vehicle Dynamics

Vehicle dynamics calculations utilised during railway vehicle engineering can be divided into the following categories, depending on the activities supported by the analyses:

- Risk assessment
- Fulfilment of customer specification
- Vehicle acceptance tests
- Support of other specialists during the design process

Typical dynamic analyses worked out during vehicle engineering are listed in Figure 17.22, together with the aim of the analysis according to categories mentioned above. A possible method to formulate each of the analyses is also provided.

Computer simulations that examine running safety, track loading, ride characteristics and ride comfort of the vehicle are the primary items of investigation. These are the characteristics that are tested during the on-track test as part of the vehicle acceptance procedure. Fulfilment of the vehicle performance according to the customer specification and required standards has to be proven and demonstrated (see Chapter 18). Simulations to investigate and optimise these issues are also part of the risk assessment and risk management process.

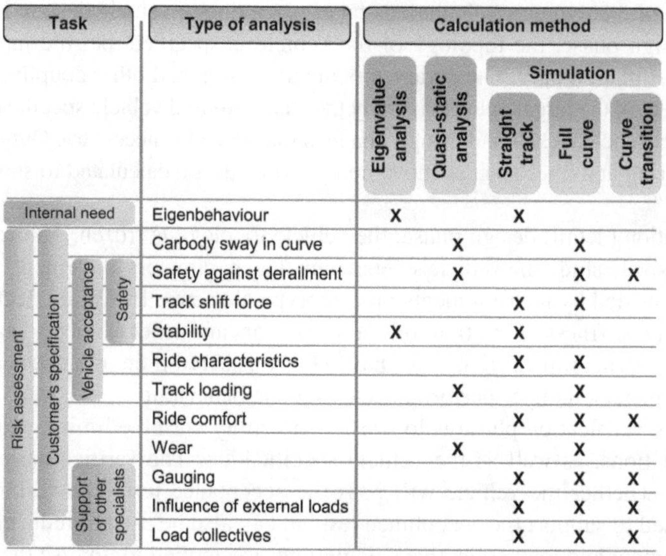

Task	Type of analysis	Eigenvalue analysis	Quasi-static analysis	Straight track	Full curve	Curve transition
Internal need	Eigenbehaviour	X		X		
Customer's specification / Vehicle acceptance / Safety	Carbody sway in curve		X		X	
	Safety against derailment		X		X	X
	Track shift force			X	X	
	Stability	X		X		
	Ride characteristics			X	X	
	Track loading		X		X	
	Ride comfort			X	X	X
	Wear		X		X	
Support of other specialists	Gauging			X	X	X
	Influence of external loads			X	X	X
	Load collectives			X	X	X

FIGURE 17.22 Typical dynamic analyses and calculation methods applied in railway vehicle engineering. (From Iwnicki, S. (Ed.), *Handbook of Railway Vehicle Dynamics*, CRC Press, Boca Raton, FL, 2006. With permission.)

Moreover, multibody simulations are used to investigate several other topics such as vehicle gauging (see Chapter 10), running through switches and crossings, longitudinal dynamics due to traction and braking (see Chapter 13), interactions between vehicles (e.g., regarding gangways and buffer contact) and influence of external loads acting on the vehicle or train configuration such as crosswind (Section 17.6.6.1). The simulations also fulfil an in-house requirement to support other specialists during the design process and to deliver necessary data for specification of components such as forces and displacements of suspension components and data for analyses of structural mechanics such as fatigue load spectra and load collectives [51].

Simulations successfully support solving of issues related to vibration and noise; examples are vibrations of running gear in tight curves, as presented by Kurzeck in [52], or curve squeal, as investigated by Glocker et al. [53]. Issues concerning dynamic analyses related to risk assessment or problem solving could be questions of wheel out-of-roundness [54], rail corrugations [55] and RCF of wheels [56–58] and rails [59,60]. The importance of assessing and counteracting these risks is increasing owing to the trend towards higher speeds, increasing axle loads and smaller wheel diameters associated with a higher risk of damage to wheels and rails.

Wear prediction and optimisation of wheel and rail profiles are further topics in which multibody simulations play an essential role (see Section 17.6.5.3 for more details). Recent trends to establish track access charges related to marginal contribution of vehicles to track damage, as discussed by Nerlich [61], and to monitor and assess track quality based on measurement of vehicle response, as illustrated by Kraft et al. [62], also rest upon the use of multibody simulations.

Multibody simulations are increasingly used to support and partly also to replace physical testing for the acceptance of running characteristics of railway vehicles in the authorisation process (see, e.g., [46,63,64]). There is a long experience with use of simulations in the context of vehicle acceptance in the UK, as shown by Evans and Berg in [46]. This experience was used during the preparation of the revision of UIC Code 518 [65], where simulation was introduced as an alternative to replace physical testing under certain conditions.

Simulation of Railway Vehicle Dynamics

The procedure and conditions for application of simulations for vehicle acceptance in European countries have been further developed in the framework of the revision of the European Standard EN 14363:2016 [66]. This standard allows the use of simulations for the following four applications:

- Extension of the range of test conditions where the full test programme has not been completed
- Approval of vehicles after modification
- Approval of a new vehicle by comparing it to a similar, already approved vehicle
- Investigation of dynamic behaviour in case of failure of the components

The range of modifications permitted in the case of use of simulations is related not to the alteration of the parameter but to the effect on the vehicle behaviour. The simulation can be applied to any modification of a vehicle parameter, as long as the influence of the modification improves the dynamic behaviour or the investigated quantities comply with the specified margin to the limit value or the deterioration is within a limited range of the existing margin to the limit value [66].

In the USA, in contrast to Europe, the application of simulations in the certification process is recently required for all vehicles accepted for operation with high speeds and at high cant deficiencies. The aim of these simulations is to identify vehicle dynamic performance issues prior to service. The simulations have to be carried out using 'Minimally Compliant Analytical Track' (MCAT), as described in Appendix D of Federal Railroad Administration (FRA) Regulation 49 Part 213 [67]. The MCAT track model contains defined geometry perturbations at the limits that are permitted for a specific class of track and level of cant deficiency.

Simulations are also used for investigations regarding the side wind stability and the impact of crosswind on vehicle safety according to EN 14067-6 [68]. According to EN 15273-2 [69], the simulations are used for the calculation of vehicle body displacements in the dynamic gauging.

The prerequisite for the use of vehicle dynamics computations in the context of a vehicle acceptance and certification process is, however, that the trust in the simulation results is substantiated by a rigorous validation of the applied simulation model.

17.5.3 Model Validation

17.5.3.1 Introduction

What does the term model validation mean? The American Society of Mechanical Engineers Standards Committee on verification and validation in computational solid mechanics describes validation of computer simulation model as a two-step process [63]:

- *Verification*: The process of determining that a computational model accurately represents the underlying mathematical model and its solution.
- *Validation*: The process of determining the degree to which a model is an accurate representation of the real world from the perspective of the intended uses of the model.

Using this definition, the verification considers predominantly the simulation code and is performed by the code's developers. The model validation, conducted by the model builder, assesses the accuracy of the simulation model in regard to the intended application. The model validation as discussed here considers a particular vehicle state (e.g. empty or laden) and a particular application of the simulation model. It includes the selection of model building, the identification of input parameters as well as model calibration with regard to unknown or uncertain parameters, if necessary.

A full definition of the model validation method should contain a unique quantitative specification of:

- Validation and application domains
- Validation quantities
- Validation metrics
- Validation limit values

The validation methods as well as validation requirements depend on the particular application of the simulation [70]. For example, as the comfort assessment is based on the measurement of acceleration in the vehicle body, the model validation has to prove the accuracy of simulations of vehicle body accelerations, including the vibration of its elastic structure. The investigation regarding safety in the case of crosswind, on other hand, requires an accurate simulation of the effect of lateral body movement on the wheel unloading. Thus, the model validation is based on comparisons of vertical wheel forces. In contrast, the vehicle gauging assessment necessitates accurate simulation of vehicle body displacements, and the model validation is consequently required to comply with the matching requirements regarding the primary and secondary roll angle evaluated in a stationary sway test.

The following model validation considerations refer to the simulations of on-track testing for approval of running characteristics, which represent the typical vehicle dynamics analyses in terms of safety assessment during the railway vehicle engineering.

17.5.3.2 Features of Testing and Model Validation in Railway Vehicle Dynamics

The model validation in railway vehicle dynamics mainly uses the tests for acceptance of running characteristics that are carried out in Europe according to the procedure specified in EN 14363 [66], cf. Chapter 18. This vehicle acceptance test consists of two steps:

- Stationary tests (first stage)
- On-track tests (second stage)

The stationary tests contain the assessment of safety against derailment on twisted track, conducted either on a test rig or during a run with low speed in a test centre. Other tests such as bogie rotational resistance and vehicle sway test are carried out on request.

The on-track tests are carried out on operational railway lines under specified target test conditions representative of operating conditions in European networks. The analyses of the test results are separated into test zones according to track curvature: straight track and very large radius curves (test zone 1), large radius curves (zone 2), small radius curves (zone 3) and very small radius curves (zone 4). Measurements in each test zone have to contain a specified number of track sections with the length between 70 m and 500 m (dependent on the test zone). The quantities measured during an on-track test are wheel-rail forces and accelerations when using the normal measuring method using force measuring wheelsets and accelerations only when using a simplified method, respectively. The measurements are evaluated per track section, and the results from track sections are statistically evaluated per test zone.

The validation method has to consider all these peculiarities when selecting the quantities to compare and the metrics to apply and matching requirements to fulfil. The components of the model validation can be divided into comparisons in the time domain, comparisons in the frequency domain and comparisons with stationary tests.

Validation in the time domain by comparing the time or distance dependent functions of measured and simulated quantities represents the most frequently considered validation exercise, as illustrated in Figure 17.23 by an example of guiding force in a curve. The comparison can either directly assess the matching of both functions, which, however, requires an excellent synchronisation of

Simulation of Railway Vehicle Dynamics

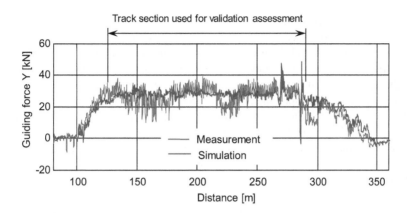

FIGURE 17.23 Validation by comparing the simulated and measured signals for guiding force over distance in a curve.

both signals, or it compares representative values of the particular track section as, e.g., maximum, quasistatic value and rms value [70]; see the marked part in Figure 17.23.

The quantitative definition of validation limits is rather difficult. Even the selection of the type of the matching errors is far from easy. Should they be relative or of absolute size? The validation limits are usually given as relative values in percent, similarly to the measurement errors. Such a definition, however, requires very high accuracy where only very small forces or accelerations occur but allows considerable deviations for large forces and accelerations. However, this is rather the opposite of the expectation that a high accuracy should be provided for the results close to the limit values regarding the vehicle acceptance test. Thus, a definition of the permissible deviation between simulation and measurement as a fixed amount seems to be more appropriate, but this requires a separate adjustment of the validation limit value for each quantity, also considering its filtering and processing, as well as the vehicle type, test conditions, etc.

Moreover, the validation limit values (matching requirements) have to consider the uncertainties of measured values and simulation values used for validation. The measurement results are affected not only by the measurement uncertainties but also by the assessment uncertainties such as the variability of friction conditions, track irregularities and wheel-rail contact geometry; the selection of valid track section samples for the test zones; the selection of statistical processing method; the choice of prototype vehicle, etc. Uncertainty of the simulation of on-track tests is affected by the uncertainties of the input data due to uncertainties of measurement of track layout, track irregularities, rail profiles and wheel profiles, as well as due to uncertainties regarding the parameters that cannot be measured, e.g., friction coefficient between wheel and rail. These manifold interrelations reveal why there are not yet common, quantitative matching requirements for the validation in the time domain.

Validation in the frequency domain represents another feature of model validation. Comparisons of Fourier transform (FFT) or spectral power densities of the simulation and measurement quantities provide the ability to check the correspondence of the main frequencies of the model and the real vehicle and thus validate the dynamic behaviour of the model. The advantage of this comparison is a low sensitivity to the synchronisation of simulation and measurement signals. However, the length of the evaluated section and the number of values considered influence the scatter of the points of the diagram.

The matching requirements between simulation and measurement regarding the frequency contents are rather difficult to justify. Fries et al. [71] recommend an agreement within 10%–15% for comparison of resonance frequencies. The agreement of amplitudes is recommended to be within 10%–15% for linear spectra and 20%–30% for spectral densities.

Validation using stationary tests by comparing the simulation and measurement of stationary tests is recognised as a possibility to check the vehicle model parameters and to point to their inaccuracy. However, this type of validation also has disadvantages. Namely, the dynamic behaviour of the vehicle during the on-track test may differ from the quasistatic behaviour in a stationary test, e.g., due to the dependence of the stiffness on frequency and amplitude, as can be observed, e.g., in rubber-to-metal suspension components. Polach and Evans [63] point out that the validation by stationary tests also holds other potential weaknesses. The measurement of static vertical wheel forces often shows an insufficient reproducibility and leads, especially in vehicles with friction damping, to a large dispersion of the measurement results. As another example, the measurement of the turning resistance of the bogie may be called for. The mostly unknown friction of the test rig influences the results and can be mistakenly interpreted as inaccuracy of the simulation [63].

17.5.3.3 Model Validation in the Context of the Acceptance of Running Characteristics

The application of simulations instead of on-track tests in the process for the acceptance of the running characteristics in European countries is described in Annex T of EN 14363:2016 [66]. This standard contains two equivalent methods to demonstrate the simulation model validity.

Validation Method 1 according to EN 14363:2016 represents an evolution of the methodology outlined in UIC Code 518 [65]. This procedure lists possible comparisons of stationary tests and on-track test runs with simulation results that can be used for validation and illustrates the assessment with examples. The scope and criteria to be met shall be determined by the model developer with regard to the application of the simulation model and the available test results. The validation result has to be assessed by an independent reviewer, who, according to the model quality assessment, can limit the scope of application or reject the validation. This methodology is thus suited for validation of a model intended for simulation of modifications, the impact of which can be well estimated.

Validation Method 2 according to EN 14363:2016 [66] was developed within the framework of the EU research project DynoTRAIN, carried out during 2009–2013. This project represents an extensive activity consisting of complex tests and simulations, comparisons with measurements and evaluations. The on-track test runs were carried out in four European countries, using a test train consisting of several vehicles equipped with a total of 10 force-measuring wheelsets as well as with simultaneous recording of track irregularities and rail profiles [72]. The validation simulations were performed using several vehicle models, built with the use of different simulation tools by different partners (see [73,74] for details).

The validation method developed within the DynoTRAIN project and implemented as Method 2 in EN 14363:2016 [66] will be introduced here in more detail. This method requires experimental data from the normal measuring method using force-measuring wheelsets. The validation assessment is based on comparisons between simulation and measurement for 12 quantities, filtered and processed by analogy with EN 14363:

- Quasistatic values and maximum values of wheel-rail contact quantities (Y, Q, Y/Q and ΣY)
- Rms and maximum values of vertical and lateral vehicle body accelerations

The simulation and measurement results of these quantities have to be compared on at least 12 track sections from all 4 test zones according to EN 14363, i.e., at least 3 track sections per test zone. Each quantity should be evaluated using at least two signals, e.g., vertical acceleration above the leading and trailing bogie, so that at least 24 simulated values can be compared with the corresponding measured values for each quantity.

The validation process is illustrated in Figure 17.24. The differences between the simulated value and the corresponding measured value have to be determined and transformed by mirroring the differences corresponding to negative measured values. Afterwards, the difference is positive if the

Simulation of Railway Vehicle Dynamics

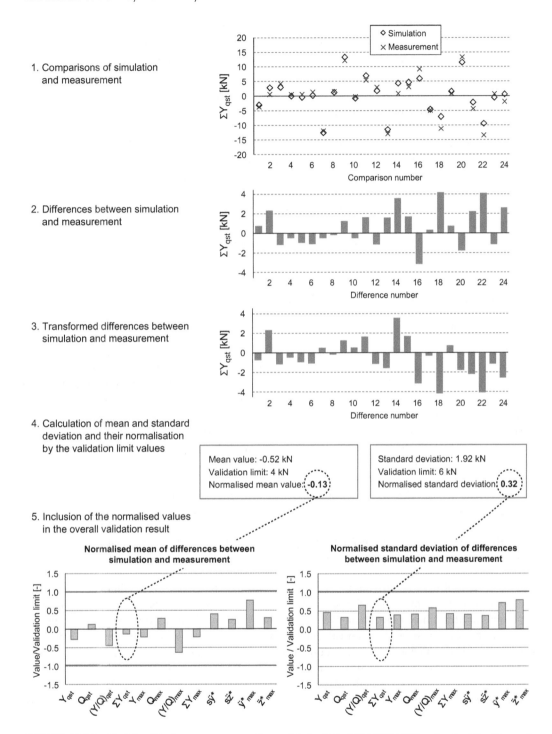

FIGURE 17.24 Model validation process according to Method 2 in EN 14363:2016, using 1 of the 12 quantities to be evaluated as an example.

magnitude of the simulation value is higher than the magnitude of the measurement (simulation overestimating the measurement), and vice versa. The following values have to be determined for the whole set of differences between the simulation and measurement for each quantity:

- Mean of differences between simulation value and measurement value
- Standard deviation of the same set of differences

The validation requirements are fulfilled if all mean values and standard deviations of differences between simulation value and measurement value do not exceed the corresponding validation limits specified in EN 14363:2016 [66]. Figure 17.24 illustrates the calculation of differences between the simulation value and the measurement value for the quasistatic values of the sum of guiding forces between wheelset and track, their transformation as well as the calculation of mean value and standard deviation, which are then normalised by the corresponding validation limits to allow easier assessment. The bottom diagram in Figure 17.24 displays an example of a successful model validation result.

17.6 TYPICAL RAILWAY VEHICLE DYNAMICS COMPUTATION TASKS

17.6.1 INTRODUCTION

There are no standards and recommendations available that specify how dynamic simulations of railway vehicles should be carried out. Specialists, companies and institutes dealing with railway dynamics have usually developed their own methods for vehicle modelling, analysis procedure and results assessment. A guideline is provided by the standards and specifications for measurements and acceptance tests of vehicles (see Chapter 18). The simulations must prove whether the vehicle performance required by the specification can be achieved and that the limit values will be fulfilled. However, on the one hand, it is difficult to apply the procedure for measurements in simulations, on the other hand, some additional possibilities in simulations are not available in measurements. Therefore, the simulations constitute a combination of the conditions specified in the standards for the measurements, together with a feasible and most efficient calculation method.

The simulation results can be basically structured in two possible manners:

- By the vehicle dynamics performance specified in the standard or specifications
- By the methods used in the calculations

The advantage of the first structure is a clear and easy comparison with requirements, while the second structure provides a simpler and better overview of the methods, input values and other conditions used in simulations. In an overview of the typical analyses used in the engineering process, the second structure based on calculation methodology is followed. Section 17.6.2 presents calculations of eigenbehaviour, and Section 17.6.3 shows examples of linearised and non-linear stability analyses. Section 17.6.4 deals with simulation of running behaviour on track with irregularities, Section 17.6.5 deals with the analysis of curving behaviour and Section 17.6.6 provides two examples from the variety of other simulation applications: effect of crosswind and interaction of vehicle and traction dynamics.

17.6.2 EIGENBEHAVIOUR

17.6.2.1 Eigenvalue Analysis

The eigenvalue analysis (cf. Section 17.3.2.1) allows the vehicle-track model to be examined and initial information concerning suspension properties to be obtained. This should be done at the commencement of the dynamic calculations as a first optimisation step. The application of linearised

Simulation of Railway Vehicle Dynamics

calculations for vehicles with strong non-linearities (e.g., friction damping between car body and bogie frame) is incorrect. A simulation of eigenbehaviour, as described in Section 17.6.2.2, would provide a suitable alternative.

The eigenbehaviour should be investigated for tare (empty) as well as for full load. The calculation can be done for any speed, but as kinematic (speed dependent) oscillations should be excluded, it is advisable to apply very low speed, e.g., 1 m/s (zero speed is often not allowed in simulation tools).

The eigenfrequencies, eigendamping and eigenmodes can be used for testing the vehicle model. An asymmetry of the modes or implausible values can provide an indication concerning incorrect or missing parameters. All eigenvalues should possess sufficient damping (as a minimum, 5% of critical damping is recommended). For correct modelling, all eigendamping properties of rubber elements, as well as other parasitic damping, have to be considered in the model parameters.

The modes and nomenclature used for car body eigenbehaviour are shown in Figures 17.25 and 17.26. The sway mode, as the combined lateral movement and rotation about the longitudinal axis, is present in two forms, with different heights of the rotation centre (see Figure 17.27): lower sway mode and upper sway mode. As an example, Table 17.1 shows the car body eigenmodes of the passenger coach from the Manchester Benchmark [44].

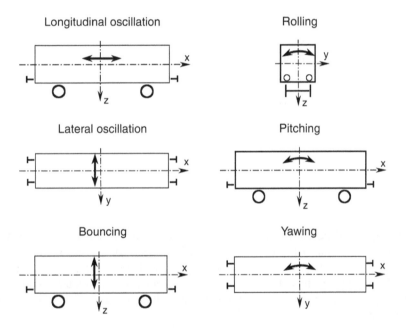

FIGURE 17.25 Basic car body modes. (From Iwnicki, S. (Ed.), *Handbook of Railway Vehicle Dynamics*, CRC Press, Boca Raton, FL, 2006. With permission.)

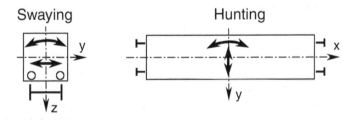

FIGURE 17.26 Combined car body modes. (From Iwnicki, S. (Ed.), *Handbook of Railway Vehicle Dynamics*, CRC Press, Boca Raton, FL, 2006. With permission.)

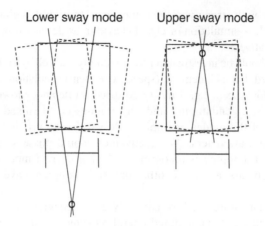

FIGURE 17.27 Two modes of car body sway movement. (From Iwnicki, S. (Ed.), *Handbook of Railway Vehicle Dynamics*, CRC Press, Boca Raton, FL, 2006. With permission.)

TABLE 17.1
Eigenfrequency and Eigendamping of a Railway Passenger Coach in the Manchester Benchmark

Mode	f [Hz]	D [%]
Bouncing	1.07	13.4
Pitching	1.28	16.0
Lower sway	0.58	21.1
Upper sway	1.10	45.2
Yawing	0.73	53.7

The bouncing eigenfrequency of the car body should be approximately 1 Hz. The frequency of the lower sway mode should be higher than 0.5 Hz; otherwise, there is a risk of motion sickness arising. The lowest eigendamping of car body modes should lay between 15% and 25%. Eigendamping lower than 15% but higher than 10% can also be acceptable, but this requires more detailed ride comfort investigations. Higher damping of remaining car body modes is not critical from a running behaviour perspective.

17.6.2.2 Simulation of Eigenbehaviour

The eigenbehaviour can also be assessed by using time integration. This is mainly suitable for vehicles with strong non-linearities, where a linearisation would be an unacceptable simplification.

The investigation can be carried out in a similar manner to measurements, simulating a reaction of the vehicle on a single-track excitation or on non-zero initial conditions. Figure 17.28 shows the simulated eigenbehaviour of a metro vehicle in comparison with measurement results. The car body sway oscillation was measured by a so-called wedge test. At the beginning of the test, wedges having a height of 15–25 mm are positioned on the top of one rail in front of each wheel at one side of the vehicle. The vehicle starts to move slowly, and the wheels roll over the wedges and excite car body swaying (see description of wedge test procedure in [75]). The sum of the measured lateral accelerations on the ceiling and on the floor of the car body provides the signal characterising the

Simulation of Railway Vehicle Dynamics

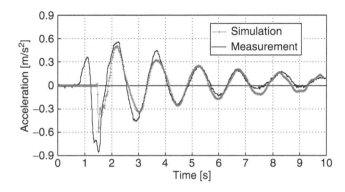

FIGURE 17.28 Comparison of lower sway mode oscillation observed during wedge test and detected by simulation. (From Iwnicki, S. (Ed.), *Handbook of Railway Vehicle Dynamics*, CRC Press, Boca Raton, FL, 2006. With permission.)

lower sway mode shown in Figure 17.28. Pitch, bounce and yaw modes can be measured and investigated by numerical simulations in a similar manner.

17.6.3 STABILITY ANALYSIS

17.6.3.1 Introduction

A self-excited oscillation of bogies and running gear with conventional wheelsets starts from a certain speed, called critical speed, as mentioned in Section 17.3.2.1. The frequency of this wheelset and bogie motion is related to wheel-rail contact geometry. If only the wheelsets and bogie or running gear are involved in this movement, this is referred to as bogie instability or bogie hunting. If the wheel-rail contact conditions lead to self-excited oscillation with low frequency, the car body sometimes starts to move together with bogies. This case is called car body instability or car body hunting.

Experimental stability investigations are exacting due to an important influence of non-linear contact geometry and friction conditions between wheel and rail. In the vehicle type tests, the vehicles are checked concerning the fulfilment of the stability up to the specified test speed, but the margin of critical speed at which the vehicle becomes unstable is normally not investigated during type tests. It is therefore seldom possible to prove the margin of stability up to the critical speed and to compare the stability calculation with measurements during vehicle design.

Owing to the wide range of input conditions and limited experimental experience with speed runs over the critical speed, the stability simulations provide the most diversified type of analysis. Methods such as non-linear and linearised calculations can be applied in various versions. The methods can vary depending on whether they are based on the theory of mechanics or on the experience from measurements. The following sections present several feasible methods as they are or may be used for stability analysis in the industrial application. The description is divided into sections concerning linearised analysis by using eigenvalue calculation and non-linear time domain simulations. Section 17.6.3.2 describes the linearisation of the contact between wheelset and track. Following this, bogie stability and car body stability investigations are discussed individually. Furthermore, the effect of the most important design parameters on vehicle's stability is mentioned. In Section 17.6.3.3, the non-linear stability analysis is limited to the investigations of bogie stability. For non-linear simulations regarding car body stability, methods and limits for assessment of ride characteristics and comfort can be used, as described in Section 17.6.4.

17.6.3.2 Linearised Stability Analysis

Despite the non-linearities in contact between wheel and rail, linearised stability calculations can be used to assess stability in the preliminary design stage. However, the contact of wheelset and track has to be linearised differently to the other coupling elements. The parameters of a linear wheel-rail contact model depend on the lateral amplitude of wheelset movement used for linearisation. In parallel, they depend on the shape of the wheel and rail profiles, track gauge and back-to-back distance of wheels. The characteristic parameter is the equivalent conicity (cf. Chapter 7).

The most widely used quasilinear wheel-rail contact model (see Mauer [76] and Wickens [77]) uses three parameters:

- Equivalent conicity λ
- Contact angle (contact slope) parameter ε
- Roll parameter σ

These parameters can be determined by:

- Linearisation of non-linear profiles for specified linearisation amplitude
- Variation of equivalent conicity, setting the other parameters as its function

Although only the first method can fully take into account the characteristics of particular contact geometry, the second method is often used to vary the conicity in a wider range, without requiring sets of wheel and rail profiles to represent those conditions.

Equivalent conicity is widely used in railway practice to describe the properties of the contact geometry between wheel and rail. It is a function of the amplitude of the wheelset lateral displacement; however, the value for wheelset amplitude of 3 mm is usually considered when presenting the value of equivalent conicity. Determination of equivalent conicity is a topic of European Standard EN 15302 [78]. Different methods to calculate the equivalent conicity are compared by Polach in [79].

As the equivalent conicity is the only parameter usually available to characterise the wheel-rail contact geometry, possibilities were evaluated to describe the other linearisation parameters in dependency on the conicity. The contact angle parameter as a function of conicity was investigated by the ERRI (formerly Office de Recherches et d'Essais (ORE)) for common combinations of wheel and rail profiles. The formula derived in the report ORE B 176 [80] for the combination of S 1002 wheel profile with 60E1 rail (formerly UIC 60), inclination 1:40, is the most widely used option for the calculation of the contact angle parameter as a function of conicity:

$$\varepsilon = 85\,\lambda \qquad (17.2)$$

For roll parameter, the following linear function of conicity is often used:

$$\sigma = 0.2\lambda \qquad (17.3)$$

These linear dependencies of contact angle parameter and roll parameter on equivalent conicity are only very rough approximations. Linearisation of different combinations of new as well as worn wheel and rail profiles and track gauge values for a lateral wheelset amplitude of 3 mm provide partly significantly different results compared with these formulas (see [81] for details).

A set of eigenvalue calculations (cf. Section 17.3.2.1) with speed as the vehicle parameter is called the root locus curve. It allows the influence of speed on the stability of the vehicle to be assessed (see Figure 17.29). The eigenmode is unstable if the real part of the eigenvalue is positive. For engineering applications, the speed at which the vehicle achieves 5% of critical damping of the lowest damped mode can be used for setting the critical speed.

A set of root locus calculations for a varying parameter is called a stability map or stability diagram. A typical parameter used as an independent variable is equivalent conicity; see Figure 17.30. In most cases, two areas of low critical speed can be recognised in this type of stability map. For high values of equivalent conicity, the limiting mode is the bogie hunting. In the range of very low conicity,

Simulation of Railway Vehicle Dynamics

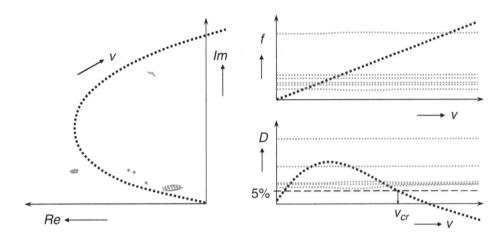

FIGURE 17.29 Root locus curve in complex plane (left), and frequency f and relative damping D as functions of speed v (right). (From Iwnicki, S. (Ed.), *Handbook of Railway Vehicle Dynamics*, CRC Press, Boca Raton, FL, 2006. With permission.)

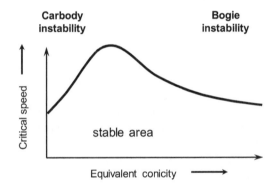

FIGURE 17.30 Example of typical stability map (stability diagram).

the limiting mode is car body instability – a combined movement of car body and bogies [51,77]. The vehicle is unstable for all speeds higher than the critical speed of bogies, whereas the car body instability sometimes disappears with increasing speed. Although the form of the stability diagram can vary within a wide range dependent on the input parameters, there is typically a sector limited by the bogie instability and, in most cases, also a sector limited by low damped car body modes.

As the bogie stability decreases with increasing conicity, the stability should be mainly investigated for the upper range of equivalent conicity anticipated in operation. Since the worst-case scenario of wheel-rail contact geometry is often not known during the design process, the calculations are done for a parameter range based on experience. In EN 14363:2016 [66], the minimum equivalent conicity λ as a function of vehicle maximum speed V in km/h is specified, for which the vehicle should be verified by measurements to run without instability:

- V ≤ 120 km/h: $\lambda \geq 0.40$
- 120 km/h < V ≤ 300 km/h: $\lambda \geq (0.534 - V/900)$
- V > 300 km/h: $\lambda \geq 0.20$

However, it is well known that the real situation varies for different railway companies and that equivalent conicity can reach higher values in service due to wear of wheels and rails as well as variation of track gauge.

TABLE 17.2
Typical Conditions for Linearised Bogie Stability Analysis

Input Parameter	Recommended Value or Conditions
Wheel-rail contact geometry	Variation of equivalent conicity up to the maximum value expected in service, representing various combinations of worn wheel and rail profiles as well as track gauge variation
Wheel-rail creep-force law	Full creep coefficients of Kalker's linear theory (dry rail)
Vehicle state	1. Intact
	2. Failure mode: failure or reduced effect of yaw dampers
Vehicle loading	1. Tare (empty)
	2. Full (crush) load
Vehicle speed	Speed variation from low up to high speed, in minimum up to the maximum test speed (according to EN 14363 [66] with maximum service speed increased by 10% or by 10 km/h for maximum speed below 100 km/h)

The typical conditions for linearised bogie stability analysis are summarised in Table 17.2. As the critical speed reaches the lowest value for dry wheel-rail contact, the linearised analysis should be done for full creep coefficients of Kalker's linear theory. For the lowest damped kinematic bogie mode, a minimum 5% of critical damping is recommended.

Linearised bogie stability computations are a very helpful instrument during concept investigation and design optimisation. The influence of several parameters on the critical speed can be investigated and an optimum range identified. However, for detailed analysis, non-linear calculations have to be used, allowing simulation of real running conditions. A comparison of linearised and non-linear methods of running stability assessment, as they are used in railway engineering, is presented by Polach in [81].

Figure 17.31 shows an example of parameter variation for longitudinal and lateral axle guidance stiffness of a four-axle locomotive. Critical speed increases with an increase of both longitudinal and lateral stiffness of wheelset guidance, whereby the influence of longitudinal axle guidance stiffness plays a more important role. However, during design optimisation, the trade-off between stability and curving requirements has to be solved (see Section 17.6.5).

Wheelset guidance stiffness, together with yaw damping between bogie and car body, is the most important parameter influencing running stability. A proper modelling of wheelset guidance stiffness and of yaw dampers (longitudinal dampers between car body and bogie), including a realistic value of their series stiffness, is thus a necessary condition for a correct stability assessment. Considering traction vehicles, a reliable stability assessment requires mature modelling of traction equipment suspended in the bogie, because the mass properties of traction motor and gear as well as the stiffness and damping of their suspension have significant influence on a vehicle's stability. Alfi et al. [82] demonstrated that lateral stiffness of motor suspension with a resonance near to the limit cycle oscillations and rather low motor damping leads to a critical speed higher than that with the motor fixed to the bogie frame. Another investigation in [83] shows that the optimum natural frequency of traction equipment is significantly affected by the primary and secondary suspensions, as well as by the wheel-rail contact conditions.

At speeds for which the frequency of the bogie oscillations approaches the natural frequency of the vehicle car body on the suspension, the possibility of considerable interaction may arise, leading to vibrations during which the amplitude of the car body is large relative to that of the wheelsets. The bogie movement is coupled with the car body movement, usually by the yaw or lower sway car body modes (or sometimes a combination of both), or also with car body pitching. Kinetic energy is transferred from bogie to car body, so that the damping of the bogie eigenmode increases, whereas the damping of car body decreases, as explained in [51,84] (see Figure 17.32).

Simulation of Railway Vehicle Dynamics

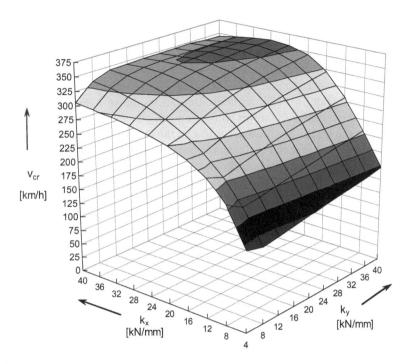

FIGURE 17.31 Example of stability map of a four-axle locomotive – diagram of critical speed v_{cr} (speed at 5% of critical damping) as a function of longitudinal and lateral axle guidance stiffness for equivalent conicity of 0.45. (From Iwnicki, S. (Ed.), *Handbook of Railway Vehicle Dynamics*, CRC Press, Boca Raton, FL, 2006. With permission.)

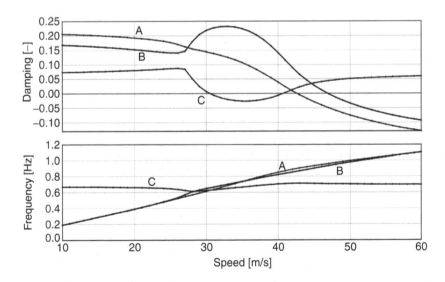

FIGURE 17.32 Example of car body instability (C) at very low conicity – damping of the sinusoidal movement of trailing bogie (B) increases, whereas the damping of car body decreases for increase in speed between 27 m/s and 45 m/s (A and B are sinusoidal movement of leading or trailing bogie, respectively, and C is car body sway movement). (From Mahr, A., *Abstimmung der Rad-Schiene-Geometrie mit dem lateralen Fahrzeugverhalten: Methoden und Empfehlungen*, Thesis, Shaker Verlag GmbH, Aachen, Germany, 2002. With permission.)

For two-axle vehicles or for freight wagons with laterally stiff coupling between bogie and car body, car body instability with a frequency of approximately 2 Hz can lead to a derailment risk, as illustrated by Stichel [85]. For vehicles with secondary lateral suspension, the car body instability does not usually lead to the limits used for assessment of stability in measurements being exceeded, as the sum of lateral forces between wheelset and track is usually smaller than in the case of bogie instability. The car body oscillations cause a deterioration of the lateral comfort behaviour. A significant comfort worsening can occur for lightly damped eigenmodes despite the fact that the instability limit according to the standard [66] is not exceeded. For this reason, it is recommended that the eigendamping of car body modes should reach more than 10% of critical damping for the whole range of speeds expected in service. Owing to the influence on running comfort, the low damped car body oscillations are often investigated from a comfort analysis's standpoint [86]. On occasion, the coincidence of self-excited bogie wave motion with the eigenfrequency of the car body is referred to as resonance. However, as the motion is self-excited by wheelset movement, it constitutes low damping or instability of the car body, as also explained in [51,77]. The car body instability is promoted by low values of equivalent conicity and sometimes also by a low creep coefficient. In service, the unstable or lightly damped car body eigenbehaviour leads to increased oscillations for a certain speed range, mainly on smooth track, whereas, on a bad quality track, the eigenbehaviour is rather disturbed by track irregularity.

The appearance of lightly damped or unstable car body eigenmodes in the presence of low equivalent conicity is dependent on all parameters of the vehicle, so that it is not easy to assess the risk of this phenomenon. The secondary suspension parameters are usually mentioned as being of influence. To avoid car body instability, according to Wickens [77], the lateral stiffness and lateral damping between bogie and car body may not exceed a certain limit. Linearised analyses of a high-speed passenger vehicle presented in [87] lead to the conclusion that the occurrence of car body instability in the presence of low conicity can be promoted by values of secondary lateral stiffness that are too large, values of equivalent stiffness and damping provided by yaw dampers that are too large or values of secondary lateral damping that are either too small or too large. An excessive lateral damping between bogie and car body arising from yaw dampers installed inclined into the car body centre was identified in [88] as the cause of car body hunting of the Korean next-generation high-speed train HEMU-430X.

The axle guidance can also influence the risk of car body instability. According to Wickens [89], vehicles with forced steering are more sensitive to car body instability than conventional vehicles. According to [87], bogies with stiff wheelset guidance promote car body instability.

In articulated vehicles, car body hunting with the yaw mode can result in a wave motion of the entire vehicle, with amplitudes increasing in the rear part. Such a phenomenon can be avoided by introducing damping between car bodies. As a design solution, two parallel longitudinal dampers between the car bodies are used. The vertical position of these inter-car dampers can be at floor level or roof level. Inter-car dampers are also implemented on high-speed vehicles, e.g., on Shinkansen trains in Japan [90], where the car body oscillations are likely to be promoted by aerodynamic phenomena in tunnels.

It is of advantage to investigate the risk of unstable or lightly damped car body oscillations with linearised calculations. Typical conditions and parameters used are shown in Table 17.3. In this manner, a set of speeds and parameter variations can be investigated very quickly, and the coincidence of frequencies can be observed. Chen and Shen [91] developed a numerical method for evaluating the risk of car body hunting by tracing the rigid-body modes of a railway vehicle and assessing the coincidence between the frequency of bogie hunting and the eigenfrequency of car body based on the similarity recognition.

TABLE 17.3
Typical Conditions for Linearised Car Body Stability Analysis

Input Parameter	Recommended Value or Conditions
Wheel-rail contact geometry	Low equivalent conicity (<0.1)
Wheel-rail creep-force law	1. Full creep coefficients of Kalker's linear theory (dry rail)
	2. Reduced creep coefficients of Kalker's linear theory, reduction factor 0.2–0.6 (wet rail)
Vehicle state	Intact
Vehicle loading	Tare (empty)
Vehicle speed	Speed variation between very low speed and maximum service speed

17.6.3.3 Non-Linear Stability Analysis

An important difference of a non-linear dynamic system compared with a linear system is the possible coexistence of multiple solutions in dependency of system parameters. Considering the running stability of railway vehicles, the multiplicity of solutions arises with variation of vehicle speed at so-called bifurcation points, as explained by True in [92,93].

To visualise the multiplicity of solutions, bifurcation diagrams are used usually displaying the amplitude of wheelset lateral oscillation versus speed. The bifurcation diagrams can be complex, including multiple solutions and chaos. Figure 17.33 presents two basic forms of bifurcation diagrams occurring in railway vehicle dynamics. The diagram in Figure 17.33a shows an example of so-called subcritical Hopf bifurcation. The solid thick line on the horizontal axis represents the stationary solution without oscillation. The Hopf bifurcation point $v_{cr\,lin}$ corresponds to the linear critical speed. The branching solution is on the same side of the bifurcation point as the stationary solution, and the new branch is unstable and therefore drawn as a dashed line. The non-linear critical speed $v_{cr\,nl}$, which is defined as the lowest speed for which a self-excited oscillation exists, corresponds to the saddle-node bifurcation point at which the unstable solution branch gains stability. For vehicle speeds between $v_{cr\,nl}$ and $v_{cr\,lin}$, two solutions exist; the vehicle can run either stably or with a limit cycle. The actual solution will depend on the size of initial disturbance or initial condition compared with the unstable saddle cycle, as shown by arrows. The diagram in

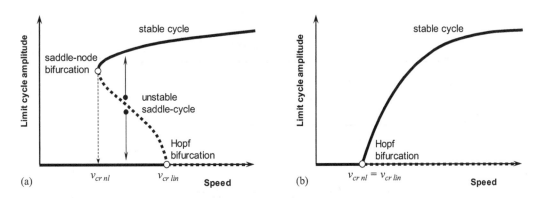

FIGURE 17.33 Amplitude of limit cycles as a function of speed in a bifurcation diagram: (a) subcritical Hopf bifurcation and (b) supercritical Hopf bifurcation.

Figure 17.33b displays an example of a supercritical Hopf bifurcation where the new branch is stable. The non-linear critical speed $v_{cr\,nl}$ and linear critical speed $v_{cr\,lin}$ are identical; the solution is unique for all speeds.

Computation of a complete bifurcation diagram by the path-following method is the topic of several scientific papers (see, e.g., [94,95]). Although it is not the state of the art in multibody simulation tools, the method has been developed by Schupp [96].

Methods commonly used for non-linear stability assessment in industrial applications are presented by Polach [97,98]. They can be classified according to the analysed values and assessment criteria as follows:

- Decay of wheelset displacement (lateral or yaw displacements)
- Quantities used in on-track testing (sum of guiding forces between wheelset and track or lateral acceleration on the bogie frame, respectively) and comparison with stability criteria according to the standard for vehicle acceptance [66]

The first group of methods is based on the theory of non-linear dynamics, and the second is based on the safety criteria regarding the instability in the vehicle acceptance tests. Because of different methods and criteria, results can obviously differ (see [97]).

Stability assessment considering decay of wheelset oscillation is carried out by simulating a run on ideal track (without excitation). The recommended method is to start at a high speed with limit cycle oscillations and then reduce the speed continuously [19,93]; see example in Figure 17.34.

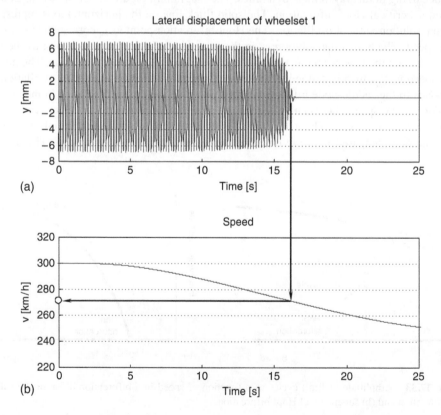

FIGURE 17.34 Non-linear stability analysis by simulation of a run on ideal track with continuously decreasing speed. (a) Lateral wheelset displacement, (b) Vehicle speed (From Iwnicki, S. (Ed.), *Handbook of Railway Vehicle Dynamics*, CRC Press, Boca Raton, FL, 2006. With permission.)

Simulation of Railway Vehicle Dynamics

As an output value, lateral or yaw wheelset movement can be considered. The critical speed is then the speed at which the wheelset oscillation stops. Applying a very small rate of speed decrease, the resultant critical speed will be in a good agreement with the non-linear critical speed identified from a bifurcation diagram.

Another possibility is to start a set of simulations with constant speed and to simulate a reaction after an initial disturbance or after a short track section with irregularities. A similar type of test with excitation by initial disturbance is used for stability investigations on a roller rig. In this case, a single lateral disturbance in the form of a half-cosine with wavelength of 10 m and excitation amplitude of 5–10 mm is used. Examples of such simulations are given in Figures 17.35 and 17.36. In Figure 17.35, there is a decay of wheelset oscillation at a speed of 275 km/h but a limit cycle with large amplitude at speed of 280 km/h. A sudden appearance of large oscillations corresponds to the subcritical bifurcation shown in Figure 17.33a. In Figure 17.36, a limit cycle with a small amplitude far away from the flange contact starts at significantly lower speed and increases slowly with increasing speed. This behaviour corresponds to the supercritical bifurcation (Figure 17.33b). It is important to stress that the simulation of a reaction to an initial disturbance can provide an incorrect stability assessment with an overestimation of non-linear critical speed. As the stable state after the transition depends on the amplitude and wavelength of the excitation, the vehicle can run stably, even though there would be a limit cycle when using another excitation.

In experiments on a roller rig in Munich, it was demonstrated that, during speed increase and decrease, the sinusoidal wheelset movement does not start and stop at the same speed. As explained in [51,99], on a smooth track, the instability commences at the linear critical speed $v_{cr\,lin}$ and stops at the non-linear critical speed $v_{cr\,nl}$. This phenomenon corresponds to the subcritical Hopf bifurcation

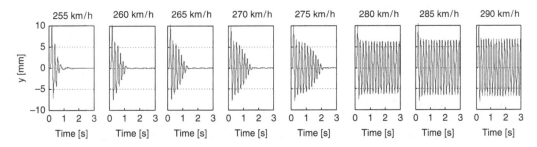

FIGURE 17.35 Example of a multi-simulation to calculate reactions after an initial disturbance – wheelset limit cycles with large amplitude commence suddenly above 275 km/h (vehicle and simulation parameters are identical to the example in Figure 17.34). (From Iwnicki, S. (Ed.), *Handbook of Railway Vehicle Dynamics*, CRC Press, Boca Raton, FL, 2006. With permission.)

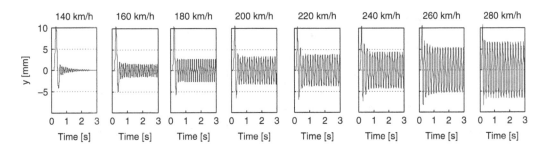

FIGURE 17.36 Example of a multi-simulation to calculate reactions after an initial disturbance – wheelset limit cycles commence for speeds above 140 km/h (the amplitude is initially small, then increases with speed and only reaches large flange to flange movement for speeds higher than 260 km/h). (From Iwnicki, S. (Ed.), *Handbook of Railway Vehicle Dynamics*, CRC Press, Boca Raton, FL, 2006. With permission.)

(Figure 17.33a) with two solutions between the speeds $v_{cr\,nl}$ and $v_{cr\,lin}$. This explains why it is incorrect to evaluate the critical speed by simulation with increasing speed [93].

Investigations to explain the differences in a vehicle's behaviour at the stability limit, when considering different wheel-rail contact geometries with the same equivalent conicity for a wheelset amplitude of 3 mm, showed that the differences observed in Figures 17.35 and 17.36 are mainly related to the shape of the conicity function. Polach therefore proposed in [79] to introduce, in addition to the equivalent conicity, a second parameter called the non-linearity parameter, with the aim of improving the characterisation of the wheelset-track contact geometry. The non-linearity parameter N_P according to [79] is defined as the slope of the equivalent conicity function between the equivalent conicity value λ_2 for a wheelset displacement amplitude of 2 mm and the value λ_4 for a wheelset displacement amplitude of 4 mm (see Figure 17.37):

$$N_P = \frac{\lambda_4 - \lambda_2}{2} \quad [1/\text{mm}] \tag{17.4}$$

Simulation investigations presented in [79,97] showed that the vehicle's behaviour at speeds around the instability limit is usually related to the shape of the equivalent conicity function, which can be characterised by the non-linearity parameter (see Figure 17.38). If the wheelset-track contact geometry possesses a positive N_P, the wheelset limit cycle appears suddenly and the wheelset oscillation achieves large amplitudes, often up to flange contact. In contrast, if the wheelset-track contact geometry results in a negative N_P, the wheelset limit cycle starts at a lower speed, with a small wheelset amplitude far away from flange contact, so that the instability safety criteria are usually not exceeded. The amplitude of wheelset oscillation grows with increasing vehicle speed, whereby the instability safety criteria are exceeded at a speed similar to the speed for other wheelset-track contact geometries with the same equivalent conicity for an amplitude of 3 mm.

Evaluations by Polach and Nicklisch [100] demonstrate that increasing conformity of wheel-rail contact due to wear of wheels and rails results in higher equivalent conicity, together with a decreasing non-linearity parameter N_P, particularly on high-speed lines, with wheel wear predominantly in the tread. Former stability studies used theoretical, new profiles, which usually led to the contact geometry with a positive N_P. Therefore, for a long time, the opinion prevailed that the subcritical bifurcation represents the most frequently occurring case. Recent studies [79,101,102] demonstrate that both subcritical and supercritical bifurcations are present in railway vehicle dynamics.

Simulation on an ideal track with an initial disturbance can be used to identify a rough shape of the bifurcation diagram. Comparisons of this brute-force method with the path-following method on a complex vehicle model in [103] demonstrate that this method provides reliable results regarding engineering applications.

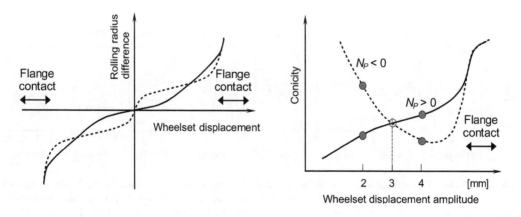

FIGURE 17.37 Definition of non-linearity parameter.

Simulation of Railway Vehicle Dynamics

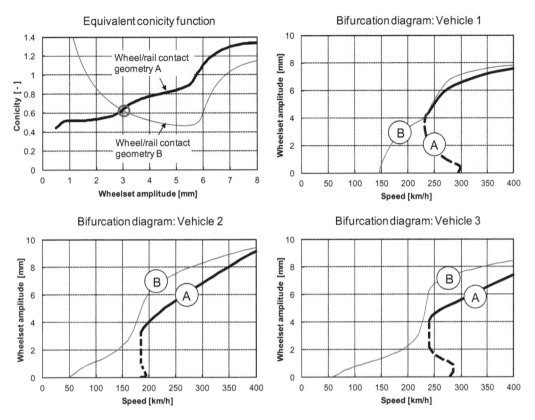

FIGURE 17.38 Comparison of three vehicles showing that the wheel-rail contact non-linearity affects the shape of the bifurcation diagram in the same way regardless of the vehicle type – Vehicle 1: Articulated EMU, Vehicle 2: Double-decker coach without yaw dampers, and Vehicle 3: Double-decker coach with yaw dampers. (From Polach, O., *Vehicle Syst. Dyn.*, 48, 19–36, 2010. With permission.)

Another possibility for stability assessment is simulation of on-track testing. A run on straight track with irregularities is simulated and criteria for vehicle acceptance testing are used for assessment. According to EN 14363 [66], the sum of guiding forces (track shifting force) is used for stability assessment in an on-track test when using force measuring wheelsets. The measured signal has to be filtered with a band-pass filter ($f_0 \pm 2$) Hz, where f_0 is the frequency of unstable bogie oscillations. Thereafter, the rms value is processed as an average over 100 m distance in 10 m increments and compared with the limit value, which is dependent on the static wheel load Q_0:

$$\left(s\Sigma Y\right)_{lim} = \frac{\left(\Sigma Y_{2m}\right)_{lim}}{2} = \frac{1}{2}\left(10 + \frac{2Q_0}{3}\right) \quad [\text{kN}] \tag{17.5}$$

The simplified measuring method according to EN 14363 [66] uses lateral acceleration on the bogie frame. The investigated signal, filtered and processed the same as the sum of guiding forces (band-pass filter, rms value over 100 m distance with steps of 10 m), should be compared with the limit value specified as a function of bogie mass m^+ in tons as:

$$\left(s\ddot{y}^+\right)_{lim} = \frac{1}{2}\left(12 - \frac{m^+}{5}\right) \quad [\text{m/s}^2] \tag{17.6}$$

Lateral acceleration on the bogie frame is traditionally used to check the bogie stability, also without the calculation of rms value. To fulfil this criterion, the amplitude of acceleration must not exceed the limit value of 8 m/s² more than six times in succession.

Figure 17.39 shows an example of stability assessment using simulation results from a run on a straight track with measured irregularities. All the signals mentioned can be used to prove the stability by simulation. However, the above limits can lead to differing results. In the example of simulation runs in Figure 17.39, one of the criteria is exceeded for the speed of 280 km/h, whereas the other criteria are still fulfilled (however with a small margin only). Compared with stability assessment by the simulation of a run on an ideal track, the stability limits mentioned above will be achieved first when the limit cycles with large amplitudes occur.

For the non-linear stability analysis, the worst wheel-rail contact conditions have to be used. The sole application of nominal parameters and nominal wheel and rail profiles is definitely insufficient. Wear of the wheels usually leads to a reduced curve radius of the wheel tread, top-of-rail wear on straight track leads to reduced curve radius of the rail corner and increased radius of the rail head crown. Both wear phenomena lead to increased conicity. The equivalent conicity also increases if the track gauge is reduced compared with the nominal value. To cover the worst-case conditions, the stability investigations should be carried out for worn shapes of wheel and rail profiles and possibly

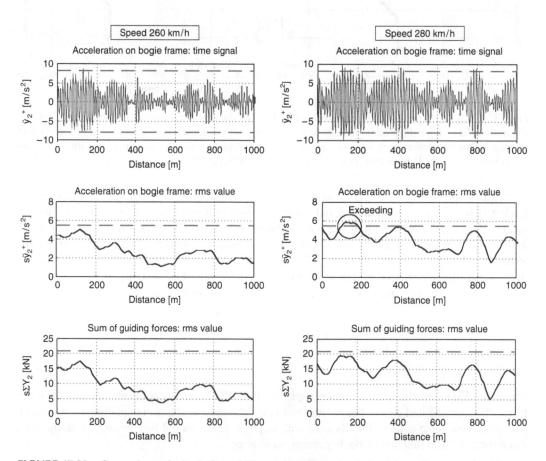

FIGURE 17.39 Comparison of criteria for stability analysis from simulation of running on track with measured irregularities – dashed lines indicate limit values (vehicle and simulation parameters identical to those in Figures 17.34 and 17.35). (From Iwnicki, S. (Ed.), *Handbook of Railway Vehicle Dynamics*, CRC Press, Boca Raton, FL, 2006. With permission.)

TABLE 17.4
Typical Conditions for Non-Linear Bogie Stability Analysis

Input Parameter	Recommended Value or Conditions
Track design	Straight track
Track irregularity	1. Ideal track (without excitation) after an initial lateral disturbance
	2. Measured track irregularities
Wheel-rail contact geometry	Variation of wheel and rail profiles, including worn wheel and rail profiles and gauge variation, representing conditions expected in service
Wheel-rail creep-force law	Non-linear theory, friction coefficient 0.4–0.5 (dry rail)
Vehicle state	1. Intact
	2. Failure mode: failure or reduced effect of components used to increase stability; usually yaw dampers
Vehicle loading	1. Tare (empty)
	2. Full (crush) load
Vehicle speed	1. Speed variation up to the critical speed
	2. Maximum test speed (according to EN 14363 [66] maximum service speed increased by 10% or by 10 km/h for maximum speed below 100 km/h)

also for narrowed gauge. Another important parameter is the friction coefficient between wheel and rail. As the critical speed decreases with increasing friction coefficient, dry conditions should be used for analysis. The friction coefficient applied is usually between 0.4 and 0.5. Typical conditions and parameters used for non-linear bogie stability analysis are summarised in Table 17.4.

17.6.4 Running on Track with Irregularities

17.6.4.1 Definition of Running Behaviour, Ride Characteristics and Comfort

Simulation of running on track with measured irregularities allows the prediction of the vehicle running behaviour, which can be anticipated during the type test. According to EN 14363 [66], the running behaviour constitutes the characteristics of the vehicle or running gear with regard to interaction between vehicle and track. The assessment of running behaviour covers the following parts:

- Running safety
- Track loading
- Ride characteristics

For the running safety on straight track, the sum of the guiding forces provides a criterion. However, the limit value for this criterion is normally fulfilled if the bogie and vehicle run stably. This criterion together with the simulations related to stability were explained in Section 17.6.3.3.

For the track loading, the term 'track fatigue' is also sometimes used, because the limit values are related to the fatigue of the track components. Investigation of track loading is important for the assessment of running on curved tracks and will be presented in Section 17.6.5.

The ride characteristics criterion provides an assessment of the dynamic behaviour of the vehicle by analysing the accelerations at the vehicle body, whereas the ride comfort criterion assesses the influence of vehicle dynamic behaviour on the human body. Although both criteria use acceleration signals, the analysis and limit or target values differ. Simulations related to the analysis of ride characteristics and ride comfort are carried out predominantly on straight track, as explained in the following sections. However, curves and transitions generally also have to be considered in order to take the same conditions into account as during measurements.

17.6.4.2 Ride Characteristics

To simulate a run of a vehicle on track, measured irregularities are usually applied. An overview concerning definition and properties of measured track irregularities can be found in several references, [104] for example. Another possibility is the use of synthesised irregularities with specified spectral density. Track irregularity data often used in European countries are the 'low level' and 'high level' spectral density functions, described by ORE in [80].

The typical conditions for simulations related to ride characteristics, as well as ride comfort, are described in Table 17.5. It is understandable that track irregularity has a significant influence on the ride characteristics. However, it is not easy to make a definite assessment of the track quality, as the same vehicle can demonstrate different ride characteristics on differing tracks. A change from one track to another does not always demonstrate a clear tendency with an overall improvement or deterioration of the vibration behaviour. In fact, the occurrence of opposing trends at various points in the vehicle is possible.

The ride characteristics are also influenced by the parameters of the wheel-rail contact. The best ride characteristics are usually achieved at medium conicities, i.e., 0.10–0.20. In the case of detailed investigations concerning ride comfort, the wheel-rail contact may play an important role, as detailed in Section 17.6.4.3.

Figure 17.40 shows an example of ride characteristics simulation for a four-axle locomotive with a maximum service speed of 140 km/h running on synthetic 'low level' irregularities, according to ORE B176 [80], at a speed of 154 km/h. Several variants of non-linear wheel-rail contact geometries were applied in the simulations. A wide range of equivalent conicity variation was achieved by applying a new wheel profile together with either variation of track gauge or variation of profile and inclination of rails, respectively. Vertical and lateral accelerations in the driver's cab filtered with a band-pass filter of 0.4–10 Hz are presented in Figure 17.40 as a function of equivalent conicity evaluated for the lateral wheelset amplitude of 3 mm.

In the case of vehicles with air springs, running on emergency suspension should also be simulated, as required during acceptance tests, in addition to the intact conditions (see example in Figure 17.41).

TABLE 17.5
Typical Conditions for Simulation of Ride Characteristics and Comfort

Input Parameter	Recommended Value or Conditions
Track design	Straight track, curves with typical radius; optional comfort analysis in curve transitions (tilting trains)
Track irregularity	According to the specification and conditions on the railway network; measured track irregularity if possible
Wheel-rail contact geometry	Nominal profiles of wheel and rail, nominal gauge; sensitivity analysis of gauge narrowing and widening, analysis of influence of worn wheel and rail profiles
Wheel-rail creep-force law	Non-linear theory, friction coefficient 0.4 (dry rail); analysis of influence of reduced friction coefficient 0.1–0.3 (wet rail)
Vehicle state	1. Intact 2. Failure mode: air springs deflated (for ride characteristics only)
Vehicle loading	Tare (empty)
Vehicle speed for ride characteristics analysis	Maximum test speed (according to EN 14363 [66] maximum service speed increased by 10% or by 10 km/h for maximum speed below 100 km/h)
Vehicle speed for ride comfort analysis	1. Maximum service speed 2. Speed of car body pitch and bounce resonance, see Section 17.6.4.3

Simulation of Railway Vehicle Dynamics

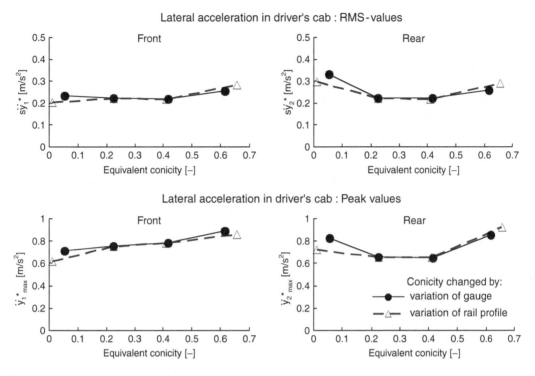

FIGURE 17.40 Ride characteristics of a four-axle locomotive as a result of simulations with non-linear wheel-rail profiles and presented as a function of equivalent conicity. (From Iwnicki, S. (Ed.), *Handbook of Railway Vehicle Dynamics*, CRC Press, Boca Raton, FL, 2006. With permission.)

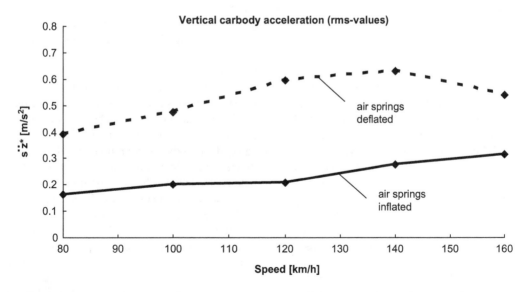

FIGURE 17.41 Influence of vehicle speed on vertical ride characteristics of a four-axle motor coach for intact and deflated air springs. (From Iwnicki, S. (Ed.), *Handbook of Railway Vehicle Dynamics*, CRC Press, Boca Raton, FL, 2006. With permission.)

Loading of the vehicle usually leads to lower natural frequencies and consequently to an improvement in ride comfort in the vertical direction. Vehicles will therefore be mainly examined without loading (tare).

The vibrations grow with increasing running speed, which in turn results in a deterioration of the ride characteristics. The ride characteristics are assessed during the acceptance test at the maximum test speed, and it is thus investigated at this speed during the engineering process. A negative influence on the ride characteristics will be caused through a resonance by harmonic components of the excitation with natural frequencies of the vehicle (see also Section 17.6.4.3), so that the maximum test speed may not necessarily provide the worst ride characteristics values. Figure 17.41 shows a comparison of the ride characteristic of a railway vehicle in intact condition (with air springs inflated) and in the failure mode with deflated air springs. Whereas the acceleration values increase over the whole range concurrent to the speed in the case of the vehicle with air springs inflated, a maximum occurs in the resonance frequency range of the emergency suspension during a run with air springs deflated, followed by a slight reduction in the values.

17.6.4.3 Ride Comfort

Ride comfort characterises passenger well-being in relation to mechanical vibrations, thereby taking the physiological characteristics of the human body into consideration. The manner in which the human body experiences ride comfort differs depending on the frequency and amplitude of the vibrations. In order to take this influence into account, frequency weighting filters are applied. The filters vary for the vertical and lateral directions and also differ depending on the standard applied.

The most widely used ride comfort assessment methods are the comfort index (N-value) according to EN 12299 [105] and the rms method specified in ISO 2631-1 [2], but other standards and evaluation methods are also used. In the simplified method according to EN 12299, accelerations in vertical, lateral and longitudinal directions are applied on the floor of the car body. Another option is a measurement on the seat and seat back in conjunction with the human body. The evaluation is based on signals measured in 5-second sequences and analysed for 60 blocks. The comfort index N is calculated as a 95% value from the statistical histogram of accelerations in all three directions. In the assessment according to ISO 2631-1 [2], the rms values are evaluated separately in the vertical and lateral directions.

The influence on ride comfort is similar to that on the running characteristics. The simulation conditions are therefore the same as for simulation of ride characteristics, as can be seen in Table 17.5. However, as the specification regarding ride comfort is usually stricter than the standards' requirement for ride characteristics, the simulation investigations and prognoses of the ride comfort are more demanding than investigations of the ride characteristics.

In contrast to other calculations, a very important role in the comfort calculations is played by an appropriate modelling of the car body (cf. Section 17.2.2). The elastic car body structure, the connection between the car body and the bogie and the distribution and suspension of the apparatus play a decisive role. Whilst the influence of the flexible car body structure above the bogies is relatively small, the vibrations at the vehicle centre and at the vehicle ends are strongly influenced by the flexibility of the structure. In a vehicle model with a rigid car body, the vibration behaviour in the vehicle centre is typically better than that above the bogies. If the flexibility of the car body structure is taken into consideration, the behaviour at the vehicle centre achieves higher vibration values, and these are usually higher than the values above the bogies, particularly in the case of lightweight vehicles possessing low car body structural stiffness. In order to achieve realistic simulation results, it is therefore necessary to model the car bodies as elastic structures (see Section 17.2.2 and [106,107]).

Vibration comfort is strongly influenced by the position in the vehicle. Above the bogies, the low frequency vibration forms of the car body (e.g., bouncing, pitching and swaying) prevail. In the centre of the car body, a frequency of approximately 8–12 Hz typically dominates, which indicates the eigenmode of vertical car body bending. Specifications relevant for comfort tests usually indicate that the limit values or target values are to be fulfilled at all positions in the vehicle. This signifies that, by way of simulations, the most unfavourable position is to be located. Practically, the ride comfort is examined

Simulation of Railway Vehicle Dynamics

at nine locations in the car body: above front bogie, transversally to the left, in the centre and to the right in the running direction; longitudinally in the centre of the vehicle, transversally to the left, in the centre and to the right; above rear bogie, transversally to the left, in the centre and to the right.

Some unfavourable conditions that are particularly detrimental for comfort are as follows:

- Low damping or even instability of the car body mode, initiated through the coupling of the self-excited wave bogie movement with an eigenmode of the car body, as described in Section 17.6.3.2
- Resonance of a vehicle component eigenmode with a periodically excited mode

To study the influence of vehicle speed, the excitation by track unevenness must be considered. The track irregularity can be broken down into the harmonic components with the aid of an FFT. Of these, the wavelengths possess special significance for the ride comfort, which are in a particular relationship to the bogie pivot pin distance p (see also [86,106]). As can be seen in Figure 17.42, bouncing, pitching or bending of the car body is excited at a particular speed by some wavelengths. These wavelengths can be described as being 'critical wavelengths' for the speed in question. In the case of car body bouncing and the first bending mode, those critical wavelengths l_m are:

$$l_m = \frac{p}{m} \qquad m = 1, 2, 3, \ldots \tag{17.7}$$

The resonance occurs when, at a particular speed v_m, the car body is excited by the critical wavelength of a car body eigenfrequency f_i.

$$\frac{v_m}{l_m} = f_i \qquad m = 1, 2, 3, \ldots \tag{17.8}$$

The critical resonance speeds for bouncing and for the first bending form of the car body can be derived from Equations 17.7 and 17.8 as:

$$v_m = \frac{p}{m} f_i \qquad m = 1, 2, 3, \ldots \tag{17.9}$$

In a similar manner, the resonance speeds for pitching or the second bending form of the car body are obtained (see Figure 17.42) as:

$$v_n = \frac{p}{n - \frac{1}{2}} f_i \qquad n = 1, 2, 3, \ldots \tag{17.10}$$

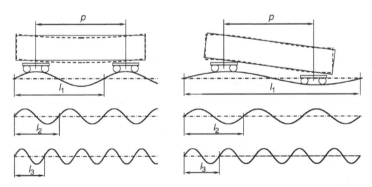

FIGURE 17.42 Wavelengths exciting vertical car body modes: bouncing and first bending mode (left), and pitching and second bending mode (right). (From Iwnicki, S. (Ed.), *Handbook of Railway Vehicle Dynamics*, CRC Press, Boca Raton, FL, 2006. With permission.)

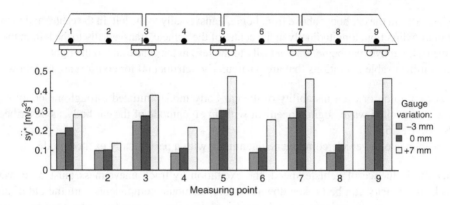

FIGURE 17.43 Lateral accelerations on the floor of the car bodies of an articulated commuter train at a speed of 160 km/h (rms values filtered according to ISO 2631). (From Iwnicki, S. (Ed.), *Handbook of Railway Vehicle Dynamics*, CRC Press, Boca Raton, FL, 2006. With permission.)

The resonance speeds for the bouncing and pitching movement of the car body as a rigid body are low, which is why the influence on ride comfort is not very large. In particular, the first bending mode of the car body structure is liable to strong excitation. While the first resonance speed for the bending of the car body usually lies above the maximum speed, the second or third resonance speeds are often critical and should be examined with the flexible car body. The resonance with car body pitching movement leads to high accelerations during emergency running on deflated air springs in the speed range of 80–140 km/h, which sometimes necessitates a speed limitation.

The parameters of the wheel-rail combination influence the comfort results, whereby these influences are strongly dependent on the vehicle design. At low conicity, the self-excited low-damped eigenmode of the car body can occur, as described in Section 17.6.3.2. If the vehicle tends to such vibrations in a specific speed range, the situation will worsen in the case of widening of track gauge (decrease of conicity) and improve in the case of gauge narrowing (increase of conicity); see the example of an articulated commuter train in Figure 17.43. Higher conicities and the allied increasing tendency of the bogie to hunting lead to forces and moments acting on the car body structure. A corresponding eigenmode of the car body structure will be excited and may lead to poor ride comfort. Ofierzynski and Brundisch [86] mention an example where an increase of the equivalent conicity from 0.20 to 0.55 leads to a tripling of the calculated rms values at the sides in the middle of the car body, whilst the influence of the conicity in the central longitudinal axis remains slight.

17.6.5 Curving

17.6.5.1 Assessment of Curving Properties

During curving, the vehicle is guided by the forces between wheel and rail in the lateral direction. These guiding forces can reach very high values, particularly when heavy vehicles run through narrow curves. For example, Figure 17.44 shows lateral forces between wheel and rail at a locomotive bogie when running with uncompensated lateral acceleration of 1.1 m/s² through a curve with radius of 300 m. The vehicle is guided into the curve direction mainly by the lateral force on the flange of the outer leading wheel on each bogie. An opposite, smaller lateral force acts on the inner wheel of the leading wheelset. The lateral forces of the trailing wheels are relatively small and can have the same or the opposite direction to each other. This is a typical force distribution for a bogie with stiff wheelset guidance running in a curve with small radius. Considering an ideal track curvature without irregularities, the simulation results are in accordance with Figure 17.44. The steady-state

Simulation of Railway Vehicle Dynamics

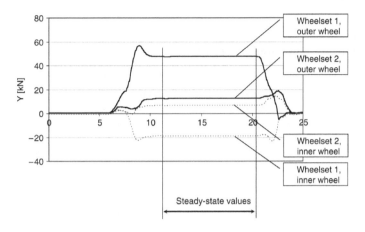

FIGURE 17.44 Curving simulation: lateral forces on leading bogie of a four-axle locomotive running on an ideal track through a curve with 300 m radius. (From Iwnicki, S. (Ed.), *Handbook of Railway Vehicle Dynamics*, CRC Press, Boca Raton, FL, 2006. With permission.)

curving can be solved in an efficient way by a special routine for quasistatic analysis, if available in the particular simulation package. Another possibility is usage of time integration, simulating a run on an ideal track without irregularities from a straight through a curve transition and into a curve with constant radius. However, the steady-state values can also be calculated from a simulation with measured irregularities, calculating 50%-values from the dynamic forces in the same way as they are evaluated in measurements. Taking track irregularities into account, the forces are superimposed by dynamic effects, as can be seen in Figure 17.45.

The following presentation of the curving analyses is oriented on the criteria according to standards where the limits for both dynamics as well as steady-state values are specified. The typical conditions used for simulations of curving are shown in Table 17.6.

The curve radii selected for simulation runs should range from the smallest radius present in the network, or used during the testing, to the largest curve radius through which the vehicle can run

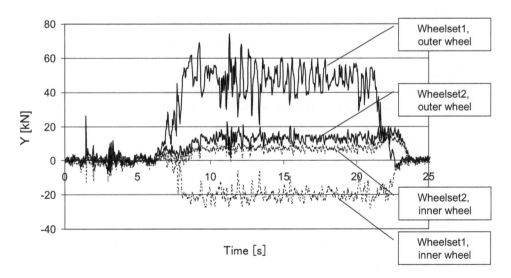

FIGURE 17.45 Curving simulation: lateral forces on leading bogie of a four-axle locomotive running on track with measured irregularities through a curve with 300 m radius. (From Iwnicki, S. (Ed.), *Handbook of Railway Vehicle Dynamics*, CRC Press, Boca Raton, FL, 2006. With permission.)

TABLE 17.6
Typical Conditions for Simulations of Curving

Input Parameter	Recommended Value or Conditions
Track design	1. Typical curve radii including transitions
	2. The smallest curve radius on the network (outside of depot area)
	3. The largest curve radius which vehicle can run with its maximum speed and maximum cant deficiency
Track irregularity	According to the specification and conditions on the railway network; measured track irregularity if possible
Wheel-rail contact geometry	Nominal wheel and rail profiles, nominal track gauge, gauge widening in tight curves according to the specification; influence analysis of worn wheel and rail profiles
Wheel-rail creep-force law	Non-linear theory, friction coefficient 0.4 (dry rail)
Vehicle state	Intact
Vehicle loading	1. Full (crush) load
	2. Tare (empty); relevant for derailment safety investigation
Vehicle speed	Speed variation in function of curve radius and cant deficiency

with its maximum speed and maximum cant deficiency (maximum uncompensated lateral acceleration). The highest values of safety criteria and track loading are usually achieved at the maximum uncompensated lateral acceleration, but low uncompensated lateral acceleration may lead to higher risk of derailment. It is thus recommended to vary the lateral acceleration between its maximum allowed value and zero. The vehicle speed v can be determined dependent on the curve radius R, uncompensated lateral acceleration a_{lat} and track superelevation h_t (expressed as the height difference between left and right rail) according to the formula:

$$v = \sqrt{R \cdot \left(a_{lat} + g \frac{h_t}{2e_0}\right)} \quad (17.11)$$

where e_0 is half the tread datum (tape circle) distance (for normal track gauge, $e_0 = 0.75$ m), and g is the gravitational constant.

Since the wheel-rail forces increase with increasing friction coefficient between wheel and rail, dry conditions of wheel-rail contact are used for simulations. Depending on the bogie and vehicle design, the results can be sensitive to the wheel-rail contact geometry. To test the 'worst case' conditions with regard to wheel-rail contact, worn wheel profiles as well as worn rail profiles from curves should be used. For investigations into track loading, a fully loaded vehicle should be considered, whereas an empty vehicle is the most critical case for investigations of derailment safety.

17.6.5.2 Running Safety

To assess running safety in curves, the following criteria are used:

- Sum of guiding forces (track shifting force)
- Quotient of guiding force and vertical wheel force Y/Q (L/V), which is covered in detail in Chapter 11

For the sum of guiding forces, the limit used in EN 14363 [66] for an average value over 2 m distance is the value according to Prud'homme:

$$\sum Y_{2m} = \alpha \left(10 + \frac{2Q_0}{3}\right) \quad [\text{kN}] \quad (17.12)$$

Simulation of Railway Vehicle Dynamics

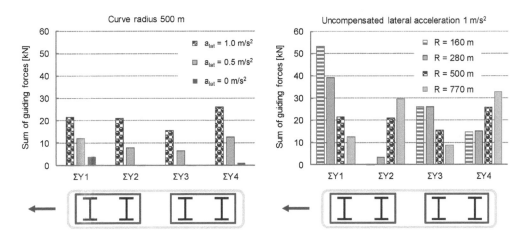

FIGURE 17.46 Distribution of the sum of quasistatic guiding forces on a four-axle locomotive in circular curves.

with α taking account of greater variations in geometrical dimensions and of the state of vehicle maintenance ($\alpha = 0.85$ for freight wagons and 1.0 for other vehicles). The criterion is valid on straight track as well as in curves. However, in curves, the dynamic value excited by irregularities is superimposed by the steady-state value, which makes this limit more critical, mainly when running at high speed through curves with a large radius. To reduce the sum of guiding forces in curves, both steady-state as well as the dynamic values of the track shifting force have to be as low as possible. The sum of all steady-state track shifting forces acting on the vehicle is defined by the product of vehicle mass and lateral acceleration, which is a function of curve radius and cant deficiency. Figure 17.46 shows a distribution of the sum of quasistatic guiding forces at a four-axle locomotive in a curve. The values of the sum of guiding forces grow with increasing uncompensated lateral acceleration, as demonstrated in the diagram on the left in Figure 17.46 giving an example of a curve radius of 500 m. The diagram on the right of this figure shows the effect of curve radius when running with uncompensated lateral acceleration of 1 m/s². In tight curves, the highest value of the sum of guiding forces occurs on the leading wheelset, while the highest value in large radius curves is achieved on the trailing wheelset. Figure 17.47 gives an example of simulated steady-state and dynamic track shifting forces of a regional vehicle running through a curve with a lateral acceleration of 1.1 m/s². One can see that the values are high for a vehicle with maximum (crush) load and for curves with a large radius.

Besides the vehicle design, curve radius and cant deficiency, the distribution of the track shifting forces is dependent on the friction coefficient and contact geometry between wheel and rail. The criterion is thus challenging mainly for tilting trains running with large cant deficiency, as possible optimisation is limited to an equal distribution of the forces on wheelsets.

17.6.5.3 Track Loading and Wear

For track loading, the individual forces between wheel and rail in vertical and lateral directions are assessed: lateral guiding force Y and vertical wheel force Q. Their limit values refer to both the steady-state and dynamic values. With regard to the dynamic forces between wheel and rail, the limit value is only given for the vertical force in EN 14363 [66] and only considering frequencies up to 20 Hz. For the steady-state forces, the following limit values apply: guiding force $Y \leq 60$ kN and vertical wheel-rail force $Q \leq 145$ kN. The given limit values refer to railway tracks with a maximum allowable load of 22.5 t per wheelset and take account of rails with a weight of 46 kg/m and the minimum value of rail tensile strength of 700 N/mm². For tracks designed for higher axle loads and greater load capacity, higher values may also be accepted.

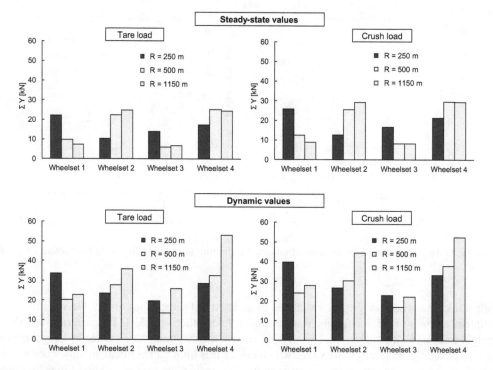

FIGURE 17.47 Distribution of the sum of guiding forces on a four-axle motor coach running through a curve with uncompensated lateral acceleration of 1.1 m/s².

The vertical wheel force Q achieves high values mainly on outer wheels of heavy vehicles, such as locomotives, when running with large uncompensated lateral acceleration. The risk of exceeding the limit value of the dynamic vertical wheel force occurs in curves with a large radius, in which the maximum vehicle speed, and, at the same time, the largest uncompensated lateral acceleration is exploited. For vehicles with a high car body centre of gravity and soft lateral secondary suspension, a large uncompensated lateral acceleration results in a large lateral displacement of the car body centre of gravity, which increases the transfer of load on the outer wheels. Consequently, high values of vertical wheel forces can be expected for heavy vehicles with a high centre of gravity, such as fully loaded double-decker coaches.

The guiding force Y achieves high values, particularly in small curve radii, during dry weather and with heavy vehicles such as locomotives. Figure 17.48 gives an example, with a four-axle locomotive, of the steady-state guiding force Y as a function of curve radius and uncompensated lateral acceleration. The guiding force increases significantly concurrent with the decreasing curve radius. Increasing uncompensated lateral acceleration leads to a slight increase of the guiding force. In Figure 17.48, two bogie versions are compared:

- Conventional bogie design with longitudinal stiff axle guidance
- Self-steering bogie with longitudinal very soft axle guidance and cross-coupled wheelsets

The self-steering bogie demonstrates a good efficiency down to a small curve radius. The design developed for a standard gauge locomotive significantly reduces the guiding force in large, medium and small curve radii. In very small curve radii below approximately 250 m, the creep forces cannot overcome the restoring moment of the axle guidance, and the guiding force is practically the same for both versions.

Simulation of Railway Vehicle Dynamics

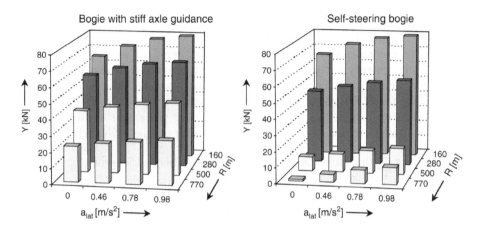

FIGURE 17.48 Guiding force Y of the outer leading wheel in a curve as a function of lateral acceleration and curve radius, with a four-axle locomotive as the example – comparison of stiff and self-steering bogie designs. (From Iwnicki, S. (Ed.), *Handbook of Railway Vehicle Dynamics*, CRC Press, Boca Raton, FL, 2006. With permission.)

Wheel wear and rail wear in curves represent an important aspect with respect to vehicle and track maintenance. In order to enable a simplified wear assessment, the friction work A_R affected over 1 m of distance (the so-called wear index) is calculated as the sum of products from creep forces and creepages as:

$$A_R = |F_x s_x| + |F_y s_y| + |\Theta \omega_s| \tag{17.13}$$

The last term mentioned, caused by spin ω_s and moment Θ around the normal to the contact surface, is very small during wheel tread contact and can usually be neglected. Most subject to wear is the wheel flange surface of the outer leading wheel. Wear is extremely dependent on the curve radius and increases significantly in very small curve radii (see Figure 17.49).

The life cycle of the wheels is not only influenced by the total friction work but more significantly by the wear distribution over the lateral wheel profile. The alteration of the wheel profile can be simulated by determining the material removal in the individual profile sequences based

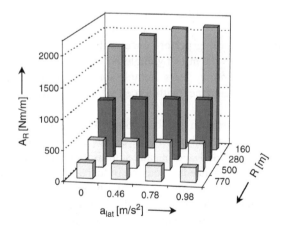

FIGURE 17.49 Wear index of the outer leading wheel in a curve as a function of lateral acceleration and curve radius, using a four-axle locomotive with stiff axle guidance as the example. (From Iwnicki, S. (Ed.), *Handbook of Railway Vehicle Dynamics*, CRC Press, Boca Raton, FL, 2006. With permission.)

on a wear theory. As the wear depends on the momentary form of the wheel and rail profile, a new simulation must be carried out with the newly determined worn wheel profile. In order to make a wear development prognosis, the simulations and analyses of profile form must be repeated several times in cycle, whereby a sequence of statistically representative running conditions such as vehicle loading, running speed, curve radius, track profile, wheel-rail friction coefficient, tractive effort and braking forces must be taken into consideration. The procedure and examples of such extensive investigations can be seen in [18,108–111]. Similar procedures can also be applied for simulation investigations regarding the alteration due to wear of rail profiles [112,113].

17.6.5.4 Curving Optimisation by Using Bogies with Wheelset Steering

During the design of wheelset guidance, contradictory objectives arise of improving the curving performance while achieving running stability at high speed. This trade-off can be solved using different advanced axle guidance solutions. One possibility is to exploit the self-steering ability of wheelsets. This, however, requires a soft wheelset guidance stiffness regarding the wheelsets' bending mode, i.e., an out-of-phase rotation of wheelsets around the vertical axis to ensure a radial wheelset position in curves. Conversely, a stable run at a high speed requires stiff wheelset guidance having regard to the bogie hunting mode.

A recently used solution of these contradictory requirements on wheelset guidance properties is a frequency-dependent guidance, as realised, for example, in the HALL axle guide bush [114,115]. The hydraulics integrated in the bush lead to frequency-dependent changes of the longitudinal guidance stiffness: a low guidance stiffness of only 2.5–7.5 kN/mm is available during steady adjusting movements when entering or exiting a curve, while it increases by 3–10 times at higher speeds during dynamic excitation with frequencies above 1.5–3 Hz.

A design solution that has been introduced in several vehicle types with the aim of solving the trade-off between stability and curving requirements is a combination of soft wheelset guidance with cross-coupling of wheelsets. Independent of the form of the wheelset guidance and suspension design in the horizontal plane, the stability and curving performance can be described by two axle guidance stiffness parameters: the shear stiffness and the bending stiffness. To improve the stability properties without increasing the bending stiffness, the wheelsets have to be restrained by an increase in shear stiffness. For conventional bogies, a limit exists to the shear stiffness that can be provided in relation to the bending stiffness. This can be improved if the wheelsets are connected to each other directly or by a mechanism fitted on the bogie frame, as shown by Scheffel [116] in his review of design options and realised examples. Suda et al. [117] present the history of research on steering bogies and running gears together with an innovative solution of an asymmetrical steering bogie with soft longitudinal guidance stiffness on the leading wheelset and a stiff one on the trailing wheelset, as used in Japan.

Another way of improving curving performance is the so-called forced steering of bogies or wheelsets by a car body yaw movement or by angle between the car bodies [116]. A design solution with coupled single-axle running gears suitable for articulated vehicles [118] combines the self-steering ability of soft axle guidance with forced steering by the car body through secondary suspension.

The forced steering of wheelsets can also be realised as an active system, with actuators steering a softly guided wheelset into radial position in curves as well as stabilising it at high speeds. A development of such an active radial steering and stabilisation (ARS) is described in [119], and its effect on the infrastructure maintenance is discussed in [61].

Next, let's take a closer look at the influence of the bogie design with self-steering wheelsets on the performance of a railway vehicle. The stability and curving characteristics of a four-axle locomotive with four different versions of a modular axle guidance system were evaluated in [120]:

- Longitudinal stiff axle guidance (ST)
- Longitudinal soft axle guidance (SO)

- Longitudinal very soft axle guidance combined with wheelset coupling shaft (CW)
- Longitudinal very soft axle guidance combined with wheelset coupling shaft and dampers of coupling shaft (CWD)

The proposed versions were compared based on the equivalent longitudinal axle guidance stiffness k_{xB} for the bending (steering) mode of the wheelsets. The advantage of the soft axle guidance combined with cross-coupling of the wheelsets becomes apparent during curve negotiation, as can be seen in Figure 17.50a. Both the guiding force of the leading wheel and the wear index decrease with reducing stiffness k_{xB}. The stability declines with lessening stiffness of the axle guidance, but a stability comparable with the stiff axle guidance version can be achieved in the case of soft axle guidance with interconnected wheelsets and damping of the coupling shaft; see Figure 17.50b.

As the radial adjustment of the self-steering wheelsets in curves is achieved through creep forces in the contact between wheel and rail, the running characteristics are influenced by the conditions in the wheel-rail contact. The influence of wheel-rail contact geometry, tractive effort and flange lubrication on the self-steering ability of a high-power locomotive was analysed in [120] and a comparison made between:

- Self-steering bogie with longitudinal very soft axle guidance and cross-coupling of the wheelsets by coupling shaft (CWD)
- Conventional bogie design with longitudinal stiff axle guidance (ST)

A full non-linear locomotive model has been used to simulate runs for curve radii $R = 300$ and 500 m with an uncompensated lateral acceleration $a_{lat} = 0.98$ m/s^2. In order to assess the curving performance, values on the outer wheel of the leading wheelset are presented for steady-state guiding force Y and wear index A_R.

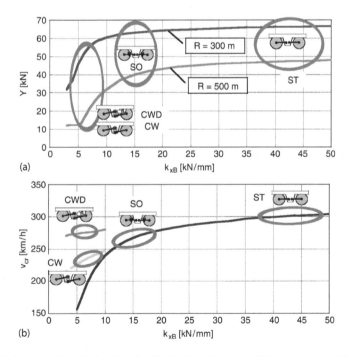

FIGURE 17.50 Influence of equivalent longitudinal axle guidance stiffness on: (a) guiding force in curves and (b) critical speed – marked areas display recommended equivalent stiffness for the proposed versions of modular axle guidance. (From Polach, O., Curving and stability optimisation of locomotive bogies using interconnected wheelsets, in Abe, M. (Ed.), *Proceedings 18th IAVSD Symposium held in Atsugi, Kanagawa, Japan*, Taylor & Francis, London, UK, 53–62, 2004. With permission.)

The lateral rail profile and rail inclination has an influence on the self-steering ability of the wheelsets. Figure 17.51 demonstrates the guiding forces between the S 1002 wheel profile and the 60E1 rail (formerly UIC 60). The guiding force is lower in the case of a rail inclination of 1:40 because the S 1002 wheel profile is optimised for this rail inclination. In order to estimate the influence of rail wear on self-steering, calculations were carried out on rail profiles that were identified from measurements as characteristic of worn rail profiles in curves and used for calculation of wheel wear by Jendel [110] (see Figure 17.52). The heavily worn outer rail and small rolling radii differences lead to a reduction in the self-steering ability (see Figure 17.53). Even though the

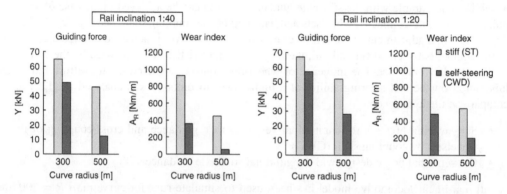

FIGURE 17.51 Influence of rail inclination on guiding force and wear index. (From Iwnicki, S. (Ed.), *Handbook of Railway Vehicle Dynamics*, CRC Press, Boca Raton, FL, 2006. With permission.)

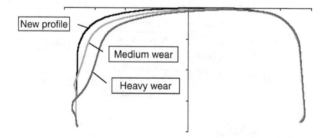

FIGURE 17.52 Worn rail profiles of high rail in curve, as used in sensitivity analysis. (From Iwnicki, S. (Ed.), *Handbook of Railway Vehicle Dynamics*, CRC Press, Boca Raton, FL, 2006. With permission.)

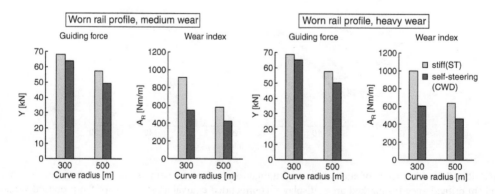

FIGURE 17.53 Influence of medium and heavy worn rail profiles in Figure 17.52 on the guiding force and wear index – results for new rail profile are shown in the left diagram of Figure 17.51. (From Iwnicki, S. (Ed.), *Handbook of Railway Vehicle Dynamics*, CRC Press, Boca Raton, FL, 2006. With permission.)

effectiveness of the self-steering is significantly lower, the self-steering bogie is still more favourable, particularly with regard to wear.

Figure 17.54 illustrates the influence of tractive effort on the examined values Y and A_R. With increasing tractive effort, the creep forces between wheel and rail reach saturation point. The longitudinal creep forces incurred by the varying rolling radius difference reach lower values, and self-steering is reduced. Wear at full tractive force is mainly caused by tractive creep and is therefore hardly influenced by self-steering.

The radial adjustment of the wheelsets will be slightly reduced by the influence of wheel flange lubrication, so that the guiding force achieves a higher value than without lubrication. However, the wheel flange lubrication definitely has a positive effect on wear. As can be seen in Figure 17.55, the wear index demonstrates values that are approximately five times lower than that without lubrication. The utilisation of flange lubrication on self-steering bogies can lead to a further reduction in flange wear if a slight increase in the guiding force can be accepted.

The sensitivity analysis demonstrates the influence of operating conditions such as rail inclination, rail wear, tractive force and wheel flange lubrication on the self-steering of the wheelsets. Despite the sensitivity to the effective operating conditions, the self-steering bogie generally achieves better characteristics and provides significant potential for savings in connection with the maintenance of vehicles and infrastructure when compared with bogies with stiff axle guidance.

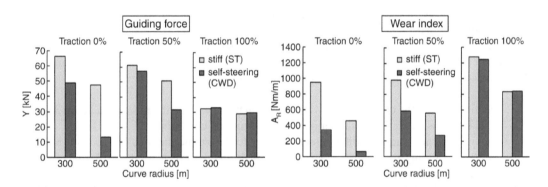

FIGURE 17.54 Influence of tractive effort on curving performance ($a_{lat} = 1.1$ m/s²). (From Iwnicki, S. (Ed.), *Handbook of Railway Vehicle Dynamics*, CRC Press, Boca Raton, FL, 2006. With permission.)

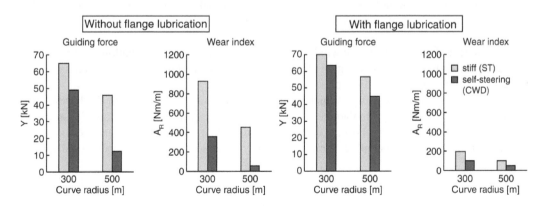

FIGURE 17.55 Influence of flange lubrication on curving performance (wheel-rail friction coefficient: tread 0.4, flange 0.1). (From Iwnicki, S. (Ed.), *Handbook of Railway Vehicle Dynamics*, CRC Press, Boca Raton, FL, 2006. With permission.)

17.6.6 EXAMPLES OF OTHER INVESTIGATIONS

17.6.6.1 Influence of Crosswind

Crosswind safety of trains is a multidisciplinary subject that embraces the topics of aerodynamics and vehicle dynamics. A review of research work on this topic as well as of risk analyses and methodologies used for prediction of accident risk provided Baker et al. in [121].

The evaluation of the limiting wind speed that leads to the vehicle overturning can be carried out both through a quasistatic approach and through a dynamic approach [121]. The time domain simulation is, however, the only one able to account for the effects of the non-linearities in the vehicle model as well as in the contact between wheel and rail.

The crosswind behaviour of a railway vehicle can be determined by the following points, which must be considered jointly and not individually:

- External vehicle design, characteristics of the running gears, track parameters such as radius, cant and track quality
- Meteorological marginal conditions, in particular the local occurrence of strong winds on the railway line
- Aggravating circumstances running on high embankments and viaducts, where winds are normally greater, and sudden gusts, for instance, when exiting a tunnel

Complex procedures for the evaluation of the wind safety limits of rail vehicles in crosswind have been developed with different peculiarities in Germany, France, Italy, UK and China [121]. The procedure developed in Germany [122] involves a comparison with reference vehicles, followed by assessment and derivation of measures in order to achieve equal or better performance than that of the reference standard. The following steps of the procedure are considered:

- Identification of the characteristic wind curves, which describe the maximum permissible wind speed as a function of the train speed and running conditions
- Analysis of the assigned railway lines in order to determine the local permissible wind speed for running the vehicle at each point of the track, which results in the so-called operating curve
- Definition of wind occurrence frequency; meteorological investigations in order to evaluate the occurrence frequency of strong winds at each point on the line
- Calculation of cumulated occurrence frequency, i.e., an integration of the local frequency of occurrence of the permissible wind speed over the complete line, and comparison with the reference value determined by the identical method for the statistical frequency of exceeding of the characteristic wind curves for the reference vehicle

The assessment of the characteristic wind curves is carried out by means of multibody simulations, where the aerodynamic forces acting on the vehicle are either measured in a wind tunnel or calculated with computational fluid dynamics.

Crosswind assessment of railway vehicles in European countries is specified in EN 14067-6 [68]. An application of multibody simulations along with a gust scenario and evaluation of wind occurrence frequency, as described above, is one of the assessment possibilities. The most critical vehicle of the train has to be modelled, running in empty condition considering operational mass in

Simulation of Railway Vehicle Dynamics

working order. The assessment criterion is the average value of wheel unloading of the most critical bogie or single axle in the case of single axle running gear. This unloading shall not exceed 90% of the average static wheel load.

17.6.6.2 Interaction between Vehicle and Traction Dynamics

The investigations of traction system dynamics are usually limited to the rotational masses of the drive systems. However, high adhesion utilisation and sophisticated vehicle dynamics design of modern traction vehicles demand complex simulations, which, at the same time, take into consideration the mechanical, electro-technical and control system fields. To carry out this kind of simulation, together with the complex modelling of systems mentioned, differing creep force models used in traction dynamics and vehicle system dynamics have to be combined into one model. A suitable option represents the creep force model for large creep applications proposed by Polach in [123] and extended by recommended parameters for modelling of typical dry and wet conditions in [32]. The proposed modelling is based on the simplified method published in [124] and uses one parameter set to simulate realistic tractive forces with adhesion maximum, while at the same time considering the influence of vehicle speed and longitudinal, lateral and spin creep on the creep forces. This model is meanwhile widely used for simulation investigations such as development and optimisation of traction systems [125], impact of traction control dynamics on fluctuations in traction forces that may cause damage to the track and vehicle components [126] and prediction of train brake system performance at low adhesion conditions [127], as well as for investigations of topics such as vibration of bogies [52] and curve squeal [53]. Spiryagin et al. [128] proposed an extended version of Kalker's code FASTSIM (cf. Chapter 7) developed to achieve results at large creepages in good agreement with the Polach method, without modifying the proven FASTSIM modelling at small creepages.

In the following examples, an application of co-simulation of vehicle dynamics and traction control under adverse adhesion conditions is shown [32]. The vehicle model is represented by a locomotive modelled in the simulation tool Simpack (see Section 17.4.2.2). The multibody model of the locomotive and one wagon (representing the hauled train) was supplemented with the complete model of the traction system. The controller was modelled in the computer code MATLAB-SIMULINK. During co-simulation, both programmes are running in parallel while only exchanging a few channels.

Figure 17.56 demonstrates the simulation of traction control reaction following a sudden worsening of adhesion conditions (see the input function of friction coefficient between wheel and rail in Figure 17.56). In addition to the sudden reduction of friction coefficient after approximately 30 m of track, small stochastic oscillations are superimposed. The torque on the rotor is unable to reach the set value due to low wheel-rail adhesion. The maximum achievable value is obtained by means of an adaptive traction controller. Following the sudden reduction of friction coefficient, the creep increases at first, but the controller stabilises the working point at a new adhesion optimum with low creepage after a short transition period.

Another co-simulation study shows a starting and acceleration of a locomotive hauling a train on a curved sloping track. Figure 17.57 presents a comparison of measurement and simulation of longitudinal forces in wheelset linkages. The observed forces on the straight track and on the left and right curves are very close in measurement and simulation. The comparison confirms the proposed method as suitable for computer simulation of traction vehicles running at adhesion limit.

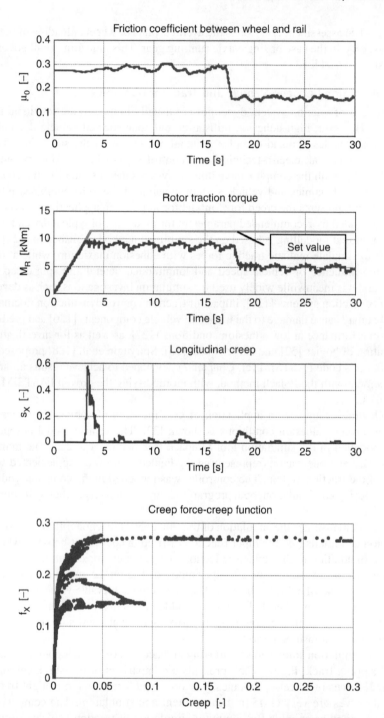

FIGURE 17.56 Simulation of impact of a sudden drop in the adhesion conditions modelled by reduction of wheel-rail friction coefficient – time diagrams of rotor torque and longitudinal creep between wheel and rail, and diagram of creep force (normalised by vertical wheel-rail force) as a function of creep. (From Polach, O., *Wear*, 258, 992–1000, 2005. With permission.)

Simulation of Railway Vehicle Dynamics

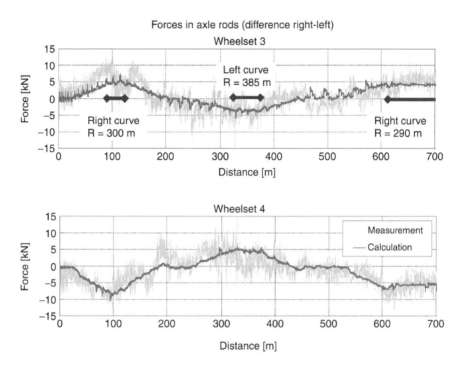

FIGURE 17.57 Comparison of calculated (dark line) and measured (non-filtered) (light line) longitudinal forces in wheelset linkages during starting and acceleration of a locomotive hauling a train on a curved sloping track. (From Polach, O., *Wear*, 258, 992–1000, 2005. With permission.)

17.7 CONCLUSIONS

The development of mathematical models and software tools for simulation of wheel-rail contact and the dynamic behaviour of railway vehicles on track have grown with the development of computing power. From the earliest analogue computers to the modern powerful digital processors, the equations that govern the contact location, pressure distribution and tangential creep forces have been developed and then coded into computer programs. These programs are now combined into a number of powerful and reliable computer simulation packages, and three examples from those currently in widespread use have been described previously.

Vehicle dynamics simulations now constitute a significant and indispensable part of railway vehicle engineering and are recently increasingly recognised as a replacement for some parts of testing. During the design period, a prognosis of the running characteristics can be carried out with the aid of computer simulations, which reduces the vehicle design period.

Further developments of dynamic analyses in railway vehicle engineering arise on the one hand due to new challenges and on the other hand by the broadening of objectives and possibilities. During the development of new railway vehicles, the range of technical solutions is being continuously extended and the technical limits exploited. From a dynamics standpoint, this tendency is accompanied by challenges of, for instance, higher demands on material and energy savings, wear reduction, avoidance of damages such as out-of-roundness of wheels, rail corrugation and RCF of wheels and rails. Several new methods to solve these new challenges and to reduce the risks have been developed, and many further research projects are in progress.

The dynamic simulations support a larger scope of development in other technical sectors. For instance, it is possible to take the whole dynamic behaviour of the vehicle or train composition into consideration for tasks such as investigation of loads acting on vehicle components or clearances between the components during running on the track, loading and risk of damage to railway tracks, vehicle's crash behaviour and aerodynamic analyses. The typical tasks of running dynamics will evolve further, with the result that, in future, more complex running tests can be virtually simulated and evaluated.

NOMENCLATURE

a_{lat}	Uncompensated lateral acceleration (cant deficiency)
A_R	Wear index
c	Viscous damping
$[C]$	Damping matrix
D	Damping normalised by critical damping
e_0	Half of the tread datum (tape circle) distance
f	Frequency
f_i	Car body eigenfrequency, $i = 1, 2, 3...$
f_0	Frequency of unstable bogie oscillations
F	Force
$\boldsymbol{F}$	General input force vector
F_x, F_y	Creep force in the contact plan between wheel and rail
g	Gravitational constant
h_t	Track superelevation (cant), expressed as height difference between the top of left and right rails
k	Stiffness
k_x, k_y	Axle guidance stiffness
k_{xB}	Equivalent longitudinal axle guidance stiffness for bending wheelset mode
k_s	Series stiffness
$[K]$	Stiffness matrix
l_m	Wavelength of track irregularity critical for excitation of a car body mode, $m = 1, 2, 3...$
m^+	Bogie mass
$[M]$	Mass matrix
N	Normal force
N_P	Non-linearity parameter of wheel-rail contact geometry
p	Bogie pivot pin distance
q	Vehicle motion vector
Q	Vertical wheel-rail contact force
Q_0	Static vertical wheel-rail contact force
R	Curve radius
s_x, s_y	Creep (slip) between wheel and rail
$s\ddot{y}^+$	Rms value of lateral acceleration on the bogie frame
$s\ddot{y}^*$	Rms value of lateral acceleration in the car body
$s\ddot{z}^*$	Rms value of vertical acceleration in the car body
$s\Sigma Y$	Rms value of the sum of guiding forces (track shifting force)
v	Vehicle speed in m/s
V	Vehicle speed in km/h
v_{cr}	Critical speed
$v_{cr\,lin}$	Linear critical speed
$v_{cr\,nl}$	Non-linear critical speed
x	Displacement

y	Lateral wheelset displacement
$\ddot{y}^+$	Lateral acceleration on the bogie frame
$\ddot{y}^*$	Lateral acceleration in the car body
Y	Guiding force (lateral wheel-rail contact force)
Y/Q	Quotient of guiding force and vertical wheel-rail force
$\ddot{z}^*$	Vertical acceleration in the car body
ε	Contact angle (contact slope) parameter
λ	Equivalent conicity
μ	Coefficient of friction (μN = friction break-out force)
σ	Roll parameter
ω	Angular frequency
Ω_s	Spin creep between wheel and rail
Θ	Moment around the normal to wheel-rail contact
ΣY	Sum of guiding forces (track shifting force)

Indices:

x	Longitudinal direction
y	Lateral direction
qst	Quasistatic value
max	Maximum value
lim	Limit value

REFERENCES

1. E. Andersson, M. Berg, S. Stichel, *Rail Vehicle Dynamics*, Royal Institute of Technology (KTH), Stockholm, Sweden, 2014.
2. ISO 2631-1, Mechanical vibration and shock—Evaluation of human exposure to whole-body vibration—Part 1: General requirements, 2nd ed., International Organization for Standardization, Geneva, Switzerland, 1997.
3. P. Carlbom, Structural flexibility in a rail vehicle car body—Dynamic simulations and measurements, TRITA-FKT Report 1998:37, Royal Institute of Technology (KTH), Stockholm, Sweden, 1998.
4. P. Carlbom, M. Berg, Passengers, seats and carbody in rail vehicle dynamics, *Vehicle System Dynamics*, 37(S1), 290–300, 2002.
5. N. Chaar, Structural flexibility models of wheelsets for rail vehicle dynamics analysis—A pilot study, TRITA-FKT Report 2002:23, Royal Institute of Technology (KTH), Stockholm, Sweden, 2002.
6. N. Chaar, M. Berg, Experimental and numerical modal analyses of a loco wheelset, In: M. Abe (Ed.), *Proceedings 18th IAVSD Symposium held in Atsugi, Kanagawa, Japan*, Taylor & Francis, London, UK, 597–606, 2004.
7. B.M. Eickhoff, J.R. Evans, A.J. Minnis, A review of modelling methods for railway vehicle suspension components, *Vehicle System Dynamics*, 24(6–7), 469–496, 1995.
8. M. Berg, An airspring model for dynamic analysis of rail vehicles, Report TRITA-FKT 1999:32, Royal Institute of Technology (KTH), Stockholm, Sweden, 1999.
9. P.-A. Jönsson, E. Andersson, Influence of link suspension characteristics on freight wagon lateral dynamics, *Proceedings 6th International Conference on Railway Bogies and Running Gears*, Budapest, Hungary, 253–262, 2004.
10. P.S. Fancher, R.D. Ervin, C.C. MacAdam, C.B. Winkler, Measurement and representation of the mechanical properties of truck leaf springs, Society of Automotive Engineers, Paper No. 800905, 1980.
11. M. Berg, A non-linear rubber spring model for rail vehicle dynamics analysis, *Vehicle System Dynamics*, 30(3–4), 197–212, 1998.
12. J. Nicolin, T. Dellmann, Über die modellhafte Nachbildung der dynamischen Eigenschaften einer Gummifeder, *ZEV-Glasers Annalen*, 109(4), 169–175, 1985.
13. J.R. Evans, Rail vehicle dynamics simulations using VAMPIRE, In: S. Iwnicki (Ed.), *The Manchester Benchmarks for Rail Vehicle Simulation*, Swets & Zeitlinger, Lisse, The Netherlands 1999.
14. P. Sundvall, Comparisons between predicted and measured ride comfort in trains—A case study on modelling, M.Sc. Thesis, TRITA-FKT Report 2001:19, Royal Institute of Technology (KTH), Stockholm, Sweden, 2001.

15. F. Xia, Modelling of wedge dampers in the presence of two-dimensional dry friction, *Vehicle System Dynamics*, 37(S1), 565–578, 2002.
16. S. Bruni, J. Vinolas, M. Berg, O. Polach, S. Stichel, Modelling of suspension components in a rail vehicle dynamics context, *Vehicle System Dynamics*, 49(7), 1021–1072, 2011.
17. S. Claesson, Modelling of track flexibility for rail vehicle dynamics simulations, M.Sc. Thesis, Report TRITA AVE 2005:26, Royal Institute of Technology (KTH), Stockholm, Sweden, 2005.
18. R. Enblom, Deterioration mechanisms in the wheel-rail interface with focus on wear prediction: A literature review, *Vehicle System Dynamics*, 47(6), 661–700, 2009.
19. H. True, Does a critical speed for railroad vehicles exist? *Proceedings ASME/IEEE Joint Railroad Conference*, Chicago, IL, American Society of Mechanical Engineers, 125–131, 1994.
20. R.M. Goodall, S.D. Iwnicki, Non-linear dynamic techniques v. equivalent conicity methods for rail vehicle stability assessment, In: M. Abe (Ed.), *Proceedings18th IAVSD Symposium held in Atsugi, Kanagawa, Japan*, Taylor & Francis Group, London, UK, 791–799, 2004.
21. H. Hertz, Über die Berührung zweier fester elastischer Korper, *Journal für Reine und Angewandte Mathematik*, 92, 156–171, 1882.
22. K. Knothe, L.T. Hung, Determination of the tangential stresses and the wear for the wheel-rail rolling contact problem, In: O. Nordstrom (Ed.), *Proceedings 9th IAVSD Symposium held at Linköping University, Sweden*, Swets and Zeitlinger, Lisse, the Netherlands, 264–277, 1986.
23. J.P. Pascal, G. Sauvage, New method for reducing the multicontact wheel/rail problem to one equivalent rigid contact patch, In: G. Sauvage (Ed.), *Proceedings 12th IAVSD Symposium held in Lyon, France*, Swets and Zeitlinger, Lisse, the Netherlands, 475–489, 1992.
24. W. Kik, J. Piotrowski, A fast approximate method to calculate normal load at contact between wheel and rail and creep forces during rolling, In: I. Zobory (Ed.), *Proceedings 2nd Mini Conference on Contact Mechanics and Wear of Rail/Wheel Systems*, Technical University of Budapest, Hungary, 52–61, 1996.
25. S.D. Iwnicki (Ed.), *The Manchester Benchmarks for Rail Vehicle Simulation*, Swets & Zeitlinger, Lisse, the Netherlands, 1999.
26. J.B. Ayasse, H. Chollet, J.L. Maupu, Paramètres caractéristiques du contact roue-rail, INRETS Report No. 225, 2000.
27. F.W. Carter, The electric locomotive, *Proceedings of Institution of Civil Engineers*, 221, 221–252, 1916.
28. P.J. Vermeulen, K.L. Johnson, Contact of non-spherical elastic bodies transmitting tangential forces, *ASME Journal of Applied Mechanics*, 31(2), 338–340, 1964.
29. J.J. Kalker, On the rolling of two elastic bodies in the presence of dry friction, Doctoral thesis, Delft University of Technology, Delft, the Netherlands, 1967.
30. D.J. Haines, E. Ollerton, Contact stress distributions on elliptical contact surfaces subjected to radial and tangential forces, *Proceedings of the Institution of Mechanical Engineers*, 177, 95–114, 1963.
31. Z.Y. Shen, J.K. Hedrick, J.A. Elkins, A comparison of alternative creep force models for rail vehicle dynamics analysis, In: J.K. Hedrick (Ed.), *Proceedings 8th IAVSD Symposium held at Massachusetts Institute of Technology*, Swets and Zeitlinger, Lisse, the Netherlands, 728–739, 1984.
32. O. Polach, Creep forces in simulations of traction vehicles running on adhesion limit, *Wear*, 258(7–8), 992–1000, 2005.
33. A. Minnis, L. Purcell, The wheel rail interface—A common language, *Maintaining the UK Rail Infrastructure IMechE Seminar*, Institution of Mechanical Engineers London, UK, 2012.
34. C. Grimes, G. Hunt, S. Wilson, A new tool for planning vehicle dynamics based track maintenance, *Railway Engineering 2007–2009th International Conference and Exhibition*, University of Westminster, London, UK, 2007.
35. V. Sakalo, A. Sakalo, S. Tomashevskiy, D. Kerentcev, Computer modelling of process of accumulation of rolling contact fatigue damage in railway wheels, *International Journal of Fatigue*, 111, 7–15, 2018.
36. R. Kovalev, A. Sakalo, V. Yazykov, A. Shamdani, R. Bowey, C. Wakeling, Simulation of longitudinal dynamics of a freight train operating through a car dumper, *Vehicle System Dynamics*, 54(6), 707–722, 2016.
37. R. Kovalev, N. Lysikov, G. Mikheev et al., Freight car models and their computer-aided dynamic analysis, *Multibody System Dynamics*, 22, 399–423, 2009. doi:10.1007/s11044-009-9170-6.
38. Q. Wu, M. Spiryagin, C. Cole et al., International benchmarking of longitudinal train dynamics simulators: Results, *Vehicle System Dynamics*, 56(3), 343–365, 2018.
39. D. Pogorelov, V. Yazykov, N. Lysikov, E. Oztemel, O.F. Arar, F.S. Rende, Train 3D: The technique for inclusion of three-dimensional models in longitudinal train dynamics and its application in derailment studies and train simulators, *Vehicle System Dynamics*, 55(4), 583–600, 2017.

40. ERRI: B176/3, Benchmark problem—Results and assessment, B176/DT290, Utrecht, the Netherlands, 1993.
41. W. Kortüm, R.S. Sharp, The IAVSD review of multibody computer codes for vehicle system dynamics, *Proceedings 3rd ASME Symposium on Transportation Systems, ASME Winter Annual Meeting*, Anaheim, CA, 1992.
42. W. Kortüm, R.S. Sharp (Eds.), Multibody computer codes in vehicle system dynamics, *Supplement to Vehicle System Dynamics*, 22(S1), Swets & Zeitlinger, Lisse, the Netherlands, 1993.
43. J.G. Giménez, W. Kik, J.P. Pascal, G. Sauvage, Simulation of the IAVSD Railway Vehicle Benchmark # 2 with MEDYNA, SIDIVE and VOCO, *Vehicle System Dynamics*, 22(S1),193–214, 1993.
44. S.D. Iwnicki, Manchester benchmarks for rail vehicle simulation, *Vehicle System Dynamics*, 30(3–4), 295–313, 1998.
45. W. Kortüm, W.O. Schielen, M. Arnold, Software tools: From multibody system analysis to vehicle system dynamics. In: H. Aref, J.W. Phillips (Eds.), *Mechanics for a New Millennium*, Kluwer Academic Publishers, New York, NY, 225–238, 2001.
46. J. Evans, M. Berg, Challenges in simulation of rail vehicle dynamics, *Vehicle System Dynamics*, 47(8), 1023–1048, 2009.
47. C. Weidemann, State-of-the-art railway vehicle design with multi-body simulation, *Journal of Mechanical Systems for Transportation and Logistics*, 3(1), 12–26, 2010.
48. M. Ofierzynski, Simulationstechnik bei der SIG Schweizerische Industrie-Gesellschaft, *ZEV+DET Glasers Annalen*, 118(2–3), 105–126, 1994.
49. M. Ofierzynski, Industrieller Einsatz der Simulationstechnik, In: H. Hochbruck, K. Knothe, P. Meinke, *Systemdynamik der Eisenbahn, Fachtagung in Hennigsdorf am*, Hestra-Verlag, Darmstadt, 171–180, 1994.
50. O. Polach, Simulations of running dynamics in bogie design and development, *European Railway Review*, (3), 76–81, 2008.
51. K. Knothe, S. Stichel, *Rail Vehicle Dynamics*, Springer International, Cham, Switzerland, 2017.
52. B. Kurzeck, Combined friction induced oscillations of wheelset and track during the curving of metros and their influence on corrugation, *Wear*, 271(1–2), 299–310, 2011.
53. C. Glocker, E. Cataldi-Spinola, R.I. Leine, Curve squealing of trains: Measurement, modelling and simulation, *Journal of Sound and Vibration*, 324(1–2), 365–386, 2009.
54. J.C.O. Nielsen, A. Johansson, Out-of round railway wheels—A literature survey, *Journal of Rail and Rapid Transit*, 214(2), 79–91, 2000.
55. P.A. Meehan, P.A. Bellette, R.D. Batten, W.J.T. Daniel, R.J. Horwood, A case study of wear-type rail corrugation prediction and control using speed variation, *Journal of Sound and Vibration*, 325(1–2), 85–105, 2009.
56. S. Stichel, H. Mohr, J. Ågren, R. Enblom, Investigation of the risk for rolling contact fatigue on wheels of different passenger trains, *Vehicle System Dynamics*, Supplement 1, Vol. 46(S1), 317–327, 2008.
57. R. Enblom, S. Stichel, Industrial implementation of novel procedures for the prediction of railway wheel surface deterioration, *Wear*, 271(1–2), 203–209, 2011.
58. S.H. Nia, P.-A. Jönsson, S. Stichel, Wheel damage on the Swedish iron ore line investigated via multi-body simulation, *Journal of Rail and Rapid Transit*, 228(6), 652–662, 2014.
59. J.R. Evans, M.A. Dembosky, Investigation of vehicle dynamic influence on rolling contact fatigue on UK railways, In: M. Abe (Ed.), *Proceedings18th IAVSD Symposium held in Atsugi, Kanagawa, Japan*, Taylor & Francis Group, London, UK, 527–536, 2004.
60. C. Casanueva, B. Dirks, M. Berg, T. Bustad, Track damage prediction for universal cost model applications, In: M. Spiryagin, T. Gordon, C. Cole, T. McSweeney (Eds.), *Proceedings 25th IAVSD Symposium held at Central Queensland University, Rockhampton, Queensland, Australia*, CRC Press/Balkema, Leiden, the Netherlands, 537–543, 2018.
61. I. Nerlich, Benefits of "track gentle rolling stock"—Expected track maintenance costs of new services in the future, In: I. Zobory, (Ed.), *Proceedings 10th International Conference on Railway Bogies and Running Gears*, Budapest University of Technology and Economics, Budapest, Hungary, 191–200, 2016.
62. S. Kraft, J. Causse, F. Coudert, Vehicle response-based track geometry assessment using multi-body simulation, *Vehicle System Dynamics*, 56(2), 190–220, 2018.
63. O. Polach, J. Evans, Simulations of running dynamics for vehicle acceptance: Application and validation, *International Journal of Railway Technology*, 2(4), 59–84, 2013.
64. B.P. Marquis, A. Tajaddini, H-P. Kotz, W. Breuer, M. Trosino, Application of new FRA vehicle-track system qualification requirement, *Proceedings of IMechE Stephenson Conference: Research for Railways*, London, UK, 581–592, 2015.

65. UIC Code 518, *Testing and Approval of Railway Vehicles from the Point of View of Their Dynamic Behaviour—Safety—Track Fatigue—Running Behaviour*, 4th Edition, International Union of Railways, Paris, France, 2009.
66. EN 14363, *Railway Applications—Testing and Simulation for the Acceptance of Running Characteristics of Railway Vehicles—Running Behaviour and Stationary Tests*, European Committee for Standardization, Brussels, Belgium, 2016.
67. United States Department of Transportation, Code of Federal Regulations, Title 49—Transportation, Part 213—Track Safety Standards, Federal Railroad Administration, Washington, DC, 2011. Available at: https://www.ecfr.gov.
68. EN 14067-6, *Railway Applications—Aerodynamics—Part 6: Requirements and Test Procedures for Cross Wind Assessment*, European Committee for Standardization, Brussels, Belgium, 2010.
69. EN 15273-2, *Railway Applications—Gauges—Part 2: Rolling Stock Gauge*, European Committee for Standardization, Brussels, Belgium, 2013.
70. G. Götz, O. Polach, Verification and validation of simulations in a rail vehicle certification context, *International Journal of Rail Transportation*, 6(2), 83–100, 2018.
71. R. Fries, R. Walker, N. Wilson, Validation of dynamic rail vehicle models, In: *Proceedings 23rd IAVSD Symposium*, Qingdao, China, 10.1-ID416, 2013.
72. M. Zacher, R. Kratochwille, Stationary and on-track tests using various vehicles, *Journal of Rail and Rapid Transit*, 229(6), 668–690, 2015.
73. O. Polach, A. Böttcher, D. Vannucci et al., Validation of simulation models in the context of railway vehicle acceptance, *Journal of Rail and Rapid Transit*, 229(6), 729–754, 2015.
74. O. Polach, A. Böttcher, A new approach to define criteria for rail vehicle model validation, *Vehicle System Dynamics*, 52(S1), 125–141, 2014.
75. H.-L. Shi, P.-B. Wu, R. Luo, J. Zeng, Estimation of the damping effects of suspension systems on railway vehicles using wedge tests, *Journal of Rail and Rapid Transit*, 230(2), 392–406, 2016.
76. L. Mauer, The modular description of the wheel to rail contact within the linear multibody formalism, In: J. Kisilowski, K. Knothe (Eds.), *Advanced Railway Vehicle System Dynamics*, Wydawnictwa Naukowo-Techniczne, Warsaw, Poland, 205–244, 1991.
77. A.H. Wickens, *Fundamentals of Rail Vehicle Dynamics: Guidance and Stability*, Swets & Zeitlinger, Lisse, The Netherlands, 2003.
78. EN 15302, *Railway Applications—Method for Determining the Equivalent Conicity*, European Committee for Standardization, Brussels, Belgium, 2008.
79. O. Polach, Characteristic parameters of nonlinear wheel/rail contact geometry, *Vehicle System Dynamics*, 48(S1), 19–36, 2010.
80. ORE B 176, Bogies with steered or steering wheelsets: Volume 1: Preliminary studies and specifications; Volume 2, Specification for a bogie with improved curving characteristics, Office of Research and Experiment (ORE) of the International Union of Railways, Utrecht, the Netherlands, 1989.
81. O. Polach, Comparability of the non-linear and linearized stability assessment during railway vehicle design, *Vehicle System Dynamics*, 44(S1), 129–138, 2006.
82. S. Alfi, L. Mazzola, S. Bruni, Effect of motor connection on the critical speed of high-speed railway vehicles, *Vehicle System Dynamics*, 46(S1), 201–214, 2008.
83. C. Huang, J. Zeng, S. Liang, Influence of system parameters on the stability limit of the undisturbed motion of a motor bogie, *Journal of Rail and Rapid Transit*, 228(5), 522–534, 2014.
84. A. Mahr, Abstimmung der Rad-Schiene-Geometrie mit dem lateralen Fahrzeugverhalten: Methoden und Empfehlungen, Thesis, Shaker Verlag GmbH, Aachen, Germany, 2002.
85. S. Stichel, On freight wagon dynamics and track deterioration, *Journal of Rail and Rapid Transit*, 213(4), 243–254, 1999.
86. M. Ofierzynski, V. Brundisch, Fahrkomfort von Schienenfahrzeugen—Die Zuverlässigkeit moderner Simulationstechnik, *ZEV+DET Glasers Annalen*, 124(2/3), 109–119, 2000.
87. C. Huang, J. Zeng, S. Liang, Carbody hunting investigation of a high speed passenger car, *Journal of Mechanical Science and Technology*, 27(8), 2283–2292, 2013.
88. C.-S. Jeon, Y.-G. Kim, J.-H. Park, S.-W. Kim, T.-W. Park, A study on the dynamic behavior of the Korean next-generation high-speed train, *Journal of Rail and Rapid Transit*, 230(4), 1053–1065, 2016.
89. A.H. Wickens, Steering and stability of the bogie: Vehicle dynamics and suspension design, *Journal of Rail and Rapid Transit*, 205(2), 109–122, 1991.
90. H. Fujimoto, M. Miyamoto, Measures to reduce the lateral vibration of the tail car in a high speed train, *Journal of Rail and Rapid Transit*, 210(2), 87–93, 1996.

91. D. Chen, G. Shen, Analysis of railway vehicle carbody hunting based on similarity identification of fuzzy mathematics, In: M. Spiryagin, T. Gordon, C. Cole, T. McSweeney (Eds.), *Proceedings 25th IAVSD Symposium held at Central Queensland University, Rockhampton, Queensland, Australia*, CRC Press/Balkema, Leiden, the Netherlands, 1019–1024, 2018.
92. H. True, On the theory of nonlinear dynamics and its application in vehicle systems dynamics, *Vehicle System Dynamics*, 31(5–6), 393–421, 1999.
93. H. True, Multiple attractors and critical parameters and how to find them numerically: The right, the wrong and the gambling way, *Vehicle System Dynamics*, 51(3), 443–459, 2013.
94. J.C. Jensen, E. Slivsgaard, H. True, Mathematical simulation of the dynamics of the Danish IC3 train, *Vehicle System Dynamics*, 29(S1), 760–765, 1998.
95. M. Hoffmann, H. True, The dynamics of European two-axle railway freight wagons with UIC standard suspension, *Vehicle System Dynamics*, 46(S1), 225–236, 2008.
96. G. Schupp, Computational bifurcation analysis of mechanical systems with applications to railway vehicles, In: M. Abe (Ed.), *Proceedings 18th IAVSD Symposium held in Atsugi, Kanagawa, Japan*, Taylor & Francis, London, UK, 458–467, 2004.
97. O. Polach, On nonlinear methods of bogie stability assessment using computer simulations, *Journal of Rail and Rapid Transit*, 220(1), 13–27, 2006.
98. O. Polach, Application of nonlinear stability analysis in railway vehicle industry, In: P.G. Thomsen, H. True (Eds.), *Non-smooth Problems in Vehicle Systems Dynamics: Proceedings Euromech 500 Colloquium*, Springer-Verlag, Heidelberg, Germany, 15–27, 2010.
99. K. Knothe, F. Böhm, History of stability of railway and road vehicles, *Vehicle System Dynamics*, 31(5–6), 283–323, 1999.
100. O. Polach, D. Nicklisch, Wheel/rail contact geometry parameters in regard to vehicle behaviour and their alteration with wear, *Wear*, 366–367, 200–208, 2016.
101. X. Wu, M. Chi, Parameters study of Hopf bifurcation in railway vehicle system, *Journal of Computational and Nonlinear Dynamics*, 10(3), 031012, 2015.
102. L.L. Xing, Y.M. Wang, X.Q. Dong, Effect of the wheel/rail contact geometry on the stability of railway vehicle, *IOP Conference Series: Materials Science and Engineering*, 392, 062134, 2018.
103. O. Polach, I. Kaiser, Comparison of methods analyzing bifurcation and hunting of complex rail vehicle models, *Journal of Computational and Nonlinear Dynamics*, 7(4), 041005, 2012.
104. F. Frederich, Die Gleislage—aus fahrzeugtechnischer Sicht, *ZEV Glasers Annalen*, 108(12), 355–336, 1984.
105. EN 12299, *Railway Applications—Ride Comfort for Passengers—Measurement and Evaluation*, European Committee for Standardization, Brussels, Belgium, 2009.
106. G. Diana, F. Cheli, A. Collina, R. Corradi, S. Melzi, The development of a numerical model for railway vehicles comfort assessment through comparison with experimental measurements, *Vehicle System Dynamics*, 38(3), 165–183, 2002.
107. M. Götsch, M. Sayir, Simulation of riding comfort of railway vehicles, *Vehicle System Dynamics*, 37(S1), 630–640, 2002.
108. T.G. Pearce, N.D. Sherratt, Prediction of wheel profile wear, *Wear*, 144(1–2), 343–351, 1991.
109. I. Zobory, Prediction of wheel/rail profile wear, *Vehicle System Dynamics*, 28(2–3), 221–259, 1997.
110. T. Jendel, Prediction of wheel profile wear—Comparisons with field measurements, *Wear*, 253(1–2), 89–99, 2002.
111. J. Pombo, J. Ambrósio, M. Pereira, R. Lewis, R. Dwyer-Joyce, C. Ariaudo, N. Kuka, Development of a wear prediction tool for steel railway wheels using three alternative wear functions, *Wear*, 271(1–2), 238–245, 2011.
112. R. Enblom, M. Berg, Proposed procedure and trial simulation of rail profile evolution due to uniform wear, *Journal of Rail and Rapid Transit*, 222(1), 15–25, 2008.
113. I. Persson, R. Nilsson, U. Bik, M. Lundgren, S. Iwnicki, Use of a genetic algorithm to improve the rail profile on Stockholm underground, *Vehicle System Dynamics*, 48(S1), 89–104, 2010.
114. Trelleborg, Antivibration Solutions, Webpage available at: http://www.trelleborg.com/en/anti-vibration-solutions.
115. D. Cordts, B. Meier, Hydraulisches Achslenkerlager zur Anwendung im Schienenfahrzeugbereich, *EI-Eisenbahningenieur*, 3, 69–73, 2012.
116. H. Scheffel, Unconventional bogie designs—Their practical basis and historical background, *Vehicle System Dynamics*, 24(6–7), 497–524, 1995.
117. Y. Suda, Y. Michitsuji, H. Sugiyama, Next generation unconventional trucks and wheel-rail interfaces for railways, *International Journal for Railway Technology*, 1(1), 1–26, 2012.

118. O. Polach, Coupled single-axle running gears—A new radial steering design, *Journal of Rail and Rapid Transit*, 216(3), 197–206, 2002.
119. R. Schneider, G. Himmelstein, Active radial steering and stability control with the mechatronic bogie, In: *Proceedings 7th World Congress on Railway Research*, Montréal, Canada, PB011066, 2006.
120. O. Polach, Curving and stability optimisation of locomotive bogies using interconnected wheelsets, In: M. Abe (Ed.), *Proceedings 18th IAVSD Symposium held in Atsugi, Kanagawa, Japan*, Taylor & Francis, London, UK, 53–62, 2004.
121. C. Baker, F. Cheli, A. Orellano, N. Paradot, C. Proppe, D. Rocchi, Cross-wind effects on road and rail vehicles, *Vehicle System Dynamics*, 47(8), 983–1022, 2009.
122. G. Matschke, P. Deeg, B. Schulte-Werning, Effects of strong cross winds on high-speed trains: A methodology for risk assessment and development of countermeasures, In: *Proceedings 5th World Congress on Railway Research*, Cologne, Germany, PB011827, 2001.
123. O. Polach, Influence of locomotive tractive effort on the forces between wheel and rail, In: *Selected papers from 20th International Congress of Theoretical and Applied Mechanics held 28 August–1 September 2000 in Chicago, IL*, Swets & Zeitlinger, Lisse, The Netherlands, 7–22, 2001.
124. O. Polach, A fast wheel-rail forces calculation computer code, In: R. Fröhling (Ed.), *Proceedings 16th IAVSD Symposium held in Pretoria, South Africa*, Swets & Zeitlinger, Lisse, The Netherlands, 728–739, 2000.
125. N. Kuka, C. Ariaudo, R. Verardi, A. Dolcini, Development and validation of a methodology for closed loop simulations of the driveline in railway vehicles, *Journal of Rail and Rapid Transit*, 232(6), 1625–1649, 2018.
126. S. Liu, Y. Tian, W.J.T. Daniel, P.A. Meehan, Dynamic response of a locomotive with AC electric drives to changes in friction conditions, *Journal of Rail and Rapid Transit*, 231, 90–103, 2017.
127. H. Alturbeh, J. Stow, G. Tucker, A. Lawton, Modelling and simulation of the train brake system in low adhesion conditions, *Journal of Rail and Rapid Transit*, OnlineFirst, doi:10.1177/0954409718800579, 2018.
128. M. Spiryagin, O. Polach, C. Cole, Creep force modelling for rail traction vehicles based on the Fastsim algorithm, *Vehicle System Dynamics*, 51(11), 1765–1783, 2013.

18 Field Testing and Instrumentation of Railway Vehicles

Julian Stow

CONTENTS

18.1	Introduction	724
	18.1.1 Reasons for Testing	724
	18.1.2 Sources of Test Data	725
18.2	Common Transducers	726
	18.2.1 Displacement Transducers	726
	18.2.2 Accelerometers	727
	18.2.2.1 Piezoelectric Accelerometers	727
	18.2.2.2 Capacitive Accelerometers	730
	18.2.2.3 MEMS Sensors	731
	18.2.3 Strain Gauges	731
	18.2.3.1 Bridge Circuits	733
	18.2.4 Force-Measuring Wheelsets	735
	18.2.4.1 Measuring Lateral Forces between Wheelset and Axle Box	735
	18.2.4.2 Measuring Lateral and Vertical Wheel-Rail Forces – Axle Method	736
	18.2.4.3 Measuring Lateral and Vertical Wheel-Rail Forces – Wheel Methods	737
	18.2.4.4 Compensation for Undesired Parasitic Effects	738
	18.2.5 Vehicle Speed and Position Measurement	739
	18.2.5.1 AC Tachogenerator	739
	18.2.5.2 Hall Effect Probes	739
	18.2.5.3 Ground Speed Radar	740
	18.2.5.4 Determining Vehicle Position	741
	18.2.6 Wireless Sensors	741
18.3	Test Equipment Configuration and Environment	743
	18.3.1 Transducer Positions on Vehicles	745
18.4	Data Acquisition	746
18.5	Measurement of Wheel and Rail Profiles	748
	18.5.1 General Requirements	748
	18.5.2 Contacting Measuring Devices	749
	18.5.3 Optical Measuring Devices	750
18.6	Track Geometry Recording	751
	18.6.1 Manual Survey	751
	18.6.2 Track Geometry Trolley	752
	18.6.3 Track-Recording Vehicles	752
	18.6.4 Unattended Track Geometry Measuring Systems	754

18.7	Laboratory and Field Testing for Validation and Acceptance		755
	18.7.1	Static/Quasi-Static Tests	755
		18.7.1.1 Wheel Unloading Test	755
		18.7.1.2 Bogie Rotational Resistance Test	755
		18.7.1.3 Sway Test	757
		18.7.1.4 Body Modes Tests	758
	18.7.2	Dynamic Tests	758
References			760

18.1 INTRODUCTION

An understanding of testing and instrumentation methods is essential for the accurate construction and validation of railway vehicle dynamic models. The dynamics engineer may need to produce specifications for test work, understand the applicability and limitations of data produced and manipulate test results to provide comparisons with modelling work. This chapter provides an overview of the situations in which the engineer may require test data, together with an introduction to common techniques and equipment used, both in the laboratory and for conducting vehicle testing on track. The examples given largely relate to vehicle testing, which forms the most relevant body of work for the vehicle dynamics engineer. However, dynamic simulation is increasingly used in rail-/track-related investigations, where many of the same techniques may be applied.

18.1.1 REASONS FOR TESTING

During the development of a new vehicle or modification of an existing one, requirements for test work may arise for a number of different reasons:

- *Component characterisation*: Tests may be required to establish the properties of the various components that make up the suspension in order to allow the initial construction of a model; such tests are normally carried out in the laboratory, using small to medium-sized test machines or dedicated test rigs.
- *Determination of parasitic or secondary effects*: Once assembled, vehicles (particularly modern passenger vehicles) can exhibit behaviour that is difficult to predetermine from the individual suspension components. These parasitic effects typically arise from the summation of a number of small stiffness contributions from components such as anti-roll bars, traction centres and lateral and yaw dampers in directions other than those in which they are mainly designed to operate. Other effects that may need to be quantified arise from flexibilities in mounting brackets or similar, as well as internal flexibility in dampers. Such tests are normally carried out statically or quasi-statically on a complete vehicle in the laboratory.
- *Structural testing*: The testing of vehicle body structures and bogie frames for strength, fatigue life, and crashworthiness is a complete subject in itself and beyond the scope of this chapter. However, the dynamicist may need to obtain parameters to enhance models, particularly with respect to simulation of ride and passenger comfort. Typical examples include the vertical and lateral bending modes of vehicle bodies and the torsional stiffness of bogie frames. Tests are most often carried out in the laboratory by using bare body shells or bogie frames mounted in dedicated structural test rigs.
- *Validation testing*: It is generally necessary to increase confidence in the correct operation of models by comparing the results with those from a series of tests. Such tests may be on bogies or complete vehicles. At a basic level, these may be carried out quasi-statically in the laboratory, but any extensive validation is likely to require on-track tests under a range

Field Testing and Instrumentation of Railway Vehicles

of conditions to fully understand the dynamic behaviour of the vehicle. The level of validation required will ultimately depend upon the intended use of the models.

- *Acceptance tests*: All railway administrations require new or modified vehicles to undergo a series of tests to demonstrate safe operation for various conditions. Such tests may be specific to an individual company or country or, as in the case of European Standards [1,2], may allow a vehicle to operate across a number of countries. The exact requirements for these can vary widely but will usually comprise a mixture of laboratory and field tests. Many administrations now allow some of these requirements to be met by simulation of the test procedure, using a suitably validated vehicle dynamics model. In any case, simulation of these tests forms a common part of vehicle development to ensure that proposed designs meet the required standards. As such, tests will be carried out on all vehicles accepted for service, and they may also provide a useful source of information to validate models of existing vehicles. In addition, the dynamics engineers may also be involved with testing to assess performance against specified criteria such as passenger comfort or to investigate problems with existing rolling stock.
- *Reproducing track geometry*: Many simulation tasks will require the use of 'real' track geometry measured by a high-speed recording vehicle or hand-operated trolleys, and, although such data is generally presented as 'ready to use', experience has shown that an appreciation of the measuring systems and instrumentation used is vital to ensure that an accurate reconstruction of the track geometry can be obtained.
- *Measuring wheel and rail profiles*: Accurate representations of worn wheel and rail profiles are vital to understanding vehicle (and track) behaviour; a number of proprietary devices are available to measure profiles; however, as with track geometry, accurate results will be aided by an understanding of the principles behind their operation.

18.1.2 Sources of Test Data

The dynamics engineer may obtain test data from one of three sources:

- *Laboratory tests*: These include dynamic tests of individual components and sub-systems, static or quasi-static tests of complete vehicles or, in the case of full-size roller rigs, dynamic tests of complete bogies or vehicles. All provide a greater degree of control over the test conditions than can normally be obtained by field testing.
- *Wayside measuring systems*: These collect data about many trains passing discrete locations or the same train passing those locations repeatedly. These data typically include the forces imparted by the vehicle onto the track, sound pressure, ground borne vibration, etc.
- *On-board measuring systems*: These allow continuous collection of data about the dynamic performance of a vehicle under a range of operating conditions. The data may be collected on special test trains or using instrumentation fitted to a service train. Wide variability in test conditions such as track quality, rail profile, load and wheel-rail friction coefficient are inherent in such tests. The test engineer can exercise some control over these, for example, by selecting track sections with the desired track quality for testing or adding weights to achieve the required vehicle load conditions.

Note that, although field tests may cover a wide range of conditions, they often do not include the worst cases, as these are rarely encountered in operation. It is therefore common to use a blended approach, that is, a mix of laboratory tests, simulations and field tests to obtain running safety acceptance.

This chapter mainly deals with laboratory and on-board measuring techniques. Laboratory roller rigs are discussed in detail in Chapter 19.

18.2 COMMON TRANSDUCERS

This section provides a brief overview of the range of transducers commonly encountered to measure displacement, acceleration and force.

18.2.1 DISPLACEMENT TRANSDUCERS

These are used for measuring linear or rotational displacements. The most common type of transducer is the linear variable differential transformer (LVDT). This comprises a transformer with a single primary coil and two secondary coils wound onto a hollow cylindrical tube, as shown in Figure 18.1. Within this tube, a ferromagnetic core can move up and down. The primary coil at the centre of the tube is excited with an AC signal, and this induces a voltage in the secondary coils. The secondary coils are normally connected as shown in Figure 18.2. This arrangement, known as 'series opposition' [3], has the effect of producing zero output voltage, with the core in its central or zero position. As the core is moved, the coupling between the primary and one of the secondary coils increases, whilst the coupling with the other secondary coil decreases in direct proportion. With correct arrangement of the coils and core, the resulting output voltage will be linear over the majority of the stroke. It should be noted that, as the core moves past the zero position (central on the primary coil), the output voltage undergoes a 180° phase shift.

In practice, a transducer that requires AC input and produces AC output is inconvenient, so a signal processing module is used in conjunction with the LVDT. This senses the zero-passing phase shift described above and uses this to distinguish between AC signals of equal amplitude on either side of the zero position. The resulting conditioned output is therefore a positive or negative DC voltage on either side of the zero position. The signal conditioning module usually also converts a DC supply voltage into the required AC excitation for the primary coil. The signal conditioning may be in a separate module but is more often incorporated within the transducer casing itself.

The LVDTs have the advantage of being inherently non-contact devices and therefore have no wearing parts. They typically achieve better than ±1% linearity over their specified range and are

FIGURE 18.1 LVDT example. (From RDP Electronics Ltd. catalogue. With permission.)

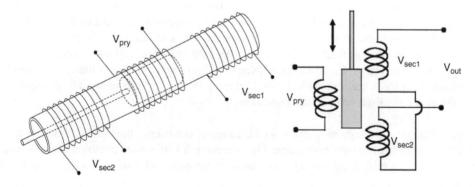

FIGURE 18.2 Schematic of primary and secondary coils in an LVDT. (From Iwnicki, S. (Ed.), *Handbook of Railway Vehicle Dynamics*, CRC Press, Boca Raton, FL, 2006. With permission.)

Field Testing and Instrumentation of Railway Vehicles

available commercially in measuring ranges from a few millimetres up to approximately 0.5 m. They generally operate on input voltages up to 24 V DC and may be obtained with floating cores or with a sprung loaded plunger.

18.2.2 Accelerometers

Accelerometers are electromechanical transducers that convert vibration into an electrical signal. Unlike displacement and velocity, acceleration can be measured as an absolute, rather than relative, quantity. This factor, combined with the accuracy, robustness and good frequency response/sensitivity of modern accelerometers, makes them ideal for use in vehicle dynamics test applications.

Figure 18.3 shows a simplified accelerometer. A mass (the seismic mass) is mounted within a rigid casing on a spring and damper. Accelerometers are designed such that the natural frequency of the seismic mass is high compared with the desired measuring frequency range. In such an arrangement, the amplitude of displacement of the seismic mass will be directly proportional to the acceleration exciting the transducer. It follows, therefore, that accelerometers work by sensing the relative displacement of the seismic mass with respect to the transducer casing. It can be shown, mathematically [3], that the maximum useful frequency range of an accelerometer, around 20%–30% of the transducer's natural frequency, is achieved with a damping ratio of 0.7. This damping ratio also provides almost zero phase distortion.

18.2.2.1 Piezoelectric Accelerometers

The most commonly used type of accelerometer is the piezoelectric accelerometer. The sensing element in such devices is a slice or disc of piezoelectric material. Such materials develop an electrical charge when they are subjected to mechanical stress. A number of naturally occurring materials exhibit this effect (e.g., quartz), but transducers typically employ man-made materials of a family known as 'ferroelectric ceramics' [4].

Practical accelerometer designs typically employ a seismic mass resting upon, or suspended from, a number of slices of the piezoelectric material. The vibration of the seismic mass within the accelerometer exerts a force on the piezoelectric material, and a charge is developed that is proportional to the force exerted. Three common designs of accelerometer are illustrated in Figure 18.4.

The centre-mounted compression design is a relatively simple arrangement where the mass is mounted on a centre pillar with a spring to provide the preloading. The mass acts in compression on the piezoelectric element. These designs have the advantage of good useable bandwidth.

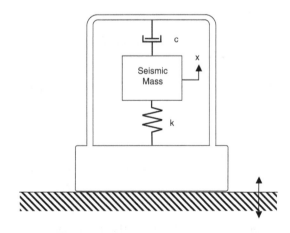

FIGURE 18.3 Simplified accelerometer. (From Iwnicki, S. (ed.), *Handbook of Railway Vehicle Dynamics*, CRC Press, Boca Raton, FL, 2006. With permission.)

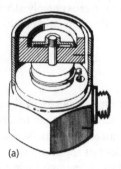 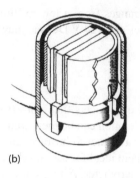 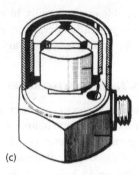

FIGURE 18.4 Common accelerometer designs: (a) centre-mounted compression, (b) planar shear, and (c) delta shear®. (From Serridge, M. and Licht, T.R., *Piezoelectric Accelerometers and Vibration Preamplifier Handbook*, Copyright © Brüel and Kjaer Sound & Vibration A/S, Denmark, 1987. With permission.)

However, as the base and centre pillar act as a stiffness in parallel with the piezoelectric element, any bending of the base or thermal expansion can cause erroneous readings.

Planar shear designs feature two slices of piezoelectric material either side of the centre post, each having a seismic mass attached to it. The masses are held in place by a clamping ring that preloads the piezoelectric elements and results in a high degree of linearity. The charge induced by the shear forces acting on the piezoelectric elements is collected between the housing and the clamping ring. In this design, the sensing elements are effectively isolated from the base, and these designs therefore have good resistance to base strains and temperature variations.

The delta shear design is similar to the planar shear version described previously. In this case, three masses and piezoelectric slices are mounted radially to the centre pillar at 120° to each other. Once again, a clamping ring preloads the elements. In addition to good resistance to base strain and temperature changes, these designs also have high resonant frequency and sensitivity.

An understanding of the useable bandwidth of a device is vital when selecting the correct accelerometer for a vehicle test application. The typical frequency response of a piezoelectric accelerometer is shown in Figure 18.5. As described previously, the upper frequency limit for the device will be dictated by the resonant frequency of the accelerometer. A commonly used rule of thumb

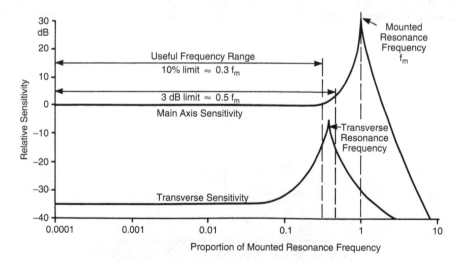

FIGURE 18.5 Frequency response of an accelerometer. (From Serridge, M. and Licht, T.R., *Piezoelectric Accelerometers and Vibration Preamplifier Handbook*, Copyright © Brüel and Kjaer Sound & Vibration A/S, Denmark, 1987. With permission.)

is that the upper frequency limit should be no more than one third of the resonant frequency. For general-purpose piezoelectric accelerometers, the resonant frequency may be of the order of 20 kHz, putting their upper useable limit way above anything likely to be required for a vehicle dynamics application. Piezoelectric accelerometers have one very important limitation with regard to vehicle dynamics test applications. The lower frequency limit is determined, not by the accelerometer itself, but by the time constant of the charge amplifier used to condition the signal from the transducer. Whilst it is possible to sense very low frequencies using preamplifiers with very high impedance, general-purpose equipment may limit the lower useable frequency limit to 1–3 Hz. Clearly, this has important implications for the vehicle dynamics engineer. A general-purpose piezoelectric accelerometer may be quite acceptable for mounting on unsprung masses, such as the axle box, but may be operating near its lower limit when mounted on a bogie frame. Vehicle body modes may occur at 0.5 Hz or less and therefore would be below the lower limit of a general-purpose piezoelectric accelerometer. Piezoelectric accelerometers are generally not capable of measuring the quasi-static accelerations due to curving. In the case of bogie and body measurements, the capacitive accelerometer, described in Section 18.2.2.2, will provide a solution to this problem. An alternative type of device, the piezoresistive accelerometer, can also provide a DC response, although these are typically used in applications where large accelerations must also be measured, such as crash tests.

An ideal accelerometer design would only respond to vibrations applied to the main sensing axis. However, in practice, most accelerometers will exhibit some sensitivity to excitations at 90° to the main axis, known as the transverse sensitivity. These are caused by small irregularities in the piezoelectric material, causing the axis of maximum sensitivity to be slightly misaligned with the operating axis of the accelerometer. It can be imagined, therefore, that the transverse sensitivity will not be constant, and, consequently, there will be directions of maximum and minimum sensitivity at 90° to each other. Some accelerometer designs will indicate the direction of minimum transverse sensitivity on the accelerometer body to aid correct mounting of the device. It is generally found that the transverse resonant frequency is lower than the main resonant frequency and therefore falls within the useable bandwidth of the accelerometer. However, at the relatively low frequencies of interest to the vehicle dynamics engineer, the maximum transverse sensitivity is usually less than 4% of the main axis sensitivity [4].

To achieve reliable results, care should be taken when mounting and cabling piezoelectric accelerometers. The device may be mounted using either a stud screwing into a hole tapped directly into the test component or by gluing it in place onto a clean flat surface. However, in many railway vehicle applications, the accelerometer will be mounted on its stud to a bracket, which will in turn be bolted or clamped to the vehicle. In this case, the mounting bracket and clamping arrangement should be as rigid as possible to ensure that the measured data is not degraded by vibrations or deflection of the mounting itself. A thin mica washer may be placed between the base of the accelerometer and the mounting, which, when used in conjunction with an insulated stud, increases the electrical isolation of the accelerometer from the vehicle. It should be noted that dropping an accelerometer onto a hard surface, such as a workshop floor, can cause a shock load that exceeds the maximum design limit and damages the device permanently.

As described previously, the piezoelectric sensor generates a small charge when subjected to mechanical stress. Any noise generated between the accelerometer and the signal conditioning/charge amplifier module can therefore adversely affect the accuracy of the results obtained. Flexing of the accelerometer cables can induce a charge as a result of the separation of the layers within the co-axial cable, known as the triboelectric effect. These charges can be sufficiently large to induce significant 'noise' when measuring low levels of vibration. It therefore follows that accelerometer cables should be securely clamped or taped in position to prevent flexing of the cables that induce such charges (see Figure 18.6). In addition, cable runs between accelerometers and charge amplifiers should be as short as possible, as the signal conditioning unit will generally output a strong DC voltage, which will be less sensitive to noise than the incoming signal from the transducer. An alternative (which may be preferable in many vehicle test applications) is to use an accelerometer with an

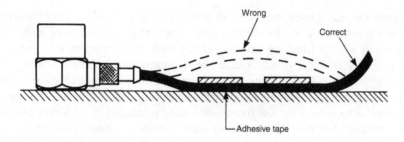

FIGURE 18.6 Accelerometer mountings and cabling. (From Serridge, M. and Licht, T.R., *Piezoelectric Accelerometers and Vibration Preamplifier Handbook*, Copyright © Brüel and Kjaer Sound & Vibration A/S, Denmark, 1987. With permission.)

in-built preamplifier that performs some or all of the required signal conditioning. For piezoelectric accelerometers, this is usually in the form of a miniature 'charge amplifier', which produces an output voltage proportional to the charge generated by the accelerometer. As with all test equipment, cabling runs should avoid sharp bends and be routed away from sources of electrical and magnetic interference, such as traction equipment and current collectors (third rail shoes/pantographs).

18.2.2.2 Capacitive Accelerometers

Although piezoelectric accelerometers have very-high-frequency upper useable limits, they can be limited to around 1 Hz at the lower end of their frequency range (dependent on the charge amplifier employed), as described in Section 18.2.2.1. In contrast, capacitive accelerometers have no lower limit on their useable frequency range, as they are capable of giving a DC or static response. They also have a number of other attributes that make them attractive for railway vehicle test applications. They generally exhibit no phase shift at low frequencies, are insensitive to thermal effects and electro-magnetic interference and have a high signal-to-noise ratio and a low transverse sensitivity, typically around 1%.

The sensor comprises a tiny seismic mass etched onto a slice of silicon, which is interposed between two further silicon plates that act as electrodes. The plates are arranged as a capacitive half-bridge, and the small space between the plates is filled with a gas that provides the necessary damping to the seismic mass [5]. The arrangement is shown in schematic form in Figure 18.7. A useful feature of the design is that the plates provide a mechanical stop for the seismic mass, preventing damage by shock loadings. When the accelerometer is stationary, the mass is central between the plates. Applying an acceleration causes the mass to move towards one of the plates and unbalances the capacitive half-bridge. This results in a charge that is proportional to the applied acceleration. Devices of this type normally include the signal processing within the accelerometer package. Therefore, although the half-bridge is excited by a high frequency AC voltage, the accelerometer requires only a low current (a few milliamps) DC input and provides a DC output that can be fed directly to a data logger.

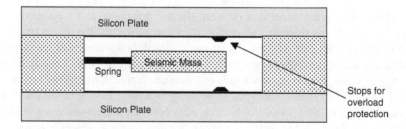

FIGURE 18.7 Schematic arrangement of a typical capacitive accelerometer. (From Iwnicki, S. (Ed.), *Handbook of Railway Vehicle Dynamics*, CRC Press, Boca Raton, FL, 2006. With permission.)

Field Testing and Instrumentation of Railway Vehicles

Owing to the modest upper frequency limit, mounting methods are less critical than for piezoelectric accelerometers, with the device being glued or bolted to the test components. Once again, however, it is essential to avoid any additional vibrations that may result from insufficiently stiff mounting brackets. Typical capacitive accelerometers suitable for body or bogie mounting may have accelerations of ±2 g with a frequency response of 0–300 Hz or of ±10 g with a frequency response of 0–180 Hz. It should be noted that, in general, increasing the acceleration range is achieved at the expense of lower sensitivity. As static devices, capacitive accelerometers also measure acceleration due to gravity. The output from the accelerometer will therefore be the sum of the vibration being measured and the component of the acceleration due to gravity acting on the main sensing axis.

18.2.2.3 MEMS Sensors

A wide range of miniaturised sensors is now available, known as micro-electromechanical systems (MEMS). The MEMS sensors are etched onto a single micro-chip and typically contain both mechanical micro-sensors and micro-electronics. The sensor elements are typically very small, ranging from a few microns up to 2 mm^2 and have very low power consumption. Readily available MEMS sensors include accelerometers, gyroscopes and temperature and pressure transducers, and many commercial off-the-shelf transducers now incorporate a MEMS sensor element. The MEMS accelerometers often work on the capacitive principle described in Section 18.2.2.2 and give a DC response, which is useful in measuring the low frequencies (typically <1.5 Hz) found in rail vehicle carbodies. In general, the same considerations apply to sensor selection for MEMS sensors as to other types of transducers. These include the sensor range, frequency response, input and output voltages, sensitivity and resolution.

The most common application of MEMS sensors are the inertial measurement units (accelerometers and gyroscopes) used within smart phones and for the stabilisation of drones. Useful applications for vehicle dynamics engineers include the ability to package sensors on printed circuit boards in small spaces. This makes them suitable for use in individual components or for wireless sensing in difficult locations (axlebox, load measuring wheelsets, etc.). It is possible to use smart phone accelerometers, together with a data logging app, as a quick and easy means of recording carbody accelerations, for example, to assist in diagnosing ride comfort issues or collecting validation data. It should be noted that the specification of accelerometers in smart phones may be difficult to determine, and data logging time and frequency range may be limited.

18.2.3 STRAIN GAUGES

The science of force and strain measurement is a complex one, and it is not possible to provide more than an introduction to the subject in this context. Strain gauges operate on the principle of measuring the change of resistance of a conductor when it is subjected to a strain. This change of resistance is generally measured using a bridge circuit as described later. The most common form of strain gauge is the foil gauge in which the required pattern is etched onto a thin metal foil, a simple example of which is shown in Figure 18.8. A good strain gauge will have two apparently conflicting requirements. It must have a short 'gauge length' in order to provide a point measurement of strain on the test specimen, whilst having the longest possible conductor to give the maximum change in resistance per unit strain. It is for this reason that most foil gauges use a folded or 'concertina' pattern, as illustrated in Figure 18.8.

The change in resistance of the gauge is related to the strain (i.e., the change in length of the gauge) by a constant known as the gauge factor given by:

$$k = \frac{\Delta R/R}{\Delta L/L}$$

and, since strain is defined as $\varepsilon = \Delta L/L$

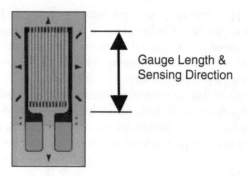

FIGURE 18.8 Foil strain gauge. (From Micro-Measurements, a Vishay Precision Group brand, www.Micro-Measurements.com. With permission.)

$$k = \frac{\Delta R/R}{\varepsilon}$$

where k = gauge factor, ΔR = change of resistance, R = unstrained resistance, ΔL = change in gauge length, L = unstrained gauge length and ε = strain (normally quoted in terms of micro-strain).

The higher the gauge factor, the higher the sensitivity of the gauge. Good linearity is also a key requirement for accurate measurement; foil strain gauges typically have gauge factors around $k = 2$ and linearity varying from ±0.1% at 4000 $\mu\varepsilon$ to ±1% at 10,000 $\mu\varepsilon$. Many configurations of foil strain gauges are available for a variety of strain measuring applications, a selection of which is shown in Figure 18.9. Thick-film and semi-conductor strain gauges with $k = 10$ to 20 and $k \approx 50$ have considerably higher sensitivity than foil gauges, but are less commonly used.

Foil gauges are particularly delicate items, and considerable care is required when mounting them on a test component if accurate and reliable measurements are to be achieved. Mounting surfaces must be polished to a good surface finish and then cleaned with specialist solvents and cleaning agents to remove any oil contamination and oxides and ensure that the surface is at the optimum pH for bonding. An adhesive will then be applied to the surface of the part, and the gauge, complete with its backing tape, will be pressed onto the surface. Finally, the backing tape is peeled

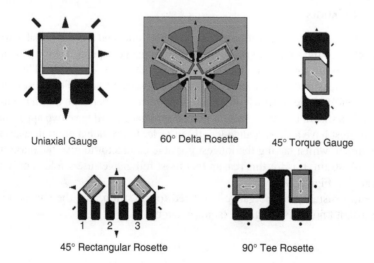

FIGURE 18.9 Examples of various strain gauge configurations. (From Micro-Measurements, a Vishay Precision Group brand, www. Micro-Measurements.com. With permission.)

away, leaving the gauge bonded to the test specimen. Gauges may be supplied with leads already attached to them, or the leads may be soldered in place once the gauge is mounted. In either case, it is good practice to bond a terminal in place adjacent to the gauges, from which the main leads can be led away to the bridge circuit. Once the installation is complete, the gauges should be tested before being finally encapsulated in a protective coating to prevent ingress of water and other contaminants. Strain gauges are now available with various mounting systems, including weldable gauges that are fixed to the component by using a small spot welder.

Strain gauges have a number of applications in the fields of rail vehicle testing. They are commonly the basis of force-measuring devices such as load cells and force-measuring wheels. Strain gauges may also be attached directly to components that are being tested under laboratory conditions. In both these instances, the load cell or component can be mounted in a test machine, and the resulting strain can be calibrated against a known force input. Strain gauges are also widely used in structural test applications such as determining the strain regimes present in bogie frames or vehicle bodies for testing performed either in the laboratory or on track. The data generated may be used to provide validation for finite element models or as the basis of fatigue calculations. However, as gauges measure strain at singular points, considerable care and skill are required to ensure that the critical elements of a structure's behaviour are captured.

18.2.3.1 Bridge Circuits

The change of resistance generated by strain gauges is very small, typically of the order of a few hundredths of an ohm. The most convenient means of measuring such changes is with a Wheatstone bridge. This is composed of four resistances connected to a DC power supply, as shown in Figure 18.10. If $R_1 = R_2 = R_3 = R_4$, the bridge is said to be balanced, and a voltmeter connected across the bridge as shown will read 0 V. It can be shown [3] that $R_1/R_3 = R_2/R_4$.

This equation highlights two important factors about the Wheatstone bridge. If more than one strain gauge is connected in the measuring circuit, the sensitivity of the bridge can be increased. It is also apparent that changes in resistance on one half of the bridge may 'balance' by changing the resistance of the other half. As described later, this provides a useful method of compensating for the temperature sensitivity of strain gauges.

In order to measure strains, the resistors shown in Figure 18.10 are replaced by one or more strain gauges (which are of course variable resistors whose resistance changes with applied strain). At the start of the test, the balancing potentiometer is used to balance the bridge, giving 0 V at the voltmeter. Applying the test load will then cause the resistance of the strain gauge to change and unbalance the bridge again, producing a voltage output that is proportional to the applied strain. The actual

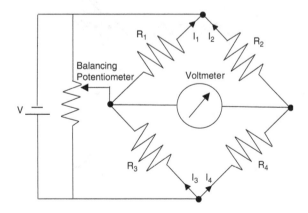

FIGURE 18.10 Wheatstone bridge (with balancing potentiometer). (From Iwnicki, S. (Ed.), *Handbook of Railway Vehicle Dynamics*, CRC Press, Boca Raton, FL, 2006. With permission.)

strain can then be calculated. It is not uncommon to find situations where it is not possible to calibrate the strain measurement system using a known test load such as, for example, when gauging a large structure such as a vehicle body. In these cases, it is possible to undertake an electrical calibration by placing a high resistance in parallel to the active arm(s) of the bridge. This method is known as shunt calibration and assumes that the surface strain in the test component is fully transmitted to the strain gauge in which it produces a linear response. The fact that the active gauge itself is not a part of the calibration is clearly a drawback, however, providing sufficient care is exercised, results from such set-ups may be used with a reasonable degree of confidence. Where long cable runs are used, it is essential that the shunt resistance is applied across the gauge, with the cable in place, to ensure that the effects of cable resistance are accounted for in the calibration.

A bridge may contain one, two or four strain gauges in the arrangements shown in Figure 18.11, and these are known as a ¼ bridge, ½ bridge and full bridge, respectively. A ¼ bridge will have the lowest sensitivity of these three arrangements. If no precautions are taken, it may also produce errors if the gauges used are sensitive to thermal effects. In order to prevent this, one of the resistors adjacent to the active gauge may be replaced with a 'dummy' gauge. This will be an identical strain gauge to the active one that is subject to the same environmental conditions but is not subject to loading, achieved, for example, by mounting it on an unstressed part of the test component. Both gauges will be exposed to any temperature changes, and the effect is to cancel out any resulting unbalance

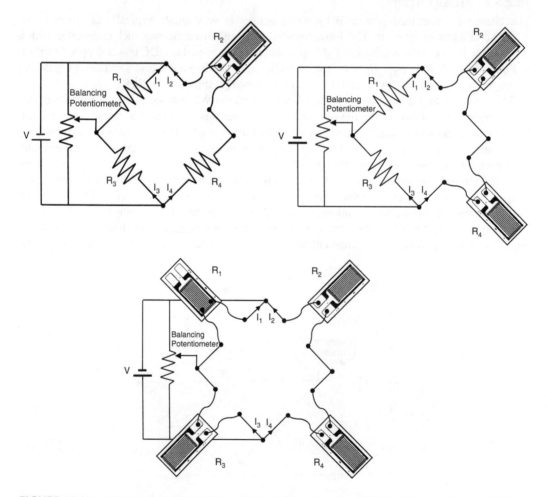

FIGURE 18.11 Arrangements of the ¼, ½ and full bridge. (From Iwnicki, S. (Ed.), *Handbook of Railway Vehicle Dynamics*, CRC Press, Boca Raton, FL, 2006. With permission.)

Field Testing and Instrumentation of Railway Vehicles   735

on the bridge due to thermal effects. A ½ bridge will have a higher sensitivity than a ¼ bridge, as the additional strain gauge will produce a larger unbalanced voltage across the Wheatstone bridge. The presence of two active gauges will also cancel out any thermal effects, as described previously. A full bridge will have the greatest sensitivity of the three arrangements and will similarly be self-compensating for temperature changes (Figure 18.11).

Strain gauges that are self-compensating for temperature changes are available, and the need for dummy gauges is therefore eliminated. However, changes in temperature can also affect the resistance of the lead wires and connectors, and, if no dummy is present, such changes may unbalance the bridge, resulting in errors in the strain measurement. Such errors can be minimised by the use of a 'three-wire' arrangement such as that described in [6].

It should be noted that the output voltage changes from strain bridges are usually very small and therefore should be amplified as close to the bridge as possible. Once again, cabling should be fully screened and carefully installed to prevent unwanted noise from interfering with the test data. It may be advisable to include dummy gauges in the system that are subject to the same environmental conditions, wiring and connection arrangements as the active gauges but are not subjected to strain. These can be used to assist in determining the level of noise present.

18.2.4 FORCE-MEASURING WHEELSETS

European vehicle acceptance standards [1] call for the assessment of wheel-rail forces (i.e., track forces) in newly developed or essentially modified main line rail vehicles, particularly for those operating at higher speeds. National standards and practices often call for track force assessment for acceptance testing. The data generated by force-measuring wheelsets is, obviously, also useful for direct validation of vehicle dynamics simulations. With force-measuring wheelsets, run over appropriate sections of track, it is possible to evaluate the vehicle-track interaction continuously under different operating conditions. Force-measuring wheelsets are often called instrumented wheelsets.

With the more advanced wheelsets, it is possible to measure lateral and vertical and sometimes even longitudinal wheel-rail forces, continuously during 'on-track' tests. With correct wheelset design and instrumentation, measuring precision can be good (of the order of 5%–10%). However, the instrumentation processes, as well as the procedures during the tests, are sometimes tedious and time consuming.

The sort of measuring device to be used is partly regulated in the current European standards. For low-speed vehicles with conventional running gear, with modest axle load and modest cant deficiency, it is not mandatory to measure Y and Q forces. In such cases, a simplified method with instrumented wheelsets measuring just the lateral forces between wheelsets and the axle boxes is considered as sufficient. In some cases (e.g., freight wagons with standard running gear at ordinary speed), instrumented wheelsets are not required. However, above a certain speed (usually >160 km/h for passenger vehicles or >120 km/h for freight wagons), Y and Q forces must be assessed. This is also the case for higher axle load or cant deficiency. In the European standard assessment of Y and Q forces, this is referred to as the 'normal measuring method'. This requires a more advanced technique for the instrumented wheelsets than for the simplified method described previously.

It should be pointed out that there is no known technique available for measuring the forces at the wheel-rail interface directly. Instead, reactions such as strains or accelerations must be measured in structures affected by the wheel-rail forces, that is, in wheels, axles and axle boxes. A brief overview of each of the techniques available is given later, whilst a detailed description of the different methods available can be found in [7]. Historically, strains in the track structure have also been measured, but such techniques are outside the scope of this chapter.

18.2.4.1 Measuring Lateral Forces between Wheelset and Axle Box

The simplest form of a force-measuring wheelset is to install strain-force measuring devices between axle journals and the axle boxes. The lateral force can then be estimated by the calibrated strain-force relationships. The operating principle of the device is shown in Figure 18.12. In this

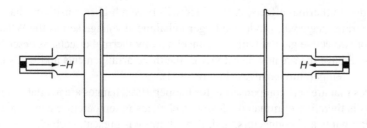

FIGURE 18.12 Lateral force, H, measured between axle journal and axle box. (From Iwnicki, S. (Ed.), *Handbook of Railway Vehicle Dynamics*, CRC Press, Boca Raton, FL, 2006. With permission.)

case, the lateral axle force is referred to as the H-force. It is similar, but not identical, to the track shifting force S. The difference is due to the wheelset mass force.

This simple H-force method can be further developed by attaching an accelerometer to the wheelset, measuring the lateral wheelset acceleration. This makes it possible to calculate and compensate for the lateral mass force of the wheelset. With this technique, it is possible to achieve a fairly good idea of the total lateral track shift force S between the wheelset and the track. However, with this method, it is only possible to measure the total lateral force on the wheelset or on the track; it is not possible to separate the lateral force between the two wheels, that is, the Y-forces, or to measure vertical Q-forces.

18.2.4.2 Measuring Lateral and Vertical Wheel-Rail Forces – Axle Method

Through the measurement of bending moments in the axle, on four cross-sections, it is possible to estimate approximate vertical and lateral forces on the wheels if mass forces generated by the wheelset are neglected. By additionally measuring two torques, approximate longitudinal forces can be calculated. Thus, with six measured moments and torques, it is possible to determine six forces (two longitudinal, two lateral and two vertical). The principle is shown in Figure 18.13. Moments and torques are measured by strain gauge bridges. Signals are transmitted to and from the axle through slip-ring devices inserted at one of the axle journals or by radio transmission.

This principle of measuring axle moments and torques seems, at first sight, to be fairly simple, efficient and accurate. A further advantage is that wheels can be changed on the instrumented axle. However, this method has two major disadvantages:

- Forces on the wheels may be applied at various positions. For example, the lateral position of the contact area may change by as much as ±35 mm over the wheel tread, thus the position of the vertical force application will also change. The changing positions will also change the moments measured in the axle, thus introducing errors that cannot be compensated because the actual position at which the force is applied is not known.

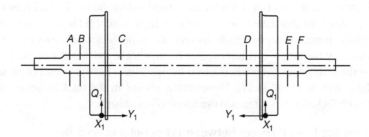

FIGURE 18.13 By measuring six bending moments and torques, approximate measures of forces x, y and q on the wheels can be determined. (From Iwnicki, S. (Ed.), *Handbook of Railway Vehicle Dynamics*, CRC Press, Boca Raton, FL, 2006. With permission.)

Field Testing and Instrumentation of Railway Vehicles

- Moments in the axle are dependent, to a small degree, on the vertical mass forces due to the unsprung mass of the axle and other unsuspended parts of the wheelset. Thus, it is not possible to fully assess the effects of the unsprung mass on the vertical dynamic forces.

Owing to the deficiencies described previously, this method has been further developed. By applying strain gauges on the wheel webs, the effects of varying positions are reported to be compensated. However, this makes the method more complicated and approaches the 'wheel methods' described in the next section.

18.2.4.3 Measuring Lateral and Vertical Wheel-Rail Forces – Wheel Methods

'Wheel methods' can be divided into two different techniques, either measuring strains in the spokes of spoked wheels, or measuring strains in the wheel web of ordinary railway wheels, that is, in the web between the axle and the outer wheel rim.

The 'spoked wheel method' is not frequently used nowadays, mainly due to the need to design and manufacture special wheels. In addition, the calibration procedures are tedious and time consuming. However, with properly designed and calibrated instrumented spoked wheels, this method is reported to produce a good accuracy. The mass forces of the unsprung mass are, to a large extent, included in the measured quantities.

The most frequently used method today (besides the simplified 'axle box method' described in Section 18.2.4.1) is the wheel web method. Within this method, a number of different technologies are used. The basic principle is that strains are measured at various locations on the wheel web as a result of the applied forces on the wheel, as shown in Figure 18.14. A number of strain gauges are applied on the same web, usually in the radial direction on the inside as well as on the outside of the wheel. However, these locations must be carefully selected.

Figure 18.15 shows an example of measured strains in single strain gauges of one wheel web as the wheel rotates and the wheel is loaded by lateral forces Y or vertical forces Q. In order to achieve signals proportional to the applied load, the strain gauges must be combined in Wheatstone bridges in an intelligent and precise way. Separate bridges are required for the lateral Y forces and the vertical Q forces. Sometimes, two bridges are used for the same force on the same wheel, installed at different wheel angles. In this case, additional data processing is needed to combine the two bridge signals. In a few cases, forces are measured in all the three directions: longitudinal, lateral and vertical. Signals are usually transferred to and from the wheels via slip rings, although radio transmission may also be used.

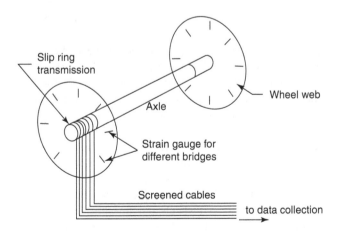

FIGURE 18.14 Schematic arrangement of force-measuring wheelset using the wheel web method. (From Iwnicki, S. (Ed.), *Handbook of Railway Vehicle Dynamics*, CRC Press, Boca Raton, FL, 2006. With permission.)

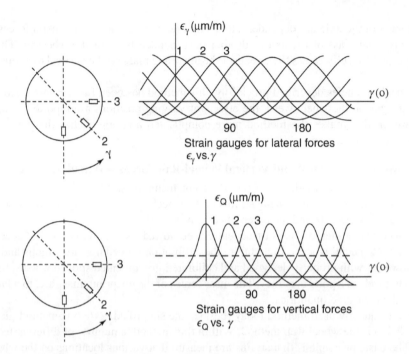

FIGURE 18.15 Example of measured strains in single strain gauges of one wheel web as the wheel rotates – y and q applied at the lowest part of the wheel. (From Iwnicki, S. (Ed.), *Handbook of Railway Vehicle Dynamics*, CRC Press, Boca Raton, FL, 2006. With permission.)

18.2.4.4 Compensation for Undesired Parasitic Effects

An important issue for force-measuring wheelsets is how to compensate for 'parasitic' effects and possible cross-talk between forces in the longitudinal, lateral and vertical directions. The parasitic effects include the influence of wheel rotation, temperature and temperature distribution and, finally, the location of the forces on the wheels. As described in the previous section, the lateral position of the vertical force application will change by as much as ±35 mm over the wheel tread, which may generate errors in the output signal. Also, electro-magnetic noise must be carefully considered, as very strong currents (1000–2000 A) will sometimes pass just some 50–100 mm away from the wheels and the cabling. Effects of water and humidity, temperature and mechanical impact must also be considered.

The overall goal is to achieve output signals proportional to the applied loads, with a minimum of cross-talk and parasitic effects. Several techniques are used for this purpose, for example, in France, Germany, Sweden, the USA and China. The selection of locations for the strain gauges, their connection in bridges and the additional data processing may vary considerably from one laboratory to another. The design, calibration and operation of this type of equipment are a highly specialised subject, and a detailed description is beyond the scope of this chapter.

The principal advantage of the 'wheel web method' is that it is possible to measure the Y and Q forces continuously quite close to the wheel-rail interface, and hence, most of the dynamic effect from the unsprung mass is included in the measured data. It is possible to measure quite high-frequency forces (up to at least 100 Hz). The measuring accuracy may be good or at least acceptable (within 5%–10% under normal conditions) if wheels and the whole system are properly designed and calibrated. The major drawback is the volume of work required for system design and calibration, requiring specialised knowledge, which generally makes the technique very expensive. A further drawback is that the instrumented wheelsets must very often be specifically designed for the type of vehicle to be tested.

18.2.5 Vehicle Speed and Position Measurement

A prerequisite of almost all on-track testing applications will be the ability to determine vehicle speed and position. This section discusses the most commonly used approaches to this problem.

18.2.5.1 AC Tachogenerator

The most commonly encountered means of generating a vehicle speed signal is the AC tachogenerator. This is, effectively, a two-phase induction motor, comprising a rotating magnet with a pair of stator coils arranged at 90° to each other and to the axis of rotation, as shown in Figure 18.16. One coil is excited with a constant frequency AC signal. The resulting eddy currents in the core induce an AC voltage in the sensing coil, which is proportional to the rotational velocity of the core. The direction of rotation can also be determined from the device, as the output voltage phase will change by 180° when the rotation direction is reversed. AC tachometers are generally used as they are less susceptible to noise and 'ripple' of the signal than their DC equivalents. They can also be fairly robust, an important consideration as axle box mounted equipment may be exposed to very high accelerations.

In order to determine actual vehicle speed and distance travelled from devices of this sort, the vehicle's wheel diameter at the nominal rolling position must be measured. Inevitably, the accuracy is limited by the lateral movement of the wheel from the nominal rolling position. If measurements are carried out over extended periods, the wheel diameter must be remeasured to compensate for the effects of wheel wear.

18.2.5.2 Hall Effect Probes

An alternative means of measuring wheel rotational speed is with a Hall effect probe. If a conductor with a current flowing through it is placed in a magnetic field whose direction is normal to the direction of the current, a voltage (the Hall voltage) will be induced across the width of the conductor. This is due to the magnetic field causing the electrons to take a curved path through the conductor. The effect is found most strongly in semiconductors, and these are therefore the basis of commercially available Hall effect probes [8]. Rotational speed measurement is achieved by combining the probe with a ferrous toothed wheel, as shown in Figure 18.17. As the ferrous tooth passes the probe, it causes the reluctance of the probes internal magnetic circuit to change, and this change produces a varying Hall voltage, where the frequency is proportional to the rotation speed. Once again, if the resulting speed signal is used to estimate distance travelled, the wheel diameter must be known and appropriate corrections made for wear.

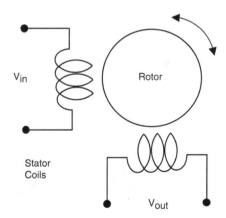

FIGURE 18.16 AC tachogenerator. (From Iwnicki, S. (Ed.), *Handbook of Railway Vehicle Dynamics*, CRC Press, Boca Raton, FL, 2006. With permission.)

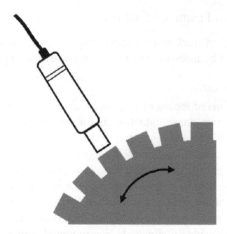

FIGURE 18.17 Hall Effect rotational speed sensor. (From Iwnicki, S. (Ed.), *Handbook of Railway Vehicle Dynamics*, CRC Press, Boca Raton, FL, 2006. With permission.)

18.2.5.3 Ground Speed Radar

This is a non-contact device that relies upon the Doppler effect to measure the vehicle speed. The Doppler effect is based upon the frequency shift that occurs when energy waves radiate from, or are reflected off, a moving object. A familiar example is the change in pitch in the noise from a train passing at speed. Owing to the Doppler effect, the pitch increases as the train approaches and then lowers as it departs. For a ground speed radar device, a high-, known-frequency signal is transmitted from the radar, aimed at a point on the track beneath the vehicle. The reflected signal will be detected by the sensor and the phase shift from the original transmitted signal will be calculated, allowing the velocity of the vehicle relative to the stationary target (the track) to be determined. The Doppler frequency shift is given by:

$$F_d = 2V \left(F_0 / c \right) \times \cos\theta$$

where F_d = Doppler shift Hz, V = velocity, F_0 = transmitter frequency Hz (typically 25–35 GHz), c = speed of light, and θ = offset angle.

A simplified arrangement of a Doppler effect speed sensor is shown in Figure 18.18. It should be noted that the sensor is usually mounted at an angle, as shown. This represents a compromise between

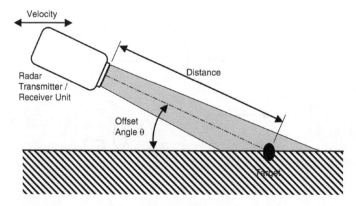

FIGURE 18.18 Simple Doppler effect speed sensor. (From Iwnicki, S. (Ed.), *Handbook of Railway Vehicle Dynamics*, CRC Press, Boca Raton, FL, 2006. With permission.)

the strength of the return signal, which is greatest with the sensor vertical, and reducing the sensitivity to vertical motion due to the vehicle/bogie bouncing and pitching on its suspension. Larger offset angles can introduce a 'cosine error' [9], as targets at the edge of the conical beam will be at a slightly different angle to those in the centre of the beam. Commercially available ground speed radar devices may have two sensors, one of which faces forwards while the other faces backwards. This arrangement can be used to automatically correct mounting or vehicle pitch errors. Devices of this type are typically mounted between 0.3 m and 1.2 m from the ground and have an accuracy for speed measurement of better than 1% above 50 km/h. Below this speed, the accuracy reduces somewhat (say within 0.5 km/h below 50 km/h), and these devices are therefore less suitable when accurate low-speed measurements are required.

18.2.5.4 Determining Vehicle Position

Before the advent of global positioning systems (GPS), a vehicle's position was recorded in the test data by means of marking events such as mile per kilometre posts in a separate data channel on the logger. This could be carried out manually by observations from the test coach, or alternatively, a known position at the start of the route could be synchronised with a data file containing a list of features and their locations for the chosen test route (known in the UK as a 'route setting file'). These would then be written to the test data based on the distance calculated from the vehicle speed. When such a method is used, it is normally necessary to resynchronise periodically against an observed position to remove the effects of measurement errors from the speed-/distance-measurement device.

Modern GPS systems make establishing the location of a vehicle relatively straightforward, provided an asset register is available to relate the logged GPS position to the location of stations and other infrastructure features. The GPS is based upon a network of satellites orbiting around 20,000 km above the Earth. They are arranged so that a GPS receiver should be able to see signals from four of these satellites at any given time. Each satellite transmits low-power radio signals, which can be detected by a GPS receiver on Earth. The signals contain information that allows the GPS receiver to determine the location of each satellite it is tracking and how far it is from the receiver. Knowing this information for three satellites allows the receiver to calculate a two-dimensional position (latitude and longitude) whilst adding the position of a fourth satellite produces a three-dimensional (3D) position (latitude, longitude and altitude). A more detailed description of the system may be found in [10]. Typical accuracies for GPS systems are in the order of 6–12 m. A system known as differential GPS (DGPS) uses Earth based 'reference stations' at known locations to determine corrections to the satellites transmitted positions. Using DGPS, accuracies of 1–5 m or better may be achieved.

For railway test applications, the GPS antenna is mounted on the roof of the vehicle to ensure that the maximum number of satellites is visible. However, since GPS is a 'line-of-sight' system, deep cuttings, tunnels, high buildings and other obstructions will prevent the system from working. In addition, it may be considered advisable to confirm the logged location by 'marking' the logged data, either by a manually activated signal against known locations (mile/km posts, etc.) or by automatically recording signals from trackside balises or signalling devices (e.g., Automatic Warning System (AWS) and Train Protection and Warning System (TPWS) loops in the UK).

18.2.6 WIRELESS SENSORS

Developments in wireless communication technologies and low-power-consumption MEMs sensors have provided an alternative to conventional wired sensors. Wireless transducers can be connected together to form a wireless sensor network (WSN). This allows data to be transmitted over short distances to a central hub. The WSNs are composed of three major elements:

- Sensor nodes that comprise the transducer itself, a microcontroller, a power source that may be a battery and/or energy harvester and a radio to transmit the data. A full range of transducers can normally be incorporated in nodes, for example, to measure acceleration, temperature and displacement.

- A gateway that is a radio receiver that collects the data from each sensor in the network. The data is then transmitted to a logger using a conventional network or internet connection. A variety of wireless communication standards are available for WSNs. These include Zigbee, which has a very low power consumption, a range of around 10 m and a data transmission rate up to 20 Kb/sec. Bluetooth radio has a range up to 100 m and a higher data transmission rate of 3 Mb/sec but, as a result, has a higher power consumption [11]. Other possibilities include Wi-Fi, using IEEE Standard 802.11-2007 [12], or the railway GSM-R radio network.
- A conventional data logger that receives the data from the gateway and stores and processes it, as described in Section 18.4.

The WSNs can be configured using a variety of different arrangements, as shown in Figure 18.19. In a star network, each node sends its data directly to the gateway, whilst, in a cluster tree, data is sent via the higher nodes in the tree until it reaches the gateway. A mesh network aims to achieve better reliability by allowing nodes to connect to a number of other nodes in the network to send data via the most reliable path. Both cluster tree and mesh WSNs have the advantage of extending the range of relatively low-power sensor nodes by allowing them to send their data via other nodes rather than direct to the gateway.

An important consideration when designing a test setup using WSNs is the battery life of each sensor node. Higher radio data rates, more frequent transmission of data and higher processing speed within the node itself, all reduce the battery life. Where continuous transmission of data is not required, sensor nodes are usually put to sleep to conserve power.

The advantages of WSNs are largely their ability to be installed in locations where providing wired sensors and power are difficult. These could include rail vehicle wheelsets and distributed networks of lineside sensors. The sensor nodes are typically compact, relatively cheap and have plug-and-play functionality; this allows new sensors to be added easily. However, WSNs also have disadvantages, which may include relatively low communication speed, limited memory and narrow bandwidth. Rail vehicle testing applications typically require continuous transmission of large volumes of data at relatively high sample rates and, due to the cost of organising track testing, with a very high degree of data transmission integrity. Conventional wired solutions are usually more suitable for these applications. However, they are several examples where the advantages of WSNs can be exploited. Figure 18.20 shows a load measuring wheelset developed by SNC Lavalin, which uses WSNs to avoid the need for slip rings to transmit data to the non-rotating parts of the system. The wheelset also uses an inductive power transfer system to boost the available battery power. Figure 18.21 shows an axlebox mounted vibration monitoring system developed by Perpetuum Ltd., which uses WSNs together with a vibration energy harvesting device to provide the necessary power. The system is used to measure accelerations on the axlebox, which are processed to diagnose the condition of bearings, gearboxes, wheels and track to avoid failures and provide data for predictive maintenance.

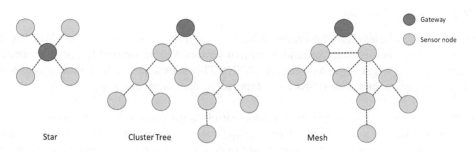

FIGURE 18.19 Alternative wireless sensor network configurations.

FIGURE 18.20 IWT4 wireless load measuring wheelset. (Courtesy of SNC Lavalin Ltd.)

FIGURE 18.21 Axlebox-mounted vibration monitoring system. (Courtesy of Perpetuum Ltd: From www.perpetuum.com. With permission.)

18.3 TEST EQUIPMENT CONFIGURATION AND ENVIRONMENT

The elements of a typical test arrangement are shown in Figure 18.22. The example shown might be appropriate for gathering data on the suspension behaviour or ride comfort of a vehicle, but the principles apply equally to many on-vehicle or laboratory test tasks.

Data from the selected transducers is passed through the appropriate conditioning electronics and transmitted to the analogue side of a data logger. Such signals are usually in the form of DC voltages, although some devices utilise AC voltage or varying current with a steady DC voltage. The signals are passed through an analogue-to-digital converter (ADC) and stored in digital format in the data logger memory. Many systems allow real-time display of the incoming data, and this is very useful for checking that the measurement system is performing correctly and that sensitivity settings for different channels are correctly configured. It is essential when vehicle acceptance or safety tests are being undertaken.

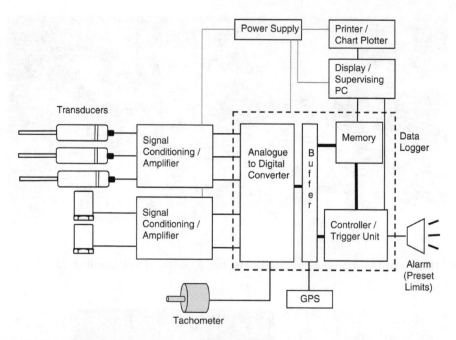

FIGURE 18.22 Typical test equipment configuration. (From Iwnicki, S. (Ed.), *Handbook of Railway Vehicle Dynamics*, CRC Press, Boca Raton, FL, 2006. With permission.)

The railway represents an aggressive environment in which to perform test measurements, and careful attention to detail is required if reliable test setups on vehicles or track are to be implemented. Transducers and cabling may be subject to high levels of transient vibrations, particularly if mounted on unsprung components (e.g., axle boxes and rails). Acceleration levels of the order of 30–50 g are not uncommon on axle boxes, and peaks of up to 100 g have been recorded. Extremes of temperature and weather must be catered for, and all externally mounted components should resist the ingress of moisture and dirt. The test data must not be affected by electrical noise from a wide variety of sources, including high-voltage AC or DC traction systems. Conversely, the test setup must preclude the possibility of generating electro-magnetic interference that could adversely affect train control or signalling systems. Fortunately, most commonly used transducers are low-voltage, low-power devices, and hence, the problem does not arise.

The following list highlights typical requirements for a reliable installation:

- All brackets should be rigid and robust and should prevent unwanted backlash between the bracket and transducer.
- Cabling should be secured to prevent (insofar as possible) movement; this is particularly important in the case of connections to transducers or joints between cables.
- Cables should be routed to avoid sources of electrical noise (traction motors, pantographs/collector shoes, wiring looms, generators etc.).
- Where possible, signal conditioning should be carried out within the transducers; where this is not possible (e.g., with some types of accelerometer requiring a separate charge amplifier), the distance between the device and the conditioning unit should be as short as possible, as very-low-power signals may be rendered useless by even modest amounts of electrical interference.
- Cables and connectors should be shielded to the highest available standards.
- All transducers, cables and enclosures should be sealed to a recognised standard such as IP66/IP67.

Field Testing and Instrumentation of Railway Vehicles

- Transducers should be selected to prevent the likely peak vibration transients exceeding the 'shock loading' specification for the device.
- Cabling lengths should be minimised and be routed inside vehicles at the earliest opportunity; cables should be arranged tidily, and long coils of spare cable should be avoided at the end of cable runs.
- Care should be taken to ensure that expensive transducers and data loggers cannot be exposed to damaging voltage 'spikes'.

Sources of power are an important consideration when conducting field testing. It is generally recommended that, when using power other than from the mains (e.g., from generators or vehicle sources), suitable voltage stabilisation and surge protection devices are used. In many cases, field power supplies cannot be guaranteed, and an alternative battery backup should be arranged to guard against the loss of important test data.

18.3.1 Transducer Positions on Vehicles

One of the most common instrumentation applications that the vehicle dynamics engineer will encounter is the fitting of accelerometers and displacement transducers to the body, bogies and suspension, for ride test, passenger comfort or dynamic response track tests. When choosing the locations for instrumentation, it is important to clearly understand the vibration modes that each transducer will 'see' to ensure that the correct data is gathered and that the desired information can be derived from it.

It is evident from Figure 18.23 that a vertical accelerometer mounted on the vehicle floor in line with the vehicle centre of gravity ($A1v$) will largely sense the bounce mode responses to the track input. As it is not practical to actually mount the accelerometer at the centre of gravity, some accelerations due to the change in floor height as the vehicle rolls will also be detected. Vertical accelerometers mounted on the vehicle floor above the bogie centre pivots ($A2v$ and $A3v$) will sense both the bounce and pitch mode responses. However, whilst accelerations due to body bounce will be in phase at both transducers, the accelerations due to pitching will be 180° out of phase. Assuming that the carbody is perfectly stiff, that is, no flexible modes occur, the pitch and bounce components of the signal at accelerometer $A2v$ can therefore be separated; thus:

$$A2_{bounce} = (A2v + A3v)/2$$

$$A2_{pitch} = (A2v - A3v)/2$$

Similarly, accelerometers $A5l$ and $A6l$ will sense body lateral and yaw responses in and out of phase, respectively. $A4l$ will likewise sense a combination of body yaw, lateral and roll modes. Providing that sufficient transducers have been provided, and their locations chosen carefully,

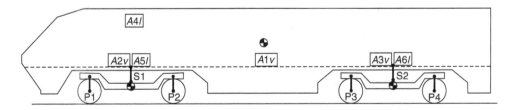

FIGURE 18.23 Simple transducer layout. Axv – vertical accelerometers, Axl – lateral accelerometers, Px – primary suspension LVDTs, and Sx – secondary suspension LVDTs. (From Iwnicki, S. (Ed.), *Handbook of Railway Vehicle Dynamics*, CRC Press, Boca Raton, FL, 2006. With permission.)

it should be possible to reliably establish the natural frequencies of the various modes of vibration of the vehicle body. The sensed accelerations on the body will also include the effects of flexible body modes. These may also be of interest to the dynamics engineer (e.g., when considering the effect of the first body bending mode on passenger comfort). Accelerometers will also detect inputs from body mounted mechanical equipment such as internal combustion engines and compressors. However, these will often occur at constant frequencies, somewhat higher than the frequencies of interest to the dynamics engineer, and may, if desired, be easily removed by filtering.

Similar considerations to those described previously will also apply to accelerometers mounted on the bogie frame or displacement transducers fitted across the primary or secondary suspension. Accurate records should be made of the mounting positions of all transducers on the vehicle to allow for later correction of geometric effects.

18.4 DATA ACQUISITION

Test data must be collected and stored in a suitable form for later analysis. Modern data loggers are usually either in the form of a PC with suitable additional hardware cards and software or as a standalone device with a PC compatible up-link. In either case, the analogue signals from the test devices must be converted to digital form to allow the data logger to store them. In addition to logging varying voltage or current signals from transducers, loggers may also have additional hardware inputs for digital signals and serial data (e.g., an RS232 connection to a GPS, CANbus, etc.).

As computers/data loggers store information in digital format, and most transducers provide an analogue signal, an ADC is required to convert the signal. The process of sampling and converting the signal leads to the possibility of two forms of inaccuracy in the digitised signal, known as aliasing and quantisation errors.

A key decision when setting up a data logging system is the sampling rate. This is the time interval at which the logger takes a 'snapshot' (sample) of the incoming analogue signal. If the sampling rate is too low, a high-frequency signal may appear incorrectly as a lower-frequency one, as illustrated in Figure 18.24. Although the aliased frequency shown here is one third the actual signal frequency, a whole series of aliases are possible, depending upon the sampling frequency chosen. Theoretically, the chosen sampling rate should be at least twice the highest frequency (of interest) in the sampled signal. However, in practice, it is normal to sample at between 5 and 10 times the highest frequency required in order to ensure good representation of amplitude as well as frequency.

An additional aliasing problem can arise if the sampled signal is degraded by an unknown, high-frequency noise component. A sampling rate of, say, 10 times the highest frequency of interest, could cause this noise to be aliased into the measurement frequency range, giving

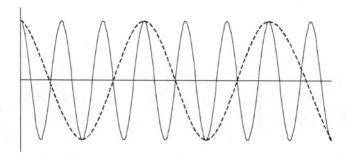

FIGURE 18.24 Aliasing due to insufficient sampling rate. (From Iwnicki, S. (Ed.), *Handbook of Railway Vehicle Dynamics*, CRC Press, Boca Raton, FL, 2006. With permission.)

incorrect results. The solution in this case is to low pass filter the signal by using an analogue filter prior to sampling to remove the unwanted component. This precaution is known as anti-alias filtering.

Quantisation errors are introduced by the ADC, forcing a continuous analogue signal into a limited number of discrete levels (binary digits or 'bits'). The error will be present in all digitised signals and has the effect of restricting the dynamic range of the signal. The magnitude of the error will be proportional to the resolution of the ADC. Quantisation errors are not generally a serious restriction for modern data loggers. However, care should be taken to ensure that the full range of the ADC is used. For example, consider an ADC using an 8-bit conversion, which allows 256 discrete states of the converted signal. If the ADC range is set to ±1 and a ±1 V signal is converted, the digitised signal will consist of 256 discrete values and will give a good representation of the original analogue signal. However, if the same signal is passed through with the ADC range set to ±10 V, the ±1 V signal may only have 256/10 discrete values, and a significant quantisation error results. It also follows, therefore, that the potential for quantisation errors are larger when a signal has a wide dynamic range.

Once the signal has been converted to digital format, further operations can be performed easily, such as filtering, bias or offset removal, amplification, etc. Detailed discussion of digital signal processing techniques is beyond the scope of this chapter, but numerous texts exist to guide the interested reader.

Important considerations when selecting a data logger include the following:

- The number of transducers to be used in the test and hence the number of channels required by the data logger.
- The rate at which each transducer is to be sampled – modern data loggers are generally capable of sampling at very high frequencies (of the order of kHz), many times higher than required for most railway vehicle dynamics applications; many loggers allow different sample rates to be set for different channels, though these all must generally be exactly divisible into the highest sample rate chosen.
- The duration of the test and hence the volume of data to be stored – assuming each sample for one transducer represents 1 byte, this can be easily calculated by multiplying the sample rate, the number of channels sampled and the required test duration to give the storage volume needed.
- The resolution of the ADC – this must be high enough to allow the transducer data to be collected to the required precision.
- The construction of the data logger – a PC may be appropriate for laboratory test work, but on-train applications may require a more rugged construction, and this will depend partly upon whether the logger is to work in a test vehicle on a temporary basis or to remain in a service vehicle over extended periods; it should be noted that traditional computer hard disk drives are not reliable when subjected to substantial or prolonged vibration, and, for on-train systems, consideration should be given to using more rugged components (such as solid-state hard drives and flash memory).
- The method of sampling employed – most high-end data loggers sample all the required channels simultaneously, ensuring that the data is perfectly synchronised; however, modern logging devices are available very cheaply, which, whilst only sampling each channel sequentially, can do so at high frequencies, and these loggers may be adequate where a small number of channels is only required to be sampled at low frequency.
- The user interface required during the tests – no interface is required if the data logger is to be unattended during the test. More commonly, a real-time display of all the data being logged may be needed allowing, for example, alarm thresholds to be set providing warnings if predetermined levels are exceeded on any channel. Logging tasks that also require extensive real-time calculations to be performed on the logged data may require the development of dedicated software.

It is always advisable to have the ability to view measured data online when conducting on-track testing work where the ability to repeat tests is limited. This is particularly important on the first day of an extended test programme in order to ensure that all data is as expected and to ensure that errors do not occur in the complete measuring system due to faulty scaling factors, electrical noise, etc.

18.5 MEASUREMENT OF WHEEL AND RAIL PROFILES

18.5.1 GENERAL REQUIREMENTS

The accurate measurement of wheel and rail profiles is critical to many vehicle dynamics simulation tasks. The general requirement is for a high degree of measuring accuracy, probably better than 0.02 mm. Early measuring devices for such tasks relied on moving a stylus over the profile, the shape being transferred via a linkage with a pen attached to a piece of tracing paper. An alternative mechanical measuring method used an indexing plate to allow a dial-test indicator to be moved to a number of known positions around the wheel or rail, a reading being taken at each. Such devices tended to be laborious to use and required the data to be manually entered into a computer for use in dynamics simulations. Several proprietary devices are available for making such measurements. These devices either use a mechanical linkage or a laser beam to electronically record the cross-sectional rail or wheel profile in terms of Cartesian coordinates. Vehicle dynamicists will usually require a high degree of accuracy to be maintained when taking profile measurements, as wheel-rail forces may be significantly affected by variations in shape of a few tenths of a millimetre. In general, measuring systems to provide this level of accuracy rely on readings being carried out manually during track walks or depot visits. Caution should be exercised if using data from automated in-track or on-train inspection systems, as the need to carry out measurements at speed may limit the accuracy of the recorded profile.

The device must be capable of measuring not only the profile, but also its correct orientation in the lateral-vertical plane. Small errors causing rotations of a wheel or rail cross sectional profile will appreciably alter the resultant contact conditions and the resulting equivalent conicity and must be avoided for realistic simulations. Several devices use the flangeback as the datum position for wheel profile measurements. This can generate significant errors when measuring profiles where the flangeback is worn, for example, by contact with check rails and the measuring device is 'tipped' as a result of being mounted on the worn flangeback. Experience shows that this simple approach therefore has limitations. For measurement of both wheel and rail profiles, the device should have an outrigger that bears on the opposite wheel or rail to ensure that the measuring head remains in the plane of the track/wheelset. Such equipment is often available for rail profile measurement but is less commonly used when measuring wheels. In all cases, however, a system using two measurement heads capable of recording a pair of wheel or rail profiles and their relative orientation simultaneously will provide superior results.

As the requirement is generally to gather profiles that are representative of a given location or situation, the exact position at which profiles are measured should be chosen with some care to avoid localised pits or defects. Profiles should always be cleaned prior to taking the measurement, as contamination on the wheels or rails (e.g., dirt, wear debris, or grease) may prevent accurate recording of the profile shape. Similarly, it is particularly critical that the measuring head of contacting devices is also kept clean.

Commonly available equipment for measuring wheel and rail profiles may be either of a contacting type, where a stylus or wheel is moved over the surface of the profile, or of a non-contacting type, where an optical sensor is employed, and it is convenient to consider these separately.

18.5.2 Contacting Measuring Devices

An example of a widely used device is the MiniProf (Figure 18.25). The measuring head in this case (which is similar in both wheel and rail devices) consists of two arms and two rotary optical encoders. These are used to determine the position of a magnetic measuring wheel. The device is connected to a laptop or rugged computer, with either a wired or Bluetooth radio connection. As the operator moves the measuring wheel around the profile, the signals from the encoders are logged by the computer. The profile of the wheel or rail being measured is then calculated from the position of the two arms, as obtained from the encoders, with suitable correction being applied to allow for the varying contact position on the measuring wheel itself. The use of a wheel rather than a stylus has a filtering effect upon the measured profile, as the measuring wheel cannot follow very small surface irregularities. However, providing that the radius of the measuring wheel remains significantly smaller than the smallest wheel or rail profile radius, this effect is of little importance. Software supplied with the MiniProf allows profiles to be viewed and provides the facility to undertake various geometric wear calculations against 'reference' profiles (Figure 18.26).

(a)   (b)

FIGURE 18.25 MiniProf measuring instruments: (a) for rail and (b) for wheels. (Courtesy of © Greenwood Engineering, Denmark.)

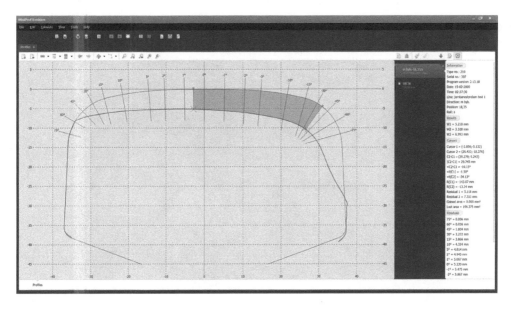

FIGURE 18.26 Viewing measured profiles using MiniProf software. (Courtesy of © Greenwood Engineering, Denmark. With permission.)

18.5.3 Optical Measuring Devices

A range of optical devices are available for profile measurement, almost all based on laser profile sensors. A laser beam is passed through a lens to enlarge it into a line that is projected onto the wheel or rail profile. The reflected light is in turn collected by a receiver and projected onto a high-resolution CMOS sensor matrix. The position of the laser line on the CMOS array allows the distance to each point on the line to be determined using the laser triangulation principle (Figure 18.27). The profile shape can then be output as a series of 2D or 3D coordinates. Modern sensors have a scanning rate greater than 1 kHz and can achieve very high accuracy, even on highly polished surfaces such as a railway wheel.

An example of a commercially available system known as Calipri is shown in Figure 18.28. This device uses a handheld laser scanner that is moved slowly over the wheel or rail. The measured profile is logged on a rugged PC or tablet in a similar manner to the MiniProf, and a software application allows the profiles to be viewed, manipulated and overlaid and carries out calculations for wear and/or equivalent conicity.

Non-contacting devices have the advantage that they are relatively easy to use and can often collect measurements more quickly than a contacting device. Conversely, they may not achieve

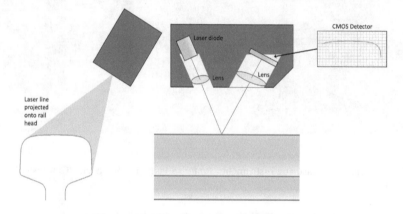

FIGURE 18.27 Optical measurement based on laser triangulation.

FIGURE 18.28 Calipri wheel profile measuring instruments. (Courtesy of Nextsense; From www.nextsense-worldwide.com/en/industries/railway.html. With permission.)

Field Testing and Instrumentation of Railway Vehicles   751

the absolute accuracy in both profile shape and orientation that can be obtained from a contacting device. Particular care is required when using optical devices to ensure that the wheel or rail is cleaned to remove dirt, debris and lubricant deposits before measurements are taken, as the laser sensor will include these in the measured profile. Any such errors will prevent the achievement of accurate wheel-rail contact calculation conditions.

Similar optical systems exist, which are designed to give 'real time' measurements of wheel profiles, either track based automated systems in depots or train-based systems. In the past, these have not achieved sufficiently high accuracy levels for use in vehicle dynamics simulations, and their use has primarily been for measuring wheel and rail profile wear. This is because some of the accuracy is sacrificed for the speed of measurement. However, with advances in scanning laser technology, this may well change in the future.

18.6 TRACK GEOMETRY RECORDING

The vehicle dynamics engineer frequently requires data describing real, representative track geometry as a basis for simulating vehicle behaviour. It is convenient, for both maintenance engineers and vehicle dynamicists, to separate the long wavelength features that represent the design layout of the track from the short wavelength features that form the variation from the design (i.e., the track irregularities). This usually results in a description based on the following five geometrical terms:

- *Curvature*: The lateral design layout of the track radii (long wavelength) and is often defined as the inverse of the curve radius in units of rad/km; however, curvature may also be quoted as a 'versine' measurement in millimetres, this being the distance from the centre of a chord of known length to the rail
- *Cant*: The vertical difference in height of the left and right rails (long wavelength)
- *Lateral alignment*: The short-wavelength lateral track irregularities
- *Vertical alignment*: The short-wavelength vertical track irregularities
- *Gauge*: The distance between the rails measured at a specified distance below the crown of the rail, and this is typically 14 mm in the UK and Europe, and 5/8 inch in North America; gauge may be given as an absolute value or as a variation from a nominal gauge (e.g., 1435-mm standard gauge)

The geometry of railway track may be recorded using one of several techniques as described in Sections 18.6.1 through 18.6.4.

18.6.1 Manual Survey

The surveyor establishes a datum (or several datums) position on the site to be surveyed, usually by placing a marker in the ground. A theodolite is then used to record the position of the left and right rail with reference to the datum position. Considerable care is required to produce accurate results from these techniques. Good results have been achieved by using a high-accuracy 'autotracking' theodolite measuring to a target placed on the fixed end of a cant and gauge stick above the rail gauge corner. The theodolite is then used to measure the position of one rail, and the position of the adjacent rail is determined from the cant and gauge measurements displayed on the stick (see Figure 18.29).

In the absence of other methods, useful results may be obtained for track design curvature by conducting a versine survey. In this case, a chord (wire) of fixed length is stretched along the high rail of the curve, and the distance between the centre of the chord and the gauge corner is measured. Chord lengths of 10, 20 or 30 m are common, depending on the curve radii to be measured. The chord length is normally chosen so that the measured versine does not exceed 150 mm on the tightest curve to be surveyed. Successive versine measurements are taken at frequent intervals,

FIGURE 18.29 Cant and gauge stick. (Courtesy of Abtus Ltd: From Abtus Ltd. catalogue. With permission.)

with the maximum recommended interval being half the chord length. Increasing the measurement frequency will increase the detail contained in the survey results. The radius of curvature at any mid-chord position can then be calculated as follows:

$$R = \frac{C^2}{8V}$$

where R = curve radius (m), C = chord length (m), and V = versine (mm).

However, the ability of such techniques to 'see' short wavelength lateral irregularities is inherently limited, as the position of the datum (the wire itself) depends upon the track irregularities at either end of the wire. This survey method is simple and cheap to carry out and requires limited equipment, but has the disadvantage of needing three people to undertake the survey.

Manual surveys are generally fairly slow to carry out and are therefore limited to short sections of track.

18.6.2 Track Geometry Trolley

A number of proprietary recording trolleys are available. These carry instrumentation and a data logging system to allow track irregularities to be measured. They commonly measure vertical irregularities cross-level and gauge and (less commonly) lateral irregularities and curvature. An electronic record of the geometry is stored in the on-board data logger for later retrieval. Some feature on-line calculations allowing alarms to be set for exceedances in, for example, track twist over a given distance. Trolleys are generally pushed at walking pace by an operator, though some are self-propelled at low speed. The length of line that can be surveyed by this method is generally greater than for manual surveys but is limited by the slow recording speed and the capacity of the on-board data logger and power supplies (Figure 18.30).

18.6.3 Track-Recording Vehicles

Most railway administrations operate dedicated track geometry-recording vehicles. These vehicles are equipped with measuring systems, often based upon the inertial principle described later, which allows data to be gathered at high speeds (up to 350 km/h). Such specialist vehicles have extensive data storage and analysis capabilities, which allow regular surveys of entire routes to be undertaken at normal running speeds. These vehicles provide the most commonly used and highest quality source of track geometry data for vehicle dynamics engineers.

The following description of the inertial measurement is based upon the track-recording systems used in the UK and described in [13–15]. The general principles are, however, common to all systems of this type. The signal from an accelerometer mounted on the vehicle body is double integrated to provide a displacement measurement. This is then low pass filtered to remove the long wavelength design information, effectively creating a moving average datum for the measurement of the shorter-wavelength features.

FIGURE 18.30 Track geometry recording trolley. (Courtesy of Abtus Ltd: From Abtus Ltd. catalogue. With permission.)

Vertical (track top) measurements are made using one wheelset on the vehicle as the sensor. Displacement transducers are fitted across the primary and secondary suspensions, as shown in Figure 18.31, with an accelerometer mounted directly above them. Subtracting the suspension displacements from the body displacement (double integrated from the acceleration) gives the track top profile. As the suspension movements are removed from the final answer, the system is effective, regardless of suspension type. It is now common practice to mount the inertial measurement unit (IMU) containing the accelerometers on the bogie frame rather than on the vehicle body, as shown in Figure 18.31. This removes the need to obtain secondary suspension displacements, which are inconvenient to measure because of the large rotational movements between bogie and body. This change, which has been partly enabled by the availability of miniaturised IMUs, has led to the development of unattended track geometry measurement systems (UGMS) fitted to in-service vehicles.

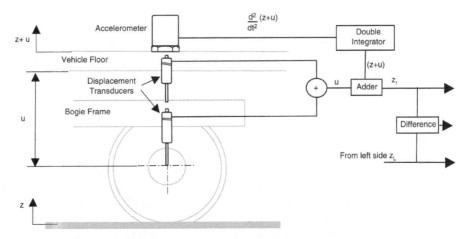

FIGURE 18.31 Vertical measuring system schematic. (From Iwnicki, S. (Ed.), *Handbook of Railway Vehicle Dynamics*, CRC Press, Boca Raton, FL, 2006. With permission.)

Non-contact measurement of the track gauge at high speed presents a considerable challenge. Modern systems usually use the laser triangulation method described in Section 18.5.3 and shown in Figure 18.27. The reflected laser light is used to measure the position of left and right rails, and these are combined to obtain the gauge.

The lateral irregularity of the track is obtained by subtracting the rail position, measured by the laser displacement sensors described above, from the inertial datum produced by a body-mounted lateral accelerometer.

The cross-level is determined by subtracting the difference in the vertical suspension displacements from the body roll angle obtained from an on-board gyroscope. The gyroscope also provides the plan view rate of turn of the body and this, together with the vehicle velocity, allows the curvature to be calculated.

The foregoing description is, of necessity, a somewhat simplified version of what is a sophisticated and complex measuring system. It is worth noting that the lateral irregularity and curvature channels are effectively short- and long-wavelength parts of the same signal. Track geometry data is normally supplied at 0.2–0.25 m intervals in the UK and will not therefore adequately capture very-short-wavelength features of less than 1.5 m, such as dipped joints. The system also does not capture wavelengths greater than 70 m and, being inertially based, will only provide data at above 30 km/h. Other systems capable of measuring longer wavelengths up to at least 100 m are now becoming available, for example the Swedish STRIX system developed by Banverket Production.

It can be seen that all the data required by the dynamics engineer to reconstruct the track geometry is available from this type of track geometry vehicle. However, several operations must be carried out to ensure it is suitable:

- As transducers are mounted on different parts of the vehicle, there will be an offset or lag in some raw data channels depending upon the position of the transducers used to obtain them, and this offset may also vary depending upon the direction in which the vehicle is running; these offsets are normally removed from the data at source, but this should be confirmed.
- Filtering data introduces both phase and amplitude distortion in the filtered signal compared with the original, and as the cross-level and gauge channels are not filtered, the filtering will also introduce an offset between channels; clearly, this offset is not a realistic representation of the real track geometry, and, as it can significantly affect simulation results, it must be corrected so this is done by re-passing (backfiltering) the data through a filter of the same design as that originally used to restore both the phase and amplitude distortions caused by filtering

Track recording vehicles may also record a range of other parameters such gradient and cant deficiency and may derive other measures, for example, track twist or cyclic top from the raw data.

18.6.4 Unattended Track Geometry Measuring Systems

The increasing availability of compact rugged electronic sensors and a move to bogie mounted inertial measurement has led to an increasing trend to fit fully featured inertial track geometry measurement systems, as described in Section 18.6.3 to service trains. The use of such UGMS brings both challenges and opportunities. The limited number of dedicated measurement trains in turn limits the frequency with which track geometry can be measured on a particular route. This frequency is typically a function of traffic tonnage and line speed and might vary between bi-weekly on a busy mainline to yearly on a lightly trafficked route. The UGMS allows the measuring frequency to be increased greatly, potentially up to many times per day. This in turn brings the possibility of monitoring track geometry deterioration close to real time. However, this benefit can only be realised if systems are in place to handle (i.e., download, store, validate and analyse) the very large amount of

Field Testing and Instrumentation of Railway Vehicles   755

data produced by UGMS. Positional accuracy (i.e., precise geo-location of the data) also becomes very important to allow accurate run-on-run overlays. This is necessary so that repeated measurements of the same route can distinguish between a single geometry fault being reported multiple times and multiple faults in close proximity. Poor positional accuracy can lead to the same fault being reported in multiple locations, causing confusion when entered into the track engineers work bank. Frequent maintenance of UGMS is also important if accuracy is to be maintained.

The UGMS systems provide a potentially useful source of track geometry data for the vehicle dynamics engineer, for a range of applications from vehicle approval to accident investigation. As described in Section 18.6.3, care must be taken to remove the effects of instrumentation offset and phase/amplitude distortion introduced by filtering to ensure that the track shape is accurately reconstructed. The UGMS may be slightly more prone to spurious spikes and other measurement errors than a system fitted to a dedicated measurement train and such errors should be removed before the data is used for simulation purposes.

18.7 LABORATORY AND FIELD TESTING FOR VALIDATION AND ACCEPTANCE

This section provides examples of laboratory and field tests that may be commonly encountered by, or provide useful information for, the vehicle dynamics engineer. Whilst they are based specifically on European practice, similar tests are employed by many railway administrations worldwide.

18.7.1 STATIC/QUASI-STATIC TESTS

These tests are normally carried out on a vehicle in a specialist laboratory. The results may be used to gain confidence in the general behaviour of a vehicle model and also to estimate the additional (parasitic) stiffness present in the completed vehicle. However, as the dynamic behaviour can vary considerably from the static behaviour, some comparisons against dynamics tests (such as ride tests) are required to enable a vehicle model to be fully validated.

18.7.1.1 Wheel Unloading Test

The test is detailed in Sections 6.1.5.2 and 6.1.5.3 of [1]. The test may be carried out by running the vehicle at slow speed over twisted track with the defined characteristics. Alternatively, it may be carried out statically in the laboratory by placing packings under the wheels on one side of the vehicle to reproduce the track twist feature shown in Figure 18.32. In the laboratory test it is permissible to apply the twist by lifting rather than lowering the wheels (Figure 18.33), provided that the defined twists are imposed over the bogie and body. The resulting wheel loads are measured using a load cell and expressed in terms of the change from the static load as $\Delta Q/Q$. The test is repeated to place each corner of the vehicle in turn at the bottom of the dip. The test provides a relatively straightforward means of testing the vertical suspension behaviour. Difficulties can be encountered if the vehicle is not placed on level track at the start of the test. In this instance, it is important that the complete load-displacement cycle is recorded (jacking up and then down to the starting point) to obtain the full hysteresis loop. The initial error can then be estimated and included in the simulation. The limiting $\Delta Q/Q$ value is normally specified as 0.6.

18.7.1.2 Bogie Rotational Resistance Test

This test is detailed in Section 6.1.5.3.3 of [1]. One bogie is placed on a turntable and rotated both clockwise and then anticlockwise to an angle that represents the minimum operating curve radius for the vehicle. The test is usually performed at 0.2 and 1°/sec rotation speeds, and the torque required to rotate the bogie is measured (Figure 18.34). Where yaw dampers are fitted, they may or may not be disconnected during the test depending on the test requirements. If yaw dampers with positional control (i.e., designed to blow-off at a particular bogie rotation angle) are included, the results from the test will reflect both the velocity and displacement dependent nature of such an

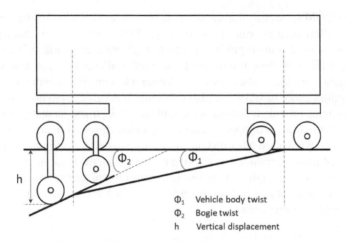

Φ_1 Vehicle body twist
Φ_2 Bogie twist
h   Vertical displacement

FIGURE 18.32 Combined carbody and bogie twist for wheel unloading test. (Adapted from BS EN 14363:2016, *Railway Applications—Testing and Simulation for the Acceptance of Running Characteristics of Railway Vehicles—Running Behaviour and Stationary Tests*, British Standards Institution, London, UK, 2016.)

FIGURE 18.33 Wheel unloading test. (Courtesy of Bombardier Transportation, Montreal, Canada.)

arrangement. It follows that the vehicle model must be simulated in the same condition if comparable results are to be achieved. The resulting X-factor is calculated as follows:

$$X = \frac{\text{Body to bogie yaw torque}}{\text{Wheelbase} \times \text{axle load}}$$

The limiting value is 0.1, except for freight vehicles where the limits are dependent on axle load, as shown in Section 6.1.5.3.4 of [1].

Field Testing and Instrumentation of Railway Vehicles   757

FIGURE 18.34 Bogie rotation test. (Courtesy of Bombardier Transportation, Montreal, Canada.)

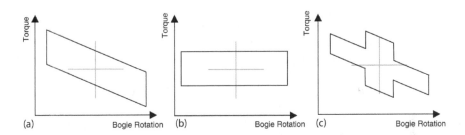

FIGURE 18.35 Typical bogie rotation test results for common suspension types: (a) air spring/flexicoil suspension, (b) friction sidebearers, and (c) air spring/flexicoil suspension plus yaw damper with positional control. (From Iwnicki, S. (Ed.), *Handbook of Railway Vehicle Dynamics*, CRC Press, Boca Raton, FL, 2006. With permission.)

The measured bogie rotation torques are particularly useful when confirming the behaviour of a vehicle model with friction sidebearers or air spring secondary suspensions. Typical examples of results from such tests for various vehicles are shown in Figure 18.35.

The bogie rotation test rig itself is likely to include a certain amount of load dependent friction. This must be quantified separately by running a test using a dummy load equivalent to the bogie pivot load of the vehicle to be tested. The resulting forces can then be subtracted from the test results to obtain the true body-bogie yaw torque.

Before a bogie rotation test is carried out, the bogie is normally moved slowly to it maximum rotation position to check clearances between all body and bogie mounted equipment.

18.7.1.3 Sway Test

Such tests are normally carried out to generate the input data for the kinematic gauging process or to verify that the vehicle will remain within a predetermined static envelope. Their usefulness to

FIGURE 18.36 Sway test. (Courtesy of Bombardier Transportation, Montreal, Canada.)

the dynamics engineer is in enabling a reasonable estimate of the parasitic stiffnesses (particularly in roll and lateral directions) to be made. An overview of the test method is given in Appendix D of [1]. A number of targets are fixed to the end of the vehicle at cantrail, waistrail and solebar level. One side of the vehicle is raised in stages (Figure 18.36) to the maximum operating cant deficiency/excess +100 mm (typically approximately 10° or 250 mm of cant). The vehicle is then lowered back to level and the second side raised. A theodolite placed some distance from the end of the vehicle is used to measure the displacement of the targets as the vehicle is raised. Additional measurements of primary and secondary vertical and lateral suspension movement are very useful when comparing test and model results. It is important that measurements are made over the full range of lifting/lowering to complete the hysteresis loops. Care is required to avoid over-jacking the suspension at each stage of the lift, that is, raising it significantly beyond the desired point and lowering back onto packings, as this can make interpretation of the hysteresis behaviour difficult.

18.7.1.4 Body Modes Tests

Such tests not specified in connection with running safety but are specifically commissioned to provide the dynamics engineer with confirmation of the fundamental vibration modes of various parts of the vehicle. They are useful for comparison with the eigenvalue analysis used at the model checking stage as they will include the effects of all the parasitic stiffnesses. Tests are usually performed by disconnecting the vehicle dampers and then applying a sine sweep displacement signal in the direction of interest via an electric or hydraulic actuator. The response of the vehicle body is measured using both vertical and lateral accelerometers mounted at various positions along the length of the vehicle. These tests may also be used to confirm the first body bending mode.

18.7.2 Dynamic Tests

Vehicle ride tests provide the means to validate the dynamic behaviour of a model. EN 14363 [1] provides a detailed methodology for carrying out on-track testing. For acceptance, this may be according to the full or 'normal' method in which load measuring wheelsets are used to measure the lateral and vertical forces at the wheel-rail contact or the 'simplified' method in which lateral axlebox forces are measured. Both methods require measurement of accelerations at various positions on the bogie frame and carbody. The requirements of EN 14363 are relatively complex and the testing required, which involves running over a wide range of track quality and alignments at the full range of speeds and cant deficiencies, is typically costly to undertake. For vehicle model validation purposes, considerable value can be obtained from simpler tests using accelerometers placed at various positions on the body, bogie frame, and axle boxes. These may be complemented with displacement transducers to measure primary and secondary suspension displacements. It is essential that an accurate measurement of vehicle speed is included in the measurements. To be of maximum value for model validation purposes, the data should also be accompanied by recent track recording coach data for the test route and measured wheel profiles from the vehicle at the time of the test. Measured track geometry will, in any case, be required for EN 14363 tests in order to determine the test zones and test sections used for the statistical analysis of the test results.

The measured wheel profiles, track geometry, and speed data can be used to set up a time stepping integration (transient analysis) for the model in question, simulating the test run and outputs. The outputs from this simulation will be accelerations (and displacements) from specified positions on the vehicle such that they replicate the signals seen by the real instrumentation. Test data and model results can then be compared both in terms of time histories and power spectral densities. It is important to note that, if the test data is filtered, the characteristics of the filter must be known to allow the simulated data to be treated in the same way.

For comparison with simulation results raw acceleration data from ride tests is normally expressed in terms of the power spectral density (PSD) of the signal versus its frequency range, as shown in the example in Figure 18.37. The test data PSD plots are likely to include peaks representing the rigid body modes, flexible modes, and rail length passing frequencies. They may also include peaks associated with the traction unit, originating from vibration of the engine or driveline. The results may be further sorted into speed banded PSD to assist in identifying speed dependent effects such as excitation of various body/bogie modes and response to cyclic track geometry. In addition, the test data may be analysed in sections according to track type to determine the vehicle's response to

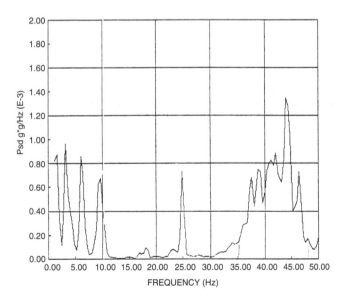

FIGURE 18.37 PSD of body vertical acceleration, mainline diesel locomotive. (From Iwnicki, S. (Ed.), *Handbook of Railway Vehicle Dynamics*, CRC Press, Boca Raton, FL, 2006. With permission.)

changes in track construction (welded, jointed, etc.) and track quality. Care should be taken to ensure that a full understanding of the vehicle behaviour is reached under a range of conditions, including any evidence of hunting or of body modes being driven by other modes such as bogie pitch.

Wheel-rail forces themselves can also be validated if test data is available from force-measuring wheelsets. Their use may be restricted due to their high cost, unless full testing using the EN 14363 'normal' method is carried out.

REFERENCES

1. BS EN 14363:2016, *Railway Applications—Testing and Simulation for the Acceptance of Running Characteristics of Railway Vehicles—Running Behaviour and Stationary Tests*, British Standards Institution, London, UK, 2016.
2. UIC Code 518, *Testing and Approval of Railway Vehicles from the Point of View of Their Running Behaviour Safety—Track Fatigue—Ride Quality, 4th Edition*, Union Internationale des Chemins de Fer, Paris, France, 2009.
3. J. Turner, M. Hill, *Instrumentation for Scientists and Engineers*, Oxford University Press, Oxford, UK, 1999.
4. M. Serridge, T.R. Licht, *Piezoelectric Accelerometers and Vibration Preamplifier Handbook*, Brüel and Kjaer Sound & Vibration A/S, Denmark, 1987.
5. T. Berther, G.H. Gautschi, J. Kubler, Capacitive accelerometers for static and low-frequency measurements, *Journal of Sound and Vibration*, 30(6), 28–30, 1996.
6. A.L. Window, G.S. Holister (Eds.), *Strain Gauge Technology*, Applied Science Publishers, Barking, UK, 1982.
7. O. Gabrielson, Analysis of a new method for measuring wheel-rail forces, Report TRITA-FKT 1996: 24, MSc Thesis, Royal Institute of Technology (KTH), Stockholm, Sweden, 1996.
8. I. Sinclair, *Sensors and Transducers*, 3rd edition, Butterworth-Heinemann, London, UK, 2001.
9. GMH Engineering, Non-contact speed measurement using Doppler radar, Application Note 1000, 2017. (Available at: http://www.gmheng.com/application_note_1000.pdf).
10. Garmin Corporation, GPS guide for beginners, Part Number 190-00224-00 Rev. A, 2000. (Available at: https://cmnbc.ca/sites/default/files/Gps4beginers_0.pdf).
11. M. Kocakulak, I. Butun, An overview of wireless sensor networks towards internet of things, *Proceedings IEEE 7th Annual Computing and Communication Workshop and Conference*, Las Vegas, NV, IEEE Press, Piscataway, NJ, Paper ID 1570326303, 2017.
12. V. J. Hodge, S. O'Keefe, M. Weeks, A. Moulds, Wireless sensor networks for condition monitoring in the railway industry: A survey, *IEEE Transactions on Intelligent Transportation Systems*, 16(3), 1088–1106, 2015.
13. R.B. Lewis, Track recording techniques used on British Rail, *IEE Proceedings B—Electric Power Applications*,131(3), 73–81, 1984.
14. R.B. Lewis, A.N. Richards, A compensated accelerometer for the measurement of railway track cross-level, *IEEE Transactions on Vehicle Technology*, 37(3), 174–178, 1988.
15. Railtrack, Track Recording Handbook, Railtrack Line Code of Practice RT/CE/C/038, Issue 1, Rev B, February 1996.

19 Roller Rigs

Paul D. Allen, Weihua Zhang, Yaru Liang, Jing Zeng, Henning Jung, Enrico Meli, Alessandro Ridolfi, Andrea Rindi, Martin Heller and Joerg Koch

CONTENTS

- 19.1 Introduction 762
- 19.2 The History of Roller Rigs 763
- 19.3 Roller Rig Facilities and Analysis of Capabilities 764
  - 19.3.1 Research Topics 765
    - 19.3.1.1 Wheel-Rail Contact 765
    - 19.3.1.2 Adhesion 765
    - 19.3.1.3 Traction and Braking 767
    - 19.3.1.4 Wear and Rolling Contact Fatigue 767
    - 19.3.1.5 Derailment and Curving 767
    - 19.3.1.6 Noise and Vibration 767
    - 19.3.1.7 Component Performance 768
    - 19.3.1.8 Environmental Conditions 768
  - 19.3.2 Overview of Rig Research Areas 768
    - 19.3.2.1 Rig Research Capabilities 768
    - 19.3.2.2 Roller Unit Issues 769
  - 19.3.3 Global Summary of Roller Rig Facilities 771
    - 19.3.3.1 RTA Large CWT Roller Rig 773
    - 19.3.3.2 China Academy of Railway Sciences (CARS) Roller Rig 773
    - 19.3.3.3 Chengdu Roller Rig 774
    - 19.3.3.4 Výzkumný ústav kolejových vozidel (VUKV) Roller Rig 775
    - 19.3.3.5 Test Rig A: Deutsche Bahn (DB) Wheelset Test Rig 776
    - 19.3.3.6 Test Rig C: Deutsche Bahn (DB) Wheelset Axle Test Rig 777
    - 19.3.3.7 Advanced Test Laboratory of Adhesion-Based Systems (ATLAS) Roller Rig 778
    - 19.3.3.8 Florence Railway Research and Approval Centre (RRAC) Roller Rig 779
    - 19.3.3.9 BU 300 – Roller Rig 780
    - 19.3.3.10 Ansaldo Roller Rig 781
    - 19.3.3.11 Railway Technical Research Institute (RTRI) Roller Rig 782
    - 19.3.3.12 Huddersfield Adhesion and Rolling Contact Laboratory Dynamics (HAROLD) Roller Rig 783
- 19.4 Roller Rig Case Studies 784
  - 19.4.1 The Chengdu Roller Rig 784
    - 19.4.1.1 Technical Overview 784
    - 19.4.1.2 Operational Modes 788
    - 19.4.1.3 Applications 788
  - 19.4.2 Huddersfield Adhesion and Rolling Contact Laboratory Dynamics Rig 793
    - 19.4.2.1 Technical Overview 793
    - 19.4.2.2 Operational Modes 798
    - 19.4.2.3 Applications 800

  19.4.3 The ATLAS Roller Rig, Knorr-Bremse, Munich, Germany 801
    19.4.3.1 Technical Overview 801
    19.4.3.2 Operational Modes 804
    19.4.3.3 Applications 805
  19.4.4 Florence Railway Research and Approval Centre Roller Rig 808
    19.4.4.1 Technical Overview 808
    19.4.4.2 Operational Modes 811
    19.4.4.3 Applications 812
19.5 Roller Rig Experimental Methods and Errors 815
  19.5.1 Experimental Methods 815
    19.5.1.1 Status of Vehicle 815
    19.5.1.2 Status of Roller Rig 815
    19.5.1.3 Stability Test 816
    19.5.1.4 Dynamic Simulation Test 817
    19.5.1.5 Curve Simulation Test 817
    19.5.1.6 Modal Analysis Test 818
    19.5.1.7 Storage Security Test 819
  19.5.2 Experimental Errors 819
    19.5.2.1 Diameter of the Rollers 819
    19.5.2.2 Gauge of the Rollers 820
    19.5.2.3 Cant of the Rollers 820
    19.5.2.4 Coefficient of Friction on the Contact Surface 821
    19.5.2.5 Vehicle Position on the Roller Rig 821
    19.5.2.6 Yaw Angle between Wheelset and Roller Axle 821
19.6 Conclusions 822
Acknowledgements 823
References 823

19.1 INTRODUCTION

Experience teaches us that the complete design, development and production cycle for a new railway vehicle that features significant levels of systems and technological innovation is very time-consuming. The duration and efficiency of prototype experimentation activities are key elements in the development of rolling stock and are instrumental in terms of technical and economic success. An awareness of this technological and competitive development has led the railway industry and research institutes to commit extensive financial and technical resources to the creation of test facilities. The main objectives are to reduce the time (and therefore the cost) of testing new vehicles and to make as wide a range of tests available as possible, in order to achieve maximum levels of performance, reliability and availability in the shortest possible time.

  A roller rig is a type of railway vehicle test equipment. Firstly, it is a system capable of testing a vehicle in a running condition without line tests, and secondly, it allows the study of interaction between a railway wheel and the rail.

  The application of roller rigs to the study of vehicle system dynamics and the development of high-speed trains and other railway vehicles have become more widespread in recent decades. Roller rigs are used by researchers and railway organisations around the world to assist in understanding the behaviour of vehicles and developing faster, safer and more efficient railways. Roller rigs have contributed to many current designs of railway vehicles and have been proved useful for both fundamental research and development of innovations in suspensions and vehicle components. Full-scale roller rigs offer the advantages that the experiments are independent of weather conditions, individual phenomena can be investigated, and the experiments and constraints, as well as the particular conditions, are reproducible.

Full-scale roller rigs have been proven to be powerful tools, not only for the demonstration of vehicle dynamics for students but also for the validation of theoretical work and the test of new concepts of innovative vehicle designs and train sub-systems.

19.2 THE HISTORY OF ROLLER RIGS

Roller rigs were originally used for the investigation of performance of steam locomotives over 100 years ago. One of the earliest such test rigs was built at the Swindon works of the Great Western Railway in 1904 (see Figure 19.1) [1–3]. The rollers were moveable and could be adjusted so that the centre of each driving wheel was exactly about the centre of each roller. High speeds could be attained while the engine remained stationary, and a braking arrangement on the rollers measured the traction power of the locomotive at various speeds.

In 1957, a full-scale roller rig with two axles was used at the Railway Technical Research Institute, Japan, which used an eccentric roller to create a sinusoidal excitation. Circa 1960, tests of bogies commenced on the newly built full-scale roller rig. The roller rig was put into use for about 30 years and played a very important role in studies related to protection against freight car derailment, regenerative braking and Shinkansen bolsterless bogies amongst others. In order to meet the demand for high-speed vehicle tests, a new four-axle, full-scale roller rig with the facility for roller lateral and vertical excitations began construction in 1987 and has been successfully applied since then.

A roller rig was built in Vitry, France, in 1964, which allowed lateral and vertical motions of the roller on each axle, simultaneously using simple hydraulic control methods. By using the roller rig, the vertical, lateral and yaw vibration frequencies, amplitudes and resonance could be measured.

FIGURE 19.1 One of the earliest roller rigs for steam locomotives at the Swindon works of the Great Western Railway. (From Iwnicki, S. (Ed.), *Handbook of Railway Vehicle Dynamics*, CRC Press, Boca Raton, FL, 2006. With permission.)

In particular, the influence of the change of the vertical and lateral excitation forces, due track induced disturbances, on the vehicle running performance could be studied, and the running safety and ride performance could also be investigated.

The roller rig in Berlin, Germany, was built in 1967. This roller rig allowed evaluation of traction equipment, acceptance tests for vehicle springs and assessment of braking systems.

The construction of a roller rig at the BR Research Centre in Derby began in 1959 and was completed in 1971. Whilst now de-commissioned, this roller rig had the capacity to assess braking power, resonant vibration and vehicle stability. Later, the roller rig was modified to a modal analysis test stand mainly used for the vibration analysis of vehicle suspension systems.

In 1977, a full-scale roller rig was built in Munich, Germany, at Deutsche Bahn AG. The rollers had four degrees of freedom, including vertical, lateral, inclination and rotation. The servo hydraulic excitation control system was adopted for the roller rig and could accurately simulate track conditions for the dynamic simulation of a vehicle operating on tracks. The rig was mainly utilised for the measurement of the dynamic performance of vehicles and determination of the effects of vehicle modifications on the system performance. The Munich roller rig has played a very important part in the development of Intercity-Express (ICE) high-speed trains but has now been decommissioned.

In 1978, a roller rig, called the roll dynamics unit, with vibrations applied through the wheels to simulate track conditions, began operation at Pueblo, Colorado, USA. The rig consisted of two separate test stands, one for roller-based testing and another used as a vibration stand. The rolling stand could be used for hunting stability and traction power simulation tests and the vibration stand used for studies of suspension system features, vehicle system natural frequencies, fatigue strength and freight load reliability, etc. The rig was designed for speeds of up to 480 km/h. This rig has now been decommissioned.

A roller rig called the curved track simulator was set up at the National Research Council in Ottawa, Canada, in 1979. It consisted of two pairs of rollers in a flexible frame that permitted the yawing motion of the roller axle to simulate curving. The frame floated on hydrostatic bearings. Unfortunately, the roller rig has since been dismantled.

In 1995, a four-axle roller rig was built at the State Key Laboratory of Traction Power (Southwest Jiaotong University) at Chengdu, China. This roller rig was built for the optimum design and testing of railway vehicles. Each roller can vibrate in lateral and vertical directions. In 2002, two new sets of rollers were added, to form six axles, allowing locomotives to be tested on the rig.

19.3 ROLLER RIG FACILITIES AND ANALYSIS OF CAPABILITIES

The main aim of building roller rigs is to provide controlled and repeatable conditions for the investigation and optimisation of railway vehicle performance and development of subsystems; a roller rig can be used to perform different tests under a variety of research topics. Jaschinski et al. [1] stated five principal differences between the roller rig and the track; these differences must be considered during the comparison of results from a test rig and a track:

- The longitudinal creepage is modified due to the variation of the rolling radius of the rollers as the point of contact is displaced laterally.
- The spin creepage is modified because the roller angular velocity has a component resolved along the common normal at the point of contact.
- As the rollers possess curvature in the longitudinal vertical plane, the geometry of the Hertzian contact areas is different.
- When the wheelset yaws on the rollers, the contact plane rotates in both pitch and yaw relative to the rollers.
- The lateral and vertical stiffness of rollers may be different from the track stiffness.

Therefore, the design of the roller unit is of particular importance.

TABLE 19.1
Full-Scale Roller Rig Availability around the World

No.	Rig Name	Operator	Location
1	RTA	RTA Rail Tec Arsenal Fahrzeugversuchsanlage GmbH	Austria (Vienna)
2	CARS	Nat. Engineering Laboratory for System Test of High-Speed Railway	China (Beijing)
3	Chengdu	State Key Laboratory of Traction Power, Southwest Jiaotong University	China (Chengdu)
4	VUKV	University of Pardubice, Faculty of Transport Engineering	Czech Republic (Pardubice)
5	Test Rig A	DB Systemtechnik GmbH	Germany (Brandenburg-Kirchmöser)
6	Test Rig C	DB Systemtechnik GmbH	Germany (Brandenburg-Kirchmöser)
7	ATLAS	Knorr-Bremse Systeme für Schienenfahrzeuge GmbH	Germany (Munich)
8	RRAC	Rete Ferroviaria Italiana S.p.A.	Italy (Firenze-Osmannoro)
9	BU 300	Lucchini RS S.p.A.	Italy (Lovere)
10	Ansaldo	Ansaldo Transport Research Centre	Italy (Naples)
11	RTRI	Railway Technical Research Institute of Japan	Japan (Tokyo)
12	HAROLD	University of Huddersfield, Institute of Railway Research	United Kingdom (Huddersfield)

This chapter provides an overview of 12 full-scale roller rigs (Table 19.1), which are able to test specimens at 1:1 scale, and compares their key features and capabilities. Figures 19.2 and 19.3 illustrate the global distribution of the rigs; it is apparent that most are located in Europe, whilst three can be found in Asia.

19.3.1 Research Topics

In order to develop faster, safer and more efficient rail systems, many different research areas require consideration. The common goal is to make savings under the economic and ecological operation of railway vehicles and their sub-systems. The focus within this section lies in the comparison of available rigs under a number of key research areas and the resulting requirements for the rigs themselves.

19.3.1.1 Wheel-Rail Contact

Wheel-rail contact is a key area of vehicle dynamics. It is the complex and compromised relationship between the vehicle and the track. Wheel-rail forces have a large influence on the dynamics of a rail vehicle, and, with a contact area of approximately 1.5 cm^2 per wheel, it is clear that this surface is highly stressed. Therefore, knowledge about the mechanisms of wheel-rail contact is of high importance and roller rigs provide an essential tool to investigate contact forces under different operating conditions in a controlled environment. The main requirement of the roller rig is the measurement of the normal and creep forces. But be careful; in contrast to the real track where the rail curvature in the longitudinal direction is infinite, the rollers have a finite curvature, and the measured contact forces therefore cannot be directly compared to vehicles on flat tracks without due consideration of the differences.

19.3.1.2 Adhesion

In contrast to a tribological point of view, where adhesion is defined as the force required to separate two surfaces that have been brought into contact, in the wheel-rail research community, adhesion is defined as the transmitted tangential force between rail and wheel in the longitudinal direction. The adhesion coefficient therefore is defined as the ratio between the adhesion force and the normal force.

FIGURE 19.2 Roller rigs in Europe (numbers from Table 19.1).

FIGURE 19.3 Roller rigs in Asia (numbers from Table 19.1).

For traction and braking purposes adhesion is necessarily required. On the one hand, adhesion must be high enough to ensure safe braking and traction. On the other hand, too high adhesion leads to wear and fatigue of the rail and wheels. The adhesion characteristics depend on many different environmental conditions, together with the roughness and geometry of the wheels and rails themselves. These conditions can be investigated effectively on a roller rig, with many rigs providing auxiliary devices to adjust operational conditions, e.g., adhesion, lubrication and temperature.

19.3.1.3 Traction and Braking

The traction and braking process is closely related to the adhesion phenomenon and is connected to the longitudinal train dynamics. In contrast to the adhesion research topic, traction and braking deals with the study of a vehicle's traction and braking control systems to avoid wheel slide or wheel slip. Therefore, a roller rig provides a perfect opportunity to test and develop wheel slide protection (WSP) systems that automatically detect, control, and prevent wheel slides. It is necessary that the roller rig is equipped with adequate traction and braking units, which supply the required braking and traction moments.

19.3.1.4 Wear and Rolling Contact Fatigue

As discussed previously, the contact patch between rail and wheel undergoes very high stresses. Therefore, the material must be resilient enough to resist these heavy loads. If the material cannot withstand such stresses, damage occurs. One can distinguish between wear and rolling contact fatigue (RCF). Wear is the process where material loss and change in profile shape becomes apparent. Many different wear mechanisms can occur. Two simple models are the 'mild' and the 'severe' wear. Severe wear is characterised by a rough surface that is often rougher than the original surface, whereas mild wear results in a smoother surface than the original one. A measure for wear is the wear rate, which is the volume loss per sliding distance. Rolling contact fatigue can occur on rails and wheels. Owing to the high stress levels in the contact patch, a continuous plastic deformation of the steel's microstructure occurs. The first stage of RCF involves work hardening of the surface layer. The hardness increases. Such recurring deformation processes lead to structural changes of the material with a high relative dislocation density in the crystalline structure and formation of fine cracking. A roller rig must be able to provide a controllable forcing unit and must be able to withstand such high loadings.

19.3.1.5 Derailment and Curving

Derailment leads to a loss of vehicle guidance, and hence, the avoidance of derailment is vital to the safe operation of railways. Reasons behind derailment include component failure or high lateral forces developed at large angles of attack between the wheel and rail, which are typically present in smaller-radius curves. Such forces can lead to the wheel flange climbing out of the rail, gauge widening, rail rollover and track panel shift. In addition to the measurement of contact forces, the roller rig must be able to adjust the wheel-rail angle of attack to study the effects of derailment and curving behaviour.

19.3.1.6 Noise and Vibration

In railway systems, noise is mostly produced by structural vibrations and aerodynamic disturbances. The aerodynamic noise produced by an unsteady airflow dominates the overall noise for speeds approaching 300 km/h and is therefore predominant in high-speed operations. Structural vibrations are therefore the most important aspect for velocities less than 300 km/h. The main source of noise is the wheel-rail contact, where the rolling noise is the result of vibrations of the structure induced by wheel and track roughness and irregularities. The wheel and track irregularities are also cause for ground-borne vibrations. Squeal noise occurs in sharp curves and is the result of unsteady friction force either on the wheel tread or on the flange. The vibrations that generate the noise can be transmitted through the structure and also into the vehicle itself, affecting the passenger comfort. To provide the basis for research in this area, a roller rig must be able to generate different rail and wheel roughnesses and induce track irregularities in combination with different lubrication scenarios. The test facility must also fulfil the required standards for noise measurements such as those provided by an acoustic chamber.

19.3.1.7 Component Performance

Railway vehicles are composed of many different components such as wheelsets, bogie frames, bearings, springs, dampers, brakes, etc. Every component has its own mostly non-linear characteristic, which makes it difficult to estimate the overall behaviour of an assembled railway vehicle. Therefore, it is necessary to test the final assembled unit. This allows the study of suspension behaviour and validation of vehicle dynamic computer models. The roller rig must provide flexible adjustments of auxiliary devices to adapt to the different components.

19.3.1.8 Environmental Conditions

Rail vehicles are exposed to the most adverse conditions. They have to withstand large temperature differences, between −40°C and +85°C, or different weather conditions such as rain, ice, snow and sun radiation. To test these environmental conditions, the roller rig must be able to work as a climate chamber.

19.3.2 Overview of Rig Research Areas

19.3.2.1 Rig Research Capabilities

Table 19.2 summarises the capabilities of the various roller rigs to accommodate the different areas of research discussed in Section 19.3.1.

The common feature of all rigs reviewed in this chapter is their full-scale nature; however, there is a wide variation in their configuration and hence their ability, as set out in Table 19.3, to test specimens of:

- Full vehicle (includes wheelsets, bogies, suspensions and carbody)
- Bogies (includes bogie frame and primary and secondary suspensions)
- Single wheelsets
- Single wheels

TABLE 19.2
Overview of Rig Research Capabilities

	Wheel Rail Contact	Adhesion	Traction and Braking	Wear and RCF	Derailment and Curving	Noise and Vibration	Component Performance	Environmental Conditions
ATLAS	●	●	●	●	○	●	●	●
Chengdu	●	●	●	●	●	●	●	○
Test Rig A	●	●	●	●	●	●	●	○
BU 300	●	●	●	●	●	●	○	○
CARS	●	●	●	●	●	○	○	●
RTRI	●	●	●	●	●	○	○	●
HAROLD	●	●	●	●	○	○	●	○
RRAC	○	○	●	○	○	●	●	○
RTA	○	○	●	○	○	○	●	●
Ansaldo	○	○	●	○	○	○	●	○
VUKV	○	●	●	○	○	○	○	○
Test Rig C	○	○	○	○	○	○	●	○

Note: ● indicates capability available; ○ indicates not available.

TABLE 19.3
Overview of Possible Test Specimens

	Full Vehicle	Bogie	Single Wheelset	Single Wheel
Ansaldo	●	○	○	○
Chengdu	●	○	○	○
RTA	●	○	○	○
RTRI	●	○	○	○
RRAC	●	●	●	○
ATLAS	○	●	●	●
CARS	○	●	●	●
Test Rig A	○	●	●	●
HAROLD	○	●	●	○
BU 300	○	○	●	○
Test Rig C	○	○	●	○
VUKV	○	○	○	●

Note: ● indicates capability available; ○ indicates capability not available.

19.3.2.2 Roller Unit Issues

As mentioned in the beginning of this section, the design of the roller unit is of particular importance. If one is confronted with the word roller rig, different kinds of roller arrangements are available [4–6]. The main roller designs are depicted in Figure 19.4, where ω_R and ω_W are the angular velocity of the roller and the wheel, respectively. The simulated direction of the driving velocity is given by v.

19.3.2.2.1 Vertical Plane Roller

This configuration is the most common design. The wheel of the test specimen is located on the top of the roller unit. The lower rig roller simulates the endless track. Normally, the heads are profiled similar to the real track and wheel profiles.

19.3.2.2.2 Tangent Roller

The wheel of the test specimen is inside the rig roller. This concept is widely used in the automotive industry for testing tyres and brake systems.

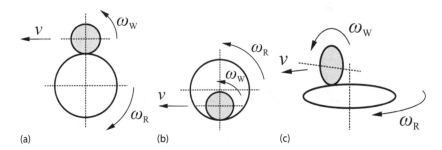

FIGURE 19.4 Different roller arrangements: (a) vertical plane roller, (b) tangent roller, and (c) perpendicular roller.

19.3.2.2.3 Perpendicular Roller

The perpendicular roller acts as a bevel gear. The axis of the rig roller and the test specimen are perpendicular to each other.

19.3.2.2.4 Key Requirements for Rollers

A roller rig acts as a track simulator; the rollers with rail profiles form an endless track. The common design is the vertical plane roller. Not only can it simulate the running of vehicles on straight track, but, by the rotation of the rollers about the longitudinal (roll) and vertical (yaw) axis, it is also possible to simulate canted and curved track conditions. Also, track irregularities can be simulated by the excitation of rollers in the different directions. When applying rotational resistance to the rollers, the roller rig can provide the traction force to simulate traction and braking effort of a vehicle. A modern roller rig should be capable of simulating track irregularities and also curve negotiation. The track irregularity shown in Figure 19.5 can be considered to have four components: gauge, cross level, lateral alignment and vertical profile. When the vertical and lateral disturbances of the left and right rails are indicated as z_L, y_L, z_R and y_R, these four types of track irregularities can be described as:

- Gauge = $(y_L - y_R)/2$
- Lateral alignment = $(y_L + y_R)/2$
- Cross level = $(z_L - z_R)/2$
- Vertical profile = $(z_L + z_R)/2$

It is possible to transfer these four types of irregularities to the specimen under test, using the roller arrangement on the roller rig. Similarly, transiting a curve and experiencing cant in a curve can also be replicated. Table 19.4 shows how to arrange the rollers.

19.3.2.2.5 Overview of Installed Roller Units

Table 19.5 summarises the roller diameter (critical for an accurate representation of contact conditions), the maximum driving velocity v_{max}, the degrees of freedom (related to the ability to represent track design features) and the arrangement of the roller unit (related to the ability to accommodate test specimens). Here, the x-, y- and z-directions relate to the longitudinal, lateral and vertical directions, respectively. Rotations about the x-, y- and z-axis are the roll, pitch and yaw movements, respectively.

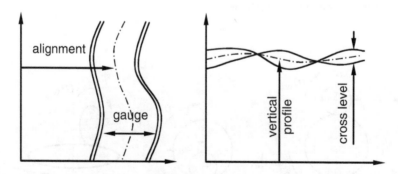

FIGURE 19.5 Track irregularity inputs. (From Iwnicki, S. (Ed.), *Handbook of Railway Vehicle Dynamics*, CRC Press, Boca Raton, FL, 2006. With permission.)

TABLE 19.4
Relationship between Status of Rails and Rollers

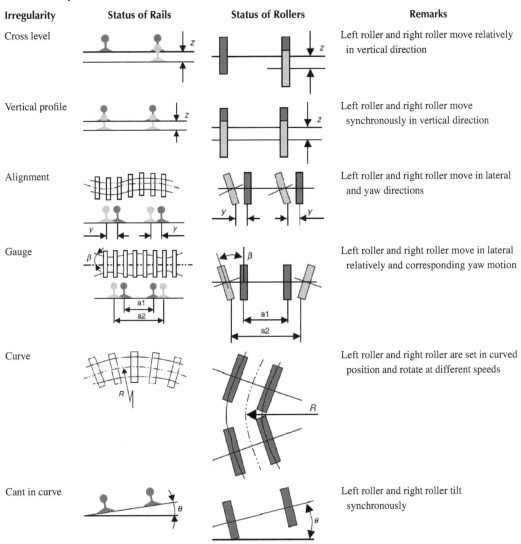

Source: Iwnicki, S. (Ed.), *Handbook of Railway Vehicle Dynamics*, CRC Press, Boca Raton, FL, 2006. With permission.

19.3.3 GLOBAL SUMMARY OF ROLLER RIG FACILITIES

This section provides an overview of roller rigs available around the world. The list does not claim to be exhaustive, but all technical data has been updated by the relevant test institutes for the purpose of this chapter. Owing to the complex design of each test facility, the overview highlights key facts and acts as a first selection guide for specific research tasks. All summaries presented in Tables 19.6

TABLE 19.5
Comparisons of Installed Rollers

	Roller Diameter	Velocity v_{max}	Translation[a]			Rotation[a]			Roller Unit Arrangement[b]
			x	y	z	x	y	z	
Chengdu	1800 mm	600 km/h	o	●	●	●	●	●	3x
RTRI	1500 mm	500 km/h	o	●	●	●	●	o	2x
VUKV	905 mm	60 km/h	o	●	o	o	●	●	
Test Rig A	2100 mm	300 km/h	●[c]	o	o	o	●	o	
Ansaldo	1500 mm	300 km/h	o	o	o	o	●	o	3x
RRAC	1590 mm	220 km/h	o	o	o	o	●	o	2x
RTA	1000 mm	280 km/h	o	o	o	o	●	o	1x
CARS	3000 mm	500 km/h	o	o	o	o	●	o	
ATLAS	3000 mm	350 km/h	o	o	o	o	●	o	
BU 300	2000 mm	300 km/h	o	o	o	o	●	o	
HAROLD	2000 mm	200 km/h	o	o	o	o	●	o	
Test Rig C	2100 mm	160 km/h	o	o	o	o	●	o	

Notes: [c] Rig can be converted from a roller rig to a linear test rig with a straight rail.

[a] Translation and rotation degrees of freedom by axis per roller pair are marked by ● indicating available and o indicating not available

[b] Roller unit arrangement, axle configuration, single roller or bogie pair, with 1x = (bogie only), 2x = (four-axle vehicle) and 3x = (six-axle vehicle).

through 19.17 are arranged in the same layout to ensure easy comparability and begin with information about the operator and location of the rig, followed by the type of test specimen that can be accommodated. The wheelbase and bogie pivot spacing is provided as an estimation regarding the minimal and maximal dimensions of the test specimens. In alignment to the section research topics, typical applications of the rigs are listed. Also, the capacity for possible curving scenarios, hardware-in-the-loop (HIL) features and maximum axle loads are presented. Also, the central key numbers of the roller unit are presented. Here, n_{Roller} represents the number of driven roller pairs, the roller diameter is given by d_{Roller} and the possible gauge range is given by g. The maximum driving velocity is denoted by v_{Max}. To estimate the possible traction and braking capacity, the maximum continuous motor power P_{Max} and maximum torque M_{Max}, per each roller pair, are given.

19.3.3.1 RTA Large CWT Roller Rig

TABLE 19.6
Key Specification for Rail Tec Arsenal (RTA) Large Climatic Wind Tunnel (CWT) Roller Rig

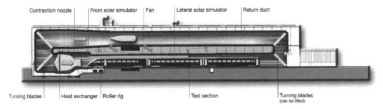

Large Climatic Wind Tunnel

Operator	RTA Rail Tec Arsenal Fahrzeugversuchsanlage GmbH				
Location	Vienna, Austria				
Test specimen	Full vehicle				
Wheelbase	1100–3000 mm				
Typical applications	Climatic test (solar radiation, rain, snow and ice), traction and braking				
Curve simulation	No				
HIL	Real-time mass simulation, emergency brake and slope braking				
Load simulation	Up to 196.2 kN				
Max. axle load	245 kN				
n_{Roller}	1 (1 DoF)	d_{Roller}	1000 mm	g	1435 mm
v_{max}	280 km/h	P_{max}	850 kW (1.5 MW overload for 90 s)	M_{max}	12.5 kNm

Source: © Rail Tec Arsenal. With permission.

19.3.3.2 China Academy of Railway Sciences (CARS) Roller Rig

TABLE 19.7
Key Specification for CARS Roller Rig

(Continued)

TABLE 19.7 (*Continued*)
Key Specification for CARS Roller Rig

China Academy of Railway Sciences

Operator	Nat. Engineering Laboratory for System Test of High-Speed Railway
Location	Beijing, China
Test specimen	Bogie, single wheelset; single wheel
Wheelbase	–
Typical applications	Adhesion, wear, RCF, derailment, braking and weather influence
Curve simulation	Yes (+/− 4 degree yaw angle of the test specimen)
HIL	N/A
Load simulation	2 × 245 kN vertical; 1 × 150 kN lateral
Max. axle load	490 kN

n_{Roller}	1 (1 DoF)	d_{Roller}	3000 mm	g	1435 mm
v_{max}	500 km/h	P_{max}	2400 kW	M_{max}	N/A

Source: © China Academy of Railway Sciences. With permission.

19.3.3.3 Chengdu Roller Rig

TABLE 19.8
Key Specification for Chengdu Roller Rig

Operator	State Key Laboratory of Traction Power (Southwest Jiaotong University)
Location	Chengdu, China
Test specimen	Full vehicle
Wheelbase	1600–3500 mm (4000–22000 mm bogie pivot spacing)

(*Continued*)

TABLE 19.8 (Continued)
Key Specification for Chengdu Roller Rig

Typical applications	Adhesion, curving, derailment, wheel-rail contact and braking				
Curve simulation	Yes (min. curve radius 200 m)				
HIL	N/A				
Load simulation	Vertical (+/− 15 mm@ max. 30 Hz); Lateral (+/− 10 mm@ max. 30 Hz)				
Max. axle load	245 kN				
n_{Roller}	6 (5 DoF)	d_{Roller}	1800 mm	g	1000–1676 mm
v_{max}	600 km/h	P_{max}	1600 kW	M_{max}	N/A

Source: © Southwest Jiaotong University. With permission.

19.3.3.4 Výzkumný ústav kolejových vozidel (VUKV) Roller Rig

TABLE 19.9
Key Specification for VUKV Roller Rig

Výzkumný ústav kolejových vozidel (Research Inst. for Rail Vehicles)

Operator	University of Pardubice, Faculty of Transport Engineering				
Location	Pardubice, Czech Republic				
Test specimen	Single tram wheel				
Wheelbase	–				
Typical applications	Adhesion, PMSM drives, traction and braking				
Curve simulation	Yes (+/− 5 degrees angle of attack)				
HIL	N/A				
Load simulation	50 kN vertical				
Max. axle load					
n_{Roller}	1 (3 DoF)	d_{Roller}	905 mm	g	–
v_{max}	60 km/h	P_{max}	55 kW	M_{max}	0.892 kNm

Source: © University of Pardubice. With permission.

19.3.3.5 Test Rig A: Deutsche Bahn (DB) Wheelset Test Rig

TABLE 19.10
Key Specification for Test Rig A

Operator	DB Systemtechnik GmbH				
Location	Brandenburg-Kirchmöser, Germany				
Test specimen	Bogie, single wheelset; single wheel				
Wheelbase	–				
Typical applications	Wheel-rail materials, RCF, traction, braking, wheelset bearings and acoustic inspection				
Curve simulation	Yes (min. curve radius 350 m)				
HIL	N/A				
Load simulation	2 × 170 kN vertical, 30 kN lateral				
Max. axle load	340 kN single wheelset				
n_{Roller}	1 (1 DoF)	d_{Roller}	2100 mm	g	1435 mm
v_{max}	300 km/h	P_{max}	500 kW	M_{max}	N/A

Source: © DB Systemtechnik GmbH. With permission.

19.3.3.6 Test Rig C: Deutsche Bahn (DB) Wheelset Axle Test Rig

TABLE 19.11
Key Specification for Test Rig C

Operator	DB Systemtechnik GmbH				
Location	Brandenburg-Kirchmöser, Germany				
Test specimen	Single wheelset				
Wheelbase	–				
Typical applications	Wheel and wheelset constructions, fatigue crack growth of wheelset axles and adjustable gauge wheelsets				
Curve simulation	No				
HIL	N/A				
Load simulation	2×170 kN vertical; 80 kN lateral				
Max. axle load	340 kN				
n_{Roller}	1 (1 DoF)	d_{Roller}	2100 mm	g	1435–1668 mm
v_{max}	160 km/h	P_{max}	100 kW	M_{max}	N/A

Source: © DB Systemtechnik GmbH. With permission.

19.3.3.7 Advanced Test Laboratory of Adhesion-Based Systems (ATLAS) Roller Rig

TABLE 19.12
Key Specification for ATLAS Roller Rig

Advanced Test Laboratory of Adhesion-Based Systems

Operator	Knorr-Bremse Systeme für Schienenfahrzeuge GmbH
Location	Munich, Germany
Test specimen	Bogie; single wheelset; single wheel
Wheelbase	–
Typical applications	Brake systems, adhesion, slip, wear, wheel-rail contact and environmental impact
Curve simulation	Yes
HIL	Real-time train braking model
Load simulation	2 × 150 kN vertical; 1 × 80 kN lateral; 2 × 80 kN longitudinal
Max. axle load	300 kN

n_{Roller}	1 (1 DoF)	d_{Roller}	3000 mm	g	1000–1670 mm
v_{max}	350 km/h	P_{max}	2 × 1400 kW	M_{max}	62 kNm

Source: © Knorr-Bremse Systeme für Schienenfahrzeuge GmbH. With permission.

19.3.3.8 Florence Railway Research and Approval Centre (RRAC) Roller Rig

TABLE 19.13
Key Specification for Florence RRAC Roller Rig

Florence Railway Research and Approval Center

Operator	Rete Ferroviaria Italiana S.p. A.
Location	Firenze-Osmannoro, Italy
Test specimen	Single wheelset; bogie; full vehicle
Wheelbase	1600–3500 mm (12000–20000 mm bogie pivot spacing)
Typical applications	Traction, dynamics, electromagnetic compatibility and electric braking
Curve simulation	No
HIL	Reproduce the motion resistance characteristics, simulate slope variation and verification of traction performance
Load Simulation	–
Max. axle load	235 kN
n_{Roller} 4 (1 DoF)	d_{Roller} 1590 mm g 1435 mm
v_{max} 220 km/h	P_{max} 1900 kW M_{max} 70 kNm

Source: © RFI SpA. With permission.

19.3.3.9 BU 300 – Roller Rig

TABLE 19.14
Key Specification: BU 300 Roller Rig

Operator	Lucchini RS S.p. A.				
Location	Lovere, Italy				
Test specimen	Single Wheelset				
Wheelbase	–				
Typical applications	RCF, wheel-rail contact, adhesion, braking, stability, curving				
Curve simulation	Yes				
HIL	N/A				
Load simulation	2 × 250 kN vertical; 1 × 150 kN lateral				
Max. axle load	500 kN				
n_{Roller}	1 (1 DoF)	d_{Roller}	2000 mm	g	1435 mm
v_{max}	300 km/h	P_{max}	500 kW	M_{max}	N/A

Source: © Lucchini RS SpA. With permission.

19.3.3.10 Ansaldo Roller Rig

TABLE 19.15
Key Specification for Ansaldo Transport Research Centre Roller Rig

Operator		Ansaldo Transport Research Centre			
Location		Naples, Italy			
Test specimen		Full vehicle			
Wheelbase		1400–3500 mm (5200–22,000 mm bogie pivot spacing)			
Typical applications		Traction simulation, start-up test, braking			
Curve simulation		No			
HIL		N/A			
Load simulation		No			
Max. axle load		245 kN			
n_{Roller}	4–6 (1 DoF)	d_{Roller}	1500 mm	g	600–1700 mm
v_{max}	300 km/h	P_{max}	1500 kW	M_{max}	1500 kNm

Source: S. Iwnicki (Ed.), Handbook of Railway Vehicle Dynamics, CRC Press, Boca Raton, FL, 2006. With permission.

19.3.3.11 Railway Technical Research Institute (RTRI) Roller Rig

TABLE 19.16
Key Specification for RTRI Roller Rig

Operator		Railway Technical Research Institute of Japan					
Location		Tokyo, Japan					
Test specimen		Full Vehicle					
Wheelbase		1600 mm					
Typical applications		Derailment, braking, track conditions					
Curve simulation		Yes					
HIL		N/A					
Load simulation		Vertical (+/− 12 mm@ 0 ~ 1.8 Hz; +/− 10 mm@ max. 3 Hz; +/− 2 mm@ max. 10 Hz); lateral (+/− 12 mm@ 0 ~ 1.8 Hz; +/− 0.4 mm@ max. 25 Hz); and rolling (+/− 11 mrad@ 0 ~ 2 Hz; +/− 0.06 mrad@ max. 15 Hz)					
Max. axle load		200 kN					
n_{Roller}	4 (4 DoF)	d_{Roller}	1500 mm	g	1000–1676 mm		
v_{max}	500 km/h	P_{max}	2000 kW	M_{max}	N/A		

Source: Iwnicki, S. (Ed.), *Handbook of Railway Vehicle Dynamics*, CRC Press, Boca Raton, FL, 2006. With permission.

19.3.3.12 Huddersfield Adhesion and Rolling Contact Laboratory Dynamics (HAROLD) Roller Rig

TABLE 19.17
Key Specification for HAROLD Roller Rig

	Huddersfield Adhesion and Rolling Contact Laboratory Dynamics Rig				
Operator	University of Huddersfield Institute of Railway Research				
Location	Huddersfield, United Kingdom				
Test specimen	Bogie; single wheelset				
Wheelbase	1500–3000 mm				
Typical applications	Adhesion, wheel-rail contact, braking and suspension behaviour				
Curve simulation	Yes				
HIL	Train-braking model				
Load simulation	2 × 245 kN vertical; 1 × 245 kN lateral				
Max. axle load	245 kN				
n_{Roller}	1(1 DoF)	d_{Roller}	2000 mm	g	1435 mm
v_{max}	200 km/h	P_{max}	450 kW	M_{max}	110 kNm

19.4 ROLLER RIG CASE STUDIES

19.4.1 THE CHENGDU ROLLER RIG [7–9]

China has constructed more than 120,000 km of railway lines; among them, newly built high-speed lines now account for 28,000 km of these. The service speed of passenger trains on main lines has risen from approximately 50–70 km/h in the 1980s up to 300–350 km/h in the 2010s. This rise in vehicle speed has been mainly attributed to the use of test facilities, especially roller rigs. The successful application of these rigs within China has resulted in a growth in their use. There are now six roller rigs in service, but only the roller rig in Chengdu has the functions of both rolling dynamics and vehicle vibration testing combined in a single test rig.

19.4.1.1 Technical Overview

19.4.1.1.1 Primary Research Objectives and History

The roller rig in Chengdu was developed by the State Key Laboratory of Traction Power at Southwest Jiaotong University, with the aim of being heavily utilised in the development of new railway vehicles. Design of the roller rig began in 1989, and it came into service in 1995. From 1995 to 2018, more than 100 railway vehicles have been tested. Each roller can move in the vertical and lateral directions independently under servo hydraulic control. The original roller rig had four roller sets (allowing testing of up to four-axle vehicles), with the two rollers of each set constrained to have the same rotational speed. These constraints meant the roller rig could only simulate a four-axle vehicle running on straight track, with a maximum gauge variation between 1435 mm and 1676 mm. As required by the testing, the rig was extended to six roller sets and the structure improved during 2002. Four roller sets of the new modified rig have the ability of gauge variation between 1000 mm and 1676 mm, and the two rollers of each roller set can be run at different rotational speeds. In 2014, the roller rig underwent renovation for high-speed vehicle tests, with the radius of each roller increased to 900 mm, the maximum motor shaft power increased to 1600 kW and the maximum test speed up to 600 km/h, which provides future proofing for even higher speed tests for Chinese high-speed vehicles. The latest configuration of the roller rig is shown in Figure 19.6.

FIGURE 19.6 The roller rig in Chengdu.

19.4.1.1.2 Configuration of the Roller Rig

The main parameters of the roller rig are shown in Table 19.18. The rig can test conventional four-axle and also six-axle railway vehicles and can run in either active or passive modes, depending on whether testing a locomotive or an unpowered trailer car.

The degrees of freedom of the Chengdu roller rig are shown in Figure 19.7 and comprise the following:

- Movement of the two rollers independently in the Y direction to simulate the track irregularities of gauge and lateral alignment
- Movement of the two rollers independently in the Z direction to simulate the track irregularities of cross level and vertical profile

TABLE 19.18
Main Characteristic Parameters of Chengdu Roller Rig

Parameter	Value	Parameter	Value
Exciting in vertical		Exciting in lateral	
Maximum frequency f_{vmax}	30 Hz	Maximum frequency f_{vmax}	30 Hz
Maximum amplitude A_{max}	±15 mm	Maximum amplitude A_{hmax}	±10 mm
Maximum acceleration a_{vmax}	±4 g	Maximum acceleration a_{hmax}	±5 g
Maximum traction force per axle F_e	10 t	Maximum axle load Mw	25 t
Maximum motor power W	1600 kW	Maximum speed V	600 km/h
Maximum cant angle ϕ_{max}	7°	Distance between bogies L	6~22 m
Bogie wheelbase l	2000~3500 mm	Range of gauge A_0	1000~1676 mm
Minimum curve radius R	200 m	Maximum wheelset numbers Nz	6

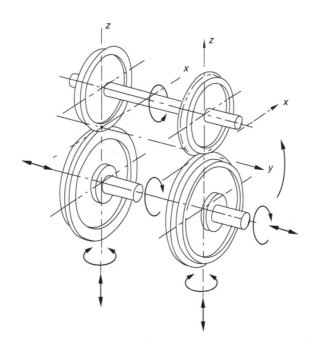

FIGURE 19.7 Degrees of freedom of rollers. (From Iwnicki, S. (Ed.), *Handbook of Railway Vehicle Dynamics*, CRC Press, Boca Raton, FL, 2006. With permission.)

- Rotation of the two rollers about the X axis to simulate the cant angle in curving
- Rotation of the two rollers at the same speed about the Y axis to simulate the forward speed of a vehicle on straight track and with different speeds to simulate curving
- Rotation of the two rollers about the Z axis to allow the development of an angle of attack between wheel and rail and hence simulation of the track curvature

The linear motions of the two rollers in the Y and Z directions and also the rotation of the two rollers about the Y axis are controlled during rig operation. The rotation about the X and Z axes is applied only for curve simulation and is preset before the test. That is, this roller rig can simulate straight track and circular curved track with track irregularities.

19.4.1.1.3 Specification

Hydraulic actuators provide the movements of the two rollers in the Y and Z directions, and the rotation about the Y axis is driven by the motor. The rig is composed of several subsystems including the test unit, driving system, hydraulic system, monitoring system, and data acquisition and processing system. The whole test system of the rig is shown in Figure 19.8a. The main power supply for the rig comes from the railway power supply with 25 kV at 50 Hz or a low-level power supply of 380 V at 50 Hz.

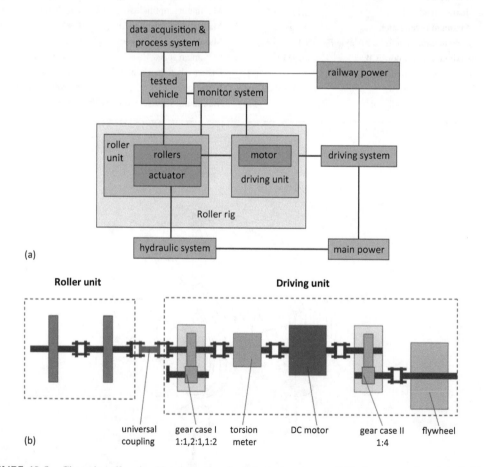

FIGURE 19.8 Chengdu roller rig: (a) schematic of whole test system and (b) the setup of each individual test cell. (From Iwnicki, S. (Ed.), *Handbook of Railway Vehicle Dynamics*, CRC Press, Boca Raton, FL, 2006. With permission.)

The roller rig has six test cells. These are independent and can be moved according to different vehicle configurations. Each test cell consists of a roller unit and a driving unit, as shown in Figure 19.8b. The driving unit can provide different rotational speeds and torque to the rollers via a double-articulated universal joint, according to the task required. Within the driving unit, there is a direct current motor, two gear cases, a flywheel and a torsion meter, all fixed to a welded frame. The motor can work as a driving motor or a generator according to the requirement of driving or braking. The flywheel is used to maintain the running stability of the roller rig and to simulate the inertia of the vehicle. Gear case II is used to accelerate the flywheel, while gear case I is used to apply different rotational speeds and torques by setting the transmission ratio as 1:1 and 2:1 (for high torque) or 1:2 (for high speed).

The roller rig has the ability to measure the output torque of the driven motor, but the roller cannot measure the applied force on the wheels of the vehicle. The rotational speed of the roller can be easily checked by an additional mounted rotational speed sensor or use of the output signal from the driven motor. The motor control system can also provide the power consumed during the test.

Driving System: The driving system consists of a control computer, digital controller, converter, motor excitation, resistance and motor. Using a feedback control technique, the operator can control the motor operation according to the defined running speed or operating torque through the control computer. The maximum difference between the rotation speeds of the six test units can be controlled to within 0.5%. Therefore, the driving system can ensure that the six roller sets are rotating almost synchronously without any mechanical connection.

Hydraulic System: The movements of the roller in the vertical and lateral directions are realised by lateral and vertical hydraulic actuators. The roller rig is a complicated system with 24 actuators for 12 rollers. By using a digital controller, the motion of the actuators is controlled by a displacement Proportional–Integral–Derivative (PID) feedback control.

Control System and Data Acquisition: The operation of the roller rig is governed by a control system. The control system can display the roller rig running speed and torque, plus the temperature of bearings and of lubricating oil. Through a 12-channel video system, the status of the roller rig, the test vehicle and the contact conditions of roller and wheel can be monitored. The system also has the function of overload protection and safety interlocking.

The terminal operating computers for the driving system, hydraulic system, monitoring system, data acquisition and processing systems are arranged within the control room as shown in Figure 19.9. During a test, the rolling and vibration states of all rollers can be controlled through several computers, such as the rotation speed and the track irregularities.

According to the railway vehicle test evaluation standards, the response of the tested vehicle should be recorded. The data acquisition and processing system can measure the signals of displacement, velocity, acceleration, strain, pressure, temperature, voltage, current, etc. All signals can be measured, conditioned and sent to the acquisition computer via a network link, with up to 200 channels of data acquired.

Supporting Infrastructure: The following describes the supporting infrastructure of the roller rig:
- *Test shed* – the test shed with a length of 72 m and a width of 24 m is divided into two sections, one section being for the roller rig and the other for test preparation and locating facilities for component tests.
- *Component test stands* – vehicle suspension parameter measurement, fatigue tests, etc.
- *Power supply* – there are two power supply systems. The civil power system with three-phase Alternating Current (AC) 10 kV is used for the driving motors of the roller rig. The hydraulic system and other systems use 380 V. The railway power system of AC 25 kV is to power electric locomotives under test.
- *Cranes* – there are two gantry cranes in the test shed, with capacity of 50 t each.

FIGURE 19.9 Control room at Chengdu.

19.4.1.2 Operational Modes

The Chengdu roller rig is used not only to verify the performance of railway vehicles but also for basic research duties, such as wheel-rail contact mechanics, wear and noise.

The following describes the primary modes of operation, thereby defining the rig's key test capabilities:

- Tangent track running stability and ride comfort
- Curving performance and safety
- Adhesion investigations
- Driving or braking power tests and optimisation of train operation
- Basic research on wheel-rail creep theory
- Wheel-rail wear, adhesion and control
- Wheel-rail noise and noise reduction
- Modal tests of the suspension vibration
- Derailment mechanisms

The following are the secondary modes of operation for the rig that are not primary design features:

- Static and dynamic parameter measurement of railway vehicle systems
- Accelerated fatigue testing
- Goods load safety
- Vibration and dynamic stresses of vehicles components

19.4.1.3 Applications

19.4.1.3.1 Stability and Ride Index Tests of High-Speed Freight Car [9]

The high-speed freight car shown in Figure 19.10, manufactured by Qiqihar Railway Rolling Stock Co., Ltd., was tested on the roller rig in 2014. This flat car was designed to be loaded with three containers of 20-feet length and has a distance between bogie pivots of 13.2 m and a bogie wheelbase

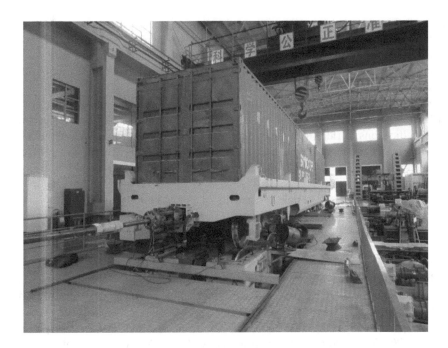

FIGURE 19.10 High-speed freight car on the roller rig.

of 2.5 m. The intended operational speed was 220 km/h on a standard gauge track; to achieve this speed, the bogies were equipped with anti-yaw dampers. The main purpose of the test was to evaluate the hunting stability and ride performance of the vehicle to validate and improve the design of the suspension system. In addition, the test cases also considered suspension failure modes.

Methodology: Different types of track irregularities were employed that were dependent on the running speed of the vehicle. The American Class 5 track irregularity spectrum was used when the test speed was between 80 and 160 km/h, while the measured Chinese Qinshen track spectrum was used for speeds of 160 ~ 260 km/h. Lower excitation amplitudes were employed in the case of higher operational speeds. Different sets of suspension parameters were examined, considering the combination of the damping of the anti-yaw damper and lateral damper, stiffness of the rubber spring as well as the steering stiffness of the joint bush mounted in the swing-arm. A total of 27 cases were tested at various speeds. In order to evaluate the dynamic performance of the vehicle, a total of 46 channels of acceleration and displacement were recorded.

Test of Hunting Stability: The actual critical speed was measured on the roller rig (the definition of critical speed is given later in Section 19.5.1.3). For these lateral stability tests, the measurement of lateral displacement of the wheelset and acceleration on the bogie frame is used to identify the hunting motion of the bogie at certain speeds when a regular harmonic motion of the wheelset or bogie frame occurs.

In the testing of the critical speed, the vehicle was excited by the track irregularities for a period of time, and the track excitation was then removed. With only pure rolling at a constant speed, the critical speed can be identified by observing if the oscillation of the vehicle restores to an equilibrium state without any harmonic movements.

If the lateral motion of the bogie frame is monitored, the peak and r.m.s values of the frame acceleration can be used to assess the stability. When evaluating the peak value, different filtering bands can be applied; for example, the European standards UIC 515 and TSI 2008/232/CE use 4 ~ 8 Hz and 3 ~ 9 Hz band-pass filtering, respectively. Furthermore, the limit value is exceeded when the peak reaches 8 m/s² during more than 6 to 10 consecutive cycles. While the stabilities determined by these criteria are close to reality, their application depends on the vehicle type and track conditions. A peak limit of 8 m/s² for six consecutive cycles with a frequency band of 0.5 ~ 10 Hz is used for Chinese high-speed trains concerning the operational speed and track conditions [10].

Ride Index: The acceleration on the carbody was used to evaluate the ride performance at each running speed. For each speed, the sampling duration must be 60s, according to the specifications of ride index analysis in railway vehicle criteria. The vibrations on the carbody, bogie frame, and wheelset were examined along the lateral and vertical directions. In addition, the relative displacements of the primary and second suspensions were recorded.

Results and Conclusions: Example test results of bogie stability are listed in Table 19.19. In the case of the original set of suspension parameters, the vehicle has an actual critical speed of more than 260 km/h and can meet the requirements for an operational speed of 220 km/h on track. It can be seen that, within the range studied, the critical speed is affected by the damping coefficient of the lateral dampers in the secondary suspension, with the critical speed reducing to 240 km/h. Removing the yaw dampers leads to the severest instability of the bogie; in this case, the vehicle has an actual critical speed of 160 and 120 km/h in the case of fully loaded and empty conditions, respectively.

The corresponding Sperling ride index on the front and rear carbody is presented in Table 19.20. Compared with the Sperling index of case A, the suspension failure or non-optimised damping leads to a larger index and vibration on the carbody, which deteriorate the vehicle ride quality.

TABLE 19.19
Test Results of Bogie Stability of the High-Speed Freight Car

Test Cases	Actual Critical Speed(km/h)
A: Original parameters	260
B: +30% damping in lateral dampers	260
C: −50% damping in lateral dampers	240
D: Remove all anti-yaw dampers, loaded	160
E: Remove all anti-yaw dampers, empty	120

TABLE 19.20
Test Results of Ride Index of the High-Speed Freight Car

Test Cases	Speed	Front of Carbody (Sperling Index)		Rear of Carbody (Sperling Index)	
		Lateral	Vertical	Lateral	Vertical
A	260	2.836	2.720	2.932	2.714
B	260	3.043	2.729	2.821	2.673
C	240	2.794	2.730	3.034	2.740
D	160	2.793	3.824	2.860	3.670
E	120	3.116	3.416	2.727	2.942

19.4.1.3.2 Stability Test and Theoretical Analysis of a High-Speed Passenger Vehicle

A stability test was also performed on a widely used high-speed electrical multiple unit (EMU) manufactured by CRRC Changchun Railway Vehicle Co., Ltd., which was designed to have a maximum continuous operational speed of 380 km/h. It has a distance between the two bogies of 17.375 m and a bogie wheelbase of 2.5 m. Currently, the vehicle has been put into service on many railway lines in China, including the Beijing-Shanghai, Wuhan-Guangzhou and Harbin-Dalian high-speed lines. Figure 19.11 shows the vehicle under test on the roller rig.

Methodology: Similar to the test methods described in the freight vehicle experiments, the testing cases considered suspension failure and various sets of suspension parameters, especially the guiding stiffness of the wheelset and damping ratio of the yaw damper. In addition to the hunting stability tests under various wheel-rail interaction conditions, the ride comfort of the carbody was tested. The test results were also compared with numerical simulations.

Results and Conclusions: The vehicle stability was analysed for three cases, including the car with original suspension parameters and without yaw dampers, as discussed in [11]. As shown in Figure 19.12a, the stability results from the roller rig and three numerical models share a similar increasing trend with respect to the vehicle speed for the original car case. While there exist obvious differences in the critical speed identification amongst different creep force models, it is seen that FASTSIM has the lowest critical speed and the Polach and Shen-Hedrick-Elkins (S.H.E.) models give almost the same results. The critical speed identified on the rig is between the results of the three simulation models, but it is closer to the result of FASTSIM when the vehicle speed is higher than 400 km/h. In addition, the simulation and test conditions are not exactly the same, and thus, the comparisons cannot determine which model is more accurate.

The yaw damper adds a strong non-linearity in the dynamic force outputs with respect to the excitation and has a significant effect on the vehicle system dynamics. In order to examine this effect, the critical speeds of the vehicle from the rig test and simulations were compared when all eight

FIGURE 19.11 High-speed Electric Multiple Unit (EMU) car on roller rig.

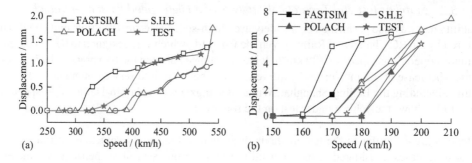

FIGURE 19.12 Limit cycles of wheelset in test and simulations: (a) original car and (b) after removal of yaw dampers.

yaw dampers mounted on the test car were removed. As shown in Figure 19.12b, the bogie hunting appears at quite a low speed, with large lateral movement of the wheelset. The simulated limit cycle of lateral movement of the wheelset resulting from the Polach model is much closer to the test results, followed by the S.H.E. model, while the FASTSIM model predicts a lower critical speed. The solid markers in Figure 19.12b present the results when increasing the vehicle speed, while the hollow markers correspond to the cases when slowing down the vehicle. It is seen that the vehicle experiences similar stability performance at a certain speed regardless of using the increasing speed and slowing down method, which means the hollow and solid markers merge with each other at some speeds.

As shown in Figure 19.13, the simulated Sperling index is larger than the test results on the roller rig, but they have the same trend with respect to the increase of vehicle speed. This difference becomes severe when the running speed is higher. In addition, the simulation results change smoothly with the increasing speed, while the test results fluctuate. This may be caused by not fully considering the structural vibration of the carbody in the numerical simulation. Moreover, the lateral Sperling ride index predicted by the S.H.E. model has the smallest value among the simulation results, especially when the running speed is higher than 300 km/h. Meanwhile, the vertical Sperling ride index predicted by the three models share nearly the same amplitude.

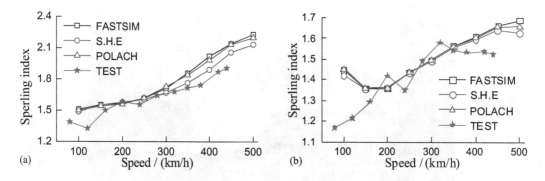

FIGURE 19.13 Sperling index in test and simulations: (a) lateral and (b) vertical.

19.4.2 HUDDERSFIELD ADHESION AND ROLLING CONTACT LABORATORY DYNAMICS RIG

19.4.2.1 Technical Overview

19.4.2.1.1 Primary Research Objectives

Commissioned in 2017 at the University of Huddersfield's Institute of Railway Research (IRR), the primary research objectives of the Huddersfield Adhesion and Rolling Contact Laboratory Dynamics (HAROLD) rig are defined by the scientific fields of railway rolling contact, adhesion and train braking. Fundamentally, such applications dictate a test rig with a large roller diameter to minimise resultant errors in the shape and size of the contact patch between wheel and roller, relative to that of a conventional flat rail.

Consequently, the HAROLD rig's key research objectives can be defined as follows:

- *Adhesion*: In railway terms, adhesion or the adhesion coefficient can be defined as the ratio between the longitudinal force, F_x and vertical force, Q, between wheel and rail. It forms an expression of the available friction and is critical in governing the tractive or braking effort that can be transmitted between wheel and rail. The study of adhesion and maximising the performance of a vehicle's traction and braking system has become a key area of research in the UK and supports a long-term vision to reduce signalling headways (the spacing between trains), hence maximising capacity and route utilisation. HAROLD is being applied to study the effect of varying adhesion conditions within the wheel-rail contact, with particular focus on how low-adhesion conditions form. Experimental studies are being undertaken to study the third-body layer formed when 'mulch' is generated as a result of leaf-fall and other contaminants within the wheel-rail contact. The rig can also be used to investigate the efficacy of top-of-rail friction modifiers or flange lubricants on wear and other wheel-rail damage modes.
- *Contact*: With an ability to re-machine the rail roller profiles in situ, using an integral lathe facility, HAROLD can represent the wheel-rail contact conditions for any rail system in the world. This capability facilitates the study of new wheel and rail profiles and the evolution of contact derived degradation modes such as RCF and wear. With an ability to measure wheel-rail forces and creepage conditions, the rig can be used to gather relevant information for a particular bogie or wheelset under test.
- *Braking*: As a complimentary activity to research around adhesion and wheel-rail contact, further primary research objectives of HAROLD are the analysis of train braking systems and the development of advanced adaptive WSP systems to improve braking performance. Currently, in the design phase, the capability of HAROLD is being extended to include an HIL simulation of the whole train braking system. This allows a prototype or service bogie with attached braking systems and controllers to be evaluated at full-scale, with other bogies in the train consist represented through real-time co-simulation alongside the bogie hardware on test.
- *Bogie performance*: As a final primary research objective, the ability of the HAROLD rig to accommodate a complete bogie assembly allows for the study of suspension behaviour. Of particular interest is the characterisation of primary yaw behaviour (steering), especially for suspensions that are complex to model, such as those with friction elements or other non-linearities. This type of study can help validate computer models, useful in the UK, where such models are used to help establish vehicle-track access charges with regard to propensity to drive track degradation.

As a secondary research objective, HAROLD's 11 m × 4 m, 50 t capacity strong floor can be utilised as a general-purpose advanced dynamic test cell. Typical applications in this non-rolling configuration might include mechanical and accelerated fatigue testing of a trackform or other multi-axis dynamic testing at high loads.

19.4.2.1.2 Configuration of the Roller Rig

The HAROLD test rig was primarily developed around the concept of investigating rolling contact, adhesion and braking, utilising a test bogie. An overview of the HAROLD test rig is shown in Figure 19.14.

HAROLD is built around the concept of a testbed that can be moved laterally and rotated in yaw to allow manipulation of a bogie relative to a fixed roller. This approach allows for either direct control of a wheelset (constrained wheelset mode) via mechanical attachments to a bogie's axleboxes (see Figure 19.15), or indirect control whereby the wheelset is free to move within the primary suspension. In this configuration, the yaw angle and lateral position of the wheelset is governed by the position of the bogie itself, which is manipulated via movement of the test bed.

Vertical load is applied to the bogie centre pivot via a loading beam and two 25 t capacity hydraulic actuators (see Figure 19.16). Load can be varied dynamically to facilitate wider studies such as the effect of cant deficiency (body induced bogie roll moment) on wheelset steering performance or vertical suspension characterisation.

Longitudinal restraints are provided to secure the bogie and ensure the leading wheelset is correctly positioned on the crown of the rollers (see Figure 19.17); these restraints can also be used to facilitate bogie rotational resistance testing.

The key components of HAROLD are described in further detail in the sections that follow.

Roller Unit: The roller unit (see Figure 19.18) is rigidly mounted to the test rig's portal frame and has a single degree of freedom in pitch (rotational driving and braking directions). Full technical details are presented in Table 19.21 of the specification section below.

FIGURE 19.14 Overview of HAROLD test rig.

FIGURE 19.15 Constrained wheelset mode of operation.

FIGURE 19.16 Universal bogie loading beam and vertical actuators.

The unit has a number of novel features to support the research objectives described above. Fundamentally, the unit provides a single 2 m diameter roller with two curved rail sections attached. Each rail is made up of four segments. Segmentation of the rails facilitates measurement of rail reaction forces within three directions, namely longitudinal, lateral and vertical, together with moments around these axes. The measurements are acquired within one adjacent segment on each of the left and right rails. A further benefit of segmenting the rails is the ability to test different rail steels within one revolution of the roller.

FIGURE 19.17 Longitudinal restraints (rig bed to loading beam).

FIGURE 19.18 Rigidly mounted roller unit with load cells.

TABLE 19.21
Summary of Rolling Unit Mechanical Specification

Parameter	Value
Roller diameter	2.0 m
Maximum axle load	25 t
Maximum running speed	200 km/h
Maximum wheelbase	3.0 m
Minimum wheelbase	1.5 m
Gauge	1435 mm
Maximum torque	110 kNm
Continuous power rating	0.45 MW

- The stiffness of the mounting between rail and roller can also be varied to allow different rail support conditions to be considered.
- The roller unit has been specified with a high torque capacity; this facilitates traction and braking studies, whereby wheel slip can be encountered at up to 25 t axle load at a friction coefficient of 0.4. This capability allows WSP algorithms to be tested and refined and new drivetrain concepts to be evaluated.
- Supporting investigations at high slip, the rail surface can be re-machined in situ; this allows both restoration of the rail following a slip event and a complete change of the test rig's rail profile shape and effective inclination.
- High-precision encoders and control systems allow the levels of creep (slip) between wheel and rail to be accurately controlled.
- The roller unit is electrically driven via a three-speed gearbox. Inverter drives are combined with high-precision sensors to provide precise control of roller speed and torque.

Portal Frame and Testbed: Structurally, HAROLD is based around a bespoke fabricated, large section portal frame. The portal frame supports not only the roller unit, but also the testbed itself. With a total weight of 120 t, HAROLD is mounted on eight pneumatic air springs for vibration isolation and is effectively 'free-floating' when in operation.

- The testbed is constructed in two sections, one section fixed in place providing a general-purpose strong-floor and the second section mounted on a slewing ring and translational sub-frame to allow for lateral and yaw displacement of the bed relative to the fixed roller unit.
- Servo-hydraulic movement of the test bed facilitates both quasi-static movement for wheel-rail contact and curving studies and dynamic movement, whereby a wheelset can be subjected to wheel-rail relative shifts, as calculated from actual route-based vehicle dynamics simulations. Such dynamic tests can be used to investigate wheel-rail wear and other damage modes.
- The testbed has embedded rails and a matrix of threaded holes to facilitate loading and constraint of the test bogie. The bogie is typically constrained to the testbed at its trailing axle, with the leading axle positioned over the crown of the rollers.

Loading Beam: The loading beam provides the means by which vertical forces are imparted to a test bogie. It consists of a reaction beam, with a universal mounting arrangement to accommodate a variety of secondary suspension arrangements. HAROLD is currently configured with a UIC freight bogie centre bowl and adapter plate arrangement, together with constant-contact coil sprung sidebearers.

The loading beam is mounted via two-axis gimbals, thereby allowing fore-aft and lateral displacement of the loading beam. This arrangement effectively creates a pendulum stiffness in the lateral direction, whilst fore-aft the bogie is typically constrained.

To allow for varying bogie wheelbases, the loading beam is connected to the portal frame via two linear traversing mechanisms; these allow the centre point of the loading beam to be moved fore-aft and laterally relative to the testbed.

Constraints: A test bogie can be constrained in a number of configurations to suit the purpose of the test work. For wheel-rail contact related studies, where direct control of the wheelset's position relative to the rollers is typically required, adapter flanges are fitted to the axlebox to provide mechanical connection to the testbed. This results in testbed translations and rotations being directly imparted to the wheelset (constrained wheelset mode). The mechanical connections are instrumented to provide longitudinal forces at each axlebox and net lateral force generated by the wheelset.

When a test requires the wheelset to remain unconstrained within the primary suspension of a bogie, then bogie constraint is provided at the trailing axle to the testbed. This is supplemented by longitudinal constraint by means of restraints mounted between the testbed and loading beam. In this mode of operation, wheelset steering and kinematic behaviour within the suspension can be observed, with wheel-rail forces measured within the roller unit.

19.4.2.1.3 Specification

The detailed specifications of the HAROLD test rig systems are presented later; these include mechanical parameters in Table 10.21, kinematic performance in Table 19.22 and instrumentation parameters in Table 19.23, and details of the control system then follow.

The HAROLD test rig has a control system that is constructed within the National Instruments Labview environment. This reconfigurable architecture controls all elements of the test rig, including safety interlocks, manual and test file-based actuator control, data logging and a HIL real-time simulation capability. This latter feature will enable train braking models to be run in parallel with HAROLD's full-scale wheel-rail interface, facilitating advanced research into adhesion and train braking technologies.

19.4.2.2 Operational Modes

19.4.2.2.1 Primary Modes

The following section describes the primary modes of operation of the HAROLD rig, thereby defining its key capabilities.

Contact Studies: In constrained wheelset mode, HAROLD can facilitate wheel-rail contact studies whereby the wheelset position is directly controlled based upon inputs from either the manual control screen, or an input file expressing a time history of

TABLE 19.22
Summary of Kinematic Performance Envelope

Parameter	Value
Testbed yaw displacement range	±0.10 rads
Testbed yaw velocity range	±0.04 rads/s
Testbed lateral displacement range	±30 mm
Testbed lateral velocity range	±55 mm/s
Loading beam actuator displacement range	±400 mm
Loading beam actuator velocity range	±250 mm/s

TABLE 19.23
Measurement Range and Resolution of Primary Instrumentation Systems

Measurement Parameter	Range (Accuracy)
Roller unit rail force (z)	±450 kN (1 kN)
Roller unit rail force (x and y)	±225 kN (1 kN)
Roller unit speed	±56 m/s (0.003 m/s)
Roller shaft torque	±120 kNm (0.03 kNm)
Roller vertical force (× 2)	±500 kN (0.25 kN)
Testbed lateral displacement	±30 mm (0.01 mm)
Testbed lateral force	±200 kN (0.1 kN)
Testbed yaw displacement	±0.105 rads (0.1 × 10^6 rads)
Testbed yaw force	±200 kN (0.1 kN)
Loading beam displacement (× 2)	±800 mm (0.01 mm)
Loading beam force (× 2)	±500 kN (0.25 kN)
Test bogie restraint load (× 3)	±200 kN (0.1 kN)

wheelset lateral and yaw displacements established from simulations. This enables precise control of lateral and rotational wheel-rail relative shift, together with creepage and normal load.

Wheel-Rail Profile Design and Material Investigations: Operating in constrained wheelset mode or utilising a full-scale test bogie, HAROLD can be used to evaluate new wheel and rail profile designs. This allows studies of materials, contact mechanics and wear, together with evaluation of curving and derailment performance for a given wheel-rail geometry.

Braking Studies: Evaluation of train braking systems can be performed in two modes of operation. In constrained wheelset mode, precise control of longitudinal creepage can be achieved; this approach has benefits when investigating low-adhesion braking conditions and development of advanced adaptive WSP algorithms using HAROLD's HIL functionality.

Study of braking systems can also be extended to full-scale bogie testing, whereby a bogie and its complete braking sub-systems can be tested under a range of wheel-rail operating conditions.

Adhesion Studies: Coupled with the development of braking technologies, HAROLD can be applied in better understanding the formation of wheel-rail contact third-body layers, whereby contamination of the contact results in a reduction in available adhesion. Research work is ongoing in this area to develop and validate low-adhesion creep curves, helping in establishing more precise train braking simulations.

Suspension Optimisation: With the capability to impart lateral shift and yaw to a test bogie, HAROLD can be used to characterise the primary and secondary yaw suspension of a bogie. This has applications in validation of complex bogie models and verification of steering performance, particularly where a steering index is applied in track access charging, such as the case in the UK.

19.4.2.2.2 Secondary Modes

The following section describes the secondary modes of operation for the HAROLD rig; these describe test capabilities that are not primary design features of the rig.

Bogie Dynamic Performance: Through design of suitable constraints and fixtures, it is possible to use the HAROLD rig to perform bogie rotational resistance testing, evaluate twist performance (wheel unloading) and perform dynamic stability assessments with respect to propensity for wheelset and bogie kinematic hunting modes. There are limitations on stability studies because only a single wheelset can run on a roller; however, useful results have been obtained under this mode of operation.

Trackform and General Mechanical Testing: Utilising the 11 m × 4 m, 50 t rated strong-floor, combined with the advanced servo hydraulic testing capability of the HAROLD rig, it is possible to carry out accelerated fatigue testing of trackform elements such as concrete track slabs, level crossing panels, sleeper assemblies, rail fixings and other high-load and high cycle mechanical component testing.

19.4.2.3 Applications

19.4.2.3.1 Testing of a Remanufactured Wheelset

The IRR carried out experimental testing of a wheelset that had been remanufactured using an additive welding process. The primary objective of the test work was to establish the effectiveness of the remanufactured wheel material when benchmarked against a conventional wheel steel, specifically in the context of wear performance.

Methodology: The HAROLD rig was configured in constrained wheelset mode, whereby the test wheelset was mounted in a bogie frame, and axlebox constraints were fitted to connect the wheelset directly to the test bed, allowing direct control of wheelset lateral and yaw displacement relative to the rail roller.

Vehicle dynamic simulations of a typical route were carried out and the resultant wheel-rail shifts output to the HAROLD test rig to construct a route-based test profile, thereby ensuring a realistic distribution of contact positions throughout the test cycles run on the rig. The tests were performed at a constant running speed and representative axle load. Following verification of the wheelset response relative to the simulated outputs, a number of test cycles were completed, following which, wheel wear measurements were taken using digital profile recording equipment (Miniprof). These were supplemented by wheel profile RCF crack detection tests using magnetic flux leakage techniques.

Results and Conclusions: The experimental work provided wear and RCF data for a new wheel material based on accurate route-based simulations of wheelset dynamic position change. Based on the experimental work, a further programme of work into additive weld techniques is being developed.

19.4.2.3.2 Freight Bogie Stability Testing

The IRR carried out experimental work to investigate kinematic instability within a freight bogie. The primary objective of the test work was to establish the potential cause of instability on two different wheel profile geometries.

Methodology: The test bogie was constrained to the test bed of the HAROLD rig via its centre pivot using a loading beam and two longitudinal constraints to ensure that the lead axle remained on the crown of the rail roller. The rear axle was unconstrained; this provided the greatest degree of freedom within the limitations of running a single rail roller test (see Figure 19.19).

Tests were performed at a range of operating speeds up to 120 km/h and axle loads from tare to fully laden (25 t). A lateral step input was imparted to the lead axle to initiate

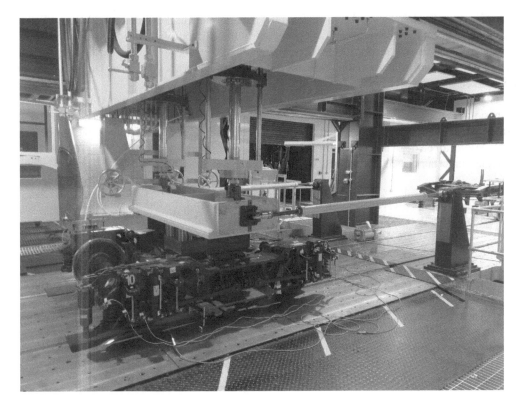

FIGURE 19.19 Bogie stability testing.

wheelset instability. In addition, the nested primary vertical suspensions were subjected to characterisation tests to assist in developing the friction breakout behaviour, aiding in validation of a vehicle dynamics model. The bogie was fully instrumented to assess how the suspension system performed during the test programme; this included lateral, vertical and rotational displacements and accelerations across key elements of the primary and secondary suspensions.

Results and Conclusions: The experimental work provided critical data for validation of vehicle dynamics models, which were subsequently used to investigate bogie stability issues. The test programme itself allowed bogie stability to be experimentally tested, and ultimately, modifications to the bogie design developed.

19.4.3 THE ATLAS ROLLER RIG, KNORR-BREMSE, MUNICH, GERMANY

19.4.3.1 Technical Overview

19.4.3.1.1 Primary Research Objectives and History

The Atlas test rig is specifically developed for adhesion and brake system research and development. Following from Newton's law, it can be found that the integral of the utilised coefficient of adhesion directly determines the stopping distance when braking. Hence, the ability to control this coefficient (in particular its available maximum) provides a brake system with far better and more predictable control of the train's stopping performance, independent of the environmental influences on adhesion. Under the current state of the art, many brake systems still control a clamping force or brake cylinder pressures. Others control the slip, but without information on whether the actual slip is causing an improvement or worsening of the adhesion, hence closed-loop optimisation of braking performance cannot be achieved. This aspect forms the primary target for research undertaken with the ATLAS rig.

The motivation to develop and construct the 1:1 scale roller rig Advanced Test Laboratory of Adhesion-based Systems (ATLAS) was based on the concept of creating innovative principles for brake system management, based on the results of experimental research around how the coefficient of adhesion can be detected and controlled to optimise stopping distance.

Other important aspects of the research programme are as follows:

- To test complete brake systems, comprising brake control (including control of electrodynamic braking), WSP, pneumatics, bogie equipment and friction materials
- To accommodate deviations in the coefficient of friction due to the characteristics of the friction material but also due to environmental influences (water, snow and ice)
- To investigate and control (if possible) the adhesion factors that influence wheel slide
- The influence of the rail-wheel contact and slip on the roughness of the running surface, and its influence on the coefficient of friction of a tread brake block
- The influence of a tread brake block on the running surface, and its influence on the adhesion levels
- Wear and other types of material damage, caused by slip, rolling, braking
- The heat impact when braking, and cooling due to wheel-rail contact and headwind
- Noise

It is also possible to test running characteristics, wear and RCF of wheels and rails, but this is not the main purpose of the test rig. Dynamometer tests of friction materials and discs can be performed, though the effort to set up the ATLAS rig is higher than a specialised brake dynamometer and hence is not the most efficient option.

The design of the rig is specifically targeted at the investigation of the slip-adhesion characteristics with original equipment (OE) brake gear as installed in a production railway vehicle; hence:

- Brake equipment, a wheelset or single lead wheelset of a bogie or a vehicle are capable of being tested.
- Axle load, driving torque and inertia are adjustable according to the properties of the vehicle being simulated.
- Besides using brake equipment, the use of a counter-driven motored wheelset provides a second option to generate controlled slip as an input to achieve variable adhesion; to obtain the required precision, the motors drive the wheels directly without a gearbox.

Conception, calculation, design, construction (a hall with a special 566 t damping foundation included) and commissioning were realised between 2011 and 2017.

19.4.3.1.2 Configuration of the Roller Rig

The basic setup of the ATLAS roller rig is shown in Figure 19.20, and the available degrees of freedom during testing are displayed in Figure 19.21.

Roller Rigs

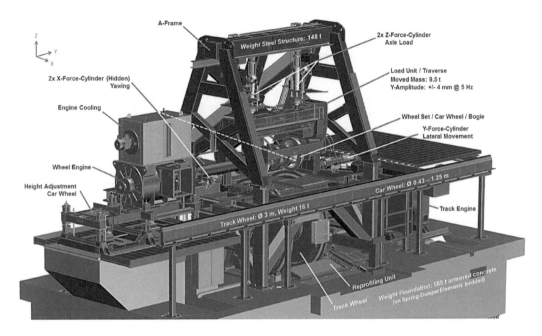

FIGURE 19.20 ATLAS roller rig, prospect and components.

FIGURE 19.21 ATLAS roller rig showing degrees of freedom of unit under test.

TABLE 19.24
ATLAS Rig Specifications

Parameter	Value
Roller diameter	3000 mm
Maximum axle load	30 t
Vertical actuator capacity	2 × 150 kN
Longitudinal (driving) and lateral actuator capacity	80 kN
Maximum running speed	350 km/h
Gauge	1000–1676 mm
Maximum torque	62 kNm
Continuous power rating	2 × 1.4 MW
Test temperature range	−20°C to +50°C

19.4.3.1.3 Specification (Table 19.24)

A summary of the ATLAS rig specification is presented in Table 19.24.

19.4.3.2 Operational Modes

19.4.3.2.1 Primary Modes

Adhesion Investigation: For adhesion investigations the test rig offers two main types of tests:

- Braking or traction of the unit under test (wheelset or single wheel) whilst the motor of the track wheel simulates the inertia of the vehicle; this means that the brake torque is controlled by electric or pneumatic means (see case study 1).
- Controlling the slip to a defined value using the counter-torque of the motors driving the wheelset relative to the track wheel (see case study 2).

For both types of tests, intermediate layers at the track may be applied (e.g., water, leaves, soap, oil and sand), and the axle load and the lateral and yaw movement of the wheelset can be controlled.

The adhesion investigation may include the following:

- Improvement of the adhesion by advanced slip control of the WSP system
- Testing of sanding and friction modifiers

Brake Tests: Besides the adhesion, the coefficient of friction of the brake pad/shoe elements may also be tested. In this case, the influence of the adhesion may be excluded.

Interference of adhesion and friction: In case of special environmental conditions (water and snow), the friction of the brake pads/shoes as well as adhesion may be affected, in that they may increase or decrease.

Interference of Electrodynamic and Pneumatic Braking (blending): Normally, the electrodynamic braking is used as much as possible for service brake application. However, the brake force of the electrodynamic brake may not be sufficient, owing to several reasons, such as thermal effects and limited capacity of the electric network. In this case, brake force generated by the friction brake can be added. The objective is to obtain a constant overall brake force to retard the train in a consistent manner. The aim of the tests is to validate the brake control algorithm to achieve this objective.

Traction: From the point of view of mechanics, the traction is no different from the electrodynamic braking, but the direction of the torque is inverted. The first motor is used for driving the wheelset, the motor at the track wheel represents the rolling inertia of the train. The control of the test rig allows adjustment of slip or a combination of maximum slip and maximum driving torque.

19.4.3.2.2 Secondary Modes

Test capabilities that are not primary design features of the rig are as follows:

- Wear test of the wheel, related to the energy dissipated by braking or due to free running
- Measuring of the mechanical load to the wheelset and the bogie equipment during movement
- Fatigue tests of elements of the bogie equipment

19.4.3.3 Applications

19.4.3.3.1 Case Study 1: Pneumatic Braking

The key test mode the rig was designed for is the braking process, utilising actual brake equipment and full-scale wheel-rail contact. The test comprises the following:

- Original brake equipment, consisting of brake control and bogie equipment
- Test specification, defining a sequence of movements, brake applications and environmental conditions such as temperature, humidity, intermediate layer at the track, wind, gradient of the track and braked weight per axle

Operational Mode

Figure 19.22 shows how such testing is performed, and what kind of results can be expected.

The test procedure is as follows:
- Installation of the brake equipment, comprising pneumatic brake control and disc brakes; the brake control includes an adjustable slip control algorithm.
- Adjustment of the mass simulation (inertial effects) at the track wheel corresponding to the vehicle being simulated.
- Contamination of the track with an intermediate layer that corresponds to the condition of the tracks due to leaf fall and wet weather in autumn.
- Acceleration of the system by the track wheel to an initial speed (the wheelset under test is not driven).
- Performing varying levels of brake applications, adjusting the slip controller to different parameters, as shown in the examples in Figure 19.22.

Results and Conclusions
- The wheel-rail adhesion can be influenced by the parameters of the slip controller.
- A defined parameter set that leads to an optimum value of the adhesion at a specific intermediate layer may fail at another type of intermediate layer due to the fact that the characteristics of the adhesion at different layers (humidity, water, oil, leaves, dust, etc.) are significantly different; subsequently, the range of slip to achieve the optimum adhesion must be adjusted for each specific use case.

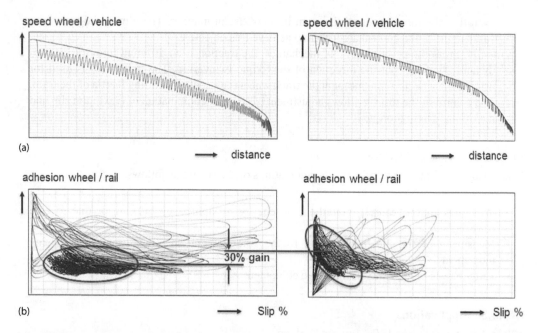

FIGURE 19.22 Speed and coefficient of adhesion at stopping brake applications with low adhesion: (a) slip control, parameter set 1 and (b) slip control, parameter set 2.

19.4.3.3.2 Case Study 2: Electrodynamic Braking of Traction with Fixed Slip

Figure 19.23 shows the speed characteristic of a stopping brake application with fixed slip.

The start of the brake application is marked with '1'. The control of the motor is adjusted to keep a constant slip during the test, until the end of the braking cycle at '3'. The example was executed with a contamination layer applied to the track that caused low-adhesion conditions. Figure 19.24 shows the corresponding adhesion in phases 1, 2 and 3.

The marks 1, 2 and 3 refer to those defined under Figure 19.23. When starting the braking ('1'), the slip is increasing to 20%, which is the set value of the control. During the braking phase ('2'), the relative slip is constant, and the adhesion (τ) is getting three times higher. When the braking torque is removed at the end ('3'), the slip again is reducing till zero.

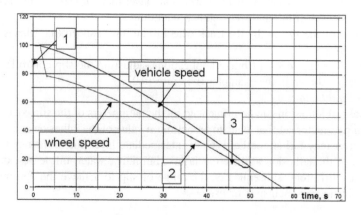

FIGURE 19.23 Stopping brake application, speed over time, with fixed slip.

Roller Rigs

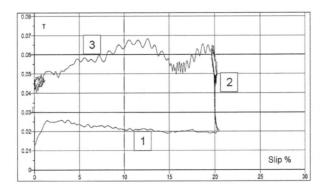

FIGURE 19.24 Adhesion for an electrodynamic stopping brake application, phases 1, 2 and 3 (see Figure 19.23).

Operational Mode: The test procedure is as follows:
- Adjustment of the mass simulation at the track wheel motor control, corresponding to the vehicle that shall be simulated
- Connecting of the wheelset under test to its electric motor, simulating the traction motor
- Application of an intermediate layer to the track that corresponds to the condition of the tracks due to leaf fall and wet weather in autumn
- Performing of varying brake applications, using the braking torque of the motor connected to the wheelset, while the motor of the track wheel is simulating the inertia of the vehicle

Results and Conclusions: The slip during electrodynamic braking can lead to a substantial increase or decrease of the adhesion level. A defined slip, which leads to optimal adhesion at a specific intermediate layer, may lead to a decrease in adhesion with another type of intermediate layer.

19.4.3.3.3 Case Study 3: Comparison of the Behaviour of a Tread Brake and a Disc Brake at a Track with an Intermediate Layer

Objective of the study is to compare the influence of the type of brake equipment (disc or tread) on the adhesion when the track is contaminated with an intermediate layer.

Operational Mode: The track is contaminated with an intermediate layer and a number of stopping brake applications, accelerations and periods of free running were performed. The initial intermediate layer disappeared due to roller rotation, the slip and the friction (in case of block brake). The procedure was repeated seven times with tread brakes and a further seven times with a disc brake.

Results and Conclusions: Figure 19.25 shows the resultant stopping distances. Immediately after the contamination of the track, they are longer, because the layer is lowering the adhesion as well as the brake friction (in case of the tread brake). After four brake applications and the corresponding running distance, the wheel with tread brakes has recovered its initial brake performance at a stable level. Due to the missing cleaning effect of the tread block, a disc brake is more sensitive to intermediate layers at the track.

Note: The example case studies are provided to explain the function of the roller rig. The results may not be generalised for other materials and different test procedures.

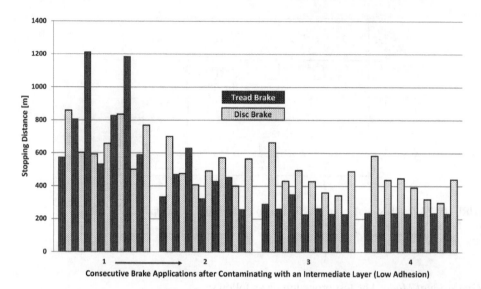

FIGURE 19.25 Improvement of the stopping distances due to consecutive brake applications. Disc brake (lighter grey) and tread brake (darker grey) in comparison.

19.4.4 Florence Railway Research and Approval Centre Roller Rig

19.4.4.1 Technical Overview

19.4.4.1.1 Primary Research Objectives

The Florence Railway Research and Approval Centre (RRAC) roller rig was commissioned in 2011 by Rete Ferroviaria Italiana SpA (RFI) and is currently located at the RRAC in Osmannoro, Florence, Italy [12–15]. The main research goals of the new roller rig are related to the study of railway vehicle dynamics (both on straight and curved tracks), braking and traction under different adhesion conditions and electromagnetic compatibility.

The capabilities of the new RRAC roller rig will enable research and industrial activities in the following fields:

- *Vehicle dynamics* – thanks to the roller rig's actuation system (four motorised axles, each one including two independent rollers actuated by two independent motors), sensor system and carbody constraint system, the lateral, vertical and longitudinal dynamics of many different railway vehicles can be investigated. The rig can be used as a fundamental tool to design and test vehicle components and on-board subsystems (OBS) and to develop and validate physical models (including wheel-rail interaction models).
- *Vehicle electric braking* – the rig enables the modelling, design and testing of braking systems. By using an advanced HIL architecture of the RRAC, many operating conditions can be reproduced, including degraded adhesion conditions.
- *Vehicle traction* – various traction systems can be developed and tested by exploiting the RRAC rig capabilities and its HIL architecture. The advanced actuation system allows reproduction of almost all relevant operating conditions.
- *Electromagnetic compatibility* – finally, applying an advanced sensor system, the RRAC roller rig can be used for electromagnetic compatibility tests and for the development of related equipment.

19.4.4.1.2 Configuration of the Roller Rig

The RRAC roller rig was designed for the study of railway vehicle dynamics, braking and traction under different adhesion conditions and electromagnetic compatibility. The rig is able to reproduce vehicle dynamics on both straight and curved tracks. An overview of the RRAC roller rig is shown in Figure 19.26. The key components of the RRAC roller rig are described in further detail in the following sections.

- *Roller Unit*: Each roller unit (one single roller) is rigidly mounted to the roller rig portal frame and has a single degree of freedom in pitch (rotational driving and braking). Each roller unit is electrically directly driven by one independently controlled motor. The spacing (gauge) of the two rollers and the roller profiles can be modified depending on vehicle and track dimensions. Right and left rollers of the same axle can rotate independently to reproduce curved tracks and can also be locked by means of a specific locking system (based on high torsional stiffness) to reproduce straight tracks.
- *Actuation System and HIL Architecture*: Each roller unit is directly actuated by an independent internal permanent magnet (IPM) synchronous motor with high performance characteristics (see Figure 19.27). The advanced actuation system and a HIL architecture that exploits sliding-mode based control techniques [13,14] enables a faithful reproduction of vehicle dynamics on straight and curved lines and under degraded adhesion conditions (without excessive wear of the wheel and roller contact surfaces). Sliding-mode control

FIGURE 19.26 Overview of the RRAC roller rig, placed in the semi-anechoic room of the Research and Approval Centre of Osmannoro, Florence, Italy. (From Allotta, B. et al., *Int. J. NonLin. Mech.*, 57, 50–64, 2013. With permission.)

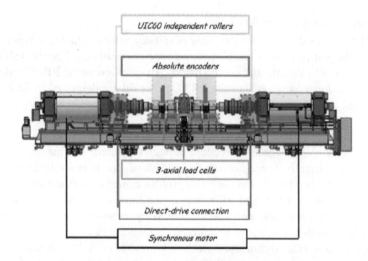

FIGURE 19.27 Scheme of the actuation system of the RRAC roller rig (one axle, including two independent rollers actuated by two independent motors). (From Conti, R. et al., *Mechatronics*, 24, 139–150, 2014. With permission.)

is a powerful robust control technique, quite suitable for this kind of applications, where a desired system dynamics has to be reproduced in presence of many unknown external disturbances.

Portal Frame: From a structural point of view, the portal frame of the RRAC roller rig is characterised by moving supports (one for each roller). This way, the bogie wheelbase and the carbody length can be modified according to the dimensions of the railway vehicle.

Loading Beam: The loading beam provides the required load both at the rear and at the front of the carbody. Static and dynamic three-dimensional (3D) loads can be applied, in longitudinal, lateral and vertical directions. General external loads are very useful to investigate the static and dynamic performances of the vehicle and its 3D forced dynamics on the rollers.

Constraints: Two constraint systems are implemented to properly constrain the front and rear of the carbody. Each constraint can limit the longitudinal, lateral and vertical motions of the carbody, depending on the dynamics to be investigated. A complete system of constraints is fundamental to study the dynamics of the railway vehicle on the roller, with and without external forces applied to the carbody.

Sensors: Each roller is equipped with a torque sensor (for the detection of the torque applied to the rollers), an encoder (for the detection of the roller position, speed and acceleration), and four triaxial load cells (for the detection of longitudinal, lateral and vertical forces). The rear and front constraints systems are equipped with further triaxial load cells to measure the external forces applied to the railway vehicle. Finally, the RRAC and the semi-anechoic room of the research centre are equipped with specific sensors for electromagnetic compatibility tests and for the development of related equipment.

19.4.4.1.3 Specification

The detailed specification of the RRAC roller rig is presented later, paying particular attention to the physical characteristics of the rig and actuation system. Mechanical parameters are shown in Tables 19.25, 19.26 and Figure 19.27 outlines the main mechanical specifications of the actuation system of the RRAC roller rig (each roller directly driven by a specific IPM synchronous motor).

TABLE 19.25
Summary of RRAC Roller Rig Mechanical Specification

Parameter	Value
Roller diameter	1.590 m
Maximum axle load	25 t
Maximum running speed	220 km/h
Minimum–maximum wheelbase	1.6–3.5 m
Minimum–maximum carbody length	12–20 m
Gauge	1.435 mm
Maximum torque per single roller	35.0 kNm
Continuous power rating (each motor)	950 kW

TABLE 19.26
Summary of the Mechanical Specifications for the RRAC Roller Rig Actuation System

Parameter	Value
Maximum torque per single roller	35.0 kNm
Continuous power rating (each motor)	950 kW
Motor inertia	200 kg/m^2
Number of poles	6
Inductance on the q-axis	0.326 mH
Inductance on the d-axis	1.800 mH
Stator resistance	15.9 mΩ

Note: Each roller directly driven by a specific IPM synchronous motor.

19.4.4.2 Operational Modes

19.4.4.2.1 Primary Modes

The following section describes the primary modes of operation of the RRAC roller rig, defining its key capabilities. These are also explored within related publications [12–15].

Vehicle Dynamic Tests on Straight Track: In this mode, the left and right rollers of each axle are locked by means of a specific locking system based on adding a high torsional stiffness between the rollers. The dynamics of the railway vehicle on straight track can then be studied in detail, especially concerning stability and safety. Furthermore, bogie and carbody dynamics can be accurately reproduced to develop and design new primary and secondary suspensions systems.

Vehicle Dynamic Tests on Curved Track: In this second operational mode, the left and right rollers of each axle are free and can independently rotate, as they are independently actuated by two different motors. The dynamics of the railway vehicle on curved tracks can be studied in detail, especially concerning derailment and safety. Furthermore, as in the previous case, the dynamic behaviour of bogie and carbody on curved lines can be accurately reproduced to develop and design new primary and secondary suspensions systems.

Braking and Traction Tests under Good-Adhesion Conditions: The dynamic behaviour of railway vehicles during braking and traction manoeuvres under good-adhesion conditions can be easily reproduced on the roller rig by means of simple HIL architecture and standard control systems. This operational mode is important to design and test new OBS such as WSP for braking and anti-skid (AS) for traction.

Braking and Traction Tests under Degraded Adhesion Conditions: The dynamic behaviour of railway vehicles during braking and traction manoeuvres under degraded adhesion conditions is much more complex and difficult to reproduce. These critical scenarios can be obtained on the roller rig by means of a new HIL architecture and more sophisticated sliding-mode based control systems. The new RRAC roller rig and its HIL architecture are able to reproduce such scenarios (without wearing the wheel and roller contact surfaces). As in the previous point, this operational mode is fundamental to design and test new OBS such as WSP systems for braking and AS systems for traction.

Electromagnetic Compatibility Tests: The RRAC roller rig and the semi-anechoic room of the Research and Approval Centre of Osmannoro (Florence, Italy) can be equipped with the sensors and instrumentation required to perform a large variety of electromagnetic compatibility tests. Thanks to the equipment of the RRAC roller rig and of the related facilities, electromagnetic compatibility tests can be carried out accurately and according to current approval standards.

19.4.4.2.2 Secondary Modes

The following section describes the secondary modes of operation for the RRAC roller rig (capabilities that are not primary design features of the roller rig).

Study of Wheel-Rail Contact: If properly equipped in terms of sensors, the RRAC roller rig can be exploited to study in detail the wheel-roller contact dynamics and to better understand the geometrical and physical connections between wheel-rail contact and wheel-roller contact.

Study of Wheel-Rail Wear, Rolling Contact Fatigue, Polygonal Wear and Corrugation: If properly equipped in terms of sensors and by using suitable HIL architectures and control systems, the RRAC roller rig can be used to investigate typical wheel-rail interaction phenomena such as wear, RCF, polygonal wear and corrugation. Such work would also consider the related effects caused by the use of rollers, as opposed to rails, as essential part of any contact-based study when using a roller rig.

19.4.4.3 Applications

19.4.4.3.1 Braking Tests under Degraded Adhesion Conditions

By way of example, in this section, a specific braking test under degraded adhesion conditions will be briefly described [12–15]. The railway vehicle under test is the UIC-Z1 coach, equipped with a fully operational WSP system (see Figure 19.28).

Methodology: The experimental conditions characterising the braking under degraded adhesion are reported in Table 19.27.

The RRAC roller rig control architecture, based on a robust sliding mode approach, performs a simulation of the effective rolling resistance through precise independent control of the roller motors. Explicitly, the roller motors are controlled to recreate a specific wheelset's angular velocity, applied torque and tangential effort, as observed in reality on rails but at a much higher coefficient of adhesion. This novel approach allows the proving of WSP systems but avoiding gross sliding (and consequently wear) between the wheelsets and rollers. In fact, since the real adhesion coefficient between the rollers and wheelset surfaces is

Roller Rigs

FIGURE 19.28 The UIC-Z1 coach. (From Conti, R. et al., *Mechatronics*, 24, 139–150, 2014. With permission.)

TABLE 19.27
Experimental Testing Conditions for the Braking Test under Degraded Adhesion

Parameter	Value
Adhesion coefficient	0.05
Type of vehicle	UIC-Z1
Wheel radius	0.460 m
Starting speed	120 km/h
Braking torque	9500 Nm

far higher than the simulated one (greater than 0.40), negligible sliding occurs, and almost pure rolling conditions are always present between them.

Results and Conclusions: Some results of the experimental braking test under degraded adhesion conditions are reported in Figure 19.29 as an example.

Considering Figure 19.29, the estimated train speed V and the axle speeds $V_i = \omega_i * R_i$ are reported, where ω_i and R_i are the angular velocity and the radius of the i-th axle, respectively. The adhesion recovery during the second phase of the braking manoeuvre can be observed, owing to the action of the WSP system that tries to optimise the adhesion conditions.

19.4.4.3.2 Traction Tests under Degraded Adhesion Conditions

In parallel to the previous paragraph, in this section a specific traction test under degraded adhesion conditions will be briefly described as an example of experimental results [12–15]. In this case, the tested railway vehicle is the E402 B locomotive, equipped with a fully operational AS system (see Figure 19.30).

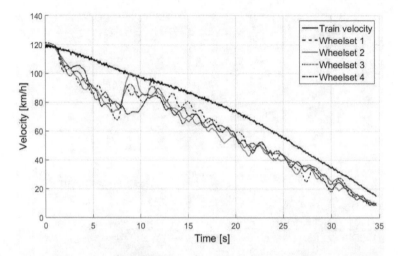

FIGURE 19.29 Experimental braking test under degraded adhesion conditions. (From Allotta, B. et al., *Int. J. NonLin. Mech.*, 57, 50–64, 2013. With permission.)

FIGURE 19.30 The E402 B locomotive.

Methodology: As in the braking test, specific degraded adhesion conditions have been considered for the traction test. The experimental conditions characterising the traction test are summarised in Table 19.28.

Also, under this test, the RRAC rig control architecture performs a simulation of rolling resistance through precise control of the roller motors. The roller motors are again controlled to recreate the angular velocities, applied torques and tangential efforts that would normally be observed between a wheelset and the rail. The new control architecture allows the achievement of this goal by only controlling the roller motors and without having sliding (and consequently wear) between the wheelsets and rollers.

Results and Conclusions: Some example results of the experimental traction test under degraded adhesion conditions are reported in Figure 19.31.

TABLE 19.28
Experimental Testing Conditions for the Traction Test under Degraded Adhesion

Parameter	Value
Adhesion coefficient	0.06
Type of vehicle	E402 B
Wheel radius	0.625 m
Starting speed	5 km/h
Braking torque	12300 Nm

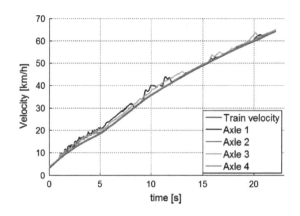

FIGURE 19.31 Experimental traction test under degraded adhesion conditions. (From Allotta, B. et al., *Int. J. NonLin. Mech.*, 57, 50–64, 2013. With permission.)

As shown in Figure 19.31, the estimated train speed V and the axle speeds $V_i = \omega_i * R_i$ are reported again, where ω_i and R_i are the angular velocity and the radius of the i-th axle, respectively. The action and the typical behaviour of the AS system, trying to optimise the adhesion conditions, can be repeatedly observed during the traction manoeuvre.

19.5 ROLLER RIG EXPERIMENTAL METHODS AND ERRORS

19.5.1 Experimental Methods

The following test methods used for Chengdu roller rig are formed through many years of test experience.

19.5.1.1 Status of Vehicle

In order to ensure the test vehicle is in a good running condition, the vehicle should run on the roller rig for about 10 hours or 500 km before the test starts. During the trial running, the running speed should cover the design speed of the vehicle.

If using a dummy carbody instead of a real carbody, the mass, moments of inertia and centre of gravity of the dummy carbody should be controlled in the error range of 15% compared with the real carbody.

19.5.1.2 Status of Roller Rig

In order to ensure the test precision, the conditions of the roller rig should be as follows:

- The wear of the rail profile should be less than 0.5 mm.
- The roller diameter difference should be less than 1 mm for the same roller unit, 2 mm for the bogie and 3 mm for the vehicle.

- The displacement error of the actuator used on railway test facility should be less than 5%, and the phase error relative to command signal at 30 Hz should be less than 60°.
- Two longitudinal fixation bars for the middle car are positioned in the place of couplers with ball joints, and the length of the bar should be longer than 1 m; one longitudinal fixation bar for the locomotive is positioned in the place of the coupler with a ball joint, and the length of the bar should be longer than 1.5 m.
- The errors of roller altitude for the same bogie should be less than 2 mm and for the whole vehicle should be less than 3 mm.

19.5.1.3 Stability Test

Motion stability of a railway vehicle is the most important factor in its vehicle dynamic behaviour. The main objective of performing stability tests is to find out the vehicle's hunting critical speed. Before introducing the stability test method for a roller rig, we first briefly review the concept of vehicle stability.

A typical example of a limit cycle diagram of wheelset motion is shown in Figure 19.32. The solid line indicates a stable limit cycle and the dashed line indicates an unstable limit cycle. When the vehicle speed is lower than V_{C2}, the vehicle system is always stable under any track disturbances. When the vehicle speed is between V_{C2} and V_{C0}, the system equilibrium position is stable for small track disturbances and unstable at larger track disturbances, whereby a limit cycle oscillation appears. When the vehicle speed is between V_{C0} and V_{C1}, the system equilibrium position is unstable and a small limit cycle oscillation emerges for small track disturbances and a large limit cycle oscillation for a large track disturbance. Finally, when the vehicle speed is larger than V_{C1}, the system jumps to large limit cycle oscillation at any track disturbance and flange contact may occur. Therefore V_{C0}, which is the Hopf bifurcation point, can be defined as the linear critical speed. V_{C1} and V_{C2} can be defined as the non-linear critical speeds. The non-linear critical speed V_{C2}, where the first stable limit cycle appears, is normally lower than linear critical speed V_{C0}, thus it should be taken as the speed limit for a vehicle running on tracks.

The aim of a stability test is to establish the three speed points V_{C0}, V_{C1} and V_{C2}. The method of stability testing on pure rolling rigs (RTU) and rolling and vibrating rigs (RVTU) is as follows:

- Gradually increase the roller rig speed under pure rolling conditions and, when a small limit cycle oscillation appears, then the speed V_{C0} is found; this test is facilitated by the fact

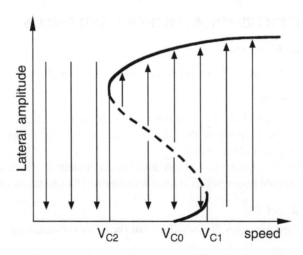

FIGURE 19.32 Limit cycle diagram of wheelset motion: V_{C0} is linear critical speed, V_{C1} is non-linear critical speed and V_{C2} is non-linear critical speed. (From Iwnicki, S. (Ed.), *Handbook of Railway Vehicle Dynamics*, CRC Press, Boca Raton, FL, 2006. With permission.)

that, even when the roller rig is in a pure rolling condition (no actioned roller excitations), there will always be small disturbances due to roller and wheel irregularities or surface roughness.
- Increase the roller rig speed continuously until the hunting motion of vehicle system jumps to large amplitude oscillation or even flange contact, then the speed V_{C1} is found.
- Reduce the speed of the roller rig slowly; when the severe hunting motion dies out to an equilibrium position, then the speed V_{C2} is found.

Since the RVTU facilitates both rolling and track excitation capabilities, it can be used to establish the critical speed of the vehicle under actual track irregularity inputs. Therefore, the critical speed found between V_{C2} and V_{C1} can be accurately influenced by track conditions.

In field line tests, the vehicle stability is estimated by the bogie acceleration, which is band-pass filtered between 0.5 Hz and 10 Hz. If the peak values of acceleration have exceeded 10 m/s² six times continuously, then the vehicle is declared as being unstable.

For the stability test, not only should the critical speed be measured but also the mode shapes of the hunting motion need to be determined. So, the lateral displacements of the carbody, bogie frames and wheelsets are examined.

19.5.1.4 Dynamic Simulation Test

The RVTU can simulate the running of the vehicle with track irregularity inputs. Normally, the responses of the vehicle, such as accelerations or displacements on the carbody or bogie, are measured. According to the applicable test standard, for instance, UIC518, the ride index is calculated according to the acceleration response on the carbody.

Let z_L, y_L, z_R, y_R indicate the irregularity inputs of left and right rails in time domain for the first roller unit; the inputs of other roller units are subsequently delayed by defined time intervals. The time delays, assuming the six-axle vehicle of Figure 19.33, can be calculated as follows:

$$t1 = l/v \qquad (19.1)$$

$$t2 = 2l/v \qquad (19.2)$$

$$t3 = L/v \qquad (19.3)$$

$$t4 = (L+l)/v \qquad (19.4)$$

$$t5 = (L+2l)/v \qquad (19.5)$$

where l is the bogie wheelbase, L is the distance between bogie centres and v is the running speed.

In the dynamic tests, the measured parameters normally include: the accelerations and displacements of the carbody, bogie frame and wheelsets (axle box); the relative displacements of primary and secondary suspensions; the relative displacements of the motors; the wheel-rail forces; stresses of key parts; temperature of bearing or gear case; etc.

19.5.1.5 Curve Simulation Test

In order to simulate vehicle curving performance, the roller rig should have the following features:

- Simulating the curve geometry – set the rollers in a radial position.
- Simulating the speeds of inner wheel and outer wheel – attain different speeds of inner roller and outer roller by using differential driving system.
- Simulating the superelevation and centrifugal force – set the cant angle of roller unit to simulate the unbalanced centrifugal force caused by superelevation and centrifugal force.

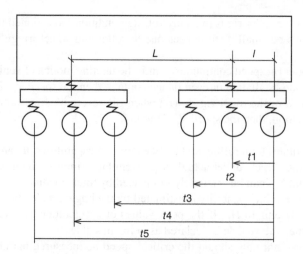

FIGURE 19.33 Time delay of input signals. (From Iwnicki, S. (Ed.), *Handbook of Railway Vehicle Dynamics*, CRC Press, Boca Raton, FL, 2006. With permission.)

The steps for a curve simulation test are as follows:

- Set the rollers in the radial position of the curve, dependant on vehicle geometry
- Widen the gauge of rollers according to the curve radius.
- Set up the cant angle of the roller unit according to the unbalanced centrifugal force.
- Lift the test vehicle onto the roller rig and locate it by fixation bars in the longitudinal direction; the fixation bars are set at an angle with respect to the carbody centre line according to the simulated curve radius.
- The roller rig runs at the prescribed speed.
- Adjust the speed difference of inner and outer rollers according to the curve radius and running speed.
- For steady-state simulation, the roller rig is in a pure rolling condition and the wheel-rail interaction forces can be measured.
- For dynamic simulation, the track irregularity inputs are considered; the wheel-rail interaction forces and the responses such as accelerations, displacements of carbody, bogie frame and wheelset can be measured.
- Then, the derailment ratio Q/P, lateral force H, wheel load reduction rate $\Delta P/P$ and ride index W can be obtained.

19.5.1.6 Modal Analysis Test

When the carbody, bogie frame, and wheelsets are considered as rigid bodies and are suspended by primary and secondary suspensions, the vehicle is a typical multibody system. Thus the natural vibrations of the vehicle system will appear at some frequencies. By using the RVTU or a test rig using a short vibrating rail under each vehicle wheel to reproduce track irregularities without wheel rolling motion (a VTU), the vibration modes (self-vibration frequency and modal shape) can be determined.

To perform the modal analysis test, the lateral, vertical, roll, pitch and yaw motions are normally considered. The rollers (or movable short rails) should be excited in separate modes. A swept sine wave or white noise is used as the roller rig inputs. Through analysing the responses of carbody and bogie frames, the resonance points under each mode will be determined and the modal frequencies can be obtained.

Roller Rigs

19.5.1.7 Storage Security Test

Loading methods are very important for some special goods, such as columned goods (pipes, cans and wood), destructible goods (glass and apparatus) and explosive goods (nuclear material and detonators). The use of RVTU or VTU can validate the goods behaviour, such as:

- The security of loading under vibration, impact, and lateral force (superelevation and centrifugal force)
- The stability of goods after long-distance travel
- The behaviour of a new loading method
- The dynamic environment of goods during transportation, such as vibration, acceleration, temperature, pressure and force

19.5.2 Experimental Errors

A roller rig should have the ability to accurately evaluate the dynamic behaviour of the vehicle system, but various setup errors always exist due to wear, incorrect installation and different environmental (friction) conditions. The following shows the influence of various setup errors of a roller rig on vehicle stability, using the example of a Chinese high-speed passenger car shown in Figure 19.34 [7].

19.5.2.1 Diameter of the Rollers

Wear and machining errors on the roller surface are practically unavoidable. The nominal diameter of the rollers of the Chengdu Roller Rig was originally 1370 mm, with an allowable minimum diameter of 1300 mm. A smaller roller radius will cause larger negative gravitational angular stiffness $K_{s\varphi}$, which may cause the hunting critical speed to become lower. Figure 19.35 shows the limit cycle of the passenger car on the roller with diameters of 1370 and 1300 mm. It is seen that the smaller the roller diameter, the lower the critical speed.

FIGURE 19.34 Passenger car on a roller rig. (From Iwnicki, S. (Ed.), *Handbook of Railway Vehicle Dynamics*, CRC Press, Boca Raton, FL, 2006. With permission.)

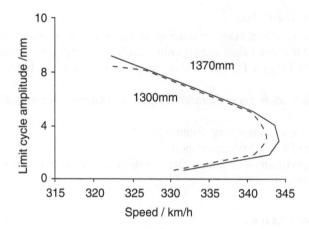

FIGURE 19.35 Influence of diameter of roller on critical speed. (From Iwnicki, S. (Ed.), *Handbook of Railway Vehicle Dynamics*, CRC Press, Boca Raton, FL, 2006. With permission.)

19.5.2.2 Gauge of the Rollers

For the RTU, the gauge of rollers is fixed during the test. But, after adjusting the rollers to meet different gauges, an error in gauge may occur. For the RVTU, the gauge can be changed arbitrarily. Theoretical results indicate that the influence of gauge is small when using the Chinese standard 'TB' conical type wheel profile. But when using the 'LM' worn type wheel profile, the influence of gauge on critical speed is obvious according to the results shown in Figure 19.36. When the gauge is widened, the critical speed increases.

19.5.2.3 Cant of the Rollers

In the early era of railways there was no installed inclination angle between the rail and the sleeper, but it is now common practice to apply an angle of 1/20, with the currently inclination in China now being 1/40. The influence of differing rail inclination is shown in Figure 19.37. It is evident that, when the railhead profile has zero inclination, the critical speed is at its lowest; with an increase in inclination, the critical speed increases, but within the range of 1/40 to 1/10, the influence is relatively small. This effect is due to the fact that increasing rail inclination typically reduces the equivalent conicity between wheel and rail.

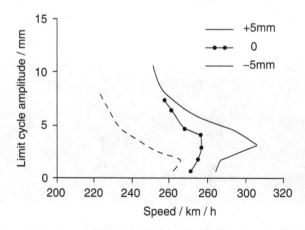

FIGURE 19.36 Influence of gauge variation on critical speed. (From Iwnicki, S. (Ed.), *Handbook of Railway Vehicle Dynamics*, CRC Press, Boca Raton, FL, 2006. With permission.)

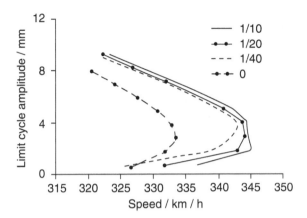

FIGURE 19.37 Influence of cant of roller on critical speed. (From Iwnicki, S. (Ed.), *Handbook of Railway Vehicle Dynamics*, CRC Press, Boca Raton, FL, 2006. With permission.)

19.5.2.4 Coefficient of Friction on the Contact Surface

A wheel rolling on rollers makes the contact surface smooth and contamination makes the friction characteristic change. The influence of friction coefficient on stability is shown in Figure 19.38. It is seen that the lower the friction coefficient, the higher the critical speed.

19.5.2.5 Vehicle Position on the Roller Rig

During testing, the vehicle sits on top of the rollers and is fixed in the longitudinal direction. Since errors exist in the wheelbase, bogie centre distance, and position of roller unit, it is difficult to accurately position the wheels on top of the rollers (see Figure 19.39a). Such errors will influence the critical speed, as shown in Figure 19.39b. When all the wheels of the vehicle are forward of the centreline of the rollers by 10 mm, the critical speed reduces; contrarily, when all wheels are rearward by 10 mm, the critical speed is increased. Such interesting results are validated by the stability tests.

19.5.2.6 Yaw Angle between Wheelset and Roller Axle

For an optimal reproduction of wheel-rail interaction, the wheels of the test vehicle should remain on top of the rollers and be parallel with the roller axle. However, the wheelsets within a bogie are

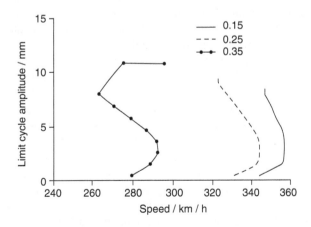

FIGURE 19.38 Influence of friction coefficient on critical speed. (From Iwnicki, S. (Ed.), *Handbook of Railway Vehicle Dynamics*, CRC Press, Boca Raton, FL, 2006. With permission.)

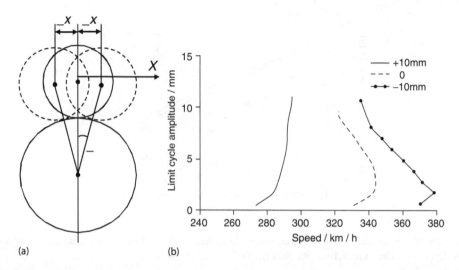

FIGURE 19.39 Effect of variation of positioning on the rig: (a) wheel position and (b) resulting influence on stability. (From Iwnicki, S. (Ed.), *Handbook of Railway Vehicle Dynamics*, CRC Press, Boca Raton, FL, 2006. With permission.)

difficult to keep parallel with each other. There is a yaw angle between the wheelset and roller axles, which results in the wheelset moving to one side of roller unit or even into flange contact. It is known that, when wheelset one is set a yaw angle, it moves from the central position, which in turn causes wheelset two to move from the central position. When wheelset two is set to a yaw angle, again, both wheelsets one and two leave the central position. Thus, if one wheelset has a yaw angle with respect to roller axle, it will cause the bogie to go to one side. Normally, this yaw misalignment of the wheelsets cannot be corrected during the test. In order to allow the wheelset to align centrally on the roller unit, it is necessary to adjust the yaw angle of the roller axle to match the wheelset yaw angle.

19.6 CONCLUSIONS

From an initial historical review, it can be seen that railway roller rigs have played an important role in the history of vehicle development, especially in the evolution of high-speed trains in Europe and China. Their application has become wide ranging, diversifying from initial test objectives based around the analysis of traction and vehicle dynamic performance, to researching wheel and rail degradation modes such as wear and RCF, adhesion and braking, whilst facilitating the development of leading edge wheel-slide protection and rail head conditioning brake controllers that help trains to operate safely under a wide range of environmental conditions.

The inherent errors in wheel-rail contact must always be considered as these will always result in some deviation in results for wheel-rail interaction related studies from actual field test data. For this reason, roller rigs cannot replace field testing but will continue to provide a valuable research and development tool for vehicle manufacturers, sub-system suppliers and Universities.

Whilst some of the famous historical full-vehicle roller rigs such as those at British Rail Research in Derby, Transportation Technology Center, Inc. (TTCI), in Pueblo, in the USA, and Deutsche Bhan (DB) in Munich, Germany, have been decommissioned, these rigs have taught us much about the fundamental behaviour of railway vehicles. It could be said that the golden age for roller rigs has passed, but, with new test rigs being developed across the globe, evidence would suggest that this is not the case. Whilst simulation techniques become ever more advanced, the most challenging aspects of vehicle-track interaction and running dynamics such as wheel-rail adhesion, braking, new hybrid drivetrains and energy efficient traction packages still require roller rig testing to help develop these systems and validate computer models; hence, the prospect of full-scale roller rigs seems secure for the foreseeable future.

ACKNOWLEDGEMENTS

We would like to thank Chongyi Chang (China Academy of Railway Sciences), Steven Cervello (Lucchini RS SpA), Eugenio Fedeli (Rete Ferroviaria Italiana SpA), Thorsten Geburtig (Deutsche Bahn Systemtechnik GmbH), Gabriel Haller (RTA Rail Tec Arsenal Fahrzeugversuchsanlage GmbH), Martin Heller (Knorr-Bremse Systeme für Schienenfahrzeuge GmbH), Patrick Hoffmann (DB Systemtechnik GmbH), Jörg Koch (Knorr-Bremse Systeme für Schienenfahrzeuge GmbH), Dirk Lehmann (DB Systemtechnik GmbH), Katrin Mädler (DB Systemtechnik GmbH) and Petr Voltr (University of Pardubice) for supporting and updating the data sheets.

REFERENCES

1. A. Jaschinski, H. Chollet, S. Iwnicki, A. Wickens, J. Von Würzen, The application of roller rigs to railway vehicle dynamics, *Vehicle System Dynamics*, 31(5–6), 345–392, 1999.
2. Anon, Locomotive testing plant in Swindon, UK. *The Engineer*, 100, 621–622, 1905.
3. D.R. Carling, Locomotive testing stations, parts 1 and 2, *Transactions of the Newcomen Society*, 45, 105–182, 1972.
4. M. Naeimi, Z. Li, R.H. Petrov, J. Sietsma, R. Dollevoet, Development of a new downscale setup for wheel-rail contact experiments under impact loading conditions, *Experimental Techniques*, 42(1), 1–17, 2018.
5. S.Z. Meymand, M.J. Craft, M. Ahmadian, On the application of roller rigs for studying rail vehicle systems, *Proceedings ASME 2013 Rail Transportation Division Fall Technical Conference*, 15–17 October 2013, Altoona, PA, Paper RTDF2013-4724, 2013.
6. S. Myamlin, J. Kalivoda, L. Neduzha, Testing of railway vehicles using roller rigs, *Procedia Engineering*, 187, 688–695, 2017.
7. W.H. Zhang, Dynamic simulation study of railway vehicle, PhD Dissertation, Southwest Jiaotong University, Chengdu, China, 1996. (in Chinese)
8. S.J. Ma, W.H. Zhang, G.X. Chen, J. Zeng, Full scale roller rig simulation for railway vehicles, In: P. Lugner, J.K. Hedrick (Eds.), *Proceedings 13th IAVSD Symposium*, Chengdu, China, 23–27 August 1993, Swets and Zeitlinger, Lisse, the Netherlands, 346–357, 1994.
9. W.H. Zhang, J. Zeng, H.Y. Dai, Dynamic performance test for the high-speed freight truck using roller testing rig, Research Report No. TPL-BG-2014-Z009, State Key Laboratory of Traction Power, Southwest Jiaotong University, Chengdu, China, 2014.
10. H.L. Shi, J.B. Wang, P.B. Wu, C.Y. Song, W.L. Teng, Field measurements of the evolution of wheel wear and vehicle dynamics for high-speed trains, *Vehicle System Dynamics*, 56(8), 1187–1206, 2017.
11. R. Luo, W.L. Teng, J. Zeng, L. Wei, Comparison of creep-force models for dynamic simulation of high-speed train, *23rd IAVSD Symposium*, Qingdao, China, 19–23 August 2013, IAVSD, Qingdao, China, 2013.
12. B. Allotta, R. Conti, E. Meli, L. Pugi, A. Ridolfi, Development of a HIL railway roller rig model for the traction and braking testing activities under degraded adhesion conditions, *International Journal of Non-Linear Mechanics*, 57, 50–64, 2013.
13. R. Conti, E. Meli, A. Ridolfi, A. Rindi, An innovative hardware in the loop architecture for the analysis of railway braking under degraded adhesion conditions through roller-rigs, *Journal of Mechatronics*, 24(2), 139–150, 2014.
14. B. Allotta, R. Conti, E. Meli, A. Ridolfi, Modeling and control of a full-scale roller rig for the analysis of railway braking under degraded adhesion conditions, *IEEE Transactions on Control Systems Technology*, 23(1), 186–196, 2015.
15. RFI Internal Report, Full-scale roller rig: technical documentation, Rete Ferroviaria Italiana SpA, Rome, Italy, 2011.

20 Scale Testing Theory and Approaches

Nicola Bosso, Paul D. Allen and Nicolò Zampieri

CONTENTS

20.1	Introduction	826
20.2	A Brief History of Scaled Roller Rigs	827
20.3	Roller Rigs: The Scaling Problem	827
	20.3.1 The Scaling Strategy of UoH	828
	20.3.1.1 Principles	828
	20.3.1.2 Materials	829
	20.3.1.3 Equations of Motion	829
	20.3.1.4 Scaling and Wheel-Rail/Roller Forces	830
	20.3.2 The Scaling Strategy of DLR	832
	20.3.3 The Scaling Strategy of INRETS	834
	20.3.4 Tabular Comparison of Scaling Strategies	836
20.4	Scaling Errors	837
20.5	Survey of Current Scaled Roller Rigs	838
	20.5.1 The Scaled Rig of DLR	838
	20.5.1.1 Design Overview	839
	20.5.2 The Scaled Rig of UoH	839
	20.5.2.1 Design Overview	839
	20.5.3 The Scaled Rig of INRETS	841
	20.5.3.1 Design Overview	841
	20.5.4 The Scaled Rig of Politecnico di Torino	843
	20.5.4.1 Design Overview	844
	20.5.5 The Scaled Rig of VT-FRA	847
	20.5.5.1 Design Overview	847
	20.5.6 The Scaled Rig of CTU	849
	20.5.6.1 Design Overview	849
	20.5.6.2 Test Bogie	850
20.6	Scaled Prototype Testing on Track	851
20.7	Scaled Prototypes: Typical Applications	853
	20.7.1 Dynamic Track Simulator	854
	20.7.1.1 Hunting Stability	854
	20.7.1.2 Contact Force Estimation	855
	20.7.1.3 Other Vehicle Dynamic Studies	856
	20.7.2 Tribology Studies	856
	20.7.2.1 Adhesion	857
	20.7.2.2 Wear	860
	20.7.3 Hardware in the Loop Applications and Control Systems	862
20.8	Conclusions	863
Acknowledgements		863
References		864

20.1 INTRODUCTION

The use of roller rigs for the investigation of railway vehicle dynamics has been discussed in Chapter 19. Their operation, application and the changes in vehicle response introduced due to geometric and kinematic differences between running a rail vehicle on rollers, as opposed to track, were described in detail. Full-scale rigs have been a useful tool when assessing the dynamic performance of a prototype vehicle, especially in the days when numerical simulation of vehicle dynamics was not as well developed as it is today, but their frequency of use for prototype design is in decline, as computer techniques become more popular.

The main obstacles to the widespread use of full-scale test rigs lie in the high costs of construction, maintenance and management of these installations. Furthermore, their use is substantially limited to tests carried out on the first prototypes of real vehicles, before their approval or placing into service.

Analogous with what happens in other areas of research, such as aeronautical or naval engineering, even the railway sector long ago decided to carry out tests using small-scale prototypes, and in this context, the use of roller rigs is very effective.

The testing of small-scale prototypes has several advantages, including the possibility of extending the research scope to a much wider range of cases. In fact, the lower costs of testing at a reduced scale allow testing even on prototypes of experimental vehicles or analysis of phenomena that could not easily be reproduced on real vehicles.

Among the other advantages, there are also the lower cost of construction of the test rig and of the prototype in scale, the reduced dimensions, the reduced security problems and the ease of use. The management of the tests is much simpler and can also be carried out in research institutes or universities, and this has allowed its wide diffusion throughout various research centres.

It is also far easier to change a large number of vehicle parameters without great effort. However, these advantages must be offset by a number of negative factors. These are primarily concerned with the effect of scaling down the vehicle dimensions. From a scientific viewpoint, it is not acceptable to reduce the dimensions of the vehicle, without giving due consideration to the effect of these changes. It is of great importance, if reliable scaled results are to be obtained, to adopt a scientifically based scaling strategy. The outcome of this strategy will dictate how well the roller rig test results will relate to the full scale, whether this is in terms of vehicle dynamics, wheel-rail forces or even wear.

This chapter describes a number of scaled roller rigs, used as research tools, and how each of the institutions involved has handled the issues related to scaling. Examples are given of the errors that can be introduced by different types of scaling strategies.

The fundamental ideas of similarity, that is, maintaining correlation between a scale model and the full scale, can be traced back to the work of Reynolds [1,2] or even earlier. Analogous to Reynolds' approach, similarity of mechanical systems with respect to dynamic behaviour and elastic deformation can be defined.

Small-scale testing of railway vehicles on roller rigs has been carried out for different purposes, including the verification and validation of simulation models, the investigation of fundamental railway vehicle running behaviour (nonlinear response, limit cycles, etc.), the development and testing of prototype bogie designs with novel suspensions, in order to support field tests and computer simulations, and last, but not least, teaching and demonstration of railway vehicle behaviour. Small-scale tests at various institutions have proven that, under laboratory conditions, influences of parameters can be revealed, which often cannot be separated from stochastically affected measurements of field tests; this is of course also true for full-scale rigs.

From a constructive point of view, the scaled roller rigs are similar to those in full scale and include one or more rollers adapted to support the wheels of the prototype. For the realisation of material characterisation tests, twin-disc systems are often used, consisting of a single pair of rollers. Twin discs can be considered precursors of roller rigs, but they have some substantial differences. The twin discs are limited to reproducing a cylinder-cylinder contact condition with the presence of only one creepage in the longitudinal direction. In contrast, the roller rigs allow the replication, even with differences due to intrinsic errors, of the contact conditions that normally occur on a railway vehicle, including an elliptical contact area and the presence of creepages in the three different directions (longitudinal, lateral and spin).

20.2 A BRIEF HISTORY OF SCALED ROLLER RIGS

Investigations using scaled models of railway vehicles on scaled tracks were performed by Sweet et al. [3,4] at Princeton University in 1979 and 1982. Experiments were concerned with the mechanics of derailment of dynamically scaled 1:5 models of a typical three-piece freight truck design widely used in North America. Careful attention was given to the scaling of clearances. Forces were scaled according to similarity laws, including the effects of inertia, gravitation, spring stiffness, creep and dry friction. These methods have been widely adopted and are used in many of the currently adopted scaling strategies.

One of the first investigations in Germany on a scaled roller rig was performed by Rheinisch-Westfälische Technische Hochschule (RWTH) Aachen [5]. Other designs of scaled roller rigs followed, in 1984 at DLR Oberpfaffenhofen [6–8], in 1985 at the Institut National de Recherche sur les Transports et leur Securite (INRETS) in Arcueil [9] and in 1992 at the Rail Technology Unit of the Manchester Metropolitan University (MMU), now the Institute of Railway Research at the University of Huddersfield (UoH) [10–12].

Because the realisation of a scaled roller rig is much cheaper than full-scale rig, scaled roller rigs have become widespread in research centres and universities all over the world. Some of these test benches will be analysed in detail, illustrating the main construction features (Politecnico di Torino rig built at the end of the 1990s, the joint Virginia Polytechnic Institute and State University and Federal Railroad Administration [VT-FRA] rig built in 2000 and the Czech Technical University [CTU] rig built in the mid-80s). Other roller rigs will be analysed by illustrating their peculiar applications in the final part of this chapter.

20.3 ROLLER RIGS: THE SCALING PROBLEM

Similarity laws and the correlated problem of scaling are of interest for the transformation of experimental results from a scaled model to the full-scale design. There are various possible approaches to scaling, including using the methods of dimensional analysis to establish several dimensionless groups from which the scaling factors can be derived; workers include Jaschinski [6], Illingworth [13] and Chollet [14]. Other methods include first deriving the equations of motion and then calculating the scaling factors required for each term to maintain similarity.

This latter method is known as inspectional analysis and requires a sound understanding of the equations of motion, which is achievable in this particular field.

Choice of material properties is also a factor in the scaling method used, particularly if the simulation work requires the preservation of the levels of strain at the contact point. British Rail used aluminium wheels and rollers [15], while Matsudaira et al. [16] used steel and Sweet et al. [4] used plastic.

The scaling factors are used to define, starting from the characteristics of the real vehicle, the dimensions and characteristics of the small-scale prototype to be tested on the roller rig. Therefore, from a theoretical point of view, a roller rig can be used to perform tests by using different scaling methods, provided it is able to withstand the loads, speeds and torques required for the prototype. In reality, however, existing roller rigs are normally used with a single similitude method for which they have been designed.

The similarity law can be obtained by choosing the length scaling factor and then defining the general terms. These general terms are outlined below; it is from these that the scaling strategies of the three research institutions are developed. The length scaling factor is:

$$\varphi_l = \frac{l_1}{l_0} \tag{20.1}$$

where l_1 is a characteristic length of the full scale and l_0 that of the scaled model. In the same way, a time scaling factor can be derived as:

$$\varphi_t = \frac{t_1}{t_0} \qquad (20.2)$$

With these definitions, scaling factors of φ_A for cross-section, φ_V for volume, φ_v for velocity and φ_a for acceleration follow:

$$\varphi_A = \varphi_l^2 \qquad (20.3)$$

$$\varphi_V = \varphi_l^3 \qquad (20.4)$$

$$\varphi_v = \frac{\varphi_l}{\varphi_t} \qquad (20.5)$$

$$\varphi_a = \frac{\varphi_l}{\varphi_t^2} \qquad (20.6)$$

When the density scaling factor φ_ρ is known as:

$$\varphi_\rho = \frac{\rho_1}{\rho_0} \qquad (20.7)$$

then the scaling factors φ_m for mass, φ_I for moment of inertia and φ_F for inertial force can be derived as:

$$\varphi_m = \varphi_\rho \varphi_l^3 \qquad (20.8)$$

$$\varphi_I = \varphi_m \varphi_l^2 \qquad (20.9)$$

$$\varphi_F = \frac{m_1 a_1}{m_0 a_0} = \varphi_m \varphi_a = \frac{\varphi_\rho \varphi_l^4}{\varphi_t^2} \qquad (20.10)$$

Once these general definitions have been developed, the scaling strategies of each institution can be used to derive the following quantities used in studies of wheel-rail interaction: scaling factor for creep forces φ_T, scaling factor for the elliptical size of the contact patch φ_{ab}, scaling factor for Young's modulus φ_E, scaling factor for Poisson's ratio φ_v, scaling factor for strain φ_ε, scaling factor for stress φ_σ, scaling factor for the active coefficient of friction φ_μ, scaling factor for stiffness φ_c, scaling factor for damping φ_d and scaling factor for frequency φ_f.

20.3.1 THE SCALING STRATEGY OF UoH

20.3.1.1 Principles

The important aspects of the behaviour that are being studied in a dynamic analysis are the displacements, velocities and acceleration of the various bodies and the forces between these bodies and at the wheel-rail/roller interface.

As the most common measurements in dynamic studies are made in the form of time histories or frequency spectra, the scaling factor for time, and therefore, frequency should be unity.

Scale Testing Theory and Approaches

$$\varphi_t = 1 \tag{20.11}$$

The roller rig has been built to 1:5 scale to give suitable dimensions for construction and laboratory installation, hence:

$$\varphi_l = 5 \tag{20.12}$$

Following through from Equation 20.1 to Equations 20.5 and 20.6 gives rise to the following expressions:

$$\text{for displacement} \quad \varphi_l = 5 \tag{20.13}$$

$$\text{for velocity} \quad \varphi_v = 5 \tag{20.14}$$

$$\text{for acceleration} \quad \varphi_a = 5 \tag{20.15}$$

and

$$\text{for frequency} \quad \varphi_f = \frac{1}{\varphi_t} = 1 \tag{20.16}$$

which is convenient for comparison of these values.

20.3.1.2 Materials

Various options were available for the material used in the construction of the roller rig, but for ease of construction and to allow a reasonably practical wear life of the wheels and rollers, it was convenient to use steel for these bodies. This is not a great disadvantage, as the roller rig is not used to perform wear investigations, which would require correlation with the full-scale case. The material properties are then similar for both scale and full size: $\varphi_p = 1$ for density, $\varphi_E = 1$ for Young's modulus, $\varphi_\nu = 1$ for Poisson's ratio and $\varphi_\mu = 1$ for coefficient of friction.

Therefore, the scaling factor for mass considering Equation 20.8 is:

$$\varphi_m = 5^3 \tag{20.17}$$

For rotational inertia, due to Equation 20.9, the scaling factor is:

$$\varphi_I = 5^5 \tag{20.18}$$

20.3.1.3 Equations of Motion

The equations of motion for a dynamic system govern the relationship between force and acceleration (and therefore velocity and displacement). In general terms, the basic equation is expressed in the form of a force balance, and all force terms in the equation, for similarity, should equate to the force scaling term, φ_F.

$$m\ddot{x} + c\dot{x} + kx = F \tag{20.19}$$

and in the angular form:

$$I\ddot{\theta} + c_T\dot{\theta} + k_T\theta = T \tag{20.20}$$

where m is the mass; I is the moment of inertia; c, c_T are the damping coefficients; k, k_T are the stiffnesses; F is the applied force; and T is the applied torque.

Therefore, for the scale model, Equations 20.19 and 20.20 become:

$$m\ddot{x}\left(\frac{\varphi_m \varphi_l}{\varphi_t^2}\right) + c\dot{x}\left(\frac{\varphi_c \varphi_l}{\varphi_t}\right) + kx(\varphi_k \varphi_l) = F(\varphi_F) \qquad (20.21)$$

$$I\ddot{\theta}\left(\frac{\varphi_I}{\varphi_t^2}\right) + c_T\dot{\theta}\left(\frac{\varphi_{c_T}}{\varphi_t}\right) + k_T\theta\left(\frac{\varphi_{k_T}}{\varphi_t}\right) = T(\varphi_T) \qquad (20.22)$$

For the translational case, from Equation 20.21 and for similarity:

$$\left(\frac{\varphi_m \varphi_l}{\varphi_t^2}\right) = \left(\frac{\varphi_c \varphi_l}{\varphi_t}\right) = (\varphi_k \varphi_l) = (\varphi_F) \qquad (20.23)$$

Therefore, using the previously derived scaling factors for φ_l, φ_m and φ_t:

$$\varphi_l^4 = \varphi_c \varphi_l = \varphi_k \varphi_l = \varphi_F \qquad (20.24)$$

giving $\varphi_c = 5^3$ for the translational damping coefficient, $\varphi_k = 5^3$ for the translational stiffness constant and $\varphi_F = 5^4$ for the applied force.

For the rotational case, from Equation 20.23 and for similarity:

$$\left(\frac{\varphi_I}{\varphi_t^2}\right) = \left(\frac{\varphi_{c_T}}{\varphi_t}\right) = \left(\frac{\varphi_{k_T}}{\varphi_t}\right) = (\varphi_T) \qquad (20.25)$$

Therefore, using the previously derived scaling factors for φ_I and φ_t:

$$\varphi_I = \varphi_{c_T} = \varphi_{k_T} \qquad (20.26)$$

giving $\varphi_{c_T} = 5^5$ for the rotational damping coefficient, $\varphi_{k_T} = 5^5$ for the rotational stiffness constant and $\varphi_T = 5^5$ for the applied torque.

The above terms of power x^5 are validated by considering that a translational spring of stiffness k will give a torsional stiffness of kl^2, hence giving rise to the power raise of two. Therefore, similarity is maintained in all equations with forces scaling at 5^4 and torques scaling at 5^5.

20.3.1.4 Scaling and Wheel-Rail/Roller Forces

A complete study of the scaling methodology must also include the effect of scaling on the equations governing the wheel-rail/roller interaction. A complete derivation of the equations of motion for a railway vehicle is not required for this study, as the wheel-rail forces act through the wheelset alone. Therefore, the equations of motion for a single bogie vehicle, which includes some simple suspension forces, are sufficient. The creep forces are derived from Kalker's linear theory.

The lateral equation of motion for a simple linear vehicle model can be represented by the following expression:

$$m\ddot{y}_w + 2f_{22}\left(\frac{\dot{y}_w}{v} - \psi_w\right) + 2f_{23}\left(\frac{\dot{\psi}_w}{v} - \frac{\varepsilon_0}{l_0 r_0}\right) + \frac{w\varepsilon_0 y_w}{l_0} + d_y(\dot{y}_w - \dot{y}_b - a\dot{\psi}_b + h\dot{\theta}_b) \\ + c_y(y_w - y_b - a\psi_b + h\theta_b) \qquad (20.27)$$

and the terms influencing the yaw of the wheelset:

$$I_z\ddot{\psi}_w + 2f_{11}\left(\frac{l_0^2 \dot{\psi}_w}{v} + \frac{l_0 \lambda y_w}{r_0}\right) - 2f_{23}\left(\frac{\dot{y}_w}{v} - \psi_w\right) + 2f_{33}\left(\frac{\dot{\psi}_w}{v}\right) + c_\psi(\psi_w - \psi_b) \qquad (20.28)$$

Scale Testing Theory and Approaches

where m is the wheelset mass; y_w is the wheelset lateral displacement; y_b is the bogie lateral displacement; ψ_w is the wheelset yaw angle; ψ_b is the bogie yaw angle; θ_b is the bogie roll angle; d_y is the wheelset-bogie lateral damping (per wheelset); c_y is the wheelset-bogie lateral stiffness (per wheelset); c_ψ is the wheelset-bogie yaw stiffness (per wheelset); w is the axle load; λ is the effective conicity; l_0 is the semi gauge; a is half the bogie wheelbase; h is the height of the bogie centre of gravity above the wheelset axis; v is the forward speed of the vehicle; ε_0 is the rate of change of contact angle with y_w; r_0 is the rolling radius with the wheelset central; and f_{11}, f_{22}, f_{23} and f_{33} are Kalker's linear creep coefficients.

The equations governing the linear creep coefficients are:

$$f_{11} = (ab)GC_{11} \quad f_{23} = (ab)^{3/2}GC_{23}$$
$$f_{22} = (ab)GC_{22} \quad f_{33} = (ab)^{2}GC_{33} \tag{20.29a}$$

where C_{ii} are Kalker's tabulated creep coefficients, G is the modulus of rigidity and a and b are the contact patch semi-axes.

Hertz theory governs the size of the contact patch and the relevant equations are requoted as follows:

$$ab = mn[3\pi N(k_1+k_2)/4k_3]^{\frac{2}{3}} \tag{20.29b}$$

where

$$k_1 = \frac{1-v_R^2}{E_W} \quad k_2 = \frac{1-v_W^2}{E_R} \tag{20.29c}$$

and

$$k_3 = \frac{1}{2}\left[\frac{1}{r_1}+\frac{1}{r_1'}+\frac{1}{r_2}+\frac{1}{r_2'}\right] \tag{29.29d}$$

where m and n are the elliptical contact constants, N is the normal force and the other parameters are as previously quoted.

The scaling factors can then be calculated:

$$\varphi_{k_1} = \varphi_{k_2} = \frac{1}{\varphi_E} = 1 \tag{20.30}$$

$$\varphi_{k_3} = \frac{1}{\varphi_l} = 5^{-1} \tag{20.31}$$

If the scaling factor for the normal force, φ_N, is 5^4, as with all other forces, then the scaling factor for the contact patch area, $\varphi_{(ab)}$ will be:

$$\varphi_{(ab)} = \left(\varphi_N \frac{\varphi_{k_1}}{\varphi_{k_3}}\right)^{\frac{2}{3}} = \left(\frac{\varphi_F}{\varphi_{k_3}}\right)^{\frac{2}{3}} = 5^{3.33} \tag{20.32}$$

From Equations 20.29a and 20.32, we can evaluate the scaling for the linear creep coefficients. Therefore:

$$\varphi_{f_{11}} = \varphi_{f_{22}} = \varphi_E \varphi_{(ab)} = 5^{3.33} \tag{20.33}$$

$$\varphi_{f23} = \varphi_E(\varphi_{(ab)})^{\frac{3}{2}} = 5^5 \qquad (20.34)$$

$$\varphi_{f33} = \varphi_E(\varphi_{(ab)})^2 = 5^{6.66} \qquad (20.35)$$

With a normal force scaling factor of 5^4, we have a conflict with the vehicle weight scaling factor due to its mass multiplied by the acceleration due to gravity:

$$\varphi_w = \varphi_m \varphi_g = 5^3 \qquad (20.36)$$

conflicting with $\varphi_N = 5^4$.

This conflict can be resolved by the use of support wires, with incorporated spring balances connected to each axle box, to remove the required amount of weight.

Considering the describing equations of motion given in Equations 20.27 and 20.28, each term can be evaluated, including the scaling factors derived above for the linear creep coefficients, to check for the required scaling factor. This gives creep coefficient scaling factors of 5^4 when considering a force term and 5^5 for a torque. All terms agree with the scaling strategy and give perfect scaling apart from those listed below:

Force terms (required $\varphi_F = 5^4$)

$$2f_{22}\frac{\dot{y}_w}{v}\psi_w \quad \text{gives a force scaling } \varphi_F = 5^{3.33} \qquad (20.37)$$

$$\frac{w\varepsilon_0 y_w}{l_0} \quad \text{gravitational stiffness term gives } \varphi_F = 5^3 \qquad (20.38)$$

Torque terms (required $\varphi_T = 5^5$)

$$2f_{11}\left(\frac{l_0^2\dot{\psi}_w}{v} + \frac{l_0\lambda y_w}{r_0}\right) \quad \text{gives a torque scaling } \varphi_T = 5^{4.33} \qquad (20.39)$$

$$2f_{33}\left(\frac{\dot{\psi}_w}{v}\right) \quad \text{gives a torque scaling } \varphi_T = 5^{5.66} \qquad (20.40)$$

In practice, the value of f_{33} is much smaller than f_{11} and f_{22}, and the gravitational stiffness term in Equation 20.38 is of the same order as the f_{33} terms during normal tread running of the wheel. Therefore, the major error sources with respect to scaling the simulated scaled forces with those of a full-size vehicle are those described by Equations 20.37 and 20.39.

20.3.2 The Scaling Strategy of DLR

DLR was involved in the development of simulation software for railway vehicle dynamics and, in particular, the nonlinear lateral dynamics, which leads to the instability known as hunting. This instability is caused by a bifurcation in describing differential equations into a limit cycle, and the strategy for the scaling of the roller rig was developed with respect to this.

Therefore, much like the UoH group, the starting point for the DLR scaling strategy focussed on the nonlinear lateral behaviour of a single wheelset suspended to an inertially moving body. An example equation is described in [17] for a wheelset with conical treads. The first component of this system of two coupled equations of motion is:

$$\frac{m}{\chi}\ddot{y}_w = \frac{I_y\Gamma v}{\chi r_0}\dot{\psi}_w - \frac{mgb_0}{\chi}\frac{c_y}{\chi}y_w + T_y + T_x\psi_w \qquad (20.41)$$

Scale Testing Theory and Approaches

The symbols used above denote the same quantities as described in the previous section, with the addition or replacement of the wheelset's rotational moment of inertia I_y; the longitudinal creep force T_x; the lateral creep force T_y; $\Gamma = \delta_0/l_0 - r_0\delta_0$; the cone angle δ_0; $\chi = \Gamma l_0/\delta_0$; $b_0 = 2\Gamma + \Gamma^2(R_R + r_0)$; and the transverse radius of the rail head R_R.

Multiplying the scalable parameters and variables in Equation 20.41 with the previously defined scaling factors and re-arranging:

$$\frac{m}{\chi}\ddot{y}_w = \frac{I_y\Gamma v}{\chi r_0}\dot{\psi}_w \frac{mgb_0}{\chi}y_w \frac{\varphi_l^2}{\varphi_l}\frac{c_y}{\chi}y_w \frac{\varphi_c\varphi_l^2}{\varphi_m} + (T_y + T_x\psi_w)\frac{\varphi_T\varphi_l^2}{\varphi_m\varphi_l} \qquad (20.42)$$

Dynamically, the scale wheelset behaves similarly to the full scale if Equations 20.41 and 20.42 coincide. This requires that the following conditions hold:

$$\frac{\varphi_l^2}{\varphi_l} = 1 \Rightarrow \varphi_v = \sqrt{\varphi_l} \qquad \text{velocity scaling}$$

$$\frac{\varphi_c\varphi_l^2}{\varphi_m} = 1 \Rightarrow \varphi_c = \varphi_\rho\varphi_l^2 \qquad \text{stiffness scaling} \qquad (20.43)$$

$$\frac{\varphi_T\varphi_l^2}{\varphi_m\varphi_l} = 1 \Rightarrow \varphi_T = \varphi_\rho\varphi_l^2 \qquad \text{creep force scaling}$$

It can be seen that, for similarity to be maintained, the scaling factors for Equation 20.43 above cannot be freely chosen and are a function of the principle scaling terms derived in Section 20.3.1. This result is identical to that found by Matsudaira et al. [16] from investigations carried out in 1968 at the Railway Technical Research Institute of the Japanese railways. From the constraint equations (a relationship between the normal forces, gyroscopic, gravitational, applied, and creep forces, see [6]), together with the scaling method described above, the scaling factors for the constraint forces, the mass and creep forces can be derived as:

$$\varphi_N = \varphi_m = \varphi_T = \varphi_P\varphi_l^3 \qquad (20.44)$$

This results in a scale factor for friction coefficient μ of $\varphi_\mu = 1$.

Assuming that Kalker's nonlinear theory is used for calculation of the contact forces, then similarity is required for the dimensions of the contact ellipse if the calculated Kalker's creep coefficients, and hence creep forces, are to be correct. This requires that $\varphi_E = \varphi_v = 1$.

If this condition is adhered to, then the scaling factor for density is derived as follows:

Kalker's theory requires that $\varphi_T = \varphi_{ab}$ and $\varphi_{ab} = (\varphi_N\varphi_l)^{2/3}$ and $\sqrt{\varphi_{ab}} = \varphi_e$ (the contact ellipse mean radius), then the scale of this radius becomes:

$$\varphi_e^3 = \varphi_N\varphi_l = \varphi_P\varphi_l^4 \qquad (20.45)$$

Assuming geometric similarity for the contact ellipse, $\varphi_e = \varphi_l$, and Equation 20.45 results in the definition of the density scaling factor as:

$$\varphi_\rho = \frac{1}{\varphi_l}$$

This scaling factor would result in perfect scaling for the contact ellipse and Kalker coefficients, which, when considering a length scale factor of $\varphi_l = 5$, requires a density that is very difficult to achieve. It was considered by DLR that exact scaling of the contact patch was only necessary at low levels of creepage and not so important during the analysis of limit cycles, where saturation of the creep forces occurs and the exact shape and size of the contact patch does not influence the creep forces (gross sliding within the contact region). Considering the above practical limitations, the density scaling factor was chosen as:

$$\varphi_\rho = \frac{1}{2}$$

which can be easily achieved and has proven through testing to give good experimental results. With $\varphi_l = 5$, and considering the above-mentioned limitation with respect to density, the other scaling factors can be determined as follows:

$$\varphi_v = \sqrt{\varphi_l} = \sqrt{5} \quad \text{velocity}$$

$$\varphi_t = \frac{\varphi_l}{\varphi_v} = \sqrt{5} \quad \text{time}$$

$$\varphi_a = \frac{\varphi_l}{\varphi_t^2} = 1 \quad \text{acceleration}$$

$$\varphi_m = \varphi_T = \varphi_N = \varphi_F = \varphi_\rho \varphi_l^3 = 62.5 \quad \text{mass and force}$$

$$\varphi_I = \varphi_\rho \varphi_l^5 = 1562.5 \quad \text{moment of inertia}$$

$$\varphi_c = \varphi_\rho \varphi_l^2 = 12.5 \quad \text{spring stiffness}$$

$$\varphi_d = \frac{\varphi_\rho \varphi_l^3}{\varphi_v} = \varphi_\rho \varphi_l^{5/2} = 27.95 \quad \text{viscous damping}$$

$$\varphi_f = \frac{1}{\varphi_t} = \frac{1}{\sqrt{5}} \quad \text{frequency}$$

$$\varphi_\mu = \frac{\varphi_T}{\varphi_N} = 1 \quad \text{coefficient of friction}$$

$$\varphi_e = (\varphi_N \varphi_l)^{1/3} = 6.79 \quad \text{contact ellipse}$$

Table 20.1 shows some typical parameters for a generic test vehicle using the DLR scaling strategy.

20.3.3 The Scaling Strategy of INRETS

Within the INRETS institution, the main area of research focus was the experimental validation of Kalker's creep coefficients. The vehicle scale of 1:4 at INRETS is large compared with other rigs; this, coupled with the very large roller diameter, means the rig is suitable for the analysis of wheel-rail contact. The validation of Kalker's theory requires exact representation of the contact patch and its elasticity, to allow accurate measurement of the quasistatic creepage and creep force relationships. Therefore, the basis of the scaling strategy was obtained by adopting a stress scaling factor of

$$\varphi_\sigma = \frac{\varphi_F}{\varphi_l^2} = 1.$$

This means that the stresses in the scale and full-scale test vehicles are the same. In addition to the advantages in investigating Kalker's theory, this stress scale factor results in a spring stiffness scaling factor φ_c, which is proportional to the length factor. This helps in the design of suspension components, as size and internal stresses are the same as the full scale:

$$\varphi_c = \frac{\varphi_F}{\varphi_l} = \varphi_l = 4 \qquad (20.46)$$

TABLE 20.1
Generic Test Vehicle Parameters

Parameter	Full Size	1:5
Bogie		
Bogie frame mass	487.50 kg	7.8 kg
Wheel mass	281.25 kg	4.5 kg
Axle mass	275.00 kg	4.4 kg
Bogie roll inertia	218.75 kg/m²	0.14 kg/m²
Bogie pitch inertia	103.13 kg/m²	0.066 kg/m²
Bogie yaw inertia	192.19 kg/m²	0.123 kg/m²
Wheel rotational inertia	51.56 kg/m²	0.033 kg/m²
Axle rotational inertia	3.13 kg/m²	0.002 kg/m²
Vehicle Body		
Body mass	2037.50 kg	32.6 kg
Body roll inertia	1403.13 kg/m²	0.898 kg/m²
Body pitch inertia	1339.06 kg/m²	0.857 kg/m²
Body yaw inertia	2342.19 kg/m²	1.499 kg/m²
Wheel Dimensions		
Wheel diameter	1.0 m	0.2 m
Gauge	1.435 m	0.287 m
Primary Suspension		
Longitudinal stiffness	8.30×10^5 N/m	6.64×10^4 N/m
Lateral stiffness	8.30×10^5 N/m	6.64×10^4 N/m
Vertical stiffness	5.90×10^7 N/m	4.73×10^6 N/m
Normal force	11,496 N	183.94 N

When similarity of elastic forces, together with similarity of gravitational forces, is required, then the following is true:

$$\varphi_c \varphi_l = \varphi_m \varphi_g \tag{20.47}$$

where φ_g is the scaling factor for gravity. Equation 20.46 shows that, for the requirement of a valid frequency scaling factor, the frequency of a mass spring system should be the same as that of an equivalent gravitational oscillator, such as a pendulum, and this condition yields that:

$$\varphi_w^2 = \frac{\varphi_c}{\varphi_m} = \frac{\varphi_g}{\varphi_l} \tag{20.47a}$$

Assuming the density scaling factor $\varphi_\rho = 1$, Equation 20.47 leads to a gravity scaling factor of

$$\varphi_g = \frac{1}{\varphi_l} = \frac{g_1}{g_0} \tag{20.48}$$

The above equation essentially results in a different scaling factor for forces generated through gravitation and those generated from inertia, and, in a similar way to the UoH strategy, this can be achieved by application of external forces, which add to the effective weight, without increasing mass. Considering Equation 20.42, this results in a scaling factor for weight of

$$\varphi_w = \frac{m_1 g_1}{m_0 g_0 \varphi_l} \Rightarrow \varphi_w = \varphi_l^2 = 16 \tag{20.49}$$

whereas the inertial force scaling factor with $\varphi_g = 1$ is

$$\varphi_m \varphi_g = \varphi_p \varphi_l^3 \Rightarrow \varphi_w = \varphi_l^3 = 64 \qquad (20.50)$$

Using this strategy, increasing the weight through an external force, which does not change the mass of the body, allows the derivation of scaling factors for velocity, time and acceleration to be formed from the frequency, using Equations 20.5 and 20.6.

$$\left(\frac{\varphi_v}{\varphi_l}\right) = \frac{\varphi_c}{\varphi_m} \Rightarrow \varphi_v = 1$$

$$\varphi_t = \varphi_l \qquad (20.51)$$

$$\varphi_a = \varphi_l$$

The above scaling strategy results in similarity of vertical dynamics, together with elastic contact, normal and tangential stresses, which in turn allows the lateral dynamics to be accurately represented.

20.3.4 Tabular Comparison of Scaling Strategies

To summarise the strategies discussed in the previous section, the scaling parameters have been listed in Table 20.2.

TABLE 20.2
Comparison of Scaling Strategies

Scaling Factors	UoH	DLR $\rho = 0.5$	DLR $\rho = 1$	INRETS
Geometry				
Length	5	5	5	4
Cross-section	25	25	25	16
Volume	125	125	125	64
Material				
Density	1	0.5	1	1
Mass	125	62.5	125	64
Inertia	3125	1562.5	3125	1024
Elasticity	G, E, cij	Approximate	Approximate	G, E, cij
Parameters				
Time	1	$\sqrt{5}$	$\sqrt{5}$	4
Frequency	1	$1/\sqrt{5}$	$1/\sqrt{5}$	1:4
Velocity	5	$\sqrt{5}$	$\sqrt{5}$	1
Acceleration	5	1	1	1:4
Stress	25	2.5	5	1
Strain	25	2.5	5	1
Stiffness	125	12.5	25	4
Forces				
Inertial forces	625	62.5	125	16
Gravitational forces	Reduced by 1/5	62.5	125	Multiply by 4
Spring forces	Modified	62.5	125	Scaled
Viscous damping forces	Modified	62.5	125	Not considered

Considering the scale factors used in the different methods, it can be noted that depending on the objective of the experimentation, some methods may be more suitable than others, depending on the case considered.

For example, if tests in which accelerometric measurements are involved are carried out, the DLR method is easier to use, because the accelerations are not scaled. The method used at INRETS, on the other hand, is particularly suitable for analysing problems related to wheel-rail contact or the design of suspension components, since the stresses and deformations are not scaled. The UoH method, having a unitary factor for time and frequency, is particularly suitable for frequency analysis, acoustic analysis, hardware-in-the-loop (HIL) tests or algorithms that need to work in real time.

20.4 SCALING ERRORS

As discussed, a scaling strategy is selected based on the type of analysis work to be carried out on the rig; this type-specific selection of the strategy has to be performed, as perfect scaling cannot otherwise be achieved. The example below, using the scaling strategy of the University of Huddersfield, illustrates the level of error that can be encountered.

The errors caused by the scaling of a vehicle can be expressed by the following equations that have been reproduced from Section 20.3.1.4 for convenience:

Force terms (required $\varphi_F = 5^4$)

$$2f_{22}\left(\frac{\dot{y}_w}{v} - \psi_w\right) \quad \text{gives a force scaling factor of } \varphi_F = 5^{3.33} \tag{20.52}$$

Torque terms (required $\varphi_T = 5^5$)

$$2f_{11}\left(\frac{l_0^2 \dot{\psi}_w}{v} + \frac{l_0 \lambda y_w}{r_0}\right) \quad \text{gives a torque scaling factor of } \varphi_T = 5^{4.33} \tag{20.53}$$

$$2f_{33}\left(\frac{\dot{\psi}_w}{v}\right) \quad \text{gives a torque scaling factor of } \varphi_T = 5^{5.66} \tag{20.54}$$

The expected level of error from the terms highlighted above can be quantified by performing an analysis of a two-degree-of-freedom wheelset model. The results of the theoretical scaling strategy are plotted against the same model but simulated with a perfect scaling strategy.

The plots shown in Figures 20.1 and 20.2 quantify the errors due to the scaling factors derived in Equations 20.52–20.54. The plots were produced with a two-degree-of-freedom model, excited with a sinusoidal disturbance, with an amplitude of 0.5 mm and a frequency of 2π radians per second. The forward speed of the vehicle was a 1:5 scale speed of 2 or 10 m/sec at full scale. The results have been scaled from the 1:5 scale values of the roller rig to the full scale; all other terms achieve perfect scaling.

Although perfect scaling has not been achieved for these terms, experimental testing has shown that the adopted scaling strategy gives good agreement with the full scale, particularly when considering stability. For the purpose of relative studies between vehicles on the roller rig, the errors in creep forces illustrated in Figures 20.1 and 20.2 are not of great significance, but modifications to the scaling method, to reduce this error, may be required if absolute values between scaled and full-size creep forces were required.

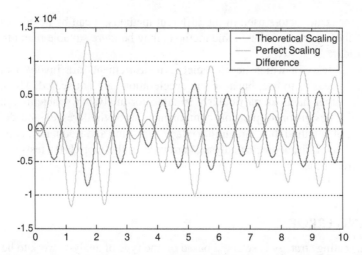

FIGURE 20.1 Lateral creep force scaling error. (From Iwnicki, S. (ed.), *Handbook of Railway Vehicle Dynamics*, CRC Press, Boca Raton, FL, 2006. With permission.)

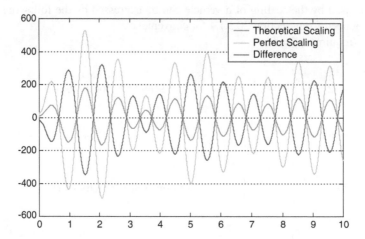

FIGURE 20.2 Wheelset creep torque scaling error. (From Iwnicki, S. (ed.), *Handbook of Railway Vehicle Dynamics*, CRC Press, Boca Raton, FL, 2006. With permission.)

20.5 SURVEY OF CURRENT SCALED ROLLER RIGS

There are several scaled roller rigs which are used for research and demonstration purposes. The design and operation of some of the scaled roller rigs in use today are described in the following section.

20.5.1 THE SCALED RIG OF DLR

The Institute for Robotics and System Dynamics of DLR has been involved in the development of simulation software for railway vehicle dynamics, based upon multibody modelling techniques, since the early 1970s.

The institute was interested in the nonlinear running behaviour of railway passenger vehicles, and experiments became very important for the validation of modelling work, which was carried out to predict the dynamic response of the vehicle. In wheel-rail dynamics, the nonlinear forces involved play a dominant role in the onset of vehicle hunting, a phenomenon that is caused by a bifurcation of the system's equations of motion into a periodic solution, or limit cycle as it is commonly referred to. For this reason, DLR developed a scaled roller rig with a single bogie vehicle running on the rollers. The primary functions of the rig were not only to perform the above validation but also to assist in the verification of parts of DLR's dynamic simulation software, SIMPACK.

The bogie was a scaled-down version of the MAN bogie [7,8]. The emphasis of the first series of tests was the fundamentals of modelling and experimental methods in wheel-rail dynamics. Once this first stage of work was completed, including investigations of limit cycle behaviour of the bogie [9], DLR adapted the rig to concentrate on the development of unconventional wheelset concepts and contributed to fundamental research in this field [18].

20.5.1.1 Design Overview

The roller rig at DLR is a 1:5 scale rig consisting of two rollers, each of which is composed of a hollow cylinder with a wall thickness of about 20 mm. A disc that has a 1:5 scale UIC60 rail profile formed around its circumference is attached at each end of the cylinder. The diameter of this part of the roller is 360 mm, and the separation of the discs is 287 mm, which is 1:5 of the standard track gauge of 1435 mm. The advantage of this type of roller construction is that the design provides a very high torsional stiffness, which is important in maintaining a true creepage relationship between the wheel and the roller. This stiffness is coupled with a large rotational inertia, making the rollers insensitive to disturbances of their rotational velocity, meaning the arrangement is well suited to simulating tangent track behaviour. A plan view drawing of the rig is shown in Figure 20.3.

The distance between the rollers can be varied to accommodate various bogies with different wheelbases. A feature of this arrangement is the inclusion of a 'Schmidt-Coupling'; this device is a parallel crank mechanism and allows the change of wheelbase, without disruption to the drive arrangement. As can be seen in the sectional view in Figure 20.3, the rollers are mounted on cones, allowing easy removal of the rollers for changes to rail profile or gauge.

The rollers are interconnected using a toothed belt with a specified longitudinal stiffness to maintain the synchronisation of the roller speeds at all times. The roller speed can be varied from zero up to 168 km/h, depending on the rolling resistance of the vehicle being modelled. Figure 20.4 shows the general arrangement with a MAN bogie being tested on the rig.

Recent applications of the roller rig at DLR are oriented to various active and mechatronic applications, including active systems to improve steering ability and stability of railway bogies [19,20]. This approach, which was also validated using a comprehensive model of the complete vehicle in SIMPACK, was used to define a new concept innovative vehicle based on a controlled pair of independently rotating wheels (IRWs) [21].

20.5.2 THE SCALED RIG OF UoH

A 1:5 scale roller rig was set up at the University of Huddersfield in 1992 for use in the investigation of railway vehicle dynamic behaviour and to assist in research, consultancy, and teaching activities. Research activities then focussed on the evaluation of a novel design of differentially rotating wheelset and the quantification of errors inherent in roller rig testing [10]. The roller rig is currently being used to investigate the behaviour of independently driven wheelsets for light rail applications.

20.5.2.1 Design Overview

The roller rig at the University of Huddersfield is of 1:5 scale and consists of four rollers supported in yoke plates incorporating the rollers' supporting bearings, with the interconnection between the roller pairs being provided by the use of splined and hook jointed shafts. While these shafts do not offer the degree of torsional stiffness given by the DLR arrangement, they do allow the simulation of lateral track irregularities by enabling rotational movement of the rollers about a vertical axis (yaw), coupled with a lateral movement of the rollers as a pair. The roller

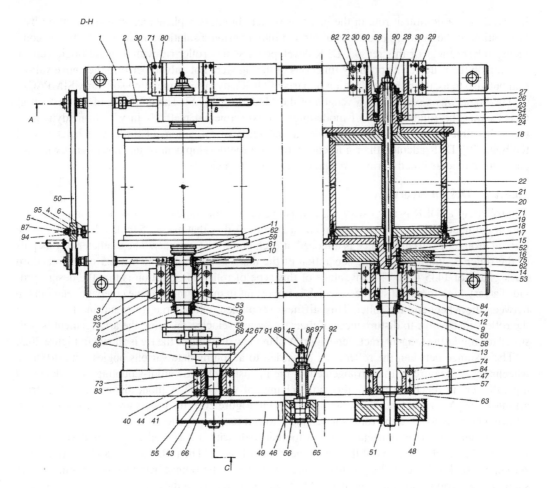

FIGURE 20.3 Plan view drawing of the 1:5 scale DLR roller rig. (From Iwnicki, S. (ed.), *Handbook of Railway Vehicle Dynamics*, CRC Press, Boca Raton, FL, 2006. With permission.)

motion is provided by servo hydraulic actuators, which are connected directly to the rollers' supporting yoke plates, these actuators being controlled by a digital controller, which allows the inputs to follow defined waveforms or measured track data. The longitudinal and lateral position of the rollers can be adjusted by means of a system of linear bearings for changing the wheelbase and the gauging of the rollers. Drive is supplied to the rollers via a belt, with pulleys on each roller drive shaft, allowing the rig to operate at scaled speeds of up to 400 km/h. The bogie is modelled on the British Rail Mark IV passenger bogie [22], but it can be easily modified. The purpose of the rig was to demonstrate the behaviour of a bogie vehicle under various running conditions and acquire nominal data from the vehicle responses. A plan view drawing of the roller rig is shown in Figure 20.5.

The bogie vehicle parameters were selected to represent those of a typical high-speed passenger coach (the BR Mk IV passenger coach). The wheel profiles are machined scaled versions of the BR P8 profile, and the rollers have a scale BS110 rail profile with no rail inclination. A bogie running on the rig can be seen in Figure 20.6.

More recently, the test rig was also used to perform wear [23] and adhesion tests [24] and to investigate the behaviour of vehicles with independent rotating wheels [25]. Further studies are related to the effect of wheel flats and track irregularities [26] and to the analysis of innovative acoustic techniques to identify the defects [27].

Scale Testing Theory and Approaches

FIGURE 20.4 The MAN bogie model under test at DLR. (From Iwnicki, S. (ed.), *Handbook of Railway Vehicle Dynamics*, CRC Press, Boca Raton, FL, 2006. With permission.)

20.5.3 THE SCALED RIG OF INRETS

INRETS is the French national research institute, and within this is a group specialising in wheel-rail interaction, with particular interest in the novel variations of freight bogies. The test facility was originally commissioned in 1984 and was used intensively until 1992.

The first railway vehicle to be tested on the rig was the Y25 type, a UIC bogie [14,28] that is very common in Europe. This bogie was selected, as it is particularly difficult to model with conventional computer software, as there are several dry friction dampers within the suspension. Studies focussed on optimising the stability of the bogie under varying vertical loads and differing suspension parameters.

Later, the rig has been used for the quasistatic measurement of the Kalker coefficients, and experiments dealing with squeal noise and braking performance are currently being carried out.

20.5.3.1 Design Overview

The test rig of INRETS was originally designed as a large flywheel of 13 m diameter to test linear motors for the Bertin Aerotrain transport vehicle, and this is the reason that the wheel is of such large diameter. Weighing 40 t, the wheel is driven by a linear 2 MW motor, which can power the wheel to a periphery speed of 250 km/h.

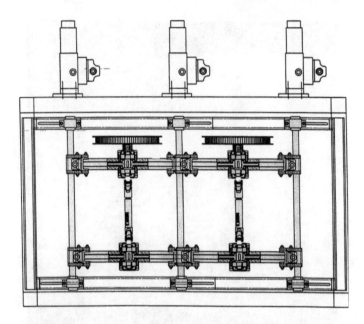

FIGURE 20.5 Plan view drawing of the roller rig at UoH. (From Iwnicki, S. (ed.), *Handbook of Railway Vehicle Dynamics*, CRC Press, Boca Raton, FL, 2006. With permission.)

FIGURE 20.6 Bogie vehicle on the UoH roller rig platform. (From Iwnicki, S. (ed.), *Handbook of Railway Vehicle Dynamics*, CRC Press, Boca Raton, FL, 2006. With permission.)

The flywheel was not designed to support very high vertical loads, and a scaling factor of 1:4 was therefore chosen for the rig. INRETS were already familiar with similarity laws used in scale models and developed a specific strategy for dynamic similarity with respect to the preservation of the elasticity of the bodies, especially for the wheel-rail contact area. The rig is illustrated in Figure 20.7.

Scale Testing Theory and Approaches 843

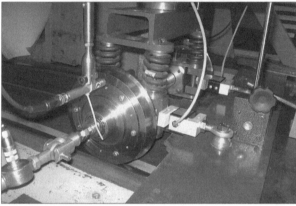

FIGURE 20.7 The INRETS rig at Grenoble. (From Iwnicki, S. (ed.), *Handbook of Railway Vehicle Dynamics*, CRC Press, Boca Raton, FL, 2006. With permission.)

The large diameter of the test wheel made the rig at INRETS particularly suitable for investigating the contact between wheel and rail, as the radius of curvature was far closer to approaching that of conventional track when compared with any other roller rig in existence, resulting in the size and shape of the contact patch being closer to reality.

The wheel can only rotate about the horizontal axis, and an angle of attack can therefore only be generated by yawing the vehicle wheelset relative to the track. A hydraulic ram is fitted at the top of the wheel to allow the variation of the vertical load on the tested bogie.

20.5.4 THE SCALED RIG OF POLITECNICO DI TORINO

A scaled roller rig was designed and installed in 2002 at the railway research group of Politecnico di Torino as shown in Figure 20.8. The aim of this roller rig was to support the simulation of the dynamic behaviour of different railway vehicles, performed by means of multibody codes. Initially, the main goal was to study the contact forces and the stability of the vehicle. The investigations were later extended to other aspects: adhesion, wear and traction control. The design of this roller rig was carried out considering the existing studies and designs (Manchester, DLR, INRETS, Delft), and it was optimised to allow investigations on various types of vehicles. At Politecnico di Torino, in order to carry out different types of studies, different variants of the roller rig were subsequently

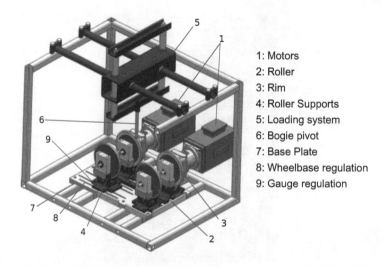

FIGURE 20.8 Schematic of test rig at Politecnico di Torino.

developed [29–37]: a test rig for a single suspended wheelset, two test rigs for 2-axle bogies and, more recently, a multi-wheelset roller rig.

The prototypes were designed according to the DLR similitude law with $\varphi_\rho = 1$.

20.5.4.1 Design Overview

The design of the roller rig of Politecnico di Torino was carried on using a modular approach in order to allow the use of different prototypes. The base module consists of a single roller and its supporting structure. This includes the bearing housings and a base structure that can be fixed to a reference plate in different locations, depending on the vehicle gauge and wheelbase. For this reason, both the base structure and the reference plate are provided with slots. The rollers are designed in order to allow the replacement of the external rims, where the rail profile is reproduced. This is useful to reduce the costs when different materials or profile shapes need to be used or when the profiles are worn and need to be replaced.

The rings mounted on the rollers have been made with different diameters between 160 mm and 185 mm. Combining the different possibilities of adjustment of the bench and considering prototypes both at 1:4 scale and 1:5 scale, the configurations that can be used on the roller rig are indicated in Table 20.3.

TABLE 20.3
Possible Configurations of Roller Rig/Vehicle Prototypes Tested at Politecnico di Torino

Vehicle Type	Wheelbase (mm)		
	Full Size	Scale 1:4	Scale 1:5
Minimum	1440	360	–
Light Rail	1500	375	–
Freight (Y25)	1800	(450)	360
Eurofima (Y32)	2400	(600)	480
Regional narrow gauge [31]	2400	600	(480)
ETR 460 [29]	2700	(675)	540
ICE 2 (maximum at 1:4)	2800	(700)	560
ETR 500	3000	–	600
TGV	3500	–	700

Each roller can be motorised independently, or, alternatively, a pair of rollers can be rigidly connected using a joint and a single motor can actuate both rollers.

All motors are controlled in real time, with the possibility of operation in different modes: electric shaft, gearing and simulated inertia. This makes it possible to carry out different types of tests by imposing velocities that vary over time according to predefined programmes, with the possibility of simultaneously running different parameters for each wheel.

The test rig is equipped with a set of sensors, which may be changed according to the test to be carried out. These include high-resolution encoders for measuring roller and wheelset rotation, torque transducers to measure the torque provided by the motors, high-resolution laser measurement sensors to measure the movement of the wheelsets and accelerometers and load cells.

The acquisition system can operate in real time and can be synchronised to the motor control system using different protocols (Ethercat, Profibus).

The simplest configuration of the test rig is that in which the prototype consists of a single wheelset, where only a couple of rollers are used, as shown in Figure 20.9. This type of arrangement is particularly suitable for carrying out studies on contact forces, adhesion and wear. In this case, the wheelset must be connected directly to the structure of the bench by means of suspension elements that allow the reproduction of the desired test conditions. The test rig of Politecnico di Torino is equipped with a suspension system that allows the adjustment of the load and alignment of the wheelset and the measurement of the forces exerted on the suspensions during the test.

The most common configuration is that in which the prototype consists of a two-axle bogie placed on two pairs of rollers. In this case, the bolster of the bogie is fixed to the structure of the test rig by means of a drag system, which reproduces that of the real vehicle, and the vertical load is applied by means of a load beam, which is vertically movable on cylindrical guides, as shown in Figure 20.10a. This type of arrangement has mainly been used to perform stability tests or validation of numerical models [29,35]. To complete the modularity of the roller rig, a modular bogie at 1:5 scale was also built, with which it is possible to simulate different types of vehicles by varying the wheelbase and the characteristics of the suspensions. Further tests were carried out on scale prototypes of new vehicles in their design phase, as in the case of the narrow-gauge bogie (950 mm) for the Circumvesuviana railway [31], which was simulated with a 1:4 scale prototype,

FIGURE 20.9 Single wheelset arrangement on roller rig.

 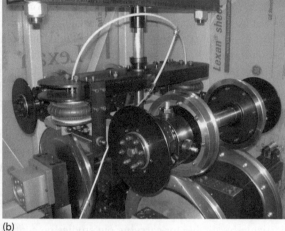

FIGURE 20.10 Tests on roller rig at Politecnico di Torino: (a) 1:5 modular bogie and (b) 1:4 narrow gauge bogie.

shown in Figure 20.10b. This bogie is characterised by various innovative design aspects, such as the articulated bogie frame, the primary suspension made of toroidal rubber elements and air springs in the secondary suspension.

The multi-wheelset test bench [34], shown in Figure 20.11, has been recently installed to carry out studies on the adhesion recovery phenomenon and to test anti-skid systems in the laboratory. The test bench is the result of an evolution of the single wheelset roller rig and is based on the same base module with rollers of radius 185 mm. The test bench, which is designed considering a 1:5 scale, is basically composed of a pair of rollers supporting four wheelsets, which are connected to a rigid frame. The position of the wheelsets in the lateral and longitudinal directions is provided by rigid connection elements that connect the wheelsets with the main frame. Considering the vertical direction, the wheelsets are elastically suspended with respect to the frame by means of helical springs. Since both

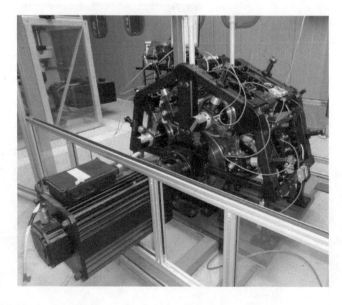

FIGURE 20.11 Multi-wheelset test bench developed at Politecnico di Torino.

wheelsets are in contact with the same pair of rollers, an unrealistic longitudinal component of the constraint force is generated. In the specific case, this component of the constraint force is transferred to the frame of the test bench by means of rigid links, and it has no influence on the experimental results, which neglect the longitudinal dynamics of the vehicle. The normal load can be independently regulated on each wheel by means of a special preloading system that includes a load cell, and this allows compensation to be made for the normal load decrease caused by the longitudinal component of the constraint force. The test bench was designed to install wheelsets with wheels of diameters in the range 184–220 mm, which corresponds, in real scale, to the range 920–1100 mm. Each wheelset is equipped with a braking system, so that an independent braking torque can be applied. The braking system consists of two pneumatic braking callipers for each wheelset, acting on disc brakes. The braking force is regulated by a proportional pneumatic valve, which acts directly on the pressure of the braking circuit. Each brake calliper is equipped with a load cell that allows precise measurement of the effective braking force. The pneumatic circuit of the brake is supplied with compressed air to reproduce the braking system of railway vehicles. In this way, it is also easy to interface the test bench with the electro-pneumatic valves of the anti-skid system and the relative control logic of a real vehicle. The inertia of the vehicle during braking operations is reproduced using the motors connected to the rollers, which can provide a torque that opposes the braking torque. The test bench can operate in two different ways, namely one mode in which the rollers are mechanically connected by means of a rigid joint and driven by a single motor, and a second mode in which the rollers are independent and driven by two motors controlled on a single electric shaft. The second configuration is used when the torque generated by the single motor is not enough to simulate the inertia of the vehicle.

20.5.5 The Scaled Rig of VT-FRA

The VT-FRA roller rig is located in Blacksburg, Virginia, USA. It was built as a collaborative partnership between Federal Railroad Administration (FRA) and Virginia Polytechnic Institute and State University (VT).

The main purpose of the rig is to study the wheel-rail contact patch mechanics and dynamics in a consistent and controlled laboratory environment [38–40]. The roller rig is an ideal tool for engineering analysis of passenger and freight trains.

Conceptualisation and design of the rig started in 2010, and the rig was successfully commissioned in 2016. Figure 20.12 shows a photograph and a schematic of the VT-FRA roller rig test setup. This design is an example of a single wheel on roller test rig with the functionalities of a roller rig [41].

20.5.5.1 Design Overview

The rig comprises a wheel and a roller in a vertical configuration that simulates the single-wheel-rail interaction in a one-fourth scale setup. The roller is five times larger than the wheel to keep the contact patch distortion to a minimum. The rig is equipped with two independent, direct drivelines that provide the required power to drive the wheel and roller, while controlling the differential speed at the contact. In addition, the setup is equipped with positioning mechanisms to replicate boundary conditions of real railway vehicles. Linear actuators are employed to actively control various parameters: the simulated load, the angle of attack, the rail cant angle and the lateral displacement.

The roller rig is equipped with an electromechanical actuation system capable of simultaneously controlling a total of six degrees of freedom, including the rotational motion of wheel and roller, as well as the motion of the positioning systems including angle of attack, cant angle, normal load and lateral displacement.

Two independent drivelines power the wheel and roller and are capable of commanding creepage at wheel and rail contact in increments as low as 0.01%. The positioning systems are capable of

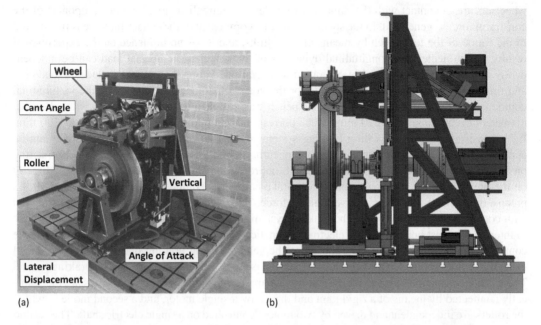

FIGURE 20.12 VT-FRA roller rig: (a) photograph showing adjustment capabilities and (b) schematic side view. (From Meymand, S.Z. and Ahmadian, M., *Measurement*, 81, 113–122, 2016. With permission.)

moving in increments of 1 micron. These capabilities ensure the VT-FRA roller rig is capable of simulating the field conditions with a very high level of accuracy.

A side view of the rig is shown in Figure 20.13. The VT-FRA roller rig is equipped with an instrumentation system, including a custom design force-measuring system and torque-measuring sensors. The rig is equipped with two custom-built force platforms, each consisting of four tri-axial load cells capable of measuring forces in the longitudinal, lateral and vertical directions. Moments are calculated by knowing the distance of wheel-rail contact from each load cell.

One of the particular features of the rig is its unified communication protocol (SynqNet) between actuators, drives and data acquisition system that eliminates data conversion between these units and hence facilitates servo update rates of up to 48 kHz.

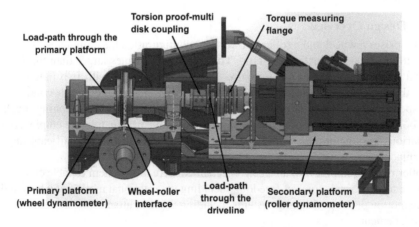

FIGURE 20.13 Side view of VT-FRA roller rig showing the two load platforms, load paths and torque-measuring units. (From Meymand, S.Z. and Ahmadian, M., *Measurement*, 81, 113–122, 2016. With permission.)

The peculiar characteristics of this test rig make it suitable to carry out tests concerning the study of wheel-rail contact, accurately controlling in real time creepages, relative position and adhesion. Its applications include experimental validation of different creep-creepage models; studying the effect on creep forces and moments of changes in angle of attack, cant angle, wheel tread position, normal load, linear velocity, contact surface condition of wheel and roller and creepage at wheel-rail contact; wheel wear testing; hollow wheel testing; third body layer analysis; testing effects of top-of-rail friction modifiers; flange contact studies; and wheel-rail vibration studies.

20.5.6 The Scaled Rig of CTU

The history of roller rig testing at the Czech Technical University (CTU) in Prague began towards the end of the 1980s, when the first 1-axle scaled roller rig was built at a 1–3.5 scale with a roller diameter of 0.5 m, and those parameters remained unchanged until now.

The rig has been improved and updated many times in accordance with objectives of the projects for which it was used [42–44]. The first major adjustment came during the first half of the 1990s, when the rig was completely rebuilt to a 2-axle type configuration. The current state of the rig, after more than two decades of modifications, is shown in Figure 20.14.

20.5.6.1 Design Overview

The CTU roller rigs consist of two pairs of rollers of diameter 0.5 m each. They have a peripheral profile corresponding to a R65 rail profile with a cant of 1:20. One of the roller pairs is fixed to the roller rig chassis, whereas the second roller pair is fixed to the subframe, which can be moved in the yaw direction towards the rig chassis. The position of the subframe is controlled by a servo motor, so it is possible to change the mutual yaw angle of the roller pairs during experiments. Each roller is individually driven by its own asynchronous electromotor. The rig is capable of carrying out experiments with maximum rollers speed up to 700 min^{-1} that corresponds to a maximum real vehicle speed of 230 km/h. The CTU roller rig is not restricted to performing experiments only on a straight track but is also capable of simulating the negotiation of curved track or track consisting of an arbitrary number of straight, transition and constant curvature sections.

A system for measuring the lateral component of wheel-rail contact forces, which is based on strain gauge measurement of roller disc deformation, is applied on all four rollers.

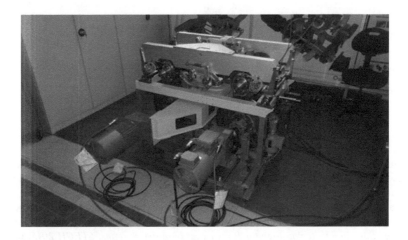

FIGURE 20.14 CTU roller rig in 2018.

20.5.6.2 Test Bogie

The CTU roller rig is equipped with an experimental two-axle bogie (Figure 20.14). The scale bogie does not correspond to any specific bogie of a real vehicle; it was newly designed in order to fulfil the requirements of experiments and measurements at the CTU roller rig. The bogie wheelbase is 714 mm, track gauge is 410 mm and wheel diameter is 263 mm. This corresponds to a wheelbase of 2500 mm and 920 mm wheel diameter of a standard gauge vehicle in full scale. Most of the parts are made of aluminium alloy manufactured via computer numerical control machining. For the connection of the mutually movable components, roller and linear roller bearings are used.

The bogie frame consists of two side-beams and the central cross-beam. The cross-beam is connected via a rigid connection to one side-beam and by a cylindrical joint on the opposite side. Thus, such a design enables these side-beams to mutually pitch, and a vertically flexible frame with equalised vertical components of wheel-rail contact forces may be achieved. The bogie frame is connected to the roller rig mainframe by a Watts linkage. This mechanism provides longitudinal fixation of the bogie with respect to the roller rig mainframe, while bogie lateral and yaw motion is not restricted.

Each wheelset is independently actuated in the yaw direction by an active control mechanism. Referring to Figure 20.15, the actuator is a permanent magnet synchronous servomotor (item 1) with a rated torque of 2.5 Nm. It can be controlled to a desired magnitude of yaw torque acting on the wheelset, or to a desired value of yaw angle between the wheelset and the bogie frame. The actuator torque is transmitted via a toothed belt (item 2) to the steering rod (item 3) and finally to the wheelset by two pairs of linkages (item 5).

The bogie has no vertical primary suspension; however, the design of the axleboxes permits variation of the lateral stiffness of the connection between wheelsets and bogie frame.

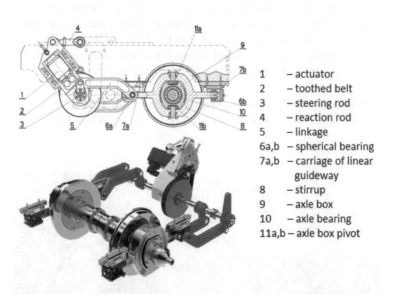

FIGURE 20.15 Design of the wheelset steering mechanism and the connection between the axle box and bogie frame. (Schematic from Kalivoda, J. and Bauer, P., Mechatronic bogie for roller rig tests, in: Rosenberger, M. et al., (Eds.), *The Dynamics of Vehicles on Roads and Tracks: Proceedings 24th Symposium of the International Association for Vehicle System Dynamics*, Graz, Austria, 17–21 August 2015, CRC Press/Balkema, Leiden, the Netherlands, 899–908, 2016. With permission.) & (Model from Kalivoda, J. and Bauer, P., Design of scaled experimental mechatronic bogie, in: *Proceedings of International Symposium on Speed-up and Sustainable Technology for Railway and Maglev Systems, STECH 2015*, Japan Society of Mechanical Engineers, Tokyo, Paper 1F34, 2015. With permission.)

Scale Testing Theory and Approaches

The wheelsets can be quickly setup with conventional or IRW types. Wheel profiles with tread tapering in the range from 1:40 to 1:5 are available. The wheelsets are designed to accommodate individual wheel drives.

The bogie is equipped with a system for measurement of axlebox forces, i.e. the forces acting between the axleboxes and the bogie frame. The system is based on the stirrup (item 8 in Figure 20.15), a specifically designed component that connects the axlebox and the bogie-frame. The bogie is equipped with four stirrups, one for each axlebox. Each stirrup is instrumented with 36 strain gauges connected to three full bridges. Measurement of x, y and z components of a force transmitted between each axlebox and the bogie frame is thus achieved.

The CTU roller rig and the test bogie were designed to perform various types of tests, allowing simulation of the running of a 2-axle railway bogie on an arbitrary shaped track within the laboratory environment. The typical applications include the following:

- Research into running dynamics of railway vehicles with conventional wheelsets and IRWs.
- Curving behaviour of railway vehicles with conventional wheelsets and IRWs.
- Improvement of curving behaviour by active wheelset steering.
- Improvement of vehicle stability by active controlled yaw torque acting upon the wheelset.
- Research into control algorithms for wheelset drives and IRW drives.
- Improvement of vehicle running dynamics and decrease of vehicle-track force interaction by advanced control of IRW drives.

20.6 SCALED PROTOTYPE TESTING ON TRACK

The use of roller rigs as track simulators involves the introduction of inherent errors due to the curvature of the rollers. As mentioned previously, such errors can be minimised with various strategies such as the adoption of a very large roller radius [14] and the correction of the transverse profile of the roller [35]. A selection of roller rig setup types is shown in Figure 20.16, with the most commonly

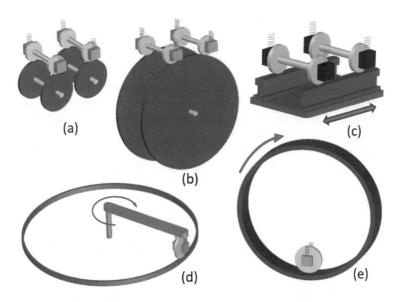

FIGURE 20.16 Types of scale prototypes: (a) roller rig with a roller for each wheelset, (b) single roller rig, (c) alternative tangential rig, (d) circular path test rig and (e) internal roller rig.

used option shown in Figure 20.16a. Roller rigs with a large roller diameter (Figure 20.16b) do not allow use of one roller for each wheelset, so all the wheelsets of the prototype are placed on the same roller, generating a longitudinal component of the constraint force that is unrealistic and must be taken into account or compensated.

Various researchers have investigated alternative strategies to overcome these limitations with the aim of making the behaviour of the prototype similar to that of the real rail vehicle, even though using a small-scale prototype. Obviously, the problems related to the scaling of the prototype and the need to use a method of similarity cannot be avoided.

Among these alternative methods, the most intuitive one is the use of a scale prototype on a real test track, also built on a small scale, in similitude with the real track. It is evident that, even when using a reference track of reduced dimensions, achieving the layout will still result in relatively large dimensions and high costs. The realisation of the track is particularly problematic because of the need to maintain the precision of various parameters (e.g., plano-altimetric defects, cant and geometry of the curves) with a reduced scaling. There are some examples of this type of approach [45] in which the geometry of the actual layout has been used to create a calibration system useful for the development of monitoring systems for on-board installation.

In the case of wear and adhesion tests in a straight line (tangent track), it is possible to use a test bench composed of a vehicle prototype and a section of rail moving along the longitudinal direction with an alternative motion. The same effect can be achieved by fixing a section of rail and moving the wheel with an alternative motion. This type of test bench (Figure 20.16c) was utilised at the University of Sheffield in both full scale and reduced scale [46] for tests of adhesion. The limitation of this bench type is related to the fact that, due to the limited length of the rail, the maximum speed obtainable for the prototype is equally limited. The test is normally carried out with an initial phase of acceleration a_x, of a duration t_1, followed by a phase with a constant speed of duration t_2, and finally a deceleration phase equal to the initial one but of opposite sign. The maximum speed that can be reached during the test can be calculated according to Equation 20.55.

$$v_{max} = a_x \cdot t_1 \tag{20.55}$$

The distance s is travelled during a cycle, and so, the length L of the rail required in the test bench, can be calculated according to Equation 20.56:

$$L = s = a_x \cdot t_1^2 + v_{max} \cdot t_2 = a_x \cdot t_1^2 + a_x \cdot t_1 \cdot t_2 \tag{20.56}$$

Considering, as an example, an equal duration of the acceleration phase and constant speed phase ($t_2 = t_1$), if the length of the rail is limited to 4 m and the acceleration is limited to 1 m/s², a maximum speed of 1.4 m/s is obtained, very small even in the case of scale models. These benches are therefore suitable for analysing friction phenomena and wear at low speed.

For the simulation of the behaviour of a vehicle on a curve, it is possible to use a test bench consisting of a circular track on which the prototype runs continuously, such as the one shown in Figure 20.16d. An example of this type of bench has been studied [47,48] and realised on a smaller scale [49] at the University of Delft. The bench has a single circular rail and four wheels arranged every 90° and fixed by suspension on a cross-shaped structure constrained in the centre of the circular track so as to be able to rotate freely. The rail is fixed to the ground by a series of connections, sleepers and junctions that can be designed to simulate the dynamic behaviour of a variety of railway track types.

This type of test bench allows accurate simulation of the contact forces between the wheel and the rail in a curve and facilitates the study of the evolution of wear of the rail and the wheel after repeated test cycles.

The limitations of the bench are related to the fact that the setup allows the simulation of only a single curve radius. However, to reduce costs, it can be significantly smaller than the radius of a real railway curve. The 1:5 scale bench currently available at the University of Delft [49] is characterised by a curve radius of 2 m, corresponding to 10 m in full scale.

Another type of roller rig constructed at the Fraunhofer Institute [41] involves the use of a hollow roller (Figure 20.16e), inside which the prototype is placed.

The hollow roller is analogous to the solution with a single large roller (Figure 20.16b), with the difference that the roll curvature is negative rather than positive. As a consequence, the contact area in the longitudinal direction will change, increasing, instead of reducing, its longitudinal length. If several axles are used on the same roller, the constraint forces generate longitudinal forces, which tend to bring the wheelsets closer together rather than distancing them, as in the case of the outer roller. This type of bench, derived from the automotive sector where similar benches are used to test vehicle tyres, has the advantage of being able to contain the dimensions of the bench (with the same curvature of the roller) and to offer greater safety during the tests because the whole prototype is contained inside the hollow roller. The projection of objects due to malfunctions is therefore prevented. However, the drawbacks lie in the difficulty of producing complex or multi-axle prototypes due to the space constraints imposed by the hollow roller. The test rig that was developed is limited to a single wheel.

Researchers have hypothesised other strategies for carrying out tests on scale prototypes [50,51], including the use of flexible rails mounted on two or more rollers, in order to reproduce a tangential direction rail in the contact zone with the wheel of the prototype. However, there have been no real applications of these configurations at this time.

20.7 SCALED PROTOTYPES: TYPICAL APPLICATIONS

Tests on scale prototypes in the railway field have been used to carry out various types of experiments. Depending on the purpose of each experiment, it is important to select the appropriate types of test, test rig, instrumentation, method of scaling and test procedure, if significant results are to be achieved. As we have seen previously, even the configuration of the test rig itself can be significantly different depending on the tests to be carried out.

From an operational point of view, tests on prototypes can be used to analyse a wide range of aspects such as:

- Validation of models (numerical or phenomenological)
- Tests of innovative solutions in a controlled and safe environment
- Calibration of algorithms, measurement systems, control methods and diagnostics
- Experimental tests related to the behaviour of materials and vehicle components (e.g., tribological aspects and design of suspension components)
- Development of HIL/software-in-the-loop (SIL) co-simulation environments

Multiple aspects are often analysed simultaneously, making the possible applications very numerous and different from each other.

In the following, the main applications put into practice by different researchers are exemplified by being macroscopically divided into the following three main representative categories:

- Use of the roller rig as a dynamic track simulator
- Tribological studies
- HIL/SIL applications and control and diagnostic systems

20.7.1 Dynamic Track Simulator

The use of the roller rig as a dynamic track simulator is the most traditional configuration, which involves the use of rollers to reproduce wheel-roller contact forces on the prototype, similar to those that would occur with vehicles running on rails. In this configuration, the test rig is typically used for dynamic tests on the prototype, but other applications are also possible.

20.7.1.1 Hunting Stability

Hunting stability was one of the first phenomena to be studied using a reduced-scale roller rig [16]. In fact, a roller rig is particularly suitable for simulating the dynamics of a vehicle on a straight (tangent) track, even at very high speeds, a fundamental requirement for running stability tests.

The hunting phenomenon occurs on a roller rig in a qualitatively similar manner to how it occurs on rails in the sense that, even on a roller rig, the vehicle presents a lateral-yaw coupled motion, which becomes unstable above the critical speed. From the quantitative point of view, the intrinsic errors present on roller rigs related to the finite curvature of the rollers cause a difference in the critical speed value, which is slightly lower than for vehicles running on rails, and result in different oscillation amplitudes when the hunting becomes unstable [52–54].

The differences in the value of the critical speed are related to the different shape of the contact area and the creepages, which generate different tangential forces with respect to the case on rail. Some differences, such as the ones on creepages, are greater in the small-scale models, and this is true in particular for the spin creepage that is not dimensionless.

The amplitude of oscillation is also altered by the phenomenon of decrowning, which makes the behaviour of the wheelset positioned on the top of roller more unstable, with the wheel being translated from its top position due to the longitudinal components of the constraining forces.

All the effects that differentiate stability on a roller rig can be minimised by increasing the ratio between the radius of the rollers and the radius of the wheel, and this explains why some test rigs were created with a large single roller to support all the wheelsets [14]. Nevertheless, the difference found on a scaled roller rig, compared with the real case once the scaling factors have been correctly applied, is normally less than 10% according to studies conducted by various researchers.

Scaled roller rigs have been used to carry out tests both to determine the critical speed [29,31,52,55], and tests that allowed determination of the instability limit cycles [35,56]. These tests can be conducted on a prototype that includes at least a single suspended wheelset [35]. Tests on a single wheelset are easier to compare with numerical models because of the simplicity of the prototype, and are therefore of greater interest for studies related to the understanding of the phenomenon or related to the study of contact forces, even when the shape of the profiles and the axle load vary.

Bogie stability tests are instead of greater interest for the study of vehicle dynamics and can be compared with numerical models or used in the pre-design phase of a real vehicle [31].

From a technical point of view, stability tests are normally carried out in the steady state, varying the rotation speed of the roller and then leaving it constant for a certain period of time, during which the data are collected. If only the critical speed is considered, it is not normally necessary to provide an external excitation to the wheelsets because, when the critical speed is exceeded, the hunting mode is self-excited due to the constructive irregularities inevitably present in the rig. When the instability limit cycles are to be determined instead, it is necessary to excite the wheelset with lateral displacements of different amplitudes, normally imposed by impulsive load.

Figure 20.17 shows an example of the evolution of the stability on a scaled roller rig using realistic profiles. When the critical velocity was reached or exceeded (at 150 km/h in the proposed case), the lateral accelerations measured on the axlebox increased. The lateral and yaw motions became synchronous, and this can be observed from the diagrams showing the lateral displacement of the axlebox versus the longitudinal displacement.

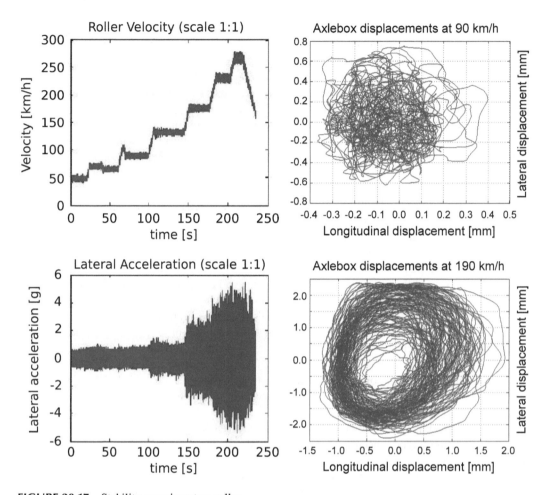

FIGURE 20.17 Stability experiment on roller.

20.7.1.2 Contact Force Estimation

Another type of test that can be conducted using scaled roller rigs is the study of wheel-rail contact forces. This makes it possible to validate numerical contact models, and as far as the results are concerned on the wheel-roller case, their validity can also be extended to the wheel-rail case.

Contact studies can be performed using simple prototypes consisting of a single wheelset or a single wheel. In this way, the vehicle model is particularly simple, and the results can be interpreted more easily.

Different methods can be used for the estimation of contact forces; the most direct among them is to place load cells under the roller support [38]. As an alternative, it is possible to instrument the rollers using strain gauges arranged on their surface [42]. The contact forces can also be evaluated indirectly by measuring the loads on the suspensions of the prototype used during the tests and considering the position and accelerations acting on the wheelset [33].

The estimate of contact forces is normally also performed by varying the axle load and the kinematic parameters of the prototype.

The contact tests can be conducted by varying the yaw angle between the wheelset and the rollers, thus changing both the angle of attack and the spin creepage.

The roller rig tests can also be used to perform analysis of the vehicle's behaviour with damaged wheel or rail surfaces. For example, it is possible to study the behaviour of a vehicle whose wheels are damaged by wheel-flats [26,57], and this allows the development of different diagnostic techniques to identify the phenomenon.

Even the presence of rail defects can be simulated on roller rigs, but the maximum wavelength is limited by what can be reproduced on the circumference of the roller. The determination of the forces exchanged between vehicle and track by a roller rig can also be used to determine actions on the infrastructure under particular conditions. In [58], for example, a roller rig is suspended on the ground by a set of stiffness to reproduce the passage of the vehicle on a bridge.

Finally, if tests are carried out by modifying the longitudinal creepage, it is possible to undertake tests on the adhesion. This topic is of particular relevance and will be analysed in detail in the following sections.

20.7.1.3 Other Vehicle Dynamic Studies

The use of roller rigs has been extended to other types of studies in the dynamic field, which include, for example, the simulation of behaviour in curves. Except the obvious case where the test rig directly reproduces a curved track [47,48,59], also a roller rig can simulate the curve by imposing a yaw angle between the rollers and the wheelsets of the prototype and at the same time by varying the angular velocity of the roller that simulates the outer rail compared to the one that simulates the inner rail. An interesting application is illustrated in [60], where an articulated vehicle with three bogies is tested on a six-axis roller rig. The vehicle's trajectory in the curve is simulated by rotating and translating the supports of the rollers, this being done in order to verify the self-steering capabilities of the vehicle.

Another possible application involves the use of a roller rig to simulate derailments [61–63]. In [64], the derailment condition is reproduced by acting on the angle of attack and creating a lifting of the rollers that simulates the outer side in the curve. For this type of simulation, the test rig must be designed in order to allow several movements to the supports of the rollers: a movement in the vertical direction and a yaw rotation need to be applied to each pair of rollers.

If the vehicle prototype is accurate enough, and it also includes the secondary suspensions, comfort tests can also be performed on roller rigs [65–67]. In [66], a 1:5 scale prototype, representing a half vehicle, was used to develop an active secondary suspension to improve vehicle comfort.

A further type of study that has been attempted on a roller rig is related to the analysis of the noise emitted by the vehicle under different conditions. Of particular interest for tram vehicles are the squeal noise simulations generated during curving [59,60]. The squeal noise can also be simulated with special twin-disc systems if these are appropriately designed to reproduce the contact forces on the side of the wheel [68]. In general, other types of noise generated by the contact due to the surface corrugation of the wheel and roller profiles (which simulates the rail) can also be studied. In this case, the finite size of the roller circumference does not impose a limitation, because the wavelength of the acoustic vibrations is small enough to be reproduced on a roller rig. The study of acoustic phenomena can conveniently be carried out using similitude models that maintain constant scaling of time and frequency, e.g., the method proposed by Iwnicki [27], where a roller rig was used to validate a technique to detect wheel and rail surface defects based on acoustic measurements.

20.7.2 TRIBOLOGY STUDIES

Tribology is one of the fundamental aspects for the dynamic design of railway vehicles and has been the subject of study by many researchers. Tribological studies in the railway sector can also be supported by the use of scaled roller rigs and can deal with different aspects: wheel-rail adhesion [33,69], use of friction modifiers [70,71] and wear of wheel and rail profiles [37,72]. The study of these phenomena is very complex and difficult to achieve on the track because of the costs due to the occupation of the line and the poor reproducibility of the results, mainly due to changes in environmental conditions. For this reason, researchers have designed test benches that permit the reproduction (in the laboratory) of the typical operating conditions of the vehicle running on track. In this way, it is possible to carry out experimental tests in controlled and easily reproducible environmental conditions. The test benches that are commonly used for tribological studies are pin-on-disc test rigs [69,72–74], twin-disc test rigs [75–78] and roller rigs, both at full [79] and reduced scale [23,33,80]. The first type of test rig consists of a rotating disc with a vertical rotation axis and a specimen that is pressed on this disc. This device allows the simulation of pure sliding conditions and is typically used to evaluate the friction coefficient and the wear of materials in the presence of high sliding values. Twin-disc rigs, on the other hand, consist of two rotating discs that are pressed against one another with a controlled load. The sliding between the two rollers can be precisely regulated by changing the rotation speed of the rollers, which are controlled in velocity. In this case, it is possible to reproduce the case of rolling contact and, therefore, to consider both low sliding values, in which the contact area is mainly in adhesion condition, and high sliding values, in which all or a large part of the contact area is in sliding condition. For this reason, this type of experimental device is mainly used to study the adhesion curve behaviour versus the sliding, considering different environmental conditions such as the presence of contaminants [33] and of friction modifiers [71,73]. The same test benches have also been used to evaluate the wear behaviour of the materials typically used for the manufacturing of wheels and rails [37]. The advantage of these systems consists in the simplicity of construction and in the possibility to precisely control the sliding, but they have the disadvantages of using simplified roller profiles, which are not able to reproduce the actual contact conditions, and of completely neglecting the dynamic effects. To overcome these problems, roller rigs are a valid alternative, and they allow obtaining results that are directly comparable to those obtainable on track by means of suitable scaling techniques.

Roller rigs allow the performance of adhesion and wear tests, considering the dynamics of the vehicle, as, e.g., during traction [24,36,80] or braking [30] actions. An alternative to roller rigs are wheel-rail test benches that are able to reproduce realistic contact conditions. These devices are composed of a railway wheel (in full scale [81] or reduced scale [46]) that is in contact with a section of rail that can move in the longitudinal direction by means of a linear actuator. These systems are typically used to evaluate the efficiency of the lubricant that is released by the wheel flange lubrication systems but do not allow the study of the adhesion curve, as they only allow simulation of low rolling speed and the wheel moves back and forth with a reciprocating motion. The wheel-rail test benches, however, are widely used for wear studies and studies related to the rolling contact fatigue (RCF) phenomenon.

20.7.2.1 Adhesion

Wheel-rail adhesion is of fundamental importance to ensure efficient braking and adequate traction to railway vehicles. The study of this phenomenon has the aim of identifying the characteristic that describes the behaviour of the adhesion coefficient, calculated as the ratio between the longitudinal friction force and the normal load, as a function of the sliding, considering different contact conditions. Traditional test methods for the evaluation of the adhesion curve involve the use of pin-on-disc and twin-disc systems.

Pin-on-disc testers are the most traditionally used tribological systems to determine the value of the wheel-rail friction coefficient in the presence of pure sliding. This system has been used to investigate the effect of air humidity [74], of iron oxide [69] and of the use of a friction modifier [70] on the friction coefficient.

Regarding the twin-disc systems, there are studies in the literature concerning the evaluation of the adhesion curve in the presence of contaminants [76] and friction modifiers [77,82].

An important contribution to the study of adhesion is certainly given by the use of roller rigs (both at full and reduced scale) that, unlike the previously mentioned systems, allow taking into account the effect of the shape of wheel and rail profiles, the effect of the three creepages and the dynamics of the vehicle, which are essential to evaluate adhesion during braking and traction operations. These two conditions are completely different because, in the braking case, the angular velocity of the wheelset decreases until eventually stopping (total sliding), causing positive sliding, while in the traction case, the angular velocity of the wheelset increases and generates opposite sign sliding. The roller rigs allow the simulation of both operating modes, and different strategies are possible to generate sliding between wheel and roller.

One method to investigate adhesion consists of the use of a motorised prototype to whose wheelsets it is possible to provide a traction torque. This type of configuration is clearly interesting when the goal is the evaluation of the adhesion conditions during traction. An example of this solution was described by Zhao et al. [80], where a 1:5 scaled roller rig was used to experimentally test an algorithm, based on the Kalman filter, which is able to evaluate the wheel-rail friction coefficient from the behaviour of the traction motor. The prototype of the bogie has both the wheelsets powered by AC motors and controlled by inverters with indirect field control, which is the typical control adopted in modern locomotives. Instead, the rollers are connected by means of pulleys to two DC generators, providing the external resistance that can be regulated during the tests to simulate different external loads. Another example of a scaled roller rig with a configuration similar to this is described in [83]. In this case, the test bench is used to test traction control algorithms to be used for wheel slide protection systems for railway locomotives. The single wheelset prototype installed on the rollers is powered by an asynchronous electric motor, while the resistances of the vehicle and of the track (inertia, slopes, etc.) are simulated with an electric motor combined with a mechanical brake.

A second method that is adopted to generate and control wheel-roller slip on roller rigs is to apply a braking torque on the wheelsets of the prototype installed on the rollers. In this case the roller motors are used to accelerate the wheelset and, if necessary, to simulate the inertia of the vehicle during braking. This strategy is described in [30], where a 1:5 scaled model of a single wheelset, equipped with a disc braking system, has been installed on a pair of motorised rollers. The rollers are driven by a synchronous permanent magnet motor (brushless), which has the task of accelerating the rollers (and therefore the wheelset) up to a reference velocity. Subsequently, a braking torque, which is regulated by controlling the pressure acting on the brake pads using a proportional pneumatic valve, is applied to the wheelset. In this case, the vehicle inertia is simulated by means of the roller motor and a pair of additional discs, which are connected to the rollers by means of a gearbox which has the task of increasing the inertia acting on the rollers. The aim of the experimental tests is to evaluate the adhesion characteristic during braking operations, considering different vehicle speeds and different adhesion conditions. Another example in which the braking torque acting on the wheelset is controlled to regulate the wheel-rail sliding is described in [34], in which the phenomenon of adhesion recovery is evaluated by means of the multi-wheelset test bench described in Section 20.5.4.

An alternative to the two configurations described above to control wheel-rail sliding is presented in [36]. The work describes the use of a single wheelset (1:5 scale) suspended on a pair of rollers, which are independently controlled, according to the scheme proposed in Figure 20.18.

The system also allows the independent regulation of the vertical load acting on each wheel. The tests are carried out by controlling both motors in velocity or by controlling one motor in velocity and the other one in torque [37]. In the first case, the angular velocity of the roller on the side where the wheel is more loaded, and therefore in adhesion (due to the greater normal load), is kept constant, while the angular velocity of the other roller is decreased in order to obtain the desired sliding value between the wheel and the roller. In the second case, the speed of the roller located on the adhesion side is kept constant, while a resistant torque is applied to the other roller. The value of the resistant torque is calculated in such a way as to maintain a reference slip value.

The strategy of controlling both motors in velocity is suitable when the aim is the investigation of the adhesion characteristic for high sliding values, where a torque control would be less effective since a small increase in resistant torque would generate a significant increase in slip. In contrast, the control of one of the rollers in torque allows the investigation of the adhesion characteristic for low sliding values, where a velocity control of the roller would not be effective since a small decrease in the angular velocity of the roller would result in a huge increase in friction force.

Scale Testing Theory and Approaches

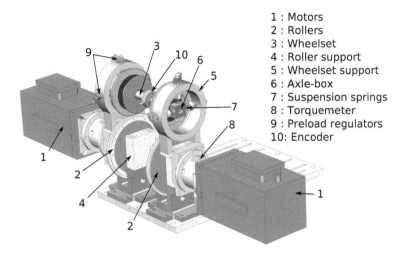

FIGURE 20.18 Scheme of the roller rig with single suspended wheelset.

Figure 20.19 shows an adhesion curve under dry conditions, obtained by controlling both motors in velocity (curve indicated with '+'). Analysing the results, it is possible to observe that this control method allows the acquisition of numerous experimental points in the region corresponding to high slip, while few points are available for low values of creepage. The results shown in Figure 20.19 for the torque control mode under both dry ('x') and wet ('*') conditions show an opposite behaviour, with many experimental points near the origin and few points in the region corresponding to higher slip values. These results were, in fact, obtained by controlling one motor in velocity and the other one in torque.

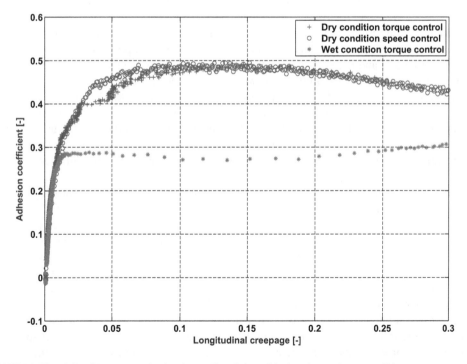

FIGURE 20.19 Adhesion curves obtained on roller rig considering dry and wet conditions.

Another application of this architecture is described in [33], where a single wheelset roller rig was used to carry out a study on adhesion recovery. This phenomenon comprises the recovery of adhesion resulting from the removal and/or destruction of the contaminant by the passage of wheelsets that progressively encounter a contaminated section of rail. The re-adhesion process is of great interest for the development of the algorithms used in wheel slide protection systems, which clearly can be optimised if the variation of the friction coefficient is known. Experimental tests were carried out by contaminating a section of the roller and using both of the control strategies previously described. During the tests, a specific cleaning system was adopted to remove the contaminant deposited by the roller on the wheelset. This is necessary to simulate the real case in which a 'cleaned' wheelset encounters the contaminated rail section. The use of a single wheelset roller rig and independent control of the roller motors was demonstrated to be effective to perform adhesion studies. The main advantage of this solution is the design simplicity. This configuration, in fact, allows the performance of the tests using a simple wheelset without the need to install a traction motor or a braking system on it. Clearly, this configuration does not allow the faithful reproduction of the longitudinal dynamics of the vehicle, which in any case would require at least the use of a scaled prototype of the whole bogie, including the traction system.

20.7.2.2 Wear

Another important tribological aspect in regard to wheel-rail interaction is certainly wear. The most common approach to perform wear tests on railway wheels and rails is the use of pin-on-disc [70,72] and twin-disc systems [76]. These systems have been largely used to carry out experimental tests to validate wear estimation algorithms and to evaluate the wear behaviour of the materials commonly adopted for manufacturing wheels and rails. The tests performed with the pin-on-disc system cannot be directly compared with those of a railway vehicle because the contact geometry and creepages are completely different. Therefore, this method requires a numerical model to take into account the consequences of effective slip and the effective distribution of pressure that occurs between the wheel and the rail. For this reason, results obtained on a pin-on-disc system cannot be directly compared with the wear process occurring on a railway wheel. The results obtained on traditional twin disc systems are also oriented to define the material properties during the wear process but are not able to predict the wear evolution on real wheel-rail profiles, since the pin-on-disc contact is cylindrical and not elliptical, and the creepage is purely longitudinal. When a twin-disc test rig is equipped with the capability of adopting real wheel and rail profiles, it works as a roller rig. An example of this procedure is proposed by Shebani and Iwnicki [23]. That work concerns the development, verification and training of a numerical code, based on neural networks, for the prediction of wear of wheels and rails. In order to evaluate the wear of the profiles during the associated experimental tests, the Alicona profilometer was used. This device allowed the wheel and roller profiles to be digitised, in order to evaluate the volume of material removed from the profiles at the end of each test.

In [37], Bosso et al. propose the use of a roller rig test bench to validate a numerical code able to estimate in real-time the wear of the wheel and roller profiles. The RTCONTACT [84] contact algorithm, which includes a wear calculation module, was compiled into a real-time system to calculate the wear of the profiles during the experimental tests, directly using the data measured on the roller rig. The test bench was used with the same mechanical arrangement which was adopted for the adhesion tests [33]. The test bench consists of a single wheelset supported by two rollers that are independently driven by electrical motors controlled in velocity. This configuration allows the operator to impose a reference creepage value.

The test bench includes a specially designed laser profilometer, which allows the acquisition of the wheel and roller profiles with a resolution of 0.1 μm. In this way, it is possible to compare the profiles estimated by the wear algorithm with those measured after the experimental tests. Before starting wear tests, the wheel and roller profiles were measured on the roller rig, and the measured profiles were used by the contact algorithm as initial profiles. During the test, the wear algorithm modified the geometry of the profiles on the basis of the calculated worn volume, and, after a selected period of time, the test was stopped, and the wheel and roller profiles were measured again on the roller rig. These profiles were then compared with those predicted by the wear algorithm.

The wear tests were carried out starting with the rollers rotating at the same speed, then the speed of one of the two rollers was decreased to a certain value and then kept constant for the rest of the test. The tests were carried out using softer materials than those commonly adopted for the manufacturing of wheels and rails in order to reduce the time required for the experimental test. The goal of the authors was to validate the wear algorithm and the use of a different material therefore did not compromise the results. The wheel profiles were conical since the wear algorithm was developed and tested only considering tread wear. In fact, the wear rate is completely different when considering tread and flange. The conicity of the wheel profile was 0.036 rad and it was manufactured of Fe360 steel with a surface hardness of 100 HB. The rollers had a profile that exactly reproduced the 1:20 canted UIC60 profile, manufactured of Fe510 steel with a surface hardness of 160 HB. The wear tests are summarised in Table 20.4, where the parameters of each test are also reported. Observing the table, it can be noticed that the creepage and the vertical load were maintained constant during each test. Figure 20.20 shows the evolution of the geometry of the wheel and roller profiles during the experimental tests. The profiles were acquired at the end of each test and compared with those estimated by the wear algorithm.

TABLE 20.4
Summary of Wear Tests Carried Out on Roller Rig Test Bench Considering Conical Wheel Profiles

Test	Creepage	Reference Roller Velocity (rpm)	Vertical Wheel Load (kg)	Distance Travelled (km)
1	0.25	500	53	21.8
2	0.05	500	50	17.0
3	0.05	500	50	23.1
4	0.1	500	48	35.47
5	0.15	250	43	10.2

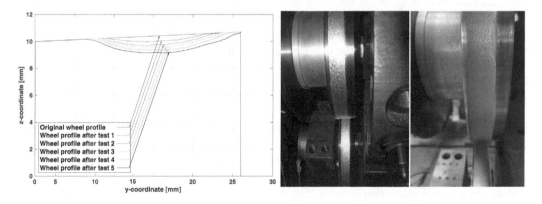

FIGURE 20.20 Wheel-roller worn profiles obtained on scaled roller rig.

20.7.3 Hardware in the Loop Applications and Control Systems

Roller rigs are particularly suitable for experiments with an HIL approach and for testing new mechatronic solutions with active control of devices mounted on the vehicle.

In fact, as well as the first tests carried out on vehicles on scaled roller rigs referred to in the discussion of hunting stability, even the first tests of active systems were aimed at developing and testing systems for active stabilisation of the hunting motion. These tests were conducted on the roller rig at the DLR [19,20]. The vehicle prototype used for this purpose consisted of a two-axle bogie with independent motorised wheels. The motors were controlled with an algorithm capable of improving the behaviour in curves and, at the same time, reducing the occurrence of the hunting phenomenon. Similar tests have been carried out, for example, using the roller rig at the CTU, where active control was instead used to impose and control a steering angle on the wheelsets of the vehicle [42,43]. The wheelsets in this case were connected to each other by a kinematic system connected to the axleboxes, which is actuated by a single electric motor. In any case, the device, actively controlled, allows the improvement of both curving behaviour and running stability.

Another application concerning vehicle dynamic control on roller rig consist in the active control of the vehicle suspension to improve both comfort level and vehicle stability [85,86]. In this case, a roller rig developed at the Korea Railroad Research Institute was used to test a semi-active lateral damper located in the secondary suspension of the vehicle. The results of experiments were used to define a skyhook control method for the semi-active damper in order to reduce the acceleration measured on the coach of the vehicle.

Roller rigs were also used for the development and optimisation of algorithms used in traction control systems (anti-slip) and braking control systems (anti-skid). The algorithms that control wheel slide protection systems (WSPs) cannot be directly tested on track owing to the risk of extensive damage to the rails and the need to occupy the track during the tests. Furthermore, roller rigs allow testing the algorithms in the laboratory under controlled environmental conditions and with the possibility of immediately testing proposed modifications of the control algorithms. A further advantage that arises from the use of scaled roller rigs is the possibility of testing, with relatively simple modifications of the test bench, the control systems considering different types of vehicles with different architectures. An example of the application of a roller rig for developing a traction control system is described in [24]. The authors developed a traction control, based on the extended Kalman filter, which considers the voltage, current and velocity of the traction motor to determine the sliding and the coefficient of friction that correspond to the maximum of the adhesion curve. Another example of using a scaled roller rig test bench for the study of a traction control system is described in [36]. That work concerns the implementation of a traction control that includes a real-time wheel-roller contact algorithm. On the basis of the data measured on the roller rig, the algorithm is able to estimate the sliding to be imposed between the wheel and the roller in order to exploit the maximum adhesion force. The experimental tests were carried out considering both the dry and the wet conditions. Another aspect that, more recently, involved the use of roller rig test benches is the development of HIL systems, which allow the testing of a real component simulating the rest of the vehicle through the test bench. In this way, it is possible to evaluate the operation of a specific vehicle subsystem considering operating conditions that typically cannot be reproduced on track for reasons of cost and safety. The use of scaled roller rig test benches as HIL devices is still limited, but recent publications in the literature demonstrate the research and industrial interest in this field. An example of the use of a roller rig as a HIL system is described in [87], where the test bench was designed to perform tests on anti-skid and anti-slip systems. Another example of a test bench explicitly designed as a HIL system is the multi-wheelset test bench [34], which was developed to install and test wheel slide protection systems.

20.8 CONCLUSIONS

This chapter discussed the history and application of scaled roller rigs, outlined their uses and described the construction and operation of six scaled roller rigs.

The first three roller rigs, developed at the University of Huddersfield, INRETS and DLR, are the rigs in which the three possible methodologies of scaling used on this type of test bench have been historically defined. The three scaling strategies of the institutions mentioned previously have been described in detail and the differences tabulated.

It is important when designing a new roller rig to first consider the primary use of the rig as this will help form the basis of the scaling strategy. The scaling strategy is the most important aspect of the rig development as it will ensure that the measured parameters are correctly related and obey laws of similarity.

In summary, the scaling strategy of the rig at UoH was developed using a comparison of the linearised differential equations for the scale and full size, the main purpose of the rig being the study of vehicle stability and general dynamic behaviour. Frequency was preserved at 1:1 for this type of analysis. The scaling method for the rig at DLR, Oberpfaffenhofen, was derived from a study of the full set of nonlinear equations of motion to give precise results for study of limit cycle behaviour and early validation of the dynamic multibody simulation software, SIMPACK. The large single wheel rig at INRETS in Grenoble allows suspension parameters to be evaluated, and the almost exact treatment of the contact conditions is allowed by the very large radius of the roller. The rig has been used extensively for the validation of Kalker's theory and development of in-house contact mechanics software.

As has been detailed in Chapter 19, there are errors inherent in roller rig testing, and these of course apply to scaled rigs. It must be realised that scaled rigs also have additional errors introduced by the scaling strategy, as perfect scaling for all parameters cannot be achieved. An example of possible scaling errors is given and the errors analysed using a typical two-degree-of-freedom wheelset model. This analysis illustrates the importance of selecting a scaling strategy, which suits the desired use of the rig.

It is sensible with a scaled roller to use the largest possible roller diameter irrespective of the scale defined for the bogie, as this will preserve the contact conditions with respect to running on conventional track. Results have been presented in Chapter 19 as to the influence of roller diameter on various parameters, and these should be considered at the design stage of a scaled rig.

The other three roller rigs described in this chapter were most recently developed at the Politecnico di Torino, the VT-FRA and the CTU and illustrated design examples of roller rigs used to analyse different types of problems. The first of these rigs, thanks to its versatile and modular concept, was used to analyse the behaviour of different types of vehicles, also analysing problems of adhesion and wear. The VT-FRA rig consists of a single roller and has been designed to perform accurate analysis of contact forces. The CTU rig is instead used to carry out tests on prototypes of mechatronic vehicles in active control condition.

The second part of the chapter analysed the different applications of roller rigs that have been applied over time, divided between applications where the roller rig is used as a dynamic track simulator and for tribological studies.

ACKNOWLEDGEMENTS

The authors wish to acknowledge the contribution of Alfred Jaschinski and Hugues Chollet, whose work has been presented, in part, under information provided regarding the DLR and INRETS scaled roller rigs, respectively [88].

Recognition also goes to Mehdi Ahmadian and Jan Kalivoda for the material provided on the test benches at VT-FRA and CTU.

REFERENCES

1. O. Reynolds, An experimental investigation of the circumstances which determine whether the motion of water shall be direct or sinuous, and of the law of resistance in parallel channels, *Philosophical Transactions of the Royal Society of London*, 174, 935–982, 1883.
2. O. Reynolds, On the dynamical theory of incompressible viscous fluids and the determination of the criterion, *Philosophical Transactions of the Royal Society of London*, A, 186, 123–164, 1895.
3. L.M. Sweet, A. Karmel, S.R. Fairley, Derailment mechanics and safety criteria for complete railway vehicle trucks, In A.H. Wickens (Ed.), *Proceedings 7th IAVSD Symposium held in Cambridge*, UK, 7–11 September 1981, Swets & Zeitlinger, Lisse, the Netherlands, 481–494, 1982.
4. L.M. Sweet, J.A. Sivak, W.F. Putman, Non-linear wheelset forces in flange contact, part 1: Steady state analysis and numerical results; part 2: Measurement using dynamically scaled models, *Journal of Dynamic Systems, Measurement, and Control*, 101, 238–255, 1979.
5. M. Cox, H. Nicolin, Untersuchung des Schwingungsverhaltens von Schienenfahrzeugen mit Hilfe des Modellprüfstands am Institut fur FSrdertechnik und Schienenfahrzeuge der RWTH Aachen, *Leichtbau der Verkehrsfahrzeuge*, 23(4), 91–95, 1979.
6. A. Jaschinski, On the application of similarity laws to a scaled railway bogie model, Doctoral thesis, Delft University of Technology, Delft, the Netherlands, 1990.
7. M. Jochim, Konstruktion eines Versuchsdrehgestells, Term study, in Lehrstuhl B fur Mechanik, Technical University of Munich, Germany, 1984.
8. M. Jochim, Analyse der Dynamik eines Schienenfahrzeuges, Diploma thesis, in Lehrstuhl B fur Mechanik, Technical University of Munich, Germany, 1987.
9. C. Heliot, Small-scale test method for railway dynamics, In O. Nordstrom (Ed.), *Proceedings 9th IAVSD Symposium held in Linköping*, Sweden, 24–28 June 1985, Swets and Zeitlinger, Lisse, the Netherlands, 197–207, 1986.
10. P.D. Allen, Error quantification of a scaled roller rig, Doctoral thesis, Manchester Metropolitan University, Manchester, UK, 2001.
11. S.D. Iwnicki, Z.Y. Shen, Collaborative railway roller rig project, SEFI World Conference on Engineering Education, 20–25 September 1992, Portsmouth, UK, 1992.
12. S.D. Iwnicki, A.H. Wickens, Validation of a MATLAB railway vehicle simulation using a scale roller rig, *Vehicle System Dynamics*, 30(3–4), 257–270, 1998.
13. R. Illingworth, Railway wheelset lateral excitation by track irregularities, In A. Sliber, H. Springer (Eds.), 5th VSD-2nd IUTAM Symposium held in Vienna, Austria, 19–23 September 1977, Swets & Zeitlinger, Amsterdam, the Netherlands, 1978.
14. H. Chollet, Essais en Similitude a l'Echelle 1/4 de Bogies de Wagons de la Famille Y25, INRETS Report 781988, 1988.
15. A.H. Wickens, The dynamics of railway vehicles on straight track: Fundamental considerations of lateral stability, *Proceedings Institution of Mechanical Engineers*, 180(6), 29–44, 1965.
16. T. Matsudaira, N. Matsui, S. Arai, K. Yokose, Problems on hunting of railway vehicle on test stand, *Journal of Engineering for Industry*, 91(3), 879–885, 1969.
17. A. Jaschinski, F. Grupp, H. Netter, Parameter identification and experimental investigations of unconventional railway wheelset designs on a scaled roller rig, *Vehicle System Dynamics*, 25(S1), 293–316, 1996.
18. A. Jaschinski, H. Netter, Non-linear dynamical investigations by using simplified wheelset models, *Vehicle System Dynamics*, 20(S1), 284–298, 1992.
19. M. Gretzschel, L. Bose, A mechatronic approach for active influence on railway vehicle running behaviour, Proceedings 16th IAVSD Symposium held in Pretoria, South Africa, 30 August – 3 September 1999, Swets and Zeitlinger, Lisse, the Netherlands, 418–430, 2000.
20. M. Gretzschel, A. Jaschinski, Design of an active wheelset on a scaled roller rig, *Vehicle System Dynamics*, 41(5), 365–381, 2004.
21. B. Kurzeck, L. Valente, A novel mechatronic running gear: Concept, simulation and scaled roller rig testing, *9th World Congress on Railway Research* (WCRR), Lille, France, 22–26 May 2011.
22. P. Meinke, L. Mauer, Koppelrahmen-Laufdrehgestell fur ICE-Mittelwagen, *VDI-Berichte*, 634, 203–219, 1987.
23. A. Shebani, S. Iwnicki, Prediction of wheel and rail wear under different contact conditions using artificial neural networks, *Wear*, 406–407, 173–184, 2018.
24. Y. Zhao, B. Liang, S. Iwnicki, Friction coefficient estimation using an unscented Kalman filter, *Vehicle System Dynamics*, 52(S1), 220–234, 2014.

25. B. Liang, S.D. Iwnicki, Independently rotating wheels with induction motors for high-speed trains, *Journal of Control Science and Engineering*, 2011, 968286, 2011.
26. B. Liang, S. Iwnicki, G. Feng, A. Ball, V.T. Tran, R. Cattley, Railway wheel flat and rail surface defect detection by time-frequency analysis, *Chemical Engineering Transactions*, 33, 745–750, 2013.
27. B. Liang, S. Iwnicki, A. Ball, A.E. Young, Adaptive noise cancelling and time-frequency techniques for rail surface defect detection, *Mechanical Systems and Signal Processing*, 54, 41–51, 2015.
28. L. Mauer, P. Meinke, Requirements of future high-speed running gears, *RTR Special Railway Technical Review*, 1993.
29. N. Bosso, A. Gugliotta, A. Somà, Dynamic identification of a 1:5 scaled railway bogie on roller rig, *WIT Transactions on the Built Environment*, 88, 829–838, 2006.
30. N. Bosso, A. Gugliotta, A. Somà, Design and simulation of railway vehicles braking operation using a scaled roller-rig, *WIT Transactions on the Built Environment*, 88, 869–883, 2006.
31. N. Bosso, A. Gugliotta, A. Somà, Simulation of narrow gauge railway vehicles and experimental validation by mean of scaled tests on roller rig, Meccanica, 43(2), 211–223, 2008.
32. N. Bosso, A. Gugliotta, N. Zampieri, Study of adhesion and evaluation of the friction forces using a scaled roller-rig, *Proceedings 5th World Tribology Congress*, 8–13 September 2013, Politecnico di Torino (DIMEAS), Torino, Italy, 2640–2643, 2014.
33. N. Bosso, A. Gugliotta, N. Zampieri, Strategies to simulate wheel-rail adhesion in degraded conditions using a roller-rig. *Vehicle System Dynamics*, 53(5), 619–634, 2015.
34. N. Bosso, A. Gugliotta, N. Zampieri, A test rig for multi-wheelset adhesion experiments, In J. Pombo (Ed.), *Proceedings 3rd International Conference on Railway Technology: Research, Development and Maintenance, Civil-Comp Proceedings*, 110, 223, 2016.
35. N. Bosso, M. Spiryagin, A. Gugliotta, A. Somà, *Mechatronic Modeling of Real-Time Wheel-Rail Contact*, Springer-Verlag, Berlin, Germany, 2013.
36. N. Bosso, N. Zampieri, Real-time implementation of a traction control algorithm on a scaled roller rig, *Vehicle System Dynamics*, 51(4), 517–541, 2013.
37. N. Bosso, N. Zampieri, Experimental and numerical simulation of wheel-rail adhesion and wear using a scaled roller rig and a real-time contact code, *Shock and Vibration*, 2014, 385018, 2014.
38. S.Z. Meymand, M. Ahmadian, Design, development, and calibration of a force-moment measurement system for wheel-rail contact mechanics in roller rigs, *Measurement*, 81, 113–122, 2016.
39. S.Z. Meymand, M. Hosseinipour, M. Ahmadian, The development of a roller rig for experimental evaluation of contact mechanics for railway vehicles, *Proceedings ASME 2015 Joint Rail Conference*, 23–26 March 2015, San Jose, CA, JRC2015-5721, V001T10A007, 2015.
40. S.Z. Meymand, M. Taheri, M. Hosseinipour, M. Ahmadian, Vibration analysis of a coupled multi-body dynamic model of a contact mechanics roller rig, *Proceedings ASME 2016 Joint Rail Conference*, 12–15 April 2016, Columbia, SC, JRC2016-5813, V001T10A005, 2016.
41. M. Kieninger, A. Rupp, T. Gerlach, Ein Neuer Radsensor zur Erfassung des multiaxialen Belastungsgeschehens an schienengebundenen Nahverkehrsfahrzeugen, *ZEVrail Glasers Annalen*, 131, TB102–TB109, 2007.
42. J. Kalivoda, P. Bauer, Scaled roller rig experiments with a mechatronic bogie, In: J. Pombo (Ed.), *Proceedings 2nd International Conference on Railway Technology: Research*, Development and Maintenance, Civil-Comp Proceedings, 104, 317, 2014.
43. J. Kalivoda, P. Bauer, Roller rig tests with active stabilization of a two-axle bogie, In: J. Pombo (Ed.), *Proceedings 3rd International Conference on Railway Technology: Research, Development and Maintenance*, Civil-Comp Proceedings, 110, 96, 2016.
44. J. Kalivoda, P. Bauer, Mechatronic bogie for roller rig tests, In: M. Rosenberger, M. Plöchl, K. Six, J. Edelmann (Eds.), T*he Dynamics of Vehicles on Roads and Tracks: Proceedings 24th Symposium of the International Association for Vehicle System Dynamics held in Graz*, Austria, 17–21 August 2015, CRC Press/Balkema, Leiden, the Netherlands, 899–908, 2016.
45. J.F. Aceituno, R. Chamorro, D. García-Vallejo, J.L. Escalona, On the design of a scaled railroad vehicle for the validation of computational models, *Mechanism and Machine Theory*, 115, 60–76, 2017.
46. P.D. Temple, M. Harmon, R. Lewis, M.C. Burstow, B. Temple, D. Jones, Optimisation of grease application to railway tracks, *Journal of Rail and Rapid Transit*, 232(5), 1514–1527, 2018.
47. M. Naeimi, R. Dollevoet, Preliminary results on multi-body dynamic simulation of a new test rig for wheel-rail contact, *Advances in Railway Engineering*, 3(1), 27–37, 2015.
48. M. Naeimi, Z. Li, R. Dollevoet, Scaling strategy of a new experimental rig for wheel-rail contact, *International Journal of Mechanical, Aerospace, Industrial and Mechatronics Engineering*, 8(12), 1787–1794, 2014.

49. M. Naeimi, Z. Li, R.H. Petrov, J. Sietsma, R. Dollevoet, Development of a new downscale setup for wheel-rail contact experiments under impact loading conditions, *Experimental Techniques*, 42(1), 1–17, 2018.
50. S.Z. Meymand, M.J. Craft, M. Ahmadian, On the application of roller rigs for studying rail vehicle systems, in *Proceedings ASME Rail Transportation Division Fall Technical Conference*, 15–17 October 2013, Altoona, PA, RTDF2013-4724, V001T01A015, 2013.
51. S. Myamlin, J. Kalivoda, L. Neduzha, Testing of railway vehicles using roller rigs, *Procedia Engineering*, 187, 688–695, 2017.
52. P.D. Allen, S.D. Iwnicki, The critical speed of a railway vehicle on a roller rig, *Journal of Rail and Rapid Transit*, 215(2), 55–64, 2001.
53. N. Bosso, A. Gugliotta, A. Somà, Dynamic behavior of a railway wheelset on a roller rig versus tangent track, *Shock and Vibration*, 11(3–4), 467–492, 2004.
54. J.P. Meijaard, The motion of a railway wheelset on a track or on a roller rig, *Procedia IUTAM*, 19, 274–281, 2016.
55. H. Yoshino, T. Hosoya, H. Yabuno, S. Lin, Y. Suda, Theoretical and experimental analyses on stabilization of hunting motion by utilizing the traction motor as a passive gyroscopic damper, *Journal of Rail and Rapid Transit*, 229(4), 395–401, 2015.
56. H.M. Hur, J.H. Park, W.H. You, T.W. Park, A study on the critical speed of worn wheel profile using a scale model, *Journal of Mechanical Science and Technology*, 23(10), 2790–2800, 2009.
57. N. Bosso, A. Gugliotta, N. Zampieri, Wheel flat detection algorithm for onboard diagnostic, *Measurement*, 123, 193–202, 2018.
58. C. Mizrak, I. Esen, Determining effects of wagon mass and vehicle velocity on vertical vibrations of a rail vehicle moving with a constant acceleration on a bridge using experimental and numerical methods, *Shock and Vibration*, 2015, 183450, 2015.
59. W. Wang, Y. Suda, Y. Michitsuji, Running performance of steering truck with independently rotating wheel considering traction and braking, *Vehicle System Dynamics*, 46(S1), 899–909, 2008.
60. H. Kono, Y. Suda, M. Yamaguchi, K. Takasaki, Y. Hironaka, K. Tsuda, Curving simulation and scaled model test for LRT vehicles, *Transactions of the Japan Society of Mechanical Engineers, Part C*, 72(724), 3899–3904, 2006.
61. B.G. Eom, B.B. Kang, H.S. Lee, A study on running stability assessment methods for 1/5 small scaled bogie of saemaul using small-scaled derailment simulator, *International Journal of Precision Engineering and Manufacturing*, 14(4), 589–598, 2013.
62. Y. Nagumo, K. Tanifuji, J. Imai, Basic study of wheel flange climbing using model wheelset, *Transactions of the Japan Society of Mechanical Engineers, Part C*, 74(738), 242–249, 2008.
63. K. Tanifuji, H. Kutsukake, J. Imai, Experimental study on flange-climb derailment using a scaled truck with single wheel-set, in Z. Zobory (Ed.), *Proceedings 12th Mini Conference on Vehicle System Dynamics, Identification and Anomalies*, 8–10 November 2010, Budapest University of Technology and Economics, Budapest, Hungary, 89–98, 2010.
64. B.G. Eom, B.B. Kang, H.S. Lee, Design of small-scaled derailment simulator for investigating bogie dynamics, *International Journal of Railway*, 4(2), 50–55, 2011.
65. Y.J. Shin, W.H. You, H.M. Hur, J.H. Park, H∞ control of railway vehicle suspension with MR damper using scaled roller rig. *Smart Materials and Structures*, 23(9), 095023, 2014.
66. Y.J. Shin, W.H. You, H.M. Hur, J.H. Park, G.S. Lee, Improvement of ride quality of railway vehicle by semiactive secondary suspension system on roller rig using magnetorheological damper, *Advances in Mechanical Engineering*, 2014, 298382, 2014.
67. W.H. You, J.H. Park, H.M. Hur, Y.J. Shin, Comparison of simulation and experimental results of railway vehicle dynamics by using 1/5 scale model, *19th International Congress on Sound and Vibration held in Vilnius*, Lithuania, 8–12 July 2012, International Institute of Acoustics & Vibration, Auburn, AL, 3348–3354, 2012.
68. S.S. Hsu, Z. Huang, S.D. Iwnicki, D.J. Thompson, C.J.C. Jones, G. Xie, P.D. Allen, Experimental and theoretical investigation of railway wheel squeal, *Journal of Rail and Rapid Transit*, 221(1), 59–73, 2007.
69. Y. Zhu, Y. Lyu, U. Olofsson, Mapping the friction between railway wheels and rails focusing on environmental conditions, *Wear*, 324–325, 122–128, 2015.
70. A. Khalladi, K. Elleuch, Tribological behavior of wheel-rail contact under different contaminants using pin-on-disk methodology, *Journal of Tribology*, 139(1), 011102, 2017.
71. R. Galas, M. Omasta, I. Krupka, M. Hartl, Laboratory investigation of ability of oil-based friction modifiers to control adhesion at wheel-rail interface, *Wear*, 368–369, 230–238, 2016.

72. H. Liu, Y. Cha, U. Olofsson, Effect of the sliding velocity on the size and amount of airborne wear particles generated from dry sliding wheel-rail contacts, *Tribology Letters*, 63(3), 30, 2016.
73. O. Arias-Cuevas, Z. Li, R. Lewis, E.A. Gallardo-Hernández, Laboratory investigation of some sanding parameters to improve the adhesion in leaf-contaminated wheel-rail contacts, *Journal of Rail and Rapid Transit*, 224(3), 139–157, 2010.
74. U. Olofsson, K. Sundvall, Influence of leaf, humidity and applied lubrication on friction in the wheel-rail contact: Pin-on-disc experiments, *Journal of Rail and Rapid Transit*, 218(3), 235–242, 2004.
75. G.W.G. Poll, D. Wang, Fluid rheology, traction/creep relationships and friction in machine elements with rolling contacts, *Journal of Engineering Tribology*, 226(6), 481–500, 2012.
76. O. Arias-Cuevas, Z. Li, R. Lewis, A laboratory investigation on the influence of the particle size and slip during sanding on the adhesion and wear in the wheel-rail contact, *Wear*, 271(1–2), 14–24, 2011.
77. O. Arias-Cuevas, Z. Li, R. Lewis, E.A. Gallardo-Hernández, Rolling-sliding laboratory tests of friction modifiers in dry and wet wheel-rail contacts, *Wear*, 268(3–4), 543–551, 2010.
78. O. Arias-Cuevas, Z. Li, R. Lewis, Investigating the lubricity and electrical insulation caused by sanding in dry wheel-rail contacts, *Tribology Letters*, 37(3), 623–635, 2010.
79. W. Zhang, J. Chen, X. Wu, X. Jin, Wheel/rail adhesion and analysis by using full scale roller rig, *Wear*, 253(1–2), 82–88, 2002.
80. Y. Zhao, B. Liang, Re-adhesion control for a railway single wheelset test rig based on the behaviour of the traction motor, *Vehicle System Dynamics*, 51(8), 1173–1185, 2013.
81. R. Stock, D.T. Eadie, D. Elvidge, K. Oldknow, Influencing rolling contact fatigue through top of rail friction modifier application – a full scale wheel-rail test rig study, *Wear*, 271(1–2), 134–142, 2011.
82. K. Matsumoto, Y. Suda, T. Fujii, H. Komine, M. Tomeoka, Y. Satoh, T. Nakai, M. Tanimoto, Y. Kishimoto, The optimum design of an onboard friction control system between wheel and rail in a railway system for improved curving negotiation, *Vehicle System Dynamics*, 44(S1), 531–540, 2006.
83. M. Covino, M.L. Grassi, E. Pagano, Traction electric drives: An indirect identification method of friction forces, *International Electric Machines and Drives Conference Record*, 18–21 May 1997, Milwaukee, WI, IEEE, Piscataway, NJ, TA2/5.1-TA2/5.3, 1997.
84. N. Bosso, A. Gugliotta, N. Zampieri, RTCONTACT: An efficient wheel-rail contact algorithm for real-time dynamic simulations, *Proceedings ASME 2012 Joint Rail Conference*, 17–19 April 2012, Philadelphia, PA, JRC2012-74044, 195–204, 2012.
85. Y.J. Shin, W.H. You, H.M. Hur, J.H. Park, Semi-active control to reduce carbody vibration of railway vehicle by using scaled roller rig, *Journal of Mechanical Science and Technology*, 26(11), 3423–3431, 2012.
86. J.S. Oh, Y.J. Shin, H.W. Koo, H.C. Kim, J. Park, S.B. Choi, Vibration control of a semi-active railway vehicle suspension with magneto-rheological dampers, *Advances in Mechanical Engineering*, 8(4), 1–13, 2016.
87. B. Allotta, L. Pugi, M. Malvezzi, F. Bartolini, F. Cangioli, A scaled roller test rig for high-speed vehicles, *Vehicle System Dynamics*, 48(S1), 3–18, 2010.
88. A. Jaschinski, H. Chollet, S. Iwnicki, A. Wickens, J. Würzen, The application of roller rigs to railway vehicle dynamics, *Vehicle System Dynamics*, 31(5–6), 345–392, 1999.

21 Railway Vehicle Dynamics Glossary

Tim McSweeney

CONTENTS

21.1 Vagaries of Railway Terminology and Jargon 869
21.2 Glossary 869

21.1 VAGARIES OF RAILWAY TERMINOLOGY AND JARGON

There appears to be as much national and regional variation in the technical and operational terminology used by different railway systems as there is in the multitude of track gauges that have been utilised around the globe. One prominent difference is in the use of the word 'railway' by those whose systems originated from British-backed development versus the use of the word 'railroad' for most systems using American-based technology.

There are many similar British/American equivalent terms that an internationally engaged railway/railroad person will need to be aware of. Prominent examples of these include driver/engineer, wagon/car, bogie/truck, sleeper/tie, turnout/switch, vee/frog and fishplate/joint bar. However, every railroad company seems to have its own peculiarities in the way it refers to items of hardware, safeworking systems, etc. In addition to the following glossary, readers may find the Railway Technical Website at http://www.railway-technical.com useful in searching out the appropriate terminology relevant to an issue of interest.

21.2 GLOSSARY

60103: the ultimate example of railway vehicle dynamics, London and North Eastern Railway Class A3 4472 steam locomotive 'Flying Scotsman', which was built in 1923. It was the first steam locomotive to reach a speed of 100 miles per hour in 1934 and made a record non-stop run of 679 kilometres in Australia in 1989.

acceptance (of rollingstock): a requirement on most railway networks for safety-related field/laboratory tests and/or simulations using validated vehicle dynamics models prior to the introduction of new types of rollingstock or performing significant modifications to existing rollingstock.

active steering: improvement of the curving behaviour of bogies by means of a variety of mechanical and mechatronic control strategies to minimise wheelset lateral displacement and angle of attack so as to reduce lateral forces and wear at the wheel-rail interface.

active suspension: use of microprocessor control to detect changes in track geometry such as curves and gradients or responses to track irregularities and activate appropriate changes in suspension configuration. Lateral movements initiate stiffer hydraulic resistance in dampers or induce hydraulic jacking adjustment of body tilt.

adhesion coefficient: percentage or ratio of the total weight on the driving wheels of a locomotive that is available for traction or braking. The adhesion coefficient is dependent on the construction and operational characteristics of rail tracks and railway vehicles, for example, the difference between wheel diameters of wheel pairs, conicity and eccentricity of wheels, track curvature, reallocation of loads between wheels, irregular loads of wheels for a wheel pair or a bogie, vibrations and unaccounted for slipping motion. It can vary from as low as 10% (0.1) on wet rail to as high as 40% (0.4) on dry sanded rail.

aerodynamic resistance: the resistance to train/vehicle motion along the track due to aerodynamic drag and the effect of wind strength and direction relative to the direction of travel of the train/vehicle, including consideration of pressure variations when transiting tunnels and slipstream effects when passing other trains and platforms.

air suspension: modern passenger vehicles often employ an air suspension system, also referred to as air bags, air cushions or air springs, with the vehicle body supported on each bogie bolster by a set of hollow rubber bags. Compressed air fed into the bags under the control of a levelling valve maintains the correct body level. Air bags usually contain a solid rubber core on which the body will rest if the bag bursts. Air suspensions are also designed to allow for shear during movement through curves.

alignment of track (horizontal): the geometrical layout of a railway track in its plan view basically specified by the location/length of sections of tangent track (i.e., straights) and the radii/length of circular curves (with or without transition spirals) that connect them.

alignment of track (vertical): the changes in the elevation of a railway track along its longitudinal profile, specified by the gradients of uphill/downhill sections and the radii of the vertical curves that connect changes in gradient.

angle of attack: the angle that the direction of wheelset trajectory makes with the tangential direction of the running rail alignment.

automatic brake: see **train brake**.

autocoupler: see **coupler**.

axle: the part of a wheelset on which the two wheels are mounted. Wheels are usually press-fitted onto shoulders/seats machined near the ends of the axle.

axlebox: the housing that attaches each end of an axle to the bogie frames; each housing contains a bearing that allows the axle to rotate. They are usually box-shaped to fit the guides/openings in the bogie frames and constrain longitudinal and lateral movement in the **track plane**, while allowing vertical freedom.

balance speed: see **equilibrium speed**.

ballast: graded crushed stone that provides vertical, longitudinal and lateral support to the sleepers (ties), transfers rail vehicle wheel loads to the subgrade with a degree of elasticity and allows the track geometry to be adjusted by **tamping** and permits drainage of the track structure.

bogie: an assembly comprising wheels, axles, bearings, side frames, bolsters, brake rigging, springs and connecting components used to support rail vehicles (usually near their ends) and capable of rotation in the **track plane** to provide guidance along the track. A bogie may hold one, two or more wheelsets and may also provide support to adjacent ends of an articulated vehicle. Also referred to as a **truck**.

bogie bolster: the main transverse member of a conventional three-piece freight bogie (truck) that transmits rail vehicle body loads to the side frames through the suspension system/s. The ends of the bolster fit loosely into the side frames and are retained by the gibs that contact the side frame column guides. Bogie bolster contact with the vehicle body is through the bogie centre plate, which mates with the body centre plate, and through the side bearers when the vehicle is tilted, as in a curve.

bogie centre plate: the circular area centrally placed in the top surface of a bogie bolster that provides the principal bearing support to the vehicle body on the bolster via the vehicle body centre plate. Bogie centre plates are often fitted with a horizontal wear plate and a vertical wear ring to improve wearing characteristics and extend bogie's bolster life.

bogie hunting: lateral instability of a bogie (truck) at high speed, characterised by wheelsets oscillating from side to side, with the flanges striking the rail gauge face. The resulting motion of the wagon causes excessive wear in wagon and bogie components and creates potentially unsafe operating conditions.

bogies, radial: rail vehicle bogies whose interconnected wheelsets have low yaw constraint links with the bogie frame so as to allow each of the wheelsets to individually align itself to the radius of a track curve.

brake pipe: on trains fitted with **pneumatic brakes**, the brake pipe connects the driver's brake valve on the locomotive with the brake cylinder on each of the vehicles in the train. It includes the branch pipes, valves, angle cocks, cut-out cocks, hoses and hose couplings used to distribute compressed air to charge those brake cylinders and control the release of air to apply braking. Also referred to as a **train line**.

broad gauge: railways with running rails spaced at more than the 1435 mm (4 ft 8½ in) standard track gauge.

buff forces: a term used to describe compressive coupler forces in a train, caused by run-in of slack from the rear end. The term 'buff' means the opposite of the term 'draft'.

bumpstop: a flexible but relatively stiff fitting to provide a limited level of impact absorption through progressive compression when the normal suspension approaches the full limit of its travel.

cant (of rail): rail cant involves tilting the tops of running rails in, towards each other, to assist rail vehicles to self-centre as they move along the track. Rail cant is usually expressed as a rate of inclination (commonly 1 in 20 or 1 in 40).

cant (of track): the cross level of track on a curve used to compensate for lateral forces generated by the train as it passes through the curve. Track cant is specified by the vertical difference in height of the outside (high) rail and the inside (low) rail measured at right angles to the centreline of the track. Track cant on a curve is also referred to as **cross level** or as **superelevation**. On straight sections of track, the cross level should be zero, except near the start and end of non-transitioned curves.

cant deficiency: the difference between the cant actually present on a curve and the equilibrium cant that should apply for speed higher than the balance speed at which a train is travelling as it passes through the curve. Similarly, cant excess occurs when trains travel through a curve at a speed lower than the balance speed.

cant rail: longitudinal vehicle member that forms the junction between the side of the vehicle body and the roof.

catenary: see **overhead traction wiring equipment**.

centre beam/sill: the central longitudinal member of a rail vehicle underframe structure that transmits draft and buff shocks from one end of the vehicle to the other.

centre pin: a large steel pin that passes through the centre plates on the vehicle body bolster and bogie bolster. The bogie rotates about the centre pin, and stress is taken by the centre plates.

coefficient of friction: see **friction coefficient**.

conicity (effective): new wheels have their tread tapered towards the field side at an inclination of 1 in 20 to 1 in 40, and new rails have a corresponding slope (also referred to as rail cant) of their railhead top surface. The effective conicity on each wheel of a wheelset varies depending on the actual worn shape of wheel and rail profiles and the lateral offset of the wheelset from the centreline of the track. Depending upon this latter factor, the wheel tread (with conicity as low as 1°–3°) or the wheel flange (with conicity as high as 70°) can be in contact with the railhead.

contact patch: the area of contact between a wheel tread and the railhead, generally considered to be elliptically shaped.

contact wire: the overhead traction power supply wire, sometimes referred to as trolley wire, against which the pantograph of an electric locomotive slides to collect its electrical current.

corrugation (of rails): wave-like undulations that develop and propagate along the top running surface of the head of the rail. Causes of this condition are not thought to be necessarily the same at individual locations. Short-wave corrugation, also known as 'roaring rail', has wavelengths of 25–75 mm and is most common on light axle load and high-speed operations. Intermediate wave corrugation with wavelengths of 75–600 mm is most common in heavy freight operations. Long-wave corrugation has wavelengths greater than 600 mm and is most common in very-high-speed operations.

coupler: the device at both ends of a rail vehicle to allow vehicles to be connected together in a train. Modern designs allow vehicles to be attached to each other simply by pushing them together; this arrangement is referred to as an automatic coupler or **autocoupler**.

creep: see **wheel creep**.

critical speed: the lowest speed at which a vehicle commences **bogie hunting** on tangent track that does not then decay away as speed increases.

cross level: see **cant (of track)**.

crosstie: see **sleeper**.

curving resistance: the additional rolling resistance imposed on the vehicles in a train as they transit a curve or curves in the track. Many parameters affect this resistance, including rollingstock design and condition, curve length/radius/cant, rail profile and degree of curve lubrication.

damper: a device used to control oscillations in the primary or secondary suspensions of a rail vehicle by energy dissipation.

draft forces: a term used to describe forces resulting in tension in the couplers of a train. The term 'draft' means the opposite of the term 'buff'.

draft gear: the term used to describe the energy-absorbing component of the draft system. The draft gear is installed in a yoke that is connected to the coupler shank and is fitted with follower blocks that contact the draft lugs on the rail vehicle centre sill. So-called 'standard' draft gear use rubber and/or friction components to provide energy absorption, while 'hydraulic' draft gear uses a closed hydraulic system with small ports and a piston to achieve a greater energy-absorbing capability. Hydraulic draft gear assemblies are generally referred to as 'cushioning units.'

draft system: the term used to describe the arrangement on a rail vehicle for transmitting coupler forces to the centre sill. On standard draft gear, the draft system includes the coupler, yoke, draft gear, follower, draft key, draft lugs and draft sill. On a vehicle with cushioning units, either hydraulic cushion units replace the draft gear and yoke at each end of the vehicle or a hydraulically controlled sliding centre sill is an integral part of the vehicle underframe.

drawbar force: the force exerted through the couplers by the locomotive/s on coupled wagons and by one wagon upon another along the train. This force is usually greatest at the coupler between a locomotive and the first wagon behind it.

dynamic braking: slowing a train by reconfiguring the powered-vehicle electric traction motors, so that they act as generators, with the energy thus produced then dissipated in on-board resistor grids (**rheostatic braking**) or returned to the external supply system via the catenary or third rail for use by other trains or the supplier (**regenerative braking**).

equilibrium speed: the speed of a train on a curve at which the wheel loads are equally distributed between the high and low rails, and the vehicle and its load (including passengers) experience no lateral force. It is dependent on the amount of cant installed on the curve. Also referred to as **balance speed**.

friction coefficient: a dimensionless scalar value, often symbolised by the Greek letter μ, which describes the ratio of the force of frictional resistance between two bodies and the force pressing them together. It is a system property that depends upon the materials involved, relative velocity of the bodies and interface issues, including geometric properties, temperature and lubrication state.

gauge corner: see **rail gauge corner**.

gauge, track: distance measured at right angles between the inside running (gauge) faces of the two rails of a track at a specified distance below the top of the railheads.

gauge, wide: any track gauge greater than a nominal design standard as a result of installation deficiencies, track component deterioration or wear of the rail. Some rail systems deliberately widen gauge slightly in small-radius curves to ease the passage of vehicle bogies and minimise rail and wheel wear.

grade or gradient: the percentage rise or fall of track over the horizontal longitudinal distance or the rate of inclination of track in relation to the horizontal. For example, a rise of 1 metre in 50 metres equals a grade of 2 percent and can also be specified as a gradient of 1 in 50.

hunting: see **bogie hunting** and **wheelset hunting**.

journal bearing: the general term used to describe the load-bearing arrangement at the ends of each axle of a rail vehicle bogie (truck). Modern designs involve roller bearings that are sealed assemblies of hardened steel rollers, races, cups and cones pressed onto axle journals and generally lubricated with grease to reduce rotational friction. Vertical loads are transferred from the journal bearing to the bogie sideframe through a device known as a roller-bearing adapter that fits between the bearing outer ring and the sideframe pedestal.

kinematic envelope: see **swept envelope**.

L/V ratio: the L/V ratio is defined as the ratio of the lateral force to the vertical force imposed by a rail vehicle wheel when flanging on a rail. When the ratio is greater than 1.0, there is a significant potential for the wheel to climb onto the railhead and derail. Also referred to as the **Y/Q ratio** in Europe.

MGT an abbreviation for Million Gross Tonnes, representing the tonnes of traffic load (including the rail vehicles' mass) that have passed over a railway section, expressed in millions and usually calculated on an annual basis.

narrow gauge: railways built to less than the 1435 mm (4 ft 8½ in) standard track gauge.

overhead traction wiring equipment: the system of support structures and wiring erected beside and over a track to provide electricity to electric locomotives, electric multiple units or trams. Also referred to as the **catenary**.

pantograph: a device fitted to the top of a locomotive or power car with which to collect traction current from the **overhead traction wiring equipment**. A spring or air pressure system is used to keep the pantograph raised and in constant contact with the catenary contact wire at any speed.

pitch (of vehicle body or bogie): rotation of the vehicle body or bogie about the lateral axis due to track irregularities or longitudinal dynamic action caused by in-train forces.

pneumatic brakes: braking systems that use compressed air to push blocks/shoes onto wheels or pads onto discs to slow or stop trains.

primary suspension: a system that connects the wheelsets to their bogie frames, designed to provide springing and damping in vertical, lateral and longitudinal directions so as to maintain the stability and ride quality of the vehicle as it is guided along the track. Primary suspension elements are usually stiffer than those in the **secondary suspension**. Many freight wagons do not have an effective primary suspension.

propulsion resistance: the sum of **rolling resistance** (excluding **curving resistance**) and **aerodynamic resistance** (excluding head or tail wind components).

radial steering: connecting the wheelsets into a bogie frame such as to allow them to maintain their alignment more easily with the radius of track curves.

rail: a rolled steel shape, most commonly a flat-bottom section, designed to be laid end to end in two parallel lines on sleepers (ties) or other suitable supports to form a track for the guidance of railway rollingstock.

rail creep: intermittent longitudinal movement of rails in track, caused by temperature changes and/or the adhesion forces imposed by trains during acceleration/braking.

rail gauge corner: the curved transition on the inner (gauge) side of a railhead joining the top surface of the rail and the rail gauge face.

rail gauge face: the side of the railhead that is located immediately below the gauge corner and contacts the wheel flanges to provide guidance along the track.

rail lubrication: the application of lubricant onto the rail gauge face and/or wheel flange to reduce the friction between them.

rail neutral temperature: the optimum temperature at which continuous welded rail is installed and anchored with no axial force in the rail so as to minimise the stresses that occur at the extreme ends of the ambient temperature range.

rail web: the vertical section of a rail that joins the railhead to the foot and provides beam strength to support rail vehicle loads between adjacent sleepers.

railhead: the top of the rail on which rollingstock wheels are guided. The railhead also accepts the weight from rollingstock in a very small area at each wheel-rail contact point.

roll (of vehicle or vehicle body): rotation about the longitudinal axis due to designed track geometry variations during curving, track geometry irregularities at any location, the effects of cross-winds and lateral forces due to lateral components of in-train forces on curves.

roller bearing: a sealed assembly of hardened steel rollers and associated components lubricated with grease and pressed onto the axle journals to transfer the loading imposed by the sprung mass of the vehicle onto the wheelsets while allowing them to rotate.

rolling contact fatigue (RCF): the process whereby the high forces in the contact patch cause the development of cracks that penetrate into the railhead or into the wheel tread surface. The cracks can grow just underneath the surface, causing flaking and spalling, or they can 'turn down' and lead to rail breaks or severe railhead and wheel damage.

rolling resistance: resistance to vehicle motion along the track due to wheel-rail contact friction and journal friction. This excludes the additional resistance experienced on curves, defined as **curving resistance**.

regenerative braking: see **dynamic braking**.

rheostatic braking: see **dynamic braking**.

roadbed: see **subgrade**.

rolling contact fatigue: the process whereby extreme contact pressures at the wheel-rail interface initiate the development of surface cracks in the railhead and/or the wheel tread that can grow at shallow depths and result in head checks, shelling, spalling, rail squats, etc., or that penetrate deeper and can result in crushed railheads, broken rails or severe wheel defects.

rolling radius differential: the difference in the effective wheel radii at contact points between the wheel tread and railhead on the low rail versus the high rail, accomplished by tapered wheels. When in curves, the wheel flange on the high rail is up against the gauge face, with the wheel flange on the low rail pulled away from the low rail gauge face. This action results in a larger radius contact point on the wheel contacting the high rail, thereby inducing a steering effect of wheelsets through curves. In addition, wheel wear and rail wear are minimised due to a reduction in wheel slip.

rollingstock: a general term for any wheeled item of equipment that operates on a railway track.

running gear: a general term used to describe the components that facilitate movement of a rail vehicle. Running gear includes the wheels, axle, bearings, suspension system and other components of the bogies (trucks).

secondary suspension: connects the bogie frames to the vehicle body (H-frame bogies) or connects the sideframes to the bolster (three-piece bogies) and provides stiffness and damping to the vertical, yaw and roll motions of the vehicle body.

self-steering (bogie): bogies designed to allow the wheelsets to align to an almost radial position when transiting a curve.

side bearer (of wagon): a component located on the vehicle body bolster and interacting with its complementary bogie side bearing to absorb vertical loads arising from the rocking motion of the body. Side bearings vary from simple flat pads to complex devices that maintain constant contact between the bogie bolster and the vehicle body.

side bearer (of bogie): a plate, block, roller or elastic component attached to the top of a bogie bolster on both sides of the centre plate and interacting with its complementary vehicle body side bearing to provide support when variations in track geometry cause the body to rock transversely on the centre plates.

sideframe: in the conventional three-piece freight bogie (truck), the heavy cast steel side member that is designed to transmit vertical loads from the wheels through either journal boxes or pedestals to the bogie bolster.

side sill: longitudinal member/s placed along both sides of a rail vehicle load-bearing frame structure to provide support for the vehicle body.

sleeper: the component of the track structure generally placed perpendicular to the rails to hold track gauge, distribute the weight of the rails and rollingstock and hold the track onto its correct surface and alignment in the ballast. Materials commonly used in the manufacture of sleepers include timber, concrete and steel. A sleeper is also referred to as a **tie** or **crosstie**.

standard gauge: the standard distance between rails used by about two-thirds of the world's railways, being 1435 mm (4 ft 8½ in), measured between the inside faces of the railheads.

structure gauge (dynamic): see **swept envelope**.

subgrade: the natural formation material and placed soil (fill material) upon which, once these are compacted, the railway track structure is constructed. Also referred to as the **roadbed**.

superelevation: see **cant (of track)**.

suspension: the resilient system through which a rail vehicle body is supported on its wheels. Suspension systems involve the use of hydraulic devices, friction elements and coil, elliptic, rubber or pneumatic springs. See also **primary suspension** and **secondary suspension**.

sway: the combination of lateral displacement and roll displacement of a vehicle body on its suspension.

swept envelope: also referred to as a kinematic envelope; this defines the maximum extent of the three-dimensional space around a railway track that, allowing for dynamic effects, can potentially be occupied by a rail vehicle, including pantograph/s where installed, travelling at a nominated speed, given its static dimensions (including bogie centres and related centre and end overhangs), the bounce, sway and lateral displacement allowed by the vehicle components and suspension system, and the curvature/cross level of the track and any allowable irregularities in same at the given location.

swing bolster: a bogie (truck) bolster suspended by hangers or links so that it can swing laterally in relation to the bogie and thus lower the effects of lateral impact received through the sideframes and wheels. Bogies equipped with swing bolster are known as swing motion bogies.

switch: see **turnout**.

tamp/tamping: the process of compacting ballast under sleepers (ties) to provide uniform load bearing under the rails and to correct horizontal and/or vertical track alignment deficiencies.

third rail system: traction current supply system using an additional rail beside or between the two running rails to transmit a direct current electrical supply via collector shoes attached to the train.

tie: short for **crosstie**. See **sleeper**.

track defect: an anomaly in any part of the track structure requiring repair or other action such as an operating speed reduction.

track plane: the plane along the longitudinal track alignment formed by joining the top of the two running rails. This plane will be canted on curved track that has **superelevation** applied but will otherwise be a horizontal plane.

track-train dynamics: the study of the motions and resulting forces that occur during the movement of a train over a track under varying conditions of speed, train makeup, track and equipment conditions, grades, curves and train handling.

track twist: the difference in track cant or cross level measured over a specified distance along the track. Also referred to as **warp**.

train brake: the combined brakes on locomotives and wagons that provide the means of controlling the speed and stopping the entire train. Also referred to as **automatic brake**.

train configuration: the composition of vehicles that make up the complete train, including the locomotive/s.

train line: see **brake pipe**.

tread: see **wheel tread**.

tribometer: a device for measuring the level of adhesion between wheel and rail.

truck: see **bogie**.

turnout: the junction where tracks diverge or converge, comprising a pair of switch blades (also called points) and a crossing (also called a frog or vee because of the shape when viewed from above) with guard rails. The frog or crossing vee allows wheels to cross from one track to another and can either be fabricated from rails and blocks or manufactured from a casting. North American railroads refer to the complete **turnout** simply as a **switch**.

vertical bounce: instability at high speed, where the vehicle oscillates vertically on the suspension system.

warp: see **track twist**.

wheel: the cast or forged steel cylindrical element that rolls along the rail, carries the mass and provides guidance for rail vehicles. Pairs of railway wheels are mounted on a steel axle and are designed with flanges and a tapered (or occasionally cylindrical) tread to provide for operations on track of a specific gauge.

wheel burn: damage to the rail and/or wheel resulting in metal flow and/or discoloration due to heat from their frictional contact.

wheel creep: an operating condition where the wheel is neither purely rolling on the rail nor purely slipping on the rail. The friction coefficient between wheel and rail is greatest at this transition between purely rolling and purely slipping.

wheel flange: the tapered projection extending completely around the inner rim of a railway wheel, the function of which is to keep wheelsets on the track by limiting their lateral movement between the inside gauge faces of the running rails.

wheel flats: flat spots on rollingstock wheels resulting from sliding along the rail, generally found on all wheels on a wheelset or bogie due to severe braking or failure to release handbrakes.

wheel profiling: the process of restoring the desired contour of rail vehicle wheels by rotating the wheelsets in a wheel lathe to remove metal under precise control.

wheelset: a pair of wheels mounted on an axle.

wheelset hunting: lateral instability of a wheelset characterised by oscillating from side to side, with the flanges striking the rail gauge face.

wheel slide: where the wheel does not rotate on its axis, and motion exists at the area of contact between the wheel and the rail, usually caused by over-braking during poor-adhesion conditions. It is a common cause of **wheel flats**.

wheel slip: where a wheel rotates on its axis, but relative motion occurs between the wheel and rail at their point of contact. Wheel rotation speed during wheel slip is greater than that during rolling. This phenomenon is caused on a powered rail vehicle by over-application of power to the drive system relative to the available adhesion. Modern creep-control systems using microprocessors permit some limited degree of slip, as this has been proven to improve acceleration efficiency.

wheel tread: the normally slightly tapered (but occasionally cylindrical) circumferential external surface of a railway wheel that bears on the top surface of the rail and also serves as a brake drum on rail vehicles with conventional bogie-mounted brake rigging.

yaw: rotation of a vehicle body, bogie or wheelset about the axis perpendicular to the **track plane**.

Y/Q ratio: see **L/V ratio**.

Index

Note: Page numbers in italic and bold refer to figures and tables, respectively.

A

Abbreviated Injury Scale (AIS) Code, **512**
absolute damping, 592, *593*
absolute gauging process, 346
accelerometer(s)
 -based control strategies, 588
 capacitive, *730*, 730–731
 centre-mounted compression, *728*
 MEMS sensors, 731
 mountings/cabling, *730*
 piezoelectric, 727–730, *728*
acoustics basics, 523–524
AC tachogenerator, 739, *739*
active lateral damping, 602
actively controlled bogies, 84
active primary suspensions
 Bombardier mechatronic bogie, 607
 concepts and requirements, 600–601
 configurations and control strategies, 601–605
 independently driven wheels, 607–609
 technology, 605–606
active secondary suspensions
 concepts, 591
 configurations, 591–592
 control strategies, 592–597
 servo-hydraulic active lateral suspension, 598, *599*
 Shinkansen/Sumitomo bogie, 599–600
 Swedish Green Train Project, 600
active suspensions
 concepts, 580–581
 controllers, 606
 design and simulation, 581–583, *583*
 primary, 600–609
 secondary, 591–600
 semi-active approach, 581
 tilting trains, 584–591
active yaw damping, 601, *602*
actuation system, 809–810
actuator(s), 591, 597–598
 configurations, active steering, *601*
 control, 606
 response, 596
adhesion, 765, 767, 793, 857–860
 coefficient, 765, 793
 loss, 294
 traction control systems, 157–158, *158*
aerodynamic(s)
 drag coefficient, 446–447, **447**
 forces, 423
 ground effect, 422, 430, 434
 head pressure wave, 443
 moving-model test, *424*, 424–425
 numerical simulation, 426–430, *431*
 pitching/yawing/overturning moment, 448, 450, **450**
 pressure wave, trains crossing, 425, 436, *438*, 443–445, *445*, 452
 real-vehicle test, 425, *425*
 tail pressure wave, 443
 wind tunnel test, 422–424
aerodynamic loads, 416, *416*, 421
 distributive load on vehicle surface, 442–445
 forces and moments, 446–450
 influence factors, 439–442
 open air, 452, **452**
 on pantograph, 646
 railway tunnels, **452**, 452–453, **453**
 safe distance, human body, 453, **453**, 454
aerodynamic noise, 524, 767
 control of, 548
 measurement methods, 547, *548*
 prediction methods, 547–548
 sources, 547
air bag compensation, 358
airborne particles emission from wheel-rail contact, 298–299
air brake systems, *152*, 152–153
air-conditioning, noise, 549
air springs, 209
 active, 221
 auxiliary chamber, 212, *213*
 connection pipe, 212, *212*
 dynamic stiffness, oscillation amplitudes, 211, *211*
 equivalent mechanical models, 214–216
 emergency spring, 199, 209, *209*, 211, 215
 horizontal plane, 218–219
 installation, *209*
 orifice plates, connection pipe, 212–213, *213*
 parameters, 219–221
 semi-active, 221
 suspensions, 210–211, *210*
 thermodynamic models, 216–218, *217*
 vertical stiffness, wavelengths, *661*
air suspension, 350, 358
AIS (Abbreviated Injury Scale) Code, **512**
Alicona profilometer, 860
alternating current (AC) electric locomotive, 125, *126*
American non-rigid type E coupler, 93, *94*
American type F semi-rigid automatic coupler, 93, *94*
Ansaldo roller rig, **781**
antialias filtering, 747
anti-roll mechanism, *570*
arcing detection, OCL, 638
arm linkage, bogie, *571*
articulated trains, 29–30
articulated vehicles, 44–47
ATLAS roller rig, **778**, 801–808
autocouplers, 473–474
automatic couplers, 51–53, *52*, 92–95
autonomous rolling stock, 116

879

auxiliary suspensions, 83–86
axle
    loads, *159*, 159–160, *160*
    method, 736–737
    sum L/V ratio, 381
    traction control, 157
axlebox(es)
    advantages, 77
    and bogie frame, cylindrical guides, *229*, 229–230, *230*
    cartridge-type roller bearings, 77, *78*
    horn guides, 229, *229*
    -mounted vibration monitoring system, *743*
    plain bearings, *75*, 75–76
    roller bearings, 76, *76*
    spherical bearings, 76, *77*

B

ballasted track, 365
    ballast motion, 314–315, *315*
    rail motion, 312–313, *313*
    and slab track, parameters, **557**
    sleeper motion, 313–314, *314*
    three-dimensional dynamic model, 311, *312*
ballastless track, *308*, 308–309, *309*, 316, 319, 329–330, 332
ballast mats, 566
Barber S-2 bogie, 79, *80*
BASS 501 method, 350, 355–356
    vs. MBS, 362–363, *363*
beamforming, 547
beam links, 230, *231*
Berg model, 215, *215*, *216*, 218–219, *219*
bifurcation diagram, subcritical/supercritical Hopf, *691*
body build tolerance (BOD), 358
body modes tests, 758
bogie constraints
    axlebox, horn guides, 229, *229*
    beam links, 230, *231*
    cylindrical guides, axlebox, *229*, 229–230, *230*
    radius links, *231*, 231–232
    traction rods, 232, *232*
    trailing (radial) arms, Y32 bogie, 232, *232*
bogies, 136–138, 839
    car body connections, 110–112
    characteristics, 51
    classification criteria, 157
    development, 10
    dynamic performance, 800
    forced steering, 27–28
    frame modelling, 656
    functioning, axles, 156
    generic configuration, *27*
    hunting, 11, 13, 243, 253, 275, 685
    instability, 685
    Jacob shared bogie, 30
    lateral stiffness, 406
    oscillation, 11
    powered, 19
    rotational resistance test, 755–757, *757*
    three-piece, 28–29, 79–81
    and two-axle vehicles, 26–27
    two-axle wagon, 49, *50*
    vehicles, 12, 568–569, *569*
    warp/rotational resistance, 407

bolster-based tilting, 586
bolsterless bogie, 236, *237*
Bombardier mechatronic bogie, 607
brake systems, 151, 799
    adhesion, 152
    air, *152*, 152–153
    dynamic, 153–154, *154*
    electrodynamic braking, fixed slip, 806–807, *807*
    electromagnetic, 155, *155*
    emergency application, 471, *471*
    non-adhesion, 152
    rail, 156
    types, 152
braking equipment, freight wagon, 53, *53*
braking test
    degraded adhesion conditions, 812–813
    good-adhesion conditions, 812
BU 300 roller rig, **780**
buffers, 98–100, *101*, *102*
buff strength, 118
bumpstop, 198, 228–232, 658–665

C

Calipri, 750, *750*
car body(ies), 45, 135–136
    accelerations, influence, *656*
    carcass type, 57–58, *59*
    double deck passenger coach, 48, *48*
    eigenbehaviour modes, 683
    elastic behaviour, 48, *49*
    hunting, 685
    instability, 685, *689*
    passenger coaches, 48, 54–59
    structure, 572
    sway movement, *684*
    types, 47, *47*
car body to bogie connections
    bolsterless bogie, 236, *237*
    bolster with centre plate, 28, 110, 233
    centre column, 234–235, *235*
    flat centre bowl and side bearings, *233*, 233–234
    pendulum linkage, 236, *236*
    spherical centre bowl and side bearings, 234, *234*
    Watts linkage, 235, *235*
CARS roller rig, **773–774**
Carter
    bogie hunting, 13
    creep, 14–15
    equations of motion, 14–15, 17
    theoretical contributions, 17
    theory of dynamic stability, 15
    wheel-rail forces, 13
Cartesian grid, 429, *430*
catenary models
    damping, 625–626
    droppers, 624–625, *625*
    FEM, 624
    finite difference models, 622–624, *623*
    lumped parameter, 622, *622*, *623*
    static position, 626
catenary wear and damage, 642
    crack propagation, 642
CFD, *see* computational fluid dynamics (CFD)

Index

Chengdu roller rigs, **774–775**, 784–792
circle theory, 266, 275–276
circular path test rig, *851*
clearance analysis, 347, 366–367
    probability distribution, *370*
closed-circuit wind tunnel, 422, *423*
coil compression helical springs
    analytical/numerical solutions, 201, *202*
    compact spring elements, 205, *206*
    definition, 199
    deformations, 201
    linear stiffness, 201
    parameters, 199
    PtP elements, 205, *206*
    shapes, 201, *201*
    steels, material properties, 201, **201**
    stiffness calculations (spring rate), 202–205
computational fluid dynamics (CFD), 426–428, *427*, 547, 646
concave wheel profiles, 266, *267*, *269*
conformal wheel-rail contact, 405
conical wheel profiles, *267*
coning, 6–8
constant contact side bearings, 234, *234*
constrained layer damping of wheels, 534, *535*
contact
    filter, 527, *527*
    jump, 251, 267, 271–272
contact angle function (CAF), *268*, 268–269
    multi-Hertzian contact, 271–272, *272*
contact forces
    CAF, *268*, 268–269
    dicone to wheelset, 266
    flange, 267, *268*
    friction and spin, 269–270, *270*
    gravitational stiffness, 266–267
    Hertzian multiple contacts, 271–272, *272*
    independent wheels, 270–271
    Nadal's formula, safety criterion, 270
    non-Hertzian contacts, 272–273, *273*
contact pressure
    CONTACT/Hertz, 283
    maximum, 283, *284*
    sliding velocity, 283, *284*
contact wire (CW), 614, 616
    and contact strip, wear of, 638–641, *639*
    uplift/overhead contact line motion, 638
controlling surface roughness, 534
co-simulation approach, 464–465, *465*
Coulomb's model, 253, *253*
couplers, 51–53; *see also* autocouplers; automatic couplers
    angles, 497–501, *498*, **500**
    forces, simulation results, *510*
cowl unit locomotive design, 135, *136*
creepages
    -creep force law, 274, *274*
    damping terms and stability, 256
    dynamic formulation, 256
    non-dimensional spin, 256
    quasi-static, 254–256
    reduced, 259
creep forces, 252, *252*
    C110, C220 stiffnesses, 259
    linear expressions, 253
    load dependence, 257
    reduced parameter, 259
    transversal force, 259
critical speed, 685
cross-level error, 366
cross level track irregularity, 770, *770*, **771**
crosswind
    safety test, pressure sensors, 425, *425*, *426*
    simulating influence, 712–713
current collection quality, OLC, 619–620, *620*
curve radius, 364
    vs. bogie pivot offset, 364, *365*
    vs. wheelset movement, 364, *364*
curve squeal
    noise generation, *543*, 543–546
    reducing, 546
curving, 767
    APT-E, 25, *25*
    equations of motion, 23
    forces, 355, *355*
    optimisation, 708–711
    properties, 702–704, *703*
    Redtenbacher's formula, rolling of coned wheelset, 8
    resistance, 469–470
    simulation, *703*, **704**
    track superelevation, 9
    two-axle research vehicle HSFV-1, 24, *24*
cylindrical contact, 248

D

dampers, 222–223
    friction, 225–228
    viscous, 223–225
damping
    coefficient, side bearing, 111, *111*
    draft gear, 53
    hydraulic/orifice, 87
    Lenoir link friction damper, *81*, 82
    oscillations, 52
    suspension, 108–110
data acquisition, *746*, 746–748, 787
dead bands, 198
decrowning, 854
delta shear design, 728
delta suspension, 616, *617*
density scaling factor, 828, 833
derailment, 30, 403–404, 767, 856
    avoidance, 2
    guardrails/restraining rails, installation, 406, *406*
    independently rotating wheels, 405
    lubrication and wheel-rail friction modification, 408
    process, *378*
    rates in United States, *375*
    switches and crossings, 409
    system monitoring, 409
    track geometry inspection/maintenance, 408–409
    track tests, 402–403
    train marshalling and handling, 408
    vehicle dynamic response/suspension, 406–407
    vehicle loading, 407
    vehicle-track interaction, 402
    wheel climb duration limit, 383–384, *384*
    wheel-rail contact parameters, 401–402
    wheel-rail profile optimisation, 404–405

design geometry, 352
Design Guide BASS 501, 350
Deutsche Bahn (DB)
    wheelset axle test rig, 777
    wheelset test rig, 776
diesel-electric locomotive, 460, *461*
diesel-electric rail traction vehicles
    AC-DC-AC topology, 149, *149*
    AC-DC topology, 147, *147*
diesel locomotives
    auxiliary equipment, 131
    design scheme, 130, *130*
    electric transmission system, 130
    hydraulic transmission systems, 131
    main frame, 129
    mechanical transmission systems, 130
    with two cabs, 130, *131*
diesel multiple unit (DMU), 118, 131–132, *132*
differential-algebraic equations (DAEs), 628
digital systems, 369
dipole-type sources, 547
direct current (DC) electric locomotive, 125, *125*
direct tilt, 585–586
discrete elastic rail-wheel contact (DEC), 669
distributive load, aerodynamic pressure
    open air with environmental wind, 444–445, *446*
    open air without environmental wind, *442*, 442–444, *443*
    vehicles running in tunnel, 444, *445*
DMU, *see* diesel multiple unit (DMU)
Doppler effect, 740
    speed sensor, *740*
double deck passenger coach, 48, *48*
double-span girder, 174, *174*, *175*, 176, *176*, *181*
double-span simple beam, *181*, 181–182
    multiple moving forces, *189*, 189–190
    single moving force, 189, *189*
double-stage draft gear, *488*
double-suspension bogies, 82–83
draft gear, 53, 95–98, 117
    characteristics, *493*, 493–494, *494*
    double-stage, *488*
    performance indicators, 95
drawbar systems, 474
draw hook/buffer systems, 473–474
driving motion forces, 10–11
driving system, 787
droppers model, OCL, 624–625, *625*
DVA (dynamic vibration absorber), 337–340
dynamic braking system, 153–154, *154*
dynamic considerations, vehicle, 354–355
    critical speeds, 356
    curving forces, 355, *355*
    loading effect, 357, *357*
    time factors, 358
    track roughness, 355–356
    vehicle height, 358
dynamic creepages, 256
dynamic displacement, track structures, 331, *332*
dynamic envelopes, 350
dynamic equations, girder, 179
dynamic interaction problem, 587
dynamic response, impact loads, 331–332
dynamic simulation test, 817
dynamic tests, 758–760
dynamic track simulator, 854
    contact force estimation, 855–856
    hunting stability, 854, *855*
dynamic vibration absorber (DVA), 337–340

E

earth-fixed reference system, 654
ECF (Extended Creep Force) model, 290–291
ECP brakes, *see* electronically controlled pneumatic (ECP) brakes
eddy current brakes, 155, *155*
EDS maglev, *see* electric dynamic suspension (EDS) maglev
effective conicity, 404
eigenbehaviour, simulation of, 684–685
eigenfrequencies, *657*, *658*
eigenmodes, 656, *657*, *658*
eigenvalue analysis, 664–665, 682–684
elastic behaviour, freight flat wagon, 48, *49*
elastic contact, 628
elastic elements
    air springs, 209–221
    coil compression helical springs, 199–206
    leaf springs, 221–222, *222*
    principal types, 199, **200**
    rubber springs, 206–209
elastic foundation beam model, 310
elastic shakedown, 288
elastic wheel, 142
elastomeric draft gear ZW-73, 97–98, *99*
electric arc welding, 70
electric dynamic suspension (EDS) maglev, 166, 169, 171
    levitation forces, 185–186, *186*
electric locomotives
    AC, 125, *126*
    DC, 125, *125*
    layout scheme, 125–126
    multi-system, 125–126, *127*
    with two cabs, 125, *126*
electric multiple unit (EMU), 118, 126–128
    high-speed train, 128, *129*
    mixed equipment locations, *128*
    speeds, 128
    train configurations, 127, *127*
electric power type, 117
electric rail traction vehicles
    AC-DC-AC topology, 149, *150*
    AC-DC topology, 148, *148*
    advantages, 122
    DC-AC topology, 149, *150*
    DC-DC topology, 147–148, *148*
    electric locomotives, 125–126
    EMU, 126–128
    hydraulic system, 122
    light rail vehicle, 122–124
    operational principle, electrified railway power supply, 121, *121*
    pneumatic system, 121
    production, 122
    traction motors, 122
    types, 121
electrodynamic braking, fixed slip, 806–807, *807*

Index

electromagnetic brakes, 155, *155*
electromagnetic compatibility tests, 812
electromagnetic levitation, 169
electromagnetic suspension (EMS) maglev, 166, 171, 187, 193
    levitation forces, 184–185, *185*
electro-mechanical wear, 638
Electromotive Division of General Motors (EMD), 379, 383, *383*
electronically controlled pneumatic (ECP) brakes, 471–472, *473*
electronic lubricators, 297
EMD, *see* Electromotive Division of General Motors (EMD)
EMS maglev, *see* electromagnetic suspension (EMS) maglev
EMU, *see* electric multiple unit (EMU)
end wall, passenger coach, 58, *58*
engine noise, 549
envelopes, gauging
    dynamic, 350
    kinematic, 349–350
environmental noise, 522
equation-based approach, 463, **463**, **464**
equations of motion, 182–183
    scaling problem, 829–830
equivalent conicity, 18, 22, 266, 686
    and kinematic hunting, 275
equivalent longitudinal axle guidance stiffness, *709*
equivalent mechanical models, air springs, 214–216
estimation technique, 605–606
ETH hydraulic combination shock-absorber, 100, *102*
ETR-460 bogie, 89, *89*
ETR-500 bogie, 87, *88*
European Standard BS-EN15273, 367
evacuated tube technology (ETT), 193
Extended Creep Force (ECF) model, 290–291

F

fan noise, 549
FASTSIM algorithm, 274
    discretised ellipse, 261–262, *263*
    extensions, 265
    linear contact forces, elastic coefficients, 263
    reduced creepages, 264–265
    stresses, 262–263, *265*
fatigue-decreased proficiency boundary (FDPB), 516–517
fault-tolerance studies, 606
feedback control system methods, 31–32
filtering process, 352
Finite Element Analysis (FEA) software, 671
finite element method (FEM), 528, 624
first buckling mode, OCL, *618*
flange climb derailments, 376
    application, 386–390
    coefficient, friction, 387, *387*
    flange angle, 386, *386*
    flange length, 386, *387*
    independently rotating wheels, *388*, 388–389
    low-speed, 389
    safety criteria, 378–386
    vertical wheel unloading criteria, 389–390
    wheel climb process, 376–378, 501

flange contact, 267, *268*
    location, 379, *379*
flange lubrication, influence, *711*
flange wear, 74, *74*
flexible bodies, modelling of, 664, 671
flexible track model, 672
floating-slab track (FST), 566
    and DVA, 337–338
    and VAT, 337–340
Florence RRAC roller rig, **779**, 808–815
flow field
    around high-speed train, *432*
    distribution and generation, *434*
    distribution law and development characteristics, 431–433, 436–439
    rail vehicles crossing, 436, *437*
    single train passing through tunnel, 436, *438*
    trains passing in tunnel, *438*, 438–439
    transient wake structure, *435*
    vehicles under crosswind, 436, *437*
    vortex structure, 433–434
    wake flow characteristics, 434–436, *435*
    zones, 432–433
flutter analysis, 15, 545
force analysis
    ballast motion, 314–315, *315*
    rail motion, 312–313, *313*
    sleeper motion, 313–314, *314*
    track slab, 315–317, *316*
force-displacement curve
    Coulomb friction model, 226, *226*
    quasi-static loading and unloading, *198*
forced steering, 27–28
    mechanisms, 84, **84**
force-measuring wheelsets, 735–738
force *vs.* displacement graphs, harmonic displacement excitation, *660*
forward translational platform, *192*
four-axle locomotive, *689*
free slack, 458
freight bogie stability testing, 800–801
freight locomotives, 119
freight traction rolling stock, 119
freight train derailments, 375, **375**
freight wagons
    bogies design, 79–86
    car body structure, *60*, 60–61
    combined structure car body, 66–68, *67*
    connections, 70
    enclosed body designs, *61*, 61–62
    load-carrying underframes, 62–63, *64*
    loading devices, 68
    load-securing devices, 68–69
    materials, 69–70
    open body designs, 62, *63*
    tanks, 64–66, *65*, *66*
    transport, 59–60
frequency response function (FRF), 627, *633*
friction
    adhesion loss, 294
    coefficient, 54
    damping, 662
    ECF model, 290–291
    force, definition, 282, 290

friction (*Continued*)
  increasing adhesion, 295–296
  model, 659, *660*
  modification, 292–293
  modification factor, 489, *489*
  sliding bodies, 290
  traction and creep, 290, *291*
  wedge factor, 484, *485*
  wheel-rail friction conditions, 291–292
friction dampers
  advantage, 225–226
  classification, 226, **227**
  Coulomb model, force-displacement curve, 226, *226*
  disadvantages, 227
  force characteristics, 228, *228*
  linear spring in series, 226, *226*
  planar and spatial, 227
  telescopic, 227
friction draft gears, 481–489
  double-stage, *486*
  force characteristics, *96*
  Miner Crown SE, *97*
  wedge thrust, *96*, *97*
FST, *see* floating-slab track (FST)
'full-active' suspension, 581
full-scale roller rigs, 762–764, **765**, 826

G

Galerkin method, 23
gas turbine traction, 132–133, *133*
Gauge Commission, 11, 29
gauges
  geometric/swept, 347
  kinematic, 347–348
  pseudo-kinematic, 347
  static, 347
  track irregularity, 770, *770*, **771**
  UIC, 348–349
gauge widening derailments, *390*, 390–392
  hollow-worn wheels effect, 393–395, *394*, *395*
  rail gauge wear, *393*
  wheel and rail geometry, *392*
gauging
  accuracy, measurement, 350–351
  advanced analysis, 368–369
  BASS 501 *vs.* MBS methods, 362–363, *363*
  ClearRoute™, 350, 368
  curvature, 351–353
  digital systems, 369
  dynamic considerations, 354–358
  geometric considerations, 353–354
  measurement, 352
  measuring system, choice, 351–352
  philosophy and history, 346–347
  risk analysis, 369–371
  shape, 350
  software, 350
  standards, 367–368
  structure-vehicle interaction, 366–367
  tilting trains, *359*, 359–361, *360*
  track/ballast settlement allowances, 366
  track geometry, 352–353, *353*
  track position, 352
  track-structure interaction, 365–366
  vehicle tolerances, 358–359
  vehicle-track interaction, 364–365
  vertical curvature, 349
general-purpose piezoelectric accelerometers, 729
generic bogie configuration, *27*
generic test vehicle parameters, **835**
GENSYS software, 283
global positioning system (GPS), 741
graphical user interface (GUI), 663
  UM, 672, *672*
  VAMPIRE, *668*
gravitational forces
  conical/concave profiles, 269, *269*
  flange contact, 267, *268*
gravitational stiffness effect, *21*, 21–22
gravity scaling factor, 835
ground-borne vibration/noise, 553–554
  controlling, 563–566, *564*, *565*
  ground-based measures, *562*, 562–563
  low-frequency vibration, 560–563
  surface railways, 554–558, *558*
  trains in tunnels, *558*, 558–559, *560*
ground improvement methods, 563
grouped drive design, 138, *139*, 156
group traction control, 157
guardrails, 406, *406*
guidance control, 603–604
guiding force/wear index, rail inclination, *710*

H

Hall effect probes, 739, *740*
hardware-in-the-loop (HIL)
  architecture, 809–810
  simulation, 632–634, *634*
harmonic excitation
  rail corrugations, 326, *326*
  wheel polygon, 327
HAROLD roller rig, **783**, 793–801
head checks, 289
head injury criteria (HIC), 512, **512**
heavy-haul freight train, 44, *45*
heavy-haul locomotives, 119
helical springs, *see* coil compression helical springs
Hertzian contact
  contact pressure, 248
  convexity/concavity/radius sign, 247
  curvature ratio A/B, relation with b/a, 246
  curvatures, semi-spaces, 245
  elastic bodies, 244
  general case, *245*
  railway case, *245*, 246
  semi-axis lengths, 246–247
  2D contact, 248
Hertz theory, 666, 831
  coefficients, **247**, **248**
  of elastic contact, 15
  of elliptical contacts, 282
  n/m values, approximation of, 249, **249**
H-force method, 736
HIC (head injury criteria), 512, **512**
higher buckling mode, *618*

Index

high-floor light rail vehicle, 122, *123*
high-pass filter, 352
high-speed freight car, 788–790
high-speed maglev, 168–169
high-speed passenger vehicle, 791–792
high-speed train, 119, 128, *129*
high-temperature superconducting (HTS) maglev, 166, 169, 171
    high-temperature superconducting bulk (HTSCB), 187
    levitation forces, 187, *187*
H-infinity controller, 599
hollow-worn wheels, gauge widening/rail rollover derailments, 393–395, *394*, *395*
homogeneous ground, 562
hood unit locomotive design, 135, *135*
hornblocks, 49
hot riveting, 70
HTS maglev, *see* high-temperature superconducting (HTS) maglev
hunting, 243, 253, 832
    limit-cycle, 21–23
    stability, 854, *855*
hybrid locomotives, internal energy storage, *134*, 134–135
hybrid pantograph models, 627
hybrid traction systems, 150–151, *151*
hydraulic dampers, 108, 661
    coaxial rubber-metal spring, 224, *225*
    force characteristics, 224, *225*
    telescopic, 223, *224*
hydraulic-gas draft gears, 96
hydraulic lubricators, 296
hydraulic system, 787
hysteresis, 198, 223
hysteretic damping model, 563

I

ideal friction conditions, 291, *292*
impact excitations
    accelerations, track structures, 331–332, *332*
    dynamic displacement, track structures, 331, *332*
    rail joints, 323, *323*
    turnouts, 323–325
    vertical wheel-rail force, 331, *331*
    wheel flats, *325*, 325–326
impact noise/vibration, 538
    rail joints/switches/crossings, 541–542, *542*
    reduce, 543
    wheel flats, *539*, 539–541
inclined suspension, *570*
'incompressible algebraic' pipe model, 218
independently driven wheels, 607–608, *609*
independently rotating wheels, 29, 603–604
    drive design, 141, *141*, 156
independent wheels, 270–271
individual drive design, 138, *139*, 156
industrial revolution, 167
inertia effect, 661
innovations, improved steering, *12*, 12–13
in-plane suspension stiffness, 106–108
in-service draft gear, 493
inside critical speed, 356

inspectional analysis, 827
integrated control design, 604
inter-car connections
    automatic couplers, 92–95
    buffers, 98–100
    draft gear, 95–98
    functions, 91
    inter-car gangways, 100–103
    screw couplers, *91*, 91–92, *92*
inter-car gangways
    hermetic, 101, *102*
    sliding frame, rubber tubes, 100–101, *102*
    Talgo train, 101, *103*
intercity transport, 166
interior noise, prediction, 553
internal roller rig, *851*
inter-wagon connections, 474, *475*
inter-wheelset connections, 27

J

jackknifing derailments, 400, *400*
Jacob shared bogie, 30
Joule effect, 641

K

Kalker's coefficients, 257, **258**, 688
    linear creep coefficients, 253, 831
kinematic envelopes, 349–350
kinematic gauges, 347–348
kinematic hunting, 275
kinematic oscillation, 7, *7*, 29
kinetic energy, 688
Klingel's formula, 275

L

laden/crush inflated, 357
Lagrange multipliers, 628
land-based transport speed, 167, *167*
laser sensors, 750–754
lateral alignment track irregularity, 770, *770*, **771**
lateral creepage, 255
lateral creep force scaling error, *838*
lateral force, *736*
    undesired parasitic effects, compensation for, 738
    and vertical wheel-rail forces, *736*, 736–737, *737*
    between wheelset/axle box, 735–736
lateral force to vertical force ratio (L/V or Y/Q), 376
    angle of attack, *385*
    duration-based criteria, 383–386
    Nadal single-wheel limit criterion, 379–381
    single wheel criterion, *385*
    Weinstock criterion, 381–383, *382*
lateral wheel-rail forces, curved track, 322, *322*
Lattice Boltzmann methods, 548
leaf springs, 221–222, *222*
length scaling factor, 827
levitation forces
    and drag forces, 186, *186*
    EDS, 185–186, *186*
    EMS, 184–185, *185*
    HTS, 187, *187*

light rail vehicles
    configurations, 122, *124*
    definition, 118
    high-floor, 122, *123*
    low-floor, 122, *123*
limit cycle, 839
linear eddy current brake cross-section, 155, *155*
linearised stability analysis, 685–690
    bogie, 688, **688**
    car body, **691**
linear synchronous motors (LSMs), 169
linear variable differential transformer (LVDT), 726, *726*
line tests, 643
loading effect, 357, *357*
loading gauge, 117
load measuring wheelsets, 742, *743*
local shielding/barriers, 538
locked-in suspension movement, 358
locked stiffness, vehicle connections, 479, 487–488
locomotives
    axle configuration, 11–12
    diesel, *see* diesel locomotives
    double-bogie, 9
    electric, *see* electric locomotives
    4-6-0, 16, *16*
    independently rotating wheelsets, 29
    Jervis's 4-2-0, 10
    and passenger trains, 118
    steering, 12–13
    two-axle, 10
    unsymmetrical, 16
locomotive traction and dynamic braking, 460–463, *461*
    characteristics, *462*
    co-simulation approach, 464–465, *465*
    equation-based approach, 463, **463**, **464**
    lookup table approach, 463
    performance curve interpolation approach, 463, *464*
    thermal derating curve, 461, *462*
longitudinal creepage, 255
longitudinal in-train forces, 400–401
longitudinal passenger comfort, 514–517, **516**
    acceleration limits, *516*
    displacement limits, *517*
longitudinal train dynamics, 457–458
    gravitational components, 470, *470*
    input forces, 460–473
    and train crashworthiness, 512–514
    train models, 458–460, *459*
    vehicle connections, 473–494
    wagon dynamic responses to, 494–511
lookup table approach, 463
Lorentz force, 144
Low Adhesion LABRADOR train braking model, 291
low-bandwidth control, 595
lower sway mode oscillation, *685*
low-floor light rail vehicle, 122, *123*
low-noise track, 535–537, *537*
low-pass filter, 352–353
low-speed flange climb derailments, 389
lozenging effects, 81
LP model, *see* lumped parameter (LP) model
LTF25 bogie, 82, *82*
lubricants, 293

lubrication, 408
    benefits, 296
    lubricator system selection/positioning, 297–298
    methods, 296–297
    problems with, 297
lubricity, 298
lumped mass, 190, *190*
lumped parameter (LP) model, 622, *622*, *623*, 627
LVDT (linear variable differential transformer), 726, *726*

M

magnetic levitation (maglev) vehicles
    classification, 170–171, *171*
    defined, 166
    dynamic impacts, 169
    dynamics analysis, 178–188
    higher speed, 168–169
    moving loads, flexible girder, 172–178
    simulation, 188–193
    speed ranges, 166
    system dynamic problems, 169–170
    technological characteristics, 171–172
    trend, higher-speed rail transport, 167
magnetic suspension, 31
MAN bogie model, 839, *841*
Manchester benchmarks, 673, *673*
Matsudaira, research contribution, 19–21
maximum adhesion/traction coefficient, 161
maximum rail displacement *vs.* load speed, *556*
maximum speed, 162
MBS, *see* multibody systems (MBS)
mechanical lubricators, 296
mechatronic model, multiple-span flexible girder, *192*
medium/heavy worn rail profiles, *710*
Medyna software, 283
mesh network, 742
micro-electromechanical systems (MEMS), 731
mild wear, 285
Miner Crown SE friction draft gear, 97
MiniProf profile measuring instruments, *749*
mobile lubricators, 296
modal analysis, 664
    test, 818
modal control, 595–596
modal frequencies, 664
mode-coupling mechanism, *545*
model-based estimator, *605*
model calibration, 627
model validation, 677–682, *681*
mode shapes, 664
monocoque construction design, 135, *136*
motion sickness, 567–569, 585, 684
motive power energy principles
    diesel traction, 128–132
    electric traction, 121–128
    energy transformation, 119, *120*
    gas turbine traction, 132–133
    hybrid traction, 134–135
motor control system, 787
moving-model test, *424*, 424–425
multibody (MB) models, 627
multibody simulation (MBS), 350, 356
    and BASS 501, 362–363, *363*

benchmarking, 673, *673*
maglev simulations, 188–193
Simpack, 669–671
UM, 671–673
VAMPIRE, 667–668
multibody systems (MBS), 627
    direct modelling method, SIMPACK, 190, *191*
    EMS levitation control loop, 190, *191*
    equations of motion, 663–664
    moving loads, flexible beam, 188–190
    vehicle-girder interaction, 192–193
multi-Hertzian methods, 666
multi-objective design optimisation process, *583*
multiple moving forces, flexible girder, 175–178
multiple pantograph operation, 620–621
multiple-span girder, *175*

N

Nadal's criteria, Y/Q analysis, 276–277
natural frequencies, 527, *528*
Navier-Stokes (N-S) equations, 426
network rail high-speed-structure, *351*
Newton-Euler method, 183–184
Nishimura model, *214*, 214–215, 219
nodes, 664
noise, 298–299
    A-weighting curve, *524*
    generation, 34
    and vibration, 522–523, 767–768
nominal vehicle model, parameters, **561**
non-adhesion brakes, 152
non-autonomous rolling stock, 116
non-dimensional spin creepage, 256
non-Hertzian contacts, 272–273, *273*
non-linear bogie stability analysis, *697*
non-linear elastic force characteristics, suspension, *105*
non-linearity parameter, 694, *694*
normal contact, 244–251, 272
    contact jump, 251, 271
    contact plane angle, conicity, 249, **249**
    contact point, wheel/rail profiles, 250, *251*
    contact pressure, 248
    longitudinal curvature, 250, *250*
    minimal distance determination, 250, *251*
    normal load, 250
    variable conicity, 266
normal/lateral tangential forces, wheelset, 20, *21*
normal wear rate (NWR), 641
nose-suspended traction motor design, *140*
no-slip reduced friction forces, 259
numerical integration method, 321–322

O

OCL, *see* overhead contact line (OCL)
octave band sound reduction index, *552*
octave band spectra, *550*
Oda-Nishimura model, *214*
Office for Research and Experiments (ORE) competition, 19–21
on-board lubricators, 296–297
on-track tests, 678
open-circuit wind tunnel, 422, *422*
optical measuring devices, 750–751
orifice damping, 661
overhead contact line (OCL), 614, 616
    conductor bar overhead contact system, 619, *619*
    continuous monitoring, 646
    heavy road vehicles in, 647
    simple CW, 616–617
    standard catenary wire, 617–619
overthrow effect, 353–354, *354*

P

panel car body designs, 56, 59, *60*
pantograph-catenary contact, 628
pantograph-catenary interaction, 628
    HIL simulation, 632–634
pantograph-catenary system, 614
    condition monitoring, 642–644
    issues in design, 614–621, *615*
    life cycle costs, 642, 644
    numerical simulation, 621–629
    track testing, 635–638
    wear/damage/condition monitoring, 638–644
pantograph damage, 642
pantograph models, *626*, 626–627
    hybrid, 627
    LP, 627
    MB, 627
pantographs, laboratory testing
    contact force, measuring system, 634–635, *635*
    dynamic characterisation tests, 631–632, *632*
    HIL simulation, 632–634
    test rigs, *630*, 630–631
    type/routine tests, 631
pantographs, long term trends
    active/semi-active pantograph control, 644–646, *645*
    current collection, 647
    instrumented pantographs/continuous monitoring, OCL status, 646
    pantograph aerodynamics, CFD, 646
passenger
    rolling stock, 118–119
    trains, 118
passenger coaches
    aluminium alloys, 55
    bearing structures, 55–56
    car body parts, 54–55, *55*
    carcass-type designs, 57–58, *59*
    end wall, 58, *58*
    frame, 56, *56*
    panel type designs, 59, *60*
    roof, 57, *57*
    side wall, 57, *57*
    unpowered bogies, 86–89
passenger ride comfort, 48, 356, 700, 725, 745, 767, 791
passive damping, 592–593, *593*
passive-support EDS levitation vehicles, 193
passive suspension components, 658
passive tilt, 585
pendulum linkage, 236, *236*
performance curve interpolation approach, 463
periodical excitations, vehicles/girders
    elevated girder piers, 173, *174*
    TR08 train on flexible girder, 172–173, *173*

PhX rail™, 350–351
piecewise linear vehicle connection model, *478*
piezoelectric accelerometers, 727–730, *728*
pin-on-disc test rigs, 857
pivot assemblies, 142, *143*, *144*
planar shear designs, 728
plastic ratchetting, 288, *288*
pneumatic brakes/braking, 53, 53–54, 470–472, *472*, *473*, 805, *806*
point-to-point force element (PtP), 205, *206*
Polach model, 274, *274*, 791–792
polymer draft gears, 96, *98*, 490–492, *491*, *492*
power braking, 458
powered rolling stock, 117
power output, 161–162
power spectral density (PSD), 759
    track irregularity, *see* track irregularity PSDs
    vertical/lateral car body acceleration, 332, *333*
    vertical/lateral rail acceleration, 334–335, *335*
    vertical/lateral slab acceleration, 335, *336*
    vertical/lateral wheel-rail force, 333–334, *334*
primary suspended H-frame bogies, *81*, 81–82
primary suspension, vibration issues, 571
profile measurements, 266
proportional-integral (PI) controller, 465
propulsion resistance, 466, *470*
    freight rolling stock, **466**, *468–469*
    passenger rolling stock, **467**
protective emergency devices, tanks, 66, *66*
PSD, *see* power spectral density (PSD)
pseudo-kinematic gauges, 347
P-SV waves, 554, *555*, *556*

Q

quantisation errors, 747
quasistatic guiding forces, *705*
quasistatic solution method, 666

R

radially steered bogies, *83*, 83–84
radiation efficiency, 532
radius links, *231*, 231–232
rail brakes, 156
rail corrugations, 326, *326*
rail joints, 323, *323*
rail rollover derailments, 390–391, *391*
    AAR criterion, *391*, 391–392
    hollow-worn wheels effect, 393–395, *394*, *395*
Rail Safety and Standards Board (RSSB), 367–368
rail vehicle aerodynamics, 416, 420
    coordinate system, *419*
    numerical simulation, *429*
    open air with crosswind, 441–442
    open air without environmental wind, *439*, 439–440, *440*
    passing via tunnel, 441
    symbols and units, **417–419**, 419
    terminologies in, 416–417
rail vehicles
    components, 768
    connection model, 480, *481*
    drag, 446–447, *447*, **447**

lateral force, 448, *449*, **449**
lift, 447–448, **448**
pitching/yawing/overturning moment, 448, 450, **450**
pitch models, *505*
stability, 454
testing applications, 742
Railway Group Standards (RGS), 367
railway noise/vibration, 524–525
RTRI roller rig, **782**
railway track system
    ballast bed, 309
    ballastless, *308*, 308–309, *309*
    booted sleepers, 566
    dynamic properties of components, 331–336
    rail pads and fasteners, 309
    rail pad stiffness, 529, *537*, **557**
railway traction rolling stock
    classification, 116–118
    freight traction rolling stock, 119
    passenger rolling stock, 118–119
    shunting locomotives, 119
    special purposes, 119
railway vehicle derailments, 374, 398
    history and statistics, 374–376
    longitudinal in-train forces, 400–401
    mechanisms and safety criteria, 376–398
    vehicle body resonance, 399
    vehicle lateral instability, 398–399, *399*
    vehicle overspeed, 401
railway vehicle dynamics, 5–6, 674–682
    active suspensions, 30–32
    Carter, 13–17
    computation tasks, 682–715, **684**
    computer simulation, 32–33, 666–673
    coning and kinematic oscillation, 6–7
    creep, 18–19
    curving, 8–9, 23–25
    derailment, 30
    engineering process, *674*, 674–675
    expanding domain of, 33–34
    hunting problem, 21–23
    improved steering, *12*, 12–13
    ORE competition, 19–21
    response, hunting and bogie, 9–12
    simulation methods, 663–666
    suspension design concepts/optimisation, 26–30
    testing and model validation, features, 678–680, *679*
    wheel-rail geometry, 17–18
random track irregularity, *see* track irregularity PSDs
Rayleigh waves, 554
RC25NT bogie, 83, *83*
RCF, *see* rolling contact fatigue (RCF)
real-vehicle test, 425, *425*
red-line kinematic gauge, 347
reduced comfort boundary (RCB), 516–517
reference point (RP), 451, *451*
regression process, 353
reinforced sheets, 55
relative humidity, 294
resilient track systems, 563–564
    rail dampers, 536, *537*
    soft resilient fastening systems, 564–565, *566*
resilient wheels, 77–79, *78*
resonance, 690

Index

restraining rails, 406, *406*
retentivity, 298
ride characteristics, 698, **698**, *699*, 700
ride comfort, **698**, 700–702
rigid automatic Voith Turbo Scharfenberg couplers, 94, *95*
rig stability test, *608*
risk analysis, 369–371
roll dynamics unit, 764
roller bearings, 281
roller rig(s), 762, 826, 857; *see also specific roller rigs*
    application, 762
    arrangements, *769*
    in Asia, *766*
    ATLAS, 801–808
    capabilities analysis, 765–771, **768**
    Chengdu, 784–792
    creepages, 254, *254*
    in Europe, *766*
    experimental methods and errors, 815–822
    facilities, 771–783
    full-scale, 762–764
    HAROLD, 793–801
    history, 763–764
    installed roller units, **772**, 770
    perpendicular roller, *769*, 770
    requirements, 770
    RRAC, 808–815
    scaling errors, 837, *838*
    setup types, *851*
    with single suspended wheelset, *859*
    steam locomotives, *763*
    tangent roller, 769, *769*
    vs. track, 764
    vertical plane roller, 769, *769*
roller unit, 809
rolling behaviour, wheelset, 7
rolling contact fatigue (RCF), 288–290, 672, 767
rolling friction model, 253, *253*
rolling noise, *525*
    generation, 525–526
    radiation, 532–533, *533*
    reducing, 533–538
    surface roughness, *526*, 526–527
    track dynamics, 529–530, *530*, *531*
    wheel dynamics, 527–529
    wheel-rail interaction, 531, *532*
rolling stock modifications, 560–562, *561*
root locus curve, *687*
root-mean-square (RMS) acceleration levels, 338
roughness, 352
RouteSpace®, 369
RSSB, *see* Rail Safety and Standards Board (RSSB)
RTA Large CWT roller rig, **773**
RTCONTACT contact algorithm, 860
rubber-metal draft gears, 96, *98*
rubber springs
    excitation frequency, 207–209
    layered rubber-metal spring, 206, *207*
    multi-dimensional models, 209
    one-dimensional models, 207–209
running behaviour, 697
running gears, 49
    and components, 70–79
running safety, 704–705

S

sanding apparatus, 295, *295*
saturation laws, 259
    exponential, 260–261
    heuristic expressions, 260, *261*
    Kalker's empirical proposition, 260
    Vermeulen and Johnson's law, 260
scaled prototype
    applications, 853–862
    dynamic track simulator, 854–856
    HIL applications, 862
    testing on track, 851–853
    tribology, 856–861
    types, *851*
scaled roller rigs
    CTU, *849*, 849–851
    DLR, 838–839, *840*
    history, 827
    INRETS, 841–843, *843*
    Politecnico di Torino, 843–847
    UoH, 839–840, *842*
    VT-FRA, 847–849, *848*
    wheel-roller, worn profiles on, *861*
scaling errors, 837, *838*
scaling factors, 827–828
scaling strategies, roller rigs, 827–828, **836**
    DLR, 832–834
    INRETS, 834–836
    UoH, 828–832
Scheffel bogie, 84–86, *85*, *86*
Schmidt-Coupling, 839
screw couplers, *91*, 91–92, *92*
screw coupling design, 51, *52*
secondary suspension, vibration issues, *569*, 569–571
self-locking relationship, 483, *484*
semi-active control, 597
semi-active suspension approach, 581
semi-Hertzian methods, 666
semi-trailers, 48
sensing technique, 597, 605–606
Series 300 Shinkansen bogie, 89–90, *90*
Series E2 Shinkansen bogie, 87, *88*
series opposition, 726
servo-hydraulic active lateral suspension, 598, *599*
severe wear, 285
shark's fin dampers, *535*
shelling, 288–289
Shen-Hedrick-Elkins (S.H.E.) model, 791–792
Shinkansen/Sumitomo active suspension, 599–600
shoe force, 54
shunt calibration, 734
shunting locomotives, 119
SH waves, 554
side bearings, 80, *81*, 87
    centre column, 234–235, *235*
    constant contact, 111
    critical speed, longitudinal stiffness, 112, *112*
    damping coefficient, 111, *111*
    flat centre bowl, *233*, 233–234
    spherical centre bowl, 234, *234*
    vertical stiffness, probability, 111, *111*
side wall, passenger coach, 57, *57*
signal conditioning module, 726

similarity law, 827
Simpack software, 669
    bogie models, *670*
    FE83 model, *214*, 215, 219
    FlexTrack module, 669, *670*
    models, *669*
    rail wear, *671*
simple double-elliptical contact region (SDEC), 265
simple transducer layout, *745*
simplified accelerometer, *727*
simplified bogie models, 257
simulation modelling process, *583*
single-axle bogies, 137
single non-oscillating axle load, *557*
single-point contact, 405
single roller rig, *851*
single-span girder, 174, *174*, *175*, 176
single-span simple beam, 180–181, *181*
    single moving force, *188*, 188–189
single strain gauges, *738*
single trolley wire, *616*
single-wedge spring system, 482, *483*
skidding, 54
sky-hook damping, 592–594
slab tracks, 315, *316*, 537–538, **557**
slackless packages design, 474
sleepers, 308
    structure, 308, *308*
    subgrade, 310
slip control principles, 157–158, *158*
small-scale testing, 826
soft girder, 176
software simulation, pantograph-catenary interaction, 628, **629**
solid-axle wheelset, 601
    stability control, 601–602
    steering control, 602–603
solid wheels, 77, *78*
sound
    emission from wheel-rail contact, 298–299
    power level, 523, *540*, *541*
    pressure level, 523
    reduction index, *552*
spalling, rails, 289
spalling, wheels, 325
speed limitation
    contact wires, *168*
    wheel-rail adhesion, *168*
speed ranges, maglev trains, 166
Sperling index, 322, *322*, 792
spin creep, 22
spin creepage, 255–256
Spiryagin model, 291
spoked wheel method, 737
spring
    draft gears, 95
    modelling, 659
    rate, 202–205
    resting, rubber-metal cylindrical joints, *237*
    sub-system, 487
squats, rails, 289, 326
squeal noise, 767

stability
    chart, *20*
    diagram, 686, *687*
    test, 816–817
stability analysis, 685
    linearised, 685–690
    non-linear, 691–697, *692*, *693*, *695*, *696*
stability control
    independently rotating wheelset, 603
    loop, *608*
    solid-axle wheelset, 601–602
standard catenary wire, *616*
static forces, 658
static gauges, 347
static/quasi-static tests, 755
stationary tests, 678
Statistical Energy Analysis (SEA), 553
steady-state curving, 666
steam locomotive, 6, 9
steel springs, 10
steering control, 602–603
Stephenson, George, 7
Stephenson, Robert, 10, 29
stepping distances, 367
stitch wire suspension, 617
stochastic analysis, 665
storage security test, 819
strain gauges, 731–733, *732*
    bridge circuits, 733–735, *734*
    foil strain gauge, *732*
    Wheatstone bridge, *733*
Stribeck effect, 544
stringlining derailments, 400, *400*
STRIPES method, 274
structural stiffness model, *480*
structural vibrations, 524, 767
structure-borne sound transmission, 553
structure-vehicle interaction
    advanced analysis, 368–369
    clearance, 366–367
    digital systems, 369
    risk analysis, 369–371
    standards, 367–368
    stepping distances, 367
surface coating, rail, 290
surface damage mechanisms
    plastic deformation, 288
    RCF, 288–290
    wear, 285–288
suspension(s)
    car body to bogie connections, 110–112, 233–237
    characteristics, vertical direction, 103–106
    constraints and bumpstops, 228–232
    dampers, 222–228
    damping, 108–110
    elastic elements, 198–222
    in-plane stiffnesses, 106–108
    primary suspension, 198, 201, 206–207
    secondary suspension, 198, 201, 206–207
    systems, 141–142
swap bodies, 68
sway mode, 683

Index

sway test, 757–758, *758*
Swedish Green Train Project, 600
swept envelopes
 dynamic, 350
 kinematic, 349–350
swept gauges, 347
swing link secondary suspension, 87, *87*
switches and crossings, 409
system excitations
 harmonic loads, 326–327
 impact loads, 323–326
 track irregularity PSDs, 327–331

T

tache ovale, rails, 288–289
Talgo Pendular passenger coach, 101, *103*
Talgo train, 30, 59, 101
tangential rig, *851*
tangent problem
 CONTACT algorithm, 261
 creepages, 254–256
 FASTSIM algorithm, 261–265
 forces/couples, wheelset, 251–252
 linear expressions, creep forces, 253
 pre-tabulated methods, 265
 rolling friction model, 253, *253*
 saturation laws, 259–261
tape circle, 71
tare deflated, 357
tare inflated, 357
Technical Specifications for Interoperability (TSI), 635
technological characteristics, maglev vehicles, 171–172
telescopic hydraulic dampers, 223, *224*
test cell, roller rig, 787
test data sources, 725
test equipment configuration/environment, 743–745, *744*
testing reasons, 724–725
tethered tests, 636, *636*
three-axle bogie, 49, *50*, 137, *138*
three-axle vehicles, 13, 28
three-dimensional dynamics model, slab track, 315, *316*
three-dimensional vehicle-track coupled dynamics models, 310
 passenger vehicle and ballasted track, *318*, 318–319
three-piece bogies, 28–29, 79–81, *80*
tilt controller
 command-driven, *587*
 nulling, *587*
 precedence, *588*
tilting trains, *359*, 359–361, *360*
 concept and equations, 584–585
 control approaches, 586–590
 history of development, 30–31
 mechanical configurations and requirements, 585–586
tilting trains, control approaches, 587–588
 actuator response, 596
 assessment of, 588–590, 604–605

low-bandwidth control, 595
modal control, 595–596
semi-active system, 597
sky-hook damping, 592–594
suspension stiffness softening, 594
tilt lag, 360–361, *362*
tilt precedence effect, 361
time scaling factor, 828
time-stepping integration, 665
top-of-rail friction management, 408
top-of-rail lubricants/products, 292–293, *293*
track
 buckling, 397
 fatigue, 697
 gauge, 51, 117
 geometric unevenness, 560
 geometry, dynamic response, 26
 irregularity inputs, 770, *770*
 loading/wear, 705–708, *706*
 models, 662, *662*
 vs. roller rigs, 764
 simulator, 770
 slab, 315–317, *316*
 tolerance, 365–366
track-based noise solutions
 local shielding/noise barriers, 538
 low-noise track, 535–537, *537*
 slab tracks, 537–538
 TSI noise limits, 526, *526*, 549
track dynamics modelling
 ballasted track, 311–315
 elastic foundation beam model, 310
 slab track, 315, *316*
 track slab, 315–317, *316*
track geometry, 352–353, *353*
 inspection/maintenance, 408–409
track geometry recording, 751
 manual survey, 751–752
 track-recording vehicles, 752–754
 trolley, 752, *753*
 UGMS, 754–755
track irregularity PSDs
 American track spectrum, 327–328
 Chinese track spectrum, 329
 German track spectrum, 328–329
 high-speed railway, 329–331, *330*, **330**
track panel shift, 395, *395*
 causes, 395–397, *396*
 criterion, 397–398, **398**
traction
 and braking process, 767
 mechanisms, 571
 motors, 122
 rods, 143, *144*, 232, *232*
traction drives
 advantages, 140
 bogie frame, 140, *141*
 design and parameters, 140
 grouped drive design, 138, *139*, 156
 independent rotating wheel bogie, 141, *141*, 156
 individual drive design, 138, *139*, 156
 nose-suspended traction motor design, *140*

traction systems
  AC traction power, 149–150
  air-gap flux strength, 145
  current carrying conductor, rotor, 144–145, *145*
  DC traction power, 147–148
  electrical power, 147
  hybrid, 150–151, *151*
  induced voltage, 146
  Lorentz force, 144
  magnetic field strength, 145
  mechanical power, 146
  torque speed curve, generalised electrical machine, 145, *146*
traction test
  under degraded adhesion conditions, 813–815
  under good-adhesion conditions, 812
tractive/dynamic braking efforts, 161
tractive effort, influence, *711*
traditional high-speed railways, 167
train crashworthiness, 512
  end car crumple zones, 512, *513*
  locomotive/power car crumple zones, 512, *513*
  vertical collision posts, 512, 514, *514*
train marshalling and handling, 408
train stop distance, 54
transducers
  accelerometers, 727–731
  displacement, 726–727
  positions, 745–746
tram bogie, 140
transverse spring rate, 202, *205*
tread wear, 74, *74*
tribology, wheel-rail contact, 281–299
  contact mechanics, 282, *284*
tribology studies, 856–857
  adhesion, 857–860
  wear, 860–861
TriboRailer, 292
tuned resonant devices, *535*
turbulent boundary layer theory, 427, *427*, *428*
turnbuckle, 92, *92*
twin-disc rigs, 826–827, 857
two-axle bogie, 49, *50*, 137, *137*
two-axle vehicles, 10, 26–27
two-point contact, 405
type 18–100 three-piece bogie, 79, *80*
typical creep force-creepage relationships, *544*
typical stability map, *687*
tyred wheels, 77, *78*

U

ultrasonic reflection technique, 282
unattended track geometry measurement systems (UGMS), 754–755
under-sleeper pads, 566
Universal Mechanism (UM), *671*, 671–673
unloading devices, 68
unpowered rail vehicles, 43
  car bodies, 47–49, 54–70
  freight wagons, *44*, 79–86
  heavy-haul freight train, 44, *45*
  inter-car connections, 91–103
  number of axles, 44, *46*
  passenger coaches, *44*, 86–91
  running gears and components, 70–79
  suspension design, 103–112
unsymmetrical vehicles, 28

V

VAMPIRE model, 215, *215*, *216*, 218–220, *219*, 667–668
VAT, *see* vibration-attenuation track (VAT)
vehicle
  design, 568–572
  load distribution, 178, *178*
  loading, 407
  pitch model, *510*
  stop distance, 54
  tolerances, 358–359
  and traction dynamics, 713, *714*, *715*
  wheel base, 45, *46*
vehicle connection modelling, 473–474
  draft gear overview, 475, *475*, *476*
  equipment overview, 474–475
  fundamentals, 476, *477*, *478*
  measuring draft gear characteristics, 493–494
  response, *486*
  softening behaviour, *489*
  for wagon connections, 477–492, *479*, **482**
vehicle dynamics
  task/applications, 675–677, *676*
  tests, 811
vehicle interior noise
  air-borne transmission, 550–553
  interior noise, prediction, 553
  levels, 549
  measurement quantities, 549–550, *551*
  structure-borne transmission, 553
vehicle models, 654, *655*
  body components, 655–658
  suspension components, 658–662
vehicle speed/position measurement
  AC tachogenerator, 739, *739*
  GPS, 741
  ground speed radar, 740–741
  Hall effect probes, 739, *740*
vehicle system dynamic, equations, 183–184
vehicle-track coupled dynamics model, 311, 317–319
  design scheme, 336–338
  dynamic simulation and evaluation, 338–339
  experimental validation, 339–341
  Zhai numerical solution method, 322
vehicle-track interaction, 653–654
  computer simulation, *653*
  dynamic simulation, 402
  modelling, 653–663
  track models, 662, *662*
  vehicle models, 654, *655*
  wheel-rail contact models, 663
vibration
  comfort in trains, assessment, *567*, 567–568
  mode/frequency, 180–182
vibration-attenuation track (VAT)
  vs. FST, vibration level, 340, *340*
  full-scale dynamic tests, 339, *340*
  prototype, *337*
  slab acceleration, time domain, 338, *338*

Index

transfer loss, 340, *341*
vibration transmission, subgrade, 338, *339*
viscous dampers, 223–225
VUKV roller rig, **775**

W

wagon dynamics to longitudinal train dynamics, *505*
    body/bogie pitch/jackknifing stability, *496*
    coupler forces, components, **501**, 501–504, *503*, *504*
    on curves, *495*
    lateral coupler angles, 497–501
    overview, 494–497
    rail vehicle body and bogie pitch, 504–506, *507*, *508*, *509*
    vehicle lift-off, coupler forces, 509–511, *511*
    vehicle rollover on curves, coupler forces, 501
    wheel unloading, coupler forces, 500–501, **501**, 510
warping effects, 81
Watt mechanism, bogie, *572*
Watts linkage, 235, *235*
wave-impeding blocks, *563*
wayside lubricators, 296–297
wear, 282, 285–288, 767, 860–861
    index, 707, 709, *710*, 711
    model, 641
    rate, 767
wedge
    friction coefficient, *485*
    test, 684
wheel
    base, 45, *46*, 51
    climb process, 376–378, *378*
    damping, 534, *535*
    design, 535, *536*
    false flange, 74, 393–395, *394*, *395*
    flange angle, 404
    flats, *325*, 325–326
    hollowing, 393
    polygon excitation, 327
    surface roughness, 387, *387*
    taper, 404
    types, 77–79, *78*
    unloading test, 755, *756*
wheel-rail adhesion, 857–860
wheel-rail contact, 765
    models, 663
    parameters, 401–402
wheel-rail coupling model, 319–321, *320*
wheel-rail forces, 13
wheel-rail friction
    conditions, 291–292
    modification, 408
wheel-rail geometry, 17–18
wheel-rail interface, *670*
    contact conditions, 282–284, 405
    research and development, 292, *293*
wheel-rail profiles
    contacting measuring devices, 749
    optical measuring devices, 750–751
    optimisation, 404–405
    requirements, 748
wheel-rail test benches, 857
wheel-rail *vs.* maglev vehicles, pier supporting forces, 169, *170*
wheelset angle of attack, 376, *377*
    on L/V distance limit, 384, *385*
    on L/V ratio limit, 380, *381*
wheelsets
    acceleration, frequency contents, 339, *339*
    benchmark, 673
    creep torque scaling error, *838*
    in curves, radial position, *83*, 83–84
    degrees of freedom, 243, *243*
    design, 71, *72*
    forces/couples, 251–252, *252*
    hunting, 601
    and rail, contact situations, 74, *75*
    rail and contact frames, 244, *244*
    steering mechanism, *850*
    tread and flange wear, 74, *74*
    wheel profiles, 71–72, *72*, *73*, 74
wind tunnel test, 422–424
wireless sensor network (WSN), 741–742, *742*
worn rail profiles, *710*

X

X-2000 bogie, 90, *90*

Y

Y25 bogie, *81*, 81–82
Y32 bogie, trailing (radial) arm suspension, 232, *232*
yaw angles, 447, **447**
yaw stiffness, 406

Z

Zhai vehicle-track dynamics numerical solution method, 321